The period between the repeal of the Corn Laws and the First World War is crucial to the understanding of contemporary rural issues. The unifying theme of this monumental volume is the changing role of agriculture and the countryside in national life, and the impact of social and economic forces unleashed by industrialisation and the growth of towns. Although science and technology had promised unprecedented advances in agricultural output and productivity, English agriculture had by 1914 lost both its 'headship among industries' and its technical supremacy among nations. Agriculture now produced less than one-tenth of national income and employment, and less than half the nation's food. Only one-fifth of the population was classified 'rural'; most traditional rural crafts had long since decayed; and the social and political position of the landowning classes was already seriously weakened.

The book is organised into seven parts. It begins with a critical view of the course of the 'Golden Age' and Great Depression, together with an analysis of output, income and employment. Separate elements are then considered: in Part II, farming regions, systems, techniques, and institutions; in Part III, three principal social and economic classes – landowners, farmers, and labourers. Part IV explores the broad industrial context of agricultural marketing, and the servicing and processing trades; and Part V the social dimension – demography, community, culture and domestic life, and social institutions. Part VI offers an assessment of the urban impact on the countryside and a survey of the main ecological regions. In conclusion, Part VII contains a statistical review of agricultural performance, with appendices.

THE AGRARIAN HISTORY OF ENGLAND AND WALES

GENERAL EDITOR

JOAN THIRSK

C.B.E., F.B.A., HON.D.LITT., HON.DU., HON.D. AGRIC.,

M.A., PH.D., F.R.HIST.S.

*Sometime Reader in Economic History in the
University of Oxford*

VII
(PART II)

1850–1914

The Agrarian History of England and Wales

General Editor: Joan Thirsk

This series initiated under the general editorship of the late H. P. R. Finberg in eight volumes, covers the agrarian history of England and Wales from the Neolithic period to the twentieth century. Series volumes marked ★ remain in print.

Volumes in the series

THE AGRARIAN HISTORY OF ENGLAND AND WALES

VOLUME VII 1850–1914

(PART II)

EDITED BY
E. J. T. COLLINS,

B.A., PH.D., F.R.HIST.S.

*Professor of Rural History in the Department of Agricultural and Food Economics, and
Director of The Rural History Centre, University of Reading*

CAMBRIDGE
UNIVERSITY PRESS

PUBLISHED BY THE PRESS SYNDICATE OF THE UNIVERSITY OF CAMBRIDGE
The Pitt Building, Trumpington Street, Cambridge, United Kingdom

CAMBRIDGE UNIVERSITY PRESS
The Edinburgh Building, Cambridge CB2 2RU, UK   http://www.cup.cam.ac.uk
40 West 20th Street, New York, NY 10011–4211, USA   http://www.cup.org
10 Stamford Road, Oakleigh, Melbourne 3166, Australia

© Cambridge University Press 2000

First published 2000

Printed in the United Kingdom at the University Press, Cambridge

Typeface Monotype Bembo 11/12 pt   *System* QuarkXPress™   [SE]

A catalogue record for this book is available from the British Library

Part I   ISBN 0 521 32926 4 hardback
Part II   ISBN 0 521 32927 2 hardback
Two part set   ISBN 0 521 66214 1

CONTENTS

Introduction   1

By E. J. T. COLLINS, B.A., PH.D., F.R.HIST.S., Professor of Rural History in
the Department of Agricultural and Food Economics, and Director of the Rural
History Centre, University of Reading

PART I · AGRICULTURE IN THE INDUSTRIAL STATE

CHAPTER 1

Food supplies and food policy   33

By E. J. T. COLLINS

CHAPTER 2

Rural and agricultural change   72

By E. J. T. COLLINS

CHAPTER 3

Agricultural output, income and productivity   224

By M. E. TURNER, B.SC., PH.D., Professor of Economic
History, University of Hull

CHAPTER 7

Farming techniques   495
Edited by Paul Brassley

CHAPTER 8

Agricultural science and education   594
By Paul Brassley

PART IV · TRADE, COMMERCE AND INDUSTRY

Edited by JOHN CHARTRES, M.A., D.PHIL., Professor of Social and
Economic History, School of History, University of Leeds and
RICHARD PERREN, B.A., PH.D., Senior Lecturer in Economic History,
University of Aberdeen

PART VI · THE URBAN IMPACT ON THE COUNTRYSIDE

By the late GORDON E. CHERRY, B.A., F.R.I.C.S., Emeritus Professor of
Urban and Regional Planning, University of Birmingham and JOHN SHEAIL,
B.A., PH.D., D.LITT., Senior Principal Scientific Officer in the NERC
Institute of Terrestrial Ecology, and Deputy Head of its Research Station at
Monks Wood

PART VII · THE STATISTICAL BASE OF
AGRICULTURAL PERFORMANCE IN ENGLAND
AND WALES, 1850–1914

By BETHANIE AFTON, B.A., PH.D., Research Associate, University of Hull
and MICHAEL TURNER, B.SC., PH.D., Professor of Economic History,
University of Hull

PART IV

TRADE, COMMERCE AND INDUSTRY

INTRODUCTION

BY JOHN CHARTRES AND RICHARD PERREN

The middle years of the nineteenth century saw the beginnings of profound changes in the context within which the agriculture of England and Wales operated. Many of the trends visible before the 1840s and 1850s came to full maturity in the second half of the century, and impacted upon farming, its commercial linkages and the wider rural community. Changing industrial organisation, domestic and international patterns of trade, transport systems and commercial infrastructure combined to determine new patterns of farming, the adoption of new techniques and technologies, shifting patterns of rural employment, food marketing and processing, and so completing the shift to a fundamentally urban and industrial society.

Although it has long been conventional to date the completion of industrialisation to mid-century, and the share of national product generated by industrial activity had certainly assumed dominance by the 1840s and 1850s, the full impact of this social and economic transformation had yet to be felt in village and countryside. Those leading manufacturing industries on which the industrial supremacy of the third quarter of the century was based were already firmly ensconced in the great industrial towns as early as 1840, where almost half the population was also to be found.[1] Industry, needless to say, was not the sole economic activity of the towns, because at least half the urban workforce was engaged in providing services. But only a quarter of the labour force in 1840 was engaged in primary production and in addition to agriculture this also included another rural occupation, that of mining. If we allow for the numbers of urban industrial workers and those rural workers who were not engaged either in primary production or in services, this means at least a third of manufacturing activity in 1840 was rural either in location or in the source of its workforce, and some of these industries

---

[1] Estimates based on N. F. R. Crafts, *British Economic Growth During the Industrial Revolution* (Oxford, 1985), pp. 62–3, 67.

947

remained rural to the last decades of the century. Regional and local experiences of these processes differed greatly, but affected even textiles, where the beginning of our period coincided, in woollens, with the core of the transition to the steam-powered integrated plant, not its conclusion, and elements of rural and domestic production endured even in cotton, worsteds and linens. In such trades as stocking manufactures, lace, straw plait and shoes, rural manufacture was still a major factor in 1850, and countless industrial activities were mapped into the rural districts of England and Wales by Augustus Petermann on the basis of the 1851 Census. The deindustrialisation of the countryside, especially in the south and west, certainly began long before 1850, indeed, as early as 1700. It had its roots in the shift of capital towards intensive commercial farming. But the bulk of the process, and its completion, were concentrated into the second half of the nineteenth century.

Similar considerations applied to the critical inputs to agriculture itself. Although fertilisers exogenous to the agricultural system were known before 1840, it was the subsequent period that saw their full growth and development, and the gradual decline of supplies of urban night soil under the impact of hygiene and gentility.[2] Labour remained largely endogenous to farming in the 1840s, and it was only after 1850 that mechanised substitutes for agricultural workers, and the agricultural engineering industry they created, emerged and matured. Livestock food was also very largely supplied from within the cycle of domestic farm production to mid-century, and the impact of interregional and international trade in displacing fodder crops, and in so doing creating a new network of suppliers and merchants, was primarily a phenomenon of the years 1850–1914. It was mainly the second half of the century that saw Britain apply the full gains from trade in world markets to the supply of factors for farm output.

The commercial framework of farming, market organisation and transport, were perhaps rather different, and had already distinguished Britain from its neighbours long before 1850, and even 1800, and the high degree of complex articulation of the markets was clearly critical to the processes of change that took farming to such high levels of output without fundamental changes in technology. Its success as a structure was proven by the endurance of many of its systems and institutions beyond 1850. Yet the drover, still a central figure in the livestock trades in 1850 bringing animals into English markets from as far as Wales and Scotland, disappeared rapidly in the 1860s under the impact of the railway, and it was still later that steam power in this form and in international shipping

---

[2] A comprehensive list of the minerals and industrial and urban wastes used for this purpose is to be found in a publication by The Society for the Diffusion of Useful Knowledge, *British Husbandry* (London, 1834), pp. 267–437.

expressed its full force for structural change in agricultural marketing. Well into the 1870s and 1880s, patterns of food processing and trade that could be found a century earlier still characterised such businesses as malting, hops, milling and tanning. Vertical integration into raw material supply was largely uncharacteristic of industries like brewing for the whole of this period, and in terms of the disposal of the products of English and Welsh farming, the full processes of 'modernisation' were concentrated into the second half of our period.

British industrialisation certainly helped to transform the trade, commerce and industry of agrarian society, but it had not completed these processes by 1850, and though the fuse had undoubtedly been lit, it burned slowly and sputtered periodically as it did so. Flexible commercial frameworks of supply and marketing had much still to offer to agriculture in the second half of the century, and it was only with the completion of such innovations as the branch line network of railways, and the new systems of collection and distribution which it spawned, largely after 1870, that these were changed, notably in the milk trades, all of which responded to the premiums available for fresh produce.[3] International developments were arguably more important still in inducing such change. Shifts, such as that towards Russian or North American hard wheats, induced and reinforced structural change in both milling and bread manufacture; malting too restructured as brewers adopted overseas supplies; and a whole raft of food processors and packagers, and the makers of containers themselves, shifted towards the city and the port as international supplies grew in importance. Certainly free-trading, world-supplied Britain was very different from fortress Germany by 1914, notably in the capacity of its farmers to focus through commerce onto higher value-added products rather than the basic means for subsistence,[4] but it is important not to antedate this change, which may have seen a turning point nearer to 1880 than 1850.

These are among the issues explored and analysed in the present volume. The chapters that follow describe and analyse the systems through which farm produce was marketed, and the nature of the price regimes of the maturing industrial and globalising economy. Richard Perren's analysis evaluates the successive impact of coastal shipping and inland waterways, roads and the railway in the domestic market for farm products, and explores their contribution to its changing framework of institutions. New structures of wholesale markets, displacing the old regulated markets, evolved, and the institutional inheritances from the Middle Ages, the open market and the seasonal fair, declined. New urban

---

[3] J. Simmons, *The Railway in Town and Country 1830–1914* (Newton Abbot, 1986), pp. 45–52.
[4] A. Offer, *The First World War: An Agrarian Interpretation* (Oxford, 1989).

wholesale markets were created both to cope with the congestion of the growing cities and to deal with the new trades in foodstuffs created by changing sources of supply, notably in foreign livestock and meat, the processed imported foods, and the extensive rail-borne traffic in fresh milk.

From mid-century, farming increasingly drew inputs from outside the traditional cycle of nutrient and power supplies. The agricultural engineering industry emerged to displace human and animal labour, to apply new power sources, such as steam, to farming, albeit on a relatively modest scale, and increasingly to supplant the established supply networks provided by smithy and wheelwright's shop. These processes were slow and uncompleted by 1914: UK markets for the engineers were distinctly limited, and David Grace's assessment is of an industry handicapped by US competition and the scale of the home market, driven to seek markets abroad for engines and larger items of machinery, while retaining a dominant position in the home market for more basic implements.

Chemical inputs to farming were also provided by industries effectively 'new' to England and Wales from the middle years of the nineteenth century, from the fertiliser suppliers analysed by Michael Thompson to the producers of veterinary products, whose uncertain but patent remedies are assessed by Richard Perren. The former is dated here only to 1842, and had completed the full cycle of birth to senescence by the First World War, drawing heavily upon riverside and port locations, determined by access to waterborne coals and raw materials, and making a genuinely urban industrial source for farm nutrients. Akin to some parts of engineering, veterinary products adopted and branded elements of traditional remedies, many of which continued in use, and drugs of genuine prophylactic power, for the cosseted horse as for humans, were a later twentieth-century development, owing much to European science. Europe too came to supply increasing shares of animal feedstuffs, with port-based mills crushing oilseeds for cattle cake, with American and Indian cottonseed added from the 1860s, and port-located mills also providing offals for animal feed, increasingly from overseas grains. The analysis of these trades and industries by Perren and Thompson thus demonstrates two layers of externalisation in supplies: inputs from abroad, and inputs from coastal and estuarine industrial plant.

At the opposite end of farming, food processing is also shown to have undergone major transformation in the second half of the century. With the increasing adoption of hard foreign wheats for domestic bread, Hungarian and German technology was recruited from the 1870s in the form of the roller mill, and essential steam power applied to it, the by-product of this change being the technological redundancy of both stone miller and millwright. In milling, and in baking, as in so many other fields discussed in this volume, change concentrated into the second half of the period, with some large-scale urban bread factories developing to bake the new and lighter loaf made from high-gluten foreign flours, and to produce the first branded products, but still leaving a large traditional industry in place. Perhaps surprisingly, with patterns of supply of malting barley mirroring those of bread grains, and the later years of the century seeing major organisational changes in brewing, malting remained relatively more traditional in its methods, if not its scale. The industrialisation of food processing in other fields was also limited, and much change concentrated into the last two decades before the war. Led by Alfred Bird, branded new and convenience foods developed late, as did canning in which the technology to deliver a low-priced consumer-oriented size only became widely available after 1890. Late development was also a feature of jam manufacturing, while creameries provided a further putative case-study of British entrepreneurial failure.

These themes of external and urban competition to traditional processes and supply systems, and the relative lateness of the impact of these changes on traditional structures, were perhaps surprisingly replicated in industry. Even in textiles, it is suggested here, many elements of manufacturing endured in rural locations well beyond 1850, and clothing, leather and other manufactures survived and thrived for long in the rural context, and not as mere relics of past modes of production. The paradox of this aspect of industrialisation lay in its partial success in transforming manufacture to an urban and power-based process: in such industries as lace or shoemaking partial mechanisation stimulated large, if dependent, rural employment, with the changing share of women in the workforce often a telling indicator of the nature of the change. In many such industries, the perfection of machine tools and manufacturing systems came towards the end of our period, and the different timings of such changes played an intimate interactive role in that of rural depopulation. Contrastingly, the very forces of Victorian technological success, coal and steam, implanted new industries into rural England and Wales, above all in the advance of deep mining and rail-borne transport of coal. As the main coal fields were scattered widely north of a line from Bristol to the Wash, and London remained the largest single domestic market, a network of rail lines running through the agricultural districts of the south and the home counties converged on it.[5] That rural England and Wales were largely deindustrialised by 1914 is true to the extent that so many of the industrial activities of Petermann's 1851 map had moved to the cities and towns; to the extent that large-scale mining and brickmaking had appeared it was not; and there were many trades, such as quarrying, in which modes and

[5] R. A. Church, *The History of the British Coal Industry*, vol. III, *1830–1913: Victorian Pre-eminence* (Oxford, 1986).

patterns remained mixed, with large-scale and relatively regular industry coexisting with the seasonal, the handicraft, and the residual elements of the 'poor man's gain'. Rural industry is shown here to have had elements of flexibility, seasonal responsiveness, and reservoirs of skills that enabled it still to compete in some industries on bases other than the low overheads and labour costs.

The final areas of activity covered in the present volume are those of a wide range of trades, the providers of services, and the professions serving agriculture and the country districts of England and Wales. Here again, patterns of change were mixed, and interacted with processes analysed in earlier chapters. Some crafts experienced the firm if late imperative of centralised production, and were potentially moribund by 1914: wheelwrights, smiths, millwrights and saddlers remained in the market towns and larger villages, but after the 1880s represented increasingly sales, repair and maintenance trades, not primary providers; in other cases, the timing of decline may have been disguised by the durability of the product, and here thatching provides an intriguing case-study; in fencing, new materials only displaced traditional crafts late, and had the additional benefit of economising in farm labour; and others, such as the rural shopkeeper, vanman and carrier also enjoyed a long period of expansion and relative prosperity, before the full impact of transport, product and population changes were expressed to peg them back towards the end of the period. In the context of the great Victorian glory, the development of the exclusive and self-defining professions, few impacted directly upon many of the rural population, although the conflict of schoolteacher and compulsory education with child labour was a notable exception. A constant supportive infrastructure was provided by the rural publican, affected by the growing concentration of the brewing industry, but it was too small in turnover to be the major target of direct acquisition, and its ageing clientele perhaps withstood changes in taste to 'light and bright' that determined the broader fortunes of the brewing trades.

The five lengthy chapters that follow explore these issues in greater depth, and in so doing provide an intriguing, if sometimes oblique image of the grand themes of the economic and social history of the years 1850–1914. Internationalisation, transport change, the generalisation of urban industry, changing tastes and the beginnings of branding, depopulation, endogenous and exogenous technical change in farming and supply and the processing industry, speedier communication, and the spread of machine technology with its corollary deskilling and gender shifts recur through the following chapters. They effectively assess the range of super- and infra-structures of English and Welsh farming in a period in which it experienced profound change, in the context of the coming of the industrial world-trading state to full maturity.

CHAPTER 15

THE MARKETING OF AGRICULTURAL PRODUCTS: FARM GATE TO RETAIL STORE

BY RICHARD PERREN

This chapter continues some of the themes that were explored in Chapter 3 of the preceding volume in this series.[1] The marketing of farm produce in a time of generally falling prices and the widening of networks of supply continued to exert pressures for change. These pressures were probably more powerful after 1850, and especially after 1870, than they had been in the previous 100 years, or at any time before that. In 1850 the great bulk of farm produce supplied to markets was still home produced, whereas by 1914 this was often outnumbered by imports for many important items. The size of the market mainly grew between 1750 and 1850 chiefly in response to the increase in aggregate demand generated by the growth of population. Demographic growth continued at a slower pace after 1850 so this force still contributed a lesser share of the expansion up to 1914, but now the really dramatic increase was in supply and sheer growth in the quantities coming to market. The organisation and institutional framework of markets in 1850 was still comparatively unregulated by either central or local government. But by 1914 both central and local government were much more willing to intervene at all levels of marketing. In some cases these were concerns that directly affected the consumer, as over matters of public health, but in others like the suppression of fairs it was a case of public order and a recognition of the redundancy of certain earlier forms of market organisation.

A PRICES

1. Sources and data

Information on agricultural and food products can be divided into three groups depending on the distance between producer and final consumer.

---

[1] R. Perren, 'Markets and marketing', in G. E. Mingay (ed.), *Cambridge Agrarian History of England and Wales*, vol. VI, *1750–1850* (Cambridge, 1989), pp. 190–274. However, it should be noted that in that volume prices are covered separately by B. A. Holderness in Chapter 2, 'Prices, productivity, and output', pp. 84–126.

In the first instance the farm-gate prices were those actually paid to the producer. After that the products passed into the wholesale distribution network and the next group of prices to appear were those charged by wholesalers to their retailer customers. At the final stage came those paid by consumers to the retailers.

Much more is known about the middle group than either farm-gate or retail prices.[2] The absence of information about the first stage is not surprising as most transactions between farmers and individual traders neither required nor left any public record. Details of individuals' sales are sometimes found in farm records but they are isolated in their occurrence and most do not provide information over a long series of years. Wholesale prices are recorded because this information was sought and published in local, national and trade newspapers. The *Economist* index of wholesale prices was based on twenty-two commodities, including basic foods and raw materials, beginning in the 1840s, and Sauerbeck's index of forty-five commodities appeared in the *Statist*, also from the 1840s.[3] Both these series include prices of imported agricultural produce as well as home produced. As British agricultural prices were largely determined by world prices in the era of free trade after 1846, imports cannot be excluded from a study of the general level of farm product prices in England and Wales.

In many cases wholesale prices can provide an accurate indication of the movement, if not the level, of farm-gate prices, but they were less reliable as the distance between farm and market increased involving dealers in the expense of collecting from a wide geographical area. Farm gate and wholesale prices could also diverge if traders needed to hold stocks over a period of time, and had to add appreciable interest and storage charges, as did wool dealers. The only indication of the differences between farm gate and wholesale prices was for potatoes between 1909 and 1914 when English main crop growers received 80 per cent of the wholesale price.[4] Before the 1870s isolated instances of retail prices exist, like those recorded by Caird in 1850 covering the main items of food.[5] But these were only for Rutland and there is no indication how far they were typical of the country as a whole and the major cities in

---

[2] The best general summary of annual price variations from 1849 to 1914 is found in M.A.F, Economic Series No. 2, *Report of the Committee on Stabilisation of Agricultural Prices*, 1925. This contains prices of pigs and pork (1867–1914), pp. 22–3; potatoes (1867–1914), pp. 24–6; beef and mutton (1849–1914), pp. 29–31; cereals (1821–1923), pp. 100–1.

[3] *The Economist, 1843–1943*, Oxford, pp. 139–40; A. Sauerbeck, 'Prices of commodities and precious metals', *JRSS*, 49, 1886, pp. 581–646. Annual prices from the *Statist* were reprinted in the *JRSS* after 1887.

[4] R. F. George, 'Potato prices in England and Wales, 1909/10–1913/14 and 1926/7–1930/31', *Journal of the Royal Statistical Society*, 94, 1931, pp. 583, 588.

[5] J. Caird, *English Agriculture in 1850–51* (London, 1851), p. 407.

particular. The earliest series of retail food prices appeared in 1903, giving London prices of bread, flour, potatoes, beef, mutton, bacon, butter, tea and sugar going back to 1877.[6] The regular collection of retail food prices was started in 1904 by the Board of Trade, and annual figures were given thereafter in the January number of the *Labour Gazette*. The weakness of all these series, both wholesale and retail, is that they depend heavily on London prices, and other regions of the country are under-represented. *Broomhall's Corn Trade Yearbook* for 1904 contains a ten-product price series for Liverpool for 1870–1902, furnished by the Medical Officer of Health. These are for animal products and potatoes but with the exception of beef and mutton, which are apparently home produced as they are described as 'carcasses of animals slaughtered at the Liverpool Abattoir', the rest are based on average values of imported products.[7]

Regional price variations were present in 1850 and still existed in 1914: they can be explained in terms of transport costs and the strength of local demand. There appears to have been no general rule about these variations, although prices in London and the conurbations seem to have been rather higher than in other areas which may have been a reflection of higher income levels. In 1900 the national average retail price of milk was 3d. a quart, but it could range from as little as 2½d. in the Yorkshire town of Keighley and the Lancashire towns of Nelson and Colne to 4d. a quart in Liverpool, Manchester and London. Between October 1905 and October 1912 bread prices in London rose from 5½d. to 6d. per quartern loaf, but in Bristol they were 5½d. at both dates, and in Liverpool a loaf cost only 4½d. in 1905 but this had risen to the London level by 1912.[8]

There is little systematic information about the relationship between wholesale and retail prices of food in 1850, but in a study published in 1913 A. L. Bowley found a strong positive correlation between whole-sale and retail price changes. But wholesale prices changed more fre-quently and by smaller amounts than retail prices. Retail food traders preferred to keep prices unchanged for as long as possible and would absorb the small weekly increases in wholesale prices by reducing quality or accepting lower profits. Only when wholesale price increases were large and sustained did retailers pass them on to their customers, and then by more than the latest wholesale increase to compensate for lower profits in the past. So although wholesale prices would move in weekly steps of 1 or 2 per cent, retail food prices changed in monthly or quarterly jumps

[6] *Memoranda Statistical Tables and Charts . . . by the Board of Trade with reference to . . . Industrial Conditions*, BPP, 1903, LXVII Cd. 1761, p. 233.

[7] *Broomhall's Corn Trade Yearbook* (Liverpool, 1904), p. 260.

[8] D. Taylor, 'The English dairy industry, 1860–1930: the need for reassessment', *AHR*, 22, 1974, p. 158; A. L. Bowley, 'The relation between the changes in wholesale and retail prices of food', *Econ. Journ.*, 23, 1913, p. 522.

of 6 to 10 per cent. Wholesale prices were of course only a part of the retailer's total costs and there was no reason why things like transport, commissions, interest on capital, and processing costs should either rise or fall at the same time as wholesale prices. In the case of flour and bread a sustained increase of 10 per cent in the price of flour would only bring about a 6 per cent increase in the price of bread.[9] This relationship held true for most other foodstuffs and had important implications for the effects of changes in farm gate and wholesale prices on the cost of living. It meant that the full impact of rises and falls in wholesale prices was not felt by the consumer as the other elements of cost in the structure of retail prices provided a cushion for the consumer.

The structure of all retailing became more complex after 1850, and more intermediaries were placed between the producer and consumer, so it is likely that the gap between wholesale and retail prices widened and retail food prices were more stable by 1914 than they had been in 1850. Certainly by 1914 in some branches of food distribution, the dealer networks were performing the function of absorbing seasonal price changes. The London retail prices of milk held steady at 1s. 4d. a gallon from March 1892, but contract prices arranged twice-yearly with farmers were 2d. a gallon higher in winter than in summer, giving dealers a larger margin in winter.[10]

Corn prices are available in the greatest detail. Those for wheat, barley and oats in the principal market towns of England and Wales were collected to form the official series published in the *London Gazette* from 1772. This series was continued after 1815 to establish when prices were high enough for the ports to be opened under the Corn Laws, and after the 1836 Tithe Commutation Act they were used to regulate tithe rents. The absence of a nationally accepted uniform system of weights and measures meant regional comparisons were not always easy. As late as 1893 there were 46 different measures for wheat, 26 for barley, and 36 for oats in the 187 towns from which information was collected.[11] There was also frequent complaint about the quality of reporting by the inspectors from some of the towns, and the number of places included in the official Corn Returns was changed a number of times. After the Repeal of the Corn Laws in 1846 it was apparent that British prices were driven by world prices, and in the 1880s and 1890s dealers in home markets saw no point in collecting information which had no practical value except to farmers and the Church. As the grain acreage declined the quantities sold in some markets dwindled to insignificance.[12] But although bread was still

[9] Bowley, 'The relation between the changes in wholesale and retail prices of food', pp. 518–20.
[10] Ruth L. Cohen, *The History of Milk Prices* (Oxford, 1936), pp. 19–20.
[11] J. A. Venn, *Foundations of Agricultural Economics*, (Cambridge, 1933), p. 279.
[12] The number of towns was: 148 in 1821; 150 in 1828; 290 in 1842; 150 in 1864; 187 in 1883; and

regarded as the 'staff of life', and great interest in the price of wheat and other grain remained in the era of increasing abundance after 1870, no specific collection was made of the official prices of imports. Here, as with other agricultural imports, they are obtained by dividing the ascertained or declared values of goods by their quantities. They are also supplemented by information collected from merchants and reprinted for the benefit of traders in the specialised newspapers like the *Mark Lane Express* and the *Liverpool Corn Trade News*.

The collection of livestock prices at an official level did not emerge until the 1890s. Before then newspaper reports, frequently irregular and always imprecise, are the only publicly available source. The 1887 Markets and Fairs (Weighing of Cattle) Act directed all markets to erect weighbridges to provide more accurate information, but the majority paid little attention to the siting, suitability or care of these machines. Farmers' complaints about the poor quality of information available led to a second Act in 1891 which allowed them to appeal to the Board of Agriculture where market owners did not provide 'suitable and sufficient accommodation' for weighing cattle. This Act compelled authorities at the selected markets of Ashford, Birmingham, Bristol, Leeds, Leicester, Lincoln, Liverpool, London, Newcastle, Norwich, Salford, Shrewsbury, Wakefield, York, and certain markets in Scotland and Ireland to provide returns to the Board. These were published in the *Agricultural Statistics of Great Britain* from 1892.[13] But the measures were never popular with farmers or livestock dealers and before 1914 only 30 per cent of fat cattle and 15 per cent of stores entering these markets were sold by weight.

Price movements after 1850 were determined by a number of interrelated factors. Some of these, such as overall trends in supply and long term shifts in the pattern of demand, were decided by forces essential to long term economic development. Without digressing into the fundamentals behind these forces, it is still possible to disaggregate price series into their various components, and that approach has been followed in the rest of this chapter.

Prices of individual products were a combination of four elements: long term secular trends, medium term cyclical fluctuations, seasonal variations, and random shocks. Their analysis is complicated by the fact that the first two in particular need to be considered against general or average price movements. The importance of the fall in cereals between 1873 and 1896 was that they declined more than the general level of prices. In addition, most other agricultural goods fell by a smaller amount

196 in 1890. W. Vamplew, 'A grain of truth: the nineteenth century corn averages', *AHR*, 28, 1980, pp. 1–17.

[13] R. H. Rew, 'English markets and fairs', *JRASE*, 3rd ser., 3, 1892, pp. 112–15.

than cereals, and many livestock products declined by less than the average price level. In the long and medium term, the movement of prices was only important to farmers, traders or consumers when the goods they produced, dealt in or bought, rose or fell in value against other commodities. Although this rule held for trends and fluctuations it was not particularly relevant when applied to seasonal and random variations. As relative prices of commodities did not undergo violent shifts within the space of a single year, the perspective changed, and individuals were more concerned with the effect of sudden shocks upon the prices of a single product considered alone. This was reinforced by the fact that a particular event influencing the price of one commodity often had no effect upon others.

While these four components can give a mechanical account of the features of individual commodity price movements, it must be remembered that different groups of agricultural products reacted on each other. For example, the changes in cereal prices had profound effects on the prices of animal feeds, and this in turn linked back into the prices of livestock and forward into those of meat and dairy products.

2. Trends

Between 1850 and 1914 three sub-periods can be identified within each of which agricultural prices moved in a clear direction. In 1849 to 1873 the average of all prices rose by 25 per cent, between 1873 and 1896 it fell 40 per cent, and from 1896 up to 1914 was once again a time of rising prices. The real importance of these movements to the farmer was not so much in generally rising or falling prices but in those of his own output relative to general prices and also to changes in his costs. This is illustrated in Table 15.1 summarising the price movements of sixteen classes of agricultural output within each sub-period. They cover most agricultural commodities, but the important dairy sector is under-represented as milk and cheese are not included. Fourteen can chiefly be regarded as output prices, but linseed and maize, though not grown in England and Wales, were animal feeds. Oats and barley occupy a dual position for, although outputs of arable farms, their use as provender made them inputs into livestock and arable farms. In this table the movement of agricultural commodities is compared with average price changes of all commodities taken from Jevons's index of wholesale prices before 1860 and Sauerbeck's index thereafter.[14]

Up to 1873 it is important to note that livestock products were generally more buoyant than were arable. The smaller than average rise in

[14] W. T. Layton, *An Introduction to the Study of Prices* (London, 1920), pp. 149–52.

wheat was associated with the Repeal of the Corn Laws, and rising imports of foreign grain. The tendency of animal products to remain above the general level of prices was because the home market relied mainly on domestic supplies and here the growth of output failed to match that of population. The effects of tariff reform were more muted with respect to animal products, and most imports came from Europe where supply constraints were broadly similar to those in Britain. In 1870 Britain relied on imports for at least 38 per cent of its wheat, 21 per cent of its oats, 18 per cent of its barley, but only around 12 per cent of its meat, in spite of the widening price disparity which might have been expected to stimulate imports as well as home production.[15] As money wages rose from 1850 to 1870 the higher income elasticity of demand for the more expensive animal products meant that all items in this group, except tallow, rose faster than wheat. Indeed, the average rise in oats and barley was a reflection of an increased demand for animal feedingstuffs.

Although the years of falling prices from 1873 to 1896 were clearly identified at the time as ones of agricultural depression, only three items, wool, wheat and maize, declined by more than the average 40 per cent of all wholesale prices. In fact, the ranking of farm prices in this group of years did not change markedly from the previous one. Only wool and tallow altered substantially. The downward movement of wool was caused by increased world production, led by Australia and New Zealand, but this was the only animal product to fall by more than the average. The fall in wheat prices has conventionally been blamed upon the increase in United States output, but in the early 1880s wheat from Russia and India played almost as great a part in hammering down prices as supplies from the New World.[16] When the performance of all agricultural prices is judged against the average of wholesale prices it is hard to see that agricultural goods fared markedly worse than other commodities. The low ranking of maize, barley and oats provides the clearest illustration of the benefit the fall in cereal prices conferred on the livestock farmer in cheaper feedingstuffs. Even the fall in wheat helped here as in some years, like 1874–5, prices were so low that quantities of the major bread grain were fed to livestock.[17] Up to the mid-1880s although the

[15] Cereals are average figures for 1869–71. J. B. Lawes and J. H. Gilbert, 'On the home produce, imports, consumption, and price of wheat 1852–3 to 1879–80', *Journal of the Statistical Society*, 43, 1880, p. 330; HMSO, *A Century of Agricultural Statistics, Great Britain 1866–1966*, London, 1968, pp. 99, 100, 110, 112; *Annual Statement of Trade*. Meat figures are interpolations from estimates for 1861–70 and 1870–4. R. Perren, *The Meat Trade in Britain 1840–1914* (London, 1977), p. 3.

[16] W. Fream, 'Canadian agriculture. Part II', *JRASE*, 2nd ser., 21, 1885, pp. 459–60; W. E. Bear, 'The Indian wheat trade', *JRASE*, 2nd ser., 24, 1888, pp. 50–80; *Return of Food Supplies (Imported) Since 1870*, BPP, 1903, LXVIII, Cd. 179, pp. 4–5.

[17] J. A. Clarke, 'Practical agriculture', *JRASE*, 2nd ser., 14, 1878, p. 469.

price of wheat had gradually been falling under the pressure of foreign competition, those of other farm produce – meat, butter, eggs, cheese – had been rising and were continuing to do so, notwithstanding the effects of foreign imports.[18] In spite of the more than fourfold increase in imports, Friesland butter prices in England, and the declared value of cheese imports, rose relative to general prices up to 1896.[19] The resilience of livestock products was due to the stronger demand resulting from increased real wages, emanating largely from the additional purchasing power released by the fall in wheat and cereal prices.

In the period of rising prices after 1896 there was some advance in those of wheat, at a time when new areas were still being opened to cultivation. As the development of regions like the Canadian prairies was more costly than new wheatlands in India, Russia and the United States in the 1880s, the price of wheat was maintained, and it improved its ranking in Table 15.1. The United States also consumed more of its own output in the twentieth century: in 1901 it exported over £55 million worth of breadstuffs but by 1912 this had fallen to £33 million.[20] Two animal products – mutton and beef – declined in the ranking to appear for the first time among the group of commodities that rose less than the average. This was explained by improvements in transport enlarging the area which supplied the British market. The growth of the frozen meat trade permitted Australia, New Zealand and Argentina to do for the meat industry what the United States had done for cereals a generation earlier. This increased volume of meat exports was mostly absorbed by Britain because European governments preferred to protect their own farmers with high tariffs on meat and other livestock products. A further factor limiting any rise in livestock product prices was the failure of money wages to keep pace with the rise in retail prices between 1896 and 1909.[21]

Although the second sub-period from 1873 to 1896 is seen as the one when British agricultural prices were under greatest pressure, their overall performance was better than in either the preceding or following sub-periods. The majority fell by less than the average of 40 per cent, and only wheat and wool were in the bottom group, compared with five before 1873 and nine after 1896. Therefore, on general price performance alone it is hard to justify these as years of depression for the majority of farm products. Additional reasons why farmers felt threatened were because real labour costs increased at the same time as they were having to reappraise both their methods and enterprises. Ó Gráda shows that

[18] H. P. Dunster, 'England as a market garden', *The Nineteenth Century*, 16, 92, 1884, p. 605.

[19] Cohen, *The History of Milk Prices*, p. 53.

[20] C. E. Solberg, *The Prairies and the Pampas* (Stanford, 1987); *The Times Food Number*, 1915, p. 212.

[21] C. H. Feinstein, *National Income, Expenditure and Output of the UK 1855–1965* (Cambridge, 1972), Table 65.

Table 15.1. *Price changes for agricultural products, 1849–1913*

1849–73 average prices rose by 25 per cent		1873–96 average prices fell by 40 per cent		1896–1913 average prices rose by 39 per cent	
Commodity	Per cent change	Commodity	Per cent change	Commodity	Per cent change
Hides	+75	Leather	−22	Hides	+72
Wool	70	Tallow	24	Maize	59
Leather	66	Mutton	25	Pork	57
Beef	55	Butter	25	Bacon	53
Butter	40	Bacon	26		
Mutton	34	Beef	29	Potatoes	43
		Hides	31	Wool	42
Barley	25	Pork	33	Linseed	48
Bacon	25				
Oats	24	Oats	38	Leather	33
Linseed	22	Barley	39	Beef	33
		Potatoes	39	Mutton	31
Pork	10			Oats	28
Wheat	6	Linseed	41	Wheat	23
Tallow	4	Flour	41	Butter	22
Potatoes	1			Flour	22
Flour	−3	Maize	47	Barley	13
Maize	−4	Wool	50	Tallow	−1
		Wheat	51		

Sources: Based on W. T. Layton, *An Introduction to the Study of Prices*, (London, 1920), pp. 53, 88. Data for 1896–1913 extracted from the *JRSS*, 60, 1897, pp. 188–94; 77, 1913–14, pp. 564–70.

between 1870/2 and 1890/2, real income arising from agriculture rose by 9.6 per cent, but that farmers' share of that income fell from 34 to 28 per cent, whereas labour's rose from 43 to 49 per cent.[22]

3. Cycles

Within and spanning the three sub-periods there were cycles of shorter duration lasting between three and seven years. These were mainly animal production cycles, and were largely determined by the time taken for livestock to reach maturity. This meant that the smaller the animal the

[22] C. Ó Gráda, 'Agricultural decline 1860–1914', in R. Floud and D. McCloskey (eds.), *The Economic History of Britain Since 1700*, vol. II (Cambridge, 1981), p. 177.

shorter the cycle so that pigs, which are the most prolific and rapidly maturing farm animals, had cycles of shortest duration. However, such cycles were also influenced by production costs, represented mainly by the price of animal feed, as well as the level of demand which influenced the price of the final output. This interaction between largely invariable biological reproduction requirements and, at times, highly variable economic factors meant that these cycles were never entirely predictable. Certain commodities were highly vulnerable to price movements, and cycles of pork prices and pig production are the best-known examples of this phenomenon. Here the time-lags in adjusting output to prices affected the way supply and demand were balanced. In the nineteenth century the predominant unit of pig production in England and Wales was small and producers' reaction to a rise in price (and profit) was to increase output. Eventually increased supplies glutted markets, reducing prices and prompting the withdrawal of marginal producers until prices stabilised at a new level. Farmers took their decision to produce all agricultural products on the basis of current prices, but in the interval before they had goods to sell, prices may well have altered to make this production unprofitable.

A number of studies of this phenomenon were carried out by Keith Murray at the University of Oxford Agricultural Economics Research Institute between the wars, but some of his data were pre-1914.[23] In his analysis he measured the prices of the specific commodity under discussion as deviations from the average of all prices over the whole period so as to eliminate changes in price due to purely monetary causes. In the case of fat sheep he found that prices at the Metropolitan Market from 1874 followed a regular cyclical movement averaging about seven years' duration with peaks in 1878, 1883, 1889, 1895, 1904 and 1913 and troughs in 1874, 1880, 1885, 1893, 1899 and 1909. The numbers of sheep received at this market also fluctuated but lagged behind prices by about a year. As the same regular cyclical movement was present in the numbers of sheep in the country as a whole, and followed the same seven-year cycle, it is likely that national prices followed the same course as London prices.[24] For pigs it took only one or two years to increase supply beyond immediate requirements, whereas more time was needed to increase the stock of sheep, which explains why the pig cycle was four years and the

---

[23] K. A. H. Murray, *Factors Affecting the Prices of Livestock in Great Britain: A Preliminary Study* (Oxford, 1931). This has been supplemented by reference to brief articles by Murray in the *University of Oxford Agricultural Research Institute Occasional Notes* between Jan. 1930 and July 1932 (cited hereafter as *Occasional Notes*).

[24] Murray, 'The effect of price on the subsequent marketing of sheep on the Metropolitan Market, London, 1874–1900', *Occasional Notes*, I, 9, Jan. 1930, pp. 5–7; Murray, *Prices of Livestock*, pp. 79, 111.

sheep cycle seven. Low points in pig prices were reached in July 1904, June 1908, July 1911 and July 1914, and high points in March 1906, September 1910 and December 1913.[25]

Strong regular price cycles were a feature of small stock but were much weaker in fat cattle. For these Murray measured the purchasing power of fat cattle and beef at the Metropolitan Cattle Market and the London Central Meat Market from 1870 to 1914. Low points occurred in 1873, 1887 and 1914 with high points in 1884 and 1897, but these cycles are so long they almost coincide with the trend sub-periods considered in section 2. For beef and cattle the change in real wages was the main factor determining relative prices, and this was powerful enough to dwarf all other causes, including the supply of cattle, and effectively removed the medium-term price and production cycles that were present for smaller animals. In isolated instances producers employed local knowledge of demand factors to plan production. One large-scale cattle feeder in the north of England used information of ship-building orders on the Tyne to determine the number of cattle he would fatten. When orders rose he knew the demand for beef would increase, due to the increased purchasing power of consumers, whereas during 'quiet' times the demand for beef declined.[26] Interestingly enough the fortunes of the shipbuilding industry also affected many local industries, like the chain and anchor makers of the Dudley and Cradley districts of the West Midlands.[27]

4. Seasonal variations

Noticeable seasonal differences in the prices of most agricultural output still remained in 1914, despite developments in food processing, transport and cold storage. Their character varied for livestock and arable products, and for different commodities within these two sectors of farming. For livestock the supply of feedstuffs and seasonal demand for livestock products were the most powerful factors. In the case of the pig cycles mentioned in the previous section, it is not surprising that lowest prices were in June and July, as the consumption of pork, in the age before the domestic refrigerator, fell very sharply in hot weather. After reaching their lowest at midsummer, pig prices increased up to Christmas and remained around the December level before declining after April. There was a similar, but less pronounced, seasonal pattern for sheep and mutton. Lowest prices for fat sheep were reached in October, and then rose up to April before they followed the pattern of the pigs and declined. There

[25] Murray, 'The price of porkers in England and Wales', *Occasional Notes*, 1, 10, April 1930, pp. 3–6.
[26] Murray, *Prices of Livestock*, pp. 54–5, 65, 68, 75.
[27] G. C. Allen, *The Industrial Development of Birmingham and the Black Country, 1860–1927* (London, 1929), pp. 226, 254.

were stronger seasonal patterns for fat cattle than for beef. Cattle had peaks in June and November, but there was a marked September dip in cattle prices that was not present in beef. Rising beef prices followed the rise in winter fed animals sold from March to May and grass fed ones between September and November.[28] The increased demand for all meats and poultry at Christmas made a seasonal peak for this branch of the livestock sector.

The movements of store cattle and sheep, based on first quality short-horns and first quality longwools, are shown in Figure 15.1(a). Store cattle prices followed the spring upsurge in beef prices as replacements were bought in to fill the stalls vacated by fattened beasts. They remained low between June and August and revived again in September and October when more were needed to replenish the yards for winter feeding. The high point for store sheep came after the success of the root crops was known in September and October.[29]

Dairy cattle prices declined after December until June and then rose again during the summer and autumn to a peak in December. Highest milk production was in the spring, the chief calving time and the best grass season, hence the low price of dairy cows in April and May. As the season progressed, milk production declined when animals advanced into their lactation period and grass became more scarce. Therefore, the demand for dairy cows increased until the end of the year when the shortage of grass meant a falling price for cows in milk.[30] As the retail demand for liquid milk was stable throughout the year, all the annual variation in dairy cattle can be ascribed to availability of feed.

London retail milk prices, as mentioned in section 1, remained fixed from 1892, and any seasonal surpluses were used for butter or cheese processing. But although retail milk prices were held stable, the peaks and troughs in production had a marked effect on wholesale milk prices and the lowest level in June was half that of December and February. The relationship between the three dairy products is shown in Figure 15.1(b). The monthly variations in wholesale butter and cheese prices mirrored those of fresh milk but the peaks and troughs were less pronounced for butter and even smaller for cheese. Here the explanation was the longer production time and storage life of these products than for their raw material, which was also purchased when the price was low.

The seasonal movements in British corn prices in 1896–1902 and 1907–13, shown in Fig. 15.1(c)–(e), reveal clear seasonal patterns revolving around the harvest. Wheat rose steadily up to harvest time before

---

[28] Murray, *Prices of Livestock*, pp. 58–9, 108–9, 140–1.   [29] *Ibid.*, pp. 108–9.

[30] Murray, 'Seasonal variations in the price of milking cows in England and Wales, 1907–13 and 1923–28', *Occasional Notes*, I, 11, July 1930, pp. 1–3.

(a) Stores

Cattle 1907–13
Sheep 1907–13

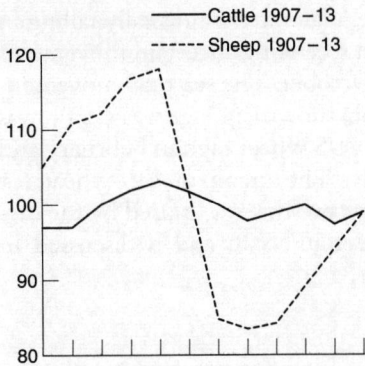

(b) Dairy products

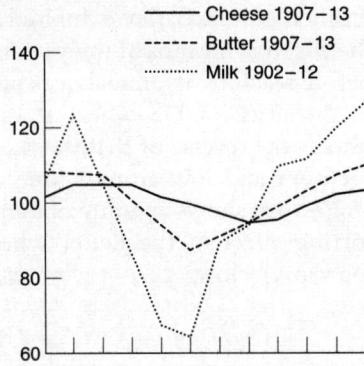

Cheese 1907–13
Butter 1907–13
Milk 1902–12

(c) British oats

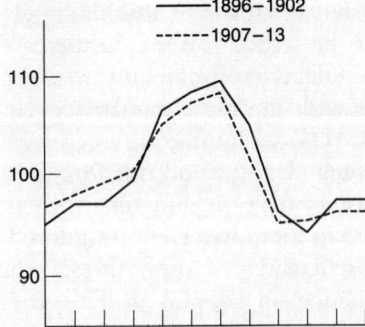

1896–1902
1907–13

(d) British barley

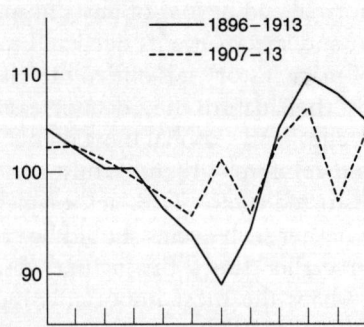

1896–1913
1907–13

(e) British wheat

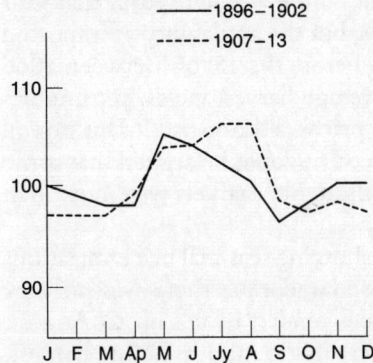

1896–1902
1907–13

(f) US wheat at Liverpool

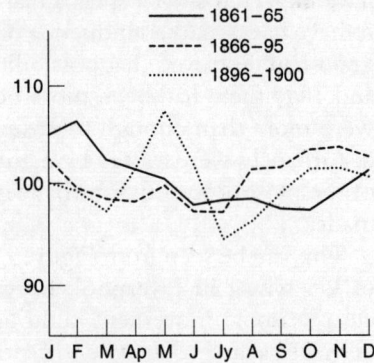

1861–65
1866–95
1896–1900

15.1 Seasonal price variations

Sources: (a)–(e), *Agricultural Returns, GB*; (f) *Annual Review of the International Grain Trade*, No. 3, 1895; *Broomhall's Corn Trade Year Book, 1901–2*

falling back in September. Oats followed a similar pattern but its peak was reached in July. Barley exhibited a less regular pattern normally falling in the first five months of the year, but then rose with uncertain movements before reaching its annual peak around October. The seasonal movement in the price of US wheat at Liverpool, shown in Figure 15.1(f), was usually the reverse of British wheat, with US wheat high in February and October and low around June and July. The graph of US wheat for 1896–1900 shows an unusual high in May, but this was caused by the distorting effect of the Leiter wheat corner in 1898, and is discussed in section 5 below.

5. Random shocks

The discussion of seasonal variations is based on the assumption that somehow conditions and circumstances during the years in question were normal and devoid of any unusual occurrence. However, this was rarely so and special events occurred to distort the settled pattern. In the case of price history it is difficult to classify the effects of random disturbances. By their nature they do not easily fit in with the more regular secular, cyclical and seasonal movements, because on occasion they merged with and reinforced them, whilst on others they clearly stood out from and distorted them. This lack of clear outcome makes it difficult to know whether such events should be considered in their own right or ignored. In earlier times the natural disaster, particularly of the harvest, was perhaps the most important single influence on people's welfare. It is undeniable that crop failures could still, even in the twentieth century, have a devastating effect on producers, even though consumers were now largely shielded from their consequences. Before 1879 observers might have expected such a spectacular harvest failure as occurred in that year to have had a marked influence on prices, but the availability of imported grain had destroyed that possibility long before the 1870s. Between 1866 and 1877 there had been more below average harvest yields, yet imports were more than enough to ensure that prices did not rise.[31] But not all disturbances were caused by nature, indeed, it could be argued that some of the most powerful disturbances to commodity markets were now man made.

This can be seen from Figure 15.1(f), showing seasonal price variations of US wheat in Liverpool. Between 1866 and 1894 there was no very clear seasonal movement, although prices tended to fall up to August, perhaps balancing the rise in British wheat in the months before harvest. But the prices for 1896–1900 are markedly out of line, especially for May

---

[31] Clarke, 'Practical agriculture', pp. 471, 474.

which shows a very marked rise. This was the effect of the Leiter wheat corner, named after a speculative operation run by a Chicago dealer in wheat futures which pushed world wheat prices over 80 per cent above normal in May 1898.[32] It affected British wheat prices as well, raising the *Gazette* price that year from 34s. 9d. in January to 46s. in May. The emergence of a more sophisticated marketing system, with a greater flow of information, does not seem to have done much to reduce such short term fluctuations, and may actually have encouraged speculative operations. The Chicago market was particularly vulnerable, and corners had been run there by traders eight times between 1867 and 1888, in 1867, 1871, 1872, 1880, 1881, 1882, 1887 and 1888.[33] This susceptibility of late nineteenth-century wheat prices to random shocks was acknowledged by the *Corn Trade Year Book* in 1902 which listed the principal causes of world wheat market fluctuations since 1890. In most cases favourable or unfavourable crop reports were enough to reduce or advance prices. But among the monotonous recitals of harvest prospects and wheat deliveries were President Cleveland's and the Kaiser's messages in January 1896, the emergence and collapse of the Leiter corner between March and July 1898, and the Fashoda incident in September and October 1898.[34] In the early nineteenth century war had been a random factor raising wheat prices, and the same effects had been experienced from the Crimean War of 1854–6 and the Franco-Prussian War in 1872–4.[35] But in the absence of major conflicts after 1875 an international crisis carrying the threat of war, albeit remote, could arouse anxieties to raise prices for a few weeks.

The random disturbance was not just confined to wheat although, because of the commodity's importance, it was likely to receive the greatest publicity when it did occur. Extraordinary weather could disturb the normal seasonal pattern of livestock and livestock-product prices. The failure of root crops over much of southern England in 1885, followed by a cold autumn and severe winter, forced so many half-fattened animals onto the markets that it was only after January 1886 that fatstock prices became remunerative once again. In this case over-supply was reinforced by a decline in demand brought about by depression in trade and manufacturing.[36] Year-to-year fluctuations in milk supplies and prices naturally affected those of cheese and butter. Milk yields fell markedly during severe winters, as well as in exceptionally cold summers and summer droughts. Butter was usually at its cheapest just before the hay harvest

[32] *Broomhall's Corn Trade Year Book* (Liverpool, 1901–2), pp. iii–iv, xvii–xix.

[33] *Broomhall's Corn Trade Year Book* (Liverpool, 1894–5), p. 91.

[34] *Broomhall's Corn Trade Year Book* (Liverpool, 1901–2), pp. 206–7.

[35] *Economist*, 10 Jan. 1857, p. 33; R. E. Turnbull, 'Farming in Shropshire in 1875 and 1895', *J. Bath and West*, 4th ser., 7, 1896–7, p. 88.

[36] H. F. Moore, 'The winter of 1885–86', *JRASE*, 2nd ser., 22, 1886, pp. 377–9, 413, 423–5.

from the latter half of May to the beginning of June when pastures were at their greenest, and it was most expensive in midwinter. Extended drought in late summer could prolong the high prices to around Michaelmas, as in 1887, 1892, 1895 and 1911. And in February 1895 particularly, high milk and butter prices were caused by a very severe and long-continued frost.[37]

B TRANSPORT

With the completion of the railway network, the transport of agricultural goods to market underwent significant change after 1850. Although the railways became the chief long distance carriers, road transport performed a complementary role in providing feeder services, and in certain instances road and water continued as effective competitors to rail.

1. Roads

The railways ensured that the average length of road journeys conveying agricultural produce became markedly shorter. In some cases, of course they entirely replaced roads, as evidenced by the rapid decline in the long-distance livestock-droving trade. This was a change initiated by the steamships and the growing use of sea transport for stock from 1820, and gathered pace in the pre-1850 period of railway construction. The long distance droving from Wales and Scotland was still a viable method of transport in 1850, but the days when a single drove of 1,500 Scottish cattle passed through Carlisle on their way to Norfolk were numbered. In 1868, 50 miles was given as the upper limit that cattle were ever sent by road. In Wales a similar process took place. The drover was not entirely extinguished – this only occurred after 1914 with motor transport – but by the end of the nineteenth century the local drover played an important role mainly in the local delivery of animals to fairs and in loading them into railway trucks.[38] Like the drover the commercial road carrier was able to survive by specialising in short-distance traffic, either carrying feeder traffic to the country station, or by becoming the final link from the urban terminus to wholesale markets or retail outlets.

On short journeys the carrier could still offer a superior total service. In the 1860s hay from Hertfordshire was seldom railed to London as

---

[37] D. Taylor, 'Growth and structural change in the English dairy industry, c.1860–1930', *AHR*, 35, 1987, p. 64; C. H. D'E. Leppington, 'The cost of food to the consumer', *Econ. Journ.*, 22, 1912, p. 136.

[38] A. R. B. Haldane, *The Drove Roads of Scotland* (Edinburgh, 1952), p. 220; G. Menzies, 'Report on the transit of stock', *Transactions of the Highland and Agricultural Society of Scotland*, 4th ser., 2, 1868–9, p. 463; R. J. Colyer, *The Welsh Cattle Drovers* (Cardiff, 1976), p. 113.

carters, starting in the afternoon, could reach the market the next morning, so the railways did not offer any important saving of time, besides involving an extra set of handling costs.[39] The question of single handling costs was an important consideration for goods where small quantities were marketed locally like poultry, rabbits, game, eggs and dairy produce. In the vicinity of Hull vegetables were brought in by growers each market day on carts or 'rulleys', and a similar practice was adopted in the Derby area where many carriers came with produce from the market-gardening parishes along the Trent.[40]

Roads remained the chief means used by market gardeners around all towns and cities. Growers in Essex, Hertfordshire, Middlesex, Surrey and parts of Kent sent in supplies of fruit and vegetables to London by road more or less the whole year round.[41] Even here changes were under way by the early twentieth century. Urban growth, with its consequent rise in land values, forced growers further away from their markets. This was a process constantly at work, but its effects were heightened during the boom years in the building industry such as the cycle that reached its peak in 1876 and the upturn of the 1890s which ended in 1903.[42] Every few miles of road added in this way increased transport time and costs. In the 1890s there was an improvement in the condition of the roads as annual investment on highways and bridges rose from £2 million in 1893–96 to £4.7 million in 1897–1903.[43] Against this was the fact that growers in the neighbourhood of the conurbations were forced so far out as to reach the limit for horse transport. Experiments with motor traction using steam engines gave encouraging results around London. A steam tractor cost more than twice as much as the four horses it replaced, but operating costs were much lower. With journey distances around 20 miles farmers on good roads and with large loads of over two tons obtained the greatest benefits from this method. In 1897, the year after the repeal of the 1865 and 1878 Parliamentary legislation limiting horseless vehicles to a top speed of 4 miles per hour and requiring a man carrying a red flag to walk 60 yards ahead of them, it was reported that one Kent farmer, using a Foden steam wagon and trailer, was able to replace six horses and carry loads of nine tons up considerable gradients. On good roads the usual London horse-drawn wagon, a heavy ponderous vehicle with raised sides, could only carry two or three tons and took four or five hours to cover a dozen miles. In other districts, particularly in Worcestershire, growers still used the

[39] H. Evershed, 'The agriculture of Hertfordshire', *JRASE*, 25, 1864, p. 283.
[40] A. Everitt, 'Country carriers in the nineteenth century', *JTH*, new ser., 3, 1975–6, pp. 182–3.
[41] E. A. Pratt, *The Transition in Agriculture* (London, 1906), pp. 99–101.
[42] Feinstein, *National Income, Expenditure and Output of the United Kingdom,* Table 39.
[43] *Ibid.*, Table 41.

lighter flat-topped horse lorry, which took less time and covered greater distances, but carried much smaller loads.[44]

2. Canals and coastal shipping

Like the railways, the canals were built primarily to serve the needs of industry and not farmers, so any agricultural traffic was of minor importance compared to their mineral and coal traffic. In 1853 Andrew Wynter did not list any internal water traffic, either river or canal, among the means by which significant amounts of food were brought into the capital, although in the 1850s Huntley and Palmer used the Thames to convey biscuits into London. In the absence of a comprehensive national railway system they sent everything to London and beyond that by sea. Their biscuits were carried on hoys sailing from Hambro's Wharf to Faversham, the Kent coast, and the Medway towns.[45]

Canals were not particularly convenient for the transport of crops, and not at all suited to livestock or dairy products. In the first half of the nineteenth century the short-distance transport of grain was by river or canal, but most long-distance traffic in cereal products used coastal shipping. Some of London's flour was sent down the Thames on river barges from mills on the western outskirts, mostly at Reading, with smaller amounts from Hertfordshire and Surrey.[46] The benefits of the canal age were more apparent as a way of supplying urban refuse and manure to selected farming districts than for any general transport of agricultural produce. Fruit was one item that could be sent, and Worcestershire growers used the Staffordshire Canal in years of glut to send their produce to Lancashire for table use, but in normal seasons it was disposed of locally and travelled by road. The Bridgwater Canal opened up the Liverpool market for potatoes produced in some parts of Cheshire. The specific marketing advantages conferred by canals were thus highly local and confined to single specialist crops.[47]

When this traffic was exposed to the competition of railways it usually disappeared quickly, even in those cases where greater speed was of no particular benefit. The opening of the Manchester, Sheffield and Lincolnshire railway in 1848 was followed by the disappearance of the wool traffic that had formerly travelled along the Witham canal from

---

[44] R. L. Castle, *The Book of Market Gardening* (London, 1906), pp. 139, 140–4, 146; T. C. Barker and D. Gerhold, *The Rise and Rise of Road Transport, 1700–1990*, (Cambridge, 1995), pp. 53, 55, 61.

[45] [A. Wynter], 'The London Commissariat', *Quarterly Review*, 95, 1853, pp. 271–308; T. A. B. Corley, *Quaker Enterprise in Biscuits: Huntley and Palmers of Reading*, (London, 1972), pp. 42–3, 63.

[46] M. D. Freeman, 'A history of corn milling c.1750–1914, with special reference to south central and south eastern England', PhD thesis, University of Reading, 1967, pp. 17–20.

[47] R. Perren, 'Markets and marketing' in Mingay (ed.), *AHEW*, vol. VI, p. 221.

Lincoln to Boston by 1854. In 1900 some agricultural traffic was still present on the Witham, but although grain, roots and potatoes were still carried, this was mainly on account of the poor state of the roads. Lincolnshire was about the only county where the lack of roads encouraged the railways to retain the canals for feeder traffic. Until 1903 the railways still used about 100 barges on the fen waterways to collect grain and bring it to Peterborough, where it was warehoused prior to distribution throughout the country. This method survived and was convenient because it was the practice of East Anglian merchants physically to assemble a large part of the local crop before dispatching it to manufacturers. But other Lincolnshire canals like the Horncastle, the Sleaford and the Kyme Eau were allowed to fall into disuse, and most agricultural produce found its way to Boston or Lincoln by rail.[48]

The amount of agricultural produce, apart from foreign grain, carried on English waterways as a whole was insignificant. The Grand Union Canal passed across 100 miles of agricultural country from Leicester, via Rugby, Hemel Hempstead and Watford, to London, yet in 1905, out of a total tonnage carried of 1.8 million, English agricultural produce, mainly potatoes, grain, hay and straw, came to no more than 6,000. It had a greater importance as a carrier of urban manure, taking 170,000 tons from London and other towns. Because farmers bought and sold in small quantities, railway trucks were more convenient than barges. As the greater part of England ceased to be arable, and much of the midlands which had the greatest concentration of canals were converted to permanent pasture, there was a drastic reduction in grain, the one bulky crop which the canals were most suited to carry. Hull millers made some limited use of its position at the terminus of this system of canals, not so much to transport flour from home-produced wheat but to forward flour milled from imported wheat. This could be sent into Lincolnshire, Lancashire, Yorkshire, and even as far as Nottingham, at rates low enough to compete with the railways, although traffic further afield was carried by the North Eastern and the Hull and Barnsley Railway companies. But despite these instances, by 1914 the whole transport system was geared towards the railways and the main roads led, not towards the largely marginalised canals, but to the stations.[49]

In the history of transport over the whole of the nineteenth century far more attention has been devoted to the railways than to coastal shipping

[48] J. Boyes and R. Russell, *The Canals of Eastern England* (Newton Abbot, 1977), pp. 266–7; *Second Report of Royal Commission on Canals and Inland Navigations of the UK, Minutes of Evidence*, Vol. III, BPP, 1907, XXXIII, Cd. 3718, pp. 269–70; Ministry of Agriculture and Fisheries, *Report on the Marketing of Wheat, Barley and Oats in England and Wales*, Economic Series, No. 18, 1928, p. 63.

[49] *Fourth and Final Report of the Royal Commission on Canals and Inland Navigations of the UK*, Vol. VII, BPP, 1910, XII, Cd. 4979, p. 62; *The Times Shipping Number*, 1913, p. 147.

because they have been perceived as more of a discontinuity than the advent of the steam coaster. But between 1860–4 and 1910–13 the net tonnage of vessels engaged in the UK coastal trade rose from 11.79 million tons to 21.87 million, and by 1914 they were still carrying some wheat, flour, rice, sugar, oil-cake, fertiliser, agricultural machinery and live-stock.[50] This was in spite of the railways' attempts to attract certain types of agricultural (and other) traffic from coastal shipping by charging lower rates, but at the same time charging higher rates on inland routes where no alternative was available.[51] H. M. Jenkins believed that by 1870 steam-ships no longer carried livestock between Scottish and English ports, although cattle boats were still employed in the growing traffic, mostly in stores, between Ireland and Britain and bringing livestock from Continental ports.[52] However, recent work by John Armstrong has shown that shipping companies continued to carry cattle, sheep and pigs from Aberdeen to London throughout the 1870s. As late as 1889 the railways were still concerned at the large weight of meat and other perishable traffic carried on this route by shipping companies. Coasters also claimed a share of the growing imports of agricultural goods after they had reached the major ports. In 1901–3 they took about half the wool arriving at the Port of London and shipped it from there to the manufacturing towns of Yorkshire, while the other half went by rail.[53]

This was achieved both by co-operation as well as by competition between the railways and coastal shipping companies, and in this way a certain amount of domestic agricultural traffic, mainly in cereal products, still remained with the shipping companies up to 1914. Records of coastal imports of grain into London in 1880, 1885, 1890 and 1895 varied between 30,312 and 54,244 tons.[54] In the 1890s there was something of a coastal trade in home-milled flour between English ports. Hull, with its principal mills situated on the waterside, made some use of coastal transport and the port management encouraged the loading of small coastal steamers by giving them free access to the dock to pick up cargoes. This was not just a one-way traffic: the direction it took depended on which flour-milling ports had surpluses at any time as well as on the rel-ative price of flour in other places.[55] Grain and flour were commodities

[50] Anon., *The Coastwise Trade of the United Kingdom* (London, 1925), pp. 47–9, 52.

[51] S. Marriner, *The Economic and Social Development of Merseyside* (London, 1982), p. 28.

[52] H. M. Jenkins, 'Report of the trade in animals . . . of the farm', *JRASE*, 2nd ser., 9, 1873, pp. 187–245.

[53] J. Armstrong, 'Railways and coastal shipping in Britain in the later nineteenth century: coopera-tion and competition', in C. Wrigley and J. Shepherd (eds.), *On The Move: Essays in Transport History Presented to Phillip Bagwell* (London, 1991), pp. 80–9, 97–100.

[54] P. S. Bagwell and J. Armstrong, 'Coastal shipping', in M. J. Freeman and D. H. Aldcroft (eds.), *Transport in Victorian Britain*, (Manchester, 1988), pp. 190–2.

[55] *Milling*, 7, 13 June 1896, p. 429; 24, 11 June 1904, p. 471; *The Times Shipping Number*, 1913, p. 147.

Table 15.2. *Livestock sent to London by different means in 1853*

	Cattle %	Calves %	Sheep %	Pigs %
By rail	57	14	57	47
By sea from Europe and the rest of Britain	22	25	15	15
Driven in by road and from the neighbourhood of the metropolis	21	61	28	38
	100	100	100	100
Total number of animals	322,188	101,776	1,630,793	127,852

Source: [Andrew Wynter], 'The London Commissariat', *Quarterly Review*, 95, 1855, pp. 284–5.

that were not highly perishable, but where time was important and speed essential the railways had the greatest advantage.

3. Railways

The great bursts of railway construction up to 1854, connecting the main towns and establishing the framework of a national transport system, made relatively little impression on the countryside. It was only in the twenty years from the end of the Crimean War to 1876, when the railways were extended to cover the remoter and less populated districts, that agriculture felt their full impact.[56] Part of this change was already in progress in 1853 when Andrew Wynter surveyed the London food supply, and commented on the major part played by the railway in supplying animal food to the metropolis and the trifling amount that now went in by foot. It will be seen in Table 15.2 that only the majority of veal calves were still driven in by road, and this was only because a regular surplus of these animals was produced by dairies on the outskirts of London. Not all animals came in live, and in addition to these, the railways (and ships) supplied London with a growing proportion of country-killed meat. In 1853, 36,500 tons of railway meat was received at the Newgate and Leadenhall meat markets, the largest amount (13,000 tons) was from the Great Northern Railway and this included a lot of Scotch beef. Before the advent of refrigeration there was probably a greater amount of seasonal variation in the traffic carried by the railways. In the summer a

[56] J. Simmons, *The Railway in Town and Country, 1830–1914* (Newton Abbot, 1986), p. 312.

premium was placed on freshness, and butchers favoured live animals which they could kill exactly when the meat was required. But in winter, as the fear of taint receded, more country-killed meat appeared.[57]

For fat animals the problem of weight loss and lowered carcass quality as a result of long-distance droving was particularly severe. On the 100 miles from Norfolk to London anything from 28 to 64 lb were lost by cattle and even yearling sheep lost 6 lb of mutton and 4 lb of tallow on the journey. Whereas in the 1840s cattle and sheep destined for sale at Smithfield on Monday morning left Norfolk the previous Wednesday, by 1858 their departure by rail could be delayed until the Saturday. They reached the Metropolitan Cattle Market that evening with 36 hours to rest before the Monday sale. Although rail freight was more expensive, the net saving conditional upon reduced weight loss and improved condition was as much as 20s. Apart from this extra profit the farmer could also gain because he could respond immediately to a dear market.[58]

The effect of the railways was felt first by farmers in the neighbourhood of London. As lines radiated out from the capital they changed, and in some cases destroyed, the geographical advantages enjoyed by producers nearest the nation's largest market. In the 1840s Middlesex farmers complained about the Great Western Railway bringing large numbers of fat cattle from the west of England to Southall station. Previously they had done well out of renting land to salesmen who needed somewhere for a few days to keep their stock when Southall's weekly cattle market was glutted. Being close to the market, local farmers could also put their own animals in for sale as soon as prices rose. But as the railway allowed farmers in the west country to send large numbers of animals at very short notice, Middlesex farmers alleged that this forced prices down, compelling them either to withdraw their animals or to sell them at a loss.[59] No doubt it removed the demand for accommodation land as well. As the network was extended its effects were felt in the home counties. Market gardeners in the Vale of Bedford used the Great Northern Railway in the 1850s, not only for sending their produce to London but to obtain manure in return.[60] In Hertfordshire some farmers found it more profitable to send their hay and corn to the London omnibus stables and their root crops to the London dairies than use them to produce meat on their farms. But the railways preserved and even extended other branches of livestock farming. In 1864 there were several dairies in the neighbourhood of Hatfield where cows were kept simply to supply milk to the

---

[57] [Wynter], 'The London Commissariat', pp. 287–9.

[58] C. S. Read, 'Recent improvements in Norfolk farming', *JRASE*, 19, 1858, pp. 290, 296–7.

[59] H. Tremenheere, 'Agricultural and educational statistics of several parishes in the county of Middlesex', *Journal of the Statistical Society of London*, 6, 1843, p. 122; *VCH Middx.*, IV, 1971, p. 47.

[60] W. Bennett, 'The farming of Bedfordshire', *JRASE*, 18, 1857, p. 18.

London market. This was conveyed there by rail in tin cans night and morning.[61]

The railways also removed some of the advantages enjoyed by market gardeners in Middlesex and the home counties in supplying the London market. This process was completed in two stages. In the first, other districts within the British Isles could now exploit any special climatic and geographical advantage they enjoyed. This was the case in Cornwall where small farmers were quick to exploit the trade in early vegetables. In the mid-1850s farmers at Marazion and Perranuthnoe were growing early broccoli, but as the railway did not extend as far as Penzance they still had to cart it six miles to Hayle and load it aboard a coaster to Bristol where it joined the train to London.[62] Secondly, railways allowed established centres of production outside the home counties to develop further. In 1850 distribution was the most serious problem for market gardeners in the Vale of Evesham as produce could only go by road to Birmingham, Worcester, Tewksbury, Gloucester and Warwick markets. At that date around 600 acres were devoted to the market-garden crops but by 1905 this had expanded to 10,000, largely as a result of better transport. By the twentieth century through-trains filled with goods ran from Evesham during the season to all parts of the country. Produce loaded at seven o'clock in the evening could be in London, Glasgow or Edinburgh for markets early next morning. The development of Evesham was not simply a matter of increasing the acreage of crops, it also involved the introduction of new and more perishable ones. In the 1840s the main crops were radishes, onions, peas, cucumbers, plums and kidney beans. After the building of the railways strawberries were first grown in 1870 and tomatoes were introduced in 1884.[63]

Although some of the smaller provincial towns had to wait to the 1860s, Liverpool felt the impact of the railways as early as London. In 1850 large numbers of Scottish cattle were brought by train, as well as potatoes, turnips, carrots, cabbages and lettuces from Lancashire, Yorkshire and further afield.[64] By the 1860s the markets of the industrial centres of the West Riding, that is Leeds, Wakefield, Rotherham, Doncaster and Pontefract, were supplied with corn, cattle and sheep by rail from the most distant farms of the other two Ridings. The railway companies went to some trouble to stimulate this traffic and provided extensive facilities for both livestock and produce.[65] By 1900 they allowed Sheffield, which had problems in obtaining sufficient supplies of perishables in the 1840s,

[61] Evershed, 'The agriculture of Hertfordshire', pp. 283–5.
[62] Pratt, The Transition in Agriculture, p. 117.   [63] VCH, Worcs., II, 1906, pp. 305–6.
[64] B. Poole, The Commerce of Liverpool (Liverpool, 1854), pp. 178–80.
[65] W. Wright, 'On the improvements in the farming of Yorkshire since the date of the last reports in this journal', JRASE, 22, 1861, p. 105.

to become something of a distributing centre for such foodstuffs in Yorkshire, supplied from more distant sources than the small market gardens on the outskirts on whose output the city had formerly relied.[66]

Some of the benefits of increased agricultural trade were experienced by the smaller as well as the larger towns of a district. The ease of railway transport to Norwich was cited as one reason why the sales of wheat and barley at its corn market rose by 35 per cent between 1843 and 1857. But the sales in other corn markets also rose and this, along with cheaper building materials, iron and glass, allowed corn exchanges to appear in every good-sized town in the county.[67] Some of these were built during the later stages of building the main railway trunk-lines up to 1856, and more were erected during the period of rapid branch-line building from 1856. At the national level, two thirds of the corn exchanges in existence in 1914 were established between 1847 and 1870 and very few of them were in places without a railway connection of some sort.[68]

C MARKET STRUCTURE AND ORGANISATION

At the retail level home-produced and imported foodstuffs were sold together, but before they reached that stage each class of goods was handled by separate groups of traders with their own distribution networks, even though both sometimes passed through the same wholesale markets, originally established to handle home-produced goods alone. The marketing of imports was highly capitalised and organised to shift large quantities quickly and cheaply from the dockside to their final destination with the minimum of handling. Home-produced goods reached the retailer through higglers, merchants, factors, dealers and commission-agents, each of whom handled much smaller quantities. The reason for this was that the network of middlemen dealing in British produce had to collect small amounts from as many separate sources of supply as there were farmers engaged in the production of each item. With each product having its own marketing requirements it is difficult to identify any overall marketing structure, except in the very widest sense. The requirements for the sale of cereals, livestock, eggs, milk and dairy products, all of which were sold throughout the year, were quite different from those for wool which became available just once a year. Between these two extremes fruit and vegetables had seasonal periods of sale of varying lengths. Products with the greatest continuity of supply generally had the most efficient and highly organised marketing structures.

[66] Janet Blackman, 'The food supply of an industrial town: a study of Sheffield's public markets 1780–1900', *Business History*, 5, 2, 1963, pp. 92–5.
[67] Read, 'Recent improvements in Norfolk farming', p. 300.
[68] Simmons, *The Railway in Town and Country*, p. 274, nn. 26 & 27.

1. The farmer as salesman

In sales on the farm itself the producer played the smallest part in the process of distribution. He did not need to visit the market, either with a sample or with the entire consignment of his produce, as the dealer visited him. It is difficult to know whether this form of sale became more popular after 1850. In the absence of any system of sales tax, there is no way of measuring the total value of produce sold on and sold off the farm. Large farmers had relied on the middleman for the distribution of their produce long before the mid-nineteenth century. It was only the small occupier who lived near to a centre of population who came into direct contact with the consumer of his produce. Although output per acre was high on small family farms, output per worker was low and this left their occupiers with time available for selling. In Carlisle and Preston on the eve of the First World War market gardeners, or more likely their wives, still brought in eggs, live poultry, cheese, butter, fruit and vegetables which they sold in the covered market on market days.[69] However, such transactions only accounted for a minor part of British agricultural output. In 1914 there were markets in many small towns throughout the country where producers brought in small amounts on a seasonal basis and where buyers were retailers or consumers. These local markets were outside the network of larger collecting and distributing markets that had grown up and formed the framework of the national system that supplied the large centres of population.

In the milk supply of towns the producer-retailer remained an important link in the distribution network. Those farmers in the vicinity of small towns, and in the outer suburbs of large ones, had a favourable location to develop a dairy business. Probably the best example was Lord Rayleigh's branch of dairy shops in London's West End. Between 1886–7 and 1895–6 the Rayleigh estate in Essex adjusted to falling corn prices by taking farms in hand and making milk production its chief enterprise. At first the estate supplied dairy shops with milk, but in 1900 it acquired the Bloomsbury shop of a bankrupt customer. The estate decided to manage the shop, and by 1913 had acquired a further eleven premises in Bayswater, Kensington, Belgravia, Marylebone and Hampstead, as well as one at Westcliff-on-Sea in Essex. This last one was to provide a retail outlet for surplus milk in the late summer months when London demand fell after the end of the season.[70] The landless milk producer-retailer with a town cowshed was a feature of the large cities like London, Liverpool

---

[69] R. H. Rew, 'The middleman in agriculture', *JRASE*, 3rd ser., 4, 1893, pp. 61–2; Ministry of Agriculture and Fisheries, *Report on Markets and Fairs in England and Wales, Part III, Northern Markets*, Economic Series, No. 19, 1928, pp. 64–5, 70–1.

[70] Sir William Gavin, *Ninety Years of Family Farming* (London, 1967), pp. 85, 112–21.

and Manchester in the 1850s. But everywhere this type of producer was declining in the face of supplies brought in by rail, although the importance of this source and its pace of retreat varied from place to place.[71]

For some products the wholesaler calling on the farmer was the usual practice. The cheese factor had played a pivotal part in the marketing of cheese since the seventeenth century in those cases where the farmer did not market the product himself. He dealt with a regular number of suppliers, gaining over the years a close knowledge of the particular products of individual farms, and visiting those on his books at intervals of a few weeks to a few months, depending on the movement of cheese from his warehouse and his forecasts of future demand. His trade called not only for business ability but for personal skill in the recognition of quality and the fine shades of flavour that would satisfy the requirements of his various customers. But we know that the proportion of cheeses sold in this way almost certainly declined after 1850 with the rise of the specialist cheese markets held at towns in the cheese-producing areas.[72] Farmers incurred expense and trouble in taking cheeses to these markets, but they believed the greater competition where a number of buyers congregated compensated for this. In the case of wool the situation was more similar to cheese, with some sold to country buyers who visited the farm, but the rest sold away from the farm either at country auctions or in the big wool dealing centres like Bradford, Liverpool and London. Such farm sales could be hurried affairs where the farmer had no precise knowledge of prices or the state of the market, and where the wool dealer might visit between thirty and fifty farms in a day.[73] Farmers' attitudes depended on how important the product was in their total output. Some would be keen followers of the market, whereas others might regard wool as just a by-product and the return from it as no more than pin money for their wives and daughters.

Farm-gate livestock sales could be small or large affairs. Small purchasers could be jobbers, or farmers who were part-time livestock dealers, and some butchers in country districts might buy from neighbouring farmers as and when they needed. But it was the large public sale, conducted for the farmer by a firm of local auctioneers, that attracted public notice. Besides managing the sale, the auctioneer provided valuable pre-

[71] R. Scola, *Feeding the Victorian City: The Food Supply of Manchester, 1770–1870* (Manchester, 1992), pp. 72–3, 77; Joan E. Grundy, 'The origins of the Liverpool cowkeepers', MLitt thesis, University of Liverpool, 1982, pp. 31–9; P. J. Atkins, 'London's intra-urban milk supply', *TIBG*, new ser., 2, 1977, pp. 383–99.   [72] See below, p. 985.

[73] R. B. Westerfield, *Middlemen in English Business* (New Haven, 1915), pp. 206–7; Ministry of Agriculture and Fisheries, *Report on the Marketing of Dairy Produce in England and Wales, Part I, Cheese*, Economic Series, No. 22, 1930, pp. 94–7; Ministry of Agriculture and Fisheries, *Report on Wool Marketing in England and Wales*, Economic Series, No. 7, 1926, pp. 20–1.

sale publicity with printed catalogues and sale notices. Only substantial farmers with hundreds of sheep and cattle, whose buyers might come from as far away as 50 miles, found it worthwhile to hire this professional for the day.[74]

2. Producers' co-operatives

Attempts at co-operative selling by farmers' organisations in England and Wales represented attempts to set up an alternative marketing system, or at least to bypass a part of the existing system. But such enterprises never had the same success in this country as their Irish and Continental counterparts. This was because the ordinary commercial network of jobbers, dealers, factors and commission agents paid higher prices and gave a better service. The earliest co-operative selling organisations involving British farmers only appeared in the 1890s, not from the initiative of farmers, but promoted by agricultural lobbyists and wealthy landowners. A national organisation, the British Produce Supply Association, was started by Lord Winchelsea in 1896, 'to assist the producer in the disposal of his produce at every stage from the farm to the market'. After struggling for some time with debts as high as £250 a week it was wound up in 1900. The first proper farmers' co-operative trade organisation was the British Agricultural Organization Society started in 1900 at Newark. In 1901 it merged with the National Agricultural Union to form the Agricultural Organization Society. Its aims were to promote co-operation in general but although 'co-operation for the sale of farm produce' was one of its objects, the belief before 1914 was that it was much more valuable for buying than marketing produce. Given this outlook, it is not surprising that the Society made little impact on the national marketing of farm produce before 1914.[75]

Some limited progress was made on a local level with individual products such as eggs, vegetables, fruit, seeds and potatoes. By 1914 there were thirty-five dairy societies, some making cheese and butter, but most engaged in retailing liquid milk. It was said that they improved prices for local producers where the only alternative was sale to higglers who travelled from farm to farm and paid what they liked. Few attempts at co-operative fruit and vegetable marketing were made before 1900, but some societies were formed after that date. The first of a group of important

[74] R. Perren, 'The effects of agricultural depression on the English estates of the Dukes of Sutherland, 1870–1900', PhD thesis, University of Nottingham, Oct. 1967, pp. 329–32.

[75] J. Long, 'The co-operation of producers in the sale of farm produce', *J. Bath and West*, 4th ser., 1, 1890–1, pp. 63–81; Ministry of Agriculture and Fisheries, *Co-operative Marketing of Agricultural Produce in England and Wales (A Survey of the Present Position)*, Economic Series, No. 1, 1925, pp. 17–18, 20–1.

organisations began in the west midlands in 1909. In 1914 there were nine of these societies, including one or two running auction marts.[76] It is significant that farmers' co-operative marketing was most successful with the minor agricultural products. Although output of these items grew rapidly between 1850 and 1914, partly as an alternative to cereals, they still remained relatively unimportant and the lack of an existing strongly organised marketing network allowed farmers' enterprises scope for development.[77] But in the case of the major agricultural products of meat, corn and milk, the commercial marketing networks were already in existence, or in the case of milk evolved with the expansion of output, effectively excluding producers from the business of distribution.

3. Markets and fairs

After 1850 the seasonal agricultural fair, formerly the chief medium for the public sale of livestock, lost ground to the auction market held at regular and frequent intervals. In the 1920s the Ministry of Agriculture and Fisheries' survey of markets and fairs remarked: 'The number of fairs . . . has gradually decreased with the rise of the properly constituted market.' But a number of fairs were held on the same day as regular markets that had effectively superseded the function of the fair.[78] The big autumn fair for the sale of livestock, usually store animals, declined steadily. By easing their movement, and travel arrangements for dealers, the railways allowed smaller but more frequent store sales, and eliminated the need to concentrate large numbers of animals in a particular place at a particular time. Livestock fairs held their ground best in the remoter hill districts of the west country and the north, but even here they were eclipsed. In Cumberland the notable sales in 1850 were the great store fairs at Carlisle, and the weekly markets at various towns were of secondary importance. But by 1874 the establishment of auction marts for the sale of fat cattle and sheep was seen as a great boon to buyers and sellers. Dealers knew where to go to obtain their weekly supplies, and farmers were sure of their money at the fall of the auctioneer's hammer. Horse fairs had the best survival rate, and here there was an economic justifica-

[76] Lord Ernle, *English Farming Past and Present* (6th edn., London, 1961), pp. 389–92; C. S. Orwin and E. H. Whetham, *History of British Agriculture, 1846–1914* (London, 1964), pp. 266–74.

[77] Ministry of Agriculture and Fisheries, *Co-operative Marketing of Agricultural Produce in England and Wales (A Survey of the Present Position)*, Economic Series, No. 1, 1925, pp. 57, 81–2, 103–4, 149; The Horace Plunkett Foundation, *Agricultural Co-operation in England: a Survey* (London, 1930), pp. 14–15.

[78] J. Brown, *The English Market Town* (Marlborough, 1986), pp. 40–2; Ministry of Agriculture and Fisheries, *Report on Markets and Fairs in England and Wales, Part I, General Review*, Economic Series No. 13, 1927, p. 7.

tion. Unlike cattle, horses could not be sold by weight, and the fair ground provided sufficient space for putting the horse through its paces where it could be closely observed by a prospective buyer.[79]

The auction market was another force acting against sheep and cattle fairs, and even fairs that survived tended to fall under the control of auctioneers. In comparison with the auction the traditional livestock fair, where deals were made by private treaty, presented a disorganised and unregulated rabble of buyers and sellers. Local residents found them a disruptive nuisance, especially when livestock fairs in decline adopted the cloak of pleasure fairs, bringing with them the seasonal crop of rogues, drunks, petty thieves and a surfeit of noise whose presence cancelled the benefits of any extra trade.

Not only fairs, but livestock treaty markets diminished in popularity in the second half of the nineteenth century in the face of the rise of auctions. Such sales for livestock first appeared in the mid-1830s, but until 1845 a tax was imposed on movables and heritable property sold at auctions except when they were held on a farm. The Act was introduced in 1776, partly to curtail the spread of public auctions, and to preserve the existing arrangements in public markets where goods were sold by private treaty between individual dealers and farmers. The measure was successful, as it made the arrangements for public livestock auctions in town so cumbersome that little progress was made in this direction until its repeal by Sir Robert Peel.[80] With the removal of the tax, and the need to notify the Excise of an impending sale, the numbers of municipally owned and private livestock auctions, some at sites close to railway stations, increased in spite of extra charges sometimes placed on them by the owners of existing markets. In Leicester beasts sold at auction in the town's cattle market paid three times as much in tolls as beasts sold by treaty. At Wolverhampton auctioneers who sold cattle on their own premises in the neighbourhood of the cattle market paid the same tolls as if they had used the market.[81]

In the second half of the nineteenth century a complicated system of markets existed in England and Wales. Wholesale and retail markets were a combination of older common-law and statutory markets which in some instances dated back to medieval times, but in others were created by Private Acts, some passed comparatively recently. In addition there

---

[79] W. Dickinson, 'On the farming of Cumberland', *JRASE*, 13, 1852, pp. 261–2; T. Farrall, 'Report on the agriculture of Cumberland, chiefly as regards the production of meat', *JRASE*, 2nd ser., 10, 1874, p. 428; Ministry of Agriculture and Fisheries, *Report on Markets and Fairs in England and Wales, Part III, Northern Markets*, Economic Series, No. 19, 1928, p. 20.

[80] W. Addison, *English Fairs and Markets* (London, 1953), pp. 184–5.

[81] *Royal Commission on Market Rights and Tolls, Minutes of Evidence*, BPP, 1890–91, xxxviii, Cd. 6268, pp. 44, 148.

were markets, like livestock auction marts, which had come into being after 1850. An almost chaotic variety of different arrangements governed the control of markets in each locality, no two places being exactly the same. Most public markets were owned and run by local authorities, but in some cases private individuals even had control of these, although they were gradually bought out in the process of nineteenth-century municipal reform.[82] The situation as it existed in England and Wales, but excluding London, in 1890, is shown in Table 15.3. At that time 44 per cent of market rights were owned by local authorities, 38 per cent by private persons, and the rest by trading companies, other persons and institutions, and a few whose ownership was not clearly established.

A broad division can be made according to function, which applied to all types of market, both for livestock and for produce, depending on their geographical location. A noticeable distinction could be seen between markets in producing areas and those in consuming centres. In many cases the markets in producing areas were used by farmers to sell their output to country dealers who exported it to markets near consuming centres where the traders were middlemen or local producers, all selling their goods for consumption in the district. The markets in the producing areas thus acted as concentrating points where output was assembled before being sent to the consuming centres. Also, sometimes a further and final process of concentration took place within some of the markets of the consuming centres. This could occur when a single very important market, like Covent Garden, or Smithfield, in London, received goods from several producing regions, in their case from both Britain and overseas.

4. Wholesale markets in producing areas

Some of the livestock-assembling markets collected animals from a much wider area than others. Such distant supplies were generally sent by dealers rather than by farmers. This was because only dealers, who made a preliminary collection by visiting several farms, could acquire enough animals to justify the expense of rail freight to the collecting markets. The farmer, who had just a few animals to sell at a time, was unable to take advantage of the lower rates for larger numbers and complete truckloads. Shrewsbury, Rugby, Northampton, Gloucester and Banbury had a greater proportion of supplies from a distance – sometimes from as far as Ireland, Scotland or the borders – than markets like those in the West Riding where fatstock was collected from within a 15-mile radius. Although all markets handled both stores and fatstock, certain ones were

[82] Perren, 'Markets and marketing', in Mingay (ed.), *AHEW*, vol. VI, pp. 224–5.

Table 15.3. *Various authorities under which market rights were exercised in England and Wales, and owners of those rights in 1890*

Alleged title or authority for markets	Owners					Total
	Local authorities	Trading companies	Private persons	Bodies other than trading companies	Ownership of rights uncertain	
1 By royal grant, charter, letters patent, etc.	90	6	110	18	–	224
2 By prescription	17	8	43	6	–	74
3 By charter or prescription confirmed or regulated by statute	41	–	4	1	–	46
4 By statute (general)	40	–	–	–	–	40
5 By statute (special) local and private Acts	42	20	5	7	–	74
6 By purchase or grant	79	–	–	1	–	80
7 Particulars not ascertained	1	14	97	3	4	119
8 No market rights claimed	3	16	15	3	18	55
Total	313	64	274	39	22	712

Source: Final Report Royal Commission on Market Rights and Tolls, BPP, 1890–1, XXXVIII, p. 47

noted for stores. Shrewsbury was the chief store market, but at times large numbers of stores appeared at certain markets in the producing areas to replace finished animals exported to the towns.[83]

Before the advent of large imported supplies, most cereals were sold and processed locally. In the inland districts the markets were farmers' markets, frequented by either millers or travelling corn merchants. For the farmer, the advantage of these markets was that trade was done on samples, and so he did not have the trouble of taking in all he had to sell. Once a sale had been arranged the usual practice was for the farmer to deliver the required quantity to the local miller or, with the decline of country milling after 1880, to the nearest railway station leaving the dealer to make the arrangements thereafter. Merchants would use these markets as collecting markets, visiting several and assembling their purchases in their warehouses for resale either in the industrial centres of the north and west, or else in London. The largest quantities of grain were shipped from east coast ports like Hull, King's Lynn, Yarmouth and Lowestoft where merchants had extensive premises for both imported and home-produced corn.[84]

The national centre of the grain trade was the London Corn Exchange at Mark Lane. The influx of foreign grain, much of it landed at Liverpool, and the declining importance of home-produced, changed the nature of the trade at Mark Lane but it still remained an important commercial centre up to 1914. The first corn market was established there in 1747, but a larger one was built in 1828 and enlarged in 1850. In 1879 a third and more extensive building was erected as the volume of business at Mark Lane responded to the ever-increasing trade in North American grain, although by then both London and Liverpool shared this trade more or less equally. Subsequently Liverpool gained pre-eminence over London in the international grain trade, although merchants in both cities worked together using similar forms of contract and common marketing arrangements. The Liverpool Corn Trade Association was started in 1853 and incorporated as a limited company in 1886 and again in 1897. It consisted of about 300 members, representing half that number of firms. As early as the 1880s a system of forward dealing was devised, borrowed from the Manchester Cotton Exchange, which allowed options to be traded some months ahead of delivery. The London Corn Trade Association was formed in 1878, but did not adopt the Liverpool system of forward contracts.[85]

[83] *VCH Shropshire*, IV, 1989, p. 247; Ministry of Agriculture and Fisheries, *Report on Markets and Fairs in England and Wales, Part I, General Review*, Economic Series, No. 13, 1927, p.42; *Part III, Northern Markets*, Economic Series, No. 19, 1928, p. 13.

[84] S. B. L. Druce, 'Sale and delivery of corn', *JRASE*, 3rd ser., 4, 1893, pp. 392–4; *Mark Lane Express*, 19, 15 April 1850, p. 1.

[85] R. C. Michie, 'The international trade in food and the City of London since 1850', *Journ. of European Economic History*, 25, 1996, p. 379; *The Times Food Number*, 1915, p. 71; *Minutes of Evidence*

At the ports the marketing arrangements for grain were inevitably dominated by imported supplies, whereas in country districts and inland towns home-grown corn continued to be traded from local corn exchanges. These were often built by private companies who provided the building and charged a rent to the merchants who used the stands inside. With the decline in home-grown corn it is doubtful whether there was enough business to justify all the exchanges, as in some cases, to help defray costs, the buildings were let for private or public use outside market hours.[86]

In the producing areas the specialist wholesale produce markets, which existed for a number of items, did not serve as wide a catchment area as some of the livestock markets. Corn markets were numerically the most important, and after these came those concentrating on some of the animal products, the most notable ones being wool and cheese. The cheese market has already been mentioned.[87] They were known as cheese fairs but were not really fairs at all but specialist markets held at regular intervals, usually every three weeks. From the producer's point of view the cheese fair had the advantage over the older method of the solitary factor visiting his farm that a greater number of bidders could see his cheese. By 1914 there were twelve regular fairs, associated with four varieties. The oldest was at Chester and this, along with others at Whitchurch, Nantwich, Wem, Market Drayton, Ellesmere and Shrewsbury, was for Cheshire cheese. Lancashire cheese was sold at Preston and Lancaster, Caerphilly cheese at Highbridge in Somerset, and Stilton cheese at Leicester and Melton Mowbray. Fairs were not always popular with cheese factors who disliked the competitive environment, but they could not be ignored as they had an important national influence on prices of their particular cheese.[88]

As fruit and vegetable growers marketed their output in short seasonal bursts, they often had larger quantities to dispose of than other producers. For this reason the dominant marketing method was consignment by the grower to distant markets. The majority of this produce was ungraded, either by the grower or by the country wholesaler who just handled a minor part of this commodity. The only collecting markets were in the west midlands. In Gloucestershire, Worcestershire, Warwickshire and Herefordshire the industry consisted of a large number of small growers who raised a variety of crops such as apples, pears, plums,

taken before Commissioners on Agriculture, Vol. III, PP, 1882, XIV, p. 345; J. G. Smith, Produce Exchanges, (London, 1924), pp. 499–500; J. Kirkland, The Modern Baker, Confectioner, and Caterer, vol. IV, (London, 1924), pp. 186–7.   [86] Simmons, The Railway in Town and Country, p. 274.

[87] See above, p. 978.

[88] Ministry of Agriculture and Fisheries, Report on the Marketing of Dairy Produce in England and Wales, Part I, Cheese, Economic Series, No. 22, 1930, pp. 93–102.

strawberries, beans, peas, cabbages, asparagus, lettuce and celery. The output of each grower in this region was too small to pass through the normal commercial channels, so assembly markets were required. The biggest was at Evesham. Practically all the produce sold was grown within a radius of ten miles of the town, and brought in by the growers. The method of sale was by auction to local dealers, some of whom had premises in the town. They acted as distributing agents, making telephone sales to manufacturers, wholesalers and retailers in England, Wales and even Scotland.[89]

In Kent and the eastern counties the output of individual growers was usually large enough to justify direct consignment to the distant centres of demand. From there individual growers sent large amounts of soft fruits, like strawberries, currants, raspberries and cherries, directly to wholesale dealers in London and, after 1870, the large urban markets of the north. In the potato-growing areas of Lincolnshire most supplies were sold at the point of production by private treaty between grower and vegetable dealer.[90]

5. Wholesale markets in consuming centres

Wholesale markets were generally found in towns of 100,000 or more inhabitants, except where these were served by wholesale markets in large towns nearby. Most of these markets were governed by local authorities and, with the exception of some of the London markets, they mainly served the district in which they were situated. At times pressure mounted on city wholesale markets as urban populations expanded and the trade done in these markets outgrew the original accommodation.

Some found it difficult to sustain their livestock markets because of the problems of traffic congestion, dirt, smell and general inconvenience caused by animals in heavily built-up areas. It is not surprising that London, being the largest city, was the first to experience this, and after years of complaints from local residents, the livestock market at Smithfield was closed in 1855 and moved further out into the suburbs at Islington. Eventually even this became less popular and dealers turned to supplying London with meat as processing industries for animal by-prod-

---

[89] Ministry of Agriculture and Fisheries, *Report on Markets and Fairs in England and Wales, Part II, Midland Markets*, Economic Series, No. 14, 1927, pp. 82–38; Pratt, *The Transition in Agriculture*, pp. 136–7; J. Udale, 'Market gardening and fruit growing in the Vale of Evesham', *JRASE*, 69, 1908, pp. 102–3.

[90] C. Whitehead, 'Fifty years of fruit farming', *JRASE*, 2nd ser., 25, 1889, p. 175; Ministry of Agriculture and Fisheries, *Report on Markets and Fairs in England and Wales, Part I, General Review*, Economic Series, No. 13, 1927, pp. 30–1; *Report on Fruit Marketing in England and Wales*, Economic Series, No. 15, 1927, pp. 47–8.

ucts were established in the provinces and closed in the metropolis on the grounds of nuisance. In 1855–9 the Islington market received an annual average of 302,000 British cattle and 1,441,000 British sheep; by 1911–13 this had shrunk to 51,000 cattle and 289,000 sheep.[91] In the 1850s Birmingham had a flourishing cattle market but by 1914 it only received small numbers of fat cattle, and only pigs after 1919. Some of the trade was transferred to the smaller country markets surrounding the city but, as with London, Birmingham became supplied with an increasing amount of meat slaughtered in the countryside.[92]

As an accompaniment to the decline of livestock markets in some of the large cities the wholesale meat markets expanded. A new Smithfield in London was opened in 1868 but as a meat market only. Although not particularly big it was, at three acres, five times larger than the previous inadequate Newgate Meat Market. Smithfield was supplied with underground rail links connecting it to the lines of the main companies bringing meat into London, relieving the road congestion that would have resulted had the increased trade relied on surface transport from the rail termini. But the limited area of the market itself, with no room for any enlargement, eventually created problems by the 1890s when Charles Booth found it 'none too large for the trade, and the shops are now all let'.[93]

Other important wholesale produce markets in London were Covent Garden, the Borough Market and Spitalfields Market, all dealing in fruit and vegetables. There were two potato markets, at King's Cross and Somers Town, the first opened by the Great Northern Railway Company in 1865 and the second by the London Midland and Scottish in 1892. Each of these handled other bulky vegetables besides potatoes, and both were a natural development of the advantages of rail transport for this class of goods. Poultry and game were traded at Leadenhall Market in the City. Although new premises were built here in 1881 they were, like Smithfield, on a congested central site with no room to expand further. In spite of this disadvantage, Leadenhall Market was conveniently situated to supply the large London hotels and eating houses.[94]

---

[91] Smithfield Market Removal Act, 14 and 15 Vict., c. 16; *Return Relating to the Past and Present Supply of Live and Dead Meat to this Country and the Metropolis*, BPP, 1867–8, LV, pp. 9–10; *Agricultural Returns, 1911–13*; Perren, *The Meat Trade in Britain*, pp. 32–41, 100–5, 151–5.

[92] Ministry of Agriculture and Fisheries, *Report on Markets and Fairs in England and Wales, Part II, Midland Markets*, Economic Series, No. 14, 1927, pp. 12, 90.

[93] *Country Gentleman's Magazine*, Oct. 1868, p. 346; *Illustrated London News*, 10 Oct. 1868, p. 349; C. Booth, *Life and Labour of the People of London*, Vol. VII (London, 1896), p. 197.

[94] *The Times Food Number*, 1915, pp. 65–72; Ministry of Agriculture and Fisheries, *Markets and Fairs in England and Wales, Part IV, London Markets*, Economic Series, No. 26, 1930; P. J. Atkins, 'The production and marketing of fruit and vegetables, 1850–1950', in D. J. Oddy and D. S. Miller, eds., *Diet and Health in Modern Britain* (London, 1985), pp. 114–18.

A number of London's wholesale markets shared the problem of cramped accommodation. Nineteenth-century decisions to rebuild some of them on existing central sites were not helpful as these tied them to locations with no room for future expansion.[95] The pressure on London was reduced by London salesmen setting up branches in other towns and cities, as well as by growers consigning their produce directly to the provincial markets. This occurred in the fruit and vegetable trade where an extra annoyance was caused by the tolls at Covent Garden, levied by the Duke of Bedford, which many people regarded as a tax on food. In 1895 one of the largest firms in the market, W. N. White & Co., who had for a long time quarrelled with the Duke over paying toll, sold the freehold of their premises to the Duke and moved elsewhere.[96] When Kent fruit growers from the 1870s onwards ceased to rely solely on the London wholesaler, and started to send fresh strawberries, cherries, currants and raspberries directly to provincial wholesalers, this allowed traders outside London to acquire a larger variety of direct supplies without relying on London merchants and adding to the congestion by coming into London themselves. Over-reliance on London was further reduced by the extension of fruit growing in the midlands and the north west after 1885.[97]

The use of wholesale markets in the suburbs also eased some of the problems that applied specifically to London. In 1850 there were long-established cattle markets at Southall in Middlesex and Romford in Essex, the latter described in 1876 as the largest cattle market near London.[98] After 1849 the opening of Greenwich fruit and vegetable market removed some of the pressure on Covent Garden. In 1891 the Brentford and Chiswick Urban District Council purchased about two acres of land from the Rothschild family on which they built a formal market that was opened in 1893.[99]

The process of market enlargement and modernisation occurred in provincial towns as well, but they usually had more land available for

[95] *Covent Garden Gazette and Market Record*, 2, 20 March 1886, p. 135; Ministry of Agriculture and Fisheries, *Markets and Fairs in England and Wales, Part VI, London Markets*, Economic Series No. 26, 1930, p. 176.

[96] R. Webber, *Covent Garden: Mud Salad Market* (London, 1969), pp. 135–6; *The Fruit Grower, Fruiterer, Florist and Market Gardener*, 10, 2 Aug. 1900, p. 75.

[97] T. J. Sharp, 'The development of the fruit industry in Kent, 1860–1914 with reference to the mid-Kent area', PhD thesis, University of Canterbury. 1982, pp. 176–7; *The Fruit Grower's Circular and Market Grower's Mail*, 42, 29, 7 April 1906, p. 160; *The Northern Counties Grocers' Review and Provision Trades Journal*, 14, 1 Jan. 1895, p. 21.

[98] *Report from the Select Committee on Smithfield Market*, BPP, 1849, XIX, QQ. 2101–14; *VCH Middlesex*, IV, 1971, pp. 46–7; *VCH Essex*, VII, 1987, pp. 75–6.

[99] *The Gardeners' Chronicle*, 3rd ser., 7, 1 March 1890, p. 257; 10, 17 Oct. 1891, p. 460; 10, 26 Dec. 1891, p. 760; *The Green Grocer, Fruiterer and Market Gardener*, 22 June 1896, p. 121; *Report of the Departmental Committee on the Fruit Industry of Great Britain*, BPP, 1905, Cd. 2589, XX, p. 572.

future expansion. This occurred in Bradford between 1866 and 1901, in Leeds between 1863 and 1869, and in Manchester after 1844.[100] Modernisation affected all categories of markets, both wholesale and retail, but just as there were examples of improvement, there were also cases of neglect where combinations of local vested interest and civic inertia left towns saddled with markets that were either too old, too small, or inconveniently sited. Livestock street markets were perhaps the worst example of unsatisfactory arrangements.[101]

Wholesale markets in the larger consuming centres could also operate as collecting markets for surplus produce that was not required by domestic consumers. This can be seen in the case of the London markets for fruit; while jam factories encouraged the expansion of fruit growing, the fresh fruit markets afforded high prices. Many growers were attracted by the high prices in London, and preferred to send their fruit there, rather than consign it to jam-makers who paid lower prices. In the 1870s the jam-makers of Liverpool, Manchester and Birmingham would use agents from both local and London firms to buy surplus fruit in the London markets. As the jam market expanded, more fruit growers were prepared to cut out the London markets and sold on contract to jam-makers, and this lessened the likelihood of low prices for fresh fruit. Growth of the demand for fresh fruit in the northern towns from the 1870s meant that more fruit was consigned to those places, and they in turn acted as collecting centres. From Manchester fruit was distributed in 1896 to Oldham, Bolton, Stockport, Blackburn, Ashton, Hyde and other towns of the north-west. In addition, jam manufacturers in these places could now draw on supplies from the Manchester market and its satellites, and so reduced their reliance on London.[102]

6. Retailing

The final stage of marketing was shared between shops, market stall holders and street traders. Although it is impossible to measure how trade was divided between the three outlets, there is no doubt that the share of the retail shop increased. The general development of retail trading meant not only an increase in the number of shops but changes in the

---

[100] Ministry of Agriculture and Fisheries, *Report on Markets and Fairs in England and Wales, Part III, Northern Markets*, Economic Series, No. 19, 1928, pp. 77–81, 92–5, 107–8; 'Fruit in Manchester', *Journal of Greengrocery, Fruit and Flowers*, 1, 27 June 1896, pp. 611–12.

[101] Ministry of Agriculture and Fisheries, *Report on Markets and Fairs in England and Wales, Part II, Midland Markets*, Economic Series, No. 14, 1927, pp. 31–2.

[102] Sharp, 'The development of the fruit industry in Kent', pp. 174–9; Atkins, 'The production and marketing of fruit and vegetables', pp. 105–6, 111; 'Fruit in Manchester', *Journ. of Greengrocery*, 1, 27 June 1896, p. 611.

nature of the shops themselves with the appearance of the department store, the multiple chain and the co-operative store.[103] All were large businesses whose bulk buying favoured imported food which came from the docks in huge quantities often packed, marked and graded for ease of selling, rather than British produce available only in small amounts and poorly presented in comparison with its competing imports. Indeed, without the appearance of imports it is impossible to envisage the rise of the multiple food store. The fact that most British produce was retailed alongside imported substitutes gave rise to problems of distinguishing between home-produced and foreign foodstuffs. Imported meat was first displayed only in shops specialising in it, but then was increasingly offered for sale by butchers selling home-produced meat.[104] Complaints about imports sold as home produce were made against a wide range of items and exacerbates the difficulty of identifying features that applied specifically to the retailing of British produce. Other products where foreign goods were sold as British were fruit and vegetables, cheese and butter.

Imported food had its greatest market share within the large cities. In London, Liverpool, Manchester, Birmingham and Leeds the whole marketing network, not just the part concerned with retailing, made the distribution of imported produce cost-effective. But in villages and small towns home-produced foodstuffs were afforded the protection of lower transport costs, but only if they did not require long-distance transport for processing. Thus the slaughterhouses attached to butchers' shops in the small towns were used to supply home-produced meat to the local population, while the conurbations obtained home-produced meat from large slaughterhouses close to country stations.[105]

By the First World War the small-town system of retail distribution was based on the shopkeeper whose food came mostly from local farmers or dealers. A good example was English eggs. In 1850 these were shipped by coastal steamers from Newcastle to the metropolis, but by the 1890s they did not normally pass through the hands of big city wholesalers because they were available only in small quantities, and their poor packaging made handling expensive. The London grocer found it easier to sell Danish eggs, whereas the provincial man could obtain British eggs with little trouble, either from the producer or through the local higgler.[106] In London comparatively little British cheese was consumed by 1914. Up to

[103] J. B. Jefferys, *Retail Trading in Great Britain, 1850–1914* (Cambridge, 1954); P. Mathias, *The Retailing Revolution* (London, 1967).

[104] Ministry of Agriculture and Fisheries, *Report on the Marketing of Cattle and Beef in England and Wales*, Economic Series, No. 20, 1920, p. 121.

[105] Anon., 'The food of London', *Quarterly Review*, 190, 380, 1899, pp. 479–80.

[106] G. Dodd, *The Food of London* (London, 1856), p. 318; E. Brown, 'The marketing of eggs', *Journ. of the Board of Agriculture*, 6, Sept. 1899, p. 149.

1890 there was a considerable sale of Cheshire cheese, but then farmers in the north-west switched production from a long-maturing to an early-ripening type in response to a specific demand from northern consumers. This crumbly sharp-flavoured cheese was liked by manual workers in the mining districts of the north. Londoners preferred mild close-textured cheese, qualities which a number of imported cheeses could supply in high volume. On the other hand many of the foreign cheeses sold in London met with little demand in the provinces.[107]

In spite of the rise of the large retail enterprise, urban food retailing was still based on the small shop and the retail market in 1914. But even here methods changed. As the process of urban growth advanced and the population of a district grew, so did a tendency for more shops to open. When a single shop served a large area there was little need for its owner to take much trouble over such things as product display, price, or customer service. As competition increased traders had to take greater care. Except in poor areas, shops needed to be clean, well lit, and goods efficiently displayed. More use was made of vans to take goods from the shop to the customer's doorstep. This was not just as a delivery service, as house calls were made with no prior invitation. The 'cold call' posed threats to shopkeepers in the countryside when tradesmen's vans from neighbouring towns appeared in the villages, carrying a wide variety of goods, including foodstuffs.[108]

In midland and northern towns regular retail markets were more common than in London and the south. Both home-produced and imported foods, as well as other goods, were sold in large, well-equipped covered market-halls – often municipally owned and run. This type of retail market was probably at its best in Lancashire where large sums were spent by local authorities in providing suitable accommodation. In 1914 there were over 150 in the county and nowhere else was a greater volume of business done at such centres. They varied in size from the large covered market-hall at Bolton, to ordinary street markets. But even here home-produced food gradually gave way to imported meat and fruit and vegetables from the wholesale markets at Manchester and Liverpool.[109] London had no such markets for the sale of food. Of the twenty-eight London boroughs only Woolwich controlled any retail markets; it had two, both of them held in the open. In London the place of the municipal market was

[107] *Departmental Committee on Distribution and Prices of Agricultural Produce, Interim Report on Milk and Vegetables*, BPP, 1923, IX, Cmd. 1892, p. 110 (paragraph 169); D. Taylor, 'Growth and structural change in the English dairy industry, c.1860–1930', *AHR*, 35, 1987, p. 49, n. 10.

[108] *The Grocer*, 77, 10 Feb. 1900, pp. 388–9; 77, 3 March 1900, p. 607; 77, 7 April 1900, p. 977; 77, 19 May 1900, p. 1356.

[109] Ministry of Agriculture and Fisheries, *Report on Markets and Fairs in England and Wales, Part III, Northern . . . Markets*, Economic Series, No. 19, 1928, p. 36.

filled by street markets that were congregations of retail stalls where food was just one of the commodities on sale. It was not the custom of London consumers to walk long distances for their food, or any other goods. As a result of this, and the inability of the London County Council to establish a single authority to control existing markets and establish properly regulated new ones when the need arose, the irregular street market set up in densely populated districts was a feature of the capital. In 1891 there were 112, all unauthorised, and containing 5,292 stalls, of which 65 per cent were set aside for the sale of perishable commodities. Some 13 of them were sufficiently important, and did such a large and general trade, that they performed a similar function to some of the big municipal markets of the north, the only difference being that they lacked a fixed market area and buildings.[110]

The retail street market, wherever it was held, was heartily detested by local shopkeepers who periodically formed associations to rid their districts of this nuisance. Generally prices on the market stalls were lower because stallholders had fewer overheads. Traders in municipal markets paid a regular weekly rent for their booths, which some even tiled and decorated. The shopkeeper also paid rent, and in addition rates, but the street trader paid neither and always added to the street litter. While the peripatetic vendor of fruit and vegetables thronging the thoroughfares was a convenience for passing shoppers, nearby residents were not quite so appreciative of the thunderous roar with which he announced his wares, right through from Monday morning to Sunday night. Food contamination in the street markets was a greater problem than in covered markets or shops, and almost impossible for the local Medical Officer of Health to police.[111]

7. The milk market

Although certain features of the trade in liquid milk are mentioned elsewhere in this book, the changes that came about here are important enough to be considered on their own. By 1911–13 milk production had become the second largest item of British agricultural output, after meat.[112] The development of the milk trade was partly a consequence of changes in transport technology, but at the same time it was influenced by changes in thinking about animal health and public health.

[110] The Gardeners' Chronicle, 3rd ser., 11, 27 Feb. 1892, p. 272; 15, 31 March 1894, p. 404; W. W. Glenny, 'The fruit and vegetable markets of the metropolis', JRASE, 3rd ser., 7, 1896, pp. 65–6.

[111] Mary Benedetta, The Street Markets of London (London, 1936); The Journal of Greengrocery, Fruit and Flowers, 1, 7 March 1896, pp. 357–8; 1, 20 June 1896, p. 602; W. Robertson (ed.), Encyclopedia of Retail Trading, Vol. VI (London, 1911), pp. 322–3

[112] E. M. Ojala, Agriculture and Economic Progress (Oxford, 1952), p. 209.

A number of studies show the substantial changes in the marketing arrangements for liquid milk, based mainly on the supply to London.[113] In 1850 the major part of the British milk market was supplied by producer-retailers. In the villages and small country towns all milk was produced locally, and its delivery to the consumer was the responsibility of the producer. The supplies for larger towns came from either farms in the immediate vicinity or stall-fed cattle kept in urban cowsheds. Only in these, and in London, had a distinction begun to appear between the producer and the retailer. In the metropolis a number of the milksellers did not own or manage cows themselves, but purchased milk directly from producers. This they hawked around the streets, calling on householders and measuring out whatever quantities their clients required into the customers' own vessels. Other milksellers were directly employed by the producers. Some milk in the 1850s was brought in to London by rail from a distance of up to 45 miles, but this did not amount to any more than 5 per cent of the total supply.[114] Retailers complained that railway milk was not as fresh as town milk, and a difference in price reflected this fact. Up to the mid-1870s it was regarded as the poor man's beverage.[115] As milk was the most perishable of all agricultural products the impossibility of long-distance transport, and the inability to store it for any length of time, meant that there could be no organised wholesale market in this commodity.

As distribution had to be rapid the distribution network needed to be simple, with one individual at most intervening between the producer and consumer. In the absence of a proper wholesale market, just an informal arrangement was made between retailers to transfer milk that was surplus to the requirements of one dairy to another where there was a shortage. In the 1850s George Barham, founder of the Express Dairy in 1881, gained his early experience in the London milk trade working as one of the 'balancers' who performed this hard manual task.[116] A proper

[113] E. R. Duldyke, 'The economic geography of the milk supply of London', MA Thesis, University of London, 1937; F. A. Barnes, 'The evolution of the salient patterns of milk production and distribution in England and Wales', *TIBG*, 15, 1958, pp. 167–95; E. H. Whetham, 'The London milk trade, 1860–1900', *EcHR*, 2nd ser., 17, 1964, pp. 369–80; G. E. Fussell, *The English Dairy Farmer, 1500–1900* (London, 1966), chapter 6; D. Taylor, 'London's milk supply, 1850–1900: a reinterpretation', *Agricultural History*, 45, 1971, pp. 33–8; P. J. Atkins, 'The milk trade of London, c. 1790–1914', PhD thesis, University of Cambridge, 1977; Atkins, 'The growth of London's railway milk trade, c. 1845–1914', *JTH*, new ser., 4, 1977–78, pp. 208–26.

[114] Jefferys, *Retail Trading in Britain,* p. 226; Atkins, 'The growth of London's railway milk trade', p. 210.

[115] [Wynter], 'The London Commissariat', p. 293; A. E. Baxter, 'Milksellers', in C. Booth (ed.), *Life and Labour of the People in London*, Vol. VII, (London, 1896), p. 173.

[116] P. J. Atkins, 'Sir George Barham, 1836–1913', in D. J. Jeremy, et. al. (eds.), *Dictionary of Business Biography*, Vol. I, *A–C*, (London, 1984), pp. 157–8.

wholesale market only emerged after 1860, as the volume of milk brought into London by the railways increased, and it ceased to be merely a supplement to urban production. The outbreak of cattle plague in 1865 caused problems for the milk supply of all large centres, but they were particularly serious in the case of London. The disease caused heavy mortality in the metropolitan cowsheds and forced milksellers to use more milk brought in by rail.

The reliance upon railway milk allowed some cowkeepers to abandon production altogether and concentrate all their efforts on wholesaling. By the late 1860s London had several large wholesalers each of whom received milk from several farms.[117] Initially, trading took place at the railway stations, although lack of platform space for those using the London and North Western Railway forced wholesalers and retailers to move into the surrounding streets, until this was stopped by the police. About ten dealers received milk via the Great Western Railway and they were sufficiently organised to reach agreement over selling prices.[118]

As business expanded from the 1870s, wholesaling firms obtained their own premises to cool and store and bulk the milk. In addition to those in London, others were established at country railway stations to receive milk delivered by farmers. Some were organised by country firms, in touch with the London trade, while others were direct branches of London firms.[119] By 1900 milk could pass through as many as five hands after leaving the producer. There were rural milk agents who collected milk from all the farms in an area which was then delivered in bulk to the railway companies. On arrival in London it was collected from the station by the wholesaler who then supplied the retailer. But it became common for large retail dairymen who bought their milk at the standard wholesale price to resell it to those who sold just a few pints a day. This allowed numerous outlets such as itinerant vendors, chandlers, confectioners, fruiterers and general dealers to be supplied while freeing large wholesalers from the trouble of delivering to thousands of back-street shops.[120] The responsibility of the middleman was to balance supply to meet demand, and he could only do this efficiently by issuing farmers with long-term contracts to supply a regular amount of milk at a fixed price.

By the 1890s there were around 20 wholesalers in London, supplying

[117] J. C. Morton, 'On London milk', *Journ. of the Society of Arts*, 14, 1865, p. 66; *idem.*, 'Town milk', *JRASE*, 2nd ser., 4, 1868, pp. 96–8; *Select Committee on Adulteration of Food Act (1872), Minutes of Evidence*, BPP, 1874, VI, Cd. 262, QQ. 2601–8.

[118] G. E. Fussell, 'Early days of railway milk transport', *Dairy Engineering*, 83, 1956, pp. 208–9.

[119] Whetham, 'The London milk trade', pp. 374–5.

[120] F. L. Dodd, *The Problem of the Milk Supply* (London, 1904), p. 13; Robertson (ed.), *Encyclopedia of Retail Trading*, vol. VII, p. 274; Atkins, 'The milk trade of London', p. 233.

about 2,400 retailers. A large London dairy company would buy milk from perhaps 400 farms scattered over the whole country. Without binding contracts, and sometimes even with them, it was not unknown for a firm that was over-supplied to telegraph some forty farmers and cancel deliveries for the next few days. Over the years the growth of liquid milk production outpaced the expansion of demand in London. This made it a buyers' market and the contract price declined accordingly. The wholesalers became the controlling force in the expansion of the liquid milk trade. They increased their influence over supplies by insisting on farmers using refrigeration and providing clean milk, both of which extended its keeping time.[121]

In none of the other large urban centres did the proportion of the milk trade in the hands of wholesalers become as important as in London. Their smaller populations meant that milk distribution did not require the same degree of organisation. None of the provincial towns saw the emergence of a class of dealers who did an exclusively wholesale trade in milk. In Manchester and Birmingham there were wholesalers, but they also carried on a retail trade.[122] Birmingham, Liverpool and Manchester all lay in the milk-producing counties, and although Newcastle did not, none of them needed to have their supplies transported over such long distances as London. Birmingham, even by the 1920s, did not draw its railway milk supplies from further than 50 miles, while in Leeds, Bradford and Sheffield the larger dairymen bought 'distance' milk, but still dealt with local producers in order to ensure a proportion of their supplies for early delivery. Also the West Riding towns still had some of their milk supplied from urban byres still in existence, although not on such a scale as Liverpool.[123]

Liverpool presented the greatest contrast to London because of its heavy reliance on town-produced and local milk. In 1910, 50 per cent of Liverpool's milk came from town-fed cows, and a further 30 per cent from Cheshire.[124] In London the supply from urban cowsheds had declined to less than 3 per cent and the railways were supplying over 96 per cent. About a half of London's milk supply in 1914 came from more than 50 miles distant while its entire milkshed extended as far as 200 miles.[125] There were two reasons why Liverpool relied so heavily on the urban cowshed. Its riverine site limited land to two sides of the town,

[121] Atkins, 'The growth of London's railway milk trade', pp. 220–2; Lord Vernon, 'Dairy farming', *The Nineteenth Century*, Feb. 1896, pp. 275–6.   [122] Pratt, *The Transition in Agriculture*, p. 21.

[123] R. B. Forrester, *The Fluid Milk Market in England and Wales*, Ministry of Agriculture and Fisheries, Economic Series, No. 16, 1927, pp. 14–15.

[124] *Annual Report on the Health of the City of Liverpool, 1910* (Liverpool, 1911), p. 135.

[125] Atkins, 'The growth of London's railway milk trade', pp. 210, 224; Atkins, 'The milk trade of London', p. 203; G. Newman, *Report on the Milk Supply of Finsbury*, 1903, p. 6.

making it too valuable for dairy farming, so even in the early nineteenth century the city was unable to obtain as much milk from its immediate hinterland as Manchester. Liverpool's early reliance on urban dairies persisted into the twentieth century because the Medical Officer of Health conducted an energetic campaign to clean up the urban cowsheds and enforce strict standards of hygiene on their owners.[126]

In London the disease and dirt associated with so many of its cowsheds in the 1860s meant their demise was not greatly lamented. But as close control of these dangers was less easy in country cowsheds, initially the growth of the railway milk trade simply removed them from close public scrutiny. The anxiety became more acute after 1882 when there was shown to be a relationship between the human tuberculosis bacillus and that of the dairy cow. In 1895 the Royal Commission for Public Health expressed the opinion that milk containing tuberculosis bacilli contributed to the disease in humans. To remedy this, by the twentieth century some cities had obtained local Acts of Parliament to exclude milk from country cowsheds where bovine tuberculosis was known to exist. But with milk passing through so many hands contamination was possible at any stage. Concern over the cleanliness of the urban milk supply usually meant more frequent inspections, most of these being either at the producer's or at the retailer's premises.[127] At the retail level most milk in 1914 was still sold 'loose', and poured into the customer's own container, although one or two of the larger retailers, like the Newcastle Co-operative Society, were trying to popularise bottled pasteurised milk.[128] The disadvantage, as one company discovered, was that the cost of supplying, delivering, and cleaning the bottles made the trade unprofitable.[129]

CONCLUSION

Although the growth in demand and the volume of produce passing through the market network imposed changes in the existing methods of business and the conduct of institutions, traditional elements existing in 1850 were by no means entirely swept away by 1914. Transport develop-

[126] J. Holt, *General View of the Agriculture of the County of Lancaster* (London, 1795), pp. 149–54; *Annual Reports of the Health of the City of Liverpool* (Liverpool, 1900–1919).

[127] G. Dodd, *The Problem of the Milk Supply* (London, 1904); *Journal of the Newcastle Farmers' Club*, 1909, p. 88; Hilary Darragh, 'Consumer protection and the development of public law in the nineteenth century: the regulation of food quality', MLitt thesis, University of Oxford, 1982, Appendix 8.   [128] Pratt, *The Transition in Agriculture*, pp. 15–16.

[129] *Third Interim Report of the Committee on the Production and Distribution of Milk*, BPP, 1919, xxv, Cmd. 315, p. 37.

ments brought about an increase in the variety of goods and an improvement in the range of choice available to the individual retail customer. The increased variety and quantity of produce meant that a larger number of wholesalers and other middlemen intervened between the producer and the consumer, yet in certain cases British farmers could still act as their own marketing agents; although even by 1850 the majority had decided not to do so, those who still did this in 1914 were either very large producers like big livestock farmers who might organise their own annual sales, or some very small producers on the urban outskirts. One benefit which the larger and more sophisticated organisation of dealers and middlemen conferred on the consumer, largely through the flow of better information, was a greater stability of retail prices. Such things as trade newspapers, the telegraph system, and latterly the telephone, improved the flow of market news and information, and in doing so they complemented the improved flow of goods brought about by railways and steamships. Although these benefits were available to all production areas, the sheer increase in quantities of imported goods that they facilitated did mean that certain markets, like some of the inland corn exchanges, were being increasingly bypassed before 1914.

The general rule was that the large town or conurbation required a more sophisticated supply and marketing system than the small towns. This was not just the simple mechanics of a huge population needing to draw on a wider hinterland, as there was also pressure on market space within the boundaries of large cities. This caused problems when some markets, as in London, Liverpool, Birmingham and Manchester, acted as collecting centres which received produce from all over the country, and from where it was redistributed to other regions. In part this could be accommodated by relocating some wholesale markets away from city centres and towards the suburbs, and establishing any new markets in the suburbs. Another solution was for the conurbations to place a greater reliance on imported produce than small towns and rural districts. Imported meat, grain, fruit and vegetables were landed in bulk and the initial decisions about splitting consignments and further distribution were made at the dockside by the importing firms. They could employ their knowledge of what each market could absorb and only consign a part of each cargo to the nearest group of large central markets, while the remainder was diverted to other outlets in the smaller provincial towns. Although this by no means eliminated the need to have collecting markets in the great cities, it did at least relieve some of the congestion involved in the system. Greater reliance on imports did give rise to some complaints that their sources of origin were kept from the consumer and they might be passed off as home-produced. This was generally of greater concern to the British farmer than to the final consumer, or, for that matter, the local authority who might regulate the market where the product was sold. Consumer protection, even by 1914, was still comparatively underdeveloped and was mainly confined to ensuring purity in processing, and full weight of goods at the point of sale, rather than to ascertaining a product's exact origin.

CHAPTER 16

AGRICULTURAL SERVICING TRADES AND INDUSTRIES

INTRODUCTION

The industries covered in this section provide something of an indication of just how far, and to what extent, farming in Britain entered into the industrial age between 1850 and 1914. At a time of complaints of economic hardship, depression and declining incomes, the outputs of the agricultural engineering, fertiliser and chemical, veterinary pharmaceuticals, and feedingstuffs industries all presented opportunities for farmers to take a positive action to improve profits by widening their range of purchased inputs. The agricultural engineering industry offered a broad group of products which allowed farmers to substitute capital for increasingly expensive labour, whereas chemicals and fertilisers, veterinary goods, and factory-made feedingstuffs all contributed towards improving the productivity of land, livestock, or both. The extent to which farmers availed themselves of the benefits from these industries depended on their output mixes, and their own willingness to undertake extra expenditure in times of depression. In their different ways, technical refinement of the products of these industries offered a challenge to the forces of agricultural conservatism. In some cases, like mechanical harvesters, chemical fertilisers and factory-made feedingstuffs, some entirely new products appeared. But whether or not they were adopted depended almost entirely on the decision of individual farmers; as they represented current outlays rather than investment in fixed capital, landlords had hardly any say in these decisions.

A

THE AGRICULTURAL ENGINEERING INDUSTRY

BY DAVID GRACE

By 1850 Britain had a clearly discernible and autonomous agricultural engineering industry, meeting the needs of the country's farmers. In the course of the next sixty years, this industry was forced to adapt to the changing fortunes of agriculture by restructuring its product mix, management and commercial structure. The survivors from the first phase of development had already seen major changes to cope with the post-1815 depression, among which the most noticeable was diversification into non-agricultural work, especially railway building.[1] The onset of farming's 'golden age', however, had restored the balance to a large extent by mid-century and the years up to the early 1870s saw major progress in agricultural technology and the establishment of a definable leading sector of large firms. Buoyant demand also enabled the continued survival of many small, local businesses.

1. The expansion of the home market, 1850–75

A great deal of the ethos of 'high farming' was based on a naive trust in technology which did not always fulfil its promise but home demand was great enough to encourage the larger firms to concentrate more fully on agricultural work. Ransomes of Ipswich, for instance, transferred all railway work to a separate company, while in 1869 Ransomes and Rapier, in order to expand agricultural capacity, and other firms, which had sustained themselves in lean years with general engineering, similarly returned to greater emphasis on farm implements and machinery.[2]

There were other factors governing such moves, particularly the unpredictability of railway work, culminating in a major slump in 1866, and the replacement of iron by steel for parts, which seems to have favoured the more purely 'industrial' manufacturers.[3]

Nevertheless, greater confidence among farmers seems to have been the major factor in encouraging specialisation. Ploughs and field imple-

---

[1] D. R. Grace, 'The agricultural engineering industry', in G. E. Mingay (ed.), *AHEW*, VI, 1750–1850, (Cambridge, 1989), pp. 520–44.

[2] D. R. Grace and D. C. Phillips, *Ransomes of Ipswich*, (Reading, 1975), pp. 4–5; R. A. Whitehead, *Garretts of Leiston*, (London, 1964), pp. 58–74; L. T. C. Rolt, *Waterloo Ironworks*, (Newton Abbott, 1969), p. 65.   [3] Grace and Phillips, *Ransomes of Ipswich*, p. 5.

ments, together with simple barn machinery, were the bread and butter of the industry, but growing enthusiasm for steam power boosted the fortunes of many firms and brought into being a second phase of new businesses. In the long run, threshing proved the most successful process to receive the application of steam, but other areas, such as ploughing, enjoyed a considerable level of enthusiasm. The manufacture of steam engines and attendant farm machinery required a more sophisticated approach to production, which launched old firms on new paths and led to the emergence of new, specialised firms. Richard Hornsby, founded in 1815, achieved national importance with steam threshing machines, Marshalls of Gainsborough commenced similar work in 1848 and, by 1885, were employing nearly 2,000 workers. Clayton and Shuttleworth began in 1842 and quickly became internationally known for steam engines with an annual output of 566 in 1861. Robey and Co. of the same city was established in 1852 and William Foster gave up milling to manufacture agricultural engines and threshers. These developments reflected a notable upsurge in interest, initially from local farmers.[4]

Similar moves can be traced by Ransomes, who expanded factory space considerably in the 1860s and by Garretts of Leiston, whose workforce increased from 60 in 1837 to 600 in 1862, largely to cope with demand for steam engines and threshing machines after 1849.[5] The connection between steam engines, both portable and traction, and agricultural engineering is remarkably close. A survey of three East Anglian counties – the pre-eminently agricultural region of England – identified thirty-four separate significant manufacturers of steam engines active within the period 1850–1914, almost all with direct links with agriculture.[6] While the steam threshing machine was, perhaps, the great success story of British agricultural engineering, ploughing by steam, despite its high profile, was less readily accepted by farmers, largely because of economic considerations. Technically ingenious, it was most appropriate to large fields in heavy-soil regions – a combination not common in Britain. The Royal Agricultural Society recognised the limited appeal as early as 1867, but a number of firms led by Fowlers of Leeds committed themselves to manufacture. Fortunately, international demand compensated for lack of interest at home.[7]

[4] VCH, Lincolnshire, II, 1906, pp. 394–5; J. A. Clarke, 'On the farming of Lincolnshire', JRASE, 12, 1851, p. 412; R. H. Clarke, Steam Engine Builders of Lincolnshire (Norwich, 1955); J. Timbs, The Industry, Science and Arts of the Age (London, 1862), p. 115; The Ironmonger, 34, 25 July 1885, pp. 136–46.

[5] University of Reading, Rural History Centre, formerly Institute of Agricultural History (hereafter RHC), Trade Records (hereafter TR), Ransomes Collection (hereafter RAN) SP1/1 F11.

[6] R. A. Clarke, Steam Engine Builders of Suffolk, Essex and Cambridgeshire, (Norwich, 1950).

[7] 'Reports of the Committees appointed to investigate the present state of steam cultivation', JRASE, 2nd ser., 3, 1867, pp. 97–427; J. Haining and C. Tyler, Ploughing by Steam (Hemel Hempstead, 1970); C. Spence, God Speed the Plough (Illinois, 1959); H. Bonnett, Saga of the Steam Plough (London, 1965).

The general trend, therefore, in the third quarter of the century, was towards specialisation, against a background of solid domestic demand. General engineering contracts became less important and, within the area of agricultural work, many firms concentrated on particular lines. Ransomes not only shed railway work, but transferred the whole of the firm's manufacture of food-preparing machinery to Hunts of Earls Colne in order to concentrate on field implements and threshing equipment.[8] Smaller firms were able to gain a wider reputation by becoming identified with a single product, such as Smyths of Peasenhall for seed drills and Cooch of Northampton for winnowing machines.[9] By restricting product range, these smaller firms were able to modify their designs regularly to keep up with technical changes in farming. Smyths increasingly produced drills to deliver the new chemical fertilisers with the seed, and the records of Reeves in Wiltshire show similar adaptations.[10]

One result of product identification, among firms of all sizes, was the breakdown of local markets for implements and machines. The Royal Agricultural Society's 'prize essays', and similar accounts of improved agriculture, show a growing awareness of 'foreign' suppliers in their counties as well as a greater complexity in the range and function of the farmer's equipment. A contemporaneous account of Dorset and Somerset emphasises the importance of specific areas for the manufacture and supply of ploughs, harrows and cultivators.[11] Local preferences, however, were still persistent, and farmers' unwillingness to adopt a standard solution remained a problem for agricultural engineers for the rest of the century. Ransomes made great efforts to encourage the use of standardised plough parts, and achieved some success, but there remained a 'bewildering variety of ploughs from different firms' as late as 1919.[12] The situation ensured the continuance of many small businesses and local craftsmen beyond what might have been expected and may have held back the introduction of the most efficient production methods. In contrast, where a firm specialised intensively, major changes could be realised. As early as 1859, Burgess and Key's factory at Brentwood was exclusively manufacturing reaping

[8] RHC TR RAN SP1/1 F6.

[9] RHC TR COO, Cooch and Co. Accounts; East Suffolk Record Office, H C 23, J. Smith and Sons, Ltd; G. E. Fussell, *The Farmer's Tools* (London, 1952), pp. 111–12, 163–5.

[10] East Suffolk RO, H C 23, J. Smith and Sons, Ltd; RHC TR REE, R. and J. Reeves, List of drills sent out, 1848–1936, MP1/1.

[11] W. Bearn, 'On the farming of Northants', *JRASE*, 13, 1852, p. 101; W. Bennett, 'On the farming of Bedfordshire', *JRASE*, 18, 1857, pp. 21–4; H. Tanner, 'The agriculture of Shropshire', *JRASE*, 19, 1858, pp. 29–61; C. S. Read, 'Recent improvements in Norfolk farming', *JRASE*, 19, 1858, pp. 280–3; W. W. Fyfe, 'Note on the application of farm implements in Dorset and Somerset', *J. Bath and West*, 10, 1862, pp. 321–61.

[12] W. Dickinson, 'On the farming of Cumberland', *JRASE*, Vol. 13, 1852, pp. 207–300; *Report of the Departmental Committee on Agricultural Machinery*, BPP, 1919, VIII, Cmd. 506, p. 783.

machines, using division of labour 'based on the American model'.[13] Demand, up to the 1870s, appears to have been sufficiently buoyant to support many degrees of efficiency. However, there is some evidence that the proliferation of small-scale businesses was being viewed as a weakness in the agricultural engineering sector, with competitors striving for unnecessary 'novelty' in their products before depression struck in the final quarter of the century.[14] The continuation of a large number of firms, of all sizes, may have been helped by the increasing publicity given to implements and machinery by the agricultural society publications and shows which were such a prominent feature of mid-Victorian England. Equally, farmers found themselves faced with a tightening of the labour market, as the agricultural workforce increased only slowly between 1831 and 1851, and then fell by about 20 per cent between 1851 and 1881.[15] Interest in 'novelty' was combined with a genuine need to save labour, particularly in the arable sector, in the years before serious levels of grain imports began. The 'Royal' shows, especially, increased their emphasis on mechanical aspects of farming and publicised not just the firms of national repute, but many small local businesses around the country. An element of showmanship may have crept into the displays at times but farmers were, at least, made aware of what was available to them.[16] In the long run, this sort of publicity was probably of greatest benefit to the large firms, which were able to establish themselves as household names by regularly exhibiting. Between 1850 and 1880, the greatest interest seems to have been in advanced threshing machinery and harvesting equipment (although in the latter case, it was American technology which dominated), indicating that labour saving was the prime objective.

The achievement, by 1870, of the agricultural engineering industry was the ability to supply the farmer with the technology to carry out all the essential processes of food production in an efficient way. Firms, like Ransomes, were making ploughs and fittings suitable for use on a wide variety of soil types and topography and iron had replaced wood on all but the most remote claylands. Root and chaff cutters and other types of barn machinery were available, powered by hand, horse and steam. Iron was being used for harrows, cultivators and land rollers, and seed drills existed in a bewildering variety of forms. This was in addition to the comparatively new technology of harvesting and steam threshing.[17] Not

---

[13] *Bell's Weekly Messenger*, 7 Feb. 1859, in RHC TR RAN ET3/188 Press cuttings; *Farmer's Magazine*, 61, July 1860, pp. 1–2.   [14] *The Engineer*, 26, 30 Oct. 1868, pp. 329–30.

[15] E. J. T. Collins, 'The age of machinery', in G. E. Mingay (ed.), *The Victorian Countryside*, vol. 1 (London, 1981), p. 201.

[16] K. Hudson, *Patriotism With Profit* (London, 1972); pp. 80–87; Grace, 'The agricultural engineering industry', p. 542.

[17] E. J. T. Collins, 'The age of machinery', p. 205; Grace and Phillips, *Ransomes of Ipswich*, pp. 4–5.

all farmers were able, or willing, to make use of the full range of equipment, as costs were comparatively high for some machinery, and there were still some social constraints on the substitution of machinery for labour. However, there is little doubt that the farmer of 1870 was more technically aware than his predecessor of, say, 1830. Against this background, agricultural engineering as a discrete industry flourished. By the 1860s most of the big firms were heavily involved with steam. Indeed, they dominated the production of steam vehicles nationally.[18] In newly expanded factories, and using some of the latest machine tools, they were active in experimentation, both with products and with manufacturing techniques.

2. The home market in decline, 1875–99

After 1875, the industry seems to have lost momentum, although, as will be seen, many businesses continued to grow. The main reason for this seems to have been problems in the home market. Demand for tillage equipment declined as the arable sector contracted, and the frenzy of activity of the previous twenty-five years had probably left the domestic market saturated. Consequently, there was little room for new business and slow demand even for replacement machinery and parts. Many small businesses ceased manufacturing and turned to either maintenance work or non-agricultural production. Some became mere agencies for the larger concerns. In small towns demand was often sufficient for domestic and municipal ironwork to sustain a local foundry, and some reverted to the survival tactics which had been used by an earlier generation between 1815 and 1835.[19]

Among the larger firms, there was a similar trend towards reduced dependence on agriculture. Those producing steam engines looked for wider markets in industry, road haulage, and pumping and drainage. Ransomes produced many road engines and much fixed industrial equipment as well as experimenting with oil engines. Fowlers, always vulnerable to agricultural depression, diversified by expanding early railway locomotive manufacture and by making engines for electrical generation and mining. Garretts's home production of steam engines (a small proportion of output) took on a non-agricultural flavour in this period, and Taskers of Andover, who in the 1860s had consciously built up their own brands of farm machinery, moved noticeably into produc-

---

[18] See W. J. Hughes, *A Century of Traction Engines* (Newton Abbott, 1968).

[19] D. R. Grace, 'The agricultural engineering industry', pp. 528–32. For examples see: D. R. Grace, 'Notes on Alexanders of Cirencester', typescript, Corinium Museum, Cirencester, 1980; Berkshire RO, D/EWr, Wilder Collection; V. F. M. Garlick, *The Newbury Scrapbook* (Newbury, 1970), pp. 176–78.

tion of traction engines for road haulage and road rollers. In 1885, it was reported that Marshalls of Gainsborough had been doing 'a very large business' in engines for electric lighting for towns, railway stations, picture galleries and industrial establishments.[20] The trade literature of the time aimed at the home market has a distinct emphasis on non-agricultural products.

The nature of the 'great depression', when arable farming bore the brunt, meant that agricultural engineering was particularly hit by falling demand at home. There is evidence that firms producing a wide range of products found themselves sustained by greater concentration on servicing the livestock sector. Listers of Dursley, indeed, saw considerable growth in the 1890s, based on dairy equipment.[21] Elsewhere, more subtle changes in emphasis prevented disaster. Samuelsons at Banbury, one of the few specialist British firms making reaping machines, faced major difficulties in the 1880s, and may well have relied to a large extent on producing turnip cutters to survive in the worst years. Garretts had slimmed down its range of implements and machines to steam engines and threshing equipment, and Ransomes relied heavily on a reputation in plough production.[22] It was now, however, not as easy as it had been in an earlier depression for firms to diversify within the agricultural sector. Specialists already existed to provide for livestock farmers and Ransomes may well have regretted selling off the food-preparing machinery business.[23] Similarly, it was more difficult to move back into general engineering where there was significant competition and where there had been major technological change in the preceding years. Steam technology, however, did offer wider markets. The self-moving (traction) engine had much broader application and could be used for road haulage, and most manufacturers preferred these to the less adaptable fixed and portable engines.[24] Steam road transport provided an alternative outlet for products, but it suffered from technical problems and legislative restrictions and, eventually, replacement by the petrol engine. Nevertheless, traction engine manufacture appeared in the activities of many firms. Two erstwhile agricultural engineers, Savage of Kings Lynn and Burrell of Thetford, moved heavily into the specialised area of road engines for fair-

[20] RHC TR RAN AD1/1; M. R. Lane, *The Story of the Steam Plough Works* (London, 1980), pp. 176–78; Whitehead, *Garretts of Leiston*, pp. 79–133; Rolt, *Waterloo Ironworks*, pp. 127–49; *Implement and Machinery Review*, 11, 1 May 1885, p. 7167.

[21] D. E. Evans, *Listers: The First Hundred Years* (Gloucester, 1979), pp. 43–54.

[22] A. Potts, 'Ernest Samuelson and the Britannia Works', *Cake and Cockhorse*, 4, 12, 1972, pp. 187–93; *VCH, Oxfordshire*, II, 1907, pp. 268–70; Whitehead, *Garretts of Leiston*, pp. 79–133; Grace and Phillips, *Ransomes of Ipswich*, p. 6.

[23] RHC TR HUN, R. H. Hunt and Co., Summary Books, 219–22.

[24] Hughes, *A Century of Traction Engines*, pp. 149–74; *Implement and Machinery Review*, 6, 3 Jan. 1881, pp. 3331–42.

ground showmen in this period.[25] Perhaps the most dramatically successful diversification was achieved by Ransomes in the manufacture of lawn
mowers for horticultural and domestic use, enabling the expansion of the
factory, against economic trends, in 1876.[26]

3. Reliance on exports, 1875–1914

A much more successful overall response to problems at home was to
build up an export trade. This was not new in the 1870s but the level of
exporting and its significance within the major firms (and some minor
ones) is remarkable in the period up to 1914.[27] Where figures are available, it is clear that the export trade dominated production in most areas
of farm machinery. Some firms appear almost to have written off the
home market. Between 1875 and 1894, up to 90 per cent of Garretts's
production of engines went abroad and the engine register of Burrells
shows a distinct upturn in exporting after 1875. Analysis of Ransomes's
output reveals export-led statistics across most of the firm's range of products in the final years of the century.[28] Around the same period, we find
some previously purely domestic producers seeking export outlets for the
first time. Bentalls of Maldon, founded at the end of the eighteenth
century, attempted to break into exporting only after 1870, and
Blackstones considered exporting seriously in the 1880s.[29] The nature of
the depression helps to explain this emergence of exporting. Falling
home demand was a reflection of rising foreign competition and, clearly,
the market for implements and machinery lay with the new competitors.
Indeed, there was some criticism of British agricultural engineers for
encouraging further competition by their activities. Looking back, in
1906, J. H. McClaren told the Farmers' Club that in the previous twenty
years it would have been impossible for agricultural engineering to have
survived without foreign trade. His listeners may have had mixed feelings
about this.[30]

It would be oversimplistic to see exporting as a straightforward solution. The growth of foreign trade from 1870 to 1914 was complex, and
required some major readjustments to cope with constantly changing

[25] R. H. Clarke, *Savages Ltd., Engineers: A Short History* (Norwich, 1964); D. Braithwaite, *Savages of Kings Lynn* (Cambridge, 1975); M. R. Lane, *Burrell Showmans Road Locomotives* (Hemel Hempstead, 1971); Hughes, *A Century of Traction Engines*.

[26] C. and M. Weaver, *Ransomes, 1789–1989: A Bicentennial Celebration* (Ipswich, 1989), p. 111; RHC TR RAN AD2/1 463.   [27] Grace, 'The agricultural engineering industry', p. 537.

[28] Whitehead, *Garretts of Leiston*, p. 120; R. H. Clarke, *Chronicle of a Country Works* (Thetford, 1952), Appendix A; RHC TR RAN AD2/1.

[29] P. K. Kemp, *The Bentall Story*, (London, 1955), pp. 19–20.

[30] J. H. McClaren, 'Farm implements in relation to labour', *Implement and Machinery Review*, 32, 1 Dec.1906, pp. 911–17.

circumstances. Initially, the prospects for establishing foreign markets were promising because, of the newly emerging 'world granaries', only the USA and Canada had their own developed agricultural engineering industries. Russia's wheat production rose by 60 per cent between 1870 and 1913, almost entirely as a result of increasing arable acreage, and this created exceptional demand for tillage, harvesting and threshing equipment, which is reflected in a distinct rise in British exports after 1875.[31] Throughout western Europe, as industrialisation placed additional pressures on domestic agriculture, there was a growing demand for machinery. In France the number of holdings using threshing machines more than doubled between 1862 and 1882, and in Germany the rate of progress was much faster.[32] The Romanian census of machinery reveals that, in 1865, only 752 units of agricultural equipment were in use. By 1875, this had risen to 2,895, nearly half of which were threshing machines bought from British firms, such as Ruston and Proctor, William Foster, and Aveling and Porter.[33]

The activity of the European market was clear to see, but the important point for British engineers was how much they could capture of the potential trade. Both contemporary observers and later historians have done much to establish the view that British industry in general was losing its grip on world markets and lacking in competitive edge. Agricultural engineering had its own critics who emphasised slower technical progress and capitulation in the face of foreign competition. Much of this, however, arose from the record in one particular field – harvesting machinery – which was firmly in the hands of American manufacturers. British makers had lagged behind and their share of world trade was very small. Indeed, by 1913, only 30 per cent of British home demand for self binding reapers was being supplied by domestic production.[34] This does not, of course, imply an overall failure in the depression years as British engineers had never been significant in this area. American levels of production had constantly been higher from the mid-1850s, when harvesting machinery first became popular. Output of Wood's American reapers, for example, in Britain, rose from 500 p.a. in 1853, to almost 21,000 in 1873, with an earlier peak of 23,000 in 1869. By comparison, Samuelsons produced no more than 18,000 per annum for both home and overseas markets, and Burgess and Key, another British firm, made only 3,000

[31] V. Timoschenko, *Agricultural Russia and the Wheat Problem* (Stanford, 1932), pp. 138, 211–26.

[32] F. Dovring, 'The transformation of European agriculture', *Cambridge Economic History of Europe*, Vol. VI, (Cambridge, 1965), p. 644.

[33] E. Mewes, *The Influence of the Relations of Production upon the Level of Techniques in Romanian Agriculture During the Nineteenth Century*, 1971, pp. 11, 19 (privately published by the author; copy in the Rural History Centre, University of Reading).

[34] *Report of Engineering Trades (New Industries) Committee*, BPP, 1918, Cmd. 9226, VIII, pp. 642–4.

machines in total in 1862.[35] By the late 1880s, American domination was firmly established and, even where concentrated effort was made to improve production techniques, British firms had insufficient base upon which to build a significant export market. Harrison McGregor of Leigh (Lancs.) made attempts to equip itself for mass production of 'Albion' reapers in the 1880s but, by 1906, annual production was only 6,409 machines, entirely for the home market. In fact, by 1906, only two British firms were producing self-binding reapers.[36] American makers, by contrast, had used an expanding home market to introduce mass production and economies of scale to achieve price advantage over any British products and this gradually eclipsed any opposition. Americans were admitting to 'dumping' excess production on foreign markets in the 1890s.[37] Thus, even in the domestic market, the last quarter of the century saw the disappearance of many home-produced models. A survey of East Lothian in 1872 noted 656 reaping machines in use, mostly on the British 'Bell' principle and supplied by 33 different firms.[38]

By 1900, the situation had changed dramatically. Contemporaries noted a major contrast between British and American firms in both organisation and output. The most obvious difference was that, while American firms produced all year round, with a concentration on a narrow range of items, their British counterparts saw harvesting machinery as a seasonal product which had to be supplemented by other work in slack times. Harrison McGregor concentrated on barn machinery and, in Banbury, it was said that it was possible to work out the time of year by observing activity at the Samuelson factory. A visitor to Bamlett's works in Thirsk, in 1880, noted that, while capacity existed to turn out thousands of machines annually and to employ many more men, the plant and machine tools were underused. Similar frantic, but short-term rushes for raking machines earned the name 'klondyke' for a section of Bamford's foundry at Uttoxeter.[39]

[35] Oxfordshire RO, Stockton and Fortescue Collection, Samuelson Papers, Box 14; *Walter A. Wood, Trade Catalogue*, 1874; *Bell's Weekly Messenger*, Supplement, 24 May 1869 in RHC TR RAN ET3/18 Press cuttings; B. S. Trinder, 'An Early Description of the Britannia Works', *Cake and Cockhorse*, 4, 4, 1969, pp. 60–1.

[36] *Implement and Machinery Review*, 11, 1 Aug. 1885, pp. 7423–4; *Engineering and Machinery Chronicle* (supplement to *Agricultural Chronicle*, 6), 15 Nov. 1906, p. 8; *Tariff Commission*, vol. 4, 1904, evidence of M. W. Harrison, para. 640.

[37] *Bell's Weekly Messenger*, 24 May 1869, in RHC TR RAN ET3/18 Press cuttings; *Implement and Machinery Review*, 25, 2 Aug. 1899, p. 24432; 17, 1892, pp. 15090–3; 18, 1 Nov. 1892, pp. 15994–5; Helen M. Kramer, 'Harvesters and high finance: the formation of the International Harvester Co.', *Business History Review*, 38, 1964, pp. 283–301; M. Denison, *Harvest Triumphant* (London, 1949), p. 105.

[38] Anon., 'Reaping machines in East Lothian', *Gardeners' Chronicle and Gazette*, 3 Feb. 1872, p. 161.

[39] *Implement and Machinery Review*, 17, 1 Feb. 1892, pp. 15090–3; 18, 11 Nov. 1892, pp. 15994–5; Bamfords Ltd, *Bamfords Ltd., 1871–1971* (Uttoxeter, 1971).

Failure to compete in harvesting technology, however, gives a false picture of the position of British firms, compounded by the dismal view of domestic farming in the 'great depression'. In other fields, the export market was exploited successfully. Official figures for all classes of agricultural machinery after 1873 show an upward trend in value at a time of falling prices. Internal records, meanwhile, seem to indicate a distinct emphasis being placed on foreign trade by the larger firms. This is easiest to trace in agricultural engines, as most firms kept detailed registers of customers. Of those produced by Burrells, between 1869 and 1913, at least 30 per cent went abroad, including all of their steam ploughing engines. Garretts sent 65 per cent of their main class of agricultural engines to foreign markets between 1858 and 1913, with periods when up to 90 per cent were exported. Both firms increased the proportion of exports throughout the period. Similarly, over 50 per cent of Savage's 'Agriculturalist' engines (used in steam ploughing) up to 1885 went to foreign buyers. Among the smaller firms, The Wantage Engineering Company was exporting 68 per cent of its engines in the 1880s and 1890s.[40]

The threshing machine appears to represent the extreme case in export domination. Government reports suggested that between 90 and 98 per cent of production was being exported in 1913. As early as 1888, James Howard, the Bedford manufacturer, was showing concern that the great switch into exports was leading to a neglect of the home market, and thus contributing to failure to recover from the depression. He emphasised the potential of British agriculture in order to 'remove the tendency to almost sole dependence which some English implement makers seem to place on export sources of demand'.[41] The Agricultural Engineers Association, using a survey of twenty-four large firms, suggested that in 1913 exports accounted for 70 per cent of overall sales of implements and machinery.[42]

A further indication of the importance of foreign trade after 1870 is a noticeable upgrading of marketing methods. Up to the 1850s, foreign sales had been achieved by haphazard means. Emigrants were encouraged to buy familiar implements, and individual foreign customers themselves sometimes took the initiative after visiting one of the agricultural shows or an established factory. Firms like Ransomes and Garretts positively encouraged potential overseas customers but did not always follow up in

[40] Clarke, *Chronicle of a Country Works*; Whitehead, *Garretts of Leiston*; D. C. Phillips, 'The Wantage Engineering Company and its predecessors', *Road Locomotive Society Journ.*, 33, 2, 1980, p. 44.

[41] *Report of Engineering Trades (New Industries) Committee*, BPP, 1918, Cmd. 9226, VIII, pp. 642–4; *Committee on Industry and Trade, Survey of Industries, Pt IV (Metal Industries)*, 1928, pp. 159–68; *Implement and Machinery Review*, 13, 1 Mar. 1888, pp. 10367–8.

[42] *Committee on Industry and Trade, Survey of Industries, Pt IV (Metal Industries)*, 1928, pp. 159–68.

the field.[43] Major exhibitions, such as 1851 and 1862 in England and 1867 in Paris, helped to gain wider recognition for major firms like Ransomes, although the direct financial gains seem to have been limited.[44] However, positive efforts were made in the 1850s and 1860s to produce trade literature in an increasing range of languages. Ransomes had a French catalogue by 1853, adding German and Dutch by 1857 and Spanish and Russian by the early 1870s. Howards had a French catalogue by 1856 and Garretts, French, German, Danish and Spanish by the early 1860s.[45] Such examples can be multiplied many times and show an active and organised interest in establishing foreign links, challenging the assumption that British firms were slow to develop marketing strategies.

Firms also despatched representatives to assist in setting up complicated machinery, giving an opportunity to canvass for further business. In 1853, Ransomes sent John Head to Warsaw to supervise the erection of two pumping engines, and he remained throughout the Crimean war, using freedom of movement granted by the Russian government to assess commercial possibilities in the area.[46] Manufacturers also looked on overseas trading trips as a type of 'grand tour' for their sons who used the experience to good effect on inheriting the business. Others set younger sons the specific task of cultivating the foreign market.

In the 1850s and 1860s, there was a trend towards establishing 'branch factories' (more properly, maintenance depots) or overseas trading warehouses, with a London warehouse to link manufacturing and retailing elements. Ransomes developed this arrangement, and Clayton and Shuttleworth had a branch factory in Vienna in 1857 which evolved into a fully fledged manufacturing establishment, employing a workforce of 700. Later, this firm had similar, if less ambitious, centres at Pest, Prague, Warsaw and other central European cities. Fowlers had their Magdeburg repair depot in operation, probably, by 1870, to handle German and Austro-Hungarian orders and to service a network of repair shops. Ransomes, meanwhile, established a branch at Odessa in 1867 and directly controlled operations in Moscow (1870–89), Prague, Pest and Seville (1867). The German catalogue for Picksley, Sims and Co. for 1866 claims branches at Constantinople, Melbourne and Smyrna.[47] The exact status of some of these branches may, of course, have been agencies.

---

[43] W. F. Hobbs, 'Report on the exhibition and trial of implements at the Carlisle meeting', *JRASE*, 16, 1856, p. 523; G. Biagioli, 'Agrarian changes in nineteenth century Italy', *Institute of Agricultural History, Research Papers*, No. 1 (Reading, 1970), pp. 12–13; RHC TR RAN SP/4; *The Engineer*, 16, 19 July 1861, p. 217.   [44] RHC TR RAN SP1/1.

[45] RHC TR RAN P1/A, B3; Whitehead, *Garretts of Leiston*, p. 74.

[46] RHC, TR RAN SP1/26.

[47] RHC TR RAN SP/1 D40, F49–50, G100; S. B. Saul, 'The market and the development of the mechanical engineering industries in Britain, 1860–1914', *EcHR*, 2nd ser., 20, 1967, p. 119; *Implement and Machinery Review*, 16, 1 Feb. 1891, pp. 13855–6; Haining and Tyler, *Ploughing by*

Except where products were locally manufactured – and this appears to have been the case only with Clayton and Shuttleworth in Vienna and Robeys of Lincoln in Pest[48] – foreign branches tended to carry the stock dictated by the home base. This may have led to some inappropriate products being held, but it does appear that branches were able to feed back valuable information on specific needs and give regular reports on local trading prospects. In addition, Britain pioneered the commercial use of the consular service, although the full potential of this system was never realised. Reports were available only occasionally through parliamentary 'blue books', and their content was often superficial, largely because consuls were poorly paid and lacked incentive. American rivals were better served by their officials, and British agricultural engineers would have found little of direct use in government sources. They, therefore, resorted to their own investigations of foreign markets.[49] These, however, were of a distinctly *ad hoc* nature and limited in their success. Ransomes, for instance, lavishly entertained Japanese visitors in 1866 but failed to follow up on initial orders.[50]

Despite these weaknesses, British manufacturers were remarkably successful in the twenty years after the Great Exhibition and this laid the foundations for much greater achievement in the period up to 1914. After 1870 there was a noticeable change from a small number of overseas branches to a much wider network of agencies, largely because of the expense and inefficient coverage of potential markets. Ransomes wound up all foreign branches, except the one at Odessa, between 1874 and 1914, and replaced them with carefully chosen agents. These often acted for a number of different firms, although they generally avoided competition in individual lines. Stefan Vidats, himself a machine maker in Hungary, stocked Howards harrows, Crosskills clod crushers, Barrett, Exall and Andrewes hay rakes, Garretts seed drills, Hornsbys dressing machines and Samuelsons chaff cutters.[51] The great benefit of using agents was that they sent their orders in blocks which enabled manufacturers to set up machine tools and production lines for long runs of the same article. Both Garretts and Ransomes took advantage of this in engine production. In 1893, one Argentine agent sent Ransomes an order for 137 engines and 143 threshing sets, and in 1908 an order was received for 338 threshers.[52] In general, Ransomes' agents appear to have been well informed and of great value to the firm in keeping up to date with changing demands – sufficiently

Steam, p. 255; T. Davis, *John Fowler and the Business he Founded*, (privately published, 1951), p. 252; RHC TR RAN P1/B5 (Picksley Sims and Co., Trade Catalogue, 1866).

[48] *The Engineer*, 24, 18 Oct. 1867, p 345.

[49] D. C. M. Platt, *The Cinderella Service*, 1971, pp. 43, 56–7; *Implement and Machinery Review*, 5, 2 May 1879, p. 2186.   [50] RHC TR RAN SP1/1 G 47, K60.

[51] RHC TR RAN AD7/49 (export book).   [52] RHC TR RAN SP/1 H12.

important, indeed, for J. E. Ransome to produce notes on how to treat agents in the course of travelling on behalf of the company.[53] Travellers were another important source of information and agricultural engineers seem to have used them effectively, again questioning the notion that British engineers were dilatory in their business practices. Marshalls of Gainsborough and Ransomes had semi-permanent representatives in Russia, and the latter sent a prominent family member on several overseas visits. Elsewhere, Ransomes used a journalist, James Drew Gay, to send back reports on prospects in India and Ceylon.[54]

Exporting, then, was efficiently organised and relatively successful. However, a common criticism of British industry was that it emerged from the nineteenth century with a much greater and unhealthy reliance on colonial markets.[55] Examination of export figures for agricultural engineering does not support this view. The first official statistics (for 1853) show some 65 per cent of agricultural implements being exported to colonial possessions, and into the 1860s the colonies continued to take around 50 per cent of output. Thereafter, while definitions of 'machinery' and 'implements' complicate the issue, there is no noticeable swing to colonial preference. In the implements sector, colonial markets account for a steady 40 per cent between 1873 and 1913, with only minor annual fluctuations. Sales of agricultural machinery remained around 15 per cent after 1875, with an exceptional 31 per cent in 1894. A similar pattern is discernible for agricultural steam engines with, if anything, a trend away from colonial markets after 1891.[56] Trading tariffs and foreign competition did affect individual countries, but restrictions in one area seem to have been counterbalanced by expansion in another. In overall terms, it seems to have been the firms which were unable to obtain more distant foreign markets which were forced into colonial dependence when European demand fell. Johnson and Son of Leeds claimed a rise of 5–35 per cent in output going to colonial areas between 1890 and 1909, while foreign markets collapsed completely from a level of 35 per cent. Among the engineers giving evidence to the Tariff Commission, it was the smaller firms which relied most heavily on the Empire.[57]

For the major engineering companies, a much more balanced devel-

[53] R. Munting, 'Ransomes in Russia', *EcHR*, 2nd ser., 31, 1978, pp. 260–3; RHC TR RAN AD7/48, 49, 64.

[54] S. B. Saul, *Studies in British Overseas Trade, 1870–1914* (Liverpool, 1960), chapter 3; P. J. Cain and A. G. Hopkins, 'The political economy of British expansion overseas, 1750–1914', *EcHR*, 2nd ser., 33, 1980, pp. 488–9; R. C. Floud, 'Britain 1860–1914: a survey', in R. C. Floud and D. N. McCloskey, *The Economic History of Britain,* vol. II, *1860–1939*, Cambridge, 1994, p. 20.

[55] Munting, 'Ransomes in Russia', *EcHR*, p. 263: RHC TR RAN AD7/63 (correspondence. 1875–6).   [56] Trends based on *Annual Trade and Navigation Returns* (in BPP from 1854 to 1914).

[57] *Tariff Commission*, vol. 4, 1909, Engineering Industries, paras 117–1226.

opment is apparent. European markets flourished initially with notable flushes of demand following the abolition of serfdom in the east, but then began to close down under the impact of import duties, foreign competition and industrial development in the receiving countries. These were replaced by major expansion in new areas, especially South America. In the colonial sector, Australasia took less from Britain as American competitors moved in and as a native engineering industry developed towards the end of the century. However, South Africa filled the gap, together with some growth in India.[58] While this apparent stability conceals adjustments in markets, it also covers major changes in the products being exported. As a general rule, exports tended to become more sophisticated over time, beginning with the simpler implements and progressing to more elaborate machinery. Larger firms, again, were able to compensate for losses in one product by gains in another. The gains were mainly in the higher value products, accounting for growing prosperity in the export sector. In addition, agricultural development was staggered in the nineteenth century; so, for instance, the loss of the plough trade in one country might be offset by growth in another. This can be seen quite neatly in the trade figures for Australia and South Africa, where growth in the latter around 1890 compensated for a significant fall in the former, at least as far as implements were concerned.[59]

Here, South African trade filled the gap left by the drop in demand from Australia as the latter developed its own, indigenous industry and also began taking increasing amounts from US and Canadian manufacturers. A similar pattern is discernible in relation to Western and Eastern Europe, where the development of indigenous production in the former, protected by increasing tariff barriers, was counterbalanced by a rapid growth in demand in the latter, as the abolition of serfdom led large landowners to purchase machinery to replace lost labour. Clayton and Shuttleworth began exporting agricultural engines to Hungary almost immediately on abolition, and within five years had some seventy machines at work in the country. By 1863, the firm had nearly 400 engines, most of which were used for threshing and, for which, it can be safely assumed, British machines were sold as an accompaniment. The main purchasers were the great landowners of Bohemia and Silesia.[60] In Russia, Lenin, using official statistics, claimed that the initial phase of emancipation was followed by feverish importation of machinery to replace freed serfs up to the mid-1860s.[61] In Eastern Germany, it was

---

[58] Grace and Phillips, *Ransomes of Ipswich*, p. 7.

[59] Based on *Annual Trade and Navigation Returns*, in BPP.

[60] *Guide to the Royal Hungarian Agricultural Museum* (Budapest, 1914); *Report on the Agricultural Exhibition at Vienna*, BPP, 1867, Cmd. 3828, LXX, p. 711.

[61] V. I. Lenin, *Works*, vol. III (Moscow, 1960), p. 219.

rising demand for the grain harvest which lay behind significant sales of Fowler steam ploughs.

In the latter part of the nineteenth century, some other growth points were noticeable. Egypt, for instance, became a major buyer of British machinery during the years of the American civil war, when it benefited from windfall demand for its cotton and a corresponding disastrous loss of draught cattle through disease. Fowlers again (along with Howards of Bedford and Ransomes) were the main beneficiaries and, although demand dropped off after 1865, these firms were sufficiently established to continue supplying other types of machinery – e.g. for irrigation – for the rest of the century.[62] The other area of dramatic activity was South America, although here political unrest and fiercer competition from the USA made progress less certain. Nevertheless, some spectacular successes were recorded, as, in 1893, when a single customer ordered 143 threshing machines and 137 engines from Ransomes. South America was one of the areas which adopted the straw burning process pioneered by Ransomes for areas where the availability of fuel for steam engines was a problem.[63] Adaptations of traditional machinery also brought some further success in the native coffee industry, where British firms were able to compete effectively.[64] Throughout this period, however, there were criticisms that firms were not offering sufficient technical or financial aid to prevent American dominance. As became general, it was in the field of steam engines and threshers that Britain was best able to hold its own.[65]

There is little doubt that in the final quarter of the nineteenth century it was the export trade which sustained the industry. This continued to be a major feature up to 1914. In some classes of machinery, Ransomes exported up to 98 per cent of production in 1913. Overall, it was claimed, the export element in the output of major firms was between 50 and 90 per cent in the years immediately prior to the First World War and these proportions appear to have been remarkably stable from the 1880s.[66]

The onset of the twentieth century brought with it a revival of domestic agriculture which made these 'golden years' for the big manufacturers, many of whom became limited companies and invested in new plant

[62] Spence, *God Speed the Plough*, p. 151; Davis, *John Fowler and the Business he Founded*, pp. 81, 261, 265; *The Engineer*, 20, 12 Aug. 1865; p. 101; 20, 5 May 1865, p. 277; 20, 23 June 1865, p. 395; E. R. J. Owen, *Cotton and the Egyptian Economy, 1820–1914* (Oxford, 1969), pp. 89–121.

[63] Grace and Phillips, *Ransomes of Ipswich*, p. 7; Weaver and Weaver, *Ransomes, 1789–1989*, pp. 51–2.

[64] RHC TR RAN SP1/1 G121; *Consular Report on Central America*, BPP, 1913, XLVIII, Cd. 6969, pp. 469–91.

[65] *Report on the Trade of the Argentine Republic for 1893*, BPP, 1894, LXXXV, p. 33; D. C. M. Platt, *Latin America and British Trade, 1806–1914* (London, 1972), pp. 233–38; *Implement and Machinery Review*, 23, 1 Oct. 1897, pp. 22052–4; 36, 1 Dec. 1910, pp. 988–9.

[66] RHC TR RAN CO2/1, 2, 7; *Report of the Departmental Committee on Agricultural Machinery (Summaries of Evidence)*, BPP, 1919, VIII, Cmd. 506, p. 813.

and machine tools. Evidence to the Tariff Commission, which concentrated on the difficulties faced by businesses, indicates that it was only in the area of harvesting machinery that foreign competition was, as yet, significant. Elsewhere, and especially in food-preparing machinery and ploughs, home demand was met predominantly by British firms.[67]

4. Limited revival of the home market, 1900–14

Ransomes's fortunes seem to have begun to improve from the late 1890s, with an increase in home demand. Initially, this was for non-agricultural products, such as industrial engines, but, eventually, farmers appear to have featured more prominently among British customers. Sales reached record levels, thanks to exceptionally large home demand and corresponding improvement in exports. Thereafter, with only minor setbacks, overall profits increased steadily up to 1914. Between 1901 and 1910 net profits increased from £50,000 to £85,000, when the company had assets of some £850,000.[68] This success was due to the coincidence of healthy home and overseas markets. From 1900 to 1914, Ransomes's home trade improved by about 40 per cent while exports more than doubled.[69] In 1909 Clayton and Shuttleworth declared gross profits of around £84,000 – 25 per cent increase on the previous year – and average pre-War profits of other prominent firms were similarly healthy – Blackstones (£40,000); Garretts (£20,000); Aveling and Porter (£24,000) and Howards (£18,000) being good examples.[70] Analysis of production registers shows a similar pattern of recovery between the 1890s and 1914 in both large and smaller firms. Those engaged almost entirely in the home market confirm the revival very clearly.[71]

The revival in home demand appears to have been largely the result of a further tightening of the agricultural labour market which had taken place, especially in the 1880s, when the 'drift from the land' had been accelerated by declining arable acreage, long run depression and alternative employment opportunities. By 1900, it is possible that serious labour shortages affected many farmers, despite adjustments in favour of the less labour intensive livestock and dairying sectors.[72] It may also have been the case that a long period of depression had discouraged new purchases and replacement of existing equipment was now essential. Whatever the reasons, agricultural engineering seems to have benefited from increased home interest, although some developments, such as the smallholdings

[67] *Tariff Commission*, vol. 4, 1909, Engineering Industries, paras. 117–1226.
[68] RHC TR RAN AD 1/1; CO2/7.   [69] RHC TR RAN SP1/1 K4.
[70] *Implement and Machinery Review*, 36, 1 May. 1910, p. 62; RHC TR FOW 49.
[71] RHC TR RAN AD 2/1; RHC TR REE.
[72] C. S. Orwin and E. H. Whetham, *History of British Agriculture* (Newton Abbott, 1971), pp. 342–4.

and allotments movements, offered little extra business. Here only the basic implements were needed and there was a tendency to hire or share equipment. Nevertheless, ever hopeful, Kell and Co. of Gloucester were offering 'allotment' ploughs and seed drills in 1902.[73]

In the home market a distinct shift in marketing methods had also occurred. By 1900, the larger firms were sufficiently well known to have begun to question the value of agricultural shows. Trade literature, agency agreements (often with what had been small independent manufacturers in the 1870s) and the greater mobility of farmers themselves, meant that now agricultural societies were 'in every way beholden to the exhibitor for the success of the show'. The need to develop sophisticated sales techniques had probably been the result of experience in the export market. It now enabled the large firms to increase their hold over the reviving British trade.[74] There is some evidence, in addition, that British agricultural engineers were also doubting the value of international shows, concluding that the expense rarely justified the business gained and probably also indicating that they were fairly confident in their own alternative sales methods.[75]

5. The industry in 1914

By 1914, the agricultural engineering industry seemed in good fettle. Home demand was healthy and exports booming. Since 1850 a clear group of leading firms had emerged, thanks, in no small measure, to the weeding out process during the 'great depression'. Geographically, it was still concentrated in eastern England, perhaps more so than before. Determining the precise size and structure of the industry is difficult as there is no clear definition of agricultural engineering in any of the statistical surveys. William Kent's list for 1867 records 540 firms, to which can be added another 100 likely omissions, while the Post Office Directory of Engineers records 724 and 889 for 1870 and 1890 respectively.[76] How many of these firms were agricultural engineers in the strictest sense is another matter. E. J. T. Collins suggests that, in 1900,

[73] E. Thomas, *The Economics of Small Holdings* (Cambridge, 1927), pp. 42, 58–61; F. E. Green, *The Small Holding* (London, 1908), pp. 18, 21; A. W. Ashby, *Allotments and Small Holdings in Oxfordshire* (Oxford, 1917), pp. 136, 163, 170; *Agricultural Chronicle*, 2, 1 Feb. 1902, p. 23.

[74] *Agricultural Chronicle*, 2, 15 Mar. 1902, pp. 18, 27; *Machinery Chronicle* (formerly *Agricultural Chronicle*), 7, 15 Sept. 1907, pp. 303–4; 7, 15 Nov. 1907, pp. 495–6.

[75] *Minutes of Evidence Taken by the Committee Appointed by the Board of Trade to Make Enquiries with Reference to the Participation of Great Britain in International Exhibitions*, BPP, 1908, XLIX, Cd. 3773, pp. 133–47.

[76] W. Kent, *The Agricultural Implement Manufacturers Directory of England* (London, 1867); Post Office Directory, *Iron and Metal Trades*, 1870 and 1890.

some 900 firms might have claimed a connection with agricultural engineering but, of these, the vast majority were very small scale and less than
50 per cent were engaged in manufacturing.[77] It is clear that there was an
obvious set of leaders who were distinct from the majority. In 1923, estimates suggested that 90 per cent of total capital in the industry was owned
by 26 firms and in 1913 29 firms employed 27,110 workers.[78] The
number of truly large firms was comparatively small. Bamfords of
Uttoxeter had 400 employees in 1893, Marshalls of Gainsborough 4,500
in 1900, Clayton and Shuttleworth of Lincoln 2,300 in 1907 and
Ransomes of Ipswich 2,500 in 1911. Of the remainder, a workforce of
around twenty was probably more typical. Location in 1913 was still commonly in market towns where individual influence on the labour market
was significant.

What is more readily discernible is the continued heavy reliance on the
export trade, with the twenty-four largest firms exporting 70 per cent of
production.[79] Ransomes's export figures for 1913 summarise well the
condition of the industry and reveal some interesting features. While
Europe accounted for the greatest market by value, 81 per cent of this
went to Russia. Western Europe was closed down by tariff barriers. South
Africa was the next largest market, followed closely by South America.
Trade in North America was virtually non-existent. 'Agricultural
machinery', primarily steam engines and threshers, accounted for well
over 50 per cent of the export total with ploughs (always a strong element
in this particular firm) also important.[80]

Herein lay some weaknesses, which similarly affected all the major
firms. The East European market was lost as the result of war and revolution in Russia and was never regained. The reliance on the threshing
machine and steam engine – the only area in which British manufacturers retained a clear lead over other countries by 1914 (with 90 per cent
of production exported) – left firms wedded to a technology which was
to be replaced by the petrol engine and combine harvester.[81] The battle
for harvesting machinery had long been lost to the Americans, who compounded their dominance by amalgamations and massive economies of
scale. Not only could British firms not compete abroad, but (with the
exception of mowing machines) they supplied less than 50 per cent of

[77] Collins, 'The age of machinery', p. 205.
[78] *Committee on Industry and Trade, Survey of Industries, Pt IV (Metal Industries)*, 1928, pp. 159–68.
[79] *Committee on Industry and Trade, Survey of Industries, Pt IV (Metal Industries)*, 1928, pp. 159–68; Collins, 'The age of machinery', p. 204.   [80] RHC TR RAN AD7/18.
[81] *Committee on Industry and Trade, Survey of Industries, Pt IV (Metal Industries)*, 1928, p. 642; For a specific example see I. M. Rubunov, 'Russia's wheat surplus', US Department of Agriculture, *Bureau of Statistics Bulletin, No. 42* (Washington, 1906), pp. 55–62, (reprinted in S. Pollard and C. Holms, *Industrial Power and National Rivalry, 1870–1914* (London, 1972), pp. 25–30).

the home market. Of an estimated annual demand for 12,000 self binding reapers, only 30 per cent were British made in 1913. A similar position prevailed in the field of dairy equipment. With one or two exceptions, such as Listers of Dursley, this fell into the hands of Scandinavian manufacturers who took advantage of the gradual shift into dairying which took place in many countries (Britain included) which were unable to compete with the great world grain producers.[82]

Elsewhere, the other great development in farming technology, the tractor, was more advanced in the USA and, in Britain, was being produced by a separate motor industry. Ransomes had experimented briefly with early tractors but, like the rest of the agricultural engineers, remained committed to steam. Thus, when the petrol tractor emerged as a major, and eventually indispensable piece of equipment, it was other countries and other industries that benefited.[83]

This neglect was compounded during the four years of the war, when all Ransomes's spare capacity, like that of the rest of the agricultural machinery industry, was switched to government orders for war-related contracts, thus removing the opportunity to establish a viable British tractor industry. In 1918, when Britain came out of the war, lost markets, technical backwardness and organisational weakness led to a period of major upheaval. Ill-fated amalgamations and the drift of companies into what seemed more promising, and more clearly industrially based, products, left a much smaller sector to fight off increasing foreign competition at home and abroad. Only a handful of businesses, notable among which was Ransomes, remained primarily committed to agriculture.[84]

[82] *Report of Engineering Trades (New Industries) Committee*, BPP, 1918, VIII, Cmd. 9226, pp. 642–4; Evans, *Listers: The First Hundred Years*, pp. 43–82.

[83] C. L. Cawood, 'The history and development of farm tractors', *Industrial Archaeology*, 7, 1970, pp. 264–91.   [84] Grace and Phillips, *Ransomes of Ipswich*, pp. 7–9.

B

AGRICULTURAL CHEMICAL AND FERTILISER INDUSTRIES[85]

BY F. M. L. THOMPSON

1. Origins of the industry

The establishment of the fertiliser industry may be precisely dated to 23 May 1842, when John Bennet Lawes obtained a patent for 'chemically decomposing for purposes of manure by means of sulphuric acid of Bones, or Bone Ash or Apatite or Phosphorite or any other substances containing phosphoric acid'. With this patent in his pocket Lawes bought an old mill at Deptford Creek, on the south bank of the Thames, began production, and in the July 1843 issue of the *Gardeners' Chronicle* advertised his patent manure, composed of Superphosphate of Lime, for sale at the factory at 4s. 6d. per bushel, the equivalent of about £7 per ton. This was the world premiere of artificial, meaning manufactured, manures on a commercial scale.[86]

As with any other new industry, the fertiliser industry did not spring full-grown from virgin soil. On the supply side it had antecedents in agricultural chemistry and in the mechanical treatment of bones; on the demand side it depended on the familiarity of a sufficient number of farmers with the advantages of buying in manures from outside their farms, a practice that had been slowly spreading in the previous half-century or more. Many substances were involved – rape cake, rape dust, untanned leather scraps, hoof and horn meal, waste wool, shredded rags, fish refuse, dried blood and waste hair were some of them – but bones were central to both sides of the supply and demand schedules. Bones contain phosphate (and a small percentage of nitrogen), and many years earlier chemists had discovered that mixing bones with sulphuric acid produced superphosphate of lime which in turn could be used to produce a moderately pure form of phosphorous, and this formula did have a tiny experimental and industrial use before 1842.[87] In addition, bones were already being used in a fairly extensive, though localised, way in farming well before the 1840s, and were found to answer best if ground

---

[85] This article is a commentary on, and interpretation of, the statistics of chemical fertiliser production and consumption tabulated in the Appendix Tables 16.1 to 16.6.

[86] E. J. Russell, *A History of Agricultural Science in Great Britain, 1620–1954* (London, 1966), p. 94; J. C. Morton, 'Lawe's manure works', *Agricultural Gazette*, New ser., 23, 2 Jan. 1888, pp. 8–10.

[87] For example in D. B. Reid, *Elements of Chemistry* (3rd edn, London, 1838).

into a fine meal, so that appropriate grinding and sieving had been devised. This was not elaborate, and a farmer using bone dressings on a large scale might well do his own grinding; but particularly where bones were imported – and over 40,000 tons a year were being imported in the late 1830s – it was convenient and profitable for importing merchants at the chief ports of entry, of which Hull was the largest, to set up their own grinding works, which may be identified as the earliest natural manure factories.[88]

Bones and bone meal were, above all, one of the ingredients in the expansion of lightland farming on the East Riding and Lincolnshire Wolds in the quarter-century after 1815, but already by 1829 experience had shown that while bones were most effective in increasing the yield of turnips, in particular, on dry sands, limestone and chalk soils, they were utterly ineffective on clays and strong loams. It was this puzzle of the inertness of bones on certain soils which Lawes set out to solve in a series of field experiments on his small estate at Rothamsted in Hertfordshire, experiments which ran for six years from 1836 and in which he established, largely by trial and error, that either bones or mineral phosphate treated with sulphuric acid produced striking results on turnip growth, while untreated bones in whatever quantity had no effect at all. The explanation came later, that bones contain calcium phosphate which is inert and is only released by contact with acid, which makes the phosphate soluble: some soils are naturally acidic and make this conversion 'spontaneously', but many are not. Any acid will do the trick, but in practice sulphuric acid was the one most readily and cheaply available for commercial application. Working farmers were not much bothered about the scientific explanation: what counted was the visible evidence that superphosphate worked wonders, so much so that even those who farmed land that was naturally capable of making use of untreated bones often came to prefer to buy manufactured superphosphate.

The well-established habit of purchasing bones and bone meal, and the frustration of those farmers who had found bones useless, had by the early 1840s created a latent and potential demand for superphosphates. The commercial success of the Lawes enterprise was, however, also helped by the coincidental introduction of Peruvian guano into the British market in 1841. Guano was virtually the Growmore of its day, a balanced fertiliser (though not many farmers appreciated this) containing perhaps 24 per cent phosphate, 17 per cent ammonia and 4 per cent potash, and enjoyed instantaneous success.[89] Annual consumption grew

---

[88] F. M. L. Thompson, 'The second agricultural revolution, 1815–80', *EcHR*, 2nd ser., 21, 1968, pp. 68–70.

[89] Thomas Ways's analysis in 1849, quoted by W. M. Mathew, 'Peru and the British Guano market, 1840–70', *EcHR*, 2nd ser., 23, 1970, p. 113. Guano being a variable, not a homogeneous commod-

from nothing in 1840 to over 100,000 tons a year in the mid-1840s, and after a slight drop in 1847–50 climbed to a peak of 230,000 tons a year in the late 1850s. Application rates varied according to the soil type and the crop which was in view, from 2 cwt an acre for potatoes up to 5 cwt an acre for turnips; it is a reasonable guess that around one million acres were being dressed with guano every year by the late 1850s.[90] The extraordinary mushrooming of the guano trade was undoubtedly mainly due to its intrinsic merits as a fertiliser, capable of raising the yields of some crops, especially roots but also wheat, contemporaries thought, by as much as 50 per cent; but it was also due to the intensive advertising and publicity campaigns mounted by the two importing merchants, W. J. Myers of Liverpool and Antony Gibbs & Sons of London.[91] To some extent superphosphate rode on the back of guano even though the two were also rivals for the attention of farmers. For one thing, the guano craze familiarised many more farmers than before with the concept of buying fertilisers. For another, being accustomed to applying manures to the root crop the shrewder farmers discovered that turnips responded to phosphates but made little or no use of the nitrogen which guano contained, so that at £10 to £12 a ton guano seemed an expensive source of phosphate when superphosphate could be had for £7 a ton.[92]

2. Superphosphate

Contemporaries thought of guano as an 'artificial fertiliser', which indeed it was in the sense that it did not grow on the farm and was only reproducible in the very long run; but it required no manufacturing and only the basic processing of breaking bulk, sorting and bagging. The guano trade was essentially a matter of merchanting, marketing and distribution. Some was sold direct to farmers, but most was handled by local agents, usually by those who were established agricultural merchants with trades in corn milling, in seeds and in animal feedstuffs. Superphosphate, by contrast, had to be manufactured. It is true that Liebig suggested in his *Agricultural Chemistry* (1840) that superphosphate could be made on the farm by stirring up a mixture of bones and sulphuric acid and spraying the fertiliser as a liquid on to the land before ploughing: this hazardous

---

ity, other analyses of 'average quality guano' have slightly different results: e.g. Thomas Brown, fertiliser manufacturer of Lynn, in 1892 estimated 31 per cent phosphate, 14 per cent ammonia, and some potash: *Agric. Gaz.*, new ser., 35, 16 May 1892, p. 454.

[90] Application rates from J. R. McCulloch, *Dictionary of Commerce* (London, 1871 edn), p. 652.

[91] Mathew, 'Guano', p. 114. After 1848 Gibbs had a monopoly on imports of Peruvian guano, until 1861.

[92] Properly made superphosphate had a phosphate content similar to that of good guano, about 27 per cent.

and unreliable procedure found few takers. The manufacturing process was extremely simple, not to say crude; it was also extremely smelly and could be dangerous. The phosphatic material, either in mineral form or as animal bones, was broken into small pieces and placed in a brick-lined mixing pit; sulphuric acid was added and produced a violent chemical reaction with much heaving, bubbling, stench and gas, leaving when it had finished a solid cake of superphosphate of lime; this cake was then broken up with picks and shovels, pressed through a large-mesh sieve, and crushed by rollers into a usable form of fine granules or powder. All that then remained was to weigh and bag the product and, if the manufacturer aspired to a good reputation, to take a sample from each batch for a chemical analysis which would be entered on the sales invoice to give the purchaser some indication of what he was buying.

The open mixing pits were quite quickly replaced by enclosed, brick-lined, chambers, known as dens. These were charged with the materials through a top opening, and the pick-and-shovel men entered through end doors to break up the cake. The advantage of the den was that the gas generated by the reaction could be vented through a chimney, and larger batches, of 20–30 tons, could be made; but the den did not do much to improve the working conditions of the men. This, the central core of the manufacturing process, scarcely changed until after the end of the century, although the capacity of individual dens was enlarged, sometimes to as much as 200 tons. After 1900, but only gradually, a few firms installed mechanically operated dens, bearing names like the Milch den, the Wenk den, or the Svenska den, which proclaimed their origins in the technologically more advanced industries of Germany or Sweden: such dens had rotating knives, screws or discs, which broke up and scraped out the superphosphate cake and deposited the pieces on a conveyor belt, dispensing with much hard, dirty and dangerous manual labour. But if the handling of the basic chemical reaction remained simple, labour-intensive, and undemanding in terms of fixed equipment, the attendant and ancillary processes lent themselves to mechanisation and to economies of scale. In the preparation of the raw materials the crushing of the bones or mineral phosphates by steam-powered machinery was an early development, and similar gear was adapted for grinding the superphosphate and for powered sieving. Another development, under way by the 1850s, was in the mechanical handling of materials and finished product, on trolley tracks within the factory and by conveyor belts and elevators which carried the superphosphate to bagging plants. Above all, there was the acid question.

The chemical recipe for making sulphuric acid was simple, and the equipment required was neither elaborate nor expensive. To make H_2SO_4 take some source of sulphur – pyrites, a little mined in Britain, but the bulk imported, mainly from Spain, were the common commercial source – roast it, pass the fumes through steam into a lead condensing chamber, and reduce this liquid by further heating until the desired concentration is obtained, that is stopping before the acid begins to eat away the lead. The acid could be further concentrated by heating at higher temperatures in platinum stills, but at that point it became a fine chemical suited for use in small quantities, rather than a heavy chemical available for bulk use. The crude acid could be stored in glass vessels, and it could be transported in large carboys protected by basketwork and straw, but that was expensive and risky. In any process using acid continuously and in sizeable amounts there were pressing economic and technical motives for making it on the spot, or locating the process alongside an acid plant.

This Lawes grasped from the start, and his Deptford factory was an integrated chemical manure works with its own acid plant as well as mixing pits and dens for making superphosphate, and grinding apparatus. The location had the additional advantage that it was alongside sugar refineries, and Lawes had discovered that spent bone charcoal from the refineries was an excellent phosphatic raw material which the sugar refiners regarded as waste. The Thames-side site was ideal for obtaining cheap supplies of seaborne coal and imported pyrites and bones, and handy for discharging effluents; it was not particularly close to the agricultural markets for the product, but many of the main turnip- and grain-growing regions which he targeted could be reached, cheaply, by the coastal trade, and a fair part of London's market-gardening district, regarded as important and responsive potential customers, was within economic range of land transport. This siting of the first chemical manure factory set the pattern for the nascent industry which had been formed within ten years by Lawes's imitators and competitors. Access to coal, pyrites and bones pointed to river and port locations: Humber, Tyne, Clyde, Mersey and Severn joined the Thames as foci of the growing industry. The discovery of coprolites in the 1840s, at first in Cambridgeshire and later in parts of Suffolk, Essex and Hertfordshire, provided a different locational factor. Coprolites were in effect a completely mineralised and fossilised kind of guano, hard nodules of ancient dung, excavated or won by shallow mining, containing concentrated phosphate that could be used untreated, once ground to a powder, as an effective fertiliser, but was more efficient when converted into superphosphate by the acid treatment. By 1850 annual production of coprolites had reached 10,000 tons; thereafter it grew to 80,000 tons a year in the mid-1860s, then increased rapidly to the peak output of 258,150 tons in 1876 before dwindling to less than 2,000 tons a year in the 1890s; by 1900 coprolite mining had ceased.[93]

[93] J. C. Morton in *Estates Gazette*, 7, 23 July 1864, pp. 358–9.

While it lasted this home-produced supply of mineral phosphate was an obvious inducement to locate some of the fertiliser industry in the eastern counties, particularly close to ports where reasonably cheap supplies of coal could be procured. This locational influence would have been at its strongest in the early phase when new enterprises were selecting their sites: once established the locational inertia of existing fixed equipment would become stronger, and raw materials would be moved to existing plant rather than the other way round, unless transport costs were prohibitive.

Besides proximity to coprolites, eastern counties sites, notably in the Ipswich region but also in King's Lynn, had the advantage of being on the doorstep of one of the largest markets for the finished product, the progressive turnip-and-barley and turnip-and-wheat farmers of East Anglia. During the currency of Lawes's patent, however, the charms of being close to coprolites were somewhat dimmed. While aware of the potential of mineral sources of phosphate, in the early days of the Deptford factory Lawes in practice used nothing but bones. He was, however, quickly alerted to the discovery of coprolites; and when production outgrew the Deptford site and he built a new factory on the north bank of the Thames on a 100-acre site at Barking Creek in the late 1840s, he turned increasingly to coprolites as the phosphatic raw material of the future. At first Lawes had assumed that his patent gave him a monopoly in the production of superphosphate as a fertiliser (as distinct from its use as an intermediate product for fine chemicals), from whatever materials it was made, but by 1848 he had conceded that because Liebig had published the bone–acid recipe in 1840, two years before the issue of the patent, he had no enforceable exclusive rights over the bone-based process. Phosphatic minerals were a different matter, and Lawes was determined to assert his rights over them and to pursue pirate firms for infringements of his patent. Challenged in 1848, the patent was upheld. Challenged again in 1852 by a consortium of technically illicit superphosphate manufacturers, the patent was again substantially upheld in the case of Lawes v. Batchelor, in November 1853: the judgment provided that anyone could manufacture superphosphate, but if they used mineral phosphate they had to pay Lawes a royalty of 10s. a ton.[94] By then the patent only had some four years still to run, but nevertheless by the mid-1850s Lawes should have been receiving about £10,000 a year in royalties if all the manufacturers using coprolites, or other mineral sources, had paid up.

[94] Russell, *Agricultural Science*, pp. 143–5. Although the basis of the patent was shaky because the chemical reaction was public knowledge, there was much goodwill towards Lawes as it was widely known that he was spending a great deal of his own money on the Rothamsted experiments, and was making their results freely available to the agricultural world.

The more important point is that, patent or no patent, by 1852 a considerable group of businessmen with interests in chemical manure manufacture, coming from several different parts of the country, had come into existence and were able to challenge Lawes. It is possible that by this date these other manufacturers were producing, in aggregate, something of the order of 50,000 tons a year of superphosphate, perhaps three-quarters of it made from bones, while the Lawes factory was turning out 10,000–15,000 tons a year.[95] In short, between 1843 when the Lawes factory started production and 1852 when a group of provincial and London manufacturers met to plan the 1853 law suit, a number of fertiliser firms had been established, many of which were to become big names in the industry and to survive into the twentieth century.[96] Some of these entered the industry gradually and initially as a side-line, stepping sideways from long-established, agriculturally based businesses as millers, maltsters, corn merchants, seedsmen and the like: such were the Ipswich and Suffolk merchant firms of the Fison brothers, Packard, and Prentice, who all began superphosphate manufacturing at this time and were destined to become the 'big three' in the East Anglian fertiliser industry.[97] Some came from farming, like the Devonshire farmer, Norrington, who built a superphosphate works at Plymouth in 1845. Some, like the London Portable Manure Co., founded in 1840, moved across from being manure merchants to being manure manufacturers. A few, like the Richardson partnership which in 1848 adopted the title of the Blaydon Manure and Alkali Co., were direct offshoots of the chemical industry.[98]

The thirty years after 1850 were looked on in retrospect, within the trade, as the great age of the British fertiliser industry when it led the world and supplied much of it as well. In size and in fertiliser technology the British industry had no serious rivals before the 1890s, although some of the work in German chemical laboratories and in German agricultural research should have been causing concern before then. An export trade in manufactured fertilisers developed from the 1860s, with markets in Germany, France and Russia, as well as in parts of the empire, notably

---

[95] These guesstimates for annual average production in 1851–3 rest on back-projection of T. Anderson's estimate in 1860 that 55 per cent of total available bone supply was made into superphosphate: cited by Mathew, 'Guano', p. 121, n. 6. This suggests an output of 65,000 tons a year of superphosphate from bones, with in addition about 17,000 tons a year produced from coprolites. An unknown part of all this was made in back-yards and farmyards. The Appendix Tables 16.1 to 16.6 use a different basis of estimation.   [96] Russell, *Agricultural Science*, p. 144.

[97] M. S. Moss, unpublished typescript, 'History of Fisons Ltd', pp. 41–55. I am most grateful to Dr Moss, of the Business Archives Centre, University of Glasgow, for allowing me to see his typescript and to quote from it.

[98] *Agric. Gaz.*, 30 Mar. 1863, p. 297 advertisement of Blaydon Co.; Benwell Community Project, *The Making of a Ruling Class* (Newcastle-upon-Tyne, 1978), p. 106.

Australia. Some of the exports were of straight superphosphate, but much was of compound fertilisers, made up to the particular recipes of different manufacturers or merchants and in general tailored to the needs of specific crops by the inclusion of proportions of nitrates (usually nitrate of soda or 'cubic nitre' from Chile, but sometimes sulphate of ammonia and potash, either 'natural' from wood ash or kelp, but more often 'mineral' from muriate), sometimes helped out with additions of a bit of guano, or bone meal. In addition, some guano was re-exported, but even without that fertiliser exports had grown to around 140,000 tons a year at the end of the 1870s, valued at over £1 million, and amounted to about 20 per cent of the total output of the industry.[99] Although this was not an especially remarkable export performance it was evidence of a strongly expanding industry which had no serious competitors in the other agricultural economies which took its products.

The expansion of the industry was evident from the rapid growth in the output of coprolites, doubling every five or six years throughout the fifties, sixties, and seventies, and from the beginnings of an import trade in phosphate rock. The existence of phosphatic rocks had long been known, such as apatite in Norway or the Extremadura deposits in Spain, but as long as supplies of home-produced raw materials remained plentiful there was no incentive to import them. An import of phosphate rock was first recorded in 1868, when 900 tons came in; by 1880 imports were running at around 160,000 tons a year. In 1860 an agricultural chemist, T. Anderson, estimated that the total production of superphosphate was about 120,000 tons, of which at least 60 per cent was made from bones; in 1864 Lawes thought that output was between 150,000 and 200,000 tons, 40 per cent of it from bones. By 1880 production was around 675,000 tons, less than a quarter of it from bones.[100] Such an expansion was clearly not achieved simply by the firms that were already established by 1850, although most of those did enlarge their operations considerably: for example, the Lawes factory which was producing 18,000 to 20,000 tons a year in the early 1860s was producing 40,000 tons or more annually by 1880.[101] The home market was obviously extremely buoyant and had taken off after the initial phase of bombarding farmers with advertisements and leaflets to convince them of the merits of artificials. And the profits to be made from selling superphosphate for £6 to £7 a

---

[99] *Annual Trade and Navigation Returns*, in BPP. Exports were recorded by value only and under the single classification of 'unenumerated manures' until 1889, after which quantities also were returned; from 1908 exports were separately recorded for superphosphate, sulphate of ammonia, basic slag, and a residual category of unenumerated manures. For years before 1889 values have been converted into estimated tonnages at the ruling price per ton of superphosphate.

[100] T. Anderson, reported in *Agric. Gaz.*, 19 Jan. 1861, p. 59; J. B. Lawes, *Agric. Gaz.*, 23 Aug. 1864, p.796.   [101] *Ibid.*; Morton, *Agric. Gaz.*, new ser., 27, 2 Jan. 1888, pp. 8–9.

ton, when the ingredients cost perhaps a third of that, were attractive, perhaps especially for those close to the acid supplies and acid expertise of the major alkali districts of Merseyside and Clydeside.

In principle it remained possible to make superphosphate virtually in a back yard, and certainly in the farmyard. Indeed, farmers and the farming press were so bothered about adulteration of fertilisers and frequent stories of buying rubbish – sand, chalk, clay, sawdust, even powdered glass – when they thought they were getting powerful fertilisers, that serious advice was given that the sole way of being sure of getting a pure fertiliser was to make it oneself.[102] However much manufacturers might explain that it was necessary to dilute superphosphates, or compound fertilisers, with a proportion of inert powdery material to assist evenness of spreading on the land, and disclaim ability to control whatever manure merchants might get up to after delivery from the factory, there is no doubt that quite a lot of fraudulent fertiliser was on the market. How many farmers followed the advice to make their own, and how many farm labourers were injured in the process by the uncontrollable violence of the chemistry, is unknowable. The proliferation of very small-scale 'factories', often operating in the local railway station goodsyard, is undoubted. From the crudity of their equipment and their lack of skill and scientific knowledge these may have been a primary source of substandard products. At the same time they may also have been one response to farmers' suspicions of adulteration, providing a locally made fertiliser inspiring local confidence. The main thrust of expansion, however, came from the large firms, Lawes, London Manure, and Anglo-Continental, all on Thames-side, Nitro-Phosphate & Odams on Merseyside, Langdale's and Blaydon on Tyneside, Packard & Co. at Bramford near Ipswich, to name a few. A few years later, in 1892, it was stated that ninety firms accounted for three-quarters of British fertiliser manufacture; and three, probably the three largest – Lawes, Nitro-Phosphate and Langdale's – produced about 20 per cent of national output.[103]

No doubt the large firms made quite a lot of money in the fertiliser boom times of the sixties and seventies: J. B. Lawes decided to retire from direct management and ownership of his Deptford and Barking factories in 1872, when he sold them to the Lawes Manure Co. for £300,000; Packards, with a somewhat smaller capacity, made a peak profit of £30,000 in 1872, and could usually rely on at least half that in the years

---

[102] 'Adulteration of manures', *Agric. Gaz.*, 5 Sept. 1863, p. 854, and many similar comments and letters from farmers in the 1860s.

[103] *Report of the Departmental Committee on the Adulteration of Artificial Manures and Fertilizers, and Feeding Stuffs used in Agriculture*, BPP, 1892, xxvi, Cd. 6742, QQ. 1751, 1757 (T. Elborough, Managing Director of Lawes Chemical Manure Co.).

before 1880.[104] J. B. Lawes continued to own, and run through a managing director, a tartaric and citric acid factory in Barking which he had acquired as a bad trade debt, but that had little agricultural significance: he spent the rest of his active life (he died in 1900) on the Rothamsted researches and fishing on Loch Etive. All the same, what was profitable for the entrepreneurs was almost certainly good for farmers. The large firms were not only technically superior to the back-yard operators, but also they were jealous of their reputations and regularly sampled and analysed their products; if they did not exactly guarantee the precise percentage of soluble phosphate in their superphosphates, they did at least pass on the information on chemical content to the wholesale merchants with whom they dealt, and to those farmers who purchased directly in bulk. The Lawes factory, above all, was widely regarded as the model of best practice and the home of innovations in manufacturing methods. Visitors from abroad came to Barking to see how a fertiliser factory should be organised; Packard in fact went one better, and set up a factory in Germany (at Wetzlar, near Koblenz) in 1873.[105]

It was said of the Barking factory that 'the engineering and mechanical equipment of the entire works can only be described as the perfection of a plant outfit for such an establishment . . . Here is now concentrated one of the most remarkable industrial systems of its kind extant; and it is safe to affirm that there is not in the world an establishment devoted to this particular branch of chemical manufacture which equals in magnitude or influence that of the Lawes Chemical Manure Co.'[106] Paradoxically, one of the claims to fame advanced in this panegyric was the beneficent sanitary effect on the surrounding district of the enormous factory chimney which was allowed to vent a small percentage of acid fumes that had, it was thought, purified the atmosphere and greatly reduced the local incidence of fevers and agues and even of smallpox. Inhabitants in the downwind vicinity of other chemical works took a less benign view of chimney emissions; although admittedly it was the hydrochloric acid generated as a waste gas in the sulphuric acid–salt–soda process (Leblanc), killing the pastures and defoliating the parks of the Cheshire gentry, that produced the first statutory reaction against pollution, the 1863 Alkali Act. The Alkali Inspector appointed to enforce this Act was initially restricted to patrolling the emissions of alkali works; but in the emerging tradition of Victorian inspectorates he and his team rapidly built up a body of knowledge and expertise in ways of detecting and reducing the polluting effects of industrial chemical processes more

[104] Russell, *Agricultural Science*, p. 145; Moss, 'Fisons', p. 78.
[105] *Ibid.*, p. 79 (it was in partnership with a friend, Carl Muller).
[106] Morton, *Agric. Gaz.*, 1888, pp. 8–9.

generally, and pressed for amending and enlarging Alkali Acts in 1874 and 1881. The 1881 Act extended inspection to all 'emitting' industries, including chemical manure works.[107] Control of atmospheric discharges obliged manufacturers to install equipment to cleanse their flue gases, something which was already best practice since it recovered commercially valuable residues, notably dilute acid. But partially purifying the flues also led to the discharge of more strongly polluted liquid effluents into the brooks and rivers on which most chemical works were sited, and this added to the strength of the long-running campaign against river pollution and helped in the passage of the 1876 River Pollution Act. This measure did not have the teeth of a special inspectorate, but it did enhance the awareness of manufacturers that they could be prosecuted and penalised for creating an environmental nuisance.[108]

For the historian the result of this legislation was to make the Chief Alkali Inspector's annual reports a useful source for the history of the fertiliser industry.[109] For the industry the result was to accelerate the installation of flue washing and scrubbing towers and tanks in both the acid-making and fertiliser-making processes, of a type already employed in the Lawes factory in the 1860s, and eventually to encourage the diversification of single-product works making straight superphosphate in order to make use of the by-products which were recovered – for example, phosphoric acid, which was present in the gas given off during the making of superphosphate. At the end of that road fertiliser manufacture largely ceased to be a distinct industry and became one element in a complex, multi-product, chemical industry. That point was only coming into sight in the interwar period. Meanwhile in the late Victorian and Edwardian industry one of the effects of the converging influences of the economics of by-product recovery and the regulations of the Alkali inspectorate was to make it harder for the very small, back-yard, producers to survive, and thus to benefit the larger-scale integrated works even though some manufacturers grumbled at being forced by state intervention to invest in what they regarded as unproductive plant. Thus, in reporting on chemical manure works in 1889 the Alkali Inspector said that 'some of these factories still use the ancient method of mixing ground mineral phosphate in open troughs, generating dense acid gases

---

[107] J. H. Clapham, *An Economic History of Modern Britain*, vol. II, *Free Trade and Steel, 1850–1886* (Cambridge, 1952 edn.), p. 105; P. Brimblecombe, *The Big Smoke* (London, 1987), pp. 137–40; R. M. MacLeod, 'The Alkali Acts administration, 1863–84', *Victorian Studies*, 9, 1965, pp. 86–112.

[108] *Reports of the Commissioners on River Pollutions*, BPP, 1866, XXXIII [3634] [3634–1]; 1867, XXXIII [3835] [3835–1] [3850] [3850–1]; 1870, XL [C. 37][C. 109] [C. 181]; 1871, XXV [C. 347]; 1872, XXXIV [C. 603–1]; 1873, XXXVI [347–1]; 1874, XXXIII [C. 951] [C. 951–1] [C. 1112].

[109] The series of *Annual Reports of the Chief Alkali Inspector*, in BPP, starts in 1864, and begins to contain material relating to fertilisers from 1881.

injurious to the workers', but stated that through the influence of the inspectorate there had been gradual improvements so that 'this now belongs to the past'. 'The universal practice', his report concluded, 'is to mix with acid in a closed vessel with a steam-driven stirrer, the mixture being discharged into a den in which the chemical reaction is completed, and the gases are led through flues to a wash tower, where they are scrubbed and then discharged up a chimney assisted by fans.'[110] The 'open trough' men were on the way out, if not yet quite extinct. Their extinction no doubt explains much of the decline in the number of registered chemical manure works known to the inspectorate from over 300 in the mid-1880s to less than 200 in 1913.[111]

Although the squeeze on the minnows from technological, economic and administrative pressures was causing a reduction in the number of businesses in the industry, the industry leaders were more conscious of depressed trade and depressed profits, conditions that set in initially in reaction to the exceptional boom in 1872–3 but then seemed to have become continuous in the 1880s and much of the 1890s. In a sense this was simply the story of British industry at large. Looking for specific fertiliser explanations the manufacturers pointed, as did many other industrialists, to foreign competition: mainly this squeezed them out of their European markets as Belgian and German manufacturers, in particular, took over their own domestic markets, but the Belgians also began to develop exports to Britain especially of a new concentrated 'double super' made by treating 'first growth' superphosphate with phosphoric acid. They pointed also to the weakness of the home market caused by agricultural depression, and the plight of cereal farmers (who included most of the turnip growers) was indeed likely to have had a direct impact on the demand for fertilisers, at least in the short run as farmers strove to reduce whatever costs could be easily cut. In the longer run, depending on the relative movements in the prices of labour, land, machinery and fertilisers, it could have made economic sense to increase the use of fertilisers in order to reduce the unit costs of producing grains, or roots (or grass, where the benefits of fertilisers only began to be popularised from the late 1890s), by producing more for the same amount of labour. The figures suggest that the quantities of fertilisers of all kinds used by British farmers fell by a quarter between 1872–6 and 1882–6 and then began to recover, slowly, just about regaining the previous level for most years between 1892 and 1906 but not increasing strongly until the last half dozen years before 1914.[112]

The superphosphate manufacturers were also apprehensive about the

---

[110] *26th Annual Report of Chief Alkali Inspector* (for 1889), BPP, 1890, xx, Cd. 6026, p. 12.
[111] *Annual Reports of Chief Alkali Inspector*, 1881–1914.   [112] Appendix, Table 16.3.

effects on their costs and profits of existing and threatened pollution legislation and regulation, and it was chiefly with that in mind that several of
the leading manufacturers came together in 1875 to form the Chemical
Manure Manufacturers Association so that the industry should have a recognised body to speak for its interests. That objective at least had been
achieved by 1892, when the Association had ninety members and its president, T. Elborough (managing director of the Lawes Manure Co.), was
called as a leading witness to the Departmental Committee on the
Adulteration of Artificial Manures and Fertilisers and Feeding Stuffs. It is
less clear that the views of the Association carried much weight in influencing the Committee's report or the subsequent Fertilisers and Feeding
Stuffs Act of 1893. The trade view was that the reputable manufacturers,
meaning the members of the Association, were blameless, exercised
careful quality control, sold a pure product, and made no more than
modest profits; though it was admitted that it was 'a season trade, and
therefore we have to prepare our manures weeks or months beforehand,
and send them to various depots; and during this time chemical changes
occur, without reducing the actual fertilizing value but changing soluble
phosphate into precipitated phosphate'.[113] While the manufacturers were
against sin and disapproved of fraud and adulteration which they held,
possibly correctly, were committed by others – petty manufacturers and
unscrupulous merchants – they argued that the 1893 Act, which obliged
them to warrant the chemical analysis of their products, might expose
them to vexatious prosecution by the newly-appointed county analysts,
who might be inexperienced and ignorant, and to fraudulent prosecutions
by devious farmers, who were enabled to take unsupervised samples to
send for analysis.[114] In truth the Act was passed largely as a sop to farmers,
who were casting round in all directions to claim that they were being
overcharged or cheated by their suppliers, of rail freight as well as of fertilisers; and it did provide a useful remedy for small farmers who did
not buy direct from the major manufacturers with established brand
names. Those manufacturers, the members of the Association, in practice
lived easily with the Act, only to surface again in 1905 when another
Departmental Committee reviewed the working of the 1893 Act. This
time the Association was represented by Edward Packard, who argued
against the official proposal to tighten the rules by obliging manufacturers to warrant fixed percentages of active chemicals instead of the ranges
of guaranteed percentages required under the 1893 Act, on the grounds
that this would taint honest manufacturers as criminals for some slight

---

[113] *Report: Committee on Adulteration*, 1892, QQ. 1751–7.

[114] Alexander Cross, MP, owner of Cross & Co., Glasgow, fertiliser manufacturers, *Agric. Gaz.*, new
ser., 38, 20 Nov. 1893, p. 441; Alfred Firby, agricultural analyst, *Agric. Gaz.*, new ser., 38, 4 Dec.
1893, p. 487.

variation in chemical content due to accidental or natural causes rather than to intentional deception. He also argued that it was impossible to brand fertiliser bags with the guaranteed analysis because 'after lying a time in the store the marks would be obliterated or rendered illegible' and because farmers often returned old bags with other makers' names on them, to be refilled, and to refuse to accept those would add to farmers' costs.[115] Nevertheless, in 1906 the law was amended to include the fixed analysis guarantee, and the industry lived with that too.

While Alkali inspectors, county analysts and adulteration incidents occupied quite a bit of the Association's time, they did not pose any serious problems for the industry. Early in its existence, in the 1880s, the Association identified excess capacity as the industry's real problem. The evidence was there in the fall in sales on the home market. The symptoms were there in the steep fall in the price of superphosphates from the £6 to £7 a ton range of the sixties and seventies to around £5 a ton at the beginning of the 1880s and £3 a ton by the mid-1890s; by 1908–13 superphosphate fetched little more than £2. 10s. a ton.[116] The classic responses to the symptoms were there, in price wars, in occasional buyouts of ailing firms in order to scrap their plant, and in attempts to form a 'manure ring' by using the Association to operate a price-fixing cartel. Such attempts, seriously made in 1889 and again in 1894, all collapsed. Pretty well all the large producers were members of the Association and agreed to regulate prices; but while, in 1892, it was reckoned that twelve firms on Thames-side and nine on Merseyside between them accounted for about half of total British output, there always remained too many independent producers for a ring to be effective.[117] It is true that the cost of the basic raw material, phosphate rock, also slumped, from nearly £3 a ton in the mid-1870s to under £1. 10s. a ton in the 1890s, as massive new supplies entered the international market from mines in Tunisia and Morocco to compete with, and soon surpass, the previous dominant source in Florida.[118] Nevertheless, a selling price of £3 a ton was reckoned, in 1890, to leave a margin of only 2s. 6d. (or 4 per cent on turnover) over the costs of superphosphate production, meagre profits indeed when in the good times manufacturers had regularly been making 10 to 15 per cent on turnover.[119]

Excess capacity among the established fertiliser manufacturers was,

[115] *Report of the Departmental Committee on the Working of the Fertilizer and Feeding Stuffs Act, 1893*, BPP, 1905, xx, Cd. 2386, QQ. 3087, 3107–9 (Packard).

[116] These were factory-gate prices (and for 1908–13 when exports were separately identified, export, i.e., f.o.b. values) and the farmer would have paid perhaps 30 per cent to 50 per cent, sometimes nearer 100 per cent, more, to cover distribution and retailing costs.

[117] Moss, 'Fisons', pp. 96–8, 101–2; *Report: Committee on Adulteration*, 1892, Q.1868 (Hermann Voss, managing director of Anglo-Continental Guano and Manure Co. London).

[118] A. N. Gray, *Phosphates and Superphosphates* (London, 1944, edn), pp. 22–32.

[119] Moss, 'Fisons', p. 97.

however, also a symptom of a larger problem which, largely coinciden-
tally, confronted the industry at the same time as the decline in demand
from British consumers. The alkali industry had grown up using the
Leblanc process, consuming large quantities of sulphuric acid. The
chemically more elegant and less wasteful Solvay (or ammonia) process,
using no sulphuric acid, had been invented in 1863 and by the mid-1870s
was the basis of a rapidly expanding industry in Belgium which comfortably undersold Leblanc soda. The British industry was slow to switch to
Solvay methods and reluctant to have to scrap its heavy investment in
Leblanc plant, but by the late 1880s international competition was forcing
the British industry to adapt. One result of this technological battle of
chemical formulae was to create a flood of surplus sulphuric acid, and
fertiliser manufacture was the only available alternative use for bulk supplies of acid. It was no accident that the United Alkali Co., a combine of
practically all Britain's Leblanc capacity, formed in 1890, rapidly became
one of the largest manufacturers of superphosphates. Surplus acid capacity from obsolescent Leblanc plant was by nature a transitory phenomenon, but it took a further twenty-five years to work out of the system.
While that was happening, fresh sources of surplus acid appeared, before
1914 notably from the construction of zinc spelter works which generated vast quantities of sulphuric acid that was easily recovered from their
waste gases. Looking a little ahead, large new sulphuric acid plants were
built during the First World War, for acid was a key ingredient in making
explosives: planning for the return of peace, the specialist committees of
the Ministry of Reconstruction could see no possible use for all this acid
plant except in the manufacture of fertilisers, and in the heady and unreal
atmosphere of 1917 and 1918 imagined not only that British farming
ought, on grounds of technical efficiency, to use at least double its pre-
war quantities of fertilisers, but also that British farmers could be compelled to purchase enough fertilisers to keep all the acid works (which
were state-owned) in business, through the 'good husbandry' provisions
of the 1917 Agricultural Production Act.[120] Abundant and cheap acid,
whose producers were desperate to find an outlet for it in fertilisers, was
thus in effect a permanent feature of the industrial scene from the mid-
1880s onwards.

3. Sulphate of ammonia and basic slag

The fertiliser industry did have another use for sulphuric acid besides
the original, and major, one of making superphosphate; that was for

[120] D. S. Landes, *Unbound Prometheus* (Cambridge, 1969), pp. 272–3; Gray, *Phosphates*, pp. 175, 188; *Ministry of Reconstruction, Report of Agricultural Policy Sub-Committee*, Part I, BPP, 1917/18, XVIII, Cd. 9079, p. 16; Part II, BPP, 1918, V, Cd. 8506, pp. 124–5; *Report of the Departmental Committee on the Post-war Position of the Sulphuric Acid and Fertilizer Trades*, BPP, 1918, XIII, Cd. 8994, p. 6.

making sulphate of ammonia. Essentially this was a waste product from gasworks, but it did require processing before it became a product which could be handled, transported, distributed and spread on the fields. The chemistry was as simple as for superphosphate. Ammoniacal gasworks liquor was mixed with slaked lime, dropped down a tower containing steam jets, and the resulting vapour was passed through sulphuric acid baths where it turned into sulphate of ammonia in granular form suitable for handling, compounding and spreading. Both the chemical process and the fertilising properties of sulphate of ammonia as a good source of soluble nitrogen were known from the 1840s, and small quantities were regularly produced. Manufacturers such as the Lawes company, being virtually alongside the giant Beckton gasworks complex, were ideally placed for getting cheap supplies of the waste liquor. Nevertheless, production of sulphate of ammonia grew only slowly during the great mid-Victorian early flowering of chemical fertilisers. On the one hand the cost pressures on gasworks – let alone on the coking industry which produced identical waste liquors – to recover commercially valuable by-products from their wastes were weak: when pressure did become stronger it was as much administrative, from the Alkali Inspector, as it was economic. On the other hand, both British farmers and fertiliser manufacturers became wedded to imported Chilean nitrate as their accustomed and convenient form of nitrogenous fertiliser. While guano contained nitrogen as well as phosphate and was in itself something of a chemically balanced fertiliser, manufacturers preferred Chilean nitrate as the source of nitrogen to be mixed with superphosphate, a little kainit (for potash), and sundry other substances of doubtful fertilising value, in the variety of compound fertilisers which were marketed under various brand names as being specially tailored for specific crops.[121] At the same time some farmers, probably those with large barley-and-turnip farms, applied Chilean nitrate direct, or conceivably made up their own compounds on the farm. One way and another, consumption of imported nitrate climbed from a little over 10,000 tons a year in the 1850s to a peak of over 100,000 tons a year in the mid-1870s. Meanwhile production of sulphate of ammonia, which had a somewhat lower usable nitrogen content per ton, had slowly climbed to about 46,000 tons a year.

Then, a run of half a dozen years from 1876 during which the price of Chilean nitrate rose by a third and the quantity imported fell by two-thirds provided the economic trigger for domestic producers to take sul-

---

[121] Mixing compound fertilisers to individual recipes provided a mystique which kept many local manure merchants in business: some called themselves 'manufacturers' but in fact they were mixers and distributors. Any issue of *Agricultural Gazette* from the 1860s onwards carried numbers of advertisements for many different brands of special turnip, potato, barley and wheat fertilisers.

phate of ammonia more seriously: even though they were facing a weakening agricultural market they had the chance to capture a larger share of the total nitrogen market. Prices of sulphate of ammonia sagged somewhat, from around £12 a ton in the early 1870s to £10 a ton or so by the early 1880s; but production increased as gasworks, ironworks and shale distilling works (all in Scotland) were forced by pressure on profits from their main products to pay increasing attention to the recovery of by-products. That pressure was reinforced from 1881 onwards by the attentions of the Alkali Inspector, whose requirements for the control of flue gases and effluents acted as an incentive to cover the costs of washing, scrubbing and filtering plant out of the proceeds of by-product recovery. Production doubled in the ten years after 1875, reached 150,000 tons a year in the early 1890s, passed 200,000 tons by 1900, and rose to over 400,000 tons by 1913: by that time imports of Chilean nitrate were running at around 140,000 tons a year. This was not, however, a straightforward case of import substitution. Already by 1889 the Alkali Inspector, remarking that sulphate of ammonia was produced in every gasworks in every market town, found it strange that 'the British farmer does not buy it' and that the bulk was exported to Germany.[122] The conservatism of British farmers was no doubt part of the explanation for the comparatively small home market. But a more important reason seems to have been that German farmers (and some others in western Europe) were developing a heavily subsidised appetite for large quantities of nitrogen to feed their greedy, even damagingly exhaustive, sugar beet crops. Sulphate of ammonia thus acquired a reputation of being primarily a sugar beet fertiliser, a perception which did not begin to dissolve until the manufacturers found it worth while soon after the turn of the century to mount a vigorous publicity and advertising campaign to persuade British farmers of its versatility.[123] Meanwhile, the comparative neglect of the substance by British farmers, who did not grow sugar beet, was understandable; they continued to obtain the bulk of their 'artificial' nitrogen supplies from imported nitrate, although on the eve of the War the proportion supplied by home-produced sulphate of ammonia had crept up to nearly one half.

The numbers of producers of sulphate of ammonia proliferated steadily. Not only did every local gasworks, and in Scotland every shale plant, come in on the act, but also from the mid-1890s onwards coke smelters began to install recovery ovens and these too added to the supply. There were 291 works with a sulphate of ammonia process attached to them in 1886, the first year in which the Alkali Inspector began to count them,

---

[122] *26th Annual Report of Chief Alkali Inspector*, BPP, 1890, xx, Cd. 6026, p. 12.
[123] Moss, 'Fisons', pp. 118, 127; *Committee on Post-war Sulphuric Acid*, 1918, p. 5.

399 in 1901, 504 in 1907 and 624 in 1914.[124] Such an expansion in numbers was to be expected in what was essentially a by-product industry, and contrasted with the equally steady contraction in the number of 'chemical manure' (sc. superphosphate) works and the concentration of production of that fertiliser in the hands of a few large firms. It was also an export industry. Hermann Voss's estimates suggest that in the early 1890s something like three-quarters of total output was being exported, declining to about two-thirds by the early 1900s; exports began to be separately recorded in the official statistics from 1908, and from then until 1914 they hovered around 70 per cent of total output.[125] The sulphate of ammonia producers formed their own trade association in the early 1900s, but since for all of them this was somewhat peripheral to their main producer interests it did not excite the commitment or carry the weight of the Chemical Manure Manufacturers Association (renamed the Fertiliser Manufacturers Association in 1904), whose main members' business interests were wholly concentrated on the fertiliser trade.[126]

The other late Victorian addition to the range of chemical fertilisers followed a very similar path. Basic slag, the residue from making steel from phosphoric ores by the Thomas and Gilchrist, or basic, process, was instantly recognised to have useful properties as a phosphatic fertiliser. Like sulphate of ammonia it was a by-product, produced at the steelworks; unlike sulphate, it required no further chemistry, but simply the mechanical processing of crushing and grinding to reduce the hard lumps of waste into a usable fine grey powder. Again like sulphate of ammonia, it was neglected by British farmers, perhaps in this case because bulk supplies of basic slag first became available in Germany, Belgium and France where the steel industry adopted the basic process in the 1880s, while the British industry remained tied to Bessemer (non-phosphoric) steel. Nevertheless, when basic steel production did begin to grow in Britain, alongside acid steel, the bulk of the basic slag was exported. In 1892 home consumption was estimated at 30,000 tons a year, about one-third of total output, and although the home market absorbed roughly 100,000 tons a year in 1900–3, that fluctuated between 28 per cent and 42 per cent of output; the exports were largely to the empire, since Germany had ample

[124] *23rd Annual Report of Chief Alkali Inspector*, BPP, 1887, XVII, Cd. 5057, p. 26; *38th Report*, BPP, 1902, XI, p. 6; *44th Report*, BPP, 1902, XI, p. 6; *50th Report*, BPP, 1914, XIV, p. 6.

[125] Hermann Voss, Anglo-Continental Guano and Manure Co., evidence to *Committee on Adulteration*, 1892. Q. 1892, and to *Committee on Fertilizer . . . Act*, 1905, Q. 3205; *Annual Abstract of Statistics of U.K.*, 1908–14.

[126] The Chemical Manure Manufacturers Association was the only trade body formally represented by witnesses to the *Committee on Fertilizer . . . Act*, 1905. See also *Committee on Post-war Sulphuric Acid*, 1918, pp. 4–5, and *Ministry of Reconstruction, Committee on Chemical Trades*, BPP, 1917/18, XVIII, Cd. 9079, p. 3.

supplies of her own.[127] It was a cheap fertiliser, selling at less than £2 a ton, and it was lucky for the superphosphate manufacturers that the inertia or prejudices of British farmers prevented a wholesale switch to basic slag from the more expensive, but well-tried, superphosphate. As it was, British farmers treated basic slag largely as complementary to superphosphate, not as a substitute for it, applying it on acid soils which did not respond well to superphosphate. This happy situation changed from about 1907 when the associated basic slag producers mounted a vigorous campaign to popularise the benefits of using their product on grasslands, particularly run-down grasslands, and the Fertiliser Manufacturers Association responded in kind with a similar campaign making identical claims for superphosphate. Down to 1914 basic slag seemed to be winning this fight, with home consumption increasing to over 250,000 tons a year, about 60 per cent of total output.[128]

4. The industry in 1914

All in all trading conditions for the superphosphate manufacturers remained generally tense and keenly competitive, though not entirely unprofitable, throughout the period from the early 1880s to 1914. Awash with surplus sulphuric acid, they faced competition at home from basic slag, and from abroad in the shape of imports of high-quality 'double' and 'treble' super; and while not in direct competition in a fertilising sense with nitrate or sulphate of ammonia, they were in competition with them for the farmers' money, for these were expensive fertilisers at £13 a ton against £2 12s. (£2.6) a ton for superphosphate (1913 export prices). Moreover, their former export markets in continental Europe had been virtually wiped out in the 1890s by the rise of continental producers, and they had to work hard with promotion campaigns and travelling salesmen extolling the benefits of superphosphates for rubber, tea and coffee plantations to establish new export markets in the empire and South America. These efforts achieved some success, with exports of 159,000 tons in 1911, though they declined to only 66,000 tons in 1914 – figures which represented from 7 to 15 per cent of total production.[129]

---

[127] Voss gave the estimate in 1892: *Committee on Adulteration*, 1892, p. 129. G. Goetze, chairman of H. & E. Albert Chemical Works, gave detailed annual figures of total production and home consumption for the eight years 1896–1903 in evidence to *Committee on Fertilizer . . . Act*, 1905, p. 204: these differ from the estimate for 1900 in J. Hendrick, 'The growth of international trade in manures and foods', *Trans. Highland and Agric. Soc. of Scotland*, 5th ser., 29, 1917, p. 22, but appear to be more reliable. Hendrick's statement that British slag was entirely exported to Germany is definitely inaccurate.

[128] Exports of basic slag peaked at 231,277 tons in 1910 and then fell away to 132,679 tons in 1914, while total output reached 404,000 tons in 1913: *Annual Abstract of Statistics of the U.K.*, 1908–14.

[129] *Annual Abstract of Statistics of the U.K.*, 1908–14; Moss, 'Fisons', pp. 129, 140.

The gross value of that output in 1914, at factory-gate prices, was about £2.4 million (and the basic raw material, phosphate rock, cost about £0.9 million): it was dwarfed by the gross value of sulphate of ammonia production, about £6 million, and of nitrate imports, about £1.5 million, although the total production of the direct rival, basic slag, was worth little more than £0.5 million.[130] All the same, the superphosphate manufacturers were the core of the fertiliser industry, and their association, the Fertiliser Manufacturers Association, was the leading body in the trade: the others were by-product processors, for whom fertilisers were a sideline, or they were simply importers and distributors. The associated but inconsiderable business of other agricultural chemicals – sheep dips, insecticides and weedkillers, for which sulphur and arsenic were common ingredients – seems to have been largely in the hands of the superphosphate men: but the 1907 Census of Production put the gross output of 'disinfectants, insecticides, weedkillers, and sheep and cattle dressings' at no more than £585,000, or £729,000 when allowance was made for some production by firms outside the 'fertiliser schedule'.[131]

In the space of seventy years the fertiliser industry, headed by the superphosphate manufacturers, had moved through the whole cycle from an infant industry to an elderly industry with antiquated equipment and technology, handicapped in keeping abreast of international competition by a home market whose buoyancy had vanished in the 1880s and only revived from the mid-1900s. Fresh problems, technological, structural and financial, were on the horizon in 1914, with the likelihood that the Germans would succeed in developing an economical, but highly sophisticated, method of fixing nitrogen from the atmosphere. The War brought a respite: imported raw materials (phosphate rock) were difficult to get, but the demand for fertilisers was strong, and the manufacturers made easy profits. It was no more than a respite. The future lay with more complicated chemistry and with the integration of fertiliser production into the heavy chemical complex, not with the old-fashioned stinking 'dens' with which Lawes had started the whole process going.

---

[130] Export values (and import values) have been taken as rough proxies for factory-gate prices.

[131] *Census of Production (1907), Preliminary Tables*, Part III, BPP, 1910, CIX, Cd. 5162, Schedule no. 27 'Fertilizers, glue, sheep dip, and disinfectant manufacturers', pp. 28–9. Under this schedule the Census estimated the *net* value of the output of the industry – basic slag, superphosphate, 'other manures' (chiefly compounds), and insecticides, etc. – at about £2 million a year: production of all fertilisers (including sulphate of ammonia which was returned under Gas Undertakings in Part IX, BPP, 1911, CI, Cd. 5813, Schedule no. 116, pp. 17–18) was, however, seriously under-recorded in the Census. Much of the market for insecticides and weedkillers was horticultural (both domestic and commercial), not agricultural. Agricultural scientists did not turn their attention to this field until the 1920s and 1930s, when they came up with derris dust and pyrethrum: Russell, *Agricultural Science*, pp. 317–18.

APPENDIX

Table 16.1. *Total production of the British fertiliser industry, 1851–1913*
(000 tons per annum)

Period	Superphosphate made from				Sulphate of ammonia	Basic slag	Total tonnage
	home bones	coprolites	imported rock	total			
1851–3	40	17	–	57	–	–	57
1854–8	41	35	–	76	–	–	76
1859–63	42	85	–	127	–	–	127
1864–7	45	180	–	225	–	–	225
1868–71	48	200	21	269	40	–	309
1872–6	51	270	143	464	46	–	510
1877–81	56	283	227	566	60	–	626
1882–6	61	46	390	497	97	–	594
1887–91	65	10	500	575	134	–	709
1892–6	66	3	577	646	171	100	917
1897–1901	69	2	617	688	206	247	1,141
1902–6	72	0	706	778	253	293	1,324
1907–11	74	0	843	917	353	340	1,610
1912–13	76	0	916	992	410	404	1,806

Table 16.2. *Exports and UK consumption of British-made fertilisers, 1851–1913 (000 tons per annum)*

Period	Exports				UK consumption			
	Unenumerated manures (chiefly superphosphate and compounds)	Superphosphate	Sulphate of ammonia	Basic slag	General	Superphosphate	Sulphate of ammonia	Basic slag
1851–3	–				57			
1854–8	–				76			
1859–63	–				127			
1864–7	20				205			
1868–71	47				262			
1872–6	87				423			
1877–81	147				479			
1882–6	275				319			
1887–91	295				414			
1892–6	275			70	542			30
1897–1901	137		147	139	551		59	108
1902–6	129		170	203	649		83	90
1907–11	138	149	268	204	(−138)	768	85	136
1912–13	142	76	304	161	(−142)	916	106	243

Note: The minus quantities for UK consumption of 'general' undifferentiated fertilisers for 1907–11 and 1912–13 are balancing items, as the rows of Table 16.2 cross-add to equal the 'Total tonnage' column of Table 16.1. Probably they relate mainly to superphosphates used in compounds, whose export is recorded in the first column, and therefore should be deducted from the amounts of superphosphate used in the UK.

Table 16.3. *Total UK consumption of fertilisers, 1851–1913 (000 tons per annum)*

Period	Net Imports				Total imported tonnage	Total home-produced tonnage	Gross tonnage used in UK
	Bones	Guano	Nitrate	Unenumerated manures			
1851–3	39	165	13	–	217	57	247
1854–8	68	231	15	1	315	76	391
1859–63	66	139	25	2	232	127	359
1864–7	68	166	39	5	278	205	483
1868–71	86	209	45	15	355	262	617
1872–6	86	115	102	69	372	423	795
1877–81	79	83	53	98	313	479	792
1882–6	72	40	78	84	274	319	593
1887–91	67	19	86	86	258	414	672
1892–6	67	29	112	74	282	572	854
1897–1901	62	24	124	92	302	718	1,020
1902–6	47	28	112	123	310	822	1,132
1907–11	43	28	121	156	348	851	1,199
1912–13	41	20	132	220	413	1,123	1,536

Note: The imports of 'Unenumerated manures' included the imports of potash, mainly in the form of kainit, which were estimated at 30,000 tons a year in 1892 and 75,000 tons a year in 1905.

Source: 'Total home-produced tonnage' is from Table 16.2.

Table 16.4. *Value (factory gate or f.o.b. prices) of British fertilisers, 1851–1913 (£000 per annum)*

Period	Superphosphate	Sulphate of ammonia	Basic slag	Total
1851–3	570			570
1854–8	760			760
1859–63	1,140			1,140
1864–7	1,800			1,800
1868–71	2,152			2,152
1872–6	3,712	500		4,212
1877–81	3,962	560		4,522
1882–6	2,982	660		3,642
1887–91	3,737	1,080		4,817
1892–6	2,584	1,368	170	4,122
1897–1901	2,064	1,854	420	4,338
1902–6	2,334	2,783	500	5,617
1907–11	2,292	4,271	510	7,073
1912–13	2,580	5,617	725	8,922

Table 16.5. *Value of exports and UK consumption of British-made fertilisers, 1851–1913 (£000 per annum)*

	Exports					Total
Period	Unenumerated	Superphosphate	Sulphate of ammonia	Basic slag	Total exports	retained UK consumption
1851–3	–				–	570
1854–8	–				–	760
1859–63	–				–	1,140
1864–7	166				166	1,634
1868–71	374				374	1,778
1872–6	699				699	3,513
1877–81	1,180				1,180	3,342
1882–6	1,928				1,928	1,714
1887–91	1,903				1,903	2,914
1892–6	2,001			105	2,106	2,016
1897–1901	744		1,323	208	2,275	2,063
1902–6	1,114		1,700	325	3,139	2,478
1907–11	770	360	3,242	302	4,674	2,399
1912–13	880	198	4,164	248	5,490	3,432

Table 16.6. *Value of total UK consumption of fertilisers, 1851–1913 (£000 per annum)*

| Period | Net imports | | | | Total imports | Total British-made retained | Total value of UK consumption |
	Bones	Guano	Nitrate	Unenumerated manures			
1851–3	195	2,270	221	—	2,486	570	3,056
1854–8	386	832	259	5	1,482	760	2,242
1859–63	334	1,333	342	13	2,022	1,140	2,162
1864–7	357	1,726	487	39	2,609	1,634	4,243
1868–71	528	2,431	1,266	73	4,298	1,778	6,076
1872–6	560	864	1,302	193	2,919	3,513	6,432
1877–81	478	461	766	205	1,910	3,342	5,252
1882–6	461	216	844	159	1,680	1,714	3,394
1887–91	333	75	780	128	1,316	2,914	4,230
1892–6	293	185	782	142	1,402	2,016	3,418
1897–1901	259	126	981	194	1,560	2,063	3,623
1902–6	203	150	1,191	223	1,767	2,478	4,245
1907–11	200	139	1,184	359	1,882	2,399	4,281
1912–13	218	115	1,382	569	2,184	3,432	5,616

Sources: Tables 16.1 to 16.6.
Annual Trade and Navigation Returns
Annual Abstract of Statistics of the UK
Annual Reports of the Chief Alkali Inspector
Professor Anderson (Dumfries), 'The present consumption of artificial Manures', *Agricultural Gazette*, 28 Feb. 1863, p. 205.
H. Voss (Anglo-Continental Guano and Manure Co., and President of Chemical Manure Manufacturers Association), evidence to *Departmental Committee on Adulteration*, BPP, 1892, XXVI, p. 129, and to *Departmental Committee on Fertilizer and Feeding Stuffs Act*, BPP, 1905, XX, App. XXII, pp. 201–2.
G. Goetze (Chairman, H & E Albert Chemical Works), evidence to *ibid.*, App. XXV, p. 204, and *Ministry of Reconstruction, Agricultural Policy Sub-Committee*, BPP, 1918, V, Cd. 8506, App. XIII, 'The supply of artificial manures', pp. 124–5.
J. Hendrick, 'The growth of international trade in manures and foods', *Trans. Highland and Agric. Soc. of Scotland*, 5th ser., 29, 1917, pp. 1–36.
A. N. Gray, *Phosphates and Superphosphates*, (London, 1944 edn).

Some further details on methods of estimation are given in F. M. L. Thompson, 'The second agricultural revolution, 1815–80', *EcHR.*, 2nd ser., 21, 1968, pp. 62–77. The figures in Tables 16.1 to 16.6 may be compared with those in E. M. Ojala, *Agriculture and Economic Progress*, (Oxford, 1952), pp. 212–14. Ojala added a margin to cover marketing and distribution costs, amounting to 40–45 per cent of the prime cost ex-works: I have not attempted any similar calculation.

It has been somewhat arbitrarily assumed in Tables 16.1 to 16.6 that 80 per cent of the home-produced supply of bones was made into super-phosphate (the remainder being used for industrial purposes), and that the entire supply of imported bones was used direct, as bonemeal. This accords with the relative proportions of these two agricultural uses of bones in the mid-1900s (the entire import of Indian bones, for example, a major source of imports, came as bonemeal and was sold direct to farmers), but may well not hold true for earlier decades. Any distortion of the overall figures of tonnage and values is, however, likely to be minimal.

Natural manures, notably farmyard and town stable manures, are by definition excluded from treatment in this contribution. Imported potash is included (as an 'unenumerated' import) in Tables 16.3 and 16.6; but home-produced potash, lime, and sundry other substances such as woollen waste or shredded rags, which were used in small quantities as fertilisers, are excluded.

C
VETERINARY PRODUCTS

BY RICHARD PERREN

1. Origins of the industry

Both the range and quantity of veterinary products available in Britain were extensively enlarged in the nineteenth century.[132] Three main factors accounted for this development: the increase in the number of animals, the increase in the number of veterinary practitioners, and the marketing efforts of the product manufacturers.

Exact numbers of animals are not known: there are no figures for cattle, sheep and pigs before 1866, and no information on domestic animals. But for horses alone the market for shoes grew by 250 per cent between 1811 and 1901.[133] The growth of the veterinary profession is known in its broad outlines and by 1911 there were 2,612 veterinary surgeons, 2 of whom were women, in England and Wales, whereas the efforts of commercial firms, sometimes in alliance with members of the profession, have not hitherto been given much attention.[134] It is argued here that they were important in expanding the market for veterinary products, an expansion that relied not only upon advances in scientific and veterinary knowledge but on purely commercial marketing efforts by manufacturers and suppliers. It may also be that greater density in the enclosed housing of animals, particularly cattle and urban horses, encouraged more disease and enhanced the demand for medications. Even if this were not so, it is quite certain that increased concern about dairy cleanliness on the grounds of public health acted as an independent stimulus.

2. Animal medicines

The largest purchases of many products were by farmers, farriers and domestic householders, rather than members of the veterinary profession, so different marketing techniques were required to attract the attention of

---

[132] R. Perren, 'The manufacture and marketing of veterinary products from 1850 to 1914', *Veterinary History*, New ser., 6, No. 2, Winter 1989/90, pp. 41–61.

[133] F. M. L. Thompson, 'Nineteenth century horse sense', *EcHR*, 2nd ser., 29, 1976, p. 80; *Agricultural Returns*.

[134] *Census of England and Wales, 1911*, Vol X, Part I, Cd. 7018, p. 11; I. Pattison, *The British Veterinary Profession, 1791–1948* (London, 1984), pp. 39–146; Ruth D'Arcy Thompson, *The Remarkable Gamgees: A Story of Achievement* (Edinburgh, 1974), pp. 58–199.

these various groups of consumers. The agricultural press started as the chief medium of communication between manufacturer and farmer. The earliest advertisements appeared in the 1850s and 1860s in the columns of *Bell's Weekly Messenger* (1796–), the *Mark Lane Express* (1832–) and *Thorley's Agricultural Gleaner* (1860–3), and were aimed at farmers. They were supplemented by the specialised journal publications like the *Veterinarian* (1848–), *Veterinary Journal* (1875–) and the *Veterinary Record* (1845–50, 1888–) which carried advertisements aimed at selling to the veterinary profession. The first group, which were newspapers, were the most important medium of publicity as farmers and farriers were the chief occupations for those administering any medicines to farm livestock in 1850. Trained veterinary surgeons did most of their work in towns and amongst horses and small animals, although their numbers in the countryside and the extent of farm practices did rise towards the end of the nineteenth century.[135]

This meant that the ideal product had two requirements; it needed to be simple to administer, and to treat a wide variety of ailments in all species of animals. Harris's Mixture claimed to be a certain cure for: 'murrain and foot rot in sheep; cancer; loe in cattle; humour or yellows in cows or ewes; splints and spavins in horses; and sprains and bruises in all kinds of animals'.[136] Cupiss's Constitution Balls, at 3s. 6d. for six, were said to 'possess the power to cure and, by putting the system into good health, prevent most of the common diseases which horses and cattle are likely to contract'.[137] This product spawned a cheaper imitator, Gostling's Condition Balls, on sale at 2s. 6d. for six, to treat horses and cattle in poor condition.[138] Another way of economising in the treatment of animals was by purchasing a general medicine chest containing several specific remedies.[139]

Remedies for individual diseases were cheaper, as Gostling's Cough Balls – for horses and cattle suffering from coughs and colds – were only 1s. 6d. for six. Here the manufacturer responded to current anxiety over respiratory disease, associated with conditions in the closed urban cowsheds, by claiming: 'They are moreover a successful remedy for influenza, inflammation of the lungs, pleuro-pneumonia (lung disease), etc.' Some conditions applied only to a particular class of animals, and for sheep farmers the problems and financial losses associated with scab were probably the most persistent. But it was the domestic horse, the nineteenth-century status symbol, that received most attention, though even within

---

[135] Read, 'Recent improvements in Norfolk farming', pp. 290–1; Joan Lane, 'The English provincial veterinarian and his practice', *Veterinary History*, 6, Winter 1975/76, p. 13.

[136] *Mark Lane Express*, 19, 25 Feb. 1850, p. 16.

[137] *Ibid.*, 20, 17 Nov. 1851, p. 16; 29, 24 Dec. 1860, p. 5.   [138] *Ibid.*, 27, 27 Dec. 1858, p. 3.

[139] *Ibid.*, 34, 13 Mar. 1865, p. 13; *Johnson's Official Handbook of Cattle Fairs*, 1894, p. 168.

this species it was hunters, urban dray-horses, carriage-horses and military mounts, rather than mere farm horses, that had most individual medication lavished upon them.[140]

3. Advertising and marketing

None of these products was original as something like them had been available long before 1850. The only change was that they were now being manufactured and sold as branded goods. Veterinary works in the first part of the nineteenth century contained numerous recipes of preparations for animal use.[141] What Victorian proprietary manufacturers did was to formulate and market them on a national scale.[142]

With increasing concern over hygiene and cleanliness as a preventive measure, together with advances in the understanding of infection, a wide range of commercial antiseptics and disinfectants became available. Most contained some form of chlorine and included Sir Edward Burnett's Disinfecting Fluid, Ellerman's Fluid, Eau de Javelle and Labarque's Liquid. Other disinfecting ingredients were creosote, carbolic acid and permanganate of potash, the last of these being the basis of a preparation known as Condy's Fluid. Gases were sometimes used to fumigate farm buildings, either chlorine which was produced by heating the appropriate chemicals, or sulphurous acid gas. The last of these enjoyed something of a vogue as a cure for pleuro-pneumonia until it was exposed as being useless by John Gamgee in Glasgow in 1867.[143] By 1900 there were at least twenty-four different kinds of disinfecting and antiseptic preparations at the disposal of the veterinary practitioner.[144] Some widening took place in the treatments available for dairy cattle with the appearance of apparatus for conditions of the udder. The growing concern for clean milk after 1900 helped the sale of products like milk fever pumps, tubes with rubber caps, teat compressors, and instruments for removing obstructions from milk ducts.[145]

---

[140] *Bell's Weekly Messenger*, No. 3397, 18 Jan. 1862, p. 8; *Thorley's Agricultural Gleaner*, 3, 3 Aug. 1863, p. 8; *Mark Lane Express*, 34, 13 Mar. 1865, p. 13.

[141] See for example, R. Pearson, *Every Man His Own Horse, Cattle, and Sheep Doctor* (Leicester, 1811), pp. 377–409 and J. Pursglove, *Guide to Practical Farriery* (London, 1823), pp. 352–87. Pearson had formerly been a chemist and druggist in Leicestershire and included an appendix with twenty-six of his own remedies – an assortment of tonics, ointments, pills and horse balls – with full directions for their use. Pursglove listed thirty-six 'valuable and original recipes from the practice of an eminent veterinary surgeon' (*ibid.*, p. 352).

[142] *Veterinary Record*, 1, 14 July, 1888, pp. i–viii.

[143] F. F. Cartwright, *A Social History of Medicine* (London, 1977), pp. 143–7; G. Armatage, *Every Man his own Cattle Doctor* (2nd ed., London, 1870), pp. 209–16, 665–7.

[144] J. Wortley Axe (ed.), *The Horse: Its Treatment in Health and Disease*, vol. VII (London, 1907), p. 166.

[145] *Veterinary Record*, 23, 30 July 1910, p. x.

4. Technical development of drugs and equipment

In addition to drugs and disinfectants, instruments were the other group of veterinary products available. Besides sprays, the earliest items were glass and metal syringes, enemas, tracheotomy tubes, funnels for passing fluids into veins, stomach pumps, catheters, various devices for ventilating animals' stomach gases, bleeding, obstetric and castrating instruments. Better care of farm animals was called for in the 1860s, and in response to such appeals, as well as the growing interest in veterinary education, a larger range of treatments appeared on the market in the forty years after.[146] In the first half of the nineteenth century the manufacture of the best drugs, instruments and appliances was concentrated in London, and after 1850 it remained the centre of this industry.[147] Sheffield was a secondary centre where blades and cutting instruments were produced by specialised surgical instrument makers and ordinary cutlers alike, and Birmingham supplied various cheaper products. The leading firm was Arnold & Sons of West Smithfield who produced a variety of hernia clamps, adjustable splints, saws, forceps, knives, needles, balling guns, hoof cutters, restraining straps and even curry combs.[148]

As developments in veterinary medicine had the effect of adding to, rather than replacing, existing techniques, and with no sharp division between obsolete and current practice, aggregate demand for veterinary instruments was constantly expanding. The fleam, used for bleeding horses and cattle, was still undergoing technical refinement as late as 1891, and firing irons for treating tendon injuries of horses were still being produced by Arnold and Sons in 1907.[149]

The resort to surgery increased with the appearance of ether after 1846 and then the wider use of the more potent chloroform in veterinary practice from 1890. The changes which anaesthesia allowed in animal surgery mirrored those in human surgery.[150] It stimulated demand for a wider range of surgical instruments as it allowed complex and delicate operations to be carried out. A general concern over theatre, instrument and patient hygiene acted as a further stimulus to demand for antiseptics. Anaesthetics were also responsible for the appearance of animal operat-

[146] *Thorley's Agricultural Gleaner*, 3, 25 May 1863, p. 8 (review of *Outlines of Veterinary Homeopathy*, by James Moore).

[147] D. Blaine, *Outlines of the Veterinary Art* (London, 2nd edn, 1816), p. 638.

[148] A good selection of the firm's products shown in J. P. Sheldon (ed.), *Livestock in Health and Disease* (London, 1906), p. 596.

[149] D. W. Wright, 'John Weiss and the Spring Fleam', *Veterinary History*, New ser., 5, 1988, pp. 83–96; *Veterinary Record*, 1, 7 Dec. 1889, p. ii; C. Vogel, 'From firing to phenylbutazone in veterinary practice', *Veterinary Record*, 111, 1982, p. 523.

[150] *Veterinary Record*, 3, 3 Jan 1891, pp. 327–8; B. M. Q. Weaver, 'The history of veterinary anaesthesia', *Veterinary History*, 5, 1988, pp. 47–9; *Veterinary Record*, 25, 9 Nov. 1912, p. v.

ing tables. There was no problem with small animals as solid timber benches with straps and restraints were cheap and easily made. But for operations on horses British equipment lagged behind what was available in America and Europe. In the absence of tables nineteenth-century British practice was to carry out horse surgery on the ground. When E. W. Hoare wanted to use an equine operating table he had to communicate with a Mr Conkey of Grand Rapids, Michigan, the inventor of the 'Simplicity' pattern, who actually travelled to Britain to assemble and demonstrate it.[151] Another alternative was Vinsot's operating table, a model used in Europe.[152]

5. Links between veterinarians and manufacturers

The proliferation of journals and farming newspapers made it easy for so-called new treatments to get publicity, often authority, as drugs and surgical instruments, and even animal products like steel frost cogs to fit to horses' shoes in winter, were advertised in their pages with the endorsement of veterinary professors who had supposedly pioneered their successful use.[153] The benefits to the practitioners were obvious, even if they were nothing more than further publicity and conferred professional credence when the production models were marketed under their name. Most of the new or improved products that appeared from Arnold and Son's factory in the east end of London did so under this type of advertisement. There was no question of such advertising being considered unethical or as conflicting with the interests of professional objectivity. Not only London professors were involved in this way; it was so widespread that provincial vets were also called upon to add their seal of approval to particular items.[154] In 1891 virtually every item advertised by the firm was endorsed by some member of the profession.[155]

In the absence of any significant advances in drug chemistry it is hard to say whether the real clinical effectiveness of many of the preparations sold in 1914 was any greater than in 1850. But as the market and the demand were there manufacturers continued to enlarge and exploit them. One development from human medicine that was applied to veterinary practice, following the successful work of Pasteur against rabies in 1885, was the use of vaccines. By 1912 a number of accounts of vaccine therapy had appeared in the veterinary literature and it had become so firmly established that the *Veterinary Record* urged every clinician to

---

[151] *Veterinary Record*, 25, 14 December 1912, p. 356.   [152] Wortley Axe, *The Horse*, vol. VII, p. 158.

[153] *Veterinary Record*, 1, 11 Aug. 1888, p. v; 1, 29 Dec. 1888, p. v; 1, 19 June 1889, p. 339; 3, 14 Mar. 1891, p. viii; 3, 27 June 1891, p. vii; 23, 30 July 1910, p. x.

[154] *Veterinary Record*, 3, 29 Nov. 1890, p. 262; 3, 14 Mar. 1891, p. viii.

[155] *Veterinary Record*, 3, 27 June 1891, p. vii.

possess some knowledge of its theory and practice.[156] Some of the vaccines, like those for canine distemper, were of limited use to the farmer, but others were available for equine influenza. As with conventional medications, they were suggested for the cure or prevention of a wide range of conditions in animals, not all of them infectious diseases. But this did not help British manufacturers as the production of these biological pharmaceuticals was a European industry carried out at the Pasteur Institute in Paris. Instead, London drug houses, principally Parke Davis & Co., acted as agents and importers of European products.[157] The other area where developments in the chemical and biological sciences were shared between animal and human medicine was in the use of anaesthetics, antiseptics and aseptic techniques in surgery and nursing.

[156] *Veterinary Record*, 25, 14 Dec. 1912, p. 353.
[157] *Veterinary Record*, 23–25, 1910–12, *passim*; T. Turner and D. Lane, 'One hundred years of small animal practice', *Veterinary Record*, 111, 4 Dec. 1982, p. 520.

D

ANIMAL FEEDING STUFFS

BY RICHARD PERREN

The rise of the commercially produced feedstuffs industry freed farmers from dependence on what they could grow themselves or purchase locally. The expansion in the use of purchased feeds between 1850 and 1914 is evident in a number of ways.[158] These were: increased imports; the appearance of new products incorporating an ever-widening range of ingredients; extensive and persuasive advertising; the rise of new processing techniques and firms; and the emergence of a network of dealers and distributive services.

1. Location and growth

In 1850 the bulk of the feedstuffs industry used home-produced raw materials, the most important of these being wheat offals, followed by barley and oats. As the first item was a by-product of flour milling the locational pattern of that industry determined the siting of feed production. Flour milling is dealt with in Chapter 17A, but the picture that emerges for 1850 is that it was widely dispersed and largely rural. This suggests that much of the wheat offals and other products were sold locally, while an unknown amount passed back to producers for consumption on their own farms.[159]

Oilseeds were the chief imported raw material used in purchased animal feeds, not in their raw form but as the residue or cake that remained after the oil had been extracted for other uses. Those most often used for cattle cake in 1850 were Russian linseed and oilseed rape. After 1860 cotton seed came into use, initially from the United States, and in 1908 soya beans from Japan and Manchuria.[160] For some time after 1850 seed crushing was a seasonal industry and the Hull mills closed down

---

158 The pioneering study of this development is by F. M. L. Thompson, 'The second agricultural revolution, 1815–1880', *EcHR*, 2nd ser., 21, 1968, pp. 62–77.

159 R. T. Hopkins, *Old English Mills and Inns* (London, 1927), pp. 128–9, 199–200; W. Foreman, *Oxfordshire Mills* (Chichester, 1983), p. 61.

160 A. Voelcker, 'The adulteration of oilcakes', *JRASE*, 24, 1863, p. 592; C. M. Daugherty, 'The cotton seed industry', *Yearbook of the United States Department of Agriculture, 1901* (Washington, 1902), p. 295; J. Hendrick, 'The soy bean', *Trans. Highland and Agric. Soc. of Scotland*, 5th ser., 22, 1910, p. 258; *Transit*, 7, 15 Mar. 1909, p. 14; K. E. Hunt, 'Raw materials', p. 36 in J. A. Van Stuyvenberg (ed.), *Margarine: An Economic, Social and Scientific History* (Toronto, 1969).

in summer after the previous year's crop was processed, and before the current year's crop was harvested and shipped from northern Europe. It was only after European supplies were supplemented from other sources that the crop was large enough to justify all the year round working.[161] In addition to oilseeds there were other feedstuffs like palm kernel cake, carob nut, sorghum, maize, and cocoa nut meal. British farmers rarely purchased these directly, although from time to time considerable quantities were incorporated in various compound feeding cakes and meals.[162] Much of the palm nut and cocoa nut meal produced in Liverpool was exported to Europe, whereas all linseed, cottonseed, and most of the rape seed cake from the city's crushing mills went for home consumption.[163]

As the feedstuffs industry developed and a wider selection of raw materials became available they were subjected to a greater intensity of processing, a development that was accompanied by the appearance of mixed feeds. In the 1840s a product called Warne's Compound was known, but was mixed by the farmer himself from a recipe published by John Warnes of Trimingham in Norfolk.[164] As an alternative to such cumbersome do-it-yourself procedures, the commercial production of compound feeds began in the 1850s. Probably the first in this trade was the Kingston Cattle Food Company, established at Hull in 1853 by Joseph Thorley. In 1856 he was marketing Thorley's Food for Cattle and in 1857 moved his business from Hull to London where he set up at King's Cross.[165] By 1863 the British farmer could buy Henri's Cow-Food and Pig-Meal, Gripper's Horse-Food, and Cattle-Food, and Pig-Meal, and the British Cattle Food Co.'s Cattle-Food.[166]

In 1850, 135,151 tons of oilseeds and 65,145 tons of cake were imported; in 1914 total imports of oilseeds were 1,262,465 tons and 329,431 tons of cake.[167] Thus in sixty-four years their consumption rose ninefold and fivefold, respectively. The fact that imports of raw seeds rose faster than cake fostered the establishment and expansion of domestically

[161] G. Jackson, *Hull in the Eighteenth Century* (London, 1972), pp. 187–92; H. W. Brace, *History of Seed Crushing in Great Britain* (London, 1960), pp. 25, 28; E. Gillett and K. A. MacMahon, *A History of Hull* (London, 1980), p. 343.

[162] A. Voelcker, 'The influence of chemical discoveries on the progress of English agriculture', *JRASE*, 2nd ser., 14, 1878, pp. 837–8; *Report of the Committee on Edible and Oil Producing Nuts and Seeds*, BPP, 1916, IV, Cd. 8247; C. Crowther, 'Palm kernel cake and meal', *JRASE*, 77, 1916, p. 47.

[163] A. Voelcker, 'On the composition and nutritive value of palm-nut kernel meal and cake', *JRASE*, 2nd ser., I, 1865, pp. 177, 181; *Royal Commission on Depression in Trade and Industry, First Report*, BPP, 1886, XXI, C. 4621, Appendix A, p. 94.

[164] J. Warnes, *On the Cultivation of Flax; The Fattening of Cattle with Native Produce; Box-Feeding and Summer Grazing* (2nd edn, London, 1847), pp. 134–266.   [165] Brace, *Seed Crushing*, pp. 88–9.

[166] T. Anderson, 'On the composition of different kinds of concentrated food', *Trans. Highland and Agric. Soc. of Scotland*, 23, 1863, pp. 176–8.   [167] *Annual Trade and Navigation Returns*, BPP.

produced animal feedingstuffs. Before the middle of the eighteenth century, when oil pressing relied heavily on home-grown seeds, it was much more widely dispersed than in 1850 when it was a predominantly port-based industry relying exclusively on foreign supplies of raw materials. The centre of the industry in Britain was Hull. Its importance can be gauged from the fact that in 1856 out of 150 hydraulic seed-crushing presses in Britain 100 were in Hull.[168]

Hull had processed Baltic oilseeds since the mid-eighteenth century, although mills in Lincolnshire, Yorkshire and other parts of the country had been pressing locally grown seeds much earlier. In 1840 there were ten mills on the River Hull, but by 1878 this had increased to forty-five.[169] Up to then the principal shipping lines from countries exporting oilseeds only visited the three chief ports of Hull, London and Liverpool, with the majority going to Hull, followed by London. The natural tendency was to build factories as near as possible to the point of arrival of these heavy goods, and where the main product of oil could be consumed close at hand. If a hinterland could absorb the by-products, this was an advantage, and Hull was conveniently sited to supply the mixed-farming districts of Lincolnshire, the East Riding and Nottinghamshire with oil-cakes.[170] In the middle of the nineteenth century, although the cake from country oil mills was consumed in adjacent districts, most of the oil had to be returned by sea to Hull or London for further processing. But improvements in shipping after 1850 allowed small ports to develop a regular liner traffic with a variety of return cargoes, and this enabled them to accept supplies of oilseeds almost as cheaply as they could be purchased in Hull or London.[171]

After 1860 the expansion of the agricultural market for oil-cake was an important factor in dispersing the industry. Rail carriage was expensive for the cake, whose value was low in proportion to its weight, and this made proximity to a farming hinterland an advantage.[172] While inland centres still sent away a large proportion of oil for further processing, their proximity to the cake-consuming districts permitted them to compete effectively against Hull and London.[173] Seed crushing was a small industry in 1850, because of limits imposed by supplies of raw materials, and technology.[174] Improved machinery with the use of the hydraulic seed press after 1850, and improved supplies of raw materials, removed earlier constraints and permitted an expansion in the size of

---

[168] Brace, *Seed Crushing*, pp. 25, 28, 49.

[169] Jackson, *Hull*, pp. 187–92; Brace, *Seed Crushing*, pp. 25, 28.

[170] Thompson, 'Second agricultural revolution', p. 67.

[171] J. W. Pearson, 'The seed crushing industry', *Journal of the Society of Arts*, 12 Dec. 1919, p. 51.

[172] *Report of the Committee on Edible and Oil Producing Nuts and Seeds*, BPP, 1916, IV, Cd. 8247, p. 14.

[173] Pearson, 'Seed crushing industry', p. 51.   [174] Brace, *Seed Crushing*, pp. 42, 45, 48, 53.

plant and the number of firms, a process assisted by the growth of the agricultural market for oil-cakes.[175] An indication of the importance of this market can be seen in the fact that agricultural depression was given as one reason for a decline in the industry from 1880. After rapid growth in the early 1870s businesses in Liverpool and Gloucester complained that increased foreign competition after 1875, combined with agricultural depression from 1878, limited expansion.[176] Growth was resumed in the late 1880s, and in the 1890s the location of the industry for the early twentieth century was laid down.

In 1850 crushers used mainly European seeds but, with the adoption of cottonseed crushing after the 1860s, supplies were obtained from India and America. The diversification of sources of supply further encouraged the establishment of other crushing industries in Liverpool, London, Bristol and Gloucester. Liverpool's position as the port, through which the raw cotton for the Lancashire textile industry passed, made it a natural place to receive the cottonseed by-product, and for a time the seed crushing industry based on cottonseed came to rival and threatened to outstrip the older linseed-based industry on the east coast.[177] The industrial market of the north west was a significant user of oil-based products, and the Lancashire and Cheshire dairy industries were important consumers of cattle foods. Liverpool was held back from domination of the industry because American cotton seed was unable to compete on quality with Egyptian cotton seed. By 1900 about twenty-five mills in England specialised in pressing cotton seed, and three quarters were in Hull which had become the world's largest single centre for crushing this seed. These mills could process about 400,000 tons a year, of which 85 to 90 per cent came from Egypt and only 15,000 to 20,000 tons from the United States.[178]

By 1914 the industry was widely distributed with centres of manufacture at Bristol, Bridgwater, Colchester, Gainsborough, Gloucester, Grimsby, Hull, Ipswich, London, Lincoln, Liverpool, King's Lynn, Manchester, Rochester, Selby, Southampton, Warrington and Weighbridge, as well as the major Scottish towns. This meant that no farmer in England or Wales was likely to be more than 70 miles from a manufacturing centre for this product. But although the agricultural market did encourage outport and inland development, Hull, though threatened, retained its primacy, and in 1914 still took over half the oilseed imported into the United Kingdom.[179]

[175] Brace, *Seed Crushing*, 1960, pp. 37–79; *The Economist*, 25, 9 Mar. 1867, p. 16; 27, 13 Mar. 1869, p. 20.   [176] *Royal Commission on Depression in Trade and Industry*, Appendix A, pp. 84, 94–95.
[177] J. Bibby and C. L. Bibby, *A Miller's Tale: A History of J. Bibby & Sons Ltd.* (Liverpool, 1978), pp. 11, 27; *Royal Commission on Depression in Trade and Industry, First Report*, Appendix A, pp. 84, 92, 94–5.   [178] Daugherty, 'Cotton seed industry', p. 295.
[179] *Hull Trade and Transit*, 10, 15 June 1914, p. 15.

2. Structure

No overall indication is available of the size and structure of the entire feedstuffs industry before the appearance of the First Census of Production taken in 1907. This provided no direct measurement of certain important parts of the industry, notably the value of wheat offals and those obtained from brewers' and maltsters' waste grains, as well as the output and value of oil-cake. But it is possible, using other information – some from the Census tables – and data on seed crushing, to form estimates of these. The results of this exercise are shown in Table 16.7. The picture that emerges is that output was from three sources: grain mills; specialised manufacturers of animal feed; and seed crushers. The last two produced mainly oil-cake or oil-cake products, but together they only accounted for 25 per cent of the industry's output. The other 75 per cent of animal feed products came from maltsters, brewers and grain millers. The output of grain millers can be divided into two categories: from provender mills specialising in animal feeds, and the remainder comprising offals from grain millers producing food for human consumption. Maltsters provided malt coombs, and brewers spent hops, yeast, and brewers' grains or draff.

In the middle of the nineteenth century the component derived from grain millers was more important as the output from seed crushers was small and there were no specialised animal food manufacturers. Also the character of the output from grain millers would have been rather different as at that time over 80 per cent of their inputs were home produced. The rise in quantity and the total percentage of imported wheat meant that by 1907 the position was reversed and approximately 80 per cent of the 33,792 cwt of wheat offals originated from imported wheat. The percentage of imports was equally high in the case of barley, and all maize, the other large item of feed produced by grain mills, was imported.

The whole industry had a large number of firms, a feature that was preserved by the changes in flour milling. In 1880 there were about 10,000 flour mills in Britain, most of them small country wind- or water-mills with two to four millstones. All these would have produced some animal feed, but with the rise in wheat imports flour production became concentrated in the ports. Against this background many small flour-millers only managed to survive by turning to provender milling of barley and oats, resulting in a large number of individual producers in the animal-feeds industry as a whole. As the scale of operations increased, this favoured the larger concerns which combined a wide range of imported feeds, as well as domestically produced ingredients. One firm to do this after 1880 was the Bristol-based milling firm of Spillers who used it as a method of disposing of their offals.[180]

---

[180] *Milling*, 7, 10 Oct. 1896, pp. 212–13.

Table 16.7. *Output of animal feed in England and Wales, 1907*

	Quantity		Value	
	thousand cwt	per cent	£ thousand	per cent
I FROM GRAIN MILLS				
Wheat offals	33,792		6,758[a]	
Oatmeal	296		156	
Barley meal	6,016			
Maize meal	8,287		5,683	
Bean meal	775			
Other meal (including some oil-cake meal)	1,144			
Peas and lentils	208		104	
Ground oil-cake	115		41	
Other animal and poultry feed, and offals	2,443		853	
Chopped hay and straw	[e]		59	
Gristing done for farmers	[e]		143	
TOTAL from grain mills	53,076	64	13,797	65
II FROM ANIMAL FEED MANUFACTURERS				
Cattle feed (includes cakes and artificial feedstuffs)	3,352		1,138	
Dog food	[e]		108	
Poultry food	[e]		85	
Other animal foods	[e]		14	
TOTAL from animal feed manufacturers	3,352	4	1,345	6
III FROM SEED CRUSHERS				
Oil-cakes	17,540[b]	21	4,145[c]	20
IV FROM MALTSTERS AND BREWERS[d]				
Offals	9,040	11	1,808	9
TOTAL, from I, II, III and IV	83,008	100	21,095	100

Notes:

[a] Estimated value of £0.20 per cwt, based on the values of oat and other offals.

[b] Assuming 70 per cent of the oil and cake produced (1,253,000 tons).

[c] Assuming 35 per cent of the total value of oil and cake produced (£11,844,000).

[d] Amount for 1911–13, estimated by Paul Brassley and valued at £0.20 per cwt.

[e] Recorded by value only.

Source: Based on the *Final Report on the First Census of Production of the United Kingdom* (1907), BPP, 1912–13, CIX, pp. 482, 517, 575.

Spillers, like the other firms in the industry, produced a variety of specialised products aimed at a range of particular niche markets. This was partly because it was realised, at an early stage in the trade, that some types of oil-cakes and other feed products were harmful to certain classes of animals if given in excessive amounts. This was mainly the concern of agricultural scientists, but manufacturers realised there was a price advantage in the branded product as well as being aware of, and keen to promote the growth of, specialised markets. In the 1860s Joseph Thorley advertised some of his output as a feed supplement to be mixed with conventional animal fodder, but at £24 a ton Thorley's Condiment was decidedly expensive as Thorley's Feeding Meal was £14 a ton, whereas the unbranded linseed cake of other makers was only £10 a ton. In the late 1870s the Royal Agricultural Society of England saw all cakes and feeding compounds in terms of either fattening mature stock or rearing young animals.[181] But by the 1890s a bewildering variety was on sale of composite cakes for cattle and sheep, mixed feeding meal for young calves and grown-up cattle, compound lamb food and compound poultry food. Joseph Bibby was marketing Dairy Cow Meal specifically tailored for the dairy farmer producing milk and butter which he advertised, like Joseph Thorley twenty-five years earlier, through the firm's own news sheet, *Bibby's Quarterly*, which was distributed free to customers from 1896. By 1911 BOCM at Hull were selling their own general branded blends of linseed and cotton cakes marked 'BO & CM Pure', and citing its chemical analysis by the Royal Agricultural Society of England as a certificate of guarantee. But they also produced Robson's Compound Cake, which their advertisements claimed to be particularly suited to beef animals, and which had been used in feeding a number of prize winners for Christmas fatstock shows.[182] Further specialisation appeared in 1906 when the members of the London Corn Trade Association who dealt in animal feed grains decided that their work was so different from the regular grain trade of the capital that they formed the London Cattle Food Trade Association.[183]

The seed-crushing section of the industry in 1911 had 10,504 persons directly employed in crushing and refining in England and Wales, with another 4,200 additional workers. Hull and Liverpool employed the

---

[181] J. B. Lawes, 'Observations on the recently-introduced manufactured foods for agricultural stock', *JRASE*, 19, 1858, pp. 199–204; *Thorley's Agricultural Gleaner*, 3, 12 Jan. 1863, p. 8, 2 Mar. 1863, p. 5; Voelcker, 'The influence of chemical discoveries on the progress of British agriculture', pp. 837–8; J. Macdonald, *Book of the Farm*, vol. 1 (4th edn, Edinburgh, 1891), p. 275.

[182] *Departmental Committee on Fertilizer and Feeding Stuffs*, BPP, 1892, xxvi, Cd. 6472, QQ. 2884–6; 'Joseph Bibby (1851–1940)', in D. J. Jeremy (ed.), *Dictionary of Business Biography*, vol. 1 (London, 1984), pp. 325–7; *Transit*, 7, 15 Feb. 1911, pp. 13, 15.

[183] H. Barty-King, *Food for Man and Beast* (London, 1978), pp. 39–43.

largest numbers with 32 and 28 per cent, respectively. London came third with 11 per cent. The next largest centre was the west midlands counties where 7 per cent of the workforce were employed, most of these in Gloucester.[184] In 1914 the industry had some sixty separate firms in the whole of Great Britain, ranging from the small country concerns with a mill capacity of around 5,000 tons a year, to the much larger plants that made up the combine of the British Oil and Cake Mills Ltd – universally known as BOCM – who together encompassed somewhere between 45 and 60 per cent of the industry's capacity.[185] BOCM was established in July 1899 by the amalgamation of seventeen firms engaged in oil and cake manufacture. In its prospectus it claimed to have twenty-eight mills and twelve oil refineries. The mills had an aggregate crushing capacity of 500,000 tons per annum, which in 1898 was over half the United Kingdom's annual oilseed imports.[186] Although most of its plants were in Hull, it had mills in London, Rochester, Bridgwater, Gloucester, Gainsborough, Liverpool, Glasgow, Leith and Burntisland, and depots at Brigg, Boston, King's Lynn, Yarmouth and Southampton.[187] In the beginning the company left day-to-day running of individual mills in the hands of their former owners and used its considerable market power to purchase raw materials, and develop a single network for the joint marketing and distribution of all its finished products, including cattle cake.[188]

CONCLUSION

The industries covered in this chapter were the late nineteenth-century equivalent of 'new industries' in that they offered British farmers the chance to direct the results of recent advances in science and technology into their own production systems. But some of these industries encountered a more encouraging market than others; the manufacturers of farm machinery and agro-chemicals faced a declining arable acreage which certainly limited, even if it did not actually reduce, the potential demand for ploughs, harvesters and fertilisers. Both reacted by basing part of their growth on overseas markets. Although some agricultural machinery had been exported before 1875, most firms developed a heavy reliance on overseas markets after that date, and exports of manufactured fertilisers, undertaken from 1860, were also to become an important item up to 1914. As exports of machinery and fertilisers assisted overseas competi-

---

[184] *1911 Census*, x, Pt 1.

[185] Pearson, 'Seed crushing industry', p. 50; H. W. Macrosty, *The Trust Movement in British Industry* (London, 1907), p. 316.   [186] *Ibid.*, p. 316.

[187] Gillett and MacMahon, *History of Hull*, p. 344; *Transit*, 7, 15 Feb. 1915, pp. 13–15.

[188] Macrosty, *The Trust Movement*, pp. 314–5; *Transit*, 7, 15 Feb. 1915, p. 13.

tors, these industries did not have an entirely beneficial effect on British arable farming, although the major export market was Europe rather than the New World. However, the other two industries surveyed in this chapter faced a far more encouraging market situation at home, and needed to export hardly anything. Manufactured feedingstuffs provided an eagerly sought-after input to livestock farming, which accounted for 66 per cent of farm output in 1870–6 and rose to 75 per cent of total output by 1911–13.[189] Indeed, this industry relied on imports for a higher proportion of its raw materials which included not only foreign maize and oilseeds, but also substantial amounts of offals from domestic millers who were now grinding increased amounts of foreign small grain. In comparison with the preceding industries, veterinary products was very small in size and hardly more than a collection of small workshop producers. The size of its undoubtedly growing market is most difficult to estimate, catering mainly for around 2,500 veterinary surgeons in England and Wales before 1914, plus an unknown number of farriers and farmers who cared enough about animal health to purchase their own supplies of animal medicines.

Among the firms within these industries competition was fierce and they shared to varying degrees the usual atomistic structure of most of British industry.[190] Although hundreds of firms made agricultural implements, there were perhaps around only 30 of any significant size by 1914 and only about 4 employing more than 2,000 workers apiece. Fertiliser manufacturers were just as numerous, and this was helped by the fact that part of their output was a by-product of gas and steel manufacture – industries that were in themselves, and for their own particular reasons, highly atomistic. Unsuccessful attempts to reduce effective competition with price agreements were tried by fertiliser manufacturers in the late 1880s and early 1890s, but foundered upon the barriers to co-operation offered by the persistence of a multiplicity of small firms before the First World War. In the feedstuffs industries a rare example of successful concentration was the establishment in 1899 of the British Oil and Cake Mills combine. However, as this industry also relied on the by-products of other industries, its manufacturing plant was widely distributed, even though the largest mills were concentrated at the ports.

---

[189] E. M. Ojala, *Agriculture and Economic Progress* (Oxford, 1952), p. 209.

[190] B. Elbaum and W. Lazonick, 'An institutional perspective on British decline', in B. Elbaum and W. Lazonick (eds.), *The Decline of the British Economy* (Oxford, 1986), pp. 3–4.

CHAPTER 17

FOOD PROCESSING INDUSTRIES

INTRODUCTION

The common experience of the industries in this chapter was their transfer from small mill, malting house, bakery and workshop to the large factory, as changes in technology raised them from small batch processors to mass-producers. These new technologies involved the organisation of large amounts of labour in single industrial units and required the forward purchasing and close control of very large volumes of raw materials. The high volume of output achieved by the biggest firms raised its own particular challenges. However, not all firms and entrepreneurs participated in these changes to the same extent. These industries exhibited one of the usual features of the diffusion of new technology among a heterogeneous collection of firms in the same sector, in that they adopted new technology at different rates, and to differing extents.[1]

A variety of reasons explains these differences, but they can best be summarised in terms of the factors that shaped individuals' perceptions of the risks of, and opportunities for, technical change. For instance, corn millers far from the ports which received increased imports of grain after 1870 faced higher transport costs and were unable to gain the same advantages from bulk unloading systems as coastal millers. They were therefore slower to scrap the old system of stone-grinding and invest in roller milling, and instead often chose to fight a losing battle against the advance of new technology. Milling was an intermediate industry and the success of the large firm was mirrored in the other intermediate industry in this chapter, malting. In 1850 this was still 'a trade of overwhelmingly small-scale, even tiny, businesses', but it too saw a more radical shift in its structure with their removal by 1914, although here the changes in technology were reinforced by independent changes in the structure of the brewing industry which favoured

---

[1] P. A. David, *Technical Choice, Innovation, and Economic Growth: Essays on American and British Experience in the Nineteenth Century* (Cambridge, 1975).

the larger maltster.[2] But in the further processing of flour, there were no such strong backward linkages and pressure for change, imposed on the milling industry from the final users. Although some town bakers found it worthwhile to invest in bread factories, the bread market remained sufficiently fragmented up to 1914 to preserve those bakers choosing to soldier on in small bakehouses with a retail shop. The other portions of food processing experienced varying degrees of change. Growth of convenience foods and the canning industry required the adoption of new technology, but the British dairy products industry only managed to achieve a limited transformation.

[2] J. Brown, 'The malting industry', in G. E. Mingay (ed.), *AHEW*, vol. VI, *1750–1850* (Cambridge, 1989), p. 503.

A

MILLING

BY RICHARD PERREN

In the second half of the nineteenth century this sector of the food industry experienced major changes. Firstly there was the replacement of home-produced by imported raw materials, and secondly the rise, and eventual dominance by 1914, of a few big, highly organised milling firms using modern business methods and large power-driven factories with the latest technology. The other change was the marked geographical concentration of the industry in coastal locations.

1. The industry in 1850

In 1850 the milling industry was engaged almost wholly in processing home produced inputs because the introduction of free trade in 1846 did not result in really heavy supplies from abroad before 1860. In 1850 the industry was characterised by generally small-scale units of production. In the 1851 Census only two master millers in England and Wales employed more than 100 men, and of the 2,394 individual returns made by master millers, 403 employed no men or failed to say how many they employed, 644 employed only one man and a further 483 employed two men.[3] Flour mills were widely dispersed across the country. The map in the 1851 Census giving the distribution of the occupations of the people did not include milling – an indication that there was no strong regional specialisation. Again, an industrial map of Great Britain, dated 1881, in John Yeats's *Technical History of Commerce*, showed no particular concentrations of flour milling.[4] There was some clustering of corn warehouses in seaports as corn merchants attending inland market towns used the ports as depots to assemble their stock and then send it by coastal vessels to other centres.[5] But the concentration was not overwhelmingly coastal, and in 1850 inland milling centres like Gloucester were as important as those at ports.

In 1850 most wheat was consumed close to the farming districts where it was grown, a pattern of production and consumption consistent with

---

[3] *1851 Census, Summary Tables*, Table xxx, p. cclxxvii.

[4] *1851 Census, Summary Tables*, p. ccxi; J. Yeats, *The Technical History of Commerce* (London, 1887), folding map in pocket at the end of the volume.

[5] H. Stephens, *Book of the Farm* (2nd edn, vol. IV, Edinburgh and London, 1851), p. 757; R. Bennett and J. Elton, *History of Corn Milling*, vol. IV (London, 1904), p. 199.

a large number of small scale milling enterprises. London was supplied with wheat from Kent, Essex, and a little from Norfolk. Northumberland's wheat was consumed in the immediate mining districts. The manufacturing towns of West Yorkshire were supplied from the North and East Riding and Lincolnshire. Nottingham and Leicester's bread was grown in those counties; Warwickshire and Staffordshire supplied Birmingham and the Black Country; and the coal and iron districts of South Wales consumed the wheat grown in Wiltshire, Somerset and Dorset.[6] This pattern of geographical location was dictated by its dependence on traditional technology, particularly wind and water as the major sources of motive power. Kanefsky has estimated that as late as 1870 two thirds of the 90,000 horse power available for all grain milling was provided by watermills. At that date the majority of steam mills were found in the ports, and these accounted for only a minority of flour mills. Technology had not undergone any major change since the introduction of steam milling at the end of the eighteenth century. By freeing the industry from its former dependence on wind and water power, it enabled it to move from predominantly rural and riverine sites to urban ones, but that was all. In 1850 millstones were still used to grind the wheat, so mechanical principles were largely the same as they had been in the Middle Ages. The replacement of wooden by more efficient iron gear wheels after 1800 permitted faster working speeds and reduced the amount of servicing.[7] But only when the steam engine became the chief motive power after 1880 did the industry acquire 'locational mobility' enabling it to concentrate in particular sites.[8]

The industry, with a large number of small units, was highly competitive. Prior to the railways, and when the barrel was the most common form of packaging for the long-distance transport of flour, problems were involved in moving the product over long distances. With water transport, either inland or coastal, there was the risk of product damage from dampness or seepage. But stone milled flour did not keep in good condition for very long before it became musty and unpalatable. Therefore there was no truly national market for the product, but a series of interrelated local ones, each led by the largest firm in the area.[9] Most country mills would sell their flour in the district a few miles around, while the larger port and urban mills would probably not go more than 30 miles beyond in search of custom. There was also a number of part-time

[6] *The Economist*, 8, 23 March 1850, p. 310.

[7] See J. Kanefsky, 'Motive power in British industry and the accuracy of the 1870 Factory Return', *EcHR*, 2nd ser., 32, 1979, pp. 372–3; G. N. Von Tunzlemann, *Steam Power and British Industrialization to 1860* (Oxford, 1978), p. 122.

[8] W. W. Rostow, *How it All Began: Origins of the Modern Economy* (London, 1975), pp. 166–67.

[9] *Milling*, 16, 23 Feb. 1901, p. 107.

millers, combining this occupation with others, of which farming was a favourite, and who added to this competition.[10] In the 1851 Census, 1,855 farmers in England and Wales reported milling as a subsidiary occupation. Miller, after victualler, beershop keeper and innkeeper, was the most common occupation for occupiers of land.[11] This meant a high surplus capacity that could be brought into production on a seasonal basis if demand was high.

2. Changes in technology

The 1870s were the start of the roller milling revolution, a process developed and perfected from the 1860s, first in Hungary, then in Western Europe, and then the United States. From start to finish the new method was radically different from stone milling. In stone milling the pulverising of the wheat was performed generally in one operation producing a coarse wheat meal containing all parts of the grain mixed together. To obtain fine white flour the mixture was passed through sieves of varying mesh. But by using rollers, the process became one of gradually reducing and separating the different portions of the wheat in various stages. The first set, called the break rolls, split open the husk and separated the floury kernel, or middlings, from the bran. It then passed through further rollers, after which the stock was again sifted and remaining coarser material removed.

There were two main advantages with this system. Firstly, it was mechanically more efficient; but although it required less power to grind the same amount of corn than stones, it did require a higher initial power threshold, which precluded its successful installation in most water- and windmills where the amount of power available depended on the elements. There were also steam-powered stone mills,[12] but the second advantage of roller milling was that it produced a higher proportion of fine white 'patent' flour, the type which was most favoured by consumers, whereas millstones produced more of the less popular coarse 'household' grades. The new process was adopted most rapidly in Europe and the United States. After the Civil War the area around Minneapolis and St Paul became an important milling centre and by the early 1880s all its mills used the new system. With the new technology and sustained by an increasing supply of river- and,

---

[10] *Royal Commission on Agriculture, Minutes of Evidence*, BPP, 1881, XVII, Cd. 3096, p. 917, QQ. 56,767–73 (Evidence of T. Rigby, Secretary of Cheshire Chamber of Agriculture).

[11] *1851 Census, Summary Tables*, Table XXVII, p. cclxxxv.

[12] J. Tann, 'Corn milling' in G. E. Mingay (ed.), *AHEW*, vol. VI, *1750–1850* (Cambridge, 1989), pp. 409–13.

more importantly, rail-borne wheat from the corn lands of the mid-west, Minneapolis soon became the premier corn-milling centre of the Western Hemisphere.[13]

The initial rate of adoption of the new technology was slower in Britain than in Europe. In 1869 only 79 British flour mills out of an esti-mated total of over 10,000 were using a variation of the Hungarian system.[14] A St Louis miller who shipped the first significant quantities of American flour to Britain in 1874-5 saw nothing new in the machinery of any of the mills he visited.[15] Even after a large exhibition of American and European milling machinery held in London in 1881 had demon-strated the advantages of the new technology not all British millers were prepared to convert their machinery. The late nineteenth- and early twentieth-century literature on milling technology gives the impression that British flour milling was a lagging sector with a slow response time when compared to Continental and United States models.[16] Technical compromises were adopted even by individuals who were later to become leading figures in the industry. Arthur McDougall at the City Corn Mills in Manchester first used the Hungarian system of porcelain rollers for secondary reductions and only installed a complete roller system – the first in Britain – in 1878.[17] When stone milling was seri-ously threatened numerous experiments were carried out to see if meas-ures like improvements in stone dressing, better systems of grain feed, or perhaps more accurate balancing of the stones, could redress the advan-tages of roller milling.[18]

Looking back in 1901, Sidney Leetham, head of the successful Hull firm of Henry Leetham and Sons, Ltd, founded in 1878, recalled:

Twenty years ago our trade was threatened with extinction owing to the pres-sure of foreign competition, this was due entirely to the fact that our machin-ery had become obsolete, and that the Americans with greater perspicuity and alertness of mind had the insight to scrap theirs as they realised it was out of date. Our convictions came, as they generally do, rather late . . .[19]

---

[13] J. Storck and W. D. Teague, *Flour For Man's Bread: A History of Milling* (Minneapolis, 1952), pp. 201, 228, 234; H. A. Calton, 'Hard wheats winning their way', *United States Department of Agriculture Yearbook*, 1914, p. 391; F. A. Shannon, *The Farmer's Last Frontier: Agriculture 1860-97* (New York, 1945), p. 148.

[14] Storck and Teague, *Flour For Man's Bread*, p. 201; J. Tann and R. G. Jones, 'Technology and trans-formation: the diffusion of the roller mill in the British flour milling industry, 1870-1907', *Culture & Technology*, 37, Jan. 1996, 1996, pp. 36-69.   [15] *The Miller*, 1, 5 July 1875, p. 102.

[16] W. R. Voller, *Modern Flour Milling*, (2nd edn, Gloucester, 1892), p. 108.

[17] *Milling*, 4, 7 Oct. 1892, pp. 219-20.

[18] P. A. Amos, *Processes of Flour Manufacture* (London, 1912), pp. 135-6.

[19] *Milling*, 16, 11 May 1901, p. 312.

3. Foreign competition

An important change that started in the second quarter of the nineteenth century was the influx of foreign wheat. Before 1850 approximately 85 per cent of wheat consumed in Britain was home grown.[20] In Table 17.1 (column 2) it can be seen that by 1875 it was under 50 per cent, and by 1914 only around 20 per cent of bread consumed in Britain was from home-grown wheat. But accompanying the influx of foreign wheat came increasing quantities of foreign flour, as can also be seen from Table 17.1 (columns 5 and 6). Until the end of the 1870s flour imports had never been more than 10 per cent of the nation's consumption, but by 1880–4 they had risen to 17 per cent and by the end of the 1880s had almost reached 20 per cent. In the 1880s one of the complaints from the Liverpool corn traders to the Royal Commission on Depression in Trade and Industry was reduced profits owing to the competition from foreign manufactured flour.[21] The response of some millers to this challenge was similar to that of the farmers. The National Association of British and Irish Millers (hereafter referred to as N.A.B.I.M) was formed in 1878 in the wake of anxieties over the new technology and flour imports. At its annual general meeting in 1886, an unsuccessful attempt was made to commit the organisation to an official policy of tariff protection.[22] In 1891 the industry's leading trade paper observed: '. . . there is no reason why we should not manufacture our food as far as possible at home, even if we have to import the raw material'.[23] The next year it lamented: '. . . the volume of imported flour continues to represent rather over 25 per cent of our home receipts . . . while on the other hand mills be idle in all districts, both town and country . . .'[24]

Flour imports reached their highest point in the second half of the 1890s, but even as late as 1902 the *Mark Lane Express* issued a special number to celebrate its 70th birthday, in which it not only recorded the demise of British farming, but also prepared to deliver the last rites to the agricultural processing industries as well. It took the view that under the onslaught of competition from imported flour home milling was in decline. This verdict was applied not only to the representatives of the older technology – the dismantled windmills and the roofless watermills – but 'even some of the best equipped mills on the newest lines of roller flour milling have been obliged to close their doors and in many of these cases the machinery has been dismantled and sold'.[25]

[20] M. G. Mulhall, *Dictionary of Statistics* (London, 1892), p. 15.
[21] *Royal Commission on Depression in Trade and Industry, First Report*, Appendix A, BPP., 1886, XXI, Cd. 4621, pp. 93–4.   [22] *Broomhall's Corn Trade Year Book* (Liverpool, 1904), p. 7.
[23] *Milling*, 2, 20 Nov. 1891, p. 12.   [24] *Milling*, 3, 12 Feb. 1892, p. 59.
[25] *Mark Lane Express*, 70th Birthday Number, 31 March 1902, p. 15.

Table 17.1. *Estimated annual quantities of wheaten flour milled in the United Kingdom, imported, and consumed per head, 1850–1914*

	1	2	3	4	5	6	7	8
			Flour milled in the UK from:					
	Imported wheat		Home-produced wheat		Imports of wheat meal and flour		Total flour	Flour consumption
Years	million cwt	%	million cwt	%	million cwt	%	million cwt	lb per head
1850–4	12.3	19	47.0	74	4.3	7	63.6	259
1855–9	11.8	19	48.1	76	3.0	5	62.9	250
1860–4	21.0	31	40.6	60	5.6	9	67.2	255
1865–9	21.5	33	39.9	61	4.2	6	65.6	242
1870–4	28.1	41	34.8	51	5.1	8	68.0	237
1875–9	36.4	48	32.5	42	7.6	10	76.5	255
1880–4	40.3	52	24.4	31	13.3	17	78.0	246
1885–9	39.3	48	26.8	33	16.0	19	82.1	251
1890–4	45.8	52	24.2	27	18.8	21	88.8	258
1895–9	48.5	54	20.7	23	20.5	23	89.7	249
1900–4	56.7	59	20.3	21	19.7	20	96.7	256
1905–9	66.7	69	17.7	18	12.7	13	97.1	249
1910–14	73.2	71	19.6	19	10.7	10	103.5	257

Sources: columns 1 and 5: *Statistical Abstract and Annual Abstract of Statistics*; column 3: 1850–54, W. Schlote, *British Overseas Trade from 1700 to the 1930s*, (Oxford, 1952), p. 62, estimates annual output for 1849 to 1854 at 16.7 million quarters; 1855–84, *Broomhall's Corn Trade Yearbook* (Liverpool, 1904), p. 138; 1885–1914, *Agricultural Returns, GB*; column 7 = columns 1 + 3 + 5; column 8 = column 7 divided by UK population from B. R. Mitchell and P. Deane, *Abstract of British Historical Statistics* (Cambridge, 1962), pp. 8–10.

Using an extraction rate for columns 1 and 3 declining from 75 per cent in 1850–4 (E. J. T. Collins, 'Dietary change and cereal consumption in Britain in the nineteenth century', *AHR*, 23, 1975, p. 108) to 70 per cent by 1875–9 (J. Kirkland, *Three Centuries of Wheat, Flour, and Bread* (London, 1917), p. 17).

Some sources for columns 1 and 3 give quantities in quarters. These were converted into cwt at 1 quarter = 4.418656 cwt, derived from Mitchell and Deane, p. 86. Where necessary in column 3, 15 per cent was deducted to allow for seed and tail corn, as suggested in *The Times*, 8 June 1914, 16c.

Small quantities of flour had been imported from Europe in the 1850s and 1860s after the repeal of the Corn Laws. Mainly these were from France, Germany and Austria-Hungary. In 1878 Hungarian mills produced around 30 million cwt of flour, half of which was exported, though most within the Habsburg empire and to the German states, and only a small amount finding its way to Britain.[26] But after 1875 the main source of Britain's flour imports was the United States. These imports of American flour were first noticed on an appreciable scale in the winter of 1874–5 when some 75,000 cwt were shipped from St Louis and sold in London, Bristol, Liverpool and Glasgow.[27] In 1889 the *Bakers Times* claimed 'the English market [is] glutted with United States Flour' and by 1892 it feared that imports were large enough to threaten the UK milling industry.[28] Between 1878 and 1894 Britain's imports of flour from the United States rose, whilst those she received from the rest of the world declined. European flour was unable to compete on price with the American product. Not all this American flour went for human food. Makers of dog biscuits, cattle feeds and animal meal were known to import and use large quantities. Just how much was consumed by the non-human population is hard to say, but it was thought that perhaps 20 to 25 per cent went into animal foods. But even so, this still left about 14 million cwt a year available for bread making in the mid-1890s.[29]

Imports of foreign flour hovered around the level of 22 or 23 per cent for most of the 1890s, though in 1899 they again reached 25 per cent. They declined after 1902, as United States supplies declined and no other producer came forward to take its place. In 1904 and 1905 flour imports made up 15 per cent of Britain's wheat consumption, and had returned to the level of the early 1870s of around 10 per cent by 1910–14.[30] Changes in the quantities of American flour followed the same pattern as changes in imports of American wheat. From 1878 to 1902 was the great surplus-producing period in the development of wheat production, and from 1903 to 1913 exports were much lower than in the previous decade.[31]

4. The response of the British industry

The apparently slow initial response to technical change by British flour millers is explained by the industry's situation in this country which was

[26] M. G. Mulhall, *The Progress of the World* (London, 1880), p. 394; J. Komlos, *The Habsburg Monarchy as a Customs Union* (Princeton, N.J., 1983), p. 140.   [27] *The Miller*, 1, 5 July 1875, p. 102.

[28] *The Bakers Times*, 3, 9 March 1889, p. 2; *Milling*, 4, 9 Sept. 1892, p. 202.

[29] *Milling*, 7, 10 February 1894, pp. 41–2.

[30] J. Kirkland, *The Modern Baker, Confectioner and Caterer* (London, 1924), vol. IV, p. 175.

[31] C. R. Ball, C. E. Leighty, O. C. Stine and O. E. Baker, 'Wheat production and marketing', *United States Department of Agriculture Yearbook*, 1921, pp. 152–4, 752.

different from those of its counterparts in Europe and America.[32] Variations in supply factors and in the nature of demand explain differing initial responses to changes in technology. British millers were faced with a slowly expanding market.[33] *Per capita* flour consumption in the UK between 1850 and 1914 remained constant at something under 5 lb a week (Table 17.1 col. 8). As prices fell and real incomes rose, people chose to eat more of the higher protein foods rather than more carbohydrate in the form of bread: increases in the carbohydrates were met by higher consumption of sugar, as a sweetener in drinks and in cakes, biscuits and jams.[34]

There does not appear to have been any real difficulty in obtaining sufficient capital to undertake investment in new technology, although there is no doubt that new process milling was more capital intensive than stone milling. Joseph Rank, with little initial capital, was able to build up his milling empire by a combination of careful living, steady reinvestment of profits, and bankers' advances.[35] Nor were British millers deterred from adopting roller milling because a cheap and plentiful supply of labour made it uneconomic. In America, where labour was dearer than in Britain, the new milling process was associated with lower labour requirements. But in central Europe where wages were lowest this did not inhibit the introduction of roller milling, although there the advance of roller milling was also assisted by the availability of hard wheat, where the British farmer only grew soft wheat. In Hungary milling was an attractive subject for early industrial development, in spite of its relatively low added value, because its requirements for skilled labour were small. The new technology was quite flexible as to how it was installed and operated, and although small amounts of labour and much automatic machinery were used in America, Europe only adopted the latter after 1890.[36]

---

[32] This is explained in some detail in R. Perren, 'Structural change and market growth in the food industry: flour milling in Britain, Europe and America, 1850–1914', *EcHR*, 2nd ser., 43, 1990, pp. 420–37.

[33] B. R. Mitchell and P. Deane, *Abstract of British Historical Statistics* (Cambridge, 1962), pp. 9–10; B. E. Supple, 'Income and demand, 1860–1914', in R. Floud and D. McCloskey (eds.), *The Economic History of Britain Since 1700*, vol. II (Cambridge, 1981), p. 31.; C. H. Feinstein, *Statistical Tables of National Income, Expenditure and Output of the UK, 1855–1965* (Cambridge, 1976), T.42.

[34] Annual per capita consumption of refined sugar rose from 30.1 lb in the 1850s to 90.8 lb by 1914. G. N. Johnstone, 'The growth of the sugar trade and refining industry', in D. J. Oddy and D. S. Miller (eds.), *The Making of the Modern British Diet* (London, 1976), p. 60.

[35] R. G. Burnett, *Through the Mill: The Life of Joseph Rank* (London, 1945), pp. 24, 32–3, 104.

[36] Storck and Teague, *Flour For Man's Bread*, pp. 210, 251, 255, 262; Kolmos, *The Habsburg Monarchy*, pp. 132, 140, 144; I. T. Berend and G. Ranki, *Hungary: A Century of Development* (Newton Abbot, 1974), p. 52; J. R. Walton, 'Varietal innovation and the competitiveness of the British cereals sector, 1760–1930', *AHR* 47, 1999, pp. 29–57.

Although adaptation may have been slow and partial before 1890, there is evidence that by the last decade of the nineteenth century strong attempts were being made to catch up. In the early 1890s it was observed that:

The mills of five years ago are nearly obsolete already . . . Machines . . . have been adopted . . . with a paean of triumph . . . only to be discarded a few months later for some more favoured rival . . . while experiment in all directions seems to have no end.[37]

The driving force behind this transformation was the change in the sources of raw materials, which had a strong influence on the rate at which the new technology was adopted. This was because imported wheat was not a direct substitute for home-grown wheat, as there were important differences of quality and type between the two products. The British farmer had always grown soft wheat because it was best adapted to the damp British climate, but by the 1890s most wheat imports were from Europe, Russia and America where climatic conditions – notably their hotter, drier, summers – favoured hard wheats. It was possible to make flour from hard wheat using millstones, but for technical reasons hard wheat was best suited to processing by rollers, and it produced a different kind of flour from soft wheat. Hard-wheat flour is 'strong' flour and soft wheats produce 'weak' flour; the basic differences being in the amount of gluten they contain, hard wheats having the highest gluten content. The classification of wheat types and the kinds of flour they produce is complex, but writing in 1903 a British bakery expert classified weak flours as those containing 6 to 9 per cent dry gluten and strong flours as having 9 to 16 per cent.[38] Flour manufactured from soft wheat alone bakes into a dense, heavy-textured loaf. The majority of the British public were eating soft wheat bread in 1850 and, because none other was available, this was what they preferred. But by 1914 millers and bakers had educated an increasing number of the British public to accept, and become accustomed to expect, a light, even-textured loaf baked with a blend of flour containing a high proportion of flour from hard wheats but with some soft-wheat flour.[39]

In the United States the dramatic increase in wheat output was largely a function of the increased acreage of hard wheat. In the early 1870s a lot of American wheat was rather nondescript types of soft wheat similar to the wheat grown in Britain. Before the Civil War St Louis in the soft winter-wheat area of the middle west was the main milling centre, but the years after 1870 saw the rise of Minneapolis in the hard spring-wheat

[37] *Milling*, 3, 12 February 1892, p. 59.   [38] O. Simmons, *The Book of Bread* (London, 1903), p. 35.
[39] Kirkland, *The Modern Baker, Confectioner and Caterer*, vol. I, pp. 55–63.

area further west.[40] And as the wheat area moved further westwards into the prairie regions of America and Canada from the late 1870s, the wheat grown there was various types of hard wheat which behaved very differently from soft wheat. Initially millers in Minnesota and the Dakotas complained they could not successfully grind these wheats with their hard friable skins using millstones. The action of the millstones, involving a large amount of friction, pulverised a considerable portion of the husk, making a dark, unpalatable flour that could only be sold at a low price, if at all. The problem had been partially solved in the 1860s in Hungary, where similar varieties of hard wheat were grown. Because Britain lacked plentiful supplies of hard wheat, and because the traditional technology was quite adequate to process the wheats that she grew, there was no need to exploit new technologies. It has also been suggested that there had been gradual deterioration in the quality of British wheats from the 1840s, a problem not specifically identified until after 1900. If this were the case, it may also have been an additional factor discouraging British millers from taking an early interest in implementing new technology.[41]

There was a need for new technology in Hungary and America, since both produced large amounts of hard wheat. This was a reflection of climatic differences and the agricultural sector's greater weight in each economy, as well as the fact that flour milling was an important industry in both countries.[42] In these cases there was every incentive to develop it to cope with large and expanding volumes of grain. But in Britain flour milling was not a major industry and its pattern of geographical location did not favour investment in large flour factories, which was a feature of the early history of roller milling. In both Europe and America transport developments meant that particular sites received large deliveries thus favouring the concentration of milling at these points.[43] In 1880 the position of the industry in England and Wales was weak with regard to withstanding foreign competition, but the increasing imports of American flour were more a function of excess capacity in that country than the

---

[40] C. B. Kuhlmann, *The Development of the Flour-Milling Industry in the United States* (Boston, 1929), pp.71–155.

[41] A. E. Humphries, 'Modern developments in flour milling', *Journ. of the Society of Arts*, 55, 21 December 1906, pp. 110–11; Storck and Teague, *Flour For Man's Bread*, pp. 275–8.

[42] US Census Office, Ninth Census, 1870, *The Statistics of Wealth and Industry of the United States*, vol. 3 (Washington, D.C., 1872), pp. 329–31, Table 8 (B); L. C. Hunter, *A History of Industrial Power in the United States*, vol. II (Charlottesville, 1985), p. 35; Komlos, *The Habsburg Monarchy*, pp. 132–43; S. M. Eddie, 'The terms and patterns of Hungarian foreign trade, 1882–1913', *JEH*, 37, 1977, p. 337; Eddie, 'Agricultural production and output per worker in Hungary, 1870–1913', *JEH*, 28, 1968, pp. 197–222; J. Komlos, 'Austro-Hungarian agricultural development', *Journ. of European Economic History*, 8, 1979, pp. 37–60.

[43] Storck and Teague, *Flour For Man's Bread*, p. 206.

weakness of the British industry. British imports of American flour did not decline after 1900 because the domestic milling industry had modernised but simply because by then the United States's population had grown sufficiently to absorb domestic production.[44]

Meanwhile, increasing imports of all descriptions of foreign wheat, and declining quantities of British wheat, forced the industry to change. Most wheat arrived at Glasgow, Liverpool, Bristol, Hull and London. These concentrated supplies placed the flour millers in these ports in a similar position to millers in America and Europe. For the first time it was worthwhile for millers in this country to invest in really large mills, and the wheat itself required such investment. This can be illustrated with regard to cleaning. In the 1860s, apart from some kiln-drying to reduce the moisture content, English wheat arrived at the mill practically fit for milling directly as it was received. Farmers were glad to use their surplus labour not only to thresh, but also to clean the grain, although as prices fell after 1870 the use of labour for this purpose became too expensive and as machinery was used in harvesting, even English wheat came to contain dirt, seeds and stones.[45] But increasing quantities of foreign wheat posed more serious difficulties in the screen room. Impurities in American and Canadian wheat were between 2 and 4 per cent – almost as good as English wheat – but in Indian and Russian, impurities averaged 6 per cent, and in the worst samples could be as high as 12 per cent.[46] To cope with the requirements of a raw material increasingly drawn from all parts of the world and containing varying impurities, further technology was developed to include textile and cyclone dust extractors, electromagnets to remove ferrous metals, as well as washing and drying machinery. Different mixtures of wheat not only required different combinations of cleaning machinery, but different sized mills could provide either more or less sophistication in this section of processing.[47]

5. The effects of change

In the difficult years after 1880, there were those who had the foresight and energy not only to install new, one hundred per cent roller milling systems, but also to expand and extend them when they proved profitable. But the industry still retained an important though diminishing amount of its dual structure up to the First World War. This was in spite

[44] The same effect occurred with imports of American beef. See R. Perren, 'The North American beef and cattle trade with Great Britain, 1870–1914', *EcHR*, 2nd ser., 24, 1971, p. 442.

[45] Humphries, 'Modern developments in flour milling', p. 112.

[46] W. Jago and W. Jago, *The Technology of Bread Making* (London, 1911), pp. 284–9.

[47] *Milling*, 2, 20 November 1891, p. 1; 27, 2 Jan. 1904, p. 5; Voller, *Modern Flour Milling*, pp. 38–65; *The British Trade Review*, 32, 1 Nov. 1909, pp. 224–5.

of the fact that by 1907 the large firm had a sufficient foothold in the industry for W. H. Macrosty to include a chapter on grain-milling in his book *The Trust Movement in British Industry*.[48] But although the large firms had attempted to combine to regulate markets, fix prices and control competition, they had not been able to achieve the long-term success necessary 'to liberate them from the thraldom of competition'.[49] They failed because most of the competition was initiated by the larger firms and directed against the smaller ones. The large firms, sustained by huge supplies of raw materials, had evolved in the major ports. By 1901 three areas were responsible for the manufacture of a third of the country's flour; Hull and York produced 10 per cent, and so did the Bristol Channel ports, while Liverpool took the lead in flour milling with 13 per cent of the nation's output.[50]

Even these places, with large and growing populations in their hinterlands, had a limited market area. To justify further growth large firms needed to extend their market areas away from their home ports. Initially they had been able to do this without competing against each other, because the competition was directed at the small milling firms in the inland areas beyond the ports. In the 1880s and 1890s this strategy allowed the port millers to maintain their position when cheap flour imports were at their highest.[51] Direct competition between the large firms would not have achieved very much as they all had a similar pattern of costs: but these firms could gain valuable additional sales when they directed their competitive energies against smaller inland millers with higher costs.

By this time many small firms had invested in roller milling, and although they still existed, the stone flour mills had ceased to be an important force in the industry. In 1901 it was estimated that only 5 per cent of flour production came from stone mills. And although there were believed to be around 1,000 roller mills and 7,000 stone mills, only 1,500 of the stone mills produced flour, the rest depending largely on milling animal foods.[52] By 1910 the hourly capacity of the industry was 6,500 (280 lb) sacks with only about 100 mills capable of producing over 20 sacks an hour. The five largest firms, with several mills apiece, each had an hourly output of around 250 sacks. Even though the average mill only

[48] This chapter first appeared in a slightly different form under the title of: 'The grainmilling industry: a study in organisation', *Econ. Journ.*, 13, No. 51, September 1903, pp. 324–34; and No. 52, December 1903, pp. 536–43. This article was updated and added to when it appeared as chapter VIII, pp. 210–28 in the 1907 study.

[49] H. W. Macrosty, *The Trust Movement in British Industry* (London, 1907), p. 255.

[50] *Milling*, 16, 23 Feb. 1901, p. 117.

[51] M. D. Freeman, 'A history of corn milling, c.1750–1914, with special reference to south central and south eastern England', PhD thesis, University of Reading, 1976, p. 118.

[52] *Milling*, 17, 31 August, 1901, p. 139.

produced between seven and eight sacks an hour these five 'leviathans' still only represented a concentration ratio of 19 per cent of total output.[53] This was a long way from the position in 1930 when the three largest firms had an output concentration ratio of 62.5 per cent.[54]

The industry's national organisation, the N.A.B.I.M., could do nothing to limit competition because it was little more than a talking shop that passed vague resolutions about the desirability of a unified approach. This was because its members represented only a fraction of millers. In 1892 out of 1,500 substantial millers in Great Britain and Ireland only 280 had joined the N.A.B.I.M. and by 1910, although the Association represented over 80 per cent of the flour milled in the country, it numbered only 240 out of a total of between 800 and 850 roller milling firms.[55] Even this membership was drawn from a wide spectrum of firms with divergent interests. For instance, in 1901, its northern branch, the North of England Flour Millers' Association, tried to get its members to agree to the long term regulation of prices but failed because Joseph Rank refused to entertain the idea. Although he was in favour of consultation between competitors when it suited him, Rank argued, in a speech where he used the phrase 'survival of the fittest', that price fixing must be at a level high enough to allow 'the worst equipped and most unfavourably situated mills' to operate at a profit. But this would force the largest mills to operate at a margin of profit that would encourage large firms from outside the North of England Association to enter their area.[56] So although large millers increased their sales at the expense of smaller firms, the stage of competing for a national market had not arrived, although it was within sight.

Large firms tried to avoid each other's local market areas, as competition between companies with a similar cost structure was more bruising than a contest with weaker rivals. When large firms competed they preferred neutral ground, like London or the south-west, as did the Liverpool and Yorkshire firms. They were also encouraged into these regions by the fact that roller milling firms there had not reached the size of those of Lancashire and Yorkshire.[57]

Regional patterns of demand also encouraged the large firms to extend their markets, although – paradoxically – regional factors also placed limits on the process. This was because regions had different preferences for various grades of flour. The Liverpool area was able to consume more fine flour, but millers could only supply this by producing more of the coarse grades for which there was insufficient demand

[53] The Standard Cyclopedia of Modern Agriculture and Rural Economy (London), vol. IX, 1910, p. 35.
[54] A. H. Hunt, The Bread of Britain (Oxford, 1930), p. 28.
[55] Milling, 4, 4 Nov. 1892, pp. 239–40; 35, 18 June 1910, p. 641.
[56] Milling, 16, 11 May 1901, p. 314.   [57] Milling, 17, 12 October 1901, pp. 228–9.

in the north-west. To sell this surplus they sent travellers to the south-west where consumers had a higher preference for coarse flour. Hull millers, who had expanded their sales of fine flour along the east coast, and as a result had surpluses of second and third grades, followed the Liverpool millers into the south-west.[58] But when Liverpool millers tried to dispose of more flour in the north east they faced problems. Before the imports of hard wheat most bread was baked with the weak 'colourey' flour from soft English and European wheats; but the loaf made from hard wheat flour was lighter in texture and colour.[59]

In the first instance this product was eaten by those living around the ports, but with increased supplies of hard wheat it became known to customers further afield. In its own way, this was a change in consumer habits almost as important as the change from barley to wheaten bread in the eighteenth century. Because this bread used a high proportion of hard wheat flour it meant that inland millers, even where sufficient quantities of British wheat were still produced, had to blend in a proportion of imported wheat or flour. But changing consumer preferences required sustained effort by the port millers. They used trade and consumer advertising and more travellers to push their sales further afield.[60] Travellers offered generous discounts on forward sales, which caused the smaller firms to accuse the larger ones of conducting this trade at a loss to drive them out of business.[61] But up to 1914 some regional variations in the demand for different grades and qualities of flour remained, particularly in the north-east. When west-coast millers, who used more expensive hard wheats than their Northumberland and Durham counterparts, tried to push sales around Newcastle they found it hard to get orders except from bakers. This was because the Tyneside housewife did not value strong grades of flour any more than cheaper weak grades. The gradual demise of home baking had been detected in 1901, but in 1910 three quarters of the flour on Tyneside was still used for home baking, which meant Lancashire millers faced a limited market in this region. Consumers in the north-east used more soft home-grown and German wheat flours than other parts of the country, and the bulk of that market was supplied more cheaply by local millers.[62]

---

[58] *Milling*, 16, 23 Feb. 1901, p. 117.
[59] *Milling*, 15, 7 July 1900, p. 12; 15, 1 Dec. 1900, p. 235; *The Miller*, 27, 1 April 1901, p. 71.
[60] *Milling*, 22, 2 Jan. 1904, p. 1; 16, 27 April 1901, p. 266.
[61] *Milling*, 6, 6 June 1896, p. 386; 6, 13 June 1896, pp. 428–9; 16, 27 April 1901, pp. 265–6.
[62] *Milling*, 16, 23 Feb. 1901, p. 108; 34, 19 March 1910, p. 240.

B

MALTING

BY JONATHAN BROWN

Few industrial buildings have attained the distinction of honourable mention in Sir Nikolaus Pevsner's *Buildings of England* series.[63] Among those few are the maltings at the country town of Sleaford, Lincolnshire. They were built between 1902 and 1905 by Bass, Ratcliffe & Gretton Ltd. of Burton-upon-Trent, Britain's biggest firm of brewers. Bass spent liberally on these buildings to produce what, indeed, are amongst the most impressive monuments of any rural industry, a range of eight malthouses having a frontage of 1000 feet alongside a railway line to Boston.

These maltings at Sleaford exemplify on a grand scale many of the changes in the trade since the mid-nineteenth century: steady investment in new buildings, with an increase in their size, improvement in their design and mechanical equipment; the interest and influence of the brewers in the business; the location of malting in the heart of the British barley country.

Both brewers and independent maltsters invested heavily in new buildings: in new malthouses and in malt and barley granaries. These were large and well equipped. Malthouses of this period were typically of 75–120-quarters capacity, three to four times the size of their early nineteenth century predecessors.[64] The amount invested was considerable, although it is impossible from surviving records to quantify it. Bass's investment at Sleaford was reckoned to run to about £250,000 for eight malthouses each of 120 quarters with large granaries, together with an engine house, workmen's cottages and other small works. That was fairly lavish. The *Brewer's Journal* was quoting about £9,000 as the cost of a 120-quarter malting in the 1880s.[65] With scores of new buildings the cumulative investment was quite substantial.

1. Demand and production

It might seem odd that there should be so much investment in new maltings, for demand and production of malt were declining for most of this period. However, the changes in malting were to a considerable extent a

---

[63] N. Pevsner and J. Harris, *The Buildings of England: Lincolnshire* (London, 1964), pp. 639–40.

[64] The size of a malthouse was usually denoted by the capacity of its steeping cistern, measured in quarters. A 120-quarter malting could thus steep 120 quarters of barley every four days.

[65] For example, £8994 for one built at Faversham. *Brewer's Journal*, May 1884.

reflection of structural changes in the brewing industry. The publicans who brewed their own beer were being squeezed out by the wholesale ('common') brewers, and the common brewers, in turn, were becoming fewer in number, dominated by the large firms of London and Burton-upon-Trent. These developments had their effect on the nature and structure of malting. Fiscal changes had influence as well. The malt tax was repealed in 1880, replaced by a new duty on beer, and with that a number of restrictions on malting were lifted.[66]

One effect of the repeal of the malt tax was to make it harder to measure change in production. Statistics collected annually for excise purposes ceased in 1880. Until then production of malt followed the upward course of beer production, except that the graph for malt began to level off sooner in the 1860s. After 1880 there is no complete record of malt production until the first Census of Production in 1907. In that year the production of malt in England and Wales was recorded as 9,624,000 cwt (or 25.7 million bushels), and that was 42.8 per cent below the 44.9 million bushels recorded in 1880. It may be that 1907 was an exceptional year.[67] Even if it was, it is evident that there had been a substantial decline in the output of malt.

Consumption and production of beer were also depressed during the last decades of the nineteenth century, but that could account for only part of the decline in the production of malt. Brewers were using less malt. They were using other materials instead of it; they were brewing lighter beers that needed less malt, and more scientifically controlled processes were able to effect further savings of materials.

By 1914 malt substitutes made up 18 to 19 per cent of the brewing grist. Sugar was by far the most important of these, but there were several others, including cereal preparations such as flaked rice and maize. Sugar was a cheap substitute for malt, and as the price of malt tended to rise so brewers were more inclined to turn to sugar, especially for their light ales. Consumption of sugar by the brewers of England and Wales, nearly 40 million lb in 1867, reached 118 million lb in 1878 and more than 300 million lb by the beginning of the twentieth century. Steward & Patteson, brewers in Norwich, exemplify this trend. Between 1896–8 and 1911–14 their consumption of sugar increased by 46 per cent, while during the same period the amount of malt used per barrel was reduced by 19 per cent.[68]

---

[66] 43 and 44 Vict. c.20. 1880.

[67] G. B. Wilson, *Alcohol and the Nation* (London, 1940), pp. 369–70; Mitchell and Deane, *Abstract of British Historical Statistics*, pp. 268–9; *Census of Production*, 1907, Final Report, p. 524.

[68] *Return of Sugar used in Brewing*, BPP, 1880, LXVII, p. 881. Return of brewers' licences for 1880–1914, T. R. Gourvish, *Norfolk Beers from English Barley* (London, 1978), p. 76.

2. Changes in technology

Against this background, all the new maltings built during the late nineteenth century represented an investment not in new capacity but in efficiency of operation, and higher standards of production. Steward & Patteson, a regional brewer of medium size, operated sixteen floor maltings in 1885, most of which were coming to be regarded as too small and inefficient.[69] During subsequent decades this company, in common with others, set about modernising their maltings.

For the most part, improving the efficiency of malting meant building larger malthouses that were designed to store and to move large quantities of grain more effectively than the 'one-man' maltings. Those old maltings were often deficient in storage, especially for malt, with a consequence that the poorly stored malt was likely to be delivered in worse condition, with greater moisture content, than brewers demanded.[70] The new maltings were designed to rectify that, having generous provision for storing both barley and malt. They were also designed so that the flow of grain from intake of barley to despatch of malt was as smooth as possible. Mechanical handling equipment was installed. Bucket elevators and power-operated sack hoists lifted the barley to the stores on the upper floors. Conveyor belts and worm conveyors carried the grain along the storage floors. Power was usually provided by a gas engine, although in the largest maltings steam engines might be used. Bass used steam to drive the machinery in their malthouses at Sleaford. By the end of this period electricity also was more frequently a source of power.

In the malting process itself, steeping cisterns were more likely to be situated on the upper floors than on the ground floor, where, influenced by excise regulations, they had usually been placed until 1880. With cisterns on the upper floors, the flow of barley to the middle floors, to be steeped, thence being dropped to the lower working floors, was rendered more efficient. Scope for change was small on the working or growing floors, except perhaps for the use of overhead pulleys along which baskets could be run to carry the green malt down the floor. Nor was there much mechanisation of kilning. Power operated conveyors might take the malt from the working floor into the kiln. Automatic kiln turners were introduced in the 1880s, but they were then of limited efficiency, and remained uncommon until after 1945.

The design of kilns did undergo considerable development, less to save labour than to improve efficiency and the quality of the malt. Furnaces were redesigned in order to provide better heat chambers which would

---

[69] Gourvish, *Norfolk Beers*, pp. 54, 66.
[70] H. M. Lancaster, *The Maltster's Materials and Methods* (London, 1936), p.109.

spread the heat more evenly through the kiln floors. Attempts were made to improve the draughting of the kilns, especially to eliminate down draughts, by the use of improved designs of cowls and shutters, and occasionally by fitting fans on the roof. Similar considerations were applied to the construction of working floors in new maltings, to give a solidity, and control over ventilation that would ensure an even temperature across the piece of germinating barley.[71]

There were other technologies to which maltsters could turn, in the 'pneumatic' systems of malting. These were founded upon the principle of germinating the barley in an enclosed container in which temperature, humidity and moisture could be controlled, rather than relying upon the natural ventilation over an open malting floor. Being artificially controlled, the malting process could be made to take less time, and so increase productivity of labour and capital. Two systems of pneumatic malting – drum malting and box malting – both developed on the Continent, were introduced to Britain during the 1870s. They did not catch on, however. British maltsters and brewers, on the whole, believed that the quality of malt made by the floor malting method was better; nor did the gains in efficiency seem sufficient to justify the higher capital cost. These arguments continued to hold sway until after 1945, when conditions in the industry changed considerably. Until then pneumatic malting remained a minority interest.

The new malthouses, then, saved little labour, except possibly, in handling the grain in storage, where greater mechanisation was possible. The gains in efficiency were of a technical nature: buildings of better design that resulted in more even temperature across the floors, better control of kilning and in storage. These all added up to the ability to produce malt of higher quality, malt which would meet the specifications on colour, moisture content and enzymes which the brewers increasingly were demanding.

3. Changes in industrial structure

With the new malthouses and their storage the maltsters were better able to meet large orders. This in turn reflected the changing structure of the industry. Brewing was becoming more concentrated; so, too, was malting, while the direct involvement of brewers in malting was becoming more pronounced. The number of common brewers in England and Wales fell from 2,507 in 1880 to 1,236 in 1914, while average output increased by 130 per cent. Licensed victuallers and beer retailers declined

---

[71] H. Stopes, *Malt and Malting* (London, 1885), pp. 197–211; H. M. Lancaster, *Practical Floor Malting* (London, 1908), pp. 8–13; Lancaster, *The Maltster's Materials*, pp. 107–10.

more dramatically, from 18,493 in 1880 to 2,357 in 1914.[72] The maltsters followed the same course. Kelly's directory of the brewing and allied trades for 1902 lists just over 1,200 maltsters in England and Wales. In 1870–1 there had been 4,728 licensed maltsters, in 1855 6,352.[73]

The decline in the number of maltsters was brought about first by the very small-scale producers being forced out of the trade. There had been thousands of these, producing as little as 1,000 or 2,000 quarters in a year for sale locally to domestic customers, licensed victuallers and beer retailers and the small common brewers.[74] Malting on this scale was as often as not a part-time business, carried on by those who otherwise were farmers, millers or corn merchants, among the more common joint trades. The demise of the retail brewers, and decline of the small-scale common brewers simply took away the custom of the small maltsters.

In their place was a smaller number of larger firms of sales maltsters. Two at Ware, the malting centre in Hertfordshire, stand typical of those firms that had a few very large contracts and at the same time supplied a number of smaller clients. The smaller of these firms, Ward & Son, in the 1890s had two principal customers who each bought more than 10,000 quarters each season; these totals made up orders for about half a dozen different types of malt. These two clients accounted for about two-thirds of the firm's production. Ward's neighbours, Henry Page & Co., at this time had a turnover more than twice as large, and did about three-quarters of their business with one client, Combe & Co., who bought more than 60,000 quarters each year.[75]

The influence of the large customers thus helped the rise of large-scale malting businesses. Ward & Son and Henry Page & Co. were by no means the largest. Arthur Soames & Son, of Grimsby, were in the late 1880s capable of producing 80,000 to 100,000 quarters of malt a year, and Free, Rodwell & Co., of Mistley, Essex, had a capacity in excess of 100,000 quarters a year by 1910.[76]

It was not only the independent sales maltsters who were increasing their milling capacity, but the brewers also. Indeed, brewers often took the lead. Bass, the largest brewer in the country, was also comfortably the largest maltster, producing over 250,000 quarters of malt by 1890 and thus

---

[72] Wilson, *Alcohol and the Nation*, pp. 48–9; R. Wilson, 'The British brewing industry since 1870', in L. Richmond and A. Turton (eds.), *The Brewing Industry: A Guide to Historical Records* (London, 1990), p. 11.

[73] *Return of the Number of Licences Taken out for Making Malt*, BPP, 1862 (123), XXX; 1870 (105), LXI; 1871 (169), LXII; *Kelly's Directory of the Wine and Spirit Trades*, 1902.

[74] Total output of this amount represented working at about 10–20 quarters per steeping.

[75] Hertfordshire Record Office, D/E WD B5; D/E Pa B5.

[76] A. Barnard, *The Noted Breweries of Great Britain and Ireland* (London, 1889–91), vol. IV, pp. 532–6, 570; *ex info*. Albrew Maltsters, Ltd.

outstripping the combined capacities of Soames and Free, Rodwell.[77] Brewers had always been maltsters. The tendency during this period was for them to take a greater direct interest. Their motive was to have greater control over the cost and quality of their malt. Thus the chairman of the Lion Brewery Company, of London, justified the board's decision to build maltings at Long Melford in Suffolk on 'the desirability of having malt made in season, so as to have it of one uniform quality'. After the maltings were in operation the company declared their satisfaction with the costs of production in the first season, and with the quality of the malt, which was reckoned to be better than they had been buying that year.[78]

With these changes the brewers making their own malt were tending to take a larger share of the malting business. This was clearly so in terms of the number of firms. Of licensed maltsters in 1855, 38.4 per cent were brewers, and 39.1 per cent in 1862. Brewers made up 45 per cent of the malting firms listed in Kelly's trade directory for 1902. This was a natural consequence of the fact that the small independent maltsters were leaving the business at a rate greater than the decline in the number of common brewers. No statistics of malt production differentiated between brewers and sales maltsters until the second Census of Production, taken in 1924. Then 45 per cent of total production was made by brewers.[79]

4. Limited changes in industrial location

The changes in the structure of brewing and malting reinforced the concentration of malt production in the main barley-growing regions of the country. Most of the large businesses and most of the large new maltings were to be found in eastern England, and on the light lands of southern England. This concentration was already established by the mid-nineteenth century, but had been obscured to an extent by the scattering of small-scale maltsters throughout the country. The excise returns for 1832, for example, showed the production of malt in the Grantham Collection to be three-and-a-half times greater than it was in the Cornwall Collection. The numbers of maltsters, however, were little different: in 1855 there were ninety in the Cornwall Collection, ninety-four in the Grantham Collection.[80] In the east the businesses, and the maltings they operated, were bigger.

---

[77] T. R. Gourvish and R. G. Wilson, *The British Brewing Industry, 1830–1980* (Cambridge, 1994), pp. 188–89, 191.   [78] *Brewer's Journal*, Feb. 1878; Feb. 1879.

[79] *Census of Production*, 1924, Final Report, Part II, p. 171.

[80] The Collection was the excise authorities' local division. Boundaries were not published, but the Grantham Collection may be taken as roughly equivalent to Kesteven. *Return of the Number of Licences Taken Out for the Making of Malt*, BPP, 1862, (123) xxx; *Fifteenth Report of the Commission of Excise Inquiry*, BPP, 1835, XXXI, p. 9.

Most of the smaller-scale maltsters in Cornwall would have been producing malt to meet local demand from the publican-brewers and the small, local wholesale brewers. It was the same throughout the country, with the result that at the beginning of this period, maltsters were widely distributed throughout the counties of England and Wales. Even in areas remote from the best barley land there was likely to be enough locally grown barley, augmented by some supplies bought in, to meet the needs of small-scale brewers. Premium-quality barley was less likely to be a priority for these brewers, and possibly beyond their means. The dispersal of maltings was also determined by the same bulk, weight and value considerations as applied to corn milling, which was also widely dispersed.[81] Inland carriage of barley cost as much as the transport of wheat, but, as barley was less than two-thirds the price of wheat in the 1850s, its relative cost of transport was higher. The decline of the small brewer, and the demise of many of the small village maltsters, meant that malting was a trade practised less commonly in places away from good barley land. The large common brewers were increasingly looking for good-quality barley, and so the maltings went to where the best barley was grown. The London brewers had long ago established this pattern, with the result that the small towns close by, if not in the midst of the good barley country of southern England, and with good communications with London, were among the leading centres for malting by the late eighteenth century. These were the towns along the Thames, from Kingston in Surrey to Abingdon in Berkshire, and, above all, Ware in Hertfordshire, on the Lea Navigation.

By the end of the eighteenth century London brewers were turning increasingly to East Anglia, where barley of the highest quality was grown. The ports along the east coast, from Maldon and Colchester to Ipswich, Yarmouth and Kings Lynn, became the focal points for a large coastal traffic, mainly to London, but also to other major cities.[82] During the course of the nineteenth century the preference for eastern England was reinforced as the major brewers of London, Burton and other large cities sought the best barley for their malt. East Anglia retained pre-eminence, but Lincolnshire, Nottinghamshire and Yorkshire attracted more attention. Carriage of the malt was transferred from ship to railway with the result that malthouses of the second half of the nineteenth century were most likely to be built as close to the railway as possible, often alongside it, as were the Lion Brewery's maltings at Long Melford, and Bass's at Sleaford. The dominance of East Anglia and Lincolnshire left the

---

[81] See above, p. 1073.

[82] In 1846 *Bell's Weekly Messenger* recorded the following individual shipments of malt from towns in East Anglia to London: 70,815 quarters from Yarmouth; 50,983 quarters from Harwich; 50,539 quarters from Ipswich; 26,921 quarters from Colchester; and 9315 quarters from Maldon.

downlands of southern England of less importance, and the malt trade of such towns as Abingdon and Wallingford declined. Even Ware was not quite the place it had been, having lost its clear lead as the capital of malting, although it retained large firms such as Ward's and Page's.

One thing was not changing and that was the fact that malting remained firmly a trade of the countryside. It was the country towns – Grantham, Newark, Ipswich, Thetford, and many smaller places – that were the centres of the trade. Few maltings were built in the large cities.[83] On the whole brewers preferred to have their malt made where the barley was grown rather than have the barley brought in to the brewery. The brewers of Burton were a major exception to this rule. The preference for malting in the barley-country had a number of reasons. One was that the large breweries were usually in the large towns, and often in the town-centres, where sites, if available, were cramped and expensive. Maltings needed a large area of ground, for the building, for free circulation of air, for ease of access. There was an element of convenience for the maltster in having his raw material close to hand, allowing ease of arranging purchases and transport of barley. There was a preference for making the malt travel the greater distance rather than the barley. Partly this was a matter of convenience again, for malt was more likely to be transported in bulk, whereas barley could be delivered in small wagonloads off the farms. Savings could be made in transport costs because a quarter of barley weighed 448 lb whereas a quarter of malt weighed 336 lb. The real savings here arose from the 'malting loss', the by-products which did not have to travel – the culms[84] sold as cattle feed, and the water which was dried off in the kilns.

Malting barley was imported in greater quantity during this period. In the mid-nineteenth century maltsters were importing central European barleys of a very high quality, known in the trade as Saale barleys. The removal of the malt tax in 1880 made it worth while to buy cheaper low grade barleys. These barleys came from such places as Asia Minor, southeast Europe and California; these were dry, sunny countries which produced grain that was light, thin-skinned, low in moisture content, although low, too, in sugar. These barleys were deemed highly suitable for the light, pale ales being produced in the late nineteenth century. Imports rose markedly in 1910–14, the quantity being about twice what it had been in 1870–4.[85] These imports had little effect on the location

---

[83] Those that were to be found in urban areas had often been sited in open countryside when they were built.

[84] Malt culms were the sprouts removed from the barley by screening, also known as malt sprouts, coombs or cummins. J. F. Lockwood, *Provender Milling* (Liverpool, 1945), pp. 33–4.

[85] 20,287,200 cwt compared with 10,281,800 cwt; Mitchell and Deane, *Abstract of British Historical Statistics*, pp. 98–9.

of malting. With but few exceptions the malthouses did not follow the flour mills to the major ports. Instead, malting stayed within English barley-growing country, for there was little incentive to change. The quantities of barley for malting were much smaller than the imports of wheat for milling.[86] The small towns around the east coast, such as Ipswich, had a strong tradition in malting and were ideally situated to handle both local and imported barley. That was another consideration, influencing Bass when they located their maltings at Sleaford, close to the best barley-lands of Lincolnshire, and an easy journey by train from Boston docks for the imports.

At about the same time that the maltings at Sleaford were built, the investment in new buildings slowed down. After 1880, partly encouraged by the lifting of restrictions imposed by the malt tax, the rebuilding of the malt trade had tended to gather pace. Now, however, it came almost to a standstill. The whole of the brewing industry entered a period of difficult trading conditions after 1900, and profits were limited.[87] This had its effect on malting. Bass had plans to extend the maltings at Sleaford which were not fulfilled, and that was symptomatic of the industry as a whole. Malting entered a time of relative quiet until, after 1945, another period of great change came along.

---

[86] In the mid-1890s, 60 per cent was low-quality Russian barley for animal feed. Anon., 'The barley supply of the United Kingdom', *Journ. of the Board of Agriculture*, 3, No. 2, Sept. 1896, pp. 116, 118.   [87] Wilson, 'The British brewing industry since 1870', pp. 9, 12–13.

C

FOOD MANUFACTURING

BY RICHARD PERREN

In 1850 the food industries were generally small scale and traditional, supplying local markets in much the same way as they had done for centuries past. Some were little removed from the domestic and craft origins that characterised them in the eighteenth century and earlier. The main exception was the brewing industry, but only in London did large scale enterprises predominate. Outside the metropolis, even in the growing capitalist brewing centres of Burton, Norwich and Liverpool, the small man was still the most important character supplying local markets.[88] Before the nineteenth century most food was relatively pure, as producers and consumers were not widely separated, and a fraudulent baker or brewer would soon lose reputation and customers. But poor housing in the industrial towns and large cities which prevented the majority of consumers doing much food preparation at home meant that a great deal of food manufacturing was carried out in small workshops and even domestic commercial premises. Such sites lent themselves to the forms of preparation that became common across the food trade. The literature on adulteration, so varied around 1850, vividly portrays the stages through which virtually all products passed as their ingredients, and descriptions, were modified.[89]

Foods can be divided into two classes for the purposes of tracing the evolution of processing techniques. Each type provoked a different response from manufacturers. In the first category are those for which demand and *per capita* consumption grew only slowly, or not at all, as incomes rose. These were mainly carbohydrates, including the products of the milling industry, and some vegetables like potatoes. For these items income elasticity of demand was low or, in some cases at higher income levels, even negative. The other group had high income elasticities of demand, and included the protein foods like meat, dairy products and milk, and also most fruit and vegetables (the latter of course not being generally regarded as protein foods). Manufacturers in the first group had

[88] J. Burnett, *Plenty and Want* (London, 1968), pp. 137–8.

[89] J. Mitchell, *A Treatise on the Falsifications of Food, and the Chemical Means Employed to Detect Them* (London, 1848); A. Normandy, *The Commercial Handbook of Chemical Analysis* (London, 1850); A. H. Hassall, *Food and Its Adulterations: Comprising the Reports of the Analytical Sanitary Commission of 'The Lancet'* (London, 1855); Hassall, *Adulterations Detected: or Plain Instructions for the Discovery of Frauds in Food and Medicine* (London, 1857).

to find ways either of enticing the public to eat more of the product or of lowering production costs. Flour millers, facing this situation did the latter, but those food manufacturers who used flour also had to find ways of making the product more attractive in order to increase sales.

1. Bread and cereal products

In many parts of the country bread was still baked at home in 1850; only in the cities and large towns had the commercial establishment taken over from the housewife. Eliza Acton complained in 1857 that 'in Kent, Sussex, Surrey, Middlesex, and in many other parts of the Kingdom . . . not one woman in twenty is capable of making a loaf!'[90] The typical bakery before 1914 was a small business which provided work for, besides the owner, between one and four journeymen. In 1862 the factory inspector, H. S. Tremenheere, published a report on the industry that revealed its backward state. At that time a business using 10 sacks of flour a week, which produced around 940 loaves weighing 4 lb apiece was about the average size.[91] One advantage of home baking was that, at this stage at least, no adulterants were likely to be added to the product. The extent of this practice is difficult to determine, but at least one observer believed that by 1850 after the advent of free trade many of the grosser bread adulterations noticed by older writers like Accum and Price had disappeared.[92]

The first large bread factories appeared in the 1870s, a process that was assisted both by changes in legislation and by developments in technology. Firstly, the 1862 Companies Act made it easier to raise the capital required and, secondly, improved baking machinery reinforced the trend towards larger manufacturing units.[93]

The pioneer of the modern bread factory was William Henry Nevill (1819–1889). He was born in Hampshire and went to London as a young man where he trained and eventually opened his own bakery in Drury Lane, later moving to Holborn. In 1854 his business had expanded, and he opened the bakery in Caledonian Road that was to be the cornerstone of his enterprise. Other bread factories followed at Herne Hill (1875), Harrow (1883) and Acton (1885). The Caledonian Road premises supplied central London while the others supplied the southern, northern and western suburbs. The business used around 3000 sacks of flour a week and employed 500 workers, 150 vans, 280 horses, and 64 ovens. But

---

[90] Eliza Acton, *The English Bread Book* (London, 1857), p.85.

[91] J. Burnett, 'Trends in bread consumption' in T. C. Barker, J. C. Mackenzie and J. Yudkin (eds.), *Our Changing Fare* (London, 1966), p. 67.

[92] Anon., 'Food and its adulteration', *Quarterly Review*, 96, 1855, p. 466.

[93] P. L. Cottrell, *Industrial Finance 1830–1914* (London, 1980), pp. 45–54.

although Nevill pioneered large-scale baking, using the latest Perkins steam ovens, he retained many traditional methods. He regarded mechanical dough kneaders, available since 1858, as more of a hindrance than a help, preferring to rely on a good master baker with properly trained and supervised helpers to achieve the desired results. He also used only millstone-ground flour. Thus, despite the scale of his operations, his methods were semi-traditional and labour-intensive.[94]

He had imitators but not all of them succeeded, and those who did remained surrounded by numerous highly competitive small firms only too eager to enter into competition with a large, and possibly more cumbersome, rival. In 1889 three joint-stock concerns, the Victoria Bread Co., the People's Bread Co. and Stein's Bread Co. all stopped trading, and another, the Golden Grain Co., was forced to write off half its capital. But there was no shortage of new entrants to take their place and the London and County, the Middlesex Aerated, the Brighton and South Coast, the Midland Counties, the Metropolitan and District, the London and Westminster and the Bread Union were all new bread companies successfully floated in that year.[95]

The development of the large-scale bread factory was permitted by two advances in bread-making technology. The modern oven was the first, and pioneer work on this was carried out by A. M. Perkins, a young American engineer who settled in Britain in 1827 and founded the Peterborough firm of A. M. Perkins, Ltd, which became foremost in the manufacture of baking machinery. In 1851 he patented an oven where the heat was circulated through the baking chamber by high-pressure steam enclosed in sealed tubes. The other technical advance was the travelling oven where bread passed slowly along a moving belt until it was baked. This allowed larger amounts of bread to be baked than in the old peel ovens where the loaves were loaded and then unloaded by hand by an operative using a peel, a long-handled shovel shaped like a flat oar.[96] Another oven-making firm, started in 1873 by Paul Pfleiderer, eventually absorbed Perkins, as well as Lewis and Pointon Painification, Ltd. The new company became Werner, Pfleiderer and Perkins, Ltd and achieved a leading position in the industry. By 1908 its premises at Peterborough covered six acres, and included a model bakery where customers could see the whole range of their machinery and the latest methods in operation.[97]

Unlike flour milling, the large companies never came to dominate the

[94] *The Miller*, 9, 4 June 1883, pp. 266–7; 15, 2 Sept. 1889, pp. 307–8.

[95] *The Bakers Record and General Advertiser*, No. 1334, 4 Jan. 1890, p. 6.

[96] R. Sheppard and E. Newton, *The Story of Bread* (London, 1957), pp. 112–14.

[97] J. Kirkland (ed.), *Twenty-one Years' History of the National Association of Master Bakers and Confectioners* (London, 1908), Appendix, pp. vii–xi.

trade. Entry costs for the small baker were low and the economies of scale were by no means as substantial as they were for flour milling. As late as 1911 a bake house could be fitted out with a hand-loaded oven, flour sifter, dough mixer and tempering trough for as little as £150, with a further £35 for other sundries like tables, racks, tins and moulds. This could comfortably handle 20 sacks of flour and produce around 1,880 loaves a week. The owner could double his capacity simply by installing another oven, with no further capital expense, as he could still use all his existing equipment but more intensively, even though more labour was required for manufacturing and deliveries.[98] Although the small baker persisted, there were great changes in the manner in which his business was conducted. The cellar bakery and the unhygienic methods noted by Eliza Acton and H. S. Tremenheere in the 1850s and 1860s had largely disappeared by 1890. After 1896 in London, where the practice was most prevalent, it was made illegal to establish any new underground bakery.[99] More emphasis was placed on a scientific approach to the industry and its processes and this was reflected in a proliferation of textbooks and trade journals dealing with baking. A campaign was launched to provide technical training of bakers, and in 1899 the National School of Bakery was opened at the Borough Polytechnic in East London. The large factory bakeries afforded greatest scope for the application of these new techniques, but even the small baker with one or two assistants was not left untouched by the application of science and technology to the industry.[100]

After 1880 a number of proprietary breads appeared, usually made from a mixture of fine flour with the addition of wheat germ prepared in a particular way. This was a result of developments in milling and pressure from food activists. Unlike stone milled, roller milled flour contained no wheat germ because its high fat content rapidly turned flour rancid. As the germ contained a high proportion of protein there was a conflict between commercial pressures and the dietary lobbyists. The problem was solved by R. Smith of Macclesfield who patented a process for treating the germ with superheated steam. The heat halted the enzymatic action and converted a highly unstable compound into one with good keeping qualities, at the same time giving the germ a malt flavour. The germ was mixed with three parts of white flour to produce Hovis flour, which was a locally successful product in Stafford and Macclesfield in the 1880s. After 1898 the flour was manufactured under patent by S. Fitton and Sons, at the Hovis Mills in London and Macclesfield, and marketed

[98] W. B. Robertson, ed., *Encyclopedia of Retail Trading*, vol. IV (London, 1911), pp. 183–5.

[99] J. Burnett, 'The baking industry in the nineteenth century', *Business History*, 5, 1963, pp. 106–7.

[100] W. Jago, *Science and Art of Bread-Making* (London, 1895); Jago and Jago, *The Technology of Bread-Making*, p. 235.

on a national scale by the Hovis Bread Company. Its advantage for the baker was that he did not have to bother with advertising or publicity as that was all handled by the company. Bakers wanting to produce Hovis bread had to buy stamped tins, paper bags and the flour from Messrs Fitton, who insisted that only bread made from their flour could be baked in the tins and sold as Hovis, an insistence they were always prepared to back with court actions.[101]

By 1914 the Hovis Bread Company provided advertisement cards, paper bags, cardboard boxes, sign boards for bakers' vans and barrows, as well as press advertising in the quality magazines and national dailies.[102] Other types of bread from flour using wheat germ, treated slightly differently from Smith's patent, were Daren and Turog, which also contained small amounts of rye meal and finely ground bran. Some fancy breads contained small amounts of malt extracts, the most common being Bermaline. They were often little removed from white bread, with the addition of small amounts of colour and flavouring, although their makers mostly claimed other properties such as an enhanced nutritional content, or that they were more easily digested than white bread.[103] Besides buying a recipe and ingredients from a patentee the baker could also simply buy the ingredients and make his own recipe fancy bread.[104]

The critics of roller milled white flour founded their opposition on nutritional grounds. Complaints about the allegedly unhealthy nature of white bread pre-dated roller milling.[105] In 1881 these sentiments became formalised with the establishment of the Bread Reform League, founded on the belief that white bread was wasteful and dietetically inferior because it discarded the most nutritious part of the wheat. The League had the support of a number of eminent scientists, including T. H. Huxley, Sir Thomas Watson, W. B. Carpenter and Erasmus Wilson, but it still exaggerated and made mistaken claims, such as that its recipe for a wheatmeal bread would eliminate most symptoms of working-class malnutrition, including rickets.[106] As consumers found white bread more palatable the League was largely ignored, although it published a pamphlet in 1909 arguing that all bread should contain a standard amount of vitamins, minerals and proteins and this should be achieved by the readdition of those that had been

[101] Jago, *Science and Art*, p. 407.

[102] E. J. T. Collins, 'The "consumer revolution" and the growth of factory foods: changing patterns of bread and cereal eating in Britain in the nineteenth century', in Oddy and Miller, eds., *The Making of the Modern British Diet*, pp. 29–30.

[103] Jago and Jago, *The Technology of Bread-Making*, London, 1911, pp. 486–7.

[104] J. Kirkland, *Three Centuries of Bread* (London, 1917), pp. 25–7.

[105] Anon., 'Bread', *Quarterly Review*, 105, 1859, p. 244; H. Thompson, 'Food and feeding', *The Nineteenth Century*, 5, 1878, p. 984.

[106] L. S. Bevington, 'How to eat bread', *The Nineteenth Century*, 10, 1881, pp. 341–56.

removed in the milling process.[107] As the science of nutrition was in its infancy many of the vitamins themselves were not identified before 1914. Also the amounts in a normal human diet contained in bread were so small that they were easily obtained from other foods.[108]

Burnett has identified two trends in the amount of bread consumed per head between 1840 and 1914. Up to 1880 was a period of rising consumption, followed by a slow decline thereafter which has continued up to the present with brief interruptions during the two world wars.[109] But total sales of bread did not decline, as they were sustained by the increase in population and the substitution of the various bakery products for the home-produced article. Those employed as bakers increased in number, an occupation which was sustained by the increased sales of cakes and biscuits.

The latter was well suited to large-scale production methods which were pioneered in the production of ship's biscuits by the Victualling Office in the first half of the nineteenth century. The firms that came to prominence after 1850 like Carr's of Carlisle, Peek Frean & Co. of Bermondsey, and Huntley and Palmer at Reading were fancy biscuit makers who supplied the civilian market. Like the bread bakers they used the Perkins steam ovens, but in conjunction with specialised mixing and cutting machinery to form the production lines for specific types of biscuit. Unlike the bread industry, most of the flour they used was from English wheat. For bread making urban bakers preferred strong flour from highly milled imported hard wheats, as it yielded far more loaves per sack than flour from native soft wheats, whereas biscuit makers found that weak flour from domestic soft wheats produced a better product than strong flour. The steadily rising demand for biscuits, and for weak flour by biscuit makers, was an important factor in sustaining the British wheat grower. The large range of biscuits available, together with substantial advertising, and the substantial cost reductions following extensive mechanisation, all helped to increase total demand. In the 1850s factory-made biscuits were relatively expensive middle-class products, but the efforts of the Co-operative Wholesale Society and multiple grocers such as Liptons and the Home and Colonial Stores, all selling their own cheaper standardised ranges after 1870, suggest that by 1914 the product was already penetrating some way down the social scale.[110]

---

[107] Bread Reform League, *Standardization of Bread* (London, 1909); *Bakers' Record*, 1910, *passim*.

[108] T. B. Wood, 'Less refined bread not helpful', *Baking Technology*, 4, 1925, p. 273; R. A. McCance and E. M. Widdowson, *Breads White and Brown* (London, 1956), pp. 120–7.

[109] Burnett, 'Trends in bread consumption', p. 70; P. Maunder, *The Bread Industry in the United Kingdom* (Nottingham, 1969), p.18. C. Petersen and A. Jenkins (ed.), *Bread and the British Economy* (Aldershot, 1995), chapter 5.

[110] *The Times Food Number*, 1915, pp. xiii–xvi, 113; T. A. B. Corley, *Quaker Enterprise in Biscuits: Huntley and Palmers of Reading, 1972–1922* (London, 1972); Corley, 'Nutrition, technology and the

The other main method of selling cereals was as cereal foods including ready-to-eat breakfast cereals that needed no further preparation. Although some sixty brands of these products were marketed in Britain by the late nineteenth century only a few had a large sale. Among the popular brands were Fabona (a mixture of wheat, oats, barley, and bean flour), and John Bull (a mix of malted cereals and dried milk). Both these varieties were manufactured in this country, but also available more or less on a national scale were imports of a number of North American products, such as Quaker Oats, Puffed Wheat, Grape Nuts and Corn Flakes. However, these products only made a serious impact on the British market after the First World War; before that time their consumption was held back by their unfamiliarity to consumers as well as the fact that they were an expensive way to consume low-cost carbohydrates.[111]

2. Convenience foods

The difficulties of preparing food in their homes for some urban consumers encouraged the production of new kinds of ready-to-eat foods and others that needed less cooking.[112] This development was not new and in the 1840s Henry Mayhew detailed the activities of thirty-three classes of London street vendors selling all sorts of prepared foods from pea soup and hot eels to baked potatoes and boiled puddings, mainly catering for the daytime needs of office and other workers.[113] Probably the oldest convenience food was bakers' bread, but to this was added a steadily increasing range of pies, cakes, biscuits, meat, fruit, vegetable and dairy products. Another factor encouraging their production was the way in which the food industry responded to nineteenth-century concern over health and diet by trying to process as much nutrition as possible into as small a volume as possible. Many of the concentrated foods that emerged as a result were highly flavoured, thus increasing their marketability.

The convenience-foods industry was attractive to manufacturers because it transformed bulky and low-cost raw materials into a compact and high-value finished product, as well as saving on storage and transport costs. Condensed milk was an item that had all these features and the addition of sugar enhanced its palatability, although the less popular unsweetened form was available. Meat extracts were another favourite, founded on the pioneering work of Baron Justus von Liebig who formed

growth of the British biscuit industry, 1820–1900', in Oddy and Miller (eds.), *The Making of the Modern British Diet*, pp. 13–25.   [111] Collins, '"Consumer revolution"', p. 33.

[112] This question is dealt with by T. C. Barker in 'Urbanization and rising earnings', in Barker, Mackenzie and Yudkin (eds.), *Our Changing Fare*, p. 24, and W. H. Chaloner, 'Trends in fish consumption', *ibid.*, pp. 109–10.

[113] H. Mayhew, *London Labour and the London Poor*, vol. 1 (London, 1964 edn), pp. 166–226.

a company to process cheap South American beef in 1865. Von Liebig's success encouraged numerous imitative products, all of which stressed their value and high reputation as convalescent foods, although this rested on slender evidence. Here there were about a dozen products, Bovril being the best known but supported by other brands such as Esco, Vitalia, Liquor Carnis, Bovinine and Puro. The interest in polar exploration at the end of the nineteenth century provided further publicity for the manufacturers of concentrated foods, although most of the provisions taken by explorers were traditional foods with high energy values. Another specialist market was the one for infant foods. Hardly any of these were milk based, most using a mixture of starch flavoured with malt or sugar as their major ingredient. In this category the consumer could buy Theinhardt's Infantina, Kufeke's Infant Food, Manhu Infant Food, and Victor Baby Food.[114]

Some early products of Alfred Bird, a chemist and druggist who turned to food manufacture, can be seen as an alternative way of marketing flour, as they were cereal based. His eggless custard powder was invented before 1850 and in the early 1870s his son, also Alfred, devised and launched Bird's Blancmange Powder which he claimed was a novel cold dessert. Alfred senior's other early product, Bird's Fermenting Powder, invented in 1843 and later to be called baking powder, was designed as a yeast substitute for bread making in military field kitchens, but was also used to increase the consumption of bread, biscuits and light cakes among the civilian population. Bird's products were early examples of convenience health foods, because the custard and baking powders were designed for those allergic to eggs and yeast. In 1895 the firm moved into animal-based products with the appearance of its jelly crystals, the forerunner of tablet jellies. All the Bird's products provoked a considerable number of imitators, because a strong demand was stimulated by attractive advertising and, once the original formula had been created, no great expertise was required to work out the production methods which were relatively simple and inexpensive.[115]

From the mid-1890s a considerable expansion took place in the numbers and amounts of branded and packeted foods being advertised. But this development did not have the same impact on all groups of consumers, as the cost of packaging meant a fractional increase in prices per lb for the goods in each tin or packet. For middle-class consumers this was acceptable but for many working-class customers the extra halfpenny or so on two or three lb of cereals meant that they still purchased in the

[114] *The Times Food Number*, 1915, pp. 160–3.; W. Tibbles, *Foods: Their Origin, Composition and Manufacture* (London, 1912), p. 146.

[115] J. Foley, *The Food Makers: A History of General Foods, Ltd.*, (Banbury, 1972), pp. 1–3, 9, 11,12.

old-fashioned way at shops that carried large supplies in bulk. The convenience foods that did make an impact on working class diets were those that increased the palatability of traditional foods. As the meals of lower-income families contained large amounts of bland high-energy carbohydrates in the form of bread and potatoes, the addition of pickles, spices and sauces gave welcome variety. Most condiments relied heavily on imported ingredients, including sugar and spices, but the mass production of bottled sauces was essentially a British industry. Varieties like O.K. Sauce and A1 Sauce contained large amounts of fruity materials and were mass produced in large factories. They were not only popular at home but, like British biscuits, they built up a considerable brand allegiance that meant a substantial export business for the expatriate market.[116]

3. Canned foods

The foundations of the modern canning industry were laid in the late eighteenth century by the Frenchman, Nicholas Appert, whose experiments with heated food preserved in glass containers earned him a prize of 12,000 francs from the Bureau Consultatif des Arts et Manufactures in Paris in 1809. In 1811 a process based on Appert's was developed at the Dartford Iron Works by John Hall and Bryan Donkin, but they used iron containers instead of glass. By 1813 Hall and Donkin had established a factory in Blue Anchor Road, Bermondsey, and were supplying the army and navy on trial. In the following years their products were used by Wellesley's army in Spain and by the navy on voyages ranging from Arctic expeditions in search of the North-West Passage to ships on station in the West Indies. By the 1840s more than one firm was supplying the military with canned provisions which included meat, poultry, fish, vegetables, soups, milk, cream and desserts.[117]

Canned products gained a notoriety in the late 1840s through the activities of Stephan – later to become Stephen – Goldner, a Hungarian businessman resident in Britain, who decided to take advantage of the lower British duties on imported meat after 1842 by establishing a cannery at Galatz on the Danube. There he was able to buy cheap Moldavian cattle, and by 1845 had supplied the British navy with around 2.7 million lb of meat. However, in 1846 some of the canisters were found to contain putrescent meat. The reasons for this are not entirely clear, but in some cases Goldner used large canisters, containing from 9 to 32 lb, and upon examination it was concluded that they were not adequately cooked in

[116] *The Times Food Number*, 1915, pp. 158–9, 167, 226–8. L. Richmond and B. Stockford, *Company Archives* (London, 1986), pp. 363, 432.

[117] *Historic Tinned Foods*, International Tin Research and Development Council, Publication No. 85 (Greenford, 1939), pp. 9–24.

the centre. There were other problems as well, probably caused by small faults in the tin-plating and the navy's practice of keeping them in damp storehouses on shore for long periods, or at sea on long voyages where the sea air corroded the container, thus breaking the vacuum and allowing the entry of bacteria to spoil the contents. Other complaints were tempered with strains of xenophobia and anti-Semitism, alleging that the cans contained offal, as well as what was euphemistically described as 'other rejectamenta', possibly placed there as a protest by some of Goldner's low-paid and disgruntled cannery workers. In spite of the fact that 95 per cent of Goldner's contract had been perfectly satisfactory, and canned meat was of better quality and lower in price than the 'salt junk' which sailors were fed normally, a scandal followed. One effect was a revulsion on the part of the British public against all preserved foods in general, and canned meat in particular; another was the failure of Goldner's business and his personal bankruptcy when the Admiralty cancelled its contract. In 1852 the Admiralty set up its own canning factory in the Royal Victualling Yard in Deptford.[118]

In these early days the Admiralty had been the principal purchasers, but in the 1860s attention turned to serving the civilian market, and in 1866 Tooth's Extract of Meat Company began importing tinned meat extract from New South Wales. The small British food-canning industry established in the 1840s continued to grow, and the pioneer firm, Donkin, Hall and Gamble, eventually merged into Crosse and Blackwell. In the 1860s a London firm, Messrs Ritchie & McCall, had their own canning plant at Houndsditch, a few streets from the meat market. Their policy was to purchase particular items when markets were glutted and prices were low. As a result they lacked the regular established lines that were a feature of later firms, and their product mix tended to change unpredictably from week to week, depending on availability and relative prices. Meat was the main item processed but they also canned expensive short season vegetables like peas and tomatoes if the opportunity offered.[119] After 1860 the public's perception of preserved foods gradually improved, partly as a consequence of the greater variety available and also of changes in the canning process extending shelf life and reducing spoilage.

At this stage, before 1890, in the absence of automated production lines the average size of canisters was much larger than at later dates. The large size of container was likely to have made the product individually expensive for private consumers in the early years, and generally imprac-

[118] A. Wynter, *Our Social Bees* (London, 1865 edn), pp. 193–8, 203; *Historic Tinned Foods*, pp. 24–9.
[119] R. Perren, *The Meat Trade in Britain, 1840–1914* (London, 1978), p. 70; W. H. Chaloner, *People and Industries* (London, 1963), pp. 107–8; A. Wynter, *Our Social Bees*, 1865, p. 193.

tical for all but the largest families. However, the long storage life and low unit costs of the contents, reduced further by bulk purchasing, gave it a market among institutional consumers, such as schools, hospitals, poor law authorities, and the military. Donkin, Hall and Gamble rarely went above 6-lb containers, but other firms would go to 14 lb, and Ritchie & McCall would sometimes use tins large enough to enclose 50 lb of beef and entire carcasses of game and poultry. But there was always the danger with large canisters that the heat applied in the cooking process was not sufficient to kill bacteria in the centre, and this caused frequent spoiling. The technology was relatively simple. After initial cleaning the uncooked produce was sealed in cans leaving a small hole in the lid. The cans were then placed in vats containing a solution of chloride of calcium which had a very high boiling point. The heat for cooking was circulated through the solution as super-heated steam in sealed pipes. When the operator judged that cooking was complete and all the air had been driven out, he sealed the cans with a spot of solder. There was even a primitive testing procedure as, some days later, the cans were placed in a chamber and reheated. The sound ones would remain unchanged but faulty ones would bulge as air trapped inside expanded.[120]

These labour-intensive, batch-production methods ensured that before 1890 consumption of canned foods was limited. In addition to institutional purchasers, private consumers were mainly confined to the middle and upper classes. Only a small range of articles was treated because often the process was so expensive that the container cost as much as its contents. In the 1890s canning techniques improved and unit costs fell with the introduction of open-topped, pressed-steel cans and continuous production lines for their packing and sealing. Advances in microbiology, based on the work of Pasteur, improved theoretical understanding, and in 1895 bacteriology was directly applied to the problems of food canning by researchers in the United States investigating the spoilage of canned corn.[121] Producers were now able to pay closer attention to packing and presentation, and this had an important influence in increasing their markets; in the mid-1890s the consumer of canned salmon had a choice of flat, tall and oval cans, in sizes as small as ½ lb.[122] The introduction of fully mechanised production also allowed canners to process a wider range of cheaper foods, making them affordable to the artisan class. By the twentieth century the industry had a diversified product range catering for two distinct income groups, but both were heavily dependent on imported items. High-quality products like

[120] Chaloner, *People and Industries*, pp. 107–8; A. Wynter, *Our Social Bees*, pp. 194–5.
[121] Chaloner, *People and Industries*, pp. 107–8; *Historic Tinned Foods*, pp. 24–9.
[122] *The Grocer*, 69, 4 January 1896, pp. 15, 17.

Oregon Chinook salmon and the best kinds of foreign fruit were consumed by higher-income groups, and cheaper Alaskan and British Columbian salmon, as well as lower-quality tinned fruit, were bought as small luxuries by the working class. Although the price of a 3-lb tin of peaches had fallen from 1s. 1d. in 1869 to 7d. by 1905, it was still an expensive item for a family supported on a weekly wage of 25s. to 30s. The working-class demand for canned foods was also highly vulnerable to fluctuations in employment. This volatility was well known to merchants in all branches of the provision trade, and was particularly apparent to those in Lancashire during the trade depression of 1904.[123]

The introduction of canned food production came later in Britain than in a number of other countries, the United States being the leader in this technology. In 1892 practically the only processing of native produce was for jam, and even here the quality was poor, containing excessive quantities of sugar that deprived it of all proper fruit flavour. The processing of jam was distinct from the canning of other foods, as the containers used for jam were glass and not metal. But a number of the firms that achieved prominence as canners started as jam manufacturers in the 1870s and 1880s. The Chivers family were originally smallholders at Histon in Cambridgeshire, but were able to extend their farm so that by 1860 Stephen Chivers held 160 acres. He responded to falling corn prices by diversifying into fruit in the 1870s and, urged on by his sons, conducted his first jam boiling in a small barn on the farm in June 1873. By 1914 the factory which had replaced the original makeshift arrangements, was processing large quantities of strawberries, currants, gooseberries, raspberries, plums and greengages. The vast amounts of fruit now required came partly from the family's extended farm, but was also purchased from neighbours who had followed him into fruit growing. When both these sources were insufficient, foreign fruit was also used.[124]

By 1914 the jam industry, like the canning industry, had two distinct parts – those who, like Chivers, made a high-quality product with whole fruit; and the 'smashers' who used a variety of low-cost ingredients. This sector of the industry had the largest share of production and used sour or over-ripe domestic fruit and imported fruit pulps – mostly apricots – as well as marrow and other vegetable pulps, combined with large amounts of sugar and glucose, and artificial flavourings. Before the expansion of the fruit acreage in 1880 domestic fruit was not widely used in commercial jams, and the dramatic increase in jam consumption in the

[123] A. Watson, *My Life* (London, 1937), pp. 80, 191; *The Grocers' Review and Provision Trades Journ.*, 34, 3 January 1905, p. 13.

[124] C. Whitehead, 'New modes of disposing of fruit and vegetables', *JRASE*, 3rd. ser., 3, 1892, pp. 590–1; W. S. Mansfield, 'The farms of Messrs. Chivers & Sons, Ltd., Histon, Cambs.', *JRASE*, 92, 1931, pp. 142–4.

generation before 1914 owed more to the fall in the price of sugar than the increased availability of domestic fruit. The quality jam makers like Chivers, Beach, and Crosse and Blackwell tended to look down upon the cheaper end of the trade like Hartleys with its factories in London and Liverpool and E. T. Pink who had a factory at Bermondsey in London's east end.[125] The consumption of tinned ordinary vegetables such as peas, carrots and beans was not common in England before the First World War, although canned and bottled varieties were used in middle-class households as an alternative to the less appetising dry goods, which were mostly dried beans and peas, when the fresh item was out of season. But as there were doubts that they had the same nutritional qualities as fresh vegetables, medical authorities generally advised that they should only be used as an occasional treat.[126]

4. Dairy products

It has been suggested that improvements in dairy technology after 1870, such as the barrel churn, milk cooler and mechanical cream separator, were largely irrelevant to English dairy farmers of the late nineteenth century because the majority had abandoned cheese- and butter-making in favour of the easier and more profitable liquid milk trade.[127] The sheer superiority of the mechanical separator over hand skimming was demonstrated at the Royal Agricultural Society's annual show at Kilburn in 1879.[128] Even in Cheshire, whose farmers produced some 12,000 tons of cheese in the 1850s, the frontiers of Cheshire cheese making were pushed back to the middle and the south of the county by the 1880s. The general decline of cheese-making, and the even more pronounced decline of butter-making, was mainly caused by the railways which opened up tempting prospects of the milk trade in the customary slack winter season. American factory-made cheese was also landed at Liverpool after 1840, although in the early years of the trade the product's poor quality meant that Cheshire cheese makers faced little effective competition from this quarter.[129]

Although the technology crucial for the transfer of cream, butter and

[125] C. Whitehead, 'The progress of fruit farming', *JRASE*, 2nd ser., 19, 1883, p. 370; *Minutes of Evidence Before the Departmental Committee upon the Fruit Industry of Great Britain*, BPP, 1906, xxiv, Cd. 2719, pp. 274–5, 291, 411.

[126] Tibbles, *Foods: Their Origin, Manufacture, and Composition*, pp. 563, 565.

[127] D. Taylor, 'The English dairy industry, 1860–1930', *AHR*, Vol. 22, 1974, p. 155.

[128] A. Voelcker, 'On the composition of cream and skim-milk obtained by De Laval's centrifugal cream-separator', *JRASE*, 2nd ser., 16, 1880, pp. 160–1.

[129] R. Scola, *Feeding the Victorian City, The Food Supply of Manchester, 1770–1870* (Manchester, 1992), pp. 86–92.

cheese manufacture from the farmhouse to commercial factories was easily available, these enterprises did not have much success in England and Wales before 1914. Before 1870 British cheese was only made in farm dairies, but in America the factory-made cheese appeared in 1851. The first British cheese-makers to feel American competition were those producing lower grade and less popular varieties. The Royal Agricultural Society recommended the adoption of the American system in 1870, and the first British factory opened in Derby that year with substantial support from local landlords under an American manager. In 1874 six cheese factories were operating in Derbyshire, and later ones were established in Staffordshire, so that by 1889 sixteen were located along the line of the North Staffordshire Railway between Derby and Ashbourne. But the movement died out and the few factories that survived into the 1890s did so by processing supplies surplus to the liquid milk trade. In 1908 the total output of factory cheese in England and Wales was a mere 53,000 cwt.[130] Butter factories had even less success and were a sharp contrast to the relative strength of this branch of food manufacturing in Ireland.[131] Only a few factories were engaged in the other branch of dairy manufacturing, condensed milk, the majority of which was imported. Two Swiss companies did manufacture a small amount in Britain, Anglo-Swiss Condensed Milk which opened a factory at Tutbury in Staffordshire in 1872, and Nestlés who started a plant at Chippenham in Wiltshire in 1901.[132] Both of these were sited on major rail routes into London, and so it is likely that they also adopted the strategy of processing cheap milk surplus to the traffic in liquid milk destined for London.

The scattered nature and small throughput of British butter factories and creameries prevented them from taking full advantage of the economies of scale available to really large plants on the Irish, Continental and American models. They processed such large quantities of milk that they could offer an economic proposition for the manufacture of the whey, whereas in Britain most of this was wasted. In Derbyshire the opposition of the local cheese factors caused problems when they refused to buy at certain seasons and forced the manufacturers to carry the costs of large stocks. But as the nature of the dairy industry changed, and farmers were

[130] D. Taylor, 'Growth and structural change in the English dairy industry, c.1860–1930', AHR, 35, 1987, pp. 50–1; H. M. Jenkins, 'Report on the cheese factory system and its adaptability to English dairy districts', JRASE, 2nd ser., 7, 1870, pp. 173–203.; Report on the Marketing of Dairy Produce in England and Wales, Part 1 – Cheese, Ministry of Agriculture and Fisheries, Economic Series, No. 22, 1930, p. 24.

[131] C. Ó Gráda, 'The beginnings of the Irish creamery system, 1880–1914', EcHR, 2nd ser., 30, 1977, pp. 284–305; M. Turner, After the Famine: Irish Agriculture, 1850–1914 (Cambridge, 1996), pp. 277.

[132] G. Jones, 'Foreign multinationals and British industry before 1945', EcHR, 2nd ser., 41, 1988, p. 450.

released from the limited local demand for liquid milk by the growth of
the railway milk traffic, some cheese factories just became surplus to
requirements.[133]

The milk cooler had its application in the short-distance, as well as the
long-distance, milk trade. In 1912 small farmers just around the towns of
Lancashire began milking at five in the morning and two in the afternoon, and as soon as the milk had passed through the refrigerator the carts
went out on their rounds. But further out, because of the difficulties of
getting the milk to the towns early enough it was turned into cheese for
sale locally, and the whey fed to pigs.[134] The question of transport was
the main factor that preserved farm house dairy processing for local consumption. In 1892 one Staffordshire farmer observed:[135]

During the last few years the sale of whole milk for consumption in London and
other large towns has increased rapidly. Year by year the manufacture of cheese
is decreasing, while the production of butter is also less, though in a considerable degree its manufacture for the time being is increased or decreased in accordance with the demand and prices current in the local markets.

CONCLUSION

In some respects, the food industries mirrored the problems found at
large in the rest of British industry. They all felt the impact of foreign
competition, and in some respects lagged behind the best practice found
abroad. All branches experienced heavy exposure as a result of the
country's growing dependency on imported food after 1850. By 1911–13
some 78 per cent of wheat, 41 per cent of meat, 35 per cent of butter
and 80 per cent of cheese was imported.[136]

Corn milling in the late 1870s was particularly vulnerable to the charge
of being a lagging sector, although this was only a transitional phase as
the industry adjusted to the new technology and switched from home-
produced to imported raw material. Malting did not face the same
problem of imports, either of finished products or of raw materials, and
so was not subject to the same exogenous pressures for change. The pre-
dicament of corn milling was similar to all industries experiencing
foreign competition: the import of the competing product was comparatively quick and easy, but the installation of foreign technology to

[133] A. D. Hall, *A Pilgrimage of British Farming 1910–1912* (London, 1913), pp. 233–5.

[134] Ministry of Agriculture and Fisheries, *Co-operative Marketing of Agricultural Produce in England and Wales (A Survey of the Present Position)* (London, 1925), p. 15.

[135] R. H. Rew, 'An inquiry into the statistics of the production and consumption of milk and milk products in Great Britain', *JRSS*, 55, 1892, p. 269.

[136] R. C. Michie, 'The international trade in food and the City of London since 1850', *Journ. of European Economic History*, 25, 1996, p. 386.

counter that competition was a longer process. The British industry did, however, have some features in its favour. There were already some seaport millers, which were now the most favourable sites for installation of roller milling, and it also had an established customer base which could be educated to accept the modified, and improved, product that the new technology imposed on the industry.

Product improvement was very much the characteristic dominating the collection of sometimes unrelated trades that are grouped together under the heading of manufactured foods. In all cases these represented a wider consumer choice, as prepared and prepacked foods began the process of reducing the amount of home food preparation that was to become a continuing feature of the twentieth century.[137] The emergence of entirely new methods of manufacture, packaging and advertising gave this group the advantage, unlike corn milling, of not being tempted to improve an obsolete technology. Advertising was particularly important to create what was essentially a new market. In some cases the marketing of the factory processed foodstuff, and the marketing of graded and branded natural or traditional products of uniform quality, like imports of colonial and foreign refrigerated and chilled meat, bacon and dairy products, involved a very similar task; in both cases it was vital to build brand loyalty. To this end the advertising agency was called in, sometimes to assist and other times to replace in-house promotions teams, with differing measures of success. The new processed savoury foods like soups, sauces and meat extracts were less popular than biscuits, cakes, home-baking products, and sweets like Bird's Custard and jellies, and of course jam.[138] It is perhaps significant that the popularity of various items in this class of foodstuffs seems to have been more or less proportional to the amount of refined sugar they contained. In many cases the products beginning to appear in British grocers' shops before 1914 depended on technologies perfected elsewhere. In canning the United States was the world leader, while refrigeration techniques to preserve meat and dairy products were also developed abroad. But in the case of dairy products the British food-processing industry was never able to achieve the flexibility of the flour millers who did belatedly adapt to foreign competition. The sector of the food industry producing cheese and butter was never properly transformed, with the result that the farmhouse producer of these items was forced into a long retreat before 1914, in the face of increasing imports of branded dairy products.

[137] R. Graves and A. Hodge, *The Long Week-End* (London, 1940), pp. 175–6.
[138] M. Jubb, *Cocoa & Corsets* (London, 1984), pp. 1, 4; J. Foley, *The Food Makers*, p. 9.

CHAPTER 18

RURAL INDUSTRY AND MANUFACTURING

BY JOHN CHARTRES

INTRODUCTION

Writing in the mid-1840s, Frederick Engels identified the destruction of rural industry as one critical element in the emergence of the rural proletariat: 'the small peasantry also was ruined when the former Union of industrial and agricultural work was dissolved, the abandoned fields thrown together into large farms, and the small peasants superseded by the overwhelming competition of the large farmers'. Basing his comments predominantly on the textile industry, he solemnly proclaimed the English and Welsh countryside to be an agricultural monoculture, populated by day labourers, tenant farmers and the landlords, where immiseration characterised the lot of the farmworker, and was exceeded only by the experience of contemporary Ireland. He correctly identified the long-run tendency for 'industry' to concentrate into larger and more centralised plant during the course of the century, and, influenced in his writing by the literature on 'Swing' and the evidence of rapid decline of the worsted and woollen handloom weavers in the face of the power loom from the 1830s, made a significant point about this transition in a major industry.[1]

Yet, on balance, his judgment was premature. In 1850, the countryside was a lesser place of manufacture than it had been during the period of dispersed and cottage production that had characterised so much industrial activity before the 1820s, but it remained an important industrial location.[2] This was graphically illustrated by Augustus Petermann's map

---

[1] Frederick Engels, *The Condition of the Working-Class in England From Personal Observation and Authentic Sources*, 1845, edited with introduction by E. J. Hobsbawm (London, 1969), p. 286. The editor and many other authors have correctly criticised Engels's own research and immaturity, notably Chaloner and Henderson, but this view about the domestic division of labour in Britain remains deeply embedded in much of the literature.

[2] This chapter is indebted to a great deal of bibliographical assistance provided by the late Dr Raine Morgan, of the Rural History Centre, University of Reading.

of occupations from the Census of 1851.[3] Industry continued to be found in the English and Welsh countryside in the years up to 1914, and even beyond, although it was generally in decline, and in many spheres confined to low-value added outwork. Conscious of the pace of change from the mid-1880s, J. L. Green attempted to survey the totality of English rural history, and painted a picture of almost universal decline. When, in the early 1920s, the University of Oxford's Institute for Research in Agricultural Economics conducted its surveys of the rural industries of England and Wales, they found survivors, but little vitality, and consistently identified the decades immediately before the First World War as the critical period of terminal and generalised decline.[4] Rural industry's obituary must therefore be postponed for the best part of two generations.

The resolution of this apparent conflict lies within the specifics of individual industries, and even of sections of them, and was implicit in the earlier comments about textiles. Whereas spinning had mechanised relatively early, and with improvements in plant and licensed application of patents had tended to become concentrated successively in cotton, linen, worsteds and woollens into larger mills by the 1820s, weaving proved the slower to mechanise successfully, with the wool-based textiles lagging well behind cotton and linen in the introduction of the power loom. Steam-powered plant grew rapidly in the woollen districts between the mid-1830s and the mid-1850s, only then effectively displacing hand and water as the power sources for textile manufacture, and creating factory-style manufacture.

Thus generalisations about industrial location after 1850 need to incorporate a sensitivity to the specifics of the individual industry, and to appreciate the typology of England and Wales' industrial structure in so doing. One helpful analytical approach was proposed in a remarkable statistical exercise by Clive Day, who explored the employment data of the

---

[3] Reproduced by J. H. Clapham, *An Economic History of Modern Britain*, vol. II (Cambridge, 1932), endpaper.

[4] J. L. Green, *The Rural Industries of England* (London, 1895), esp. pp. 9–10, 56–129, 191–6, 198–9. (Note that Green's survey was limited in coverage and quality; was based upon fuzzy and impressionistic criteria; and had a clear tendentious purpose, to argue against accelerating rural depopulation, which it ascribed primarily to agricultural causes, including excess fiscal burdens on the land and the collapse in land values). K. S. Woods, *The Rural Industries Round Oxford: A Survey* (Oxford, 1921, hereafter Woods, *Oxford*); Helen E FitzRandolph and M Doriel Hay, *The Rural Industries of England & Wales*, vol. I, *Timber and Underwood Industries and Some Village Workshops* (Oxford, 1926); vol. II, *Osier-growing and Basketry and Some Rural Factories* (Oxford, 1926, reprinted East Ardsley, Wakefield, 1977); vol. III, *Decorative Crafts and Rural Potteries* (Oxford, 1927 hereafter FitzRandolph, *Rural Industries*, I–III); and Anna M. Jones, *The Rural Industries of England & Wales*, vol. IV, *Wales* (Oxford, 1927, reprinted East Ardsley, 1978 hereafter, Jones, *Rural Industries . . . Wales*).

1841 Census in association with material from 1861 (in practice little used). Day divided English industry into four groups, defined by location and by spatial concentration: 'national', with more than half of its employees located in districts containing less than one quarter of total population; 'metropolitan', where more than half its workers were located in London; 'provincial' if neither concentrated as above, nor sufficiently regular in distribution to indicate dependence upon immediate markets; and 'local', which was broadly ubiquitous in distribution. Cotton textiles, 85 per cent concentrated into Lancashire and Cheshire, is the perfect exemplar of the 'national' industry; artificial flowers, 92 per cent in London, 'metropolitan' and 'national'; brickmakers 'provincial'; and carpenters, butchers and blacksmiths characteristic of the 'local' category. Day's categorisation, though based upon data from outside the period presently under consideration, nonetheless provides a useful shorthand for the economics of industrial location, and is applied in discussions of industry below.[5]

Day's classification is not, however, satisfactory in identifying all the locational factors of significance here, and, London apart, has nothing directly to say on 'urban' or 'rural' location: hence a more complex taxonomy is required. Day's analysis was focused with deliberation upon the data of the 1841 Census for two principal reasons: occupations were presented in fine detail; and 1841 was the last to be taken before the railway had significant impact on the economics of industry and industrial location. This latter point is, of course, critical to the consideration of the industries of the countryside. The development of the trunk railway system was effectively completed between 1840 and 1870, and rapidly provided for the more effective expression of natural advantage in industry, and the intensification of industrial activity in regions thus favoured. The corollary of that was to raise the relative costs of operating on a dispersed, out-working system in businesses such as textiles, and to raise the capacity of mechanised plant to attain 'national' or 'provincial' markets. The decline of rural industries may therefore be as much the function of the development of the transport system, as the simple outcome of advances in industrial technology. Both factors clearly applied, and at different times to different industries.

There were also good reasons for the continued existence of rural industry. In many branches of textiles, for example, transport costs had long been secondary to access to labour supply at the right price in producing goods, or, more commonly, intermediate goods. Rural industries in general retained comparative advantage as low labour cost zones, especially

---

[5] Clive Day, 'The distribution of industrial occupations in England, 1841–1861', *Transactions of the Connecticut Academy of Arts and Sciences*, 28, 1927, pp. 79–235.

for industries that generated piece-rate tasks, in which access to particularly cheap female, elderly or child labour remained unconstrained, and in which the household could continue to provide intermediate supervisory and managerial functions. Motivated by 'harsh necessity', as with hand-loom weaving and, post 1850, wrought iron nailmaking, or, especially for female and elderly workers, by the desire for 'the little extras of life' or to meet short-term family crisis such as the costs of sickness, rather than pure subsistence, the potential supply of outworkers remained rich throughout the later nineteenth century.[6] One general question, therefore, is the extent to which the countryside remained *industrial* purely in a residual sense, mopping up low value-added tasks and processes at each end of manufacturing.

It also offered advantages in flexibility. There were many industries in which demand was unstable, erratic, highly and narrowly seasonalised, or hypersensitive to business cycles, in which rural and by-employed workers permitted the entrepreneur to offload the potential costs of idle plant. Many trades, ranging from the traditional clothing industries and highly fashion-oriented textiles such as lace, to box-making, some building supplies, and even chain-making, fitted elements of these patterns in whole or part. For some industries, as in lace or even in the Staffordshire earthenware manufacture, outworkers, some rural, were used throughout our period as a reservoir to be tapped at the peaks of demand. Demand-side determinants therefore sustained the continued existence of outwork in both urban and rural locations at least to around 1900.

There was also a fair degree of inertia in the later nineteenth-century industrial system. On the supply side, immobility, stemming from the specifics of locality, life-cycle, or other family circumstances, preserved pools of such labour to be tapped. It was only relatively gradually that these were eroded by the processes of rural depopulation, which gathered particular force from the 1870s.[7] Mere habit may also have retarded the transition from one mode of production to another, especially in trades in which elaborate networks of middlemen had developed, and which could adjust to changing market conditions: the nail 'foggers' of the west midlands may have shifted towards part-time and female outworkers in the middle years of the century, to cut rates and thus delay the disappearance of the putting-out trades.[8] Inflexible War Office regulations fixed the specification for some military underpants into handicraft

[6] Duncan Bythell, *The Sweated Trades: Outwork in Nineteenth-century Britain* (London, 1978); and on the wider economics of domestic production, R. Millward, 'The emergence of wage labour in early modern England', *Explorations in Economic History*, 18, 1981.

[7] J. D. Saville, *Rural Depopulation in England and Wales, 1851–1951* (London, 1957), pp. 20–9; D. E. Baines, *Migration in a Mature Economy: Emigration and Internal Migration in England and Wales, 1861–1900* (Cambridge, 1985).   [8] Bythell, *Sweated Trades*, pp. 124–6.

production until 1908, thus preserving supply from the midland framework knitters. Only at the end of the century, and from 1906 with the multi-party Anti-Sweating League, did regulation begin to end such trades.[9]

By-employment or the survival of 'dual economies' of agricultural and industrial work were a further force in the endurance of rural industries.[10] Far from disappearing as Engels suggested, many such systems endured, partly motivated by the industrial considerations discussed above, but partly also by the strong attachment of so many communities to land in its own right. The preservation of a landholding in some form increasingly appears to have motivated petty dealers and industrial outworkers in the later nineteenth-century countryside, and these groups perhaps added a fourth to the traditional 'three-class train' model conventionally applied by historians to agrarian society.[11] From the Weald to Swaledale and the Pennines to Lincolnshire, there is evidence to suggest the use of rural industrial work, or of short-term migration to industrial work even on a weekly basis, by some members of a family to preserve a stake on the land. A smallholding, unviable in itself as a source of family livelihood, and superficially therefore 'irrational' in economic terms, was often an element in a dispersed family earnings system. Good seasonal returns from the agricultural enterprise, such as from dairying or other urban fringe production, may have preserved the perception of the land unit as the cornerstone of this family economy. It may not be altogether fanciful to draw an analogy with Irish 'peasant' behaviour in the same era. Rural industrial activity, however much 'sweated' or intermittent in character, may have been one element of a strategy primarily designed to permit the family to remain in the countryside.[12] This may help to account for its longevity.

Not all of these issues can be resolved in the present study, which is confined to an assessment of the principal industrial activities of the English and Welsh countryside up to the First World War. In the sections that follow, these wider considerations will be touched on in the

---

[9] Ibid., pp. 97, 234–6.

[10] Emphasised by some contemporary studies, such as G. B. Longstaff, 'Rural Depopulation', JRSS, 56, 1893. [11] Adding to labourer, farmer, and landlord.

[12] My interpretation of the implications of a range of modern studies, including J. M. Neeson, Commoners: Common Right, Enclosure and Social Change in England, 1700–1820 (Cambridge, 1993), which, in its study of a county with rural industry, has implications for later in the century; A. Howkins, 'Peasants, servants and labourers: the marginal workforce in British agriculture, c. 1870–1914', AHR, 42, 1994, pp. 49–62; C. S. Hallas, 'Economic and social change in Wensleydale and Swaledale in the nineteenth century', PhD thesis, Open University, 1987, especially chapters 5 and 6, pp. 95–143; and A. Hall, Fenland Worker-Peasants: The Economy of Smallholders at Rippingale, Lincolnshire, 1791–1871, Agricultural History Review, Supplement Series, I, 1992.

consideration of the principal rural or outwork industries grouped successively into textiles, clothing, leather and footwear, wood and underwood, and extractive industries. For each of these groups, broad patterns of spatial distribution, structural and employment trends, and the impact of mechanisation, new processes, or the application of power are analysed before considering the state of the rural element of the industry in the decade before the First World War, and providing the basis to determine the extent to which the countryside was still a significant 'industrial' location in 1914.

A TEXTILES

To begin with the single most important industry possessing a rural history, textiles, it is clear that the experiences of each sector were divergent, with some, notably cotton and linen, relatively early subjected to full mechanisation, and others, woollens, knitting, and parts of lace manufacture, long preserving outwork, often rural in nature. As indicated above, these differences reflected a complex mix of technological and organisational factors, and combined with the specifics of landholding to determine the relative longevity of the rural mode of production. Some of the textile trades considered here went into absolute decline in employment terms during this period, and others merely reallocated employment towards urban mills, and it is useful first to assess overall national patterns of employment in the principal trades considered here, before exploring these issues in somewhat greater detail.

The Census data for this period are not by any means wholly reliable, especially for the female occupations that were increasingly marginalised after mid-century, and whose exclusion became explicit in 1881. Hence in many of the discussions below, following the practice of scholars from Charles Booth onward, occupational data are assessed largely in terms of the information about males, while acknowledging that long before 1881 female employment had been patchily covered, and was always understated.[13] Acknowledging these clear weaknesses, and the inconsistencies over time, Table 18.1 presents summary employment data for the principal textile manufactures, 1841–1911.

These figures do no more than permit the broad-brush description of

---

[13] Charles Booth, 'Occupations of the people of the United Kingdom, 1801–81', *JRSS*, XLIX, 1886, pp. 314–444; W. A. Armstrong, ' The use of information about occupation', in E. A. Wrigley (ed.), *Nineteenth-century Society: Essays in the Use of Quantitative Methods in the Study of Social Data* (Cambridge, 1972), pp. 191–310; B. Hill, 'Women, work and the census: a problem for historians of women', *History Workshop Journal*, 35, 1993, pp. 78–94; E. Higgs, *Making Sense of the Census* (London, 1989); and Higgs, 'Occupational censuses and the agricultural workforce in Victorian England and Wales', *EcHR*, 2nd ser., 48, 1995, pp. 700–16.

Table 18.1. *Employment in textile industries, 1841–1911*
(England and Wales, 000s)

Occupation	1841	1851	1861	1871[b]	1881[b]	1891[b]	1901[a,b]	1911[a,b]
Woollens, male	67.3	76.9	81.2	71.7	57.3	61.6	87.7[a]	95.5[a]
Woollens, female	21.4	45.4	48.8	56.8	58.5	61.3	122.1[a]	127.1[a]
Worsted, M	12.3	54.9	32.8	38.6	35.4	40.5	–	–
Worsted, F	15.5	55.3	50.4	62.5	63.8	69.9	–	–
Flax/linen, M	10.4	13.1	9.3	7.4	4.2	2.6	1.1	1.1
Flax/linen, F	6.4	13.2	12.8	10.6	7.8	5.6	3.3	2.9
Cotton, M	100.5	178.7	198.4	188.4	185.4	213.2	196.9	233.4
Cotton, F	116.9	194.8	259.4	279.9	302.4	332.8	332.2	371.8
Lace, M	6.5	9.4	8.9	8.5	11.4	13.0	12.6	15.2
Lace, F	20.3	52.3	45.1	40.8	32.8	21.7	23.8	25.8
Knitters, F	1.4	2.5	2.0	–	–	–	–	–

Notes:
[a] Woollens and Worsteds not distinguished in these years, and Woollens rows
cover totals of both.
[b] Knitters not separately enumerated in these years.
Sources: Figures for 1841–91 are taken from Armstrong, 'Information about
Occupation', being his revision of Booth's reworking of the data,
'Occupations of the People . . .'; 1901 and 1911 data are my attempt to
replicate Armstrong's approach, taken from Census reports, summary data.

a few key features of each industry. Charles Booth, whose methods were
largely followed but improved by Armstrong, made numerous adjustments to categories for 1841, but did not record them fully in his notebooks, and presented only aggregated results for textiles, and hence
cannot resolve much of the obvious uncertainty over whether the first
set of figures is meaningful. The 80 per cent growth in our sub-set of the
1841 and 1851 figures of textile employment far exceeds the 55 per cent
of Booth and Armstrong's pooled data, and is largely the outcome of
omitting from these data the 99,000 in the catch-all category of 1841,
'weavers, spinners and factory hands (textile)'. This vagary of 1841
categorisation may have had a particular impact upon the 'worsted'
figures, coinciding as it did with a point of particular significance in the
move to mills.[14]
Three other features stand out from these aggregates. In the woollen

[14] D. T. Jenkins, *The West Riding Wool Textile Industry, 1770–1835* (Edington, 1975); P. Hudson, *The Genesis of Industrial Capital: A Study of the West Riding Wool Textile Industry c. 1750–1850* (Cambridge, 1986).

and worsted categories, male employment was almost exactly displaced by the growth of recorded female workers, a sign of both the development of mechanisation and the deskilling of the workforce, but also perhaps indicative of the understatement of women's work in 1841. Second emerges the familiar massive growth in employment in cotton, again with a dramatic female component, and fluctuations in lace. Third, there was a striking decline of flax and linen manufacture from mid-century, as the industry largely relocated in Scotland and Ireland.[15] In all of these industries, the generality of experience was of the concentration of production into mills, and the decline of rural outwork.

Some further light on this process and its timing can be derived from closer analysis of the most detailed occupational data of the Census of 1861, applying an approach analogous to, but significantly different from, that of Day, the location quotient. This measures the extent to which particular industries were concentrated into the specific counties of England and Wales, measuring their distribution against that of the economically active workforce. It provides a crude but simple screen to pick up clusters of particular industries or trades, and a focus for discussion of locational factors and trends within them. Because of the differences of scale, which would distort the overall measures, this analysis has omitted 'London', and hence 'Middlesex', 'Surrey' and 'Kent' are assessed only in their 'extra-metropolitan' districts. For kindred reasons, 'Wales', where most county populations were small, has been analysed separately, although pooled England and Wales results were employed to check the overall magnitude of clusters of significance. Distinctive concentrations of employment have been defined, as in an earlier study, as those at or above one standard deviation from the mean quotient.[16]

Among the major textiles summarised in Table 18.2 this approach

---

[15] Bythell, *Sweated Trades*, p. 58

[16] This approach was suggested by L. A. Clarkson in a contribution to G. E. Mingay, ed., *AHEW*, vol. VI, *1750–1850*, (Cambridge, 1989), 'The manufacture of leather', esp. pp. 476–81, and adopted in my study of 'country trades, crafts and professions' in the same volume, pp. 416–66. Clarkson's analysis presented location quotients derived from the 1841 Census to identify the finer gradations of leather manufacturing, whereas my study applied 1851 data. Thus use of the 1861 data in this and the succeeding chapter here explicitly provides comparative results, subject, of course, to the differences in compilation of each of these successive Censuses, and applies this crude screen to the most detailed of Census listings of occupations. Location Quotients are estimated as:

$$L.Q. = \frac{\text{Percentage of a particular occupation in a county}}{\text{Percentage of the civilian occupied workforce in the same county}}$$

These occupational descriptions are, of course, self-descriptions, and hence contain what many historians may regard as 'aspirational' features, whereas Booth and subsequent scholars have been interested primarily in their evidence on industrial groupings, such as in the present instance 'Textiles', and rearranged returns according to those objects. In this present study, including usage in Chapter 19, a data set for ninety-three occupations (counting male and female of several trades

Table 18.2. *Location quotients, England and Wales, major textiles, 1861*

Occupation	Mean L.Q.	S.D.	County concentrations (in rank order)
England:			
Woollens, M	0.379	1.294	Yorks, WR, 7.953; Wilts, 2.255
Woollens, F	0.459	1.309	Yorks, WR, 6.722; Wilts, 4.736; Glos, 2.426
Worsted, M	0.339	1.426	Yorks, WR, 8.868; Leics, 2.610
Worsted, F	0.309	1.447	Yorks, WR, 9.233
Cotton, M	0.287	0.880	Lancs, 4.939; Chesh, 2.175; Derbys, 1.689
Cotton, F	0.271	0.744	Lancs, 3.911; Chesh, 2.004; Derbys, 1.974; Cumb, 1.064
Wales:			
Woollens, M	1.221	1.479	Montgomery, 5.428
Woollens, F	1.133	2.842	Montgomery, 10.524
Worsted, M	1.152	1.015	Anglesey, 3.444; Merionedd, 2.307
Worsted, F	0.792	2.179	Pembroke, 7.933
Cotton, M	1.831	4.188	Flint, 15.662
Cotton, F	2.579	9.070	Flint, 32.763

Source: Calculated as explained in note 16; data from *Census*, 1861.

generates no surprises. Male employment in cottons, worsteds and woollens was concentrated into a small group of English counties, as expected from 'national' industries, and the first two were already long urban in character, as was the small cluster of female cottonworkers in Flintshire, focused on Holywell. Leicestershire's cluster predominantly reflected worsted stocking manufacture, which endured in heavily industrial villages, notably Shepshed.[17] Though woollens concentrated into the old town-based industries of the West Country, the West Riding was still a very mixed zone, sprawling south-westwards from its commercial centre, Leeds, and made up of cottage-, small township-, and smaller mill-based enterprises, and even in 1914 the industry retained some of its characteristic village mills.[18] However, the third quarter of the century saw the

separately) was extracted and analysed, although, as will be seen, several of these proved so small in numbers returned as to offer little analytical value. I am indebted to Jon Knox for his assistance in the final computation of these results.   [17] Bythell, *Sweated Trades*, p. 7

[18] M. T. Wild, 'The Yorkshire wool textile industry', in J Geraint Jenkins (ed.), *The Wool Textile Industry in Great Britain* (London, 1972), pp. 211–28.

growth of large integrated plant, and this eroded the position of the smaller spinning and weaving mills.[19] Increasingly, from 1870, the industry also abandoned home wool supplies in favour of Australian merino and New Zealand cross-bred wools, many of the latter derived from Cheviot or English breeds.[20] Numbers employed in rural outwork in woollens fell rapidly during the years after 1850, mirroring the earlier shift of cottons and worsteds, although in West Yorkshire some family members may have remained on the traditional farm unit, with others working in the mill while still part of the same household.[21] By the first Census of Production in 1907, numbers of 'outworkers' recorded in both woollens and worsteds were very small.[22]

Not all rural-based woollens declined. There were significant if small clusters in Wales among both males and females: in Montgomery, where Welsh tweeds were manufactured, they endured around Rhyadyr into the present century as a village and small town mill industry, though much smaller than its 1861 level; flannel manufacture continued in and around Newtown, 'the Leeds of Wales', but lost ground to the 'real Welch flannel' made in Rochdale; and in the north, at Llangollen, there were large spinning mills.[23] In the south-western counties of Wales, new demand from the developing coalfield appears to have created some growth in woollen and flannel manufacture, based upon small rural mills, and larger factories in the Teifi valley, centring on Llandysul.[24] Total employment in woollens (including a tiny element of declared worsteds) in the three counties of Cardigan, Carmarthen, and Pembroke stood at 1,750 in 1891, double the level of 1861, and was still nearly 1,200 in 1911, two-thirds in Carmarthen. The industry remained active into the post-war period, and Jones's survey of 1922–3 identified 107 'wool factories' in the three counties, 71 per cent of the Wales and Monmouthshire total, of which 62 were still powered solely by water. Most failed to adjust to changing fashions, and declined dramatically in the inter-war years.[25] Some handloom weaving survived in small clusters where spinning and pandy (fulling) facilities were accessible, as around Lampeter, making quality goods, and Jones found a weaver declaring, 'One can make a fair living by it, but a man can never get rich at it.'[26] Worsteds, by contrast,

[19] D. T. Jenkins and K. G. Ponting, *The British Wool Textile Industry 1770–1914* (London, 1982), pp. 77–91, 166–83.

[20] J. W. Turner, 'The position of the wool trade', *JRASE*, 3rd ser., 7, 1896, pp. 67–76.

[21] Information kindly provided by Alison Knowles. The same mixed income-generating strategy endured in parts of Lancashire beyond 1870: M. Winstanley, 'Industrialization and the small farm: family and household economy in nineteenth-century Lancashire', *PP*, 152, 1996, esp pp. 181–92. FitzRandolph, *Rural Industries*, III, pp. 7–15.

[22] *Census of Production, Final Report*, BPP, 1910, CIX, pp. 294–347.

[23] J. Geraint Jenkins, 'The Welsh woollen industry', in Jenkins (ed.), *Wool Textile Industry*, pp. 288–91.

[24] *Ibid.*, pp. 292–3.   [25] *Ibid.*, pp. 296–300.   [26] Jones, *Rural Industries . . . Wales*, pp. 20–1, 26.

Table 18.3. *Location quotients, England and Wales, minor textiles, 1861*

Occupation	Mean L.Q.	S.D.	County Concentrations (in rank order)
England:			
Flax/linen, M	0.595	1.221	Yorks, WR, 6.281; Yorks, NR, 3.983; Dorset, 2.656; Westm, 2.023; Cumb, 1.923
Flax/linen, F	0.549	1.369	Yorks, WR, 6.639; Westm, 4.669; Cumb, 3.789
Lace, M	1.169	5.925	Notts, 37.916
Lace, F	1.561	3.751	Bucks, 16.395; Northants, 12.974; Beds, 9.374; Notts, 9.337; Hunts, 5.494
Knitters	1.790	3.662	Rutland, 19.943; Leics, 9.066; Dorset, 8.844; Westm, 6.499; Yorks, NR, 6.001
Wales:			
Flax/linen, M	1.313	1.128	Radnor, 3.537; Brecon, 2.903
Flax/linen, F	2.822	10.174	Flint, 36.684
Lace, M	1.473	3.663	Radnor, 12.974
Lace, F	0.634	0.760	Glamorgan, 2.378; Monmouth, 1.794
Knitters	1.045	0.704	Merionedd, 2.445; Carmarthen, 2.052

Source: Calculated as in Table 18.2 above.

were little made in Wales, and the large quotients have no analytical significance.

The lesser textile manufacturing trades, summarised in the location quotients of Table 18.3, present a more complex and varied picture. As noted earlier, linen declined in the face of Scottish and Irish competition, and had already become highly localised in England and Wales in 1861. In England it was principally an urban industry, centred on Leeds led by Marshall, with smaller clusters, the moribund residues of outwork, in Barnsley, Knaresborough and Darlington.[27] Smaller clusters were located around Kendal and Windermere, where it survived to the 1920s

---

[27] W. G. Rimmer, *Marshalls of Leeds, Flax Spinners* (Cambridge, 1960); Bythell, *Sweated Trades*, pp. 57–9.

on a minor scale, for premium-quality household linens and 'Greek lace', linen threadwork. At the opposite end of the spectrum, the Dorset quotient reflected the sailcloth- and canvas-manufacturing region around Bridport, Broadwindsor and Beaminster, which extended to Chard, Crewkerne and the Cokers in Somerset, residual traces of which again endured after 1918.[28]

Lace, however, provided a complex mix of two rural handicraft industries, the Devon and Midland, and one machine sector, that of Nottingham. The latter generated significant handicraft work, put out to largely female domestic and rural workers, and the Midland handicraft industry at least came during our period to employ machine-made nets, onto which pillow-lace *sprigs* were mounted. Within each district, a range of quite distinct products grew up, reflecting the timing of adoption of patterns and their sources: and three broad groups could be distinguished in the hand-made laces: 'Bedford', which contained Maltese, Torchon, and Cluny patterns, and was the more recent of the two Midland types; 'Buckingham', or 'point ground', the older, allegedly owing its origins to the seventeenth-century Flemish migrants and its patterns consequentially to Brussels, Mechlin and Valenciennes; and Honiton 'point' or 'stitch', the products of the East Devon industry, a Belgian-style bobbin lace unknown in the midlands. Lace is perhaps the perfect illustration of the complexity of rural industry, and its interrelationship with the mills of the industrial town.[29]

Lace also illustrated the vicissitudes of handicraft industry and rural manufacture, and its attraction for the entrepreneur. The industry was vulnerable to changes in fashion, hence the attraction of cheap and flexible outwork for both sectors; increasingly its labour intensity made it vulnerable to foreign competition, by the twentieth century including that from China, while machine lace was a major export and component of exports; and early mechanisation first by the point-net and subsequently by the bobbin-net devices intensified the trend towards factory production, while not producing a perfectly finished final product, and thus generating a secondary demand for distributed handicraft work. The clusters of lacemaking observed in Table 18.3 therefore enjoyed very different experiences during the subsequent half century. That the traditional Devon industry was already in decline in mid-century may be indicated by the absence of a significant cluster of employment although, being highly localised within a large county, the quotient may not prove a wholly adequate measure of its significance. Actual numbers declared

[28] FitzRandolph, *Rural Industries*, III, pp. 7–19; J. H. Bettey, *Rural Life in Wessex, 1500–1900* (Bradford-on-Avon, 1977), pp. 44–5.
[29] FitzRandolph, *Rural Industries*, III, pp. 50–71; Bythell, *Sweated Trades*, pp. 97–105.

for the 1861 *Census* show an industry already half the size of that of Buckinghamshire and Northamptonshire, and two-thirds that of Bedfordshire.[30] Despite efforts to turn the industry around, including the incorporation of lacemaking into the curriculum of new compulsory schooling, decline continued, and the industry was on its last legs in 1914.[31]

In the midland industries, survival meant cost-cutting inside a complex organisational structure, where lacemen dealt with village dealers, or direct with the largely female producers. Thomas Gilbert, one of the leading Buckinghamshire putters-out, admitted in 1863 to employing 'about 3000 persons', by selling patterns and thread, and accepting a general but not absolute obligation to buy the subsequent produce.[32] Pillow lacemakers responded to machine competition by design innovation and cost-cutting in a craft of high skill, in which training from early childhood was an essential prerequisite, and provided an extreme case of the trap in which so many female workers were to find themselves. A key symptom of the decline of the Honiton industry had been its loss of design and assembly skills, and the gradual suppression of the midland 'lace-schools', where children were sent to learn the complex craft from the age of six or seven, by the impact of compulsory schooling after 1870, led to an ageing workforce, and an industry in which those of lower skills effectively became adjuncts of the machine makers.[33] Symbolic of this was the 'bertha' of Paulerspury (Northants), essentially an application of point ground lace appliqué to net collars and veils, a forlorn if imaginative attempt by the vicar's wife to revive a dying industry in the 1920s.[34].

In rural Nottinghamshire, a huge dependent female and often juvenile workforce serviced factory-made lace: Mrs Barber's evidence from her own 'seaming and running' business in 1843 suggested up to a hundred mistresses in Nottingham, employing between five and ten young women directly, and others in the smaller towns and villages on a putting-out basis, and through these cascading layers, each of the 'first-hand mistresses' might ultimately employ around 300, a total outwork employment of up to 30,000. In 1865, Felkin suggested up to 100,000 such outworkers, ten times the numbers listed in the *Census* for Nottinghamshire in 1861,

---

30 Actual numbers were as follows, for female lacemakers: Devon, 4,841; Bucks, 8,459; Northants, 8,187; and Bedford, 6,714. The Welsh quotients can again be ignored, as measures of concentration in industries of no significant magnitude in the country.

31 H. J. Yallop, 'The history of the Honiton lace industry', *Textile History*, 14, 1983, pp. 195–211.

32 In evidence to the Royal Commission on the Employment of Children, 1863, cited by Bythell, *Sweated Trades*, pp. 102–3; G. F. R. Spenceley, 'The health and disciplining of children in the pillow lace industry in the nineteenth century', *Textile History*, 7, 1976, pp. 154–71.

33 Bythell, *Sweated Trades*, pp. 103–4; G. F. R. Spenceley, 'The English pillow-lace industry, 1840–80: a rural industry in competition with machinery', *Business History*, 19, 1977, pp. 68–87.

34 FitzRandolph, *Rural Industries*, III, p. 55.

double the female occupied workforce, and two-and-a-half times the total of direct factory employment.[35] The figures point therefore to a practice of very large scale, but cannot be interpreted as a precise estimate. While much of this work gradually moved inside the factory, outwork remained significant to the end of the century and beyond; directly contracted outworkers still numbered over 4000 according to the *Census of Production*, and may have been as high as 7000, according to the evidence of the Committee on Homework, also in 1907. Conditions remained poor, with exploitative piece rates paid to workers in both sectors, and led FitzRandolph and Hay to the sorry conclusion, based upon surveys of 1921–3: 'Lace-making is either a sweated industry, practised only by the most needy, or a delightful hobby for leisured people whose time is not reckoned in terms of the necessities of life.'[36]

Knitting as recorded in these Census data was a fuzzy category, and bracketed two quite distinct industries, the figures of Tables 18.1 and 18.3 certainly underestimating both. One was the framework knitting industry of the east midlands, in which the workforce was predominantly classified in 'hosiery', discussed below,[37] and the other was the long-standing hand knitting industry of the north, where some workers at least may have slipped into 'woollens' or even into 'glovers'.[38] As an industry, even allowing for underestimation by the Census, it was already largely moribund in 1850, and confined to areas such as the Dales and the Lake District, in which it provided a means to cash income where few alternatives existed. Both areas, with the coarse stocking industry of Dorset, are indicated by the figures of Table 18.3, which must be treated with some caution in the light of the evident poverty of the basic Census data.[39]

Hand-knitting, even if in terminal decline in the 1850s and 1860s, had long been a major domestic employer, using simple tools to sustain what for many was a full-time occupation. Speed was remarkable, and sustained by the use of a wooden sheath tucked into a broad leather belt to support one needle, normally the left-hand , wire curved needles or hooks, and tape employed with a top-crook inserted into the knitting and drawn back round the belt to sustain tension. Simple worksongs, which were disappearing fast with the industry, had been employed to sustain the rhythm that was critical to speedy and sustained output. As with straw-plait and lace, village knitting 'schools' were directed by older women.[40] The prod-

[35] Cited by Bythell, *Sweated Trades*, pp. 100–1, which notes R. A. Church's scepticism over Felkin's figures; *Census of Production*, 1907, BPP. 1910, CIX, p. 184.

[36] FitzRandolph, *Rural Industries*, III, p. 71.   [37] See section B, pp. 1115–16, 1122.

[38] On glovers, section B, pp. 1122–3.   [39] On Dorset, Bettey, *Wessex*, pp. 44, 50.

[40] M. Hartley and J. Ingilby, *The Old Hand-Knitters of the Dales* (Clapham, 1951; 5th edn, 1991), pp. 17–22.

ucts were predominantly heavy woollen stockings, mittens, with some caps and jackets, made largely from the coarse 'bump' yarns supplied from Kendal or Bradford by the carrier, who so often provided the middleman functions in these northern districts. One such was Ralph Fawcett, of Hawes, who still delivered bump after 1870 to communities from Keld to Gunnerside at the head of Wensleydale, and took a halfpenny commission for each pair of stockings. The coarse products were made large, with mittens going beyond the elbow, and the 'elephant stockings' a yard long and thirteen inches in each foot, all focused onto the mining, marine and Greenland fishery markets.[41] It remained no more than a residual activity, with many of the men and women 'wapping away' at their cottage doorways because they lacked an alternative. As Mrs Martha Dinsdale of Appersett (near Hawes) commented, c.1880, 'we were fain to deu it, ther' wer' nowt else', much the same experience as lacemakers a generation later.[42]

B CLOTHING

The clothing industries were, quite naturally, much smaller generators of total employment, apart from the hosiery trades of the midlands, and the vast numbers in tailoring, but were still occupations with real significance for the rural districts, most strikingly perhaps in the rise and fall of straw plait manufacture. The broad trends of employment in this group of trades are summarised in Table 18.4.

As in the earlier discussion of textiles, it must be noted that female employments were unevenly recorded, and in general poorer after 1881, and should be regarded generally as understating actual levels. This was particularly the case with the hosiery trades, where our data do reflect the transition, discussed above, into factory production, and hence to the formal recording of females as employees, but may still underestimate the extent of female outworkers, employed, as in the machine lace industry, through a hierarchy of putting-out, and hence almost certainly veiled from the Census record.[43] Daniel Angrave ran a small six-frame workshop at Gilmorton (Leics) from 1895, making grey worsted stockings for the army, and employed fifteen women outworkers toeing in the village.[44] Tailors listed here are confined to males, and this excludes the substantial growth in female employment in the industry particularly from the 1880s, but which was overwhelmingly located in urban workshops. In Day's categorisation for 1841, all these trades, bar tailoring

---

[41] Ibid., pp. 34–5.   [42] Ibid., p. 58; Hallas, '. . . Wensleydale and Swaledale . . .', II, pp. 398–9.

[43] S. O. Rose, 'Proto-industry, women's work and the household economy in the transition to industrial capitalism', Journ. of Family History, 13, 1988, pp. 181–93.

[44] R. Foster, 'The hosiery trade in a Leicestershire village', Folk Life, 11, 1973, pp. 85–7.

Table 18.4. *Employment in clothing industries, 1841–1911*
(England and Wales, 000s)

Occupation	1841	1851	1861	1871	1881	1891	1901	1911
Straw plait, M	1.5	*a*	4.1	3.6	3.0	3.4	3.9	5.0
Straw plait, F	17.7	46.5	44.2	45.3	28.0	15.0	10.9	9.9
Hatter, M[b]	11.7	9.4	7.2	10.1	9.5	12.3	12.1	12.8
Hatter, F[b]	2.7	6.7	2.0	6.2	8.0	9.5	11.9	12.5
Tailor	113.3	121.7	110.3	111.8	107.7	119.5	119.5	122.4
Buttonmaker, M	2.3	3.0	2.8	2.4	2.3	1.9	1.2	1.4
Buttonmaker, F	1.7	3.9	3.8	3.4	4.1	3.1	3.0	4.3
Hosiery, M[c]	27.5	33.6	24.4	22.4	18.9	18.2	20.7	22.5
Hosiery, F[c]	8.5	25.3	21.5	19.7	21.5	30.9	36.3	43.6
Glover, M	3.1	4.5	3.8	2.7	2.3	2.8	2.4	3.0
Glover, F	6.1	25.3	24.0	20.3	13.2	9.2	7.8	8.1

Notes:
[a] No data
[b] Excludes the much more numerous milliners
[c] Compounds stocking manufacturing and dealing
Source: Compiled as Table 18.1 above.

which was ubiquitous and hence 'local', were 'national' in distribution, and thus highly concentrated in geographical terms.[45]

The straw plait industry was highly concentrated into three English counties, although some activity existed in many others. The spatial distribution of this and the other clothing trades is summarised by the location quotients calculated from the Census of 1861, in Table 18.5. It is evident that the plait industry overlapped significantly with the south midland lacemaking counties, with Buckinghamshire demonstrating distinctive quotients for both trades. This may not have been wholly accidental, in that the skill, manual dexterity, and early training, felt to be so important to the finer lace trades, were also held to be prerequisites for straw plait, and in both industries there was an extended continuum of employment, from the full-time domestic worker to the very much part-time, and both involved very extensive child labour. 'Old ladies' were recorded in the early 1920s as testifying to the two trades being carried on by the same workforce, which switched between them according to demand and season.[46] These Census data therefore indicate the significant regional concentrations of activity, but may underestimate total numbers

[45] Day, 'Distribution of industrial occupations', pp. 223–33.
[46] FitzRandolph, *Rural Industries*, II, p. 119.

Table 18.5. *Location quotients, England and Wales, clothing trades, 1861*

Occupation	Mean L.Q.	S.D.	County concentrations (in rank order)
England:			
Straw plait, M	2.502	11.838	Beds, 73.617; Herts, 20.339
Straw plait, F	1.626	5.255	Beds, 27.624; Herts, 19.595
Hatter, M	0.654	0.938	Chesh, 5.228; Lancs, 2.492; Glos, 2.605; Warks, 1.599
Hatter, F	0.708	1.228	Herts, 5.992; Chesh, 3.546; Glos, 2.975; Yorks, NR, 2.502; Cumb, 2.269; Lancs, 2.103
Tailor	1.010	0.228	Yorks, NR, 1.609; Yorks, ER, 1.568; Cumb, 1.388; Devon, 1.325; Westm, 1.311; Northd, 1.297
Buttonmaker, M	0.724	3.534	Warks, 22.576
Buttonmaker, F	0.659	3.289	Warks, 21.100
Hosiery, M	1.332	5.306	Leics, 28.028; Notts, 19.994
Hosiery, F	1.172	4.687	Leics, 26.612; Notts, 14.742
Glover, M	1.211	2.993	Som, 13.769; Worcs, 12.579; Oxon, 7.082
Glover, F	1.193	2.721	Som, 10.589; Worcs, 8.983; Oxon, 7.848
Wales:			
Straw plait, F	0.917	0.649	Pemb, 2.699; Monmouth, 1.602
Hatter, M	1.039	0.795	Cardigan, 2.899; Carmarthen, 2.077; Anglesey, 1.837
Tailor	1.096	0.215	Cardigan, 1.553
Hosiery, M	1.028	1.601	Cardigan, 5.856
Hosiery, F	1.023	1.602	Cardigan, 5.951

Source: Calculated as in Table 18.2 above, with figures for occupations in Wales returning fewer than 100 in all omitted for clarity.

employed, and the figures of Table 18.4 correspondingly provide a guide to magnitude and to trends in employment between 1841 and 1911, but should not be regarded as precise.

Plait was therefore heavily concentrated into the countryside of Bedfordshire, Hertfordshire and, to a much lesser degree, Buckinghamshire, and was strongly associated with the chalklands in these counties:

the high quotients for the two southern counties of Wales accounted for fewer than 200 workers in all, and represent a small cluster in a trade effectively uncharacteristic of Wales in the 1860s.[47] As indicated by these figures, the industry was predominantly female, but engaged children of both sexes from as early as five, skills being inculcated as in lace through plaiting 'schools', which may have been little more than rural sweatshops. Kitteringham's study of the trade cited Miss Vaughan's recollections of Harpenden in 1893 in support of this view, and 'schooling' was held to consist of a quarter-of-an-hour a day diverted to the Bible.[48] The Factory Inspectors found corroborative evidence in 1871:

On the opposite side of Dunstable I visited the villages of Silsworth, Stanbride, and Egginton: there was no school in any of these villages, but I visited seven straw plait schools and found 85 children under 13 years of age working in them. I saw at least 20 to 30 others in the lanes and at the doors of the cottages.[49]

Child labour, masked as schooling, was compounded by long hours and poor conditions, and in that respect straw plait mirrored the other principal outwork trades.

Yet the straw-plait industry was very different organisationally. Though there was some putting-out and some middleman activity by retailers, village shopkeepers and even the straw-producing farmers, and though the trade depended to a great extent upon the increasingly concentrated stitching and hatmaking centres of Luton and Dunstable, perhaps the majority of plaiters worked independently at the beginning of our period, and sold their produce direct in the town markets. It was generally reckoned that four plaiters were needed to keep one hand-stitcher supplied.[50] Rapid growth in recorded employment indicated also that it was a business subject to a long upswing in demand, and pointed to the extensive endowment of this small region with plaiting skills: though it did not produce equivalent results for Hertfordshire, the plaiting trade was directly responsible for Bedfordshire's distinctive female activity rate of nearly 50 per cent in 1861, nearly double that of the run of English counties.[51]

[47] For more detailed locational analysis, see C. M. Law, 'Luton and the hat industry', and D. J. M. Hooson, 'The straw industry of the Chilterns in the nineteenth century', *The East Midland Geographer*, 4, 1966–9, pp. 329–41, 342–9.

[48] Miss Vaughan, *Thirty Three Years at Harpenden*, 1893, cited by Jennie Kitteringham, 'Country work girls in nineteenth-century England', in Raphael Samuel (ed.), *Village Life and Labour* (London, 1975), p. 120.   [49] *Reports of the Inspector of Factories for . . . 1871*, cited *ibid*.

[50] Bythell, *Sweated Trades*, p. 121.

[51] Census figures analysed in this and the two subsequent chapters are based upon the ratios of identified employments to civil occupied population, defined as total population net of orders II (defence), IV (children, relatives and scholars), XVII (persons of rank and property), XVIII (supported by the community or unspecified), with categories such as 'farmer's "Wife"' addition-

Both contemporary and subsequent assessments of the industry are contradictory to a degree. That there were gradations of skill required to produce different kinds of plait, from the simple 'Dunstable twist' up to the twenty-straw plait of chined straw was clear, but given the contemporary view of plaiting as 'easy' work, and the apparent ease with which young women established themselves in the trade, and thus gained an independence of living that carried with it a reputation for want of chastity, it is hard to see human capital barriers to entry, or determinants of location other than local supplies of suitable straws, and proximity to urban centres of plait usage, notably Luton, Dunstable, Hitchin and Ivinghoe. Unmarried women met the seasonal demands for hat-making by temporary migration into Luton to stitch hats and other products, returning home to the countryside for much of the summer, when hats were purchased and worn, not made. Munby certainly considered that the industry was critical to the welfare of the rural labouring families, and as in several other cases the timing of its prosperity and decline affected that of out-migration.[52]

Decline set in with changing fashions, cheaper Chinese imports (1873), and the sewing machine (1874), but the trade survived in attenuated form, with a greater concentration into the towns.[53] Contemporaries often blamed the introduction of compulsory schooling for the loss of the child labour that had been the staple of the industry. It was, however, only a marginal cause of decline, and it was rather the closures of the lace schools that led to overcrowding in the board schools from the 1880s.[54] By 1907, perhaps a third of the regular workforce remained rural, when gross output of straw hats, at 1.434 million dozens, still exceeded that of felt hats, with exports broadly comparable at 37 per cent of total.[55] When visited by the Oxford survey in the 1920s, numbers of

---

ally deducted from the female figure. This is a fairly hard definition of the occupied workforce, and correspondingly tends to understate the economic activity rate which is expressed here by this figure as a percentage of total population in the county. The results are, however, striking: whereas male activity rates thus defined ranged between a low of 57.2 per cent (Hampshire, the consequence of effectively removing Portsmouth) to a high of 67.7 (Hereford), and mean of 64.9 (S.D. = 1.95), women ranged from 14.2 in Durham, to Bedfordshire at 49.2, with a mean of 27.0 (S.D. = 6.02).

[52] Kitteringham, 'Country work girls', p. 126; Baines, *Migration*, pp. 178–249.

[53] Kitteringham, 'Country work girls', pp. 121–2, citing the *Morning Chronicle* from 1850; Bythell, *Sweated Trades*, p. 121. By their economic self-determination, plaiters fell short of Victorian expectations for dependence upon male breadwinners, and opened themselves to accusations of questionable morals: the young women of St Alban's had thus a very questionable reputation, according to Munby, cited by Kitteringham, p. 126.

[54] D. Thorburn, 'Gender, work and schooling in the plaiting villages', *Local Historian*, 19, 1989, pp. 107–13.

[55] *Census of Production, 1907*, BPP, 1910, CIX, p. 401 used to make a very rough approximation, given the consolidation of all hatmaking into the assessment of employment structure.

outworkers were put at 'well under one hundred', confined largely to women who had formerly worked in the factories.[56]

The remaining hatters were more heterogeneous, and almost certainly overlapped with the straw hat makers, as indicated by the cluster of women workers in Hertfordshire. They included two principal groups, using woollens to make felt hats, and those making hats from a wide range of skins. They were located in areas in which there had been a traditional cap- and hat-making industry, such as Gloucestershire or the knitting districts of North Yorkshire, Cumberland and parts of Wales, and while involving extensive outwork, in the principal northern clusters of Cheshire and Lancashire the occupation was predominantly urban in nature. There were few signs of even 'rural factories' in the Oxford surveys of the 1920s, with that at Brandon (Suffolk), established on the basis of processing rabbit skins, a notable exception.[57]

Tailoring was ubiquitous at the beginning of the period, being distributed down to village level, like the still more numerous shoemakers, and it compounded some putting-out manufacture with the generality of making-up, repair, and refurbishment work. It was in a real sense a direct service trade, and the closeness of the mean Location Quotient to unity and the small standard deviation in England and Wales confirm the point. In 1879, tailors were the craft businesses that appeared at the lowest population threshold in the North Riding, at one per 330 persons, ahead of blacksmiths (340) and wheelwrights (350), with a similar distribution in the village of Wortham (Suffolk) in the 1860s.[58] With the advent of the sewing machine in the later 1850s, more outwork was generated, particularly for women, in a range of activities from buttonholing to finishing, and full make-up work diffused into some villages with the machine itself later in the century. A. J. Hollington, manufacturer of clothing for the colonies, had all of his goods made up in the country districts of Hampshire, Essex and Suffolk in this way in 1888.[59] While the great centres grew in London and Leeds, female rural outworkers thus comprised an element of the massive growth of women's employment in tailoring: from 17,500 thousand in 1851, 12.6 per cent of all recorded workers in tailoring, 38,000 women accounted for a quarter by 1871, achieved parity with 117,600 employed in 1901, and with 127,100 employed, exceeded this by 1911.[60]

---

[56] FitzRandolph, *Rural Industries*, II, pp. 119–20. The Oxford survey was conducted in 1920–1.

[57] FitzRandolph, *Rural Industries*, II, p. 143.

[58] E. R. Kelly, *The Post Office Directory of the North and East Ridings of Yorkshire* (London, 1879); R. Cobbold, *Features of Wortham, 1860*, edited R. Fletcher, *The Biography of a Victorian Village* (London, 1972).

[59] Evidence to House of Lords Select Committee, cited by Bythell, *Sweated Trades*, pp. 70–1.

[60] Figures derived as in Tables 18.1 and 18.4 above.

Tailoring can thus be treated as a two sector trade, with the leading element of manufacturing subdivided between the factories and sweat-shops and outwork, and the widely distributed 'service' element occurring everywhere as small individual or workshop businesses. Many such rural tailors, in market town or village, were peripatetic, meeting the needs of the country dweller for work clothes and 'Sunday best', as recalled by Bill Partridge of Lindsey (Suffolk) for the years before 1914:

You'd git a topping suit for thirty bob, that'd last you ten years, and just as good then as that was when you bought it. There was this tailor bloke used to live at Bildeston, his name was Prentice: he'd come round and measure you one Sunday morning, and bring your suit the next Sunday morning. You didn't want to go to his shop, he'd come right the way from Bildeston on his old bicycle: he didn't hev no pony and trap – they couldn't afford to buy one, such as they.[61]

Hugh Abercrombie entered apprenticeship in a rural tailor's workshop around 1900, at St Michael's-on-Wyre in the Fylde, where he was one of five workers, supplemented in the summer months by one or two tailors 'on the tramp', and described the range of work:

. . . gamekeepers' coats, morning coats with bound edges, riding breeches with buckskin strappings and box-cloth leggings, plenty of whole-fall and split-fall trousers, also the new fly-front trousers then coming into vogue and an endless number of waistcoats, double-breasted and with collars. After five years I was able to make anything that was wanted.

In addition, the firm made women's corsetry, 'stiff terrifying garments which looked like armour-plate', all made up in the workshop to patterns and measurements collected by the proprietor, John Fisher, from the houses and farm customers he visited in his pony and trap.[62] In Yorkshire's East Riding, farm servants purchased clothes from tailors on credit, paying bills in instalments at the annual hirings, although the coming of the bicycle in the years immediately before 1914 opened access to shops at other times, at least in the summer months when the roads were passable.[63] Mass production and sales post-war by retail chains, notably 'wholesale bespoke', ultimately reduced both types of business, but both survived the Second World War.[64]

---

[61] Evidence of Bill Partridge, former 'head horseman' born in 1900, oldest inhabitant of Lindsey, near Hadleigh, when recorded in the early 1980s, in Charles Kightly, *Country Voices: Life and Lore in Farm and Village* (London, 1984), p. 23.

[62] H. Abercrombie, 'A country tailor looks back', summer 1967, in Elizabeth Seager (ed.), *The Countryman Book of Village Trades and Crafts* (Newton Abbot, 1978), pp. 172–3.

[63] S. Caunce, *Amongst Farm Horses: The Horselads of East Yorkshire* (Stroud, 1991), pp. 63, 169, 178.

[64] For the wider context, see E. M. Sigsworth, *Montague Burton, The Tailor of Taste* (Manchester, 1990), pp. 34–5; K. Honeyman, 'Montague Burton Ltd: the creators of well-dressed men', in J. A. Chartres and K. Honeyman (eds.), *Leeds City Business, 1893–1993: Essays Marking the Centenary of the Incorporation* (Leeds, 1993), pp. 187–8.

Buttonmaking requires little comment beyond the telling statistics of Table 18.5. It was very heavily concentrated into the metalware workshops of Birmingham and adjacent towns, although residual traces of fancy handicraft 'buttony' were still to be found in Dorset, around Bridport and Sturminster Newton.[65] It did generate substantial volumes of outwork in the small towns and villages of the region, but very much part-time employment at poverty rates of pay, largely confined to women carding buttons, or similar goods such as hooks-and-eyes, hairgrips and safety-pins. As in the lace trade, successive layers of formal and informal subcontracting remained as late as 1900, with perhaps 700 women engaged in carding buttons, the field most mechanised and soon to disappear, but with up to 20,000 such outworkers engaged in all carding of metalwares.[66]

As suggested above, this pattern of extensive female outwork, much in the villages of the countryside, being created by the processes of mechanisation and mill or factory production, recurred in the hosiery trades of the midlands, and persisted to the end of our period. In hosiery it produced a striking gender division of labour, when males were displaced from framework knitting, and women were left to hand tasks akin to those in lace in the same region. Work was generated for *cheverning*, or embroidering clocks onto stockings, still reported to be found in 'nearly every cottage for several miles round Belper' in the 1920s; seaming and toeing in the villages from Leicester to Nottingham; and, in a related trade, also around Belper, cottage manufacture of special orders of stockings or silk underwear.[67] Rose's study of the village of Arnold (Notts.) pointed to startlingly high activity rates as seamers among the wives of framework knitters in 1881, at over 70 per cent.[68] Mining ultimately helped to absorb the men displaced by the relative shift towards a female workforce visible in Table 18.4.[69]

The last of these clothing trades, gloving, was 'urban' in its concentration into three main centres, clearly indicated by the Location Quotients of Table 18.5, Yeovil, Worcester and Woodstock, with smaller clusters in the textile districts of Nottinghamshire and Leicestershire, and in north Devon, around Torrington, Bideford and Barnstaple. Its organisation ranged between outwork by putting out, through to factory production, supported by rural outworkers, who were used even around the long-centralised industry in Worcester.[70] The industry differed significantly from most of the others considered here in being regarded with some

---

[65] Bettey, *Wessex*, p. 50.

[66] Bythell, *Sweated Trades*, p. 141, citing the Select Committee on Homework, 1907.

[67] Foster, 'Hosiery trade', pp. 86–7.

[68] FitzRandolph, *Rural Industries*, II, pp. 117–18; Rose, 'Proto-industry, women's work', pp. 186–7.

[69] Bythell, *Sweated Trades*, p. 97.   [70] FitzRandolph, *Rural Industries*, II, pp. 123–4, 145–6.

optimism in the early 1920s, and experiencing new capital investment. With mechanisation of sewing coming late, and never complete in our period, rural labour, reached through these urban centres, long remained critical to the industry.[71] Morley's, the London glove house, reported glovemaking to be 'purely domestic work' in 1860, populated by 'whole villages of born glovemakers', and they put out the manufacture of 70,000 dozen pairs a year to up to 6,000 outworkers on their books in Somerset, Devon or Worcestershire.[72]

Putting-out characterised the Somerset trades, where hand-sewing was reputed by some to reach its highest standards, and skills were inculcated from childhood in the house of an 'overlooker', who may also have acted as putting-out agent. Women specialised in gloving because of this, and because it was felt that fieldwork was incompatible with the trade. Hence they gained economic independence from a relatively early age and, as with the straw-plaiters, this became associated with moral laxity, and attracted the opprobrium of the local clergy.[73] By the end of the century, machinery provided for part assembly of the lighter gloves, but the division of labour remained complex, and outwork remained critical to finishing, as noted by Woods' survey of the Woodstock industry. One firm in the town, surveyed *c.* 1920, employed 131, disposed as follows: 14 men, comprised of the manager, one 'sorter and puncher', and a dozen cutters; 14 women indoors; and a further 103 women, 87 of them 'pointers', working outdoors. Another firm in the town blamed shortages of female labour for missing their production targets, and for being far below their pre-war output of 1,150 dozen pairs a week. Handwork remained essential to the final assembly of part-sewn gloves, the insertion of the *forchettes* and *quirks* between the fingers, and the padding of cricket gloves, made by Bryans of Worcester at their Woodstock branch.[74] Around Worcester, a similar satellite network of glove-making villages was to be found, ranging as far as Ledbury, Evesham, Inkberrow, and Redditch.[75] A similar industry clustered in Sussex and Kent, after Duke's creation of the modern cricket ball in 1780, this and the related industries of Tonbridge and Robertsbridge building upon the skills of leather and woodworking in the region.[76] Despite apprehension about labour supply, and of competition from imports from France, the USA and even Japan, gloving endured as a rural manufacture well beyond the end of the period.[77]

---

[71] Woods, *Oxford*, pp. 134–47.   [72] Bythell, *Sweated Trades*, p. 118.

[73] Kitteringham, 'Country work girls', p. 118.

[74] Woods, *Oxford*, pp. 136–7; also on the Woodstock industry, see A. Crossley (ed.), *A History of the County of Oxford*, vol. XII, Oxford, 1990, pp. 366–9.

[75] FitzRandolph, *Rural Industries*, II, p. 123.

[76] *Ibid.*, I, p. 22; H. S. Altham, 'Dates in cricket history', *Wisden, Cricketers' Almanack*, 100, 1963, p. 179.   [77] Woods, *Oxford*, pp. 143–4.

Table 18.6. *Employment in leather, tallow, and footwear industries, 1841–1911 (England and Wales, 000s, males unless otherwise indicated)*

Occupation	1841	1851	1861	1871	1881	1891	1901	1911
Fellmonger	[a]	1.8	1.9	2.0	[b]	[b]	[b]	[b]
Tanner	[a]	7.9	8.4	8.6	10.2	10.3	9.5	10.6
Skinner	[a]	1.6	1.7	3.7	4.7	5.7	5.9	8.5
Currier	9.5	11.4	12.8	14.2	15.0	17.6	23.6	15.3
Tallow	3.1	4.8	4.5	3.7	3.0	2.5	2.2	2.9
Boots and shoes, M	176.7	211.0	211.6	197.5	180.9	202.6	174.8	160.1
Boots and shoes, F	10.8	29.3	39.5	25.9	35.7	46.1	43.8	42.4
Clog and patten	3.2	3.7	5.0	1.2	7.4	[c]	6.3	5.8

Notes:
[a] No data.
[b] Fellmongers and Tanners not distinguished from 1881, and totals listed with the larger component.
[c] Subsumed into Boots and Shoes in 1891, in both summary and county tables.
Source: Compiled as Table 18.1 above.

C LEATHER AND RELATED INDUSTRIES

The experiences of the leather and related industries echoed many of these themes of the ways in which technical change impacted at different times on particular parts of the trade, the mixtures of factory, distributed outwork and petty manufactures, and patterns of local and regional concentrations in activity. These trades fall into two groups, a mixed bag of leather and animal skin processing trades, including tallow-makers, and, in the case of curriers, also comprising product manufacture, and the major and minor footwear industries of boot- and shoemaking, clog and patten making. The footwear trades notably encompassed both a distributed national manufacture, and, like tailoring, the petty industry of repair, refurbishment and adaptation. Like several of our other businesses, footwear experienced late mechanisation, and a corresponding shift to centralised factory production, but greater exposure to foreign competition than most other rural manufactures in this period. Employment in this group of trades is summarised in Table 18.6.

Recorded employment in most of these trades was predominantly male, apart from the significant element of women's work in boots and shoes, which was still rather small by comparison with clothing. While there was a shift towards female workers in the industry, it was small, and may even have been inflated by the procedure adopted here, following Booth, of omitting such categories as 'shoemaker's wife' from the pre-

1881 data: in 1861, for example, their inclusion would add around 80,000 females to this workforce.[78] The only other group in which their omission is significant was the curriers, which appears to have comprehended the largely urban 'furriers', and women were a fairly consistent 40 per cent of the total workforce from 1851 onward. Some figures are clearly 'rogue' for reasons that remain unclear: the data for clog- and patten-makers for 1871 are inextricably inconsistent with earlier and later observations, and the employment of male curriers in 1901 was evidently inflated by the inclusion of a large number of leather goods manufacturers. More than most industrial classes considered here, these were trades in which Census categories were inconsistent.

This makes the interpretation of the spatial concentrations in the leather trades the more difficult to interpret with certainty. All bar one, the skinners, were assessed by Day's study of 1841 as 'provincial' in character, somewhat concentrated in market terms, and indicative of the technological base, which did not as yet dictate large-scale centralised production.[79] The distribution of these trades, with tallow, is described in the Location Quotient figures of 1861 in Table 18.7. These patterns broadly confirm the indications of Day's work, and fit closely with Clarkson's detailed analysis of this group of industries before 1850. The majority generated mean Location Quotients close to unity, and relatively low standard deviations, both suggestive of industries very widely distributed, securing raw materials and serving markets on a regional or county basis. The smallest trades, notably the fellmongers, and to a lesser degree tallow, may exhibit some false indicators of concentration in the smallest county units, where small aggregate changes have a large impact on these relative measures.[80] Even in 1841, most of these leather industries were urban, and Day's analysis by registration districts for four counties showed curriers and tanners in Hereford, Huntingdon, Rutland and Yorkshire as overwhelmingly concentrated into the towns.[81]

The explanation for these patterns, which intensified in the second half of the century, lay largely in their externalities: the principal factors were already the external economies to scale gained at each end of the

[78] Booth, 'Occupations of the people', and other authorities cited in note 13 above.

[79] Day, 'Distribution of industrial occupations', pp. 128–30, and, noting the imperfection of some categories, which grouped rather different trades to generate false indicators of 'national' distribution, p. 220.

[80] Monmouth's 15 Fellmongers of a Wales total of only 37 is indicative of this, and several Welsh counties, such as Radnor and Anglesey had small occupied populations within which this effect may be easily magnified. The same would apply to a greater degree to Rutland observations, though they are here 'confirmed' by their closeness to the L.Q. for Leicestershire.

[81] Day, 'Distribution of industrial occupations', pp. 146–8. In Hereford, curriers appeared at the rate of 21 per 10,000 in the urban population, and 5 in the residue of the county, and tanners at 192 and 28 per 100,000 respectively; in Huntingdon, the equivalent figures were 8 to 3 and 113 to 19; Rutland 0 to 7 for curriers; and Yorkshire 31 to 1, and 45 to 10.

Table 18.7. *Location quotients, England and Wales, leather and related industries, 1861*

Occupation	Mean L.Q.	S.D.	County concentrations (in rank order)
England:			
Fellmonger	1.305	0.795	Rutland, 3.043; Leics, 2.913; Lincs, 2.704; Hunts, 2.531; Sussex, 2.269; Dorset, 2.181; Wilts, 2.146; Notts, 2.115; Yorks, ER, 2.102
Tanner	0.992	0.558	Som, 2.386; Devon, 2.176; Yorks, ER, 2.031; Surrey, 1.884; Westm, 1.764; Cumb, 1.676; Northd, 1.598; Heref, 1.578; Worcs, 1.553
Skinner	1.115	1.392	Westm, 7.251; Cumb, 4.131; Salop, 3.135; Heref, 2.926; Notts, 2.836
Currier	0.973	0.577	Yorks, WR, 1.884; Yorks, ER, 1.736; Yorks, NR, 1.729; Westm, 1.716; Northd, 1.612
Tallow	0.945	0.388	Chesh, 2.549; Yorks, ER, 1.618; Glos, 1.487; Cumb, 1.345
Wales:			
Fellmonger	0.746	0.785	Monmouth, 2.565
Tanner	1.268	0.553	Radnor, 2.484
Skinner	1.118	1.207	Denbigh, 3.444; Merionedd, 3.088; Montgomery, 2.902
Currier	1.155	0.326	Anglesey, 1.860; Denbigh, 1.493
Tallow	0.911	0.418	Flint, 1.867; Carmarthen, 1.440

Source: Calculated as in Table 18.2 above.

industry, raw material supplies and end users and distribution. The bulk of skins, hides, and fats for the tallowchandler were obtained at the greatest concentrations of slaughter, and hence in proximity to the urban meat markets. Nearly 30 per cent of curriers and tallowchandlers in 1861 were in London, and over a quarter of tanners in London and the nearby metropolitan counties, into which hides had long been rotated from the capital for tanning.[82] Bermondsey in particular was noted as a centre of

---

[82] Clarkson, ' Leather', in Mingay (ed.), *AHEW*, vol. VI, pp. 466–83, on the longer-term background.

tanning, from which the canal network was employed to distribute the finished leather. The process of continuing urban growth to 1914 was combined with that of rural depopulation to make the smaller scale and rural leather processors increasingly unviable.

The shift of the raw material was compounded by that of the end users, from shoemakers to saddlers, all of whom adopted or intensified their urban basis in these years. By the early 1920s, this process was largely complete, and apart from a few niche markets, and small scale supply to the makers and repairers of sturdy country boots, few rural leather processors could compete with those of the town, and demand even on a county basis was insufficient to sustain the country tannery.[83] Increasing use of imported hides added to the difficulties, and helped further to concentrate processing industries at the ports. Such changes also helped to reduce the role of the fellmongers, the traditional middlemen who had purchased and traded in hides and especially sheepskins, and linked butcher and processor.[84]

Technical change both inside and outside these industries also contributed to the decline of rural locations. Though it came typically late in the century, mechanisation of both the tanning and currying processes favoured newer and largely urban plant, and opened the industry to the use of imported tanning agents, normally used in combination with the traditional oak barks, rather than their complete replacement, but critically also speeding the processes, and raising productivity. Such materials as gambier from the East Indies, valonia, an acorn cup derivative from the eastern Mediterranean, and sumach from Smyrna, joined imported barks as inputs to the industry, which were running at over 1.5 million cwt in the years 1905–13.[85] As early as 1858, the Nickols's 'Joppa' tannery in Leeds had pointed to future trends, using imported hides, covering 4 acres with 500 tanning pits, and using a 30-HP steam engine and fulling stocks to soften the leather.[86] Drum processing and greater power inputs to modern tanneries and currying plant could produce enormous gains in labour productivity in what had traditionally been a labour-intensive process: by 1920 it was estimated that even a small tanning factory with a workforce of fifty could finish 1,000 skins a week, processing each hide in a week, whereas the traditional plant might not better an output of two a week, using a process so protracted that it might not turn over its best hides in a year.[87] Some niches did survive, as in North Wales, where best cured leathers were supplied to clad the rollers of Lancashire textile

[83] FitzRandolph, *Rural Industries*, I, pp. 157–8.

[84] The location quotient for this small trade points strongly to the links with sheep country.

[85] FitzRandolph, *Rural Industries*, I, pp. 160–1, 169.

[86] W. G. Rimmer, 'Leeds leather industry in the nineteenth century', *The Thoresby Miscellany*, Thoresby Society Publications, 46, 1963, pp. 143–4.   [87] FitzRandolph, *Rural Industries*, I, p. 166.

Table 18.8. *Location Quotients, England and Wales, footwear industry, 1861*

Occupation	Mean L.Q.	S.D.	County concentrations (in rank order)
England:			
Boots and shoes	1.084	0.672	Northants, 5.130
Clog and patten	0.414	0.709	Lancs, 3.706; Chesh, 1.846; Yorks, NR, 1.718; Yorks, WR, 1.669
Wales:			
Boots and shoes	1.053	0.133	Anglesey, 1.271
Clog and patten	1.354	1.176	Pembroke, 3.627; Cardigan, 3.212; Anglesey, 2.795

Note: Calculated as in Table 18.2 above.

mills, and the remainder, as at Llangollen, used as the base for a purse manufacture.[88] Overall these industries experienced late restructuring into modern plant, but rural producers were already becoming marginal in 1850, and were little better than a relict population by 1914.[89]

The footwear industries were, of course, a major user of these products, and their demands impacted strongly upon the organisation of leather production, as did American import competition in ready-made shoes and boots which was strong for a decade up to 1905, before being repulsed by the adoption of US methods of production and machinery in the British industry. Such changes had accelerated the process of decline among traditional leather processors from the 1890s. There were also two other footwear industries, in the numerous petty shoemakers and the small industry of clog making, that combined a specialised, largely rural timber industry with urban leatherworking. The spatial distribution of these industries in 1861 is described in Table 18.8.

The two-sector nature of the making of boots and shoes is immediately evident from these Location Quotients. The mean in both England and Wales lay close to unity, and, certainly in the latter case, remarkable evenness of distribution was indicated by the low standard deviation, apparently confirming Day's classification of the industry for 1841 as 'local' in its distribution, apart from the already distinctive cluster of

---

[88] Jones, *Rural Industries . . . Wales*, pp. 90–1.

[89] Major leather industries in the northern counties, and in the textile districts, had been stimulated by industrial demand for such items as pumps, machine belting and industrial clothing. All were beginning to face competition from rubber-based products by the time of the *Census of Production* of 1907, and from the beginnings of electrical, directly-driven machinery.

Northamptonshire. The introduction of machine stitching from the mid-1850s, and of effective mechanical closing of the uppers over the next decade accelerated the flight of the wholesale trade from London, and led to the concentration of the manufacturing industry into the towns, particularly where there existed a kindred and adaptable skill base. Already heavily concentrated into Northampton and its surrounding small towns – such as Wellingborough, Daventry, Kettering, and Raunds – by the 1850s, and in Stafford, factory production was established in the 1860s and onward in Leicester, Street (Somerset), Kendal and Norwich, all mostly for the lighter shoes, and in Leeds and Bristol for the heavy products, including riveted boots. Even when fully mechanised, the industry's power requirements remained relatively low, and sustained its off-coalfield development.[90] The very extended rural manufacture that had supplied the wholesale trade before successful mechanisation partially adapted itself because, as in clothing and gloving, the sewing machine was compatible with cottage industry, and was thus, from the 1860s, an urban/rural hybrid. The impact of the final stages of mechanisation, post 1890, was the force that eliminated rural manufacture.[91]

Yet large numbers declared themselves to be employers in the Census of 1901, and others were still working on their own account. Among male shoemakers, 5 per cent were returned as employers, 8,999 in all, and nearly 40,000 claimed to be independent workers, overlapping with the 30 per cent who 'worked at home'; women by contrast were overwhelmingly employees.[92] The evidence from contemporaries suggests that these independent cobblers were already residual. Eric Powell recalled his father's one-man shop at Blakeney in the Forest of Dean before 1914, where it took half a day to make a pair of boots for his clientèle of miners and farmworkers, and where he also carried out general repairs to other goods, from gloves to harness.[93] In less isolated villages, such as South Marston, near Swindon, such village cobblers were more typically extinct as makers by 1912, when Alfred Williams recorded his village bootmaker as still working at 'past ninety', but on repairs. In his prime, he had made

[90] P. R. Mountfield, 'The footwear industry of the East Midlands', I, 'Present locational pattern and the problem of its origins'; and III, 'Northamptonshire 1700–1911', *East Midland Geographer*, 3, 1962–5, pp. 293–306, 434–453; Mountfield, 'The shoe industry in Staffordshire 1767 to 1951', *North Staffordshire Journ. of Field Studies*, 5, 1965, pp. 74–80; Mountfield, 'The footwear industry of the East Midlands', IV, 'Leicestershire to 1911', *East Midland Geographer*, 4, 1966–9, pp. 8–23.

[91] Based upon R. A. Church, 'The effect of the American export invasion on the British boot and shoe industry 1885–1914', *JEH*, 28, 1968, pp. 221–54; Church, 'Labour supply and innovation, 1800–1860: the boot and shoe industry', *Business History*, 12, 1970, pp. 25–45; V. A. Hatley and J. Rajczonek, *Shoemakers in Northamptonshire 1762–1911, A Statistical Survey*, Northampton Historical Series, 6, 1972; Bythell, *Sweated Trades*, pp. 107–19.

[92] *Census*, 1901, Table XXXV. The reliability of these declarations is normally regarded as uncertain.

[93] Eric Powell, winter 1969, writing of *c*. 1900, in Seager (ed.), *Village Trades and Crafts*, pp. 169–70.

boots and sold to his countrymen on credit of up to a year, closely paralleling the practices of many rural craftsmen and tradesmen.[94]

In employment terms, clog-makers exactly paralleled the shoemakers, with around 30 per cent describing themselves as employers or working on own account for the Census of 1901, and about a quarter working at home. The Location Quotients of Table 18.8 clearly reflect the taste for clogs in northern mills and foundries, and their final assembly near the ultimate market, but omit the blockcutters, who supplied the timber-work. These, naturally enough, tended to locate away from these manufacturing counties, and often did so as part of a wider woodworking business, making good use of the offcuts in what was a trade otherwise very wasteful of wood. By the end of the period, some of this work, in counties such as Shropshire, Durham, Norfolk and Oxfordshire, was conducted by itinerant gangs. Blocks were cut to four standard sizes – men's, women's, boys' and children's – and despatched by train to the makers.[95] Some, of course, did integrate the whole production process, like Myles Bainbridge, who was apprenticed as a clogger in Sedbergh in 1915, and who worked on his own account as an independent from around 1923, when he too adopted 'machine-made' soles for his clogs, supplied normally from Hebden Bridge.[96] Even in a traditional form of footwear, country and town were joined in an extended division of labour.

D TIMBER AND UNDERWOOD INDUSTRIES

As the clog trade illustrated, a wide range of manufacturing industries and crafts servicing agriculture could be grouped under the general head of timber and underwood trades, and any division between the two is to some extent arbitrary. Here, therefore, three industries are assessed, turnery, basket and brush manufacture, all of which in Day's terms were 'provincial' industries in 1841, and thus served more extended markets than more basic timber trades.[97] All were relatively small, although total recorded employment did more than double between 1841 and 1911, and having more than 45,000 engaged in these trades in 1911, were hardly insignificant, as indicated in Table 18.9. The figures indicate that the

---

[94] Alfred Williams (ed. M. J. Davis), *In a Wiltshire Village: Scenes from Rural Victorian Life* (Gloucester, 1981), pp. 90, 175, selected from two of Williams's books of 1912, *A Wiltshire Village* and *Villages of the White Horse*. Williams was a forgeman at the Swindon works, and remained there until ill-health forced him to leave in 1914, but he was also educated through Ruskin Hall, Oxford.

[95] FitzRandolph, *Rural Industries*, I, pp. 64–8.   [96] Kightly, *Country Voices*, pp. 206–7.

[97] Day, 'Distribution of industrial occupations', pp. 233–5. Two other trades or industries, which might be considered with this group, namely, coopers and hoopmakers and fence and hurdle makers, both elided into generic carpentry in much of the countryside, and are considered separately in Chapter 19 below.

Table 18.9. *Employment in timber manufactures, 1841–1911*
(England and Wales, 000s)

Occupation	1841	1851	1861	1871	1881	1891	1901	1911
Turner and boxmaker, M	7.9	9.1	10.1	12.4	11.4	12.7	19.2	19.5
Turner and boxmaker, F	0.6	1.5	2.0	5.4	2.6	1.9	2.2	2.5
Basketmaker, M	6.3	13.7	9.5	8.4	9.0	11.4	9.6	8.7
Basketmaker, F	0.5	1.0	1.3	0.9	2.5	2.6	1.9	2.1
Brushmaker, M★	5.2	7.6	8.5	8.5	8.7	9.7	9.3	9.8
Brushmaker, F★	0.7	1.8	2.7	3.2	4.2	6.2	7.0	7.7

Note: ★ For 1891 only, includes all Hair and Bristle workers, and is not directly comparable with other years.
Source: Compiled as Table 18.1 above.

growth of employment was uneven, being concentrated principally into only two groups, turners and boxmakers after 1890, attributable principally to the latter, and from the last quarter of the century, female brushmakers, the latter being an occupation, like many in textiles and clothing, in which women's employment, despite the uncertainties of the Census as its record, appears significantly to have displaced that of men.

All were 'provincial' industries in 1841, exhibiting some spatial concentration, and this broad pattern is confirmed by the analysis by Location Quotients from the 1861 Census, presented in Table 18.10. These industries were generally widely distributed, but contained a few distinctive county clusters.

In combining turners with boxmakers, the figures of Table 18.9 linked two distinct trades, and the latter, covering generic container-makers, distorted the profile. Their inclusion here would have displayed their concentration near the cities and ports, where the greatest demand for crates, packing-cases and so on existed. The turners were rather different, being a woodland-based occupation manufacturing chairs, bowls, plates, reels and bobbins, taps and spigots, and drawing upon a variety of timbers for the purpose, principally beech, birch, alder, ash and sycamore.

Despite the growth of the use of enamelled and galvanised iron, tinware and other metalwares, wood was still an important industrial raw material in the early years of the twentieth century and, as a point of reference for the present consideration of timber manufactures and that of other wood-using crafts described below in Chapter 19, Table 18.11 summarises the principal commercial uses of native woods, *c.* 1920.

Turnery took place wherever suitable supplies of wood were available, but the single distinctive concentration lay in Buckinghamshire, around

Table 18.10. *Location quotients, England and Wales, timber manufacturing, 1861*

Occupation	Mean L.Q.	S.D.	County concentrations (in rank order)
England:			
Turner	0.892	0.661	Bucks, 3.358; Warks, 2.745; Notts, 1.964; Derby, 1.869; Glos, 1.624; Leics, 1.585; Yorks, WR, 1.573
Basketmaker, M	1.031	0.322	Norf, 2.013; Yorks, ER, 1.650; Glos, 1.639; Middx, 1.544
Basketmaker, F	1.099	1.310	Middx, 8.207; Hants, 2.587; Heref, 2.571
Brushmaker	0.805	0.579	Warks, 2.420; Glos, 2.095; Norf, 2.003; Surrey, 1.967
Wales:			
Turner	1.211	1.094	Flint, 4.103; Cardigan, 2.374
Basketmaker, M	0.986	0.396	Brecon, 1.577
Basketmaker, F	0.736	0.741	Glamorgan, 2.098; Cardigan, 1.924
Brushmaker	1.015	1.316	Denbigh, 5.185

Source: Calculated as in Table 18.2 above.

High Wycombe and Stokenchurch, where the making of chairs with simple pole-lathes arguably attained the standing of a 'national' industry. During our period, these 'bodgers' came increasingly to supply turned parts for assembly in the factories of the towns, and by the end of the century it was a trade of poor piece rates and long hours. From the 1890s onward, power lathes in larger urban plants cut into their trade, and as fashion shifted away from traditional chair types, such as the Windsor of this area, the living standards of these turners declined, pre-war conditions being described by a Stokenchurch manufacturer as 'slavery'. It was reckoned that before 1914 it took a 7 a.m. to 7 p.m. working day, plus a half-day on Saturday, to produce three gross of chair legs, and an income of about 12s. from which the costs of tools had to be found. No young man with a family to support could by then subsist from the trade. Joint family enterprise thus characterised much of the surviving Chiltern trade on the eve of the War.[98]

Elsewhere, supplies of reels and bobbins for textile industries had been

[98] *Ibid.*, p. 105; FitzRandolph, *Rural Industries*, 1, pp. 48–53.

Table 18.11. *Principal home-grown commercial timbers and their uses, c.1920*

Wood	Principal commercial uses
Oak	Wheelwrighting; wagons; motor bodies; gates, posts, rails; barrel staves (imported oak strongly preferred); boats; baskets
Ash	Shafts; felloes; fence rails; heel trees; baskets; chairs; hammer hafts; pump buckets; bobbins; barrel hoops; heads and teeth of rakes
Elm	Wheel hubs; chair seats; coffins
Sycamore	Silk reels; dairy and small bowls; mangle rollers; butter prints; walking sticks
Beech	Factory clog soles; chair legs; brush backs; cotton reels; malt shovels
Birch	Hand-cut clogs; cotton reels; brush backs; spoons; taps; besoms; twigs in steel plate manufacture
Alder	Preferred material for clog soles; bobbins; chair legs
Willow	Legs; ditching shovels; cricket bats; crate rods; charcoal
Poplar	Willow substitute – shovels; motor bodies; cricket bats
Lime	Spoons; large taps; shovels
Maple	Egg cups; spigots
Chestnut	Coppice split for fencing
Hazel	Baskets; besom handles; hurdles; wattle
Heather	Besom heads (in the northern counties)

Other minor uses of holly, larch (boatbuilding) and crab (windmill gearing)

Sources: Based upon FitzRandolph, *Rural Industries*, vol. I, pp. 84ff; Jones, *Rural Industries . . . Wales*, pp. 44ff; Woods, *Oxford*, pp. 79–134.

important markets for the turner's trade, and clusters of activity occurred locally to serve these and other markets, not all of which are revealed by the Location Quotients of Table 18.10. Most were characterised by small-scale rural enterprise, and most like the Chiltern bodgers, were in serious decline by the beginning of the century. Notable among these small trades were the mixed urban–rural turners of the Lake District and the West Riding, where water-powered mills had concentrated production but, like that at Steeton, they were unable to compete with more modern plant much after 1914; the turners of tool handles and similar products in Gloucestershire, on the fringes of the Forest of Dean, and Warwickshire, serving demand of Birmingham and Coventry industries;

and the East Midlands, again largely small relict industries in the forest or woodland fringe. Detailed analysis of the bobbin manufacture and lesser turnery industries of Cumberland and Westmorland showed an industry reaching its peak in the third quarter of the nineteenth century, when it was supplying over half of Britain's demand, and declining thereafter in the face of changing technology within the cotton industry, and growing international competition from countries such as Norway. When the Oxford surveys were conducted in the early 1920s, their strong focus on the Chiltern chair manufactures, and on the single high-grade turnery plant at Kingscliffe (Northants), which had come to concentrate on the making of taps, spigots and pump buckets because of the disappearance of other traditional markets, demonstrated an industry in terminal decline.[99]

Basketmaking comprehended two quite different raw materials and organisational structures: the spelk makers of trugs, swills, whiskets and skips, using boiled and slit or shaved poles, and those using osiers and reed. Both were widely distributed, as the figures of Table 18.10 indicate, with few concentrations of magnitude. While their markets overlapped to some degree, they remained largely distinct, and the spelk manufactory of Furness and Cumbria, Shropshire and Derbyshire, tended to serve the heavy-duty uses of the north, notably for potatoes, coaling and loading ships, charging furnaces in Sheffield, and packing ironmongery in the West Midlands. The Sussex fruit trugs were also manufactured in Wealden villages such as Herstmonceux and Hailsham, using a kindred technique, but using mostly split willow. Spelk manufacturing was more workshop or domestic in character than that of osiers, using local raw materials and requiring few specialised tools, but relatively high skills were acquired and protected by extended apprenticeship. In Furness and Shropshire units consisted of only one or two workers, and had small masters organised into their own union before 1914, whose work routines were integrated, characteristically allocating Monday to preparing the strips for the rest of the week's making. Six 22-inch baskets could be made by the skilled worker in an eight-hour working day and, in the inflated times of the early 1920s, carried a piece rate of 2s., putting the 'wage' at roughly three times that earned by the Chiltern turner.[100]

The osier industry was very different in structure, involving typically a more extensive division of labour, with a complex series of preparatory and supply processes, workshop or small factory production, and much more extensive use of female labour. It required a relatively labour-

---

[99] Ibid., pp. 23–4, 31–2, 36, 48–9; J. D. Marshall and M. Davies-Shiel, The Industrial Archaeology of the Lake Counties (Newton Abbot, 1969), pp. 55–75.

[100] FitzRandolph, Rural Industries, I, pp. 96–8; Woods, Oxford, pp. 104–5.

intensive growing industry, since quality osiers depended upon clean cultivation, and varied significantly in quality and by variety.[101] The principal growing areas lay in Somerset, around Sedgemoor, Leicestershire, the Trent Valley, and Mawdesley (Lancs.), each with its own characteristic willows. In the Trent valley, which produced osiers of the finest quality, 'Black Maul' and the 'Wissender', varieties of the hard willow (*salix triandra*) were the most commonly grown of more than twenty types of common osier (*salix viminalis*), hard willow, and bitter willow (*salix purpurea*). The coarsest produce came from Somerset, where the 'Black Mole' was again favoured. In mid-Lancashire, the local variety was developed by, and supposedly named after, Richard Meadowes of Mawdesley, 'Dicky Meadows' or 'Dicks'. Elsewhere, production was typically for local use, and in Gloucestershire, Worcestershire, Warwickshire and Yorkshire particular varieties were favoured for local use, and basketmakers were commonly also growers.[102]

In all districts, male labour was used in cultivation and cutting, but women carried out the preparatory processes of peeling the bark, which was very seasonal, peeling white willow during April and buffed (boiled) willow during the winter, releasing labour for other agricultural work in late spring and summer.[103] Even in the season, it was often carried on part-time, and cutting was customarily a male piece-work task.[104] Processes were labour intensive, a 30-acre growing ground or 'holt' in the Trent valley producing up to 150 tons of rod, and needing three or four workers per ton week to handle it all.[105] Rod thus stripped was sold wholesale to urban manufacturers in London, Birmingham, Leeds, Liverpool and Hull, and in smaller centres such as Ormskirk, where male workers were applied to the heavy willows, and women were confined to light and cane work. This last speciality is picked up by the Location Quotients in Middlesex also, where the market gardening west of London may have produced a specific stimulus to manufacture.[106] In 1901, the average establishment employed only nine workers; 73 per cent of all males and 80 per cent of all females were employees, with only 19 per cent claiming to be working on their own account.[107] Fewer than a quarter were recorded as working at home, and the Oxford surveys of

[101] E. J. Baillie, 'Willows and their Cultivation', *JRASE*, 3rd series, 5, 1894, pp. 234–50.

[102] FitzRandolph, *Rural Industries*, II, pp. 32–4.

[103] W. P. Ellmore and T. Okey, 'Osier and willow cultivation', *Journal of the Board of Agriculture*, 18, 1911–12, p. 17. Buffing produced interesting colour variation in the osier, and was held to be a distinctively British process.

[104] W. P. Ellmore and T. Okey, 'Planting, cleaning, and cutting willows', *Journ. of the Board of Agriculture*, 18, 1911–12, p. 213.

[105] FitzRandolph, *Rural Industries*, II, pp. 46–7 – my summary of labour inputs.

[106] A large manufacture of baskets had been located at Feltham before 1914, but was noted as in major decline in the 1920s: *ibid.*, p. 82.   [107] *Census*, 1901, Table XXXV.

the 1920s regarded only the industry of Bridgwater as having an extensive tradition of putting out or sub-contracting.[108]

The industry's products met a wide range of markets, many agricultural, and the sturdiness of baskets was seen as essential in the Covent Garden trades. FitzRandolph and Hay's survey of the early 1920s noted up to ten types of 'agricultural baskets', a similar number of 'baskets for transport purposes', including 'strikes', 'flats', 'chicken-crates' and 'Rabbit Hampers', plus a wide range of industrial basketry, principally 'skeps' and wicker covers for jars and carboys; local specialities also abounded, notably in coastal districts for fish and lobster and crab pots.[109] The development of agricultural specialisms, such as strawberry-growing at Botley (Hants.), generated a new secondary business in basket-making. Despite this rich variety, for traders marketing fruit or vegetables the returnable sturdy basket was increasingly an underemployed luxury, nets and light boxes were increasingly substituted for it, and imports competed with domestic suppliers in the decades before 1914. Ellmore and Okey drew attention to the market potential, which under the right conditions offered rates of return from osiers as good as any in agriculture.[110] By creating labour shortages, cutting imports of rods, and eliminating the principal competition in finished goods, from the Netherlands and Belgium, the War ended our period providing the industry with some artificial protection, but by the early 1920s decline was general, especially among the rural makers. Firms in Ely, St Neots, Peterborough, Biggleswade and Leighton Buzzard, which had together employed 240 men before the war, were down to 13, plus 8 or 9 half-time, by 1923.[111]

Our final industry, brushmaking, also ranged across a number of different products, from besoms at one end to paint and similar brushes at the other, and combined timber, underwood, hair and bristle trades. The former was widely pursued in the woodland fringes, was predominantly rural in character, remaining so throughout the period, and engaged large numbers of part-time and seasonal workers. While the 'broom squires' of Hampshire, Sussex and Surrey, and the besom makers of the north and Wales, attracted great interest from those recording rural crafts and industries at the end of the century and beyond, its tools were simple, and production rarely that of the full-time specialist.[112] Traditional full-time broom squires, like the uncle in Hampshire recalled by Ernest Boxall, bought up birch in spring for manufacture during the winter, making heather brooms during the autumn, and went on the road to sell them at

[108] FitzRandolph, *Rural Industries*, II, p. 51.   [109] *Ibid.*, pp. 58–67.

[110] Green, *Rural Industries*, p. 82; Ellmore and Okey, 'Osier and willow cultivation', p. 17.

[111] FitzRandolph, *Rural Industries*, II, pp. 81–2.

[112] K. S. Woods, *Rural Crafts of England, A Study of Skilled Workmanship* (East Ardsley, 1975), pp. 122–4 perhaps overstated the craft skill in his retrospective assessment.

thirteen to the dozen in early spring.[113] In the fringes of Snowdonia, Derbyshire, and the northern moors and fells of England, heather besoms were made only during the summer months, very much as an income supplement for the labourer.[114] Some real concentrations did exist, as in the fringes of the Chilterns and the North Downs of Surrey (as indicated by the Location Quotient of Table 18.10), Berkshire, Hampshire and Oxfordshire, where Woods noted large quantities being made, c. 1919, for despatch by rail to the ironworks of South Wales and the Midlands. The birch besom remained the preferred tool for brushing slag off the surface of pigs of iron, and from steel plate, making Sheffield 'an unlimited market for besoms' up to and beyond the War.[115]

Although it combined timber machining with hair and bristle work, brushmaking throughout the period was largely urban in character, and concentrated overwhelmingly in London where more than a quarter of total employment was located in both 1861 and 1901. Within this small workshop and factory trade, there were regional concentrations in Warwickshire, Gloucestershire and Norfolk, as indicated by the figures of Table 18.10, and in Lancashire, which was second only to London in its share of total employment in the industry in 1901, having around 10 per cent. Although the industry was typically of modest scale, with an average twenty-one workers per employer in 1901, it was also overwhelmingly characterised by direct employment, with 84 per cent of males and 97 per cent of females in the industry described as employees in the 1901 Census.[116]

Two important changes did occur within the period: as in several other textile and clothing industries, the period saw the relative replacement of male by female workers within the industry; and with this there developed a significant proportion of outwork, principally among women. Brush-drawing engaged more than a quarter of the female workforce in homework in 1901, and Bythell has suggested a process of diffusion into districts where older textile trades had decayed.[117] The West Country, notably the former lace districts of Devon, and parts of Somerset and Dorset, were areas into which this outwork spread.[118] Three-quarters of output by volume in 1907 lay in domestic brooms and brushes, but only 64 per cent by value, and it was in this low value-added section of the trade that outwork, often for educational purposes in the 'philanthropic

[113] E. W. Boxall, 'The broomsquire', Autumn 1952, in Seager (ed.), *Village Trades and Crafts*, pp. 160–2.

[114] FitzRandolph, *Rural Industries*, I, pp. 115–19; Jones, *Rural Industries . . . Wales*, pp. 80–2.

[115] Woods, *Oxford*, pp. 94–7; Helen FitzRandolph, 'Besom-making in Derbyshire and Nottinghamshire', *Journ. of the Ministry of Agriculture*, 28, 1921–2, p. 442.

[116] *Census*, 1901, Table xxxv.   [117] Bythell, *Sweated Trades*, p. 172.

[118] FitzRandolph, *Rural Industries*, II, p. 141.

institutions', was most important.[119] Toothbrush manufacturing also involved extensive rural outwork to the end of the period, at least in the traditional bone-handled and bristle industry, where machinery could not be applied, but it was already in retreat in the face of the factory celluloid-based industry.[120]

A final underwood industry was much discussed by the Oxford surveys in the 1920s, but was already a minor residual activity in mid-century, and declined thereafter, namely, charcoal burning. Employment was only separately identifiable in the Census of 1861, and then stood at a mere 443 in England and Wales, all, bar a few, being males: in other Censuses, they were subsumed into the much more numerous coke burners and coal dealers, and even in 1861 were not differentiated within the county schedules. The ending of its use in iron-making, the increasing use of imports for gunpowder, and the growing difficulties of recruiting workers to what was seen as an increasingly unsociable and itinerant trade, were all held to be the sources of its decline. Thus the industry was found in isolated outposts in the years before 1914, representing the extreme case of niche marketing among perhaps all rural industries. The Lake District was probably the leading area of surviving production, where demand was stimulated by the specialist needs of the Backbarrow furnaces, which still employed substantial quantities; the woods of Monmouthshire, again for some small inputs to the iron and copper industries; at Stirchley in the Wrekin as the by-product of wood-spirit distillation; and in the other traditional centre, around Midhurst in Sussex, from which some supplies still went to the Faversham powder mills. Since coaling had long been seen as an integral part of the normal woodman's duties both at Windermere and in the Weald, the industry may have been somewhat larger than suggested above, but it was still small, and perhaps symbolic of the future for so many of the tradition-ally rural, timber-using industries.[121]

E MINING AND EXTRACTIVE INDUSTRIES

The final group of industries, mining and extractive, might at first glance appear not to fit with some of the patterns of change identified for more obviously 'rural' industrial activities. However, all contained very exten-sive rural elements at the beginning of the period, some parts of each retained by-employment as an element of their operations, and all

---

[119] *Census of Production, 1907*, BPP, 1910, CIX, pp. 711–13, 732–4.

[120] FitzRandolph, *Rural Industries*, II, p. 155.

[121] *Ibid.*, I, pp. 21–3, 119–23; Jones, *Rural Industries . . . Wales*, pp. 62–4. Marshall and Davies-Shiel, *Lake Counties*, pp. 75–88, provide a detailed analysis of the charcoal and local powder industry, which survived well into the 1920s.

Table 18.12. *Employment in mining and extractive industries, 1841–1911*
(England and Wales, 000s)

Occupation	1841	1851	1861	1871	1881	1891	1901	1911
Miner	168.5	256.6	329.0	371.2	438.5	599.2	672.2	908.8
Quarrier	17.0	34.1	41.0	46.7	55.1	55.3	71.8	50.6
Brickmaker	16.7	27.8	37.7	36.2	47.3	41.0	60.9	49.2

Source: Compiled as Table 18.1 above.

included seasonal patterns of activity that permitted this part-time commitment of the workforce. Each impacted significantly on the timing, direction and process of internal migration, through their growth or decline on a local and regional basis. Above all, the vicissitudes of the tin-, copper-, and lead-mining districts appear to have been a powerful force in determining the processes of rural depopulation; to a degree each of these had been characterised in 1850 by links with small-scale farming, and were the residues of the farmer-miners of early industrial development. By contrast, in many parts of Britain, it was only after 1850 that the scale and capital intensity of many coalfields developed, and the exploitation of coal reserves through deep-mining techniques created both new settlements at the pithead, and rapid growth in the villages in which collieries were located.[122] The summary of male employment trends in these industries, above all in mining, in Table 18.12, makes their significance abundantly clear. While none was the archetypal 'rural industry', all were largely rural in location, and their impact upon the countryside of later nineteenth-century Britain fundamental.

Within these figures for mining, which comprehend all of those males described as miners in the Census data, coal predominated, but in the counties in which this occupation represented a cluster of particular significance in total employment, described for 1861 in the figures of Table 18.13, it was not the sole determinant. Regionally at least, other mining activity was still able to be more significant in the employment profile of a county. This was particularly evident in Cornwall, where these data reflect the hard-rock mining of copper and tin before the collapse of world prices in 1866 provoked a crisis from which the industries never recovered. Rowe estimated that in eighteen months, over 11,000 jobs, nearly a third of the total, were lost as a result of this profound depression, and Cornwall's already high rate of migration was

---

[122] For a wider context to such new industrial entrants to the countryside, J. D. Marshall, 'Industrial colonies and the local historian', *Local Historian*, 23, 1993, pp. 146–54.

Table 18.13. *Location Quotients, England and Wales, mining and extractive industries, 1861*

Occupation	Mean L.Q.	S.D.	County concentrations (in rank order)
England:			
Miner	0.735	1.243	Corn, 5.334; Durham, 4.324; Staffs, 2.871; Derby, 2.816; Northd, 2.412; Cumb, 2.260
Quarrier	0.781	0.772	Salop, 2.801; Yorks, WR, 2.448; Dorset, 2.440; Derby, 2.207; Leics, 1.757; Corn, 1.681; Staffs, 1.648
Brickmaker	1.030	0.609	Middx, 3.659; Kent, 2.536; Staffs, 1.792
Wales:			
Miner	0.721	0.599	Glamorgan, 1.750; Flint, 1.679; Monmouth, 1.347
Quarrier	1.094	2.218	Caernarvon, 8.039
Brickmaker	0.488	0.905	Monmouth, 2.354; Carmarthen, 1.981; Glamorgan, 1.858

Source: Calculated as in Table 18.2 above.

significantly enhanced.[123] A similar experience befell the lead miners of North Yorkshire, Cumbria and the north Pennine districts, Derbyshire and Flint, in the later 1860s, in the face of the downward drift of prices, which accelerated from 1874, and continued to fall as Australian ores reached the market. By the 1880s, this had been compounded with the exhaustion of seams in many lead districts, and in Swaledale, where mining and smallholding had long been interactive occupations, land prices collapsed.[124] Recorded employment in copper, tin and lead mining almost halved between 1861 and 1871, and by 1900 stood at less than a quarter of its 1861 level.[125] With this decline went that of the many rural

[123] W. J. Rowe, *Cornwall in the Age of the Industrial Revolution* (Liverpool, 1953), p. 378; Baines, *Migration*, p. 159.

[124] Discussed extensively by Hallas, 'Wensleydale and Swaledale', I, pp. 95–115, 281–349.

[125] Figures, from Census, with percentage changes in brackets, were as follows: 1861, 50,593; 1871, 28,243 (−44%); 1901, 11,741 (−58%); and 1911, 10,372 (−12%). Losses were heavily concentrated into copper and lead. Within our generic 'miners' of Table 18.12, coal and, to a minor degree, iron miners grew in numbers over the same period. Output (UK) of tin peaked in 1871, at 16,272 tons of ore; copper, in 1856, at 320,500 tons; and lead at 98,200 tons in 1870: B R Mitchell and P. Deane, *Abstract of British Historical Statistics* (Cambridge, 1962), pp. 155, 159, 160.

communities in relatively remote parts of the country, which had depended upon these mining industries; the northern Pennines appear to have been particularly severely affected, in addition to the obvious impact upon Cornwall.[126]

The coal industry can hardly be explored in great depth in this chapter, but a number of points sustain its case for treatment as a rural industry. As is evident in the figures of Table 18.13, several counties were prominent in mining and quarrying or brickmaking, notably Staffordshire, Derbyshire, Monmouth and Glamorgan, and just outside these leading ranks, one might reasonably add Shropshire and the West Riding.[127] But the advance of deep mining techniques that was essential to the continued development of Staffordshire, for example, was also the route to the development of entirely new coalfields, and with this the promotion of small villages and dispersed hamlets into mining communities larger than traditional market towns; with the rapid growth of coal, particularly in these new fields, industry was moved dramatically into the countryside. A spectrum of relationships between agrarian society and coalmining therefore existed, ranging from the traditional patterns of many of the areas of outcropping coal, where it had been long worked, in which smallholding or even small-scale arable farming had been combined with mining, to the spread of miners to new pits from contiguous districts, up to the implanting of a new coal industry into areas historically of mixed or arable farming.[128]

Principal among these new developments were the 'concealed' coalfields, notably those of the east Midlands and south Yorkshire, and the application of deep mining and the railway system to the reserves of coal within exposed fields, such as Staffordshire and south Wales. Their impact upon rural areas can be summarised fairly briefly. In south Wales, the beginnings of deep mining in the Rhondda around 1851 symbolised the transformation of existing small settlements into large mining communities, mainly through the movement of population from the adjacent counties, and from other coalfields.[129] In the east Midlands, new pits from

---

[126] The most important modern study, by Dudley Baines, unfortunately aggregates the three Ridings into one Yorkshire, and the localised impact on the Dales of the collapse of lead and related industries cannot be seen with clarity (*Migration*). However several affected counties do appear in the top third of those losing population by migration, 1861–1901: Cornwall (1); Durham (7); Westmorland (8); and Cumberland (15), *ibid.*, p. 150.

[127] L.Q.s respectively, mining, 1.427 and 1.264.

[128] Based in part upon D. G. Hey, 'Industrialized villages', in Mingay (ed.), *Victorian Countryside*, vol. 1, pp. 360–2.

[129] R. Church, with A. Hall and J. Kanefsky, *The History of the British Coal Industry, 3, 1830–1913: Victorian Pre-eminence* (Oxford, 1986), pp. 6–11. On the complexities of the recruitment to new coalfields, and the views of Brinley Thomas on south Wales, see *ibid.*, pp. 224–6, and Baines, *Migration*, pp. 222f. Discussion of these views, and of the pioneering work of Cairncross using Welton's pre-1914 data, is impossible within the confines of the present chapter.

the 1860s and 1870s were developed around Nottingham, Derby and Leicester, the timing coinciding neatly with the displacement of men from the contiguous hosiery and knitting districts; northward to Chesterfield; and from the 1890s, the Top Hard seam around Mansfield. The development of the south Yorkshire coalfield began with the exploitation of the exposed Silkstone and Barnsley Beds, at the Oaks (1838) and Denaby Main (1866) pits, and was sustained by large-scale migration.[130] The latter was perhaps a pure example of a coal company's new town, nearer Mexborough than the old village of Denaby, and built almost in its entirety by the company, including church, schools, housing, shop and hotel.[131] Nearby Cadeby (1892) represented the first exploitation of the concealed seams, and the further intrusion of a purely industrial settlement into a largely agricultural district.[132] Cleator Moor (Cumbd) developed from the 1840s as a new iron-mining settlement, roughly a mile from the village of Cleator, and built an outstanding cooperative society, by 1883 operating its own farm and corn mill.[133] Similar impact came from the ironstone mine developments of north Yorkshire and Northamptonshire, again largely after 1870.[134]

Quarrying for a wide range of stone, slate and sands was a widespread activity in later nineteenth-century Britain, and the Census is generally considered to have understated employment in that activity. It was in many districts erratic and or seasonal, and remained so to the end of the period. Quarrying was in many parts of the country an unspecific trade, as noted by an official report in 1912: 'You may have a perfectly good quarryman working three weeks or a month in a quarry, and another time he is a farm labourer or working on some other work altogether'.[135] Thus a limited number of large quarries, serving distant markets normally by water transport, were mixed with large numbers of localised sources of critical intermediate goods, such as sand or gravel for building, or limestone for cement, iron and steel, gas, construction and agriculture, and an even greater number of minor quarries, still worked on a contract or permission basis. Samuel's taxonomy of the industry from the reports of the Mines Inspectors in 1912 made it clear that the industry was still characterised by this dramatic contrast in scale: of 7,132 quarries recorded, only 1,017 employed more than ten, and only 334 more than thirty people; and in the 1890s, Beattie Scott, inspector for the Stafford district, had excluded 1,200 of the 1,900 quar-

---

[130] Church, *British Coal Industry*, pp. 7–8.   [131] Hey, 'Industrialized villages', pp. 360–1.

[132] Church, *British Coal Industry*, p. 8.   [133] Marshall, 'Industrial colonies', pp. 151–2.

[134] Census employment data, 1861–1911 show iron miners growing in number from 1969 to 8595 in the former, and 20 to 2518 in the latter.

[135] Cited by Raphael Samuel, 'Mineral workers', in Samuel (ed.), *Miners, Quarrymen and Saltworkers*, (London, 1977), p. 4.

ries and pits returned to him as being less than 20 feet in depth.[136] The range of quarry business, and the leading products of each county, summarised for the leading producers (by volume) in 1911, are set out in Table 18.14, and include the extraction of clay for bricks, also discussed further below.

The figures for total employment in Tables 18.12–18.14 may thus tell less than for many other industries, but they do at least highlight areas of significance. The industrial demands for clay and limestone were reflected in the figures for Staffordshire, Derbyshire, and the West Riding; for building stone in Dorset, Purbeck and Portland, and for Cornish granites; Cornish china clay; and the supply to the new iron industries of Scunthorpe and Corby and generic associations with mining in the midland counties.[137] Variations in output per worker reflected end users, ores and industrial inputs against dressed stone or other end products, and the vintage of the industries. Chalk (for lime and cement) and ironstone correspondingly led the way.[138] Outstanding was the slate industry of north Wales, largely in Caernarvonshire and Merionedd, which in 1910 was employing up to 14,000 men, and which produced 80 per cent of UK slate output.[139] Penrhyn quarry, near Bethesda, was the largest slate quarry in the world around 1900, employing 2,800 men, and blasting hourly.[140] Even in this great industry, where full-time commitment of male workers inducted through an extensive quasi-apprenticeship system was the norm, by-employments survived, and within a largely inhospitable landscape, small landholdings were exploited agriculturally, much as among the quarriers of Dorset and the northern Dales.[141] The area had another quarrying industry of significance, the production of granite setts at the coastal site of Penmaenmawr, which provided ready access to the road surfacing markets of the industrial cities of Liverpool and Manchester.[142] In all of these industries, easy access to sea and rail transport was the key to their scale and intensity.

More commonly the industry was of small scale and, even within the great quarries, systems of team sub-contracting were the norm for exploitation of stone. At the bottom end of the distribution, the spirit of the 'poor man's gain' was retained, as is evident in the auctioned

---

[136] *Ibid.*, pp. 27, 76, citing HM Inspectors of Mines, *Mines and Quarries Reports for 1911*, BPP, 1912–13, XLI, and *Annual Report on Mines and Quarries*, BPP, 1896, XXII, p. 20.

[137] Based largely upon the comments of Samuel, *Miners*, pp. 1–17.

[138] Unfortunately, these data do not permit the corresponding range of comparisons by value across the set.   [139] Jones, *Rural Industries . . . Wales*, p. 11.

[140] M. Jones, 'Y chwarelwyr: the slate quarrymen of North Wales', in Samuel (ed.), *Miners*, p. 101.

[141] *Ibid.*, esp. pp. 111–12; Bettey, *Wessex*, pp. 45–8; Hallas, 'Wensleydale and Swaledale', II, pp. 350–65.   [142] Samuel, *Miners*, p. 27.

Table 18.14. *Output, employment, and numbers of quarries in leading counties of England and Wales, 1911*
(counties in rank order of output)

County	Output 000 tons	Number of quarries	Total number of workers	Average output per plant 000 tons	Mean output per worker 000 tons	Average number of workers per plant	Leading product and share of total per cent
Kent	3,921	120	1,970	32.7	2.0	16.4	Chalk, 73.2
Northampton	3,090	86	2,896	35.9	1.1	33.7	Ironstone, 87.9
Durham	2,593	139	3,054	18.7	0.8	22.0	Limestone, 22.0
Yorkshire, WR	2,555	531	6,771	3.8	0.4	12.8	Clay, 42.3
Derbyshire	2,484	242	3,476	10.3	0.7	14.4	Limestone, 81.1
Leicestershire	2,432	74	4,181	32.9	0.6	56.5	Igneous rocks, 53.0
Lincolnshire	2,419	63	1,703	38.4	1.4	27.0	Ironstone, 78.0
Lancashire	2,145	330	4,405	6.5	0.5	13.3	Clay, 45.6
Staffordshire	1,571	240	2,171	6.5	0.7	9.0	Clay, 60.6
Caernarvon	1,334	101	10,001	13.2	0.1	99.0	Igneous rocks, 46.4
Warwickshire	1,209	70	1,454	17.3	0.8	20.8	Clay, 42.3
Glamorgan	1,170	283	1,854	4.1	0.6	6.6	Sandstone, 37.5
Cornwall	1,140	384	6,336	3.0	0.2	16.5	China clay, 62.6
Devon	1,066	417	2,173	2.6	0.5	5.2	Limestone, 46.8

Source: BPP, 1912–13, XLI, reports I, pp. 17, 21; II, 25–6; III, pp. 30–1, 34; IV, pp. 17–18, 21–3; V, pp. 20–22, 29–30; VI, pp. 22, 24; VII, pp. 35–6, 41–2.

relettings of rights to work Oxford's Headington Quarry, a working-class suburb of stone workers and brickmakers.[143] Cotswold slaters of stone roofing operated on a similar basis, and all moved in and out of employment seasonally, as did those at Collyweston, on the Leicestershire–Lincolnshire borders.[144] Small specialised trades, such as the jet of Whitby and the serpentine carvers of the Lizard, also represented independent work, but were linked to workshops where finished goods were produced, often, in the case of the latter, in the costume-jewellery manufacture of Birmingham.[145] Even flint-knapping just survived the First World War: at Brandon (Suffolk), where around fifty had been at work in 1868, including the diggers, five remained in 1923. The flints met the residual demand for flintlock guns, still in use in West Africa, where old-fashioned firearms were still supplied to meet the needs of the 'natives . . . but whom it is politic to arm less efficiently than the representatives of the ruling race'.[146]

Our final industry, brickmaking, was by far the most evenly distributed of the mineral trades, although Day's analysis of the 1841 Census had found it to be 'provincial', but attributed its apparent 'irregularity' of location to a quirk of his methodology, in that brickmaking was normally to be found around, rather than in, England's towns.[147] Census data may also understate the real level of employment, since brickmaking was highly seasonal in character, and many workers drifted in and out of the industry according to the time of year, Hobsbawm noting the countervailing movement with the gasworkers, who joined brickworks off-season.[148] In the smaller brickyards that characterised much of the industry, often with bricks still burnt against a specific contract, the core workforce was relatively small, and was topped up with these spring and summer seasonal workers.[149] This seasonal fluctuation was still visible in the largest 'factories' surveyed in the *Census of Production* of 1907, which was confined to power-using plant: average employment through the censal year was 63,287, involving 3,407 salaried staff among whom seasonal variations were noted as minimal; seasonal fluctuation around the mean was lower among the small proportion of female wage-earners, roughly −4 to +3 per cent for the 4,044

[143] R. Samuel, '"Quarry roughs": life and labour in Headington Quarry, 1860–1920. An essay in oral history', in Samuel (ed.), *Village Life and Labour*, pp. 170–1; Samuel, *Miners*, pp. 60–1.

[144] Described by J. G. Jenkins, *Traditional Country Craftsmen* (London, 1969), pp. 171–5; H. Harris, 'An industry built on rock', Autumn 1975, in Seager (ed.), *Village Trades and Crafts*, pp. 93–7.

[145] FitzRandolph, *Rural Industries*, III, pp. 157–60.

[146] *Ibid.*, pp. 161–5. It is unclear whether the quoted comment is that of FitzRandolph and Hey, or an indirect quotation from the author of the 1879 monograph on the industry, S. B. J. Sketchley.

[147] Day, 'Distribution of industrial occupations', pp. 163–4, 233.

[148] E. J. Hobsbawm, 'British gas workers, 1873–1914', *Labouring Men* (London, 1964), pp. 162, 170, cited by Samuel, *Miners*, p. 5.   [149] *Ibid.*, p. 4.

involved, than among the 55,836 males, for whom the figures were −6 to +7.[150]

Noting these doubts about the coverage of Census data, which are clearly reinforced by those of 1907, the figures of Table 18.12 demonstrate the dramatic growth of employment in this industry during the years after 1841, as bricks spread in industrial and domestic usage. The evidence of spatial concentration provided by the Location Quotients of Table 18.13 also identifies known clusters of significant production which remained so to the end of the century. Middlesex confirmed the importance of brickmaking on the city fringes, as did the traveller's recurrent experience of entering London through smouldering brickfields, but it was in absolute terms a small industry employing only 1,381 males, and was perhaps at a peak in that censal year: Lancashire, where brickmaking could achieve no great proportionate 'significance' in employment, had 4,380 (L.Q. = 0.789) in the same year. More significant perhaps were the results for Staffordshire and Kent, which accounted for over a fifth of total employment in the industry in 1861, and 17.7 per cent in 1901, and made a case for both as industries on the national scale.[151] In this they clearly pre-dated the brickmaking industry of Fletton and Peterborough, normally regarded as the first plants making for a wider national market. It was developing only from the 1890s, and employment in this south-midland region remained below that of either county before the War.[152] In Wales the association of brickmaking concentrations with industrial and urban counties was clear, but in 1861 represented only relatively small numbers.

The scale of both extraction of clay and the making of bricks was thus polarised between the large integrated plant in which the pug mill, for cleansing and tempering the clay for moulding, and the mechanical excavator or 'steam navvy', were increasingly used from the 1880s, using ever larger and more productive labour forces, and the handicraft industry, extracting clay with the spade and pugging by foot. The large 'factories' of England and Wales alone produced 3,812 million bricks and 294 million tiles in 1907, and indicate something of the scale of the industry as a whole.[153] The Census of 1901 produced disaggregated figures for the

---

[150] *Census of Production, 1907*, BPP, 1910, CIX, pp. 205–6. Calculations of seasonal variation are mine, and details are as follows: January, males, 52,323, 93.7 per cent; females, 3877, 95.9 per cent; April, males, 57,348, 102.7 per cent; females, 4097, 101.3; July, males, 59,780, 107.1 per cent; females, 4170, 103.1 per cent; October, males, 53,894, 96.5 per cent; females, 4031, 99.7 per cent.

[151] Sources as Table 18.2 above.

[152] Samuel, *Miners*, pp. 73–4 makes this suggestion, but employment in brickmaking in Bedfordshire, Northamptonshire and Huntingdonshire combined was far smaller than that of either Kent or Staffordshire in 1901, as it had been in 1891, and remained in 1911.

[153] *Census of Production, 1907*, BPP, 1910, CIX, p. 246.

two leading counties, Staffordshire and Kent: in the former, the 5,643 male and 1,045 female wage-earners worked on average in units of 40, and the predominantly male workforce of Kent, heavily concentrated around Faversham and Sittingbourne, and numbering 5,037, worked in units of 70.[154] Even these were still largely unmechanised in the 1890s, and the brickmaking of Kent and Middlesex was still heavily a handicraft and summer trade.[155] For most of our period, access to clays could be rented for fixed periods, and a small workforce of diggers, puggers and molders could produce saleable products, and leave the industry if and when demand fell, either seasonally or cyclically.[156] Many brickworks may have been like those at Weymouth, established around 1859, which adopted continuous firing in the Hoffmann kiln from around 1914, and were fully mechanised only in the mid-1920s.[157] Entry and exit were therefore relatively easy right up to 1914, and the rural and urban-fringe workforce on which the industry depended, and which needed few specific skills, provided the ultimate case study in flexibility, a key factor in the survival of industry in the early twentieth-century countryside.

CONCLUSION

Industry in the rural districts of England and Wales in the years between 1850 and the First World War might easily be seen as a mere residual of past times, something left behind in pools or gullies by the changing tides of manufacturing. Yet the evidence discussed in the present chapter presents a more complex and varied picture. Certainly some of the hand-loom weavers left in the West Riding, Gloucestershire or Wiltshire in 1914 were special cases, often elderly workers, or those making special pieces of experimental patterned cloths for ubiquitous mill-based manufacture. Much the same might be argued for many rural shoemakers, and for the depressed makers of pillow lace. That there were the declining vestiges of older rural industries, now urban and mechanised, is not in doubt, and of this there were plenty of cases in the present study.

The widespread rural manufacturing activity of the years up to about 1830 that had made the countryside so 'industrial' in character certainly changed, but in a complex manner determined by the different mixes and timing of the agencies of change. The combination of urban growth, transport change, especially the spread of the railway's influence, the mechanisation of former hand processes, the application of power to industrial processes, the introduction of new materials and the impact of

---

[154] *Census*, 1901, Table xx and xxx.   [155] Samuel, *Miners*, p. 44.

[156] Comments based upon Samuel's study of Headington, *Village Life and Labour*, pp. 236–7, and Woods, *Rural Crafts*, pp. 214–26.

[157] D. Young, 'Brickmaking at Weymouth, Dorset', *Industrial Archaeology*, 9, 1972, pp. 188–96.

foreign competition can all be seen as forces for this change in the period, but they applied in differing permutations. Whereas even in the lagging sector of woollens, power-based manufacturing in mills was becoming the norm by the 1850s, in lace a two-sector industry lasted for another generation or more; it was only in the 1860s and 1870s that the first stages of integration of the production of shoes took place, and fully 'industrialised' manufacture after the onset of US influences in the 1890s; the sewing machines that, with the band knife, helped to generate the urban sweatshops of London and Leeds clothing in the years after 1880 also permitted rural glovers to continue to contribute significantly to the industry up to and beyond 1914; and it was only after 1890 that steam power, in the form of the railway and the 'steam navvy' at Fletton and Peterborough, began to invade the predominantly handicraft and seasonal brickmaking industries.

While disappearance was to be the long-term fate of much of the rural industry considered here, there is a case for treating much of this period as one of change and transition, and not writing premature obituaries. The imperfect and incomplete processes of mechanisation in lace, hosiery and gloving actually created new rural outwork, but largely for a low-wage and female workforce. Rural industry may also have permitted some lasting child work in similar trades, and introduced many intermittent new activities, such as the carding of buttons or pins, in which mechanisation was the creator. Others were in effect great industrial 'cuckoos' in the late Victorian countryside, principally the mineral trades that sustained so much urban growth. Quarrying, brickmaking and mining retained both a heavy handicraft base and a rural location, especially as the development of deep-mining techniques for the exploitation of concealed coal seams brought collieries into new areas of south Yorkshire or the east Midlands. At the very end of the period, the development of ironstone mining intruded new quarrying activity on a huge scale into rural Northamptonshire and Lincolnshire. The growth and survival of industry with a rural location thus in part depended upon the random effects of raw materials and natural endowments.

While such factors may have produced the rural industries of the greatest scale and employment, other factors also applied. Certainly, there were many sweated trades in which an immobile rural workforce was trapped, in the words of the Dales knitter, because 'there were nowt else', but this must not be allowed to obscure the flexibility that could be sought by both employee and employer, nor the extent to which much UK industry, even in 1914, remained of relatively low intensity. The synergies of brickmaker and gasworker, of slater or miner and dairyman, and of lace- and plait-maker, were signals of this feature of industrial structure, and perhaps also a sign of the capacity of a workforce in the countryside to adjust price and work regime in order to remain there. Viewing earnings as a family not an individual matter, at its extreme in the poorly paid Buckinghamshire chair bodgers, then the combination of industrial activities within the family unit with a smallholding or even, as in Caernarvonshire slate, with just enough land to sustain a cow was both rational and in many areas successful. For all of these reasons it was in general only in the last years before 1914 that many traditional rural industries were forced into full retreat, which the war and the changes of the 1920s turned into a final rout.

CHAPTER 19

THE RETAIL TRADES AND
AGRICULTURAL SERVICES*

BY J. A. CHARTRES

INTRODUCTION

Together with the many engaged in the varying forms of industrial and manufacturing activity, the crafts and trades servicing and supporting agriculture and wider rural society formed an important element of the countryside of England and Wales in the years 1850–1914. As suggested earlier, the commonplace taxonomy of rural society as the 'three-class train' of landlords, tenant farmers, and agricultural labourers, must be augmented by this fourth group, members of which clearly also over-lapped into the traditional three classes. A number of the proprietors of important servicing businesses, and retailers, certainly held land, and had been an important component of those who had acquired land through post-enclosure auctions.[1] Many others also held land as a necessary adjunct or additional component of an occupation as publican, farrier or carrier, and many who declared themselves as craftsmen for the purposes of the Census were generic labourers, for whom the designation 'carpenter' was a commonplace aspirational mask. As discussed in the present chapter, this group of people and families inhabiting the country districts and market towns of the period thus formed a distinct set worthy of separate analysis, but not one that was wholly separable from the strata of farming society. Many were very much *of* as well as *in* rural society, and their craft, trade or service lay in a continuum of specialisation.[2]

Two features in particular make their precise assessment problematic. Even at the end of the period, such were the hard realities of rural busi-

---

* Where a reference is cited in short form at the first citation in this chapter, the full reference may be found in Chapter 18.

[1] My summary of much of the literature on the secondary distributive consequences of enclosure, many studies of which note the strong presence of the small town tradesmen and shopkeepers among the purchasers.

[2] For a detailed demonstration of these points, Adrian Hall's case-study of the Lincolnshire community of Rippingale is very helpful: *Fenland Worker-Peasants: The Economy of Smallholders at Rippingale, Lincolnshire, 1791–1871, Agricultural History Review, Supplement Series*, 1, 1992.

ness that many of these trades or crafts were practised as multiple occupations, and the categorisation of occupation in such sources as the Census correspondingly created an illusion of specialisation and precision that did not match reality. Thus the comparison of businesses in rural England as revealed by directories and the patterns of employment presented by the Census produces a contrasting picture as the latter world of 'public house keeper', 'carpenter', 'coal merchant' and 'blacksmith' encountered the former in which the formulation 'publican and coal merchant', 'farmer and carter' or 'grocer and draper' was commonplace.[3] It is perhaps only in the tidy world of urban and industrial society after 1850 that specialist occupations can be conceived, and to one degree or another they have always been rare features of the country districts. Personal and family strategies for subsistence involved occupations mixed often on a seasonal basis, as in many areas of rural industry, and the disappearance of one or other element could prejudice livelihood as a whole.[4]

The second of these preliminary points is related: since Census categorisations, and for that matter declarations in directory listings, were essentially self-descriptive, and not the outcome of a precise analytical taxonomy, they were in some degree aspirational. The labourer might declare himself a carpenter, the carpenter a wheelwright, or the smith a farrier, and thus introduce an upward bias in the revealed skill/specialism endowment of the countryside. The occupational descriptions upon which much of the analysis of the present chapter necessarily depends may be best treated as a measure of central tendency in work patterns, and not a precise description.[5]

Bearing these caveats in mind, the following pages survey a range of trades and crafts that provided support in services or equipment for agrarian society. Most were little subject to technical or technological change during the period, and thus most bore a direct and dependent relationship to rural society as a whole. Most therefore were classified as occupations distributed 'locally' in Day's analysis of the 1841 Census data: markets were highly localised; there was little potential for elements of comparative advantage to induce concentration of activity; and the characteristics of local farming, building materials and population density were the principal determinants of their distribution.[6] These trades and services are the subject of the present chapter, and have been grouped for analytical purposes into the following broad groups, in the full recognition that each

[3] J. A. Chartres, 'Country tradesmen', in G. E. Mingay (ed.), *The Victorian Countryside* (London, 1981), vol. I, pp. 301ff.

[4] Discussed in Chapter 18 above, and based upon Saville, *Rural Depopulation*, pp. 20–30.

[5] Surveyed as above by J. A. Chartres and G. L. Turnbull, 'Country craftsmen', in Mingay (ed.), *Victorian Countryside*, vol. I, pp. 319–21.

[6] Day, 'Distribution of industrial occupations', pp. 87–118.

may be an imperfect bracketing for its components: building and construction; horses, vehicles and equipment suppliers; transport and distributive trades; the retail trades; and the public house and beershop trades. It is to the consideration of these successive groupings that we now turn.

A THE BUILDING TRADES

The building trades of rural England and Wales as considered here have been confined to a set of four: carpenters, masons, slaters and tilers, and thatchers. This is scarcely the occasion to attempt to write the history of building work in general, and as will be more obvious from the discussion that follows, this group exhibits some of the principal contrasting features of the building trades.[7] There was a case, at least looking back towards 1850 from the end of the period, for including the Census categories of 'bricklayers', who became increasingly numerous over the second half of the century. Bricklayers overtook masons in number between the 1861 and 1871 Censuses, and were increasingly the leading purely building workers, but while brick was very extensively made and consumed in the countryside, it was above all the material of the towns and their growth, and this relatively new mass occupation was also predominantly urban in character.[8] Brick, and later concrete, were the principal agents in the separation of building from the determinants of underlying geology which first occurred on a large scale in our period. On that perhaps somewhat arbitrary basis, bricklayers have been included in the figures of Table 19.1 for comparison, but omitted from the more detailed survey.

The four trades selected provide a range of scale, from the small thatchers to the very numerous carpenters, and all lock into the traditional patterns of building of before 1850, perhaps more generally characteristic of rural building activity. Their numbers, for England and Wales, 1841–1911, are summarised in Table 19.1 as a preliminary to further analysis.

These summary data confirm some of the suggestions made above, principally that for much of the period the most generic building crafts and trades bore a pretty constant relationship to that of the occupied population as a whole, and retained their almost completely male nature.

---

[7] C. G. Powell, *An Economic History of the British Building Industry, 1815–1979* (London, 1980), pp. 68–84; Marian Bowley, *The British Building Industry, Four Studies in Response and Resistance to Change* (Cambridge, 1966), pp. 3–35, 325–61; A. Satoh, *Building in Britain: the Origins of a Modern Industry* (Aldershot, 1995). Most of these were concerned with the larger scale of building activity, although each contains some useful material on crafts and trade processes.

[8] Comments based upon the summary occupational data in Armstrong, 'Information about occupation', p. 258.

Table 19.1. *Employment in the building trades, 1841–1911*
(England and Wales, 000s)

Occupation	1841	1851	1861	1871	1881	1891	1901	1911
Carpenter	136.6	156.1	177.8	205.6	235.0	220.7	270.7	214.4[a]
Mason	63.6	78.7	82.4	94.1	97.4	84.6	96.0[a]	63.0[a]
Bricklayer	39.6	67.1	79.5	99.9	125.0	130.4	218.8[a]	171.6[a]
Slater, tiler	4.0	4.4	5.2	6.1	7.5	6.8	9.8	8.4
Thatcher	3.7	5.9	5.3	4.1	3.7	3.6	b	b

Notes:
[a] Figures include 'mason's labourers' and similar where indicated
[b] Not separately enumerated in 1901 and 1911
Source: Compiled as Table 18.1 above.

Even allowing generously for the unknown proportions of females omitted, the nature and culture of all of these trades was overwhelmingly masculine, and these figures for males only therefore provide a reasonable plot of the course of employment. Up to 1901, employment among carpenters almost exactly mirrored the growth of the workforce, and tilers and slaters ran well ahead of this, indicating the substitution effects in roofing practice so evident in the thatchers; even more striking were the bricklayers, whose growth was double that of the workforce, 1851–1901, and the relative stagnation among masons, again perhaps indicating the impact of substitution in building.[9] Two of the four trades selected for more detailed assessment were therefore in at least relative decline during the period, one in relative growth, and the most numerous in a constant state. The consistent pattern of decline in 1911 is largely explained by the taking of this Census at or near the bottom of the house-building cycle, where all other censal years lay in the mid- or upper parts, and the figures, therefore, bear no certain interpretation.[10]

The inference drawn from these overall figures was that two at least of our building crafts were associated with traditional building, and thus more likely to be concentrated into the rural districts, and that the others were less likely to display such spatial patterns. In part this reflected a shift in the building industry in general: as Marian Bowley demonstrated, the later eighteenth and nineteenth centuries saw the emergence of a composite form of construction, in which load-bearing was shared between

[9] Compared with 1851 as 100, for the civil occupied male population, index figures stood at 111.7 in 1861; 123.8 in 1871; 136.2 in 1881; 154.5 in 1891; 171.9 in 1901; and 196.8 in 1911.

[10] Interpretation based upon J. Parry Lewis, *Building Cycles and Britain's Growth* (London, 1965), pp 196–210, and Appendix 4, pp. 301–19.

Table 19.2. *Location quotients, England and Wales, building trades, 1861*

Occupation	Mean L.Q.	S.D.	County concentrations (in rank order)
England:			
Carpenter	1.066	0.220	Devon, 1.529; Surrey, 1.486; Dorset, 1.432; Hants, 1.390; Heref, 1.336; Som, 1.303; Yorks, NR, 1.299
Mason	0.933	0.760	Devon, 2.518; Heref, 2.344; Som, 2.252; Corn, 2.075; Westm, 2.009; Dorset, 1.923; Glos, 1.900; Northd, 1.803; Yorks, WR, 1.724
Slater/tiler	0.780	0.770	Northd, 2.831; Glos, 2.276; Oxon, 2.266; Derby, 2.103; Lancs, 2.064; Chesh, 1.950; Cumb, 1.569
Thatcher	1.300	1.592	Dorset, 6.021; Suff, 4.991; Devon, 4.859; Som, 4.062; Essex, 3.652; Wilts, 3.640; Cambs, 3.002
Wales:			
Carpenter	1.042	0.330	Pembroke, 1.747; Cardigan, 1.614
Mason	0.964	0.282	Pembroke, 1.588
Slater/tiler	1.117	0.860	Merionedd, 2.575; Denbigh, 2.305; Caernarvon, 2.018; Flint, 1.983
Thatcher	1.037	1.702	Radnor, 6.418

Source: For details on this measure, and its calculation, see note 16 of Chapter 18 referring to Table 18.2 above.

walls and framing, in place of the previous timber and subsequent iron- or steel-framed form. This was rather more significant for the large construction projects, but helped to segregate crafts and skills, and meant that the maintenance or alteration of timber-framed structures, whether stone-clad and roofed, or mud-and-stud and thatched, demanded a rather different portfolio of skills than that of structures in which masonry or brick were significantly load-bearing.[11]

The figures for the Location Quotients for these trades in Table 19.2 offer some confirmation for this, notably for the carpenters, for whom a mean quotient near to unity appeared for both England and Wales, each

[11] Bowley, *British Building Industry*, pp. 7–27.

with a small standard deviation, indicative of a ubiquitous trade.[12] Even
so, there were interesting relative concentrations of carpenters, and much
more striking regional and local clusters of masons, slaters/tilers and
thatchers, indicated by their distinctly higher variations from the mean.[13]
The implications of these indicators are explored successively for each
trade below.

Two features in particular stand out in respect of the carpenters: the
general evenness of their distribution; and the pronounced concentration
of such clusters into very 'rural' areas of the country. These counties were
those where traditional building materials remained important, where
agriculture was a relatively greater source of employment than in the
average 'mixed' or 'industrial' county: Surrey here is exclusive of metropolitan London, and in 1861 perhaps the most rural of the shires immediately surrounding the capital; and the iron and steel of Middlesbrough
had only just begun, and so the transformation of North Yorkshire by the
sudden implanting of perhaps Britain's last great industrial town had yet
to take place.[14] The coincidence of the clusters of carpenters with counties in which other 'traditional' building craftsmen also concentrated,
above all in the West Country, was notable, and reinforces the interpretation of carpentry as a strongly rural craft, despite its ubiquity.

There may be several reasons for this distribution, the first of which
was identified by Barbara Kerr, and was essentially terminological rather
than real. Kerr noted the phenomenon in Dorset, and suggested that
those displaced by enclosure from heath and waste cottages moved into
carpentry as a livelihood, and were thus disproportionately represented
in the Census description of the workforce.[15] In a county with much
timber, cruck framing, wattle-and-daub and similar building, and many
rural wood crafts, such as hurdle-making, this was a sensible strategy for
a refugee, and probably represents a partial explanation of the cluster.
All labourers would perform 'hedge carpentry' and, for those with a
modicum of skill, specialisation was a real possibility.[16] Several of the
other counties in this group also had extensive experience of late enclosures, but not all: this quasi-cultural explanation of the clusters is helpful

---

[12] This measure, its calculation and meaning are explained fully in Chapter 18, especially note 16
referring to Table 18.2, but, for reference, a mean score of 1.0 and a minimal Standard Deviation
of say 0.01, would indicate perfectly even distribution across the country, with only plus or minus
½ per cent deviation by county around this.

[13] As in Table 18.2 and others, clusters have been somewhat arbitrarily defined here as those counties in which the Location Quotient lies more than one Standard Deviation above the mean.

[14] A. G. Parton, 'Parliamentary enclosure in nineteenth-century Surrey – some perspectives on the
evaluation of land potential', *AHR*, 33, 1985, pp. 51–8; D. J. Watson, 'The effect of the growth
of Middlesbrough on the surrounding rural economy, 1830–1914', MPhil thesis, University of
Leeds, 1993, *passim.*   [15] B. Kerr, *Bound to the Soil* (London, 1968), pp. 132–3.

[16] Walter Rose, *The Village Carpenter* (Cambridge, 1937), p. 61.

but not exhaustive. Rather more important were surely the dominant characteristics of the vernacular building of these districts, and the aspirational attraction of a craft or trade in a county overwhelmingly populated by agricultural labourers.

Many of the retrospective accounts of carpentry work, such as those of George Sturt or Walter Rose, dealt with supreme skills in the business, shading into the 'companion trade' of the wheelwright, and it is clear that the village 'carpenter' covered a wide range. Carpenters were diffused down to the smaller villages, and like smiths appeared at population thresholds of between 300 and 400 in the late 1870s.[17] Much of their work was clearly the 'hedge-carpentry' of the farm, described by Rose as the men who 'had never learned what is termed the higher order of the carpenter's craft', and used a limited set of tools – axe, auger, saw, claw hammer, mallet, gouge, planes, chisels, and rules – packed into a basket for transit, as such carpenters travelled from farm to farm to conduct repairs and alterations.[18] Gates and rails, cow-stalls and other farm building work, and general repairs and maintenance were their stock in trade, and were of a functional rudeness, not to be confused with the higher level work: as Rose noted,

They had never become enslaved to line and level; their minds had not been trained to revolt if their work deviated from the square, or if it was slightly on the twist and the faces of their joints not absolutely flush . . . the work they did was part of the beauty of the countryside; the cleft fence-rails and posts split from oak saplings, with the bark left on in places, and the rough knots trimmed with axe or drawing knife . . . The carpentry of the open countryside ought not to savour too much of the joiner's bench. In fact, it is a separate craft and should be kept so.[19]

Not surprisingly, the documentary records of property sales or accounts also tend to relate to the upper and more complex businesses of carpentry and joinery, but present a fairly consistent picture across the period, with a mixed range of work being undertaken, and a role as supplier of sawn and other timber also being performed. The goods of John Reavell senior, of Guilden Morden (Cambs.), sold at Royston in 1847, showed him mixing smallholding and carpentry, with around six acres of land under crops, and pit saws, adzes and planes in his shop, perhaps transported to work elsewhere in his donkey cart.[20] The more direct record from mid-century of two carpentry businesses of Mr Arnold of Wallingford, 'carpenter and coffin-maker' and of Thomas Parr, carpenter and wheelwright of Newbury, indicated the range of activity of the market towns. Both sold stakes, palings and other sawn timbers; repaired

[17] Chartres and Turnbull, 'Country craftsmen', p. 321. [18] Rose, *Village Carpenter*, pp. 61–76.
[19] *Ibid.*, pp. 61, 63. [20] Cambs RO, 296/SP 24.

timber materials from window bars to chairs, pumps and wheelbarrows; sharpened saws and other tools, and repaired spade and other tool handles.[21] In this they closely resembled the business of George Sturt, at Farnham (Surrey) in the 1880s, formally a 'wheelwright's shop', but practically performing a vast range of joinery, building, and repair work, or like Sturt's 'Bettesworth', who repaired guttering and roofing, and used a one-piece joiner's ladder of 'forty-five round' to approach but not reach the high eaves of the Crown and Bell in the town.[22] In Buckinghamshire, Walter Rose's family was running a very similar workshop recalled in his 'reminiscences' prior to 1893. Charles Cooper, of 'Bunkum' (Suffolk), around 1910, was builder, wheelwright, repairer and builder of dog-carts, tumbrils and wheelbarrows, general carpenter, smith, paint-maker and painter and, according to his son, made elm or oak coffins 'guaranteed to larst a bloody lifetime'.[23]

That these were the generic business activities of the pre-1914 carpenter was confirmed by a Welsh case. Richard Faulkner, of Llangunllo (Radnor), was a substantial self-employed carpenter, who continued the business to 1946, and his ledgers for 1905–10 indicated his range of work: making and erecting a granary, two barns, a bungalow, verandas, sheds, seven pigsties, two fowl-houses, a goose-coop, sets of 'horse-works' (gins) and oil-engine works, in addition to a comprehensive range of wood and similar repairs, including restringing a piano; yet he still had time to catch and sell 400 rabbits.[24] Such carpenters were clearly the generic supporters of much of the building and equipment of rural England and Wales. At one end lay the most numerous itinerant, jobbing and 'hedge' carpenters, using timber available on the spot, and carrying minimal overheads; at the other, the integrated businesses, like Faulkner's, with wheelwrighting equipment and perhaps sawpits, employing perhaps half a dozen journeymen, and based overwhelmingly in the larger villages and market towns, and taking the limited economies of scale. Between them were numerous variants, and this explains why they were so numerous and widespread, and why this flexible craft defied uniformity. Both types were in retreat by the early 1920s, as changing materials, equipment, and losses of skilled manpower made their cumulative impact.[25]

The distribution of stonemasons in Table 19.2 pointed perhaps more

[21] Berks RO, Ledger of Mr Arnold of Wallingford, W/Z 15; Reading, Rural History Centre, Account Book of Thomas Parr, Newbury, 1847–66, D66/Z1. I am grateful to Professor E. J. T. Collins for information on the former.

[22] George Sturt, *The Wheelwright's Shop* (1923, Cambridge Canto edn, 1993), pp. 9–20; Sturt, *The Bettesworth Book* (1901, facsimile of the second edn of 1902, Firle, 1978), pp. 102–3.

[23] C. Ketteridge and S. Mays, *Five Miles from Bunkum* (London, 1972), p. 38.

[24] Kightly, *Country Voices*, pp. 195–201.

[25] Woods, *Oxford*, p. 45; Jones, *Rural Industries . . . Wales*, pp. 100–8.

clearly than in the other trades to the enduring impact of geology on building. All of the prominent counties in England were located in the classic 'stone-belts', where traditional building had employed rubble, free or dressed stones, and where in general quarrying remained important, if not an industry on the national scale,[26] and in Wales stone was widely used, as indicated by the low Standard Deviation from the Location Quotient. Though generally a craft perceived as small in scale, the evidence on employment structure from the 1901 Census indicated only around 10 per cent of masons acting as employers or working on their own account, and having employees distributed at a ratio of 18 to each employer; carpenters, at 28 to one, were employed on average in larger units, but had a higher proportion of the self-employed, at 6.3 per cent of the total, against the masons' 5.3.[27] Large-scale work had long been known in stonemasonry, of which Rennie's use of stone, precisely prefabricated in Cornwall, for London Bridge, had been a telling case.

Overall, while much of a mason's work may have been on large structures, the trade appears to have been almost a service craft, in which numerous employees sold traditional skills to work on traditional materials, and did so overwhelmingly in narrowly defined locations. The West Riding provided a particular example of this phenomenon: though appearing at the back of our list of significant clusters, it was also a leading quarrying and brickmaking county, in which Bradford and the other textile towns of the west, and the Dales villages remained predominantly stone-built throughout the period, whereas Leeds, while surrounded by small quarrying, shifted dramatically to brick.[28]

These features tended to make the craft one that was seasonal in activity; even Cotswold masons, working perhaps in the area of England in which stone retained its greatest market share before 1914, were laid off by weather in the winter, when frost precluded work on the porous stones of the region. Masons were therefore seen as skilled workers, true artisans, and men who enjoyed wages substantially higher than those of the more intermittent quarryworkers or brickmakers, but were still subjected to frequent lay-offs. Raphael Samuel indicated this tellingly with extracts from the diary of a Headington stonemason, Charles Snow, whose work was further seasonalised by the habit of Oxford colleges confining work to the vacations, 'when the toffs was all down', making the long vacation the stonemason's 'harvest'.[29] Snow's diary for the winter of 1882 indicated the intermittent nature of his work:[30]

---

[26] See above, Chapter 18, pp. 1142–5.   [27] Census, 1901, Table xxxv.

[28] Samuel, *Miners*, pp. 13–14, citing *Leeds Mercury*, 20 March 1875.

[29] Samuel, *Village Life and Labour*, p. 239, oral evidence of Coppock and Augur.

[30] *Ibid.*, pp. 239–40.

Friday 5 January	went to Oriel to work
Monday 9 January	begin putting up chimney at Oriel Coll.
Thursday 19 January	had the sack from Oriel
Saturday 28 January	planted Mr Blake's flowers
Thursday 2 February	killed pig 9 stone 10 lb. 10/6 4–19–6
Friday 3 March	went to work at Oriel Coll. roughing steps
Monday 24 April	went to Hollywell & Jericho
Saturday 29 April	wt day had the sack
10 May	went to work Bodlyon
10 July	went to University coll. to work

In skills, the possession of the tools of his trade, and in this relatively low intensity of work, the mason was very much the representative of the traditional and rural craft, gradually coming under threat from the new building trades of urban and industrial society.[31]

Tilers and slaters, by contrast, were both distinctly representatives of the modern, as their relative growth in the figures of Table 19.1 indicated. Yet they were strongly contrasting: true slate as quarried and mined in north Wales and the north-west was a business of large-scale production, and great skills in use; stone slates, such as those of the Cotswolds or Collyweston, modest in scale of production and more traditional in its markets and uses; and the modern tile, very much the product of industrial kilns and the semi-skilled tiler. Unfortunately, the Census data make it impossible to disaggregate the two trades across the whole period, although slaters predominated in the 1861 data, and the Location Quotients of Table 19.2 provided an unsurprising picture of clusters within the ambit of the principal producers. Outside the north-west, urban demand was increasingly met by tiles, but London's share of total employment in these crafts was unchanged, 1861–1901, at just over 9 per cent, perhaps confirming that one did not have to bear the occupational status of 'tiler' thus to roof a building.

Slating, by contrast, carried all the trappings of the true artisanate: it had its own traditional measures, regional patterns and vocabulary, with an extended process of entry by at least informal apprenticeship. Thus sizes of slates in Cheshire ranged from *Haughattees* and *Widetts* to *Jenny why gettest thou* and *Rogue why winkst thou*, and North Walian slates classified by size from *Duchesses* and *Viscountesses* to *Broad Ladies* and *Narrow Ladies*; in the stone slate areas of the Cotswolds and the Pennines, *slat rules* or *wippett sticks* were employed to measure the twenty to thirty distinctively named types customarily employed.[32] Working in the smallest average units of the three crafts in 1901 (one employer to thirteen

[31] J. G. Jenkins, *Traditional Country Craftsmen* (London, 1969), pp. 159–69; K. S. Woods, *Rural Crafts of England: A Study of Skilled Workmanship* (repr. East Ardsley, 1975), pp. 182–91.

[32] Jenkins, *Country Craftsmen*, pp. 173–5.

slaters or tilers), their work was also highly seasonal, as in all building trades, and Cotswold 'slatters' expected to roof in the summer, and to plaster or limewash in the winter months.[33] The relatively high skill and labour content of slating in both quarrying and use combined with its unsuitability for mechanisation to reduce its position in the inter-war years.[34]

Perhaps because of romantic rusticity among the observers of the late-Victorian and early twentieth-century countryside, our final building tradesman, the thatcher, has enjoyed an undue prominence. The evidence of such observations is correspondingly conflicting, and the Location Quotients of Table 19.2 raise immediate questions. Since almost all corn cut in England and Wales in 1861 was stacked for some months before threshing, and since such stacks were normally thatched, how did the counties with no thatchers at all manage to preserve their crops? The problem was emphasised by the retrospective comments of Woods:

Farm thatching is not always done by a specialist craftsman: many farm-workers, including land-girls, have learned to do it well . . . the work was the pride of Harvest Home; it put the crown on the efforts of the year, and safely stored the precious grain for barren winter days. It used to be the aim of the stack-builder to pack his straw so tightly and to shave off the surface so evenly that no rat could find a foothold on the outward-sloping walls.

Extolling the quality of the stack, and by implication its roof, revealed an apparent conflict with the non-specialist thatching.[35]

Certainly in many districts the thatching of the corn stacks was no amateur business. In counties such as those of East Anglia, where thatchers were in any case clustered, the specialist came onto the large farm to secure the crop, as noted by Bill Partridge of the farm at Lindsey (Suffolk) around 1914: 'There used to be twenty or thirty stacks in the stackyards, and you hed to hev a thatcher come and thatch 'em all.'[36] In general, however, contemporaries differentiated the professional thatcher from the senior farmworker who might cover the stack, as Alfred Williams observed of pre-war Wiltshire:

The old Thatcher and his wife were a unique pair. He was at one time an agitator, and secretary of the old-time branch of the agricultural union in this district, but that has long ago died out here. He was very hearty and independent, unconquerable in spirit, and fond of a glass; 'Must ha' drap beer,' was a frequent saying of his. He did not believe in the Divinity, and never went to church but once a year, that was on Club-Day. He would not have gone then if he had not

---

[33] Census, 1901, Table xxxv; Jenkins, Country Craftsmen, pp. 171–2.

[34] Other uses too were in decline, exercise books came to replace birch-framed slates and slate pencils in British and empire schools (Jones, Rural Industries . . . Wales, p. 11).

[35] Woods, Rural Crafts, pp. 196–200.   [36] Kightly, Country Voices, p. 129.

been compelled, but on that festival each member must attend the service or be fined half a crown, and the old thatcher could not afford to lose such a large sum of money as that. He was in constant work all the year round, for, though almost every farm has its rough thatcher, they are not capable of performing skilled work, such as covering cottages, and farmhouses and buildings.[37]

The fundamental difference of course lay in the anticipated duration of these thatched roofs, despite the value of the stacked corn: stacks had to last months, whereas a well-thatched roof of best Norfolk reed was generally expected to last up to a century. The tarpaulin or 'rick cloth' also assisted in the stackyard before the introduction of Dutch barns reduced dependence on the straw thatch roof.[38] The thatcher was there-fore integral to building work, and the need for his continuing services created by the interaction with the traditional building of these counties, notably the wattle-and-daub of parts of Dorset, or the plastered clay-lump buildings of Suffolk. Changes in working patterns around 1850 may already have been eroding the future of thatch, and the use first of the scythe, then of mechanised reaping and the drum thresher, damaged the wheat straws preferred for thatch in the southern and south-western counties, and thus reduced supplies of replacement raw materials. Norfolk reed was far superior in durability, but more than twice as expen-sive, while being still unable to bear the costs of transport on a national scale. Aesthetically pleasing and romantically conjuring up the rural idyll, the thatcher naturally drew attention to his trade, but external circum-stances may already have determined his decline long before his elimina-tion as a separate category in the Census in 1901. The durability of his roofs may just have masked its real timing.

B HORSE AND VEHICLE CRAFTS

The group of trades providing equipment for the farm and the wider support of horses and vehicles has traditionally been regarded as the core of the creative rural crafts. They attracted the lion's share of attention from the surveyors of the changing British countryside after 1918, and became symbolic of the losses from the striking consequences of rural change. That many of the blacksmiths, wheelwrights, saddlers, coopers, millwrights, and fence- and hurdle-makers embodied high levels of acquired skills was not in question, but this cannot be held absolutely to distinguish these trades from many others in the later nineteenth and early

---

[37] Williams, *Wiltshire Village*, p. 105.

[38] Detailed descriptions of stacking wheat and thatching the stack were provided by Henry Stephens' *Book of the Farm*, here cited from C. A. Jewell (ed.), *A Sourcebook, Victorian Farming* (Winchester, 1975), pp. 116–18, extracted from the third edition of Stephens, *c.* 1870.

twentieth centuries. It was perhaps the ways in which the wheelwright and blacksmith symbolised the old institutions, habits and cultures of rural and horse-drawn society that helped to create this 'craftism', and elevated these craftsmen in the minds of social commentators into artisan symbols of integrated handicraft in a world increasingly of machine manufacture.

Woods, writing in 1920 of the village wheelwrights and carpenters of Oxfordshire, lamented the quality of modern Bristol-built waggons merely repaired in the village workshop, where the poor quality of materials and execution was revealed beneath the masking coat of paint. By implication, long-perfected and still viable crafts were being destroyed by the 'fashion' of buying from urban workshops, and this consequently dissuaded young men of talent from entering the extended process of acquiring these skills.[39] Yet this feeling that these creative crafts were waning in the face of relatively recent circumstances is hard to reconcile with other commentators, and the evidence of employment trends and distributions. Our period was the one that saw the generalised transition of these crafts from creative craftsmen, perhaps true artisans, making their own products to the local patterns and designs acquired through apprenticeship, to mere repairers. As FitzRandolph and Hay noted in their follow-up to Woods's study, 'A notable feature of the present-day smiths, as compared with their prototypes of fifty years ago, is that the majority have lost the art of making things'.[40] As will be suggested below, perhaps even this represented an exaggerated view of the creativity of the rural craftsman, and one must be very cautious about inferring continuity back to the eighteenth century, in trades such as edge-tool making, where, as Collins has trenchantly suggested, many of the tasks for which tools were created were themselves of very recent origin in the 1850s.[41]

The exact balance between the traditional making of tools and the proliferation of types beyond the needs of strict utility varied between these trades and crafts, but the experience of stagnation and decline during the period 1850–1914 was more generalised.[42] As in many other trades, the First World War accelerated these processes and heightened the awareness of contemporary observers, but rarely initiated change.

[39] Woods, *Oxford*, p. 45.   [40] FitzRandolph, *Rural Industries*, I, p. 183.

[41] E. J. T. Collins, 'Agricultural hand tools and the Industrial Revolution', in N. Harte and R. Quinault (eds.), *Land and Society in Britain, 1700–1914: Essays in Honour of F. M. L. Thompson* (Manchester, 1996), pp. 57–77, esp. p. 74. Jenkins, *Country Craftsmen*, illustrated the point by reference to the trade catalogue of Isaac Nash of Stourbridge, for 1899, which contained ninety patterns of billhook, sixteen slashers, forty-six axes and hatchets, and ninety-four hooks and sickles, p. 125.

[42] Collins, 'Agricultural hand tools', p. 74, citing anthropological approaches of Geertz, and defining the process of cultural involution as 'progressive internal complication which takes over from inventiveness once the limits to utility have been reached'.

Table 19.3. *Employment in equipment and horse and vehicle crafts, 1841–1911 (England and Wales, ooos)*

Occupation	1841	1851	1861	1871	1881	1891	1901	1911
Blacksmith	81.6	94.2	107.8	112.1	112.2	124.5	136.8	125.3
Wheelwright	25.1	28.0	30.0	30.3	28.6	27.8	28.8	23.7
Saddle, whip, and harness, M	15.2	16.1	18.1	21.2	21.8	24.4	26.0	20.9
Saddle, whip, and harness, F	0.4	0.7	1.3	1.8	2.0	2.9	4.7	3.5
Millwright	6.9	7.6	8.2	7.6	6.9	6.1	5.3	5.5
Cooper	15.6	15.9	19.7	19.2	18.6	17.1	15.7	14.0
Fence/hurdle	1.4	1.7	2.7	3.0	3.0	2.4	2.2	1.8
Sawyer	25.0	30.5	31.7	28.0	24.7	23.3	32.3	40.2

Source: Compiled as in Table 18.1 above.

The broad outlines of employment in the crafts considered in the present section are summarised in Table 19.3, and provide the basis for the more detailed consideration in the present section successively of blacksmiths; wheelwrights; saddle, whip and harness makers; millwrights; coopers; fence- and hurdle-makers; and sawyers. Female employment has been listed for the only craft in which significant numbers were recorded.

All of these occupations provide clear evidence of the early growth of these crafts, followed generally by stagnation and then absolute or at least relative decline, although the groupings mask the evidence in some cases. Blacksmiths, the first and most numerous of these crafts, included at least some of the farriers, that is, the shoeing smiths and horse doctors whose work overlapped at the beginning of the period with that of the veterinary surgeon, who were as yet not fully developed as a profession. But at the opposite end of the spectrum they merged with the smiths engaged in engineering and foundry work, many in railway-related construction. The absolute disentangling of the trades is impossible, but the relative weights of the traditional market town and village smith, the foundry worker, the iron manufacturer, can be indicated, and suggest that within this broad band the traditional smith shared the experiences of his fellow craftsmen.

This was indicated in the analysis of the relative concentrations in horse and vehicle services of the 1861 Census, identified by the use of the Location Quotients. The outcome of this analysis is presented in Table 19.4, and for 1861 it indicates crafts very widely and relatively evenly distributed across England, and the same for smiths in Wales, displaying in

Table 19.4. *Location quotients, England and Wales, horse and vehicle services, 1861*

Occupation	Mean L.Q.	S.D.	County concentrations (in rank order)
England:			
Blacksmith	0.997	0.242	Durham, 1.823; Northd, 1.681; Corn, 1.396; Salop, 1.294; Devon, 1.273; Cumb, 1.253
Wheelwright	1.090	0.460	Salop, 2.384; Lincs, 2.076; Heref, 1.767; Rutland, 1.696; Suff, 1.661
Saddle, whip, and harness, M	1.110	0.321	Warks, 2.232; Staffs, 1.747; Rutland, 1.520; Hunts, 1.447
Saddle, whip, and harness, F	0.796	2.252	Staffs, 12.822; Warks, 7.640
Wales:			
Blacksmith	0.941	0.175	Carmarthen, 1.259; Pembroke, 1.199; Glamorgan, 1.127
Wheelwright	1.282	1.112	Radnor, 3.298; Montgomery, 3.200
Saddle, whip, and harness, M	1.155	0.274	Montgomery, 1.541
Saddle, whip, and harness, F	1.252	2.009	Radnor, 6.685; Monmouth, 3.548

Note: Calculated as in Table 18.2 above.

all but the special case of the female saddlery workers a mean close to one, and comparatively small standard deviations. In this they appeared in general to confirm Day's findings for 1841 that smiths and wheelwrights were crafts distributed 'locally', and saddlers and their related occupations 'provincially'. In mid-century, then, the degree of dependence upon these crafts for the operation of horse-drawn society was fairly even, and their distribution shared fairly evenly between town and country, and county and county.[43]

Perhaps surprisingly, in the context of these data, and the remarkable evenness of their distribution, the true village or market-town smith can only have formed a small proportion of this total, and one that declined

---

[43] Day, 'Distribution of industrial occupations', pp. 234–5. The significance of Day's work is discussed more fully above, Chapter 18, pp. 1102–3, and in the introductory section of the present chapter, pp. 1151–2.

to the end of our period. Smith businesses appeared at the lowest population thresholds in mid-century Yorkshire and Norfolk, defined fairly crudely as the midpoint between the mean of villages with and without the trade, at 475 in the 1830s in Norfolk, at around 410 in the early 1850s in the Skipton district of the West Riding, while in 1879 they were to be found in settlements having around 350 people in the North Riding.[44] The 1870s may have seen the peak in the distribution of the rural smithy, for while the farm horse and other horse population of England and Wales did not peak until around the end of the century, other factors were already eroding their market. Factory building of implements cut into the trade of some, others failed to adapt to the new world of partial mechanisation on the farm and, after 1900, the increasing use of motor vehicles was a new source of rapid decline in rural shoeing demands.[45]

Evidence available from the 1901 Census and the 1907 Census of Production that allows the estimation of their numbers therefore comes after the peak, when decline had already set in, but it is helpful in defining the nature of the business, its scale of operation and the work undertaken. Only 10,490 of the smiths in the 1901 Census were recorded as working on their 'own account', and only 11,723 as working 'at home'; if we make the reasonable assumption that each would normally have at least one employed man, then perhaps 20,000–30,000, or around one in five or six of all 'smiths', worked in the true smithy.[46] This rough and ready estimate is supported by the evidence of the 1907 Census of Production, which confined its attention to power-using factories and workshops, where around 40,000 were employed in smithy and edge-tool work, principally the making of fences, gates, builder's ironmongery and jobbing work, and rather less than a third of this, by value, was 'agricultural' in nature.[47] By implication, the residue, at least half of recorded employment, worked as blacksmiths in transport, mines, building and other industrial or urban plant, a point suggested by the Location Quotient clusters of 1861.

The evidence of surviving accounts and direct business records of smithies indicates that, even at the beginning of our period, the work of the rural smith consisted primarily in repairs, maintenance, and in the creation of replacement parts, such as bolts, for equipment. While some

[44] Skipton estimated from White, Directory, 1853; North Riding from Chartres and Turnbull, 'Country craftsmen', p. 321.

[45] FitzRandolph, Rural Industries, I, pp. 183–4; Jones, Rural Industries . . . Wales, pp. 108–9.

[46] Ibid., and others make it clear that a smith and at least one assistant were required to work the bellows of the forge before the advent of portable machinery, largely after the War, and this fixed labour input represented a problematic overhead in the declining market of the last years of our period. Data on employment structure taken from 1901 Census, Table xxxv.

[47] Census of Production, 1907, BPP, 1910, CIX, pp. 146–53, 209–211.

blacksmiths, such as John Williams of Cellan (Cardiganshire) or the forgers at Aberaeron, were still making tools,[48] and even ovens in the 1850s and 1860s, the majority of smiths were already concerned predominantly with shoeing and repair work: such was the case with Surridge of Chipstable (Somerset), 1854–8, Webb of Linton (Cambs.), 1839–46, and the Banks family at Westonbirt Forge (Glos.), 1846–70, where shoeing, sharpening, retining, and repairs provided almost the entire work. The detailed inventory of Joseph Peace, blacksmith and shoeing smith, taken in 1864, recorded stocks of 'sundry iron, including old shoes', and seventeen new shoes at one of the two forges, and three dozen at the other,[49] of which much was paid for, as in Wales, with scrap or other payment in kind.[50] At this village level, the smith was deeply integrated into the fabric of rural society, where until late in the century cash transactions were minimised, and credit and payments in kind remained remarkably current.

Blacksmiths of course also made some tools, or made adaptations or repairs so extensive that they warranted what George Ewart Evans recalled as the ultimate in compliments – 'blacksmith made'.[51] Charles 'Clarky' Cooper of Ashdon, near Saffron Walden, around 1900, was described, perhaps fancifully, as such a miracle-worker:

There was scarcely an implement used in agriculture that Clarky did not have to make, repair or renew. Farm workers came with problems, for most of the labourers had to find their own tools. They asked for left-handed scythes, wider or narrower hoe blades to chop out weeds from wide or narrow drillings; men with big arms would demand longer scythe blades, and gamekeepers and poachers asked for long-bladed curved spades to dig out rabbits and ferrets. He sharpened billhooks, scythes, sickles, axes, and every type of edged tool, often replacing the broken-off split staves and stakes. He made pitchforks and four-tined forks, plough spuds, coulters, shears, drills and harrows; sometimes renewing the entire set of teeth on worn out harrows.[52]

Smiths like 'Clarky' also 'roughed' horseshoes on an emergency basis to cope with hard frosty weather and improve purchase, and Clifford Rose recalled this being done for 107 horses in a day by one smithy in Needham Market.[53] The farrier's shop was a warm meeting place for

[48] J. G. Jenkins, 'Rural industry in Cardiganshire', Ceredigion, 6, 1968, pp. 99–106.

[49] Bristol RO, Inventory and Valuation . . . of Mr Joseph Peace, of Maisey Hampton [Meyseyhampton] and Marston Meysey, 4 November 1864, D1070\II 1\F3.

[50] Reading, Rural History Centre, Account Book of S. Surridge, Blacksmith, Chipstable, Somerset, 1854–8, D62/32; Cambs RO, Ledger of John Webb of Linton, 1839–46, L95/17a/2; Gloucester RO, Banks Family of Westonbirt Forge, Customer Ledger, 1846–70, D775; on payment in kind, see also J. G. Jenkins, 'Rural industry in Brecknock', Brycheiniog, 14, 1970, pp. 9–10.

[51] George Ewart Evans, The Farm and the Village (London, 1969), p. 134.

[52] Ketteridge and Mays, Bunkum, pp. 41–2.

[53] Ibid., p. 43; George Ewart Evans, The Horse in the Furrow (London, 1967), p. 197.

cronies, which made the dual employment of smithy and beershop a natural partnership, if one often perceived by the upper orders of society as fundamentally dangerous.

Some such forges were effectively village 'foundries'. Alfred Williams, exceptionally knowledgeable about forge work in his Wiltshire village from his own experience in the Great Western workshops in Swindon, noted that around 1912 South Marston had such a foundry, its yard cluttered with machinery for scrap, repair or storage, and that it was a survivor of what had recently been a more common village institution, effectively the village engineering workshop and machinery contractors. The foreman reckoned that virtually every casting in the district returned once in seven years for renewal or after smelting for recycling to smiths.[54] With their trip hammers and resources, such foundries were more easily able to provide 'fire-welding' repairs, which were vital to machine repairs before the wider diffusion of the oxy-acetylene process largely after 1920.[55]

Some foundries, as with the larger urban plants, and iron stockholders, also came to supply ready-made shoes for smithies to adapt: in the decade before 1914, George Depledge of Leeds held large stocks of shoes for such sales.[56] Factory-made shoes were to become more important in inter-war farriery, but the extent of their penetration into even urban shoeing before 1914 seems to have been small, and the Census of Production of 1907 recorded only 4 per cent of total factory output under this specific head, while acknowledging that most shoeing was subsumed under 'jobbing' work, allocated by the same source a 64 per cent share of the value of annual production.[57] However, there was evidence of consumer resistance to the farrier's craft skills and prices – the full cost of a smith-made shoeing, including iron, nails, fuel and labour, was put at 12s. before 1914 – and smiths appear to have been tempted to cross-subsidise this work from other jobs, shifting but not disarming the criticism. Farmer organisations attributed these prices to restrictive practices, alleging that the prices stemmed from restrictive rings that linked members of the Master Farriers' Association with ironmonger suppliers.[58] It is impossible to disentangle these from generic complaints about costs in the context of change in rural society on either side of the war, when the relatively inflexible systems of extended credit on which rural England and Wales had long operated were being rapidly eroded, and

[54] Williams, *Wiltshire Village*, pp. 153–5.
[55] Jenkins, *Country Craftsmen*, p. 124; Woods, *Rural Crafts*, pp. 32–6.
[56] Depledge ledgers, and other information kindly supplied by Mr H. R. Sutton.
[57] Census of Production, 1907, BPP, 1910, CIX, p. 150.
[58] FitzRandolph, *Rural Industries*, I, pp. 184–5. The MFA had no effective penetration into Wales, and Jones, *Rural Industries . . . Wales*, makes no mention of such complaints.

fully charged in supplier pricing. Certainly the complaints were indicative of a strong predisposition to dispense with dependence upon the farrier smith, and a source of the competitive pressures that were already eroding numbers of rural and village smithies long before 1914.[59]

The wheelwright was in every sense the close companion trade of the farrier and smith. As was clear from the Location Quotients of Table 19.4, wheelwright businesses were very widely and evenly distributed in England, though less so in Wales, and they appeared at broadly equivalent population thresholds. Some, like Mrs Lane in Flora Thompson's 'Candleford Green', united the two trades, and the tyring platform was an essential feature of the wheelwright's shop.[60] The platform, 'shoeing hole', and smith's shop had been allocated its spot in the Sturt wheelwright yards at Farnham as early as 1706, and remained there to the 1880s.[61] In Norfolk in 1836 and the North Riding in 1879, wheelwrights were to be found at population thresholds almost identical with those of the blacksmith, 500 and 350 respectively.[62] In 1853, in the Skipton District of the West Riding, wheelwrights appeared at thresholds of 460 compared with the 410 of the smith.[63] The data for Wales indicated a significantly more 'lumpy' distribution, in part the outcome of terminology, and rather greater functional blurring between 'carpenters' and 'wheelwrights' than characterised England. Pembroke and Cardigan, for example, had significant clusters of carpenters, but were well below average in their numbers of wheelwrights in 1861, and the apparent clustering of the latter within Radnor may be largely a function of the small scale of the civil occupied population of the county; the cluster of Montgomery cannot be so explained and both counties may have been rather more intensive vehicle users than the western Welsh counties.[64]

All commentators have agreed with J. Geraint Jenkins that 'the craft of wheelwrighting may be said to be one of the most complicated of all crafts, for although in many respects it is akin to hardwood joinery it differs from it in that the wheelwright relies on tightness of joints alone to hold the work together, and unlike the joiner he never uses glue'.[65] In

---

[59] The decline in village forges from the late nineteenth century was noted by both FitzRandolph and Hay, and Jones, and evinced in several of the local sources used above: Williams's *Wiltshire Village,* for example, noted the decline of village forges and small foundries, even though there remained seven or eight within a twelve-mile radius of his village, pp. 68, 153–5.

[60] Flora Thompson, *Lark Rise to Candleford* (Oxford, 1945), pp. 407–8.

[61] Sturt, *Wheelwright's Shop,* pp. 10–11.   [62] Chartres and Turnbull, 'Country craftsmen', p. 321.

[63] White, *Directory,* 1853. For clarity, thresholds here are defined as the midpoint between the mean distribution of crafts in villages/small towns with representatives of the trade and the mean of those without.

[64] See also Table 19.2, p. 1154 above. The comments on wagon and other wheeled vehicle usage are based upon J. G. Jenkins, *Agricultural Transport in Wales* (Cardiff, 1962), pp. 42–3, 81.

[65] *Ibid.,* p. 81.

this, he appears consistently as the epitome of what has been termed above 'craftism', and it is clear that the outstanding contributions of George Sturt's *The Wheelwright's Shop* to the epistemology of artisan skills helped foster such elegiac views. The wheelwright, as Sturt himself explained, was subjected to significant changes during our period, which may be summarised principally, with Sturt, as 'the new iron age'.[66] The long transition in the craft and its products through the century from 1780 was symbolised by the changing nature and balance of use of ironwork in the building of vehicles, its impact upon the highest skill content described above, and the consequential opening of the market to urban factory-built products, and to imported raw materials.

Partly because of the taxonomy adopted by the Census, the evidence from towards the end of the period indicates clear structural differences between the two harmonious crafts of smith and wheelwright. The 1901 Census recorded roughly double the numbers of wheelwrights as working at home (15.9 per cent), as employers (10.0 per cent), and working on their own account (13.1 per cent), and correspondingly indicated a much smaller average unit of employment, after the deduction of the self-employed, at 6.3.[67] The near-contemporary evidence of the 1907 Census of Production covered a somewhat different field, and was limited to the power-using factories and workshops for all 'carriage, cart, and wagon trades', which included motor car bodies. It therefore recorded a higher figure for employment in this part of the wheelwrighting business than the Census – 31,103 males against the 28,844 of the 1901 Census – but does provide some indication of the extent to which the factory building which Sturt and others bemoaned had grabbed the market. Almost half the value of the output of these factories and workshops was described as repair and jobbing wheelwright work (45.5 per cent for England and Wales by value), with some horse-shoeing (a mere 0.8 per cent); and while, to protect confidentiality, the data were only presented for the UK as a whole, a further 8 per cent by value was attributable to wheels made for carriages, carts, motor cars and perambulators, and poles, shafts and other wooden parts for vehicles.[68] The elegy provided

[66] Sturt, *Wheelwright's Shop*, Chapter xxx, pp. 153–8 perfectly demonstrates this use of the craft as the tolling bell for a declining rural society. This and related issues are discussed in Edward Thompson's short introduction to the Canto edition, pp. xi–xiii.

[67] Census, 1901, Table xxxv. Of total employment of 28,844, 4593 were working at home; 2893 were employers; 22,005 employees; and 3767 working on their own account. In the figures above, the ratio of employees has been expressed after subtracting the last from the employees.

[68] Census of Production, 1907, BPP, 1910, CIX, pp. 708–11, 729–31. The terms under which the Census was conducted, set by 6 Edward VII, and which led to the destruction of its original returns in the 1950s, precluded the tabulation of data where disaggregation would reveal an individual firm's performance. The implication of these figures is that a single firm in each of Scotland and Ireland was engaged in this segment of the industry, and probably on a small scale.

by Sturt, and given emphasis in the Oxford surveys of the early 1920s, was on this evidence fully warranted, and decline well established in the generation before 1914.[69] Wheelwright work was thus organised into two sectors: the large urban factory or workshop, gaining an increasing share of the market; and the traditional shop of the larger village or market town, which had been the dominant mode of production in the 1850s.

The village wheelwright traditionally employed a very particular range of woods for his craft, many of which required long seasoning, and thus tied up capital for many years.[70] The wheelwright co-ordinated the purchase of standing timber of various kinds, requiring deep local knowledge to select oak or ash that was of the right growth, directed its felling and carriage to his yard, where he would superintend the work of sawyers making planks, other wheelwrights cutting ash for felloes, splitting summer-cut oaks for spokes, and cutting elm heartwood into the huge blocks for use on the fifteen-inch naves (hubs) of wheels after up to ten years of seasoning. All had to be stacked and stored for seasoning, watched and appropriately rotated, and the nave-blocks above all carefully inventoried for use in this long rotation. The older type of wheel, straked tyred, used these large naves, and was a heavy user of timber; the more modern, hoop-tyred, wheels could employ smaller twelve-inch naves, saving perhaps half the costs of seasoning, and make corresponding economies in other materials while demanding a higher level of co-ordinated skills of wheelwright and blacksmith. Involving large labour inputs of high levels of skill, using a complex range of materials that represented a heavy overhead burden on the business, and depending upon craft skills for the ultimate strength and long durability of products, wheelwrighting was thus the epitome of traditional rural industry.

It was correspondingly vulnerable to technical change that could economise on these inputs, and to changing fashion or vehicle types. Both groups of change affected the craft increasingly after 1850, and came to have particularly dramatic impact in the last twenty years or so before the war. It has been customary to regard the supersession of the straked tyred wheel by the hooped during the middle quarters of the nineteenth century as the outcome of simple technical progress. This is an incomplete analysis of the change, given the endurance of the old type in some parts of the country, notably in Shropshire, Somerset and parts of Wales, where it had advantages in terms of lateral adhesion on hills, in durability on rough ground, and in the lower levels of smithy skills which straking

[69] Sturt, *Wheelwright's Shop*, passim; FitzRandolph, *Rural Industries*, 1, pp. 173–82; Jones, *Rural Industries . . . Wales*, pp. 104–8.

[70] This paragraph is based largely upon Sturt, *Wheelwright's Shop*, esp pp. 23–82; Jenkins, *Country Craftsmen*, pp. 100–13, and Woods, *Rural Crafts*, pp. 44–56.

required, opening up the possibility of at least some repairs at farm level. In Somerset, hybrid tyring combining strakes and hoops was found. Its survival into Sturt's days and beyond was a complex outcome of local needs, available ironworking skills, and perhaps also the extent of price competition, given the greater material content of the older form of wheel.[71]

Changing vehicle types and, from the 1880s, the renewed improvement of road surfaces, first accelerated the shift towards the hooped wheel, and subsequently towards still lighter factory-made 'steam wheels', US imports and those made with imported Canadian timbers. Shifting patterns of transport demand forced changes in vehicle types, notably with the completion of the railway branch line networks after 1870, which shifted rural transport towards the local station, and induced further change in vehicle types. Above all, ironwork grew in use in providing, as in the hooped or factory wheel, an alternative route to integral strength, with lower inputs of seasoned hardwoods, and perhaps less refined artisan skills, at the same time as increasing robustness against bumping into trucks at the railway sidings. Other shifts in the patterns of rural transport, such as the development of the travelling baker's van, based on the market town, and the town brewer's dray, tended to reinforce the change, and increasingly to favour the lighter and sprung vehicle types against the products of the traditional wheelwright, very much the pattern of production reflected in the 'market shares' revealed by the 1907 Census of Production.[72]

By 1914, the country wheelwright was under severe pressure, and after the War various authorities discussed whether there was any route to survival, and in so doing provided helpful evidence about the comparative economics of the pre-war factory and village businesses. Sturt certainly considered that pricing in the village business as he knew it was traditional and haphazard in the 1880s, and that the true costs of overheads, materials and stockholding were not met: with an element of hyperbole, he noted that his books showed remarkably little difference in price for wheelwright work in 1884 from those noted by Arthur Young for 1767.[73] Yet, contrary to Sturt's elegiac views of the trade, Woods and the Oxford surveyors of the early 1920s considered it rescuable, given the higher quality and durability of the products of the traditional wheelwright.[74] An assessment by the Rural Industries Intelligence Bureau in 1924 made it clear that this was not the case: allowing for overheads (including materials) and labour costs on an equal basis, their analysis showed that

[71] Sources as above, esp. Sturt, *Wheelwright's Shop*, pp. 123–8, and Jenkins, *Country Craftsmen*, pp. 110–11.   [72] Based largely upon Sturt, *Wheelwright's Shop*, pp. 153–71.

[73] *Ibid.*, pp. 199–201, 211–12; FitzRandolph, *Rural Industries*, I, pp. 173–6.

[74] Woods, *Oxford*, pp. 44–5.

despite cheaper labour costs, the country shop had no means to compete with the factory on the basis of price.[75] Applying the same approach to the data provided for factory making in 1907, the implicit value added to labour and material costs was in excess of 40 per cent, and provided a margin for competition and profit with which the country wheelwright could never contend.[76] Even before motor transport became generalised, the market had changed irrevocably against the traditional craft, and successful strategies for survival depended upon change and diversification, often back into dual employment, marrying the vehicle maintenance business with the pub or small farm.[77]

The final group of crafts, saddlers, and makers of harness and whips, had in many senses foreshadowed the trend toward concentration, but also shared the generalised experience of decline, particularly in the last decades of the period. Day's analysis for 1841 had identified it as provincial in distribution, and the evidence of the Location Quotients for 1861 confirms the impression.[78] Saddlers were already becoming largely the repair shops and retailers of equipment by the third quarter of the nineteenth century, with the manufacture of new horse furniture highly concentrated into the west midlands, where it had established prominence before 1800. The figures of Table 19.4 suggest two aspects of their distribution: as with smiths and wheelwrights, saddlers were essential support crafts to all horse-users, and hence relatively even in their distribution through the provincial workforce; manufacturing equally had become clustered, seen in the high quotients for Staffordshire and Warwickshire, where Walsall and Birmingham were notable centres of manufacture.[79]

Evidence from mid-century confirms this twofold structure: saddlers occurred at the large village and market-town level, having populations well in excess of 500 in Norfolk (1836) and Yorkshire (1853 and 1879); manufacture was relatively concentrated, but largely into workshops of small scale.[80] Net of those working on their own account, saddlers in 1901 worked at an average 5.2 per employer, though female employment, overwhelmingly concentrated into the midland manufacturing region, was in substantially larger units (23), indicative of the more advanced division of labour in the larger firms that were developing in the later part of the century.[81] With the advent of mechanised cutting out, around the

---

[75] Rural Industries Intelligence Bureau, 'The Country Wheelwright and His Outlook', *Journ. of the Board of Agriculture*, 31, August 1924, pp. 466–74.

[76] Census of Production, 1907, BPP, 1910, CIX, p. 729, Table II, indicates a return in excess of 140 per cent over material costs in the trade.   [77] FitzRandolph, *Rural Industries*, I, pp. 178–9.

[78] Day, 'Distribution of industrial occupations', pp. 185–7.

[79] *Ibid.*; *VCH Staffordshire*, II, pp. 236–8; XVII, pp. 202–4.

[80] Chartres and Turnbull, 'Country craftsmen', p. 321.   [81] Census, 1901, Table XXXV.

1890s, this process was accelerated, and by 1914 Matthew Harvey & Co. Ltd of Walsall was employing 600, but the manufacturing industry remained generally characterised by small firms of fewer than 25.[82] Three-fifths of the output of the industry recorded in 1907 was for manufacture rather than repair, both of finished products, saddlery and harness (41.9 per cent), bits, spurs, and stirrups (4.3), and horse clothing (4.0), and whips (2.7), and of intermediate goods, such as saddletrees, hames and buckles (around 5 per cent).[83]

The village saddler was assessed in 1907 as doing mainly 'jobbing work for farmers', and appearing typically in two- or three-man units, very much on the scale of the smithy.[84] By 1900, their businesses were coming under pressure, and by the early 1920s they were in general retreat before both the factory-made products and the contraction of the market in the face of the internal combustion engine.[85] Levi Archer and his assistant Albert Bassett at Ashdon near Saffron Walden repaired and made harness, saddles, brasses, headstalls and nosebags, but also repaired rick cloths and the canvases of reapers around 1900, and most of their fellows appear to have worked on a similar basis, increasingly diversifying into small jobbing work which was the secret of survival.[86] Even businesses such as the Ragges and Lloyds of Leatherhead, with more than 250 years behind it, effectively died in 1905.[87] After 1920, the craft was generally unable to reproduce itself, and was increasingly practised by older men, working upon a range of sports and fancy goods, and orthopaedic work, as their traditional market dwindled.[88]

C EQUIPMENT CRAFTS

The remaining crafts can be dealt with more briefly, given their generally smaller scale, or declining relevance to the countryside. New materials, shifting patterns of supply towards imports, and technological change affected each to differing degrees. These issues were already clear in the data derived from the 1861 Census presented in Table 19.5.

Millwrights were very clearly closer in 1861 to the traditions of Rennie and the other engineering giants whose careers had been formed in the

---

[82] *VCH Staffs*, XVII, p. 204.

[83] Census of Production, 1907, BPP, 1910, CIX, p. 678. As in the above analysis of wheelwright work, some categories – whips, hames and saddletrees – are percentages of the value of UK output, because of the risk of disclosure of details of particular firms.

[84] *Ibid.*, p. 662.

[85] Jenkins, *Country Craftsmen*, pp. 201–14; Woods, *Rural Crafts*, pp. 82–9; FitzRandolph, *Rural Industries*, I, pp. 194–5.   [86] Ketteridge and Mays, *Bunkum*, pp. 45–6.

[87] F. B. Benger, 'The Ragge, Lloyd, and Walker families: Leatherhead saddlers and harness makers from the 17th to the 20th century', *Proceedings of the Leatherhead and District Local History Society*, II, 5, 1961, pp. 144–54.   [88] Woods, *Rural Crafts*, pp. 82, 86.

Table 19.5. *Location quotients, England and Wales, equipment crafts, 1861*

Occupation	Mean L.Q.	S.D.	County concentrations (in rank order)
England:			
Millwright	0.750	0.427	Northd, 2.198; Lancs, 1.880; Yorks, WR, 1.776; Cumb, 1.405; Yorks, ER, 1.247.
Cooper	0.949	0.415	Yorks, ER, 1.896; Glos, 1.845; Heref, 1.829; Westm, 1.740; Northd, 1.545; Staffs, 1.387
Fence/hurdle	1.506	1.706	Oxon, 6.413; Dorset, 4.649; Heref, 4.623; Beds, 4.110; Wilts, 4.070; Bucks, 4.022; Hants, 3.436; Suff, 3.288; Herts, 3.267
Sawyer	1.131	0.429	Bucks, 2.121; Heref, 1.899; Hants, 1.874; Beds, 1.672; Salop, 1.668; Oxon, 1.646; Wilts, 1.623
Wales:			
Millwright	1.035	0.400	Flint, 2.066
Cooper	1.077	0.520	Cardigan, 1.909
Fence/hurdle	0.901	1.335	Radnor, 3.891; Monmouth, 3.322
Sawyer	1.074	0.479	Radnor, 2.210; Flint, 1.637

Source: Calculated as in Table 18.2 above.

craft, and in both England and Wales were tending to concentrate into the industrial districts, though in the figures of Table 19.5 they remained widely and moderately evenly distributed in general. The rapid technical changes in the organisation and structure of grain milling after 1870, considered above, led to the subsequent decline of the traditional rural millwright, serving water- and windmills.[89] By 1901, the Census recorded a craft consisting overwhelmingly of employees (92.2 per cent of total), and with very few millwrights either working 'at home' (114, 2.2 per cent) or on their own account (216, 4.1 per cent), and their average distribution to employers in the same trade, at 27.7, indicated substantially greater units than typified most of the other crafts and call-

[89] See above, Chapter 17, pp. 1062–5; Jennifer Tann and R. Glyn Jones, 'Technology and transformation: the diffusion of the roller mill in the British flour milling industry, 1870–1907', *Technology and Culture*, 37, 1, January 1996, pp. 36–69.

ings considered here. This confirms the impression that these millwrights were primarily industrial maintenance engineers.

Walter Rose's account of the calling from the opposite perspective confirms this view. Rose's 1937 account was explicitly stated to record 'reminiscences . . . of our old carpentry business prior to my grandfather's death in the year 1893' and is not, therefore, a purely historical source, but his account does indicate that his grandfather and above all his father were servicing the last millwrighting works of the old kind.[90] To Rose this was almost a matter of relief in terms of windmills, where, while regretting the passing of the old post and tower mills which the family had serviced, he was happy to avoid the dangers involved in the setting and replacement of sails. The firm faced a more insistent demand for the replacement of the cogs cut from a variety of native woods to interact with iron gears in various parts of the mill mechanisms: the large spur wheel, for example, was located immediately under the second floor of a watermill, and drove the stones above, and this typically had 140 cogs.[91] Stocks of hardened timber were maintained for their making, since there was a universal preference for wood to engage iron in gearing, holding grease better and provided a smoother mechanism, but there were preferred timbers for each stage: for the pit wheel, where conditions were damp, oak was preferred; for the spur wheel, beech, holly or hornbeam; and for the hard-worked great crown wheel close-grained apple or pear was favoured. Rose considered these skills and demands as disappearing in the last years of the century of which he was writing, and by the 1920s and 1930s, many of these traditional millwrighting skills were already rare, and represented a barrier to the survival of remaining traditional mills. The dressing of the millstones had already become primarily the task of the miller.[92]

By the end of our period the cooper was regarded as an almost wholly urban creature, and had long been largely so, and surviving village coopers were seen typically as old men, confined to repair work if fit for anything.[93] This represented a major change from 1850, when coopered products were widely employed within agriculture, more native timber was employed in the craft, and major users were more widely diffused. Some indication of the generalised importance of the cooper in the earlier years of the period may be taken from the fact that 1861 represented the high point of employment in the craft (Table 19.3), before alternative materials, principally galvanised iron and imported timbers, and machine coopering had had their full impact. This generalised

[90] Rose, *Village Carpenter*, pp. xvii–xxi explains the provenance of his book.   [91] *Ibid.*, pp. 104–20.
[92] Woods, *Rural Crafts*, pp. 57–77.
[93] See, for example, the comments of Jones, *Rural Industries . . . Wales*, pp. 52–3; FitzRandolph, *Rural Industries*, I, pp. 61–3; Woods, *Oxford*, p. 93.

importance was also reflected in the Location Quotients of Table 19.5, at least where England was concerned: coopers were relatively evenly distributed across the English counties, and displayed a modest degree of deviation from the mean, and in Wales showed no great extremes, while being a little less evenly spread. The principal clusters were also relatively modest in their degree of difference, being characterised by counties already distinct in other woodworking crafts, by the impact of packaging demands on the great river systems of the Trent and Ouse, and the Severn, and by the demands for the packaging of meat in the north, and pottery in Staffordshire.

Even before the full impact of change, the concentration of coopers in towns was already notable. In the East Riding of Yorkshire, in 1851, just under half the coopers (49.4 per cent) were employed in Hull alone, and 84 per cent in the top three clusters in the county, Hull, York and Sculcoates or Driffield; this contrasted with the much less concentrated patterns for carpenters, 39.5 per cent in the three leading locations; smiths, 26.6; wheelwrights, 28.7; and saddlers, 26.6.[94] In our leading county concentration, then, even at the beginning of the period, the demands for packaging in the major port, where supplies of staves were also most readily available, generated a predominantly urban craft. The decline of the village cooper's business, mapped through directories, emphasised the impact of change: in 1853, there were fifteen in the villages of Bedfordshire, and fourteen in 1861; numbers fell to nine in 1871, three in 1894, and to one in 1910.[95]

The evidence of the Census of 1901 and the first Census of Production (1907) provides the basis for the assessment of the differential impact of change upon the three primary sectors of the cooper's craft. By 1901, they were overwhelmingly employees (90.2 per cent of recorded occupation), very few working either on their 'own account' (6 per cent) or 'at home' (5.2 per cent), figures that clearly reinforce the decline of village businesses.[96] Although limited in its coverage to factories and workshops, the 1907 Census of Production collected additional data on a voluntary basis, securing returns from three-quarters of the trade, which indicated the distribution of work between the three sectors, and the unit values.

The introduction of machinery from the later 1850s combined with the growth of scale in brewing to concentrate the premium wet coopering work, and by 1889 the great Bass Brewery at Burton was employing 400 coopers, covering 25 acres with staves in seasoning, and consuming over 700 tons of hoop iron per annum, with the stock of

[94] Census, 1851, Part 2, II, pp. 702–19. I am indebted to Mr R. A. Davies for his assistance in estimating these figures.   [95] Kenneth Kilby, *The Cooper and his Trade* (London, 1971), p. 152.

[96] Census, 1901, Table XXXV.

Table 19.6. *UK factory and workshop output of coopering, 1907*

	Output by volume	By value £	% by value	Unit value £
Wet or tight coopering:				
Machine-made casks for breweries and distilleries	233,000	209,000	33.2	0.90
Hand-made casks for breweries and distilleries	81,000	49,000	7.8	0.60
Herring pickle barrels	1,113,000	169,000	26.8	0.15
Other casks	288,000	74,000	11.7	0.26
Sub-total	1,715,000	501,000	79.5	0.29
Dry coopering:				
Machine-made casks	188,000	19,000	3.0	0.10
Hand-made casks	596,000	94,000	14.9	0.16
Sub-total	784,000	113,000	17.9	0.14
White coopering:				
Churns, tubs, buckets, etc.	78,000	16,000	2.5	0.21
TOTAL	2,577,000	630,000	100.0	0.24

Source: Census of Production, 1907, BPP, 1910, CIX, p. 715.

casks exceeding 518,000.[97] This, of course, was exceptional, and with capital tied up in casks often exceeding the value of the rest of the brewery, rapid turnover of casks and much smaller teams of coopers, thirty to fifty in London, were more typical of the trade.[98] Skilled men preferred brewery work: complaining of labour shortages in 1919, a Banbury cooper noted the severe losses of manpower from the War, and claimed to be able to employ forty men and six apprentices, were he able to compete with the brewers.[99] Dry coopering was hit badly by the decline of salting meat for shipment with the development of chilling and later refrigeration, from the later 1870s, and the white coopery, serving dairying and some hunting or racing stables, faced direct

---

[97] Kilby, *Cooper*, p. 163.

[98] T. R. Gourvish and R. G. Wilson, *The British Brewing Industry, 1830–1980* (Cambridge, 1994), pp. 93, 202, citing also the earlier *A Glass of Pale Ale and A Visit to Burton* (1880), on Bass's 'Babylon'. Bass, distributing to a national market via the rail network, has thus to carry much more substantial overheads of casks than a brewer distributing beers of shorter life to a concentrated market.

[99] Woods, *Oxford*, p. 93. Gourvish and Wilson, *Brewing Industry*, pp. 199–200, confirm the high level of wages and perquisites of the brewery coopering in the early 1900s, which was combined with long hours and erratic patterns of working.

competition from the new materials, enamelled and galvanised iron and tinwares.[100] In the face of this, and left with the work of lower value, the village cooper could not compete, and came to concentrate upon the repairs that could not be conducted economically in the large work-shops, or spread into dealing or other trades: Thomas Foster, of George Square, London, was also a seedsman; Thomas Wingrove a furniture broker; and Wratten of Hailsham (Sussex) claimed to make or sell baskets, yokes, rakes, shovels, churns, dishes and spoons, trenchers, tubs and casks, butter prints and milking pails.[101] Though hand coopering survived, it did so tenuously, and as a rural craft was effectively moribund by 1914.[102]

Just as the coopery of the country districts suffered through the advent of new and cheaper materials, so too did the makers of traditional fencing, who were the primary industrial victims of barbed and other wire, and of the changing role of sheep in farming. The evidence of the 1861 Location Quotients makes clear the highly localised nature of the making and use of hurdles and fences made of split ash or willow, oak and hazel wattle. Counties where sheep were folded predominated, and hurdles were primarily employed for folding or shelter in the downland farming districts: in Wales, where sheep were more hardy, their use was limited.[103] Fencing, in general, as Sturt and Rose made clear, was part of the 'hedge-carpentry' conducted on the farm by labourers, sometimes using posts and rails bought in from sawyers, and later in the century par-tially displaced by machine-made products from the urban workshops.[104] Estate woodyards had integrated production of a wide range of timber products, and were often equipped with extensive machinery, including steam-saws. After the 1890s, machine-made wired, split chestnut fencing came increasingly to be used where temporary barriers were required. The rise of these new materials, the declining use of the sheepfold as an integral feature of lowland farming, and the growth of timber imports which favoured urban supply of fencing timbers left the hurdle and fence makers as largely a relict population by 1914.[105]

As indicated in the earlier discussions of carpentry and wheelwright-ing, the sawpit and the pair of hand sawyers were also features of the traditional craft in 1850 that experienced rapid decline thereafter. The Location Quotients of 1861 in Table 19.5 indicated their concentration

[100] FitzRandolph, *Rural Industries*, I, pp. 61–2.   [101] Kilby, *Cooper*, pp. 152–3.

[102] J. G. Jenkins, 'The cooper's craft', *Gwerin*, I, 4, 1957, provides the most effective short survey of the processes involved in the craft.

[103] FitzRandolph, *Rural Industries*, I, pp. 108–14; Jones, *Rural Industries . . . Wales*, pp. 60–2; Woods, *Oxford*, pp. 97–101; Woods, *Rural Crafts*, pp. 117–22.

[104] Sturt, *Wheelwright's Shop*, p. 29; Rose, *Village Carpenter*, pp. 61–76.

[105] The 1901 Census, Table XXXV, recorded only 272 fence and hurdle makers working on their own account (12.2 per cent of total employment) and 209 (9.4 per cent) working from home.

as an occupation into the traditional counties of wood processing, where native hardwoods were applied to construction, vehicle-building and a wide range of other uses. This heavy work was highly skilled, and involved remarkable techniques for manipulating large timbers over sawpits without the use of power: rollers, levers and the ring-dog were used to handle the logs, and cutting lines were marked with chalk. For both the team, and especially for the bottom sawyer in the pit, the work was very demanding, often little more than in Rose's description, 'a monotonous slog', and the fondness of sawyers for beer was consequentially almost proverbial. Sturt's views on the trade were perhaps exaggerated, but indicative of the problems involved in co-ordinating a pair of workers who might not synchronise their taking of beer:

One sawyer was no good without his mate – he was as useless as one scissor would be. So, on a Monday morning, the one who reached his work first would loaf about waiting for the other, and then, sick of waiting, drift off to a public house – his home perhaps for a few days or weeks. His mate, coming at last, would presently find that his predecessor had begun boosing [sic]; and was likely enough to end a disgusted and wasted day by following suit. He might be, himself, in the thick of a great drink by the time that the first man was ready. And so it would go on. I have known sawyers unable to get together and start their week's work until Thursday morning.[106]

Sturt's heightened view of the decline of the craft in the face of the sawmill, informed Walter Rose's comment that he had 'never met a sawyer who has expressed regret at the passing of the work over to the machine'.[107] Mechanisation was compounded with the impact of the growing use of softwood timbers imported from Canada or Scandinavia, to induce a general shift towards large sawmills, often integrated with machine joinery and manufacture. 'Sawyers' in the 1901 Census were overwhelmingly employees (97.7 per cent), virtually none working either 'at home' (0.6 per cent) or on their own account (0.9 per cent), and the average number of employees per employer exceeded 120.[108] By 1907, net imports of timber, including manufactured goods, exceeded £27 million, when total output of timber in Great Britain was a mere £800,000, and further emphasised the forces that induced the terminal decline of the traditional sawyer's craft.[109]

D TRANSPORT AND DISTRIBUTION

Three trades provided critical transport and linking services between the consumer and the provider, sometimes the producer: the hawker;

---

[106] Sturt, *Wheelwright's Shop*, p. 39.   [107] Rose, *Village Carpenter*, p. 32.
[108] Census, 1901, Table xxxv.
[109] Census of Production, 1907, BPP, 1910, CIX, pp. 692–701, 719–28.

Table 19.7. *Employment in transport and distribution, 1841–1911*
(England and Wales, 000s)

Occupation	1841	1851	1861	1871	1881	1891	1901	1911
Hawker, M	11.2	16.5	27.3	30.6	29.4	42.4	46.9	54.0
Hawker, F	3.7	9.2	13.3	19.2	17.7	16.6	14.4	15.4
Drover	4.9	7.5	9.3	8.8	8.4	8.1	8.2	7.8
Carrier, M	26.0	43.7	67.7	74.5	124.6	169.3	272.3	278.4
Carrier, F	0.5	0.6	0.6	0.7	0.7	1.0	0.7	0.3

Source: Compiled as Table 18.1 above.

the drover; and the carrier. Each can be considered only briefly here, but these trades in many ways replicated the experiences of some of the craftsmen and other traders. The processes of urban growth, of the development of the railway and the steam coaster, the growth of the shops of the market town and larger villages, and the renewed improvement of the roads after 1890, and with this the expanded use of the bicycle, all impacted upon their trades, and produced still greater change than is immediately revealed by the broad trends in employment described by the figures of Table 19.7. While each may have been expected to have been vulnerable to such changes, perhaps to the point of technological extinction in the case of the drovers, the broad patterns of experience were those of adjustment and change, combined with relative but not precipitate absolute decline.

Although hawkers recalled the rural traditions of earlier times, their growth, and the very significant record of female participation in the trade in our period were indicators of their primarily urban location. Metropolitan London's share of the trade, omitted from the measures of spatial clustering in Table 19.8, was broadly proportionate to its population, and this, combined with the mean Location Quotient close to unity, and the low standard deviation among male hawkers in both England and Wales, is indicative of their importance in the distribution of goods, much of it food, in the cities and towns. These Census data may be very misleading, since Mayhew regarded hawking as highly seasonal, its numbers rising in his view fifteen to twenty times above the level recorded in 1841, but this was almost certainly confined as a problem largely to the capital.[110]

Despite this, they were perceived as a problem, and urban worthies,

---

[110] Roger Scola, ed. W. A. Armstrong and Pauline Scola, *Feeding the Victorian City: The Food Supply of Manchester, 1770–1870* (Manchester, 1992), p. 249.

Table 19.8. *Location quotients, England and Wales, transport and distribution, 1861*

Occupation	Mean L.Q.	S.D.	County concentrations (in rank order)
England:			
Hawker, M	0.997	0.311	Notts, 1.771; Rutland, 1.676; Lincs, 1.593; Northd, 1.537; Leics, 1.478; Yorks, ER, 1.318
Hawker, F	0.769	0.459	Northd, 2.422; Lancs, 1.952; Durham, 1.752; Yorks, ER, 1.280
Drover	1.457	1.477	Hunts, 5.572; Middx, 4.288; Norf, 4.161; Bucks, 3.132
Carrier, M	0.835	0.452	Sussex, 2.269; Hants, 1.916; Middx, 1.890; Lancs, 1.672; Northd, 1.485
Carrier, F	1.391	1.199	Hunts, 6.742; Rutland, 4.045; Northants, 3.640; Oxon, 2.746
Wales:			
Hawker, M	0.977	0.324	Denbigh, 1.369; Brecon, 1.335
Hawker, F	0.981	0.804	Flint, 3.220; Denbigh, 1.853
Drover	1.243	0.982	Cardigan, 3.889
Carrier, M	1.166	0.743	Flint, 2.889; Anglesey, 2.237
Carrier, F	1.372	0.903	Merionedd, 2.918; Flint, 2.779

Source: Calculated as in Table 18.2 above.

including many shopkeepers, flocked to give evidence to the Royal Commission on Market Rights and Tolls in 1888 on the damage done to their trade and to the tone of their town by the hawker.[111] In the smaller towns, in general, their invasive tendencies were easier to control, and it was in the sale of meat, fruit and vegetables, fish and milk that these hawkers and costermongers contributed most. Providing a contrasting case study to Mayhew's London, Scola's analysis of Manchester, employing the invaluable evidence of 'Felix Folio', the pseudonym of John Page, Clerk of the Market, from 1858, supported this view, and argued that these 'hawkers' comprised a wide range of trades. Nor, in Manchester before 1871, was there evidence to indicate a rise in numbers even in line with those of shops or population in

---

[111] *RC on Market Rights and Tolls*, 1888, BPP, 1890–1, XXXVII–XXXIX, *passim*.

general, and thus the commonplace burgher complaints of unfair dealing were not substantiated.[112]

This adds to the evidence of Table 19.8 to suggest that Scola's argument for Manchester cannot be sustained for the generality of the larger towns. On a national basis, total hawkers ran well ahead of both general employment and that in general shopkeeping, which also outran the former, indicating that they were rather more important to food distribution than Scola suggested.[113] Within a fairly even relationship to the overall profile of employment in both English and Welsh counties, some clustering occurred among both male and female hawkers, notably in the east midlands, the East Riding, and the industrial districts of both the north of England and north Wales. The comparatively high profile of recorded female hawkers in the areas of heavy industry may be a reflection of the low level of alternative employment opportunities, and unsurprisingly, in counties in which female employment was comparatively high, such as the textile districts, the quotients were correspondingly low. Overall, though, these trades were fundamental to urban distribution, and remained so throughout our period.

If the hawkers were basic elements of retail food distribution, then the drover was an element of the wholesale livestock trade, and one that evoked marketing systems of the past, whose survival is perhaps the surprising feature of the figures in Table 19.7, which record the numbers of drovers together with those of more general livestock market operatives. Declining in relative terms, their absolute numbers held up, as did the numbers of drovers within the category: numbers peaked at 3,215 in 1861, but held at around 2,900 for the rest of the century, and were still 2,882 in 1911. Concentrations of the trade in Cardigan, Huntingdon, Buckinghamshire and Middlesex in 1861 reflected differing stages in the primary production of stores, grazing and the finishing trades nearer the capital, the principal market for livestock, and Norfolk, also a major grazing centre; Lincolnshire (2.671), Leicestershire (2.033) and Essex (2.106) were also relatively high, though lying within one standard deviation of the mean.

Sheep and cattle had long been driven from the producing regions of Wales, the north and Scotland to the midland counties, where stores were brought on before being driven once more towards the capital, being rested and fattened in the home counties before sale.[114] The coming of

---

[112] Scola, *Manchester*, pp. 246–53.

[113] Taking the uncertain basis of Census employment figures, and 1851 as 100, hawkers, both male and female, grew to 194 by 1871, when shopkeepers (M and F) reached 149, and total male civil workforce 124; by 1901 the figures were, respectively, 239, 136 and 172; and in 1911, 270, 228, and 197.

[114] P. G. Hughes, *Wales and the Drovers* (London, 1943); A. R. B. Haldane, *The Drove Roads of Scotland* (London, 1952); K. J. Bonser, *The Drovers*, (London, 1970).

the railway was the primary influence in changing the nature of the business after 1850, and its impact was to restructure the pattern of droving rather than extinguish it. Auction marts gradually shifted towards the railheads, and drovers rapidly incorporated the new means of transport into their activities: the accounts of David Johnathon, a Cardiganshire drover who also held grazing land at Spratton (Northants.), showed his herd of 300 beasts being driven in 1856 to meet the new railway at Shrewsbury, which took the cattle on to sales in Rugby, Northampton, and Turvey (Beds.), before the bulk went to Essex to sales in fairs at Harlow and Ingatestone, and markets at Brentwood and Chelmsford.[115] Initially shifting the focus of the droves from the traditional centres of Ludlow, Kington and Hereford to Shrewsbury, railway extension reached to the heart of Welsh producing regions in Pembroke, Aberystwyth and Holyhead during the 1860s, and extinguished the long-distance trades. By the late 1860s, fifty miles was regarded as the economic limit to droving.[116] Branch-line development from the mid-1860s induced further adjustments. With the opening of the Wensleydale Railway, linking Leyburn with Hawes (Yorks., NR) in 1876, the former's position as market centre for the lower dale and forwarding point grew appreciably: cattle handled at Leyburn station rose from 4,717 in 1871 to 10,699 in 1891, before falling back to 7,029 in 1911; and sheep numbers handled increased from 10,301 in 1871, to 21,509 in 1891, and 22,999 in 1911.[117] Drovers thereafter became the intermediaries between railhead, auction mart and fair, little diminished in numbers but employed in very different patterns of work.

By comparison with the other trades in Table 19.7, employment among 'carriers' grew very dramatically, far exceeding even that of the hawkers. The vast majority reflected the changing nature of trunk transport, the continuing growth of coastwise and other shipping, some inland navigation, and above all the development of the railways. Trunk waggon services, which had once provided maintenance contracts for the Sturt wheelwright business, faded rapidly with the coming of the railway, but both in town and country the new modes of transport all generated huge demands for cartage and delivery services.[118] While there was a degree of overlap, the country and village carriers were rather different from the bulk of these, and perhaps numbered 25,000–30,000 in the 1880s, estimated by Everitt at the rate of one per 800 inhabitants of

[115] R. J. Colyer, 'Welsh cattle drovers in the nineteenth century – 1', *NLW Journ.*, XVII, 1971–2, pp. 401–2.

[116] George Menzies, 'Report on the transit of stock', *Trans. Highland and Agricultural Society of Scotland*, 4th ser., II, 1868–9, p. 463.

[117] C. S. Hallas, 'The social and economic impact of a rural railway: the Wensleydale Line', *AHR*, 34, 1986, pp. 35–6.   [118] Sturt, *Wheelwright's Shop*, p. 8.

England, London excluded.[119] The 1880s may have represented the peak of the trade, but numbers held up well to 1914, and some towns, notably Guildford, bucked this trend and continued to grow, and the Census numbers of 1901 may help to identify the relative share of the country carrier in total employment.[120] Of the 272,300 males listed as 'carriers' in 1901, more than 90 per cent were employees, and fewer than 16,000 – less than 6 per cent – were working on their own account, although more than a quarter of the 660 women carriers were self-employed, and the bulk of the rest the proprietors of firms.[121] If each of those working on their own account employed a further half a person to conduct the service, then absolute numbers exactly compatible with Everitt's estimates for the mid-1880s emerge, and strongly suggest a pattern of change comparable with that of the drovers; absolute stagnation, but only relative decline.

Carriers served specific functions which were largely uncompetitive with the new trains, and improved transport and retail systems, to the extent that these fostered the Victorian market town and county town, and created a new market for the expansion of the country carrier. Everitt's analysis stressed four primary functions for such carriers: shopping agents for the rural population; parcels carrier to station or halt; public passenger conveyance; and carriage of country produce to the town.[122] At least to the 1880s, the trade grew as towns like Leicester, Melton Mowbray or Guildford became important places of resort for the populations of the rural districts, and the continuing and enticing spread of manufactured consumer goods raised the importance of shopping as an activity. The outcome was the growth in services, in stops en route, and the increase in frequency, and this perhaps reflected the ease with which the business was entered at village level.[123] Among the inhabitants of Headington Quarry (Oxford) at the end of the century, Samuel noted many men deriving 'a bit of a living' from carrying, and the business was itself fissiparous, new opportunities being met by perhaps the addition of a second horse and cart, and the creation of a job for a second member of the family. Tommy Webb started up as a

---

[119] Alan Everitt, 'Country carriers in the nineteenth century', *Journ. Transport History*, 2nd ser., 3, 3, 1976, pp. 179–202, with the particular comments on numbers being pp. 186, 188.

[120] *Ibid.*, pp. 188, 193–4, noting that many performing carrier functions may not have been returned as such in the Census, since they followed multiple occupations. See also A. Everitt's earlier detailed study of Leicestershire, 'Town and country in Victorian Leicestershire: the role of the village carrier', in A. Everitt (ed.), *Perspectives in English Urban History* (London, 1973), pp. 213–40. Professor Everitt's pioneering studies have greatly informed this and the following paragraphs.   [121] Census, 1901, Table xxxv.   [122] Everitt, 'Country carriers', pp. 181–2.

[123] *Ibid.*, pp. 184–92.

carrier at the age of eleven with a pony given by his uncle, but he indicated also that the village carrier's trade was no guarantee of fortune: when a pony died, there was a collection towards a replacement among the inhabitants of the Quarry.[124] Typical in more remote districts was Mrs Knight of Milton, a Dorset 'higgler' in the 1890s who started her travels in the night, collecting eggs, chickens and butter from outlying farms for carriage to market.[125]

Evidence on their patterns of development supports that of the Location Quotients of Table 19.8, at least for males, that in both England and Wales the country carrier was already pretty evenly distributed by 1861, with little distinctive clustering. There were few distinctive clusters, Sussex and Hampshire representing the more purely rural counties, and Flint, Denbigh, Lancashire and Northumberland the strongly mixed rural/industrial, and Middlesex the rural fringe of the capital, the place of intensive food production and extensive residences. In the clothing districts of the West Riding by 1853,[126] more than 500 carrying firms were operating from 86 locations, and serving 168 places, and little more than half these firms were concentrated into the leading seven towns: Leeds (24.2 per cent), Bradford (9.0), Huddersfield (4.7), Halifax (4.4), Wakefield (3.8), Skipton (3.4) and Settle (2.4); and for the West Riding as a whole, in the Census of 1851, one male in 122, on average, was in the carrying trade.[127] Everitt's analysis of individual towns documented patterns of activity and development on a wider basis, and details for the five for which more than one observation is available are reproduced in Table 19.9.

The analysis of directory evidence cannot be taken as a completely satisfactory statistical base, but this is clearly sufficient to demonstrate significantly better access from the country districts to these towns during our period, and, looking with a longer perspective at Leicester, a massive change from the earlier years of the century. Carrying goods and people, performing shopping duties and acting as adjuncts in the marketing process, the carriers were fundamental in opening up the communications of the countryside after 1850.

---

[124] Samuel, *Village Life*, pp. 229–30.

[125] R. Best, *Powerstock in Wessex* (Bournemouth, 1970), p. 25, citing the evidence of the vicar's wife, Mrs Sanctuary.

[126] White, *Directory*, 1853. Clothing Districts were traditionally defined as north of Woolley Edge.

[127] Regression only possible by the twenty-nine unions/districts, which represented an undesirable level of aggregation, however, this produced the following equation:

$$\text{Carriers} = -29.01 + 0.011(\text{Male Adult Workforce}) \qquad R^2 = 0.906$$

significant at the 99 per cent level.

Table 19.9. *Country carriers, five towns from Everitt's weekly figures*

Town	Year	Carriers	Places Served	Services	Calls
Leicester[a]	1815	116	105	152	188
Leicester	1884	204	220	461	1100
Devizes	1890	47	91	72	242
Devizes	1903	46	70	68	158
Guildford	1854	45	66	106	261
Guildford	1894	39	70	115	339
Guildford	1914	41	77	166	553
Hull	1882	149	166	256	465
Hull	1899	151	146	280	591
Preston	1853	72	60	151	[b]
Preston	1882	60	53	158	[b]
Preston	1901	39	39	98	[b]

Note:
[a] Excluding 'long-distance' carriers
[b] Preston directories did not permit the estimation of total calls
Source: Everitt, 'Country Carriers', p. 189. Unfortunately, no Welsh data were analysed in this form.

E THE RETAIL TRADES

Earlier comments have indicated the presence of the retail trades in the countryside, and these were already well established by 1850, contrary to the view that once placed their generalised diffusion into the years after 1850.[128] The second half of the nineteenth century saw the massive growth of the service sector, nationally, in the face of which the development and work of the shopkeeping trades of the villages appeared relatively modest. This growth was characterised by the proliferation of specialisms primarily within the urban environment, and this experience was shared by the market towns. Cooperative retail societies largely shared these characteristics, although where they encountered 'political' support, as among the Northamptonshire shoemakers, they appear to have been stronger.[129] These developments were part of the background to the great expansion of the village carriers discussed above, the attraction of the maturing retail trades drawing business from the rural consumer. In the context of this massive and largely urban growth, these trades cannot be discussed in detail here, but as elements of the connec-

[128] J. B. Jefferys, *Retail Trading in Britain, 1850–1950* (Cambridge, 1954), suggests this misleading impression.

[129] Martin Purvis, 'The development of co-operative retailing in England and Wales, 1851–1901: a geographical', *JHG*, 16, 1990, pp. 314–31.

Table 19.10. *Employment in the retail trades, 1841–1911*
(England and Wales, 000s)

Occupation	1841	1851	1861	1871	1881	1891	1901	1911
Shopkeeper, M	15.3	19.9	9.6	29.9	29.1	28.0	23.5	38.7
Shopkeeper, F	10.0	17.8	15.5	26.2	25.8	25.6	28.0	47.3
Draper/mercer, M	24.6	37.5	45.7	55.3	53.5	60.7	67.2	66.4
Draper/mercer, F	2.7	6.1	12.0	19.1	28.8	46.3	68.4	84.6
Butcher, M	44.7	60.6	65.6	72.7	78.2	93.6	105.2	123.3
Butcher, F	1.3	1.6	2.5	3.2	3.5	5.3	3.9	11.9
Cheesemonger, M[b]	2.5	3.6	3.9	4.3	4.0	4.7	16.5	18.4
Cheesemonger, F[b]	0.2	nd	0.3	0.4	0.4	0.4	4.4	6.1
Poulterer, M	1.2	2.0	2.2	2.7	2.9			
Fishmonger, M	4.4	7.1	9.5	12.7	15.5	25.6[a]	28.7[a]	37.0[a]
Poulterer, F	0.2	nd	0.6	0.7	0.7			
Fishmonger, F	0.7	2.0	2.2	2.2	2.4	4.1[a]	3.5[a]	7.7[a]
Baker/confectioner, M[c]	37.7	53.4	56.3	62.8	75.9	92.2	102.2	122.4
Baker/confectioner, F[c]	5.6	10.6	12.3	14.1	20.6	38.5	46.8	76.1
Greengrocer, M	5.6	8.7	14.4	19.4	22.8	30.5	40.7	51.8
Greengrocer, F	2.8	4.7	5.5	6.9	6.8	10.5	11.9	20.5
Grocer, M	34.3	58.3	74.0	88.6	103.4	135.6	151.2	166.0
Grocer, F	7.5	13.4	19.8	22.5	26.4	46.3	43.3	53.6
Tobacconist, M	1.6	1.6	3.6	5.0	5.8	6.5	10.2	?
Tobacconist, F	0.9	0.4	1.3	2.0	3.0	8.0	6.7	?
Ironmonger, M	6.0	7.7	10.3	15.9	15.2	20.1	26.1	29.4
Ironmonger, F	0.3	nd	0.6	1.5	0.9	1.4	2.1	3.7

Notes:
[a] Returned as one unit from 1891, and incorporated with the larger trade
[b] Last two observations incorporate 'buttermen and provision dealers'
[c] Subsumes the manufacturing bakers, and biscuit and cake makers
Source: Compiled as Table 18.1 above.

tions of rural England and Wales, as the rivals of the hawkers and coster-mongers, and as the lowest fixed element in the distribution and trade of foodstuffs, analysed in Chapter 16, their growth and importance must be assessed. In the present section, therefore, the broad trends in employment in the principal retail trades are described, together with the evidence of their spatial distribution, as indicated by the Location Quotients. A final retail trade, perhaps that most widely dispersed and most immediate to the inhabitants of the countryside, the public house, is examined separately in the succeeding section.

The magnitude of the growth in these retail trades is clearly shown in the descriptive statistics of Table 19.10, which provides broad employment data derived from the Census, including in most cases the data on female as well as male employment. While recognising that the Census, especially from 1881, represents an unreliable guide to the employment of women, female activity rates were particularly high in the retail trades. In the context of particularly rapid general growth in the shop trades from 1891, the share of total employment taken by women underwent a pronounced increase. The relative stagnation of the general category of 'shopkeeper' up to the final observation provides further evidence that this growth was reflected in the proliferation of specialists.[130] The magnitude of growth among many of the food trades compares rather more favourably with that of the hawkers discussed above, and points to the roles of both shop and barrow in food distribution.[131]

Given the definitional problems involved, the analysis of the Location Quotients for 1861, to provide evidence on the distribution of these retail trades by counties, generally indicates remarkable evenness in the principal retail trades. Several of these categories have been omitted from the summary results of this analysis in Table 19.12 for Wales, because numbers involved were too small to provide a basis for meaningful assessment: female dealers in cheese, poultry, tobacco and iron were particularly subject to this problem. Their relatively small numbers in non-metropolitan England also counsels caution in employing these categories: almost half of the female tobacconists listed for 1861 were located in London, making the base for the analysis presented in Table 19.11 very much smaller. That being acknowledged, the nearness of so many mean scores to unity, combined with standard deviations that indicate a smaller variance than for most of the crafts, industries and trades already analysed, suggests a uniform strength in the development of retailing.

These quotients were particularly striking for the draper/mercers, butchers and grocers among males in both England and Wales, and indicated a remarkable consistency and evenness of distribution. All three were trades that had appeared at modest population thresholds in Norfolk in the 1830s, and Yorkshire in the later 1870s: in the former, grocers appeared at population thresholds under 500, a little behind the generic shopkeeper; and in the North Riding in 1879, shopkeepers were found at thresholds of 309, butchers at 372, and grocers at 489. In the latter, shopkeepers, grocers, butchers, and drapers, with carriers and victuallers, were to be found in each market town, and most of the larger villages, those with parochial populations in excess of 500, but were much more

---

[130] Some of the figures for shopkeepers, notably that for 1861, wholly confirm Armstrong's judgment on the 'erratic' nature of some observations when viewed in series, 'Information about occupation', p. 254. Unfortunately neither he, nor Charles Booth, 'Occupations of the people', provides a clear means to reconcile this figure for male shopkeepers with those of the printed Census report, a total of 4556 of all ages, and it clearly represents a figure that is the outcome of some reallocation from still more general categories.   [131] See above, pp. 1180–2.

Table 19.11. *Location quotients, England, retail trades, 1861*

Occupation	Mean L.Q.	S.D.	County concentrations (in rank order)
Shopkeeper, M	0.904	0.532	Norf, 2.737; Heref, 2.259; Som, 1.747; Suff, 1.539; Glos, 1.478; Chesh, 1.459
Shopkeeper, F	0.972	0.438	Northd, 1.950; Norf, 1.880; Heref, 1.816; Worcs, 1.617; Durham, 1.597; Som, 1.516; Devon, 1.469
Draper/mercer, M	1.018	0.221	Cumb, 1.701; Northd, 1.629; Sussex, 1.437; Yorks, ER, 1.359; Glos, 1.263
Draper/mercer, F	1.019	0.760	Corn, 4.121; Devon, 3.162; Norf, 2.009; Som, 1.836; Dorset, 1.783
Butcher, M	1.048	0.185	Rutland, 1.343; Yorks, ER, 1.339; Leics, 1.336; Hunts, 1.277; Som, 1.268; Middx, 1.263; Sussex, 1.237
Butcher, F	1.050	0.434	Hunts, 2.016; Norf, 1.715; Middx, 1.654; Suff, 1.651; Kent, 1.634; Northd, 1.570; Durham, 1.556
Cheesemonger, M	1.016	0.881	Dorset, 2.952; Middx, 2.845; Wilts, 2.750; Westm, 2.640; Som, 2.449; Northd, 2.319; Glos, 2.252
Cheesemonger, F	0.974	1.250	Northd, 4.872; Som, 3.914; Staffs, 3.115; Glos, 3.093; Devon, 2.735; Cambs, 2.729; Durham, 2.609
Poulterer, M	1.091	0.766	Sussex, 2.826; Lincs, 2.483; Essex, 2.074; Cambs, 2.021; Suff, 1.972; Berks, 1.969; Middx, 1.909
Poulterer, F	1.009	0.646	Yorks, ER, 3.275; Devon, 2.734; Berks, 1.778; Heref, 1.760; Northd, 1.692
Fishmonger, M	0.968	0.697	Norf, 3.608; Sussex, 2.400; Yorks, ER, 2.252; Middx, 2.133; Suff, 2.045; Kent, 1.855

(continued)

Table 19.11 (*cont.*)

Occupation	Mean L.Q.	S.D.	County concentrations (in rank order)
Fishmonger, F	0.758	0.876	Northd, 3.386; Norf, 3.154; Corn, 2.635; Devon, 2.580; Lancs, 1.811; Cumb, 1.671
Baker/confectioner, M	1.176	0.585	Berks, 2.418; Middx, 2.123; Oxon, 1.913; Hants, 1.853; Beds, 1.838; Kent, 1.834; Surrey, 1.783
Baker/confectioner, F	0.994	0.525	Westm, 3.197; Cumb, 2.687; Durham, 1.639
Greengrocer, M	0.829	0.527	Middx, 2.138; Kent, 2.085; Warks, 1.621; Sussex, 1.620; Surrey, 1.619; Yorks, ER, 1.488; Worcs, 1.362
Greengrocer, F	0.855	0.556	Corn, 2.427; Kent, 2.172; Durham, 1.781; Glos, 1.762; Hants, 1.578; Staffs, 1.554; Northd, 1.525; Devon, 1.417
Grocer, M	1.011	0.185	Kent, 1.453; Surrey, 1.401; Yorks, NR, 1.309; Sussex, 1.302; Cumb, 1.257; Yorks, ER, 1.250; Middx, 1.223
Grocer, F	1.171	0.545	Corn, 3.137; Westm, 2.484; Cumb, 2.245; Durham, 2.127; Northd, 2.079
Tobacconist, M	0.787	0.672	Northd, 2.916; Westm, 2.475; Yorks, ER, 1.751; Lancs, 1.741; Glos, 1.650; Warks, 1.559
Tobacconist, F	0.801	0.713	Glos, 2.226; Surrey, 2.140; Hants, 2.041; Middx, 1.964; Northd, 1.790; Berks, 1.766; Kent, 1.725; Sussex, 1.637; Warks, 1.590; Lancs, 1.530
Ironmonger, M	1.118	0.297	Glos, 1.686; Devon, 1.642; Hunts, 1.527; Oxon, 1.498; Surrey, 1.496; Norf, 1.451; Berks, 1.434
Ironmonger, F	1.066	0.749	Corn, 3.142; Rutland, 2.852; Heref, 2.669; Devon, 2.660; Bucks, 2.354; Sussex, 1.895; Warks, 1.847

Source: Calculated as in Table 18.2 above.

Table 19.12. *Location quotients, Wales, retail trades, 1861*

Occupation	Mean L.Q.	S.D.	County concentrations (in rank order)
Shopkeeper, M	1.260	1.028	Anglesey, 4.014; Cardigan, 2.506
Shopkeeper, F	1.157	0.786	Anglesey, 3.425
Draper/mercer, M	1.025	0.209	Pembroke, 1.531; Carmarthen, 1.270
Draper/mercer, F	0.954	0.292	Glamorgan, 1.318; Anglesey, 1.314; Brecon, 1.293
Butcher, M	1.019	0.243	Flint, 1.588; Denbigh, 1.428
Butcher, F	0.888	0.478	Glamorgan, 1.727; Anglesey, 1.590; Brecon, 1.506
Cheesemonger, M	1.135	1.146	Carmarthen, 3.821; Merionedd, 3.190
Poulterer, M	1.528	1.921	Anglesey, 6.997
Fishmonger, M	0.807	0.531	Pembroke, 1.965; Glamorgan, 1.435
Fishmonger, F	0.797	0.835	Flint, 2.457; Glamorgan, 2.161
Baker/confectioner, M	0.846	0.371	Monmouth, 1.513; Glamorgan, 1.284
Baker/confectioner, F	1.081	0.571	Anglesey, 2.029; Pembroke, 1.953; Cardigan, 1.805
Greengrocer, M	0.630	0.569	Glamorgan, 2.032; Monmouth, 1.507
Greengrocer, F	0.676	0.690	Glamorgan, 2.460; Monmouth, 1.580
Grocer, M	0.904	0.265	Flint, 1.409; Glamorgan, 1.246; Monmouth, 1.242
Grocer, F	1.003	0.289	Flint, 1.500; Brecon, 1.337
Tobacconist, M	0.767	1.172	Anglesey, 4.082; Glamorgan, 2.273
Ironmonger, M	0.926	0.303	Pembroke, 1.317; Brecon, 1.317

Source: Calculated as in Table 18.2 above.

rare in the smaller villages, which sustained only the generic shopkeeper and the victualler.[132] While the many other specific traders of Tables 19.11 and 19.12 grew in range and number in the third quarter of the century, they did not in general diffuse beyond the market towns, and indeed their development was the critical 'bait' for the personal shopper, or those shopping vicariously through the village carrier towards the end of the period. With the gradual diffusion of the bicycle, and of roads fit for its use, in the two decades before 1914, they may even have dented the trades of these older and smaller, all-purpose village shops.[133] Larger village and market-town bakers and vanmen came increasingly to travel to customers in the years before the war, and adversely affected the small shops. Contrary to the indications of the employment figures of Table 19.10, the 1880s may have seen the high point of the rural shopkeeping trade.

What should perhaps be noted is that while these most fundamental of retailers may have carried the name of 'grocer' or 'draper', and the designation of 'shop' in the trades directory, the reality may have been rather different.[134] Many, such as the small back-parlour shops of the mining districts and of Headington Quarry, developed during our period as women's activities, this being their direct and most sensible earnings strategy when family care was needed. While the quotients of Tables 19.11 and 19.12 for females cannot be regarded as entirely reliable indicators, the strong clustering of such female shopkeeping activity is noticeable in the industrial and mining counties of the north of England, south Wales, and the west country, where opportunities for female employment were otherwise limited. At Headington there were many more retailers in the Census returns than appeared in the trade directories, and to a degree shopkeeping was an opportunistic or life-cycle phenomenon: Harry Stiles was returned as a stone-digger in 1851, grocer in 1861, and baker in 1871; Harry Coleman, grocer in the directory of the mid-1880s, was blacksmith from 1896; and Emma Webb ran a sweet shop to ease her old age, and it died with her.[135]

The village butcher also came increasingly to sell dead meat from his cart or van in the years before 1914, but his trade was a hybrid between retailer and provider of a service. Such men killed on their own account, selling meat directly, along with the wide range of meat and fat products, but most rural butchers also provided the service of pig-sticking for the family friend of the cottagers. Flora Thompson's family pig was dealt with by the itinerant butcher in her Otmoor village (Oxfordshire), and

[132] Chartres, 'Country tradesmen', pp. 302–4.

[133] P. Horn, 'The rise and fall of the village shop', in Seager (ed.), *Village Trades and Crafts*, pp. 12–18.

[134] Williams, *Wiltshire Village*, pp. 103–4, notes the village bakery and stores, kept by Daniel Lewis, which also brewed ale, but which was poor, carrying great trade debts from its impoverished customers, *c.* 1912.   [135] Samuel, *Village Life*, pp. 233–4, 262 note 47.

sisters Mary Watson and Cissie Elliott (born 1898 and 1904), mine-deputy's daughters of Ovingham (Northumberland), recalled the butcher coming to kill the two family pigs in November and February.[136] Ted Bateson described his stepfather's butchery business at the remote Holderness village of Skipsea around 1900: he killed a bullock a week, on Wednesday morning, and left his wife to keep shop while he travelled in his cart to sell meat on Thursday, Friday and Saturday, still using the pole-axe for his work; and he too killed pigs for the cottagers, but at his slaughterhouse, charging 1s. 6d. for the service, killing Monday and butchering on Tuesday, and awaiting the owners who collected the fresh blood for black puddings.[137] Shopkeeper, vanman and itinerant slaughterer mixed to provide the village butcher with a living.

F THE PUBLICAN

Still more regular and central to rural life was the village pub, beershop or off-licence. Despite the gradual development of the temperance movement, the pub remained the dominant social institution of the rural districts, although the basis of its business organisation may have changed with the wider developments in the brewing industry. Its functions in the rural districts were several, and ranged from the pick-up and stopping points for the growing country carriers described above, to the supply of beer as a continuing basic element of the rural diet, to informal market or illicit exchange for poached game, and the place of resort and leisure time, strongly biased towards the male worker, but still perceived by the evangelicals as a threat to female virtue.

For the first half of the period, numbers in the trade were perhaps as great as at any time before or since, and entry under the beer house legislation of 1830 remained relatively easy in regulatory terms until the renewed restrictions of 1869 and 1872. It was estimated that around 40,000 beerhouses were added to the stock of 60,000 public houses under this legislation, and, despite acts of 1869, the last quarter of the century was generally regarded as a golden age for the drink trade and the public house. Only after 1900, with consumption of beer falling dramatically, and its distribution diversifying into shops, hawkers and off-licences, did this come to an end, and begin to reflect itself in serious falls in numbers.[138] While the data are imperfect, the employment figures of the Census indicate fairly consistent growth among males to 1901, with stagnation thereafter, with, as in several other trades, later growth in the

---

[136] Thompson, *Lark Rise*, pp. 10–11; Kightly, *Country Voices*, pp. 75–6.   [137] *Ibid.*, pp. 93–5.
[138] Comments based largely upon P. Clark, *The English Alehouse: A Social History 1200–1830* (London, 1983), esp pp. 306–43; Gourvish and Wilson, *Brewing Industry*, pp. 3–63.

Table 19.13. *Employment in inns, public houses, and beershops, 1841–1911*
(England and Wales, 000s)

Occupation	1841	1851	1861	1871	1881	1891	1901	1911
Inn, pub, beer, M*	47.9	64.0	66.1	78.1	76.0	75.8	90.3	90.6
Inn, pub, beer, F*	11.0	14.9	17.7	19.1	16.7	29.4	25.0	56.4

Note: * Includes cellar staff
Source: Compiled as Table 18.1 above.

Table 19.14. *Location quotients, England and Wales, drink trade, 1861*

Occupation	Mean L.Q.	S.D.	County concentrations (in rank order)
England:			
Inn, pub, beer, M	1.062	0.210	Bucks, 1.456; Middx, 1.430; Surrey, 1.409; Cambs, 1.403; Kent, 1.393; Herts, 1.353
Inn, pub, beer, F	1.140	0.399	Durham, 2.501; Westm, 1.949; Cumb, 1.871; Northd, 1.863; Hunts, 1.705; Salop, 1.585
Wales:			
Inn, pub, beer, M	0.946	0.242	Brecon, 1.426
Inn, pub, beer, F	1.014	0.223	Anglesey, 1.352; Pembroke, 1.313

Source: Calculated as in Table 18.2 above.

level of female employment, which in the figures of Table 19.13 was particularly striking between 1901 and 1911, strongly suggesting a substitution effect.

In the context of the industry overall, the Location Quotients for 1861, used in Table 19.14 to assess the degree to which this employment was concentrated, describe the position short of its peak, but near to the high point of distributive outlets. These data confirm the even distribution of this employment within the workforce of England and Wales, among both men and women, and the small number and scale of clusters, by comparison with most of the other trades thus analysed. There was a degree of concentration among males within the counties surrounding London and, as noted in some of the other distributional trades, relatively high quotients among women in the northern counties, where other employment opportunities appear to have been more limited than in

England as a whole. With means close to unity, and with generally low standard deviations, the figures show the ubiquity of public house employment, and its fairly consistent relationship to the general patterns of both male and female work across the country.

Publicans therefore appeared at the lowest population thresholds when analysing Norfolk and Yorkshire directories for 1836 and 1879, respectively 377 and 296, and official licensing statistics, while also somewhat problematic, suggest that the mean distribution of drink outlets of all forms, on- and off-licences in the latter county, were still more regularly distributed by the later years of the century. For the boroughs of Yorkshire in 1886, outlets averaged one per 211 persons, slipping back to one per 243 by 1896; perhaps surprisingly, the data for the Petty Sessional Divisions, which comprehended the rural districts and the smaller towns, but are complicated by the presence of growing communities not yet of borough status, indicate a very similar but more stable distribution, at 226 and 224.[139] This broad distribution held good for both England and Wales, with remarkable consistency in the pattern for the Petty Sessional Divisions of the two, and with greater differences between the boroughs, the reflection of basic differences in the degree of urbanisation between the two. Summary data, derived from the Royal Commission on Liquor Licensing of 1897, in Table 19.15, substantiate this point. Some decline in numbers, particularly among beerhouses, took place between 1890 and 1913, just over 5 per cent overall, but was probably more pronounced in the towns and in the subsequent decade.[140]

Public houses, beershops and off-licences were therefore fairly generously distributed in the rural districts of England and Wales in the 1880s and 1890s, and may even have slightly enhanced their position, and numbers remained fairly stable as the countryside depopulated. Behind this structural stability, however, much more significant changes were taking place. Two fundamental developments took place in the brewing trade in this period: the large-scale displacement of the brewing publican by the commercial brewer, and the growth of the tied trade.

For the 'country' brewery, meaning those that were neither of the great

[139] Calculated from data collected by the *Royal Commission on the Liquor Licensing Laws*, BPP, 1898, XXXVII, Cd. 8696, pp. 293, 313. The data were pooled for licensing districts, and therefore cannot be used to generate exact comparisons with the directory thresholds, but did generate stable relationships with population for boroughs, this produced an R^2 of 0.88 for 1886, and 0.87 in 1896 ($N = 18$), both significant at the 99 per cent level; and for the Petty Sessional Divisions, an R^2 of 0.57 in 1886, and 0.67 in 1896, again both significant at above the 99 per cent level ($N = 29$).

[140] Gourvish and Wilson, *Brewing Industry*, pp. 68, 412. Numbers of licensed premises fell by 7.2 per cent, 1913–23, and by a further 7.3 per cent, 1923–39. The balance between urban and rural losses before 1914 is strictly an unknown, but the evidence is that urban JPs took the opportunity to weed out unnecessary pubs after the case *Sharp v. Wakefield* (1901) clarified their power so to do.

Table 19.15. *Distribution of all Licensed Premises by population intervals, 1896*

Unit	<100	100–149	150–199	200–299	300+
England:					
Boroughs, number	12	57	54	66	31
Boroughs, %	5.5	25.9	24.5	30.0	14.1
Petty Sessional Divisions, number	14	96	168	217	121
PSD, %	2.3	15.6	27.3	35.2	19.6
Wales:					
Boroughs, number	3	8	4	1	2
Boroughs, %	16.7	44.4	22.2	5.6	11.1
PSD, number	1	18	27	36	26
PSD, %	0.9	16.7	25.0	33.3	24.1
Total, England and Wales:					
Boroughs, number	15	65	58	67	33
Boroughs, %	6.3	27.3	24.4	28.2	13.9
PSD, number	15	114	195	253	147
PSD, %	2.1	15.7	26.9	34.9	20.3

Source: Calculated from data collected by the *Royal Commission on the Liquor Licensing Laws*, BPP., 1898, XXXVII, Cd. 8696, Table II, p. 92.

industries of London or Burton, there were several potential incentives to tie: the extent of perceived threat from the big producers, notably the London 'power-loom' breweries; integration of malting and brewing business to gain greater economies of scale; distributional economies and the overhead costs of establishing sales agencies; fear of contraction in the supply of licensed premises, which was established by the restrictive legislation of 1869 and 1872, but accelerated in the 1890s and early 1900s; concern to guarantee returns in the declining beer market after 1879; the rising costs of loans to publicans in the beer bubble of the 1880s and 1890s; and the easy access to capital provided by joint stock incorporation in the same period.[141] While these aggregate to explain the transition, none suggests itself as an overwhelming imperative for the acquisition of the smallest and most dispersed rural public houses.

This is difficult to assess, given that the process showed extensive regional differences in timing and degree, but it is now regarded as a long-

[141] See also Chapter 17 B, above, pp. 1076–84.

term trend from the liberation of licensing in 1830, and not just the product of restriction and the coming of joint-stock breweries from the 1870s and 1880s.[142] For the country brewery, the pressures to tie undoubtedly rose in the latter years, and established a 'domino effect', the costs of which some were to rue after 1900. Data are available for 1892, a little before the peak in the tumbling of the dominoes, to indicate the extent of multiple ownership of public houses by brewing companies, but not the full extent of ties, given that so many properties were tied on leases, and owners of a single house were not delineated. A simple example from the North Riding illustrates the case: in the wapentakes of East and West Bulmer, rural districts south-west of Malton, eighty-two of ninety-two public houses were owned by someone other than the tenant, but only thirty-six were in the hands of those owning two or more; twelve were owned by the Riley Smiths, ten by the Tadcaster Tower Brewery Company, and the rest by Lady Ingram – two, Miss Agar – two, Gordon and Dean – three, Robert Brogden – three, H. M. Stapylton – two, and the Hon. P. Daunay – two, making explicit brewery ownership only around a quarter of the pubs in the two wapentakes.[143]

Brewery control was almost certainly much larger: the Tadcaster Tower Brewery, known as 'Snobs' because of its aristocratic owners and perhaps also for its lack of astute management, had the Hon. Geo Dawney among its partners in 1882, suggesting a family connection with the public house owners of East and West Bulmer. This brewery was listed as holding only 22 houses in 1892, but had around 150, mostly of poor quality and barrellage in and around York, though its business had refocused towards the free trade.[144] The Riley Smiths, who incorporated as John Smith's Tadcaster Brewery Co. in 1892, owned 119 fully licensed premises, but tied another 104 by lease or loan, but on the whole quality houses.[145] Only in rare cases do country brewers appear to have acquired extensive estates containing the modest rural public house. Warwicks and Richardson of Newark were one such exception: they had bought 68 freehold pubs, 1888–1907, to make a complete tied estate of 111 freehold and 91 leasehold, ranging in value from £700–£1000 for the country pubs, up to £8000 for those in Grimsby and Nottingham.[146] Still more modestly valued rural houses clearly had little to offer such empire-builders, and a greater probability of lasting independence.

The other change, the generalised displacement of home-brewing by the commercial brewery, may reinforce this view. This process occurred

---

[142] Based largely upon Gourvish and Wilson, *Brewing Industry*, pp. 27–178.

[143] *Return of the Number of On-Licenses in each Licensing District where the Tenant and Owner are different Persons*, BPP, 1892, LXVIII, p. 221.

[144] Gourvish and Wilson, *Brewing Industry*, pp. 240–1, 164–5.   [145] *Ibid.*, p. 279.

[146] *Ibid.*, p. 281.

at differing times, and for reasons that are yet to be fully explained, but the change was dramatic: brewing publicans, including keepers of beerhouses, produced around 45 per cent of beer output in England and Wales in 1832, 23 per cent by 1870, 10 per cent in 1890, and a mere 5 per cent in 1900. The changes were compounded by those of the scale of this residual domestic production, since 47 per cent of licensed victuallers, and 42 per cent of beerhouses were brewing for themselves in 1832; 29 and 21 per cent by 1870; 9 and 8 by 1890, and 4 for each by 1900, suggesting some large but many small producers.[147] For many rural public houses, the visiting drayman from the country town brewery came increasingly to supplant home production by the end of the period, though it almost certainly survived most extensively in the more remote pub and the larger farm.[148] Rural preferences for the old tastes in beers, the sweeter and vatted beers, rather than the bright pale ales and new 'running beers' of later in the century, helped reinforce the bastions of resistance, although the brewing publican had already largely disappeared from what has been described as 'the heart of agrarian England' by 1830.[149]

The pub's functions were those of resort, leisure and interpersonal exchanges, according to a correspondent in the *Birmingham Morning News* in 1870, 'when the workman's only idea of play was drinking in a public house'.[150] Jefferies certainly distinguished the 'low public', which was clearly a beershop, from the fully licensed inn in the village, attacking it as the agricultural labourer's 'club, almost his home', and the agency of his brutalisation, putting a different complexion onto Peter Clark's suggestion that the arrival of the beershop after 1830 recaptured the spirit of the old alehouse, lost in a previous half-century of gentrification.[151] Jefferies explicitly accused the beershop of being the resort for dealing in poached or stolen goods, including poultry, eggs and coals, bartered for beer left on the doorstep:

Nowhere else in the parish, from the polished mahogany at the squire's mansion to the ancient solid oaken table at the substantial old-fashioned farmer's, can there be found such a constant supply of food usually considered as almost the

---

147 Discussed extensively, *ibid.*, pp. 64–75, but noting that the question is as yet not fully resolved. Given the small-scale operations of the rural pub or beerhouse, as evident here, it is quite natural that they have as yet attracted little analytical attention among brewing historians.

148 Sturt, *Wheelwright's Shop*, p. 198; Kightly, *Country Voices*, pp. 36–7, 130; Williams, *Wiltshire Village*, p. 171.

149 Jefferies, *Hodge*, pp. 188–9, suggests, perhaps a little misleadingly, that you could call for, and get, Bass in Cairo, Bombay, Sydney or San Francisco, but not in Hodge's 'low public'; Gourvish and Wilson, *Brewing Industry*, pp. 40–7, note the trend towards new types of beer, and the disappearance of traditional regional patterns, such as the old Welsh spiced ales, abandoned by Brains of Cardiff and Soames of Wrexham around 1914. On the early disappearance from the southern counties, *ibid.*, p. 73, citing Baxter.   150 *Ibid.*, p. 75 and note 16.

151 Jefferies, *Hodge*, p. 189; Clark, *English Alehouse*, p. 337.

privilege of the rich. In brief, it is the strangest hodgepodge of pheasant and bread and cheese, asparagus and cabbage.[152]

He characterised the landlord as 'the octopus of the hamlet', sucking all into his den, with wife and daughter little more attractive, and the clientèle as including a young man, suspected thief, horse-maimer and incendiary; the indolent drunken ditch-digger; the cheeky lad; and the representatives of the 'workhouse families', perpetually bouncing into and out of relief through profligacy.[153] The reality was less extreme, but the pubs of Headington Quarry did exchange the odd rabbit for beer, and were the place of sale for watercress, blackberries, mushrooms, briars, ferns and holly collected by its 'totters', some of them used as 'latch-openers' to the pub door.[154] If the village pub was a resort for those working on the fringes of legality, it was nonetheless a tranquil place by comparison with its urban counterparts, and perhaps because the rural community was more effectively self-regulating, rates of conviction for drunkenness in the 'agricultural counties' lay far below those of all, bar the home counties, in 1891 and 1911.[155] Elements of the 'alternative society' lingered on in the rural public houses of Victorian and Edwardian England and Wales, but rather more in the social segmentation of their customers than as a source of potential disorder.[156] It was a primary and, to all but the temperance movement, generally benign social institution.

G THE PROFESSIONS

In general, the professions were less immediate in their impact and influence upon the generality of the rural community of the years 1850–1914, although doctors, veterinary surgeons, the clergy and teachers diffused more widely into the market towns and larger villages in this period. Many of the other professionals servicing land in its broader senses, the surveyors, land agents and lawyers, interacted little with the generality of country people, and had in any case already established their base by 1850. For both the law and medicine, the years before 1850 had seen the establishment of professional characteristics and power, but the task of establishing criteria of qualifications and exclusions that finally polished the true 'profession' was not yet complete. This very largely was achieved in the second half of the century.[157] Growing in number and importance

---

[152] Jefferies, *Hodge*, p. 190.   [153] *Ibid.*, pp. 194–8.   [154] Samuel, *Village Life*, pp. 214–24.

[155] Gourvish and Wilson, *Brewing Industry*, Table 2.4, p. 32, from G. B. Wilson's *Alcohol and the Nation* (London, 1940).

[156] My use of the phrase is a direct reference to Peter Clark's analysis of its earlier roles; see *English Alehouse*, pp. 166–249 for the earlier transition to respectability.

[157] On the background over the longer term to the concept of the profession, see P. J. Corfield, *Power and the Professions in Britain 1700–1850* (London, 1995), *passim*.

Table 19.16. *Employment in the professions, 1841–1911*
(England and Wales, 000s)

Occupation	1841	1851	1861	1871	1881	1891	1901	1911
Solicitor[a]	13.7	15.8	14.5	15.9	17.4	20.0	21.0	21.4
Physician[b]	17.1	19.2	14.4	14.7	15.1	18.9	22.5	23.0
Dentist[c]	0.6	NSR	1.6	2.3	3.6	4.6	5.2	7.4
Teacher, M[d]	21.1	28.0	28.6	19.4	39.8	59.4	58.7	68.7
Teacher, F[d]	30.9	66.9	27.7	38.8	94.2	153.0	171.7	183.3
Banker[e]	1.6	1.5	1.4	1.3	1.7	1.9	30.1	39.9
Auctioneer	3.0	3.5	5.3	6.3	10.0	11.7	13.9	17.6
Accountant	4.4	5.7	6.3	9.8	11.5	7.9	9.0	9.5
Vet/farrier[f]	5.1	6.1	6.8	6.7	7.5	9.5	2.9	2.6
Surveyor[g]	4.5	6.6	6.5	4.8	5.4	5.8	6.4	5.1

Notes:
[a] Includes judges, barristers and, for 1841 and 1851, law students
[b] Combines physicians, surgeons, practitioners and, for 1841 and 1851, medical students and assistants
[c] NSR indicates not separately recorded. Figures include dental assistants 1911
[d] Includes teachers, professors and lecturers in 1841 and 1851, 1891–1911
[e] All banking employees included, 1901 and 1911
[f] Veterinary surgeons only recorded for 1901 and 1911
[g] Compounds land, house and ship surveyors from 1881; includes assistants, 1911, without them, 4.0
Source: Compiled as Table 18.1 above.

after 1850, many were significant to agriculture, but not directly to the majority of country dwellers, and several, concerned principally with landownership, were relatively heavily concentrated in the capital.

The evidence of trends in employment in the professions over the second half of the century tends to confirm this picture; they were established, if not fully transformed into professional organisations, by the 1850s. The figures of Table 19.16 show several, the lawyers and doctors, bankers and veterinary surgeons/farriers, whose growth was substantially slower than that of employment as a whole. These data are subject to many problems, themselves attributable to the definitional changes associated with their reorganisation into professional institutions, and to the inclusion of entrants in training or assistants in some Census years. Surveyors represented an extreme case of the problem, discussed more extensively below. Only some displayed growth of such magnitude as to indicate the creation of a new calling: teaching was overwhelming in this respect and dominated the figures. It also saw the recruitment of women on a large scale, representing a significant shift towards women in the teaching workforce. In proportionate terms, dentists and auctioneers exceeded the growth of the male employment as a whole, and the inclusion of the full range of clerks and others in the banking figures for 1901 and 1911 seriously distorted a pattern of long-term stability.

Since so many of these professions, above all the law, were very heavily concentrated on London at the beginning of the period, the exclusion of the capital from the Location Quotient measures in Table 19.17 below is the more difficult to interpret. Solicitors, dentists, accountants and vets/farriers listed here were professions with a third or more of their agents in London in 1861, and solicitors and accountants remained so in 1901, along with those employed in banking. The remainder diffused more widely through the workforce of England and Wales in the second half of the century, or were already more evenly distributed in 1861. Noting, therefore, the partial distortion of the picture by the exclusion of London, the figures of Table 19.17 indicate a surprisingly even distribution of these professions through England and Wales, their numbers being relatively small in the latter.

Despite the removal of London from the data for analysis, its impact remains clear in many of the figures for England. The principal location for many of these activities being London, it was also the natural focus for the creation of the agencies of professionalisation, such as the Land Surveyors' Club of 1834, the engineering institutes and the Law Society, which reinforced these centripetal processes.[158] Double-banked institutional factors thus help to explain the evident clustering of so many of these professions into the home counties, London's residential corona. All of law, medicine, banking, auctioneering, and veterinary and farriery work demonstrated this in the figures of Table 19.17, only the accountants displaying a secondary zone of concentration into the west country, and the teachers and the minimal numbers of dentists breaking the pattern by their wider distribution.

Wider developments from mid-century or earlier reinforced these processes. Above all the development of the railways generated a new mass of business for the legal profession, dealing very largely with railway companies incorporated in London, with the investigative processes of parliamentary committees likewise, and with the major agencies of land-ownership which also were predominantly located there. The high transaction costs of litigation and the associated processes provided a powerful incentive for railway companies to minimise the number of landowners

---

[158] A broad survey of these trends is provided by Corfield, *Power and the Professions*, and in the early chapters of F. M. L. Thompson, *Chartered Surveyors: the Growth of a Profession* (London, 1968), esp pp. 19–108. See also J. A. Chartres, 'Country trades, crafts, and professions' in G. E. Mingay (ed.), *AHEW*, vol. VI, *1750–1850*, pp. 445–66.

Table 19.17. *Location quotients, England and Wales, professions, 1861*

Occupation	Mean L.Q.	S.D.	County concentrations (in rank order)
England:			
Solicitor	1.108	0.476	Surrey, 3.019; Middx, 2.415; Worcs, 1.894; Glos, 1.711; Sussex, 1.618
Physician	1.041	0.637	Middx, 2.981; Sussex, 2.972; Surrey, 2.279; Glos, 2.226
Dentist	0.884	0.771	Yorks, ER, 3.674; Glos, 2.454; Sussex, 2.339; Notts, 1.794; Devon, 1.657
Teacher, M	1.113	0.297	Yorks, NR, 1.972; Middx, 1.936; Westm, 1.828; Rutland, 1.462; Sussex, 1.453
Teacher, F	1.233	0.382	Suff, 1.912; Hunts, 1.899; Rutland, 1.817; Hants, 1.699; Essex, 1.652; Sussex, 1.645; Kent, 1.615
Banker	1.181	0.627	Middx, 3.006; Surrey, 2.717; Heref, 2.285; Herts, 2.165; Hants, 2.153
Auctioneer	1.098	0.349	Middx, 2.363; Surrey, 1.866; Sussex, 1.651; Kent, 1.453
Accountant	0.973	0.731	Glos, 4.016; Dorset, 2.685; Wilts, 2.235; Som, 1.926; Devon, 1.800
Vet/farrier	1.159	0.534	Middx, 3.086; Surrey, 2.024; Rutland, 1.839; Suff, 1.800; Norf, 1.788
Surveyor/land agent	1.060	0.339	Herts, 1.831; Bucks, 1.790; Salop, 1.578; Lincs, 1.526; Yorks, NR, 1.530; Westm, 1.486
Wales:			
Solicitor	1.018	0.266	Carmarthen, 1.432; Denbigh, 1.301; Caernarvon, 1.322
Physician	1.088	0.573	Flint, 1.934; Caernarvon, 1.854; Brecon, 1.774
Dentist	0.727	0.802	Brecon, 2.366; Monmouth, 1.758

(continued)

Table 19.17 (*cont.*)

Occupation	Mean L.Q.	S.D.	County concentrations (in rank order)
Teacher, M	1.133	0.299	Cardigan, 1.640; Pembroke, 1.530
Teacher, F	0.926	0.388	Monmouth, 1.764
Banker	1.108	0.574	Brecon, 2.302
Auctioneer	0.978	0.395	Carmarthen, 1.626; Pembroke, 1.547; Cardigan, 1.395
Accountant	0.723	0.468	Glamorgan, 1.754; Monmouth, 1.304
Vet/farrier	0.945	0.455	Monmouth, 1.538
Surveyor/land agent	1.114	0.425	Flint, 1.932; Brecon, 1.809; Montgomery, 1.677

Source: Calculated as in Table 18.2 above.

to be dealt with when selecting their routes, and this reinforced the tendency to deal with the legal agencies of the grandees in the capital.[159] In a rural world, until the 1870s or 1880s, dominated by the great landowners, the key legal support services followed their residential patterns, and this overwhelmingly meant that London was their focus. Solicitors did have duties in respect of property in the provincial cities and towns, including the market towns, and the even distribution of the profession in 1861 in Wales, albeit with pretty small numbers engaged in the least urbanised area under consideration, confirmed this. The country town solicitor had important functions to perform for the generality of rural England and Wales, but the growth of new agents to perform their traditional roles of financial intermediary attenuated their position, and in terms of the overall volume of business and the most lucrative processes of litigation, the large-town and London firms were the primary beneficiaries of change in the period.[160]

One such intermediary was perhaps the critical beneficiary of the development of the railway, the surveyor and land agent. While much of the early work of railway surveying, in the 1830s, was conducted by a motley group of former military men and a fair number of charlatans, by

[159] Thompson, *Chartered Surveyors*, pp. 109–27; J. R. Kellett, *The Impact of Railways on Victorian Cities* (London, 1969), explores these issues at length in a series of important case-studies.
[160] On the trustee and public duties of the country solicitor, see Jefferies, *Hodge*, pp. 129–36.

the 1850s and 1860s it had established the overwhelming need for professionalisation. Such was the power of these land agents and surveyors in these complex transactions that by the early 1860s landowners complained of being in their thrall, and prices being depressed by a conspiracy of surveyors and railway promoters.[161] This helped to concentrate the founding membership of the Institution of Surveyors in 1868 into the field of land agency, and explained the predominance of landed affairs in *The Estates Gazette*, founded by Henry Allnutt in 1858, and its transition towards the profession's journal. Well into the 1870s, they were consequentially conservative agents of the landed status quo, generally hostile to land reform, and to the landed concerns of the urban middle classes.[162] Chartering (from 1881) and a common professional examination (from 1880), assisted the spread of membership, as they diversified into building, mine and quantity surveying. At the same time, recruitment of practitioners of surveying and land agency to membership became largely complete.[163]

Early records of membership confirmed the predominance of London interests: sixteen of the original twenty members of the Institution were London based, one was partially so, and three only came from provincial locations, Epperstone (Notts), Birmingham and Bristol; and only twelve of the next twenty-nine recruited were based outside the capital, in Salisbury (two), Bath, Birmingham, Cambridge, Exeter, Harrogate, Reigate, Manchester, Rochester, Rugby, and Newark.[164] Records do, however, survive for the Oxley Parkers, land agents and surveyors, founder member of the Club in 1834, of Woodham Mortimer, near Maldon in Essex, covering the years 1826–87, to illustrate the day-to-day work of the land agent. The Oxley Parkers dealt with land survey, including the tithe work of the 1840s; the letting or sale of farms; lease covenants and their observation; buildings and repairs; railway work; coastal land reclamation; some trustee duties; country banking; and combined all of this with farming themselves, both as landowners and as tenants. Not untypical of the leaders of the profession, represented by the élite membership of the Club, the Oxley Parkers demonstrate the full evolution of the land agent and surveyor into a central position as an intermediary in later nineteenth-century rural society.[165]

[161] Thompson, *Chartered Surveyors*, p. 126.

[162] *Ibid.*, pp. 170–2. Land agents comprised more than a quarter of the first 200 elected members, to 1873.

[163] *Ibid.*, Chapters VIII–XI, pp. 148–255, and Appendix I (ii), p. 349. Appendix II, pp. 342–6, reproduced the first examination papers, and sustained the case.   [164] *Ibid.*, pp. 130–1.

[165] A very brief summary of the remarkable evidence of J. Oxley Parker, *The Oxley Parker Papers from the Letters and Diaries of an Essex Family of Land Agents in the Nineteenth Century* (Colchester, 1964).

Closely linked in the service of landed society were the auctioneers, whose role had become established in property sales from the mid-eighteenth century, and whose mart, focused upon Garraways in Bartholomew Lane from 1810, removed to Tokenhouse Yard in 1866. They shared the ambitions of the surveyors for full professional status and organisation, with whom many overlapped, but did not establish their own Institute until 1886.[166] Sales of land by auction took place in the provinces too, the figures of Table 19.17 indicating a fairly even spread across both England and Wales, with modest standard deviations, and the shifting patterns of the livestock trade created from mid-century a new field of activity, namely, sales by auction of herds at the new railheads. As town and city governments faced the growing problems of congestion and the soiling of central streets, many removed their livestock markets to new locations, although most meat remained slaughtered and traded by small butchers in cities such as Manchester. Most retained their traditional systems of jobbers and dealers in this transition, but sale by auction now enjoyed a larger role, if one performed by salesmen outside the property-dealing élite who were ultimately to form the professional association.[167] Property, houses and furniture remained the stock business of the provincial auctioneer, who remained predominantly a professional of the market town and larger urban centre, travelling out to conduct the farm sale, or holding the auction in his town salesrooms, after buyers had viewed the property in advance.[168]

At the beginning of the period, banking was still a complex mix of the private bank, as yet not fully stabilised by central banking functions, and the beginnings of larger joint-stock banks and branch networks. The bulk of landowners with capital still managed their own finances, and the increasing diversification of their investment portfolios, through the private and merchant banks. Failures among the small networks of provincial banks remained endemic, and it was only in the second half of the period, notably from the late 1870s, that the spread of more stable networks of provincial banks and branches delivered the benefits to the financial system that had been anticipated in the reforms of the 1840s. While these predominantly small-scale banks had already become distributed fairly widely across England by 1851, the Location Quotients of Table 19.17 indicate considerable concentration of 'bankers', into London and its wider hinterland, combined with significant variation

---

[166] Thompson, *Chartered Surveyors*, pp. 144–5.

[167] Scola, *Feeding the Victorian City*, pp. 186–8; Richard Perren, *The Meat Trade in Britain 1840–1914* (London, 1978), pp. 143–4.

[168] The full history of auctioneering, subsequent to the era covered by F. M. L. Thompson, 'The land market in the nineteenth century', *Oxford Economic Papers*, 9, 1957, is as yet unwritten. On their earlier history and development, see also Chartres, 'Country trades, crafts, and professions', pp. 445–51.

around the mean distribution. The evidence from the principal trade directory, *The Banking Almanac*, for 1851 indicated the weaknesses of these employment data as indicators: by 1851, most of the major towns of England and Wales had their bank, the greater centres having their banks, but provided by different systems, the joint-stock branch networks being more significant providers to the north and west of the country, and the private country banks remaining predominant in the south and east.[169] By the beginning of the period, then, access to the British banking system for those who sought it was relatively easy, either directly or, as remained normal for the landowner, through attorney or land agent. It was through this network of connections that the rural capital was channelled into the railway system and other industrial developments, and offered compensation through equities for the 'conspiracies' by surveyors to depress land prices in sales to railways of which the landed interest had complained so much in the 1860s.[170]

In terms of the delivery of financial services to the land, the banking system was, therefore, largely developed by the third quarter of the nineteenth century, but it was only from the 1870s that branch banks became general in the market towns, and choice confronted the user. The course of development by Gilletts, the Banbury bankers, illustrated a pattern commonplace in the south: based predominantly at Banbury, to which they drew remittances from much of north Oxfordshire in the 1850s, and connected with London, they expanded to branches in Woodstock by 1860, with a further agency at Witney. From 1877, they opened branches in Oxford and Witney, the latter by merger with the existing Witney Bank, Clinch's, and in the 1880s spread to Bampton, Chipping Norton, and Abingdon. By the time of their acquisition by Barclays in 1918, Gilletts had established branches at Brackley and East Oxford, and offices in eight other locations. The base clientèle remained agricultural, and depressed conditions in farming from the later 1870s placed the bank under some stress, but it survived, unlike many others.[171] John Oxley Parker, the Essex land agent, had been involved in unsuccessful attempts to save Thomas Johnson & Co., Romford bankers in 1844, and subsequently joined Sparrow, Tufnell & Co. of Chelmsford, to help stabilise its capital base and improve management, but his diaries regrettably tell little of the business of their country banking.[172] Gilletts retained their

---

[169] *Ibid.*, pp. 458–66.

[170] For an exemplary analysis of one grandee and his investment portfolio in the period, see D. Cannadine, 'The landowner as millionaire: the finances of the Dukes of Devonshire, *c*. 1800–1926', *AHR*, 25, 2, 1977, pp. 77–97, 26, 1, 1978, p. 47. See also above, pp. 1203–4 on surveyors.

[171] A. M. Taylor, *Gilletts: Bankers at Banbury and Oxford* (Oxford, 1964), pp. 121–218.

[172] Oxley Parker, *Oxley Parker Papers*, pp. 292–4. The family had also been involved in attempts to rescue Crickett & Co. of Chelmsford in 1826.

note issues to the end, and these, with their Oxford sheep emblem, reinforced their strength with the farmers against the threat of Birmingham-based joint-stock bankers, 'Gie I a shep' becoming the Banbury manner of expressing the preference for a Gilletts' note.[173] More practically, this was reinforced by the bank's policy of considerate lending to confirm its strength in the farmer market, to which it added some local industrial interests, such as the Witney blanket makers and W. R. Morris.[174]

Banking services thus grew for the support of the agricultural community and its supporting retailers and craftsmen, but of course left most residents of rural England and Wales untouched, at least in any direct sense. The coming of banking options to the market towns from the 1880s and 1890s was significant in the long term, but this was a business for long without immediate competition, given the intimate symbiotic relationship of banker and client, although the arrival of joint-stock bank branches and the spread of cheque accounts among the farming interest led to its expansion. When the 'new bank' came to Hodge's market town, it left him untouched, and, apart from market day, neither bank was busy, though the 'old' had to pick up its socks and embellish its premises:

On an ordinary day the customers that come to the bank's counter may be reckoned on the fingers. Early in the morning the Post-office people come for their cash and change; next, some of the landlords of the principal inns with their takings; afterwards, such of the tradesmen as have cheques to pay in. Later on the lawyer's clerks, or the solicitors themselves drop in; in the latter case for a chat with the manager. A farmer or two may call, especially on a Friday, for the cash to pay the labourers next day, and so the morning passes. In the afternoon one or more of the local gentry or clergy may drive up or may not – it is a chance either way – and as the hour draws near for closing some of the tradesmen come hurrying in again. Then the day, so far as the public are concerned, is over. To-morrow sees the same event repeated.[175]

Country banking was less extensive in practice, its critical functions lay in finance and the clearing of cheques, what Jefferies described as the 'indirect profit', but the new institution of the country branch bank remained in general a quiet place before 1914.[176]

Medical provision by the trained profession, for either humans or animals, also showed limited diffusion into the rural districts in this period. Coverage of the country was clearly patchy in 1861, as indicated by the variation around the mean of the Location Quotients, and the country doctor or vet was predominantly the figure of the market town,

---

[173] Taylor, *Gilletts*, p. 188.

[174] *Ibid.*, esp pp. 199–201, which shows how this policy recruited the Morris Motors account. A. Plummer and R. E. Early, *The Blanket Makers 1669–1969: A History of Charles Early & Marriott (Witney) Ltd* (London, 1969), pp. 58–9, 157.   [175] Jefferies, *Hodge*, p. 145.

[176] *Ibid.*, pp. 145–8.

not the village. Institutional change in Victorian England, notably in the poor law institutions and the workhouse hospital, did create salaried employment for doctors in rural districts, and later concerns for rural sanitary problems and education further assisted the process of diffusion.[177] Country doctors were already widely distributed at the beginning of our period, and remained so, the south-eastern segment of the country becoming emphatically better endowed by 1911. Their business was generally more stable and less competitive than those of the towns, and bolstered by proportionately greater access to poor law union and other appointments: in both 1877 and 1899, Anne Digby estimates that two-thirds of rural doctors held such appointments. By 1918, physicians held over 30,000 public appointments of various kinds, though many, of course, were pluralists. Despite this, country practices produced smaller average incomes than the better urban practices, £400–£800 per annum against £1000 or more, but tended to avoid the overcrowding of town medicine, which generated an extended tail in revenues. Country practitioners also tended to enjoy greater non-monetary rewards in status.[178]

Even so, medical services for many country people continued to be provided by apothecaries' remedies, by sales of patent medicines, and by the continuing use of folk remedies and the informal counsel of neighbours.[179] Wise women and local untrained midwives, like 'Granny Scott' of Muker in Swaledale around 1905, continued to attend the confinements of neighbours, and to minister their remedies.[180] Like the country doctor, and so many other trades and professions, they were paid in kind, a practice which Alfred Williams saw as continuing to the eve of the War in his Wiltshire villages.[181] In a world with very few effective medicines, traditional applications of goose-grease, a panacea in rural Wales, purges, herbal remedies and the charming of warts were the staples of country folk, whose best remedies lay still in the fundamental benefits of country air and food.[182]

The same applied to the veterinarian, whose skills were heavily concentrated into the towns, and to horse work, in such institutions as the

---

[177] On the provision of medical services before 1850 by those outside the formal profession, see the discussion by Corfield, *Power and the Professions*, pp. 157–9.

[178] A. Digby, *Making a Medical Living: Doctors and Patients in the English Market for Medicine, 1720–1971* (Cambridge, 1994), pp. 20–4, 120–5, 143–8.

[179] V. Berridge, 'Health and medicine', in F. M. L. Thompson (ed.), *The Cambridge Social History of Britain, 1750–1950*, vol. III, *Social Agencies and Institutions*, (Cambridge, 1990), esp. pp. 186–91.

[180] Kightly, *Country Voices*, pp. 98–101, oral evidence of 'Maggie Joe' Chapman, born 1899.

[181] Williams, *Wiltshire Village*, p. 133.

[182] Kightly, *Country Voices*, pp. 218–29, collects together a wonderful range of these from oral sources across the country. On diet and its quality, see W. A. Armstrong, *Farmworkers: A Social and Economic History, 1770–1980* (London, 1988), pp. 99–101, 139–40.

central Leeds 'horse hospital', in which the distinctions between the skilled farrier or smith, and the veterinary surgeon or farrier remained blurred to the end of the century.[183] The figures for their employment in Table 19.16 provide a helpful indication of the balance in the new century, and suggest that most horse work was carried out by the farrier, not the professional vet. With perhaps 30,000 smiths working in the early 1900s, and a proportion of these no doubt being hidden farriers or 'horse-doctors', qualified vets or those in training probably provided less than a quarter of 'medical' services to all livestock, and a very much smaller proportion to those still on the farm.[184] Indeed, there was every incentive for farmers, fearful of the detection of disease among their herds, to avoid using the professional veterinarian, for whom the duty to public health was an imperative, although numbers of vets formally employed by government to meet these tasks remained small.[185] It was hardly before 1900 that stronger controls on tuberculosis, creating work for qualified veterinarians, emerged. Control was incomplete in 1914, but the golden years of bovine tuberculosis, and much work for the vet, were yet to come, and developed rapidly with the tanking of milk in the inter-war years.[186]

All of these professions were, as noted, numerically insignificant by comparison with the booming teaching profession. Stimulated by the legislation of 1846 providing government grants for the training of pupil teachers, educational provision in the rural districts at elementary level was in general as efficient as that of the largest cities which had more to gain from the changes of the 1870s. Some information on the distribution of schooling was collected in the Census of 1851, and this emphasised the role of the churches, above all the Anglican Church, in its provision. Of the total of 44,836 day-schools in England and Wales, with 2,108,592 registered pupils, the public day-schools comprised 15,411 (34.3 per cent), but had two-thirds of the registered pupils, and over two-thirds of these, with three-quarters of the pupils, were in church schools.[187] Three of these schools were described as 'agricultural schools', having 203 male and 61 female pupils.[188] Evidence from a study of Derbyshire suggested that this elementary education established itself the

---

[183] Information kindly supplied by Mr S. J. Lingard – the 'horse hospital' was located near Albion Street. Pickfords maintained a similarly named institution for their horses in London (information from Dr G. L. Turnbull). See also Chapter 16 C above, pp. 1062–5.

[184] On the problems of evaluating the numbers of smiths, and the nature of their work, see above, pp. 1164–5.   [185] Perren, *Meat Trade*, pp. 63–7, 84–91.

[186] P. J. Atkins, 'Sophistication detected: or, the adulteration of the milk supply 1850–1914', *Social History*, 16, 1991, pp. 317–39, and Atkins, 'White poison? The social consequences of milk consumption, 1850–1930', *History of Medicine*, 5, 1992, pp. 207–27.

[187] J. M. Goldstrom, *Education: Elementary Education 1780–1900* (Newton Abbot, 1972), pp. 109–18.

[188] *Ibid.*, p. 117.

more easily in the central limestone plateau and the south-east of the county, in other words, the agricultural districts, but was resisted in the textile and mining areas, to which elementary education only penetrated with compulsion after 1870 and 1876.[189]

Rural education thus gained relatively less from the direct impact of the Forster Act of 1870, which most forcefully attacked the areas that were worst provided, shown by a survey of 1869 to have been cities like Birmingham, Leeds, Liverpool and Manchester.[190] More important in some areas was the capacity to compel attendance, only reinforced by the Sandon Act of 1876 imposing penalties on employers: even this was limited for the rural districts, for it exempted children who lived more than two miles from the nearest school, allowed free employment outside school hours, and freed six weeks a year to the 'necessary operations of husbandry'.[191] In areas of entrenched child domestic labour, as in the lace and straw-plait industries of Buckinghamshire and Bedfordshire, this was a source of continuing conflict.[192]

Rural districts also faced rather greater problems in supporting the bureaucratic structures of compulsion, in finding the money and the people to serve the boards. A survey of 1903 revealed the extent of these problems: 2,865 civil parishes in England and 607 in Wales represented the non-urban districts, and between them had 2,363 separate school boards, and 578 independent school attendance committees; these represented a double pressure on the rate revenues and human resources of the rural districts. These deficiencies were only gradually tackled as county councils took over responsibilities from 1902.[193] Although the syllabus was reformed to appeal more directly to country children from 1900, secondary education remained dependent upon the distribution and support policies of the grammar schools, and continued seriously deficient in the country districts to 1914.[194] For these reasons, the comparative standing of rural and urban education may have shifted somewhat in favour of the latter from the 1870s, where in any event the vast majority of the new teachers were employed. Country schools retained the characteristics of the rural districts in which they were located, and as late as 1912 Alfred Williams noted the closure of Wiltshire village schools for gleaning during the harvest holidays.[195] In education, as in a number of other

[189] Marion Johnson, *Derbyshire Village Schools in the Nineteenth Century* (Newton Abbot, 1970), pp. 51–5.

[190] Frank Smith, *A History of English Elementary Education 1760–1902* (London, 1931), pp. 284–7.

[191] *Ibid.*, pp. 297–8.   [192] See above, Chapter 18, pp. 1113, 1118.

[193] Charles Birchenough, *A History of Elementary Education in England and Wales from 1800 to the Present Day* (3rd edn, London, 1938), p. 152.

[194] Smith, *English Elementary Education*, p. 341; Birchenough, *Elementary Education*, pp. 144–5.

[195] Williams, *Wiltshire Village*, p. 127.

fields, the impact of reform on the rural districts of England and Wales was distinctly limited.

H CONCLUSION

Even at such length, the present chapter has been able to do little more than sketch the broad trends of the subject, and there remains a need for many more detailed local and regional studies on the topics it has covered. Taking that cautionary note, a number of themes recur which marry effectively with many of those raised by earlier chapters. The first is the relative buoyancy of so many traditional crafts and trades in the face of a late industrialising society. Faced by the combined challenges of the coming of new materials, new technologies, new patterns of trade and internationalisation, and the deskilling of so many traditional craft employments, combinations of institutional inertia and flexibility helped to delay the full impact of change into the second half of our period.

However, from the 1880s and 1890s these long-term changes were reflected in rates of rural depopulation that made the rural strategies of adaptation and survival increasingly difficult to sustain. Many of the traditional employments analysed here were already mortally wounded by 1900, and the cataclysm of the First World War finished them off. War killed many members of the successor generations of craftsmen, and irreversibly changed the experiences and expectations of others. When the Oxford reporters of the 1920s, in their very well-observed and sensitive surveys, called for their revival, and discerned hope, they were largely whistling in the wind or describing relict populations largely preserved by their insulation from these forces.

International trade represented a fundamental force for the whole of rural Britain in these years, that in timber providing an excellent exemplar for the indirect but profound impact on rural crafts, in reinforcing the shift of coopering to the towns and ports, eliminating the dry-coopering of meat, and hitting both the building trades and wheelwrighting with new access to Norway or Canadian deals. The millwright too suffered from the impact of imported wheat, and the consequential adoption of roller-milling and relocation of the milling industry to the towns.

Domestic transport changes also eroded traditional callings, or shifted them fundamentally. Railways ended the long-distance droving trades, and focused transport on the rural railway station and halt, especially through the intensive developments of branch-line networks between the 1870s and the late 1890s. The buffeting of the station yard combined with the use of softwoods and iron to erode traditional vehicle building, a process which was intensified with the improvement of rural roads after the 1890s, and the introduction of new vehicle types, even before the arrival of the motor vehicle.

The retail trade had a long-established importance in 1850 in the country districts, but expanded to the 1870s or 1880s before its position was also eroded by the vanmen of the market town, new tastes and branded goods, and the removal of people into towns. The village beer-shop and pub remained, and perhaps resisted contraction and engross-ment into large tied estates that characterised the brewing industry as a whole in this period. Both the empirical data that are available and the logic of the economics of the industry suggested that licensed premises per head of rural population may even have increased over the period, and were at best tied informally by suppliers, but from a position of rel-ative weakness not sturdy independent strength, although home-brewing undoubtedly survived better in the country districts, and was sustained by a lingering resistance to the new tastes of the wider market.

If the great feature of the later Victorian economy was the creation of a massive service sector, the establishment of élite, or would-be élite, pro-fessions at its apex, this was a development that clearly affected the coun-tryside less than the towns and cities. The bulk of country dwellers were uninfluenced directly by these newly self-defined professional classes, apart from the teacher and the parson, and were more likely to encounter them in selling milk or eggs as they took 'cottage' holiday homes in the summer, than as part of everyday rural life. Combined with the long term impact of rural deindustrialisation, this suggests that in these fields the differentiation of rural and urban lifestyles was as great or greater in 1914 than it had been in the 1850s.

CHAPTER 20

CONCLUSION

BY JOHN CHARTRES AND RICHARD PERREN

By the mid-1820s, Britain had reached the limits of the first major phase of its economic growth into an urban and industrial nation. This had been based upon the processes of reallocation of resources within a newly integrated national economy to nearer optimal use: the full development of wind and water power in industry and transport systems, extensive supply of human and animal power fuelled from home food output, and the large-scale provision of raw materials, capital, and land from the farm sector for urban and industrial development. Farming had operated to that point largely within a closed economy: international competition had as yet little place in English and Welsh food markets, although Ireland and Canada had established and growing positions, and the threat from the corn surpluses of Continental Europe was limited by the Corn Laws.[1] Within agriculture, systems of nutrition, labour supply, and fertilisation remained largely endogenous, applying the traditional circular flows and linkages within the farm system and domestic regulated market structures. Having approached or reached these limits in applying best-practice and established technologies, subsequent changes were more fundamental, and radically altered the context in which agriculture operated.[2]

The analysis of these processes of change in the later phase of industrialisation in many ways points up the somewhat paradoxical roles of the farm sector in Britain's transition. Relatively modest technical and organisational changes under the stimulus of advancing urban systems and population growth in the long eighteenth century had generalised the application of best practice, the intensification of the division of labour within the British farm, and contributed very significantly to economic growth. Crafts's estimates certainly highlighted agriculture as the principal sectoral contributor to net growth.[3] By contrast, within the larger and

[1] S. Fairlie, 'The nineteenth-century Corn Law reconsidered', *EcHR*, 2nd ser., 18, pp. 562–73.
[2] M. Overton, *Agricultural Revolution in England* (Cambridge, 1996).
[3] N. F. R. Crafts, *British Economic Growth during the Industrial Revolution* (Oxford, 1985).

more diversified economy from the middle years of the nineteenth century, rather greater technical changes took place in the infrastructure and superstructures of English and Welsh farming, but with a necessarily smaller net economic impact as a consequence of the shrinking relative size of the farm sector. The structural and organisational changes discussed in the chapters of this present section, notably the marketing of farm produce and the supply of farm inputs and rural industry, were unparalleled in their magnitude, yet sum to relatively small effects on national income. These radical changes thus impacted upon the countryside of England and Wales profoundly in terms of human experience, but in an increasingly open economy disturbed the economic aggregates rather less.

The secular decline in the relative standing of agriculture within the British economy therefore masks some of the significance of these changes within the sector. The point can be illustrated clearly by reference to the modern historiography of the railway. While details can be debated, few dispute Hawke's findings that the net social savings attributable to the railway as an innovation in Britain are rather smaller than those identified by Fogel and others for the USA, and the empirical results of studies of other economies, both European and 'frontier'.[4] Nor does modern research on the principal feature of railway development after around 1865, city suburban services and the creation of the branch-line networks, do all that much to dispel the view that much investment was poor and some irrational in net economic returns, and would not have been made with better appraisal. Pre-1870, then, Hawke's allocations of social savings derived from agriculture in his appraisal of the impact of the railway were necessarily modest, and, despite Irving's recent more positive evaluation of branch-line performance from the 1870s, so too were the economic benefits from the completion of the network.[5] Yet for the farmer, on the local and regional basis if not in aggregate, their impact was massive: trunk-line development had an immediate and significant impact upon the droving trade and upon the farm income of grazing regions; the coming of branch lines to the remoter parts of the country opened areas such as south Wales and the Yorkshire Dales to the lucrative fresh milk traffic; and both combined to complete the development of an integrated

[4] G. R. Hawke, *Railways and Economic Growth in England and Wales, 1840–1870* (Oxford, 1970); R. W. Fogel, *Railroads and American Economic Growth: Essays in Econometric History* (Baltimore, 1964); P. K. O'Brien (ed.), *Railways and the Economic Development of Western Europe, 1830–1914* (Oxford, 1983).

[5] J. Simmons, *The Railway in England and Wales 1830–1914*, I, *The System and its Working* (Leicester, 1978), pp. 103–12; R. J. Irving, 'The branch line problem in British railway history: the financial evidence from north-east England', *Journ. Transport History*, 3rd ser., 14, 1993, pp. 27–45.

urban system, and thus to permit the fullest expression of external forces for change on the countryside and farm production, principally, from the mid-1870s, those of international trade.[6] The specific, local and regional consequences of broader changes in the economy for farming and the rural districts of the country were often very significant in terms of income, mode of production, community or individual experience, if not in terms of national economic aggregates.

This illustrated the recurrence of certain themes throughout the chapters of this book, and the complex ways in which change on the grand scale in the later stages of industrialisation interacted with the structures and institutions of the primary industry, agriculture, that had forced so much of the early pace of the process. These grand themes can be summarised as those of economic integration, the completion of the development of modern and efficient urban systems, internationalisation, institutional change, the development of new tastes and products, shifts in the relative shares of capital and labour and in the nature of the labour process, continuity and discontinuity in the relationship of town and country, power sources, materials and their usage, and in modes of production, and demographic change, in which population growth slowed and combined with the large-scale depopulation of the rural districts. At the end of the period, there was the impact of war scares, and of war itself, arguably representing a watershed for many of the occupations, structures and systems analysed here.

As Richard Perren indicated in the present and earlier studies, the extent of integration of the domestic market was already very advanced before 1850, and was indicated by the nature of the price mechanism and the consequential allocations of resources to productive ends.[7] Agriculture, which was of prime importance in the eighteenth century, remained sufficiently important as a sector in the 1830s and early 1840s for the harvest cycle still just to predominate as a determinant of fluctuations in the British economy, but thereafter its influence, and the magnitude of the agricultural fluctuations themselves, diminished.[8] This was a powerful indicator both of the growth of the non-agricultural economy, and of the declining significance of the home farm in British food supply.

---

[6] D. W. Howell, 'The impact of railways on agricultural development in nineteenth-century Wales', *Welsh Hist. Rev.*, 7, 1974–5; C. S. Hallas, 'The social and economic impact of a rural railway: the Wensleydale line', *AHR*, 34, 1986, pp. 29–44.

[7] See above, Chapter 15; R. Perren, 'Markets and marketing', in G. E. Mingay, ed, *AHEW*, vol. VI, 1750–1850, Cambridge, 1989, pp. 190–274.

[8] T. S. Ashton, *An Economic History of England: The Eighteenth Century* (London, 1955), pp. 55–62; Ashton, *Economic Fluctuations in England 1700–1800* (Oxford, 1959), Chapter 1, *passim*, pp. 31–49; W. W. Rostow, *British Economy of the Nineteenth Century* (Oxford, 1948), pp. 18, 50–2, 55–6, 109, 169, 209, 210.

However the British farmer was shown here to have suffered no more from falling prices than the producers of many other primary goods, and the responses of farm production in the highly integrated markets of Victorian Britain produced the classic pig ('two years up, two years down') and sheep (seven-year) cycles, though neither attained perfect regularity much before the 1890s.[9] Seasonality was also attenuated during the period by the evolution of complex systems of adjustment and arbitrage, notably in the retail milk trade from 1892, storage, and wholesaling facilities, and provided a further indication of fully integrated market structures.[10] This was confirmed by the consistent reaction to external shocks, which from the second half of the period, as with the Leiter wheat corner in 1898, became those of the Atlantic market place.[11] Changed patterns of prices and price fluctuations reflected the new world of internationally integrating markets.

The impact of all this was expressed through what was, by the beginning of our period, an urban system of unparalleled spatial efficiency. Already displaying patterns of population distribution by 1801 comparable with the highly urbanised Netherlands, by 1851 England and Wales had established a hierarchy of cities and towns that displayed an almost perfect log-linear distribution. This pattern, measured by de Vries on population data, provided a short-hand statement of the complex economic efficiency of the later Victorian state, and helps to explain the evolving new structures of urban markets and distribution, of the delivery of trades, services and professions, and the stratification of the retail trades.[12] However, the degree of sophistication achieved by mid-century can be over-stressed. An earlier study of the meat marketing and distribution system suggests that in 1850 the urban portion of this network had marked deficiencies. This was part of the pattern of low investment in urban social overhead capital before 1870, a view that fits in with other work before and since then.[13] Although already relatively advanced by the 1850s, this hierarchy was further refined in the subsequent half-century by the impact of transport change, in the form of the railway at one level and, at the base of the pyramid, by the proliferation of country carrier networks, both of them completed to village level by the 1890s.

After 1870 further significant improvement was made to the system,

[9] HMSO, *A Century of Agricultural Statistics, Great Britain 1866–1966* (London, 1968), pp. 50–4; see above, Chapter 15 section (A)3.   [10] See above Chapter 15 section (C)7.

[11] See above Chapter 15 section (A)5.

[12] Jan de Vries, *European Urbanization 1500–1800* (London, 1984), esp. pp 84–120.

[13] R. Perren, 'The meat and livestock trade in Britain, 1850–1870', *EcHR*, 2nd ser., 27, pp. 385–400; J. G. Williamson, 'Did England's cities grow too fast during the Industrial Revolution?', in P. Higgonet, D. S. Landes and H. Rosovsky (eds.), *Favourites of Fortune: Technology, Growth, and Economic Development since the Industrial Revolution* (Cambridge, Mass., 1991), pp. 385–90.

but not in response to any stimulus offered by the output of Welsh and English farmers. The main reason was the arrival of bulk consignments of foreign foodstuffs. Foreign grain, dairy products, fresh fruit and refrigerated meat, all required rapid and efficient transport from port to wholesale market and thence to the consuming centres. This was largely possible because imports arrived in a form that allowed for easy handling, whilst British farm products continued as a lagging sector in this respect, dominated by small quantities of diverse products and differing quality, despatched from a large number of farms. Unlike imports there was no mechanism to assemble, grade, repackage, and then send them on in such a form that made bulk handling even remotely possible. As the volume of imported produce grew, it did so through a streamlined distribution system, while both domestic foodstuffs and the output of rural industry continued to pass through a whole network of higglers, merchants, factors, dealers and commission agents, most of whom handled only small quantities before sending them on to the appropriate retailer.[14] However, this did not mean that farmers and craftsmen were untouched by the general improvement in the distributive network, as even the system handling domestic produce was, to some extent at least, forced to modify and rationalise. For rural districts this provided for the full expression of comparative advantage in the supply of goods and services, the most articulated market for produce, and meant that though they were perhaps physically remote, they were fully integrated into the national economic system. For the market gardeners of Evesham or those around Bedford, this meant a chance to seize new opportunities, but for village smith or saddler, the invasion of his market for local products by the factory-made meant relegation to repair and maintenance work.[15]

As indicated above, it was transport change that engendered these final elements of economic integration and the polishing of the urban hierarchy. Transport innovations had a profound impact on rural patterns of production, the marketing and processing of agricultural commodities, and the supply of farm inputs, their timing successively reshaping the domestic market. Within Britain, the successive benchmarks of these changes came first in the 1830s and 1840s, with the advent of steam coasters linking Aberdeen and Kincardine with London, and Ayrshire with Liverpool; the completion of trunk railway lines by the mid-1850s, with the consequential displacement of the Welsh droving trades; and the building of the rural branch-line network from the 1870s, local economic conditions being altered as they opened. Late in the period, from the 1890s, renewed improvement of roads further enhanced access to and

---

[14] See above Chapter 15 sections (C)1, (C)3.
[15] See above Chapter 15 sections (C)4, (C)5; Chapter 19 section (B).

from the rural districts, and individual mobility, and laid the basis for the impact of motor transport which was to have the greatest impact after the War.[16] Domestic transport changes, though superimposed upon an already well-developed system, successively altered the shape of the home market within which farming operated.

Arguably more dramatic were the international dimensions of transport change. If the primary impact of the development of the canal system to *c.* 1830 had been to invert the British economy, in respect of coal above all turning it 'outside in', then changes from the 1850s onward reversed the process, where rural England and Wales were concerned. The familiar elements of this were the delayed impact of free trade on cereal producers, with the advent of large quantities of American, Indian and Russian grain from the mid-1870s, and the knock-on effect on millers and feedstuff manufacturers, leading to the introduction of new roller-mill technology in primarily port locations, and the destruction of the traditional millers and their supporting services.[17] The impact of international trades in meat had come somewhat earlier, first in the European and US livestock trades, and then in the chilled and frozen meat traffics, and these too had extensive consequences for traditional methods of trade, notably the ending from the 1880s of dry-coopering for despatch of meat in barrels. This was hastened by the international timber trades which also shifted cooperage to the ports, and introduced Norway and Canadian deals, increasingly to displace English woods from building and the construction of vehicles.[18] Leather and hides, tanning materials, dairy products, straw plait, lace and osiers all shared the experience of sharp exposure to international competition before 1914, strongly reinforcing this as a dominant theme of the history of rural England and Wales after 1850.

Inevitably, these factors combined to engender institutional changes, which were most pronounced in marketing but extended more widely. With the radical alteration of the domestic livestock trades, the terminal decline of the fair accompanied the end of droving, and although the exceptional lowland fair of Weyhill, established around 1225, endured to 1957, it had been deteriorating from the 1880s, and barely survived the First World War as a commercial institution.[19] Urban livestock markets as at Islington, auction marts at railheads and stations, and others, as in Deptford, serving the seaborne and international trades displaced the fair as the fundamental institution of the cattle and sheep trades.[20] The

---

[16] See above Chapter 15 sections (B)1, (B)3.

[17] See above Chapter 17 section (A)1–5; Chapter 19 section (C).

[18] See above Chapter 15(A)2; Chapter 18 sections (C), (D); Chapter 19 section (B).

[19] A. C. Raper, *Weyhill Fair '. . . the Greatest Fair in the Kingdom'* (Buckingham, 1988), pp. 15–16, 75–82, 95–6.   [20] See above Chapter 15 section (C)3.

weekly or more frequent town market disappeared as the instrument of all but petty trades, and faced increasing competition even here from retail shops and, in the larger towns and cities, from the petty traders hawking foodstuffs in the streets and selling at the door in the expanding residential districts. Wholesale trades had long abandoned market day as even the occasion for transactions, and with new transport and communications systems and sources of supply, the plethora of middlemen and intermediaries that had characterised the integrating processes of the previous centuries lost their critical role, and declined. Even institutions of relatively recent origin, such as the village shop, which entered the period strongly diffusing into the country districts, found its role being eroded by the vans and itinerant traders who spread from the country towns from the 1880s and 1890s.[21] Long-standing institutional frameworks, which had withstood and adapted to earlier change, largely succumbed to these new forces, or were irrevocably transformed by them. The process of this institutional change began before 1850, and its working through continued beyond 1914, but it concentrated powerfully into the second half of the nineteenth century.

Food processing industries and the development of the consumer market in the context of largely rising real incomes and expanding trade opened the way to new tastes, new products, or the associative values added to goods by branding. These constituted a massive break with the past, and from mid-century the Burton brewers, led by Bass, encapsulated the new order of centralised production on a large scale of a consistent good, distributed on a national basis by the railway system and consequentially establishing a huge need for barrels and coopers to service them. They established a series of qualitative associations with the name, developing a brand image, which was reinforced by the use of their red triangle trade mark.[22] Through its processing of agricultural produce, and of branding more widely, brewing led the nation, and followers within the fields considered here were only to be found in the last third of the period; this was the classic phase in the development and marketing of branded goods in the United Kingdom.[23] But for foodstuffs such as bread and flour, many of the animal feeds and fertilisers, and veterinary products, the application of names such as Hovis or Turog, Thorley's Food, Lawes Manure or Harris's Mixture, added consistency of manufacture and advertising promotion to what was at root a generic product,

---

[21] See above Chapter 15 section (C)6; Chapter 19 sections (D), (E).

[22] K. H. Hawkins and C. L. Pass, *The Brewing Industry: A Study in Industrial Organisation and Public Policy* (London, 1979), pp. 20–1; Gourvish and Wilson, *Brewing Industry*, pp. 146–78; see above Chapter 17 section (B).

[23] For the wider context, C. Wilson, 'Economy and society in late Victorian Britain', *EcHR*, 2nd ser., 18, 1965, pp. 190–2.

and signalled the path to added value taken more widely in the agricultural product industries after the War. To these were added new products, symbolised by the triumph of Bird's custard and the invention of 'blancmange', and together they represented a major discontinuity in the history of the sale of food and related products.[24]

In his work on industrial enterprises Alfred Chandler identifies the food sector as one of the four groups (the others are chemicals, metals and machinery) in which the large scale multi-functional enterprise first appeared and tended to cluster in the last quarter of the nineteenth century. These were all industries where new techniques offered unprecedented cost advantages both in scale and in scope of production. But it was not only on the production side that such economies of scale and scope existed; they were also to be found in distribution. In addition, these enterprises required a managerial hierarchy to oversee the day-to-day process of production and distribution and to allocate resources for future operations; in short, new products needed the new techniques of business management. The extent of this change can be judged when we remember that in 1919, according to Chandler's calculations, 61 out of the 200 largest industrial enterprises in Britain were in the food and drink industry.[25]

More familiar as the element of discontinuity with the past was motive power, first steam and subsequently electricity and the internal combustion engine, the primary impact of which was largely exogenous to agriculture itself. Although, as David Grace's study indicates, UK agricultural engineers had succeeded by the 1850s in providing relatively lightweight and economic high-pressure steam engines, and displayed continuing ingenuity in the application of steam to such heavy field work as ploughing or land drainage and reclamation, much of the last was primarily an export industry, and had modest applicability to the home farm.[26] By contrast, new steam-powered systems of transport, first at home and subsequently across the Atlantic, and the transition of so many industries to powered factories and workshops, using powered machine tools to displace handicraft processes, strongly reshaped farming in England and Wales and profoundly affected the nature of labour. Widespread mechanisation in many of the industries assessed in Chapter 18 combined, as in footwear, to shift location towards the towns, de-skill the trade and eliminate the 'craft' element, often opening the way to the displacement of

---

[24] See above Chapter 16 sections (B)3, (C)2, (D)1, (D)2; Chapter 17 sections (C)1, (C)2.

[25] A. D. Chandler, Jr, *Scale and Cope: The Dynamics of Industrial Enterprise* (Cambridge, Mass., 1990); appendices B.1–B.3; Chandler, 'Creating competitive capability: innovation and investment in the United States, Great Britain, and Germany from the 1870s to World War I', in Higgonet, Landes and Rosovsky (eds.), *Favourites of Fortune: Technology, Growth, and Economic Development since the Industrial Revolution*, p. 437.   [26] See above Chapter 16 section (A).

male by female labour, and thus radically altering the labour process, eroding the practice of controlled labour supply inducted through formal or informal apprenticeship systems. Power machine manufacture had established its position by 1850, but only in a limited range of industries, notably textiles, but it consolidated its place in the second half of the century to complete the break with past modes of production.

This severance of industry from rural locations broke one of the strands holding the population in the country districts. Other links had lain in the older mining industries of the west and north, where decline in tin, copper and lead precipitated large-scale migration. However, up to the 1850s, farming had remained relatively extensive in its use of labour, and by several systems effectively hoarded labour sufficient to meet all but the peak load of harvest. This hoarding strategy became increasingly untenable in the next thirty years or so in the face of the shift in the share of farm income towards labour documented by Ó Gráda, and combined with the increasing availability of effective and reliable machinery to accelerate the release of workers from farming.[27] The gradual adoption of new materials for farm tasks, such as barbed wire, enamelled or galvanised buckets and churns, furthered this process, economising on the farm and support labour associated with traditional craft methods. Changing factor prices thus combined with the facilitating contributions of engineering to make the period after the 1850s the critical phase in rural depopulation. This early and plentiful release from the land provided the means for the labour-extensive urban industrial and service trades that characterised so much of the British economy in the pre-war period.

There were further elements of discontinuity in the traditional relationship of town and country in this period. Varying significantly by locality, and the relative standings of improvers and economisers in town politics, high Victorian England experienced the full impact of street improvement, the rationalisation of old street markets, or their in-filling under the pressures of enhanced land values, and the triumph of the sanitary engineer.[28] These combined to remove the 'night-man' from urban streets, and to break at least some of the traditional linkages of town muck and farm output in the cause of defeating cholera and satisfying the sensitivities of the urban middle classes.[29] As Michael Thompson shows, new industrial sources more than supplanted these supplies of fertility, or those garnered within the farm by the intensive techniques for gathering

---

[27] C. Ó Gráda, 'Agricultural decline 1860–1914', in R. C. Floud and D. N. McCloskey (eds.), *The Economic History of Britain since 1700*, vol. II, *1860 to the 1970s* (Cambridge, 1981), p. 177.

[28] Wales, apart from Merthyr, Cardiff and Swansea, was effectively not an urbanised society.

[29] A. Briggs, *Victorian Cities* (London, 1963), pp. 19–21, 144–7, 217–19; E. P. Hennock, *Fit and Proper Persons: Ideal and Reality in Nineteenth Century Urban Government* (London, 1973), pp. 107–11, 188–9.

dung and urine advocated by Mechi between 1860 and 1880 at Tiptree Hall in Essex.[30] While it should not be overstated, it was the period after 1842 that broke the cycle of pasture/fodder crop–manure–arable within which farming had previously operated, and began the processes that led to the growing of crops increasingly with 'artificial' or industrial sources of nutrition. Indirectly, mediated through farm livestock, imported feedstuffs added to this new nutritional balance.[31] As the comparative experiences of war demonstrated, British farming was still far from Germany's state of dependence upon chemistry and mineral imports for the successful cultivation of its fields, but the basic and harmonious manure cycle of the past had been irrevocably broken.[32]

The discontinuities discussed above in farm labour and the supportive crafts, and rural industries, by approaching the issues on the basis of the individual worker perhaps understates the complexity of subsistence in the rural districts in 1850, and the extent to which these structures were fragmented thereafter. The family rather than the individual was clearly the unit of earning, and the mixed earnings strategies of small land-holdings, coupled with a mix of industrial or other employments, some perhaps founded in temporary or cyclical spells, lived in towns, often indicated the premium placed by so many of the rural population upon staying put, and against migration. Circumstances eroding the viability of such strategies thus helped rural depopulation: the reduction of child labour by compulsory education appears to have been significant in some of the midland textile districts after 1870; mechanisation in shoemaking, completed in the 1890s, provided a late incentive to migrate; and, also in the east midlands, the deep mining of coal in the late years of the century provided a countervailing force, mopping up male labour that was displaced from traditional framework knitting by full mechanisation and the shift towards a female workforce.[33] These discontinuities in family earning strategies appeared at different times in different regions, with even the resistance to universal primary education being varied, but over time became general in their impact, making family 'bits-and-pieces' incomes more difficult to sustain in 1914 than in 1850, and weakening the hold on the land of the smallholder or petty trader. Male employments, as the best rewarded, were clearly fundamental in determining family income, but women's work in such industries as lace, straw plait, or gloving, had great regional or local significance. Child labour, always a marginal factor, was helpful to the cause but never sufficient in itself.

---

[30] C. S. Orwin and E. H. Whetham, *A History of British Agriculture 1846–1914*, (London, 1964), pp. 126–9, 130.   [31] See above Chapter 17 section (D).

[32] T. H. Middleton, *Food Production in War* (Oxford, 1923); P. E. Dewey, *British Agriculture in the First World War* (London, 1989); A. Offer, *The First World War: An Agrarian Interpretation* (Oxford, 1989).

[33] See above Chapter 18 sections (A), (B), (C), (E).

Decline in the first was often sufficient to precipitate change, at least for the young, but the need to gather cash meant that others in such trades, or in hand-knitting, having no capacity or inclination to move, adjusted to new circumstances, and reduced the effective wage, there being no alternative in the circumstances. At its extreme this was exemplified by the elderly bedridden widow, gloving or lacemaking to stay out of the workhouse.

The crafts and service trades were clearly less subject to the dramatic discontinuity of mechanisation, and were the more generally resilient. However the cumulative impact of the erosion of their market base in the rural population, new systems of and approaches to distribution, such as the urban vans from the 1890s, and changing materials meant that at least by the last twenty years of our period, many were under pressure, and in decline.[34] Contrastingly, the great growth of the professions and service industries of Victorian and Edwardian England and Wales, while not leaving the rural districts untouched, was primarily concentrated into the larger towns, and in the case of the law and banking above all into the capital.[35] This element of the shifting structure of employment towards services and the wide tertiary sector predominantly left the country districts behind, and even the rural retailer, craftsman, or publican, still numerous and widely distributed in 1914, were already in comparative decline, and vulnerable to the shock changes of war and its aftermath.[36]

While such workers were not Ben Gunns, abandoned on rural islands as the world passed them by, neither were they coping with the world of new materials and widening markets. Materials affected trades and crafts widely, as brick, slate and tile came to replace stone and thatch, and ultimately, at the great London brick works at Fletton after 1900, initiated large-scale and relatively continuous manufacture in place of the intermittent and the seasonal. A range of material sources of breaks with the past have been identified here, from tarred roads, easing access to the urban retailers, ironwork in vehicle- and equipment-making allowing the displacement of traditional timbers and perfect joints, to the celluloid toothbrush, which at the very end of our period was beginning to erode the skilled women's domestic work of bristle and bone manufacture. Osier and spelk basketmaking survived, but was not to endure long after the war, and well before 1914 crafts working flint or making charcoal were relics of dying trades bypassed by new processes and materials.[37]

---

[34] See above Chapter 19 sections (C), (D), (E).   [35] See above Chapter 19 section (G).

[36] See above Chapter 19 sections (E), (F); R. Perren, *Agriculture in Depression 1870–1940* (Cambridge, 1995), pp. 62–65.   [37] See above Chapter 18 section (D); Chapter 19 sections (B), (C).

The market also shifted against the costs of crafts, expressing a preference for the cheaper but less durable carts and waggons of town manufacture, and becoming increasingly intolerant of the shoeing charges of the traditional smithy, especially as alternatives to horsepower became practicable. As Sturt pointed out, the true costs of the wheelwright's business had never been fully charged: high real costs were borne in large-scale and long-term stockholding of hardwoods for construction, and suppliers of iron for their work or for farriery came increasingly into conflict with the informal and extensive credit basis on which so many rural crafts and services operated.[38] This spread of economic rationality to the crafts was encapsulated in the complaints, on either side of the War, of the excess charges for shoeing in England, and the perception among consumers that it was founded in a ring of iron. Traditional approaches to rural business, not just materials and processes, were therefore in growing conflict with the changing market.

Perhaps, in the absence of war as a final element of discontinuity, many crafts and trades would have experienced more protracted decline, but the War compounded decline in recruitment, and produced a pronounced ageing of the workforce of many rural crafts, implanting the certainty of decline in the 1920s and 1930s. Table 20.1 compares the age-structures of employment in a number of crafts in 1901 and 1921 with that of the male workforce at large to support this point. Each craft of wheelwright, saddler, clogger and mason exhibited this profile of a strongly ageing workforce between these dates, displaying small numbers in the younger age groups, and high proportions of older workers. Even the more buoyant crafts, which, as we have seen, incorporated a dominant urban element, aged as a result of the war, by comparison with the structure of overall employment. While these data are not to be taken as definitive proof of the structural weaknesses of the crafts, the preponderance of high index figures to the right of the 35–44 age-group, and low figures to its left, provides a clear diagnostic indication of decline.

This ageing workforce was symbolic of the decline of so many of the industries and crafts of rural England and Wales from their position of 1850 or earlier. It reflected part of the extensive processes of change experienced by farming and rural society during the period, as an already highly integrated rural economy encountered new infrastructure, international influences, new processes and new materials. Shifting patterns of industry and population reflected and compounded these changes, but neither wholesale innovation nor the decline of traditional structures and support systems was an evident certainty in the 1850s. Change came piecemeal, industry by industry, sector by sector, and locality by locality,

---

[38] See above Chapter 19 section (B).

Table 20.1. *Index of employment in selected crafts, compared with overall age distribution of the male workforce, England and Wales, 1901 and 1921 (Share in each age group, 1901 and 1921 = 100)*

	<20	20–4	25–34	35–44	45–54	55–64	65+
England & Wales, 1901 (per cent)	18.1	14.1	24.0	18.6	13.2	8.0	3.9
England & Wales, 1921 (per cent)	14.9	11.6	21.2	20.1	17.0	10.5	4.7
Smith, 1901	91.2	99.3	94.2	98.9	114.4	118.8	100.0
Smith, 1921	79.9	87.1	94.8	104.0	110.6	118.1	123.4
Cooper, 1901	63.5	75.2	85.0	99.5	101.1	170.0	114.9
Cooper, 1921	73.8	67.2	75.0	127.9	117.6	154.3	197.9
Wheelwright, 1901	92.8	90.8	81.7	101.1	84.6	128.8	159.0
Wheelwright, 1921	91.3	56.0	72.9	100.5	98.8	140.0	200.0
Saddler, 1901	95.0	92.9	97.5	98.9	106.8	122.5	110.2
Saddler, 1921	43.0	46.6	70.3	110.9	138.2	164.8	217.0
Clogger, 1901	96.7	82.3	83.8	101.1	124.2	137.5	120.5
Clogger, 1921	91.9	62.1	82.1	92.5	104.1	150.5	204.3
Mason, 1901	75.1	101.4	86.7	109.1	129.5	120.0	107.7
Mason, 1921	47.7	50.0	60.8	132.3	163.6	163.8	193.6

Source: Estimated from the occupational tables of Census, 1901 and 1921, which are not strictly comparable in respect of the overall age distribution, employment data covering those aged ten and over in 1901, and twelve and over in 1921. Proportions in the younger age group were consistently small, and so the distortion is modest, but small differences between index figures for some trades between 1901 and 1921 must therefore be treated as of no significance.

and was attributable to factors both endogenous and exogenous to agriculture. Much can be traced to the years before 1850, and few of these processes of change were exhausted by 1914, but many only began to bite from the 1880s or later, and the years 1850–1915 retain the central place in the long process of modernisation of institutions, supply networks, services and trades.

PART V
RURAL SOCIETY AND COMMUNITY

INTRODUCTION

BY ALUN HOWKINS

Recent, previous volumes of this work have been criticised for their lack of social history and their overemphasis on the purely economic or econometric history of rural England. This was not the case with the first volumes. For example the volume covering the period from 1540–1640 is rightly seen as one of the 'founding documents' of modern social history. The section that follows is in part an attempt to redress that balance and answer some of those criticisms. However, we should begin at least with some consideration of what is meant here by social history.

In the world of the common senses social history is still seen as the 'history of everyday things'. That element remains important and is present in the following chapters. This may be unfashionable, and even seen as purely antiquarian, yet the simple description of the pasts of family life, community institutions or popular beliefs and practices is of fundamental importance to our understanding of the past. Yet clearly we must move on from this, and most here would consciously or unconsciously follow some version of the view that social history is some form of an historical 'social science'. This assumes methodological as well as subject-based specialisms. Put simply it is concerned not only with 'everyday things' but with particular ways of looking at them. As F. M. L. Thompson puts it in his introductory chapter to the *Cambridge Social History of Britain*, 'social historians draw widely on concepts from historical demography, social anthropology, sociology, social geography and political science, as well as from economics'.[1] Again, the influence of all, or most, of these disciplines will be obvious in the next chapters.

Yet adopting these methodologies uncritically is fraught with dangers. Tony Judt argued some years ago that taking concepts from the social sciences like 'modernisation' or 'urbanisation', or the wholesale adoption of 'systems' of explanation derived from the social sciences, not only often

---

[1] F. M. L. Thompson, 'Editorial preface' in F. M. L. Thompson (ed.), *The Cambridge Social History of Britain, 1750–1850*, vol. I, *Regions and Communities* (Cambridge, 1990), p. xiii.

makes for bad history but marks a loss of faith in the idea of history itself.[2] As Lawrence Stone puts it with characteristic majesty:[3]

Economic and demographic determinism have collapsed in the face of the evidence . . . Structuralism and functionalism have not turned out much better. Quantitative methodology has proved a fairly weak reed which can only answer a limited set of problems.

More recently David Cannadine from history and John H. Goldthorpe from sociology have issued similar warnings, Goldthorpe in particular stressing the near impossibility of generalising in, or from, historical evidence except in very unusual circumstances.[4] Yet it is not that simple. Very often these criticisms turn out to be 'attacks' on particular pieces of work, usually, although not always, those associated with left or *marxisant* work of the 1960s and 1970s – a real case of babies and bath water even if many would argue that the latter did need changing.

We hope, as always, that we have avoided these pitfalls one and all! In the chapters that follow, it is not mainly that 'social science' methods are followed slavishly, rather that the 'questions' that some social scientists might ask of a society are the ones which structure the material presented. But it is not a social science of one school. The chapter on rural population change by Brian Short draws on the methods and approaches of quantitative historical geography, while those by myself come much more out of a humanistic strain in the social sciences. What is interesting is the extent to which both approaches rely on regionalism, a concept of the social sciences in many respects, to explain different patterns of behaviour.

There is also an important sense in which aspects of social history, probably because of their initial interest in the everyday, have become synonymous with the history of the poor and the underclasses. Again there are aspects of this in the following chapters, although a good deal of space is also given to the 'wealthy and great', who interestingly have largely escaped the view of social historians. This was (historically) very much a legacy of social history's social science parent. However, less easy in these changed times is social history's lifting of the stone to reveal the squalor and harshness of the lives of most people in the past, and growing from that, attempts to create models of society which are driven by the relative positions of different groups. In the most general sense the chap-

---

[2] Tony Judt, 'A clown in regal purple: social history and the historians', *History Workshop Journ.*, 7, 1979, pp. 66–94.

[3] Lawrence Stone, 'The revival of narrative: reflections on a new old history', *PP*, 85, 1979, p. 19.

[4] David Cannadine, 'British history: past, present, and future?', *PP*, 116, 1987, pp. 169–91; John H. Goldthorpe, 'The uses of history in sociology: reflections on some recent tendencies', *British Journ. of Sociology*, 42, no. 2, 1991, pp. 211–230.

ters that follow see a connection between political/cultural power and socio-economic position, although none of them would see that relationship as unproblematically determinist. I shall return to some of these questions in the final chapter of this section.

The chapters that follow, although written by different hands, present a logical, if not complete, account of the social history of rural England and Wales in the period 1850–1914. We begin with a discussion, by Brian Short, of the complexities of the demographic history of rural England and Wales. This is followed by an account, by Alun Howkins, of the spatial and community structures of rural England, and then a largely descriptive account of the social, family and cultural structures of the different classes in rural England and Wales (the aristocracy, the farmers and the farmworkers). The last two chapters, by Anne Digby, look at institutional social history of the period – the poor law, religious behaviour, local government, crime and education.

Aspects of these chapters overlap with one another showing a remarkable degree of accord – other parts are more of a problem. Here, a difference of emphasis, or a difference of materials used, can result in different accounts of similar processes. However, these are few and do not centrally alter the main areas of argument. On the other side, many areas which, had this project been a different one, would have been included, are not. For example, much material on labour, wages and working conditions is included under the section on the farmworker. Similarly a lot of the material on the rural élites which might have found its way into a 'social history' account will be found elsewhere. Nevertheless, what follows can be read and understood as a whole.

CHAPTER 21

RURAL DEMOGRAPHY, 1850–1914

BY BRIAN SHORT

I. INTRODUCTION: DEFINITIONS AND APPROACHES

We still await a full demographic history of rural England and Wales in the late nineteenth century. Incorporated in several texts which concentrate on population increases and urbanisation as part of broader demographic trends, or in texts on rural issues as part of the changing socio-economic complexity of the countryside at this time, many questions remain unanswered. While most attention has been held by the burgeoning towns and cities attending England's premier position in industrialisation and urbanisation, the rural issues are highly interdependent.[1]

It is unavoidable for studies of population in the nineteenth century to wrestle with problems of definition and data. What was rural? Who was a rural inhabitant? How to incorporate areas which were rural in 1850, but which had become urbanised by 1914? The difficulties faced by successive writers in attempting to define the 'urban' population of England and Wales also affect, by extension and inversion, definitions of 'rural' between the census of 1851 and 1911. There is also the spatial unit problem: can one work with units as large as a county? Is the Rural District a better prospect (after its 1894 appearance), or should all analysis remain at the level of the census authorities' Registration District?

It should also be noted that the later nineteenth century witnessed people moving into the countryside, or into its suburban fringes, who had otherwise no 'agrarian' connection other than their purchase of living space in an environment which was becoming a fashionable consumer choice. The impact of these newcomers' demographic behaviour on local rural inhabitants is another theme which requires attention. A class element is thereby an inevitable ingredient of any detailed analysis of demographic trends, at a time when one class, urban and bourgeois, faced another, rural and artisan-labouring in the countryside, often with

---

[1] R. Lawton and R. Lee (eds.) *Urban Population Development in W. Europe from the late-eighteenth to the early-twentieth Century* (Liverpool, 1989).

mutual incomprehension. The old rural certainties of mutual class suspicion or distanced respect now gave way to a less certain dimension as 'new men' and their families moved into close village proximity, bringing demographic behavioural differences with them.

Definitions and data have bedevilled much past work on rural demography. One of the most sophisticated, near-contemporary analyses of rural change was that by Bowley, whose work has influenced later writers. He based his 1914 examination of population change on the relatively new Rural Districts, and defined his rural data base in terms of population density (under 30 persons/100 acres in 1911, together with a judgmental number with between 30–50/100 acres, and with individual parishes with higher densities omitted). Adna Weber had previously based his 1899 study of settlement change on Rural Sanitary Districts, precursors of the Rural Districts, for the British element of this analysis, but on the other hand Welton in 1911 used the Census registrar's Registration Districts as of 1891, together with some rural parishes from within otherwise urban Registration Districts, referring to the whole as the 'rural residue'. Cairncross later followed Welton, with some modification but no real change.[2]

Unfortunately far less illuminating are those texts based on county-level statistics, which inevitably fail to capture the complex interpenetration of town and country. Thus the North Riding of Yorkshire in the decade 1871–81 contained the growing towns of Scarborough and Middlesbrough, whose growth was sufficiently large to mask the exodus from the rural Registration Districts covering the rest of the county. Lancashire, with Liverpool, Manchester and the coalfield and textile towns, also contained many more remote moorland parishes. The controversial paper by Ogle in 1889, which appeared to refute claims of contemporary rural depopulation, was based on such county-level data, excluding only the towns of 10,000 inhabitants and more. Even less satisfactory, of course, are studies which arbitrarily group counties, such as those of Wales, into northern and southern classes, whether the study area is the British Isles or Europe.[3]

---

[2] A. L. Bowley, 'Rural population in England and Wales: a study of the changes of density, occupations and ages', *JRSS*, 77, 1914, pp. 597–645; A. F. Weber, *The Growth of Cities in the Nineteenth Century: a Study in Statistics* (1899, reprint New York 1963); T. A. Welton, *England's Recent Progress: an Investigation of the Statistics of Migrations, Mortality etc. in the Twenty Years from 1881 to 1901 as indicating Tendencies towards the Growth or Decay of Particular Communities* (London, 1911); A. K. Cairncross, 'Internal migration in Victorian England', *The Manchester School*, 17, 1949, pp. 67–87; A. K. Cairncross *Home and Foreign Investment 1870–1913* (Cambridge, 1953).

[3] J. Saville, *Rural Depopulation in England and Wales 1851–1951* (London, 1957), p. 46, fn. 1; W. Ogle, 'The alleged depopulation of the rural districts of England', *JRSS* 52, 1889, pp. 205–14. Much of the work of the Princeton European Fertility Project is presented at the county or amalgamated county level, as illustrated in M. S. Teitelbaum, *The British Fertility Decline: Demographic Transition in the Crucible of the Industrial Revolution* (Princeton, N.J., 1984); and in the summary volume, A. J. Coale and S. C. Watkins, *The Decline of Fertility in Europe* (Princeton, N.J., 1986).

A relatively satisfactory compromise takes the Registration District as the prime data source. England and Wales was initially divided into 624 Registration Districts, based on the existing Poor Law Unions (with later changes at different census dates) in the latter half of the nineteenth century, of which about 360 were rural, as defined crudely by population density.

These finer units deal with the localised variations across rural England and Wales, although they generally still included at least one country or market town, whose demographic character might differ from that of its surrounding hinterland. The parish is even more useful, but no aggregate work based on rural parishes has yet been presented for England and Wales as a whole, although we have many valuable case studies. It is also salutary to remember that in 1851 the parish of Leeds, one of the largest parishes in the north, included out townships which were more truly rural rather than suburban in character, so that even this spatial unit can be internally split between town and country.[4]

More sophisticated delineations of rurality are also available, one used in recent years being the mirror image of that used to define urban areas by Law. In dealing with nineteenth-century urbanisation he attempted to provide a uniform data set by defining urban areas as reaching a minimum size (at least 2,500 people); reaching a density of at least one person per acre; and relating to a contiguous built-up area.[5]

Using his definition, the residual figures are obtained from his data for urban populations, which can be compared with those for Bowley, Welton and the Census (Table 21.1).

Clearly, there is no one definition of 'rural' (let alone 'agrarian') which can be used. It will be seen that Law's definition excludes more population from being rural than does the official census, since he does have a low threshold in terms of aggregate population per spatial unit. This should be counterbalanced however by his other two components, and

---

[4] R. Lawton, 'Census data for urban areas' in R. Lawton, ed., *The Census and Social Structure* (London, 1978), p. 95. For studies using Registration Districts see, for example, R. I. Woods 'Approaches to the fertility transition in Victorian England', *Pop. Studs*, 41, 1987, p. 283–311; R. Lawton 'Rural depopulation in nineteenth century England' in R. W. Steel and R. Lawton, eds., *Liverpool Essays in Geography: a Jubilee Collection* (London, 1967); M. Anderson 'Marriage patterns in Victorian Britain: an analysis based on registration district data for England and Wales, 1861', *Journ. Family Hist.*, 1, 1976, pp. 55–78; and Dov Friedlander, 'Demographic responses and socio-economic structure: population processes in England and Wales in the nineteenth century', *Demography*, 20, 1983, pp. 249–72. For valuable case studies using parish-level data, see for example, J. Saville, *Rural Depopulation* for South Hams, Devon parishes; and P. R. A. Hinde, 'The marriage market in the nineteenth century English countryside', *Journ. European Economic Hist.*, 18, 1989, pp. 383–92 for 31 parishes in Derbyshire, Yorkshire, Norfolk and Shropshire.

[5] C. M. Law, 'The growth of urban population in England and Wales, 1801–1911' *TIBG*, 41, 1967, pp. 125–43.

Table 21.1. *The rural population of England and Wales, 1850–1914*

	Total pop.	1 Law's		2 Census dftn[a]		3 Bowley's		4 Welton's	
		Rural	%	Rural	%	Rural	%	Rural	%
1851	17.9	8.2	45.96	8.9	49.8			7.6	42.60
1861	20.1	8.3	41.27	9.1	45.4	4.9	24.6	7.9	39.28
1871	22.7	7.9	34.83	8.7	38.2	4.9	21.8	8.3	36.57
1881	25.9	7.7	29.87	8.3	32.1	4.8	18.3	8.5	32.95
1891	29.0	7.4	25.52	8.1	28.0	4.6	15.9	9.1	31.25
1901	32.5	7.2	22.00	7.5	23.0	4.5	13.7	6.6	20.30[b]
1911	36.1	7.6	21.08	7.9	21.9	4.6	12.7	7.9	21.95[c]

Notes:

[a] Based on places recorded as towns in the 1851–71 censuses; and for 1881 onwards for Rural and Urban Sanitary Districts (later Rural and Urban Districts) and County and Municipal Boroughs.

[b] Welton's towns over 1,000 taken out

[c] Welton's towns and 'populous districts' taken out.

thus the numbers may well be a more realistic appreciation of the rural proportions, showing between 1 and 4 per cent fewer of the total population being rural than the official statistics, but with the figures coming more into line in the twentieth century.

Bowley's definition is more stringent, both in density terms (for the most part just 0.3 per acre, going up to 0.5 in some areas, compared with Law's 1 per acre) and in excluding large suburban and mining populations. In 1911, for example, he maintained that the official figure of 21.9 per cent rural contained many who were not rural at all, but urbanised and living in administrative Rural Districts. Occupations were thus included to define further his rural area, but unfortunately he did not include wives and children in the populations so distinguished, and it is clear that the distinction between 'agricultural' and 'rural' populations is by no means an easy one to make. Bowley actually took no fewer than 3.1 million acres out of the 'rural' area, since this was countryside where population was actually growing faster between 1891 and 1911 than in the towns themselves. He was left with a more truly rural area of 30.2 million acres, about 81 per cent of the total area of England and Wales. Welton's is an erratic measure, even though by 1911 he was basing his work on the 1891 Registration Districts (see Part VII, Chapter 40, Table 40.1 for a different interpretation of the size of Welton's rural population).

The fundamental problem is, then, one of synthesising from a barrage of data which is available for this period. The lack of agreement means that there is no agreed area which can be seen as 'rural' (Fig. 21.1). Different writers used different areas, with varying populations, for different purposes. This was anyway a time of enormous change as the urban areas expanded in size and number across England and Wales, thereby rendering even a satisfactory definition at one time possibly insufficient at another. As Lawton has written, 'One must deal with various formal definitions – all more or less unsatisfactory.'[6]

Depending on definition therefore, the rural population of England and Wales declined from a Census definition maximum of 49.8 per cent of the total population in 1851 (8.9 million), down to 21.9 per cent (7.9 million) by 1911 (Table 21.1). The absolute maximum was 9.1 million in 1861 (Census) or in 1891 (Welton). In terms of Rural and Urban District boundaries, the period saw a steady decline from 1861 to 1901, with a recovery from 1901 to 1911, although Bowley perceived little change between 1861 and 1871. Whatever definition is used however, the fundamental point is clear: the rural population from 1850 was in the minority, and it shrank from then until 1914 from just under one-half to just over one-fifth of the total population of England and Wales. It reached its lowest point in 1907 in absolute terms, recovering to 1911, although continuing to decline relative to the total population. Within that broad compass, the following analysis attempts to unravel the complexities of the demographic structures and processes.

Demographic structure, process and place

Demographic structures (the patterns of age and sex differentials); the distribution and timing of vital events (births, marriages and deaths); and the processes of population change (natural growth or decline, migration and emigration) are closely interrelated. Analytically, they must be separated to allow a closer understanding of the rural changes operating, and separate sections will therefore follow on nuptiality, fertility, mortality and migration. But all gave to, and drew from, particular local traditions and economies. Indeed, just at a time when the countryside and its cultures was so threatened by industrialisation, great cities and suburban sprawl, the age brought its own collectors of local curios, customs and folk memorabilia, many of which provided tangible symbolism of the demographic processes being experienced. Local wedding and courting customs, local baptismal practices or their avoidance, local customs connected with bereavement and mourning thus all gave colour and local

---

[6] R. Lawton, 'Census data for urban areas', in R. Lawton, ed., *The Census and Social Structure*, p. 84.

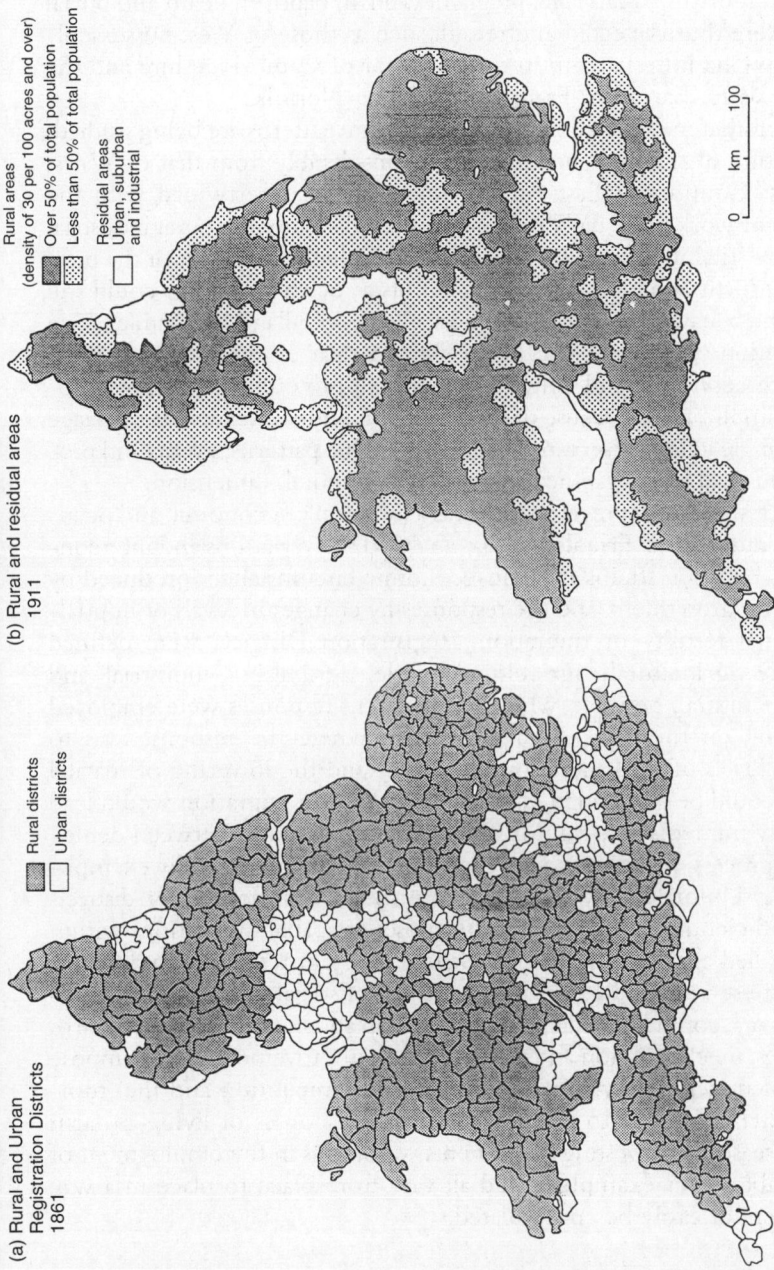

(a) Rural and Urban
Registration Districts
1861

    ■ Rural districts
    □ Urban districts

(b) Rural and residual areas
1911

Rural areas
(density of 30 per 100 acres and over)
    ■ Over 50% of total population
    ▨ Less than 50% of total population

Residual areas
    □ Urban, suburban
      and industrial

0   km   100

21.1 Competing evaluations of population density in late nineteenth-century rural England: (a) urban and rural areas 1861, by registration district, based on population densities above and below 100 persons per km² in 1861, adapted from Woods (*Pop. Studies* 41, 1987, p. 306, Fig. 8; (b) rural and residual areas 1911. Registration districts are also classified by population density (based on Bowley and Lawton, 1967, p. 232)

dignity to population processes, especially as they affected the rites of passage. Courtship, marriage, pregnancy, birth, baptism, death and burial rituals were thus collected and retold, such as those of West Sussex collected by Charlotte Latham in the 1870s, or of Cambridgeshire and the Fens by W. H. Barratt of Framingham Pigot, Norfolk.[7]

Therefore it matters where the demographic patterns are being studied. The culture of the rural north differed considerably from that of Wales, the West Country or East Anglia. The cultures, intertwined with the rhythms of work – the daily routines of the agricultural or rural industrial calendar – still influenced, for example, the timing of marriage or the incidence of birth. The local economy and its associated customs could still define the behavioural norms for most people in village and hamlet. The continuation of living-in service provides a clear example of an institution both economic and cultural, which had profound regional, demographic implications because of its association with a later age of marriage for the single living-in servants. All demographic patterns, events and processes thus had a geographical, as well as a temporal, dimension.

In part, the geographical variation is a function of economic and occupational differences. Friedlander has analysed the variations and interconnections between strains in socio-economic circumstances produced by population growth etc. and the responses by changes in levels of nuptiality, marital fertility or migration. Registration Districts were defined as purely agricultural, agricultural-textile, agricultural-industrial and non-agricultural, between which the different responses were employed depending on the type of district. Thus where the response was to increase levels of out-migration, nuptiality and the lowering of marital fertility could be postponed (and vice versa), since migration would lead to a fairly immediate release of stress.[8] Clearly the links between demographic process, place and socio-economic structure are vividly exemplified here. Unfortunately this otherwise useful classification of districts was insufficiently sensitive for our purposes to spatial distinctions within the so-called agricultural districts, where different responses to different sets of stresses, depending on a wide variety of social, economic and cultural factors, could be made. Changes in crop and livestock patterns, farm structures, mechanisation and transportation improvements could impose strains of underemployment due to foreign competition and the 'tumbling down' of arable to pasture. Again the responses of living-in farm service, seasonal harvesting movements, variations in the employment of family labour, for example, could all vary from place to place in a way which cannot easily be encapsulated.

---

[7] E. Porter, *Cambridgeshire Customs and Folklore* (London, 1969), pp. 1–36; J. Simpson, *The Folklore of Sussex* (London, 1973), p. 11.   [8] Friedlander, 'Demographic response', pp. 249–72.

Demographic structures and processes thus range in scale from the local to the international, linked to trade fluctuations and political decisions, for example, via changes in localised cropping or farm structures with their implications for labour requirements. Sweeping demographic changes came to rest in particular environments, which transmuted their message into a localised context, as specific historical moments. This reiteration between the local and the national will constantly inform this analysis.

II. NUPTIALITY

Age at marriage exerted a powerful influence on the formation and structure of the rural household. Indeed nuptiality is now regarded as the main variable in accounting for fertility changes in the population as a whole, at least until the onset of general fertility decline within marriage in the 1870s. Certainty for the 'long' eighteenth century:

Marriage was the hinge on which the demographic system turned, and given the crucial importance of the tension between production and reproduction which affected all preindustrial societies, its significance was far wider than the purely demographic. Many aspects of English social and economic life influenced, and were influenced by, marriage behaviour[9]

Even for late nineteenth-century England and Wales there are many questions to be posed about the conditions for marriage and how such conditions varied from one agrarian region to another or between occupational groupings. We need to know more about the resources needed before a separate household might be established; how such resources were obtained (inheritance, loans, savings); about courtship patterns; or about the role of sex in the timing of marriage. There is also some debate over the extent to which age at marriage varied between localities, and the relative influence of different factors which may have accounted for the observed patterns. Such investigations are specialised but one thing is clear: the interconnections between rural production and reproduction come strongly to the fore when considering nuptiality.[10]

[9] E. A. Wrigley, 'The growth of population in eighteenth-century England: a conundrum resolved', *PP*, 98, 1983, pp. 121–50, reprinted in E. A. Wrigley, *People, Cities and Wealth* (Oxford, 1987), pp. 239–40.

[10] Anderson, 'Marriage patterns', pp. 55–78; Anderson, 'Historical demography after "The Population History of England"', in R. I. Rotberg and T. K. Rabb, eds., *Population and History: from the Traditional to the Modern World* (Cambridge, 1986), pp. 46–7; D. Friedlander and E. B. Moshe, 'Occupations, migration, sex ratios, and nuptiality in nineteenth century English communities: a model of relationships', *Demography*, 23, 1986, pp. 1–12; R. I. Woods and P. R. A. Hinde, 'Nuptiality and age at marriage in nineteenth-century England', *Journ. Family Hist.*, 10, 1985, pp. 119–44.

Of the available measures which can be employed to give the essential data on nuptiality, the most useful perhaps is the Singulate Mean Age of Marriage (SMAM). SMAM for males fell from 26.94 in 1851 to 26.43 in 1871 and then rose to 27.65 by 1911. The SMAM for females was 25.77 in 1851, 25.13 in 1871 and 26.25 in 1911. Thus the two series moved in concert.[11] It has been demonstrated that the mean age at marriage of women fell during the first half of the nineteenth century to 22.9 in the late 1850s, and then rose to 24.4 at the time of the First World War. That of men fell to 24.4 and then rose to 26.7, the latter figures bringing the levels back to those of around 1700.[12]

The proportion remaining unmarried also affected overall demography. One measure of this is the marriage rate (married people per 1000 population). In 1850 this was 17.2 per 1,000 for England and Wales, falling to 14.9 in 1880, and rising to 15.9 by 1914. The proportion of women aged 20–24 who were ever married was 34.8 per cent by 1871, but by 1911 this had fallen to 24.3 per cent, as larger numbers were deferring marriage. The numbers of permanently unmarried women rose at the same time from 12.2 per cent in 1871 to 16 per cent in 1911, and the effect of these changes alone was to reduce the Gross Reproduction Rate (GRR) by 10 to 15 per cent between these years. It would appear that the percentage of women married in the population increased to a peak in 1871 or 1881, but showed an overall decline by 6 per cent between 1851 and 1911.[13]

Clearly therefore the English countryside from 1851 to 1871 must be seen in the context of an overall increase in marriages, a lower age at marriage and lower proportion of celibates, followed by a reversal around the 1870s which ushered in higher ages at marriage, higher numbers remaining celibate and fewer marriages.

Within rural areas the lowest mean age at marriage and the highest proportions of married women, standardised by age (Im), and thus fewest spinsters, were to be found in areas of, or adjacent to, mining activity.[14] Anderson suggested that 'a clear relationship existed between living in

---

[11] E. A. Wrigley and R. Schofield, *The Population History of England, 1541–1871: a Reconstruction* (2nd edn, Cambridge, 1989), p. 437.

[12] M. Anderson, 'The social implications of demographic change' in F. M. L. Thompson, ed., *The Cambridge Social History of Britain*, vol. II, 'People and their environment' (Cambridge, 1990), p. 32.

[13] B. R. Mitchell and P. Deane, *Abstract of British Historical Statistics*, Cambridge, 1971, p. 45; D. Levine, *Reproducing Families: the Political Economy of English Population History* (Cambridge, 1987), p. 185; Teitelbaum, *British Fertility Decline*, pp. 97–100; Wrigley and Schofield, *Population History*, pp. 436–7.

[14] The Princeton Study of fertility in Europe initiated several indices which have become standard in demographic analysis, and this contribution uses the several standardised indices as appropriate. Im represents the proportion married, standardised by age. For the full derivation of these indices, see A. J. Coale and S. C. Watkins, eds., *The Decline of Fertility in Europe* (London, 1986).

areas dominated by textiles, coal mining, engineering and metal manufacturing, or shoemaking and one's statistical chances of marriage and of
an early marriage age'.[15] In 1851 mining districts were less urbanised than
the national average, and although they had become more urbanised than
average by 1871, such districts can be regarded as rural at least during the
early part of the period under review. Such districts were rural but with
a non-agrarian character.

Thus in 1851 the counties of Durham and Staffordshire had the largest
proportion of married women of child-bearing age, together with Essex,
Cambridgeshire, Northamptonshire and Huntingdonshire which ranked
with the mining counties.[16] By 1861 Im by Registration District clearly
distinguishes the Durham coalfield and the coal-mining areas of
Yorkshire–Derbyshire–Nottinghamshire, southern Lancashire, southern
Staffordshire and south Wales as having the highest values. Military districts also had high values, as still did East Anglia and the rural south of
Midlands from the Wash to the Bristol Channel. By 1891 there were
fewer Districts with such high values because of the general fall in Im,
but the coalfields were still prominent, whilst much of eastern and central
rural England also showed above-average values, although in a less continuous belt.

In the relatively isolated mining villages there was an abundance of
bachelors, with an earnings peak showing early in life, and this relative
wealth at an early age was combined with little available work for women
in mining (especially after Lord Shaftesbury's Act of 1842) to produce a
low opportunity cost (in terms of income forgone) for women to marry
and have children. It was the cheapness of a wife's labour that accounted
for early marriages in colliery districts. One old collier said that he had
'been obliged to get a woman early' to avoid paying away all his profits.
Some secondary work in farming or handicraft production might also be
available, and under these circumstances new households could be established with some ease. The practice of employing members of one's own
family underground in the ironmining districts of Tankersley, in the West
Riding of Yorkshire, would also have been a powerful inducement to
early marriage.[17]

---

[15] Anderson, 'Marriage patterns', p. 64.
[16] *Census of Great Britain 1851*, Population Tables II. Ages, civil condition, occupations and birth-
places of the people (vol. I, 1854), xxxii–xxxiii, xlii–xliii; Teitelbaum, *British Fertility Decline*,
p. 110.
[17] I Pinchbeck, *Women workers and the Industrial Revolution 1750–1850* (London, 1981), p. 264; M.
Haines, *Fertility and Occupation: Population Patterns in Industrialisation* (New York, 1979); Woods
and Hinde, 'Nuptiality and age at marriage', pp. 119–44', M. Jones, 'Combining estate records
with census enumerators' books to study nineteenth century communities: the case of the
Tankersley ironstone miners c. 1850', *Local Pop. Studs.*, 41, 1988, pp. 13–27.

However the availability of industrial or other non-agricultural work and higher wages in themselves were not necessarily related to early marriage, since the sexual division of labour and organisation of production were important intervening variables. Thus in rural districts with industrial activities, the work available to women might be far greater, and nuptiality might be suppressed or delayed. The best-known example is that of textiles, but again a differentiation should be made between cottage or putting-out work, and work in the mills. In mid-Victorian Swaledale and Wensleydale, for example, many female knitters were engaged in cottage work, especially in the 1850s and 1860s. Only with the 1905 closure of the last mill, which had been providing outwork to cottages, did the textile era end. Although remaining subsidiary in terms of earnings to lead mining and agriculture, there can be little doubt that the diversity of employment was strongly associated with a marked rise in population before 1851, and that the subsequent decline of this diversity adversely affected population growth. Similarly, in the West Country women and children spun for men to weave in the family cottage. Work in the mills, on the other hand, generally led to a higher female age at marriage. In the mills of Lancashire, girls left school at fourteen to go straight into the weaving sheds, and marriage was thereby postponed.[18]

The picture is complex. In many Bedfordshire and Buckinghamshire communities women's and girls' employment was still available after 1850 in lace-making and straw-plaiting for the hat industries of Dunstable. In the straw-plaiting parish of Ivinghoe in 1871, 275 women and 76 men were earning a living as plaiters; altogether over 60 per cent of women and girls were so employed. Here, in contrast to the Dales, the numbers of women marrying early were low, and many children left home early, 'hardened and intractable', to gain independent incomes, and escape family earning pressures. The availability of domestic plait- and lace-making in the Buckinghamshire–Bedfordshire area, in association with male agricultural employment, could arguably have been an inducement to marriage, rather than a deterrent, but this appears not to have been the case. Thus in the Coggeshall area married women, working at lace-making at home, could make between 6s. and 10s. per week in the early 1920s, a substantial help with the domestic budget, but for the unmarried this equally gave a high opportunity cost to marriage.[19]

---

[18] C. Hallas, 'Cottage and mill: the textile industry in Wensleydale and Swaledale', *Textile History* 21, 1990, pp. 203–21; Hallas, 'Migration in nineteenth-century Wensleydale and Swaledale', *Northern History*, 27, 1991, pp. 139–61; E. Roberts, *A Woman's Place: an Oral History of Working-Class Women, 1890–1940* (London, 1984), p. 59.

[19] C. A. and P. Horn, 'The social structure of an "industrial" community: Ivinghoe in Buckinghamshire in 1871', *Local Pop. Studs*, 31, 1983, pp. 9–20; P. Horn, 'Child workers in the pillow lace and straw plait trades of Victorian Buckinghamshire and Bedfordshire', *Hist. Journal,*

Table 21.2. *Contrasting marriage rates in Durham and rural Wales, 1891*

Age at end decennium	Durham			Carmarthen etc.		
	Single 1891	Married in next 10 yrs	%	Single 1891	Married in next 10 yrs	%
20–25	88,006	31,000	35.2	16,397	3,346	20.4
25–35	119,070	80,500	67.6	25,861	11,187	43.3
35–45	26,259	11,000	41.9	9,417	3,248	34.5
45–55	8,541	1,300	15.2	3,905	623	16.0

Source: T. A. Welton, *England's Recent Progress* (London, 1911), p. 63.

Clearly the availability of female non-agricultural employment requires careful examination before any causal inferences can be drawn. Many sometimes opposing factors were at work, counteracting each other in different localities, and it is always vital to remember local individuality.[20]

More commonly in rural areas commentators noted the lateness of marriage. The proportions of married women in 1851 between the ages of twenty and forty were lowest in Wales and the rural border counties, such as Herefordshire and Shropshire, in rural Westmorland and Cumberland and the south-west. The numbers of bachelors were similarly high in such areas, together with Kent and Hampshire, with their naval and military establishments. Welton demonstrated the difference in marriage rates by a comparison of the Durham colliery districts and the Carmarthen, Pembroke and Cardigan area in 1891. The rates of the Welsh area were constantly lower, except for the oldest age groups shown where later ages at marriage produced higher figures (Table 21.2).

Nuptiality, as noted above, was one of several demographic responses relieving strain within rural England and Wales.[21] One key factor which had general importance in retarding marriage was the extent and survival of 'life cycle servants' within living-in farm service and the availability of employment in domestic service. Both were common pursuits for the young and single, the latter being almost totally reserved for females. The extent of the survival and decline of living-in in its various forms and its replacement with day- or wage-labour may be debatable but the

---

17, 1974, p. 791; M. Hewitt, *Wives and Mothers in Victorian Industry* (London, 1958), pp. 40–1; H. E. Fitzrandolph and M. D. Hay, *The Rural Industries of England and Wales* 3 (London, 1927), pp. 67–9.   [20] Levine, *Reproducing Families* p. 1.

[21] Friedlander, 'Demographic responses'. For the concept of nuptiality as a 'strategic control', see also E. A. Wrigley, 'The local and the general in population history', *16th Harte Lecture*, Univ. of Exeter (Exeter, 1983), p. 5.

impact of its occurrence on rural demography should be noted. In particular, there is the significance of the division between traditional (living-in) and non-traditional (wage-labour) relationships within agriculture. For Anderson, 'Almost half the variance of the married fertility potential in agricultural areas is "explicable" in terms of the variables "traditional agriculture" and "wage labour agriculture."'[22] In the most traditional farming areas the proportion of men unmarried in the 25–34 age group was 40 per cent higher than in the least traditional, and the mean age at marriage was almost two years later. For women the comparable figures were 30 per cent higher, and eighteen months later. In the twenty years after 1850 it is likely that living-in actually increased in the north, as prices of animal products relative to grain rose. In such areas potential marital fertility was almost 20 per cent lower. In ways similar to this the role of female domestic service also depressed nuptiality, although this was perhaps more of an urban or suburban middle-class than a rural phenomenon. Certainly there was a negative association between Im and numbers of females in service in 1861.[23]

Analyses of farm servants in 1851 demonstrate the northern and western distribution of the farm servant (approximately 74 per cent north of a line from the Mendips to the Wash), together with pastoral outliers in the south, as in the Weald, for example, whilst the situation *c.* 1900 shows the same broad features. The difference between pastoral Atcham (Shropshire) and arable-based Mitford (central Norfolk) is relevant here, with the former still relying by 1861 on a large number of living-in farm servants compared with the latter where they had become relatively unimportant. The SMAM for the former was high (between 26.96 years and 28.44 years 1851–81), for the latter low (24.21 to 25.5 years). Again, there is overwhelming evidence for the significance of employment for an understanding of nuptiality. Opportunity to marry, or even to meet eligible partners, was restricted for those young people living with their employers, learning a trade and saving for the future. Many married within one year of leaving service.[24]

In broad terms this can help to explain why the largest numbers of unmarried rural adults were to be found in the smaller family and pasto-

[22] Anderson, 'Marriage patterns', p. 64.   [23] *Ibid.*, pp. 65–9.

[24] A. Kussmaul, *Servants in Husbandry in Early-modern England* (Cambridge, 1981), pp. 20–1, 131–2; B. Short, 'The decline of living-in servants in the transition to capitalist farming: a critique of the Sussex evidence', *Sussex Archaeological Collections*, 122, 1984, pp. 147–64; K. D. M. Snell, *Annals of the Labouring Poor: Social Change and Agrarian England, 1660–1900* (Cambridge, 1985), pp. 96–7; A. Howkins, *Reshaping Rural England: a Social History, 1850–1925* (London, 1991), pp. 52–5; Howkins, 'The English farm labourer in the nineteenth century: farm, family and community', in B. Short, ed., *The English Rural Community: Image and Analysis* (Cambridge, 1992), p. 88; P. R. A. Hinde, 'Household structure, marriage and the institution of service in nineteenth-century rural England', *Local Pop. Studs*, 35, 1985, pp. 43–51.

ral farms of the northern counties of England and in Wales, whereas the
lowest numbers were on the larger cereal farms of East Anglia and the
Midlands. This feature of rural demography has frequently not been made
sufficiently clear, and the distinction between the earlier marriages of the
eastern and midland areas compared with the later marriages of the northern and western counties represents a fundamental demographic, as well
as social, cultural and economic divide. These same regions of lower age
at marriage and higher nuptiality in East Anglia and the Midlands were
also those enclosed most thoroughly in the fifty to seventy-five years
before 1850. They most effectively showed the relationship between the
enclosure of commons, the decline of farm service, the ending of prudential motives to delay marriage, and the great population surge in the
first half of the nineteenth century.[25]

Welton saw the general distribution of nuptiality but explained it in
terms of the higher wages and higher living standards in the rural north
which helped to create a 'comparative indisposition to marry' compared
with other parts of rural England where there was a lower standard of
living and 'less thought for the future'.[26] An analysis by occupational
structure would have helped his commentary which gave insufficient
attention anyway to what he termed the 'rural residues', focusing as it did
on urban and industrial centres. He would have done well to have read
Hasbach's 1908 English edition of *A History of the English Agricultural
Labourer* where it was noted that in Wales and the west the older system
of farm service prevailed in the mid-nineteenth century, unmarried farm
servants living in the farmhouse since 'pasture farming required uninter-
rupted work. Elsewhere the servants were married. In the corn-districts
of the South East, where large farms, close villages, and an extensive use
of machinery were found, all regular connection between employer and
employee tends to disappear. Here was the regime of gangs and casual
labour.'[27]

The generally accepted hypothesis of Hajnal that service acted as a
mechanism in pre-industrial north-west Europe for delaying marriage
and thereby population growth can thus be amplified and demonstrated
by broad regional variations within nineteenth-century England and
Wales.[28] But at a more localised level, the living-in system also becomes
more complex. The living-in practices in Northumberland, for example,
were totally different to those of other northern counties, since cottage
accommodation was provided for the hinds and whole families were

[25] Snell, *Annals*, pp. 213–7, 345; Kussmaul, *Servants, passim.*

[26] Welton, *England's Recent Progress*, pp. 70–71.

[27] W. Hasbach, *A History of the English Agricultural Labourer* (Engl. edn, 1908), pp. 62–3.

[28] J. Hajnal, 'Two kinds of pre-industrial household formation system', *Population and Development
Review*, 8, 1982, p. 481.

hired, while few bachelors were hired at all, especially in the northern part of the county. This was an area of isolated 'clachan' dwellings and the system fitted the environment and farming practices. It is noteworthy that in 1851 Northumberland had the highest proportion in any county of females who were agricultural labourers and a higher proportion of married women between the ages of twenty and forty than nearby Cumberland and Westmorland. The marriage services themselves were different, being of the 'border variety' or similar to the Scottish systems, and more likely after the 1850s to be civil marriages rather than being solemnised in church.[29]

Nevertheless, apart from these traditional areas of farming practice, the social and economic components associated with agrarian capitalism were undoubtedly affecting rural demography by 1850. New working practices, losses in use rights, a shift in the balance of the labour force between male and female as innovations were developed (such as the use of the heavy scythe), changes in the amount of seasonal work available, and the fluctuations in seasonal migration in search of work all affected SMAM as households became the scene of reproduction rather than production. Thus the loss of female employment in arable regions such as East Anglia during the nineteenth century left the way open for earlier marriages. Overall, such changes in rural marriage patterns during the second half of the nineteenth century were anyway overshadowed by the national decline in Im. In many areas change did not depart radically from the norm, but the numbers of women married fell particularly fast between 1861 and 1891 in mid-Wales, the northern Pennines, much of Yorkshire, and in parts of Essex, Sussex and Surrey. Districts where Im increased were either in Norfolk, or near the coalfields.[30]

However, it would be misleading to claim that rural England at this time presented few opportunities for female employment, and that as work in the fields became scarcer as the century progressed, so female migration from the villages grew. It may well be that in areas such as rural Kent or Surrey women's farmwork opportunities did become more restricted, at the same time as there were more opportunities to work in market gardens, or as urban or suburban domestic servants. Gertrude Jekyll noted in 1904 that women were no longer being employed even to glean, and George Bourne noted the low wages and declining numbers in Surrey field-work in 1909. But on the other hand, many

[29] Howkins, *Reshaping Rural England*, pp. 50–2; J. Gielgud, 'Nineteenth-century farm women in Northumberland and Cumbria: the neglected workforce' DPhil thesis, Univ. of Sussex, 1992; *Census of Great Britain 1851*, Population Tables II. Ages, civil condition, occupations and birth-places of the people (vol. I, 1854), pp. xxxii–xxxiii; O. Anderson, 'The incidence of civil marriage in Victorian England and Wales', *PP*, 69, 1975, pp. 50–87.

[30] Woods and Hinde, 'Nuptiality and age at marriage', pp. 119–44.

farmers complained that women no longer wished to work in the fields, and that in Kent it was only the labourers' wives who would do so. There was certainly far more work for women than was once believed, and by 1921 the census noted large increases in women 'farmworkers' in Middlesex. Previously in many areas the yearly labourer was hired almost as much for the strength of his wife and children, who had to work when required, as for any intrinsic merit of his own.[31]

Such broad regional differences in farming systems, each with their different seasonal rhythms of labour requirement in terms of numbers, demand for specialist and seasonal workers, and relations with rural non-agricultural industries, similarly affected such variables as the sex balance, sex-selective migration and the supply of marriage partners. The particular social relations involved in production, and the complexities of dual employment and cottage production, similarly engendered their own demographic regimes, which in turn acted and reacted on the rate of stability or change in local socio-economic structures. The textile-producing family working a small area of land in the Pennines, the lead miner with his cottage, the man who worked as a gardener in the suburban garden or *ferme ornée* of the southern *parvenu*, all consciously or unconsciously adapted strategies of reproduction to fit their particular circumstances. Within agricultural districts, reduced nuptiality resulted directly from domestic living-in service and the need to postpone marriage and to save. But at the same time the occupational structure, yielding high rates of female out-migration and a young-male surplus, contributed towards a higher female nuptiality for those left. Either way, occupational structure operated in ways both direct and indirect, whose relative strength in any one District could tip the nuptiality levels in a positive or negative direction. In these different regions, not only I_m but marital fertility (I_g) and overall fertility (I_f) would have been affected. Every spatial and temporal change in rural productive capacities thus triggered demographic shifts.[32]

One strong link between nuptiality and locality is provided by the seasonality of marriages. In the nineteenth century it was still the case that most arable areas witnessed a peak in marriage ceremonies in October/ November, while pastoral areas tended towards peaks in the months of May/June. Such an early summer peak could thus occur in pastoral areas even in the south-east, as in the Weald, for example which therefore contrasted with neighbouring corn-producing areas in Kent, where harvest

[31] E. J. T. Collins, 'Harvest technology and labour supply in Britain, 1790–1870', *EcHR* 22, 1969, p. 470; J. Connell, *The end of tradition: country life in central Surrey* (London, 1978), p. 21; E. Higgs, 'Women, occupations and work in the nineteenth century censuses', *History Workshop Journ.*, 23, 1987, pp. 59–88; Pinchbeck, *Women Workers*, pp. 84–110.

[32] Friedlander and Moshe, 'Occupations, migration, sex ratios, and nuptiality', pp. 1–12.

wealth and cessation of the exhausting fieldwork gave opportunity and time for marriage. Where hiring fairs still existed the wedding peaks frequently coincided with their dates as a time of payment and change for many young people.[33]

One variable which is frequently overlooked by demographers is the provision of rural housing. Marriage normally meant setting up a separate household, and because of overcrowding in much of the countryside, the availability of cottages was another constraint on marriage. Certainly the lack of accommodation was a delaying factor for marriage in much of rural Shropshire, where in 1869 it was stated that 'the general state of labourers' cottages . . . was worse than in any other English county except Dorset'.[34] Landownership, land, accommodation and nuptiality, were strongly linked although a continuing move from a degree of reliance on self-sufficiency to a money wage weakened that link in the course of the nineteenth century.

The conscious limitation of accommodation by large landowners in 'close' settlements therefore had a negative effect upon nuptiality, and conversely, the relatively greater access to cottage accommodation in 'open' settlements had a positively encouraging effect. Certainly in the Lincolnshire Wolds the 'open' parishes had a higher proportion of smaller households (fewer than six people) than the 'close' ones, which had greater percentages of large households. The operation of the Poor Laws, the type of farming, and the nature of the workforce all help to determine these circumstances, but it might well be that the smaller nuclear households in the 'open' settlements here were the product of earlier marriages, as one might expect from a system of partible inheritance and smaller landowners. The converse, associated with impartible inheritance, was the stem family household comprising some married children with their parents, giving larger households as observed in the 'close' villages of the Lincolnshire Wolds. From a southern viewpoint, Richard Jefferies wrote: 'When a young man does marry, he and his wife not uncommonly live for a length of time with his parents, occupying a part of the cottage.'[35]

In a north-west Essex village such as Elmdon, where living-in service

[33] Wrigley and Schofield, *Population History*, pp. 302–5; B. Short, 'The geography of local migration and marriage in Sussex, 1500–1900', *Univ. of Sussex Geography Research Papers*, 1983, pp. 9–11.

[34] P. R. A. Hinde, 'The marriage market in the nineteenth-century English countryside', *Journ. European Economic History*, 18, 1989, p. 383–92; and in a thought-provoking contribution, Wall has drawn attention to the different modes of household formation, and the consequently different structures of those households: R. Wall, 'Introduction', in R. Wall, J. Robin and P. Laslett, eds., *Family Form in Historic Europe* (Cambridge, 1983), pp. 1–63.

[35] C. Rawding, 'A study of place: the North Lincolnshire Wolds, 1831–1881', DPhil thesis, Univ. of Sussex, 1989, p. 245; *RC on the Housing of the Working Classes. First report*, BPP, xxx, 1884–85, pp. 24–7; R. Jefferies, *Hodge and his Masters* (London, 1880, 1966 edn., 2 vols.), vol. II, p. 68.

had disappeared by the 1860s, there were no multi-family households since the principal landowner allowed the same opportunity to work in the parish as had been enjoyed by previous generations. An heir waiting to take over a farm could work in a shop, with a blacksmith, or in a pub, etc. Although migration was another choice, the acquisition of sufficient goods to contemplate earlier marriage here would have been easier than in villages with few trades or alternatives to waiting to inherit property. Of all males aged 20–29, 54.1 per cent were heads of household in 1861 – a high headship rate when compared with available pre-industrial listings, and comparable with headship rates of between 34.2 and 42.4 per cent in Mitford, central Norfolk, between 1851 and 1881. In contrast the headship rates in the pastoral Shropshire District of Atcham grew from 15.5 to 25 per cent over the same period, as living-in farm service lingered.[36]

The 'open/close' system in England and Wales clearly affected nuptiality and family formation. The link between population increase before the changes in the Poor Law in 1865 and the 'open' communities has frequently been demonstrated, but there is no clear evidence yet of population growth because of earlier marriages or greater propensity to marriage in these communities, as opposed to growth through larger numbers of in-migrants finding niches within the open parish. This ability to settle in some parishes was helped by the operation of the irremovability legislation of 1846 (9th and 10th Vict., c.66), giving settlement to those continuously resident for five years in a parish, a practice condemned by some observers. Capt. Robinson, in his report to the Poor Law Board, noted of the labourer that:

He has no direct interest that his work should be of the best quality; there is no inducement for him to be thrifty, even if he had the power; an early marriage is rather to his advantage than otherwise. If there is a scanty demand for employment, a preference will be given to the married over the single man, and he will often receive higher wages.[37]

Any continuation of outdoor relief within the New Poor Law would also have favoured the unemployed married man at the expense of the bachelor, and various links between unemployment, interpretations of the law and nuptiality can therefore be posited. It cost twice as much to relieve a married man and his family in Norfolk than to employ him,

---

[36] R. Wall, 'The household; demographic and economic change in England 1650–1970', in R. Wall, J. Robin and P. Laslett, eds., *Family Forms*, pp. 494–5; Hinde, 'Household structure', pp. 48–9.

[37] 'Report by Cap. Robinson, RN, to the Poor Law Board, on the operation of the Laws of Settlement and Removal of the Poor in the Counties of Surrey and Sussex', in BPP, *Reports and a Memorandum to the Poor Law Board on Settlement and Poor Removal with an Appendix, 1850–54*, p. 83.

and farmer-Guardians would therefore seek to employ married men at the expense of single ones.[38] Although such interpretations varied between different localities, it is certain that given the framework of the Poor Laws and Laws of Settlement, many a young labourer would have married early 'in his own defence'. In 1851 George Coode wrote that young single people had become:

in rural places, mere outcasts, the last to be employed and the first to be pauperised, as they still remain, whenever and wherever work is scarce for the heads of large families.[39]

Clearly, linkages can be established between earlier marriage in parts of the south and east and farming type, the decline of living-in and other facets of agrarian capitalism. Since the control of fertility within marriage can normally be dated from the 1870s for most rural working families, it would appear that before this time no 'conventional check' on growth operated in these areas. It is to be hoped that a greater understanding of the unemployment, underemployment, low wages and out-migration from mid-century in the lowlands associated with a high density of rural population can now be achieved. In 1850–1 average agricultural wages in southern England were 8s. 5d., which was 26 per cent lower than that of the north; and the incidence of pauperism 12.1 per cent of the population compared with 6.2 per cent in the north.[40] But such analyses are insufficient without consideration being given to local interpretations of the Poor Laws, which affected so many families in precisely these areas. If early marriage was sought as a way to combat underemployment and the harshness of the system of poor relief, then this too must be seen as another powerful factor in any explanation of regional patterns of nuptiality.

III. RURAL FERTILITY

Rural fertility varied less between communities than did either nuptiality or mortality. Working after 1850 for the most part through a capitalist wage relationship, rather than a 'low pressure' peasant system, it can be discussed under four main headings: firstly, fertility differentials by occupation and social class; secondly (and strongly interlinked), the differences between rural and urban fertility; thirdly, the fertility decline in rural England and Wales; and finally, rural illegitimate fertility.

Differences between the fertility of occupational and social groups in the countryside can be clearly seen. Much depended on the presence of mining or industrial activities, and on the organisation of those industries

---

[38] A. Digby, *Pauper Palaces* (London, 1978), pp. 118, 145.   [39] Snell, *Annals*, pp. 350–2.

[40] Brinley Thomas, 'Escaping from constraints: the industrial revolution in a Malthusian context', in Rotberg and Rabb, *Population and History*, pp. 169–93.

by factory or cottage. Most fertile were the mining villages, especially those where a son was seen as an earner to supplement the father's wage as both grew older, or where kin were hired as a group to work in tasks handed down within families, as in some Yorkshire iron-mining villages.[41] Family limitation was practised least under such circumstances. The next closest group in terms of high fertility, however, was agricultural labourers. And of the middle-class families, it was those of the farmers who showed least interest in family limitation at this time. So clearly fertility remained relatively high in rural England and Wales, at least among farmers and labourers, quite the opposite behaviour of those families with women working in the textile industries of Yorkshire or Lancashire, where low fertility and the spacing of children within marriage was an early and general tendency.

Malthusian claims that the New Poor Law had led to prudential restraint by the labourer are difficult to justify. Instead fertility rates remained high, with prizes to be won for raising numbers of children without recourse to the Poor Law, such as those, for example, given by the Norfolk Launditch Agricultural Association. Where the income generated by women and children remained significant, as for example in the Ivinghoe straw-plaiting industry, different household strategies might emerge, depending on individual circumstances and the life-cycle of the individuals, but clearly any loss of earnings by the wife through repeated pregnancy and childcare could be made up by setting the youngsters to work, often in overcrowded households.[42]

Thus by the time of the 1911 Census of Fertility, an excellent snapshot of fertility by social class and occupation, the highest parities were returned from miners, farmers and graziers. This confirmed findings of the Local Government Board whose 1909 statistical memoranda had compared the high birth rates but high infant mortalities of mining districts, with the high effective parities and fertility of the rural Registration Districts.[43]

---

[41] Jones, 'Combining estate records'.

[42] Digby, *Pauper Palaces*, pp. 21–6 for the case of Norfolk; Horn and Horn, 'Ivinghoe in 1871', pp. 11–12.

[43] R. I. Woods, 'Approaches to the fertility transition', p. 288; Local Government Board, *Statistical Memoranda and charts prepared in the Local Government Board relating to Public Health and Social Conditions* (HMSO, 1909, Cd. 4671). The two volumes of the 1911 Census of Fertility included responses from women currently married and with their husband present on the total numbers of children ever born to them, numbers surviving, the duration of the marriage, and the age of the wife at marriage. (*Census of England and Wales* 1911, Fertility of Marriage, Pts I and II). For difficulties of working with the Census see Teitelbaum, *British Fertility Decline*, p. 49. See also J. W. Innes, *Class Fertility Trends in England and Wales, 1876–1934* (Princeton Univ., Press, 1938); M. R. Haines, *Fertility and Occupation*; and Haines, 'Social class differentials during fertility decline: England and Wales revisited' *Pop. Studs.*, 43, 1989, pp. 305–23.

Sidney Webb's observation in 1907 that family limitation was most likely 'in places inhabited by the servant-keeping class' forms a link between studies of occupation and social class differences in fertility on the one hand and studies of fertility in rural as opposed to urban areas on the other. Clearly the overlap is potentially great, but studies of rural fertility 1850–1914 can point to higher overall fertility as a valid generalisation, together with the generally later acceptance of family limitation in the countryside, and a correspondingly slower rate of fertility decline.[44]

Certainly, whatever the occupational mix and whatever the administrative area, higher fertility was to be observed in the earlier part of the period at least. But when the impact of differential out-migration came later in the century, birth rates over much of northern and western England and in Wales would fall. Thus between 1861 and 1911 the rate of natural increase in Wensleydale and Swaledale was consistently below the national average. In the country as a whole, the fall in the rate of natural increase was 8.8 per cent: in Upper Wensleydale 40.4 per cent, 25.9 per cent in Lower Wensleydale, and 64.8 per cent in Swaledale.[45]

Data on rural fertility has been given by Rural Districts, Counties and Registration Districts, and the empirical findings will necessarily reflect such variations in spatial unit. Thus the Rural Districts had legitimate Birth Rates per 1,000 married women aged 15–45 of 204 in 1911, compared with 199 in London, 195 in the County Boroughs and 192 in the Urban Districts. Within the Rural Districts a great arc of higher fertility in 1911 existed from Monmouthshire and Glamorgan through southern and eastern Wales, Staffordshire and the north-west Midlands to west Yorkshire and the North and East Ridings, northwards to the highest of all in Durham, and then southwards through Lincolnshire, the Isle of Ely, Huntingdonshire and Norfolk.[46]

At the less discerning county level, marital fertility between 1851 and 1931 was consistently highest in the mining counties of Durham and Monmouthshire, together with, at different times, Staffordshire, Glamorgan, Derbyshire and the West Riding of Yorkshire.[47] In its extreme case, in 1851 it could be claimed that married women in

[44] Woods, 'Approaches to the fertility transition', p. 298; A. Sharlin, 'Urban–rural differences in fertility in Europe during the demographic transition', in Coale and Watkins, The Decline of Fertility in Europe, p. 236.   [45] Hallas, 'Migration in Wensleydale and Swaledale', p. 146.

[46] National Birth Rate Commission (NBRC), The Report of and the Chief Evidence taken by the National Birth Rate Commission, instituted, with Official recognition, by the National Council of Public Morals – for the Promotion of Race regeneration – Spiritual, Moral and Physical (1916), p. 3; E. Charles and P. Moshansky, 'Differential fertility in England and Wales during the past two decades', in L. Hogben, ed., Political Arithmetic: a Symposium of Population Studies (London, 1938), pp. 108–17, 142.

[47] D. V. Glass, 'Changes of fertility in England and Wales, 1851 to 1931', in Hogben, Political Arithmetic, p. 174.

Cornwall were 28 per cent more fertile than their counterparts in London, though not because of any family limitation techniques in the capital, but because of some combination of infant and foetal mortality, the health and nutrition of the mothers and infant feeding patterns.[48]

And finally, at the level of the Registration District, calculations in 1909 showed that legitimate birth rates per 1,000 married women aged 15–45 in 1881, 1891 and 1901 were consistently higher for 112 entirely rural Registration Districts (population 1.33 million) compared with aggregates of 21 large towns (population 9.8 million). Age-standardised fertility levels were higher in the rural Registration Districts of every English and Welsh county than for England and Wales as a whole.[49]

However, the basic distinction between a pastoral north and west, where living-in lingered, and an arable south and east, where paid labour predominated, was again important. The former had lower nuptiality which kept overall fertility low although marital fertility remained high. In the latter, earlier marriage was more possible without the constraints of living-in, but the pressure to reduce marital fertility may therefore have been greater.

Ideas of higher rural fertility did not go completely unchallenged. In fact, ten years before the Local Government Board published its findings, Adna Weber's statistical analysis of urban growth, using Charles Booth's figures, noted that 'in England the natural increase in town and country is almost precisely the same'.[50] While such conclusions are no longer considered valid, the issue of residence in a rural or urban environment is not an absolute, since relative proximity to towns obviously varies. Indeed it could be argued that rather than the areas of highest rural fertility being in remote parishes, they would be in closer proximity to towns because of the ease of earning a good income for produce, and the easier urban access for labour mobility. But although urban influence might well stimulate the economy, it might also influence reproductive behaviour, thus damping down fertility, since access to urban values and knowledge, as well as the increased incomes, tended to give a lifestyle by the end of the nineteenth century which could see children more as economic liabilities than as assets. But this lack of any clear relationship between incomes and fertility in rural areas certainly undermines a strictly economic explanation.[51]

The Victorian practitioner of family planning would be more likely to

[48] Anderson, 'Social implications of demographic change', p. 42.

[49] Local Government Board, *Statistical memoranda*, p. 4; W. A. Armstrong, 'The influence of demographic factors on the position of the agricultural labourer in England and Wales, *c.* 1750–1914', *AHR* 29, 1981, p. 73.   [50] Weber, *The Growth of Cities*, p. 243.

[51] Friedlander, 'Demographic response', pp. 260–1; E. H. Hunt, *Regional Wage Variations in Britain, 1850–1914* (Oxford, 1973), pp. 232–7.

have come from the professional or semi-professional, and therefore urban or suburban, classes. Arguably the knowledge and availability of effective contraception would have been lower in rural areas. The increased sales of the Knowlton pamphlet and the publicity surrounding the Bradlaugh-Besant trial in 1876 would have been a more urban phenomenon, and sales of condoms, pessaries and suppositories were through urban barber shops and pharmacies. Local and national newspaper advertisements for 'surgical stores' multiplied in the first quarter of the twentieth century but these supplies were still generally available only in sizeable towns or through mail order. Their price too would have made it unlikely that farmworkers could afford them.[52]

Perhaps the most obvious feature of late nineteenth-century fertility in the countryside, and the most disturbing to contemporaries, was its decline. There has been discussion as to whether English and Welsh Gross Reproduction Rates (GRR) in the period 1850–70 were stagnating or rising (with many claiming this to be the result of improving birth registration), but there is general agreement about the fall thereafter. Marital fertility (Ig) in England and Wales may have risen from 0.675 in 1851 to 0.686 in 1871, but it fell thereafter to 0.467 by 1911 (and to 0.292 in 1931). Crude Birth Rates peaked in 1876 at 36.3 per 1,000, falling thereafter through to the 1930s, and female age-specific fertility declined from the 1880s. Birth rates (births per 1,000 women aged 15–44) stood at 141 in 1850, increasing to 152.9 by 1870, but falling to 127.9 by 1890, 100.7 by 1910, and 96.2 by 1914.[53]

This decline in fertility was of great concern partly because it was perceived to operate unequally across both space and society. Signs of 'behavioural inertia' were seen in the slower fall in completed marital fertility in rural areas between the 1850s and 1880s. Any earlier ideas that a social diffusion of knowledge of contraception spread through rural working populations via female servants from their mistresses has now been effectively dismissed. On the other hand, it is accepted that female servants and farm servants, by the very nature of their occupations, necessarily had low fertilities. It has been estimated that a change from a system based on living-in farm servants to one based on wage labour could have a significant impact on potential fertility: a shift involving 20 per cent of the agricultural labour force in this way would have increased

[52] J. Peel, 'The manufacture and retailing of contraceptives in England', *Pop. Studs.* 17, 1963–4, pp. 115–19; J. A. Banks, *Victorian Values: Secularisation and the Size of Families* (London, 1981), pp. 29, 109.

[53] Glass, 'Changes of fertility'; Innes, *Class Fertility Trends*, p. 1; B. Werner, 'Fertility statistics from birth registrations in England and Wales 1837–1987' in General Register Office, '1837–1987: 150 years of the General Register Office', *Population Trends* 48, 1987, pp. 4–10; Mitchell and Deane, *Abstract*, pp. 29–33.

potential marital fertility in agricultural areas by about 6 per cent, with a fall in SMAM of seven months for women.[54]

Social class differentials in changing fertility after 1870 were small for marriages entered into before the 1860s, but reached a maximum difference for marriages of the early 1890s when professional and middle-class fertility fell most rapidly. While all social classes showed a decrease in numbers of children born per 100 wives for those marrying between 1850 and 1886, Social Class I showed the greatest decline, followed by Class II and so on down to Class VIII (agricultural Labourers) whose slower speed of decline was only matched by that of the miners. This inverse relationship between fertility and socio-economic class or status was also manifest in rural areas as a difference between the continuing rel-atively high fertility of agricultural labouring families (Class VIII) com-pared with the lower fertility of farmers' and graziers' families (Class II), although the latter still had appreciably higher fertility levels than those pertaining in contemporary urban groups such as cotton spinners and weavers, clergy, teachers and doctors, and social groups of similar status (clerical, mercantile and lower white-collar workers in the 1911 Census of fertility). And marriages between 1890 and 1920 continued to show farmers with fertility a little below the national average, and agricultural workers rather more above average.[55]

The change to a smaller family and a higher mean age at first marriage must also be seen in the later nineteenth-century context of the increas-ing costs of child education, wider knowledge of birth control methods, and a greater degree of decision-making within the family for women. In rural Oxfordshire in the 1880s, 'the babies did not pour so quickly into these new homes as into the older ones. Often more than a year would elapse before the first child appeared.'[56] Constraints on such decisions related to class and access to employment in the locality, while effective state intervention in education directly affected the earning capacity of children who before 1870 had been extensively employed within agri-culture, especially at times of harvest, either casually or in gangs. Not

[54] Armstong, 'The influence of demographic factors', p. 73; Roberts, *A Woman's Place*, p. 95; Anderson, 'Marriage patterns', p. 76.

[55] The class references are those of the 1911 Census of fertility. T. H. C. Stevenson, 'The fertility of various social classes in England and Wales from the middle of the nineteenth century to 1911', *JRSS*, 83, 1920, pp. 401–32; Innes, *Class fertility changes*, pp. 1–69; Haines, 'Social class differentials'; NBRC, *Second Report of and the Chief Evidence Taken by the NBRC, 1918–20* (London, 1920), p. 15; D. A. Coleman, 'Population' in A. H. Halsey ed., *British Social Trends since 1900* (London, 1988), p. 89.

[56] F. Thompson, *Lark Rise to Candleford* (Oxford, 1945), p. 160. Other demographic material in Thompson's account has been criticised, but this passage appears to be correct. See B. English, '*Lark Rise* and Juniper Hill: a Victorian community in literature and history', *Victorian Studies*, 29, 1985, p. 30–31.

until 1873 was it illegal to employ children under eight years of age, and not until 1876 under ten years of age. Although thereafter school holidays were fashioned around farmers' needs, and absenteeism was difficult to counter, the flow of intergenerational wealth within labouring families was reversed, for the cost of children to parents became greater than the income generated by children.

Under such circumstances the principal agent of demographic change ceased to be nuptiality, as it had been for perhaps 300 years up until the 1870s, and became marital fertility. This turned downwards in the 1870s and sharply downward again from the 1880s, with signs of definite fertility control by the 1890s.[57] Such a decline amounted to at least a 7 per cent reduction in completed fertility between the 1870s and the First World War. Once marital fertility had plunged like this, nuptiality could rise again, and the age at marriage could fall, with far fewer demographic consequences. Completed family sizes (for women married once only) fell within two generations from 6.2 in the 1860s to 2.8 by 1911, those women born in the 1880s having only half the number of children, on average, that their own mothers had raised, and their daughters again having only about two-thirds as many. Small families had rapidly become the norm, and within fifty years most of rural England had passed from a state of 'natural fertility' to one of 'controlled marital fertility'. The average age of the mother at the birth of the last child also came down from about forty years at the beginning of the nineteenth century to thirty by the 1920s.[58]

It is clearly necessary to relate decisions taken within the family on such matters to the strength of capitalist relations and patterns of employment, and the class formations and cultures in particular localities. But when earlier macro-level work on fertility at the county level is supplemented by using the more sensitive 620 or so Registration Districts, it becomes apparent that declining fertility must no longer be seen on a regional, rural–urban, or other ecological basis, since the pattern of rural decline shown by Registration Districts is fragmentary: 'neither overwhelmingly urban nor rural, northern nor southern'.[59]

---

[57] Teitelbaum, *British Fertility Decline*, pp. 118–19; A. J. Coale and R. Treadway, 'A summary of the changing distribution of overall fertility, marital fertility and the proportions married in the provinces of Europe', in Coale and Watkins, *The Decline of Fertility in Europe*, p. 47.

[58] Friendlander, 'Demographic response', p. 265. The changeover from 'natural fertility' to 'controlled marital fertility' is defined by a change in the Princeton Fertility Index of I_g >0.6 to I_g <0.6. See also Anderson, 'Social implications of demographic change', p. 39; J. Simons, 'Developments in the interpretations of recent fertility trends in England and Wales' in J. Hobcraft and P. Rees, eds., *Regional Demographic Development* (London, 1977), p. 118; Woods, 'Approaches to the fertility transition', pp. 299–300.

[59] R. I. Woods, *Theoretical Population Geography* (London, 1982), pp. 111–29; R. I. Woods and C. W. Smith, 'The decline of marital fertility in the Late Nineteenth century: the case of England and Wales' *Pop. Studs*, 37, 1983, p. 215; Woods, 'Approaches to the fertility transition', p. 300.

Changes at the level of the Registration District in the early stages of declining fertility showed a wide variation in Ig. In 1861 most of the rural Districts with high marital fertility were to be found in the north and west, but linked with low values of Im to give only moderately high overall fertility (Figure 21.2). In the English lowlands marital fertility was lower, while the small market towns showed even lower levels. By 1891 much of the rural South showed Ig<0.6, as in north Norfolk, east Kent, Bedfordshire and Huntingdonshire, Surrey, western Sussex, the Downs and Hampshire chalklands, Dartmoor and Dorset. Parts of the Yorkshire Dales, Lincolnshire and north Wales showed similar levels.

The decline in rural fertility between 1861 and 1891, by some combination of change in Ig and Im to produce overall fertility levels, was most evident in northern and central Wales, the northern Pennines, Lincolnshire, and in a broad band from Suffolk, Essex and the east Midlands west and south to Cornwall and east Kent and the Channel coast. In some rural areas marital fertility (Ig) increased but this was outweighed by a reduction in nuptiality to produce lower overall fertility, as in the Lleyn Peninsula, coastal Pembrokeshire, a large part of Norfolk, the Lincolnshire Wolds, parts of Somerset, east Devon and Exmoor, the Dorset Downs and parts of the Cotswolds. Conversely in some rural localities at this time the decline in Ig was outweighed by an increase in nuptiality (Im) as in Furness, parts of Norfolk or Carmarthenshire. In rural Norfolk, for example, the relative strengths of Ig and Im acting in concert or in opposite directions, could serve to mask or emphasise overall population change, producing quite different outcomes in neighbouring Registration Districts. Nevertheless, by 1911 there were very few areas of rural England and Wales with Ig>0.6, and none at all in the lowlands (Fig. 21.2): the largest island of high rural fertility was now limited to Carmarthenshire and Cardiganshire in south Wales.[60]

The reasons for this concentrated decline in fertility are complex and debatable, and take us beyond the confines of this volume. They have been the subject of the Princeton Fertility Project, established in 1963, essentially to explore the decline of fertility in Western Europe between about 1870 and 1940. As Watkins has stated of the decline: 'If levels of marital fertility before 1870 were shaped by longstanding social customs, it would appear that the cake of custom crumbled rather rapidly' – in a period of perhaps sixteen years in England and Wales.[61]

The factors surrounding the decline may be grouped into four types: biological, technological, socio-economic, and cultural. Surrounding

[60] Woods and Smith, 'The decline of marital fertility', pp. 214–15, Woods, 'Approaches to the fertility transition', p. 300.

[61] S. C. Watkins, 'Conclusions' in Coale and Watkins, *The Decline of Fertility in Europe*, p. 433. The Princeton Fertility Project is therein reviewed.

1861   Variations in *Ig*
  ■ ≥ 0.70
  ▨ 0.60–0.69
  ▦ < 0.60

  ☐ urban districts

1891   Variations in *Ig*
  ■ ≥ 0.70
  ▨ 0.60–0.69
  ▦ < 0.60

  ☐ urban districts

1911   Variations in *Ig*
  ■ ≥ 0.60
  ▨ 0.50–0.59
  ▦ < 0.50

  ☐ urban districts

0   km   100

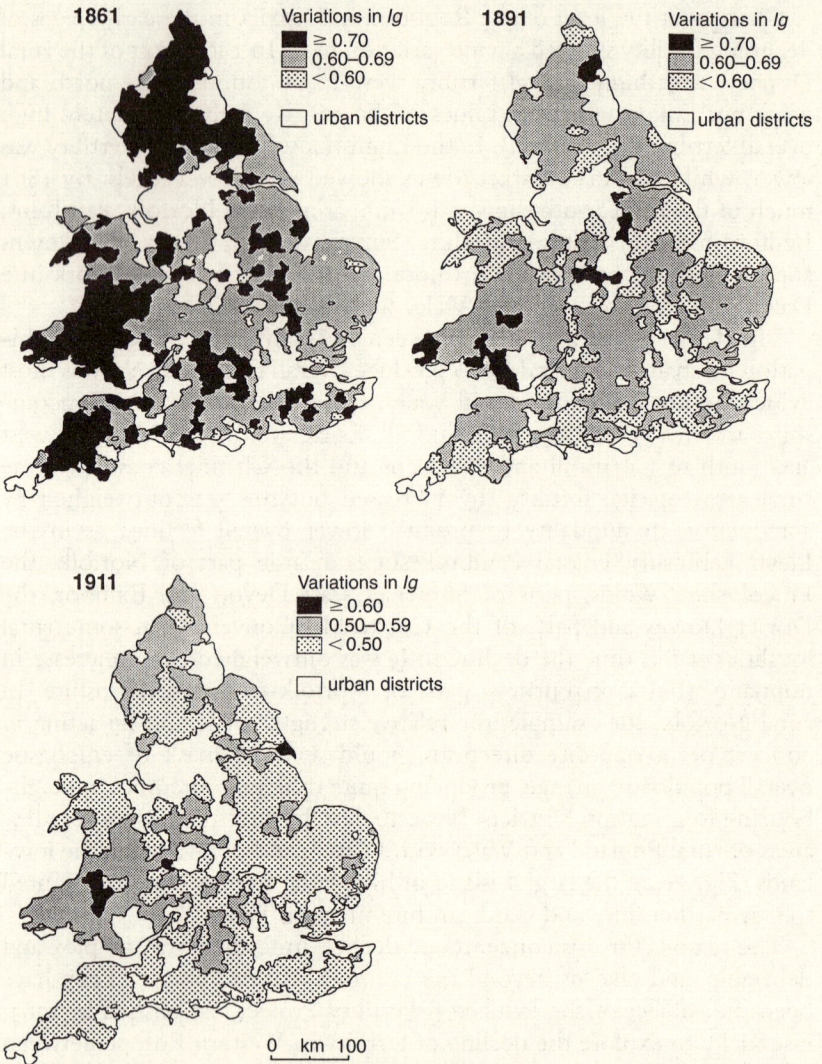

21.2 Rural England: levels of marital fertility (I*g*) in 1861, 1891 and 1911
Source: Woods, 'Approaches to the fertility transition, Fig. 8, p. 306; Fig. 6,
p. 300; and Woods and Smith, 'The decline of marital fertility',
Figs. 4 and 5, p. 213

them all are the macro-concepts of 'modernisation' and urbanisation, but other interacting factors include economic cycles, the knowledge and availability of improved methods of family limitation, mass education and the perceived costs and returns of children, the availability of female employment, the 'socially determined allocation of food' among the family and improved nutrition for women, the symbolic importance of child rearing, relative deprivation among the middle classes, and the increasing secularisation of Victorian society. Declining child and infant mortality helped confirm the downward trend, especially after 1900. Some of these factors have a rural and agrarian component, but most have wider dimensions, and apply in various ways to much of Western Europe at this time.[62]

Alternatively the decline has been seen as a 'strain-relieving response' and as a substitute for migration. The latter might delay fertility decline if it were possible for large enough numbers of people to leave hard-pressed countrysides, and if it enabled those left to have sufficient access to land, employment, etc. for there to be little change in fertility. On the other hand, migration of the younger people within the community might seriously affect fertility. Either way, it is clear that migration, as well as nuptiality, needs to be taken into account in any explanation of declining fertility, and with them, the particularities of the locality which would affect all such variables. In the Norfolk arable-farming area of Mitford, for example, birth control is detectable within marriage among farm workers and general labourers marrying young. Conversely in pastoral Atcham, Shropshire, couples married later and, with more resources, had less stimulus to control fertility.[63]

Kingsley Davis's concept of 'multi-phasic demographic responses' incorporating the idea of the interchangeability (to some extent) of emigration, rural–urban migration, nuptiality and marital fertility to cope with population growth and societal strain, clearly does apply to the English rural situation. Fertility is socially determined, and further refinement would therefore be needed to incorporate distinctions between different agrarian regimes, with their differing requirements of labour, farm servants, female employment, etc.; and to recognise the varying degrees of local political and economic power exerted over in-migration, as in the 'close' parish where the exclusion of those who might threaten the socio-economic stability of the community established a quite different context for reproductive behaviour within the family from that in the 'open' community, where population pressure was frequently

---

[62] Teitelbaum, *British Fertility Decline*, *passim*; Woods, 'Approaches to the fertility transition', p. 298.

[63] Friendlander, 'Demographic response', pp. 265–8; P. R. A Hinde, 'The fertility transition in rural England', PhD thesis, University of Sheffield, 1985, pp. 291–316 and *passim*.

severely felt by the 1850s. Demographic restraint in the 'close' parish under such protective conditions could be practised without fear that newcomers would flout this restraint and swamp the community. The near-static populations of many 'close' communities have at least partially resulted from such thought processes, as seen in South Lindsey, Bedfordshire or Leicestershire.

The possibility of conscious choice itself was new to many: as one elderly inhabitant from an Oxfordshire hamlet said in the early 1930s, 'If they knew what it meant to carry and bear and bring up a child themselves, they wouldn't expect the woman to be in a hurry to have a second or a third now they've got a say in the matter . . .' Such feelings echoed those as found in the singing of Harry Cox of Catfield, Norfolk:

> And what if you should have a child, would make you laugh and smile,
> And what if you should have another, would make you think a while.
> And what if you should another, another another un too.
> Would soon put a stop to your follish young tricks
> And make you think of the foggy dew.

The young man, the narrator, had made love to a 'serving maid'. He is identified as a weaver in some versions of the song, although Harry Cox, himself a farm worker, sang, 'When I was a bachelor early and young, I followed a weary trade.'

And alongside changes in the family itself, exogenous change, stemming from the decisions of estate functionaries or landowners, as well as the state itself, was at work, imposing 'institutional determinants' on fertility.[64]

Whatever the causes of fertility decline, the disparity between urban and rural rates was clear to many contemporaries, as was that between different social groups. For many people, the practice of birth control represented a threat to the quality of the population. As Dr Millard, Medical Officer of Health for Leicester, told the National Birth Rate Commission, 'Birth control is practised by what we may call the A1 classes and neglected by the C3 classes', and thus the net result of birth control was dysgenic, i.e. detrimentally affecting later generations. Furthermore: 'at present the only class which is not practising it is the lowest and least desirable class'. The wealthier social groups were failing

---

[64] J. Obelkevich, *Religion and Rural Society: South Lindsey, 1825–1875* (Oxford, 1976), pp. 12–13; D. Mills, *Lord and Peasant in Nineteenth-Century Britain* (London, 1980), pp. 84–5; R. M. Smith, 'Transfer incomes, risk and security: the roles of the family and the collectivity in recent theories of fertility change', in D. Coleman and R. Schofield, (eds.), *The State of Population Theory* (1986), p. 190; G. McNicholl, 'Institutional determinants of fertility change' *Population and Development Review*, 6, 1980, pp. 441–62; F. Thompson, *Lark Rise*, p. 128; 'The Foggy Dew' from the singing of Harry Cox, Catfield, Norfolk in Ralph Vaughan Williams Memorial Library, Cecil Sharp House, London.

to reproduce themselves, whether in town or country, while the poor, not using the newer methods of contraception, were 'overproducing'. The question became a broad one, in which social Darwinian, imperialist and eugenic issues were discussed. An overriding concern for the health of the Nation was expressed in the face not only of a declining birth rate, but also of continuing high infant mortality, poverty and the poor physical condition of recruits for the Boer War, when between 40 and 60 per cent of men presenting themselves for enlistment were found to be physically unfit for service.[65]

Town life was widely held by the 1880s to be the cause of a degenerating working class, and as Freeman–Williams noted in 1890, 'It is a matter of common observation that most of the best workmen are men from the country, not born and bred in London.'[66] It may be that the belief in the innate superiority of the sturdy countryman was actually a preference on the part of employers for a rural-bred, pliable and docile labour force. Nevertheless, countrymen were certainly deemed superior in the skilled trades and better unskilled ones, making up for example 70 per cent of the metropolitan police force and dominating the building, printing and iron-founding trades. In 1893 Longstaff wrote:

That the town life is not as healthy as the country is a proposition that cannot be contradicted . . . The narrow chest, the pale face, the weak eyes, the bad teeth, of the townbred child are but too often apparent. It is easy to take an exaggerated view either way, but the broad facts are evident enough; long life in town is accompanied by more or less degeneration of race. The great military powers of the continent know this well enough, and it may be surmised that with them agricultural protection is but a device to keep up the supply of country-bred recruits.[67]

The worries continued, and Jephson's *The Sanitary Evolution of London* (1907) still contained the concept that once the rural population was sucked into London, degeneration of the race would follow, while Greenwood's *The Health and Physique of School Children* (1913) noted that the rural child tended to be heavier and larger than the urban one. The Report of the Interdepartmental Committee on Physical Deterioration had been published in 1904, but lack of data bedevilled their investigations, and an anthropometric survey was felt to be needed to measure changes in physique, since 'the condition of the rural population as a reservoir of national strength is of first-rate importance'. Various measures of social engineering were recommended to help retain this population in the countryside 'with a view to combating the evils resulting from the

---

[65] NBRC, *Second Report, 1918–20*, pp. 276, 282.

[66] In G. Stedman Jones, *Outcast London: a Study in the Relationships between Classes in Victorian Society* (London, 1984), p. 127.   [67] G. B. Longstaff, 'Rural depopulation', *JRSS*, 56, 1893, p. 416.

constant influx from country to town', and to arrest the processes leading to what Weber referred to as the 'survival of the unfittest'. The Eugenics society was established in 1907 and the journal *Biometrika* followed together with a battery of statistical tests devised by eugenicists such as Galton and Pearson.[68]

Turning finally to illegitimate fertility, we might begin by invoking the concept of a woman's 'procreative career' to indicate that the fertility period should be seen as lasting between the ages of fifteen and forty-five approximately, and that births took place before marriage, during first marriage, in widowhood or in subsequent remarriage. Pre-nuptial pregnancy and illegitimate births were highly correlated spatially, and both appreciably higher in rural parishes than in urban areas before the 1930s.

At the county level, Norfolk, Cumberland, the North Riding of Yorkshire, Nottinghamshire and Shropshire had high illegitimacy levels between 1870 and 1900. Little attention was paid at the time to this, probably because it was felt to be a phenomenon associated with the more peripheral areas of the country, and that it would consequently disappear. By contrast urban areas had the lowest illegitimacy rates, London and the southern towns having the lowest of all. This urban/rural difference may have arisen through the greater practice of illegal abortion in towns, the migration of pregnant women from town to countryside, or through the difficulty of evading the registration of bastard children in the more close-knit village communities.[69]

In common with marital fertility, illegitimate fertility (Ih) also declined, from a high point in the 1840s and at 6.7 per cent in the five years around 1851 to 3.9 per cent in the five years around 1901, continuing to decline thereafter. But by 1911 'locality persistence' ensured that Ih was still higher in rural east Yorkshire, the east Midlands, Fenland and East Anglia.[70]

[68] *Report of the Inter-departmental Committee on Physical deterioration*, vol. 1. Report and Appendix (Cd 2175, HMSO London, 1904), p. 87; Weber, *The Growth of Cities*, p. 444; D. J. Oddy, 'The health of the people', in T. Barker and M. Drake, eds., *Population and Society in Britain 1850–1980* (London, 1982), pp. 121–2.

[69] Wrigley and Schofield, *Population History*, p. 438; P. Laslett, K. Oosterveen and R. M. Smith, *Bastardy and its Comparative History* (London, 1980), pp. 34–5; A. Macfarlane, *Marriage and Love in England, 1300–1840* (London, 1986), p. 26; P. Laslett, 'Illegitimate fertility and the matrimonial market', in J. Dupaquier, E. Helin, P. Laslett, M. Livi-Bacci and S. Sogner, eds., *Marriage and Remarriage in Populations of the Past* (London, 1981), p. 461. R. M. Smith, 'Marriage processes in the English Past: some continuities' in L. Bonfield, R. M. Smith and K. Wrightson, *The World we have Gained: Histories of Population and Social Structure* (London, 1986), p. 98. Flora Thompson instances a woman of thirty returning to the hamlet to live with her sister to have a baby, and more than one house had women also bringing up grandchildren (F. Thompson, *Lark Rise*, pp. 129–30).

[70] Laslett *et al.*, *Bastardy*, pp. 13–81; E. Shorter, J. Knodel and E. Van De Walle, 'The decline of non-marital fertility in Europe, 1880–1940', *Pop. Studs*, 25, 1971, pp. 375–93; Woods 'Approaches to

In the 8th Report of the Registrar General (1849) the rural area of Brampton and Longtown in Cumberland was highlighted as having an illegitimacy rate of 17.4 per cent, with a cluster of villages in Radnorshire not far behind. Cohabitation and prenuptial pregnancy, along lines very different from those of most middle-class Victorian English, allied with civil rather than church or chapel weddings, in the northern borders, north and south Wales (in places such as Bala or Newcastle Emlyn), gave a wholly different cultural perspective. The Welsh 'experimental unions' or the practice of 'bundling' undoubtedly lay behind the higher numbers of pregnant rural brides in the north and west. The unlegalised unions of Denbighshire, the *byw tali*, or the Carnarvonshire broomstick wedding, the *priodas coes ysgub*, were inimical to clerics, although as one writer commented dubiously in 1930, 'the habit represented standards not essentially less moral than those reflected in our own time in summer holiday resorts in most European countries'.[71]

Many illegitimate births were to women in farm service, waiting for marriage, but perhaps the informal unions or 'domestic concubinage' among farmworkers accounted for the greater part of the children born, and it is certainly noteworthy that on average the age of the mother at the birth of her first child was virtually the same, whether inside or outside marriage. It has been suggested that rural illegitimacy was in general higher where parental control over courtship was low, and where access to resources for supporting themselves was open to single mothers. It was also noted of the Hunston area of Norfolk in 1843 that a lack of employment during much of the year increased the bastardy list, but further research is clearly needed here, perhaps along the lines of that on late nineteenth-century Colyton in Devon. Here, between 1851 and 1881 only 38.5 per cent of all known births were conceived inside wedlock, 41 per cent were conceived prior to marriage, and 20.5 per cent were illegitimate, the latter cases being used as evidence of a 'bastard-prone sub-society', where unions were entered into with others from similar backgrounds.[72]

Many Anglican clergymen suspected that 'they simply don't care much about the marriage service' and figures ranging from 40 per cent to 75

the fertility transition', pp. 292–3; N. F. R. Crafts, 'Illegitimacy in England and Wales in 1911'. *Pop. Studs.*, 36, 1982, pp. 327–31; T. H. Hollingsworth, 'Illegitimate births and marriage rates in Great Britain 1841–1911' in J. Dupaquier *et al.*, *Marriage and remarriage*, pp. 437–51.

[71] Laslett *et al.*, *Bastardy*, p. 63; Macfarlane, *Marriage and Love*, pp. 305–6; T. Gwynn Jones, *Welsh Folklore and Folk Custom* (London, 1930), pp. 184–7.

[72] Hollingsworth, 'Illegitimate births', pp. 437–51; Anderson, 'Social implications of demographic change', pp. 37–8; J. Robin, 'Prenuptial pregnancy in a rural area of Devonshire in the mid-nineteenth Century: Colyton, 1851–1881', *Continuity and Change*, 1, 1986, pp. 113–24; Robin, 'Illegitimacy in Colyton, 1851–1881', *Continuity and Change*, 2, 1987, p. 324.

per cent of brides being pregnant were commonplace. Many were simply acting upon the old saying of 'no child, no wife', and quite clearly the incidence of cohabitation and pregnancy, with the assumption that marriage would follow at some date, was widespread. Certainly the practice of marrying after the birth of the baby was commonplace in Oxfordshire or Buckinghamshire, where 40 per cent of brides 1837–87 were pregnant. Marriage itself might follow for a variety of reasons, including the greater ease of gaining poor relief, employment or charity; and the settlement of man and wife together being assured. While the genteel thought illegitimacy a regretable outcome of overcrowded accommodation and mixed field work, rural labouring classes lived it as custom. Little recrimination would follow, unless the mother-to-be found herself in need of poor relief, in which case, at least in Norfolk, she would be forced to go into the work house, and even at the end of the nineteenth century might be put to hard labour in the work-house laundry a few days after her confinement. This indicated the moral censure contained in the disciplining of the labouring poor, which was encapsulated in much of the New Poor Law legislation.[73]

IV. RURAL MORTALITY

In contemporary discussions the subject of rural mortality surfaced regularly, and the level of differences in urban–rural mortality rates was one of the leading research questions in demographic studies before the First World War. And since that time, the question of the fall in mortality rates has continued to incorporate differences between town and country.[74]

Mortality was more intensely localised, however, than nuptiality or fertility, and even within rural areas there were wide variations in its incidence, and the relationships between observed mortality levels and measures of occupation and overcrowding form a subset of questions. Infant and child mortality and the sexual distribution of mortality forms another subset. Also relevant is the fact that variations temporally or spatially in the rate of mortality could affect the rates at which land or work was taken up when available.[75]

Rural areas were perceived as healthier than Victorian and Edwardian towns, in common with findings throughout north-western Europe. Adna Weber noted the consistently lower rural death-rates for all age

---

[73] Snell, *Annals*, p. 354; Digby, *Pauper Palaces*, pp. 4, 54, 153; Hinde, thesis, pp. 229–35.

[74] Mortality has not attracted interest comparable with that afforded to the other main variables in rural demography. For the urban case see R. Woods and J. H. Woodward, eds., *Urban Disease and Mortality in Nineteenth-Century England* (London, 1984) and R. Schofield and D. Reher, 'The decline of mortality in Europe' in R. Schofield, D. Reher and A. Bideau, *The Decline of Mortality in Europe* (Oxford, 1991), pp. 1–17.   [75] Wall, 'The household', p. 512.

groups and for both sexes, and this followed up observations by Chadwick and by Ratcliffe who in 1850 had commented upon the greater longevity and life expectancies of rural members of the Manchester Unity of the Independent Order of Odd Fellows, compared with members from town and city districts. William Farr also estimated that the average level of mortality in urban areas was 25 per cent higher than in rural areas in the decade after 1851. Near contemporary accounts agreed with such findings. The Local Government Board, reporting on public health and social conditions in 1909, calculated both crude and age-specific rural mortality rates at 23 per cent lower than urban ones.[76]

Striking urban–rural differences in mortality for all age groups before 1914 have since become a commonplace in British demographic literature. As late as 1949–53 the male and female standardised mortality rates for rural areas were 79 per cent and 91 per cent respectively of rates for the conurbations. The twenty-six years life expectancy of a male baby born in Liverpool in 1861 compared very unfavourably with the fifty-seven years of a similar baby born in Okehampton, for example. Whereas Liverpool experienced infant death rates 83 per cent above the national average, those for rural Ripon in the West Riding of Yorkshire were 11.7 per cent below.[77]

However, the pattern was not straightforward, and many such as Greenhow in 1858 found confusing results when comparing urban with rural mortality, although the overall trend was clear. In 1875 Welton demonstrated that for women aged 15–25 in the outer ring of eight counties around London the death rate was higher (9 per 1,000) than both London (5.9 per 1,000) and the national average. His explanation was that women who had previously moved to work in domestic service in London became ill and returned home, in many cases to die. For 'such is the effect of a life of increased excitement and effort upon girls accustomed to the quiet habits of a village population, that I feel sure many of these emigrants lose their health after one or two years, and not a few become consumptive'.[78] Many towns were also relatively healthy, the southern resort towns and smaller towns and suburbs for example, but there was a clear difference between most rural areas and the large towns, textile centres and heavy industrial centres.

[76] Weber, *The Growth of Cities*, pp. 355–9; H. Ratcliffe, 'Observations on the rate of mortality and sickness . . .' (Manchester, 1850), reprinted in R. Wall, *Mortality in mid-nineteenth-century Britain* (London, 1974), p. 27; Local Government Board, *Statistical memoranda*, pp. 13–15.

[77] Anderson, 'Social implications of demographic change', p. 36; R. Woods, 'The structure of mortality in mid-nineteenth century England and Wales', *JHG*, 8, 1982, p. 376; B. W. Benson, 'Mortality variation in the North of England, 1851–60 to 1901–10', PhD thesis, Johns Hopkins University, 1981, pp. 77–84.

[78] E. H. Greenhow, *Papers relating to the Sanitary State of England* (1858), in R. Wall, ed., *Pioneers of Demography Series* (Farnborough, 1973); T. A. Welton, 'The effect of migrations upon death-rates' *J. Stat. Soc. London*, 38, 1875, p. 326; Welton, *England's Recent Progress*, p. 45.

Clearly, it mattered greatly where people lived as to how soon they would die. And it was not just an urban–rural dichotomy, since many rural areas too were considered unhealthy. By the mid-nineteenth century, for example, infant mortality rates were high in the East Riding of Yorkshire, south Lincolnshire and Norfolk, and the rates for the area around the Wash and north Norfolk were equivalent to those pertaining in parts of London. Rural Norfolk was certainly not a salubrious environment, and in areas of poverty typhus, diphtheria and scarlet fever were found, together with rheumatism and ague in the marshlands, and widespread consumption and scrofulous diseases, allied with outbreaks of smallpox or cholera. Freebridge Lynn Union, where the Sandringham Estate was purchased by the Prince of Wales in 1862, was a notably unhealthy district. The Cornish and Durham mining districts also had high male mortality rates, together with record numbers of widows; of women over twenty years of age, 14.5 per cent were widows. Here there were urban problems but without urban amenities to cope with them. Much collier housing was built on poorly drained land, or was lacking in light or ventilation, such as that, for example, at Earsdon, Tynemouth.[79]

By contrast the mortality rates in the south and south-west and in the remoter Pennine valleys of Swaledale and Upper and Lower Wensleydale were lower than the national average, although age-selective migration could at times leave the dales with a less healthy population.[80] Overall, with their better wage levels, better diets and more reliable employment, rural families from the north might be expected to show signs of better health and greater longevity. However, Welton remarked on the lower mortality of those in the fifty-year and over age groups in the eastern counties, and the lower infant mortality in the south and south-west. In general something over 50 per cent of the variation in infant mortality rates over the whole of England and Wales in the later nineteenth century has been attributed to variations in population density.[81]

While certain rural areas experienced lower death rates for all ages, it is important to analyse the death rates by age group, sex and socio-economic status within those areas. Child mortality was greater in the families of agricultural labourers, for example, than among farmers.[82]

High rural infant and child death rates could still be demonstrated in

[79] R. I. Woods, P. A. Watterson and J. H. Woodward, 'The causes of rapid infant mortality decline in England and Wales, 1861–1921 Part 1', *Pop. Studs.*, 42, 1988, p. 360; Digby, *Pauper Palaces*, pp. 24, 175; Digby, 'The rural Poor Law' in D. Fraser, ed., *The New Poor Law in the Nineteenth Century* (London, 1976), p. 164; *Census of Great Britain 1851*, Population Tables II. Ages, civil condition, occupations and birthplaces of the people (vol. I) (1854), xxxviii; Benson, 'Mortality variation', p. 15.   [80] See p. 1252 above.

[81] Welton, *England's Recent Progress*, pp. 49, 57; Woods, Watterson and Woodward, 'Infant mortality decline', p. 356; and see also J. Burnett, 'Housing and the Decline of Mortality' in Schofield *et al.*, *The Decline of Mortality*, pp. 158–76.

[82] *Census of England and Wales 1911*, XIII (Pt 2), 1923, pp. cv–cvii.

the mid-nineteenth century in certain areas. Indeed, the counties of Bedfordshire, Cambridgeshire and Norfolk had infant mortality rates comparable with those of London or the industrial north in 1861. Elsewhere, pockets of high mortality were to be found, such as the hoppers' huts of the south-east, where cholera prevailed, or in the Black Fens and south Lincolnshire where gangs of women field workers were important but where it was alleged that children were neglected, left in poor supervision or 'stilled with laudenum and often, no doubt unintentionally, poisoned to death'.[83]

The sixth Report of the Medical Officer of the Privy Council in 1864, cited extensively in the House of Lords debate on 'Agricultural women and children' in 1865, contained many observations on the interrelationship of systems of agricultural practice and infant and child mortality:

With wonderful accord the cause of the mortality was traced by nearly all these well-qualified witnesses to the bringing of the land under tillage – that is, to the cause which has banished malaria, and has substituted a fertile though unsightly garden for the winter marshes and summer pastures of fifty and 100 years ago. It was very generally thought that the infants no longer received any injury from soil, climate or malarious influence, but that a more fatal enemy had been introduced by the employment of mothers in the field . . . It appears that the recently reclaimed 'black lands' are very light indeed, and may be submitted to womens' work to a far greater extent than anywhere else in the Kingdom . . . a party of women will often come from several miles off to work at a village . . . heedless of the fatal results which their love of this busy independent life is bringing on their unfortunate offspring, who are left pining at home.[84]

One consequence of the female gang system, it was claimed, was that infant death rates in Ely or North Witchford were the same as those of Salford or Blackburn; those of Spalding the same as Whitechapel; and those of Wisbech comparable with Manchester.[85]

The interaction between mortality and local economic and social institutions was strong. Infant and child mortality was a feature of many rural areas where housing was poor or 'cottage herding' manifest, and where food- or waterborne diseases could spread easily. As Weber observed: 'for every ill-kept city tenement, there is at least one rural shanty in as bad or worse condition'. In *Arcady: for Better for Worse* the author notes that 'the little rookeries in our open parishes are a blighting curse'.[86] Scarlet fever,

---

[83] C. H. Lee, 'Regional Inequalities in Infant Mortality in Britain, 1861–1971: Patterns and Hypotheses', *Pop. Studs.*, 45, 1991, p. 57; *Hansard* CLXXIX, May–June 1865, col. 177.

[84] *Hansard* CLXXIX, May 1865, col. 1471.

[85] V. Berridge, 'Opium in the Fens in nineteenth-century England', *Journ. of the History of Medicine and Allied Sciences*, 34, 1979, pp. 293–313; M. Hewitt, *Wives and Mothers*, p. 112.

[86] Weber, *The Growth of Cities*, p. 417; A. Jessopp, *Arcady: for Better for Worse* (London, 1891), p. 134; J. Montgomery, 'On Overcrowded Villages', *Trans National Association for the Promotion of Social Science (1860)*, 1861, pp. 787–9.

typhoid, diarrhoea, diphtheria and respiratory diseases were the main killers. Amongst a horrifying litany of degrading housing reports mustered in 1866 was one from the incumbent of Builth:

The Welsh are very dirty. I found a house in Builth, where in the bedroom, down stairs, were two pigs in one corner, and two children ill with scarlet-fever in the other. The dung-hills are placed in front of the houses in some parts of the town.[87]

By the 1860s the quality of rural housing was a matter of increasing concern. Dr H. J. Hunter noted in 1864 that in a survey of 821 rural parishes, in which population had increased by 5.4 per cent between 1851 and 1861, the number of cottages had actually fallen by 4.5 per cent, and of these over 40 per cent had just one bedroom.[88]

Nevertheless, in common with most areas of Europe, infant mortality rates fell from 162 deaths per 1,000 live births in 1850 to 105 per 1,000 in 1914. But in many rural counties the rates were far lower – under 90 in Dorset, Wiltshire. Hereford, Cambridgeshire or Sussex, for example. Child mortality (1–4 years) had fallen earlier down to about 1850, then fell a little until the 1880s and was then little changed until the late nineteenth or early twentieth century. Young-adult mortality rates had also fallen earlier since the epidemic infectious diseases (smallpox, malaria, typhus, plague) had largely gone by c. 1850, and the endemic infectious diseases (tuberculosis and other respiratory disorders, and diarrhoea), accounting for about 60 per cent of deaths c. 1850, were diminished by improvements in living standards. The better distribution of milk and its processing, and general sanitary improvements especially affected the 1–4-year age group, and it was this group which therefore showed most improvements; a particularly rapid decrease in mortality occurred after 1900. Infant mortality was calculated in 1909 to have been about one-third heavier in urban than rural districts.[89]

Among the older age groups the rural–urban distinction in mortality was less marked. In the north this may have resulted from the effects of industrial occupations spreading out beyond the urban boundaries to give an 'urban-industrial' mortality pattern in the surrounding countryside.

---

[87] C. Edwards Lester, *The Glory and Shame of England* (2 vols, London, 1866), vol. II, p. 334.

[88] *7th Report of the Medical Officer of the Privy Council*, 'Inquiry into the State of the dwellings of Rural Labourers', BPP, 1865, XXVI, 1865, Appendix 6.

[89] Mitchell and Deane, *Abstract*, p. 36; Local Government Board, *Statistical Memoranda*, pp. 15–16; Anderson, 'Social implications of demographic change', p. 16; W. Brass and M. Kabir, 'Regional variations in fertility and child mortality during the demographic transition in England and Wales', in J. Hobcraft and P. Rees, eds., *Regional Demographic Development* (London, 1977), p. 74; A. J. Swerdlow, '150 years of Registrar Generals' medical statistics', *Population Trends*, 48, 1987, p. 22. See also P. A. Waterston, 'The role of the environment in the decline of infant mortality', *Journ. Biosocial Science*, 18, 1986, pp. 457–70.

The complications for women in the 15–24 age group around London, noted by Welton, have been referred to above. It becomes clear, therefore, that age groups fared differently, but that these differences were also dependent upon the social and economic interrelationships within differing localities. In the northern counties, for example, the mortality rates of those in the 1–24 years categories declined first, followed by female and then male adults, then infants and finally the elderly (55 years and over).[90]

For the rural elderly, conditions improved only slowly after 1850, and mortality rates for the 55-year-and-over group were little changed before the twentieth century. In their sixties the elderly now paid the 'physical and mental penalties of lives of poverty, poor housing and heavy and dangerous work in appalling conditions'.[91] A majority of those living in the countryside into old age would have become paupers, and being a single woman or widow in old age entailed almost certain pauper status. In the workhouse it only became compulsory to have medically and surgically qualified personnel in attendance after 1858. Between 1850 and 1870 poor law provision in the form of both outdoor and indoor relief was generally extended as the aged were removed from a heavily overstocked workforce, but after 1870 a 'return to the principles of 1834' cut back outdoor relief until the coming of the interventionist state at the end of the century. Much came down to the 'prejudices, parsimony and ignorance of the Boards of Guardians', and although standards of care may actually have been better inside the workhouse than within village cottages, better nursing care was not generally available until the 1890s, by which time the workhouses were mostly populated by the aged and infirm. Aged paupers took no share of any improved living standards at the end of the century.[92]

Mortality differences were apparent by occupation. Of rural adult males the lowest mortality in the 1860s was that of the middle-class country clergymen. They were followed by seedsmen, nurserymen and gardeners; gamekeepers; farmers, graziers' and farmers' sons; stockmen; agricultural labourers and farm servants. By 1900–2 the rankings were little changed. Although occupation-related mortality could hit farm-workers, especially at harvest time when arduous work could lead to deaths from sunstroke, accidents to children, exhaustion, falls and fights, such as those between villagers and visiting reapers, groups in rural occupations were generally reckoned to have the lowest mortality rates. By the 1880s the mortality rates of agricultural labourers were just 66 per

[90] Benson, 'Mortality variation', p. 106.
[91] D. Thomson, 'Welfare and the historians', in Bonfield et al., The World we have Gained, p. 361.
[92] Digby, Pauper Palaces, pp. 167–72; Thomson, 'Provision for the elderly', pp. 200–3.

cent of the age-standardised rates for all adult males, and by the 1890s mortality from TB was at 62 per cent of standard levels. The mortality rate for agricultural labourers was lower, for example than that for the doctors or tradesmen of Manchester. Infant mortality too was lower in the families of agricultural labourers (although higher than in the families of their farmer employers): by 1911 their mortality rate was 97 per 1,000, compared with the national average of 125 per 1,000, and this rate, as noted above, could be even lower in several localities.[93]

Female mortality differed in important respects from that of males in country areas. Average life expectancies about 1850 were approximately equal (women 41.8 years, men 39.9 years) with little change through the 1860s, but a more rapid decline in female mortality became clearly apparent in the 1880s. But in terms of age-specific mortality, that for females was higher than for males, for example in the 12–19-years age group by 1911, and in 1901–2 was greater in the 3–14-years groups. In 1838–54 female mortality was higher in the 10–39-years groups, and this continued in rural areas (though not in urban) through to 1914. In general female mortality exceeded males in the year groups 10–45.

How was this excess in certain age groups to be explained? Some undoubtedly resulted from what William Farr in 1876 referred to as the 'deep, dark and continuous stream of mortality' incurred by childbirth trauma (accounting for about 5–10 per 1,000 nationally in the nineteenth century), but this could not account for the excess in the younger age groups.[94] The most important factor throughout the age groups was the incidence of TB, accounting for perhaps half of all female deaths at this time. Only after the 1880s did female mortality from TB fall below that of males, and thereafter female death-rates all fell below those of males, with girls aged 5–19 years the last to achieve this.[95]

Also influencing the female age groups was the additional factor that in many agricultural households, dependent upon the wages of the men, the most nutritional food went to the men rather than women, thereby increasing the risk of TB for women. A comparison of mortality rates in 1851–61 between Cornish women living in a purely agricultural District (Stratton) and one dependent upon copper mining (Redruth) showed that for women it made little difference in which community they lived. Indeed for women aged 5–14 the mortality rates in industrial Redruth

[93] Local Government Board, *Statistical Memoranda*, pp. 15–24; Anderson, 'Social implications of demographic change', pp. 21–2; D. H. Morgan, *Harvesters and Harvesting 1840–1900: a Study of the Rural Proletariat* (London, 1982), p. 144; Armstrong, 'The influence of demographic factors', p. 72; Registrar General, 65th Annual Report, Decennial Supplement *BPP*, 1905, XVIII, p. xv.

[94] R. Schofield, 'Did the mothers really die? Three centuries of maternal mortality in "The World we have Lost"', in Bonfield *et al.*, *The World we have Gained*, pp. 231–2.

[95] M. Vicinus, ed., *A Widening Sphere: Changing Roles of Victorian Women* (Indiana, 1977), pp. 164–78.

were lower than in agricultural Stratton. The explanation almost certainly involves the greater possibility of paid female employment in Redruth, where women were employed from an early age in surface work at the mines or in stores, etc., compared with Stratton where only domestic service offered itself. Miners' family diets were more nutritious, and girls got a larger share of the better food than did their agricultural counterparts, whose lower economic status might result in covert neglect. The high rates of female mortality for rural Wales (almost on a par with textile manufacturing towns and colliery districts for females aged 15–45), especially remote north Wales in 1861, and in the Pennines and parts of the south-east, may be explicable in the same way.[96]

A fall in mortality rates within rural areas after about 1870 mirrored those in the nation as a whole, and urban and rural rates began to converge. Infectious disease was attacked vigorously, and since this was an important cause of high mortality rates which was, to some extent, class-free, no great difference was to be observed in reduced mortality rates across the social spectrum. Mortality from TB was halved between 1850 and 1914, with improved nutrition and hygiene. The impact of endemic diseases too was lessened as living standards improved.[97]

Mortality rates for children, followed immediately by young adults, fell first after about 1870, but those for infants did not fall irreversibly until after *c.* 1900. Those for the 55-year-plus group also showed little change before the twentieth century. At the local level, many rural areas did not show steady declines in mortality, since rural depopulation was age selective and the older residual populations would have been correspondingly less healthy during the course of the later nineteenth century. But overall mortality rates fell during the 1880s and more dramatically from the 1890s onwards; mid-nineteenth-century life expectancy of around forty years rose to fifty-two (males) or fifty-five (females) by 1911–12.[98]

V. MOVEMENT AND MIGRATION IN THE COUNTRYSIDE

If before 1850 out-migration from rural areas had been prompted by a lack of farming capital and problems of employment, after 1850 a new

[96] Anderson, 'Social implications of demographic change', pp. 18–19; Vicinus, *Widening Sphere*, pp. 176–9; S. R. Johannson, 'The demographic transition in England: the economic, social and demographic background to mortality and fertility change in Cornwall, 1800–1900', PhD thesis, Univ. of California, Berkeley, 1974, pp. 346–79; Welton, *England's Recent Progress*, p. 55; Woods, *Theoretical Population Geography*, pp. 389–90.

[97] Calculated from Swerdlow, '150 years of Registrar Generals' medical statistics', p. 24 (Fig. 4); D. Friedlander, J. Schellekens, E. Ben-Moshe and A. Keyser, 'Socio-economic characteristics and life expectancies in nineteenth-century England: a district analysis', *Pop. Studs.*, 39, 1985, pp. 137–51.   [98] Anderson, 'Social implications of demographic change', pp. 15–16.

set of circumstances combined to accelerate the process. The demographic strains in rural communities between 1850 and 1914 could be met by a variety of strategies, and changes in nuptiality, marital fertility or migration could, to some extent, be seen either as alternatives or as links in an interdependent chain of reaction, depending on the time and locality under discussion. But out-migration was the most likely response in purely agricultural communities.

Between 1851 and 1911, the labour force in agriculture, horticulture and forestry fell from 1,788,000 men and 229,000 women to 1,436,000 men and 60,000 women, a fall of 19.7 per cent for men and a staggering 73.8 per cent for women: in all a fall of over half a million workers. When the families of these workers and non-agricultural workers such as tradespeople and their families are taken into account, it has been calculated that between 1851 and 1911 the countryside of England and Wales witnessed a loss of 4,063,581 people.[99]

Altogether, between 1841 and 1911 a natural increment in rural areas of 5.3 million people (+86 per cent) was actually offset by a migrational loss of 4.5 million (−73.2 per cent). This represented a total which absorbed 84.9 per cent of the natural increment during these years, and this of course was an average figure which concealed still greater losses in some upland areas.

However the calculations were done, the impact on rural areas was undeniable. Only one contemporary analysis seriously suggested a different view, that by Ogle who questioned fears of a large depopulation. He took the 15 leading agricultural counties (where more than 10 per cent of persons were employed in agriculture), and defined his rural areas as a residual population after subtracting those urban districts with 10,000 people or more. He then looked separately at areas with populations of 5,000 to 10,000, and at rural areas with fewer than 5,000 people. He found growth in the small towns of 15 per cent between 1851 and 1881 at the expense of villages and hamlets which had decreased by 2 per cent. However, his starting assumption that all rural areas could be classified as having below 5,000 people, when most rural parishes actu-

---

[99] The figures are problematic, and relate to the way in which the census figures were collected and analysed. The fall in women's employment is certainly highly exaggerated, and in some regions the fall was negligible. See Gielgud, 'Nineteenth-century farm women in Northumberland and Cumbria'. The standard figures are from Mitchell and Deane, *Abstract*, p. 60. The migration statistics were initially from Welton, *England's Recent Progress*, later revised by Cairncross, 'Internal migration in Victorian England', Table II, p. 83; and Cairncross, *Home and Foreign Investment*, Table 15, p. 15; and accepted by R. Lawton, 'Population and Society, 1730–1914' in R. A. Dodgshon and R. A. Butlin, eds., *An Historical Geography of England and Wales* (2nd edn, London, 1990), pp. 285–321.

ally had populations less than 500, served to disguise the real extent of depopulation. His work is cited by Weber in 1909, but surely Longstaff's 1893 view, based partly on Ogle's work, that the intensity of rural depopulation 'has been greatly exaggerated: in the few spots where it is at its worse, it only amounts to a thinning of the people such as should be viewed with reasonable equanimity', was quite misleading.[100]

Between 1851 and 1871 rural populations were very nearly stationary. Then, with employment opportunities becoming scarcer, between 1871 and 1901 they fell by 10 per cent, and finally between 1901 and 1911 they increased by 3 per cent.[101] Rural migration appears to have been as high in periods of economic growth (e.g. the 1860s or 1890s) as in more depressed periods (e.g. the 1880s). In general the third quarter of the nineteenth century has been characterised as a time of some prosperity, but it also coincides with a time of peak out-migration from the countryside. And by the last quarter of the century the volume of out-migration could not have been as great, despite the boom in urban building 1894–1903, since so many of the young had already gone. Only in 1901–11 was out-migration exceeded overall in rural areas by in-migration, although many more remote areas did not share in this rise. Clearly 'the ebb and flow of economic conditions' were part of, though not the whole story.[102]

Local differences of timing also complicated the overall picture. Cornish out-migration was at its peak in the late 1860s and 1870s, but that of Wensleydale/Swaledale peaked in the late 1870s and 1880s. The parish of Corsley (Wilts.) had experienced its greatest losses between 1841 and 1861 when its clothing factory and last vestiges of manufacturing closed down, but there were also losses in the 1870s as labour on arable farms was displaced by a move into pasture for dairying. Once again, it is important to see the processes of out-migration working within very different physical, economic and socio-cultural environments, and at different scales of analysis. Superimposed upon the general regional trends between high-wage north and low-wage south and pastoral west and arable east were the inter-parish differences. Thus many

---

[100] Ogle, 'The alleged depopulation of the Rural Districts', 205–32; Weber, *The Growth of Cities*, p. 45; G. B. Longstaff, 'Rural depopulation', *JRSS*, 56, 1893, p. 412.

[101] Bowley, 'Rural population in England and Wales', pp. 597–645.

[102] Cf. Cairncross, 'Internal migration in Victorian England', pp. 70–71 who cites the 1880s as a time of peak out-migration. J. Saville, 'Internal migration in England and Wales during the past hundred years', in J. Sutter, ed., *Les déplacements humaine: aspects methodologiques de leur mesure* (Entretiens de Monaco en sciences humaines, 1962), p. 10. The coincidence of the 'Golden Age' of agriculture with high out-migration rates was pointed out by Lord Eversley, 'The decline in number of agricultural labourers in Great Britain', *JRSS*, 70, 1907, pp. 267–8.

could report that 'open' villages might be growing at mid-century, with consequent overcrowding, while 'close' villages were the scene of stationary or declining populations. After the 1865 Union Chargeability Act and the relaxation in the laws of settlement, the larger 'open' parishes might lose population but the smaller 'close' ones gain, as on the Lincolnshire Wolds.[103]

Inevitably then, the impact of out-migration fell unevenly across the countryside (Figure 21.3). In the period 1851–1901 as a whole, areas of persistent loss included the large belt of countryside stretching from East Anglia to the south-west peninsula, rural Wales, the northern Pennines, east Yorkshire and the Vale of York. After 1901 the decline was less severe but still continued in remoter Wales, the south-west and the Pennines. The plotting of natural change and migrational change (Figure 21.4) shows swathes of countryside over which out-migration exceeded natural gain, compared with the smaller areas in which natural growth exceeded out-migration.[104]

Serious out-migration affected most of rural Wales, where population declined absolutely in Cardigan, Montgomery and Radnor, 1841–1911; and losses occurred 1871–81 in Brecknock; 1881–91 in Anglesey, Carnarvon, Flint, Merioneth and Pembroke; 1891–1901 in Merioneth and Pembroke; and 1901–11 in Carnarvon and Merioneth. Such growth as there was in Wales focused almost entirely on the southern coalfield and coastal area, which attracted agricultural workers from as far away as Somerset. By 1851 Merthyr Tydfil, the largest town in Wales, had declining rural counties in its hinterland where 'agricultural labourers are less numerous than they formerly were owing to their being drained off to Merthyr and the manufacturing districts'. Overall between 1861 and 1911 the population of rural Wales fell from 304,000 to 238,000, a decline of 21.9 per cent, a figure only exceeded by Cornwall.[105]

Although the northern counties as a whole demonstrated a more stable rural population, there were large pockets of decline here too. Rural

---

[103] M. F. Davies, *Life in an English Village* (London, 1909), pp. 83–8; D. R. Mills, 'Percentage native-born as an index of rural population mobility, 1851–81', unpublished paper, Oxford (July, 1985), p. 4.

[104] R. Lawton, 'Rural depopulation in 19th century England' in R. W. Steel and R. Lawton, eds., *Liverpool Essays in Geography: a Jubilee Collection* (Liverpool, 1967), pp. 227–55; Lawton, 'Population changes in England and Wales in the later nineteenth century: an analysis by registration districts', *TIBG*, 44, 1968, pp. 55–74; Lawton, 'Regional population trends in England and Wales, 1750–1971' in Hobcraft and Rees, *Regional Demographic Development*, pp. 29–70; Lawton 'Population and Society, 1730–1914', pp. 285–321.

[105] H. Rider Haggard, *Rural England: being an Account of Agricultural and Social Researches carried out in the Years 1901 and 1902* (2 vols., 2nd edn, London, 1906), vol. I, p. 250; Haines, *Fertility and Occupation*, p. 162; V. C. Davies, 'Some geographical aspects of the depopulation of rural Wales since 1841', PhD thesis, University of London, 1955.

Rural out-migration 1851–1911

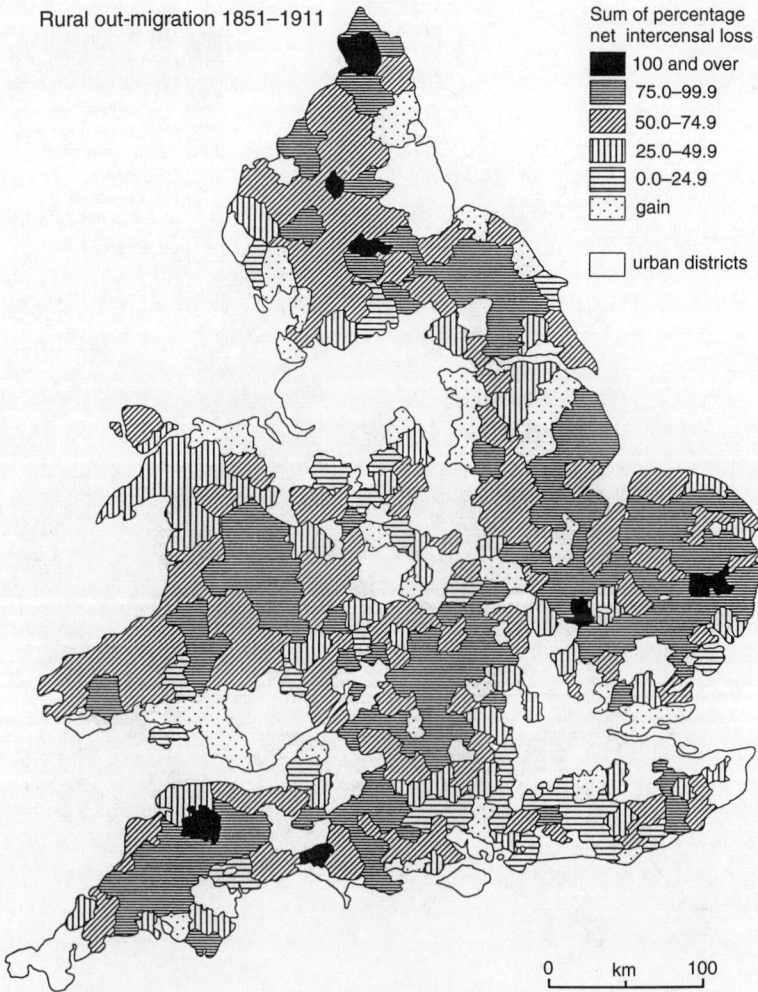

Sum of percentage
net intercensal loss

▉ 100 and over

▤ 75.0–99.9

▨ 50.0–74.9

▥ 25.0–49.9

▬ 0.0–24.9

⬚ gain

☐ urban districts

0   km   100

21.3 Rural out-migration 1851–1911
The data represent the summation of percentage net migrational change in
each census decade, i.e. the differences between natural and total change.
Source: Lawton, 'Rural depopulation'

Northumberland and Pennine Yorkshire showed steady losses, with
increased out-migration from the 1850s absorbing more than the natural
increase, and with massive decreases in the 1880s. Collieries and ironworks were easily reachable alternatives here, though many commentators felt that mining districts mostly reproduced their own labour force,
because of their higher fertility regimes and family recruitment patterns.
Even where land was fertile and wages relatively high, as in Glendale or

Population Trends 1851–1911

INCREASE

■ Net in-migration exceeds natural gain

▤ Natural gain exceeds net in-migration

▨ Natural gain exceeds net out-migration

DECREASE

▒ Net out-migration exceeds natural gain

□ urban area

0   km   100

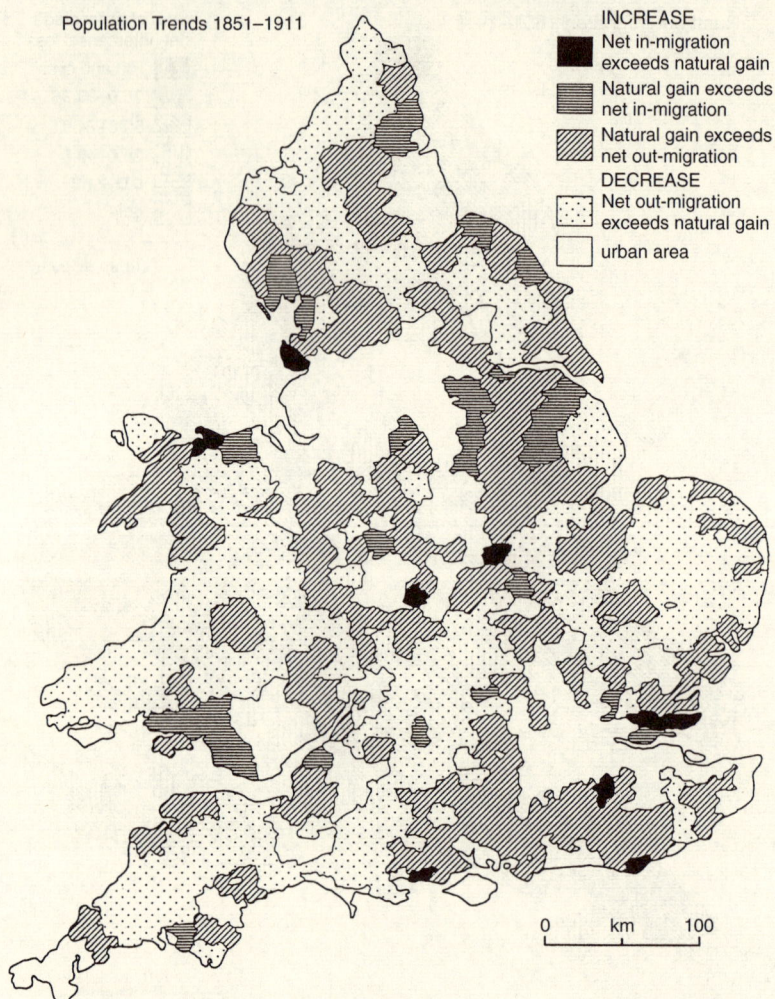

21.4 Rural England and Wales: Natural change and migrational change
1851–1911
Source: Lawton 1977, Fig. 2.6, p. 51

in the Till valley near Wooler, Northumberland, with wages at 15s.
weekly plus 6s. in kind, depopulation continued.[106]

In the south–west, Cornwall in particular showed absolute losses after
the collapse of the mining industry in 1865. Between 1865 and 1867 in

---

[106] Jones, 'The Tankersley ironstone miners *c.* 1850', pp. 13–27; P. A. Graham, *The Rural Exodus*
(London, 1892), pp. 4–7; A. Redford, *Labour Migration in England, 1800–1850* (3rd edn, London,
1976), pp. 56–7.

the parish of Uny Lelant 5 mines were closed throwing 360 men out of work. Over 100 found jobs locally, but 200 more left without their families, to seek work elsewhere. Such husband-only migration was common in Cornwall, and was already well established by 1850. But the exodus after 1865 reached vast proportions; over 11,000 left permanently or temporarily, and about 3,000 found work in the St Austell China Clay works. By 1891 about 25 per cent of all Cornish-born people lived in some other county in England. In the more purely agricultural parishes of eastern Cornwall, such as the Stratton area, absolute losses also were common after 1850. Overall between 1861 and 1911 Cornwall's population fell by more than any other county, 45,000, representing a decline of 27.1 per cent. For the other south-western counties (Devon, Somerset, Dorset, Wiltshire) the figure was a reduction by 130,000 – 17.1 per cent. Although one Dorset witness to the Committee on Settlement and Poor Removal averred in 1847 that 'In Dorset we very much vegetate where we are born, and live very close indeed', the ties that bound most people to their localities were weakening by mid-century, if indeed they had ever been as strong as mythology would have it.[107]

In East Anglia the population of Norfolk as a whole fell by 28,000 between 1861 and 1901, the county having already lost 2 per cent of its population since 1851. Overall Norfolk's losses between 1851 and 1901 accounted for about 50 per cent of its natural increase. The 1851 Census reported that '4,521 of the youth of Norfolk, Suffolk and Essex leave their native counties every year to reap elsewhere the fruits of the education, skill and vigour which they have derived, at great expense, from their parents at home'. It was duly acknowledged, however, in classic economic terms that a 'free circulation of the people is now necessary in Great Britain, to meet the varying requirements of the public industry'. The eastern counties (including Huntingdon, Lincolnshire and Cambridge) lost 10.9 per cent of their rural population between 1861 and 1911.[108]

Parts of the rural Midlands, such as Staffordshire and Rutland, showed absolute declines also. In rural Bedfordshire every census from 1841 to 1881 showed more parishes returning a decline of population; with the attractions of London, the Northampton iron industry or boot and shoe making, or local industrial urban centres making themselves felt. By 1861 Rutland, with no rural industries left, had retained just 60.8 per cent of its native population (the equivalent figure for Lancashire was 91 per cent).[109]

Higher net migration losses were recorded in the southern countryside (2.86 million) than in the north (1.64 million) between 1841 and

[107] Johannson, 'Mortality and fertility change in Cornwall, 1800–1900', pp. 262–6.
[108] Digby, *Pauper Palaces*, p. 28; *Census of Great Britain 1851*, vol. 1 (1854), p. cviii.
[109] Welton, *England's Recent Progress*, pp. 41–7.

1911. Typical perhaps was the parish of Cerne Abbas (Dorset) whose 1851 population of 1,343 was reduced to a miserable 585 by 1912, including 46 people in the workhouse.[110] This experience was repeated across a broad band of the southern countryside. On the other hand, areas in the south also showed signs of rural population growth, such as the suburban Home Counties and around London. Adventitious urbanites mingled (sometimes uneasily) with stage migrants and increased numbers working in the horticultural and market-gardening sectors, and their combined strength outnumbered the decline in the agricultural labour force here. Even in the south-east though, there were areas of loss: between 1851 and 1911 in the Weald of Kent and Sussex, for example, 80 per cent of the population growth was centred in Hastings and Tunbridge Wells, while losses were counted in the Battle, Ewhurst and Tenterden areas. The eastern Registration Districts of Rye, Tenterden and Hollingbourne, together with Petworth in the west, showed overall losses at this time.[111]

In 1881 the Census listed Registration Counties in which 'the stationary proportion of the numerated natives are less than 65%', in an attempt to show the counties which had lost the most of their native populations. All were agricultural counties, and included Rutland, Huntingdon, Buckingham, Oxford, Berkshire, Hertford and Cambridge from the English lowlands; Radnor and Brecknock from Wales; Hereford and Shropshire from the Welsh borders; Westmorland from the North; and Wiltshire and Dorset from the south-west.[112]

Bowley, in his seminal paper in 1914, noted that 'however thoroughly we purify the population of urban and mining influences, we still find that the remaining population falls less or increases more in the neighbourhood of industry or residence'. This 'radiating urban influence' varied. Thus in Hampshire the rural population north of a line from Fordingbridge to Farnborough was practically stationary at this time, whereas to the south it had increased. Similarly the Berkshire districts east of Reading had increased compared with stationary areas such as Bradfield, Newbury, Wallingford and Abingdon, and declining areas in western Berkshire, as in neighbouring Wiltshire. In the north of England too, it was noticeable that while moorland populations fell, coastal populations increased. As Glendale and the Till Valley became depopulated, so Newcastle grew. In 1911, of the 581 rural parishes which had exhibited the largest amount of growth in the improved conditions of the 1901–11 decade, 296 (51 per cent), accounting for 43.3 per cent of the

---

[110] H. D. Harben, *The Rural Problem* (London, 1913), p. 4.

[111] P. Brandon and B. Short, *The South East from AD 1000* (London, 1990), pp. 270–71.

[112] D. J. Davies, 'The condition of the rural population in England and Wales 1870–1928', PhD thesis, Univ. College of Wales, Aberystwyth, 1931, p. 20.

total population increase in the parishes, owed their growth to residential development because they were near towns or enjoyed improved communications. A further 142 (24.4 per cent), accounting for 36.7 per cent of the total population increase, occurred because of colliery development; while other growth was because of manufacturing development, the erection of hospitals and other institutions, etc. In only 16 of the 581 parishes (2.75 per cent), accounting for less than 1 per cent of the increase, was it due to agricultural development such as smallholdings, fruit farming or market gardening. The statistics demonstrate very well the importance of towns for growth nearby.[113]

Some well-worn rural–urban routes were part of community lore, such as the strong links between south London's labour market and the rural communities of Kent via a north Kent route. In some cases people clustered in towns from linked rural trades. London gardeners, smiths, millers, saddlers and building craftsmen were predominantly rural in their origins, while Welsh milk sellers and cowmen came from Pembroke or Cardiganshire. Indeed, the latter county furnished enough migrants to London for them to be seen as 'a little colony', and drapers filled their own shops with Cornish or Welsh assistants. Wensleydale migrants settled in the Liverpool area where they set up in business as cowkeepers; tradition had it that many were younger sons of farmers, and thus unlikely to inherit within the primogeniture system in operation. Close family ties were maintained and remittances sent back to the village.[114]

The processes producing this movement of 4 million people in 60 years cannot easily be separated. But clearly some major influences were at work, whether contingent or causally related, and the reasons given vary according to the time and place under examination, and the scale of analysis being used. Some analysts have looked to the workings of the economy, some more specifically to agricultural change – both technical and structural, some to the social and cultural context of village life and the changes wrought by increasing state intervention in education, some to exogenous factors such as the mechanisation of craft production or the spread of efficient communications. The de-industrialisation or 'pastoralisation' of the countryside at this time was also significant, as large populations engendered in the lead mines of Cardiganshire, or Wensleydale and Swaledale, or lead mines, slate quarries and woollen manufactures of Montgomery, could not be sustained when lead-mining

[113] Bowley, 'Rural population in England and Wales', pp. 607–8; *Census of Great Britain 1911*: General Report, pp. 38–9.

[114] Snell, *Annals*, p. 61; E. J. Hobsbawm, 'The nineteenth-century London labour market' in Centre for Urban Studies, ed., *London: Aspects of Change* (London, 1964), pp. 8, 15–16; Stedman Jones, *Outcast London*, p. 137; Davies, 'The condition of the rural population in England and Wales 1870–1928', p. 40; Hallas, 'Migration in Wensleydale and Swaledale', pp. 152–3.

collapsed, and a new equilibrium between people and resources had to be attained.[115]

Trades- and craftspeople also were caught in a double trap as their products were more cheaply produced in industrial, urban manufactories and workshops at the same time as their rural customers declined in number. In Huntingdonshire between 1851 and 1881 the number of wheelwrights fell by 21 per cent, sawyers by 51 per cent, coopers by 50 per cent, shoemakers by 49 per cent and tailors by 41 per cent. Lace-making declined at the same time by 62 per cent, and was symptomatic of the inability of rural cottage industry to compete in such areas as glove-making, button-making, hosiery or net-making. Lucy Luck, a straw-plait worker from Hertfordshire, was married to a farmworker who was given a month's notice when he asked for a rise from his 13s. weekly. They moved to London, where her husband got a job as a horse-keeper for a railway company, and she kept at her strawwork. The work was sent to her once a week, until she changed to work for shops in Westbourne Grove because the pay was far better. Thousands of other couples and individuals replicated this story in various ways.[116]

If, as Rew calculated, 3 or 4 men were required for every 100 acres of arable but only one man for every 100 acres of pasture, then the loss of 2 million acres of arable during the depression also had profound effects. Harben calculated in 1913 that the loss of arable involved throwing between 60,000 and 80,000 labourers out of work between 1881 and 1901, while Lord Eversley calculated a loss of two adult males per 100 acres laid to pasture (40,000 to 60,000 based on the above figures). Wholesale restructuring of local agricultural systems led to areas such as north Wales increasing their emphasis on grassland products on estates such as Pengwern or Bodrhyddan in the Vale of Clwyd (Flintshire), entailing a demand for fewer workers, and especially fewer seasonal harvesters. And as mechanisation became a possibility in the second half of the century, so still fewer workers were required at peak harvesting periods. Many farmers, however, maintained that mechanisation replaced the migrating workers, rather than forcing redundancies. Others may have turned increasingly to the employment of younger and cheaper males, so that two boys might be employed at 3s. 6d. week rather than one man at 9s. By 1911 the number of males in agriculture aged 15–20 stood at 55 per cent of those aged 25–45, whereas in all occupations the figure was 29.8 per cent. But increasingly such work was perceived as a

[115] Hallas, 'Migration in Wensleydale and Swaledale', pp. 139–61.

[116] Saville, 'Internal migration in England and Wales during the past hundred years', p. 20; J. Burnett, ed., Useful Toil: Autobiographies of Working People from the 1820s to the 1920s (London, 1974), pp. 67–77.

'blind alley' for boys, and their discontent was expressed in depopulation.[117]

The differentials between urban and rural wage levels are often cited as an important factor, but this was a complex issue. Nationally agricultural wages were about 50 per cent of wages for comparable skills in urban areas in 1911–14. At the regional level it is true that wages generally remained higher in the north than south throughout this period, and that regional migration levels were also higher in the lower-wage south. But at the more local level, wages were frequently highest near towns and industries in an effort to retain workers on the land. Between 1904 and 1911 young men were leaving southern Warwickshire, more remote from towns and railways, and paying wages of 12s. or even 10s., far more quickly that in the east near Birmingham or west of the county where wages ran to 15s. or 16s. But elsewhere it was also precisely in such areas that migration was at its height at this time. By 1913 Sutherland noted that 'the rate of emigration is greatest at present not from those localities where the agricultural wages are lowest but from the areas where they are highest and where there is the greatest demand for labour generally'. Clearly either the high wages were still not high enough to compete with urban levels, or the wage differential was only one of a number of important causal factors. Thus between 1871 and 1921 low-wage Dorset lost 42 per cent of its agricultural workers, but high-wage Northumberland also lost 39.7 per cent.[118]

In many areas in the later nineteenth century migration actually led to higher wages, to combat increasing labour shortages. Between 1860 and 1880 money wages for agricultural labourers rose on average by 30 per cent, and the regional variation in agricultural wages was reduced from a maximum–minimum difference of 44 per cent in 1867–70 to 28 per cent by 1907. Allied to this, it is argued that the cost of living fell considerably at the end of the century, giving a higher real income for farm workers. Set against this, however, were the reduced opportunities for seasonal and casual work by family members, the decline once more in real wages experienced after 1900, and the fact that in most areas the cash

---

[117] Harben, *The Rural Problem*, p. 4; Eversley, 'The decline in number of agricultural labourers in Great Britain', p. 289; C. Thomas, 'Seasonality in agricultural activity patterns: examples from estates in the Vale of Clwyd, 1815–1871', *Publications of the Flintshire Historical Society*, 26, 1974, pp. 110–13; Davis, 'The condition of the rural population in England and Wales 1870–1928', pp. 55–8.

[118] J. R. Bellerby, 'Distribution of farm income in the United Kingdom, 1867–1938', *J. Proceedings Agricultural Economics Society*, 10, 1953, p. 127–44; Bowley, 'Rural population in England and Wales', Appendix III; Lawton, 'Rural depopulation in 19th-century England'; Land Enquiry Committee, *The Land: the Report of the Land Enquiry Committee* (4th edn, London, 1913), p. 34; W. Sutherland, *Rural Regeneration in England* (London, 1913), p. 25.

wage was only part of the whole earnings, since garden plots, food, milk or tied cottages (for better or for worse) were also included as payment in kind. In Durham, for example, cash wages in 1907 were 18s. and total earnings 21s. 9d., compared with Norfolk where wages were 12s. 7d. and earnings 15s. 4d., although estimates of the value of the payments in kind were always a source of dispute.[119]

The relative social and cultural inwardness of village society compared with the perception of town life, was also a strong factor, as was the 'lack of outlook and prospects for the future'[120] within farming. A very few might become smallholders, bailiffs, foremen or even small farmers but most would be faced by hard and tedious work for little return. Because of the heated debate over the land issue, many county councils by the early twentieth century were providing smallholdings to attempt to stem the outflow, and were encouraging the provision of allotments. But alternative occupations were distinctly lacking compared with the possibilities in urban areas, and this was especially the case for women in the grain-growing south, where many could find alternative work in domestic service in the south coast resorts.

Insecurity of cottage tenure, as well as the appalling condition of many rural cottages, was also cited by contemporaries. Overcrowding and cottage shortages were general, although alleviated somewhat where out-migration left more room than before, and although Graham thought that housing conditions were of little relevance to the 'rural exodus', others, such as Jessopp or Rider Haggard, disagreed. The Land Enquiry Committee felt that in the low-paid counties the farmworker 'is housed in a way in which no up-to-date farmer would dream of housing his prime stock', and reported a shortage of accommodation in 50 per cent of the 2,759 parishes investigated.[121]

What Longstaff referred to as 'sentimental' reasons were also important. In the search for a less constrained and more varied lifestyle, the interpenetration of urban-based education into the villages was an undoubted aid. Its gradual acceptance into village culture from the 1870s coincided with some of the worst years for rural depopulation, confirming the fears of many farmer-employers, who saw the educated youth leaving the villages and farm employment. As Jesse Collings put it: 'By it new ideas, totally dissociated from the localities in which they live, are instilled in the minds of the young, who believe that such ideas can

[119] Eversley, 'The decline in number of agricultural labourers in Great Britain', p. 280; E. H. Hunt, *Regional Wage Variations in Britain, 1850–1914* (Oxford, 1973), pp. 58–9, 244–8; Land Enquiry Committee, *The Land*, pp. 1–29; Harben, *The Rural Problem*, pp. 7–9. See also p. 1258.

[120] Land Enquiry Committee, *The Land*, p. 34.

[121] See pp. 1267–8 above; Jessopp, *Arcady*, p. 113; Rider Haggard, *Rural England* e.g. Warwickshire (vol. I, pp. 410–11 and *passim*); Land Enquiry Committee, *The Land*, p. 36.

only be realized elsewhere.' If children were trained as clerks, was it surprising that they wanted to leave the villages to become clerks? Collings and others argued for a more relevant rural education which laid more emphasis on nature study and practical agrarian topics. However, the extension of rural education under the aegis of the county councils after 1902 actually came during one decade of rural growth at the beginning of the twentieth century, so once again the issue was not clear cut.[122]

As migration and movement widened, so too did the extent of exogamous marriages increase, and the sharp breakdown of localism in marriages accompanying greater movement and easier communications during the nineteenth century has often been noted, although most movement remained localised. The latter point also holds for the remnants of farm service across the country. Although farm service continued far longer, especially in pastoral areas, than once believed, its decline also meant that many young people in their teens and early twenties also left the countryside altogether. Taken together, the institutions of marriage and living-in service promoted a constant groundswell of movement, in which the former particularly stimulated the migration of younger women. Nevertheless, once married and with families, migration was less likely. There was anyway discrimination in favour of married men for receipt of poor relief, who were given parish work in preference to the single, who might consequently migrate. Such selective outmigration affected in turn the marriage market because of the changed sex ratio among courting age groups in the rural locality now deprived of its unmarried women.[123]

The most widely cited contemporary study of migrations was by Ravenstein. His findings can briefly be summarised: that most migrants moved only short distances, producing a 'universal shifting' and 'currents of migration' to industrial and urban centres; that migration resulted in stepwise movements, whereby the spaces left by migrants moving into nearby towns were filled by people from a little further away, producing a distance–decay effect away from the towns; that dispersion was 'the inverse of absorption'; that each current of migration produced a compensating

[122] Longstaff, 'Rural depopulation', p. 413, differentiating between two main factors causing depopulation: 'sentimental' and 'economic'; J. Collings, *Land Reform* (London, 1908), p. 23; *BPP, Interdepartmental Committee on the Employment of Schoolchildren*, 1902 xxv (Cmd. 894), p. 15.

[123] R. F. Peel, 'Local intermarriage and the stability of rural population in the English Midlands', *Geography*, 27, 1942, pp. 22–30; P. J. Perry, 'Working class isolation and mobility in rural Dorset 1837–1936: a study of marriage distances', *TIBG* 46, 1969, p. 121–41; B. M. Short, *The Geography of Local Migration and Marriage in Sussex, 1500–1900*, Univ. of Sussex Research Papers in Geography, 1983; J. Robin, *Elmdon: Continuity and Change in a North-west Essex Village, 1861–1964* (London, 1980), p. 185; Hinde, 'The marriage market in the nineteenth-century English countryside', p. 384; for unbalanced sex ratios, see also H. V. Musham, 'The Marriage Squeeze', *Demography*, 11, 1974, pp. 291–9. See also pp. 1248–9.

counter-current; that long-distance migrants generally went to 'great centres of commerce and industry'; that urban dwellers were less migratory than rural ones; and that females were more migratory than males.[124]

These findings have been confirmed by much subsequent empirical research, with the result that we now have a much clearer idea of localised rural movements. Samples from the 1851 census indicate that 52 per cent of the rural population lived more than 2 km from their place of birth, although over two-thirds moved less than 25 km. A sample of nine case studies for 1851 yields an average of 51.5 per cent being born within the parish of residence. Between 1851 and 1861 Brenchley, in the Kentish Weald, showed a turnover rate of 6 per cent per annum. Another sample of eight parishes from lowland England in 1871 still found between 70 and 79 per cent of male agricultural labourers and farm servants working in the parish of their birth. Apart from the farmers themselves, these were, however, the least mobile sectors of the labour market since by definition they did have jobs, but there were few who were not potential migrants. Many others, especially younger women, left, although Ivinghoe (Bucks.) with its straw-plaiting manufactures retained many unemployed men who could rely on wives and children for support at times of unemployment. If everyone is included, the percentages living in the parish of their birth fall to levels typically between 40 and 72 per cent. Finally, one piece of empirical work on 'countercurrents' suggests, however, that much of the migration out of London would have been destined for urban areas rather than rural ones.[125]

Given the importance of different regional experiences, it is not surprising that some have attempted to relate migration to agricultural types. Roxby in 1912 set out to display 'the direct relationship which can be shown to have existed between a particular type of agricultural organisation and a particular movement of population',[126] noting the relationship between the arable-based, large-farm, low-waged eastern counties and the high rate of out-migration. But in the poorer, heavy clay, arable

---

[124] E. G. Ravenstein, 'The Laws of Migration' *J. Stat. Soc. London*, 48, 1885, pp. 167–227; E. G. Ravenstein, 'The Laws of Migration. Second Paper', *JRSS*, 52, 1889, p. 241–305; and see D. B. Grigg, 'E. G. Ravenstein and the "Laws of Migration"', *JHG*, 3, 1977, p. 41–54.

[125] Anderson, 'Social implications of demographic change', pp. 11–12; M. Kitch, 'Population movement and migration in pre-industrial rural England' in Short, *The English Rural Community*, p. 75; B. Wojciechowska, 'Brenchley: a study of migratory movements in a mid-nineteenth century rural parish', *Local Pop. Studs.*, 41, 1988, p. 30; Horn and Horn, 'Ivinghoe in 1871', pp. 15–16; Horn, 'Child workers in the pillow lace and straw plait trades of Victorian Buckinghamshire and Bedfordshire', pp. 779–98; D. R. Mills, ed., *Victorians on the Move* (Mills Historical and Computing, Lincs, 1984), p.v.; W. A. Armstrong, 'Some counter-currents of migration: London and the South in the mid-nineteenth century', *Southern Hist.*, 12, 1990, pp. 82–113.

[126] P. M. Roxby, 'Rural depopulation in England during the nineteenth century', *The Nineteenth Century*, Jan. 1912, p. 182.

districts such as a group of thirteen parishes in Kimbolton district, west Huntingdonshire, the loss of population 1861–1901 was 35.9 per cent, compared with a group of fourteen parishes in the richer and better placed arable belt of High Suffolk, which lost just 15.6 per cent. In the central pastoral counties he noted a contrast between the extensive large-scale stock-farming in twenty-four parishes in the Barrowden and Great Easton districts of Rutland and Leicestershire, with a population decrease of 21.6 per cent, and thirteen parishes within the Vale of Evesham where intensive fruit and market gardening prevailed, and which showed an increase in population of 33 per cent. Such contrasts in demographic experience were related to soil and farming type. The Biggleswade (Bedfordshire) and Wisbech (Cambridgeshire) districts were also cited as intensive production areas where population had similarly grown. They were held up by Roxby as offering examples of areas where 'the prospects of a more numerous and contented peasantry are brighter than is often supposed'.[127]

Roxby's findings have been criticised by Lawton, who analysed the population changes for the whole Registration District from within which Roxby's arable examples were drawn. Apart from the clayland arable area of Kimbolton (St Neot's Registration District) which did show heavier out-migration, the other arable areas were remarkably consistent in their pattern of loss. Lord Eversley similarly demonstrated that groups of counties with varying proportions of arable land, from under 20 per cent of the cultivated area (e.g. Derbyshire, Westmorland and Monmouthshire) to over 60 per cent (Norfolk, Suffolk, Lincolnshire and others) also displayed great uniformity in their losses of farm labour between 1861 and 1901, 'irrespective of any differences in their systems of cultivation'.[128]

However, neither counties nor Registration Districts were by any means uniform in their natural or agrarian characteristics. For that matter, neither were parishes, and it remains extremely difficult to correlate land use or other agrarian variables with population change. Population densities and rates of change were also examined by Vince, who showed that only on the best quality arable land (level or gently undulating, and comprising deep, fertile, easily worked loams, silts or mild peats) which constituted L. D. Stamp's Grade 1A land, often intensively cultivated by 1939, were losses of agricultural population markedly lower than on other types, i.e. at 11 per cent between 1831 and 1931. On good, general-purpose farmland, such as western ley farming areas, there were losses of 42 per cent; and on shallow light soils of downland arable losses were

[127] Roxby, 'Rural depopulation', p. 188.
[128] Lawton, 'Rural depopulation in 19th-century England', pp. 245–7; Evesley, 'The decline in number of agricultural labourers in Great Britain', p. 285.

56 per cent. On grassland losses ranged from 43 per cent on first class land, given over by 1939 to fattening and dairying, to 52 per cent on fertile but heavier pastures. On poorer land losses ranged from 45 per cent on productive, medium-quality land which was often under long leys in the west, to 49 per cent on mountain moorlands or rough pasture.[129]

It is probably true to say that it was not so much the soil type or land use which directly determined population movements as much as factors such as holding size, tenure and inheritance, labour requirements, or access to markets or communications for produce. Such factors varied with little regard for the boundaries of the units of demographic data. And Lawton's insistence on the primacy of urban proximity in explaining population trends is similarly missing such intermediate factors as these. For example, farmers' sons tended to follow their fathers in business on larger farms, where there was work to be had for family members, but this pattern did not hold to the same extent on smaller ones, where out-migration was more likely.[130]

If it is difficult to correlate out-migration with farming types, it was certainly selective by age, sex, marital status and occupation, and the first two were by far the most prominent. Age selectivity varied once again by time and place: in 1881–1901, for example, all working ages from 15 years upwards showed heavy losses in the 'unprogressive' rural areas, but losses of females were greater at ages 15–20 and after 35 years, and males at between 20 and 35 years. The 1911 Census found heavy concentrations in the 25–35 age groups in urban and mining districts, whilst Bowley put most agricultural labourers' out-migration at between 17 and 25 years, and migration from rural Essex, was most common for those aged 15–30. Clearly, it is difficult to separate the significance of age and sex here, and in analysing the out-migration of young people in relation to the decline of living-in, the impact on males would have been felt in the age groups, 10–14 and 20–29 years, and females at 20–29 years.[131]

The sex differential in out-migration was widely noted. With falling rates of female employment in the arable lowlands, many young women left to work in domestic service or, if available, in industrial manufacturing, as in the Pennine textile mills, or to marry. These 'superfluous women' created a flurry of interest which emerged in both social research

---

[129] S. W. E. Vince, 'The rural population of England and Wales, 1801–1951', PhD thesis, Univ. of London, 1955, pp. 217–18 and *passim*. Vince found, for example, 'general agreement between the 1831 agricultural population density and the quality of agricultural land' (p. 223).

[130] B. Preston, 'Occupations of father and son in Mid-Victorian England', *Reading Geographical Papers*, 56, 1977, p. 36.

[131] Welton, *England's Recent Progress*, pp. 2, 7–9; Saville, 'Internal migration in England and Wales during the past hundred years', pp. 7–8; Hinde, 'Household structure, marriage and the institution of service in nineteenth-century rural England', pp. 48–9.

and literature (the governess, the spinster aunt, the domestic staff, the marriage market) at this time. At Berwick St James (Wilts.) a large reduction of young females in the parish between 1851 and 1871 left just one native-born female aged between 15 and 29 in the parish at the end (1871 population 248). But a warning about facile generalisations comes from research on Elmdon (north-west Essex), where female out-migration was active in the age group 15–29 but also at later years, while male out-migration was more concentrated in the 30–44 age group.[132]

As Graham wrote: 'Country women look upon town as a kind of Eden.' The overall extent of female out-migration was such that by the middle of the twentieth century a lower female:male ratio in all rural areas was to be found at all ages below fifty years, despite the gradual increase in the ratio in England and Wales from 104.2 in 1851 to 106.8 in 1911. In purely agricultural areas such as Rutland, however, a slight male surplus by 1851 was maintained into the twentieth century with a ratio of females:males of 103 being lower than the national average. In 1851, with a female:male ratio of 104.4 in England and Wales, the ratio for the Dorset parish of Hinton Martell was 102.4 whereas that of neighbouring Poole, receiving young female migrants, was as high as 118.3. Such a reversal of sex ratios between town and country has also been noted for parts of the South Hams and Totnes.[133] In mining communities a male surplus was found, and this applied whether the collieries were set in towns or amongst rural surroundings. The implications of this sex ratio for earlier marriage and higher fertility have been noted above.[134]

The most controversial aspect was whether migration was selective by physical and mental qualities, since this could not be so easily quantified. Many, following Galton in 1873, accepted unquestioningly that the towns drained off the fittest and most able, to replace those urban dwellers whose working life had been foreshortened by the poorer urban environments: 'the more energetic of our race, and therefore those whose breed is the most valuable to our nation, are attracted from the country to the towns'. The deterioration of the English breed would follow.[135]

---

[132] Snell, *Annals*, p. 404; P. R. A. Hinde, 'The population of a Wiltshire village in the nineteenth century: a reconstitution study of Berwick St James, 1841–71', *Annals of Human Biology*, 14, 1987, pp. 475–85; Robin, *Elmdon*, pp. 184–5.

[133] Graham, *The Rural Exodus*, p. 99; Saville, 'Internal migration in England and Wales during the past hundred years', pp. 7, 15, 21; Mills, *Victorians on the Move*, p. v.; D. Bryant, 'Demographic trends in South Devon in the mid-nineteenth century', in K. J. Gregory and W. L. D. Ravenhill, eds., *Exeter Essays in Honour of Arthur Davies* (Exeter, 1971), pp. 128–32.

[134] See pp. 1241, 1252. For the distribution of age and sex ratios at the county level in 1911, and changes 1851–1911 see R. Lawton, 'Population', in J. Langton and R. J. Morris, eds., *Atlas of Industrialising Britain, 1780–1914* (London, 1986), pp. 10–29.

[135] F. Galton, 'The relative supplies from town and country families to the population of future generations', *J. Stat. Society*, 36, 1873, p. 19.

Augustus Jessopp, rector of Scarning (Norfolk), was in no doubt. From his parish in the thirty years before 1883, thirty-one men had gone to enrol as London policemen:

the very pick of the parish – men not only of splendid physique but of approved character; men above the average in intelligence and education . . . We retain the sediment; the vicious, the immoral, the men whose character is not above suspicion, the sickly, the depraved, the dissipated and profligate, the roughs who have been poachers in the days when poaching paid.[136]

Wilson Fox, investigating 'Country-born men in large towns' in 1906 for the Royal Commission on the Poor Law, found that out of 5,657 men in the Inner Divisions of the Metropolitan Police who were questioned, 66 per cent were country-born, while 25 per cent had been accustomed to farm work; and taking 12,558 workpeople employed by 16 large municipal corporations in England, he found 37 per cent country-born and 22 per cent who had been farm labourers. The Land Enquiry Committee concluded that 'it is the most energetic of our agricultural labourers who migrate to the towns, and . . . especially in the lower paid districts, it is the underfed, ill-nourished labourers, who remain'. Rider Haggard wrote of 'People . . . deserting the villages wholesale, leaving behind them the mentally incompetent and the physically unfit'; and from Lincolnshire: 'only the old and the dullards [were] left behind'. A careful analysis of rural Essex in 1925 similarly concluded that migration selected the 'brighter and stronger (mentally if not physically as well) of the young men and women in the rural districts . . . in the country villages there remains the weaker element'.[137]

Although most, including the 1881 Census, accepted that the mentally and physically strongest were most attracted to the towns, not everyone agreed. Sir James Caird, in a November 1881 letter to *The Times*, wrote that 'I cannot agree . . . that, although their numbers have been diminished by the attractions of more lucrative employment, they have become less skilful or effective.' Kebbel noted in 1887 that the same points about quality, which were made in 1880, had also been made in 1870 and in 1834. But 'The difference between the three periods is this, that in 1834 and 1870 the inferior work complained of was due rather to want of will rather than to want of skill on the labourer's part. Now it is due to both. Then the skilled workmen were still there, now they are not.' Charles Booth thought that the countryside sent both its 'dregs' and its 'cream' to

[136] Jessopp, *Arcady*, pp. 117–18.
[137] Land Enquiry Committee, *The Land*, pp. 33–4; Haggard, *Rural England*, vol. II, p. 566; J. Thirsk, *English Peasant Farming* (London, 1957), p. 322; A. B. Hill, *Internal Migration and its Effects upon the Death Rates with Special Reference to the County of Essex* (Medical Research Council, London, 1925), p. 123.

London, while Ashby, writing in 1935, thought that migration was not sufficiently selective, except in a few isolated areas, to have lowered rural physical or mental capacities. The issue was debated hotly, but by the mid-twentieth century most seemed to dismiss the notion of selection.[138]

Graham thought that by the early 1890s:

it is no longer only the choice few, the cream of the rural youths, the one clever boy or girl of a family, who come to town. The movement has outgrown all proportion, and those who are adapted for the struggles of town life, as well as those who are not so, are laying down the implements of agriculture and hastening to compete for places at desk and counter, in the dockyard and the railway station, in the public-house bar and the police force.[139]

Finally, migration was related to occupation by many commentators. Again, there are the obvious correlates with age and sex, but some compared the great mobility shown by agricultural labourers with the relatively static nature of the farmers' lives. The 'divorce of men from the soil' particularly affected the former, whose numbers fell from a ratio of 5.1 to farmers and graziers in 1851 to 2.9 by 1901. Certainly from 1871 to 1901, and again from 1911–21, the fall in numbers of farm workers was faster than that for the rural population as a whole, although there is evidence that the brunt of the reduction was borne by seasonal or casual workers, including women. At the same time, farm workers themselves were staying put rather longer where they had employment, although this again is also a function of employees' ages and marital status.[140]

Seasonal migration was not as important by the last half of the nineteenth century as previously, for hired men took on the remaining tasks that could not be met by machines, such as reapers or horse hoes. But high-season labour demands in summer in the eastern arable counties were as much as 100 per cent above low season by the late 1860s, and so long-distance movements of travelling harvest bands continued, and hence were noticed in spring/early summer in census enumerators' returns. As late as 1912 men from eastern Oxfordshire scythed parks and fields in London and hoed for market gardeners, harvested on the Wantage Downs, came home for the harvest at Filkins, and then moved on to Northleach. Harvesters from Buckinghamshire went 'uppards' to the Middlesex hayfields regularly. The overall peak years for Irish harvesting gangs were

[138] Jessopp, *Arcady*, p. 103; J. Saville, *Rural Depopulation in England and Wales*, p. 127, fn. 1; A. W. Ashby, *The Sociological Background of Adult Education in Rural Districts* (British Institution of Adult Education, 1935), p. 9; B. S. Bosanquet, 'The quality of the rural population', *Eugenics Review*, 42, 1950, pp. 73–92.   [139] Graham, *The Rural Exodus*, p. 2.

[140] Weber, *The Growth of Cities*, p. 160; Armstrong, 'The influence of demographic factors', p. 80; Davies, 'The condition of the rural population in England and Wales 1870–1928', pp. 7–8; C. Thomas, 'The Vale of Clwyd, 1815–1871', pp. 112–13.

in the late 1870s, when many flocked to the north and Midlands rather than the south. By then the movements had been channelled along more acceptable routes and the harvest migrants were far removed from the famine-stricken and diseased Irish who had roamed the English country-side in the early 1850s.[141]

In the south-east and west Midlands, hop picking attracted many migrant workers. From Birmingham and the Black Country seasonal pickers went to the orchards and hop gardens of Herefordshire and Worcestershire. London hoppers were perhaps the most notorious, con-stituting a large number of the Kentish pickers: perhaps 70,000 or 80,000 pickers out of a total of about 250,000 by the 1880s or 1890s. 'Very dark and destitute in spiritual things', the hoppers often terrorised whole par-ishes and while earnings were high, their insanitary and overcrowded living conditions created a scandal. In 1849 a cholera outbreak at East Farleigh claimed forty victims, and deaths from disease, accident or expo-sure were by no means uncommon. The Society for the Employment and Improved Lodging of Hop Pickers issued guidelines for accommo-dation in 1867, and the Local Government Board was active in the early 1870s on the question of overcrowding, as was the Church of England Missionary Association for Hop-pickers from 1877. But the three-week period concentrated vital events dramatically, as 'hopper marriages', births and deaths occurred with great frequency among both 'home dwellers' and 'foreigners'.[142]

All these movements, stimulated by differences between regional spe-cialities in wood/pasture compared with arable, or differences in the timing of farming tasks between regions, or by differences between low-productivity peasant subsistence farming and large-scale capitalist farming, had demographic implications. They afforded supplementary income to buy food or clothing, or reduce dependency on poor relief in more marginal communities, and they explain seasonal absences of males (and females and children before the 1860s when the 1867 Gangs Act and the Education acts of the 1870s curtailed their exploitation) from upland, pastoral and small-farming regions. Probably the only group which increased mobility in the later nineteenth century was that of the older vagrants, since it would seem that more elderly tramps were admitted to workhouses by the beginning of the twentieth century, as

[141] E. J. T. Collins, 'Migrant labour in British agriculture in the nineteenth century', *EcHR*, 29, 1976, p. 39; Morgan, *Harvesters and Harvesting*, pp. 48–9, 81–3; Redford, *Labour Migration in England*, pp. 158–9.

[142] C. Baker, 'Home dwellers and foreigners: the seasonal labour force in Kentish agriculture with special reference to hop-picking', MPhil thesis, Univ. of Kent, 1979; J. H. Laws, 'Some aspects of hop picking in Kent, 1860–1890', unpub. paper Univ. of Kent, May 1968; C. Whitehead, 'Fifty years of hop farming', *JRASE*, 3rd ser., 1, 1890, p. 336.

the elderly were displaced from field-work or marginalised in the depression.[143]

The impact on the countryside itself of these movements was demographically severe. Rates of natural growth were slowed down in rural areas compared with towns, and by 1900 many rural areas, especially in lowland England, were experiencing low marital fertility and nuptiality as differential out-migration of the young reduced the potential for growth. High male:female ratios resulted where heavy female out-migration had occurred, such as in northern and central Wales, Lincolnshire, the south-east Midlands or rural south-east, and indeed the rural female surplus was much smaller by 1914 than for England and Wales as a whole. Elsewhere, as in western Cornwall, out-migrations of young males, either temporarily or permanently, on the closure of the mines in the 1860s, left equally unbalanced communities. In 1871, 18 per cent of married women aged 25–29 were not living with husbands (for wives of all ages comparable figures were 9 per cent in 1851, and 12 per cent in 1871), while crude marriage rates dropped more rapidly in the 1870s than in England as a whole.[144]

The role of migration in producing inter-district variations in age and sex ratios, and hence affecting nuptiality and marital fertility, was of wide demographic import in rural England. The availability of partners was both a consequence and a further cause of out-migration from isolated rural communities as sex ratios between town and country were inverted.[145] By 1851 many rural areas already had long histories of out-migration, resulting in ageing populations, and the majority of English and Welsh elderly were thus now to be found in the countryside. Southern England as a whole had two or three times as many aged paupers as did the north, while two to three times as many aged paupers lived in agricultural areas as in industrial areas. The situation for elderly women was particularly difficult since 'being a single or widowed woman in old age meant almost certain pauper status'.[146]

Thus the lessened rural out-migration by 1914 was anyway from a much reduced and ageing population. All aspects of local life were affected by the out-migrations. School attendances and church and chapel congregations suffered, as did custom for the trades and crafts as hamlet populations left, thereby stimulating still more out-migration in a series of spiralling downward decline. Therefore, despite the relatively low rural mortality, there was a sharp downturn in natural increase after 1891. Rural births had peaked already in the 1860s in the south and in the

---

[143] Thomson, 'Provision for the elderly', pp. 290–1.
[144] Johansson, 'Mortality and fertility change in Cornwall, 1800–1900', pp. 262–4, 284, 401.
[145] Friedlander and Moshe, 'Occupations, migration, sex ratios, and nuptiality', pp. 8–9.
[146] Thomson, 'Provision for the elderly', pp. 20, 41.

1870s in the north, and by the early twentieth century rural birth rates were 30 per cent below their nineteenth-century peak. Out-migration left an older, less fertile population from the 1860s until the 'quiet revival' experienced in many rural areas in the first decade of the twentieth century.

Alongside migration from rural to urban areas was that destined for foreign soil. The general pattern of overseas emigration is complex to chart but average levels of out-migration were 11 per 10,000 population for the last six decades of the nineteenth century, with a high point of 23 per 10,000 in the 1880s and a low point of 2 per 10,000 in the 1890s. The first decade of the twentieth century saw a still larger volume of emigrants, only slowing down with the onset of the First World War. They went mainly to Canada and the USA, together with New Zealand and Australia, and to a lesser extent South Africa and South America.

There is probably insufficient evidence to know whether emigration and internal migration were regarded as substitutes by country dwellers, or whether the two processes were to some extent complementary.[147] Unfortunately, a decline in governmental interest in emigration in the later Victorian period entailed a lack of statistical evidence on the migrants themselves. We do not know, for example, how many were from rural backgrounds, how many had moved to the towns before emigrating, or how many were agricultural labourers and their families as opposed to farmers, craftsmen and the slightly better-off. The in-migration of Irish and Scots, as well as return migration later in the century, also complicated the statistics, and net rather than gross figures need to be kept in mind. Instead, much research has tended to focus on the 1830s, a time of poor law stress and assisted emigration, both of which left documentation more ample than that for the later part of the century.

Developing from earlier ideas of the interrelation of colonisation and the relief of the poor, propaganda for overseas migration ensured that the countryside became integrated with that huge European movement of 44 million people between 1821 and 1915, which took 21 million from the British Isles. Their emigration rates were amongst the highest in Europe. In 1850–70 alone there were probably as many emigrants from Great Britain as in the whole of the seventeenth and eighteenth centuries combined, and this in a period generally thought to represent a lull in overseas movements, and at a time of some prosperity in much of the English countryside. Nevertheless, as many as 21.4 per cent of all emigrants by 1851 were farmers. The lower emigration figures were, in fact,

---

[147] D. Baines, *Migration in a Mature Economy: Emigration and Internal Migration in England and Wales, 1861–1900* (Cambridge, 1985), pp. 7, 213–49; B. Thomas, *Migration and Economic Growth* (Cambridge, 1954, reprinted 1973), pp. 124–5.

such that in 1850–1 the Emigration Commissioners had difficulty in meeting the required numbers to fulfil their Australian contracts, despite the increasing sophistication of transport. Emigration was sponsored by the Poor Law Commissioners or the Colonial Commissioners during the late 1830s to the 1850s, for the relief of social ills, but it was seemingly poorly supported by agricultural workers. But between 1841 and 1852 this sponsored emigration from rural Bedfordshire, for example, relieved local unemployment. In 1844, 118 had been assisted, of whom 48 came from the parish of Stevington alone. At least 71 went in 1849, and 130 in 1852. With the cotton famine affecting northern industrial villages and the crisis in Cornish mining, emigration expanded in the later 1860s, and by 1869 emigration from England exceeded that from Ireland for the first time on record, prompting parliamentary debates but no government action. Instead, some forty societies promoted emigration, but these were primarily (as were the few government schemes) aimed at urban rather than rural audiences.[148]

Nationwide, the 1870s were low-level years for emigration, but they were important for the funding of emigration by rural trade unions which anticipated improved wage bargaining for those remaining. Some individual villages were dramatically affected. Arch's National Agricultural Labourers' Union and the Kent Union (later the Kent and Sussex Labourers' Union) were active: by the end of the winter of 1873–4 the former union had sent about 1,000 from the Warwickshire area to Brazil, and the latter had dispatched about 1,000 to New Zealand, mostly farm workers. In 1874 the great lock-out affecting the grain-producing eastern counties produced another flood of emigrants to New Zealand from Lincolnshire, East Anglia and Kent; and in 1878–9 the Kentish lockout produced a similar response. These New Zealand emigrants were predominantly southern villagers: the open villagers of the Wychwood area of Oxfordshire featured prominently, while 5,000 or more from Cornwall actually constituted more than 10 per cent of all the colony's assisted migrants in the 1870s. In his evidence to the Richmond Commission in 1882, Arch estimated that about 700,000 had emigrated through his Agricultural Labourers' Union in the past eight or nine years – an exaggeration, but one which hinted at the magnitude of the rural

---

[148] W. A. Carrothers, *Emigration from the British Isles* (London, 1929), p. 185; W. Van Vugt, 'Running from ruin? The emigration of British farmers to the USA in the wake of the repeal of the Corn Laws', *EcHR*, 2nd ser., 41, 1988, p. 417. See the advocacy of overseas emigration for the poor in *BPP*, 'Report by Cap. Robinson, RN, to the Poor Law Board, on the operation of the Laws of Settlement and Removal of the Poor in the Counties of Surrey and Sussex', *Reports and a Memorandum to the Poor Law Board on Settlement and Poor Removal with an Appendix* (1850–54), p. 87. For the Bedfordshire figures see L. M. Marshall, 'The rural population of Bedfordshire, 1671–1921', *Bedfordshire Historical Record Society*, 16, 1934, p. 41.

movement to Canada and New Zealand in particular. On the other hand, very few came from the north or from Wales. Most were also labourers, in contrast to those heading for North America, three-quarters of whom were farmers and graziers. A wave of emigration also occurred from the Pennine lead-mines as the industry collapsed, and direct links between Swaledale and Dubuque (Iowa) were built up, reinforced by enthusiastic letters from emigrants back to their former neighbours, describing the prospects of similar lead mining and agricultural work on offer. Overall, in the 1870s over 40,000 adult farm workers emigrated, compared with fewer than 18,000 in the 1860s.[149]

The 1880s witnessed another boom, now financed privately or by societies, rather than by unions. Overall the numbers were now composed of 15.4 per cent agricultural labourers (constituting 11.4 per cent of the national labour force at that time). About 33,000 of the emigrants (one-sixth of the total in the decade) were agricultural labourers and their families. More now were single people, more were liable to return, and most made rational decisions about the relative states of the economies of England and Wales and their country of destination. It was clear that the earlier rural emigrants, mostly farmers, skilled tradesmen and craftsmen, were now being superseded by a broader social mix. Thus one successful scheme, that of the Fielding settlement in New Zealand, took 3,000 colonists, the families of agricultural labourers from Buckinghamshire and Middlesex, with passages paid for by the New Zealand government. Not all such schemes were as successful, however, and those of Sir James Rankin, Lord Brassey, or the Barr colony in Saskatchewan failed or were only partially remunerative.[150]

After a lull in the 1890s, followed by a gradual build-up in the new century, the peak year for emigration was reached in 1912 when it topped 315,000. In this year, according to the Land Enquiry Committee '. . . about one in every 50 of our male agricultural population found their prospects in the United Kingdom so poor that they decided to leave the country altogether'. Numbers of emigrants had increased in the first decade of the twentieth century, for travel was now easier and safer. Between 1911 and 1921 net losses from England and Wales by migration

---

[149] R. Arnold, *The Farthest Promised Land: English Villagers, New Zealand Immigrants of the 1870s* (Wellington, NZ, 1981), pp. 69–235, 348; P. Horn, 'Agricultural unionism and emigration 1872–1881' *Hist. Journ.*, 15, 1972, pp. 89–102; Hallas, 'Migration in Wensleydale and Swaledale', pp. 155–60. For the substantial demand for immigrants to New Zealand in the 1870s see J. E. Martin, *The Forgotten Worker: the Rural Wage Earner in Nineteenth-Century New Zealand* (Wellington, 1990), pp. 16–31.

[150] Baines, *Migration in a Mature Economy*, pp. 3, 77; C. Erickson, 'Who were the English and Scots emigrants to the United States in the late nineteenth century?', in D. V. Glass and R. Revelle, eds., *Population and Social change* (London, 1972), pp. 347–81; Carrothers, *Emigration from the British Isles*, pp. 36–7.

amounted to 620,000: about 1.7 per cent of the total population. This was the highest absolute loss, although in relative terms the 1880s, with losses of 601,000 (2.3 per cent), made more impact.[151]

Relatively little work has been undertaken on the social and geographical origins of the emigrants in the later nineteenth century. Between 1861 and 1900 it has been suggested that the rural counties lost about 12.6 per cent of their population by migration in each decade (net of those returning) of which about 10 per cent was through internal migration, and 2.6 per cent was overseas. Of the younger populations, however, there was a loss of 62.1 per cent, composed of 49.8 per cent internal and 12.3 per cent overseas. When statistics on the sex ratio of the migrants become available in the late 1870s it is seen that nearly twice as many men as women emigrated, compared with the greater tendency of women to move short distances within England and Wales. The ratio shifted somewhat by 1914, however, as more women seized the opportunity to move to the colonies and to North America – now 'tamed' and as emigration became a movement for betterment rather than survival, as it had been in many respects before 1850.[152]

What work has been done seems to suggest that the emigrants in the 1880s, for example, came from urban rather than rural backgrounds. Only about 38 per cent of emigrants between 1861 and 1900 were 'stage migrants', born in rural areas but emigrating from a last address in a town. Probably less than one-third of emigrants overall during this period came from country areas, a finding which certainly undermines much of the earlier work which pinned explanations of rural depopulation and emigration to rural stress and to the agricultural depression in particular. Emigration flows differed, however. Thus some 72 per cent of male immigrants to New Zealand between 1854 and 1876 were in rural-related occupations.[153]

Among the rural areas particularly prone to lose population through emigration was the south-west. Between 1861 and 1900, 323,000 emigrants left Cornwall, Devon, Somerset, Dorset and Wiltshire – nearly 14 per cent of all English and Welsh emigrants and nearly half of the total of rural emigrants. Cornwall lost about 40 per cent of its young adult males and 25 per cent of its females, making this one of the most important emigration regions of Europe at this time. In the crisis after the mid-1860s 11,400 jobs were lost in the Cornish mines in 18 months as miners went to North and South America, the Transvaal, Australia and elsewhere.

[151] Land Enquiry Committee, *The Land*, p. 31; N. H. Carrier and J. R. Jeffery, *External Migration: a study of the available Statistics, 1815–1950* (General Register Office Studies on Medical and Population subjects no. 6, London, 1953), pp. 33, 47.

[152] Baines, *Migration in a Mature Economy*, pp. 227–37; Erickson, 'Who were the English and Scots emigrants?', pp. 347–81.   [153] Martin, *The Forgotten Worker*, p. 203, fn. 6.

Husband-only emigration, with remittances now sent home, became crucial to the survival of many west Cornish rural communities. Above-average rates of emigration also came from rural Wales and the borders, although many younger people initially went to the Black Country, south Wales or Merseyside. In Brecon, Herefordshire and Pembrokeshire the losses averaged about 25 per cent or more of young adult males between fifteen and twenty-four years of age. Brecon, like Cornwall, lost the equivalent of 25 per cent of its females aged between fifteen and twenty-four years. Also significant were losses from Cumberland and Westmorland to Scotland as well as to the New World, and from Durham, where many left older pit villages.[154]

On the other hand, the Home Counties, the south-east, East Anglia, the east Midlands, Lancashire and Yorkshire had below-average emigration rates. Emigration increased from east to west, the causes being largely related to the state of the regional economy. But the timing of emigration was probably still more related to the possibilities of betterment within periods of boom in the receiving countries. Such decisions to leave were not taken blindly but were based on sometimes considerable knowledge. What has been referred to as 'chain migration' played a large role. The experience of previous emigrants, as revealed personally by those who returned, or in their letters home, was active in stimulating interest among their relatives and friends. Thus 'emigration was a self-reinforcing process'.[155]

---

[154] Baines, *Migration in a Mature Economy*, pp. 147–59; Johansson, 'Mortality and fertility change in Cornwall, 1800–1900', pp. 262–4.   [155] Baines, *Migration in a Mature Economy*, pp. 26, 255–7.

CHAPTER 22

TYPES OF RURAL COMMUNITIES

BY ALUN HOWKINS

INTRODUCTION

The demographic changes discussed in the previous three chapters took place within other and different social structures and were mediated through them. England and Wales in the late nineteenth century was a diverse and deeply regional economy and society. The great farming regions, discussed elsewhere in this work, are seen primarily as economic entities yet, as we shall see, they produced their own social and communal structures which profoundly shaped the lives of those who lived and worked within them. A very basic division, for example, is that between areas of clachan settlement, largely the north and west and much of Wales, and the areas of nucleated village settlement, mostly in the south and east. This division is based, in the first instance, on the nature of the land and the farming and economic systems which grew up as a result of those forces, yet the different social structures which emerged cannot simply be reduced to these differences.

Sociologists and social historians have often talked about these differences in terms of community. However this concept is nebulous and one fraught with difficulties. At the most basic level, the one it should be said where much history begins and ends, community is seen simply in spatial or numerical terms. There is sometimes a nod in the direction of social theory, often seen, as Dr Mills says, as 'a time wasting diversion',[1] then it is back to the hard facts of space, landownership or the nature of the vestry.

Important as these factors are, and we shall return to them below, taken alone they are not only unhelpful to an understanding of the past of rural Britain, they simply do not accord with what other disciplines, or for that matter the man on the Suffolk omnibus, think of as community. Many classic sociological studies of community, for instance Stacey's of Banbury, or Williams's of Gosforth, often took a purely local government

---

[1] Dennis R. Mills, *Lord and Peasant in Nineteenth Century Britain* (London, 1980), p. 22.

definition of parish or borough, to start with at least.[2] However it quickly became clear, to sociologists and anthropologists if not historians, that these divisions are seldom sufficient. In her study of Elmdon, an Essex village, Marilyn Strathern remarks that 'administrative parish boundaries are of little relevance'.[3] Against them she sets the notion of the 'real Elmdoner', essentially an internal, village description based on a group of families long associated with the village. The 'local history' of the town where I was born puts this view graphically:[4]

It was in the 1920s. The appropriate Committee was considering an application for one of the earliest Council Houses. 'Who are these applicants?' asked one member. 'They came to Bicester,' said the Town Clerk, 'twenty six years ago, and have been living in King's End ever since.' The member shook his head disapprovingly. 'Then,' he remarked, 'they are not really Bicester people.'

Even these kinds of definition of community have problems since, as Strathern writes, 'which families are actually recorded as real Elmdon depends . . . on the status and knowledge of the speaker'.[5]

A further problem with this kind of internal description, as Colin Bell and Howard Newby pointed out some years ago, is that it is often overlaid with a set of subjective and emotional ideas linked into myths of a lost, organic and ideal past. 'The concept of community . . . was not a cold analytical construct. On the contrary the ties of community, real or imagined, came from . . . images of the good life.'[6] Or as Raymond Williams puts it in *Keywords*:[7]

Community can be the warmly persuasive word to describe an existing set of relationships, or the warmly persuasive word to describe an alternative set of relationships. What is most important, perhaps, is that unlike all other terms of social organization . . . it seems never to be used unfavourably, and never to be given any positive opposing or distinguishing term.

For our purposes it is important to stress that the notion of community became increasingly identified with an almost mythical idea of 'the village' and was reinforced both by popular ideas and by academic work. These operated very much within a paradigm laid down in the late 1930s by studies of Irish life and developed very importantly in the 1950s in a series

---

[2] Margaret Stacey, *Tradition and Change. A Study of Banbury*, (Oxford, 1960); W. M. Williams, *The Sociology of an English Village: Gosforth*, (London, 1956). For a general discussion of this work, see Susan Wright, 'Image and analysis: new directions in community studies', in Brian Short (ed.), *The English Rural Community. Image and Analysis*, (Cambridge, 1992).

[3] Marilyn Strathern, *Kinship at the Core. An Anthropology of Elmdon, a Village in North-west Essex, in the Nineteen-sixties*, (Cambridge, 1981), p. 7.

[4] Sid Hedges (S.G.), *Bicester Wuz a Little Town* (Bicester, 1968), p. 7.   [5] Strathern, *Kinship*, p. 15.

[6] Colin Bell and Howard Newby, *Community Studies* (London, 1971), pp. 21ff.

[7] Raymond Williams, *Keywords. A Vocabulary of Culture and Society* (London, 1976), p. 66.

of studies of English and Welsh village communities. These stressed the local world and its strengths as against the national. They seem deliberately to have chosen remote or 'backward' areas, mid-Wales, Cumberland, the West Country or the Scottish Highlands where the personal, 'traditional' and especially face-to-face relationships dominated the social structure.[8] This has seriously skewed not only the area of community studies but the sociology of rural areas since it has tended to foist onto the social history or historical anthropology of rural Britain a set of ideas about 'community' derived from a relatively limited geographical and social area. Even Ronald Frankenberg's influential and occasionally critical synthesis of some of these community studies, *Communities in Britain*[9] of 1966, reproduced these problems. This began from an earlier working definition of community describing it as 'an area of social living marked by some degree of social coherence. The bases of community are locality and community sentiment.'[10] This definition reinforces many of the localist preconceptions of the community studies of the 1950s and 1960s, creating an ideal of community which is small and personal which in turn idealises a particular version of the rural, for example, 'in truly rural society the network may be close knit, everybody knows and interacts with everybody else'.[11]

Such a view may be tenable in a hill farming community, although I doubt even that; it certainly would not be applicable to the highly stratified and class-divided villages of East Anglia. This was the starting point for the study by Newby *et al.* of 'capitalist' farming, *Property, Paternalism and Power.*[12] Nevertheless the imbalance in terms of numbers of studies remains striking. Further, concepts derived from the early community studies continue to structure popular and even academic ideas of the rural communities of the recent past.

Alternative definitions would stress class or ethnicity as the basis of community. Especially in the 1950s and 1960s a number of important studies, for example Brian Jackson's *Working Class Community*,[13] stressed class position as a key determinant of community structures and community relations. This kind of work however seems to have had very little effect on rural community studies where older definitions remained paramount. The exceptions here are more recent studies, especially of

---

[8] See for example Ronald Frankenberg, *Village on the Border* (London, 1957); W. M. Williams, *A West Country Village Ashworthy* (London, 1963); Alwyn D. Rees, *Life in the Welsh Countryside* (Cardiff, 1960); Arthur Geddes, *The Isle of Lewis and Harris*, (Edinburgh, 1955). There are many other such studies both monographic and article length.

[9] Ronald Frankenberg, *Communities in Britain* (Harmondsworth, 1966).   [10] *Ibid.*, p. 11.

[11] *Ibid.*, p. 19.

[12] Howard Newby, Colin Bell, David Rose and Peter Saunders, *Property, Paternalism and Power. Class and Control in Rural England* (London, 1978).

[13] Brian Jackson, *Working Class Community* (London, 1968).

urban/rural relations like Pahl's influential *Urbs in Rure* where class or social status appear as dividing the community into different groups and where questions of access to power and control have become more central.[14]

Yet this imprecision also points to the strengths of the term community in other respects, that is its internal and experiential nature. As Anthony Cohen argues, the 'sense of "community" in industrialised, mass societies [is] largely a matter of sentiment and conceptualisation . . . made coherent through the creation of symbolic defences or boundaries'.[15] Seeing things in this way helps to get behind the categories of explanation often adopted by historians and sociologists and into the mental world of the past. Here we reach the polar opposite of the administration description of village or parish. In this sense communities are created not by administrators (though they might coincide with their boundaries) but by those who live in the communities by their own usage, practice and belief. As Cohen writes elsewhere, 'such boundaries are not "natural" phenomena; they are relational, they may be contrived and their very existence is called into being partly by the purpose for which one group distinguishes itself from another'.[16]

Looked at in this way 'community' becomes a complex idea which embraces not simply the 'village' or 'parish' of much historical modelling. Quite different units could generate a sense of belonging. In this an individual may have a series of identities, like the layers of an onion, or the different lines of a postal address, which relate to different levels of community and different involvements. What David Jenkins writes at the beginning of his very important study of south-west Wales applies to the whole of these islands:[17]

In this society people stood related to one another in various and characteristic ways. As members of the staffs of individual farms they stood related as farmers on one hand and farm servants and labourers on the other. People stood related too as farmers and cottagers in their capacities as members of the work groups which were connected with every farm. They stood again related to one another as members of places of worship . . .

Since these different communities co-exist, on occasion they may well come into conflict with each other. For example a trade union member

---

[14] Ray Pahl, *Urbs in Rure. The Metropolitan Fringe in Hertfordshire* (London, 1964).

[15] Anthony P. Cohen, 'Of symbols and boundaries, or, does Ertie's greatcoat hold the key', in Anthony P. Cohen, ed., *Symbolising Boundaries. Indentity and Diversity in British Cultures* (Manchester, 1986), p. 7.

[16] Anthony P. Cohen, *Belonging. Identity and Social Organisation in British Rural Cultures* (Manchester, 1982), p. 3.

[17] David Jenkins, *The Agricultural Community in South-West Wales at the Turn of the Twentieth Century* (Cardiff, 1971), p. 5.

could belong both to a national community of the union and to the local community of the village or farm; a landowner could belong both to the national community of class and to the local community of 'his' village. Again we shall return to some of these issues.

I. VILLAGE COMMUNITIES

Leaving aside the family, which is dealt with in the next chapter, the base unit of rural society, the inner layer of the community onion, is usually a small-scale, spatially delimited unit of human settlement. This could be a village, but it could easily be something else depending on what part of England and Wales is being discussed. Christopher Taylor points out in his study *Village and Farmstead*[18] that we can see two major physical types of settlement in England and Wales. The settlement pattern of the north and west is broadly speaking farmstead-based while that of the midlands, the south and east is village-based. 'Village England', writes Taylor, 'is really a broad zone stretching from the south coast through the Midlands to the north-east.'[19] The rest of England, the north and west plus mid-, west and north Wales, is either based on farm or hamlet or on a mixture of village and other settlement types. This needs treating with some caution since parts of northern England are dominated by large 'townships', nevertheless these divisions remain useful and do point to the complexities behind many apparently unproblematic definitions. They also accord roughly with those used in Mills's *Lord and Peasant in Nineteenth Century Britain*, and described by him as hamlet and champion England. Mills, however, also points to areas of former wood pasture, hamlet settlement in the south and east especially the Weald, the Dorset Heaths, the Downs and some parts of East Anglia.[20] According to Brian K. Roberts, 'True' village England is mainly in the lowland areas of mixed farming identified by Thirsk for the sixteenth century, but laying down a pattern for later periods. Around these are areas of pasture farming and 'wood pasture' regions, 'generally, but by no means invariably, associated with hamlet and single farm settlement'.[21]

The physical nature of these settlement patterns produced quite different communities and using Taylor's slightly simplified idea of two basic regions 'village' and 'farmstead' England I want to look at them in turn, to begin with the area dominated by nucleated or 'true' village settlement – 'village communities'. The notion of the village tends to be an unspoken axiom in much writing about rural history. This is largely a

[18] Christopher Taylor, *Village and Farmstead. A History of Rural Settlement in England* (London, 1983).
[19] *Ibid.*, p. 125.   [20] Mills, *Lord and Peasant*, pp. 16–20.
[21] Brian K. Roberts, *Rural Settlement in Britain* (London, 1977), p. 21.

product of the emphasis, in terms of both many of the ninteenth-century sources and twentieth-century historical writing, on the south and east of England. As Alistair Mutch writes in the 'Introduction' to his study of farm labour in Lancashire, 'there is a bias towards the southern experience in much rural history'.[22] In view of this it is important to stress that considerably less than half the land area of England and Wales was settled in nucleated villages in the nineteenth century. Against this these areas were, with some exceptions in the north-east, the most advanced in terms of farming techniques and organisation and the most densely populated rural areas. In 1851[23] the lowest densities of males per square mile employed in agriculture lay predictably upon the mountains and moorlands of northern England, and, to a lesser extent, in the south-west. High densities occurred in south-eastern England – particularly in parts of Essex, Hertfordshire, Kent and Surrey – where the rural population as a whole was most numerous.

Moving in a little closer we can recognise distinctive types of village in spatial terms. Conventionally these are 'agglomerated', that is clustered around a central point, often a church; 'linear' running along a road; and 'green', built around or along a stretch of common land or grazing.[24] These different arrangements have socio-cultural dimensions but these tend to be a result of other than the purely spatial factors. A key factor here, and one which has dominated discussion of community types in village areas, is the notion of 'open' and 'close' socio-economic structures.

The idea that villages can be divided into 'open', crudely where settlement was free from restraint, and 'close', where it was controlled by one or at best a few landlords, is present in contemporaneous discussions of the 'problems' of the poor since at least the seventeenth century. There is, for example, a great deal of similarity between seventeenth-century writing on squatter and woodland parishes and later writing on 'open' villages.

It was, however, in the years after 1830 that the actual use of the terms 'open' and 'close' emerged and had attached to them particular categories which appeared to have a firm basis in empirical social observation. In 1843, for example, when Mr Dennison reported for the Poor Law Commissioners on the employment of women and children in agriculture in Norfolk, he specifically identified Castleacre as 'what is called an "open" parish': that is, in the hands of a considerable number of proprietors'.[25] However, it was the investigations into the laws of settlement and removal of poor in the late 1840s and early 1850s, and the series of reports

[22] Alistair Mutch, *Rural Life in South-West Lancashire, 1840–1914* (Lancaster, 1988), p. 1.

[23] J. B. Harley, 'England *circa* 1850', in H. C. Darby, ed., *A New Historical Geography of England after 1600* (Cambridge, 1976), p. 235.   [24] Roberts, *Rural Settlement*, pp. 17–18.

[25] *Reports of the Special Assistant Poor Law Commissioners on Employment of Women and Children in Agriculture*, BPP, 1843, XII, p. 221.

on the employment of women and children in agriculture, which fixed the categories of 'open' and 'close', and gave them their most detailed nineteenth-century definitions.

Nevertheless, as Sarah Banks has pointed out, these definitions remain remarkably slippery. At the most basic level commentators both modern and contemporary often confused the idea of 'parish' and 'village'. This is a problem which can only be dealt with in detailed local studies but should always be borne in mind in what follows, as well as in much writing on this problem. In relation to the definition of 'close' a central confusion rests on the difference between cause and effect, with some commentators stressing landownership while others stress the restriction of settlement and labour supply.[26] However, although this causes problems, this conceptual distinction is much less important than the ideological assumptions made by most nineteenth-century observers. The material from the 1830s and 1840s is very much 'part of a propaganda campaign to reform the settlement laws'. As such, 'the "open" and "close" parish system was often spoken of in connection with immorality, ill-health, insanitary conditions and poverty',[27] while that from the 1860s and 1870s was concerned very much to link the 'immoral' conditions in 'open' villages with women's work and crime.

These in a sense reverse the dictum of earlier observers, who tended to blame landlords for creating close villages, and shift the fault onto the poor who lived in open villages or on landlords who neglected their community duties. The greatest faults, always attached to open villages, were immorality and lawlessness. Two examples from the 1860s reports on the employment of women and children in agriculture will serve to illustrate this. The former vicar of Castleacre in Norfolk, a village which became a byword for the problems of 'open' villages, wrote to the Commissioners on children's employment in 1867:[28]

The parish is usually regarded as the one blot in the diocese of Norwich, and I have no hesitation in saying that it exceeds anything of which I have any experience in the moral degradation of its poor. I have been to Sierra Leone, but I have seen shameless wickedness in Castleacre such as I never witnessed in Africa.

Two years later the Rev. James Fraser reported from Poulton, an open village in Gloucestershire, that 'the people' of the village although 'hard working' were 'lawless, and the morality of the parish is of a low type, drunkenness being very prevalent'.[29]

---

[26] Sarah Banks, 'Nineteenth-century scandal or twentieth-century model? A new look at "open" and "close" parishes', *EcHR*, 2nd ser., 42, 1988, pp. 51–3.   [27] *Ibid.*, p. 63.

[28] *RC on the Employment of Children in Trades and Manufactures not regulated by Law, Sixth report, Appendix. (Agriculture)*, BPP, 1867, XVI, p. 91.

[29] *RC on Employment of Children, Young Persons and Women in Agriculture. First Report. Appendix (Evidence of Assistant Commissioners)*, BPP, 1867–8, XIII, pp. 112–13.

Descriptions like this created what Sarah Banks has called 'a nine-teenth-century scandal' out of conditions especially in open villages, which has obscured real social relations and social structures by imposing on them essentially moral judgments derived from contemporary debates.[30] This is very similar to what has been called in a different situation a 'moral panic'. Stan Cohen writes:[31]

Societies appear to be subject, every now and then, to periods of moral panic. A condition, episode, person or group of persons emerges to become defined as a threat to societal values and interests . . . socially accredited experts pronounce their diagnoses and solutions; ways of coping are evolved or (more often) resorted to; the condition then disappears, submerges or deteriorates.

Something like this clearly happened about open villages between the 1840s and the 1870s. As Banks argues, 'increasing numbers' of parliamentary enquiries in this period were concerned with the problems of the rural poor and the terms 'open' and 'close' provide a convenient and emotionally charged shorthand in which to discuss them.[32] The terms also hinted at both the origin and solution to the problems (they came about because of 'openness', or indeed 'closeness', which could be rectified by taking the opposite course of social action).

Despite this, and probably because of the all-pervasiveness of the use of the terms in the mid-century, the categories of 'open' and 'close' have found widespread use and acceptance among historians. The core of these discussions, in the work of Dennis Mills especially, and to a lesser extent in the earlier work of B. A. Holderness, is the relationship between restriction of settlement and population growth in the 'close' village and the supply of labour.[33] A contemporary view is a good starting point. James Caird wrote in 1852:[34]

There is another evil with regard to the labourer . . . the system of 'close' and 'open' parishes, by which the large proprietors are enabled to drive the labourer out of the parish where he works, to a distant village, where property being more divided, there is not the same combination against poverty . . . Nor is this the sole evil of the practice, for the labourers are crowded into villages where exorbitant cottage rents frequently oblige them to herd together in a manner destructive of morality and injurious to health.

This relationship between landownership and 'closing' a parish, either by pulling down houses or by restricting settlement, is the basis of Mills's

---

[30] Sarah J. Banks, 'Open and close parishes in nineteenth-century England', PhD thesis, University of Reading, 1982, pp. 68ff.

[31] Stanley Cohen, *Folk Devils and Moral Panics. The Creation of the Mods and Rockers*, new edn (Oxford, 1980), p. 9.   [32] Banks, 'Nineteenth-century scandal', p. 113–14.

[33] B. A. Holderness, '"Open" and "close" parishes in England in the eighteenth and nineteenth centuries', *AHR*, 20, 1972, pp. 132–3.

[34] James Caird, *English Agriculture in 1850–51* (London, 1852), p. 516.

categorisation of the closed village. However, Mills goes further than simple restriction of settlement as a defining feature of 'open' or 'close', as does another historical geographer, Brian Short.[35] Their elaboration centres in the first place on other characteristics of 'open' and 'close' parishes in which, openly for Mills, and by implication, in other work like Samuel's 'Quarry roughs',[36] there is a relationship of determinacy between landownership and socio-cultural structure. As Mills puts it, 'the model is taken to be not merely a schematic view of rural society but also explanatory and predictive'.[37] This position is best seen by reproducing Brian Shorts 'reworking' of Mills's tables of causal links in different village types (Table 22.1).[38]

Looking at this table, crudely speaking, an open village was likely to be more lawless, radical and socially various than a close one, producing a 'freer' culture and social structure: precisely the kind of community described so brilliantly in Samuel's work. 'Quarry', he writes, 'was a village which had grown up singularly free of gentlemen. For centuries it had enjoyed what was virtually an extra-parochial existence, a kind of anarchy in which the villagers were responsible to nobody but themselves.'[39] Samuel's work also points up another key element in the classic 'open' village – the variety of non-agricultural occupations. Partly encouraged by its nearness to Oxford, Quarry supported a population who, in Samuel's powerful phrase, were 'men of many occupations and often none'.[40] These craftsmen, small tradesmen, 'free' labourers and 'penny capitalists', who were free from the restraints of master and squire, provided Quarry with much of its social independence.

Similar employment patterns were present in woodland and heathland communities. Leafield in Oxfordshire, although surrounded by the firmly controlled lands of the Marlboroughs, grew up as a squatter settlement on the forest waste, with 'no persons of quality' to lord over it, and without even a church until the 1850s. Here a mixture of woodland trades, smallholding and poaching supported a community which was thought of as wild and lawless until the 1930s.[41] At Tadley in north Hampshire a firm underpinning of small holdings, the majority of which were between 0.25 and 4.75 acres even in 1910, combined with woodland occupations, abundant commons and migration to support a

---

[35] Denis R. Mills and Brian M. Short, 'Social change and social conflict in the nineteenth-century; the use of the open-closed village model'. *Journ. of Peasant Studies*, 10, 4, 1983, pp. 253–62; see also Peter Brandon and Brian Short, *The South East from AD 1000* (London, 1990), pp. 316–22.

[36] Raphael Samuel, 'Quarry roughs', in Raphael Samuel, ed., *Village Life and Labour* (London, 1975).

[37] Mills, *Lord and Peasant*, p. 116.

[38] Brian Short, 'The evolution of contrasting communities within rural England', in Short, *English Rural Communities*, p. 30.   [39] Samuel, 'Quarry roughs', p. 155.   [40] *Ibid.*, pp. 227ff.

[41] See, for example, John Kibble, *Historical and other Notes on Wychwood Forest, and Many of its Border Places* (Charlbury, 1928).

Causal links in the close village

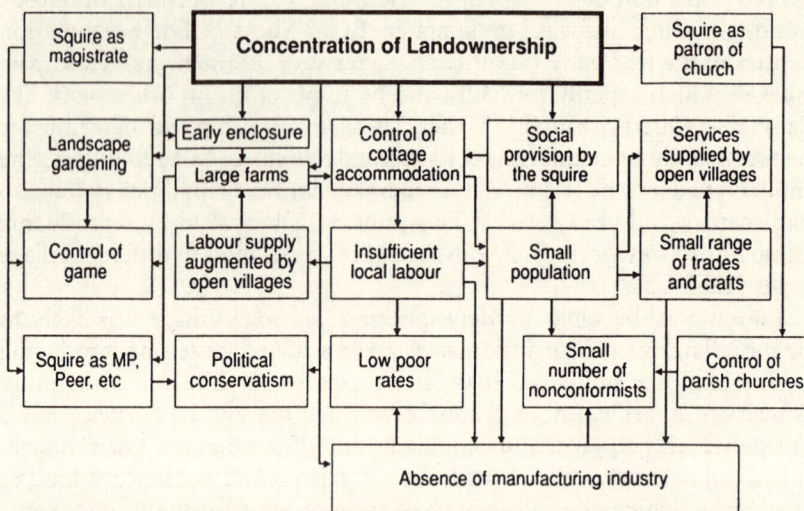

```
Squire as          Concentration of Landownership          Squire as
magistrate                                                  patron of
                                                            church

Landscape      Early enclosure    Control of      Social        Services
gardening                         cottage       provision by   supplied by
               Large farms       accommodation   the squire    open villages

Control of     Labour supply     Insufficient    Small         Small range
game           augmented by      local labour    population     of trades
               open villages                                   and crafts

Squire as MP,  Political          Low poor        Small         Control of
Peer, etc      conservatism       rates          number of      parish churches
                                                nonconformists

                          Absence of manufacturing industry
```

Causal links in the open village

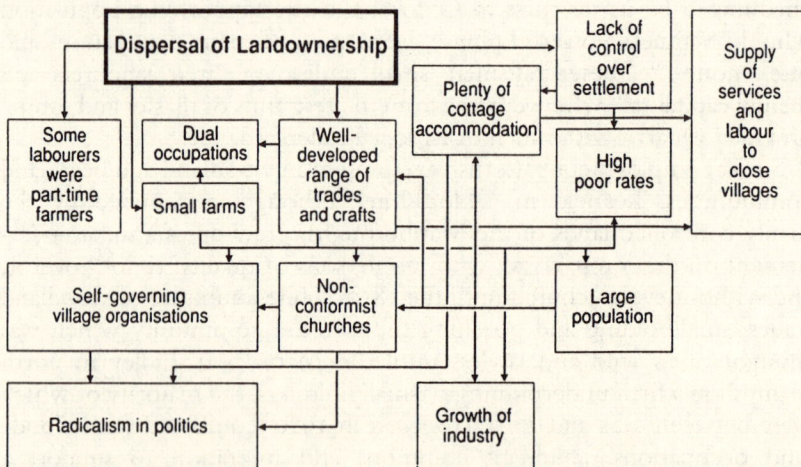

```
        Dispersal of Landownership              Lack of        Supply
                                               control         of
                                               over          services
                              Plenty of        settlement     and
         Dual      Well-       cottage                         labour
Some     occupations  developed  accommodation   High          to
labourers            range of                   poor rates     close
were     Small farms  trades                                  villages
part-time            and crafts
farmers

Self-governing        Non-                      Large
village organisations conformist                population
                      churches

Radicalism in politics            Growth of
                                  industry
```

22.1 Causal links in 'open' and 'close' villages. Reproduced from Brian Short, 'The evolution of contrasting communities within rural England', in Brian Short (ed.), *The English Rural Community. Image and Analysis* (Cambridge, 1992).

flourishing, almost peasant economy until the inter-war period. This was also true of Ashdown Forest in Sussex and parts of the Kentish Weald.[42] Tadley, like Leafield, retained a lawless reputation. 'Tadley was notorious for its insobriety and general unruliness . . . Tadley folk were reckoned a race apart – secretive, clannish and inbred.'[43]

However, it is the causal nature of the argument by Mills which is most criticised by Banks on the basis of a re-examination of his figures but more centrally in relation to her own work on Norfolk. She argues that Mills's figures do not show a direct relationship between landownership and village size or population growth, nor do her own for Norfolk. Additionally she has been able to find 'little clear-cut evidence demonstrating the widespread practice of settlement restriction'.[44] However, we should set against this Holderness's figures which do show evidence of settlement restriction in other counties although again we must move with some caution.[45] Even so problems remain. Holderness uses Lincolnshire as an example of a county where deliberate 'clearance' took place, but Richard Olney has argued, about the same county, that although the numbers of cottages and inhabitants in some villages decreased in the middle years of the nineteenth century this was not necessarily directly a result of a wish to restrict settlement. 'Some landlords did fail to pursue an enlightened cottage-building policy. Some even cleared old dwellings away and failed to replace them. But how far this was due to a desire to discourage the settlement of paupers or potential paupers is very hard to establish.'[46]

Next Banks looks at alternative bases for village typologies. Firstly, she looks at the idea of the 'classification of parishes'. This takes a number of variables: in her case these are landownership, population density, percentage change in the population, poor law expenditure and occupational structure, which are also part of Mills's definition of 'open'; she then relates them together via cluster analysis. The results are not very surprising. What emerges is a number of different parish types, not two mutually exclusive groups of open and closed, nor 'is landownership concentration the key distinguishing variable' in this classification.[47] A version of this explanation is used by Short and Mills when they insist that although there are 'pure' types of open and close it is more helpful

[42] E. J. T. Collins, 'Farming and forestry in central southern England in the nineteenth and twentieth centuries', in Helmut Brandl (ed.), *Geschichte der Kleinprivatwaldwirtschaft, Geschichte des Bauernwaldes, Tagungsvorträge*, Mitteilungen der Forstlichen Versuchs- und Forschungsanstalt, Baden-Württemberg, Heft 175, 1993, pp. 290–306.　　　[43] Collins, 'Farming and forestry'.

[44] Banks, 'Nineteenth-century scandal', pp. 57–62.

[45] Holderness, '"Open" and "close"', pp. 128–9.

[46] Richard Olney, *Rural Society and County Government in Nineteenth-Century Lincolnshire* (Lincoln, 1979), p. 73.　　　[47] Banks, 'Nineteenth-century scandal', pp. 64–5.

to see them as the extremes and to see other villages as a 'continuum (from wide open Headington Quarry – to firmly shut Cotesbach), or at least as subdivisions of each of the two main categories'.[48]

Banks goes on to look at the notion of an interrelationship between 'open and close' of the kind alluded to by Caird. This argues essentially that there is a mutually dependent system of villages with 'close' villages restricting settlement and labour supply and 'open' providing both these things for a migratory labour force. A classic example here is examined in more detail by Banks in relation to Castleacre, also in Norfolk, and its surrounding 'close' parishes. She concludes that there was such a relationship of mutual dependency although the relationship is not as clear as many historians, or indeed contemporaries, have suggested. Open and close become here not categories which constitute a 'model with predictive powers' for the internal socio-political structure of the village, but descriptive terms for different parts of a socio-spatial relationship.[49] Again, it must be said that this point is already present in the Mills and Short article where it is argued that 'individual parishes or townships cannot be studied in isolation from their neighbours', rather they have to be seen as part of a system of related entities dependent on one another but crucially lacking in determinant linkage.[50] However, what neither of these takes up is that the experience of work for the labourer in this situation was similar if not identical whether she or he came from an open or close parish. In other words the experience of labour unified those from different communities even if their relationships within the rural élite were different and potentially divisive, regardless of the type of community from which they came. This is not simply to 'reassert class' but to point out that 'work' creates communities which are just as important as where people live. At different times these may be the same, and that was more likely to be the case in nineteenth-century rural England than at other times and in other places. But that is not inevitably the case. If 'open' and 'close' have any predictive value it is in terms of the origins of the behaviour of the poor – were they rebellious or quiescent? Where in all this did a male labourer of Dersingham 'learn' his place – in the rough and tumble of the pubs and radical politics of the large village in which he lived, or in the smaller, controlled and respectable world of his work?

These kinds of questions and formulations have many virtues. Crucially they begin to deal with the uncomfortable fact that many villages which have 'open' characteristics nevertheless display none or very few of the socio-cultural manifestations which Mills's predictive model suggest should follow. This is the kind of situation which Wells suggests

---

48 Mills and Short, 'Social change', p. 254.   49 Banks, 'Nineteenth-century scandal', p. 71.
50 Mills and Short, 'Social change', p. 255.

existed in Burwash. Here, 'a large and typically "open" village in the East Sussex Weald' was in fact controlled by 'the village's select vestry, dominated by a handful of the greater farmers and the largest resident landowner'. This group exercised control over charity, outdoor relief, housing, work and drinking. As a result, 'the contemporary and historical view that "open" parishes were a haven for the footloose and displaced poor seeking a home finds little support from Burwash'.[51]

Conversely, there were 'closed' villages where an 'underworld' of radicalism existed despite the dominance of a powerful and resident élite. Hockham in Norfolk was a closed village, but Michael Home's memoir of his boyhood shows a powerful social and political underworld of radical politics and poaching.[52] In many such apparently close villages there was a sometimes overt, sometime covert struggle between the local élite and those who opposed them. North Elmham, also in Norfolk, was a close village dominated by the seat of Lord Sondes. In 1874 when the whole of the East Dereham area was aflame with trades unionism North Elmham stood out because 'of good allotments and charitable gifts' as well as the more prosaic fact that there was nowhere that the union was allowed to hold a meeting.[53] Nevertheless the 'agitators' persisted; by July that year they were able to hold 'a monster meeting' in the open air and by the autumn were holding regular meetings at the George Inn.[54]

A key issue here was the nature of the local élite, not in terms of their direct economic power, but in their ability or willingness to use any kind of power they did possess. In North Elmham we saw how 'outside' influences, in this case trades unionism coupled with the more general economic problems of the 1870s and 1880s, curtailed the ability of the élite to control the village situation. Elsewhere the will of the local élite was more important. The presence of resident and interventionist élite members, whether they were landowners, farmers or simply clergy, had a direct effect on the social, cultural and political nature of a village. In Slinfold in Sussex, for example, although the parish had 'several proprietors' and was thus in landownership terms 'open', the fact that 'two of the chief landowners' were resident led to the village demonstrating many of the socio-cultural aspects of a close community.[55] However, even this simple equation did not always work. In the Hailsham District of Sussex in the 1860s most of the villages in the area were open and 'the majority

---

[51] Roger A. E. Wells, 'Social conflict and protest in the early nineteenth century: a rejoinder', *Journ. of Peasant Studies*, 8, 4, 1981, pp. 516–20.   [52] Michael Home, *Winter Harvest* (London, 1967).

[53] *Eastern Weekly Press* (Norwich) 10 Jan. 1874, p. 3.

[54] *Ibid.*, 25 July 1874, p. 1; 7 November 1874, p. 1.

[55] *RC on the Employment of Children, Young Persons and Women in Agriculture. First Report*, BPP, 1867–8, XVII, p. 77.

of the landowners non resident'. But only in 'some of the parishes' was non-residence 'found to operate disadvantageously'.[56] This points to the fact that while landownership was an important element in who 'controlled' the village community, it was not the only criterion and in many cases a lieutenant class or group acted instead of the landowner. This was the case at Burwash and in the Hernhill – Dunkirk area of Kent in Barry Reay's important study of the 1830s.[57] Sometimes this created situations where 'open/close' modelling suggests one outcome but the reality of village life produced quite another. Saham Toney in Norfolk was described as an 'open' village and certainly supplied surplus labour to parishes around. However, in the 1860s an active vicar created within the village a range of social institutions, like schools, charities and clubs usually associated with a 'close' community.[58] There is also the question of groups of villages owned by the same landowner, as was often the case. Here it was not possible for all to be 'close' in a demographic sense, some would have to be 'feeder' villages, suggesting a wide variety of practice even within one area of landlord control.

Even given these kinds of modification problems remain, both Banks and Mills are locked into a system which seeks essentially measurable criteria of a socio-economic kind based on an assumed unit of social organisation – 'the village'. As a unit of social organisation though, the village was only one of a number of possible ways of dividing the world which might have a key role in structuring and ordering the experiences of those who lived in it. The most obviously different category was that of 'farmstead' settlement.

II. FARMSTEAD COMMUNITIES

In the north and west and most of Wales settlement was sparse, scattered and mixed in character. As Mills says, 'both true villages and scattered farmsteads could be found there in the nineteenth century'.[59] Parishes tended to be large and frequently not linked to one village or settlement. In some areas, north Northumberland or the Yorkshire Wolds for example, as well as large stretches of Wales, there was little village settlement at all. In these areas boundaries were complex and often relied not on any notion of the village or parish as a community but on the farm or the work group. John Coleman said of north Northumberland in 1881, 'each large farm represents a small colony in itself, with accommodation provided for labour requirements for all but extraordinary occa-

[56] Ibid., p. 85.
[57] Barry Reay, The Last Rising of the Agricultural Labourers. Rural Life and Protest in Nineteenth-Century England (Oxford, 1990), pp. 18–27.   [58] RC on Children (1867–8), p. 59.
[59] Mills, Lord and Peasant, p. 19.

sions, such as hay or corn harvest'.[60] In the small farming areas of south-west Wales it was more complex:[61]

Groups of cottages were connected with each farm in this organisation of overlapping farm groups, and membership of these farm-cottage groups might overlap in the same way that groups of farms overlapped. Hence no boundaries can be drawn that would isolate one community from all others.

As well as the spread of settlement other factors came into play to influence the nature of community structures in areas of farmstead-hamlet settlement which, while they were present in 'village' England, were more intense in the farmstead areas. On the borders between England and Wales and England and Scotland a combination of geology and custom created boundaries which while recognised by locals simply did not fit with administrative ideas. On the borders between Radnorshire and Herefordshire in 1870, Mr Price of Maesgwyn, a Radnorshire farmer, said that:[62]

he considered the boundary of Radnorshire and Herefordshire to be quite arbitrary, the real natural boundary was a line drawn across the summits of the Begws and Clyro Hills . . . On the east side of that range the farming is very similar to that of Herefordshire. To the west we have almost entirely stock farms, with the Welsh peculiarity of the labourers, men, women and boys, living in the farm-house. Where the Valley of the Lug, however, runs up to Bleddfa and Llangunllo an arm of England runs up between the Welsh mountains.

On the Anglo-Scottish border in Northumberland shepherd and sheep migrated backwards and forwards from England to Scotland, their boundaries defined essentially by hiring practices.[63] Nor was it only the shepherd. As Michael Robson writes:[64]

upland and lowland divisions of agriculture are much more distinct from each other than are the two countries on either side of the border. In considering the life of those who have worked the land, the shepherd and ploughmen, dykers, drainers and many others, it is of greater use to treat the English and the Scottish sides as part of the Border country.

There were other general factors which marked off and created aspects of 'community' in farmstead areas. Although much of these areas was held from large landowners, and was even in some areas held in large

---

[60] *RC on the Depressed Condition of the Agricultural Interest*, BPP, 1882, xv, Mr J. Coleman's Report, p. 6.   [61] Jenkins, *Agricultural Community*, p. 6.

[62] *RC on the Employment of Children, Young Persons and Women in Agriculture, Third Report*, BPP, 1870, xII, p. 58.

[63] Northumberland Record Office (hereafter NRO) NRO T 62, acc.no. 1055, Interview with Mr Murray, shepherd, b. Harwick 1891.

[64] Michael Robson, 'The Border farm worker', in T. M. Devine, ed., *Farm Servants and Labour in Lowland Scotland, 1770–1914* (Edinburgh, 1984), p. 71.

units, the remoteness and the lack of nucleated settlements mean that the rural élite were not the ever-present force they were elsewhere in the British Isles. F. M. L. Thompson has shown that although there is no obvious regional pattern to large aristocratic estates, the distribution of great houses does show a marked tendency to cluster in the south and east regardless of where the majority of an individual owner's land was situated.[65] Even where the main house was in an upland area simple difficulty in travelling across rough and open country made regular visits to remote farms impractical, and the presence of the gentry, and other sections of the élite, was often restricted to the immediate area of the great house. These could be well away from the body of the estate since the rural élite were not inclined to live in remote and backward areas. John Walton has recently shown that in Lancashire, 'large estates were concentrated disproportionately into the low-lying and generally relatively fertile south and west', as were the residencies of the gentry.[66] Even where there were large estates Alistair Mutch's work on the same county shows that the larger and even the medium landowners seldom lived in rural Lancashire for more than part of the year, leaving the running of their estates to stewards.[67] Although this pattern was common in many counties it seems to have been more widespread in the upland areas. In the East Riding of Yorkshire, the Wolds, colonised in the eighteenth century, were left all but devoid of great houses, although the lower slopes were a favourite area among the urban élites of Hull.[68] This distancing was often symbolised within the estates themselves by what Williamson and Bellamy call 'distance decay', 'the tendency for estate buildings and other features to be less ornamented the further they were from the house'.[69]

These problems were much worse in Wales. D. W. Howell writes of the 'Welsh' landowners that, as the nineteenth century progressed, 'their falling numbers . . . together with a growing incidence of absenteeism among the better-off families . . . indicates [that] their role as leaders of their communities was in decline'.[70] Owen Hughes, a tenant farmer of 191 acres in north Wales, told the 1894 *Royal Commission on Land in Wales*

[65] F. M. L. Thompson, *English Landed Society in the Nineteenth Century* (London, 1963), pp. 29–34.

[66] J. K. Walton, *Lancashire. A Social History 1558–1939* (Manchester, 1987), p. 127; see also J. K. Walton, 'The north-west', in F. M. L. Thompson, ed., *The Cambridge Social History of Britain 1750–1950*, Volume 1, *Regions and Communities* (Cambridge, 1990), p. 372.

[67] Alistair Mutch, 'Rural society in Lancashire, 1840–1914', PhD thesis, University of Manchester, 1980, pp. 104–5.

[68] K. J. Allison, *The East Riding of Yorkshire Landscape* (London, 1976), pp. 181–2.

[69] Tom Williamson and Liz Bellamy, *Property and Landscape. A Social History of Landownership and the English Countryside* (London, 1987), pp. 128–9.

[70] D. W. Howell and C. Barber, 'Wales', in Thompson, *Cambridge Social History*, Volume 1, p. 287.

and Monmouthshire, 'though I have been at Brynefail 13 years now, I never saw my landlord except once, when he was pointed out to me at a cattle show'.[71] David Jenkins's local study of Cardiganshire identifies four 'gentry houses' in his two parishes only one of which was occupied by anything resembling an active and interventionist gentry family on the English model in the years between 1850 and 1914. Even in this case the additional problem of 'Englishness' verses 'Welshness' set the family apart. 'The gentry families of the locality', writes Jenkins, 'were members not so much of nation-wide as an empire-wide stratum of society connected with the armed forces on the one hand and the church on the other.'[72] This dimension produced in much of rural Wales a quite different set of 'problems' for the notion of community, especially after the 1880s, in which festering grievances about rents and tenant right produced a conscious split within Welsh farming communities based on class and nationality. As Michael Jones, a tenant farmer of 76 acres near Pwllheli, said in 1893, 'the landlords are, generally speaking, alien in religion, language, politics and some in race, even from the tenants, and there is but little direct communication between one class and the other'.[73]

There were other important factors which influenced the nature of community in the areas of farm-hamlet settlement. A key variable was the historical development of settlement in the area. Two important areas of farm-hamlet settlement in the late nineteenth century, the Wolds of Lincolnshire and Yorkshire and north Northumberland, were both the product of relatively late agricultural colonisation. Much of the settlement pattern of north Northumberland was a result of a 'vigorous programme of rural development undertaken by landlords in the eighteenth and nineteenth centuries'[74] which created a countryside, as we have already noted, dominated by large farmhouses surrounded by groups of cottages. In the Yorkshire Wolds mid- and late-eighteenth-century enclosure created a landscape of isolated farmsteads,[75] a pattern which was also present in Lincolnshire.[76]

In these areas there developed, not as a 'survival' of farm service but, as a new form, a community based clearly on the farm and the work group. However, the different areas reacted differently. In north Northumberland, where family hiring was the norm, a particularly close and integrated community of the farm and work group seems to have developed. Family hiring meant that a head of household, either a 'hind'

[71] *RC on the Land in Wales and Monmouthshire*, BPP, 1894, XXXVI, p. 416.

[72] Jenkins, *Agricultural Community*, p. 31.   [73] *RC on Land in Wales*, p. 488.

[74] Robert Newton, *The Northumberland Landscape* (London, 1972), p. 138.

[75] Alan Harris, *The Rural Landscape of the East Riding of Yorkshire, 1700–1850* (Oxford, 1961), pp. 98–9.

[76] Joan Thirsk, *English Peasant Farming. The Agrarian History of Lincolnshire from Tudor to Recent Times* (London, new edn, 1981), pp. 267–9.

(a male farm worker) or a 'cottar' (a woman) was hired with their whole family for a year and lived in a cottage very close to the farmhouse and farm buildings around it. At the Lammas Hirings at Hexham in 1865 James Nixon hired himself as a hind to Sir Edward Blackett at Hatton Hall Farm. He was given a house, coals led (that is brought to the door) and 16s. 6d. a week. His son was to have constant employment 'for the summer half year' at 9s. per week and his two daughters were 'to work when wanted and to be paid at the discretion of the farm steward'.[77] Further north Castle Heaton Farm near Coldstream employed between seven and ten families on this kind of hiring in the last half of the nineteenth century and the first years of the twentieth. In 1890 for example there were seven families hired on the farm at the end of October comprising twenty-three people, fourteen men and nine women. There were also four non-family hired workers.[78] However, women could also be hired with their families as head of household. In 1881 Castle Heaton Farm had five out of ten families hired and living on the farm headed by women.[79] These were 'cottar' hirings. For example in 1891, Margaret Jane Miller and her two daughters and two sons were hired for a year with Mr Hindmarsh at Ilderton in the Glendale Union, and had a house, coals led and 1,200 yards of potatoes, as well as 1s. 6d. per day for the women all the year round and 3s. per day at harvest. The two boys had 7s. per week for 'the whole year'.[80] Most cottars were widows although a few were single women.[81]

This system of family hiring with the families living on the farms produced a particularly socially stable community. John Coleman put this very clearly to the 'Richmond' Commission in 1881: 'there is a sort of clannish feeling between the labourer and his employer, from their living together on the farm, and being so much dependent on each other'.[82] This dependence had a material base. Given the remoteness of many farms and the custom of yearly hiring a farmer found it difficult to dismiss and replace a worker once hired especially in periods of labour shortage.[83] A more extreme case was that of shepherding where the farmer was frequently totally reliant on his shepherd who took the flocks away to the hills for weeks at a time literally out of sight or control. This dependence was symbolised and reinforced by the importance of payment in kind. At a symbolic level, payment in kind marked a rejection of a purely cash nexus of relationships between master and man; it made the labourer

---

[77] NRO. NRO ZBL/78 'Hiring Agreements', Hatton Hall Estate.

[78] NRO, NRO 302/24 Wood (Castle Heaton) 'Mss. Wages Books'.

[79] NRO. Ms Census Returns, Berwick-on-Tweed Union, Cornhill.

[80] RC on Labour . . . Glendale, BPP, 1893–4, XXXV, p. 120.   [81] Ibid., p. 102.

[82] RC on the Depressed Condition of the Agricultural Interest. Minutes of Evidence, Part III, BPP, 1882, XIV, p. 392.   [83] RC on Labour . . . Glendale, p. 103.

'to a certain limited extent a co-operator with the farmers'.[84] It also had a direct effect, as Coleman argued in 1882, in that the labourer 'shared with his master in the vicissitudes of the season as affecting the quality of the produce' which gave him a direct interest in its quality.[85] If this was true for the hinds and their families it was more so for shepherds. Northumberland shepherds worked almost entirely for wages in kind in the form of a share in the flock. As Mr Pringle, a shepherd from Middleton, told Arthur Wilson Fox in 1892: 'Shepherds prefer to be paid in kind . . . If a man has not enough principle to work, then self steps in, and he knows it is in his own interest to look after the sheep.'[86]

There was however another side. If a Northumberland farm was like a village, it was a very 'close' village indeed:[87]

This concentration of labour is not only economical, as allowing the whole of the physical energy being devoted to the production effort and not wasted in walking to and from the village, but the workpeople are more or less under the influence and direct supervision of the employer.

This led, in its 'model form', to something approaching a Victorian moralist's idea of heaven with everybody in their place, each carefully graded by rank and status. At the head of the 'community' were the 'master and mistress' who, according to John Coleman, had a 'grave responsibility' as 'an influence for good or evil' on their 'colony'.[88] Below them in houses often graded according to status were the regular workers beginning with the shepherds, then the hinds, then the spade hinds and then the byre men. At the bottom were the casuals and the Irish admitted to the community on sufferance and temporarily, and sleeping in the barns. However, even these were included by being given a 'plentiful supply of plain food'.[89] Finally, of course, there were no problems of drink or a rough sub-culture, a fact which delighted the Parliamentary commissioners:[90]

At Carham, a parish of over 1,000 inhabitants, there is not a single public-house, and among the rural population drunkenness is practically unknown, except perhaps on hiring day . . . This fact alone accounts for much prosperity, and the absence of squalid and unhappy homes.

Nevertheless such settlements were not totally isolated. The larger townships like Wooler or Coldstream provided much casual labour in the summer months, even drawing on industrial workers and their families in some cases. Nor did the poor necessarily share the idyllic view of their betters. Bob Hepple of Cowpen said in an interview in the early 1970s,

[84] RC . . . Agricultural Interest, p. 393.   [85] RC . . . Agricultural Interest, p. 6.
[86] RC on Labour . . . Glendale, p. 129.   [87] RC . . . Agricultural Interest, p. 6.   [88] Ibid., p. 7.
[89] RC on Labour . . . Glendale, p. 103.   [90] Ibid., p. 110.

'Oh 'twas quiet, there was no road for two miles to Greenlea and I walked to Haltwhistle every Saturday night, seven miles and seven back . . . for the company of other lads.'[91]

None of this should suggest that the communities of the great farms of Glendale were conflict free, rather that such conflict took place outside their bounds. Although shepherds frequently stayed with one master for many years, most hinds did not, and moving each year was one certain way of expressing discontent. The hiring fair was also the point at which more formal disagreements could come forward. As Dunbabin has shown for Northumberland, Caunce for the East Riding, and Carter for Scotland, hiring time was the point at which semi-formal and formal agreements among hinds not to hire could bring real pressure on a potential master and could raise wages or alter conditions.[92] In both Northumberland and Aberdeenshire, for example, there were attempts to form Unions in the early 1870s, in common with much of rural England and Wales, yet in both cases they failed. They failed not because, as Joseph Arch thought, the labourers of these areas were slaves because of yearly hiring and the hiring fair, but because of the exact opposite. What Carter says of Aberdeenshire needs little modification to fit Northumberland. 'Not only did servants have a strong market position . . . but the feeing markets (hiring fairs) served a crucially important function . . . they established prevailing wage levels.'[93]

If we move south into the Wolds of Yorkshire and Lincolnshire a different kind of community on the farm is seen – that of male, youth hiring. In these areas the remoteness of farm settlements was dealt with by hiring by the year usually young, unmarried men. These then lived on or near the farm either in 'barracks' or cottages, or with a farm bailiff or steward. A similar system, the 'lloft stabal', was present in areas of north Wales. While not a community in a conventional historical sense, this group culture presents an interesting contrast to the standard views mediated through village structures.

As in Northumberland there was in these areas a close physical relationship between master and man in that they lived close to each other, but there the similarities end. The 'Richmond' Commission put it succinctly in 1881:[94]

Day labourers are scarce and often reside at long distances from the farm; hence the system of hiring team and cattle men by the year is universal, and barracks,

---

[91] NRO. NRO T.70. Interview with Mr Bob Hepple, Shepherd, Cowpen, Northumberland.

[92] J. P. D. Dunbabin, *Rural Discontent in Nineteenth Century Britain* (London, 1974), Chapter VII; Stephen Caunce, *Amongst Farm Horses. The Horselads of East Yorkshire* (Stroud, 1991), Chapter 5; Ian Carter, 'Unions and myths: farm servants' unions in Aberdeenshire, 1870–1900.' in Devine, *Farm Servants.*   [93] Carter, 'Unions and myths', p. 223.

[94] *RC on the Depressed Condition of the Agricultural Interest. Minutes of Evidence,* BPP, 1881, XV, p. 143.

under the eye of the hind, who is practically their master, are necessary. Formerly the hired servants boarded at the tenant's house and were more immediately under his control and supervision.

These servants, like their equivalents in Lincolnshire and parts even of Nottinghamshire, were all young men. In the East Riding as a whole in 1871 40 per cent of male workers lived in, but of these 79 per cent were aged between fifteen and twenty-four. For Lincolnshire in the same year the figures were 49 per cent and 76 per cent respectively.[95] Living outside family control but also away from the normal structures of settlement, the farms of the Wolds developed a rough and masculine community and culture which found its expression in the male bonding of the hiring fair and the 'stable' loft:[96]

We spent most of our nights in the stable until nine o'clock . . . Sometimes other farm lads dropped in for an hour, and other times we walked across to their stables – there being two or more farms near ours. Usually one of them would bring a melodeon, and he was considered a poor gawk who couldn't knock a tune out of a mouth-organ or give a song to pass away the evening.

As community forms, the farmstead-based settlements of the highland zones were quite different from the nucleated villages of the south and east. However, they were no more of one type than more conventional villages were or are. Although, for example, the physical appearance of a Northumberland and a Wolds farmstead was superficially similar, they concealed totally different communities. The Northumberland farmstead was multi-generational and family-based. Even if hinds changed every couple of years it was internally stable, patriarchal and respectable. The Wold farms of Yorkshire were quite different in that as they were based on one generation they therefore lacked the patriarchal family-based control of those further north. Their culture and their community was therefore much rougher although conflict on the farms remained personal and was not institutionalised.

Different again were the smaller farm-based and hamlet settlements of the north-west and parts of Wales. Here, some village forms were present as well as farmstead settlements, although usually on a smaller scale. However, in the small farming communities of north Lancashire and in the fells of Cumberland and Westmorland and Wales quite different organisational forms emerged, based partly on physical boundaries, but partly on work group and family, and we shall return to them below. Firstly, however, we will turn to settlement and community types which,

---

[95] *Census of England and Wales, 1871. Population Abstracts*, BPP, 1873, LXX, pt II, pp. 376ff.

[96] Fred Kitchen, *Brother to the Ox* (new edn, West Firle, 1981), pp. 59–60. For a more detailed discussion of this community and its culture see Alun Howkins and Linda Merricks, 'The ploughboy and the plough play', *Folk Music Journal*, 6, 2, 1991.

while they may have had the characteristics of a nucleated village or an upland hamlet, because of their origins existed in both upland and lowland zones. These were estate villages, industrial villages and utopian communities.

III. CREATED COMMUNITIES: ESTATE VILLAGES, INDUSTRIAL VILLAGES AND UTOPIAN COMMUNITIES

Throughout the nineteenth century, and indeed from much earlier, communities had been created within rural areas for particular purposes. We saw above how, for instance, aspects of the settlement patterns of both Northumberland and the Wolds of Yorkshire and Lincolnshire were created in response to a specific economic need at a particular historical moment. This process was taken further with the creation of village communities for specific reasons. The most obvious of these were estate villages. In his Introduction to Havinden's classic study of Ardington and Lockinge in Berkshire, Andrew Jewell describes the estate village as 'a distinct type with its own special characteristics'.[97] For our purposes here an estate means a large compact landholding unit with the landowner resident for at least part of the year, and farming or working at least some part of the land and estate with a directly employed workforce. Crucial was the fact that a large part of the workforce actually worked the estate either in the house, the woods or the gardens. It is this combination of factors which distinguishes an estate village from what otherwise would simply be an extreme version of a 'close' village.

Lord Wantage's estate at Lockinge and Ardington shows this process clearly. The core of the estate was purchased between 1854 and 1860. Between then and 1890 small areas of land were added, and another major purchase was made in 1890. By that date the estate represented a unit of over 20,000 acres. It was also an estate which, even at the time of purchase, was let mostly in large units and the few small freeholders present in the 1850s were bought out by 1890.[98] For example, in March 1861, the same month that the Ardington Manor estate was added to Lockinge, increasing its size by over 50 per cent, the family 'tidied up' the business by buying a 'messuage, orchard, barn, etc. in Ardington' from William Goodwin and others, a total area of 2 rods 29 polls. These kinds of purchases continued until the Great War, some of them being simply a cottage with a few polls of land attached.[99] By the 1890s Wantage not only owned practically all of the land in the parishes of Lockinge and Ardington but also more or less every building in them. In addition to

---

[97] M. A. Havinden, *Estate Villages. A study of the Berkshire Villages of Ardington and Lockinge* (Reading, 1966), p. 12.   [98] *Ibid.*, p. 207.   [99] *Ibid.*

this compact estate, which was the family's seat, Lord Wantage owned another 15,000 acres in Berkshire.

As a result of the agricultural depression and falling rents, Lord Wantage 'took in hand' nearly all the farm land on the estate; 'by December 1893', writes Havinden, 'he was farming 4,427 acres himself, most of which lay within a ring fence in the parishes of Ardington and Lockinge'.[100] By that date he was effectively employing the whole population of those parishes directly. This directly employed and housed labour force was added to when he built an estate yard at Ardington which serviced all Wantage's land in Berkshire, and which by 1907 employed over 100 men.[101] The compact nature of the estate and the fact that by the 1890s it was farmed as a unit gave Lockinge a particular economic coherence. This was supplemented at the socio-spatial level by the rebuilding of the estate and parts of the estate villages. Beginning in 1860 a substantial part of Lockinge village, which clustered around the manor house and church, was demolished and replaced with model cottages. These were designed by Lord and Lady Wantage and built with direct labour. They were (and are) certainly attractive, brick and tiled, and were a great improvement in sanitary terms on what they replaced. But they also give a sense of regulation and control which was characteristic of many estate villages. Flora Thompson's father commented sardonically on this aspect of estate villages when they drove through Hardwicke, rebuilt in the 1850s and 1860s by the Earl of Effingham:[102]

The village was so populous and looked so fine, with its pretty cottages standing back on each side of an avenue of young chestnut trees, that Laura thought at first it was Candleford. But, no, she was told; it was Lord So-and-So's place . . . It was what they called a model village, with three bedrooms to every house and a pump to supply water to each group of cottages . . . Only good people were allowed to live there, her father said. That was why so many were going to church.

Lockinge was perhaps extreme in some respects, particularly in the fact that Lord Wantage farmed it all after the 1890s, but in many other ways it was reproduced throughout England and even in Wales; and in areas of farmstead settlement as well as in areas of village settlement. In Northumberland, for example, there are a number of estate villages dating from the nineteenth century, such as Cambo, Simonburn, Rock, Wallington, Etal and, most famously, Ford 'a study in Victorian Gothic, a mannered suburb set down near the border'.[103] West Ella in the East Riding of Yorkshire was rebuilt in the 1860s like Lockinge, in a consistent estate

---

[100] *Ibid.*, p. 77.   [101] *Ibid.*, p. 69.
[102] Flora Thompson, *Lark Rise to Candleford* (Harmondsworth, 1973), p. 302.
[103] Newton, *The Northumberland Landscape*, p. 134.

style with 'slate roofs . . . distinctive gabled dormers and porches, as well as decorative hoodmoulds and bargeboards'. Several other East Riding villages, for example, Scampston, Bishop Burton, North Cliffe, Warter and Langton, were also rebuilt in the nineteenth century.[104] In Flora Thompson's Oxfordshire there was a good deal of rebuilding, much of it like Lockinge, but on a less grand scale. Sandford St Martin for instance was enclosed in 1868 and in the next sixty years all the small land owners were swallowed up into two principal estates. In the 1850s and 1860s these came into the hands of Dr Edwin Guest, Master of Gonville and Caius College, Cambridge, and the Rev. Edward Marshall. Between them they rebuilt the village and completely 'altered its appearance'.[105]

Even where rebuilding did not take place, high farming and an increased emphasis on a rationalised and profit-based estate management produced elsewhere similar conditions to those found in Lockinge. West Firle in Sussex, for example, has been in the Gage family, and the main family seat since the sixteenth century. In 1843 the Gages owned 3,320 acres out of 3,392 in the parish.[106] The village itself, apart from four or five outlying farms and the Old Workhouse, was effectively surrounded by the Park of Firle Place and the village street was a cul-de-sac. At the end of the nineteenth century the Gage family had 'a monopoly of land and houses', and the estate, although most farms were tenanted, was still the largest single employer of labour, employing over 100 workers.[107] In Firle, as elsewhere, cottages were let directly to those who lived in them rather than being let with the farms.

Estate villages were not just economic entities but had distinctive social structures. Many were the subjects of social experiment, becoming the laboratories of a particular landlord's philosophy. Even where this was not the case, as in West Firle, the particularly close relationship between the big house, the estate and the workforce created special kinds of social relationships. At West Firle this was symbolised very clearly in the memory of one old inhabitant by the seating 'plan' in church.[108]

The Gage family sat in front of us – we were the largest farmers – with the Wadmans, Lord Gage and any visitors he might have. Behind him were the tenant farmers in order of seniority, then the tradesmen and behind them the farm-workers. Across the aisle sat the servants of Firle Place all dressed in black.

This symbolism could be reinforced by particular social–spatial profiles with a standardised architectural style sometimes carefully graded accord-

---

[104] Allison, *East Riding Landscape*, p. 192.

[105] Frank Emery, *The Oxfordshire Landscape* (London, 1974), pp. 173–5.

[106] N. J. Griffiths, 'Firle: selected themes from the social history of a closed Sussex village 1850–1939', MA thesis. University of Sussex, 1976, p. 15.   [107] *Ibid.*, p. 74.

[108] *Ibid.*, p. 21.

ing to the status of the tenants. It was almost always a tightly controlled social structure. All this together with the regular employment provided by estate work made for a distinctive set of social relations. Again Lockinge is a good, if extreme, example.

Lord Wantage developed as part of his response to the agricultural depression a set of ideas in relation to profit sharing on his farms, co-operative stores and small holdings. However, it is extremely important to stress that these ideas came from what was in many ways a traditional view of the responsibilities and duties of a landlord, and in that respect were similar to less radical moves made elsewhere. For instance, the co-operative store grew out of a notion that the poor lacked 'intelligent foresight', and therefore the organisation and detail of co-operative buying and selling could only be worked out by 'a man blessed with time and means'.[109] Although the store was run on Rochdale principles, Wantage himself put up nearly all the initial share capital. His profit-sharing scheme similarly came out of traditional principles urging 'that it was the duty of landlords and the clergy to help the poor by whatever practical means lay within their power'.[110]

The conservative basis of many of Wantage's ideas and, therefore, their similarity in form, if not precise content, to more mainstream estate practice was not lost on contemporaries. The Liberal *Daily News* said in 1891: 'Lord Wantage has done for the people what the people ought to be able to do for themselves – not individually, of course, but collectively and unitedly, and by their own sturdy independent and manly effort.'[111] Elsewhere and in less extreme form the management of estate villages reproduced idyllicist and settled social relationships by a careful balance of carrot and stick. George Ewart Evans's work on Lord Tollemache's village of Helmingham shows this clearly.[112] Like Lockinge, Helmingham was a consciously recruited village and estate with model cottages and even some small holdings. Cottages in Helmingham were let directly from Tollemache, and the renting agreement extracted a high price in terms of a strict code of behaviour in exchange for a good house. A cottage and allotment agreement from the 1880s laid down a cropping plan for the allotment which specifically excluded growing for seed which could be sold at a profit and enforced the 'morally beneficial' practice of spade husbandry. It was not only allotments that were controlled. All tenants had, by the agreement, 'to attend some place of worship once each Sabbath Day'.[113] There was also an unwritten code understood by all those who took cottages and remembered by people who grew up in the village. As one old villager told Evans:[114]

---

[109] Lord Wantage, 'A few theories carried into practice' (1893), quoted in Havinden, *Estate Villages*, p. 85.   [110] *Ibid.*   [111] *Daily News*, 25 Nov. 1891.

[112] George Ewart Evans, *Where Beards Wag All. The Relevance of the Oral Tradition* (London, 1970), Chapters 11 and 12.   [113] *Ibid.*, p.121.   [114] *Ibid.*, p. 112.

I recollect there was one girl who became pregnant, and they told her to leave the village. Her parents had to turn her out of the house. If they didn't they themselves and the whole family would have had to go – and would probably have finished up in the workhouse. This went on for many years until someone higher up made a mistake. Nothing much was heard of the custom after thet. [sic]

In Etal in Northumberland, as on most estate villages, there was strict control on the keeping of animals. No dogs at all were allowed because of poaching, and poultry and pigs were banned partly for aesthetic reasons, and partly because, as elsewhere, the landlord feared labourers would steal food to feed them.[115]

There was also though the carrot. The houses at Helmingham, Lockinge or Etal were, as already suggested, much superior to those in most non-estate villages, and even in villages like Firle, where there were no obviously 'model' dwellings, estate cottages were by and large better built and especially better looked after than in the villages around. Helmingham's 'double dwellers' where a pair of semi-detached cottages shared a kitchen and wash-house were well built with good, if small bedrooms, boilers and ovens. Each cottage had a good allotment, even if its cropping was controlled and, unlike Etal, a pig sty.[116] The cottages at Lockinge were similar. Most had three bedrooms, although, as at Helmingham, notions of what a cottage ought to look like meant they had very steep roofs. As a result, it was almost impossible to put furniture in them other than a bed. They also had good gardens, built-in coppers and an efficient water supply.[117] At Etal the cottages were good and given rent free to those who worked on the estate.[118]

Along with cottages the regularity of employment provided by estate villages was probably the most important carrot for working people. Unlike farm work, estate work was not subject to violent seasonal variations which characterised much arable agriculture, at least for the core group. The Firle Estate for example 'required a large retinue of regular workers all the year round to repair the estate lands and buildings and as servants and gardeners'.[119] This was also true at Lockinge, as we have seen, where the 'working centre' of all of Wantage's considerable estates was set in the estate village providing regular work for over 100 men.

In all estate villages charities and general social provision could be much better then elsewhere although they lacked the charities administered by the vestry which were, it was argued, an attraction of some large open villages. Lord Wantage's social experiments were unusual in their

[115] NRO. NRO T/120 acc no 1637. Interview with Mr Jim Tulley.
[116] Ewart Evans, *Where Beards*, pp. 126–30.   [117] Havinden, *Estate Villages*, pp. 95–6.
[118] NRO T/120.   [119] Griffiths, 'Firle', p. 33.

scope, but their form was reproduced all over England even if they were rare in Wales. Many estate villages, as we have already seen, provided allotments. Where this was the case they were frequently backed up with allotment societies, created and patronised by the landlord and large tenant farmers. At Firle a 'premium' was given for the best management of cottages, flower gardens and the best vegetables grown, given by the Gage family every year at the Flower Show held in the grounds of the Park.[120] At Helmingham, John Tollemache, who owned the estate in the late nineteenth and early twentieth centuries, went round the village inspecting the allotments and rewarding and punishing 'his people':[121]

He used to drive a coach-and-pair, sometimes a coach-and-four when he went round the villages looking at the farms and the allotments. At that time o'day all the rows had to be pointed inwards from the road so the Lord could see the rows were straight and properly weeded. There had to be at least one pig in the sty and nearly every tenant had to have a couple.

As at Firle there was an annual show with prizes for the best allotments. The judges were the local farmers, and prizes in the 1900s began at 30s. – the equivalent of two weeks' wages for a labourer on most Suffolk farms.[122]

A whole range of charitable gifts, formal and informal, permeated the estate. At Firle there are detailed accounts of the expenditure. These show that the estate and the family paid pensions to old workers, supported the village friendly society, built and supported a clothing club.[123] It was not only, however, the regular payments of this kind which made up estate charity but a whole range of ad hoc and occasional payments. These were possibly even more important since they, by their very irregularity, took on a 'special' nature, that of the 'gift'. This form of charity 'often given, significantly', as Newby says 'on a personal, localized basis', was a key element in the deferential relationship which he calls the 'deferential dialectic'. It is important to stress here, with Newby, that there was nothing natural in this relationship nor was it simple or conflict free.[124] The gift was a central element since it 'buttressed [deference] in a positive manner by the application of some substantive . . . form of reinforcement'.[125]

Personal charity was used with care and discrimination to reward the good and faithful, and this is dealt with later in a more general sense.

[120] Ibid., pp. 53, 96.   [121] Ewart Evans, Where Beards, p. 131.   [122] Ibid., pp. 131–2.

[123] Griffiths, 'Firle', p. 97.

[124] For a longer discussion of this, see Alun Howkins, Reshaping Rural England (London, 1991), Chapter 3.

[125] Howard Newby, 'The deferential dialectic', Comparative Studies in Society and History, 17, 2, 1975, p. 161.

Nevertheless, it is worth stressing how important it was in estate villages. At Firle meat was given out at Christmas to all those who worked on the estate, and flannel petticoats and blankets were given to poor women.[126] The estate records of the Ashburnham estate, also in Sussex, show the enormous care which went into the giving of charity. As at Firle there were regular payments of an institutional kind, but more frequent were endless small amounts paid to individuals and families on the estate. A few examples will serve to illustrate the range involved. In August 1865 the estate paid 'John Heads Expenses to and from London taking his daughter to the Hospital 10/-'; in April 1871, 10 s. was paid for 'Grocery for the Winchester family' and later that year Mrs Scotcher got 10 s. to provide clothing for her daughter going into service. Money was also paid out regularly for mourning and for hospital treatment of various kinds.[127]

At Ashburnham, Lady Ashburnham had a central role in giving out charity and, one assumes, deciding who was to receive it. On the Courthope estate, also in Sussex, some twenty years earlier, charity money had not only been given by the 'lady of the house', Lady Anne Courthope, but entered in 'her' accounts and set against the money she earned from the home dairy.[128] The role of the Lady of the house as 'lady bountiful' was extremely important as Jessica Gerard has shown. She writes,[129]

in the landed estates it was the women who most frequently made these personal contacts, not only as part of their traditional roles, but also because charity was seen as the special duty of the female sex; women's nature was more suited to perform acts of benevolence.

Yet we need to move with caution. The 'carrot' was in many respects as much a method of control as the stick. While we cannot doubt that many great landlords and their wives were motivated in part by Christian charity, they were also driven by the need to manage those beneath them. It could hardly have been put more clearly than in a 'guide' written for young women in 1860, 'the rich may be endeared to the poor by the blessed charities of life'.[130] Nor was this aspect missed by the villagers. 'You were under and you dussn't say anything', an old inhabitant of Helmingham told George Ewart Evans. 'The old Lord used to come round to look at the house, the garden and the allotment just as he did

---

[126] Griffiths, 'Firle', p. 97.

[127] East Sussex Records Office, (hereafter ESRO) Ash. Ms 3412. Asburnham Estate Charity Records.

[128] ESRO SAS Acc 1276 GB. Anne Courthope's General Account Book 1841–1854.

[129] Jessica Gerard, 'Lady Bountiful: women of the landed classes and rural philanthropy', *Victorian Studies*, 30, 2, p. 189.

[130] Maria Louisa Charlesworth, *The Cottage and the Visitor*, quoted in Gerard, 'Lady Bountiful', p. 189.

the farms; and the farmers were as afraid of him as we were.'[131] The 1891 *Daily News* report of Lockinge made a similar point when it quoted a man from a nearby village as saying, 'They daren't blow their noses over at Ardington without the bailiff's leave.'[132]

Industrial settlements in the countryside often shared a common origin with estate villages in that they were frequently built to order and with the single purpose of providing a workforce. There the similarity ended. Unlike estate villages which seem to have increased in numbers after 1850, they were, with the exception of pit villages, increasingly a thing of the past although there are exceptions like Leverhulme. Industry in the countryside was very much part of the phase of proto-industrial-isation and, as historical geographers have long argued, one of the most significant aspects of the nineteenth-century landscape was the continu-ing and increasing separation of urban and rural. As Dennis Mills argues, after 1850 'the rural areas became increasingly dependent on agricul-ture'.[133] Yet there remained a significant number of industrial commu-nities in rural England and Wales in these years and in some areas their numbers were increasing. David Hey writes, 'the coming of the railways, the sinking of the deep coal mines and the development of large quar-ries meant that the total number of industrialized villages . . . rose significantly'.[134] However, we need to exercise some caution about using the term 'industrial' in this context. Many villages contained a significant outwork trade of various kinds which frequently contributed to a sense of independence from any particular landowner or farmer and led in turn to a radical, or at least non-deferential culture. This was the case of the outwork boot and shoe trades of Northamptonshire which were hot beds of Chartism, and supported Bradlaugh in the 1860s and 1870s. Here, although factories never dominated, in villages like Earls Barton over half the working population was employed in the outwork boot and shoe trade by the 1860s.[135] Less obvious still were villages like Leafield (Field Town) in Oxfordshire where two potteries, brickmaking and a substan-tial outwork gloving industry, employing women and girls, gave the village a notoriously independent (or rough, depending on your view-point) character, as we have already noted.[136] Villages like this certainly had a character which was different from that of a purely agricultural set-tlement, however they were hardly industrial settlements.

At the other extreme there were settlements, products of the Industrial

---

[131] Evans, *Where Beards*, p. 123.   [132] Havinden, *Estate Villages*, p. 155.

[133] Dennis R. Mills, *English Rural Communities. The Impact of a Specialised Economy* (London, 1973), p. 15.

[134] David Hey, 'Industrialized villages', in G. E. Mingay, ed., *The Victorian Countryside*, vol. I, (London, 1981), p. 353.

[135] V. A. Hatley and J. Rajczonek, *Shoemakers in Northamptonshire, 1762–1911* (Northampton, 1972).

[136] Alun Howkins, *Whitsun in Nineteenth-Century Oxfordshire* (Oxford, 1973).

Revolution, which were almost entirely industrial in their employment patterns although they are far removed from the usual notion of a 'rural village'. The most obvious of these were pit villages. Particularly after the 1850s and 1860s in Yorkshire, the north of England and south Wales newer and deeper pits created a demand for labour where none had been before. The eastern edge of the old Yorkshire coalfield provides a number of striking examples of these kinds of settlements. Denaby Main was opened in 1868 on a site at the edge of Denaby parish where there was no existing settlement. For this reason the owners built, 'a "company town" consisting almost entirely of colliery-owned terraced houses – two up, two down brick cottages with no bath and only an outside WC'. Other villages followed spreading eastward across the formerly largely agricultural areas around Doncaster. Many, like Denaby Main, were worked by outsiders from all over the north, and they had the feel of raw frontier communities. Nevertheless they were controlled, and the pitmen and their families were never able to forget that the company owned all the houses, all the public facilities and most of the shops. Denaby was, as a result, the scene of mass evictions on no fewer than four occasions between 1869 and 1903. Elsewhere and later, perhaps as a result of lessons learned at Denaby, other south Yorkshire pit villages were better. New Edlington, for instance, although it had a bad start, had a character by the Great War which was not unlike some estate villages:[137]

The houses were erected in units of four or six with large flower and vegetable gardens to the rear. On the ground floor was a living room with a Yorkshire range and a small side-boiler, a scullery with a sink and copper to heat water for washing clothes and for bathing in a portable bath, a small pantry, a sitting room and two outshuts serving as WC and coal place. Upstairs were two bedrooms and a boxroom. The manager's house was a large brick villa set in its own grounds, and the other officials lived in semi-detached houses . . . By 1914 New Edlington had a church, school, co-operative store, pub and concert-cum-dance hall.

In West Cumberland the population of the coal-mining districts increased by 75 per cent in the thirty years after 1851 and 'a large part of it came from neighbouring agricultural areas'.[138] Joyce Neale writing in the 1970s commented on these new villages:[139]

Today these houses stand out because of their uniformity and monotonous symmetry. Of all the settlements, Moor Row is the most spectacular, insomuch as it was purpose-built for the miners . . . No gardens were provided, although allotments were, and still are, tended near by. Washing was hung on a common green out of sight of the main road.

---

[137] David Hey, *Yorkshire from AD 1000* (Harlow, 1986), pp. 280–5.

[138] T. H. Bainbridge, 'Population changes over the West Cumberland coalfield' in Mills, *English Rural Communities*, p. 139.　　[139] Quoted in Mills, *English Rural Communities*, p. 14.

In south Wales this kind of settlement gradually joined together in strings of communities up the valleys. Nevertheless, intense local feeling remained with each settlement which was reinforced by the social and political institutions of the working class, the miners' lodge, the chapel and the friendly society, which tended to be local as much as national in their cultural focus. They also remained surprisingly rural in their focus. The autobiography of George Ewart Evans, who grew up in a family shop in Abercynon, shows how close country and pit 'village' were in economic terms. His father's shop served not only the pitmen and their families but the hill farms above the valley, where farm-made cheese was bartered for factory-produced goods by the farmers' wives. The pitmen and their families worked for the farmers on the hills as casual labourers at harvest, and the farms provided new recruits for the pits. The two communities, 'industrial' and 'rural', also shared a culture based around the Welsh language. So, even though its population was above, 10,000 by 1900, Abercynon remained a remarkably 'rural' community.[140]

Coal mining, and especially lead mining, produced, or rather supported, a number of peasant communities which, while not solely industrial in themselves, could not have survived without the extractive industries. The main examples of this were in south-west Wales and the northern Pennines. David Jenkins's study of south Cardiganshire in the last years of the nineteenth century shows that here both the demand for seasonal labour on large farms and the survival of small-scale 'cottager holdings' was guaranteed by labour debt incurred by the planting and cultivation of potatoes on the large farmers' land. But this could not have worked without industry providing work for the men of the family for at least part of the year:[141]

A family might have as few as two rows of potatoes 'set out' or as many as eight, usually all on one farm but sometimes divided between two farms. At the turn of the century many South Cardiganshire men worked in the coalfields of Carmarthenshire and Glamorgan while maintaining their homes and families in Cardiganshire. They took holidays in August and September to return home to pay their families' work debts in the harvest fields.

In the north of England lead mining was similarly instrumental in the survival of the peasant communities of the north Pennines. Richard Heath met such a man near Reeth in Swaledale in the early 1870s. 'He worked in the mine in his own time, and spoke well of the masters . . . In addition to his earnings in the mine he farmed three acres, on which he kept a cow – selling the milk for a penny a pint, and sometimes

---

[140] George Ewart Evans, *The Strength of the Hills* (London, 1983), especially Chapters 1 and 2.
[141] Jenkins, *Agricultural Community*, p. 52.

"kearning a bit".[142] Further north in Allandale lead mining, occasional coal mining, small farming and 18,000 acres of stinted moorland, created a series of communities up the Dale. Mr J. H. Reed, interviewed by the Northumberland Record Office, was born on High Sinderhope Farm. His father 'was also working in the lead mines':[143]

They had a system. They would go in possibly at 2 o'clock one day and work till 10, and then they would work till 10 and stay overnight in what they called the shops, and go back at 1 o'clock the next morning, come out at 2 and come home till the next day . . . At some of the mines, one week they would finish at 2 o'clock on the Friday and would be clear till 2 o'clock on the Monday, well they had the whole weekend clear and, them that had a small holding, that was then they got a bit of work done.

Nevertheless they remained rural. As Mr Armstrong who also grew up in Allandale said of his father, 'he came back to the farm . . . 'cos he was bred to the farm . . . they all made farmers'.[144]

The lead mines and the coal mines at Consett and Stargate provided the necessary capital to improve the holding as they did for many migrant Irish peasant farmers. 'The idea was in those days to collect a little bit of stuff together. Then a bit more, and try and get a bigger farm, that's the way they stepped up in them days.' Mr Reed's father followed that route and in 1911 he moved to Sipton Shield, a farm of about 50 acres as well as 25 stints on the moor on which he kept sheep.[145] Nor was it only men. As in Wales women worked and held the land while sons worked in the mines. The schedules which accompany the 1910 'land tax' survey show many women holding small farms in Allandale.[146]

These patterns were reproduced all over the rural areas of England and Wales, especially in the upland areas. As Raphael Samuel has shown, miner farmers were also found in Cornwall, Leicestershire, the Forest of Dean, Teesdale and Montgomeryshire.[147] Industrial villages also grew up around quarries. Again we need to distinguish between communities like Headington Quarry, where quarrying was one of several occupations even if an important one, and settlements like those in north Wales where villages were created for, and by, quarrying, and contained many of the structural elements of pit villages with 'homes, gardens and pensions'.[148]

The coming of the railways also transformed and created industrial set-

[142] Richard Heath, *The English Peasant* (new edn, London, 1978), p. 95.
[143] NRO. NRO T/132. Interview with Mr J. Reed.
[144] NRO. NRO T/30. acc no 1,034. Interview with Mr J. R. Armstrong.   [145] NRO Reed.
[146] NRO. NRO 2000/8 Schedules.
[147] Raphael Samuel, 'Mineral workers', in Raphael Samuel, ed., *Miners, Quarrymen and Saltworkers* (London, 1977), pp. 63–6.
[148] Merfyn Jones, 'Y chwarelwyr; the slate quarrymen of North Wales', in Samuel, ed., *Miners, Quarrymen and Saltworkers*, p. 117.

tlements some of them of quite a small scale. Didcot in Berkshire, Wolverton in Buckinghamshire and Melton Constable in Norfolk were all changed utterly by large-scale housing built by railway companies for their employees. Hoskins showed many years ago how the railways transformed his parish of 'Midland peasants' into an industrial settlement. South Wigston was transformed in the years after 1883 by the building of 600 identical brick cottages and a new population of 2,400 working on the railways and industries brought to the area by the iron road.[149] Melton Constable was perhaps the most striking of these, a company railway town set in the middle of the great agricultural estates of north Norfolk, and created by one of the county's greatest landlords, Lord Hastings. In 1881 the parish had a population of 118 with no settlement at all on the present village site; by 1911 it had reached 1,157 with nearly all the increase coming in the village itself. The centre of the village was the railway works which was surrounded by a gas works, sewage works and water tower. Houses were built to bring workers to the spot, and built in terraces with larger ones at each end to house the supervisory staff. The company also built a hotel, an elementary school and a railway institute. By the 1890s 'a close and distinctive community came into being . . . in appearance and type quite different from any other in North Norfolk'.[150] Even where railway towns did not spring up railway settlement and their workers made an impact often within a surprisingly traditional setting. In south Cardiganshire 'around Llandysul a concentration of cottager railway workers' set out potatoes and incurred labour debt which they repaid in the same way as the miner farmers did.[151]

There were however, as well as these new communities, one significant group of more traditional villages which had a rural setting but which were in some respects separate from agriculture, and these were fishing villages. Even here there were changes in our period. The most interesting of these were in Scotland where, as Paul Thompson argues, 'several fishing villages were originally laid out by entrepreneurial landlords . . . There was little work on the north-eastern farms . . . and the people of the fishing villages had no crofts of their own. From the start, therefore, these were communities of full-time fishermen.'[152] Most nineteenth-century developments in England and Wales tended to take place in larger and already specialised port towns like Hull, Grimsby or Yarmouth, and the result of these changes was usually indirect in terms of the history of the rural world. Nevertheless the coastal areas did produce distinct if not

[149] W. G. Hoskins, *The Midland Peasant* (London, 1951), pp. 261–82.
[150] R. S. Joby, *Forgotten Railways: East Anglia* (Newton Abbot, 1979), pp. 28–30.
[151] Jenkins, *Agricultural Community*, p. 52.
[152] Paul Thompson, Tony Wailey and Trevor Lummis, *Living the Fishing* (London, 1983), p. 15.

totally separated communities which by and large fished inshore, from small boats which were based on family or work-group capital. The men who fished pilchards out of Port Isaac in Cornwall or codling and herrings off the beaches of Northumberland or East Anglia were, in some senses, literally peasants of the sea:[153]

Along the Norfolk and Suffolk coast . . . where Norwich city market had encouraged specialised villages much earlier, there were still within living memory many independent-minded men who would switch from fishing to farm labouring according to season. Smallholders would become horse-and-cart fish-hawkers around the villages for the autumn fishing season, and farm labourers take berths on boats after the hay was cut . . . These men from the farms tend to remember fishing for the fighting, spitting and dirt aboard, 'That's Hell's Playground, they say. But it toughened you up.'

In the north-east Mr Jack Stewart of Alnmouth, a fisherman, remembered his father working the cobles of Alnmouth, fishing for salmon in the Croquet and, when all else failed, 'you could always get a job at the harvest time getting the harvest in, because it was generally cut with scythe and sickle'.[154]

As with estate villages, all these industrial and semi-industrial villages developed their own social characteristics, but usually they were very different from the closed and ordered world of Lockinge or Firle. A partial exception may be railway villages. We have already seen how Melton Constable had some of the physical characteristics of a normal estate village, and this may in part be a result of its particular origin. However, as Frank MacKenna has argued, railway workers, like other sections of the uniformed working class in Victorian England, were regimented and controlled often in a paternalistic way. It would be no surprise then if railway villages, before the rise in militancy in the 1900s, were a controlled and deferential environment.[155] After that date things were different, and railwaymen were often a disruptive element in the apparently harmonious rural world. The strike of Lancashire farmworkers in 1913 owed much of its success to the support of the NUR, and the 'new' farmworkers' union of the 1900s in Norfolk had close links with the NUR which was represented by an annual demonstration at Melton Constable from 1912 onwards.[156] Finally, in many rural areas the railwaymen were a key element in the founding of the Labour Party, being one of the few groups of workers in rural areas with national contacts and structures. As Raymond Williams said of Pandy on the Welsh Borders:[157]

[153] *Ibid.*, p. 14.   [154] NRO T/95 Interview with Mr Jack Stewart.

[155] Frank McKenna, *The Railway Workers, 1870–1970* (London, 1980), Chapters 1 and 2.

[156] On Lancashire, see Mutch, *Rural Life*, pp. 56–7; on demonstrations, Alun Howkins, *Poor Labouring Men. Rural Radicalism in Norfolk 1872–1923* (London, 1985), Chapter 6.

[157] Raymond Williams, *Politics and Letters. Interviews with New Left Review* (London, 1981), p. 24.

The interesting thing is that the political leaders of the village were the railway-men. Of the three signalmen in my father's box, one became the clerk of the parish council, one the district councillor, while my father was on the parish council . . . All of the railwaymen voted Labour . . . They were in touch with a much wider social network, and were bringing modern politics into the village.

Other industrial villages or industrial groups stood more apart, even where contact was quite close. The Morayshire saying, 'the corn and the cod dinna mix', was simply an extreme manifestation of the difference felt between fisher people and farm people in many areas. In East Anglia, for example, Yarmouth and Lowestoft fishermen were traditional strike breakers, being used in farmworkers' disputes in 1894 and 1911.

Social relations in the pit villages were least like those experienced in most agricultural communities. Here, a community based unproblemat-ically on class developed. Although some early pit villages may have had a manager's house 'on site', this was set away from the rows, and often separated by a high fence. By the 1900s most were single-occupation and single-class communities, dominated by their pits. Everything grew out from the pit, the close-knit family structure created by multi-genera-tional, shared work experience; the informal work organisations of the 'gang' and the 'butty system'; the co-operative store; the Miners Welfare; the Chapel and the Union Lodge.[158] These formal and informal struc-tures were created by the pitmen and their families for themselves, although the owners did contribute on occasions. In that, they were very different from the essentially cross-class and spatially defined cultures of many agricultural villages. But because of their single-class nature many pit villages were free from class strife and antagonism. Battles were with the owners who were physically elsewhere, and with the impersonal forces of production. When conflict came to the pit villages, as it did in Denaby Main, or even to the quarrying communities of north Wales in the 1880s and 1890s, what ensued was not a battle within a community but a battle in which the outside and alien world of the 'owners' was pitched against, and sought to destroy, the very idea of community. In the pitmen's strikes of the 1890s and 1900s a community was created in opposition to the rest of the world, a community which was not to be broken until our own day.

A final group of 'created' communities were those set up by philan-thropists or reformers who sought a return to the land or the country-side, believing, for a whole range of reasons, that country life was superior to urban life. This is not the place to go into the history of these communities in detail but we need to look at them briefly. They are best

---

[158] For a good picture of a pit village in rural Northumberland, see Linda McCullogh Thew, *The Pit Village and the Store* (London, 1985).

seen under two headings. Firstly, 'philanthropic communities' – that is those created by the élite often to provide a healthy or stable environment for manufacturing industry. These clearly could have much in common with estate villages. Secondly, 'utopian communities' – these were usually the product of a group or movement which sought to transform social or economic relations by living in particular ways in the countryside. It is, of course, not always easy to separate the two. Robert Owen's New Lanark clearly partook of both, but generally speaking the distinction works well enough.

Philanthropic village communities, as distinct from simple estate villages, do not really appear before the nineteenth century. The earliest was probably New Lanark although some early millowners, like the Strutts at Styal, did build decent cottage accommodation (by the standards of the day) near their mills.[159] It was from the 1840s that the real spread of philanthropic industrial villages began. The most famous was Sir Titus Salt's Italianate fantasy at Saltaire. Begun in 1850, it consisted by the 1860s of 'five hundred and sixty houses . . . grouped around the magnificent soaring mill, and a church and temporary buildings for dining rooms, school room and a lecture hall were in use'. Yet, as Darley points out, the housing, though good, was still essentially back-to-back with no gardens, and a laundry to prevent lines of washing from disfiguring what appears today as much an aesthetic development as a social one.[160]

Salt's example certainly inspired imitation in the next few years. However, it seems more likely that the horror produced, particularly within the religious section of the élite, by uncontrolled housing development as a result of industrialisation, was a more important spur to village building. In Somerset, for example, the Clarkes, a Quaker family, built a model village around their factory in Street:[161]

From the beginning they decided to provide housing for their work people, and the roads and houses were carefully laid out to ensure that ample gardens and recreation space were left. The local lias stone . . . provided an excellent building material, and the company houses were built in a modified vernacular similar to that which is so attractively displayed in nearby towns and villages like Somerton and Compton Dundon.

In the less salubrious surroundings of Merseyside, the Wilson family, inspired by a religiously modified Owenism, built a model village at Bromborough to house their employees busily making Price's Candles.[162] Most famous of the settlements inspired by a mixture of religious zeal

[159] Gillian Darley, *Villages of Vision* (London, pbk edn, 1978), pp. 122–4.   [160] *Ibid.*, p. 132.

[161] Michael Havinden, 'The model village', in Mingay, ed., *The Victorian Countryside*, vol. II, p. 421.

[162] J. N. Tarn, 'The model village at Bromborough Pool', in Mills, *English Rural Communities*, pp. 147–52.

and genuine horror, were those of the Cadburys at Bournville near Birmingham, begun in the early 1890s, and the Levers at Port Sunlight on Merseyside, begun in 1888. It is difficult now to think of either of them as villages since they are both now surrounded by urban development, but they were built as contained units of village size with access to open land both in the villages and surrounding them. Both these communities had a very high standard of housing, with good gardens and excellent communal facilities, especially at Bournville, where workers were provided with a gymnasium and swimming baths as well as the more normal parks and public institutes.[163]

However, there was another side to this. These philanthropic communities were conceived of in moral terms; they were built as models of improvement and as such they laid upon those who lived in them a variety of obligations. At their most minimal these usually involved an idyllic notion of a united and non-class community. Colonel Edward Akroyd who built two villages in the Halifax area, Copley and Akroydon, in 1847 and 1859, deliberately mixed the status of housing and hence the status of occupiers, a policy followed elsewhere. This was, according to *The Builder* in 1863, so that 'the better paid and better educated might act usefully on the desires and tastes of others in an inferior social position'.[164] Elsewhere, at Bromborough, for example, this mixing was seen not so much as leading by example as being watched over by a management which, though benevolent, was all pervasive. It was particularly true at Port Sunlight. As Darley writes:[165]

Like a possessive parent, Lever could not, as Cadbury was able to, leave the village which he had founded any measure of independence. With the brooding presence of the factory, the sole employer of every man in the village, the constant respect required for its founder and the extraordinarily hermetic effect of the village plan, it is hardly surprising that one union official commented that 'no man of an independent turn of mind could breathe for long in the atmosphere of Port Sunlight'.

In fact it was, in every way, precisely like the 'ideal' estate villages of Lockinge or Helmingham.

The philanthropists who created villages like Port Sunlight or Akroydon were consciously responding to the problems created by industrialisation and urbanisation by going back to an older and, as they saw it, superior model – that of the village. Their view of the village was, however, a very particular one. At its root was a Carlylean notion of an ordered pre-capitalist world in which all had known their place and in which relationships between people were based not on the cash nexus of

---

[163] Darley, *Villages of Vision*, Chapter 9.
[164] *The Builder*, 14 Feb. 1863, quoted in Darley, *Villages*, p. 135.   [165] *Ibid.*, p. 141.

industrialism but on the human nexus of the small-scale and structured community. They sought to recreate a paternalist world of obligation where good conditions would morally improve the worker but only in particular ways, essentially those which tended to social harmony and cohesion. Ironically, the other form of ideal community created in our period, the utopian, shared a similar starting point in the idealisation of a lost past. There, however, the similarity ends, for utopian communities looked to the past for a model which would disrupt and overturn the social order rather than simply reproduce it, even if the effects of many such experiments were a good deal more limited.

A starting point for most nineteenth-century utopian villages and colonies can be found in what Malcolm Chase has called agrarianism.[166] Chase sees agrarianism as one of the responses of the 'new working class' to the dislocation of the industrial and urban revolutions of the nineteenth century. This response essentially sought a solution in the land. It began with an historical account of how land law, itself a Norman imposition, and enclosure, had robbed the poor of the land which was their birthright. Chase quotes an early nineteenth-century popular ballad which sums this view up neatly:[167]

> Draw near if you would understand,
> The Rights of Man art in the Land,
> Let feudal Lords say all they can,
> A Nation is the People's Farm,
> They build, they plant, 'tis their strong arm,
> That till the clod, defend their clan.

Although, as Chase argues brilliantly, agrarianism was a complex set of ideas, its essence was to reverse this process and give the land back to the people. The radical paper *The Poor Man's Guardian* put it eloquently in 1834:[168]

We look to the possession of the land by the Working Classes as the means of organising them into a perfectly distinct and wholly emancipated body . . . we consider the monopoly of land is the source of every social and political evil . . . when it is borne in mind that our national debt, our standing army, our luscious law church, our large police force, our necessity for 'pauper'-rates . . . our pampered court and the pampered menials thereunto belonging, are one and all so many fences thrown around the poor man's inheritance.

From at least the 1820s, but especially from the 1840s, a number of schemes to restore the 'peoples' farm' were tried. In the early 1830s probably the most famous was the Owenite community at Ralahine in

---

[166] Malcolm Chase, *'The People's Farm'. English Radical Agrarianism, 1775–1840* (Oxford, 1988), especially Chapter 1. [167] Quoted in *ibid.*, p. 1. [168] *Ibid.*, p. 5.

County Clare but it provided inspiration for similar, if short-lived communities in various parts of England and Wales.[169] The most important
were the 'colonies' of the Chartist Land Plan. Between 1846 and 1851
five Chartist settlements were started in Hertfordshire, Gloucestershire,
Oxfordshire and Worcestershire. Their aim was simple: to provide for
each of the settlers a house for him (or her, for there were women colonists) and his/her family, plus enough land worked with spade husbandry
to provide subsistence and a small surplus for sale to pay the rent.
Communal facilities were also provided in the shape of a school and
meeting hall. Their story is well known though, with the exception of
Chase's work, it is usually told simply as a sideline on failure.[170] It is
argued that the plan was doomed to failure because of a combination of
lack of expertise and capital on the part of the settlers and the misman-
agement of Feargus O'Connell and the other Chartist leaders. A few
accounts also mention the implacable hostility of central and local
government which liked their model villages firmly under aristocratic
control.

In fact the story is more complex. It was not, as Chase argues, a side-
line but a part of the mainstream of English radicalism, as was the agrarian impulse as a whole. At its peak it attracted the support of some 70,000
shareholders, and even if few of them were settled on the estates this was
real testimony to the appeal of the idea. Even in defeat many colonists
clung to these beliefs. T. M. Wheeler, the first secretary of the Land Plan
and a settler in O'Connorville in Hertfordshire, wrote defiantly in 1858
that far from being a failure it had 'proved a signal success', for it had made
the land 'pre-eminently a People's question', and 'had rekindled the
ancient flame, and reknit the ties which bound the labourer to his mother
earth'.[171] Nor was it just ideas. Peter Searby has shown how the colony
at Great Dodford near Bromsgrove survived until the Great War by a
mixture of soft fruit growing and outwork. In 1905 there were still ten
out of the original thirty-six Chartist families represented in the
colony.[172] Even where the colonies were dispersed as at Minster Lovell a
tradition of small holding remained. In 1889 there were in the parish 80
holdings which totalled only 252 acres and which supported in the region
of 300 people.[173]

The Chartist Land Plan was the last large-scale attempt at creating a
utopian community in England but it was certainly not the end of the

---

[169] *Ibid.*, Chapter 6.

[170] *Ibid.*, pp. 172–7; J. MacAskill, 'The Chartist Land Plan,' in A. Briggs, ed., *Chartist Studies* (London, 1959); A. M. Hadfield. *The Chartist Land Company* (London, 1970).

[171] *National Union*, 5 Sept. 1858, quoted in Chase, *The People's Farm*, p. 176.

[172] Peter Searby, 'Great Dodford and the later history of the Chartist land scheme', *AHR*, 16, 1, 1968.   [173] *Report from the Select Committee on Small Holdings*, BPP, 1889, XII, p. 139.

tradition. In an article in 1908 in the Labour Leader John Bruce Glasier wrote, 'The land calls the people, and the people unbound hasten to the land',[174] and recent years certainly had given a good deal of support to that idea. Between the 1860s and the outbreak of the Great War, and in larger and larger numbers from the 1890s, utopians, dreamers and perhaps just cranks and eccentrics of every political and religious persuasion went 'back to the land'. Jan Marsh, Peter Gould and Gillian Darley between them have outlined many of these little histories while several of the 'experiments', particularly the religious ones, have produced their own.[175] What these settlements have in common is more difficult to decide than is the case of the philanthropic villages. Fundamentally, they all sought a better life by both rejecting industrial and urban living and values and elevating the life of the country. But the mixture was not always made of equal parts. It is difficult to see the inhabitants of the Bellagio bungalow estate built near East Grinstead in 1887 as rural uto-pians but in some perverse way they were, with their desire for a 'delight-ful rural retreat' where 'hidden away amongst wooded slopes, you can be lost to the outer world as completely as you wish'. Nevertheless the mixture here was mostly country life rather than any real rejection of urban values for they had to work and Bellagio was for 'tired Londoners'.[176] More total rejection was followed by the Purleigh Tolstoyans. In 1897 they set up a colony in rural Essex. Their plans were more carefully thought out than many other colonies since they intended to grow vegetables for an urban market and eggs and goat milk. However internal divisions rather than agricultural practice proved fatal. But the remnants of the split went on to found the colony at Whiteway in Gloucestershire which still exists.[177]

In these respects they prefigure the 'incomers' who come into the countryside in such large numbers after the 1900s and who are examined in Part VII of this volume, in that they wanted to 'get away' from an urban or industrial society to a rural and, by implication, 'better' or 'purer' one. Sadly perhaps, at least after the Chartist settlements, they seem to have had little effect on the areas where they were created, usually drawing only hostility or derision from those around them.

This section has brought together a group of rural communities of a rather diverse kind around the notion that they were 'created' and were therefore in some way 'unnatural'. This is obviously false at some level.

---

[174] *Labour Leader*, 14 Aug. 1908.

[175] Jan Marsh, *Back to the Land. The Pastoral Impulse in Victorian England from 1880–1914* (London, 1982); Peter C. Gould, *Early Green Politics. Back to Nature, Back to the Land and Socialism in Britain, 1880–1900* (Brighton, 1988); Darley, *Villages of Vision*.

[176] *The British Architect*, 1888, quoted in Anthony King, *The Bungalow* (London, 1984), p. 94.

[177] Marsh, *Back to the Land*, pp. 105–11.

We saw earlier that many rural communities were created, and created in the relatively recent past, as in the upland areas of the north-east and Yorkshire, but these communities do differ in their overriding wish to 'transform' existing relations. Whether they looked back to an ideal of paternalism as some estate, model and even industrial villages did, or forward to a new Jerusalem or Utopia as did others, they shared in common an alternative ideal to that which they saw around them. Relationships within them, as we have already said, of course varied – the Whiteway colonist was about as different from the 'model' labourer of Etal or Ford as it was possible to be – yet both were different from those around – if only by degrees.

IV. MARKET TOWNS AND PAYS

When the census takers in 1851 sought to measure population density they turned not only to conventional forms like persons per square mile but to a social measure of an interesting kind. 'To the twenty-one preceding "villages"', they wrote, 'there is on average a town, which stands in the midst of 110 square miles of country, equivalent to a square 10½ miles to the side, a circle having a radius of nearly six miles; so that the population of the country round is on average about 4 miles from the centre.'[178] This social measurement had a clear material reality. If we look at the late nineteenth-century local directories we see that this area accords almost exactly with the marketing area of many country towns. For example, East Dereham in Norfolk is described in *Kelly's Directory* for 1896 as being 'in the centre of a fertile and highly cultivated district, and its distance from any other market of any consequence, cause Dereham to rank among the best markets in the county'.[179] Elsewhere the precise distances might be different but the principle remained the same. The market town created an area around it which had for its inhabitants a clear meaning. On occasions it was even given a name like Hexhamshire in Northumberland or Banburyshire in north Oxfordshire, which distinguished it from the neighbouring areas. This is what I mean by *pays*, a meaning which has more in common with the French notion in which the *pays* of the country man or woman, *mon pays* is literally 'my land' or my area, and a cultural definition of belonging, rather than the historical geographer's notion of a much larger and physically defined area.[180]

---

[178] *Census of Great Britain, 1851. Results and Observations*, BPP, 1851, XLIII, p. xlvi.

[179] *Kelly's Directory of Norfolk, 1896* (London, 1896), p. 100.

[180] For example, Alan Everitt, 'Country, county and town: pattern of regional evolution in England', *TRHS*, 29, 1979, pp. 79–108. For a more recent view which sees *pays* differently, Charles Phythian-Adams, 'Local history and national history: the quest for the peoples of England', *Rural History, Economy, Society, Culture*, 2,1, 1991, pp. 1–25.

The pays also frequently accorded with other institutional and cultural boundaries. Especially from the 1880s the 'local state', which is the subject of a later chapter, often covered the same area. In East Dereham for example the pays roughly accorded with the Poor Law Union and Petty Sessional area of Mitford and Launditch, the Rural Deanery, the Primitive and Wesleyan Circuits, the District Registrar's Office and, after 1894, the Rural District Council Area. It also provided a social focus for the same area with Conservative and Liberal Clubs, the Headquarters of the Norfolk Volunteer Infantry Brigade, a Reading Room and Athenaeum, Assembly Rooms, a Corn Exchange and, of course, its Friday market.

As the centre of the pays the market towns of rural England attracted all classes from the countryside around, although not always on the same occasions. By doing this they provided, for different groups, different senses of community. In the areas where hiring of labour by the year remained the main form of hiring, which was true for much of the north until the Great War, the annual or twice annual hiring fair provided a focus and a definition. In the East Riding, for example, although men moved farm frequently they tended to stay within one 'hiring area'.[181] In Northumberland the farm books of the Rutherfords' farm near Seaton Sluice show that the hinds from the farm always went to the same hirings at Morpeth, about six miles to the north-west.[182] Again the border was of little importance, with Scottish hinds hiring at Wooler (seven miles into England), and English hinds moving the same kind of distance into Scotland to hire at Kelso.[183]

Others moved across their pays for less utilitarian reasons. The young Joseph Ashby walked the nine miles from Tysoe, the northern edge of 'Banburyshire', to Banbury with his mother, for the annual purchases of calico and boots after harvest, a journey reproduced in every arable shire of England.[184] At Horsham in Sussex there were two fairs, one after harvest in early November and one in July between the hay harvest and the corn harvest. Then, according to Henry Burstow, 'the country people flocked into the town by hundreds and thousands'. The 'business part of the fair was confined to one day only, but the pleasure fair lasted any number of days from three to nine'.[185] In Norwich the end of harvest attracted the rural population from the area around to spend their 'largesse', the traditional if vanishing end-of-harvest collection. 'That harvest time is well nigh over may be concluded from the rustic visitors who now roam about the streets of Norwich gaily dressed, rejoicing in

[181] See Stephen Caunce, *Amongst Farm Horses*, p. 9.   [182] NRO. Rutherford Papers.
[183] Orr, 'The Border farm worker', pp. 79–81.
[184] Mabel K. Ashby, *Joseph Ashby of Tysoe* (Cambridge, 1961), pp. 26–7.
[185] Henry Burstow, *Reminiscences of Horsham* (Horsham, 1911), p. 71.

the trams, and gasping at the wonders of the Arcade and the brightest shops.'[186] Even academic Oxford forgot the ivory towers at the end of harvest and gave itself over to the county around for St Giles's Fair. Here, from the late 1850s, the railways added to the crush bringing up to 10,000 extra passengers a day to the Fair.[187] However, it was still the waggoner's cart which brought most from the villages around. Henry Taunt, the Oxford antiquarian, wrote a dialect poem published in the 1890s, which was a list of the villages in a circle of about twelve miles around. The verse about the 'eastern segment' goes:[188]

> Us cums from Head'nton Quarry, an' sum from Shotover Hill,
> An' sum on us cums fru' Whaately, an' sum all th' way fru' Brill,
> Fru' Milton and an fru' 'Aasley, from Stad'am an' Talmange Stoke,
> Frum out o' th' Otmoor country, an' t' others they cums frum Noke
> We all onus cums to Auxford, brought in by th' ould gray mare,
> 'Tis only wunce a year us cums, an' that's's to St Giles's Faier.

The market towns and county towns 'us' visited were, by the 1860s and 1870s, not backward survivals of an older urban world but successful and expanding consumer markets. Although many 'marginal' and smaller markets, for example Steyning and Petworth in Sussex, or Caistor and Market Deeping in Lincolnshire, declined, others increased in their importance, 'as the countryman looked to the towns to meet more of his needs'.[189] Although growth in terms of numbers was not as great as in the industrial areas of the north or, especially after the 1860s, in the suburbs of London, towns like Chichester in Sussex and Maidstone in Kent saw population growth of 58 per cent and 74 per cent respectively in the hundred years after 1831.[190] In Yorkshire in the same period Hey writes of the ability of the market towns 'to take advantage of improved communications and to prosper in a modest way'.[191] The real changes, though, were not in terms of population but in what the town offered. An increasing tendency away from village-based crafts in all but the most remote area like rural Wales, and a concomitant growth in factory production of necessities like shoes and clothing as well as luxury goods, brought more and more of the 'benefits' of industrialisation into the rural areas. By the 1860s Boston in Lincolnshire as well as the county town had department stores; while new, specially built shops with plate-glass windows were increasingly an ordinary part of market towns by the

---

[186] *Eastern Weekly Press*, 19 Sept. 1900, p. 4.

[187] See Sally Alexander, *St Giles's Fair, 1830–1914* (Oxford, 1970), pp. 20ff.

[188] Quoted in *Ibid.*, p. 20.

[189] Brandon and Short, *The South East*, p. 308; Neil R. Wright, *Lincolnshire Towns and Industry, 1700–1914* (Lincoln, 1982), p. 224.   [190] Brandon and Short, *The South East*, pp. 307–10.

[191] Hey, *Yorkshire*, p. 286.

1890s.[192] The Lincoln store, Bainbridges, boasted in the 1890s of its 'many departments, each being well ordered, and having its own staff of assistants, whose duties are confined to the one branch alone. These (or the principal ones) are silks, mantles, millinery, dresses, hosiery, gloves, carpets and furnishing goods, etc., etc.'[193] Even smaller towns in the eastern counties like Newmarket, Chelmsford or Colchester had a considerable range of shops both departmental and specialist. Cole and Co. of Newmarket were complete house furnishers, bedding manufacturers, carpet warehousemen, painters and decorators, house agents and removal men.[194]

For most of the rural poor and even middling sort on their annual visit to Norwich or Lincoln, shops like these were simply to be stared at during 'a crowded hour of glorious life in the city'.[195] There were, however, an increasing number of shops which aimed both to present a 'smart' façade and to supply working men and women. Edwin Durham, 'merchant tailor etc' of Chelmsford, boasted that, as well as a frontage of 'upwards of seventy feet, the whole length of which is occupied by fine plate-glass windows', his main department was the one 'devoted to the sale of ready-made clothing'.[196] Similarly in Colchester, Mr Arthur A. Dyer's did not have a smart front, indeed according to his advertisement, 'no display is made in the windows', but he was 'the cheapest house in town for new and second hand furniture' and specialised in linoleum, that still new floor covering of the respectable working class.[197] In Lincoln such an appeal went even further with C. Whear and Co., 'The People's Tailors' of Silver Street, 'an establishment where the middle and working classes could have the advantage of inspecting a large and various stock of high class goods, and purchasing at prices considerably lower than those charged in many similar places for inferior articles'.[198]

There were, however, many who visited the market and county towns to whom such considerations were irrelevant. Although the aristocracy and sections of the gentry were, by the 1850s, accustomed to shop in London, and deal direct with firms like Fortnum and Masons even for food, they did patronise local shops, and, when they did, increasingly demanded something approaching metropolitan standards. Additionally, the lower gentry and the large tenant farmers provided a regular and wealthy clientele. For groups like this the market and especially the county towns of rural England produced a varied and surprisingly sophisticated range of goods and services, which suggest that notions of the domestic simplicity of rural life need serious modification. Even in a relatively small town like Colchester, albeit a garrison town, the range avail-

[192] Wright, *Lincolnshire*, p. 217.
[193] *Industries of the Eastern Counties. Historical, Statistical, Biographical* (Birmingham, n.d. but *c.* 1890).
[194] *Ibid.*, p. 141.   [195] *Eastern Weekly Press*, 27 Sept. 1902, p. 4.
[196] *Industries of the Eastern*, p. 163.   [197] *Ibid.*, p. 181.   [198] *Ibid.*, p. 32.

able, particularly of foodstuffs, is very striking. Evatt Sanders and Son, 'The Household Supply Stores' of High Street, had direct contacts with the Continent and had warehouses in London and Harwich as well as Colchester. As a result their stock in the 1890s would do credit to a very good modern food shop. They made their own Cheddar and sold Derby and Stilton. More surprisingly, they also imported and sold American, Dutch and other cheeses, including Gorgonzola, Gruyere, Parmesan and Camembert. They imported, roasted and sold their own coffee, blended their own tea, and imported and packed their own spices.[199] The upper classes of the rural areas also found their equivalents of 'The People's Tailors' in firms like Alfred Welch, Military Tailor and Outfitter of St Botolph Street, Colchester. Here all the work was bespoke and Mr Welch informed would-be customers that 'this business has been established for many years and has, from its first inception, been extensively patronised by the elite of the town and county, the officers of the garrison, and members of the various hunts'.[200]

As well as being the centre of a rural pays, market and county towns had permanent and semi-permanent populations of their own. Even in the 1850s some market towns supported a local season when the gentry, large farmers, and even occasionally the local aristocracy came to live briefly in their own or rented town houses and created around them a small social whirl. Nathaniel Paine Blaker wrote about the end of this style in Lewes, Sussex, in the early 1850s:[201]

Lewes in former times, as the County Town, was a rather festive place. Balls and parties were frequent. The County Ball, an important function, was held at the County Hall, and there were several other balls during the season . . . There were [also] a great many dinner parties in Lewes in those days, and men drank rather freely.

In larger towns the season was more structured and survived longer. Lincolnshire Society had in the 1880s a clearly demarcated season in the autumn. During these few weeks, which began with Lincoln Races, the aristocracy and gentry travelled from their country estates to Lincoln, and organised a series of parties, plays and balls which centred on the Stuff or Colour Ball at the county assembly rooms. However, as the century progressed, even in a large town like Lincoln, the importance of this social season declined, especially since the aristocracy were tending more and more to spend autumn in Scotland or on the Continent. By the turn of the century it had all but vanished.[202]

There remained the permanent inhabitants of the market and county

[199] *Ibid.*, pp. 194–5.   [200] *Ibid.*, p. 197.

[201] Nathaniel Paine Blaker, *Sussex in Bygone Days* (Hove, 1919), pp. 115–22.

[202] Olney, *Rural Society*, chapters I and IX. See also Sir Francis Hill, *Victorian Lincoln* (Cambridge, 1974), pp. 70–9.

towns. These were in essence created to serve the pays which surrounded them, even if, as they grew, they created in turn a market of their own. Like the country around them, although more obviously, the market towns were rigidly divided creating within them communities of class power and status. The divisions of status and class tended to be reflected in the topography of the town with different areas creating different types of community. At the bottom were those communities, familiar in most nineteenth-century market towns, like the one described on the outskirts of 'Casterbridge' by Thomas Hardy:[203]

Mixen Lane was the Adullam of all the surrounding villages. It was the hiding-place of those who were in distress, and in debt, and troubles of every kind. Farm-labourers and other peasants, who combined a little poaching with their farming, and a little brawling and bibing with their poaching found themselves sooner or later in Mixen Lane. Rural mechanics too idle to mechanize, rural servants too rebellious to serve, drifted or were forced into Mixen Lane . . . Much that was sad, much that was low, some things that were baneful could be seen in Mixen Lane.

Few towns in rural England of the 1850s and 1860s lacked a version of Mixen Lane. It was, like Hardy's, often on the outskirts, a medieval squatter settlement outside the town walls. Pockthorpe outside the St Andrew's Gate of Norwich was like this. William Lee's report on sanitary conditions of Norwich gives a harrowing description of Pockthorpe. There was no lighting or water supply and the walls of over-crowded and semi-derelict houses rose from a stagnant ditch a quarter of a mile long and filled with the 'refuse from the barracks, containing when full from 200 to 300 persons, and probably 200 horses'. Cholera, influenza, typhoid and whooping cough were endemic. Even the widely experienced Lee found it 'one of the most horrible places I have ever seen'.[204]

Other areas like this were in the heart of Victorian country towns where old and formerly grand buildings had decayed and been subdivided, and then infilling had taken place around them. Paradise Square in St Ebbes in Oxford was such an area as were parts of St Thomas's parish in the same city and the courts off the central streets. Many of them lay next or close to respectable and even wealthy housing. John Moore gives a graphic account of such a 'street' in his autobiographical novel about Tewksbury, *Portrait of Elmbury*:[205]

[203] Thomas Hardy, *The Life and Death of the Mayor of Casterbridge. A Story of a Man of Character* (London, 1924), p. 307.

[204] Neil MacMaster, 'The battle for Mousehold Heath, 1857–1884: "Popular politics" and the Victorian public park', *PP*, 127, 1990, p. 123.

[205] John Moore, *The Brensham Trilogy. Portrait of Elmbury* (Oxford, 1985), p. 6.

... even among Elmbury's slums, Double Alley was something to be wondered at. Respectable women drew their skirts closer about them as they passed its nauseous opening; even the doctor and the priest were unwilling adventurers on the rare occasions when they were summoned to visit it; and policemen, who were more frequent visitors, took care to go in pairs when their duties took them there.

These rural slums were often the source of social disruption. Pockthorpe was the site of a fierce and long battle against Norwich City's attempts to 'civilise' Mousehold Heath by turning it into a public park.[206] From Cliffe, an extra-parochial settlement on the marshes beyond the eastern walls of Lewes, issued the fiercest and most independent of the 'Bonire Boys' whose organised chaos took over the whole town every year on 5 November.[207] Even 'Mixen Lane' or 'Double Alley' produced constant disturbance with their drunkenness, petty crime and insistence on old rights like the carrying out of the Skimmington Ride. For these reasons, as well as a changing social climate from the 1860s and 1870s onwards, town councils and social improvers began the clearance and improvement of the worst of these areas. In Oxford, for example, it was written in the 1900s that[208]

on the whole conditions are clearly improving ... Fifty or sixty years ago most of the present houses and streets which now give trouble to all the local relieving agencies ... were in existence, most of them in a very much worse condition, by all reports, than at present.

After the Public Health Act of 1872 even small towns were forced to ensure that new housing was built to a minimum standard, while before that date local Boards of Health had, in some towns, demolished the worst of these semi-urban slums.

Above those who lived in 'Mixen Lane' or Pockthorpe, although often mixing with them, were the craftsmen and tradesmen who served the rural community. These were, according to Chalklin, between them by far the largest group of the population of country towns. In Ashby de la Zouche in 1861, for instance, craftsmen accounted for 35 per cent of the working population while tradespeople accounted for 13 per cent. The majority of these were small producers of clothing and shoes. At Tadcaster in the West Riding a third of all those described as craftsmen were in these areas.[209] However, as we have already suggested, a huge range of goods and services was available in even the smallest towns, and

---

[206] MacMaster, 'Mousehold Heath', *passim*.

[207] James E. Etherington, 'The community origin of the Lewes Guy Fawkes Night celebrations', *Sussex Archaeological Collections*, 128, 1990, pp. 195–224.

[208] C. Violet Butler, *Social Conditions in Oxford* (London, 1912), p. 109.

[209] C. W. Chalklin, 'Country towns', in Mingay (ed.), *Victorian Countryside*, vol. 1, pp. 280–1.

this diversity increased as the century went on. It increased, though, at the cost of the local craftsmen. In the same way that village crafts tended to decline in the face of opposition from larger urban centres, so, in turn, the trades of these centres came under attack as industrialisation moved into other areas. As Chalklin writes, 'towards the end of the nineteenth century more and more factory-made boots, watches and clothes were being sold'. As a direct result[210]

Nearly all the traditional crafts of the country town still existed at the beginning of the twentieth century, but they were drawing increasingly on the factory or large scale industry to supply half finished goods – iron axles, saddler's ironmongery, lead piping, sawn planks and sewing thread – and employment opportunities for young townsmen in these trades were becoming more and more limited.

At the top of the social scale, on the other hand, there seems to have been an expansion in the years after 1850. This was in part due to the growth of the local state and the beginnings of welfarism, which is dealt with later, and which created a whole range of local government officers and employees, especially after 1870 with the growth of public education. In other professions it was not so much an expansion of numbers, although that took place, as a recognition of professional standards. As Mingay writes, 'the later nineteenth century saw a remarkable growth of education and training, the setting of standards, and the recognition of qualifications, all of which became the hallmark of the professions'.[211] The Law and the Church were of course established and recognised before the nineteenth century but the professionalisation of medicine, veterinary science, land agency and the auctioning of property, as well as commercial services like banking and insurance, were all products of the nineteenth century and many of them of the years after 1850. This created in the county and market towns of rural England a powerful middle-class élite of the kind described by Moore[212]

The others [of my family] prospered moderately and lived long and respectable lives. They were all large and substantial, rather like family portraits come to life: Uncle Reg the doctor; Uncle Jim the lawyer; Uncle Tom and my father, the auctioneers. They sat together upon the Town Council; they took it in turns to be Mayor, and Chairman of the local Conservative Association; they administered charities and trusts with meticulous care; they shared a monopoly of the post of churchwarden of the Abbey. The editor of the local paper had little trouble when they died; the same obituary notice, with a few trifling alterations would serve for all of them. 'He played a prominent part in Public Life.' And that indeed was their tradition; so long as the public was not too large.

---

[210] *Ibid.*, p. 282.   [211] G. E. Mingay, *Rural Life in Victorian England* (London, 1977), p. 167.
[212] Moore, *The Brensham Trilogy*, p. 23.

Yet the golden age of the market town, like the golden age of its characteristic product, the local newspaper, was not long. We have already noted that even in the 1850s many smaller market towns were suffering from competition from larger nearby towns and this process continued throughout the period. As Jonathan Brown writes, 'the great amongst provincial towns tended to become greater; some of the middling towns of 1801 were growing steadily by 1901; but large numbers of humbler places stagnated'.[213] In the years after 1870 this became worse. As agricultural incomes declined and the process of rural depopulation intensified, the money and the customers available to each of these towns grew less and less.

In addition, the new ranges of goods which so attracted visitors and customers to markets were increasingly produced away from their traditional centres even if they were not always 'factory made'. Further, many other trades became concentrated, like the shops, in the larger settlements. As a result regional centres like Oxford, Reading or Lincoln expanded at the cost of their smaller neighbours. To quote Brown again:[214]

The towns which continued to expand until 1914 were those at which the markets, shops and rural services became concentrated. They were usually the ones with stronger industries, but even so few country towns grew at a rate equal to the national average . . . Those market towns that matched the national rate of growth were those to which a considerable amount of new industry had been attracted during the later nineteenth century.

Market towns created a community both within themselves and with the countryside around, which was reinforced by developing social and political changes after the 1860s. Change over time was crucial here, and we shall return to that shortly, but finally I want to look at the most nebulous 'communities' of all, those defined by cultural practices.

V. BOUNDARIES AND COMMUNITIES

The 'community' forms dealt with in this section, the farmstead, the hamlet, the village and even the market town and its *pays*, retain clear similarities. Although the edges may be sometimes blurred, they are still physical entities and are recognised as such. A Northumberland hind may not have distinguished between England and Scotland, may never have lived in anything which a southerner would have recognised as a village, but north or south of the border he would still have recognised a farm town, with its comfortable farm house, its steddings and its hinds' cottages.

[213] Jonathan Brown, *The English Market Town. A Social and Economic History, 1750–1914* (Marlborough, 1986), p. 117.   [214] *Ibid.*, pp. 118–19.

There were however, especially in the upland areas, though also in the areas of village settlement, quite other ways of defining community, which, while they had a physical dimension, had more to do with relationships of inclusion and exclusion. In short, as Pahl so eloquently pointed out, 'any attempt to tie patterns of social relationships to specific geographical milieux is a singularly fruitless exercise'.[215]

In his study of Whalsay in Shetland, Anthony Cohen argues that 'the depth . . . of belonging is revealed in the forms of social organisation and association in the community'. In Whalsay the key elements which are seen by Whalsay people themselves as defining a community position are 'largely pragmatic social constructs – principally of kinship, neighbourhood and fishing crew (work group)'. However, these were, and are, active definitions; membership of a community is given by moving out from these bases into active relationship with other members of these groups, 'as providing a means whereby the individual belongs to the whole'.[216]

These kinds of category are particularly useful when we come to look at marginal areas and/or non-capitalist forms of social organisation. Mr Price of Maesgwyn in Radnorshire, whom we have already met, described how neighbourhood coupled with work group functioned as an organising principle in his area. 'We help one another too . . . often lend one another a team; and in the case of small farmers, they will often go out together, and get the harvest in on one farm first, and then off to the next.' Or, from the same area, 'a farmer in Old Radnor, who farms 90 acres, of which half is arable, with only two men living in the house, if he wants another hand in the summer gets one of the neighbouring lime-burners to come out. One of them will often borrow his team in winter, saying if he will lend it he will do a day's work for him in the summer.'[217]

This work group was defined both by 'neighbourliness', how near you were to one another, and by status. In Radnorshire there were men who were 'half farmers, half labourers . . . [who] will work a day or two on their own land, and the rest of the week at 2s. a day for some larger farmer'.[218] In Cardiganshire a complex system of obligations was based on labour debt. This was incurred by the small farmers and cottagers, setting out rows of potatoes on the land of the larger farmers. This labour debt, which was based in turn on neighbourhood, was the basis of the work group. This was, according to David Jenkins, the key group:[219]

[215] Ray Pahl, 'The rural–urban continuum' in *Readings in Urban Sociology* (Oxford, 1968), p. 293.
[216] Anthony P. Cohen, 'A sense of time, a sense of place: the meaning of close social association in Whalsay, Shetland', in Cohen, *Belonging*, pp. 21–49.
[217] *RC on the Employment of Children*, pp. 62–3.   [218] *Ibid.*
[219] Jenkins, *Agricultural Community*, p. 55.

The work group was . . . the connecting link between farmers and cottagers, and perhaps above all else it was the feature of the life of that time which caught people's imagination. Co-operative work at harvest provided occasions for good fellowship, a sense of shared endeavour and price in attainment, and a feeling of companionship and personal 'nearness' which gave a deep satisfaction to those who worked together.

Elsewhere in Wales kinship could perform a similar role in providing extra hands when needed, especially when coupled with industrial work. Mr John Jones, a small tenant farmer from the Vale of Neath, told the Royal Commission on Land in Wales, 'we are in the neighbourhood of several collieries and works, and people earn money in the works and they have four or five sons, and they take a farm up and they actually pay for it from the works'.[220] Mr Cadwaldr Roberts, also a tenant farmer but from north Wales, described a similar system:[221]

In my father's time I worked at the quarries, and used to hand over the whole of my wages to my father to assist him to live and pay the rents and rates . . . Besides this in the summer months I used to work upon the farm, after returning from the quarry, until it was too dark to work.

This coming together of kinship, neighbourhood and work group as definitions of community was recognised by the rituals of community. For example in Cardiganshire and Carmarthenshire the work group, created from neighbours and kin at harvest and haymaking, was fed communally not only during those tasks but also at Christmas. These meals served both to cement the group and to define its limitations and ranks within it. The larger farmhouses, for instance, had a board room (*rhwm ford*) where the labourers ate along with the hired servants and the children. The master and the mistress ate in a separate room, the best kitchen (*cegin orau*), along with the more superior members of the neighbourhood, for instance, a visiting seamstress. On the smaller farms status was marked by different tables. Different food was served at different tables, especially different bread, white being served to the master and mistress and 'mixed' for the servants.[222] These mutual obligations continued throughout the year. Regular supplies of buttermilk from the farmer to the poorer members of the work group throughout the year, straw for the pigs' bedding, even curds and whey to make cheese, cemented the bond created by labour debt and the setting out of potatoes.[223]

Nor was it only in Wales, in a peasant area, that these kinds of bonds

[220] *RC on Land in Wales*, p. 90.   [221] *Ibid.*, p. 432.

[222] S. Minwel Talbot, 'Liberality and hospitality. Food as communication in Wales', *Folklife. A Journal of Ethnographical Studies*, 24, 1985–6, pp. 32–44. See also Trefor M. Owen, *Welsh Folk Customs* (Cardiff, 1959), pp. 113–19.

[223] *Ibid.*, p. 43; See also Jenkins, *Agricultural Community*, pp. 57–61.

existed, although it was here that they were most developed. Mick Reeds' important work on Sussex and Surrey shows how similar work-group bonds developed in economically marginal sectors within a largely capitalist, agricultural area to create something approaching a community within a community. In his picture of Phillip Rapson, sawyer, cottage landlord, cider maker and small-scale agriculturist of Lodsworth in Sussex we have an account which sounds more like Wales or Ireland than the supposedly market- and class-orientated south-east. Rapson created around him a network of neighbourhood and kinship bound together by work and labour debt, and paid in kind more often than in cash:[224]

Debts for these goods were paid in cash, by work, or by provision of the goods. Apples were supplied for cider, dung supplied by one tenant, pigs were fattened and given to the landlord, while in 1826 Rapson received thirteen geese from his son Anthony in lieu of rent.

Similar patterns were common in northern upland areas. Donajgrodski has shown that in Swaledale well into this century a local consciousness was preserved and developed around neighbourhood, kinship and work group. Here, as in Wales, family members worked away, sending money home, or bringing it home from the coal pits, while the family holdings were worked by the women. Here, because of a very precise geographical locality ('th' dale'), the neighbourhood and work group came together with a spatial dimension:[225]

Older Swaledalians often talk of places in the dale almost entirely in terms of people who lived there, their characteristics and who they were related to – a human landscape as vivid as the geographical one. It was a conceptual structure which expressed their distinct position and reinforced solidarity.

Further north it seems possible that a similar situation existed among the lead miner/farmers of Allandale who worked part of the time in the mines and part on their small farms and almost certainly relied on family and work-group support.[226]

Even on slightly larger farms the records (which are not usually good on these matters) give tantalising glimpses of work-group and neighbourhood structures. The farm records of James Griffin of Whitehall Farm, a holding of 133 acres at Luppitt, near Honiton in Devon, show a complicated work group operating between him and other farmers in the area.

---

[224] Michael Reed, 'Social and economic relations in a Wealden community: Lodsworth 1780–1860', MA Thesis, University of Sussex, 1982, p. 62. See also his 'The peasantry of nineteenth-century England: a neglected class?', History Workshop Journ., 18, 1984, pp. 53–77.

[225] A. P. Donajgrodski, 'Twentieth-century rural England: a case for "Peasant Studies"', Journ. of Peasant Studies, 16, 3, 1989, p. 435.

[226] See for example NRO. NRO T/132 acc no 1690 Interview with Mr J. H. Reed, Allandale farmer.

At threshing, as in Wales, other farmers from the neighbourhood, especially smaller farmers, shared in the work, probably in exchange for
getting their own little harvest threshed. Larger farms also sent men to
help with the threshing and this was reciprocated by Griffin. In 1895
James Griffin had help with threshing from five extra men, two of whom
were small farmers, including his own brother who farmed 33 acres
nearby. In return, three days later, two of his men were away threshing at
Fred Daniel's farm. In June the same year one of his men was away cutting
grass for Mr Harris, a small farmer of Luppitt.[227]

There were many other boundaries which delineated the community
for those who lived in them. Many, as we saw at the beginning of this
chapter, lacked even the functional precision of neighbourhood, kinship
and work group. The 'real Elmdoner' of Strathern's studies, while recognisable to all within his or her group, is and was much more difficult for
the historian to recognise. Nevertheless, they are clearly there. A sense of
belonging, or place, was clearly a central part of the social being of
English countrymen and women in the nineteenth and early twentieth
centuries. Yet we must be cautious for belonging was exclusive as well as
inclusive and, despite idyllic fantasy, the nineteenth-century countryside
was deeply divided in a number of ways.

The most important of these divisions was class. Class clearly has
within it some of the same feelings and idea as 'belonging' or 'community', yet crucially for us its distinctive form is that it stands against community, though not necessarily against the 'local'. What E. P. Thompson
wrote years ago remains central here. 'Class happens when some men, as
a result of common experiences (inherited or shared), feel and articulate
the identity of their interests as between themselves, and as against other
men whose interests are different from (and usually opposed to) theirs.'[228]
This shared identity need not move outside a small community, and
indeed I have argued elsewhere, as has MacMaster, that a 'class consciousness' of this kind is perfectly compatible with a 'localised' world
view on many occasions.[229] However, it is still a challenge to notions of
community which are all embracing. Nor is class consciousness of this
kind the prerogative of any one class, and we shall return to this in the
next chapter. But briefly it is clear that the aristocracy and the large tenant
farmers both had identities of interest with other members of their
respective classes which cut across the local boundaries of any of the geographically based communities we have talked about.

Nevertheless class, and especially class-based organisations, challenged

---

[227] Devon Record Office (Exeter), (hereafter DRO) DRO 337 B.Add 2 SS1. Dunning Bicknell Mss.
[228] E. P. Thompson, *The Making of the English Working Class* (Harmondworth, 1968), pp. 9–10.
[229] Howkins, *Poor Labouring Men*, Chapter 2; MacMaster, 'Mousehold Heath', pp. 153–4.

the notion of community with a different perspective, which frequently became 'national'. For instance, when a farm worker joined a trade union, as many of them did after 1872, he disrupted the local world and the local view by bringing into it a national one – he had an identity both as a trade unionist (which could be a class identity) and as a villager of 'X'. This did not need to conflict but often did. Nor was it only trades unions which caused this kind of conflict. Friendly Societies, especially the national affiliated orders, did exactly the same thing. The young Joseph Ashby met this when he was urged by the vicar of Tysoe to join the Tysoe Village Club. 'It would be interesting to find out what some of his old friends would say if he were to join; he knew they were Foresters.' His 'old friends' were firm. He should not join the Vicar's Club as it was run by the trustees who were all gentry, 'as good as to say a labourer's got no sense. Why can't the members manage their own money?' And there was more, 'the boys wouldn't be in Tysoe all their lives. Go where they might, there would be a Foresters' Lodge, and their membership would be transferred.'[230] Other institutional forms could break the communities' bonds. Membership of a nonconformist sect both challenged the idyllic world view of the village and linked its members to a wider world through the district and national dimensions of the organisation.

However, none of these were unproblematically anti-local or even anti-community. The village club day, for example, could be organised by a national affiliated order like the Oddfellows or the Foresters, yet it remained, at another level, a village celebration. It was the day, for instance, when children came home from service or working away to their own village, reaffirming locality and community. As George Swinford wrote of his village, Filkins in Oxfordshire, 'Club day was when members who had left the village and gone away to work, came home to see their parents. It was a day to look forward to, as for many it was the only time in the year when they saw their friends.'[231] Even trades unions' apparently unproblematic, nationalising class forces, were more complex than that suggests. The vision of Joseph Arch was of a *national* union with a national headquarters, a national organisation, a national newspaper and three-farthings in the penny going to the headquarters at Leamington. To many members after the initial enthusiasm of 1872–4 this was less obvious. The Oxford District of the National Union voted in September 1878 that 'this committee are forced to come to the Conclusion that some steps must be taken to so alter or amend the con-

---

[230] Ashby, *Joseph Ashby*, p. 71.
[231] George Swinford, *Jubilee Boy. The Life and Recollections of George Swinford of Filkins*, ed. Judith Fay and Richard Martin (Filkins, 1987), p. 81.

stetution of the union so that the Dists. will have more controle of the funds raised therein' (as original).[232] By then Norfolk had gone even further and split with Arch on localist grounds. As 'An Old Member of the Labourer's Unions' put it in an open letter to Joseph Arch, 'we shall not forsake our long-tried and proved leaders to please your whims and fancies. We have summered and wintered them long before we heard of your existence.'[233] George Rix, the district secretary who led the split, wrote, 'centralism is Toryism rank and rife. Federation is Radicalism pure and simple.'[234] Elsewhere, particularly in Wales and the north, it was clearly the regional nature of the farming system which modified the development of class in particular ways.[235] All in all the clash of community and national cultures is clearly complex and, although we will return to aspects of it in the next chapter, is outside the scope of this work.

VI. CHANGE AND THE COMMUNITY

This section is written as if the community structures and boundaries remained constant through the period from 1850–1914. Clearly this is not the case, and in some of the sections I have shown ways in which things changed. However, some general points should be made. Throughout the period there was, even in the most remote areas, a move away from isolated and culturally separated communities. Gradually at first, but with gathering speed, a 'national' community was imposing itself, or being imposed on a variety of local and regional structures. However, this had not gone as far by 1914 as many historians (and contemporaries) thought, and local and regional distinctiveness remained, indeed to remain, vital and important.[236]

More specifically, the categories of 'open' and 'close' declined in importance after the changes in chargeability in the 1860s, though as ideological constructions they continued to dominate discussion until the mid-1890s. They survived even longer in folk memory where certain villages classified as 'rough' well into the 1950s were often former 'open' settlements. In the upland areas and in Wales the structure remained unchanged longer. It was not until the inter-war period that serious modifications of the upland community structure became apparent, and living-in continued in the north until the Second World War.

Trends towards the 'nationalisation' of cultural and economic structures were also resisted, not only on the basis of genuine local consciousness like that of George Rix or the Oxfordshire trades unionist of the

---

[232] Pamela Horn, ed., *Agricultural Trade Unionism in Oxfordshire, 1872–81*, Oxfordshire Record Society, XLVIII, 1974, p. 102.   [233] *Eastern Weekly Press*, 8 February 1879, p. 1.

[234] *Ibid.*, 22 February 1879, p. 1.   [235] Howkins, *Reshaping Rural England*, Chapters 2 and 7.

[236] Cohen, *Belonging*, Chapter 1.

1870s, but as part of an attempt to 'rediscover' a rural idyll in which the local could be counterposed as part of national identity with ideas of country and even Empire.

A key element here was the growth in country literature, and especially popular country literature after 1880. The countryside has, of course, always been a source of literary and artistic inspiration, but in the years after 1880 it took a particular form.[237] Briefly, from those years onwards there was an increasing sense of urban crisis. Booth's work on London and Rowntree's on York were followed by other city studies which showed that even towns like Cambridge, Norwich and Oxford hid 30 per cent of their populations who lived below the poverty line. In this situation many younger writers and thinkers drank deep of William Morris and country rambling, especially in southern England, and came to the conclusion that country life and country ways were superior. Additionally there were others, like Henry Rider Haggard, who linked Britain's military and industrial decline in the face of American and German competition to theories of racial decay. These ran together a crude social-Darwinism with notions of urban life and urban values to produce a theory which suggested that rural life was superior. Kipling put it clearly in his bitter poem about the Boer War, 'The Islanders', written in 1902. In this he links urban life to Britain's failures in South Africa and her reliance on the rural populations of Australia and New Zealand to save her in her darkest days.[238]

Yet ye were saved by a remnant (and your land's long suffering star)
When your strong men cheered in their millions while your striplings went to
the war.
Sons of the sheltered city – unmade, unhandled, unmeet –
You pushed them raw to the battle as ye picked them raw from the street
... And you vaunted your fathomless power, and ye flaunted your iron pride,
Ere – ye fawned to the Younger Nations for the men who could and ride.

Whatever the source of their initial inspiration, a remarkably uniform picture of rural England and rural life emerged from these writers and it is a picture which still shapes many views of late Victorian and Edwardian England. Theirs was a land of the south, of village settlement, village greens and half-timbering. It was a worked landscape, not a barren romantic one, and the work, particularly country crafts and aspects of farm labour, were given a timeless honesty which contrasted with the new and vulgar of the industrial and urban world created by the nine-

---

[237] For much of what follows, see Robert Colls and Phillip Dodd, *Englishness. Politics and Culture, 1880–1920* (London, 1986), especially, Alun Howkins, 'The discovery of rural England', and Peter Booker and Peter Widdowson, 'A literature for England'.
[238] 'The Islanders', *Kipling. A Selection by James Cochrane* (Harmondsworth, 1977), p. 117.

teenth century. Within these crafts was hidden a tradition which had held together England in the past and which could, if recognised, do so again. As George Sturt wrote:[239]

Tradition is a form of group life. It tends to composure and conservation in individuals. No industry can long go on without it. Hence it prevails more in the country than in London, where individuality most flourishes. But to name the many ancient industries is to recite the group efforts of the English from time immemorial.

Particularly in the 'Georgian' years on the eve of the Great War this idyllic view of rural life, culture and community was reproduced time and again in essays, novels, poems and prose. Through the work of W. H. Hudson, Edward Thomas, Hilaire Belloc, the contributors to the Georgian anthologies, English people came to see the 'village' as their real 'home' and the countryside of the south as the real England. No matter how critical a few of these writers really were, and no matter how complex they knew the countryside to be, it was the simplified version which survived and is still reproduced today.

We shall come back to these questions at the end of the next chapter but first we need to look at the social, cultural and domestic life of rural England – to put people into our communities.

---

[239] George Sturt, *The Journals of George Sturt, 1890–1927*, ed. E. D. Mackerness (Cambridge, 1967), 2 vols., vol. II, p. 839.

CHAPTER 23

SOCIAL, CULTURAL AND DOMESTIC LIFE

BY ALUN HOWKINS

The rich man in his castle,
The poor man at his gate.
GOD made them high or lowly.
And order'd their estate.

Mrs Alexander, 'All things bright and beautiful'
Hymns Ancient and Modern
(this verse was removed in the revision of 1950)

INTRODUCTION

When James Caird looked at the people of rural England in 1850–1 he saw them divided into three groups, what he called, 'the three great interests connected with agriculture – the landlord, the tenant and the labourer'.[1] These groups were related in precise, if regionally differentiated, ways to the ownership of the means of production, to power and to cultural forms. They were classes. By the 1850s, in southern and eastern England at least, there had developed distinctive patterns of social, cultural and domestic life which marked them off from one another just as firmly as the number of sovereigns in their respective pockets. This is not to say that in all areas and at all points in time there were no 'grey areas'. The tiny 'round frock' farmers of the Sussex Weald or the hill farmers of Cardiganshire were clearly very different from the great sheep barons of the South Downs or the 'wheatocracy' of Norfolk, yet within their environments they were usually marked off as different from the labourers beneath them and the landlords above them. In Cardiganshire, for example, David Jenkins shows how the 'diacritical use of names' indicated with precision the divisions within a society which, at first glance, seems relatively homogeneous apart from a basic division into landlords and the rest.[2] At the top stood *yr gwyr mowr* (the great

[1] James Caird, *English Agriculture in 1850–51* (London, 1852), p. 520.
[2] Jenkins, *Agricultural Society*, Chapters 1 and 2.

people or gentry), then came the *ffermwyr* (the farmers), who were further divided by names indicating the number of horses required to work the farm, hence *lle par o geffyle* (a one pair place), or, at the bottom end, the number of cows *lle buwch* (a cow place). This group had really ceased to be farmers at all and become *obol tai bach* (the people of the little houses). As Jenkins notes of this crucial division, 'well within living memory school children in some Teifi Valley schools divided "naturally" into *plant ffermwyr* (farmers' children) and *plant tai bach* (children of the little houses) for team games in the playground'.[3]

Below the *obol tai bach* (people of the little houses) but also part of them came those who had nothing to sell but their own labour. Even here there were divisions. The 'superior' workman was the *gwas* (the servant), a single man who hired by the year and lived in on the farm. As in parts of England his work was with horses, 'the highest work of the farm'.[4] Some *gwas* (servants) could be small farmers' sons waiting time and earning money until they could get their own places, but the majority were the sons of labourers (*gweither*) who would usually become labourers on marriage. This meant leaving the farmhouse and yearly hiring for a cottage, often a cow place and weekly labour at the 'lesser' work of the farm, what was called 'pick and shovel work'. Finally, there were those 'outside' the system: casual workers, often skilled men who worked at thatching, or clamping potatoes and harvest. These were called *dynion hur* (hired men). This gave a clear hierarchy of prestige with the family and servants at the top, then the regular labourers who were often known by the name of the farm, and finally the hired men. 'This order of precedence was given form on such occasions as the harvest which brought the farm's workers together.'[5] There was another element – that of gender. Women had a similar hierarchy headed by the mistress of the farm. Then there were maids who hired by the year and lived on the farm. The first maid was connected with the dairy and was the most skilled and senior. The second maid was employed outside and was known until the 1880s as *morwyn llaw sgubor* (the barn hand maid).[6]

From these examples it is clear that naming and language explained real differences in what to outsiders looked like a unitary system. Simplified, it is also clear that Caird's three classes, landlord, tenant and labourer, are present in west Wales. Although the distinctions between them, or between the smallest farmers and the largest cottagers, were very slight, it is clear that culturally there were differences which were recognised by those who lived in that society and which were based, in the end, on a relationship to economic power. That this relationship was complex, and produced not a simple threefold model of culture and social behaviour but

---

[3] *Ibid.*, p. 12.   [4] *Ibid.*, p. 83.   [5] *Ibid.*   [6] *Ibid.*, pp. 84–5.

a diverse one, should not surprise us, but nor should it obscure the realities of the divisions on lines of class.

There were also changes over time. In the most general sense the divisions into a three-tier model were becoming clearer and harder as the century went on. This was not so much a result of real changes in economic position as slight and marginal changes which were perceived to be great. It is perfectly possible, for example, for economic historians to produce wage- or profit- or rent-figures which show, for instance, that rents were dropping, or even that there was no depression at all. Yet this made little difference to the many if their perceptions were different as they clearly were.[7] A rise in wages of a few pence a week or a fall in the price of meat or bread did not bring the vast majority of labourers nearer to the farmers, let alone the landowners, in economic, social or cultural terms. Yet at this time it is clear from the range of labourers' organisations which appear that many labourers did feel a separation from those above them even if they were 'improving' their relative position.

Similar points could be made about the relationships between farmers and landlords. Again there clearly were changes, but quite how significant they were in terms of individual experience is difficult to gauge. For example, it was the larger estates and those with non-rural resources that fared best in the lean years. Those that went to the wall were often those which stood nearest to the farmers. Also there clearly developed a growing sense of solidarity among farmers, against labourers, but also against landlords, which created a clearer demarcation of farmers as a group. This found institutional expression in the formation of farmer-only organisations from the 1870s, but especially from the 1890s and 1900s. Finally, the accelerated decline, which seems to have taken place in the number of holdings under 25–40 acres after 1880, opened up the gap between the labourer and the farmer by cutting away the first rungs of the farming ladder.

This chapter will look at these groups in terms of their social, domestic and cultural history. It cannot look in any but a cursory way at the detailed relationships between them, but these remain central since it is how such groups interrelated which explains how members of these groups saw one another. As E. P. Thompson says in a different context, 'we cannot have . . . distinct classes each with an independent being and then bring them into relationship with each other. We cannot have love without lovers, nor deference without squires and labourers.'[8] Finally, we will consider the 'social' history of the 'newcomers,' those who moved

---

[7] I write here as a social historian and am aware that this raises many problems. See especially F. M. L. Thompson, 'An anatomy of English agriculture, 1870–1914', in B. A. Holderness and Michael Turner, eds., *Land, Labour and Agriculture, 1700–1920. Essays for Gordon Mingay* (London, 1991).

[8] Thompson, *Making*, p. 9.

into the countryside in ever larger numbers after 1890. We will examine each group in turn, beginning with the élite, in terms first of their domestic life, then their social relations, and finally their cultural and leisure patterns. We will return, in a final section, to relationships between the groups and to change over time.

I. A CLASS APART. THE ARISTOCRACY AND GENTRY

The key to the membership of the rural élite and, once that membership was achieved, the essential business, even the obsession, of its members, was landed property. As Beckett writes: 'almost down to the First World War land was the most important single passport to social and political consideration . . . Land represented not merely wealth, but stability and continuity, a fixed interest in the state which conferred the right to govern.'[9] Landed property was also, almost by definition, rural and created a particular set of values and ideals which dominated élite thinking and conferred territorial and national power. As Léonce de Lavergne, a French agriculturalist, observed in 1855:[10]

A country life is sought after, not only for itself – for its absence of restraint, its comfort, quiet occupation, and domestic happiness, those cherished *penates* of the English – but in addition it gives consideration, influence, power, everything that a man can desire after his first wants are satisfied . . . Everybody would be born in the country, because a country life is the mark of an aristocratic origin.

By the middle of the eighteenth century, if not earlier, those who held the great landed estates were 'an elite of wealth, status and power . . . bound together by a common culture and sociability, and ties of patronage and marriage, by an intricate network of personal, family and official relationships'.[11]

Seen in the most cynical light the family was to the élite both the purpose and the means by which and for which it was preserved and if possible extended. Strategies of marriage and inheritance seem to have been geared to these ends with a bloodless nicety. Even Alan Macfarlane's wholesale dismissal of 'fixed position allotted by birth' as the major determinant of aristocratic union stresses that marriage remained 'a contractual arrangement, which both symbolized social rank and allowed complex conversions and exchanges to take place'.[12] Two key elements

---

[9] J. V. Beckett, *The Aristocracy in England, 1660–1914* (Oxford, 1986), p. 43.

[10] Léonce de Lavergne, *The Rural Economy of England, Scotland and Ireland* (Edinburgh, 1855), pp. 124–5.

[11] Lawrence Stone and Jeanne C. Fawtier Stone, *An Open Elite? England 1540–1880* (Abridged edn, Oxford, 1986), p. 41.

[12] Alan Macfarlane, *Marriage and Love in England, 1300–1840* (Oxford, 1986), p. 260.

(or sets of elements) here were the establishment of patrilineal primogeniture often coupled with entails and settlements as the normal route of inheritance; and the social management of marriage.

The first principle was well established by the middle of the eighteenth century and was all but universal in 1850. As Stone writes:[13]

> The long-term interest of the family . . . as expressed in legal documents such as entails and settlements, demanded passage of property more or less intact to the nearest male relative. Finally, everyone was more or less agreed that the bulk of the estate should be passed intact to a single individual on the principle of primogeniture, the other children, both male and female, being looked after by annuities or cash gifts when they reached twenty-one or at their marriage.

During the eighteenth century this provision for younger sons and daughters, as well as that for the wife, was fixed by strict family settlement. This ensured, although it could be circumvented, that whatever the whims of the heir the rest of the family were to an extent protected. While such clear cultural notions dealt with some problems, they created others. The first was, of course, what happened should no male heir be born or if he should pre-decease his father. In this case, by the nineteenth century, it was established that the title and the estate should go to a younger son, then, if that failed, sideways to a patrilineal nephew. If that failed it went 'upwards' or sideways to a paternal uncle. Although these movements resolved the problem in some respects, they created others. If, for example, the heir was a powerful figure in his own right, he might quit the county, abandon the established family seat, and the name and place would be lost. More serious still was the lack of any male heir, even a 'fictive' one produced by indirect inheritance. Between 1840 and 1880 about 10 per cent of all property transfers took place through women.[14] Here the family name and the continuity of the estate were put in even greater jeopardy. However, outside romantic fiction imprudent marriages were seldom made and a judicious match with a respectable but unendowed younger son coupled with a change of name or hyphenation of two family names was a perfectly satisfactory conclusion to this problem.

The problems of inheritance and the preservation of the family name and estate were of course linked closely to those of marriage. It is clear that by our period marriage was not simply a matter of arrangement. Affective marriage in which the children chose their own partners was the rule, and romantic love was accepted by all (or most) as the key basis for pairing. Nevertheless, the tight social network which dominated aristocratic and gentry life ensured that the choice upon which romantic love could be based was restricted:[15]

---

[13] Stone, *An Open Elite?*, p. 69.   [14] *Ibid.*, p. 74.

[15] *Ibid.*, p. 75. See also the Memoirs of Mary Elizabeth Lucy, *Mistress of Charlecote* (London, 1983), for a marriage, initially at least entirely loveless and based on social and economic 'need'.

Whether marriages are determined by parents or by children, both opportunity – the chance meeting at a reception, a ball or in a hunting field – and inclination – the affinity of cultural like to like – will always make social endogamy the norm. The heir of a squire who marries a milkmaid may occur in fiction, but very rarely in real life, where he is, if carried away by sexual passion, more likely to become entrapped by a singer or actress.

In fact a small number were 'entrapped' in this way since nineteen members of the old nobility married actresses between 1884 and 1914.[16] However, the vast majority continued to marry within the group in which they were born. According to D. M. Thomas 40 per cent of heirs to titles married within the group in the eighteenth and nineteenth centuries.[17] Great efforts were made to ensure suitable marriages apart from the constraints of environment. The aunt of the heir to Lord Monson gave him the following advice to be passed on by his father in 1850:[18]

She should like to see you married to a nice girl with a good fortune and she says Miss Clara Thornhill who is about just coming out promised to be a very nice girl and has ninety thousand a year (that would do eh!) . . . there are two younger daughters of £40,000 each, not bad but the first prize is the large prize. I should be very sorry for you to marry for money but a nice wife with it would not be bad.

If no marriage was possible to a social equal, then one was sought outside, classically from families who had made money in business and wished to link themselves to 'older' wealth, although the extent to which this was actually done is a matter of some debate. However it is clear that it was not as widespread as contemporaries thought. More importantly social acceptance was not readily found in the first generation of such matches. Beckett puts it clearly, 'entry into society through the marriage market was achieved only at a cost, and according to rules laid down by those already inside the group'.[19] It was only at the beginning of the twentieth century that these barriers began to crumble and even then the ghost of Lady Bracknell with her notebook and 'questionnaire' for prospective husbands stalks any easy generalisation.[20]

An important group who do not fit easily into this view were younger sons. Although strict settlement in theory protected them to an extent they were often ill served by the system. If a marriage to a much courted

[16] J. M. Bulloch, 'Peers who have married players', Notes and Queries, 169, 1935, quoted in F. M. L. Thompson, English Landed Society in the Nineteenth Century (London, 1963), pp. 302–3.

[17] D. M. Thomas, 'The social origins of marriage partners of the British peerage in the eighteenth and nineteenth centuries', Population Studies, 26, 1972.

[18] Quoted in Thompson, English Landed Society, p. 99.

[19] Beckett, The Aristocracy in England, p. 108.

[20] Oscar Wilde, The Importance of Being Ernest, Act 1 in Richard Ellman, ed., Oscar Wilde: Selected Writings (Oxford, 1961), pp. 303–4.

heiress could be arranged then all was well, if not a younger son could find himself in reduced circumstances and dependent on either his brother's charity or even work. 'Younger sons of the landed aristocracy', Thompson writes, 'were certainly not debarred from having careers, but these were more likely to be dignified than self-supporting, and to require injections of private income supplied by allowances from the family estate.'[21] This happened even with the minor gentry. The Rolfe family of Heacham in Norfolk were beset by such problems. When Eustace Neville Rolfe succeeded to the estate in 1869 it was 'incumbered' by a stepmother, the product of a late and unfortunate marriage by his father, and younger brothers and sisters. These were a constant drain on the estate over the next year. Charley showed a marked disinclination to live on the reduced means he was now expected to. He was set up in a series of 'professions' and ended up being sent to Australia, whence he constantly wrote to his brother alternately cajoling and slightly threatening letters asking for money. The youngest son Herbert joined the Navy and followed a similar course. Interestingly both eventually 'settled down' but in 'lower stations' in life.[22] This pattern was not uncommon, even among much greater gentry. As Stone says:[23]

. . . younger sons were downwardly mobile, with few career options except the church, the army or at worst unpaid bailiffs on their fathers' or elder brothers' estates, unless they should have the good fortune to marry an heiress. Because there were no special legal privileges or hereditary titles attached to them, younger sons had to make their own way in the world.

The core of the aristocratic family, once established, was nuclear. However, that simple description conceals elements of extended and complex familial structures even in the nineteenth century. For example, although the separate 'dower house' occupied by the widowed mother was less common by the mid-nineteenth century, the continued presence of unmarried female relatives, the ubiquitous maiden aunts of Victorian fiction, remained a common feature of many gentry and aristocratic families and a financial millstone. Similarly unmarried 'younger sons' exercised their rights by visiting: 'the country house and the London club provided two havens for their bachelor existence; and the growing network of railways made it easier to get from one to the other'.[24] Just how 'complex' a family could be is shown by Jock Yorke's account of his childhood on an estate of 15,000 acres in Yorkshire in the years before

[21] Thompson, *English Landed Society*, p. 22.

[22] Veronica Berry, *The Rolfe Papers. The Chronicle of a Norfolk Family, 1559–1908* (Brentwood, 1979), pp. 75–199.   [23] Stone, *An Open Elite?*, p. 165.

[24] Mark Girouard, *Life in the English Country House. A Social and Architectural History* (New Haven and London), 1978, p. 296.

the Great War. At various times the 'family' consisted of his mother and father, his grandfather and step-grandmother, a maiden aunt, and his cousin whose parents were in Egypt, who lived in the house for five years.[25] Fifty years earlier, also in Yorkshire, the 1851 Census showed that the 'family' of Lord and Lady Wenlock consisted of 'their son and daughter-in-law, two grandchildren, two sisters, two visitors, and thirty-four indoor servants.'[26] Outside the immediate family there was the household, those who lived under the same roof but were not related to the core group. In the past these had consisted of servants, friends, pupils and even unrelated children. However, by the 1850s among the aristocracy at least the division between the family and the household was clear and increasingly demarcated, especially for the adult members of the family. Robert Kerr wrote in *The Gentleman's House* of 1864:[27]

it becomes the foremost of all maxims . . . that the Servants' Department shall be separated from the main house, so that what passes on either side of the boundary shall be both invisible and inaudible to the other . . . The idea which underlies all is simply this. The family constitute one community; the servants another.

Power within the aristocratic family was, by the 1850s, firmly patriarchal. The 'portion' brought to the marriage went to the husband, although at the time of the marriage careful parents would have made adequate provision for a widow's jointure. In 1841 Baron Parke wrote to his prospective son-in-law that:[28]

. . . the proposed jointure is what is very common in settlements, being ten per cent on the lady's fortune, yet [my solicitor] does not think it enough for the widow of a baronet of considerable property; it would not enable her to live with a separate establishment, with decent comfort, in a manner proportioned to what she will have done.

Additionally marriage settlements could be used to protect a wife's property from the common law right of a husband to all his wife's possessions, 'by special legal rules under equity, often in the form of a trust where the wife or daughter was nominal owner but the property was administered by male relatives or attorneys'.[29]

Despite these limited powers, the lives of the women of the aristocracy were increasingly bound, as were women of other classes, to a model of femininity which saw women as 'dependent, young, weak and childlike, encouraged by the widening age gap between spouses'. This was

[25] Thea Thompson, *Edwardian Childhoods* (London, 1981), pp. 193–5.
[26] Barbara English, *The Great Landowners of East Yorkshire, 1530–1910* (Hemel Hempstead, 1990), p. 218. [27] Quoted in Girouard, *Life in the English Country House*, p. 285.
[28] Quoted in Thompson, *English Landed Society*, p. 101.
[29] Leonore Davidoff, 'The family in Britain', in F. M. L. Thompson, ed., *The Cambridge Social History of Britain, 1750–1950*, vol. II, *People and the Environment* (Cambridge, 1990), p. 73.

contrasted by a view of masculinity which stressed manliness, strength and the pseudo-chivalric code derived from Scott and a romanticisation of the Middle Ages. Beneath this lay a double standard in which male sexuality was expressed in relationship with women of the lower class, but publicly social behaviour acted through the idea of separate spheres. At its most simple this gave to men the public spheres of 'economic, political, and intellectual affairs with its sordid compromise and inevitable immorality' and women 'the idea of the pure home and family'.[30] This separation came to be represented in most areas of aristocratic domestic life by the 1870s and 1880s. It is seen at its clearest in the creation of separate areas within the great country houses for men, women and children which paralleled, but followed, the creation of separate space for servants. Mark Girouard shows how particularly after the 1870s the 'male domain' of the billiards room, the smoking room and the gun room became an increasingly important part of the Victorian country house. Here, especially after dinner, the men retired, dressed in elaborate smoking jackets, to talk of 'men's' things. At Bylaugh Hall in Norfolk 'one half of the ground floor was given up to a sequence of billiard room, w.c., dressing room, gentlemen's room and library – the latter described by Kerr in a cruelly revealing phrase as "rather a sort of morning room for gentlemen than anything else"'.[31] In some houses the double standard peeked through and these rooms contained 'books and pictures of a mildly naughty nature to go with the smoking-room stories', as at Callay Castle in Northumberland.[32] Women and children also had their own space. Nurseries became an increasingly common part of great houses and were accommodated usually in the 'family' wing along with her ladyship's boudoir. This was the initial site of the women's space but by the 1850s and 1860s this had moved into the drawing room, which was also all female after dinner.[33]

However, this did not mean that women had no role or power within the families of the rural élite, rather that the different roles were carefully outlined and divided by these 'separate spheres'. At one end of the spectrum some women, even among the aristocracy, continued to perform some of the roles of the 'housewife' of earlier generations with responsibility for the domestic sphere of the estate. Even Queen Victoria, according to a suitably amazed Léonce de Lavergne, showed an interest in the domestic end of agriculture. 'The Queen herself takes great interest in her poultry-yard; and the newspapers have lately announced a cure which her Majesty has discovered for a particular disease among

---

[30] Ibid., pp. 83–6.
[31] Mark Girouard, The Victorian Country House (revised edn, New Haven and London, 1979), p. 35.
[32] Mark Girouard, Life in the English Country House, pp. 297–8.   [33] Ibid., pp. 286–93.

turkeys.'[34] More mundane, but probably more representative, was Anna Courthope, the wife of the Hon. George Courthope of Whiligh in Sussex, an estate of just over 3,600 acres. She kept separate accounts which show that she managed the dairy attached to the house, the poultry and pigeons of the home farm and the hiring of female servants. Against this 'income' was set household expenditure, the education of her children and, as we have already mentioned, charity.[35]

The range of expenditure covered by Anna Courthope's 'income' was unusual but the responsibility for the female servants and for charity was not. As noted earlier the women of the aristocracy were assigned an important role in the paternalist system as distributors of charity. They also decided who received it, assessing the needs of individuals and organisations. Anna Courthope switched her payments backwards and forwards between supporting the Irish Church, various Indian missionary societies and the London Hospital for Women, as well as payments to elderly servants and more generalised soup kitchens in winter.[36] Similarly, overseeing the female servants was seen as 'natural' to women's domestic nature. As Jessica Gerard explains, 'domestic service fitted upper-class women's view of social relations; it retained the pre-industrial personal relationship of authority and subordination while insisting on the inequalities of class'.[37] Women were also increasingly given the role of organising the important aspects of socialising, the purpose of which was both to present the family to the outside world and to enable social mixing of an approved kind to take place.

The social and cultural world of the aristocracy, and to a more limited extent the larger gentry, was twofold. Firstly, and publicly the most important, was that of country life and the great house. The great landowners saw themselves, and wished to present themselves, as the representatives of rural England and of the real and unchanging values of their class. Continuity of family name and of place were, as we have seen, vital to this self image. As de Lavergne pointed out in 1855:[38]

Look at the list of the House of Lords in official publications; it is their country residences, and not their town addresses, which follow their names . . . It is the same with the members of the House of Commons as with the Lords. All those who have country houses take care to have them indicated as their habitual residences. Everybody knows the name of Sir Robert Peel's country seat – Drayton Manor.

This view is supported by Barbara English. 'In John Bateman's survey of great landowners of 1883 only 8 English landowners out of 1,363, gave

---

[34] De Lavergne, *Rural Economy*, p. 134.

[35] ESRO. SAS Acc 1276 GB, 'Anna Courthope's General Account book 1841–1854.'

[36] *Ibid.*   [37] Gerard, 'Lady Bountiful', pp. 199–200.   [38] De Lavergne, *Rural Economy*, p. 126.

a London as opposed to a country address . . . Englishmen wished to represent themselves as living in the shires.'[39]

It was these country seats which de Lavergne saw as the real basis of aristocratic and gentry culture. 'Show and splendour are reserved for the country . . . and it is there more especially that an interchange of visits, fêtes, and pleasure parties takes place.'[40] This cannot be denied, particularly in the earlier part of our period. The great theatrical show of the birth of an heir, of a coming of age, a marriage or a funeral 'on the estate' was always a matter of celebration in the county. However it was celebrated at different levels and at different distances from the centre, depending on one's precise status. In March 1871 when an heir was born to the Duke of Northumberland the 'principal farmers' ate at Alnwick Castle while 'large spreads' were held in Tynemouth, Ovingham and Bellingham for lesser tenants. Finally in the 'small villages' 'the people were granted various sums of money . . . and had it expended however on a tea, a supper or a ball . . .'[41] Similarly when Spencer Lucy brought his new bride back to Charlecote in Warwickshire in 1865 he had already celebrated his wedding with his social equals. At Charlecote 'a booth 120 feet long was erected in the park . . . where dinner was laid for all the tennants [sic]'. Additionally 'all the poor on the estate were feasted with as much as they could eat and drink', elsewhere in the park.[42] Here we see an ostensibly 'communal' celebration which was in fact divided by class with the aristocracy a class apart from those around them. This division was constantly reproduced in day-to-day life. Diaries and memoirs show time and again the extent to which the aristocracy and gentry, even when in their county seats, only mixed with others of their class. This was becoming a national and even international phenomenon. Visiting and socialising, even at the bottom end of the gentry, was closely related to one's precise class position. In the 1890s Elinor Glynne, newly married into the aristocracy, was informed by Lady Warwick exactly who was admissible into county society, where and how.[43]

Army or naval officers, diplomats or clergymen might be invited to lunch or dinner. The vicar might be invited regularly to lunch or supper if he was a gentleman. Doctors and solicitors might be invited to garden parties, though never, of course, to lunch or dinner. Anyone engaged in the arts, the stage, trade or commerce, no matter how well connected could not be asked to the house at all.

[39] English, *The Great Landowners of East Yorkshire*, p. 199.   [40] De Lavergne, *Rural Economy*, p. 126.
[41] T. Fordyce, *Local Records; a Historical Register of Remarkable Events which have occurred in Northumberland and Durham, Newcastle and Berwick-on-Tweed*, 4 vols. (Newcastle-upon-Tyne, 1880), vol. IV, p. 131.   [42] Lucy, *Mistress of Charlecote*, p. 135.
[43] Quoted in Leonore Davidoff, *The Best Circles. Society and the Season* (London, 1973), p. 61.

This clear separation seems to have become even clearer as the century progressed. Although the ideology of the ruling élite continued to stress their county roots and place, reality was gradually changing. The demands of politics and fashion coupled with the fact that a declining proportion of their income was coming from land, while the growth of the local state was removing much of their local power, led many of the aristocracy to spend less and less time on their estates and leave the day-to-day functioning of a paternalistic regime to a lieutenant caste, particularly the larger farmers and the parish clergy. Barbara English, as we have said, notes that while the vast majority of English landowners gave their main address as their rural seat this conceals the fact that by that point in time most also had London addresses and, by the late nineteenth century, virtually all had London clubs.[44] However, the social round was changing. The Lincoln Stuff Ball, after its decline in the 1870s, altered its date to Christmas in the 1880s.[45]

At the highest social level fashion had now altered and was now spent on the Scottish moors, not in the English countryside, and continental excursions often followed the end of the London season . . . The only regular time for residence in Lincolnshire for such magnates as Lord Brownlow was the Christmas and new year season.

By the 1900s it had shifted further among the wealthiest at least. Joan Poynder, the daughter of Sir John Poynder Dickson-Poynder, High Sheriff of Wiltshire, seems to have been moving constantly during her Edwardian childhood. 'Upper class Edwardians lived much less at home than their parents had, being always on the move between London, Biarritz, Bordighera, Cowes, Goodwood, Ascot, Newmarket and each other's country houses.' The time she spent in the country was limited to a short period in winter around Christmas.[46] All this was aided by the increasing ease of travel. The railways had opened up England in the 1850s and 1860s, fast steam packets added the continental system in the 1870s and in the 1900s came the motor car. This transformed the country life of the élite, not always, the older members thought, for the better:[47]

The party is in a ceaseless state of metabolic flux. You come down to breakfast to find that your charming neighbour at dinner the night before has gone off in her car to some other country house 200 miles away. Somebody else – probably a complete stranger – arrives during breakfast and introduces a discordant note

---

[44] There seems to be no modern survey of nineteenth-century gentlemen's clubs although many have 'official histories'. A good, near-contemporary account can be found in Major Arthur Griffiths, *Club and Clubmen* (London, 1907).   [45] Olney, *Rural Society*, p. 173.

[46] Thompson, *Edwardian Childhoods*, pp. 211ff.

[47] Lord Ernest Hamilton, quoted in Clive Aslet, *The Last Country Houses* (New Haven and London, 1982), p. 51.

that does not, perhaps, even begin to blend in with the general harmony for two or three days. It is upsetting.

From the south of France to the north of Scotland the aristocracy formed, by the 1900s, an international class whose relationships were cemented by marriage across the continent. Almost as much as they were 'English' they were 'Continentals', yet the national part of their culture, and especially its presentation, had to be retained.

Some of this has already been touched upon in the description of the public show of the manifest rituals of the country house. This extended further particularly through sport, and most especially through fox hunting. Hunting was the supreme county ritual. It brought together all levels of the élite, and put them on display for their inferiors. At the first meet of the Heythrop in 1876, there were 'supposed to be 1000 persons, [to] witness the opening'[48] Hunting was also about internal relations, about the intra-class notion of the organic society. Here, on the hunting field, according to this version, all were 'equal', the real judgment was made not on wealth but on skill at handling a horse:[49]

Hunting was the activity *par excellence* which brought together local people and those involved in London Society. It had all the elements of aristocratic patronage and deference masked by a male equality in sports . . . Hunting, too, allowed a limited amount of class mixing in the field: that is, local farmers, doctors and similar people could hunt along with the great as long as they were sufficiently keen and skilful.

But it was, nevertheless, as Bagehot recognised, a show, a grand theatrical spectacle, which demonstrated the power of the rural élite, while at the same time marking out its boundaries and making clear who could cross and where. The aristocracy and the greater gentry stood as a class apart. Separated by wealth and space from those beneath them, the primacy of preserving the estate and the family restricted their social and cultural being to a small and even international group of equals. Yet, until the Great War they managed to retain the image of a county-based, organised and essentially English group. This was done through the careful manipulation of public show and charity. However, in parts of Britain this was not always so easy or so obvious.

In the upland areas, as we saw earlier, there had always been less of a landed presence and this was especially true of Wales. Here, as we said previously, resident aristocracy were very thin on the ground. More than that they had, wherever possible, become a separate group:[50]

---

[48] John Simpson Calvertt, *Rain and Ruin. The Diary of an Oxfordshire Farmer 1875–1900*, ed. Celia Miller (Gloucester, 1983), p. 29.   [49] Davidoff, *The Best Circles*, pp. 28–9.

[50] Howell and Baber, 'Wales', p. 287.

. . . increasingly from the late eighteenth century onwards they [the aristocracy and gentry] were ceasing to be an organic part of rural society; more and more so, they were in, but not of, the community. Already by the late eighteenth century, they had become largely anglicised, had adopted English fashions and values and lacked real concern for their native language and literature.

This trend continued and indeed accelerated through the nineteenth century and meant that an ideal of an organic society was not a part of Welsh rural society to any extent after the 1850s. If the aristocracy of England was a class apart, it at least felt it necessary to maintain a public face of country life, whereas the great Welsh aristocrats felt no such need. Although most had Welsh addresses in Bateman, some major landlords did not. The Marquis of Anglesey, who held 9,600 acres on the island, lived in Staffordshire, where he held 17,000 acres; Richard Arden who owned over 5,000 acres in Carmarthenshire and Pembrokeshire and less than 250 in England lived in Slough; and Lord Bagehot, who held 20,000 acres of north Wales and 10,000 of Staffordshire, chose to live in Rugeley.[51] Even if these large landowners were present on their Welsh estates for part of the year, this did little to endear them to Welsh feeling, especially after the widespread political victimisation following the Liberal landslide in Wales in 1868. In Cardiganshire, for instance, following this election, there was a Tory backlash of unprecedented proportions, with the boycotting of shops and tradeplaces of Liberal voters and, according to Thomas Harris of Llechryd, '. . . the dispatch of almost two hundred notices to quit'.[52] Although never repeated on such a scale they were fiercely etched in the folk memory and landowners never lived them down.[53]

In this situation the land agent became a central figure. Like landowners, agents varied, but they were seen, especially in the Welsh-speaking areas, as unsympathetic and out of touch at best and as hostile foreigners at worst. Mr T. Evans, a Welsh-speaking tenant of Lord Windsor, whose main seat was in Shropshire, told the *Royal Commission on the Land In Wales* that the Windsor agent was 'a Scotchman [sic], a stranger to the district, who has no sympathy with the Welsh tenantry around him and little knowledge of our customs or our methods of farming'.[54] The reality, as David Howell has argued, may well have been different in other places, but as the century progressed the revival of Welsh national feeling and growing agitation for land reform made life intolerable for even the most sympathetic.[55]

[51] Bateman, *Great Landowners*, under those names.

[52] R. J. Colyer, 'The gentry and the country in nineteenth-century Cardiganshire', *Welsh History Review*, 10, 4, 1981, p. 516.   [53] Howell and Baber, 'Wales', p. 287.

[54] *RC on Land in Wales and Monmouthshire*, BPP, 1894, XXXVI, p. 59.

[55] David W. Howell, *Land and People in Nineteenth-Century Wales* (London, 1977), pp. 44–5.

The situation was little better with regard to the middle and smaller landowners, the Welsh gentry. The onset of agricultural depression, coupled with the lessons of the Irish and the Crofter movements, led to widespread accusations that the gentry were guilty of 'economic oppression' since they refused rent reductions or alterations in the terms and conditions of tenancy. This in turn led to a further widening of the gap created by language, and emphasised by the fact that most tenants were nonconformists while most gentry were Church of England. In the 1880s and 1890s Liberal radical attacks on landlordism were fuelled still further by landlord resistance to the creation of the marginally democratic School Boards after 1870, which would have vested some control of education into nonconformist and Welsh-speaking sections of the community.[56] By the creation of the County Councils and the other local government reforms of the late 1880s and 1890s the rejection of the county gentry by the mass of their tenants was complete, as was their own recognition of that fact. Unlike England the first county council elections of 1889 marked a complete change in the personnel of local government, 'as a political institution the gentry was now a spent force. Reeling under the trauma of the inevitable breakdown of the old system, its members became increasingly disinterested in politics.'[57] Hand in hand with this, especially in the more remote areas, small estates were sold off as they 'became increasingly subject to progressively increasing economic [as well as] social pressures'. Rents declined and increased taxation affected those at the margin, but especially important was the growth in other areas of investment, 'particularly joint stock companies, which gave highly lucrative returns on invested cash'. Embittered by political and social rejection and seeing better money-making opportunities elsewhere, sections of the Welsh gentry quit the land in the 1890s never to return, or at least cut their losses, sold part of the holdings and went elsewhere.[58]

As in other ways, the experiences of England and Wales were quite different in respect of the social and cultural history of the 'wealthy and great'. Although many Welsh aristocrats shared the national culture of their English counterparts, and were indeed often effectively English themselves for all their titles, they were increasingly unable to maintain the 'organicism' which was so important in England. They were in almost every respect a class apart. Also in Wales there was, partly for reasons we shall examine below, no real lieutenant class to perform the rites and rituals of paternalism. Even the 'old' gentry were closely identified with the English, and their political failure was in part at least a result of their

[56] Howell and Baber, 'Wales', pp. 287–90.

[57] R. J. Colyer, 'Nateos: a landed estate in decline 1800–1930', Ceredigion, 9, 1980–4, p. 72.

[58] Colyer, 'The gentry and the county', pp. 529–30; Howell, Land and People, pp. 24–5.

earlier neglect in carrying out the obligations of charity.[59] In England the 'system' held much better, although, as we shall see later, it was more threatened than many have argued. Crucially, for most of the period between 1850 and 1914 the cultural power of the aristocracy and the gentry ensured that their version of themselves as paternalist, organic, resident and benevolent members of the county and the country remained dominant.

II. THE FARMERS[60]

> I have fields, I have flowers,
> I have trees, I have bowers,
> And the lark is my morning alarmer.
> So Jolly boys now, and God Speed the Plough,
> Here's health and success to the farmer.

(Traditional)

If the work of Bateman at least shows us who the wealthy and great were, we have no such guide to the farmers as a class. Beyond a most basic definition, for instance, 'those who rented or owned agricultural land and worked it either themselves or with hired labour as their main or only source of income' it is difficult to see any really common bounds. Certainly, there was little organised consciousness among farmers as a group of their common interests, as distinct from other groups at national level before the late 1880s at the earliest. However, this does not mean that there was not a substantial degree of similarity at local level between farmers, sufficient to create at least communities of interest, and concerted, even class action on occasions. Despite this, it is necessary to see farmers as divided into two main groups: firstly, those who were essentially managers, even if they worked themselves, and were employers of labour; secondly, those who worked their holdings entirely with their own and their family's labour, perhaps very occasionally hiring casual workers. The acreage attaching to these two groups is problematic and varies from area to area, but for most of our period the dividing line between the two groups on a mixed or arable farm was probably about 40 acres. The 1871 census took a sample of 17 counties (which did not include a Welsh county) by holding size, and compared 1851 and 1871. In 1851 there were, in the sample, 25,688 farms below 40 acres (37.4 per cent of total) and in 1877, 22,679 (37.8 per cent). In 1851 there were, in the same sample, 23,540 farmers who employed no labour and in 1871, 25,618.[61] The figures are sufficiently

[59] Colyer, 'The gentry and the county', pp. 500–3.   [60] Readers are also referred to Chapter 12.
[61] *Census of England and Wales for the Year 1871*, BPP, 1873, LXXI, pp. xliv–xlvii.

close to suggest that 40 acres is a fair dividing line, and rather better than the 100 acres suggested by Reed.[62]

The two groups were not absolutely separate. Occasionally, especially in the north, the west and Wales, a family might cross the line but, and this can only be impressionistic, this seems to have been unusual in most of the south and east, and to have become less and less common as the century went on. There were exceptions to this. The depression enabled many smaller farmers to 'move up' the farming ladder, especially as migrants. This was reinforced by the general movement of land sales around the Great War, and particularly in Wales in this period and in the 1870s and 1880s, although in the latter case it seems unlikely that the move up the farming ladder was of great significance for most, since farms remained very small. However, in the most general terms it is possible to make fairly clear divisions between 'management' farmers and 'peasant' farmers.

The larger farmers

The big farmers of nineteenth-century England, and to a much more limited extent Wales, had a lifestyle which in most respects was indistinguishable from the smaller gentry. John Simpson Calvertt who farmed nearly 1,200 acres of north-west Oxfordshire as a tenant lived a life which was virtually indistinguishable from that of the country gentleman, with the important exception that, since he farmed, he could not be absent for long periods of time from his 'estate'. He clearly regarded himself, and was regarded by his contemporaries among the gentry, as an equal, in many respects. He visited and drank with Mr Albert Brassey, son of Lord Brassey the railway magnate, and attended various public dinners connected with the Heythrop Hunt and the Chipping Norton Agricultural Society which were patronised by the gentry and even the aristocracy.[63] Similarly William Wood, who grew up on the low Weald of Sussex, writes of his father:[64]

For such men as this, with ample capital, farming was a good business through the major part of the nineteenth century, up to about 1876. They lived in the style of country gentlemen, were very hospitable, entertained largely, kept a good table and a good cellar, and enjoyed what sport there was going – enjoyed life to the full.

Yet even here the subtle differences of place were clear. Although Calvertt had social contact with the lower end of the gentry, this seems to have been occasional and largely related to hunting. His regular social

---

[62] Reed, 'The peasantry in nineteenth century England', p. 56.
[63] Calvertt, *Rain and Ruin*, pp. 23–30. [64] William Wood, *A Sussex Farmer* (London, 1938), p. 36.

contacts were with other large tenant farmers and the middle classes of the country towns. In 1876 he dined at Chipping Norton fair with the owners of Bliss's Tweed Mill and a farming family from Charlbury. His frequent visits around the county seem almost inevitably to have been to other large farmers, many of them connected with viewing agricultural improvements, or discussing agricultural matters. He did, on occasion, go to great houses like Blenheim but only as a visitor 'viewing' them, not as a guest or social equal.[65]

Nevertheless, in terms of social, domestic and cultural life, the largest tenant farmers lived in a style little different from the lesser gentry immediately above them. Particularly in the arable areas of the south and east, although also in parts of Northumberland and Yorkshire, there was considerable house rebuilding in the first fifty years of the century and most of the homes of large tenant farmers were well equipped and modern. In these houses, especially the fine brick and stone houses of East Anglia, there was, as in gentry houses, a complete physical separation of servants and family although this did not go to the same lengths as in the country houses of the élite. Stephens and Burns's *Book of Farm Buildings* of 1861 contains a design for 'a First-class Farmhouse in the Elizabethan style, which has now become so fashionable'. This contained a library, drawing room, dining room, nursery, bedrooms for at least two servants and, in common with the houses of the gentry, two staircases, one for family and one for servants. The essential difference though is stressed in that, even with all these 'mod-cons', the building also contained a milk house, pantry and cheese room, albeit in a separate wing. By the outbreak of the Great War many of these grander farmhouses had hot and cold water, water-closets and even electricity; while the grander farmers now had motor cars instead of the traditional trap.[66]

Even if they were, as Cobbett and successive generations of rural radicals contended, 'aping their betters', the cultural life of the great tenant farmers moved close to that of the gentry. Like them they hunted, and the well-mounted and hard-riding farmer was as much a part of Siegfried Sassoon's South Down Hunt in 1914 as of John Simpson Calvertt's Heythrop in the 1850s. Similarly education was changing the farmers as a group, but especially the most wealthy, even in the north-west. Here, the *Barrow Herald* wrote in 1875, 'a respectable farmer's daughter would be thought ignorant if she could not treat you to a selection of operatic airs on her neat instrument. In place of the village school, the sons go either to boarding school or college.'[67] A small number had always gone

[65] Calvertt, *Rain and Ruin*, pp. 24–35.
[66] Quoted in Martin S. Briggs, *The English Farmhouse* (London, 1953), p. 229.
[67] Quoted in Alistair Mutch, 'Rural Society', pp. 167–8.

to university and with university expansion this number certainly increased. Probably more important was the growth of the agricultural colleges, although it was the early twentieth century before any significant number of farmers' sons attended. However even these were divided. The pupils of the 'Royal' at Cirencester, founded in 1845, were by the end of the century, according to the Principal, mainly from 'the landed gentry, the learned professions, the army and navy, commercial men and so forth – the upper classes . . . We have some occasionally from the wealthier farmers' families.'[68]

Yet the mixing, as the last quotation suggests, was sometimes uneasy, and cultural boundaries continued to exist. Institutions like 'the farmer's ordinary', the market dinner held in most market towns on market days, were very much an institution of the large tenant farmers:[69]

These Market Ordinaries were great institutions in those more leisurely times; these men were not prepared to stand beside a counter and eat a bun or a sandwich, they wanted a good dinner and time to enjoy it, with an hour or two over the wine to digest it in cheerful company . . . When I attended the Market Ordinary at the White Hart, over sixty years ago, there was always a choice of thick or clear soup, a choice of fish, joints of beef and mutton or pork, always game when in season . . . For these dinners a chairman was elected, or selected by some natural process, and often occupied the chair for many years giving the loyal toast, and ensuring a cheerful tone to the proceedings.

Nor was it only the ordinary, for market day was supremely the farmers' day. Here they met their equals from all over their *pays* and sometimes further afield. They met at the market, discussed prices and the imminent ruin of agriculture. Their wives, who came with them on some occasions, set off to shop while the men went to the ordinary. It was at markets that the great protection meetings were organised in the 1840s, and where in the 1880s the farmers' federations were formed. It was also where informal agreements were made about the level of piece-work wages, and where in East Anglia in 1874 it was decided to lock out union members. More prosaically, it was where meetings were held to inform farmers of cattle plague regulations and the other gloomier aspects of country life.

Outside the market, there existed other farmers' institutions. A tenant farmer might attend the Hunt Ball of a subscription hunt but would seldom visit the homes of the gentry on other social occasions, while the balls held in Corn Exchanges, which were effectively public, would not have been 'graced' by the gentry. Certainly, they were more educated but

---

[68] Quoted in Barnard K. Tattersfield, '"An agricultural college in the Cotswold Hills". The Royal Agricultural College, Cirencester, and the origins of formal agricultural education in England', PhD thesis, University of Reading, 1985, p. 276.   [69] Wood, *A Sussex Farmer*, pp. 76–7.

none but the very wealthiest would have gone either to a major public school or to Oxford or Cambridge. Similarly, although tenant farmers might hunt, they seldom shot with their landlords, since this was a much more closed and private world. The son of a Devon tenant farmer born in 1900 put it clearly. 'You see, the landlord and his associates didn't mix socially with the tenants . . . And the workers that worked for all the farmers didn't mix socially with the farmers.'[70] Finally, the farmers remained much more 'county' – or *pays* – based than the élites. Even though they were managers, few could risk spending long periods away from home even in winter. John Simpson Calvertt certainly travelled a lot, and most years went to London for several days, visiting the theatre, shops and generally 'doing' the sights. He also visited Devon and Cornwall, Coventry, his native Lincolnshire, Bath and Bristol. He also went frequently to the Royal Show, and was on the council of the Royal Agricultural Society of England. Yet he never seems to have travelled abroad, and the thought of a second home, in the town or by the sea, would have been impossible for him as for most farmers, no matter how wealthy.[71]

However, one did not need to go far down the scale of farm sizes, or to different areas of England, before the difficulties of generalising even about this group became clear. Cornelius Stovin farmed 600 acres on the Lincolnshire Wolds above Louth and was certainly not a peasant farmer, yet his whole social being was totally different from that of Calvertt or of William Wood's family. A key element in Stovin's case was that he was a Methodist which shaped and created a totally different socio-cultural world. Not for him the hunt, the farmers' suppers with long pipes and port or the town balls, but Zion. The cultural gap between Stovin and Calvertt, who was also Lincolnshire in origin, is graphically illustrated by entries from their diaries for two Sundays. Stovin wrote of his spiritual experiences in 1872:[72]

I left my bed in time to walk to our Free Methodist Hill of Zion 7 o'clock prayer meeting. The loss of sleep and bed comfort has proved a great gain in the spirit of prayer, of meditation, of heavenly taste and sensibility . . . Three of us met at the altar of praise and the throne of grace . . . The dew of Hermon fell upon my spirit.

Four years later John Simpson Calvertt, while far from neglecting his soul, had a very different and much less spiritually uplifting Sunday:[73]

Miss M. Fowler, & Elizabeth Anne, accd. me to Ascott Church, the new Minister, Revd. York preached, it being only his second Sunday's Duty. Jane,

---

[70] EOHC Tape no. 352.   [71] Calvertt, *Rain and Ruin, passim.*
[72] Jean Stovin, ed., *Journals of a Methodist Farmer* (London, 1982), p. 59.
[73] Calvertt, *Rain and Ruin*, p. 20.

Alice and Grace, accd me to Leafield Church in the afternoon to hear *our* Clergyman, the Revd. Lee. Miss M. Fowler, & Elizabeth Anne, accompanied me to Witney Church in the Evening – Music, Singing and Intoning, nicely performed, the Rector is a fine Apostolic Old Gentleman . . . we returned back via Crawley and saw the Blanket Manufactories on the Windrush.

But it was not only how they spent their Sundays. Stovin was clearly separated not only from the labourers below him, although his religion gave him some contacts here, but from the gentry above him, and even it seems the other large tenant farmers in the village. The chapel provided him with a complete social life, a variety of friends, and a whole range of duties. He spent none of his social life with the gentry or landlords, and indeed sometimes showed an almost Welsh dislike of them. 'Talk of Trades Unions and other organisations, but is it not high time a grand and effective organisation was effected amongst the tenantry of this country in order to hold the selfish injustice of the landlords in check?'[74] Part of this was a result of personal experience as much as political or religious belief. His 'first' landlord, Mr Edmund Beckett Denison (knighted in 1871 as Sir Edmund Beckett), was a good landlord who lived locally and supervised his estate. His son, however, was an absentee whose political and financial dealings took him to London, and he dealt mainly through his agent, which Stovin clearly resented.[75] However, it was not only a matter of personality and religion. Stovin's farm, although above average size for Binbrook, was considerably smaller than Calvertt's and the land was much poorer. Socio-economic factors separated Stovin from the gentry as much as personal inclination.

At all levels among the 'management' farmers, family was important, although it had none of the obsessive qualities which were associated with the aristocracy. Most tended to be nuclear and there does not seem to have been the extended element which was present in some aristocratic households. However, in many areas even in the 1850s and 1860s the persistence of living-in did create complex households even among larger farmers, although it was more usual among the smaller. At the top of the farming tree the great tenant farmers and their families were, as we have already indicated, little different from the gentry in most respects. A crucial difference though lay in inheritance. Although there was little formal requirement for a nineteenth-century landlord to allow son to follow father in a tenancy, this was frequently the practice. On the Holkham Estate even tenants with annual leases enjoyed considerable security of tenure and 'frequently passed their farms to their sons',[76] a

---

[74] Stovin, *Journals*, p. 48.   [75] *Ibid.*, 'Introduction', pp. 11–12.
[76] Susanna Wade Martins, *A Great Estate at Work. The Holkham Estate and its Inhabitants in the Nineteenth Century* (Cambridge, 1980), pp. 73–7.

practice which Holderness argues was the norm for most of England.[77] This fact meant that marriage was a much less fraught affair than with the aristocracy and gentry.

Until the 1850s and 1860s at least, and even among the larger farmers, there remained the notion that the wife should be economically active, and indeed was regarded as such by the census enumerators until 1871. There certainly were among the largest tenants those who had wives like those described by Jefferies in *Hodge and His Masters*, with her carriage and pair, her furs and sealskin, and 'her maid skilled in the toilet . . . grooms, footmen, just exactly as she would have done had she brought her magnificent dowry to a villa in Sydenham'.[78] Yet we should be cautious. Jefferies, like many before and since, saw his day as the dawn of a new and much worse era, with the recent past already enshrined as a golden age. Although farmers' wives of over 1,000 acres like Mrs Simpson Calvertt did little more than their gentry sisters, as Jefferies says, 'a moderate-sized farm, of from 200–300 acres, will no more enable the mistress and the misses to play the fine lady today than it would two generations ago'.[79] On farms of this size, at the bottom end of the large tenant-farmer scale, a wife's work was not an extra but a necessity. Butter, eggs, cheese, fowls and vegetables were all part of a subsistence or at any rate a semi-cash economy.[80] In the cheese-making districts and dairying areas elements of an older 'housewifery' remained. Most importantly the persistence of living-in farm service, especially on smaller farms and in the North, until the 1900s, not only created a 'complex' family but allocated to the women particular jobs. In Cumberland and Westmorland women not only ran all the dairy and poultry operations but 'taught' and supervised the women farm servants. As in Wales the house and the farmyard were her domain. In Wales, David Jenkins writes:[81]

The mistress was responsible for the female staff, daughters and maids where there were both. She herself prepared the household's food and undertook the butter-making and the cheese-making, perhaps helped by one of the maids, but the greater part of the maids' time was occupied not in the farm-house but in the farm-yard . . . The dairying was the mistress's responsibility, and this was not confined to work in the dairy but included much of the care of the milk cattle.

The language distinguishes between the responsibilities of men and women by referring to the money earned from the women's work as *arian*

[77] B. A. Holderness, 'The Victorian farmer', in G. E. Mingay, ed., *The Victorian Countryside*, vol. 1, p. 234. [78] Richard Jefferies, *Hodge and His Masters* (2 vols., London, 1966), vol. 1, pp. 94–7. [79] *Ibid.*, p. 101. [80] See, for example, H. St G. Cramp, *A Yeoman Farmer's Son. A Leicestershire Childhood* (London, 1985), pp. 14–16. [81] Jenkins, *The Agricultural Community*, p. 76.

y closs (farmyard money) or *arian y fasged* (egg money), while the man who sold eggs was seen not only as breaking some version of the natural order but as a failure and was the butt of endless jokes.[82]

Among the 'management' farmers both family structure and the internal workings of the family varied, depending on the size of the holding and its geographical location. In the great arable farms of the south and east by the 1800s the family was simple, nuclear with the wife often making no direct economic contribution to production. However, even here, movement down the scale very quickly brought the wife into the productive system of the farm. In the north and in most of Wales there were, even in the 1900s, still many complex households. Here, especially since the farms were usually smaller, or at least less profitable, the wife had key productive and organisational roles.

Although social gradations often differentiated the larger tenant farmers from the gentry and aristocracy, in many other respects they remained extremely close. Leaving aside for the moment the smaller, but still employing farmers, and Wales, what we see is the development in the years before 1850 of what Obelkevich calls the 'new-style farmer'. Farming a large tenant farm as a 'managerial worker', the new-style farmer saw farming as a business in which the key relationships were based on cash rather than obligation. Farming became 'scientific and geared to profit and the farmer himself became more individualistic, less concerned with work, more with consumption, less with village society, more with family and with others in their class'.[83] Yet, as Holderness points out, for most of the century this group, despite the likes of Cornelius Stovin, closely allied itself with the landlords. There were, of course, good economic reasons for it so doing but it remains a crucial relationship.[84] An important part of this identification was the fact that many tenant farmers took on the role of the governors of the county, acting as lieutenants for the aristocracy and gentry, and carrying out key roles in the development of organicism. Calvertt was, for instance, a member of the Board of Guardians as were Stovin and the father and grandfather of William Wood. Calvertt played an active part in local charities and 'played the squire' to the village friendly societies around, especially in his own village, Leafield, which was a notorious 'open' forest community, as did Wood's father. In addition, Stovin entertained 'his' chapel people on many occasions.[85]

However, as the century progressed the bonds between gentry and farmers seem to have become strained:[86]

---

[82] *Ibid.*, pp. 78–9.   [83] James Obelkevich, *Religion and Rural Society*, pp. 46–61.

[84] Holderness, 'The Victorian farmer', pp. 231–4.

[85] Calvertt, *Rain and Ruin*, for example, pp. 21, 39; Stovin, *Journal*, pp. 198–9, 65–6; Woods, *A Sussex Farmer*, pp. 44–60, 89–90.   [86] Obelkevich, *Religion and Rural Society*, p. 54.

As the farmers developed into a class they took new stances towards the other major groups in rural society. This involved a dramatic rupture of their ties with the labourers; it also altered more quietly their relations with the gentry. From their traditional subservience they shifted to co-operation, and when faced on occasion with gentry indifference, they gradually learned to assert their own social ideals and their own economic and political interest.

Obelkevich argues that these social ideas were best expressed by the very largest farmers who adopted an essentially 'middle-class gospel of work and of an active life', as distinct from a more traditional 'fascination with gentility and leisure'.[87] There are certainly elements of this in Stovin's journals as we have already seen. Stovin clearly felt, as many probably did in the early 1870s, that they were between the devil of rising rents and the deep blue sea of demands for increased wages. 'The breach is opening on both sides of them [tenant farmers]. The landlord on one side and the labourer on the other are deserting him. He is met by rigid exactions on every side.'[88]

A key issue here was tenant right and, associated with it, problems of game preservation. On these issues there were clear areas of conflict between landlord and tenant. The importance of tenant right in England is a matter of some debate. It is clear that it was an issue in that it was a major concern of both the members of the Royal Commissions of the 1880s and the 1890s and of many of the witnesses. J. R. McQuiston has argued that this concern was also reproduced in the realities of rural life especially after the fall in wheat prices in the mid-1870s, and that the local battles over tenant right led to the Liberals increasingly capturing rural seats after the 1880s. They 'demonstrated that rural England was no longer bound together, if only superficially, by the deference of the tenant and the benevolence of the landlord'.[89] This view was subsequently criticised by J. R. Fisher in his study of regional politics which concluded that not only was tenant right an irrelevance but also that 'at no point in the nineteenth century did the agricultural community seek to challenge the nature of the existing social and political structure of rural England'. Fisher continues by showing that any organisations of tenants were short lived and that the recognition of a unity of interest between landlord and tenant always overcame these temporary problems.[90] However, Cox, Lowe and Winter's study of the origins of the NFU suggests that this process was less clear. For example, they argue the tenant farmers were driven out of the Central Chambers of Agriculture

[87] Ibid., p. 55.   [88] Stovin, Journals, p. 62.

[89] J. R. McQuiston, 'Tenant right: farmer against landlord in Victorian England, 1847–1883', AHR, 47, 1973, pp. 95–113.

[90] J. R. Fisher, 'The limits of deference; agricultural communities in a mid-nineteenth century election campaign', Journ. of British Studies, 21, 1981, pp. 90–105.

which became 'the preserve of the landlords', leaving a vacuum which the NFU was able to occupy.[91]

In part these differences are a matter of focus, but it seems that at their most basic the concerns expressed by contemporaries as well as feelings like those in Stovin, plus the undoubted Liberal revival in rural areas as a whole after the 1880s, cannot be easily dismissed. This has been reinforced by recent work on Lancashire by Alistair Mutch and on Devon by J. H. Porter.[92] Mutch argues very convincingly for a widely based tenant-farmers' organisation, founded in 1893 in Lancashire, whose 'distinguishing features were a stress on the need for tenant farmers to be organized in their own clubs, and the demand for judicially fixed rents'.[93] Mutch shows that although the organisation was relatively short lived, its refusal to allow landlords as members, and its insistence on the interests of the tenants, especially in relation to rents and tenure, 'reflected wider shifts in rural society', leading to the formation of the NFU in 1908.[94] Again this view is given some support by Cox, Lowe and Winter who argue that especially before the Great War the NFU 'saw itself primarily as a tenant farmers' organisation'.[95] Similarly the Farmers' Federation, founded in East Anglia in the early 1890s, which also eventually became part of the NFU, showed a distinct unwillingness to involve landlords in its deliberations. Indeed, its central aim, to provide 'blackleg' labour in industrial disputes, was often deeply problematic for the landlords who were seeking a more harmonious set of social relations.[96]

Porter's examination of the Ground Game Act produces similar conclusions. The fact that ground game had been reserved to the landlord was a serious cause of problems throughout the 1860s and 1870s, and it has been usually argued that the Act of 1880 dealt with these. However, Porter shows that the Act 'provided no remedy for the depredations of ground game where the tenant had a landlord who was a strict preserver'. Further, he argues, prosecutions allowed under the Act on the basis of the strict interpretation of leases by JPs, many of whom were landowners, 'reveal the continuing underlying tensions between landlord and tenant over ground game in Devon'.[97] However, it was not simply a matter of ground game. In some areas, especially during the depression, landlords turned to the wholesale breeding of game as a major source of income, so even where hares and rabbits were allowed

---

[91] Graham Cox, Phillip Lowe and Michael Winter, 'The origins and early development of the National Farmers' Union', *AHR*, 39, 1, 1991, p. 31.

[92] Alistair Mutch, 'Farmers' organizations and agricultural depression in Lancashire, 1890–1900,' *AHR*, 31, 1983, pp. 26–6; J. H. Porter, 'Tenant right: Devonshire and the 1880 Ground Game Act,' *AHR* 34, 1986, pp. 188–97.   [93] Mutch, 'Farmers' organisations', p. 31.

[94] *Ibid.*, pp. 34–6.   [95] Cox, Lowe and Winter, 'Origins and early development', p. 35.

[96] See Howkins, *Poor Labouring Men*, Chapters 4–6.   [97] Porter, 'Tenant right', p. 196.

to the tenants the breeding of pheasant and partridge continued to create conflict.

In Wales, as we saw briefly in the previous chapter, the situation was much more serious. About 60 per cent of the land of Wales was owned by great estates of over 1,000 acres in 1877, while only about 10 per cent was farmed by owner-occupiers. The majority of Welsh landowners seem to have been absentee for much of the year which led to a widening gap between them and their tenants, especially after the 1870s, which was aggravated by religion, politics, but above all language and culture. Owen Hughes, the tenant farmer of 191 acres in North Wales mentioned above, put it clearly to the *Royal Commission on Land in Wales*:[98]

There is very little fellow-feeling between the tenants and their landlords, who belong to an entirely different and distinct class which never mixes with the tenant's class . . . My landlord, Lord Harlech, is an Englishman, a Tory and a Churchman, whilst nearly all his tenants are Welshmen, Liberals and Nonconformists.

In this situation what have been seen as minor problems in England became undeniably major ones in Wales. Game for instance, especially the breeding of birds for battue shooting, led to a breakdown in relations between landlord and tenant in many areas, while attempts to privatise the fishing of the Wye led to the reappearance of Rebecca and her daughters in the 1860s.[99] Further, rents, tenant right and tithes were all serious problems in Wales because of an increasing perception that a question of nationality was involved. From the 1880s Welsh radicalism made the land question its great rallying cry, and although Howell has queried quite how widespread was real support for wholesale land reform, there can be no doubt that by 1914 a deep cleavage existed between the tenant farmers of Wales and their landlords. As Kenneth O. Morgan perceptively remarks, 'the land question was basically social, not economic, the product of growing cultural alienation between owners and occupiers'.[100]

Small tenant farmers

For our second category of farmers, the small family producers, questions of identification with the landlords as a class, if not irrelevant, were much less important. At any point between 1850 and 1914, according to Griggs's estimates, about 40 per cent of all holdings were under 50 acres and, therefore, likely to be worked with family labour alone. However, it has been argued more recently that this figure could be a serious underestimate,

---

[98] *RC on the Land in Wales*, p. 416.
[99] David Jones, 'The second Rebecca riots: a study of poaching on the River Wye', *Llafur*, 2 1976, pp. 32–57.   [100] Kenneth O. Morgan, *Wales. Rebirth of a Nation 1880–1980* (Oxford, 1981), p. 84.

given that the number of holdings under 50 acres was, by one reckoning, nearer 70 per cent in 1880.[101]

There clearly is a problem here as to what extent 'new' small 'farms', specialising in horticulture, are distorting this figure after about 1890, but even so the majority of these farms must have been worked by small family producers. It is also important to note that although farms of this size remain a roughly constant percentage of the total number of holdings, their actual number, especially those under 20 acres, declined rapidly in the years 1850–1914.[102] However, the opportunities to get a small farm appeared much the same from the statistics, since they remained a similar proportion of the whole number of farms. But because the actual number of such holdings declined while population increased, this meant in reality that opportunities for movement up the farming ladder were getting less and less.

Small farms were most common in 'the south-west, Wales, the industrial counties of the midlands and the north-west',[103] although they existed in most counties, especially in the 1850s and 1860s. In Sussex, for instance, a whole range of small farms on the Weald had poor soil and difficult access which meant that 'high' farming was neither feasible nor very profitable. As a result a settlement pattern laid down on sixteenth- and seventeenth-century assart tenancies persisted into the nineteenth. This pattern was supported in turn by the ability of these small farmers to travel as casual workers to the large farms of the coastal plain for harvest, functioning in much the same way as the travelling Irish, and taking back much needed cash to their subsistence holdings. Additionally, the development of 'chicken cramming' (forced fattening) and some hops and fruit enabled many of these small farmers to survive until the Great War, although there were clearly difficulties from the 1870s onwards.[104]

To the small farmer, family was vital, since in most cases the farm workforce and the household were, by definition, the same thing. Marriage was almost as important here as it was to the aristocracy, but for very different reasons. A man called 'Norfolk Jack' who was interviewed by Alice Catherine Day in the 1920s put it graphically:[105]

There was a young woman servant at my employer's house. I married her, and then my prosperity began . . . She was strong, active, industrious and a good

[101] Alun Howkins, 'Peasants, servants and labourers; the marginal workforce in British agriculture, c. 1870–1914', *AHR*, 42, 1994, pp. 49–62.

[102] David Grigg, 'Farm size in England and Wales, from early Victorian times to the present', *AHR* 35, 1987, pp. 184–7.   [103] *Ibid.*, p. 187.   [104] Brandon and Short, *The South East*, pp. 322–7.

[105] Alice Catherine Day, *Glimpses of Rural Life in Sussex During the Last Hundred Years* (Kingham, n.d. c. 1928), pp. 10–11.

manager. She would do all kinds of work . . . I heard of three drill which were to be sold for sixty-eight pounds and borrowed fifty pounds and bought the drills and before long repaid my loan.

More prosaically a small Lancashire farmer told the 1894 *Royal Commission on Agriculture*:[106]

I was a labourer up to 22 years ago. I worked very hard when a labourer and so did my wife, who was a servant on the farm where I was employed. We married when in service, 40 years ago. We had not a sixpence when we married, but then we began to save a bit. We first took a farm of seven acres, and kept cows. We sold milk and butter and also brought calves up.

Even further up the scale a hard-working wife could transform a small farm by her labour or by her own money.[107]

Once married the wife's work was vital. 'What a wife mine was! Brought up chickens? Yes, I should think so. Why she used to go leasing three miles from home by four o'clock in the morning throughout harvest.'[108] To Welsh small farmers and cottagers the wife's work in and out of the house was vital to the survival of the holding. 'She would help in the dairy to make butter and that kind of thing', and on the smaller holdings do a good deal of the outdoor work, with the potatoes especially.[109] We have also already seen how, in parts of Wales, language distinguished between the work of the husband and the wife on even middle-sized farms. Further down the scale, on the cottar holdings, it was more important still. In many areas of west Wales a system of labour debt operated, where potatoes were laid out for cottar families by the larger farmers in exchange for the work of the whole family at harvest. In many cases, especially towards the end of the nineteenth century, the wife's work became more important still as the husband often worked part of the year on the coalfields of Carmarthenshire and Glamorganshire, and left his family to carry out work on the holding.[110] On the hill farms of Northumberland the wife's work with hens and cows came back to the family in the form of groceries. On certain days higglers would go to pre-arranged points 'and they [farmers' wives] met then at Kielder Hill foot, taking the butter and eggs down and coming back with groceries if they wanted'.[111] In Bicester, in Oxfordshire, up until the Great War, the wives of small holders and small farmers brought their farmyard and garden produce to the shop of Charlie Clifton in North Street, where it

---

[106] *RC on Agricultural Depression. Reports by Mr Wilson Fox on the Garstang District of Lancashire and the Glendale District of Northumberland*, BPP, 1894, XVI, pt 1, p. 63.

[107] Davidoff and Hall, *Family Fortunes*, chapter 6.   [108] Day, *Glimpses of Rural Life*, p. 11.

[109] *RC on the Land in Wales*, p. 403.   [110] Jenkins, *The Agricultural Community*, pp. 52–7.

[111] NRO. NRO T/63. Interview with Mrs Murray, b. 1900, Barrowburn, Northumberland.

was exchanged for goods or paid off debts run up in the winter.[112] Even
on a relatively large farm in Leicestershire a version of this system operated:[113]

. . . eggs were Mother's currency, as cowrie shells are to some remote islanders.
Whenever there were minor debts to settle with local people they were given
the choice of cash or eggs. The latter was always the best buy and all the villagers knew it. So eggs paid for the knitting of socks and scarves, the making and
altering of dresses, sweeping the chimneys, and beating of carpets.

The labour of other family members was also vital to the small farmers.
A large and predominantly male family could actually work quite a large
farm as at Cowbridge in Glamorgan where Mr D. J. Jenkins told the *Royal
Commission on Land in Wales*, 'I know a farm near me of 300 acres where
it is worked by the father and four sons. They are scarcely ever employing outside labour, except for harvest purposes.'[114] But it was not
only sons, as Mrs Murray, who grew up on a farm near Kielder in
Northumberland, remembered;[115]

In these [sic] days the cows were put on the hill and had to brought in and milked
up the dyke back in the summer, and the milk carried back . . . We helped with
what ever farm work was going on, setting potatoes, or tasking up potatoes, and
hoeing and weeding potatoes . . . We made wor own butter there, then the water
was to carry [sic] . . . always plenty of work on a farm, never time to be early [sic].

Most family labour was unpaid, which represented a major gain to the
small farmer. In some cases as the 'boys' came of age, even on a relatively
large farm, regular workers were laid off and replaced by family labour.
'Well about that time when I started . . . he had to give up this Mister
Mason because he couldn't keep him any longer you see . . . and within
the next year or two I had the job of looking after the horses . . . There
was just me and my father and my Uncle then.'[116] The central importance of this was stressed by Mr Morgan, a retired tenant farmer of
Cowbridge in Glamorgan, when asked by the *Royal Commission on the
Land in Wales* if farmers' children were paid:[117]

Oh no they do not get anything. I paid no wages to my children . . . I have
brought up 10 children myself on a farm, and then, of course, I did not require

---

[112] Interview with Mrs Cherry (Charlie Clifton's daughter) b. Bicester *c.* 1905. Tape in author's possession. See also Thea Vigne and Alun Howkins. 'The small shopkeeper in industrial and market
towns' in Geoffrey Crossick, ed., *The Lower Middle Class in Britain, 1870–1914* (London, 1977),
pp. 184–210; Mick Reed, '"Gnawing it out". A new look at economic relations in nineteenth-
century rural England', *Rural History*, 1, 1990, pp. 83–94.

[113] Cramp, *A Yeoman Farmer's Son*, p. 15.   [114] *RC on the Land in Wales*, Cmd 7439, p. 36.

[115] NRO T/163 Interview with Mrs Murray.

[116] Interview with Harold Hicks, Trunch, Norfolk. Sept. 1974. Tape in author's possession.

[117] *RC on the Land in Wales*, Cmd. 7439, p. 36.

much labour at one time and that was an advantage to me to enable me to rise from a small farm and have a bigger one.

In Cornwall it was the same, though here, as one suspects in Wales, there could be long-term gains. 'I never earned a wage. We didn't in those days. Not even the boys. They were working and then they were old enough to get married. They might be set up – have a bit give to them, you know, a cow or two or something when farming.'[118]
Inevitably this lack of payment was resented by some of the children especially those who felt they had got little or nothing out of it. Many daughters felt unseen or unspoken pressures. A small farmer's daughter from Co. Durham was told by the local doctor that she should train as a nurse. 'I went to night school and I was carrying on fine, learning French and Latin . . . but me mother had too much work and I had to stay at home.'[119] Even boys who could inherit sometimes felt disadvantaged as a Devon farmer born in 1900 said, 'being the older member of the family we were expected to go onto the farm and help the family. Whereas the next son – had an opportunity to go away to a boarding school and had a better education.'[120] The experience of Henry St George Cramp in Leicestershire was very much the same although he went to a village school:[121]

Yeoman farmers down the ages have managed their farms with mainly family labour. We did likewise and rarely employed more than one outside labourer except at harvest . . . Those of us at school made our contribution to work in the evenings, at week-ends and during holidays. Fortunately my regular school-ing was not interfered with. My older brothers were not so lucky. When in later years I searched the primary school records, there seemed an all-too-obvious liaison between the local school headmaster and my father, who were friends, and it showed in the absences of my brothers at harvest time.

The social and cultural life of the small peasant farmer looked down much as that of the large managerial farmer looked up. Particularly in Wales the chapel culture and the language created a common bond between the smaller farmer, the cottager and, in many districts, the labourer. 'It was the practice for nonconformists to walk to chapel . . . People walked in the company with whom they happened to fall in, farmers and cottagers alike.'[122] This was reinforced by institutions like the work group, common eating at harvest and threshing and the fre-quency of movement up and down the (very short) farming ladder which created a shared culture of work and family. Nor was this only

[118] EOHC Tape no. 442. Woman, b. 1901, Helston, Cornwall.
[119] EOHC Tape no. 256. Woman b. 1893, Cockfield, County Durham.
[120] EOHC Tape no. 352. Man b. 1900, Staverton, Devon.
[121] Cramp, A Yeoman Farmer's Son, p. 38.   [122] Jenkins, The Agricultural Community, p. 214.

the case in Wales, though elsewhere the distinction was less clear. De Lavergne noticed that in the Weald, which he described as 'perhaps the most backward part of the whole of England in point of agriculture', there was little or no difference between the small farmers and labourers, both being 'as ignorant as they are poor'.[123] But it was more than that: here in the scattered communities of the Weald among these peasant farmers flourished the remnants of seventeenth-century rural nonconformity in chapels like Cade Street near Heathfield or Jarvis Brook near Crowborough. Religion and remoteness encouraged endogamy here, as in Wales, and again, especially on the Ashdown Forest, 'co-operative' work cemented bonds of group across the indistinct lines of class.

In terms of their physical surroundings again there might be little difference in real terms between the labourers and the small farmers, although even small farms tended to be separate and surrounded by some land no matter how little. However, as with the larger farms, there was some rebuilding in the late nineteenth century, both on estates which sought to encourage small farms, and in individual houses. In Wales the old-style small farms with byre and living accommodation under the same roof were gradually rebuilt or abandoned for newer houses. Similar changes seem to have occurred on Dartmoor and Exmoor.

Nevertheless, as with the larger farmer/gentry divisions, there were real differences. The elaborate divisions expressed in the arrangement of eating during the communal meals associated with harvest and threshing in Wales point to a wish to divide groups and to symbolise those divisions. The peasant farmer of the Weald or of Allandale may have been close to the labourers, indeed may have worked alongside them for part of the year, but he remained different in that he had his bit of land and its product. In the west of Cornwall, although threshing was done communally, only the small farmers shared the communal meals.[124]

The farmers were not a unified group at any time during the years between 1850 and 1914. However, by the 1880s, it does seem clear that in some parts of England some sort of consciousness of the needs of the larger farmers as a class had emerged which found its expression in farmers' organisations, culminating in the foundation of the National Farmers' Union. Nevertheless, they remained close to the gentry politically and socially although the number of bankruptcies among smaller landowners, and the increase in owner occupation among farmers, may have opened up a gap there. The attitudes of small farmers varied greatly from area to area and depended crucially on what kind of farm they worked. Certainly in Wales a real consciousness of a national/economic

[123] De Lavergne, *Rural Economy*, pp. 202–3.   [124] EHOC Tape, no. 442.

kind emerged in the 1880s and 1890s, which was the mark of an advanced class and local pride. Elsewhere things were not so clear cut. The agricultural trades unions of the years after 1906 tried to recruit small farmers, as is shown by the title of the 'new' union of that year – the Eastern Counties Agricultural Labourers and Small Holders Union. At the founding meeting, Richard Winfrey MP said 'he looked upon the promotion of the small holdings movement as the greatest for the future in our rural districts'.[125] There certainly were small holders in the Union but they remained a tiny minority, and the word 'small holders' seems to have been dropped from the title in less than a year. However, there were still small holders in the Union long after the Great War.

Social, cultural and domestic life in more specific senses depends again on where and whom. A central feature, though, was the use of family labour. The small farmer was, in this respect, quite distinct and the importance of the family here, though it never reached the importance it did (and does have) in mainland Europe, was vital. Over time it seems possible that differences were increasing. The number of farms as a whole were falling through our period while the population was growing. Up to the late 1880s the number of farms over 300 acres increased, thereafter they declined although the number between 100 and 300 acres continued to grow up to the 1920s. The total number of small farms fell, but as a percentage of all they remained much the same.[126] Coupled with rising costs this suggests that access to land was becoming more difficult, although there clearly was movement from the top of the small to the bottom of the middle, and (perhaps) from the bottom of the large farm downwards during the depression. Yet the real changes, and the real continuities, are hidden by these kinds of generalisation. The farmers remained in 1914, as they had been in 1850, divided crucially by those who employed labour and those who did not. Both groups had changed but in subtle and complex ways that were mediated through place as well as time. Probably the most significant long-term change of all, though, had begun. In the last years before the Great War the amount of land in owner occupation nationally began to rise for the first time in generations. At the same time the farmers founded their own national organisation, the NFU, while the number of holdings between 5 and 50 acres peaked before going into irreversible decline. Most subtle of all was the gradual destruction of many of the cultures of the small farmer. In west Wales the introduction of the reaper-binder destroyed the need for the key work-group based around the harvest work-debt. This tiny change rendered a whole social structure unnecessary and irrelevant; gone were the work groups, the co-operation between cottar and farmers and the

[125] *Eastern Weekly Press*, 28 June 1906.   [126] Grigg, 'Farm size,' pp. 184–6.

great communal eatings of harvest and threshing. But change like that does not appear in the agricultural statistics.

III. THE FARMWORKERS

There's some that say the farmer's best,
But I must needs say no
If it weren't for we poor labouring men,
What would the farmers do.
They would beat up all their own stuff,
Until some new come in.
There's never a trade in all England,
Like we poor labouring men.

(Traditional. From the singing of 'Queen'
Caroline Hughes, Wiltshire.)

As we have shown elsewhere in this volume, to see those who worked the land as one homogenous group simply does not make sense for the period between 1850 and 1914. Most obviously, regional variations in hiring patterns produced a variety of social structures and social relationships.[127] Broadly speaking, in England north of a line from the Wash to the Severn a substantial part of the farm workforce was made up of workers who lived in, or on, the farm and were hired by the year even in the 1900s. This was also true of Devon and Cornwall. Even in the west and south midlands and parts of Kent and Sussex 'living in' was still met with in 1900.[128] In Wales, with the exception of parts of the Vale of Glamorgan, Pembrokeshire and the Vale of Clwyd, living-in was dominant.[129] South of the line from the Wash to the Severn most farmworkers lived off the farm and were hired by the day or week. There were also some workers of this kind in areas of the north.

Looking first at the areas of 'living-in', it is important to stress that there was more than one form of living in and not all those who lived in were classic 'farm servants', that is young, single men and women who lodged in the farmhouse in the early part of the lifecycle. We have already seen the broad outlines of these differences in our discussion of communities in the previous chapter, but we will return to them again here more specifically in terms of social and domestic and cultural life.

To begin with the 'classic' farm servant. Again we differentiate between those, probably the majority in England at least, who were

---

[127] See Peter Dewey, 'Farm labour in England and Wales, 1850–1914', above, Chapter 12. For a detailed discussion, see Alun Howkins, 'The English farm labourer in the nineteenth century; farm, family and community', in Short, ed., *English Rural Communities*, pp. 85–104.

[128] See Howkins, 'The English farm labourer', pp. 87–9.

[129] Howell, *Land and People*, pp. 94–107.

young and single and who hired into the farmhouse, and those who hired into the foreman's house or, very occasionally, into separate, purpose-built buildings. Living in the farmhouses was the predominant form in Cumberland, Westmorland and north Lancashire and to a lesser extent in the North and West Ridings of Yorkshire. Here, on the small upland farms only one or two servants were hired, and they themselves were not infrequently the sons and daughters of small farmers who hoped one day to have a holding of their own. In this situation social and domestic relations took on a particular tone. Master and servant were often, in theory at least, social equals in origin who were distinguished from one another only by being at different stages of the lifecycle. This was put very clearly in Wilson-Fox's discussion of trade unions in his 1894 report on Garstang in north Lancashire:[130]

The labourers have no trade unions. Possibly a good rate of wages, continuity of employment, a feeling of friendliness which arises from master and man living under one roof, and a very narrow social distinction between the small farmer and the man he hires, are all elements which tend to avoid the necessity of combination to assert rights or demand privileges.

This was constantly reinforced by both domestic life and work as Mr Brian Dodgson of Catterall told Wilson Fox, 'I go to work with men and show them how. My wife and I dine with the servants. They have the same food as we have', while a labourer in the same district said, with a telling comparison, 'in a good farm-house the man becomes one of the family'.[131]

However, conditions did vary from place to place. Food in particular varied from the excellent to the appalling as did accommodation, and both these were the source of anger and division. As Alistair Mutch shows, inadequate food and bad or cold accommodation were the reasons most frequently given for servants moving at the end of the term, as well as creating a rich dialect to describe mean farmers with words like 'stiff-hefted, ettle, tight i' t'heft, nip cheese and noyas'.[132] In Lincolnshire and Yorkshire farm servants created the character of 'Yorky Watson' or 'Iron Ned' who, in different versions of the same ballad, not only wanted his servants to plough 'four acres in a day' but also fed them on 'pies made of lead and cakes made of clay' and mutton pies made of 'an owd ewe' which had been dead for 'a twelvemonth and a day' and had 'maggots crawling off her hundreds, thousands, millions thick'.[133] At the other

[130] *RC on Labour . . . Garstang*, BPP, 1893–4, xxxv, p. 167.   [131] *Ibid.*, pp. 167, 173.

[132] Mutch, 'Rural society in Lancashire', pp. 185–7.

[133] 'Four acres in a day' from the singing of Mr Maurice Ogg, Coleby, Lincolnshire in Roy Palmer, ed., *Everyman's Book of English Country Songs* (London, 1982), p. 36. See also 'Rattle mutton pie', sung by Mr Hill of Tetford, Lincolnshire on Vaughan Williams Memorial Library Tape VWML 003, *The Leaves of Life. The Field Recordings of Fred Hamer*, 1989.

extreme there were 'good tommy places', and many farmers recognised the real rewards of keeping a good table in terms of getting hard work and a stable workforce. As Mr Dodgson said of his servants:[134]

The more they eat the better I am pleased. No man can work with an empty belly. What goes in the stomach comes out in work. If you want to keep an Englishman up to his work and in good humour keep his belly full of good meat. Besides which he won't want drink in it then.

On a 'good tommy farm' a servant could get three full meals a day with meat five times a week, at one meal a day plus tea, bread, and cheese or butter morning and afternoon. Similarly, the personal character of the farmer or the 'missus' took on a relevance in these areas that was unthought of elsewhere. On a good farm a young man or woman could actually find a 'home'. As Fred Kitchen said of one of his 'places':[135]

The missus was a motherly old soul who referred to us as 'oor lads' – 'oor Fred' and 'oor Jack' – it had a homely sound the way she said it, and it was prefixed to every animal on the farm, making lads and animals into one large family . . . I verily believe had we been a bit smaller she would have tucked us in bed.

Social relations on these farms, particularly in the early part of our period, were close, and antagonism between master and man, though not uncommon, was personal and usually resolved by leaving at the end of the term, cursing the farmer and his kin. In some areas of Lancashire and elsewhere the tradition of 'runaway hirings' persisted. These were held a fortnight after the full hirings and by custom, rather than formal agreement, a servant could quit his or her place and seek another. An old servant gave Alistair Mutch a graphic account of this process:[136]

You'd get into a bad shop like, a bad tommy shop or he'd done something you didn't like, you tried it for a fortnight. They had the first week hirings, then the second week hirings, the 'runaway' hirings. 'Well, its no use stopping on here', you'd think, 'I'll run off like, you know, go and get a better place.' So just get your things in the middle of the night open the door and walk away . . . you'd lifted the binding money and you'd given them a fair trial, you see, and you had a legitimate excuse. If the tommy was bad, or bad treated or summat like that.

In the majority of Welsh counties the situation was similar. David Pretty points out how 'the tenant farmer stood only a shade higher in the social scale' than the labourers, and argues that 'small farms which employed one or two male workers formed a close, paternalistic unit, labourers who received board and lodging being generally regarded as part of the family'.[137] Yet again we should beware of oversimplification.

[134] *RC on Labour . . . Garstang*, p. 167.   [135] Kitchen, *Brother to the Ox*, pp. 142–3.

[136] Mutch, 'Rural society in Lancashire,' p. 185.

[137] David A. Pretty, *The Rural Revolt that failed. Farm Workers' Trade Unions in Wales 1889–1950* (Cardiff, 1989), p. 10.

Although trade unionism, a key indicator of cultural separation, was late in coming to Wales and never widespread, there was clearly the same kind of personalised antagonism in social and domestic relationships which appeared in other areas of living-in service. D. Roy Saer said of the songs of the *llofft stabal*:[138]

Most prominent of all, perhaps, with the satirical vein ranked the servants' own occupational protest songs, a defence mechanism against harsh or mean employers that was doubly vital in pre-Union days. They circulated surreptitiously but their practical effect among the labour force must have been decisive, since they explicitly blacklisted undesirable masters and farms.

A song collected in Caernarvonshire in the early years of the century shows just how bitter these 'harmonious' social relations could become:[139]

> Farewell to Begi the bulldog,/ Farewell to skinny Ifan,
> Farewell to the treacle butty,/ Farewell to the flea ridden bed;
> Penmorfa Fair is close − / That's good news!
> Farewell to the bloody brother, I'm singing 'Fal-dee-rah'
>
> (Trans. from the Welsh)

The other main form of 'living-in' was the one followed in the East Riding, parts of Lincolnshire and the east Midlands. This was where in an area of 'new' large farms a substantial part of the workforce was composed of young men who lived on the farm, but not in the farmhouse, rather they lived either with a foreman, in a separate wing or, rarely, in a 'barracks'. Again we have looked at aspects of this in the previous chapter, and I want to return again briefly to examine family and domestic life. In these areas on the large farms, as we saw above, the young men who hired themselves lived away from the house and hence away from the 'moderating' family atmosphere of the farmhouse and the farm family although they still ate with the family on smaller farms. Elsewhere they were fed in the foreman's (hind's) house.[140] This much clearer separation was a recognition of real social differences. Unlike their fellows in the north-west, few if any of the ploughboys or farm lads of these large upland farms would ever become farmers themselves nor were they very likely to be the sons of farmers. The closeness of social relationships which was noted in north Lancashire and parts of the West County and Wales by the *Royal Commission on Labour* in 1894 was conspicuously missing here. One farmer, for instance, told Edward Wilkinson that the young horsemen were 'sulky and ill-mannered', while Caunce makes the point that, while the farmer continued to have legal responsibility for

---

[138] D. Roy Saer, *Caneuon Llofft Stabal*, p. 8.   [139] *Ibid.*
[140] For an excellent account of the East Riding, see S. Caunce, *Amongst Farm Horses. The Horselads of East Yorkshire* (Stroud, 1991).

hiring and boarding the servants, the reality of this was increasingly delegated to the hind as the century progressed, reducing still further social contact between master and man.[141]

The other side of this was that the servants of the East Riding, and in a similar way in Lincolnshire and the east Midlands and parts of Wales, were a more coherent group than their north-western brethren in that they shared a common culture. Although the farms of these areas were isolated, as we have already seen, their larger size meant that the farm lads at least mixed with each other rather than just with the farmer or the farmer's family. In the evenings and on Sundays they gathered in the kitchen or more often the stable and talked, boasting of their skills and their horses, playing fox and geese or cards and singing or playing a melodeon or mouth organ.[142]

This culture also asserted itself at the hiring fair where, with their distinctive dress and public show of masculinity, the plough boys constituted themselves into something which looked very like a modern youth subculture. Folk customs were also associated with these areas, which were largely restricted to them, especially the plough-plays with their boisterous mixture of sexuality and toughness which occur only in Lincolnshire and the east Midlands, and the dances of the Plough Stots, associated with the East Riding.[143] In all this the farm lads of the north and east were a very different group from the more traditional farm servants of the north and west.

A final form of labour organisation practised particularly though not exclusively in Northumberland was the hiring of a whole family to live on the farm. We have already looked at this in some detail in the last chapter and little more needs to be added here. In general social relationships seem to have been close in Northumberland. Nevertheless, there was here, as in other areas of living in, signs, especially after the 1870s, that personal closeness could not overcome all the antagonisms of the wage relationship. It is clear from the work of Dunbabin, Mutch and Caunce that in all living-in areas informal combinations at hiring were increasingly common. In Northumberland these took on their most organised form with meetings held at hiring fairs. In 1872 a series of meetings held at hiring fairs urged the men not to hire unless they got a nine-hour day, with overtime and Sunday work to be paid extra and a quarter-day off on Saturdays. In the next two years a firmer trade union organisation emerged based on the National Agricultural Labourers Union.[144] However, the formal organisa-

---

[141] *RC on Labour . . . Driffield*, BPP, 1893–4, xxxv, p. 62; Caunce, *Amongst Farm Horses'*, pp. 197–9.

[142] Kitchen, *Brother to the Ox*, pp. 59–60.

[143] See Alun Howkins and Linda Merricks, 'The plough boy and the plough play', *Folk Music Journ.* 6, 2, 1991. Also Caunce, *Amongst Farm Horses*, Chapters 5, 14 and 15.

[144] Dunbabin, *Rural Discontent*, Chapter VII *passim*.

tion was weak and short-lived but informal organisation continued. Word of mouth and even in some areas songs and rhymes 'blacklisted' bad employers, as they did in the north-east of Scotland. More important though was the general sense, as Caunce argues, that the fairs were always the site of informal collective bargaining.[145]

The hiring fair can usefully be compared to an informal and temporary union directly shaped by the needs of the farm servants . . . With the hiring fairs, the farm houses emptied once a year and until the farmer hired his new quota of lads, he could get very little done with his horse – in other words his position resembled that of some who had been struck against. There was no need to picket, and such a ritualized strike generated no ill-feeling . . . Hirings were like wrestling matches between two very evenly matched opponents who continually tested each other's defences.

Before we move to those areas where the majority of workers lived off the farm we need to look at women workers in living-in areas. In all such areas women were hired into the farmhouse as well as men, but their experience was often very different. Most obviously they were paid less, even where they did the same work. On Castle Heaton farm in north Northumberland male workers were paid 2s. or 2s. 6d. per day in the late 1880s while women workers working outdoors were paid 1s. 4d. with no extra for harvest.[146] In Garstang women hired into the farmhouse were paid £19 a year while men got £26.[147] In most northern areas, with the major exception of Northumberland, which we discussed in some detail in the previous chapter, there was a distinction, in theory at least, between the work of men and of women.[148] For example, many women were, according to most contemporary accounts, hired as domestic servants. However, even the most cursory examination of the sources shows that the prevailing urban ideas of domestic service simply do not make sense in many rural areas. J. A. S. Green has shown that many women, described as domestic servants in Lincolnshire, were expected to milk cows, make butter and cheeses and attend animals, while Edward Higgs argues that the instructions given on the census schedules were misleading on this account especially after 1871:[149]

According to the instructions on their schedules 'Farm servants sleeping in the farmer's house must be described in this schedule as "Carter", "Dairyman" etc.,

---

[145] Caunce, *Amongst Farm Horses*, pp. 69–71.   [146] NRO 302/24.

[147] *RC on Labour . . . Garstang*, p. 164.

[148] For a detailed discussion of women's work in the north, see Judy Gielgud, 'Nineteenth century farm women in Northumberland and Cumberland – the neglected workforce', PhD thesis, University of Sussex, 1992.

[149] Edward Higgs, 'Women, occupations and work in the nineteenth century censuses', *History Workshop Journ.* 23, Spring, 1987, p. 69; J. A. S. Green, 'A survey of domestic service', *Lincolnshire History and Archaeology*, 17, 1982, pp. 65–9.

as the case may be.'" No specific instructions dealt with any distinction to be made between domestic service and work on the farm.

This problem was particularly pronounced on the small farms of the north and Wales. In Cardiganshire the 'maids' made butter and cheese, milked and looked after the farmyard animals as well as working at threshing, corn drying and at harvest times. In north Wales 'servant girls' had, according to D. Lleuffer Thomas, 'the longest hours of all those engaged for farm service, extending from 5 am until almost any hour of the night'. They did housework but also milked and worked in the fields at weeding and thinning crops as well as at harvest times. In Devon the 'domestic servants in the farmhouse . . . generally do the dairying and make the Devonshire cream, which in some places is a considerable industry'.[150] In many of these areas, especially where the farms were small, only one or two women would be hired into the farmhouse. Here, as with the men, relations with the employers were close, but with the 'missus' rather than the master. It was the 'missus' who 'oversaw' the work of the women servants, who trained them and who was often responsible for hiring them, creating, as we suggested above, a 'women's domain' of the house and farmyard.[151]

It is worth stressing again that for most servants, male or female, living-in was a stage in the lifecycle. In the north-west and parts of Wales some went, on marriage, to their own hoped-for, and long-awaited, farm to continue the cycle laid down at the end of the Middle Ages. More normally though they moved to day labour. The plough lads of Yorkshire and Lincolnshire became hired labourers, living in 'congested villages' in the valley bottoms or sometimes in tied houses. Here their lives, apart from the fact that they seldom worked with horses, followed a similar pattern to their southern brethren with a mixture of casual and day labour.[152] Many migrated to urban or industrial areas on marriage or went to the pits like Fred Kitchen.[153] A few went on the railways or used their horse-handling skills as carters, draymen or tramway drivers, but all but a very few followed a life of wage labour rather than taking their own farm.

In the southern and eastern areas of Britain farm service persisted in some areas but the lot of most who worked the land was that of the 'classic' farm labourer. In these areas, as Wilson Fox said, 'the term of engagement of the ordinary labourer is weekly'. However,[154]

---

[150] David Jenkins, *The Agricultural Community*, pp. 84–5; *RC on Labour. The Agricultural Labourer. Wales. Report by Mr D. Lleuffer Thomas . . . upon the Poor Law Union of Pwllhelli*, BPP, 1893–4, XXXV, p. 147. *RC on Labour . . . Crediton*, BPP, 1893–4, XXXV, p. 92.

[151] Gielgud, 'Nineteenth century farm women in Northumberland and Cumberland', Chapter 4.

[152] *RC on Labour . . . Driffield*, p. 54.   [153] Kitchen, *Brother to the Ox*, pp. 192–3.

[154] *Earnings of Agricultural Labourers*, BPP, 1900, LXXXII, p. 10.

in districts where farmers do not always give employment and wages on wet days to ordinary labourers . . . or do not pay them if absent on account of illness during the week, the engagement would appear, strictly speaking, to be a daily one. But the matter appears to be one of custom rather than law.

In this situation the social and domestic life of the labourer and his or her family was, for most of the time, quite separate from that of the farmer. The exceptions were those occasions when either work or the demands of paternalism brought them together. Payment in kind, though increasingly rare by the 1900s, could mediate the harshness of a relationship based only on cash wages. Harvest beer or cider was probably the most common form of this, but in some areas it spread much further than stone bottles in the harvest field. In Yorkshire although they seldom did harvest by the piece, labourers were paid extra wages for four weeks, and 'given three good meals a day, with extras of beer etc.'[155] In parts of the West Country and the Welsh Borders a wide range of what Wilson Fox called 'allowances in kind' were paid. In parts of Devon ordinary labourers received cider every day, not just at harvest, and it was seen as part of wages. In the Crediton area, for instance, they got two quarts a day in summer and three in winter. Again they were fed at haymaking and harvest.[156] In Wales day labourers often ate in the farm house at mid-day although they ate in a different room (*rhwm ford*) or at least at different tables from the farmers. This 'meating' of labourers extended to their families when those families had worked at the key communal tasks of the farming year. 'Wives of tenants and cottagers would help with hayharvest on the larger farms. For their service, the farmers would give them a regular supply of buttermilk during the whole year and a special measure of oatmeal.' Additionally, they often got bread and cheese at Christmas.[157]

Payment in kind could spread further than this in some areas. Potato ground was given in many western and some northern counties as were cheap cottages which were, however, often tied to the job. In Devon, it was said in 1894 that both cottages and potato grounds were ways 'of holding the men' especially if a cottage had a garden attached, as it often did, since men were unwilling to leave once they had that level of personal investment in the land.[158] Indeed, much payment in kind outside the small farming areas had this motivation. As Robert Hartley Lipscombe, a Devon estate steward, told the Richmond Commission, allotments made 'the men feel there is an advantage in living under the wing of a great landowner'. But it had another side, as he also firmly pointed out: 'if a man is guilty of any great or abominable crime in any

[155] *RC on Labour . . . Driffield*, p. 55.   [156] *RC on Labour . . . Crediton*, p. 95.
[157] Minwed Talbot, 'Liberality and hospitality,' p. 43.   [158] *RC on Labour . . . Crediton*, p. 95.

way, or behaves very shamefully, I take his allotment from him; and that has a very good effect'.[159] However elsewhere, particularly in the 'high farming' counties in East Anglia, this 'advantage' seems to have been foregone. Allotments were often disliked as preventing men from working hard or as giving them too much independence. Even the keeping of pigs was prohibited 'on most estates' in the Swaffham area of Norfolk since, a farmer told Wilson Fox, 'it is a temptation to take corn &c., and manure is too valuable a commodity for farmers to be able to part with, whereas without it of course allotments are valueless'.[160]

In most counties where hiring by the day or week was the norm, men in charge of animals, horsemen, cowmen or shepherds, received a larger part of their payment in kind although nowhere do we find examples where no money wages were paid at all, as was the case with the shepherds of the Border regions. These differences were less pronounced in the western and northern counties where ordinary labourers got relatively regular allowances. However, in the eastern counties, as well as higher wages, horsemen in particular got a range of 'perks', In the Swaffham district shepherds and horsemen (team-men) often got free or cheap housing and extra payments.[161] However, in all these cases even in Wales where allowances were widespread and regular the basic relationship remained one of wage labour. To the horseman of Suffolk, the team-man of Norfolk or even the *gweithwr* of Cardiganshire, the mark of Cain was all too real. For them, 'In the sweat of thy face shalt thou eat bread' was literally true.

Most areas where day labour dominated were areas of village settlement, and so most labourers lived in what were in many cases virtually occupational communities. A few tied cottages were on the farms but these were often unpopular. As Mr Hartley Lipscombe said of North Devon: 'cottages let to the farmers . . . are not popular with the labourer; he would much rather live in the village under Mr Rolle'.[162] This was not only, he insisted, because the men who lived near the farm were more under the farmer's control, but also because the men and their families preferred to live in villages 'where they can get a gossip and so on. There is not much cordiality of feeling between the farmers and the labourers.'[163] On the other side of England Wilson Fox found the same feeling in the Swaffham district:[164]

Although the cottages on the farms are generally superior to those in villages as regards construction and size of garden and, moreover, have the advantage of

[159] *RC on the Depressed Condition of the Agricultural Interest*, Minutes of Evidence Part II, BPP, 1881, XVII, 1, p. 725.   [160] *RC on Labour . . . Swaffham*, BPP, 1893–4, XXXV, p. 71.   [161] *Ibid.*, p. 69.
[162] *RC on the Depressed Condition of Agriculture*, p. 726.   [163] *Ibid.*
[164] *RC on Labour . . . Swaffham*, p. 69.

lower rental, still the labourers often prefer to live in a village where they get far less value for the rent they pay, in order to have the advantage of the society of their neighbours or the proximity of shops and schools.

For all but a minority of the labouring poor the basic unit of family life was, it has been widely argued, some version of the simple or nuclear family living together under one roof. However, there were still great variations in this, especially if the same family is seen over time. What Reay says of two villages in Kent is surely true for many areas.[165]

. . . the census type listing . . . can be misleading. It provides a convenient snapshot at a given point, yet it fails to capture movement over time. When households in Hernhill and Dunkirk are traced through their *life cycles* for three or more censuses, it emerges that from just under 50 to 60 per cent of families experienced an extended phase . . . Finally, while it is true that the mean household size was small, the *majority of people* . . . lived in households of six or more members.

Equally important, the bald description 'simple or nuclear' conceals a wide range of functions within the household. We have already noted the importance of family members as part of the productive unit of the small farm, and this was equally so in many labourers' homes. Many women continued to work 'a field' after marriage as did children, both boys and girls, from seven or eight years old. Also, outwork of many kinds was widespread throughout rural England and showed little sign of national decline before the 1890s. Although some industries like outwork-weaving had almost vanished others, not necessarily in the same areas, like shirt finishing, appeared to replace them.[166] Nor was this 'pin money'. Without the labour of women and children many labouring families would simply not have survived. Lucy Luck wrote in her autobiography, published in the 1920s, that her husband, a labourer, only earned 'twelve shillings a week, and at that time bread eightpence a quartern loaf; also meat, tea and sugar, and other things were very dear' so she continued to do straw plait working 'as hard myself as I had ever done'.[167] The Employment Commission noticed a similar reliance in parts of Sussex on children's labour. It was reported from the Hailsham District that 'the value of the child's labour to the parent in the case of large families is too great to be easily dispensed with'.[168]

---

[165] Reay, *The Last Rising*, p. 15.

[166] The central importance of women's (and children's) out-work has been consistently ignored by historians, but see Jennie Kitteringham, 'Country work girls in nineteenth-century England', in Raphael Samuel, ed., *Village Life and Labour*, pp. 127–33 for a brief survey; Pamela Horn, *The Victorian Country Child* (Gloucester, 1985), Chapters 6 and 7 have good material on children's work.

[167] Lucy Luck, 'A little of my life', *The London Mercury*, 13, November 1925–April 1926, p. 370.

[168] *RC on the Employment of Children, Young Persons, and Women in Agriculture*, Appendix Pt II to First Report, BPP, 1867–8, XVII, I, p. 83.

If the simple or nuclear family was the dominant form, there were always those who were excluded from it, or even may have chosen not to be part of it in the first place. The strict sexual morality, which many assume to have been central to Victorian England, is, as much historical work shows, a product of our own wishes rather than a reflection of reality. Detailed village studies of the nineteenth century are rare, but Reay estimates for his Kentish parishes that nearly 50 per cent of brides were pregnant at marriage, a figure broadly supported by Robin's study of Colyton in Devon (52 per cent), and Mills of Melbourn in Cambridgeshire (40 per cent).[169] Further, as Reay points out, this certainly underestimates the level of pre-marital sexual activity.[170]

When the chances of fertilization and spontaneous miscarriage are taken into account, it is quite clear that sexual relations before marriage were commonplace in these rural communities. Indeed, first births conceived in wedlock were outnumbered by illegitimacies and premarital conceptions.

This behaviour was sanctioned by folk wisdom and tradition. The earthy, if unpleasant saying current in Oxfordshire in my parent's generation that 'You don't buy a cow 'till she be in calf' represented at least some popular attitudes. In west Wales the custom of 'bundling' or courting in bed (*caru'n y gwely*) allowed the couple to spend nights together in bed or a stable loft, while 'night watching' (*noswaith w wylad*), a more widespread custom, allowed young couples to sit up all night in the kitchen.[171] Although they were supposed to remain chaste this was frequently not the case. As David J. V. Jones writes, 'Courting between the same couple often took place over several years, and sexual intimacy was a precursor to marriage.'[172] This was the reason that so much hostility was shown, even by respectable small farmers, to the bastardy clauses of the New Poor Law. John Rees, a small farmer of Pansod in Cardiganshire, told the Commissioners enquiring into the Rebecca Riots: 'If we think a girl to be a bad character, or if we think she gives a wrong father to the child, we have no objection to her having all the punishment. But in our country there is no such thing.'[173]

There were also those who were forced out of the nuclear family or rejected it. The manuscript census returns show many widows with small children and lodgers as well as unmarried mothers, especially where there

[169] Reay, *The Last Rising*, p. 12; J. Robin 'Prenuptial pregnancy in a rural area of Devonshire in the mid-nineteenth century: Colyton, 1851–1881', *Continuity and Change*, 1, 1986, p. 115; D. R. Mills, *Aspects of Marriage: An Example of Applied Historical Studies* (Milton Keynes, 1980), pp. 14–15.   [170] Reay, *The Last Rising*, p. 12.

[171] Jenkins, *The Agricultural Community*, pp. 126–7.

[172] David J. V. Jones, *Rebecca's Children. A Study of Rural Society, Crime and Protest* (Oxford, 1989), p. 38.   [173] *Report of the Commissioners of Inquiry for South Wales*, BPP, 1844, XVI, p. 56.

was outwork. In Leafield in Oxfordshire in 1871 Jane Shayler, a widow of thirty-one, supported herself, her son and her daughter by gloving and taking in two lodgers, while the same trade enabled Maryanne Pratley, aged nineteen and unmarried, to support her twenty-month old son. Other non-nuclear 'families' also occurred in the same village; Charlotte Eeles, an eighty-year old widow, headed a household made up of her son, unmarried at fifty-five, her widowed daughter of thirty-three and a ten-year-old grandson. Nor was it only women who headed such households. Elsewhere in the village Thomas Benfield headed a household comprising himself, his brother, his sister, who was married, and her daughter. No husband lived in the house.[174]

The housing of the rural poor varied enormously from area to area, and anything like a detailed coverage is a subject far outside the scope of this survey, but some general points should be made. Usually rural housing was at its best on estate villages. Many of these, as we saw in the previous chapter, had elements of social experiment about them which led to a high standard of building, but they were also so much more 'public' in that they were easily identified as belonging to a particular owner. A good example of this is provided by the Prince of Wales's rebuilding of Sandringham. According to his agent, admittedly not the most unbiased witness, when the Prince took the estate, 'the cottages were of the most miserable kind, containing in many cases but one bedroom'. However, he continued, in the twenty or so years after 1860 'every old one has been removed and replaced by a new one, or old ones have been enlarged and made comfortable'. The new cottages:

nearly all . . . have three bed-rooms and two living rooms, the former all separately arranged from the staircase. They also have out-house for firing and tools and rather more than 20 perches of garden . . . The closets are now made to be supplied with earth, and gradually the soil tanks will be removed . . . The cottages are built of car-stone, brick, tile and slate, and cost from £300–£400 per pair.[175]

Such cottages were rare, though, even in the 1890s when sanitary regulations were gradually being enforced. At Cradley in Herefordshire, in contrast to the Prince of Wales's luxury homes, the cottages for the industrious poor, despite 'forming a very pretty picture', had 'two [rooms] up and one down, and the upstairs rooms are rather one divided room divided by an insufficient screen, than two separate rooms'. These cottages also had no drainage, and 'a cesspool in the garden is used for the deposit of all refuse and a gutter from somewhere near the back door

[174] PRO, RG 9 889 MS Census Return for the Parish of Leafield.
[175] RC on the Housing of the Working Classes, BPP, 1884–5, XXXI, Minutes of Evidence and Appendix as to England and Wales, p. 594.

conveys the water waste to the cesspool'. The assistant commissioner, who wrote the report, added: 'it is to be feared that here – as in some other districts – the sanitary inspector is not sufficiently independent to insist on all being done which is required'.[176] Nearer to Sandringham, and ten years after the Prince's improvements, Wilson Fox found a large number of very bad cottages in west Norfolk. At Great Cressingham he found a family of six living in a one-up, one-down clay lump cottage in 'very bad repair', with a 'perpendicular ladder' instead of a stairway, and a well shared with four houses and a closet shared with three. There were, however, worse examples. At Sporle he found a family of eleven living in a brick and tile cottage with one room and an outhouse downstairs and two rooms upstairs, and no water within 100 yards. He at least adds here: 'sanitary authority has ordered this to be closed'.[177]

The interiors of cottages also varied enormously. Few had ovens, especially in the early part of the period and many lacked coppers. Again, though, cottages on estates were generally better. On Lord Leicester's estate by the mid-1890s cottage tenants were provided with ovens 'and in most cases with cottage ranges'.[178] To all this the poor added their own 'gear', although often that was little enough. The memoir of James Bowd, a Cambridgeshire labourer who married in 1849, put it clearly:[179]

As regards our Household stuf we had very Little true I had a bed and I had a very good Family Bible and I had but very Little money, Must I say Only three shillings not much to start a life was it, Well now I will tell you how we trudged along We Lived with my father and my Mother up to the following Mickalmas so we had a chance to gather a few sticks togther and were very Comfortable with them. [as original]

James Nye, a Sussex labourer, had even less. 'After I bought a wedding ring and a wedding dinner and paid the parson my money was gone and no house furnished.'[180] However, by the 1880s all but the very poorest homes (and there were still plenty of those) were beginning to show some of the benefits of the consumer revolution as town shops began sending carts round with 'materials at cheap rates'.[181] By the 1890s the *Royal Commission on Labour* reported from Maldon in Essex that 'the furniture and internal fittings of the cottages I visited was usually good and comfortable . . . even in some cases when the fabric was ruinous and dilapidated'.[182] Decent furniture, china plates, pictures on the wall, and

[176] *RC on Labour . . . Bromyard*, BPP, 1893–94, XXXV, pp. 80–1.
[177] *RC on Labour . . . Swaffham*, pp. 74, 77.   [178] *Ibid.*, p. 88.
[179] James Bowd, 'The life of a farm worker', *The Countryman*, 51, 2, 1955, p. 296.
[180] James Nye, *A Small Account of my Travels through the Wilderness*, ed. Vic Gammon, (Brighton, n.d. but 1981), p. 14.   [181] *RC on Labour . . . Crediton*, BPP, 1893–4, XXXV, p. 99.
[182] *RC on Labour . . . Maldon*, p. 80.

knick-knacks and fairings gave many a cottager a home and perhaps a sense of respectability. However, even this modest expenditure seemed to smack of class hubris to some of the élite, and the Duke of Bedford wrote in 1897:[183]

It is found that if two dwelling rooms of the same size are provided, one is often kept idle as a parlour where china dogs, crochet anti-macassars and unused tea services are maintained in fusty seclusion. The idle parlour adds nothing to the comfort of the cottage.

Unlike, one assumes, the dozens of rooms at Woburn.

The nature of the villages which brought together these basic family units varied enormously, as we saw in the previous chapter, but all, with the possible exception of the completely 'close' estate villages, provided places where the labouring poor could create a social system and a culture which was, as we shall see below, increasingly separate from that of their 'betters'. The street corner, or a bit of waste, provided a meeting or loitering place; the closeness of the cottages made communication easy; and the public house provided what was essentially a private and single-class club for the poor. Yet the cultural life of the labouring poor was more complex than that, and contained many elements of a common culture with those above it, at least in the early part of our period. I want now to look at this in more detail.

IV. A SHARED CULTURE? FOLKLORE, FOLKLIFE AND POPULAR CULTURE

There is a sense in which in many rural areas at the end of the eighteenth century there was a culture which was shared by all classes. This was the remnant of the old communal culture which marked out the year with public festivity, celebrating the changes in the seasons and functions of those who lived and worked within the community. What Peter Borsay says of the provincial, urban culture of the seventeenth and eighteenth centuries applied equally well to rural areas. This culture comprised[184]

. . . [a] corpus of beliefs, customs, recreation and festivals, concerned with local rather than national affairs, rooted in magic rather than reason, employing oral and visual rather than literary forms of expression, located in public rather than private space, and intimately tied to the seasonal and Christian calendars. Because it was shared by a broad range of social groups, though often on an unequal basis, it contributed a good deal to social cohesion.

---

[183] Duke of Bedford, *The Story of a Great Estate*, London 1897, p. 86.

[184] Peter Forsay, '"All the town's a stage": urban ritual and ceremony, 1600–1800', in Peter Clarke, ed., *The Transformation of English Provincial Towns* (London, 1984), p. 246.

The precise nature of this culture varied from region, but it centred on three periods of festivity – mid-winter (roughly Christmas to Shrove Tuesday), spring (May Day to Whit Week), and summer (usually around harvest). In addition there were large numbers of local festivals at different points outside these periods, depending both on the economy of the area and on specific local observances. As Bushaway writes, 'the calendar of customs was finely intermeshed with the specific structure of the local economy and society in the eighteenth and nineteenth centuries'.[185] Thus, in forest areas customs associated with beating the bounds, common droves and wood gathering had much more importance than in pasture areas. In the Forest of Wychwood in Oxfordshire, for example, a Whitsun 'chartered hunt' took place in which the forest 'towns' claimed the right to kill deer for one day.[186] In Wiltshire the inhabitants of Great Wishford had a right, fought for throughout the nineteenth century, to gather wood from the Forest of Groverly between May Day and Whit Monday. This right was affirmed by an annual ceremony of dancing to Salisbury and shouting, before the altar, 'Groverly, Groverly and all Groverly', and a wide range of similar customs existed in other forest areas.[187]

This customary calendar relied upon all classes for its continued existence. For instance, the celebration of Whitsun in many midland counties was characterised by several days of drinking and dancing, the erection of May Poles and the setting up of temporary 'Bowers', essentially temporary shelters made of rick cloths thrown over a wooden framework. This seems simple enough, yet it all required a high degree of gentry support. The field for the celebration, the tree for the May Pole, the frames and the sack cloths, all had to come from somewhere, and there had to be at least tacit acceptance that those who took part could have some time off from work. This was certainly the case in Woodstock, Oxfordshire, where the patronage of the Dukes of Marlborough was essential to the continuation of the Whit-Ale as these celebrations were called locally.[188] In May 1837, the Duke presided over the erection of 'one of the finest May Poles ever seen . . . set up at Woodstock, as the signal for the rural sports of a Whitsunale'.[189] Also at Whitsun, at Necton in Norfolk the local landowner, Colonel Mason, opened the grounds of Necton Hall for 'a *guild* or festival of rural sports', where, 'arranged under his immediate patronage, and conducted by his principal tenantry, it soon

---

[185] Bob Bushaway, *By Rite. Custom, Ceremony and Community in England, 1700–1880* (London, 1982), p. 35.

[186] Alun Howkins, *Whitsun in Nineteenth-Century Oxfordshire* (Oxford, 1973), pp. 15–18.

[187] Bushaway, *By Rite*, Chapter 6 *passim*.

[188] For Whitsun in Oxfordshire, see Howkins, *Whitsun*.

[189] *Jackson's Oxford Journal (hereafter JOJ)*, 6 May 1837 p. 4.

became, and still continues, the most respectable resort of Whitsuntide festivals in Norfolk'.[190] In Buckinghamshire the 'Right Honourable Earl Chandos Temple' supported his local Whit festivities by presenting prizes for Morris dancing at Brill in 1808.[191] The Whit-Hunt relied heavily on landowner support since it required the suspension of the game laws for the day. As late as 1837 the local press reported the active involvement of the rural élite in the Hunt:[192]

Our annual chartered hunt had a numerous attendance on Monday last. At an early hour in the morning the whole of the athletic population of Witney appeared to be in motion, and were seen pouring in crowds to the Forest Copses, the scene of the action. The noble stag hounds of the Lord Churchill threw off at five and by eight o'clock a brace of deer were killed. The sport was suspended for a time to refresh the hounds and another deer was shortly after killed.

The three deer were 'claimed' by three of the forest 'towns', Witney, Hailey and Crawley, each of whom had sent a 'company' to the hunt. Once killed they were taken back to the villages where they were cooked and sold. Those in at the kill were entitled to a piece of the skin 'and happy was the maiden whose lover could sport a piece of skin in his cap, for it brought good luck and ensured her marriage within the coming year'.[193]

The gentry played their part in other seasonal feasts organised on a less grand scale, requiring the support of the gentry and the élite, if only as donors of 'pence and spicy ales' as the wassail song puts it. Customs associated with Plough Monday, Shrovetide, Easter Monday, May Day, Midsummer, Harvest, 5 November and Christmas all required the support of the élite. When the wassailers sang:

> There's a master and mistress sat round by the fire
> While we poor wassail boys are here in the mire,
> So now pretty maiden with your silver headed pin
> Pray open the door love and let us come in.

This represented the reality of 'country customs' which could not survive without the charity of the rich, and as a result, few of them, in the first years of the nineteenth century at least, were restricted to the poor. Antiquarian surveys like Hone's *Every-Day Book* talk time and again of the presence of the rural élite as bystanders, observers and patrons of 'rural sports'. Henry Burstow's account of the customs of Horsham in Sussex in the 1840s and 1850s shows this in practice. May Day garlands, for instance, were taken round 'private residents and trades people' as well

[190] William Hone, *The Every-Day Book*, vol. II (London, 1827), p. 670.
[191] *JOJ*, 21 May 1808, p. 3.   [192] *Ibid.*, 20 May 1837, p. 4.
[193] Percy Manning, 'Some Oxfordshire seasonal festivals,' *Folk Lore*, 8, 1897–8, pp. 311–12.

as the gentry at the Manor House. 'We represented a well-recognised institution, and invariably got well received and patronised.' The Jack-in-the Greens, who came out later on May Day, also relied on local patronage. 'Lady Shelly used to patronise them handsomely by giving them plenty to eat and drink and a good round sum of money. She one year gave the Whiting party a new set of dresses fitting them out in a very gay manner.'[194] The value of these celebrations for social cohesion was apparent to many. Washington Irving wrote of May Day, 'I value every custom that tends to infuse poetical feelings into the common people and to sweeten and soften the rudeness of rustic manners without destroying their simplicity.'[195] Others saw the connections between the sports and social harmony even more clearly:[196]

Uncouth as some of these amusement may be deemed by our modern refined taste, they had their charms and their utility; the novelty and dexterity of them excited admiration; they did not tend to promote vice and immorality, and they afforded an opportunity for all ranks of people to assemble and spend their time in innocent mirth and hilarity.

However, from at least the 1820s élite support that existed for these kinds of customs was beginning to wane as the village community itself became more socially divided. When Hone compiled his *Every-Day Book* in the mid-1820s he was conscious of this change beginning.[197]

The sheep shearings are the only *stated* periods of the year at which we hear of festivities and gatherings together of the lovers and practisers of English husbandry; for even the harvest-home itself is fast sinking into disuse, as a scene of mirth and revelry, from the want of being duly encouraged and partaken in by the great ones of the earth; without whose countenance and example it is questionable whether eating, drinking, and sleeping would soon become vulgar practices to be discontinued accordingly.

The end of the common elements of this culture followed three routes. As we suggested earlier in this chapter, the large landowners and other sections of the élite progressively withdrew into class-specific and exclusive cultures of their own where they mixed only with 'their own kind'. This in itself was enough to bring many of the great festivals to an end, as Hone and others saw. Dr Falconer, whom we quoted earlier, wrote:[198]

The nobility and gentry before they adopted the pernicious custom of deserting their native mansions and misspending their time and substance in the levities predominant in foreign countries, thought [such customs] sufficient entertainment for themselves and their children to attend them.

---

[194] Henry Burstow, *Reminiscences of Horsham, being the Recollections of Henry Burstow* (Horsham, 1911), pp. 69–70.   [195] Quoted in Hone, *Every-day Book*, vol. II, p. 550.
[196] Quoted in Bushaway, *By Rite* p. 158.   [197] Hone, *Every-Day Book*, vol. II, 787–88.
[198] Quoted in Bushaway, *By Rite*, p. 158.

Even the better-off farmers and their families seemed to be setting themselves apart in a separate world – or at least that is how the poor saw it.[199]

> A good old-fashioned long grey coat the farmer used to wear sir,
> And on old dobbin he would ride to market or to fair sir.
> But now fine geldings they must mount, to join all at the chase sir,
> Dressed up like any lord or squire, before their landlords face sir.
>
> When wheat it was a guinea a strike, the farmers bore the sway sir,
> Now with their landlords they will ride, upon each hunting day sir.
> Besides their daughters they must join, the ladies at the ball sir,
> The landlords say we'll double the rent, and then their pride must fall sir.

More public, if no more potent than the withdrawal of support, was the active suppression of customs, or the transfer of support elsewhere. Between the 1820s and 1860s a whole range of public festivities were either totally put down or transformed beyond recognition. The great Woodstock Whit-ales as public and intra-class celebrations had largely ceased by the 1840s, though aspects of them lingered on as we shall see below. Elsewhere in Oxfordshire the Charlton-on-Otmoor garland procession was abandoned after 1863, the Kirtlington 'Lamb-Ale', held at Whitsun, was suppressed in 1858, and the Whit-Hunt vanished in 1852. Elsewhere, the Stamford Bull Running was suppressed in 1840, though illegal bull baiting continued in east Oxfordshire until the late 1840s, while Shrovetide football matches were either completely controlled and regulated or abolished in many areas in the north and east midlands between the mid-1840s and the late 1860s. Even where not suppressed, these, like many of the great communal customs, were brought under control of the magistrates and the police. Shrovetide football in Ashbourne in Derbyshire was only allowed to continue after 1862 on condition that it was moved away from the Market Place to a field outside the town. However the Shrovetide football which was common in Kingston, Richmond and Twickenham in the 1820s had long vanished by that date. The history of 'Bonfire', the celebration of 5 November in south-eastern England, shows both these approaches. In many towns, notably Guildford, the processions and disorder associated with Bonfire led to their suppression in the 1840s and 1850s. However, in some Sussex towns, especially in Lewes but also Horsham, agreements were entered into between 'respectable' inhabitants of the towns and the magistrates. These led to the forming of 'Bonfire Societies' or 'Bonfire Boys' who policed their own processions and proceedings. By the late 1850s Lewes

---

[199] 'The new fashioned farmer', broadside by Pitt of Seven Dials, University of Cambridge, Madden Collection, 9 (III). For these texts see Alun Howkins and C. Ian Dyck, '"The time's alteration"; popular ballads, rural radicalism and William Cobbett', *History Workshop Journal*, 23, 1987.

had four such societies representing different areas of the town. Not all attempts at giving Bonfire an acceptable face succeeded though. In Lewes efforts to force the societies to hold their fires outside the town centre were resisted until the 1900s and even today the event has a darker side.[200]

However, suppression or modification alone was seldom the only long-term solution to the problem of 'barbarous' pastimes. For this the gentry and élite pursued, consciously or unconsciously, what Brian Harrison has called, in relation to temperance, a policy of 'counter attraction'.[201] At its most simple this meant providing religiously, socially or morally acceptable alternatives to the 'old rough sports'. Side by side with this went more general attempts to reform the character of popular recreation by the classic mixture of carrot and stick. Again the history, and particularly the regional dimension, of these changes, is complex but some general points can be made. A key element was the 'moralisation' of the élites themselves and especially reforms in the Church of England in the years after 1840. A new village clergy whether high or low, 'popish' or 'evangelical', followed the dictum of Bishop Wilberforce of Oxford, 'live in your parish, live for your parish'.[202] In many villages throughout England these newly active and often young clergymen made a real impact in creating 'counter-attractions' and attacking 'sin' wherever it was found. The Rev. Edward Elton of Wheatley in Oxfordshire was a Wilberforce appointee who took up the living of this 'peculiarly rough and lawless place'[203] in 1849. He was confronted with a village which acted as a kind of Soho for Oxford undergraduates, as well as supporting a complex but violent popular culture. Although bull-baiting seems to have ended in the late 1840s, badger baiting, dog fighting and cock fighting, all illegal, continued openly into the 1850s. Drunkenness was commonplace and the village feasts were notorious. In 1851 he wrote in his diary, 'the Wheatley feast began today, a sad time of drunkenness. A badger baiting intended . . .'[204] Wheatley was 'run' by the Juggins family, descendants of a respectable yeoman family. When Elton arrived, John Juggins headed the family and the opposition to the reforming vicar. The picture painted of him by Elton when the old man died in October 1870

---

[200] On Lewes see James E. Etherington. 'The community origin of the Lewes Guy Fawkes Night celebrations,' *Sussex Archaeological Collections* 128, 1990, pp. 195–224 and R. D. Storch, *Popular Culture and Custom in Nineteenth Century England* (London, 1982).

[201] Brian Harrison, *Drink and the Victorians. The Temperance Question in England, 1815–1872* (London, 1971).

[202] *The Letter Books of Samuel Wilberforce, 1843–1868*, transcribed and edited by R. K. Pugh, (Oxford, 1970), p. 406.

[203] G. E. Russell, *Edward King, Sixtieth Bishop of Lincoln* (London, 1912), p. 53. King was Elton's curate for a time.

[204] 'Diary of the Rev. Edward Elton', in W. O. Hassall, ed., *Wheatley Records, 956–1956*, (Oxford, 1956), p. 106.

is a rare portrait of the mentality of those who opposed the moral reforms of their betters:[205]

He was the head of a clique who had set an evil example, and managed every-thing in the parish in their own way. This man lived on a moderate annuity and went about from one public house to another, living on the gossip which was current, drinking with the richer as with the poorer inhabitants . . . He was fond of local power and filled the office of church warden I believe. He liked to boast to his friends that he would keep me under . . . [and] every Easter vestry was a time of anxiety for me as he generally contrived to propose some hostile candi-date . . . He was a desperate old cockfighter, allowing fights to take place secretly on his premises and he was a patron of dog-fighting and other cruel sports. With him passed away much of the traditions of the place, and a great deal of the opposition.

Despite Juggins, and the initial hostility that forced him to fly the village for a period in 1854, Elton persisted. By the late 1850s he had started a night school and by 1860 was able to organise most of the village children into a school procession on feast day, while by the 1870s he had sufficient adult support to hold a Temperance Meeting and Tea on the day of the village feast. By 1881 the village had both a Church of England Temperance Society and a Band of Hope. Added to the policy of counter-attraction Elton used charity, especially in winter, to bring his erring flock to the fold and he also fought hard for any improvement in the village like bringing the railway to the village or the opening of a national school. By the 1880s he had largely succeeded in transforming his rough and drunken parish into, if not a model of respectability, at least something approximating to an ideal.

Nor was Elton alone even if his case was a particularly well-docu-mented one. In Melbourne, Cambridgeshire, the vicar ran a campaign to move the village feast out of the centre of the village and reduce its length from three days to one. The villagers fought back by rough 'musicking' him and breaking his windows while singing, to the tune of 'Who killed Cock Robin?'[206]

> Who stopped the Feast?
> I, said the Priest,
> I'm a meddlesome beast,
> I stopped the Feast.

In many areas of England, as Vic Gammon has shown, conflict arose between the church bands and reforming clergy who wanted to impose a new and 'more Godly' repertoire on village musicians and 'their'

---

[205] *Ibid.*, pp. 109–10.
[206] Enid Porter, *Cambridgeshire Customs and Folklore* (London, 1969), p. 144.

churches.[207] In village after village, especially from the foundation of The Society for Promoting Church Music in 1846, new clergy, especially high church clergy, sought to introduce organs and 'modern' hymns, frequently against the opposition of the old wind bands with their repertoires of metrical psalms and traditional carols. The choirs and many villagers fought back as best they could with deferential delegations to the new vicar, as in Hardy's *Under the Greenwood Tree*, and meetings. When this failed they took to other courses of action, as described in an editorial in *The Parish Choir*, the magazine of The Society for Promoting Church Music, written in 1847.[208]

We have heard of one instance in which a body of young men who were being trained in church music were attacked on their way home from a practice by a mob consisting of friends of the 'old quire', and the staunch admirers of the Babylonian performances with which they made the wall of the church respond for many a long day. All were ill used, and one had his eye knocked out by a stone.

There was little point, though, and by the end of the 1850s *The Church of England Quarterly Review* reported that 'the days are happily numbered in which a fiddle and a bassoon were looked upon as the appropriate accompaniments to a church choir'. The article continued, 'few churches are now without an organ, and the wives and sisters of the clergy form an excellent staff of organists'.[209] Victory for the respectable reformers in this area was marked by the publication and widespread acceptance of *Hymns Ancient and Modern* after 1861 in which, as Gammon says, 'we see the national and standard triumphing over the local and various'.[210]

The transformation of rural 'popular' culture in the years after 1850 was not only a matter for the élite. In many areas, especially the West Country, Wales and parts of East Anglia, plebeian nonconformity played a key role. Temperance, for instance, was by no means simply imposed from above but widely supported, indeed often initiated, by working people themselves. When George Rix, a Norfolk labourers' leader, urged his fellow workers to 'unite for mutual intercourse, instruction and information [and] leave off smoking and tippling' he was merely echoing ideas widespread among a large section of rural workers.[211] In west Oxfordshire, John Kibble, a methodist stone mason and historian of his native area of Charlbury, made the link from drinking to other forms of public entertainment. 'Sometimes we heard a fiddle. This was very

[207] Vic Gammon, '"Babylonian performances": the rise and suppression of popular church music, 1660–1870' in S. and E. Yeo, eds., *Popular Culture and Class Conflict. Explorations in the History of Labour and Leisure* (Brighton, 1981).

[208] *The Parish Choir*, 18, June 1847, p. 145. Quoted in Gammon, 'Babylonian performances', p. 78.

[209] *Ibid.*   [210] *Ibid.*, p. 82.   [211] *Eastern Weekly Press*, 15 May 1880, p. 8.

delightful to our young ears but generally associated with drink, dancing, shame and sorrow which made it not quite the thing for us to know.'[212] In many parts of rural Wales the nonconformist conscience was all pervasive and it produced, for its adherents and members, a total culture. Chapels were:[213]

social centres providing access to literary and musical entertainments, at first through hymn singing but later through the performance of classical choral works. Above all else, these chapels provided the means by which the Welsh people attempted to educate themselves.

As this suggests, a key element in the process of reformation of plebeian culture, whether from above or below, was the creation of alternative and more 'rational' forms of entertainment. For example by the 1870s the rough sports of Whitsun and May Day had been replaced, in many midland counties at least, with the ordered display of the village friendly society going 'to Church in a decent manner, walking two and two' behind their club banner.[214] By 1862 at Bletchington in Oxfordshire, a couple of miles from the site of the Kirtlington Lamb Ales which had only been suppressed five years earlier, the local press applauded the change of scene:[215]

It was a pleasing sight to see nearly a hundred young, fine, clean, and well-dressed labourers follow their banner to the quiet old church – it was a convincing proof of what unanimity and good feeling can effect . . . This was a meeting bearing strong contrasts to those of years gone by, when riot and drunkenness was the result.

It is important to stress here that, although a part of this transformation was 'superimposed', a good deal was not. What David Neave has written about the 'club days' of the East Riding suggests that in some parts of England local control by the gentry was far weaker than in the midlands or parts of East Anglia. 'Respectability', he writes, 'was not a product of social control by landowner and parson but it had been achieved by the members on their own terms . . . The approval and financial support from other classes was welcomed but there was great hostility to any form of direct control and subordination.'[216] As so often is the case the 'truth' lies somewhere in between these two extremes of 'self expression' or 'social control'. As Cunningham has written, 'leisure is perceived as a field of

---

[212] Kibble, *Historical and Other Notes on Wychwood Forest*, p. 54.

[213] Christopher Turner, 'The nonconformist response', in Trevor Herbert and Gareth Elwyn Jones (eds.), *People and Protest: Wales 1815–1880* (Cardiff, 1988), pp. 79–80.

[214] *Rules of the Victoria Club held at the Harcourt Arms, in Stanton Harcourt*, Bampton, Oxon., 1874, p. 8.

[215] *JOJ*, 21 June 1862, p. 8.

[216] David Neave, *Mutual Aid in the Victorian Countryside: Friendly Societies in the Rural East Riding 1830–1914* (Hull, 1991), p. 95.

contention and negotiation in which the outcome was neither the submission of subordinate groups to new standards nor an untrammelled celebration of class identity'.[217]

Nevertheless, 'respectability' became an increasingly important part of
the public culture of rural areas, whatever its source. The festivities of
Spring, Summer or Mid-winter became shorter, eventually compressed
into the one-day breaks of the newly established Bank Holidays Act,
although the Act did not apply to farm workers. Broader national changes
also had a rural edge. The 1874 Licensing Act introduced 10 pm closing
for all licensed premises on weekdays and restricted Sunday opening to
two hours at lunchtime and four hours in the evening. In Wales this was
followed by a complete ban on Sunday opening in 1881 and in England
by further restrictions by the 1900s. This was a factor in the general
decline in alcohol consumption which fell from 1.11 proof gallons of
spirits per head in 1831 to .22 in 1931, while beer consumption fell from
21.6 gallons per head to 13.3 gallons over the same period.[218] Festivity
also became less violent as nationally crimes of petty violence and drunkenness decreased, even if they never vanished.[219] By the 1900s the characteristic 'public' holiday of the rural areas was probably the fete and
sports day which, although it may have had a beer tent, was certainly a
model of decorum compared with the 'saturnalia' of the Whit-Ale. An
ironic comment on this is provided by an account in the local press in
Oxfordshire from the village of Milton-under-Wychwood in 1903. It
was reported that on Whit Monday 'a May-pole dance, conducted by
Miss M. D. Venvell, was much enjoyed, the children taking part being
picturesquely attired in quaint Old English costumes'.[220] Just under a
hundred years earlier a quite different May Pole had gone up to mark the
beginning of the week-long celebration of Milton's Whitsun Ale.[221]

Yet this account is still too straightforward, even with the gloss given
by Cunningham, for if the public world of culture was made more
respectable a darker side persisted outside the influence or gaze of the
reforming clergyman or self-improved artisan. Probably most important,
although least spectacular, was the persistence of a whole cosmology of
folk belief which, particularly in the early part of the period, had as great
an effect on how most people viewed the world as the more formal belief
systems of the educated élite. Edward Peacock, a Lincolnshire antiquarian, put it neatly: 'Those who are not in daily intercourse with the

---

[217] H. Cunningham, 'Leisure and culture', in F. M. L. Thompson (ed.), *The Cambridge Social History of Britain, 1750–1950*, vol. II, p. 335.   [218] *Ibid.*, p. 331.

[219] Clive Emsley, *Crime and Society in England, 1750–1900* London, pp. 40–2. See also V. A. C. Gatrell, 'Crime, authority and the policeman-state', in F. M. L. Thompson (ed.), *The Cambridge Social History of Britain 1750–1950*, vol. III, *Social Institutions and Agencies* (Cambridge, 1990), pp. 293–7.

[220] *JOJ*, 7 June 1903, p. 4.   [221] *Ibid.*, 14 May 1808, p. 3.

peasantry can hardly be made to believe or comprehend the hold that charms, witchcraft, wise-men, and other relics of heathendom have on the people.'[222] Throughout England and Wales up to the Great War, and indeed later, the everyday acts of working and living were accompanied by frequent recourse to charms, signs and tokens. The following examples from Staffordshire were found time and again in county after county.[223]

It was then considered unlucky to start a journey on Friday, the day of the Crucifixion, and also to turn back, once a journey had begun, or even to say good-bye at a gate and, in certain circumstances, to meet a woman on the way to work. In the north of the County, those who saw a person in a round hat thought it necessary to touch iron . . . It was lucky to have a crooked sixpence, while looking at a new moon through the window and turning a chair right round on one leg brought bad luck. It was also unlucky to have a haircut when the moon was waning. The number thirteen was unlucky . . . but other odd numbers like three, five, seven, nine, and eleven can bring good luck. It was believed that if thirteen people sat at a table the first to rise would die within a year . . . Misfortune may occur to anyone eating a double nut.

The changes in the lifecycle were also marked by beliefs and customs which took on a much more serious meaning especially where they predicted the future or influenced future events. In Lincolnshire, as elsewhere, funerals were associated with many such beliefs. 'After a funeral procession has started, *no one* must "head" [precede] the coffin until the parson comes to meet it inside the Churchyard, or the worst kind of luck and sudden death will surely overtake the person who has "headed" it.' Or from elsewhere in the same county, 'On no account must there be an *odd* number in the funeral party, or the dead will soon call out for a companion.'[224] Less grim beliefs were associated with courtship and marriage. Attempts to predict the name of a future husband or wife, for example, took many forms. In Carmarthenshire, and many other counties including Oxfordshire in my own childhood in the 1950s, an apple was peeled and the peel thrown back over the head. Then when this peel had hit the floor, particular notice was taken in what form it appeared, and whenever it resembled a letter of the alphabet, the same was supposed to be the first letter of the Christian name of the thrower's future wife or husband.[225] In Warwickshire, Roy Palmer writes:[226]

[222] *Notes and Queries*, 2nd ser., 1856, p. 415, quoted in Keith Thomas, *Religion and the Decline of Magic* (Harmondsworth, 1973), p. 798.
[223] Jon Raven, *The Folklore of Staffordshire* (London, 1978), pp. 68–9.
[224] Ethel H. Rudkin, *Lincolnshire Folklore* (Gainsborough, 1936, new edn, London, 1973), p. 15.
[225] Jonathan Ceredig Davies, *Folk-Lore of West and Mid-Wales* (Aberystwyth, 1911), p. 78.
[226] Roy Palmer, *The Folklore of Warwickshire* (London, 1976), p. 90.

The coming of a wedding might be foretold simply by the girl's cheek burning, or alternatively by the appearance of three magpies: 'one for sorrow, two for mirth, three for a wedding, four for a birth'. If the girl were curious as to who the man thus indicated would be, she scattered fern seed in a garden or wood at midnight on Midsummer's Eve, saying:

> Fern seed I sow, fern seed I hoe.
> In hopes my true love will come after me and mow.

She would then see the young man's image.

This traditional belief system was not simply, as many have maintained, basically pessimistic since it also enabled those who followed it to intervene in the processes of nature, predict their future or protect themselves from bad luck. The East Anglian faith in the 'toad's bone' which gave a man power over horses is one such set of beliefs.[227]

I had a relation got to be groom for Lord Buxton. Well he could practically do anything with horses, make them lay down and all sorts. He was supposed to be one of these men who had a running toad. That's a particular toad . . . you got to stick something through it, stick it in an ants nest, and then the ants will eat all flesh off it. Then you take the bones, break them up, and go to a running stream at midnight and put them in the stream. And you get the bones that go against the tide . . . and if you've got those two you can do anything with horses. [These old team men] they kept them, you could never get rid of them, they said, they said that was something to do with the Devil, and they always advised you not to. Them that had them, then generally died, went mad or something . . .

More common were folk remedies and medicines. Again, like 'good luck' charms, these were ubiquitous, and while they had local variations which are of considerable interest, there is a basic similarity from county to county. Cures for common ailments like warts, coughs and colds were similar in most English and Welsh counties. In many areas warts, for instance were rubbed with a snail or a slug which was then impaled on a thorn. As the slug or snail withered and decomposed so the wart would go. Elsewhere the snail was replaced by a piece of stolen meat or even an apple which was then buried.[228] Whooping cough, another common complaint, had a wide variety of cures, although again many were common to many counties. Frying a mouse in butter was recommended in Staffordshire and Cambridgeshire although the latter cure was also used for bed-wetting in Lincolnshire. An alternative cure offered in Cambridgeshire and Somerset was to let a child 'run with the sheep'.[229]

[227] Interview Alun Howkins/Jack Leeder, Knapton, Norfolk, Aug. 1974. Tape in author's possession.

[228] Palmer, *Warwickshire*, p. 67; Ralph Whitlock, *The Folklore of Wiltshire* (London, 1976), pp. 165–6.

[229] Raven, *Staffordshire*, p. 50; Porter, *Cambridgeshire*, pp. 89–90; Kingsley Palmer, *The Folklore of Somerset* (London, 1976), pp. 114–15.

To see these beliefs as simply quaint or even as 'archaic survivals' is both to misunderstand them and to underestimate their importance. Although hidden from the élite until the growth in interest in folklore at the end of the nineteenth century, they were central to the lives of the poor. James Obelkevich, in what is still one of the few serious discussions of this belief system, writes of them,[230]

This universe was the result of syncretism – the universal religion of the peasant – which combined elements from the 'higher' religion and the 'lower' religion without any regard for logical compatibility. It involved no inconsistency for a villager to attend the parish church on Sunday morning and the Methodist Chapel in the evening – and with equal conviction put up a horseshoe over the door or ask permission of the 'Old Gal' before chopping elder wood . . . If nature affected men's affairs – usually for the worse – men could still affect, and participate in, Nature. Although villagers had little control over the circumstances of their lives, they could respond to hardship in other ways than prayer and passivity, by taking advantage of favourable luck and by performing magical techniques.

There was another, and more public side to this universe which involved folk rituals and performance, particularly of song, dance and folk tale. Some aspects of these were linked to the old 'communal' festivals and suffered decline alongside them, but many lingered on among the poor, despised and ignored by their betters. The great corpus of English and Welsh traditional song, for example, was performed and preserved almost entirely by the rural poor. Its style of performance and its repertoire, like that of the church bands, was, at best patronised, and at worst mocked, by the élite, before the work of Cecil Sharp and others in the 1890s and 1900s. Yet it remained a living form adapting to change well into the twentieth century.[231] Performed largely in public houses or at home, the site of much singing was away from the view of either the gentry or the farmers, as well as the respectable working men and women of the village. This separateness was added to by its performance style which was unaccompanied and declamatory, very different from the approved modes of Victorian musical performance. Further, much of the content of the songs was considered 'shocking' or, less often, subversive. For example, when Vaughan Williams published the song 'The Long Whip' in the *Journal of the Folk Song Society* in 1906, he published only one verse, adding, 'the rest of the words are not suitable for this journal'. Similarly, the collector and composer, George Butterworth,

---

[230] Obelkevich, *Religion and Rural Society*, p. 307.

[231] The best account of the social context of traditional music performance, unfortunately, remains unpublished. It is V. A. F. Gammon, 'Popular music in rural society', PhD thesis, University of Sussex, 1985. See also Reg Hall, *"I Never Played to Many Posh Dances". Scan Tester, Sussex Musician, 1887–1972* (Rochford, Essex, 1990).

omitted five of the nine verses of 'Fourteen Years of Age', collected from Mrs Verrall of Monks Gate near Horsham, on the same grounds, which made a complete nonsense of this powerful song of seduction and desertion.[232]

However, aspects of folk culture and practice were less hidden but no less resistant to control. Throughout the nineteenth century parts of the folk culture persisted which were neither picturesque nor easily assimilated within a respectable holiday calendar. In the summer of 1910, for example, a group of men, women and children from St Faiths in Norfolk followed a strike breaker, Charles Rayner, home from work beating pots and pans and making discordant noises on musical instruments. It was, according to their defence, a 'custom not unknown in the county' to show disapproval of those who infringed the communities norms. One of those accused said that:[233]

'We heard on Monday night that he had knocked old Bridgett [an elderly man] about and we said we would 'ting' him to make him ashamed of himself' . . . Witness had taken part in their 'tingings' and on the previous evening helped to 'ting' Manes because he would not come out. They 'tinged' Mallet, Bridger and Oakes because they kept at work.

Despite (or perhaps, because of) this defence they were all fined £5 each or, in default, sentenced to two months hard labour.

What they were doing was of course sanctioned by custom and practice. It was a mild version of the practice known elsewhere as 'riding the stang', 'skimmington', 'rough music', 'ran-tanning' or, in Wales *Ceffyl pren*. All had in common a procession and the use of discordant sound, usually beating pots and pans, often accompanied by obscene or threatening verses. Usually the procession went to the offender's house for three nights running. In many areas it was more elaborate still. In 1839 the 'Constabulary Commission' was told that in many areas of Cardiganshire and Carmarthenshire:

the magistrates are greatly embarrassed by the increasing practice called the 'Ceffyl pren', or wooden horse; a figure of a horse is carried at night in the midst of a mob with their faces blackened, and torches in their hands, to the door of any person whose domestic conduct may have exposed him to the censure of his neighbours, or who may have rendered himself unpopular, by informing against another, and by contributing to enforce the law. On the horse is mounted some one, who, when the procession makes a halt opposite the residence of the person whom it is intended to annoy, addresses the mob on the cause of their assembling, and on the delinquency of the obnoxious party. When the exhibition is directed against supposed domestic irregularities, it is often accompanied with the grossest indecency.

---

[232] Gammon, 'Popular music', p. 168.   [233] *Eastern Weekly Press*, 27 August 1910, p. 3.

This was not the only problem though, since by the late 1830s the practice was directed against the magistrates themselves, employers and 'a Scotch [land] agent' who had informed against someone cutting wood. Worst of all it was also directed against 'a clergyman, who procured the restoration of a stolen sack to its owner'.[234]

In the next few years the practice was to grow more serious still when the model of the *Ceffyl pren* was used by Rebecca and her daughters in the ceremonies of gate breaking which were part of the Rebecca Riots.[235] As a result the authorities moved against the ceremony. In the early 1850s in the aftermath of Rebecca, Major General Love, the commanding officer of troops in South Wales, told the Home Office, 'the common people . . . are very prone to take the Law in their own hands, or perhaps more properly speaking having recourse to punishments of a wild and disorderly nature such as the "Cyffl-pren [*sic*] . . . but latterly the Police with the aid of the Military, (not infrequently called out for this purpose) have prevented taking place'.[236] The public and quasi-political *Ceffyl pren* seems not to have survived the authorities' concern but in much of Wales and elsewhere it remained as a public judgment on domestic behaviour. In 1893 Scan Tester, the Sussex concertina player, aged five or six at the time, took part in a rough musicking of an unpopular landlord in Horstead Keynes:

They had all sorts of instruments; some had teapots and blowed in the spout and they had anything they could get hold of, and we done it three nights running. You didn't dare do it no more . . . We marched up and down the road three times, then we had to leave off, see. Course, a lot of them knew the rules about rough musicking, and, I'll tell you, that's a tidy row that is. I should think every workman there was in the village was there, and, I can tell you, there was a tidy gang there. These police was there to see we didn't cause no trouble. Well that's the law. That was a law, you daren't rough music more than three nights.

Nor was this the last occasion in this area. As late as the 1950s a man was rough musicked in West Hoathly, also in East Sussex.[237]

Rough music was the visible manifestation of communal law and identity which made up a separate and often antagonistic culture, restricted to the rural poor at least after the 1840s and 1850s. The law of the new

---

[234] First Report of the Commissioners Appointed to Inquire as to the Best Means of Establishing an Efficient Constabulary Force in the Counties of England and Wales, BPP, 1839, XIX, p. 44.

[235] Jones, *Rebecca's Children*, pp. 195–8.

[236] Quoted in Rosemary A. N. Jones, 'Popular culture, policing and the disappearance of the Ceffyl Pren in Cardigan, c. 1837–1850', *Ceredigion*, 11, 1, 1988–9, p. 33.

[237] Hall, *"I Never Played to Many Posh Dances"*, p. 129. For general discussions of 'rough music', see E. P. Thompson, '"Rough Music"; le Charivari anglais', *Annales E.S.C.*, 27, 1972, pp. 285–310. (This is now republished in English in *Customs in Common* (London, 1991)); Alun Howkins and Linda Merricks '"We be blacke as hell"; ritual disguise and rebellion', *Rural History*, 4, 1, 1993.

police and the social structure of parson and squire which stood behind it could not, and did not, take account of the needs and ideas of the poor. Wood 'stealing', flower gathering, turf cutting, gathering furze for firing or bedding for animals, led into 'poaching' and were all sanctioned by, and protected by, custom and belief. The élite may no longer have believed in these 'rights' but to the poor they were fixed within their everyday lives. In the 1860s and 1870s on the outskirts of Oxford the 'poor' of Headington Quarry fought for the rights (which legally did not exist) to graze animals, cut furze and kill rabbits on a stretch of land called the 'Open Magdelens' and the 'Open Brasenose'. As late as the 1890s one of the 'commoners', George Webb of Headington Quarry, was charged with 'wilfully damaging certain shrubs . . . the property of Abel Bicknell'. Webb's defence was that he had taken the wood (blackthorn, white thorn and furze) for firing from 'Brasenose Common' and 'that he had cut wood there for the past fifty years, and that if the witness were there the next day he would see him have some more probably, for the wood belonged to the poor of Headington'.[238] That no such right existed was in a sense neither here nor there. The gathering of wood was a central part of the village's domestic economy and that was recognised in the corpus of tradition and oral history which passed on from generation to generation the belief in 'rights' of the poor of Headington. Cases like this could probably be multiplied time and again in most areas of England. Certainly the cases of Ashdown Forest in Sussex and Mousehold Heath in Norfolk show the extent to which very well organised resistance to enclosure existed in the 1870s and 1880s.[239] Additionally the *Report of the Select Committee on the Inclosure Act* of 1869 suggests that resistance to enclosure and insistence on rights of various kinds occurred in Surrey, Sussex, Devon, Yorkshire, Rutland, Lincolnshire, Westmorland, Glamorgan and Suffolk, and there is no doubt that if detailed local work were done more cases would be revealed.

Half-remembered customary rights could be suppressed by court and police but it was a less obvious process which gradually wore away the folk cosmology as a whole. A key element was the slow but certain penetration of national and urban cultures into the rural areas which increasingly made the old ways seem out-moded, irrelevant, and even embarrassing especially to the young. The penny post, the railways and the gradual spread of newspapers, both local and national especially after the 1870s, all contributed to this long and complex process.

A key element here, the increased pace of rural migration, is dealt with

---

[238] *JOJ*, 11 March 1895, p. 7.

[239] On Mousehold Heath, see Neil Macmaster, 'The battle for Mousehold Heath 1857–1884', *PP*, 127, May 1990. For Ashdown, see R. Cocks, 'The Great Ashdown Forest Case', in T. G. Watkin (ed.), *Legal Record and Historical Reality* (London, 1989).

elsewhere, but it is still worth noting what the young man or woman who went to 'town' sent back or brought back with him or her. In Flora Thompson's *Lark Rise to Candleford* there is a careful description of one of the effects of this in how three different groups – the old, the middle aged and the young – sang different songs in the village pub in the 1880s. Significantly only the very old then sang only folk song. The young sang music hall songs learned from urban songbooks, bought in the market town or sent by brothers and sisters 'away'. The 'old' was not yet gone but the young showed little interest in it.[240] A similar fate seems to have befallen Morris Dancing in many villages where it was increasingly difficult to get young men to take part. In Abingdon in Berkshire, for example, it was only the visit of the folk dance collector, Mary Neal, in 1909 and the interest and one assumes the status brought by her to the elderly Hemmings brothers which 'revived' the tradition in the town.[241] Elsewhere, like Leafield (Field Town), even the activities of the collectors, Percy Manning and Cecil Sharp, over a twenty-year period could not revive the side or encourage the young to take up the dances. Keith Chandler puts it well in his excellent study of the south Midlands Morris, *Ribbons, Bells and Squeaking Fiddles*:[242]

But the root of the problem of transmission of lore was more deeply entrenched. It lay in the unwillingness of young men who remained in the area to become associated with a cultural form increasingly at odds with acceptable social behaviour . . . At Leafield the old Morris foreman George Steptoe succinctly summed up the problem when he commented how 'the lads arter we gin out never seemed to get on with it'.

Another important element, especially in relation to the popular cosmology was the growth of first a local, and then a national, press. Through this the influences and ideas of the urban world were brought into the villages and the villages were often found wanting. As Flora Thompson wrote, '*Tit-Bits*, was taken by almost every family, and the snippets of information culled from its pages were taken very seriously indeed.'[243] By the 1880s and 1890s weekly papers like *Reynolds News* had reached most rural areas, bringing with them urban and London ideas and values which pushed the older ways further to the margin, both geographically and socially. 'The farm labourer', wrote Henry Rider Haggard in 1902, 'is looked down upon, especially by young women of his own class' and there was seldom much more need than that to become

[240] Thompson, *Lark Rise*, pp. 69–75.

[241] Jonathan Leach, *Morris Dancing in Abingdon to 1914* (Eynsham, 1987), pp. 21–8.

[242] Keith Chandler, '*Ribbons, Bells and Squeaking Fiddles*'. *The Social History of Morris Dancing in the English South Midlands, 1660–1900*, Publications of the Folklore Society: Tradition, 1 (London, 1993), p. 217.   [243] Thompson, *Lark Rise*, p. 499.

more modern.[244] Also in the pages of these papers, especially *Tit-Bits* and *Answers*, 'superstitions' figured alongside 'interesting facts' – the cosmology of the countryside became the object of urban patronising and hence rural shame. Gradually to admit that you believed in any of the 'old stuff' was to mark you off as a 'clod hopper', backward and stupid. Other pressures, especially education, are dealt with elsewhere, but they added to these changes, national and local, and the changing nature of public festivity outlined above, to undermine, slowly but surely, the old customary framework and the peasant cosmology. George Sturt looked at the end of this old world in 1912.[245]

. . . the 'peasant' tradition in its vigour amounted to nothing less than a civilization . . . To the exigent problems of life it furnished solutions of its own – different solutions, certainly, from those which modern civilization gives, but yet serviceable enough. Beside employment there was an intense interest for them in the country customs . . . Best of all, those customs provided a rough guidance as to conduct – an unwritten code . . . It is [in] the virtual disappearance of this civilization that the main change in the village consists. Other changes are comparatively immaterial . . . had but the peasant tradition been preserved in its integrity amongst the lowlier people; but with that dying, the village too, dies where it stands.

V. THE NEWCOMERS

This revival of our national English folk music is . . . part of a great national revival, a going back from the town to the country, a reaction against all that is demoralising in city life. It is a re-awakening of that part of our national consciousness which makes for wholeness, saneness and healthy merriment.

Mary Neal, *The Esperance Morris Book*, 1909

For generations England's urban population had looked back to the land. As Raymond Williams explained:[246]

English attitudes to the country, and to ideas of rural life, persisted with extraordinary power, so that even after the society was predominantly urban its literature, for a generation, was predominantly rural; and even in the twentieth century, in an urban and industrial land, forms of the older ideas and experiences still remarkably persist.

However, it had not been restricted to literature. The London merchant made squire by buying an estate and title is a stock figure of history as well as of imaginative writing from at least the sixteenth century. Even at

---

[244] Henry Rider Haggard, *Rural England*, 2 vols. (London, 1902), vol. II, p. 540.

[245] George Bourne (Sturt), *Change in the Village* (Harmondsworth, 1984), pp. 69–70.

[246] Raymond Williams, *The Country and the City* (pbk edn, London, 1975), p. 10.

the height of the Industrial Revolution few great industrial magnates could keep a place in society without a country house of some kind even if its estate was more for show than for profit.

In the years after 1850, but especially after 1880, this rural impulse seems to have changed qualitatively and quantitatively. At one level, as Martin J. Weiner has argued, the ideology of the 'new' élite of the industrial revolution was, after the first generation at any rate, essentially rural in its direction, looking to the 'country' values of the old aristocracy rather than new urban and bourgeois ones of the town middle classes. As a result 'new money' attempted to integrate itself into the upper reaches of county society by following its way of life particularly in relation to country living, country 'pursuits' and education.[247] Even in Wales there is some evidence of a similar pattern, where late nineteenth-century mine and factory owners like Sir Henry Hussey Vivian deserted the valleys where their wealth was made for a country house in England, as well as a 'London house, a palatial home in Scotland and a yacht'.[248]

Aspects of Weiner's account have been strongly criticised by W. D. Rubenstein, among others, who argue that very little capital was moved from industrial to landed wealth in the second half of the nineteenth century, a view challenged in turn by F. M. L. Thompson.[249] Further, Hartmut Bergoff argues that the public school system only educated a very small part of the business community.[250] Nevertheless, many of Weiner's central ideas about the ideology of the élite remain, and in fact seem to be accepted by all the protagonists even if they disagree about the 'reality' of land purchase. Indeed, Weiner's 'literary sources', so scorned by many historians, probably underestimate the extent to which a simple, even naïve but personal anti-industrialism was spreading far outside the élite in the years after 1880, and becoming a central part of the beliefs, and above all hopes, of a substantial section of the urban lower middle class and even the respectable working class. The sources and variety of this anti-industrialism are complex. We mentioned earlier the impulse of radical agrarianism as a constant factor in radical movements of the nineteenth century,

---

[247] Martin J. Weiner, *English Culture and the Decline of the Industrial Spirit, 1850–1980* (Cambridge, 1980).

[248] G. W. Roderick, 'South Wales industrialists and the theory of gentrification; 1770–1914', *Transactions of the Honourable Society of Cymrodorion*, 1987, p. 78.

[249] This is a very complex argument which is dealt with elsewhere in these volumes. For a brief guide see the debate between W. D. Rubenstein and M. J. Daunton in *PP*, 132, 1991, pp. 150–87. Most recently see F. M. L. Thompson, 'Life After Death: how successful nineteenth-century business men disposed of their riches', *EHR*, 2nd ser., 43, 1990, pp. 40–61; W. D. Rubenstein, 'Cutting up rich: a reply to F. M. L. Thompson', *EHR*, 2nd ser., 45, 1992, pp. 350–61, and Thompson's further comments, 'Stitching it together again', pp. 362–75 of the same issue.

[250] Hartmut Bergoff, 'Public schools and the decline of the British economy 1870–1914', *PP*, 129, 1990, pp. 148–67.

but a desire to go 'back to the land' was a small part of a much wider structure of ideas and feelings. In essence it concerned a rediscovery of rural England (as opposed to Wales or Scotland) and a revaluation of the 'simple', 'pure' and rural as against the corruption of the urban. At its most extreme it nurtured eugenicism and even racism, but its most normal form expressed itself in a practical sentimentality for the rural, and especially for rural life. The 'practical' part of the practical sentimentality took a number of forms only two of which will concern us here. First the literal move back into the countryside and the extent to which these 'newcomers' differed in their 'social history' from the other groups has already been discussed; so in second place have the ways in which rural life, and especially rural culture, were revalued as part of this process.

As we have already said, sections of the urban population, especially the urban élite have always moved 'back' into the countryside. Whether such movements had much effect on the community around, or indeed on the lives of those who made the move, depended very much on the where and when. In most cases, one suspects, the whole process was viewed ambiguously, as it was in Michael Home's native village of Hockham in Norfolk. Here, in the early 1890s, the Hall was bought by a member of the Baring family from a relatively long-established 'squire'. The Barings could not command the same respect as the old squire, but they brought prosperity to the village by developing the shooting rights. As a result they were accorded a grudging deference but little more.[251] However, the changing tenancy of the great house, no matter how significant at one level, was not a new phenomenon and was very different from the influx of much larger numbers of middle-class newcomers into the villages of the 'Home Counties', and around many other great conurbations, from the 1880s and 1890s. This movement took two forms: the creation of estates, be they village or suburb based, and the gradual movement by individuals or groups of middle-class, especially 'arty' people into more remote villages. The two of course were not exclusive, and the first often followed on from the second.

In some sense the most obvious incursion into the rural areas by a non-agricultural population was seen in the suburbs, the gradual outward growth of the urban areas of the nineteenth century. The clearest example, though by no means the only one, is the expansion of London. As London's importance grew as a financial and commercial centre so the relative size of its resident population declined. By 1871, '750,000 clients a day poured into the commercial heart of London, served by twelve railway stations and 170,000 employees, but fewer than 75,000 inhabitants'. Increasingly the homes of these workers were in the suburbs, the

[251] Michael Home, *Winter Harvest* (London, CBC edn, 1969), pp. 114–21.

new areas built in a ring around the metropolis, and built on former village sites. As a result the 'suburban' counties of London had the fastest population growth rates of any areas of England and Wales by 1900, with Essex at 36.3 per cent, Surrey 20.5 per cent and Kent with 16.8 per cent.[252] By the late 1860s the former village and even small-town communities of Tulse Hill, Camberwell, Peckham, Dulwich, Clapham and Streatham had been absorbed. However, it was the expansion of the suburban railway network which was crucial in pushing the town outwards and creating, between the 1850s and the Great War, a huge commuting region around London where city and country merged into an admixture which was neither one nor the other.

However, suburbanisation quickly destroyed aspects of the rural. Infilling removed the market gardens and small farms which had initially prospered under the impact of a much increased local market, and gradually and inevitably the wealth moved on. As Charles Booth wrote of south London in 1902, 'Southwark is moving to Walworth, Walworth to North Brixton and Stockwell, while the servant-keepers of Outer London go to Croydon and other places.'[253] Similarly, the cities of Lancashire generated a carefully graded band of suburbs taking over more and more of the old agricultural villages. Katherine Chorley's memoir, *Manchester Made Them*, presents a fascinating picture of the complex social geography of the south Manchester suburbs around the 'village' of Alderley Edge.[254] As John Walton says: 'At all levels of the propertied and professional classes across Lancashire . . . the flight from the city centre accelerated [after the mid-century] and the most desirable goals became more geographically remote.'[255] Even in Wales, especially around Cardiff and to a lesser extent Swansea, a similar phenomenon was beginning:[256]

If Cardiff was not yet the Athens of the west, it certainly exhibited some of the style and dignity of civic status. Its growing middle class, expanding with the progress of commerce and shipping in the port of Cardiff, began to settle in new bourgeois suburbs, assisted by the growth of municipal bus and tram services which enabled them to live at some distance from their place of work in the central dock areas . . . And the grandest citizens of all, the tycoons of trade and industry like the Cory family, fled to the rural pastures of the Vale of Glamorgan, to become aspirant gentry in their turn, socially and geographically distant from the sooty valleys, overcrowded dockyards, and overworked dockers who brought them their profits.

---

[252] P. L. Garside, 'London and the home counties', in Thompson, *Cambridge Social History*, vol. 1, p. 508.   [253] Quoted in Brandon and Short, *The South-East*, p. 286.

[254] Katherine Chorley, *Manchester Made Them* (London, 1950), Chapter 9, *passim*.

[255] John K. Walton, *Lancashire. A Social History, 1559–1939* (Manchester, 1987), pp. 226–7.

[256] Morgan, *Wales*, pp. 127–8.

Despite their essentially 'urban' character many of the suburbs, particularly the later 'outer' ones, consciously attempted to take on the character of country life. H. G. Wells may have thought that 'all effect of locality or community had gone from these places', but for many the suburbs represented a kind of village England. At its most extreme, and most respectable in the London 'village' of Bedford Park, all the accoutrements of a recently lost rural ideal were there for its inhabitants. Here amidst houses deliberately made to appear old, designed by Norman Shaw and E. W. Godwin, artistic suburbanites dressed in Tudorbethan costumes, danced around the Maypole and drank at the Tudor 'Tabard Inn'. As an editorial in the *Bedford Park Gazette* put it, 'If we claim thus to be a colony of a new and hopeful sort, we also aspire to a sort of modern revival of the very ancient conception of a village community.'[257] If aspects of Bedford Park were extreme this search for a rural and organic life was present in many other suburban developments especially where they took on a 'Garden Suburb' aspect. The work of Ebenezer Howard and Raymond Unwin at Letchworth, begun in 1903–4, was intended as a model where town and country would be brought together, but many of the institutions and the structures of the new 'Garden Suburbs' in Hampstead in London, or Rhiwbina near Cardiff, were those of an imagined village. Howard's book *Garden Cities of Tomorrow* spells this out clearly, representing as it does 'a transvaluation of values' which has replaced the 'bad dream of the industrial revolution [and] the ugly nineteenth century has been wiped off the slate'.[258]

Letchworth was, in a way, a large-scale version of what was gradually happening throughout the home counties from the 1890s onwards, that is the growth of essentially middle-class estates as enclaves within existing village communities. These were different from suburbs in that they were completely rural and, although they sometimes completely overwhelmed the original settlement eventually, initially at least that original settlement appeared to be part of them. George Sturt wrote in 1912 of Lower Bourne in Surrey:[259]

The valley has been 'discovered' as a 'residential centre'. A water company gave the signal for development. No sooner was a good water supply available than speculating architects and builders began to buy up vacant plots of land, or even cottages – it mattered little which – and what never was strictly speaking a village is at last ceasing even to think of itself one. The population of some five hundred twenty years ago has increased to over two thousand . . . In fact, the place is a suburb of the town in the next valley [Farnham], and the once quiet high-road is noisy with the motor cars of the richer residents.

[257] Quoted in Margaret Jones Bolsterli, *The Early Community at Bedford Park. The Pursuit of 'Corporate Happiness' in the First Garden Suburb* (London, 1977), p. 78.

[258] Dugald MacFadyen, quoted in Marsh, *Back to the Land*, p. 225.

[259] Bourne, *Change in the Village*, p. 9.

Michael Ferguson's study of the Bagshot Sands region shows how this kind of change operated in detail. Between the late 1880s and the inter-war period the area was transformed from farming to residential use. As in other areas the pull of rural living was reinforced by the spread of the railway system which brought the region within commuting distance of London. However, both the distances involved, and the fact that the development was consciously controlled by keeping house and land prices high, gave the area a particular character. 'Surrey pines', Ferguson writes, 'had special associations with out of town living; they stood for privacy, health and exclusiveness, and they provided what was considered a rural setting for housing . . . The pines symbolised an antithesis to the polluted, built-up environment of the metropolis.'[260] A perhaps extreme example of this kind of development, although it was reproduced else-where, was that around the Sunningdale golf course. The club was set up in 1893 on land leased from St John's College, Oxford.

A legally separate company (but controlled by the same promoters) would develop the housing. To attract the opulent, the housing had to be expensive, and to maintain exclusiveness the club too had to be expensive. Those who pur-chased a house would have the right of nomination to the club (and would be put at the top of the waiting list). Wealth was a necessary but not sufficient con-dition for election. Housing and golf were mutually beneficial.

However, these controls were not enough. Housing around Sunningdale 'was zoned applying minimum building costs' which controlled the type of housing built. 'Densities were controlled . . . Building styles and the quality of materials were carefully vetted by the ground landlord. Extra buildings in the grounds were restricted except for garden paraphernalia, and maintenance clauses were inserted which bound future assignees.'[261] The development around Sunningdale was exceptional in some ways but was frequently repeated in other areas of the south-east as Lockwood's work on the inter-war development of the Weald shows.[262]

If suburbs extended the boundaries of the great towns into the coun-tryside, and the estate development of places like Bourne or Chobham transformed villages, there was a third kind of urban exodus which delib-erately sought to keep things the same. This was the much smaller-scale movement of groups and individuals, often 'arty', into 'unspoilt' villages. These were often the 'pioneers' who paved the way for other and larger developments to follow, ironically often destroying precisely what they sought to preserve. From at least the 1860s 'colonies' of like-minded souls began to congregate especially in Surrey and Sussex. Such was the influx

---

[260] Michael Henry Ferguson, 'Land use, settlement and society in the Bagshot Sands region, 1840–1940', PhD thesis, University of Reading, 1979, p. 147.   [261] *Ibid.*, pp. 496–7.

[262] Carol Lockwood, 'The changing use of land in the Weald region of Kent, Surrey and Sussex, 1919–1939', DPhil thesis, University of Sussex, 1991.

that the 1871 census, a little optimistically perhaps, attributed the increase of population on the sandy heathlands of Surrey to 'the attractions of the scenery, many artists having taken up residence in the district'.[263] By the 1890s and 1900s it seems to have been almost impossible to visit any south-eastern village without falling over resident bohemians. Around Haslemere clustered the Peasant Arts Fellowship, attracted partly by the Allinghams and partly by the early music of the Dolmetsch workshops. Around Crockham Hill, Fabians and Russian emigrés gathered, seeking a new life. At Amberly in Sussex another group of artists and writers established themselves partly in the wake of Hilaire Belloc, while at Steep another group was created around the progressive school at Bedales. Nor was it only the south-east, though this area was favoured; rather it was a land-scape 'type' which attracted these incomers. By the 1900s C. W. Ashbee had moved his craft workshops and its workers out of the infernal wen to Chipping Campden in Gloucestershire. To a young Scots member of Ashbee's Guild of Handicraft it was a new and wonderful world, 'a mile long street with hardly a mean house . . . It was . . . as foreign as Cathay and as romantic of the architecture of fairy tale illustrations . . . Was I really in the twentieth century, or in the sixteenth?'[264] The Cotswolds also attracted the Barnsley Brothers and Ernest Gimson, again craftsmen and architects, to Sapperton, and by the late 1900s the poet Lascelles Abercrombie and his family and friends had set up at Ryton Dymock.

The 'social history' of these newcomers is as varied as the groups them-selves. For most, a vague desire to escape from the city, very much a *motif* of the times, was sufficient in itself.[265] The inhabitants of Sunningdale, Chobham, or even Bromley, continued the lifestyle of the urban élites of which they were members especially where, as in Sunningdale, they were physically quite separate. To an extent they did replace the old rural élite as employers, but those who observed this aspect of their coming usually did so with cynicism. George Sturt wrote in his journal:[266]

The Radical Press is very cock-a-whoop now-a-days, in the expectation that Feudalism will soon receive its quietus. But I think on the whole it is no change for the better, to have replaced the old-fashioned Lord of the Manor by the Resident Tripper. The former did at least know and value the countryside, as a countryman; the latter is making of it his pleasure-place, so that, here near London at any rate, country life is dying out fast, in proportion as its feudal char-acteristics disappear.

[263] Quoted in Brandon and Short, *The South East*, p. 340.
[264] Fiona MacCarthy, *The Simple Life. C. R. Ashbee in the Cotswolds* (London, 1981), p. 48.
[265] See Alun Howkins, 'The discovery of rural England', in R. Colls and P. Dodd, (eds.), *Englishness: Politics and Culture, 1880–1920* (London, 1986).
[266] *The Journals of George Sturt, 1890–1927*, ed. E. D. Mackerness, 2 vols. (Cambridge, 1967), vol. II, p. 696.

Even those who went more cautiously and, in theory at least, had the 'good' of country people at heart, often got social relations wrong. In Helen Thomas's memoir of her husband Edward, the community around Bedales, although sympathetic to the villagers, is separated from them because, 'owing to their temperance they could not hob-nob with them in the inn'.[267] Similarly, their attempts to revive village life met with gentle rebuttal; '. . . they stood in their Sunday black in a ring on the village green watching the gentry who were dancing folk dances in print dresses and sun bonnets. "Well", said my neighbour with a shrewd smile, "I suppose they've got to be up to something."'[268]

Indeed, the 'revival' of folk dance, song, and even custom was a central element in the ideas and lives of many who moved into the countryside in the years before the Great War. Organisations like the Folk Song Society and the Folk Dance Society tried actively to promote the performance of dance and song. Cecil Sharp, for instance, took groups of young mainly Oxbridge graduates, who went to villages where a dance tradition existed to learn from old performers and to try and revive the dance among the younger villagers. Sadly they met with little success in revival at least. More successful were the attempts, both formal and informal, to bring folk music and dance into schools both urban and rural. Organisations like The Guild of Play and the Esperance Morris Guild produced booklets for schools on how to run an 'Olde English Pageant', using modified versions of folk song, folk dance and calendar customs.[269] By the outbreak of the Great War Sharp had, almost singlehanded, persuaded the Board of Education to base the teaching of music in elementary schools on 'music drawn from our folk and traditional song'.[270]

Nevertheless, the newcomers did bring some material benefits to rural areas. Even on the Bagshot Sands their coming prompted agricultural readjustments which benefited the villages in the short term at any rate. Around Woking new development, as well as the nearness to London by train, produced nearly 200 acres of small fruit and orchards as well as 584 acres of nurseries by 1930.[271] They also probably raised wages by competition with agriculture and provided employment for local women workers. The attempts by the Peasant Arts Fellowship and others to revive rural industry were also not as totally foolish and patronising as some would now suggest. However, the main change that came about has elements of both a cause and an effect. This was simply a change in attitude,

---

[267] Helen Thomas, *As It Was and World Without End* (London, 1972), p. 115.   [268] *Ibid.*, p. 146.

[269] See G. T. Kimmins, *The Guild of Play Book of Festival and Dance* (London, Part I, 1907; Part II, 1909); Mary Neale, *The Esperance Morris Book*, Part I (London, 1910). See also Chandler, *Ribbon, Bell and Squeaking Fiddles*, Chapter 11.

[270] *Board of Education Circular 873*, London, 1914, repr. 1923, 'The teaching of singing', p. 106.

[271] Ferguson, 'Land use', p. 500.

a revaluation of country life, and especially southern English county life. Whether they went to Sunningdale as commuters, or to Chipping Campden for a new life, they shared a belief in the virtues of country life and country ways. Yet it was a very particular country life. As one of these new countrymen, in one of its most important books, has it:[272]

The Mole saw clearly that he was an animal of the tilled field and hedgerow. Linked to the ploughed furrow, the frequented pasture, the lane of evening lingerings, the cultivated garden plot. For others the asperities, the stubborn endurance, or the clash of actual conflict that went with nature in the rough; he must be wise, must keep to the pleasant places in which his lines were laid and which held adventure enough, in their way, to last for a life time.

George Sturt put it less kindly:[273]

. . . I don't believe that well meaning reformers . . . are ever likely to benefit the Country to any appreciable extent. It isn't much good to look on – to be an enraptured spectator. The real thing is more solid, more hard . . . its clothes, and hands, and speech, and personality, become part of 'the Countryside', very often rank, and sweaty and coarse, even as cattle keepers and ploughmen have to be.

To the Mole and his human counterparts the countryside was a place of freshness and purity where a man or woman could be themselves and get nearer to some imagined Eden. Initially they were a few, but by the 1900s many thousands had 'moved out' of the cities to the suburbs, or even to the true countryside to find that life, even if commuting meant that it was a weekend Paradise. By 1914 this idea had spread well outside that group even. A cottage in the country, or at least in Surbiton, became part of Everyman's dream. Even for those who could never hope for this, the countryside was becoming a site of leisure and temporary escape on foot or by bicycle. A whole new generation of newcomers was on the way.

---

[272] Kenneth Grahame, *Wind in the Willows* (London, 1953), p. 162.
[273] George Sturt, *Change in the Village*, p. 109.

CHAPTER 24

THE LOCAL STATE

BY ANNE DIGBY

I. INTRODUCTION*

The local state may be narrowly defined as the system of local government, and more broadly construed as a system of power relationships in
rural society. In its former sense the local state was in a process of administrative transition at the beginning of our period as a result of centralist
initiatives such as the New Poor Law of 1834, the rural police of 1839,
and the increasing powers of summary jurisdiction given to local justices
in petty sessions. An even more obvious transformation apparently came
with fundamental reforms of local government in 1894 which created
parish councils and rural district councils. Such innovations might be
thought to have entirely replaced the traditional parochial state. Here
parish overseer and parish constable had been appointed and controlled
by the parish vestry, composed of substantial farmers under the chairmanship of the local clergyman. The benign administration of these
'ancient institutions' had been idealised, and any interference attacked as
'mischievous' or 'unconstitutional' by conservative romantics – amongst
the most prolific of whom was C. D. Brereton.[1] Effectively Brereton,
the rector of Little Massingham in west Norfolk, saw the time of the
autonomous, parochial state as a golden age, and a necessary precondi-
tion for the continued economic and moral vitality of the village.

Brereton had been a little premature in lamenting the demise of the
parochial system during the 1830s, although more perceptive in seeing
the clergyman's declining influence within it. It is helpful here to distinguish between the administrative means through which policies were
delivered, and the nature of the policies themselves; the former changed
to a much greater extent than the latter. Using our second and wider
framework of the local state, the continued resilience of parochial interests and thus of the traditional power structure in the 'new' era of boards

---

* This chapter was submitted in 1992 and has not been revised subsequently.
[1] C. D. Brereton, *A Refutation of the First Report of the Constabulary Force* (n. d.), pp. 3, 17–18; *A Letter on the Proposed Innovation in the Rural Police* (Swaffham, 1839), pp. 6–8, 59.

of guardians and county police forces needs emphasis. The parish constable continued to be active in many areas until the 1850s and 1860s, and in some rural areas beyond this.[2] And the parish overseer – although his function of giving poor relief was largely superceded by that of the relieving officer in 1834 – was still to collect or supervise the collection of the poor rates in villages, as well as perform another dozen miscellaneous duties, until the end of our period.[3] The parish also remained the financial unit of the poor law until the union rating reforms of the 1860s so that an economical engine – driven by a small rating base – remained in place to influence the administration of poor relief. After 1834 each parish elected one or more poor-law guardians to represent its interests; a system of plural voting among substantial local ratepayers resulted in the 'election' (usually uncontested) of the same class of people, most often substantial farmers, to rural boards of guardians that had earlier been found in parish vestries. So in country areas these rural boards were dominated by comparable employers' interests but, since they operated over a wider area, exerted an even more powerful control over the local labour market. Richard Jeffries, that perceptive commentator of the Victorian countryside, understood the character of the farmers very well, since he was the son of a Wiltshire smallholder. His verdict was that 'the united farmers of a parish were kings of the whole place'.[4]

During the mid-Victorian period the parish vestry continued to appoint overseers and churchwardens and to administer parochial charities. Under the Parochial Rate Assessment Act of 1869, however, all *bona fide* householders had the right to attend and vote at parish vestries on these matters. But it appears that there was minimal participation by the labourers, in part because meetings were held when they were at work, and in part because it was made clear to them that any activity would be unwelcome. This was shown very clearly, for example, in Norfolk when the agricultural union leader, George Rix, attempted to get himself elected as churchwarden at Swanton Morley vestry meeting in 1875. 'He raved and stormed . . . and would not hear reason' wrote Rix of the rector's conduct in chairing this meeting. The latter's denial of the labourers' rights to participate led to successful legal proceedings being taken by the labourers, whilst ensuing publicity in *The Labourer's Chronicle* in turn led others to similar activity.[5]

Both formal and informal patterns of control continued to be exerted in mid-nineteenth century villages as is apparent if we focus on public

[2] J. J. Clarke, *The Local Government of the United Kingdom* (fifth edn, London, 1929), p. 310.

[3] W. M. Mackenzie, *The Overseer's Handbook* (eighth edn, London, 1915), p. 7.

[4] R. Jeffries, *The Toilers of the Field* (Futura edn, London, 1981), p. 44.

[5] L. M. Springall, *Labouring Life in Norfolk Villages 1834–1914* (London, 1936), pp. 106–8; F. G. Heath, *British Rural Life and Labour* (London, 1911), p. 231.

and private welfare systems. The Town Lands, allotments awarded by enclosure commissioners in lieu of common rights, and other property held by the parish were administered by trustees, usually the parson and other influential parishioners, aided by churchwardens or overseers who organised doles in money and kind. Distribution involved a selection of recipients intended to reinforce socially desirable behaviour amongst the poor. Those deemed the deserving poor, whose impoverishment was clothed in a decent and deferential respectability, stood the best chance of receiving public or endowed charity. Private charity was at times even more explicit on the conditionality of the gift relationship. For example, the allotments created from the glebe by the incumbents of Lyddington, Wiltshire had a list of rules that included not only provisions about husbandry and rent but a concern for right conduct and thinking. 'All tenants shall maintain a character for morality and sobriety and shall not frequent a public house on the Sabbathday', and 'all the tenants are requested to attend regularly at the House of God during the times of Divine Service, with their families, to the best of their abilities'.[6] A similar concern for the moral character of the recipient also informed the distribution of poor relief at the guardians' board; a decision being made as to whether applicants were rewarded by being given outdoor relief in their own homes, or punished for non-conformist behaviour by being offered only indoor relief in the workhouse. And for the more thoroughly deviant the board of guardians was also linked to petty sessions, since under the Poor Law Amendment Act of 1834, magistrates sat as *ex officio* guardians. Rather than being punished in the workhouse, refractory paupers were increasingly sent to petty sessions. And with what has been called the 'vast increase of summary powers' conferred on petty sessions by legislation between 1827 and 1914,[7] the power of local justices over the countryside was augmented. In turn, this magisterial power became more secularised in character, since numbers of clerical justices declined in mid-century.[8]

Customary outward subservience was clearly required if the labourer was to survive in this situation, but this demeanour might mask a growing internal independence. The resulting tension showed most obviously in a series of labour disputes from the 1870s onwards but such conflicts cost the labourer dear, so that ancillary, 'softer' targets constituted a more continuous battleground. Conflicts over the administration of parochial charities, the payment by dissenters of church rates, the right of nonconformists to

[6] *The Times*, 23 November 1872.

[7] L. Radzinowicz and R. Hood, *A History of English Criminal Law* (London, 1986), vol. v, p. 622.

[8] R. Quinault, 'The Warwickshire magistracy and public order, *c.* 1830–1870' in R. Quinault and J. Stevenson, (eds.), *Popular Protest and Public Order* (London, 1974), pp. 188–9: G. Kitson Clark, *Churchmen and the Condition of England, 1832–1885* (London, 1973), pp. 249–51.

be buried in the parish churchyard, or the nature of religious teaching in the village school flared up from the 1850s to the 1880s. In this context rural religious dissent was rather more an expression of class than of difference in belief. The assumption by the clergyman and his wife that charity was a gift necessarily involving deferential gratitude provoked a challenge from the more independent spirited. The public and symbolic washing out of charity from a gift of red flannel in Tysoe, Warwickshire – during a dispute over the Town Lands between the vicar and the villagers – has come to typify the new spirit of the villagers. Less often noticed, however, was that this insubordination had only been made possible by the husband of the 'washerwoman' having previously got a job *outside* the village.[9] A contemporary observed that, 'If you are not pliant . . . it is best to get out of the local village.'[10] The power of the propertied in the countryside in their control of employment, housing, poor relief and charity, was truly formidable. This was also buttressed by the penal sanctions possessed by these same groups on the magistrate's bench. A searching light on the self-interested way such sanctions might be applied was shown during the union disputes of the 1870s, for example, when justices prohibited the labourers' open-air meetings because they were said to obstruct the highway.[11]

The solidarity of propertied groups in the face of the labourer's demands for greater dignity and economic independence could also be turned against any of their own kind who too conspicuously 'broke ranks' in his defence. The case of Canon Girdlestone was the most notable example of one who, in successfully assisting farm labourers to move away from serf-like dependency on local farms, split his north Devonshire parish of Halberton in the years from 1866 to 1872, when he himself moved to a Gloucestershire parish. Rigid social roles did not permit radical activity and the bitter hostility aroused by Girdlestone's assisted migration schemes was shown in complete social ostracism of the canon and his family by neighbouring squires, clergy and farmers. The practical obstructionism of the farmers in the parish vestry prevented a church rate being made, and almost led to the election of both churchwardens of the farmers' rather than the cleric's choice, whilst farmers also attempted to empty the church school, and themselves stayed away from the parish church.[12]

Farmers were not only dominant in the parish vestry but in most rural boards of guardians and rural district councils as well.[13]

---

[9] M. K. Ashby, *Joseph Ashby of Tysoe 1859–1919* (London, 1974), p. 46.

[10] R. Jefferies, 'Primrose and Gold' quoted in *Hodge and His Masters* (Quartet books, London, 1979), p. xvi.

[11] F. E. Green, *A History of the English Agricultural Labourer 1970–1920* (London, 1920), p. 41.

[12] Heath, *Rural Life*, pp. 227–37.

[13] Under Public Health Acts of 1872 and 1875 the rural poor-law union (minus any urban parts)

The farmer class, however, almost everywhere captured and controlled the Rural District Council, which is the real executive body in rural districts. The Rural District Councils are largely the Guardians of the Poor. They decide whether cottages are to be built or not; they control the highways; they are the sanitary authority, and they are the executive body with regard to rights of way, wayside wastes, commons and water supply.[14]

Few members of the agricultural working class stood for or succeeded in getting elected to the RDCs, not least because of the prospect of giving up a day's pay in order to attend fortnightly meetings. Two activists in the Warwickshire agricultural union were successful, whilst in Norfolk the secretary of the Norfolk and Norwich Amalgamated Union, George Edwards, together with his wife, were elected to the Erpingham RDC. These were exceptional cases. So conservative and economically minded bodies could be relied on by local ratepayers to veto any reforming resolution, such that, for example, the Housing of the Working Classes Act of 1890 should be activated. These potentially expensive resolutions might originate in parish councils, should they have a popular member-ship. Labourers did make considerable gains on these bodies, since whilst it was estimated that up to a half of the seats were won by farmers, and a quarter by craftsmen, most of the rest were taken by labourers, with a few gentlemen, ladies, professional men such as doctors, and clerics.[15] Usually it needed a formal organisational backing for labourers to capture many seats on the parish council, so that representation tended to be strongest in areas where unionism had flourished, as in Norfolk or Warwickshire. In 24 parishes with branches of the Warwickshire Agricultural Labourers' Union there was a total of 140 councillors elected, of whom 91 were labourers' candidates. Of these, 54 were farm workers and the rest were sympathetic tradesmen or artisans. And in Norfolk, where anticlericalism was combined with the land issue, the months before the first election in 1894 saw crowded village meetings. Notable electoral successes were achieved in parish councils such as Swanton Morley or West Raynham, Attleborough where the labourers won every seat. This promised, in the words of the *Eastern Weekly Leader*, a 'rural revolution'.[16]

The 6,000 or more parish councils had few powers, however, whilst the parish meetings, which were elected in almost the same numbers in villages with under 300 people, had still less authority. Even with popular participation such bodies could only make very limited reforms, most

was constituted as the rural sanitary district under the authority of the board of guardians and, with minor amendments, this situation continued under the Local Government Acts of 1888 and 1894 (H. Finer, *English Local Government* (London, 1933), p. 90).   [14] Green, *Labourer*, p. 127.

[15] R. Heath, 'The rural revolution', *Contemporary Review*, October 1895, p. 190.

[16] *Eastern Weekly Leader*, 4 December 1894.

effectively through the administration of civil charities together with the exercise of a permissive power to start allotments. Changes were greatest in the 'open' villages where social controls had traditionally been few. One example was Horsford St Faith's in Norfolk, where three labourers were returned, and the parish council succeeded in pressurising the county council into making an enquiry into the appalling state of local cottages, whilst also hiring eight acres for allotments in the village. In some other villages in Yorkshire, Derbyshire and Berkshire recovery of lost common land was secured by the new councils making a thorough examination of parochial documents. (But in Angmering, in Sussex, fear of such action had led to the destruction of the parish chest and all its contents by the unreconstituted parish vestry.) Other successes of parish councils were in improving village amenities by building village halls and reading rooms, reopening rights of way, and making footbridges over streams.[17] In the longer term labourers took less electoral interest as their limitations became apparent, although continuing to carefully monitor the parish council's work in such immediate areas of interest as allotments and charities.[18]

Whilst labour historians have long recognised that the class composition of local government was widened by the introduction of working people, it has taken more recent work, much of it by feminist historians, to acknowledge the comparable contribution made by women. They had a key role in the administration of charity in the villages. For women this work was a vital stage in the transition from the private to the public spheres. Local governmental work enabled women to capitalise on knowledge and skills hard won in their charitable endeavours and to emerge into a fully public role.[19] By the end of the Victorian era there were one million women on the local electoral registers, who were as determined as men to use their vote, and whose concerns therefore helped shape the agendas of the new local councils and boards. And female political influence was not only indirect but also direct. Nationwide, by 1914–15 there were some 679 female members of school boards; 1,546 lady guardians, of whom 200 were also members of rural district councils; as well as 48 female members of town and county councils. In the first elections of 1894 at least 80, and possibly as many as 200, women were returned as parish councillors.[20] On parish and rural district councils women made a very substantial contribution to rebuilding village society – combatting the demoralisation consequent on agrarian

---

[17] Green, *Labourer*, pp. 123–8, 134–7.   [18] Springall, *Norfolk Villages*, p. 119.

[19] F. Prochaska, *The Voluntary Impulse. Philanthropy in Modern Britain* (London, 1988), pp. 23, 73; F. K. Prochaska, *Women and Philanthropy in Nineteenth- Century England* (Oxford, 1980), pp. 227–8; P. Hollis, *Ladies Elect. Women in Local Government 1865–1914* (Oxford, 1987), pp. 10–11, 47.

[20] Hollis, *Ladies Elect*, pp, 32–3, 365, 486.

depression and rural depopulation. In a real sense they formed a bridge between the old and new order of the countryside since, although they came from propertied groups, they usually had a reforming agenda. Mrs Barker, elected chair of Sherfield-on-Loddon (Hants) Parish Council in 1894, analysed her own contribution, and those of other female parish councillors in modest terms:

A polluted well, an overcrowded cottage, a barrier across a footpath, are too trivial for men to make a stir about, and perhaps offend the wage-giver into the bargain; but an independent woman, knowing that 'trifles make the sum' of things, and that these trifles if looked into will reveal further defects to remedy will be earnest for frequent meetings.[21]

Women were also active on rural district councils, usually entering such work through a prior interest in the poor law; of 875 women elected as poor-law guardians in 1894, 140 were also members of an RDC. Patricia Hollis has portrayed them, 'the wives and daughters of the vicar and squire, using their social status to intimidate parsimonious farmers into spending the rates'.[22]

How much did this new membership change the character of the local state? Did the Local Government Act of 1894 deserve its early reputation as the 'Rural Magna Charta' in creating the parish and district councils? The socially conservative *Wilts and Gloucestershire Standard* rejoiced after the 1894 elections, 'either that "Tory tyranny" existed only in the heated imagination of party grievance-manufacturers, or else that the down-trodden electors hug their chains with provoking complacency'.[23] A longer historical perspective tends to confirm this view that popular participation in parish councils only dented the influence of parson and squire in the village, whilst rural district councils were frequently little more than associations of farmers. The new county councils of 1888 swiftly became the seat of the landed gentry and large farmers, with relatively few contested elections taking place. Only occasionally were individual labourers able to go sufficiently far up the hierarchy of political organisation to challenge the entrenched power of the county gentry in rural areas, as they could if elected to the county council. But men like George Rix in Norfolk were exceptional, and had already served their apprenticeship in public life in the pulpit of the Primitive Methodist chapel, and in agricultural trade-union organisation.[24] The tenacity of the old order in the late Victorian and Edwardian countryside needs emphasising since it meant that more than changes in administrative bodies were needed if the traditional biasses of class and gender in the

---

[21] Quoted in Hollis, *Ladies Elect*, p. 368.   [22] *Ibid*, pp. 10, 359–60.
[23] Quoted in P. Horn, *The Changing Countryside in Victorian and Edwardian England and Wales* (London, 1984), p. 190.   [24] A. Howkins, *Poor Labouring Men* (London, 1985), p. 52.

local state were to be displaced. One contemporary observer of the rural scene regarded farmers as rulers of the parochial system who 'paid very little regard to the liberty' of the labourer, although they also saw themselves as the head of a parental system.[25] Beyond the farmer stood the squire. In 'close' parishes in particular the paternalistic authority of the resident squire over the life of his village was virtually complete, since directly or indirectly he controlled employment, housing and welfare. A modern study has warned that the paternalism of the Victorian landowner and justice of the peace was less about benevolence than about 'the obligation to rule firmly and to guide and superintend'.[26] From the labourers' perspective the penalties as well as rewards of these class relationships of the Victorian countryside were clear. 'If the gift and charity played a vital part so did their withdrawal and blacklisting and eviction from a tied house which often followed.'[27] Labourers, and even village schoolmasters, who played too active a role in challenging the social establishment in local elections could swiftly find themselves with neither job nor home.[28]

Diversity in the Victorian countryside meant that economic conditions varied between and within regions as well as among localities, whilst social structures also showed considerable differentiation, most obviously in the contrast between 'open' and 'close' villages. It was in the overstocked labour market of the rural south and west during the earlier part of our period that the local state could be most oppressive for the labourer, while in the north of England and in south Wales alternative urban and industrial employment acted as a corrective. Later, the economic foundations of the local state were weakened to some extent by rural depopulation which gave remaining southern labourers a little more muscle in directly influencing not just their conditions of employment, but indirectly the social circumstances of the village as well. Offsetting the effect of depopulation, however, was agrarian depression in the late nineteenth century, which in many corn-growing areas in particular provided a financial constraint on any proposed social improvement. And the enhanced independence for the rural worker that in theory should have followed the Secret Ballot Act of 1872, and the 1884 and 1918 extensions of the parliamentary suffrage, might be curtailed in practice by customary patterns of deference that were rooted in economic dependency.[29] In the early twentieth century, for example, when Lord Wimborne changed his politics, those on his Dorset estate – whether tenant-farmer or labourer – were expected to follow suit; on one occasion, an agent from

[25] Jefferies, *Toilers*, p. 44.
[26] D. Roberts, *Paternalism in Early Victorian England* (London, 1979), pp. 7–8.
[27] Howkins, *Labouring Men*, p. 35.   [28] Hollis, *Ladies Elect*, pp. 365–6.
[29] Springall, *Norfolk Villages*, p. 111.

the estate stood outside the polling station during the 1910 election to record in a notebook his observations of voters.[30]

Encompassing a range of economic power from a landed magnate like Lord Wimborne on the one hand, to a Norfolk squire such as J. P. Boileau on the other, the scale of activity by the landowner in the local state was immensely varied. The more important county families constituted the interface between the local and central state: their members traditionally represented county divisions in Parliament, and also gave service as magistrates in quarter sessions where it had been customary to give a lead in county affairs over such issues as poor law policy. The traditional semi-autonomy of the county community was diminishing during the Victorian period because of economic developments such as the railway system, and also as a result of more uniform, national bureaucracies in such social questions as education, the poor law, or criminal justice. However, in my view it would be misleading to focus too exclusively on these changes and to underestimate continuities at the local level. Public service as a local magistrate in petty and quarter sessions still took up to two days a week for the minority of landowners who were selected for the bench. Such work in the mid-Victorian period encompassed a wide range of duties from inspecting prisons, punishing petty crime, or acting as *ex officio* poor-law guardian, to inspecting weights and measures, and licensing alehouses.[31] In estate villages, too, the local squire might be very active in social questions, as was J. P. Boileau in Ketteringham, Norfolk. He provided model cottages, ensured that there was an efficient village school, and provided decent employment and wages on his estate. In return he regarded the village as his property and villagers were expected to behave in approved ways such as sending their children to school until they were twelve, and not taking in lodgers.[32] Yet as F. M. L. Thompson has reminded us, there was another category of Victorian landowner, those 'who were narrowly self-centred in their interpretation of what was required to protect their interests' and whose 'indifference' to the wider issues of social order in the countryside consequently 'preserved the capacity of the rural population for independent development'.[33]

The generalised model of the local state that has been outlined above cannot do full justice to its rich complexity. A more detailed empirical analysis of important constituents follows in an attempt to remedy this. There are two sections on the poor law and on crime and policing in this chapter; and further sections on the church and chapel, and on the

---

[30] Horn, *Changing Countryside*, p. 192.

[31] G. E. Mingay, *Rural Life in Victorian England* (London, 1976), pp. 34–5.

[32] O. Chadwick, *Victorian Miniature* (London, 1960).

[33] F. M. L. Thompson, 'Landowners and the rural community' in G. E. Mingay (ed.), *The Victorian Countryside* (London, 1981), vol. II, pp. 458–9.

school, in the following chapter. Chapter 25 on 'Social Institutions' discusses agencies with a more obviously socialising character, whilst the subjects that are discussed in the two ensuing sections in this chapter provide evidence of more formal controls in society. In both chapters the categorisation of what is rural has necessarily to be broad-based. Not only might the territorial divisions of poor-law unions, for example, include a subsidiary urban component as well as the major country area, but the growing urbanisation of England and Wales meant that such complexity increased during the period from 1850 to 1914.

II. THE POOR LAW, CHARITY AND SELF-HELP

The means to alleviate poverty or destitution arising from sickness, unemployment and old age are the main themes of this section. The agricultural worker in Victorian and Edwardian times remained poorly paid, receiving little more than half the wages of the industrial worker.[34] The farmers, who were the principal rural employers, continued to confuse the farm labour bill with public welfare payments through their office of poor-law guardian. This was particularly evident in arable areas, where the seasonality of the demand for farm labour meant that it was economical periodically to make the labourer unemployed and relieve him from the poor rate, to which other rural ratepayers had contributed.

In this period the most central of the welfare agencies for the majority of the population has generally been seen to be the poor law. Yet the complex interaction of statutory poor relief and voluntary charitable agencies for the relief of poverty needs greater emphasis than it usually receives. So too does the dynamic interplay between the relief policies of the administrators of welfare and the strategies for survival of the poor themselves. The abundance of data on the former and the paucity on the latter have tended to distort our interpretation of the use that the rural poor made of welfare services that were available to them. This has resulted in an over-emphasis on the role of formal assistance, and a corresponding neglect of the selective take-up of such help, employed in combination with familial or community/neighbourhood support mechanisms. It is necessary, too, to link these patterns of help with the lifecycle of the recipients and thus to appreciate the dynamics of poverty highlighted by Rowntree in his contemporary poverty surveys. Important aspects of the lifecycle involved such factors as marital status, family size, the ages of children (and therefore the cost of their dependency or later their earning capacity), and the changing opportunities for employment at different

---

[34] J. R. Bellerby, 'The distribution of farm income in the U. K., 1867–1938', in W. E. Minchinton (ed.), *Essays in Agrarian History*, II (Newton Abbot, 1968), p. 271.

ages. Equally it is becoming apparent that we need to take greater account of distinctions within the category of welfare recipient – not so much with respect to contemporary labelling of the poor as able-bodied or impotent, as in looking at permanent or temporary dependency. And the necessary emphasis in this chapter on short-term changes in attitudes and/or administrative practices within our period from 1850 to 1914 needs some correction by reference to longer-term continuities. S. Woolf's valuable perspective on the poor in eighteenth and nineteenth century Europe suggests that concepts of poverty have shifted more than either individual causes of poverty or the actual social composition of the poor. There were the structural poor (incapable of earning a living by reason of illness, age or handicap), the 'crisis' poor with casual earnings desperately vulnerable to poor harvests or bad winters, and the low-earners whose dependency might result from trade stoppages or adverse changes in personal/household circumstances.[35] Much more research is still needed before we can address the complexity of these interrelated issues at all adequately.

A. Indoor and outdoor relief policies

Placing a study of the Victorian and Edwardian poor law within a wider context of demographic structures and developments within the agrarian economy provides a useful perspective. Whereas one in two people had lived within a rural poor-law union at the start of our period, fewer than one in four did so before the end of it.[36] Demographically and economically this sector of the poor law was declining in importance. Its principal period of expansion had been in the first years of the New Poor Law, during the late 1830s, when most rural workhouses in southern and eastern England had been built. However, there were lesser periods of activity after 1850 when the rural west of England and Wales was more active in workhouse building.[37] But Welsh country guardians disliked the workhouse system, thinking it an alien English device that was unsuitable for their thinly populated unions.[38] In consequence, more outdoor relief continued to be given in Welsh unions than English ones; in 1881, for example, the former had more than four-fifths of their total in the form of outdoor allowances compared to three-fifths in the latter.[39]

---

[35] S. Woolf, *The Poor of Western Europe in the Eighteenth and Nineteenth Century* (London, 1986), pp. 6, 11, 17.

[36] A. Digby, 'The rural poor law', in D. Fraser (ed.), *The New Poor Law in the Nineteenth Century* (London, 1976), p. 150.

[37] F. Driver, 'The historical geography of the workhouse system in England and Wales, 1834–1883', *JHG*, 15 (London, 1989), p. 280.

[38] *Select Committee on the Andover Union*, BPP, 1846, V, Q 24681, evidence of Mr William Day.

[39] *Tenth Report of Local Government Board*, BPP, 1881, XLVI, p. 17.

What was the character of the rural workhouse? Those in Cumbria were not untypical in that they were not the bastilles of popular stereotype; there was little attempt to create less eligible conditions for inmates, as the philosophy of the New Poor Law required.[40] But, erected as general institutions designed to cater for all types of indoor pauper, these early union workhouses were deficient in the specialist facilities later thought desirable, notably those for the sick or the aged. Poor-law inspectors' reports repeatedly condemned the 'old-fashioned' nature of the buildings and, as one stated, 'I should like to see the entire premises remodelled.'[41] However, with the onset of agrarian depression in the late nineteenth century, poor-law guardians increasingly saw themselves as guardians of the rates rather than guardians of the poor, so that there was opposition to proposals to spend money on institutional improvements – such as the provision of an infirmary to replace antiquated sick wards.[42] As a result rural unions were largely missing from the third phase of expenditure on workhouse improvements that occurred from the 1870s onwards, when there was a move to create more specialist accommodation not only for the sick but also for children, for the insane, and for the vagrant. In this context the rather idealised picture painted of the country workhouse as a largely well-administered almshouse for the aged, infirm and children (in evidence given to the Royal Commission on the Poor Laws of 1905–9, and then taken up by the *Majority Report*) should be set against a more realistic view provided by H. Rider Haggard, himself a Norfolk guardian. Echoing the original debates on the deterrent character of the union workhouse, he balanced material against psychological factors involved in the lives of its inmates. He concluded that, 'In truth, to whatever extent it may be brightened and rendered habitable, one cannot pretend that a workhouse is a cheerful place.' He went on to ruminate that, despite the fact that the warmth, cleanliness and diet were better than in the previous homes of its aged inmates, yet 'how they hate it most of them'.[43] A detailed study of the rural administration of the poor law in early-twentieth century Essex also concluded that 'the rose-coloured rural workhouse of the Report of 1909 . . . was not, and is not to be found'.[44] What was found universally were large buildings with few

---

[40] R. N. Thompson, 'The working of the Poor Law Amendment Act in Cumbria, 1836–1871', *Northern History*, 15 (London, 1979), p. 135.

[41] PRO, MH 32/102, report of H. G. Kennedy in 1900, fos. 15–16.

[42] Speech by Mrs Fuller, Guardian of Chippenham Union in Wiltshire on 'The aged in rural workhouses' at the Central Poor Law Conference (*Poor Law Conference*, 1899) pp. 576–8.

[43] *Reports of Visits to Poor Law and Charitable Institutions*, BPP, 1910, LIV, p. 223: *Report of the Royal Commission on the Poor Laws* (BPP, 1909, XXXVI), vol. I, p. 174; H. Rider Haggard, *A Farmer's Year being his Commonplace Book for 1898* (London, 1899), pp. 428–9. (Haggard was a guardian in the Loddon and Clavering Union.).

[44] G. Cuttle, *The Legacy of the Rural Guardians* (Cambridge, 1934), p. 322.

inmates and much spare capacity, a situation which led some unions to consider amalgamation and closure. Proposals in the 1890s to save on heavy administrative costs by concentrating indoor provision in fewer workhouses, as in Norfolk for example, resulted only in the closure of one workhouse in the county, because of the perceived difficulty of sending country paupers a long distance from their homes.[45]

Historians have developed divergent perspectives on the impact of the New Poor Law in concentrating relief in the workhouse by cutting the outdoor relief that was given to people in their own homes during the mid-nineteenth century. Basing his conclusions on the administrative records of the central poor-law authority Karel Williams has argued that the creation of union workhouses and 'the exclusion of unemployed men from the classes obtaining relief' were the 'conspicuous discontinuities' that followed the New Poor Law. [46] Revisionist historians would find these conclusions problematic because they fail to take sufficient account of the manipulation of relief categories by local guardians in the submission of their statistics to the central board. In an appendix Williams notes, for example, both the 'obfuscating category' of the term able-bodied, and the fact that half of the temporarily sick were returned as able-bodied, but does not acknowledge the possibility that this meant that relief in aid of sickness could have been used as a means to give out-relief to the temporarily unemployed.[47] In contrast to what we might call this 'orthodox' view, the 'revisionist' view would regard such relief policies by rural guardians as being shaped to a significant extent by sectional and economic considerations. The day to day business of country boards was dominated by farmers, who were both the main employers and the principal ratepayers. Regional differentiation between north and south was important here; in rural Cumbria, for instance, adult able-bodied male pauperism was very rare,[48] in contrast to arable areas in the east and south where the pressure of surplus labour was felt in formulating relief policies. It is the south on which most recent research has concentrated. Here the conflation of private economic interests and public welfare policies has been well documented, particularly for the period before 1850, by Baugh, Blaug, Digby and – most recently – by Boyer.[49] Through econometric

---

[45] PRO MH32/104, Letter from poor law inspector, H. Preston Thomas to Local Government Board, 19 February 1896: A. Digby, *Pauper Palaces* (London, 1978), p. 61.

[46] K. Williams, *From Pauperism to Poverty* (London, 1981), p. 83.   [47] Williams, *Pauperism*, p. 203.

[48] Thompson, 'Cumbria', p. 133.

[49] D. Baugh, 'The cost of poor relief in south-east England, 1790–1834', *EcHR*, 28, 2nd ser., 1975, pp. 50–68: M. Blaug, 'The myth of the Old Poor Law and the making of the new', *JEH*, 23, 1963, 151–84: M. Blaug, 'The Poor Law report re-examined', *JEH*, 14, 1964, pp. 229–45: A. Digby, 'The labour market and the continuity of social policy after 1834: the case of the eastern counties', *EcHR*, 2nd ser., 28, 1975, pp. 69–83; A. Digby, 'The rural poor law', in Fraser (ed.), *New Poor Law*; Digby, *Pauper Palaces*, especially chapter 6.

work Boyer has convincingly indicated the reasons why outdoor relief was maintained as a means of securing an adequate and cheap peak-season labour force in arable areas. 'A combination of seasonality and a tax system that allowed farmers to be subsidised by other parish ratepayers' permitted seasonal layoffs of labour where ensuing outdoor relief was partially financed by other ratepayers.[50]

This economic influence on welfare policies continued in the period that mainly concerns us, from 1850 to 1914, although the balance of considerations altered. With prolonged agricultural depression after the 1870s the desire to economise on the rates became even more imperative, so that expenditures were pared to the bone. Meanwhile, the traditional concern in southern counties over the need to use the poor rate to relieve a surplus labour problem became less urgent after mid-century, as out-migration and a changing labour market eventually produced anxieties about labour shortages. Although some argued that out-relief was 'not given consciously . . . to retain labourers on the farms'[51] yet its wide-spread use as a dole to supplement small incomes arguably had this effect. Outdoor relief was preferred by guardians to an offer of the workhouse since it was very much cheaper.[52] Relief to the aged pauper also fitted into this framework, since allowances to an old person living with chil-dren was effectively a device for giving indirect relief in aid of wages. According to the *Majority Report* of 1909 this was used in many rural dis-tricts.[53]

Changes in rating may have facilitated the pursuit of sectional eco-nomic interests on rural boards. The traditional system of parish rating and settlement was reformed by the Union Chargeable Act of 1865, which established union rating, and made all paupers chargeable to a common fund in the union. In theory this was intended to alleviate the worst feature of the rating system whereby, in impoverished villages, poverty was rated for the relief of pauperism. In practice, however, as the poor-law inspector for Yorkshire explained, its impact in certain country unions could be rather different since it:

Retarded the diminution of pauperism in those unions which consist merely of an aggregate of purely rural parishes, without a single centre where population and consequent pauperism are to be found more than elsewhere. Each Farmer Guardian thinks that he may as well have a pull at the common purse for the benefit of his own parish; and the vigilance of the villages in watching their own cases is diminished because the cost now comes out of a general fund.[54]

---

[50] G. R. Boyer, An *Economic History of the English Poor Law, 1750–1850* (Cambridge, 1990), pp. 266–7.
[51] *Royal Commission on the Poor Laws*, BPP, 1909, XLIII, p. 615 (Report of T. Jones on Suffolk and Cambridgeshire).   [52] *Ibid*, pp. 615, 628, 630.   [53] *Report*, vol. I, p. 208.
[54] PRO, MH32/102, Report of H. G. Kennedy on the Yorkshire District, 1887.

The contemporary opinion that the 1865 act eased financial constraints has been subjected to critical scrutiny recently by Mary McKinnon. She has argued that changes in local taxation in the 1860s assisted ratepayers in reconsidering relief policies, and thus to furthering the central crusade against outdoor relief from the 1870s.[55] In her interpretation, the poor-law union differs crucially from that sketched above since it would include not only the poorer parishes (which gained from the act) but also wealthier ones whose interest was now – for the first time – to keep a close eye on general relief expenditure, and thus to monitor numbers offered expensive out-relief in poorer parishes. McKinnon's econometric analysis is persuasive on a *national* basis, but at the *local* level her findings pose some intriguing problems. She indicates that although southern rural unions had 'much higher' rates of pauperism than northern unions, yet the variations across unions, revealed that 'the differences within each region were not easily explained by the occupational structure or degree of urbanization of each parish'.[56] It is possible that we may have to fall back on the argument of historical specificity here. It is clear that under the New Poor Law individual unions created distinctive policy stances of restrictiveness or liberality on relief; these could vary over quite a small area. Well-documented rural unions such as Brixworth or Bradfield also suggest the importance of strong personalities. The chairmen of boards of guardians, in these unions were largely responsible for framing and successfully maintaining unusually restrictive policies over a long period.[57]

More generally, the administrative mechanisms of the rural poor law facilitated a continuation (or resumption) of parochial interests in relief policies. The duty of serving as guardian was usually attached to certain farms in each parish, with long service as guardian therefore being quite common.[58] And cases before the board, as was shrewdly observed, 'are very often heard by a few guardians, and are practically decided by the guardian of the parish from which each application comes'.[59] Although relief scales existed they were usually disregarded since in Suffolk and Cambridgeshire, for example, 'Guardians prefer to give small doles to many persons to thoroughly helping a few.'[60] Zeal in researching cases by the relieving officers was therefore discouraged by the guardians.[61] And criticism, or suggestions for improvements in administration, by newcomers to the board, were usually ignored.[62] Also frustrated by the

---

[55] M. McKinnon, 'English Poor Law policy and the crusade against outrelief', *JEH*, 47, 1987, pp. 608, 614.   [56] McKinnon, 'Crusade', p. 622.

[57] A. Digby, 'The rural poor', in G. E. Mingay (ed.), *The Victorian Countryside* (London, 1981), vol. II, pp. 599–600.   [58] Cuttle, *Rural Guardians*, p. 3.

[59] PRO, MH 32/100 Report of B. Fleming, February 1889.

[60] *RC on Poor Laws*, PP, 1909, XLIII, p. 630.   [61] *RC on Poor Laws*, 1909, XLII, pp. 630–1.

[62] Cuttle, *Rural Guardians*, p. 7.

farmer guardians, according to a chaplain in a country union, were reforming initiatives by magistrates who, before 1894, were *ex officio* members of the boards.[63] Traditional attitudes and policies died hard despite the fact that, before the end of our period, the financial burden of the rural poor law had been alleviated. In 1896 the Agricultural Rating Act relieved occupiers of agricultural land of half their poor rates at the expense of the Exchequer, while in 1911 relief costs were lowered considerably by the removal of the pauper disqualification from old age pensions.

B. Making ends meet: paupers and non-paupers

A vigorous debate on the relief of the old under the Victorian poor law has recently been sparked off by David Thomson's controversial thesis. He has argued – as part of a wider debate concerned with the levels of pensions before and after the modern welfare state – that the cutback on relief intended by the New Poor Law of 1834 was not enforced against the elderly until the last quarter of the nineteenth century. In Thomson's view it was the 1870s and 1880s that marked a dramatic shift downwards, when numbers of pensions were halved and children and other kin were pressurised for the maintenance of the elderly. Until that time older values were allegedly maintained; the community, rather than the family, continued to take responsibility for the old, and generous relief levels were maintained.[64] Whilst there has been support for the view that before 1834 there was good community support for the elderly,[65] other parts of Thomson's thesis have been challenged. E. H. Hunt has suggested both that there was considerable variation in the amounts of outdoor relief given to the elderly Victorian pauper, and that levels of payment did in fact decline after 1834.[66] Evidence from Norfolk unions, for example, would lend support to this argument. There payments to the old were cut as part of the general review of relief at the start of the New Poor Law, while at the same time adult children were summoned before magistrates to enforce contributions for the maintenance of parents.[67] Although much more work is needed to elucidate both levels and variations in poor-law support, it appears likely that local pressure on relatives

[63] L. Twining, *Workhouses and Pauperism* (London, 1898), p. 81.

[64] D. Thomson, 'The decline of social welfare: falling state support for the elderly since early Victorian times', *Ageing and Society*, 4, 1984, pp. 453–4, 456.

[65] R. M. Smith, 'The structured dependency of the elderly as a recent development : some sceptical historical thoughts', *Ageing and Society*, 4, 1984, pp. 411–12.

[66] E. H. Hunt, 'Paupers and pensioners: past and present', *Ageing and Society*, 9, 1989, pp. 410–12.

[67] For example, Norfolk Record Office, Blofield Union Minutes November 1835–January 1836: Depwade Union Minutes 13 May 1836; Mitford and Launditch Union Minutes 2 June 1836.

to support aged parents became less insistent in mid-century, before reviving after 1870 as part of the crusade against outdoor relief.[68] Also relevant in this discussion is Sonya Rose's work on three industrial villages in mid-nineteenth century Nottinghamshire. Her perspective is helpful in bringing the relief of the elderly into the wider context of adaptive family economies and labour-pooling. An analysis of the changing household status of the elderly indicated a much higher proportion of the elderly who were living with their children than had been suggested earlier by Thomson. Rose has also emphasised the importance of self-help in life-course transitions and has drawn attention to their continued independent earning power.[69]

Who were the old? Looking at the practices of boards of guardians and friendly societies in Victorian times the answers varied between fifty and sixty-five years old.[70] But to see this solely in temporal terms, and not in the related context of physical health, is inadequate as contemporaries appreciated with their conflation of the categories into 'aged and infirm' in relief statistics. Finding appropriate levels of relief for the aged and infirm was a dominant theme in rural policy making by the end of the nineteenth century, since the out-migration of the young had left an ageing population in the countryside. But in fixing on relief, contemporaries found that estimating diverse sources of income of the elderly was a complicated business as these might include earnings, contributions from relatives, charitable assistance, savings and poor-law allowances. Within this context poor relief seems to have been used mainly as a means of lifting the old above the poverty line, although discounting other kinds of income could well involve the creation of a poverty trap, as does modern welfare provision. The oft-repeated argument, that non-contributory pension provision would discourage thrift in the working class, was not usually seen as relevant in country areas, since agricultural wages were acknowledged to be so low that saving for old age was virtually impossible. Indeed, Lloyd George recognised the relationship between inadequate income and absence of lifecycle savings when he introduced old-age pension legislation in 1908.[71]

[68] K. D. Snell, *Annals of the Labouring Poor. Social Change and Agrarian England, 1600–1900* (Cambridge, 1985), pp. 366–7.

[69] S. O. Rose, 'The varying household arrangements of the elderly in three English Villages: Nottinghamshire, 1851–1881', *Continuity and Change*, 3, 1988, pp. 101–22; D. Thomson, 'Welfare and the historians' in L. Bonfield, R. M. Smith and K. Wrightson, (eds.), *The World We Have Gained: Histories of Population and Social Structure* (Oxford, 1986), p. 364.

[70] M. Anderson, 'The impact on family relationships of the elderly since Victorian times in governmental income-maintenance provision', in E. Shanas and M. Sussman, (eds.), *Family, Bureaucracy and the Elderly* (Durham, NC, 1977), p. 40.

[71] P. Johnson, *Savings Behaviour, Fertility and Economic Development in Nineteenth-Century Britain and America* (Centre for Policy Research, Paper 203, 1987), p. 16.

The provision of outdoor relief for old people was standard in most rural unions; such relief was sufficiently widespread as not to take away self-respect, or demean the recipient in the eyes of neighbours.[72] However, the occasional offer of the workhouse to selected old people kept customary anxieties about the stigma of the poor law alive almost to the end of our period. 'The fear of the workhouse is always haunting the minds of workers when their youth is past and old age overtakes them', remarked George Edwards, the agricultural workers' trade union leader who in 1894 had become a guardian for the Erpingham Union in Norfolk.[73] Such indoor relief was mainly reserved either as a last resort, because individuals were thought incapable of looking after themselves any longer, or as a means of forcing children to contribute to the maintenance of aged parents.[74] Interestingly, according to David Thomson's calculations, the overall redistribution of resources to the elderly was most generous in southern rural counties despite the fact that the old made up a greater proportion of the population there, and contrasted markedly with northern counties, where there was little provision made.[75]

Agricultural labourers 'could not make ends meet at all if it were not for charitable gifts – sometimes of coals, sometimes of food and clothing', concluded an authoritative survey of rural incomes in 1913.[76] A Christmas dole of coals from the squire, money from the parson or beef from the farmer was a familiar ingredient of voluntary (private) charity, as were unobtrusive acts of occasional generosity – gifts of soup and meat during times of sickness in the labourer's family. Such philanthropy consisted of benevolence tempered by self-interest in varying proportions. An extreme example of the latter would be the farmer who paid below-subsistence wages yet supplemented them by 'charitable generosity' in the form of a rent-free hovel or a free potato ground. Another instance was the offsetting of increases in wages against a diminution of 'privileges' or gifts from farmers, such as firewood or straw for the pig.[77] Charity was part of a continuing moral economy; in the inward-looking village world benevolence could be conditional on perceptions of deference, and loyalty. What was given with one hand could be taken away with the other. This was especially true in estate villages where substantial economic benefits had to be offset against a significant loss of independence and self-reliance. On Lord Wantage's estate in the villages of Lockinge

---

[72] *Royal Commission on the Aged Poor*, BPP, 1895, XIV, QQ 6725–8.

[73] *Poor Law Conferences Eastern District*, 1899, p. 330.

[74] *RC on Aged Poor*, Q 7920 (evidence of Rev. Hinds Howell); Rider Haggard, *Farmer's Year*, p. 429.

[75] Thomson, 'Social welfare', pp. 469–70.

[76] B. S. Rowntree and M. Kendall, *How the Labourer Lives. A Study of the Rural Labour Problem* (London, 1913), p. 34.

[77] F. G. Heath, *British Rural Life and Labour* (London, 1911), pp. 246–9, 270–1.

and Ardington, for instance, the provision of allotments, a reading room, a co-operative store, an excellent school and good employment had to be set against a close control of social life such that a labourer in a neighbouring village commented in 1891 that 'They daren't blow their noses over at Ardington without the bailiff's leave.'[78] More generally, however, charity could be used as a reward for a socially desirable lifestyle – most obviously a life of industry which kept oneself and one's family off the poor rates. This upright behaviour was then more likely to lead to help from parochial or private charity in hard times: food during sickness, the provision of a blanket during a cold winter, and low-rented or free housing during old age. In this way parochial charity reinforced self-help. It might also have reformist implications as, for instance, with a vicar's wife who gave gifts of furniture to non-pregnant brides.

The conventional Victorian wisdom, that charity was reserved for the deserving whilst the poor law dealt with the undeserving, appears to have had rather limited application in low-wage villages where outdoor relief in inadequate doles supplemented income, and charity and mutual self-help then attempted to bring total income up to subsistence level. The poor-law inspector for Wiltshire and Dorset fulminated in 1889 that:

Poor relief as generally administered not only does positive harm but prevents a great deal of good . . . Paupers receiving two shillings [10 new pence] can only live by obtaining charity as well. Thus the rates and charity are equally misapplied and incomplete in the relief they afford. Charity should help those deserving cases of whose merit the poor law can take no heed, and keep them above pauperism. It should not be frittered away in unavailing doles to those in chronic destitution.[79]

In these circumstances it is not surprising that there was apparently little formal co-operation between poor-law guardians and charitable organisations in rural areas, although a few branches of the Charity Organisation Society were set up in market towns such as Leominster and Bury St Edmunds.[80]

Charity was not just a matter of transfer payments from the propertied classes to the poor, however, since there were extensive donations from the working class to formal charities.[81] Jeffries called village women a 'charitable race . . . eager to help each other'.[82] Also important in this context was mutual self-help. Gifts from neighbours might include meaty bones to make soup, a cod's head to make a fish pie, fallen apples for jam,

---

[78] G. E. Mingay, *Rural Idyll* (London, 1989), pp. 31–5.

[79] PRO, MH 32/100, report of B. Fleming, 1888–9, fos 11–12.

[80] *RC on the Poor Laws. Evidence of Rural Centres*, BPP, 1910, XLVII, QQ 72971–5, appendix CLXXII; *RC on Poor Laws. Report on Endowed and Voluntary Charities*, BPP, 1909, XLII, p. 737.

[81] *Family Budgets: Being the Income and Expenses of Twenty Eight British Households* (London, 1896).

[82] R. Jefferies, *The Toilers of the Field* (Futura edn, London, 1981), p. 104.

or other garden produce to eke out a piece of bacon for a family meal. Alternatively, a sick man's friends might club together to help out in bad times. Or money be sent to a hard-pressed large family from more affluent relatives; often regular small payments of money might come from a daughter in service or from a father whose own standard of living was comfortable since he no longer had children to maintain at home. Clothing for the agricultural labourer and his family was rarely bought but came instead from cast-off clothes from a neighbour or relative, or a bundle of clothes obtained by a daughter in service. In addition, items might be supplied by a ladies' guild in the locality or be the product of weekly pennies subscribed on an instalments plan to the village clothing club, which were then supplemented at the end of the year by an equivalent amount donated by others.[83] Such hard-won and meagre resources had perforce to be divided unequally, with the breadwinner in the family having first claim on food and clothing, and wife and children taking what was left over.[84]

Charitable activity in the village was very much a female world and one in which women from different social backgrounds could join in a shared endeavour. Women were prominent in charitable activities since this was an extension into the community of feminine expertise in familial and religious practices. Visits to the poor or the sick took place with a Bible or tract in one hand and food or clothing in the other. Improving the material condition of the recipient was the immediate aim, but long-term moral or religious improvement through encouraging the participation of the women at the Mothers' Meeting or the children at Sunday School was also on the agenda.[85] The women engaged in this charitable, parochial world were pragmatic and enterprising, and soon acquired skills in organisation and fund-raising, thereby enhancing their self-confidence. From the 1850s women also undertook workhouse visiting, and improved conditions for the children and the sick there. They were eventually enabled to use their expertise in the public world of local government, where their service after 1894 included that on parish and rural district councils as well as that of poor-law guardian. Here their work often concentrated on improving the comfort and health of workhouse inmates: adding a stove to a sick ward or nursery; providing brushes and combs for every child; or organising dentistry and the provision of toothbrushes for the young.[86]

[83] Rowntree and Kendall, *Labourer*, pp. 46, 69, 93, 102, 105, 110, 115, 134, 157.

[84] *Ibid*, p. 187.   [85] Prochaska, *The Voluntary Impulse*, pp. 23, 31, 45.

[86] L. Twining, *Workhouses and Pauperism and Women's Work in the Administration of the Poor Law* (London, 1898), pp. 3, 136–8; Prochaska, *Women and Philanthropy in Nineteenth-Century England*, pp. 222–6; P. Hollis, *Ladies Elect*, pp. 10–11, 374–5, 390–1.

C. Sickness and health

The medical services of the New Poor Law were mainly curative rather than preventive; the exceptions to this generalisation being their compulsory vaccination duties (from 1853) and their cursory involvement in public health (after 1872). Although these medical services have had a generally critical press, nevertheless the poor law did improve access to a range of services that would otherwise have been either limited or unavailable to the village inhabitant: subscriptions to voluntary hospitals in neighbouring towns permitted advanced surgery to be given in a few cases; payment for violent lunatics in county asylums gave access to the specialist care of the day; less acute cases of mental or physical infirmity received more rudimentary care in the sick wards of workhouses; whilst medical relief by the district medical officer was supplied in the pauper's own home. Many aspects of this provision had undoubted defects. In terms of indoor provision the most notable deficiencies were the mixing of different kinds of patient in sick wards (the chronic mentally and physically infirm, the infectious and the venereal), and the apparently insuperable difficulties in providing trained rather than pauper nurses even by the end of the period.[87] Where outdoor medical relief was concerned the principal weaknesses were the large size of the areas that the district medical officers covered, together with the cumbersome bureaucratic procedures that had to be undertaken before access to the doctor was obtained.[88] Whilst there were undoubted inefficiencies, and some cases of horrific neglect arising from these factors, the quality of medical relief given to outdoor paupers was not so demonstrably different from that given to independent patients from other social classes, not least because in country areas the same general practitioner would usually have been called on for all social classes. In consequence, rural medical officers might be held in considerable esteem by villagers.[89]

The alleged generosity with which poor-law medical assistance was given attracted robust comment during the Victorian period. Boards of guardians shrewdly considered that poor-law medical officers effectively used their 'medical extras' of alcohol and meat to give relief in aid of subsistence rather than in aid of sickness, and hence that they usurped the role of the relieving officer.[90] But medical officers were only too aware that with families dependent on low wages in agriculture, a patient's

---

[87] C. Maggs, 'Nurse recruitment in four provincial hospitals', in C. Davies (ed.), *Rewriting Nursing History* (London, 1980), pp. 33, 38.

[88] M. Flinn, 'Medical services under the New Poor Law', in D. Fraser (ed.), *The New Poor Law in the Nineteenth Century* (London, 1976) pp. 48–57; M. A. Crowther, *The Workhouse System, 1834–1929* (London, 1981), pp. 161–3.   [89] Cuttle, *Rural Guardians*, pp. 5–6, 100–1.

[90] Crowther, *Workhouse*, pp. 157–8; Digby, *Pauper Palaces*, p. 176.

recovery partly dependent on extra nourishment. However, the rising status of the medical profession, together with the increasing social esteem that scientific expertise commanded, made it more difficult before the end of our period for parsimonious boards of guardians to challenge their medical officer's judgments on medical issues.[91] On the other hand the virtual inevitability of poor-law medical relief to low-paid rural workers was implicitly accepted. The central role of poor-law medicine was recognised by the Medical Relief Disqualification Removal Act of 1885 which stated that sick people receiving such help should not lose their right to vote. Nevertheless, the poor law medical service was only one of several alternatives; lay as well as professional medicine flourished in rural areas with bonesetters, herbalists, wise women, abortionists and sellers of patent medicines, as well as midwives and general practitioners. By the late nineteenth-century a shift was discernible towards what were considered more expert and professional forms of medical assistance. These might be channelled through friendly societies or sought independently.[92] Self-help in the form of subscriptions to friendly societies was important for their mainly male members in providing medical assistance (the attendance of a local doctor as well as the medicines he recommended), as well as in giving sickness or burial benefit.[93] On the infrequent occasions when ordinary village inhabitants consulted a doctor independently they had to struggle to pay off the resultant debt.[94]

Despite the compelling case made out by Edwin Chadwick that sickness was a major cause of poor-law expenditure, the deterrent framework of central policies for poor relief, when reinforced by the economical instincts of local administrators, gave little space for the concept of preventive measures in public health. The inadequacies of guardians (under the Public Health Act of 1872) in performing their responsibility to act as the rural sanitary authority, was unfortunately merely repeated in many cases for their comparable reponsibilities on Rural District Councils after 1894.[95] Sanitary and highway work was performed much more perfunctorily than poor-law duties; RDCs being stirred reluctantly into action by the more active parish or county councils and by their own officers. In suggesting that sanitary committees progressed from this responsive mode to a more active and investigative one, Hollis gives a more optimistic interpretation than this.[96] The key question, however,

---

[91] Cuttle, *Rural Guardians*, p. 62.

[92] V. Berridge, 'Health and medicine', in F. M. L. Thompson (ed.), *The Cambridge Social History of Britain, 1750–1950* (Cambridge, 1990), p. 190.

[93] P. H. J. H. Gosden, *The Friendly Societies in England 1815–1875* (Manchester, 1961), pp. 138–48.

[94] Rowntree and Kendall, *Labourer*, pp. 115, 156.

[95] Digby, *Pauper Palaces*, pp. 175–6; E. Thomas (ed.), R. Jefferies, *The Hills and the Vale* (London, 1911), pp. 156–7.   [96] Hollis, *Ladies Elect*, p. 384.

was rural housing and here the RDCs were notoriously inactive in remedying overcrowded, insanitary dwellings. In the Maldon Union in Essex, for example, an inspector's damning verdict that he had 'never been in a district where the housing conditions had been so appalling' met with a predictable response. The Chairman suggested that this 'outside visit' had not revealed that although 'most of the houses did not appeal to the eye . . . [they] were exceedingly comfortable to live in'.[97] Indeed, the deficiencies and shortage of rural housing had been a matter of concern and debate since the late eighteenth century and had in the mid-nineteenth century climaxed in a crescendo of allegations that proprietors in so-called close parishes had actually *destroyed* cottages in order to keep settlements and poor rates down. Sarah Banks has challenged such stereotypes of open and close parishes,[98] giving additional weight to an alternative interpretation that it was not an economic proposition to *build* cottages for low-paid labourers.

D. Dependence and independence

In these circumstances how feasible was it for the rural labourer to retain his independence? In mid-century there was abundant testimony that the condition of the independent labourer in southern counties was so bad that – if they acted with economic rationality – it would have repaid them to make themselves indoor paupers, thereby enjoying the superior material standards of workhouse diet and accommodation.[99] Even that staunch friend of the agricultural labourer, Canon Girdlestone, could remark of the rural poor law in 1869 that, 'There never was invented a more powerful instrumentality for robbing a man of his independence and making him idle and improvident.'[100] But the precariousness and economic insecurity generated by low incomes needs emphasis as a corrective to this kind of moralistic middle-class interpretation. Paul Johnson's work has emphasised the desire of the ordinary person to achieve security through diverse forms of saving, and thus to prevent dependence on the poor law or on charity. He has argued that low incomes constrained choices so that preferences for insurance or club accumulation, involving small payments and short-time horizons, were economically rational, though not always recognised as such at the time.

---

[97] Cuttle, *Rural Guardians*, p. 198.

[98] S. Banks, 'Nineteenth-century scandal or twentieth-century model? a new look at 'open'and 'close' parishes', *EcHR*, 2nd ser., 41, (1988), pp. 51–73.

[99] *The Morning Chronicle*, 22 and 26 January 1850 (Letters VII and XXIX on Devon, Somerset, Hampshire, Surrey and Essex).

[100] 'How may the condition of the agricultural labourer be improved?', *Transactions of the National Association for the Promotion of Social Science*, 1869, p. 552.

The range of working-class savings included: the nearly universal subscription for burial insurance to prevent the stigma of a pauper funeral;
very common payments-by-instalment to a village clothing club; and
subscriptions to a friendly society for some kind of sickness benefit
(where the incidence seems to have shown considerable variation both
within and between agricultural counties).[101] Saving was for good times
and borrowing for bad. Making ends meet in difficult times through
obtaining credit at the local shop or by recourse to a pawn shop were less
accessible options in the countryside or market town than in the industrial city. Mutual self-help was a common form of risk sharing that provided a useful village support mechanism, and only after this was recourse
made to charity. Old people often chose to be helped by what were
effectively informal maintenance contracts with their families, and only
then through charity.[102] The poor law was a last resort for everyone. A
woman in the Malmesbury Union, for example, did not approve of her
husband having applied for an out-relief order for some bread in 1897,
and 'indignantly refused to accept it and allow her family to become
paupers'.[103]

The Victorian and Edwardian poor law was better at penalising dependency than at rewarding efforts to achieve or maintain independence. In
1870 the Local Government Board had stated that the full value of
allowances from friendly societies needed to be taken into account by
guardians when determining outdoor relief, thus effectively penalising
earlier thrift. In practice, local boards sensibly disregarded this and only
took half of such payments into account, thus anticipating the ruling
under the Outdoor Relief (Friendly Societies) Act of 1894. A few boards
of guardians made particular kinds of relief (notably the payment of children's school fees or of medical relief) available only on loan, thus combatting any incipient tendency to dependency.[104] But such policies were
the exception rather than the rule. The poor-law inspectorate argued
despairingly that if only 'the lavish distribution of demoralizing doles
could be checked' then an 'undercurrent of good feeling' could develop
in the rural population.[105] Individual and collectivist self-help manifestly
showed that such 'good feeling' already existed. But low wages in the
rural sector meant that this was an inadequate bulwark against the dependency that sickness, old age and unemployment could soon produce. It
took the vastly, more ambitious collectivist initiatives of the so-called

---

[101] P. Johnson, *Saving and Spending. The Working Class Economy in Britain, 1870–1939* (Oxford, 1985),
pp. 71, 58–60, 217–18, 221–2, 232.
[102] M. E. Davies, *Life in an English Village* (London, 1909), p. 189.
[103] PRO MH 32/100 report of poor-law inspector B. Fleming, 1897.
[104] P. Horn, *The Victorian Country Child* (Kineton, 1974), p. 67.
[105] PRO MH 32/100 B. Fleming's report.

'social-service state' between 1908 and 1911 to begin to redress this situation, although for several decades after this the new administrative bodies operated alongside older forms of welfare. This co-existence of old and new authorities was also a noticeable feature in the field of policing and crime.

III. AUTHORITY, POLICING AND CRIME

It was urbanisation and its perceived links with increased crime and the development of a criminal class that alarmed the Victorians; except in years of exceptional distress there was little anxiety about disorder or law-breaking in the countryside. Correspondingly, attention was focused on central or urban agencies of control – the 'convict' prisons or the 'new police' in the towns – rather than on the 'local' prisons or on country constables. Yet it is historically inaccurate to view the Victorian countryside as a hinterland of virtue, or to ignore some significant developments in policing that occurred in rural areas during the mid-nineteenth century. Examining these developments from a historical perspective gives a clear insight into the tenacity of traditional values and of forms of rural jurisdiction.

Fundamental questions about the nature of this authority and the legitimacy of customary values in the countryside were raised in the period before 1850 in the partial substitution by county élites of a political, for a moral, economy. Under the latter the rural poor had been able to appeal to customary social justice in their defence of traditional rights. But this was gradually eroded and a new set of relationships developed that were increasingly dominated by the cash nexus.[106] This transition suggested that it was not the poor but rather some magistrates, landowners and farmers who might be radical (even deviant) in relation to developments in criminal and civil justice.[107] Despite lawyers' habitual assumptions that law reflects social norms, in this context there was not a universal consensus. Rather, as Hay has aptly commented, 'state law and popular belief shared important areas of agreement but also important areas of disagreement'.[108]

Amongst the areas of contention were the traditional 'use-rights' of the rural poor, such as the right to glean, to cut turves, to take dead wood, to gather fruit and nuts and, in Wales, to gather wool.[109] These

---

[106] E. P. Thompson, 'The moral economy of the English crowd in the eighteenth century', *PP*, 50 (1971); E. Fox Genovese, 'The many faces of moral economy', *PP*, 58, 1973.

[107] D. Hay, 'Crime and justice in eighteenth- and nineteenth-century England', *Crime and Justice*, 2, 1980 pp. 47, 75.

[108] D. Hay, 'The criminal prosecution in England and its historians', *Modern Law Review*, 47, 1984, pp. 6–7.   [109] D. Howells, *Land and People in Nineteenth-Century Wales* (London, 1977), p. 103.

had been drastically curtailed by enclosure well before 1850 although those relating to gleaning – the custom whereby the poor gathered the uncut or fallen grain in the fields after harvest – were more resistant. An analysis of gleaning and the growth of absolute conceptions of property ownership has led to an interesting perception of the nature of rural authority in relation to law enforcement, and of the intricacies inherent in the relationship of formal law and social practice. Although never defined as a criminal offence, a civil law decision in 1788 had made gleaning a matter of trespass. But this decision was not translated into practice, since many justices would not allow farmers to indict gleaners for theft or trespass, so that the poor usually managed to continue quietly with this traditional activity. In this context the crucial determinant for the continuance of gleaning was whether it was well established in local custom. In practice the decline of gleaning was related to technological advances in the use of the horse rake, reaper and binder during the third quarter of the nineteenth century rather than to earlier legal changes. Instead of the clear-cut line-up of the propertied against the poor which the supremacy of a political economy might have indicated, this case study suggests a multiplicity of norms among social groups. As such it is instructive in cautioning against a too functionalist view of society, and also in increasing awareness of the problematic nature of social control.[110]

The face-to-face nature of village society meant that the character of the rural magistracy was distinctive, and that informal methods of control continued for a longer time in the country than in towns. In their defence of this 'local state' Victorian county justices fought a rearguard action to preserve what were essentially eighteenth-century patterns of order and control against centralist and bureaucratic innovations.[111] One of the crucial ingredients in a pluralistic interpretation of law, legal institutions and the police in Victorian society is this dichotomy between local and central government. A later discussion both of the nature of the 'new police' in rural areas, as well as the administration of local prisons before 1877, will indicate the continued strength of the county justices in their administration.

In reflecting on the nature of authority it is also useful to see that the system of criminal justice was not just a matter of imposition but also

[110] See D. Sugarman, 'Law, economy and the state in England, 1750–1914: some major issues', in D. Sugarman (ed.), *Legality, Ideology and the State* (London, 1983) p. 235 for an interesting discussion of this general point; F. M. L. Thompson, 'Social control in Victorian Britain,' *EcHR*, second ser., 33, 1981.

[111] D. Philips, '"A just measure of crime, authority, hunters and blue locusts": the revisionist social history of crime and law in Britain, 1780–1850', in S. Cohen and A. Scull, (eds.), *Social Control and the State* (London, 1983), pp. 65–6.

involved the assent of the working classes or, at the very least, a section of them. This is indicated by their use of prosecution. Whilst it is true that this social group were disproportionately represented among the defendants and under-represented among the prosecutors, nevertheless as many as one-fifth of theft prosecutions came from this class during the eighteenth and nineteenth centuries.[112] Greater account of this kind of complexity should modify a misleading view of the state as all-powerful in disciplining disorder and punishing crime.[113]

A. Criminals and crime

The accepted contemporary wisdom was that serious crime in the countryside was 'comparatively scarce', and petty sessions usually dealt with 'drunkenness, quarrelling, neglect or absenteeism from work, affiliation, petty theft'.[114] Modern discussion of rural crime and disorder usually focuses on theft, poaching, vagrancy and varied forms of social protest. Yet historians need to be careful when interpreting the 'crime' statistics of these and other 'offences' in the period. V. A. C. Gatrell has warned of the 'high degree of imprecision in the way in which crimes are perceived and recorded'.[115] Of relevance here was the disjuncture between popular and élite views of certain rural activities, a gap which widened with the increasingly class-fragmented nature of rural society. There was also the extent to which communal sanction might be given to acts seen as social protest rather than as crime. Amongst other obvious pitfalls in interpretation was the uncertain relationship between actual and recorded crime. There may have been fluctuations in the criminal statistics which reflected not so much the real incidence of crime as contemporary administrative developments which influenced the rate of commitals recorded and thus the criminal statistics. One instance of this was the 'new rural police' which improved the public's ease in reporting crime in the countryside after 1839. Related to this was the comparative efficiency over time in the detection of crime. Another relevant factor was change in the criminal law, notably an early nineteenth-century rise in the rate of prosecutions and, by the 1850s, a transition from private to

---

112 Hay, 'Criminal prosecution' p. 9; D. Philips, *Crime and Authority in Victorian England. The Black Country, 1835–1860* (London, 1977), p. 285.

113 See Ignatieff's penetrating discussion of M. Foucault's view of the powerful disciplinary state (found in *Discipline and Punish*, London, 1977), in 'State, civil society and total institutions: a critique of recent social histories of punishment' in Cohen and Scull (ed.), *Social Control*, pp. 85–6 and 100–1.

114 R. Jefferies, *Hodge and His Masters* (Quartet Books edn, London, 1979), p. 257.

115 V. A. C. Gatrell, 'The decline of theft and violence in Victorian and Edwardian England', in V. A. C. Gatrell, B. Lenman and G. Parker (eds.), *Crime and the Law. The Social History of Crime in Western Europe since 1500* (London, 1980), p. 246.

police prosecution.[116] Also pertinent was the increase in summary justice (i.e. without the use of the jury) which rose from two-thirds to four-fifths of all cases between 1857/60 and 1911/13, and which facilitated the trial of many more cases. Finally, Victorian changes in punishment – with shorter prison sentences and a greater use of fines – meant that offenders were recycled more rapidly and thus were potentially free to commit more crimes more quickly.[117]

Given these difficulties it would be possible, but in my view misleading, to see criminal statistics only as an artificial construct of law enforcement.[118] Rather, an awareness of the difficulty of disentangling administrative and judicial changes from recorded crime, should lead us to adopt a less positivistic attitude to Victorian criminal statistics. But used selectively, and with caution, such data can be illuminating.[119] An admirable attempt to discern trends in the criminal statistics has indicated that theft was the major crime of the countryside, although rural crimes of violence remained high until a downturn in the 1860s. Using both statistics and contemporary observation this significant study has concluded that theft and crimes of violence declined nationwide during the second half of the nineteenth century, after a previous rise that ended in the late 1840s or early 1850s.[120] Yet it is also important to appreciate that more can be gained if such data are placed in their full historical context, as the new social history together with a more critical interest in the social and intellectual history, of law, has indicated. There is a 'framework of meanings within which the history of law-enforcement and law-breaking might be located',[121] enabling us to transcend what has been perceptively called 'the myopia of the criminal law'.[122]

A significant question within this wider conceptual framework is whether a distinction can be made between 'social protest' and 'real crime'. In a pre-democratic age rural attacks on property have been seen as one of the few remaining ways in which the labouring poor could express their discontent about the development of a pervasive political economy, in which a narrow cash-nexus replaced an earlier bond of paternalistic protection by the propertied towards the poor. In the countryside

---

[116] D. Philips, 'Good men to associate and bad men to conspire. Associations for prosecution of felons in England, 1760–1860', in D. Hay and F. Snyder (eds.), *Policing and Prosecution in Britain, 1750–1850* (Oxford, 1989), p. 151.

[117] *Ibid*, pp. 249, 274, 277; Hay, 'Crime and justice', pp. 55–7; V. A. C. Gatrell, 'Crime, Authority and the Policeman State' in F. M. L. Thompson (ed.), *Cambridge Social History of Britain, 1750–1950* (1990), vol. III, pp. 251, 303.   [118] Gatrell, 'The policeman state', p. 287.

[119] V. A. C. Gatrell and T. B. Hadden, 'Criminal statistics and their interpretation', in E. A. Wrigley (ed.), *Nineteenth Century Society* (Cambridge, 1972), pp. 361–2; D. Jones, 'Rural crime and protest', in G. Mingay (ed.), *The Victorian Countryside* (London, 1981) , vol. II, pp. 572–3.

[120] Gatrell, 'The decline of theft', pp. 246, 251; Gatrell and Hadden, 'Criminal statistics' p. 374.

[121] Gatrell, 'The policeman state', p. 245.   [122] Hay, 'Crime and justice', pp. 61–2.

of southern England and particularly in East Anglia, where rural capital-ism was highly developed, it has been suggested that social protest was endemic. Market-based arable farming bred economic insecurity and alienation in the farm labourer. Cattle-maiming, arson, machine break-ing, as well as more serious anti-enclosure, poor-law and bread riots were his responses. In one study, East Anglian villages up to 1850 were depicted as a 'powder keg', where Hodge 'protested all the time', and where 'every year was violent'.[123] This appears a somewhat exaggerated view of rural discontent which, if certainly widespread, was neither universal nor continuous. Alan Peacock's assertion that in East Anglia, 'An enormous amount of crime went on, most of it capable of being regarded as crimes of protest, and most of it undetected'[124] raises further problems in relation to the extent of social protest. Another more detailed study of incendiar-ism – as one form of social protest that occurred in East Anglia in 1844 – has come to more cautious conclusions, emphasising both the *lacunae* in our knowledge, and the difficulties of interpreting the data that we have.[125] Nevertheless, this study also reveals the communal sanction given by the labouring population to such fires, with labourers who not only refused to help put out the burning stacks of farmers but even actively obstructed fire-fighting operations.[126] Arson continued in the south-western counties and in certain Midland counties such as Bedfordshire, Buckinghamshire and Berkshire. It was still to be found in the early twentieth century, but it appears that its incidence decreased at a time when more peaceful and organised forms of action were developed in agricultural trade-unionism during the late 1860s and early 1870s.[127] It was not only arson that has been linked to social protest since a study of cattle-maiming in East Anglia between 1830 and 1870 has argued that this could be seen as 'a kind of symbolic murder of the farmer'.[128]

Much rural crime was linked to the poverty of the labourer, being sea-sonal in nature and related to fluctuations in employment. Winter, when agricultural work was irregular and the farmers' requirements for labour reduced, produced subsistence-related crime, with higher levels of theft of items such as food, clothing and firing.[129] 'More people stole in hard

---

[123] A. Peacock, 'Village radicalism in East Anglia, 1800–1850', in J. P. D. Dunbabin (ed.), *Rural Discontent in Nineteenth-Century Britain* (London, 1974), pp. 60, 27, 39.

[124] Peacock, 'Village radicalism', p. 55.

[125] D. Jones, 'Thomas Campbell Foster and the rural labourer; incendiarism in East Anglia in the 1840s', *Social History*, 1, 1975, p. 37.   [126] *Ibid*, p. 16.

[127] D. Jones, *Crime, Protest, Community and Police in Nineteenth-Century Britain* (1982), pp. 34–5.

[128] J. E. Archer, '"A fiendish outrage?" A study of animal maiming in East Anglia, 1830–1870', *AHR*, 33, 1985, p. 156.

[129] Howkins, *Poor Labouring Men*, p. 31; A. Howkins, 'Economic crime and class law: poaching and the game laws, 1840–1880', in S. Burman and B. H. Bond (eds.), *The Imposition of Law* (New York, 1979), pp. 285–6.

times than good'; in Victorian England there appears to have been a positive correlation between property offences and poverty.[130] And in a low-
wage economy hardship tended to dissolve moral boundaries between
petty criminality and balancing the budget.

Poaching, that quintessential – and somewhat romanticised – country
activity, also raises complex issues of interpretation in relating it to such
issues as the poverty and alienation of the labourer, as well as to the sanction of communal values and customary use-rights of the poor, and to
the contemporary debate on the criminal as amateur or professional. For
some individuals there was a strong element of social protest involved in
their defiance of the Game Laws. One Victorian poacher justified his
actions in these terms:

> We had no voice in making the Game Laws. If we had I would submit to the
> majority for I am a constitutionalist. But I am not going to be a serf. They not
> only stole the land from the people but they stocked it with game for sport,
> employed policemen to look after it . . . If I had been born an idiot and unfit to
> carry a gun – though with plenty of cash – they would have called me a grand
> sportsman. Being born poor, I am called a poacher.[131]

Poaching was a multifarious activity. Particularly in the mid-Victorian
period it could be linked to social tension and popular morality regarding the right of the poor to free food. But there were also professional
gangs who were active and these were of a very different character from
the amateur poacher.[132] By 1911, at a time when game preservation had
reached its peak, the threat of these gangs had contibuted to the growth
in the number of gamekeepers who outnumbered rural policemen by
two to one.[133] The presence of poaching gangs was also used as one
justification for rural forces (such as that in East Suffolk) being armed
with cutlasses, as well as the standard truncheon.[134] And the provisions of
the Night Poaching Prevention Act of 1862, meant that anyone walking
the roads had the onus of proving that they were *not* carrying game. As
the Chief Constable of Norfolk remarked contentedly, this meant that
'every constable now has the power to stop and search any person'.[135] In

---

[130] Gatrell and Hadden, 'Criminal statistics' p. 378. And for similar evidence for an earlier period see
D. Hay, 'War, dearth and theft in the eighteenth century', *PP*, 95, 1982.

[131] G. Christian (ed.), *James Hawker's Journal: A Victorian Poacher* (Oxford, 1961), pp. 62, 109.

[132] Jones, *Crime, Protest*, chapter 3 *passim*.

[133] F. M. L. Thompson, 'Landowners and the rural community' in G. E. Mingay (ed.), *The Victorian
Countryside* (London, 1981), vol. II, p. 459.

[134] C. Emsley, 'Arms and The Victorian Policeman', *History Today*, 934 (November 1984) p. 39. This
was permitted under the 1839 act. Police were also armed in certain rural forces during incendiarism and when railway navvies were thought to pose a threat to rural order (C. Emsley, '"The
thump of wood on a swede turnip": police violence in nineteenth-century England' *Criminal
Justice History*, 5, 1984, pp. 132–4.

[135] Quoted in C. Steedman, *Policing the Victorian Community. The Formation of the English Provincial
Police Forces, 1856–1880* (London, 1984), p. 150.

practice, however, contemporaries were clear that it was not the rate-paying, middle classes who were stopped by the police.

Equally, contemporaries were in no doubt about several basic characteristics of criminals, although the accuracy of these perceptions has rightly been subjected to considerable criticism by modern historians. But these mid-century stereotypes (particularly when unsubstantiated by hard evidence) are revealing in suggesting a much more fragmented society with intense apprehension of the so-called 'dangerous classes'. Victorians believed in a definite criminal class, as did M. Davenport Hill when he referred in 1839 to 'the existence of a class of persons who pursue crime as a calling'.[136] This concept was developed so that a prison chaplain could later write that:

There is a population of habitual criminals which forms a class by itself. Habitual criminals are not to be confounded with the working or any other class; they are a set of persons who make crime their object and business of their lives; to commit a crime is their trade; they deliberately scoff at honest ways of earning a living, and must accordingly be looked upon as a class of a separate and distinct character from the rest of the community.[137]

However, modern research into criminal statistics for Staffordshire suggests that only one-tenth of theft offences were committed by the professional burglar, pickpocket or experienced thief whilst most indicted offenders were not full-time criminals.[138] And a study of Sussex and Gloucestershire during the early nineteenth century endorses this conclusion on the lack of a distinct criminal class.[139] Other more general analyses confirm these local researches in depicting most Victorian crimes as casual offences committed by ordinary people.[140]

The second important contemporary perception of the criminal was that of a mobile villain who moved on to easier pastures as policing developed first in urban areas after 1835, and then in some (but not all) rural areas after 1839. Comparative work on rural and urban areas suggests that available statistics did not support the hypothesis of a migratory criminal class. It therefore gave little credence to the supposition that more efficient policing in the boroughs after 1835 shifted crime to the surrounding countryside and hence was an important cause of the Rural Constabulary Act of 1839. Rather, continued inefficiencies in the towns made it imperative to improve policing in both urban and rural areas.[141] Interestingly, there was also a similar Victorian hypothesis concerning

[136] Quoted in L. Radzinowicz and R. Hood, *A History of Criminal Law* (London, 1986), vol. v, p. 74.

[137] W. D. Morrison, *Crime and its Causes* (London, 1891), pp. 141–2.

[138] Philips, *Crime and Authority*, p. 287.

[139] G. Rude, *Criminal and Victim: Crime and Society in Early Nineteenth-Century England* (Oxford, 1985), pp. 125–6.   [140] Gatrell, 'The decline of theft', pp. 265–6; Jones, *Crime, Protest*, p. 6.

[141] J. M. Hart, 'Reform of the borough police, 1835–1856', *EHR*, 70, 1955, pp. 413, 427.

'hard' and 'soft' areas with respect to local prison regimes during the 1860s and 1870s, and hence the same kind of inference was drawn about shifts in the location of criminals.[142]

A conflation of these beliefs gave a very high profile to the vagrant, since self-evidently this group was mobile, and also marginal to the communities through which it travelled. However, there was a concern that the 'deserving' traveller (stereotypically the man down on his luck through lack of work) should be distinguished from the 'undeserving'. The latter might be construed as feckless and workshy or as having criminal proclivities. A poor-law official reflected contemporary assumptions when he stated in 1848 that 'they are for the most part if not criminals, at least on the verge of crime'.[143] Arguably, the increasingly intensive policing of public spaces in Victorian England and Wales did much to manufacture this criminality. Of course, many of the more respectable travellers did not have much contact with the social agencies of poor law and police that attempted to control those on the road. Indeed those travelling the roads were a very mixed group – an inconvenient truism that bedevilled social policy towards them. Numbers of vagrants have been variously estimated but we are unlikely ever to be able to reconstitute numbers on the road at all accurately. Economic fluctuations obviously affected the flows of trampers; one guesstimate was that in prosperous times the traveller in genuine search of work made up one-third of the total, whereas in times of economic depression this rose to two-thirds.[144] But, as Raphael Samuel has warned, 'The wayfaring constituency was in a constant state of flux . . . The distinction between the nomadic life and the settled one was by no means hard and fast.'[145] For example, migrant agricultural workers came to bring in the hay in June, to pick peas or hoe turnips in July, for the wheat harvest in August, and for fruit or hop picking in September. This reserve army of labour regularly spent the period from the late spring to early autumn in country areas and retreated to the towns as the weather got colder.[146]

The poor law was the main agency of social policy towards vagrancy; the vagrant had the right to poor relief but under deterrent conditions. However, the desire to produce such conditions in a bureaucratic regime dominated by economy led to under-investment in the casual wards, and hence to the indiscriminate mixing of different types of traveller. Paradoxically this then fuelled contemporary fears of moral contamination and the spread of criminality. A contemporary expressed these fears vividly when he wrote of 'the whole rabble of the vagrant and dissolute classes

---

[142] S. McConville, *A History of English Prison Administration*, vol. 1, *1750–1877* (London, 1981), p. 376.

[143] Quoted in Jones, *Crime, Protest*, p. 180.

[144] S. and B. Webb, *English Poor Law History, Part 2* (London, 1963) vol. II, pp. 403–4.

[145] R. Samuel, 'Comers and goers', in H. J. Dyos and M. Wolff (eds.), *The Victorian City* (2 vols., London, 1973), vol. I, p. 152.   [146] *Ibid*, pp. 128, 134, 153.

who labour by fits and starts and eke out subsistence by pilfering and who are ever on the verge of more serious breach of the law'.[147] The desire to discipline the work-shy and rid the roads of potential criminals led to increasingly deterrent policies. In general, however, these proved difficult to sustain so that strict regimes adopted in individual poor-law unions or groups of unions were almost invariably succeeded by laxer ones. One of the more deterrent features from the 1840s onwards was the employment of rural policemen as assistant relieving officers. In general, attempts to regulate or discourage vagrancy on a permanent basis were unsuccessful. As deterrent measures increased at the end of the nineteenth century they either became unenforceable or merely shifted the geographical incidence of vagrancy. In the longer term it was economic fluctuations rather than administrative action that determined the numbers on the tramp.[148]

Did the vagrant substantially increase crime? Wide legal powers given to the police in controlling and arresting vagrants both reflected social fears that this was the case, and allowed it to become a self-confirming stereotype. The Vagrants Act of 1824 had given the police very wide discretionary powers, and legislation on habitual criminals passed between 1869 and 1871, gave the police powers of arrest on suspicion of *intent* to commit a felony. Since the 'new police' targeted public spaces as a way of demonstrating their effectiveness in improving public order, the vagrant was necessarily in the front line of their attentions. The treatment vagrants received can be linked occasionally to subsequent criminal activity, as in damage at workhouses or, more seriously, in their involvement in arson in the Midlands and Flintshire and Denbighshire during the early sixties when farmer-guardians' stacks were fired. But such serious crimes were few in number; criminal statistics suggest that vagrants were mainly active in petty crime – being drunk and disorderly or involved in opportunistic small-scale theft of food or clothing.[149] However, the perception of the vagrancy threat to public order contributed between the 1850s and the Edwardian era to a threefold increase in the numbers of vagrants who were cautioned by the 'new police'.[150]

B. Policing the countryside

Before the end of the nineteenth century Victorian Britain had become a policed society.[151] There had developed what has been called 'the

---

[147] T. Plint, *Crime in England, its Relation, Character and Extent, as Developed from 1801 to 1846* (London, 1856).

[148] R. Vorspan, 'Vagrancy and the New Poor Law in late-Victorian and Edwardian England', *EHR*, 92, 1977; Jones, *Crime, Protest*, chapter 7 *passim*; Radzinowicz and Hood, *Criminal Law*, p. 351.

[149] *Ibid*, pp. 205–6, 193–4.   [150] Gatrell, 'Crime and the law', p. 315.

[151] Gatrell and Hadden, 'Criminal statistics', p. 377.

policeman state'; an interventionist and disciplinary, rather than paternalistic state.[152] The 'new police' were the most visible sign of this transition in rural areas. The County and Borough Police Act of 1856
introduced the principle of central inspection and regulation of local
forces. This was reinforced by the opportunities of obtaining a central
government grant if the inspectorate deemed local forces efficient in
terms of numbers and discipline; all but seven county forces qualified for
a grant in the first year and all but Rutland in the next.[153] Central pressures made the counties employ more policemen. Numbers rose from
7,829 in 1861 to 15,866 by 1911, and ratios of population to police
improved from 1:1,489 in 1861 to 1:1,192 in 1891.[154]

It is worth emphasising, however, the local as well as central nature of
this interventionist state. Both the Rural Constabulary Act of 1839 and the
County and Borough Police Act of 1856 strengthened the power of the
local magistracy in the counties.[155] Under both acts chief constables were
selected by the local bench from their own kind; in a growing number of
instances they came from the younger sons of county families, and had a
military background. The chief constable continued to be responsible for
personnel and the justices for finance from the county rate.[156] And in the
prison system county justices fought very hard for the retention of their
powers. In 1863 there were still 193 diverse, local prisons, although the
Prisons Act of 1865 later reduced the justices' autonomy by enforcing a
uniform system of separate cells for prisoners.[157] What powers remained
were defended fiercely in the counties: *The Times* referred to the localities
combatting the 1877 Prison Act 'inch by inch'.[158] However, the central
government's desire for greater uniformity in the prison service triumphed; savings were made from the administration of fewer small gaols.
This was despite the justices' complaint that the magistracy would be
undercut and their powers of patronage reduced, without substantial
economy being effected.[159] In the event the prisons were consolidated so
that by 1894 the numbers of 'local', as distinct from 'convict', prisons had
been reduced to only fifty-six. The result of a tight and efficient regulation of these institutions by the Home Office after 1877 was that local visiting justices entirely ceased to take any active interest in them.[160]

[152] Gatrell, 'The policeman state', pp. 259–62.
[153] H. Parris, 'The Home Office and the provincial police in England and Wales, 1856–70', *Public Law*, 1961, p. 238; Foster, *Rural Constabulary*, pp. 20–2.
[154] Gatrell, 'The decline of theft', p. 275.
[155] Steedman, *Policing*, p. 2; D. Foster, *The Rural Constabulary Act, 1839* (London, 1982), p. 17.
[156] Steedman, *Policing*, pp. 47–50.
[157] S. and B. Webb, *English Prisons under Local Government* (London, 1922), pp. 188–90.
[158] *The Times*, 2 March 1877.   [159] McConville, *Prison Administration*, pp. 472–3.
[160] Webbs, *English Prisons*, pp. 202–3, 216–17.

The vitality of the local bench in handling mid-century disturbances has been emphasised.[161] Their concern to regulate — as well as police — rural society was also obvious in their use of the new constabulary. Nearly all the county benches employed the rural police in one or another administrative capacity in the mid-nineteenth century. In his roles as assistant relieving officer; as inspector of nuisances, inspector of weights and measures, or inspector of common lodging houses; as market commissioner; and as impounder of stray cattle; the country constable was a useful workhorse for the local ratepaying community. As such he also visibly embodied the increasing range of bureaucratic, as well as policing, activities of Victorian government. Not surprisingly, perhaps, such activity led to a rise in newer categories of crime in rural areas. During the 1880s and 1890s, for instance, large numbers were prosecuted for offences involving breaches of the Highways Acts or the Education Acts.[162]

How did the efficiency of the new police compare with their predecessors? An older whiggish view tended to belittle the effectiveness of the elected, amateur village constables and to emphasise the impact of their successors — the professional rural police after 1839. More recently, the reputation of the former has received at least a partial rehabilitation; a study of early nineteenth-century Sussex and Gloucestershire concluded that 'the old system worked reasonably well in the case of purely local crime',[163] whilst research into Staffordshire suggests that the old system operated adequately for normal offences like thefts, or assaults where the victim had a reasonable idea who the criminal might be. Indeed, the rural constable could be an experienced man with long service to his credit rather than the annually rotating bumbler of historical stereotype. The village constable continued to be active in counties such as Staffordshire and Worcestershire, even after the creation of the new county rural constabularies.[164]

Even before the Rural Constabulary Act of 1839 there had been reforms in policing in a minority of country parishes where, during the 1830s, there had been some volunteer subscription forces, the employment by poor-law guardians of police, and private watchmen.[165] Certain areas had utilised the Lighting and Watching Act of 1833 to establish a police force in rural areas, as had Wymondham and the Blofield area in

[161] R. Quinault, 'The Warwickshire magistracy and public order, c. 1830–1870', in R. Quinault and J. Stevenson (eds.), *Popular Protest and Public Order* (1974), p. 212.

[162] Jones, 'Rural crime', p. 573.   [163] Rude, *Criminal and Victim*, p. 90.

[164] Philips, *Crime and Authority*, pp. 62, 64.

[165] R. D. Storch, 'Policing rural southern England before the police: opinion and practice, 1830–1856', in D. Hay and F. Snyder (eds.), *Policing and Prosecution in Britain, 1750–1850* (Oxford, 1989), pp. 227–8, 235.

Norfolk, Buxted in Surrey, and Horncastle in Lincolnshire.[166] An interesting study of Horncastle's police force between 1838 and 1857 suggests that local policemen dealt with the kind of minor crimes that afflicted a small market town and which were not amenable to the private prosecution process, namely those of drunkenness, prostitution and street rowdiness. These were all offences against what the local inspector termed 'decency and order'. This research is salutary, both in combatting the orthodox, minimalist view of the 1833 Act, and in stressing the nature of the community policing that this legislation encouraged, since the Horncastle police addressed problems that the local people themselves thought needed remedial action.[167]

It is unlikely that the scattered forces established under the 1833 Act did a substantial amount to help gain acceptance for police forces in rural areas, as has been claimed.[168] In looking for explanations as to why certain areas adopted the Rural Constabulary Act of 1839 it is not possible to correlate counties (or county divisions) which adopted it with localities that had been unusually disturbed previously, notably during the Swing riots or the agitation against the imposition of the New Poor Law.[169] Nor does the usual general linkage between fears of Chartism and the formulation of the Act[170] relate well to the actual chronology of the Act's inception. The genesis of the legislation was complex; it owed relatively little to the centralist ideas of Edwin Chadwick, who wrote the preceding report of the select committee on the subject, and much more to the ideas of county justices.[171] The local debates that accompanied its adoption, non-adoption or attempted abandonment in the localities showed considerable ambivalence to a rural police force. This centred both on doubts about the potential political implications of any central control of local forces and, more persistently, on the likely cost of the new force.[172] In the event, twenty counties adopted this permissive act almost immediately and a further ten counties (or divisions of counties in the case of Cumberland or Warwickshire) before 1856.[173] A few counties (Derbyshire, Buckinghamshire and Lincolnshire) failed to adopt it because of the expense, whilst others (Denbighshire or Lancashire) later attempted to abolish their force for the same reason.[174] Such parsimony

---

[166] B. J. Storey, *Lawless and Immoral. Policing a Country Town, 1838–1857* (Leicester, 1983), pp. 188–9.

[167] *Ibid*, pp. 5, 180, 187, 197; L. Radzinowicz, *A History of Criminal Law* (1968), vol. IV, p. 217.

[168] Davey, *Lawless and Immoral*, p. 196.

[169] C. Emsley, *Policing and its Context, 1750–1850* (London, 1983), p. 71; Storch, 'Policing rural England', p. 250.

[170] F. C. Mather, *Public Order in the Age of the Chartists* (Manchester, 1959), p. 128.

[171] A. Brundage, 'Ministers, magistrates and reformers: the genesis of the Rural Constabulary Act of 1839', *Parliamentary History* 5 (1986), p. 62.

[172] Foster, *Rural Constabulary*, pp. 14–15, 20–2; Emsley, *Policing*, pp. 73–4.

[173] Parris, 'The Home Office', p. 237.   [174] Foster, *Rural Constabulary*, pp. 20–2.

also led to undermanning: for example, in both East Sussex and Leicestershire early ratios of police to population were of the order of 1:6,000.[175]

Administrative reform of rural policing seems to have gradually gained more widespread acceptance during the 1840s and early 1850s. Models were, however, remarkably diverse. Some areas continued with the parish constable, and this well-tried office was invigorated by the Parish Constables Act of 1842 which allowed quarter sessions to build lock-ups, but enjoined the appointment of a paid superintending constable for them. This provision was widely adopted in Dorset and Devon. And a later act of 1850 – adopted widely in Kent, Oxfordshire, Buckingham-shire and Lindsey – permitted the appointment of a professional police officer paid from the county rate, to superintend parish constables in each petty sessional district. By 1853 fourteen of the counties that had not adopted the 1839 Act had superintending constables; a significant attrac-tion being the lower cost of policing when compared to the provisions of the 1839 Act.[176] Thus, even before the County and Borough Police Act of 1856 made it obligatory for all areas to establish a police force, many areas already had a formal paid police force.[177]

The new county forces were recruited predominantly from the agri-cultural labour force in country areas; in Buckinghamshire and Staffordshire this varied between one-third and two-thirds of total recruit-ment. There were also a few servants, and artisans, although in more industrialised counties, such as Lancashire, there were many more recruits from skilled trades.[178] The rules laid down by the Secretary of State after 1856 specified ages between twenty-two and thirty-five for recruits but in practice, in Lancashire the average age was in the mid-twenties.[179] It has been suggested that the attraction of the job lay for the majority in the relatively good rate of pay in comparison to that for the agricultural worker, although clearly the margin was greatest in the southern counties where farm wages were low. Rates of pay in the county force averaged 17s. at recruitment in 1857 and 20s. by 1877. Arguably, an attraction of the police force was that it provided an opportunity – in contrast to labouring jobs – of earning better money as service lengthened; promo-tion to second-class constable led to an additional one or two shillings a week, and the same occurred with elevation to first class rank. In addition

[175] *Ibid*, p. 25; C. R. Stanley, 'The birth and early history of the Leicestershire constabulary', *Justice of the Peace and the Local Government Review*, 108, 1954, p. 605.

[176] T. A. Critchley, *A History of Police in England and Wales, 900–1966* (London, 1967), pp. 91–3.

[177] Storch, 'Policing rural England', pp. 253–5.

[178] C. Steedman, *Policing*, p. 71; W. J. Lowe, 'The Lancashire constabulary 1845–1880: the social and occupational functions of a Victorian police force', *Criminal Justice History*, 4, 1983, p. 47.

[179] Lowe, 'Lancashire constabulary', p. 46.

there were fringe benefits such as uniform, acommodation and, after 1890, pensions for long service.[180]

It was not an easy way to make a living, however, and these long-run benefits may have paled beside the short-run disadvantages of the job. Long hours were worked, with daily beats of up to twelve hours being common; these were performed seven days a week until 1910 when a rest day was granted.[181] The work was physically demanding, requiring much walking; in Leicestershire during the 1840s each constable had to patrol 32 square miles. Chief constables thought the farm labourer particularly suited to the arduousness and loneliness of the country beat, whilst his knowledge of the local terrain was invaluable. Also contributing to the hardship of the job was the social isolation; the result of a deliberate policy of cutting the constable off from his working-class roots. The rural constable was instantly recognisable since he lived in a station house, and had to wear uniform of frockcoat and stovepipe hat (later a helmet) even when off duty. An early regulation of the Pembrokeshire force declared, 'All men of the force who shall associate, drink, or eat with any civilians, without immediately reporting the same to the superintendent, will be dismissed the force.'[182] The rural policeman was supposed to embody the ideals of respectability and sobriety whilst subjected to a hard discipline and an arduous workload. Not surprisingly the turnover in most county police forces was very high. In the early years this amounted to as much as one-third to one-half within a year of recruitment,[183] whilst fewer than one-tenth stayed the course.[184] The Chief Constable of Essex warned in 1852 of the need for officers to combat 'drunkenness, sleeping, inattention and connivance' amongst their men.[185] Dismissals, as well as resignations, were common – usually for drunkenness or insubordination. But mean lengths of service gradually improved; in Lancashire they rose from 4.5 to 6.9 years between the 1840s and 1860s.[186] And by the Edwardian era greater self-discipline and a growing *esprit de corp* were contributing to an increasing sense of professional vocation.

How effective were the new police in the countryside? The country policeman walked and watched, hoping to check thefts of sheep, wood or garden produce, partly by the pressure of his physical presence, and partly by keeping alehouses and lodging houses under close scrutiny and checking up on doubtful characters on the tramp. At night he wrote up

---

[180] Steedman, *Policing*, pp. 110–12, 115; Lowe, 'Lancashire constabulary', pp. 52–3; Critchley, *History of Police*, p. 170.

[181] Stanley, 'Leicestershire constabulary', p. 605; Critchley, *History of Police*, p. 171.

[182] Quoted in Critchley, *History of Police*, p. 151.

[183] Critchley, *History of Police*, p. 147; Steedman, *Policing*, p. 93.

[184] S. H. Palmer, *Police and Protest in England and Ireland, 1780–1850* (Cambridge, 1988), p. 535.

[185] Foster, *Rural Constabulary*, p. 32.   [186] Lowe, 'Lancashire constabulary' p. 53.

his observations in long reports for his superiors. However, much of the contemporary evidence that has been adduced to show the impact of this sort of activity on crime rates has been self-interested testimony.[187] In contrast a recent study of the Bedfordshire force before 1856 suggests that it had little success against the kind of crime that worried contemporaries such as arson or burglary, rather than the easily apprehended drunk or runaway horse and cart.[188] But it is surely unwise to discount too heavily the impact of a more effective police presence when seeking for explanations for the national decline in theft and violence during the second half of the nineteenth century. One possible indication of its effectiveness lay in increased popular hostility to the police during the 1860s and 1870s.[189] It seems likely that by the 1860s the new police were having more impact on the detection of crime than had the old parish constables.[190] Such policing was directed not so much at prevention as detection, although technical expertise was adopted very slowly; Somerset and Monmouthshire had neither photography nor a specialised detective force until after the First World War.[191]

The police also gradually helped to enforce stricter standards of decorum in public spaces. These standards were themselves both cause and effect of changing attitudes. In this context it is important to appreciate that an increasingly sober and respectable society probably tamed itself quite as much as having order imposed upon it. Policing, as Silver has reminded us, was as much an extension of the moral community and the moral assent of the people, as it was of the organising state.[192] Within this evolving set of Victorian social attitudes, it was therefore important that policing practices did not depart too much from communal expectations. A clear demonstration of this was shown in 1846 by the Eaton Bray incident in Bedfordshire, when rough music – a traditional method of showing popular disapproval – was directed at the two local constables, who were seen by the locals to have been too officious in dealing with village drinking habits.[193] Policing of popular customs, notably Guy Fawkes Nights during the 1860s and 1870s, also produced riotous resistance in agricultural market towns such as Richmond and Malton in

[187] D. V. Jones 'The new police, crime and people in England and Wales, 1829–1881', *TRHS*, 5th ser., 33, 1983, pp. 151–2.

[188] C. Emsley, 'The Bedfordshire police 1840–1856: a case study in the working of the rural constabulary Act', *Midland History*, 7, 1982, pp. 81–2.

[189] This hostility was evident in Birmingham rather than rural Warwickshire (B. Weinberger, 'The police and the public in mid-nineteenth-century Warwickshire', in V. Bailey (ed.), *Policing and Punishment in Nineteenth Century Britain* (London, 1981), pp. 66–8).

[190] Philips, *Crime and Authority*, p. 87.   [191] Parris, 'Home Office', p. 254.

[192] A. Silver, 'The demand for order in civil society: a review of some themes in the history of urban crime, police and riot', in D. J. Bordua (ed.), *The Police: Six Sociological Essays* (New York, 1967), pp. 14–15.   [193] Emsley, 'Bedfordshire police', pp. 83–5.

north Yorkshire.[194] In southern England policing of 'the Fifth' was a significant element in its transformation from a plebeian- to a bourgeois-controlled occasion.[195] And, although some accounts may magnify the degree of social control that it was possible to exercise, the more pervasive activity of policemen as 'domestic missionaries' was sufficiently intrusive to have caused resentment in rural communities. According to one perceptive commentator, the police were seen by villagers 'as a potential enemy, set to spy on them by the authorities'.[196] Only occasionally was this sort of hostility shown by middle-class farmers — as was the case during insensitive policing at a race meeting.[197] More usually the police managed to successfully balance protection against undue interference, so that the more usual middle-class criticism was that of the ratepayer anxious about the perceived high cost relative to what was seen as small-scale police activity.[198]

---

[194] R. D. Storch, 'The policeman as domestic missionary: urban discipline and popular culture in northern England, 1850–1880,' *Journal of Social History*, 9, 1976, p. 490.

[195] R. D. Storch, '"Please to remember the Fifth of November": conflict, solidarity and public order in Southern England, 1815–1900' in R. D. Storch (ed.), *Popular Culture and Custom in Nineteenth-Century England*, (London, 1982), pp. 86–93.

[196] F. Thompson, *Lark Rise to Candleford* (Oxford, 1954), p. 553.

[197] R. D. Storch, 'The plague of blue locusts: police reform and popular resistance in northern England, 1840–1857', *International Review of Social History*, 20, 1975, pp. 78, 90.

[198] Weinberger, 'Police and public', pp. 76, 81.

CHAPTER 25

SOCIAL INSTITUTIONS

BY ANNE DIGBY

INTRODUCTION

Idealised scenes of village life depicted in numerous Victorian paintings are revealing in suggesting a contemporary nostalgia for a 'paradise lost': a vanished idyll of hardworking labourers and picturesque cottages grouped round the village church, the symbol of stability and enduring virtues.[1] Urbanisation and industrialisation had fostered a pyschological need to manufacture a rural golden age whilst simultaneously denying the realities of a changing countryside. The more urgent social problems of the towns – seemingly populated by growing numbers in the 'dangerous classes', and of the pauperised, the illiterate and the godless – stimulated activity by both voluntary bodies and the collectivist endeavours of the Victorian state. In consequence the less visible and diffuse deficiencies in rural life were usually seen as secondary to urban issues. However, Benthamite models of efficiency – which were to be achieved through administrative uniformity – tended to mean that what had begun in one sphere was also extended to the other. Public social institutions in the countryside thus included the same coercive and socialising features as did their urban counterparts. It is interesting to see this kind of social philosophy embodied in concrete form in the complementary grouping of institutions. In the small market town of Downham Market in Norfolk, for example, we can see from a print of 1860 that a National School had been built on the outskirts of the town; such establishments typically were designed to socialise the children of the labouring classes into desirable virtues – cleanliness, godliness, obedience, and industry. Beyond, but immediately next to the school, stood the substantial union workhouse of the New Poor Law, reinforcing the political economy of the Victorian curriculum by suggesting the penalties of failure to achieve and maintain economic independence. And should this visibly deterrent edifice not efficiently communicate its social ideology, then the complex was com-

---

[1] C. Wood, *Paradise Lost. Paintings of English Country Life and Landscape* (London, 1988).

pleted by the Court House. Here, on the other side of the workhouse gates, and facing the school, was the meeting place of magistrates in petty sessions; they punished those who, by breaking the law, offended even more drastically against the norms of Victorian society.[2] This provides an instructive exemplar of the interconnectedness of the Victorian social policies and institutions discussed in Chapters 24 and 25; those of criminal justice, the poor law, religion and the school. Historical reality was, however, often more complex than this. The historian may impose neat but over-simplistic patterns of interpretation; areas of heavy game preservation, for example, did not necessarily promote equally large investment by landowners either in the rural police or in education for the lower orders, which a simple model of social control might have suggested would have been the case.[3] Geographically, the uneven incidence of improved elementary schooling more obviously reflected the very local activities of the clergyman or minister.

I. CHURCH AND CHAPEL

The mid-nineteenth century has been termed the period of 'high rectory culture' in England; an age of comfortably off, confident, parochially active and still predominantly rural clergymen of the Church of England.[4] On the traditional secular involvement of the Anglican clergy in their parishes was superimposed the spiritual revival associated with the Oxford Movement. Yet this outward impression of competence and security was misleading; the state was taking over many of the customary secular functions of the clergy whilst, at the same time, wider social changes were eroding its pastoral role. The underlying weaknesses of the Anglican church in its traditional stronghold of rural England were to become only too obvious during the agrarian depression at the end of the nineteenth century.

At mid-century, however, the position of the Church of England in the English countryside still appeared to be one of strength. The conspicuous weaknesses of the past had been largely remedied since pluralism and non-residency had declined markedly.[5] The traditional profit-inspired pluralism had gone; pluralism was now mainly confined to contiguous parishes where one alone gave insufficient income for the

---

[2] The print is reproduced in A. Digby, *Pauper Palaces* (London, 1978), p. 181.

[3] F. M. L. Thompson, 'Landowners and the rural community', in G. E. Mingay, *The Victorian Countryside* (London, 1981), vol. II, pp. 466–8.

[4] G. Kitson Clark, *Churchmen and the Condition of England, 1832–1885* (London, 1973), p. 182.

[5] R. A. Soloway, *Prelates and People. Ecclesiastical Social Thought in England, 1783–1852* (London, 1969), pp. 433–4.

incumbent to live on, although a few deans and canons continued to hold a country parish to which they could retire. In the remaining cases of non-residency (about two thousand at mid-century) a curate was now mandatory.[6] From the 1830s increased numbers of clergy had been ordained and these youthful incumbents gave credibility to the prospect of revival.[7] Country livings formed the vast majority of openings for these ordinands and at this time rural parishes seemed an attractive prospect to them. Samuel Wilberforce, notable Bishop of Oxford, suggested in 1853 that the clergy continued to have a rural, pastoral orientation.[8] The 'Golden Age' of agriculture was giving prosperity to the countryside in the 1850s and 1860s, so that there was ample funding for building, altering or repairing church and parsonage. In Wales there was also an Anglican revival, although its vigour was greatest in the industrialising Welsh valleys. Pluralism and non-residence were stamped out, extra clergy were appointed, and parsonages and schools were built. Despite this, however, Wales remained a predominantly Nonconformist country.[9]

In England traditional society still could find visible embodiment in some village churches, 'The squire was in his pew, his friend the parson in his stall, respectable farmers in pews, and on the benches the labourers.'[10] In an idealised rural past there had been a presumption in favour of attendance at the parish church, more perhaps as a visible sign of allegiance to a cohesive society than as a commitment to denominational membership. In these circumstances the seating plan in the church reflected the village hierarchy. At South Carlton, in Lincolnshire:

The squire had a pew in the chancel; the incumbent's family sat in the front of the nave; immediately behind them was the largest farmer in the parish (with 615 acres); in the central part of the nave, on either side of the aisle, were the other farmers, the larger (with 320 and 200 acres) in front, the smaller behind: sharing seats near the back were a shopkeeper, shoemaker, schoolmaster, and cottagers, with a special sitting reserved for 'servant maids'. There were also sittings in the north aisle; from front to rear, they were occupied by 'servant men', cottagers, a foreman . . . labourers . . . a shoemaker . . . a pauper widow, a free sitting, and finally the vestry.[11]

---

[6] G. F. A. Best, *Temporal Pillars* (Cambridge, 1964), pp. 406–7. 13 and 14 Victoria *c.* 98 virtually abolished pluralism since the maximum distance between parishes could only be three miles, and the value of the first parish could not exceed £100 for a second living to be permissible.

[7] A. Haig, *The Victorian Clergy* (London, 1984), pp. 2–4.

[8] S. Wilberforce, *'I Have Much People in this City.' A Sermon Preached etc.* (London, 1853), p. 11.

[9] D. Gareth Evans, *A History of Wales, 1815–1906* (Cardiff, 1989), pp. 219, 229–35.

[10] O. Chadwick, *The Victorian Church*, Part II, *1860–1901* (London, 1970), p. 151.

[11] J. Obelkevich, *Religion and Rural Society in South Lindsey, 1825–1875* (Oxford, 1976), pp. 109–10.

In many villages non-conformists had customarily attended the parish church in the morning or afternoon, before going on to a service in the local chapel.[12]

A. The role of the English parson

The parson's role in relation to this English rural society was, however, being revolutionised. An important reason for this was the moral lead given by the new breed of bishops, of whom Wilberforce was the most outstanding example of Victorian vigour. Wilberforce came to his diocese in 1845 and found some non-residency, a lack of church services, church buildings in poor repair, and few parsonages. He charged the clergy in his largely rural diocese in 1849 that: 'Not only should the clergyman be able to reach the people, but the people reach the clergyman: he should be accessible at all hours whenever he might be wanted: he should be capable of illustrating by his life the doctrines he taught, and affording his parishioners the example of the amenities of family life.'[13] Those old-style country clergy who tried to keep up the 'huntin', shootin', and fishin' style of life' were reprimanded; Wilberforce being sufficiently determined to root this out that he assembled lists of clerical subscribers to county balls and applications for game certificates![14] His letterbooks indicate the unremitting zeal with which he pursued those whose lifestyle he thought undesirable. 'I cannot but consider you as distinctly a sporting clergyman', he wrote to the Rev. F. Burges whose habit of hunting three times a week led to a request for his resignation, because Wilberforce considered it 'brought much reproach on the Church thereby'.[15] He was equally concerned about the spiritual calibre of those he ordained and also encouraged his existing parochial clergy to take two Sunday services and to preach a good sermon. Through annual conferences and visitations, and the use of rural deans as his local lieutenants, Bishop Wilberforce succeeded in vitalising his diocese, which covered not only Oxfordshire but Berkshire and Buckinghamshire as well. A fair indication of the measure of his episcopal energy and leadership was given by the material injection of funds into the diocese; £2,100,000 was spent on education, buildings and increments to small benefices.[16] But if much was given, much was expected and self-sacrifice, single-mindedness and service to the parish were the order of the day for the parish incumbent.

[12] Chadwick, *Victorian Church*, vol. II, p. 185.

[13] *Rules, Proceedings . . . of the Diocesan Association on the Increase of Church Attendance* (1849), p. 5.

[14] S. Meacham, *Lord Bishop. The Life of Samuel Wilberforce, 1805–1873* (Cambridge, Mass., 1970), p. 103.

[15] Letters to Rev. F. Burges, 2 July 1849 and to J. B. Sumner (Archbishop of Canterbury), 23 July 1849 in *The Letter-Books of Samuel Wilberforce 1843–68* (Buckinghamshire and Oxfordshire Record Societies, 1970), pp. 167, 174.   [16] Haig, *Clergy*, pp. 178–80.

The professional role of the incumbent was notably diffuse; parochial activity in most mid-Victorian country parishes became wide-ranging. The Norfolk parson, Augustus Jessopp, encapsulated his range of incidental parochial activities as, 'from writing a letter to making a will, and from setting a bone to stopping a suicide'.[17] Indeed, establishment assumptions in the Church of England had been largely transformed by 1850; the clergy should give value for money through service to the community.[18] The clergy took pride in its practical ministry and typically this involved a spectrum of activities from paternalistic assistance to the encouragement of self-help. Not all of these were purely clerical, some being the product of earlier lay initiative and many depending on contemporary lay involvement. Some of these activities had their origin in the late eighteenth or early nineteenth century but they became more widespread in the mid-Victorian countryside. The almost universal coal, boot or clothing clubs involved labourers paying in a penny a week and this money being matched by that of subscribers. For ill-paid agricultural labourers this was a valuable supplement to scarce household resources, and the activity of the parson's wife in administering the scheme probably brought village and parsonage closer together. Some parishes also had a blanket club, whereby small payments spread over several years, allowed a warm blanket on loan during winter months, with the promise of eventual ownership. More ambitiously, a few of the more affluent clergy financed co-operative stores for their parishes, as did C. R. Christie at Castle Combe in Wiltshire, or J. W. Leigh at Stoneleigh in Warwickshire.[19] In some parishes, more substantive undertakings were begun, such as medical clubs or savings clubs (typically with a weekly sixpenny subscription). Arguably, these were more stable, because better organised, than those run by the labourers themselves, but increasingly they seem to have been regarded as intrusive and paternalistic.

Less obviously intrusive were the traditional visits to sick parishioners, in which there was a nicely calculated blend of practical and material help on the one hand and, on the other, capitalisation of the spiritual opportunities that such an occasion offered. The parson's wife was important here, and publications for her guidance, such as *Hints to a Clergyman's Wife or Parochial Duties* advised her not only to take gifts of soap and castor oil, but also a pamphlet from the Religious Tract Society. The parson himself could initiate strategic action and administer a village dispensary or be actively involved in implementing the instructions of the local Board of Health. The *Five Papers of Advice* given to Yarnton parishioners by Vaughan Thomas during the 1853–4 cholera outbreak bear the mark of

---

[17] A. Jessopp, *Arcady: for Better for Worse* (London, 1887), p. 4.   [18] Best, *Pillars*, pp. 398–401.
[19] Kitson Clark, *Churchmen*, pp. 183–4.

the incumbent's earlier experiences as chair of the Oxford Board of Health during the outbreak of 1832. The advice ranged from liming the village houses to prayers for 'Protection against the Cholera'. In Oxfordshire clerical concern for the health of parishioners extended to service as governors of urban hospitals or asylums such as the Radcliffe Infirmary or the Warneford Hospital, where the clergyman could actively promote not only the general good of the institution but also the interests of those of his flock who were admitted to them by virtue of the parish's annual subscription.[20] Once the traditional duty of visiting the sick parishioner was undertaken, awareness that caring for the soul of the parishioner necessarily involved action to improve his material environment swiftly followed. Decent living self-evidently required decent housing, for how could a labourer and his family in an insanitary, overcrowded and squalid cottage benefit to the full from the educational activities undertaken by the clergyman, either through the National School for the children, or the lending library and night school for adults. Concern over this state of affairs was abundantly evident in clerical agitation over housing conditions, as in testimony to parliamentary commissions, or in letters to The Times. After 1866 it found most effective expression in the involvement by clergymen-guardians in the nuisance-removal functions of rural boards of guardians.

Vigorous intervention in the lives of the labourers, however elevated the motivation, might well cause a reaction in the village. This was particularly the case after about 1870 when rural social sensibilities had become almost pathologically sensitive as a result of the labourers' disputes over wages and union recognition. The social fabric of rural society was being modified; customary relationships, traditional structures and routine procedures were being insidiously eroded by a changing social consciousness. Precisely because he was expected to be all things to all men, the country vicar or rector was the unfortunate hostage to fortune in many of these developments. To the agricultural labourer humane attempts to alleviate his hardships could appear at best, as condescension or, at worst, as unwarrantable interference. Ignorance of the emerging cultural values of the working class tended to undercut the incumbents' good works so that, as E. R. Norman has aptly commented, the latter's 'remorseless goodwill was often ineffective, despite their social idealism and their hard work for the removal of social evils'.[21] In some cases it was less ignorance of, than hostility to, these values as in the largely successful attempts to morally sanitise traditional village amusements, notably the harvest frolic and crying of largesse, which was frequently replaced

[20] D. McClatchey, Oxfordshire Clergy, 1777–1869 (Oxford, 1960), pp. 166–77.
[21] E. R. Norman, Church and Society in England, 1770–1970: A Historical Study (Oxford, 1976), p. 124.

by a special harvest thanksgiving in church and by a decorous meal.[22] Repressive action was also sometimes taken against boisterous village feasts.[23]

Whilst the labourer saw the parson as one of the propertied classes, any radical action on his part could in turn appear as a class betrayal to the rural landlord or employer. To farmers, attempts by the clergyman to institute allotments, had traditionally been seen in this light. Work put in on allotments at the end of the working day implied that the labourer would have shortchanged his employer in so far as he had not exhausted his strength on his paid work. Indeed, this was part of a more general suspicion of attempts to elevate the labourer, as the incumbent of Hitcham, Suffolk, the Rev. J. S. Henslow, recorded in 1850: 'All schemes, educational, recreational, or however tending to elevate [the labourer] in the social scale, are positively distasteful to some of the employers of labour, whom nevertheless we are bound to recognise as worthy men, not wilfully opposed to the comforts of those beneath them.'[24] Even before the open hostilities of the Labourers' Dispute, certain clergymen had split their parish, leaving farmers and the local gentry implacably opposed to their social interventions on behalf of the labourers. A well-known example of this, cited in the previous chapter, was the encouragement of emigration from the village of Halberton, Devon, by its parson, the Rev. Edward Girdlestone, who between 1862 and 1872 moved over 400 labourers and their families away. His early support of the union activities of the agricultural labourers could only serve to exacerbate this situation, and the village was rancorously divided over his well-intentioned actions.

Despite the anticlerical propaganda of the labourers' trade unions during and after the Labourers' Dispute it is clear that others besides Girdlestone gave support to the labourers, as did Robert Laurence, vicar of Chalgrove in Oxfordshire, who acted as secretary to the local branch of the union.[25] Nationally, the clergy was divided on the unionisation of the agricultural labourer, although the exact proportions of those favouring, opposing, or advocating neutrality are unlikely to be discovered.[26] There was an understandable reluctance to take sides; the Church Congress at Bath in 1873 favoured clerical neutrality. Earlier, there had been some notable statements in the labourers' favour, (as by Bishop Fraser of Manchester, at the previous congress), and even some clerical

---

[22] A. Armstrong, *Farmworkers. A Social and Economic History, 1770–1870* (London, 1988), pp. 106–7.

[23] Obelkevich, *Rural Society*, p. 85.

[24] L. Jenyns, *Memoir of the Rev. J. S. Henslow* (London, 1862), p. 73. Henslow was friend to Darwin and had been Professor of Botany at Cambridge.

[25] P. Horn, 'Problems of a nineteenth century vicar, 1832–1885', *Oxford Diocesan Magazine*, October 1969, p. 16.   [26] Kitson Clark, *Churchmen*, pp. 246–60.

activism in helping to build up the labourers' unions (as by the vicar of Leintwardine). But there were also the ill-considered remarks of the Bishop of Gloucester, C. J. Elliott, in 1872, about agitators from afar who organised 'iniquitous combinations'; and the well-publicised action of two clerical magistrates in Oxfordshire, who sent sixteen women to prison and hard labour in 1873, for allegedly intimidating blackleg labour during the lock-out.[27]

Traditionally, rural society had been rooted in inequality and the Labourers' Dispute was only the most visible indication that this hierarchical state of affairs was being challenged. In 1878 the Rev. J. W. Millar of Birdham, near Chichester, commented that, 'The parish has become a Republic, instead of an Oligarchy, and socially has much deteriorated.'[28] Deference was declining and with it a sense of a hierarchical village community centred on the parish church. By the 1860s employers on the Yorkshire Wolds no longer bothered to ensure at hiring fairs that their farm labourers agreed to attend the parish church.[29] In Lindsey, in central Lincolnshire, farm servants by the 1870s were not encouraged to attend church by their employers.[30] Also in the 1870s had come demands from some radicals within the agricultural unions for disestablishment of the Church of England. However, it is far from clear whether the membership as a whole shared these views since in Lincolnshire and Norfolk, both counties that were heavily unionised during the 1870s, church attendance remained stable. Yet the clergy themselves complained that their churches were less full.[31] Visitation returns from country clergy in the Chichester Diocese during the 1870s and 1880s showed incumbents dispirited by uninterested labourers and unhelpful farmers.[32] It was perhaps no longer self-evident – either to priest or to people – that the parish church remained the natural focus for village life. These deeper and more long-term changes in the role of the Anglican church in the Victorian countryside were to be accelerated from the middle of the 1870s by agrarian economic depression.

The depression affected the income of the village clergyman who had been accustomed to sustain many parochial activities from his own pocket. Obviously speaking from experience, the Rev. Edward Girdlestone commented that:

Were it not for the kindness of the clergy in the rural districts; for the sacrifices which they have made to build and endow schools; for the many meals which they have sent from the parsonage to the cottage, when the breadwinner is ill,

---

[27] Chadwick, *Victorian Church*, vol. II, pp. 155–6; Kitson Clark, *Churchmen*, p. 247; Norman, *Church and Society*, pp. 156–8.   [28] Quoted in Haig, *Clergy*, p. 292.

[29] Armstrong, *Farmworkers*, p. 107.   [30] Obelkevich, *Rural Society*, p. 68.

[31] Chadwick, *Victorian Church*, II, p. 159.   [32] Haig, *Clergy*, p. 292.

his wife confined, the children poorly fed; the many comforts which they have denied to themselves and their families in order to clothe and feed the naked and hungry among their flock; not to mention their prayers by the sick-bed, and their loving words of comfort to the widow and fatherless in their sorrow, the condition of the agricultural labourer would have been very much worse than it is.[33]

Yet this subsidy had been wrung from a financial position that usually had been far from secure. It was generally recognised by contemporaries that 'without private means a clergyman will have a hard time of it'.[34] This was partly because in order to wield influence he needed the trappings of middle-class respectability, but partly also because the demands on his purse for charity were considerable. A Cornish vicar wrote anxiously in 1874 that, 'Money terrors, too, have reached a climax. I have so many claims upon me . . . On the school building account I am responsible for seventy pounds odd more than I have collected from subscribers.'[35] With the onset of agrarian depression these kinds of claim on country clergymen grew larger whilst their income inexorably shrank. In rural areas the bulk of their income came either from tithes, usually commuted on the basis of grain prices, or from glebe land. By 1900 income from tithes had declined by about one-third, (in line with grain prices), whilst much glebe land, (most of which was situated on cold Midland clays) had had to be taken in hand because no tenants were forthcoming.[36]

The late-nineteenth century decline of rural clerical incomes merely intensified the disadvantages of what many ordinands had begun to perceive as an unappealing sector of the Anglican church. Bishop Durnford of Chichester noted 'an increased reluctance' in young clergymen to serve in the country.[37] A country living was increasingly viewed as an ecclesiastical backwater, in contrast to the real vocational challenges that were seen as being in the towns, with their opportunities of working in the settlement movement. This unattractiveness was made more conspicuous by an ageing country clergy, who appeared to lack dynamism, and by a tendency for bishops to make this even worse by using small rural livings as a form of pension for elderly clergy no longer capable of holding down a town living.[38] Whether old or young a

[33] Quoted in Kitson Clark, *Churchmen*, p. 264.

[34] F. Davenant, *What Shall My Son Be* (London, 1870), p. 16.

[35] S. Baring Gould, *The Vicar of Morwenstow. Being a Life of Robert Stephen Hawker, M.A.* (second edn, London, 1903), p. 257. Hawker, a social eccentric, was nevertheless devoted to his impoverished parishioners, and had built the school from his own money in 1843.

[36] Haig, *Victorian Clergy*, pp. 297–302.

[37] W. R. W. Stephens, *A Memoir of Richard Durnford, DD* (London, 1899), p. 338.

[38] Haig, *Victorian Clergy*, p. 321.

country living could mean being 'put out to grass' for an individual because of a marked lack of mobility in career terms. 'The rule in country parishes is that where a man is put down at first, there he dies at last.'[39] Immobility in a small country parish could breed inertia over the years; a preference for a privatised genteel existence instead of an active public role. For example, earlier successes of rural Lincolnshire clergy in breeding champion hollyhocks, or in carrying off the majority of prizes at local horticultural shows, were probably indicative of a more generalised malaise that was not entirely removed by the pastoral revival of mid-century.[40] Such underemployment meant that time hung heavily; one parson who had held country livings in Dorset and Oxfordshire commented in 1899 that 'assuredly the work of a small country living is not enough for an active-minded man'.[41]

B. Church and village

By this time the changing nature of church membership had in any case diminished the role of the country cleric. Traditionally, membership of the established church had been coterminous with citizenship; those who lived in the parish had the right to be baptised, married and buried there. The religious census of 1851 had exposed the extent of the gap between aspiration and achievement. The 1851 census indicated that both in provision of accommodation and in attendances the southern half of England was better off.[42] Anglicanism's strength was in lowland areas where parishes were small and easily worked; in East Anglia, the south, south-east, and south midlands. Conversely, in upland areas of moorland and mountain in the extreme south-western and northern counties of England, and in Wales, it was comparatively weak.[43] Although rural adherence to the established church was stronger than urban, yet even on the most liberal interpretation of defective statistics the Anglican country church was attended by only a large minority of the population.[44] Labourers' diminishing participation was the result both of broad social developments and of narrower sectarian ones. Whether the church had in the long-term lost their adherence, or had never had it in the first place,[45] is a controversial issue that – given the lack of attendance records before 1851 – is unlikely to be resolved satisfactorily. But even within the

---

[39] A. Jessopp. *The Trials of a Country Parson* (London, 1890), pp. xvii–xix.

[40] Obelkevich, *Rural Society*, pp. 124–6.

[41] C. Kegan Paul, *Memories* (London, 1899 and 1971), p. 247.

[42] J. D. Gay, *The Geography of Religion in England* (London, 1971), pp. 57, 271–2.

[43] A. D. Gilbert, 'The land and the church' in G. E. Mingay (ed.), *The Victorian Countryside* (2 vols., London, 1981), vol. i, pp. 44–5.   [44] Obelkevich, *Rural Society*, pp. 156–7.

[45] Soloway, *Prelates and People*, p. 445.

period with which we are concerned here, it is tolerably clear that agricultural labourers' church attendance showed some decline. Clerical efforts to stem the drift to alternative worship among adults seem to have been reasonably successful, but earlier ground that had already been lost was not recovered. However, those who attended church might do so for social rather than religious reasons; farmers often appear to have come to the parish church as a social obligation that conferred respectability,[46] and other villagers attended as a mark of personal respect to the parson himself. In many cases this respect had been hard won by the incumbent's earlier generosity to individuals. It was said in the parish of Mareham-le-Fen, for instance, that the incumbent's liberality meant that there was 'no need for any relieving officer to come to the village with the half crown and the loaf'.[47] This kind of sustained subsidy to the parish was unlikely to have persisted to the end of the century since agrarian depression had hit clerical pockets hard. And attempts to increase participation in the church in other ways, as through educating the young appear to have borne a disappointing harvest. In Lindsey, for instance, there were few pupils from the National Schools in the villages who were confirmed, and of those even fewer became regular communicants.[48]

Where did the future lie for the Anglican church in this increasingly secularised society? The Archbishop of Canterbury, Archibald Tait, argued in the House of Lords in 1880 that: 'The glory of the Church of England [is] that it is a National Church, wide as the nation, ready to embrace all in the nation who are anxious to join it, and not making narrow sectarian distinctions between those who adhere very rigidly to one or another set of opinions.'[49] For Tait the enemy was not non-conformity but secularism, and through his moderate policies he let much of the steam out of the issue of disestablishment in England.[50] 'The old Christian state was dismantled by Christians for the sake of keeping the people Christian.'[51] To this end the Burials Act, passed in 1880, permitted anyone with civil rights to be buried in the parish churchyard, and ensured that this could be effected either silently or with a Christian service of choice. Another of the non-conformists' bitter grievances had been removed earlier in 1868, with the abolition of a compulsory church rate; this meant that dissenters no longer might have to pay for the upkeep of churches and churchyards yet be denied their use. In Wales, however, disestablishment came about in 1914 because of both the overwhelming predominance of nonconformity there, and grievances over the Anglican

---

[46] Obelkevich, *Rural Society*, pp. 178, 316.   [47] *Lincoln Gazette*, 25 January 1862.
[48] Obelkevich, *Rural Society*, pp. 167, 138–43.   [49] *Hansard*, 3rd series, CCLII (1880), 1023.
[50] P. T. Marsh, *The Victorian Church in Decline* (London, 1969), pp. 247, 251.
[51] O. Chadwick, *The Secularization of the European Mind in the Nineteenth Century* (Cambridge, 1975), p. 93.

church's continued privileged position. Prominent amongst contentious issues were: attempts by Anglican landlords to coerce their tenants into attending the Anglican church; disputes over tithes – especially from 1885 to 1891; and passive resistance in the early years of the twentieth century over the Education Act of 1902, which put church schools on the rates.[52]

The closing decades of the nineteenth century marked the formal separation of civil and ecclesiastical functions in the parish. The parish vestry had been the seat of parochial government under the *ex officio* chairmanship of the vicar or rector, and all villagers had had the right to attend and vote. Historically the vestry had been the seat of local government; as has been indicated in the previous chapter, the vestry had appointed the parish overseer and constable, and had had responsibilities over law and order, poor relief, highways, sanitation and water in the village. Yet from the 1830s these extensive powers had been progressively eroded, and by 1893 this process of depriving the vestry of secular functions had gone so far that the President of the Local Government Board could refer to vestries as, 'decrepit survivals of former days . . . They have the form but not the power of local government; they do not possess the confidence of the rural population.'[53] Under the Local Government Act of the following year, villages of more than 300 people could elect a parish council. The parson no longer served *ex officio* but now needed election. Some contemporaries saw this as effectively disestablishing the church.[54] A modern church historian, P. T. Marsh, has commented perceptively that 'England's way of becoming a secular society was not to break Church from State but rather to push church affairs aside.'[55] Informed but sympathetic contemporaries, such as Richard Jefferies, described how the same process had already gathered strength by 1880:

The parish seemed to have quite left the Church . . . The modern institution was introduced, championed by the Church, worked for by the Church, but when at last it was successful, somehow or other it seemed to have severed itself from the Church altogether. The vicar walked about the village, and felt that, though nominally in it, he was really out of it.[56]

One instructive example of this process occurring was the village school.

In mid-century the majority of English village schools were church schools that had been founded by the National Society. In Wales, also,

---

[52] Gilbert, 'Land and church', pp. 49, 54; D. W. Howell, *Land and People in Nineteenth-Century Wales* (London, 1977), pp. 84–5; G. A. N. Lowndes, *The Silent Social Revolution* (Oxford, 1937), pp. 57, 84–5; H. C. Barnard, *A History of English Education* (London, 1961), pp. 214–15.

[53] *Hansard*, 3rd series, x, 1893, 687.   [54] Chadwick, *Victorian Church*, II, p. 197.

[55] Marsh, *Church in Decline*, p. 252.

[56] R. Jefferies, *Hodge and his Masters* (London, 1979), pp. 159–61.

the Anglican church provided one-third of the school places; a disproportionately large amount when compared to only about one-tenth of
the population that adhered to it.[57] In England, although the findings of
the Newcastle Commission in 1861, had probably over-estimated the
proportion of education provided by the Church as being around three-
quarters of weekday pupils and half of sunday-school pupils, yet its position was undoubtedly one of dominance in educational provision.[58]
Without the parish clergy this situation could not have been achieved, as
earlier evidence to a commission on rural employment had revealed: 'But
for the zeal and activity of the clergy, and their large sacrifices, not only
of money, but of labour and time in three fourths of all rural parishes of
England there would be either no school at all, or at best only the sem-
blance of a school.'[59] In these National Schools the parson had to be a
manager, and in many schools he was the only one.[60] He intervened frequently in the administration of the school and also usually provided religious instruction to the pupils, often with the help of his wife or
daughter. The parson held powers of professional life or death over the
village schoolteacher, since in 1852 he had been given the right by the
state to dismiss any teacher 'on account of his or her defective or unsound
instruction of the children in religion, or on other moral or religious
grounds'.[61]

The teachers were normally members of the Church of England and
their pupils were given a Christian training through being instructed in
the catechism and liturgy of the Church of England. Yet the catechisms
used could be highly offensive to non-Anglicans. The popular *Some
Questions of the Church Catechism and Doctrines Involved Briefly Explained*,
which went through thirteen editions betwen 1870 and 1896, included
the following:

Q Is it very dangerous to leave the Church?
A Yes; and it is also a grievous sin.
Q Is it very wrong to join in the worship of dissenters?
A Yes, we should only attend places of worship in connection with the
Church of England.[62]

[57] D. Gareth Evans, *A History of Wales, 1815–1906* (Cardiff, 1989), pp. 219, 249.

[58] J. S. Hurt. *Elementary Schooling for the Working Class 1860–1918* (London, 1979), pp. 52–3; Kitson Clark, *Churchmen*, p. 122.

[59] *Report of the Commissioners on the Employment of Children, Young Persons and Women in Agriculture*, BPP, 1867–8, XVII, p. 79 (evidence of James Fraser, Bishop of Manchester).

[60] P. Gordon, *The Victorian School Manager. A Study in the Management of Education 1800–1902* (London, 1974), pp. 10, 14.   [61] P. Horn, *Education in Rural England 1800–1914* (London, 1978), p. 235.

[62] Quoted in Hurt, *Elementary Schooling*, p. 176. Questions were raised about this text in the House of Commons in 1876 and 1891, and Archbishop Benson condemned it in 1889, as out of tune with progressive thinking in the Church of England.

In many villages the National School was the only one available and hence was attended not only by the children of Anglicans but by those of non-conformists as well. This was potentially explosive and only needed some insensitivity by the incumbent to set denominational hostilities alight. At Littlemore, in Oxfordshire, the Baptist minister complained to the Education Department in that the children had been told that they must attend the church Sunday School if they wished to attend the National School during the week, an accusation that was denied.[63] The Cowper-Temple clauses of the Elementary Act of 1870 had attempted to defuse this situation by stating that in any school receiving a government grant, no child could be compelled to receive religious instruction. Yet very few pupils were withdrawn, perhaps because of fears of social repercussions; returns to diocesan inspectors suggested only 0.3 per cent of pupils were taken out of classes.[64] The 1870 act weakened but did not lead immediately to the dramatic decline in church schools that some had feared. Indeed, the act stimulated vigorous building activity by the established church, and attendance at church schools more than doubled between 1870 and 1895. Yet the long-term implications were clear enough; the future lay with the new Board Schools with their superior buildings and much better funding. A significant straw in the wind was that 836 Anglican schools had transferred their functions to school boards by 1880.[65] The Church of England was ceasing to be national schoolteacher.

The steady encroachment of state organisations over functions hitherto discharged by the parish clearly diminished the secular influence of the parson. Only in certain spheres was the clergyman able to participate in these new authorities – most obviously as a poor-law guardian. Long service was not uncommon: the Rev. E. B. Ellman, rector of Berwick in Sussex, was a guardian for forty years, thirty-five of them as chairman of the West Firle Board of Guardians. Many parsons were enabled to serve *ex officio* by virtue of their office as magistrate, until the local government reform of 1894 removed this privilege. Clerical justices had usually been amongst the most active on the bench, as was the case in Lincolnshire.[66] But the magisterial function of the clergy must have imposed some acute tensions between their secular and spiritual roles. The responsibilities of clergy as JPs in administering the game laws, for example, seems likely to have further underlined the labourer's perception of the class divide between the parson and his flock. In the early nineteenth century it was estimated an average of one quarter of magistrates had been clergymen,

---

[63] Horn, p. 145.   [64] Hurt, *Elementary Schooling*, p. 174.
[65] G. Sutherland, *Policy Making in Elementary Education* (Oxford, 1973), pp. 112, 350.
[66] Obelkevich, *Rural Society*, p. 32.

a proportion that could rise to two-fifths in certain rural counties such as Norfolk.[67] But in mid-century this position was challenged in certain counties such as Warwickshire, while the controversial actions of the Oxfordshire clerical justices in the Chipping Norton case of 1873, cited above, may well have led to a further decline in these appointments in the late nineteenth century.[68] In Wales, however, although the proportion was smaller, there was continuing growth in the number of clerical justices, and this despite their unpopularity.[69]

Increasingly circumscribed in relation to secular activities, and with his pastoral role narrowed, the late nineteenth-century country parson focused his energies on building up what was in reality a denominational flock. To this end there was a renewed emphasis on the fabric of the parish church in its 'beauty of holiness'. This was overdue in many places where prior neglect meant that the church participated only too fully in the agrarian character of the neighbourhood. On his appointment to Stoke Talmadge in 1855, for example, Francis Pigou recorded that 'The church was in a woeful condition. Cocks and hens roosted on weekdays in the pulpit.'[70] Typically, the layout of the Victorian church was radically altered and the old high pews and three decker pulpit were removed. At the same time the organisation of services was reformed. The old practice of an independent-minded village band in the gallery – so sympathetically depicted in Thomas Webster's painting 'The Village Choir' of 1847[71] – was replaced by a disciplined choir in the chancel, accompanied by the parson's wife or daughter on an organ or harmonium. A single Sunday service was often replaced by two, and a substantial sermon provided. Formal membership of the Anglican church was given a new importance through encouragement to take communion more regularly.[72] These initiatives made the cleric much more central to religious proceedings.

At the very end of the century bishops encouraged incumbents to create parochial church councils with a membership of churchwardens, sidesmen and elected councillors, all drawn from male communicants but elected by all resident Anglicans. Women churchwardens (who had previously served in dioceses such as Ely, Salisbury and Surrey) were explicitly excluded from these councils by votes of Convocation in 1897 and 1898.[73] In the minds of the conservative majority in Convocation the church was in danger of feminisation, precisely because girls and women

---

[67] Junius, *A Letter to Lord Brougham on the Magistracy of England* (London, 1832), pp. 15–16.

[68] R. Quinault, 'The Warwickshire magistracy and public order, *c.* 1830–1870', in R. Quinault and J. Stevenson (eds.), *Popular Protest and Public Order* (London, 1974), pp. 188–9; Kitson Clark, *Churchmen*, p. 251. [69] Evans, *Wales*, p. 235. [70] Quoted in Meacham, *Lord Bishop*, p. 114.

[71] Reproduced in Wood, *Paradise Lost*, p. 210. [72] Obelkevich, *Rural Society*, pp. 146–50, 137–40.

[73] Chadwick, *Victorian Church*, vol. II, p. 201.

came to outnumber boys and men in confirmations, church attendants, and communicants.[74] In their view women's rightful role should remain ancillary and supportive. In rural areas this meant that substantial duties – which were central to parochial activity but were sufficiently diffuse because multifarious – went largely unacknowledged by the church hierarchy. Typically, these involved singing in the church choir; Sunday School teaching; district visiting; and participation in the Mothers Union, begun in 1876 but made into a national organisation in 1893. However, the votes in Convocation so incensed the more active, articulate and responsible women in the Church of England that church feminism was created.[75] Practical gains from this activism were, however, slow to follow. In 1905 women gained the right to vote for men on the Representative Church Council; in 1914 the right to be elected in their own right to Parochial Church Councils; and four years later the right to participate in Ruridecanol and Diocesan Conferences. But only when the Parliamentary Reform Act of 1918 had given some women the suffrage did the Church follow suit; women gained participation in the National Assembly in 1919.

By the end of our period, the Church of England had become a religious denomination rather than a broad church. Yet precisely because its social role was still extensive, conflict with other religious groups on a range of issues was almost inevitable. We have already noted the problems that arose from church rates, services accompanying the burial of dissenters in the parish churchyard, and religious teaching in the village school. To these should be added the administration of parochial charities which caused recurrent ill-feeling between labourers and parson, and which was sometimes intensified by the maladroit interventions of the Charity Commissioners.[76] Friction, suspicion and resentment between church and chapel over administration of town lands, allotments and doles were a feature of the period from the 1860s to the 1890s. The changing nature of social consciousness meant that villagers regarded charity as belonging to the village, not the parson and his helpers serving as feoffees or trustees. 'The Bible talks o' charity, but there were none o' that sort, only the sort that goos wi' bootlicking.'[77] Joseph Arch recollected the deference that was imposed as *quid pro quo* for charity, and the coercive attempts to make recipients stay away from the chapel.[78] At the

[74] Obelkevich, *Rural Society*, pp. 150, 179–80, 313.

[75] B. Heeney, 'The beginnings of church feminism: women and the Councils of the Church of England, 1897–1919', in G. Malmgreen (ed.), *Religion in the Lives of English Women, 1760–1930* (London, 1986), p. 267.

[76] L. M. Springall, *Labouring Life in Norfolk Villages, 1834–1914* (London, 1936), pp. 58, 113–14.

[77] M. K. Ashby, *Joseph Ashby of Tysoe, 1859–1919* (London, 1974), pp. 46, 52.

[78] K. S. Inglis, *Churches and the Working Classes in Victorian England* (London, 1963), p. 253.

time of the creation of parish councils, and the formal separation of the civil from the ecclesiastical parish in 1894, it was problematic as to where to locate charities. And by deciding that charitable doles were secular, and hence the province of the parish council, the government silently vindicated the earlier radical stance of the agricultural labourers.

C. Non-conformity in England and Wales

English dissent had grown up in communities 'marked by an unusual degree of freedom' according to Alan Everitt's study of the micro-distribution of non-conformity in four mainly rural counties – Leicestershire, Lincolnshire, Northamptonshire and Kent – during the nineteenth century. He has argued that non-conformity was strong in communities with many independent freeholders, traders or craftsmen, usually in large, populous parishes; also that Old Dissent (Baptists, Quakers, Congregationalists, Presbyterians) was characteristic of wood and wood–pasture areas with their scattered settlement pattern.

In 1851 the religious census indicated that, overall, almost half of church goers were non-conformists. David M. Thompson has concluded that all the main non-conformist groups were 'stronger in the countryside than the towns'.[79] There was, however, considerable diversity; non-conformist sittings in England varied from 28 to 60 per cent of the total. In Wales, non-conformity was more uniformly strong. The religious census of 1851 showed that 87 per cent of Welsh attendances belonged to non-conformist chapels. The census suggested both that the hold of organised religion was stronger in north than south Wales, and that Anglicanism was less weak in English-speaking and rural areas. As in England attendances at either chuch or chapel in urban and industrial districts were relatively poor.[80]

Geographically the macro-distribution of non-conformist churches in 1851 was extremely complicated. Both Old and New Dissent grew up where the established church had been weak. New Dissent or methodism was the outcome of the Evangelical movement of the eighteenth century. Wesleyan Methodism, the most powerful group within New Dissent, was influential in eastern, northern and south-western counties. Two smaller sects within the methodist movement were particularly strong in country areas – the Primitive Methodists and the Bible Christians. Primitive Methodism, the fourth largest group in 1851, was found in rural areas along the east coast from Durham to Norfolk; from Nottinghamshire north-westwards through to Shropshire; and from

---

[79] D. M. Thompson, *Nonconformity in the Nineteenth Century* (London, 1972), p. 121;.
[80] Evans, *Wales*, pp. 218–19.

Cambridgeshire south-westwards through to Wiltshire. The Bible Christians were more localised and were to be found in the extreme south-western counties of Cornwall and Devon, where the sect had originated, and also in Kent. The circuit system of Methodism made it particularly suitable for the countryside since this meant that small villages did not have to maintain their own minister.[81] New Dissent appears to have complemented Old Dissent in its location. The second and third most numerous non-conformist groups were Congregationalists and Baptists. Their rural chapels were plentiful in Wales, the east Midlands, East Anglia, the home counties north of London, and the south-eastern counties.[82] In the 1850s Congregationalists, like other non-conformist sects, were conscious that without renewed effort stagnation or even decline was inevitable. In consequence they initiated a system of rural evangelists in the 1860s.[83]

The nature of the non-conformist ministry, particularly that in country areas, to some extent exacerbated more general weaknesses by imposing physical, psychological and financial stresses on its ministers. The itinerant character of the circuit could prove very demanding physically; in rural areas – and particularly in Wales – distances could be very considerable, and some could only be covered on foot. Charles New's Wesleyan circuit in Cornwall during the 1860s, for instance, involved weekly walks totalling seventy miles.[84] Giving pastoral care to a scattered country flock was also very demanding, particularly in Primitive Methodism which laid down a norm of five such visits per day. Endemic grumbles in religious periodicals about the inadequacy of visiting attributed falling membership to this cause. This was but one reason for tensions between ministers and lay members and which the itinerant nature of the Methodist circuit tended to exacerbate. In country areas where a ministerial presence was comparatively infrequent, lay people might establish positions of considerable influence, power and activity so that clashes of opinion with the visiting minister were made much more likely. The fact that lay preachers outnumbered ministers in the Methodist connections also tended to intensify conflicts between the laity and ministers.

It was not just within Methodism that such organisational and personal pressures led to what K. D. Brown has termed an 'unsettled ministry', a

[81] Gay, *Geography*, pp. 162–3.

[82] A. Everitt, *The Pattern of Rural Dissent in the Nineteenth Century* (Occasional Paper no. 4, second series, University of Leicester Department of Local History, 1971) pp. 11, 13, 19, 44–6, 69; Thompson, *Nonconformity*, p. 121; J. D. Gay, *The Geography of Religion in England* (London, 1971) pp. 116–17, 122, 151–3.   [83] Thomson, *Nonconformity*, p. 123.

[84] K. D. Brown, *A Social History of the Nonconformist Ministry in England and Wales 1800–1930* (Oxford, 1988), p. 139.

'ceaseless ebb and flow', since even in Congregationalism, where movement was a matter of choice for the minister, pastorates were relatively short-lived. Thus the kind of long-established relationship that was common in England between a country vicar and his parishioners was much more unusual in rural non-conformity. Another contrast was apparent in the generally greater involvement in the secular affairs of the community of the clergyman compared to the non-conformist minister. Anxieties were expressed among lay Baptists and Congregationalists that secular activity by ministers might be a diversion from their true mission of saving sinners and making souls. Despite this some ministers in Old Dissent were active and even in New Dissent, where this was less obvious, philanthropic and political involvement increased. Opinion among the laity by the 1870s became more favourable to the ministry's secular involvement, and to the exercise of the non-conformist conscience, in issues of education, temperance reform and poverty.

Poverty among non-conformist ministers themselves was not viewed particularly sympathetically by their congregations. The financial position of the non-conformist minister was usually much less favourable than that of his Anglican counterpart. There was some concern that ministers should not be placed in too comfortable a position lest it insulated them from the poor; earlier in the Victorian period some Primitive Methodist congregations had underlined this point by paying a salary equivalent to that of a farm labourer. Even among the Congregationalists, where earthly reward was the most generous among non-conformist sects, there could be a large gap between expected income and that actually paid. And among Baptists, where such rewards were notoriously low, English annual rural payment during the 1860s was of the order of £65, whilst in Wales it fell below £25. Here many had perforce to make their ministry part-time and earn their bread through farming or shopkeeping.[85] During the agricultural depression at the end of the nineteenth century declining rural population and hence shrinking congregations meant falling pew rents and a lowering even of these small ministerial incomes. At the sharp end was the minister's wife, and the 'dreadful strain of pinching and saving' experienced in trying to keep up a professional middle-class lifestyle was well attested by their letters to religious journals.[86]

Within the non-conformist chapel secular anxieties and preoccupations could be forgotten for a time and attention concentrated on rousing emotional hymns, extemporary prayers, and sermons designed to convert – and entertain – by fiery theatricality. The chapel aspired to be a family

---

[85] K. D. Brown, *A Social History*, pp. 139, 141, 144–5, 153–7, 164–5, 169, 210–17, 220.

[86] For example, *Independent and Nonconformist*, 31 January 1895.

to its members, a place where participation was encouraged, and a certain cosy domestic atmosphere was cultivated, not least by the heating that was provided. This contrasted markedly with the parish church, where the laity were increasingly onlookers in a formalised service, and were usually also beset by draughts in an unheated medieval building. Lay participation was much more prominent in non-conformity than in Anglicanism: the intense fellowship of the chapel was one of its strengths.

In Wales the non-conformist chapel had an even more central role in society. In the words of D. Gareth Evans:

After 1850 the chapels regarded themselves as the custodians of Welsh life; Welsh culture was reshaped in the mould of denominationalism. Between 1850 and 1900 the chapels developed their choirs, orchestras, drama groups, literary societies, temperance groups, bands of hope, and innumerable eisteddfodau.[87]

For a short period the chapel became the cultural centre of community life and Welsh the language of the preacher, so that the chapel seemed to embody national consciousness. By the 1880s dissenters outnumbered Anglicans by three to one. Welsh non-conformity continued to expand until 1905–6; allegiance having been boosted in the early twentieth century by struggles over the 1902 Education Act.[88]

In Wales it was noticeable that the social composition of the chapel was much broader than in England. But in time even this social cohesion and self-confidence in Welsh chapel life weakened; leisure activities migrated to secular buildings, whilst the state increasingly took over educational functions. Chapels in the English countryside rarely had the cultural dynamism of their Welsh counterparts. And different social groups were unequally represented there. Country congregations of Baptists and Congregationalists recruited strongly from village craftsmen and farmers, whereas Wesleyan and Primitive Methodists attracted more agricultural labourers.[89] Whilst such labourers appear to have become class leaders in evening meetings or, occasionally, preachers, they were under-represented in Wesleyan Methodism as officeholders in comparison with farmers, tradesmen or craftsmen.[90] Primitive Methodism was the non-conformist sect that gave most scope to the labourer; many later agricultural trade-union leaders underwent their apprenticeship in public speaking in Primitive Methodist pulpits.[91] However, denominational

[87] Evans, *Wales*, p. 242.

[88] A. D. Gilbert, 'The land and the church', in G. E. Mingay (ed.), *The Victorian Countryside* (2 vols., London, 1981), vol. 1, p. 55; R. Currie, A. Gilbert and L. Horsley, *Churches and Churchgoers. Patterns of Church Growth in the British Isles since 1700* (Oxford, 1977), p. 78.

[89] Thomson, *Nonconformity*, p. 14.   [90] Obelkevich, *Rural Society*, pp. 194–5.

[91] For example, George Edwards, farm worker and trade union organiser who later became an MP (G. Edwards, *From Crow-Scaring to Westminster* (London, 1922), pp. 32–6).

membership in the English village was defined by reference first of all to the parish church and then by the availability of alternative chapels.

English non-conformity, like English Anglicanism, had chequered fortunes during our period. Stagnation threatened in mid-century, not only because of the revival of pastoral efforts in the Church of England but also because of non-conformist weaknesses, principally the split in Wesleyan Methodism after 1849, which resulted in a loss of 100,000 members in only five years.[92] As social discrimination against dissenters eased,[93] so affluent non-conformists were brought into closer contact with Anglicans of similar background; some assimilated to this new social setting only too well and joined the Church of England. Religious allegiance was also fragmented to some extent both by the depopulation of the countryside and by the suburbanisation of the towns; each made it easier for those who moved to break with their earlier religious allegiance. However, whilst these wider social changes must have affected both non-conformity and Anglicanism, in key respects it might be expected that the former might perhaps have been less adversely affected. The circuit system of Methodism, for instance, had earlier brought country and town closer together, with the formal links of the chapel having been reinforced by familial and financial contacts, so that moving from village to town may not have been such a decisive break. And non-conformist churches might also follow their congregations when they moved to the more salubrious suburbs.

In the countryside, however, the late nineteenth-century agrarian depression weakened non-conformist churches such as the Primitive Methodists (with farmworkers central to their meetings), and Baptists and Congregationalists (where small farmers had customarily been important in their congregations).[94] In 1894 the Baptists recalled 'the fact that the tenant farmers of this country were the backbone of religious Non-conformity. That class of supporters may be looked for almost in vain among our village churches.'[95] The Primitive Methodists also reviewed their position gloomily in 1896 and concluded that: 'We are a village church. Nearly 75 per cent of our chapels are in the villages . . . during the last twenty-five years we have abandoned 516 places, and only succeeded in opening up 236 new ones.'[96] However, in the late nineteenth and early twentieth centuries came the start of an organisational response to this decline in non-conformity: the union of General and Particular Baptists occurred in 1891, and the creation of the United Methodist Church took place in 1907. Earlier, in 1892, there was a more

[92] Thompson, *Nonconformity*, pp. 14, 120.
[93] For example through the ending of discrimination against dissenters in admissions to Oxford and Cambridge Universities in 1871.   [94] Thomson, *Nonconformity*, pp. 14–15, 123, 227.
[95] *Baptist Handbook* (1894), p. 77.   [96] *Minutes of Primitive Methodist Conference, 1896*, pp. 183–4.

general initiative aimed at increasing unity among the dissenting churches in the Movement for Free Church Unity.[97] This aimed at securing 'all the practical advantages of unity without sacrificing any of the indisputable advantages of denominational organisation'.[98]

'The tide is ebbing within and without the Churches . . . Even the noisy warfare between the various denominations may be interpreted less as a sign of secure vitality than as evidence of uncertain position; a struggle excited less by confidence than by foreboding', asserted that perceptive commentator, C. F. G. Masterman, in 1909.[99] Archbishop Tait had been correct when, thirty years before, he had seen secularism as the real cause for concern. It was secularisation that was eroding the significance of religious institutions and religious personnel in town and country alike. Thus although the physical presence of the parish church remained central to the village landscape, its ministry was becoming marginal to everyday living. And the continuing depopulation of the countryside meant that both rural church and chapel suffered from ageing congregations and half-empty buildings. Yet to see rural religion solely in terms of institutional structures would be to take too narrow a view. Rural labourers adopted a selective view of organised religion. Obelkevich has aptly commented that, 'They regarded the chapel as a source of entertainment just as they regarded the parson as a source of half-crowns and confirmation as a remedy for rheumatism.'[100] Popular religion – in the sense of a complementary and pluralistic set of beliefs involving not just unorthodox Christianity, but also pagan survivals, superstition, folklore and belief in magic – was embedded in the culture of the rural labourer.[101] Cultural and social disjunctures between the classes in the Victorian and Edwardian countryside were to weaken not only the impact of organised religion, but attempts to improve popular education in the countryside.

II. EDUCATION

Education was one of the sectors in which there was considerable progress made in rural life, yet the gap in resources and standards between town and countryside had still not been eliminated by the end of our period. The main impediments to a more rapid levelling up of standards remained that of inferior resources reinforced by obstinately hostile attitudes to schooling among influential groups. Indeed, as the exodus of the labourer from country to town accelerated this tended to confirm earlier

---

[97] Thomson, *Rural Nonconformity*, p. 228.
[98] H. P. Hughes, 'Free Church unity: the new movement', *Contemporary Review*, March 1897, p. 447.   [99] *The Condition of England* (1960 edition), p. 207.
[100] Obelkevitch, *Rural Society*, p. 319.   [101] *Ibid.*, p. 262.

suspicions among substantial ratepayers that too much elementary education was disadvantageous to their interests. This then belatedly posed the
question of what was an appropriate form of schooling for the agrarian
population; a topic on which there were favourite hobby-horses but few
convincing answers. Indeed, it was arguable that the sceptics had a point
in that – in devising what had become before the end of the nineteenth
century a national system of schooling – remarkably little thought had
been given to the different circumstances of the countryside, so that
reforms tended to be driven by an urban not rural engine.

In 1850 the typical English village school was a voluntary school,
founded by the National Society in order to promote the truths of the
Church of England. At this time denominational rivalry was the spur to
educational expansion but the rival British Society, favoured by non-conformists, was much stronger in the towns than the villages, so that the
children of chapel-going parents were disadvantaged. In Wales the British
Society was stronger in the north than the south in 1850 but more vigorous activity during the next decade helped to redress this inequality. The
1850s and 1860s were years when the drive towards a 'Nonconformist
Nation' was at its peak and educational activity in establishing British
schools was a powerful engine in this.[102] The country National School
would characteristically have the local incumbent as the sole manager; he
would tend to look primarily for religious and moral, rather than academic, qualifications in appointing the school master or mistress. This was
not too disadvantageous given the common contemporary perception
that the aims of the curriculum should feature the moral and the social
more strongly than the academic. Educating village boys and girls to be
God-fearing, and to know their lowly place in society, was arguably of
greater importance than the inculcation of a very basic literacy and
numeracy. However, fulfilling even these limited aims needed more
resources than could be raised by voluntary organisations, even when
aided financially by the government: schools were still to be found in
make-shift premises; school places were fewer than numbers of potential
scholars; school attendance was variable; the age at leaving school
remained low; and educational standards were indifferent. Although the
educational problems of the towns were rightly perceived to be worse
than those of the countryside, by mid-century more radical schemes for
further state intervention were being discussed to increase educational
provision and to raise standards in both urban and rural areas.
Consequently, within our period from 1850 to 1914, there were several
major pieces of legislation and, since each fundamentally reshaped rural
education, it may be helpful to outline these initially.

---

[102] D. Gareth Evans, *A History of Wales, 1815–1906* (Cardiff, 1989), pp. 127, 249.

The Elementary Education Act of 1870 had the potential to enable a national system of schooling to be provided, by requiring a school board to be formed where there was insufficient school places provided either by the National or British Societies, or alternatively by a variety of dame and other private schools. One of the marked deficiencies of the 1870 Act was that rural school boards could be formed on the basis of a single parish, with consequent inefficiency and waste of money on replicated administration and elections. However, because of the strong influence of the Church of England, with its network of village schools, there was little initial enthusiasm to create school boards in the English country-side. Whereas half of the largest towns had created boards within a year of the act, for each of two dozen counties there were between none and three.[103] Creation of rural boards was a piecemeal and long-drawn out process. In Devon, for instance, a large and very rural county, four-fifths of the rural boards had been formed within five years yet others contin-ued to be created into the 1890s.[104] It took twelve years in Derbyshire before half the population was served by school boards.[105]

The principal reason for the formation of the boards was the inabil-ity of voluntary schools to provide adequate schooling by means of a vol-untary rate, whereas the boards could levy a compulsory rate. Once formed the overall impact of the school boards was undoubtedly beneficial; their objective was to provide school places for every child and, by the end of the century, approximately three million new places had been created. The higher standards of building, equipment and staffing which were achieved by the superior financing of board schools also gradually forced up that in voluntary schools where both existed in competition with each other, but this process was less evident in the thinly populated countryside than in urban areas. The 1870 Act did not eliminate a great diversity in educational provision but it certainly nar-rowed its range. However, since differences in standards between board and voluntary schools remained large further reform was considered necessary.

Under the 1902 Education Act county councils were to 'maintain and keep efficient all public elementary schools in their areas'. With the county as the rating unit a more equitable economic base could be estab-lished with which to finance schooling, and a professional body of administrators created to form and implement policy. In the first years after the act there was a 'prickly relationship' between school managers and the local authority and conflicts centred on appointments.[106] In the

[103] P. Gordon, *The Victorian School Manager* (London, 1974), pp. 115–16.

[104] R. R. Sellman, *Devon Village Schools in the Nineteenth Century* (London, 1967), p. 55.

[105] M. Johnson, *Derbyshire Village Schools in the Nineteenth Century* (London, 1970), p. 125.

[106] Gordon, *School Manager*, p. 273; Sellman, *Village Schools*, pp. 151–2.

longer term, however, the 1902 Act had a positive outcome. In rural Devon, for example, a levelling up took place between the 300 voluntary and 200 council (formerly board) schools. A uniform staffing and salary scale was introduced, and a backlog of necessary improvements to buildings was cleared.[107] Belatedly, the economic problems of rural education could be redressed more efficiently.

In Wales the Welsh Intermediate Education Act of 1889 was important in attempting to provide post-elementary schooling on a more widespread basis than was possible with the few endowed grammar schools in the principality. The Aberdare Report of 1881 had been sympathetic to Welsh aspirations and had recommended that these distinctive institutions should be set up. Financed with a halfpenny rate, matched by parliamentary grant, by 1902 there were ninety-three such schools. Interestingly, twenty-one were for girls, seven were mixed, forty-three were dual, and twenty-two were for boys. Pressure from the recently formed Association for Promoting the Education of Girls ensured that the needs of girls would be adequately catered for, and by 1914 they were considered to have achieved equal parity with boys.[108]

A. Financial resources

The resource issue was fundamental: not only were economic resources more limited in rural areas but, since their populations were more scattered, the actual expenditure to achieve any given standard of schooling was proportionately much greater. Figures are not available before the 1870s, but from 1873 to 1895 annual returns revealed the disparity between the average rates raised for elementary education in country and town. That in parishes in England was higher by 11 per cent than that in boroughs (excluding London), and that in Welsh parishes was 27 per cent higher than Welsh boroughs. Such figures should cause no surprise given that the Elementary Act of 1870 had permitted even the smallest civil parish to elect its own school board with the power to levy a compulsory education rate. As a result three-quarters of such boards served populations of fewer than 5,000 and one-quarter had populations of under 500.[109] Higher rates were therefore needed in these less densely populated areas and this tended to provide a rationale for inaction or delay.

The vexed issue of financing rural schooling had been even more pressing before 1870 when – in the absence of a compulsory public rate

[107] Sellman, *Village Schools*, pp. 148–52.

[108] Evans, *Wales*, pp. 269–70; W. Gareth Evans, *Education and Female Emancipation. The Welsh Experience, 1847–1914* (Cardiff, 1990) pp. 159, 168.

[109] G. Sutherland, *Policy-Making in Elementary Education 1870–1895* (Oxford, 1973), pp. 84–5, 104–5, 355.

– educational initiatives rested solely on voluntary activity. As we have seen, Anglican clergymen were in the front line here since most voluntary schools in the countryside were those under the National Society. After 1870 the probable fate of these voluntary schools might have been one of decline; the result of an unequal competition with rate-aided Board schools. In fact, in the short term at least the voluntary sector showed unexpected resilience after 1870. In the decade after the 1870 Act there was a spurt in building activity by voluntary bodies in order to take advantage of the last-available government grants for such purposes.[110] Attendance at church schools doubled in the last three decades of the nineteenth century,[111] whilst numbers of children in attendance at all voluntary schools grew by 3 per cent per annum between 1870 and 1895. But this absolute growth represented a relative decline since attendances grew at twice this speed in the new Board schools.[112] This reflected the growing differences in the facilities provided which, in turn, were the result of underlying disparities in financial resources between these two sectors. Unfortunately, it is not possible to disaggregate figures into rural and urban ones so that broad inferences have to be drawn about the financial underpinning of rural schooling, given the ubiquity of voluntary schools in villages. But it is clear, nevertheless, that an unequal race was being run. Expenditure in voluntary schools was only four-fifths of that in board schools. The difference in income per child from rates and voluntary subscriptions at board and voluntary schools also meant that the latter had little more than one-third of the former's income from these sources per child, although roughly equivalent income from government grants. As a result of this shortfall from rates and subscription, fee levels per child in voluntary schools had to be correspondingly greater; those obtained were one-quarter higher.[113] An added problem for village schools was the exodus from rural areas of young adults of child-bearing age; in consequence school rolls fell and income from school pence and government grant declined proportionately.

British Schools were particularly vulnerable. Only one-third of these non-denomination voluntary schools – founded by the British Society since 1809 – were still in existence by the time of the Voluntary Schools Act of 1897.[114] This contentious and short-lived act provided extra

---

[110] *Ibid.*, p. 88. It took the Education Department twelve years to clear the flood of applications, at a total cost of £312,000.

[111] J. Hurt, *Elementary Schooling and the Working Class, 1860–1918* (London, 1979), p. 176. The figures were 844,344 to 1,855,802.   [112] Sutherland, *Policy-making*, p. 351.

[113] Based on figures in Sutherland, *Policy-making*, pp. 356–9. The figures relate to the years 1880–95 and to numbers of children in average attendance.

[114] H. Bryan Binns, *A Century of Education 1808–1908: the Centenary History of the British and Foreign School Society* (London, 1908), p. 231.

government help once such institutions joined an Association. Inexplicably, the government made a lower rate of payment to country than to town schools under this scheme. The effect of this legislation was to improve books, equipment and furniture and to allow much-needed minor works of improvement to buildings.[115] However, such aid was not sufficient to give financial stability to the voluntary schools and their numbers continued to decline. By the turn of the twentieth century the Board of Education itself recognised in a confidential memorandum that over half the voluntary schools in the counties were 'under water' in financial terms.[116] When the 1902 Education Act gave the option of rate aid to voluntary schools in return for loss of autonomy over secular education, a single county authority also became responsible for all elementary schools.

B. Attitudes to rural education

Reinforcing the financial difficulties of country schools were antagonistic attitudes to the formal schooling of the labourers' children. Farmers were a very influential social group but, throughout the period, they saw little if any value in elementary schooling for their labourers' children. In consequence, whether in their role as ratepayers, or alternatively as managers of voluntary schools and members of school boards, their reluctance and recalcitrance did much to delay, frustrate or obstruct educational developments. A typical instance was that of Carlton Husthwaite in the East Riding, where four of the five board members were farmers and the fifth was an 'agent for manure'. In consequence the school was frequently closed so that, instead of securing an education, as the board explained to Whitehall in 1890, 'the principal part of the children assist in securing the crops'.[117] Such farmer-dominated boards were, as Sellman comments on Devonshire schools, 'at least as much concerned to keep down the rates as to keep up the school'.[118]

In the Victorian and Edwardian countryside working-class parents' own attitudes to the schooling of their children were equivocal. This ambivalence tended to magnify real issues of contention between parents and teachers over such issues as school discipline, with assault on teachers not infrequently occurring.[119] Teachers also complained that parents showed their true feelings by keeping children at home on the day of the

---

[115] Gordon, *School Manager* pp. 214, 232. The area of the Association varied according to the religious denomination with Anglicans choosing the diocese or archdeaconry, Wesleyans or British schools the geographic region, and Jews the nation.

[116] P Horn, *Education in Rural England 1800–1914* (London, 1978), p. 266.

[117] Quoted in Gordon, *School Manager*, p. 114.   [118] Sellman, *Village Schools*, p. 61.

[119] Horn, *Education*, p. 144.

HMI's annual examination so that the school lost income, and teachers might consequently suffer a drop in salary.[120] There was a more general disincentive to children attending school since, as we have seen, rural schools charged comparatively high school pence. Even those schools which levied school pence on a sliding scale appeared to have made relatively little allowance for the disparities of parents' incomes. At Bamford, Derbyshire, for instance, the children of farmers paid threepence or fourpence weekly and the labourers' children twopence or threepence according to age.[121]

A basic source of tension in the village – and one that coloured most aspects of rural education – was denominational religion. Few villages had both National and British Schools, but where they existed much bitter competition ensued with complaints about poaching of pupils.[122] Even worse than rival denominational teaching in many eyes was the 'undenominational religion' associated with the board schools. In England this was both scorned and feared, so that tremendous efforts were made by religious groups after the 1870 Act to forestall the creation of a school board by building or expanding voluntary schools.[123] When, despite all efforts, a school board was created locally, then each religious group attempted to maximise its influence on its proceedings. There were complaints that in some areas the board was 'merely a subordinate off-shoot of the church-school managers'.[124] In cases where the incumbent was virtually *ex officio* chair of the board, and a non-conformist presence on it weak, such suspicions bred bitter accusations. Were the Temple-Cowper clauses of the 1870 Act – which prescribed non-denominational teaching of religion – being adhered to in the board school? In practice there appears to have been great local variation in this,[125] but in such a controversial area the reality was much less important than social perception. In Wales there was a somewhat different situation, such that by 1886 there were fifty secular school boards. In 1899 nearly one in five Welsh school boards had no religious instruction and more than one in three taught the Bible without comment.[126]

Daily attendance and the duration of schooling were influenced strongly by more mundane factors. In arable areas particularly a farm labourer's job might also carry with it an assumption that the labour of his children (as well as that of his wife) would also be supplied when seasonal demand for labour peaked. The age structure of village schools was different from that of urban schools since more rural children left earlier. Once the school leaving age had been raised to eleven in 1893, and twelve in

---

[120] M. K. Ashby, *The Country School. Its Practice and Problems* (Oxford, 1929), p. 45.

[121] Johnson, *Derbyshire Schools*, p34.　　[122] Sellman, *Village Schools*, pp. 46–7.

[123] Johnson, *Derbyshire Schools*, pp. 125–8.　　[124] Sellman, *Village Schools*, p. 60.　　[125] *Ibid.*

[126] Evans, *Wales*, p. 265.

1899, farmers alleged that boys were ruined for farm work.[127] In addition, children were removed for long periods at a much earlier age. In mid-nineteenth century Norfolk, for example, village schools were closed for agricultural reasons from August until November, and from the early age of nine, children left school permanently to go to work in the fields.[128]

Wider socio-economic changes merely served to reinforce rural suspicions and prejudices. The exodus of rural labour from the land during the agrarian depression at the end of the century was attributed by farmers to the 'town book-learning' of the local school which made country boys into clerks not farmworkers.[129] The Lincolnshire schoolmaster, Henry Winn, noted how by the Edwardian era his pupils had grown reluctant to become a 'farmer's drudge', and 'began to look down on their parents' condition'.[130] A commentator sympathetic to rural education suggested that agriculture could be a blind alley, so that 'A teacher may, as a villager and a countryman, be never so sorry to see his finer boys join the police force or go on the railway; [but] as their friend he cannot but congratulate them.'[131] In reality, it was less the system of schooling than the poor economic opportunities of farming, reinforced by the social monotony of the village, that were the main reasons for more able and enterprising young men leaving the rural areas. And for young women also, the lack of suitable domestic situations locally too often reinforced the social attractions of the town and its more plentiful posts in domestic service.[132]

C. Attendance and school fees

Earlier, rural employment had been much more plentiful. The Reports of the Royal Commission into the Employment of Children, Young Persons and Women in Agriculture of 1867–70 showed clearly how diverse were the patterns of child employment, and hence of rural schooling. In arable corn-growing areas of southern and eastern England, where casual or gang labour was common, children were most extensively employed. Children under ten years of age were most likely to be employed where male agricultural wages were lowest, and their education correspondingly impeded.[133] The worst features of child labour were

[127] H. Rider Haggard, *Rural England* (second edition, London, 1906), vol. II, pp. 290, 302.

[128] A. Digby, *Pauper Palaces* (London, 1978), p. 194.

[129] Haggard, *Rural England*, vol. I, p. 24; vol. II, pp. 324.

[130] H. Winn, 'Some reasons for the depopulation of Lincolnshire villages in the 19th century', *Lincolnshire Historian*, 6, 1950.   [131] Ashby, *Country School*, p. 180.

[132] B. S. Rowntree and M. Kendall, *How the Labourer Lives. A Study of the Rural Labour Problem* (London, 1913), pp. 322–4.

[133] W. Hasbach, *A History of the English Agricultural Labourer* (2nd edn, London, 1908), pp. 259–67.

tackled in the Gangs Act of 1867 which prohibited the employment of children under eight years old, whilst the Agricultural Children Act of 1873 extended this prohibition to agriculture generally. It made part-time employment after this age dependent on a specified number of school attendances, and full-time work dependent on passing the standard four examination. The ablest children were therefore encouraged to leave school the soonest. The 1873 Act came into effect in 1875 but its impact was extremely limited since it did not have any enforcement agency. The Education Act of 1876 remedied this deficiency and also tightened up the earlier provisions. Now children under ten were supposed to attend school full-time, children who had not passed standard four but were aged between ten and twelve had to make 250 half-day attendances, and those between twelve and fourteen 150 such attendances. The act had a let-out clause, that those between eight and ten could work for a maximum of six weeks if authorised by the local authority, 'for the necessary operations of husbandry and the ingathering of crops'.[134] Effectively this meant that children from the age of eight upwards continued to be extensively employed in casual work to get in corn, hay, hop, apple and potato harvests. Continuity of schooling was still disrupted but perhaps less seriously than in the days of high farming. For more general economic reasons the most extensive era of child labour in arable agriculture was passing; one farmer remarked that, 'the days of neat farming are at an end. We don't pick stones or weed corn as we did.'[135] However, the view that children were an important source of casual farm labour was a resilient one. Just before the outbreak of the First World War attendance bye-laws were again relaxed; some rural authorities granted attendance exemptions to children under twelve.[136]

The efficient enforcement of legislation was clearly essential for it to be effective, yet all too often in country areas the poacher was supposed to act as gamekeeper. Poor-law guardians were given the responsibility for forming the School Attendance Committees that were instituted by Sandon's Act of 1876 to operate in districts without a school board. Until Mundella's Act of 1880 local implementation of the Act varied tremendously since it was up to the individual parish to decide whether they wished the bye-laws of the attendance committee to be imposed.[137] Even after this, however, much depended on local attitudes and initiative; in Norfolk, for example, some poor-law unions made the Act into a dead letter whilst a few enforced the Acts efficiently through the appointment of relieving officers as attendance officers.[138] Overall, however, the school

[134] 30 and 31 Vict. c 130; 39 and 40 Vict. c 79; 36 and 37 Vict. c 67.
[135] Quoted in Hasbach, *Labourer*, p. 272.   [136] Horn, *Education*, p. 270.
[137] Hurt, *Elementary Schooling*, p. 189.   [138] Digby, *Pauper Palaces*, p. 195.

attendance committees made little impression on the illegal employment of schoolchildren.[139] School boards had had responsibilities for attendance since 1870 but in rural areas usually only made a part-time appointment as attendance officer. Whether it was school board or attendance committee made little difference to the enforcement of attendance legislation since in country areas each was usually dominated by local employers. So too was the rural bench; magistrates had to deal with recalcitrant attendance cases and were supposed to impose a fine of up to 5s. on parents whose children were not at school. But in general the presumption was in favour of the poverty of the parent, or the needs of the employer, rather than the letter of the law.[140] Frustrated Devon teachers recorded that: 'compulsion, as exercised here, is a mere farce'; that 'parents seem to regard attendance as optional'; and, most damningly, that 'the Board members will persist in employing boys under age [and since] all are guilty alike . . . they cannot summon when they are the chief offenders'.[141] However, there was also some sympathetic recognition by teachers of the driving force of family poverty, especially during the depression at the end of the century. 'I earnestly want their attendance; but it is very hard to snatch a few pence from a parent's weekly income when such an opportunity offers.'[142] In these circumstances attendance legislation was slow to bite and average attendance in country areas rose very slowly.

Apart from agricultural child labour another important factor discouraging school attendance was the custom of charging school pence. These fees were usually varied according to the children's age and standard, as well as by their parents' status, but were limited by the 1870 Act to a maximum of ninepence weekly. But whilst there was concern about ability to pay in fixing school pence, this seems in practice to have been relatively unsuccessful in achieving regular payments. Managers and boards differed as to whether children were sent home if they failed to bring their pence.[143] Competitive fee levels were in any case essential, since there was considerable price elasticity of demand; raising the level of school pence could actually reduce income if the pupils then deserted to go to a cheaper dame school. It was not just parents but also some

---

[139] L. M. Davison, 'Rural education in the late Victorian era: school attendance in the East Riding of Yorkshire, 1881–1903', *History of Education Society Bulletin*, 45, 1990, p. 7. See also, L. M. Davison, 'School attendance and the school attendance committees: the East and North Ridings of Yorkshire, 1876–1880' *Journal of Educational Administration and History*, 17, 1986; L. M. Davison, 'School attendance and school attendance committees: further evidence from Suffolk, Norfolk and Cambridgeshire', *JEAH*, 20, 1988; L. M. Davison, 'School attendance and the activities of the school attendance committees in Wales and the South West, 1877–1880', *History of Education Society Bulletin*, 42, 1988.

[140] Johnson, *Derbyshire Education*, p. 150; Sellman, *Village Schools*, pp. 117–22.

[141] Sellman, *Village Schools*, pp. 119–20.   [142] *Ibid.*, p. 121.

[143] *Ibid.*, p. 131; Johnson, *Derbyshire Schools*, p. 151.

teachers that worried about school pence since their salaries could be partly dependent on it. For example, in 1873 the school log book recorded that Dunsford teachers were 'thoroughly disheartened' and in 'continual anxiety' at the irregularity with which payments were made.[144] This practice of variable salaries became less common, as the Cross Commission of 1886–7 noted with approval.[145] Financing rural education involved assumptions as to how much each social group should pay. Where voluntary schools were concerned, notional income from school pence paid by labourers had to be balanced against the product of a voluntary rate paid by local farmers and landowners.

In 1886 school fees paid by parents amounted to 97 per cent of the total with only a small amount being paid by the poor-law guardians (under legislation of 1855 and 1876) for those parents too poor to pay.[146] The attitudes of guardians varied; those in Northumberland were favourable whereas in most areas guardians were much more reluctant to do so. Some even demanded eventual repayment from the parents.[147] Although the Cross Commission of 1886–7 had seen no reason to end the fees system, there was sufficient dissatisfaction for legislation to be enacted that abolished school pence in 1891, in return for a 10s. grant instead. However, at the local level little, if any, difference resulted. 'The abolition of fees does not appear to have produced any good . . . the law respecting compulsory attendance seems to be a dead letter here', commented the vicar of Hooe in 1892.[148]

There were a multitude of other factors too which militated against a satisfactory level of attendance. Notable here were: bad weather, and inadequate boots and clothes to cope with it, sometimes leading to illness; children going to see local weddings or funerals; other visits to fairs, feasts and races; taking food to parents in the fields during haytime or harvest; and girls helping mother with household tasks or looking after younger siblings. Even after school, gender roles meant that girls tended to lose out in the educational process. Hannah Mitchell, born in 1871 on a remote Derbyshire farm, found that her evenings after school were filled with household tasks whereas her brothers might read or play.[149] Also militating against the formal educational process was childhood truancy – an eloquent protest against the boredom of school.[150]

---

[144] Sellman, *Village Schools*, p. 128.

[145] *Royal Commission to inquire into the working of the Elementary Acts*, BPP, 1886, xxv, p. 82.

[146] J. W. Adamson, *English Education 1789–1902* (Cambridge, 1930), pp. 381–2. The earlier legislation permitted, and the later, required guardians to pay the fees of children of outdoor paupers.

[147] Stephens, *Literacy and Society* p. 42; Horn, *Education*, p. 141.

[148] Quoted in Sellman, *Village Schools*, p. 131.

[149] J. Burnett, *Destiny Obscure. Autobiographies of Childhood, Education and Family from the 1820s to the 1920s* (London, 1982), p. 140.   [150] Ashby, *Country School*, pp. 53–6.

D. Schooling

The village school, then, has it seems to me a high function to fulfil . . . its work is national, not to say imperial, rather than parochial. Its business is to turn out youthful citizens rather than hedgers and ditchers; and it should, in its humble way, give a liberal rather than a technical education.[151]

Such sentiments – by an Oxfordshire HMI – would have encouraged the beleaguered village teacher but alarmed many members of the agricultural community. It gave substance to their entrenched view that the country school was subverting the supply of farm labour by educating village youth above its rightful station. In terms of the curriculum how well founded were such suspicions?

The subjects laid down by the education codes made no formal differentiation between town and country schools, so that in certain respects such criticisms had substance. Beginning with the Revised Code of 1862, and lasting until 1890, the national curriculum was in a narrow strait-jacket of the grant-earning subjects of reading, writing and arithmetic. Together with average attendance, the examination of these subjects formed the basis of the government grant to a school – the so-called payment by results system. The economics of institutional survival thus ensured that the crushing rigidities of inculcating the '3 Rs' for formal examination were dominant in every school. Daily learning could be enlivened by the well-chosen local example, or the object lesson focussed on familiar country topics – as at Ashover where stone-quarrying and haymaking were among the topics studied.[152] Such practices do not figure prominently in recollections and reports of the period, instead it appears that a generation of pupils had a daily grind composed mainly of repetitive rote-learning. Only to a minor extent was a more liberal education later officially encouraged. Government grants for so-called 'extra', 'class' and 'specific' subjects became possible; the specific subjects included both domestic economy and agriculture, whilst needlework and elementary science were listed as class subjects. These were subjects directly relevant to the rural economy and to the future lives of many of the children but only a very small minority took them. During the 1890s there was a further injection of vocational relevance for country pupils since girls could learn dairy work or housewifery and boys could take cottage gardening. And in 1900, with the introduction of the block grant system, teachers and managers could at last determine their pupils' programme of learning by reference to its educational rather than financial merit. Fond reminiscences of lively lessons tend to date from this later

---

[151] *Report of the Board of Education*, BPP, 1900, XIX, p. 254, (Report of Edmond Holmes, HMI, on Oxfordshire schools).   [152] Johnson, *Derbyshire Schools*, p. 182.

period. By this time a few intelligent rural children also won scholarships to grammar schools in country towns. Although one Shropshire lad had to milk three cows before and after school![153] But only a minority of country grammar schools attempted to give a special orientation to their curriculum that was related to their rural catchment area. Some did this by imparting a 'rural colour' to a traditional academic curriculum, whilst others had a more pronounced agricultural bias, although without possessing the separate agricultural forms that the National Union of Farmers later considered to be desirable.[154] In Wales, the small size of the Intermediate Schools made it difficult to make adequate provision for technical courses. However, in one girls' school with a special technical department the girls could learn dairywork.[155]

What should have been taught in the village elementary school? In 1897 the Board of Education revealed that it had undergone an educational conversion. In country schools 'the aim should not be to produce multitudes of clerks but multitudes of good craftsmen'.[156] Shortly afterwards the board issued a circular on *The Curriculum of the Rural School*, which directed school managers to see that the pupils' learning was 'more consonant with the environment of the scholars'. Commentators such as Rider Haggard were keen to encourage rural studies and singled out for praise those few schools that taught natural history and gardening.[157] For those who had left school, there was growing provision in the form of evening classes, and again the emphasis was on vocationally relevant topics. Some government grants had been available for this purpose since 1851 but relatively few classes were run before the Technical Instruction Act of 1889. In Cambridgeshire, for example, there had been only eleven classes before 1889 but after it there was 'a well-equipped and well-attended evening school in every second parish'.[158] About one in three young people went to these courses in the county and were taught woodwork, cooking, gardening, ironwork, basketwork, drawing and design. There were additional short courses by peripatetic teachers on such topics as laundry work, poultry-keeping, beekeeping, pig-keeping and veterinary science.[159] Nationally there appears to have been considerable variation in the formation of continuation schools: Buckinghamshire,

---

[153] Burnet, *Destiny Obscure*, pp. 159, 164.

[154] Board of Education, *Report of the Consultative Committee on Secondary Education* (London, 1938), pp. 191–2.

[155] Board of Education, *Secondary Education*, p. 344; W. Gareth Evans, *Education and the Female Experience*, pp. 173–4.

[156] Quoted in R. R. Sellman, 'The country school', in G. E. Mingay (ed.), *The Victorian Countryside*, vol. II, p. 545.   [157] Haggard, *Rural England*, vol. I, p. 17; vol. II, p. 120.

[158] M. E. Sadler, *Continuation Schools in England and Elsewhere* (Manchester, 1908), p. 211.

[159] *Ibid.*, p. 212.

Warwickshire, and the East Riding of Yorkshire were active whilst Gloucestershire was not.[160]

The assumption that local and rural topics in education were necessarily a panacea for the problems of those who lived in the countryside begged several questions. A basic issue – and one that was seldom addressed – was why, if it was wrong for children to engage in casual farm work or to stay at home to engage in domestic tasks, it was right to introduce these activities into the curriculum of day or evening schools.[161] On another important issue one HMI had the temerity to suggest that, since there was rural depopulation, 'it would be unfair to a country child to make his curriculum differ so widely from that of the town child as to disable him from competition, if his future lot should be cast in a town'.[162] Later, M. K. Ashby, in her well-judged study of the country school, argued that the presence of local topics in the elementary school was not in itself sufficient since it did not necessarily focus on the subjects that would interest the child – the parish church or manor was studied rather than the history of the labourer. Similarly, she suggested that 'knowledge of the countryside does not necessarily attach people to it for economic reasons'.[163] The substitution of a rural for an urban curriculum would not automatically result in a contented and static village population, as some had argued. And the presence of a hidden curriculum in some village schools was probably less successful in the long-term preservation of the village than in a short-term lubrication of social niceties. Acquiescence in the *status quo* and deference to rural notables were encouraged by such practices as making the girls curtsey to social superiors.[164] And it was not just pupils but also teachers who needed to know their lowly place in rural society.

Village teachers were often isolated and, in their professional lives, might see few adults save angry parents, condescending clergymen, semi-literate members of the school board and, more occasionally, that educational deity – the HMI. But they might also be respected in the village community, or more rarely, even loved as was the Cheshire school-mistress, Miss Gilchrist for whom villagers subscribed a pension when her health failed.[165] A more usual fate, however, was to be exploited through additional tasks such as taking charge of the Sunday school, playing the organ, training the church choir, or taking minutes at vestry meetings. A survey of 1891 suggested that one in three teachers was in a position where extraneous duties were demanded and, despite agitation

[160] *Ibid.*, pp. 217, 231, 234; Hasbach, *Agricultural Labourer*, p. 340.
[161] See *Report of the Board of Education for 1899–1900*, pp. 163–4 for a rare mention of this fundamental issue.   [162] *Reports of H. M. Inspectors on Elementary Schools for the Year 1902*, p. 79.
[163] Ashby, *Country School*, pp. 180–1.   [164] Horn, *Education*, p. 120.
[165] Burnet, *Destiny Obscure*, p. 157.

by teachers' professional organisations, this situation continued largely
unchanged.[166] Since economy was often the driving force of the local
school board, cheeseparing on teachers' salaries was also a hazard; payers
of the voluntary rate might withhold their contribution until managers
rescinded a proposed salary increase, as at Plymtree in 1890.[167] In other
cases assistant teachers were actually replaced by monitors or pupil teachers in order to save money.[168] Women were cheaper to employ than men,
typically earning only about two-thirds of the salaries of their male col-
leagues in comparable posts.[169] It is significant in this context that the post
of elementary teacher became increasingly feminised; this was particularly evident in smaller schools, where salaries were in any case lower. A
large rural/urban salary differential also existed, so that the highest-paid
head in a rural school often earned only as much as the lowest-paid head
in a town institution. Even after 1902, when the transfer to county councils had raised salaries, country headmistresses were still earning only half
that obtained in comparable London posts.[170] In Welsh Intermediate
Schools in 1914 male heads outnumbered female heads by three to one,
and earned £377 against the women's £324. Female assistant teachers
earned £125 compared to their male counterpart's income of £156.[171]

How much progress was made in rural education during the period
from 1850? Even before the advent of a national framework for schooling in the 1870s there had been no simple differentiation in literacy levels
between urban and rural areas. Although illiteracy was higher in the latter
and was diminishing more slowly, marked regional variations still persisted.[172] The institutional developments of the period after 1870 meant
that the improvement in literacy continued, as children in remoter areas,
or from more deprived backgrounds, learned to read. Literacy among
girls also continued to increase more rapidly than that among boys, so
that an earlier gender disparity closed, and by 1900 there was almost no
difference between the sexes.[173] Overall, the educational record in the
countryside was a chequered one but there had undoubtedly been a levelling up of standards in provision of schooling; many more children had
been educated and to a much higher standard. Given the conservative
social attitudes of the countryside that was no mean achievement.

---

[166] A. Tropp, *The School Teachers* (London, 1957), p. 132.   [167] Sellman, *Village Schools*, p. 40.

[168] Horn, *Education*, p. 80.   [169] *Ibid.*, pp. 69, 222–4.

[170] Hurt, *Elementary Schooling*, p. 179.

[171] W. Gareth Evans, *Education and Female Emancipation*, p. 168.

[172] W. B. Stephens, *Education, Literacy and Society 1830–1870: the Geography of Diversity in Provincial England* (Manchester, 1987), pp. 5, 8, 16–17, 23, 37, 41–2.

[173] R. D. Altick, *The English Common Reader* (Chicago, 1957), p. 171–2.

CHAPTER 26

OVERVIEW

BY ALUN HOWKINS

The social history of rural England and Wales discussed in the previous chapters presents if not a linear and uncomplex model at least some consistencies which in turn can provide the basis of a generalised account of the period as a whole. Firstly, and this has to be said even if it appears obvious, there was a consistency of inequality which was legitimated and even praised. Social mobility was rare, and the often vaunted farming ladder a virtual myth except in parts of the north. This rigid social structure was supported further by an almost total social separation. Of course, there were exceptional individuals, but in the huge majority of cases there was little or no personal or private contact between the classes. Publicly they occasionally met and, according to their places, acted together. At agricultural shows, village fetes, harvest festival and, above all, at the public spectacle of paternalism like the Jubilees of 1887 and 1897 the countryside seemed as one. On other occasions, especially around the giving and receiving of charity, the classes touched – but their separation remained absolute and was reinforced by this contact. Nor was there, at least before the Great War, any exception for the newcomers. Where they moved into existing communities, they moved in at their already set level. Where they created communities, be they on Alderley Edge or in Sunningdale, they produced a new and different class structure which knew neither squire nor labourer but which was no more egalitarian than the rural structure it had replaced.

Nor did the social and institutional reforms within our period do much to alter this. As Anne Digby points out, by and large the new local government structures of the 1880s and 1890s continued to be dominated by the same élite groups drawn from the aristocracy, but particularly from the gentry and the farmers. In a few areas, especially East Anglia, working men and even women were elected but they remained, at least until the Great War, an insignificant minority. Similarly education had little short-term impact. The curriculum offered was all but specifically designed to keep the poor in their place and in most areas of rural England the school boards and school management were dominated by the local élites. Just how powerful these groups were, and how they could move against teachers or parents who transgressed these norms is demonstrated by the history of Annie and Tom Higdon. The Higdons were appointed to Wood Dalling School in Norfolk in 1906 with Mrs Higdon as Mistress and Tom as assistant. In the next few years they devoted themselves to the village; however, their devotion was not that of the managers of the school who were farmers and, in at least two cases, employers of 'illegal' child labour. Tom Higdon was also a branch secretary of the agricultural labourers' union and Mrs Higdon an educational reformer and a firm believer in school attendance, even at harvest time. All these elements came together in the winter of 1910 and the Higdons were removed to Burston in South Norfolk. Again, exactly the same series of events took place and they were dismissed. Supported by the vast majority of the village they set up a 'Strike School' supported by the labourers' unions, the TUC and the Labour Party which remained as the village school until the Higdons' retirement while the 'county' school stood virtually empty.[1] The Higdons' case is extreme, but it does clearly demonstrate the strictures under which rural education very often had to function.

This inequality was structural and all pervasive. As Brian Short notes, social position is a key determinant of demographic change. This is most obvious in relation to mortality where, although country dwellers as a whole were healthier than those living in urban areas, there were significant differences between life expectancy and general health of labourers and farmers, let alone between labourers and the rural élite. However, social class also determined fertility, the poor producing larger families than the élite for at least part of the period, and, crucially, migration.

All this is not to suggest that these socio-economic positions simply determined political or cultural behaviour in a 'one-to-one' way. Class and class consciousness is complex and many faceted. Centrally, in the rural areas at least, it needs to be related to specific local and regional conditions as well as national trends. A clear example here are the differences between northern and southern England and between England and Wales. In both these contrasts, one sees how what in a simple 'model' looks very similar in terms of land holding, employment patterns and broad social relationships, is in fact very different 'on the ground'.

However, difference does not mean that generalisation is impossible or useless. In the broadest sense relations between the classes in rural society

---

[1] There are several accounts of the Higdons, the best of which by far is B. Edwards, *The Burston School Strike*, (London, 1974).

in the period 1850–1914 were neither stable nor constant. At the beginning of the period in England and to a much greater extent in Wales relationships were tense and difficult. The legacy of bitterness among the labouring poor, created by the New Poor Law and the suppression of rural unrest from 1830 to 1850, remained potent. Even in the 1860s, as John Archer reminds us, Norfolk and Suffolk alone experienced 250 incendiary fires, nearly a hundred more than in the supposedly much more disturbed 1820s.[2] This bitterness, on occasions amounting to blind hatred, and usually, though not always, individual anger was never to disappear completely although it became rarer and rarer. In its place came the more respectable and muted protests of the Trade Union and the Chapel.

Similarly, outside Wales at least, relationships between farmers and their landlords seem to have stabilised in the years after 1850. This should not, however, lead us to believe that a 'natural' harmony somehow came into being. The calmer and more paternalistic social relations which characterised the years after the mid-century, in England at least, were achieved by careful balancing of a complex dialectical relationship of the kind described so well for the later period by Howard Newby.[3] With the onset of the agricultural depression in the late 1870s this balance was disturbed and other elements (like education) combined with these economic problems to produce rifts within the apparently previously stable order.

This opens up another area where there is a measure of agreement among the authors of this section which stands in clear distinction to the writings in other parts of these volumes, and this concerns chronology. Initially at least, the 'social' history of rural England from 1850 to 1914 suggests a watershed in the 1870s and 1880s. In the past this was unproblematically attributed to the 'Great Depression' in agriculture following the disastrous harvests of the late 1870s, the cattle and sheep 'plagues' of the same period, and most importantly, the opening up of the wheatlands of North America. However, much work in recent economic history suggests that the extent of this depression was exaggerated by contemporaries, and that that exaggeration has greatly influenced historians. As a result, according to these accounts, in much earlier writing the 'Great Depression' was made to appear much more important than it actually was.

It is difficult to refute this in pure economic terms and at best we are left with a 'map' of the depression which restricts its impact to a few

---

[2] John Archer, 'By a Flash and a Scare'. Incendiarism, Animal Maiming and Poaching in East Anglia, 1815–1870 (Oxford, 1990), pp. 251–2.

[3] Howard Newby, 'The deferential dialectic', Comparative Studies in Society and History, 17, 2, 1975.

regions. Yet one has a deep sense of unease. At the bottom line the appalling years of the late 1870s had a deeply traumatic effect on the confidence of British farmers. As B. A. Holderness says:[4]

The very poor harvest of 1879 in northern Europe sounded the alert since customary compensation for scarcity in higher prices was not realised. Part of the depression in arable husbandry was psychological. Unease became foreboding because the traditional relationships between prices and output, between farmers and consumers were perceived to have broken down.

Two very different testimonies give a sense of this. The language used in the diary of John Simpson Calvertt, a gentleman tenant farmer of 2,000 acres in Oxfordshire, shows for 1879 just how traumatic even the 'first' year of the collapse was. Calvertt was normally temperate, even cold, in his daily record but the events of what he called 'the most *ruinously* ugly *seasoned* year, of *this century*' produced a very different reaction:[5]

Aug. 22. Continues the most cursed, ruinous weather, on record – cannot thrash Oats – plough Fallows – skerry Turnips, nor even manure on clover land for Wheat!!! – and this state of things been going on *all over* the *Country* since *last April*!!!

Aug. 31 . . . A terrible hindering ugly month.

Sept. 10. Commenced cutting Barley (scythes) with all hands in Witley Hill – the only day this week – showers and rains frequently prevailing – to the great loss and ruin of Farmers in general all over England, Ireland and Scotland – the accounts we read in the Papers are really alarming.

In Calvertt's view this was the beginning of ten dreadful years and although he 'benefited' by rent reductions and certainly changed his cropping to a limited extent, he never changed his mind on that.

Towards the end of that ten-year period a very different kind of farmer, Bishop Samuel Harrod, smallholder, market gardener and elder of the Essex religious sect the Peculiar People, addressed his flock with no less feeling:[6]

Tenant farmers have had to break [been made bankrupt] by tithes, parish charges and rents. Next it must go to the landlords, and then when the landlords go, where must it come to then? . . . but in the midst of the grumbling I feel we have cause to be thankful to God ('Praise Him'; 'Yes') I believe God has sent as much as the country deserves. I mean in this way. If we don't do our duty to the land, how can we expect great profits? ('No, no'). Can God give increase to that

---

[4] B. A. Holderness, 'Agricultural responses to the "Depression" of the late-nineteenth century in Britain and France: towards a comparative history', Paper Presented to IV Congresso di Storia dei Movimenti Contadini – L'agricoltura en Europa e la nascita della "questione agraria" (1880–1914), Rome, 1993.

[5] John Simpson Calvertt, *Rain and Ruin*, ed. Celia Miller (Gloucester, 1983), pp. 63–7.

[6] *Essex Herald*, 24 Sept. 1889, quoted in Mark Sorrell, *The Peculiar People* (Exeter, 1979), pp. 40–1.

which is not sown or not done? That is our part just as it is to find salvation. Man must make a start and the God will meet him ('Bless Him'). There's crops of only two quarters to the acre all parishes through. But if you don't grow crops, don't expect God to send increase. 'Tain't in noways reasonable. A shortened-crop is on the land's side and not the Lord's side.

Bishop Harrod making his appeal to the Lord, and it should be said to peasant proprietorship, was very different from Calvertt, yet what they had in common was an experience which they believed was one of profound crisis. It may have been fanned by the press, as with Calvertt, or by a near-millenarian belief system, as with Harrod, but it was real, and more importantly they and thousands of others acted upon that experience. In this sense the 'reality' of the depression may not be the only question. Put crudely, if men and women believe there is a 'depression' they will act as if there is one until and if they are convinced, usually by a different experience, that the problem does not exist or is over. Calvertt had, one assumes without reading him, almost exactly the same chronology of the 'Great Depression' as Rowland Prothero.

In these circumstances it might be better to use the mainland European expression for this period – 'agrarian crisis'. This suggests something both more far reaching and wider than 'depression' and also takes away the purely economic element of the discussion. This would fit much more with profound socio-cultural change in many levels of rural society which characterised the twenty years after 1870. This is not to say that there was a direct relationship in all cases between them, or that the relationships involved were of a determinist kind, rather than, taken together, they do mark the 1870s and 1880s as a watershed. Brian Short in Chapter 21 points to important changes in demographic patterns in those decades. The rural population as a whole began serious relative and absolute decline in the years after 1870, after a period of relative stability. The 1870s also show a clear change in marriage and fertility patterns as family size decreased. Taken together these represent one of the most significant changes of the whole period, the absolute decline of the rural population and its near extinction as a major part of the labour force, and the 'normalisation' of a much smaller rural family. Anne Digby also points to important institutional changes in these decades. Although, as we noted above, the personnel were often the same or similar, the new institutions of education, law and local government do mark the penetration of the local state into many aspects of the lives of ordinary people for the first time. Also, even before the Great War, there were glimmers of what could come. The election of women to all levels of local government marked a substantial change even if they were usually women of the rural élite carrying on by other means the role of lady bountiful. Similarly the election of 'independent working-class candidates', albeit mainly in the north and East Anglia, did show that in some areas the rural poor were willing to 'buck' the political parties of the wealthy and great.

This view is supported by the electoral politics of the rural areas. The farm labourer 'got the vote' in 1884, although registration requirements debarred many, and used it for the first time to secure a Liberal victory in the 1885 election. The split in the Liberal Party over Home Rule hit the rural Liberal vote but increasingly through the 1890s the rural areas of England and Wales became the heartland of political Liberalism. Even at the notorious 'khaki' election of 1900 some rural seats went against the national trend and returned Liberals. However, it was the landslide of 1906 which proved just how powerful a force Liberalism had become in the county districts. At the end of the count rural England was Liberal. Of eighteen rural northern counties in Cumberland, Westmorland and Durham only one returned a Conservative compared with six in 1900. In East Anglia not one of the sixteen county constituencies returned a Tory although six had done so in 1900. Perhaps more striking, if less spectacular, were victories through rural England. Hardy's Wessex went 'radical' with Liberals gaining three out of the four seats from the Tories; while even Kipling's Sussex returned a non-Tory to a county seat for the first (and only) time. Many of these gains were lost in 1910, but a large core of rural English seats remained Liberal which, added to those in Wales, gave British Liberalism, in the years before the Great War, a distinctively rural character. This was in turn recognised in the Liberal programme which, especially after 1910, had a powerful agrarianist aspect.[7]

These political changes seem to mark a disruption in patterns of social relations which was fairly general. Here the effects of an agrarian crisis (as opposed to a simple economic depression) are clearer. Falling rents increasingly took the capital of landlords elsewhere and culture seems to have followed this capital, leaving a 'power vacuum' in many villages. This was filled in part by the farmers whose ideology was less paternalistic. Evidence here is difficult to come by, but it seems possible that this, combined with the growth in power of the local state which was usually in the farmers' hands, created more instrumental relationships in rural areas. Digby notes as an adjunct to this that the 1870s mark not only the growth of trades unionism but increasing numbers of political attacks on the Church of England. Nor was it only a labour–farmer conflict

---

[7] Surprisingly there is relatively little on this aspect of rural social or political history. The standard histories of Liberalism make very little of it, and 'politics' is notably absent from agricultural history. The basic data can be found in Henry Pelling, *A Social Geography of British Elections, 1885–1910* (London, 1967). There is an account of south-west Norfolk in Alun Howkins, '"The Great Momentous Time", radicalism and the Norfolk farm labourer 1872–1923', DPhil thesis, University of Essex, 1982; and of north Norfolk in Clyde Binfield, *So Down to Prayers. Studies in English Nonconformity, 1780–1920* (London, 1977).

although this was the most obvious and organised. Again evidence is conflicting, and more work is needed, but in Wales and parts of the north of England certainly, and elsewhere possibly, there is evidence of conflict between tenant farmers and landlords.

All this was exacerbated by much broader cultural trends. We saw how the market town and its *pays* was a key 'community'. The localism of this view of the world was certainly powerful yet it was ambiguous. The 'penny press for the working classes', which was so much a feature of the 1870s and after, contained local news and stressed regional pride and identity, yet it was also a conduit for the national, and even the international, through its news and features. East Dereham or Colchester may well have been the regional seats of Toryism or Liberalism through their respective clubs, but these also linked into a national world of politics. Similarly many historians have stressed the closed and local world of the wayside Bethels of rural non-conformity, yet the most humble chapel linked its members from village to market town, region, nation and even the world through its missionary work. These links brought new ideas and new consciousness into the rural districts. To give a crude example: by 1914 both labourers and farmers (though nothing like a majority of either) were unionised into national unions, with programmes which saw their respective memberships, if not as warring classes, at least as powerful socio-economic groups with often opposed interests. In Wales this local–national move created, by the 1900s, a strong sense of national identity, and a feeling among farmers and labourers alike that they were victims of a foreign landlord class. This never reached the level of disturbance that was seen in Ireland but it was real enough.

The Welsh example though points us back on the road to caution. The chapters in this section show how all this was mediated through a powerful local consciousness which reflected the deeply diverse and regional nature of rural England and Wales well beyond the end of our period. Brian Short's materials on the changes in population show just how regionally specific these patterns could be, and stress that they were not arbitrary. What we have in the years of 1850–1914 is not one rural social history but many. The great farming regions identified first for the end of the early modern period by Joan Thirsk were often also local worlds in a social or cultural sense. Any examination of the major division between the upland–pastoral and lowland–arable areas of these islands reveals great differences not only in terms of 'factors of production', but in settlement pattern, farm size, labour market, hiring practices, type of building, diet, dress and forms of traditional ritual dance. These differences were seldom respectors of village, county or even national boundaries, but were based on a complex weave of all this and more which I would call, following Raymond Williams – culture. Even the reforms which created a nation system of poor laws, police and local government by 1900 often, as Anne Digby shows, responded to local existing practice and custom as much as to the imperatives of London. The Welsh attitude to the poor law, with its proud and stubborn refusal to accept English definitions of poverty and morality, lasts from Rebecca and her daughters at the end of the 1830s to the Royal Commission of the 1900s.

This regionality, stressed in these chapters, is ill served in many accounts. Too easily and too often 'agricultural history' becomes the history of economic performance in the arable areas. This has led in turn to a neglect of the complexities of social relations elsewhere in Britain. As a result whole groups, who are of fundamental importance to the rural history of the nineteenth and even the twentieth century, have been written out of history. This is particularly true of small family producers, worker/peasant farmers and their families, farm servants, who were much more common even in 1900 than the standard accounts would suggest, and above all women workers. Our model of the social history of rural England is still far too easily one based on a village with its manor house, church, farmhouses and cottages forming a compact and half-timbered group around a village green. Even in terms of models which look at conflict, complex nuances are reduced to a three-tier 'class structure', which may well be appropriate in many areas, but which is clearly far too simple in others.

All this should not suggest that any kind of generalisation is impossible – indeed these chapters are in part an attempt to do that – and the trends are fairly clear. As a whole the social history of rural England and Wales between 1850 and 1914 was one of change, mediated through local social, economic and cultural structures. At the beginning of the period rural life and labour were still the lot of a narrow minority of British men and women. At the end rural depopulation, falling family size and urban growth had reduced that from about 49 per cent to about 29 per cent. Chronologically, the period is divided by the agrarian crisis and a period of profound socio-cultural changes in the twenty or so years after 1870. These were the years of the watershed in terms of demographic change but they are also characterised by social disruption and institutional change which marked the beginning of the end of a certain kind of rural social order in England and Wales. The specific nature of this change is charted to an extent in the previous chapters but the main trends were towards a professionalised local government system which took power away from traditional élites, at least over the long term. Added to this were long-term changes in religious behaviour. In the rural areas at least the rate of growth of religious non-conformity slowed after the 1860s, as did the revival of the Church of England. This remains uncharted territory but it seems likely that the influence of organised religion was less in the 1900s than it had been at the beginning of the period. Of course there were exceptions. In East Anglia religious non-conformity was certainly not a 'withered branch' in 1900. In rural Wales the 'Awakening' of 1904–5 brought a new vigour to what seemed, to some at least, an ailing cause.[8] Its impact may have been short lived, but to those who experienced it, the awakening remained a fundamental religious and cultural experience.

However, this is not to suggest that the nineteenth-century countryside was Godless. Even though regular church or chapel attenders were probably a minority of the population, even in Wales, their influence was strong. Nowhere is this clearer than in the general trend towards making rural society more 'respectable'. The figures show a general decrease in casual criminality, while the consumption of alcohol, particularly spirits, began a steady decline in the mid-1870s. Side by side with this went a decline in public 'rowdiness' and festivity, especially where this was associated with 'traditional' feasts. Even in the 1860s this change was clear. At Bletchingdon in 1862, the next village to Kirtlington, formerly a site of a traditional Whit-Ale, Whit-Monday had been completely transformed. As the local paper put it,[9]

It was a pleasing sight to see nearly a hundred young, fine, clean, and well-dressed labourers follow their banner . . . to the quiet old Church – it was a convincing proof of what unanimity and good feeling can affect . . . This was a meeting bearing strong contrast to those of years gone by, when riot and drunkenness was the result.

How the countryside was viewed also changed crucially in this period. In 1850 it was still possible to think of an 'agricultural interest' of enormous power, politically, economically and socially. Although the decline of this group may have been exaggerated, it is certain that by the 1890s that power had been greatly reduced. However, just as the rural areas 'lost' their political power and became less and less important to politics and the economy, so they gained in cultural and ideological importance. Eugenics coupled with imperial problems and fears of the urban 'masses' led to a revaluation of the countryside as the 'true Heart of England'. Painters, composers and above all writers created a vision of a rural 'land of lost content' in the years between 1890 and 1914 which is still with us now. At a practical level it began the reversal of rural depopulation, which had characterised the period between 1850 and 1911, as the rich and

---

[8] See Eifion Evans, *The Welsh Revival of 1904* (London, 1969).

[9] *Jacksons Oxford Journal*, 21st June 1862. See also Alun Howkins, *Whitsun in Nineteenth-Century Oxfordshire*, History Workshop Pamphlet, Number Eight (Oxford, 1973), and Keith Chandler, "*Ribbons, Bells and Squeaking Fiddles*," Folklore Soc. Publications, Tradition, 1 (London, 1993), *passim*.

mobile moved 'back to the land'; at a cultural level it created the belief that all real 'Englishness' is rural.

I wrote at the beginning of this section that 'social history' has been a poor relation in the area of agrarian history – and this is still very much the case. Certainly many hundreds of books and articles are published annually which cover what might be called (and indeed is called) rural social history. Unfortunately, the vast majority of these are products of the spurious 'English ruralism' we have already examined. This means that the serious and rigorous social history of the rural areas remains curiously unstudied. Even the *Cambridge Social History of Britain* gives very little space in its three volumes to the rural areas.[10] This is especially true of the period after 1850, which perhaps lacks the debate and excitement for social historians as the period of enclosure, 'agricultural revolution' and Captain Swing. However, clear areas of work need study urgently. We need to know much more about the small farmers – the British peasantry – than we do now. This involves crucially moving away from the narrow confines of economic definitions to the broader ones of culture, family and labour process. Much is to be learnt here both from recent British anthropology, and from comparisons with continental Europe. Curiously, we also need to know more of the social history of almost all the rural élites. There are memoirs aplenty of the aristocracy, but we still probably know more about the social and family life of this group in the early modern period than we do in the modern. Similarly, the rural middle class is almost non-existent in the historiography of the years after 1850, although again there is excellent material on the earlier period. We still need to see much more done on these groups, and on others, including the history of women in the rural areas. This has been raised so often that there is sometimes a sense that we have gone backwards. But the work still has to be done. We also need, as we move into histories of the twentieth century, a better sense of the newcomers to the country areas than we now have; I am conscious that even in these chapters this group has been portrayed in a far too simplistic way.

Finally we need to think of these different topics in terms of regionality. This is not only a plea for more 'local studies' but also a plea for a genuine sense among all historians of the rural areas of what is 'representative'. Far too much rural history (social and otherwise) is still centred on the south and east – the areas of arable dominance and large-scale farming. There are exceptions, and I hope these have been indicated in the previous chapters, but we still need to know much more about the north and west, the Welsh and Scottish borders, and indeed Wales itself

[10] See Alun Howkins, 'Social history and agricultural history', *AHR*, 40, II, 1992.

where rural social history seems too often to be a poor cousin of the more heroic history of the industrial areas. Looking seriously at the north and the west should raise new questions about peasants, about farm servants, and about the centrality of family and women's labour which will counterbalance the crudity of the modelling involved in many if not all accounts of our rural past.

PART VI

THE URBAN IMPACT ON THE COUNTRYSIDE

CHAPTER 27

INTRODUCTION*

BY GORDON E. CHERRY AND JOHN SHEAIL

Previous sections of this volume of the *Agrarian History of England and Wales* have focused directly on the issues and circumstances of farming between the Great Exhibition and the outbreak of the First World War. Detailed accounts have been given of the land and its labour force, and of trade and society, largely from the perspective of the agricultural enterprise of the time. In this sixth section, the opportunity is taken to appraise the countryside, in which farming played so large a part, from a much broader perspective. The intention is to consider the changing visual appearance and use of the countryside in the context of the fundamental departures that occurred in the rural economy and society of that period.

The title, 'The face of the countryside', might have sufficed for a section that simply recounted *when* and *where* changes occurred in the appearance and character of the wider countryside, but this section also seeks to consider *how* and *why* those changes came about. For contemporary observers and later historians, a principal driving force for change was the urban impact on the countryside – in short, the consequences of urbanism and urbanisation. It is this impact which we have adopted as the overarching theme of the section.

It will be apparent that the title, 'The urban impact on the countryside', cannot be taken to imply a narrow focus. The urban impact was expressed most obviously through the loss of rural land to urban building, but there were many other, less tangible manifestations of the gradual shift of a nation, both economically and socially, from rural to urban. Rural England and Wales changed so dramatically during the six decades of the period that we can write in terms of complete breaks with the past. Paradoxically, however, the urban and industrial processes that were so destructive of rural features also had the effect of stimulating interest in, and concern for, what were perceived to be the intrinsic qualities of the countryside and rural life. Never before had they been so cherished, ironically by an industrialised and urbanising society that had done so much

---

* This chapter was submitted in December 1989 and has not been revised subsequently.

to destroy them. Typically, in a book on *The moors, crags and caves of the High Peak*, published in 1903,[1] the author recounted how, midway between Sheffield and Manchester, 'at the threshold of the world's most populous cluster of manufacturing towns', there was a broad area of wild countryside, where

> Through the factory smoke and the steam we have glimpses, now and again, of the dark line of the edges. Even business is lightened a little by the knowledge that an hour or two might take us clear away, on to a heathery moor that wears the same harsh, impassable face as it wore when Britain was peopled by savages.

There was in fact a deluge of publications on the countryside, resulting from an interest informed by rapid advances in the earth and life sciences, and facilitated by unprecedented opportunities to gather information and see the world at first hand.

Both the declared intention to broaden the perspective of this sixth section, and the fact that the dynamics of change were so bound up with urban, or at least urban-oriented, influences, make it especially important to place the dimensions of countryside change in the context of the period itself. The economic transformation of the country is an obvious starting point. In the second half of the nineteenth century and up to the Great War, Britain 'matured' as an industrial nation. On the criterion of the size of its labour force, the cotton industry had reached its peak by 1851, but extractive and heavy industries (coal, iron and steel) continued to surge, producing a flood of manufactured goods. The country became 'the workshop of the world', its economic vitality resting on a lively export trade, allowing Britain to import freely not only the raw materials needed in manufacturing but also foodstuffs – with direct and far-reaching consequences for the performance of home agriculture.

The locus of industrial expansion was increasingly in the cities and towns. The countryside no longer boasted any manufacturing strength, as had previously been the case, notably with textiles. By mid-century, England and Wales already formed an urban nation in that more than half its population resided in towns, a proportion that increased to around four-fifths by 1911. As a general rule, large cities grew faster than smaller towns; London became a world-giant, and manufacturing centres in industrial districts coalesced to form regional pivots of population concentration. Before the end of the century, urban territorial spread had become as emotive an issue as sheer population numbers. Suburban peripheries marched into the countryside, absorbing farmland and evoking a spectre of uncontrolled growth, as towns absorbed the relatively deprived and landless, seemingly sucking the very life blood from rural areas. The picture

---

[1] E. A. Baker, *Moors, Crags, and Caves of the High Peak* (London, 1903), pp. 9–10.

of ultimate urban dominance was completed by the rise of new industrial settlements (for example, Middlesbrough) and towns with completely new functions (such as railway nodes and leisure centres on the coast).

Variably, but generally after the 1850s, the rural population first stagnated, and then began to fall. Better living conditions in the towns, measured in terms of housing, work, wages, prospects and education, encouraged a powerful drift from the land. The promise of a new start in overseas colonies, now increasingly accessible and indeed positively encouraged, swelled the outflow, as farm work began to decline. Partly in consequence of agrarian depression, and partly through the gradual introduction of mechanised operations in farming, the drift became a sustained movement.

As England and Wales both industrialised and urbanised, the rural areas were profoundly affected from so many points of view. But the impact was all the greater because the economic transformations were on a global scale. Expansion in engineering led to a huge growth of output in railway rollingstock and steam vessels; communications both within and between nations were greatly speeded up. By the 1870s it was possible to take full advantage of a free-trade system. British agriculture was fully exposed to competition from countries anxious to pay for their imports through their agricultural produce. British wheat prices fell sharply, but a shift from arable to pasture brought little relief as the techniques of freezing and chilling enabled meat to be imported, most typically from Australia and New Zealand. Home agriculture was depressed, and the countryside fell into decline as investment, affecting housebuilding and land improvement, was curtailed.

So in the years between the Great Exhibition and the Great War, the very appearance of rural England and Wales changed as a result of national and international forces: a surge in the global economic system stimulated the inherent vitality of the capitalist order, not only of Britain but of other industrially advanced countries. The fact that this system was essentially urban based allows us to write of 'the urban impact on the countryside'. The very use and occupancy of the countryside changed: types of farming (as old practices became redundant), new urban settlements, peripheral expansion of existing towns, the changing role of small, market towns, new buildings to accommodate new functions, and the new inter-urban thread, the railway. As an old order showed every sign of being left behind, if not exactly swept away, there was, paradoxically, a resurgent interest in things pastoral. In a later chapter, we show that rural values and attributes became highly prized by those living in towns, increasingly conscious of a heritage that could be lost so easily. The beginnings of a conservation movement can be discerned well before the turn of the century.

In recounting how the physical appearance of the countryside took shape, close account has to be taken of social and political trends. Both took on an increasing urban bias. At a national scale, the essentially urban housing-reform movement had some impact on rural housing conditions. But more importantly, as the garden city and garden suburb aspirations took hold, in an attempt to bring relief to poor urban living conditions, the new Town Planning movement sought environmental solutions, which could only be obtained at the expense of rural land. Meanwhile, the 'Land Question' came to impact on British politics, significantly so by the turn of the century, as the established power of the landed 'squirearchy' was attacked. In other fields, national educational policies improved the provision of rural schooling, and political reform extended the electoral franchise, with implications for both local and central government. Overall, the countryside became less contained than before. More evidently part of a wider economy and society, it was increasingly having to adjust to circumstances, the origins of which lay well beyond the rural areas themselves.

Scientific discovery and invention underpinned much of the change reflected in nineteenth-century society. The sheer self-confidence of the Victorian period depended on it. In these circumstances, man seemed to be so pre-eminently the instrument of progress – even to the point of deciding the appearance and character of the countryside. In the scheme of things, even the wild plants and animals of the countryside had their allotted place. In fact, it is not difficult for the historian to find many instances of self-doubt and concern as to what was happening, but, whatever the perceptions of contemporary observers may have been, it is worth recalling that the natural environment of England and Wales was comparatively benign. If there was anywhere on earth where Man might feel in control of the world in which he lived, it was in the British Isles.

There were so few large-scale, natural hazards that, when one did occur, it was perceived with a mixture of drama and disbelief. In 1884 there occurred the 'Great English Earthquake', the most destructive earthquake ever to be recorded in Britain. It had its epicentre to the south-east of Colchester in Essex, and shook half the country. Fissures opened up in the ground, and enormous waves swept along rivers. Over 120 buildings were shattered, and thousands more damaged. Wivenhoe looked as if it had been bombarded, and not a single house escaped between Fingringhoe and Langenhoe, a distance of 4 miles. Sightseers came by bicycle and tricycle, and by train (a special relief had to be run). National newspapers and official commentators tended to play down the event – natural disasters on that scale just did not occur in Britain.[2]

---

[2] J. E. Taylor, 'The earthquake', *Nature*, 30, 1884, pp. 18–19; P. Haining, *The Great English Earthquake* (London, 1976).

More commonplace were severe frosts and droughts. The frosts of the severe winter of 1869 killed to the ground much of the furze on Roborough Down, near Plymouth.[3] The American word, 'blizzard', came into popular use, following the snows, easterly hurricane and bitter cold of March 1891.[4] Flooding could have an even more severe impact, but usually over a smaller geographical area. Norwich and its suburbs experienced severe floods on two occasions – the first in 1878 and the second in 1912, when 6 inches of rain over a period of 12 hours caused the highest flood level ever recorded in Norwich. John Burns, the President of the Local Government Board, 'paid a flying visit to the city' to make a personal inspection of the damage. Conditions were particularly severe in the Heigham district, where many of the boot-and-shoe operatives supplemented their earnings by breeding thousands of canaries in specially constructed sheds in their back gardens. Hundreds of aviaries were swept away, and entire strains of birds wiped out.[5]

Whatever their precise cause and extent, such natural disasters posed the question as to how far the use and management of land and water should be modified as an insurance against a recurrence of such events. Such disasters often highlighted the heavy social responsibilities borne by the civil engineer. River and sea defences, dams, bridges and other such major structures could bring major economic and social benefits, where well constructed. If deficient, they might result in the most serious of man-made catastrophes. Numerous difficulties were encountered in the construction of reservoirs in the faulted gritstone valleys above Huddersfield, for example. In the construction of the earth embankment for the Bilberry reservoir in the Holme valley, a spring emerged from the centre of the puddled-bank. Remedial work was ineffective and in 1852, the second wettest year of the century, and after heavy rains, the embankment gave way, releasing an estimated 86.2 million gallons of water. Eighty-one people perished, and the village of Holmfirth was almost completely destroyed.[6]

The loss of the dam above Holmfirth was no isolated incident – there were other instances where human error turned achievement into humiliation. Each prompted doubts as to the omnipotence and wisdom of human action. It was not simply a case of retribution for embarking on a course that was so patently misguided and wrong. Even the most praiseworthy of motives, such as the greater provision of water supplies to

[3] T. R. A. Briggs, *Flora of Plymouth* (London, 1880), pp. 79–80.

[4] C. Carter, *The Blizzard of 79* (Newton Abbot, 1971), pp. 11–15.

[5] H. H. Goose, *Norwich under Water 1878 and 1912* (Norwich, 1912).

[6] R. A. Buchanan, *The Engineers: a History of the Engineering Profession in Britain, 1750–1914* (London, 1989); H. J. Morehouse, *The History and Topography of the Parish of Kirkburton* (Huddersfield, 1861, repr. 1984), p. 229; T. W. Woodhead, *History of the Huddersfield Water Supplies* (Huddersfield, 1939).

urban populations, could have unforeseen consequences. The technological and social advances commonly associated with the industrial and urban revolutions of the nineteenth century conferred on Victorian and Edwardian England and Wales an unprecedented freedom of choice, but with that freedom went responsibility. In coming to a judgment as to the ways in which that choice was exercised, contemporary observers could make their own assessments through studying the wealth of statistical, cartographic and other published information becoming available, and by making visits to town and countryside themselves, taking full advantage of the railways and latterly the bicycle and motor car.

The broad canvas of the book will now be clear: the nature of countryside change as seen in the context of other pertinent changes – economic, technological, social and political. In our account, we too have drawn heavily on the fact that so many of these changes were measured during the years in question – in the form of maps and tables of statistics. The remainder of this chapter focuses on these records and the remarkable opportunities for writing and photography which the new forms of travel encouraged.

The next two chapters set the scene. First, there is an overview of those changes in town and country which had a particular impact on the countryside: agriculture, rural industry, rural housing, the consequences of urban growth, and other urban land-demands. This is followed by an explanatory account of the dynamics of change: the motors of economic, social and land-use transformation.

We then take note of the expert guidance given to those who lived and worked within, and visited, the countryside, particularly from the perspectives of the geologist, natural historian and antiquarian. Drawing heavily on their observations, four chapters then describe four very different expressions of the countryside: woodlands, lowlands, uplands, and the coast and rivers. A broad synoptic frame is filled with a number of cameo scenes.

We conclude with a backward and forward look. The threads are drawn together to emphasise a countryside increasingly at risk from the demands of urban man and the impact of urban circumstances. The period between the Great Exhibition and the Great War may have been no pivotal period in rural history; rather we may see it as one in which both long-standing trends were sustained and new issues arose, which were only worked out in subsequent years. Yet considerable change did take place; a set of conditions was established, from which there was to be no return.

A. THE MEASUREMENT OF CHANGE

In his *History of England*, published in 1848, Lord Macaulay commented on how it was no one's business to collect information on 'the produce of the English soil'.[7] That viewpoint was, however, beginning to change. Historians have made much of the Victorians' predilection for gathering data, particularly in statistical form. Collections of facts were increasingly published in the form of Blue Books or lectures, or were displayed, say, as museum collections. The Museum of Practical Geology was founded in 1851, and pressure on space caused the natural-history exhibits of the British Museum to be transferred to new premises in South Kensington in 1880. However presented, the various assemblages of facts were more than inert entities. To the Victorian mind, a fact was something achieved – the result of laborious and patient investigation.[8]

How far would this outpouring of effort provide any kind of detailed and comprehensive picture of the countryside? How successful were the Victorians and Edwardians in measuring and mapping the changes taking place? We will look in turn at the two major sources of statistical evidence, the progress made in mapping the countryside, and conclude by reference to the most innovative form of record-keeping, namely the photograph.

i. Agricultural statistics

The most obvious advance in the collection of rural statistics was the introduction of a census of agricultural crops and livestock in every parish, held on an annual basis from 1866 onwards. No other type of land use and farm management was so closely monitored. This was not, however, the original purpose for mounting the census. In the protracted debate that led up to the first collection, it was assumed that the overriding need would be not so much to discover how the land was used, but rather to find out how much food would be produced in the ensuing harvest. If available in time, the statistics would make it easier to meet market needs.

The scope and character of the census which emerged in 1866 reflected the muddle and confusion that surrounded the search for a means of collecting statistics and for allaying the suspicions and resentment of many landowners and tenants. Encouraged by experience in Scotland and Ireland, experiments were mounted in 1853–4, which sought to collect information from holdings, initially in parts of Hampshire and Norfolk, and then from eleven counties, using the Board of Guardians as the collecting agency. The response to the more extensive survey proved extremely disappointing. Many landowners and tenants regarded the census as 'inquisitorial and invidious in its nature and mischievous in its

---

[7] Lord Macaulay, *The History of England from the Accession of James the Second* (London, 1858), vol. I, pp. 280–2.

[8] W. T. Stearn, *The Natural History Museum of South Kensington* (London, 1981), pp. 41–54; G. Lewis, *For Instruction and Recreation. A Centenary History of the Museum Association* (London, 1989), pp. 1–26.

tendencies'. Furthermore, a considerable amount of work was devolved on the Boards of Guardians. The Boards for the Basingstoke and Hartley Witney Unions in Hampshire had refused to take part, protesting that their duties were already 'sufficiently onerous'.[9]

Whilst a Select Committee of the House of Lords supported the collection of three types of data, namely for the acreage under different crops, the number of livestock and estimates of produce to be harvested, a Bill to implement these recommendations was defeated at the second reading in the Commons by 241 to 135 votes. It seemed very unlikely that even farmers and dealers would find the census of practical use – its only value might be to the historian in fifty years' time.[10]

No further progress was made until 1864, when a resolution was moved by James Caird in the House of Commons, calling for 'the collection and early publication of the Agricultural Statistics of Great Britain'. One of the more perceptive speakers in the debate maintained that it was an illusion to believe that the returns would enable precise estimates to be made of the prospective yield of crops. All they could be expected to provide was 'an accurate amount of the various descriptions of ground under different crops every year, and the quantity of livestock, so that they might trace the increase and decrease of the agricultural wealth of the country'. Nearly every other European nation, and even America, had this information. The resolution was approved by seventy-four to sixty-two votes.[11]

In March 1865, the President of the Board of Trade announced that the 'acreage under cultivation of different crops might be ascertained by voluntary Returns and other means'. No legislation would be needed, and an allocation of £10,000 was approved. Events took an unexpected turn when a severe cattle plague broke out and, in October 1865, the secretary of the Royal Commission, appointed to investigate its origins and character, wrote to the President of the Board of Trade, stressing 'the importance of obtaining correct information respecting the number of horned cattle and sheep existing in the country'. The availability of such information in the Netherlands made it possible to ascertain not only the numbers of stock to die, but the proportionate loss in every commune. As a second leader in *The Times* recalled, it was the cattle plague that was responsible for 'getting rid of some of the unreasonable prejudice against returns'. In view of the frightening losses of cattle, and the consequent need to award compensation and remedy deficiencies in the supply of

---

[9] *Agricultural Statistics (England)*, BPP, 1854–5, LIII; Hampshire RO, PI III, 5/9; *Parliamentary Debates* (=PD), 3rd ser., CXXXVIII, 1782–3; P. Dodd, 'The agricultural statistics for 1854: an assessment of their value', *AHR* 35, 1987, pp. 159–70.

[10] Select committee, Lords, *Agricultural Statistics*, BPP, 1854–5, VIII; PD, Commons, 3rd ser., CXLIX, 1871–1919.   [11] PD, Commons, 3rd ser., CLXXV, 1362–83.

meat to the markets, it seemed only reasonable that the Government should require statistics.[12]

The census of livestock was taken in March 1866, and the results published in the following May. On the 25 June, the areas under different crops were recorded for holdings of five or more acres. When received, these returns were consolidated into parish summaries, and the original schedules destroyed. From 1867 onwards, the livestock and acreage figures were collected on the same day (June from 1877 onwards). Although responsibility for the Returns was transferred to the newly constituted Board of Agriculture in 1889, the staff of the Board of Customs and Excise continued to distribute and collect the schedules until 1919.[13]

The number of items on the form rose from twenty-five in 1866 to fifty by 1919. Data were sought for the first time on the areas under orchards, market gardens and sugar beet in 1871, 1872 and 1873 respectively. The completion of a return was not made obligatory until the Corn Production Acts of the First World War. Where the occupier declined to make one, the excise officer or local assessor of taxes submitted an estimate, after seeking the help of local magistrates, clergy, land agents and others. By 1907, however, the proportion of non-co-operative occupiers had fallen to only 2.4 per cent of the total. Of much more serious concern was the lack of consistency in completing the forms. In 1869, the wording of the schedules was changed so that it might be taken 'to mean the same thing by farmers in all parts of England, Wales and Scotland'. The decision in 1907 to print the schedules in both English and Welsh in those parts where Welsh was the more familiar language was thought to have increased the accuracy of the returns and facilitated their collection.[14]

In the event, the principal value of the Annual Returns proved to be historical. The area recorded as crop and grassland in England and Wales reached its peak of 28 million acres in 1891, and thereafter fell by 3 per cent over the next two decades – a decline being recorded in every year. The Board of Agriculture attributed the trend to two factors. The withdrawal of farming from the least profitable land, in response to economic pressures, may have been the more important reason for the decline of the 1890s, when the area under crop and grassland fell by 463,000 acres. The further decline of 268,000 acres between 1901 and 1911 was thought to be mainly due to

---

[12] PD, Commons, 3rd ser., CLXXVIII, 560, and CLXXIX, 1312; *RC on the Origin and Nature of the Cattle Plague, First Report*, BPP, 1866, XXII, p. 180; *The Times*, 18 June 1867, p. 11.

[13] J. T. Coppock, 'The statistical assessment of British agriculture', *AHR* 4, 1956, pp. 3–21, 66–79.

[14] *Agricultural Returns of Great Britain*, BPP, 1867–8, LXX, pp. 3–9; 1868–9, LXII, C. 4200, p. 4; 1880, LXXVI, C. 2727, p. 3; 1886, LXX, C. 4847, p. 2; 1908, CXXI, Cd 3870, p. 6.

the extension of the towns and the demands for more and more of the surface of the country for the residential, manufacturing, mining, railway and other purposes requisite to an ever-growing population.

As the Board commented, it was impossible to calculate with any degree of accuracy the rate at which towns encroached upon the countryside, owing to the difficulties of defining the area that might be called urban.[15]

The most striking feature to be recorded within the body of agricultural land use was the conversion of 4 million acres of arable land to grass in England and Wales between the early 1870s and the First World War. On the premise that these 'new' grasslands would support cereals, and were likely to be less valuable for livestock husbandry than those established before 1870, the aim of the ploughing-up campaign of 1917–18 was 'Back to the Seventies – and Better'. The county and district ploughing quotas were based largely on the decline in area of cereals and other arable crops, as recorded in the Returns (Figure 27.1).[16]

ii. Population statistics

An even longer continuity was afforded by population censuses, which were held at regular ten-yearly intervals after 1801. They were complementary to parish registers and (from 1837) the civil registers which recorded births, deaths and marriages. The earlier censuses of 1801–31 were largely single population enumerations. From 1841, the census was undertaken by the Office of the Registrar General, and recorded the names and members of households resident at specific addresses, and particulars in respect of each individual regarding marital status, age and sex, occupation, birthplace and infirmity. The 1911 census saw a number of important additions to the enumeration, with questions on the duration of marriage and the number of children born to each marriage, on industry as well as occupation, and on the number of rooms per dwelling.[17]

An early difficulty was that the units of census organisation and registration, the registration districts, did not themselves distinguish between urban and rural in tabulations. The creation of urban and rural sanitary districts in 1872, the further adjustment of muncipal boroughs and urban sanitary districts in 1884, and the separation of 'urban' and 'country' areas

---

[15] *Agricultural Statistics*, BPP, 1912, CVI, p. 7.

[16] J. Sheail, 'Changes in the use and management of farmland in England and Wales', *TIBG* 60, 1973, pp. 17–32.

[17] R. Lawton, 'Urbanization and population change in nineteenth-century England', in *The Expanding City*, ed. J. Patten, (London, 1983), pp. 179–224; R. Lawton, *The Census and Social Structure: an Interpretive Guide to Nineteenth Century Censuses for England and Wales* (London, 1978); R. Lawton, 'Population changes in England and Wales in the later nineteenth century', *TIBG*, 44, 1968, pp. 55–74; R. Lawton, 'Rural depopulation in nineteenth Century England', in *Liverpool Essays in Geography*, ed. R. W. Steel and R. Lawton (Liverpool, 1967), pp. 227–55.

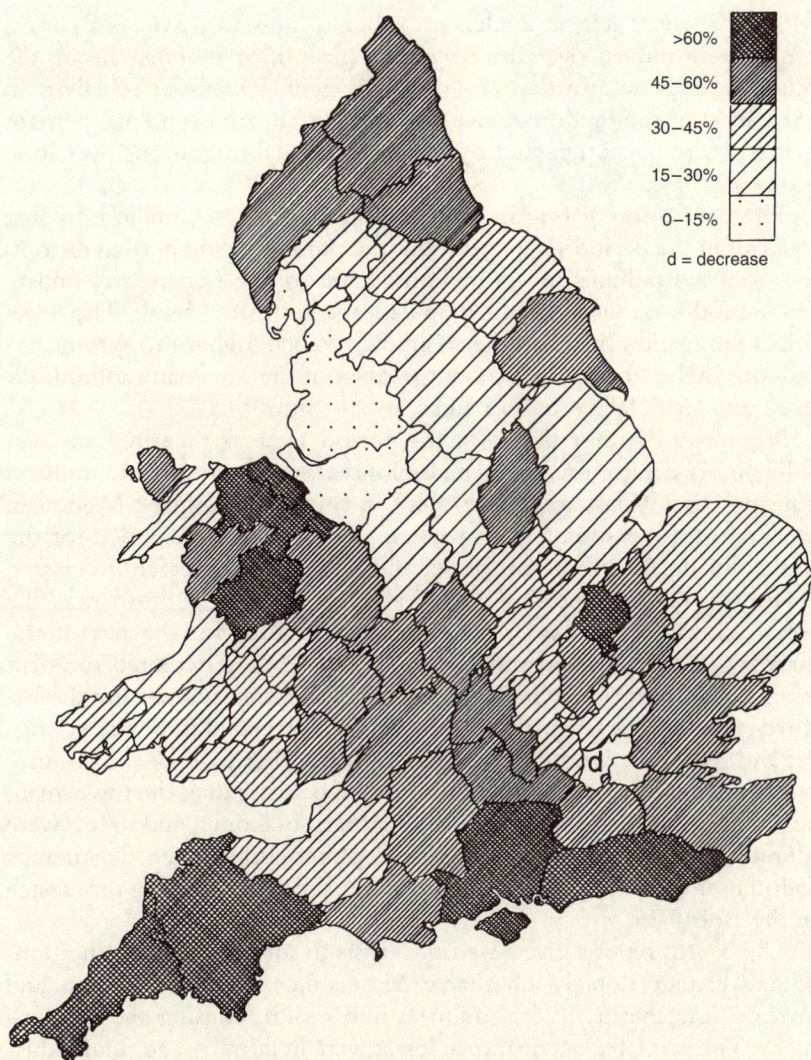

27.1 Increase in the area of permanent pasture, or grass not broken up in rotation, as recorded in the Agricultural Returns of 1870 and 1916 (BPP, 1870, LXVIII and 1917–18, XXXVI, Cd 8436).

as part of the local-government reforms of 1889, helped to clarify the situation, but the statistical problem remained, namely of how to define precisely urban and rural populations.

Whilst the census-based estimates of migration were comprehensive, there were important limitations to their accuracy and utility. The county was the smallest geographical area adopted, and migration within a county was likely far to exceed that across county boundaries.

Moreover, there was no evidence to suggest how many times a person might have moved over the decennial period, or whether he or she might have gone away, and come back again. Ravenstein's analysis in 1885 showed the predominance of short-distance movements, perhaps in two stages, accompanied by more occasional migrations over long distances.[18]

The population of England and Wales rose from 17.9 millions to 36.1 millions in the period 1851–1911, but the number living in rural districts fell from 8.9 millions to 7.9 millions. Even so, the figures may underrepresent the extent of rural depopulation and urban growth. The 'true' urban population may have been almost 700,000 higher in 1851 and 2.3 millions higher in 1911, if account is taken of the way many administrative boundaries failed to distinguish 'urban' from 'rural'.[19]

Net losses by migration over the period 1841–1911 were considerable: 79 per cent of calculated natural increase in northern and midland counties and Wales, and 89 per cent in southern counties. Moreover, the drain was persistent, with particularly high losses recorded for the south during the whole of the second half of the nineteenth century. In the north, the midlands and Wales, the decade 1881–91 was the worst, although this was a far better record than for the rest of the country in any decade from 1851 to 1901.[20] Many deplored the drift from the countryside, which was particularly noticeable during the last thirty years of the century, not least because it tended to be the youngest and most able who migrated, not just to other parts of the country but overseas. The countryman was held to be fitter than the townsman; certainly urban death rates were higher than rural ones, and so too were infant mortality rates. Those who deplored urban race degeneracy found useful data in the relative unfitness of recruits enlisting for service in the Boer War.

The census reports give a statistical basis to some of the key questions of rural change, notably suburban growth and population migration, and provide a starting point for enquiries into health, housing and employment. The principal occupational losses were in farming and rural industries. In that sense, migration was a market response to changing economic conditions and, without it, rural unemployment would have been a striking feature. The precise extent of the loss is not, however, entirely clear. Farm workers might describe themselves as general labourers, but there was certainly a reduction in demand for casual labour and harvest gangs. As a proportion of the total occupied labour force, the shift was nevertheless dramatic: in 1851, more than one person in five (21.5

[18] E. G. Ravenstein, 'The laws of migration', *JRSS* 48, 1885, pp. 167–235.

[19] Lawton, 'Rural depopulation'.

[20] W. A. Armstrong, 'The flight from the land', in *The Victorian Countryside*, ed. G. E. Mingay (London, 1981), vol. 1, pp. 118–35.

POPULATION CHANGE
1851–1911

0   80
Kms

INCREASE

▨ Net in-migration exceeds
natural gain

▨ Natural gain exceeds
net in-migration

▤ Natural gain exceeds
net out-migration

☐ DECREASE

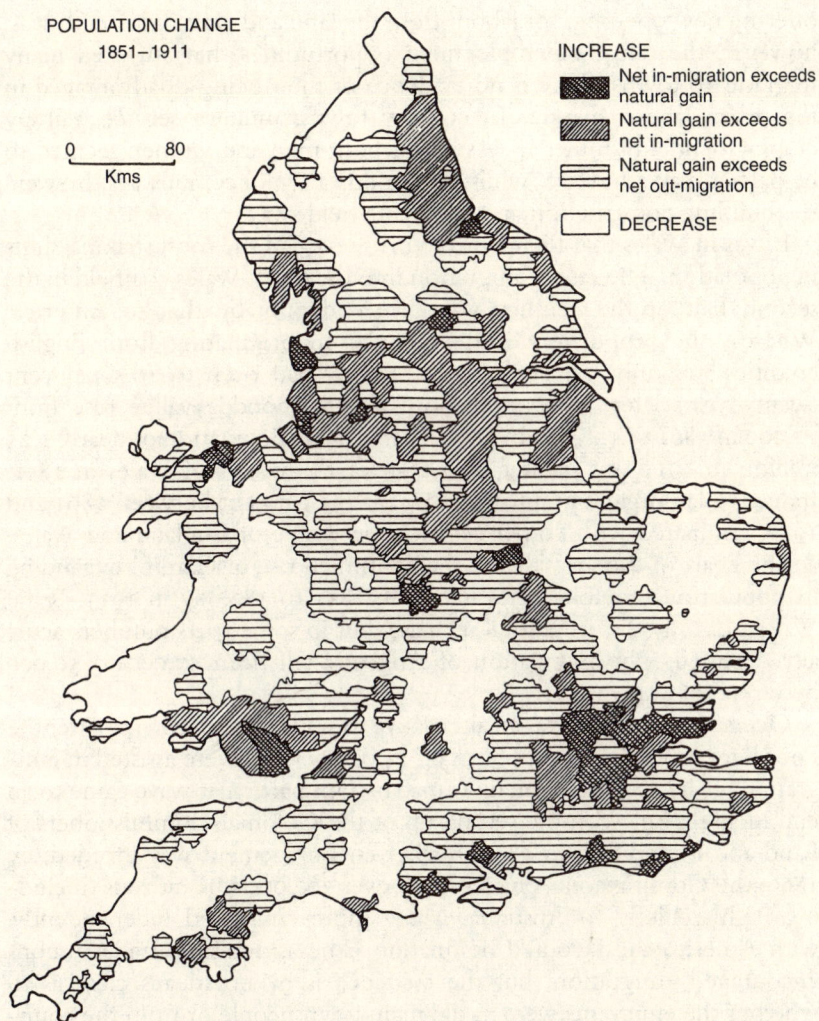

27.2 Population change 1851–1901, based on the sum of decadal natural and migrational changes in each Registration District (from R. L. Lawton, 'Nineteenth century population changes', *TIBG*, 44, 1968, p. 67).

per cent) was engaged on the land, whereas in 1911 the ratio was less than one in eleven (8.5 per cent).[21]

Many rural crafts and small industries were transferred to urban locations, accelerating the outflow of rural population. But there was some interchange of job opportunity, with rural railways and the police force

[21] J. T. Coppock, 'The changing face of England: 1850–*circa* 1900', in *A New Historical Geography of England*, ed. H. C. Darby (Cambridge, 1973), pp. 596–602.

offering new openings for labour from the land and craft industry. It was, however, the range of employment opportunities that attracted many migrants to towns; there is no evidence of their being disadvantaged in the urban labour market. In country towns, military service, railway employment and domestic service for both men and women seemed to be particularly attractive, whilst in London the police, railways, breweries, building construction and labouring held sway.

It was in Wales that the census returns revealed the most striking shifts in population. The rate of migration into the south Wales coalfield in the second half of the century was exceeded only by that to America. Whereas the proportion of the population originating from English counties was only 9.6 per cent in 1871, it had risen to 16.5 per cent twenty years later. The population of the Rhondda valley rose from 12,000 in 1861 to 128,000 in 1891. Glamorgan's population of nearly 1.25 million in 1911 was greater than that for the whole of Wales in 1851. Indeed, Glamorgan's population rose by 253 per cent between 1861 and 1911, compared with 80 per cent for the whole of England and Wales. At the heart of the growth was the continued rise of Cardiff, expanding its population ninefold from 20,000 in 1851 to 182,000 in 1911. Rural Wales not only lost its natural increase, but in some parts numbers actually declined. The population of Anglesey fell from 57,000 to 50,000 between 1851 and 1911.[22]

Overseas migration was an alternative to settlement in towns. After the Poor Law Amendment Act of 1834, 14,000 persons were assisted in emigrating to the colonies (chiefly Canada). This particular wave came to an end in 1846, but with the setting up of the Colonial Commissioners of Land and Emigration in 1842, further encouragement was afforded; by 1869, the Commissioners had assisted over 300,000 UK citizens (including Irish), chiefly to Australia. Many more emigrated independently, with America the favoured destination. The agricultural trades unions encouraged emigration, but the evidence is by no means clear as to whether the emigrants were in the main townspeople or from the countryside, although there can be no doubt that the receiving countries preferred the latter.

iii. The mapping of town and country

The origins of the Ordnance Survey lay in the later years of the eighteenth century. The primary triangulation was almost completed by 1845, and maps were published at the 1-inch scale (1:63,660) for all of England

---

[22] G. E. Jones, 'Wales 1880–1914', in *Wales, 1880–1914*, ed. T. Herbert and G. E. Jones (Cardiff, 1989), pp. 1–10.

and Wales as far north as a line between Hull and Preston. A record total of 28,000 copies of the ninety English sheets was sold. On the Cornish Lizard, the debate as to the most southerly point was finally settled.[23] Meanwhile, surveys at the 6-inch scale (1:10,560) had begun in Lancashire, Yorkshire and some Scottish counties.

It was far from easy to decide at what scale surveys should be carried out. The most cogent arguments for a scale of 25 inches to the mile (1:2,500) were put forward in a report by the Commission on Registration and Conveyancing in 1850. In 1853, the Ordnance Survey embarked on a survey of County Durham at the 25-inch scale and, in 1856, recommended that the whole country should be mapped at that scale, with the exception of uncultivated districts where 6-inch maps would be adequate. The case for a national large-scale survey was accepted by the Royal Commission on the Ordnance Survey in 1858, and, five years later, the Treasury agreed to the surveys being extended to the south of England.[24]

It was no coincidence that these decisions were taken at the same time as the Land Registry Act of 1862 introduced registration of title into England. The compilation of 'official cadastral maps' became the primary justification for the large-scale surveys. Each parcel of land within a parish was identified on the maps by a unique number and, until 1873, each 25-inch sheet was published as a parish map, with the area beyond the parish boundaries left blank within the sheet's rectangular frame. It was not only the staff of HM Land Registry who found the cadastral survey invaluable. The Director-General recounted in 1863 how a nobleman in Northumberland discovered from the Ordnance Survey maps that his estate was 10,000 acres smaller than he had supposed; elsewhere Irish reapers in Westmorland used the maps to establish the size of the fields which they had been engaged to reap.[25]

Responsibility for the Ordnance Survey passed from the Board of Ordnance to the War Office in 1855, and then in turn to the Board of Works in 1870 and to the Board of Agriculture in 1890. The published maps at a range of scales provided not merely a comprehensive topographical record of unprecedented detail and accuracy, but a national inventory of land use. From the data recorded on the 25-inch maps, it was possible to discover for the first time the area of *every* parcel of land, parish and other type of administrative unit. Where appropriate, landscape features

[23] D. Smith, *Victorian Maps of the British Isles* (London, 1985); C. A. Johns, *A Week on the Lizard* (London, 1848), p. 67.

[24] J. B. Harley, *The Historian's Guide to Ordnance Survey Maps* (London, 1964); W. A. Seymour, *A History of the Ordnance Survey* (London, 1980).

[25] Topographical Department, War Office, *Report of the Progress of the Ordnance Survey and Topographical Depot*, BPP, 1865, XXXII, p. 3.

were drawn to scale. Symbols were used to help identify such features as woodland, marsh, rough pasture, quarries, and sand, gravel and clay pits. Until 1879–80, the Ordnance Survey followed the practice of many earlier estate and parish plans, and the Tithe Commutation Survey, of compiling schedules that supplemented the information on the maps. Books of reference were published for each parish surveyed, listing the parcel numbers as given on the maps, together with the acreage of each parcel and 'remarks' on its state of cultivation. As well as corroborating much of what could be deduced from the maps, the books distinguished arable from pasture. By the time the practice ceased, about a quarter of England, Wales and Lowland Scotland was covered.

At times, progress was far from smooth. For some landowners, the right of surveyors to enter private property, and record and publish whatever they found, was a further instance of state intrusion. More seriously, doubt was at times expressed as to the accuracy of what had been recorded. In his vegetation surveys of the Pennines, and of the Bath and Bridgwater district of Somerset, Moss found inconsistencies in the distinction drawn between farmland and 'primitive pasture', even on the same individual Ordnance Survey 6-inch quarter-sheets. It was never easy to define the boundary between cultivated and non-cultivated land, especially where regional farming practices favoured a cycle of rotation grasses rather than permanent grassland. Such descriptions as 'clover fallow', 'space or enclosure', and 'mountain grass land', in some of the earlier books of reference, may have reflected the quandary of Ordnance Survey personnel, unfamiliar with local customs and faced with assigning 'a state of cultivation' to particular parcels of land.[26]

Decisions to stop publishing the parish books of reference and to reduce the amount of vegetation detail on the maps after 1880 reflected a growing feeling that the only way to speed up progress in completing the survey for the whole country was to dispense with all that was not strictly necessary for a cadastral survey. By the 1890s, the 'elaborate arboreal detail representation' had come under attack as 'a waste of public money for a cartographic superfluity'. A large majority of respondents to a circular distributed by a departmental committee in 1893 believed hedgerow trees, single trees, and trees and shrubs around houses might be 'better omitted altogether'. The outcome was a further round of 'cartographic disafforestation' in rural, suburban and town areas alike, on both the 25- and 6-inch maps.

It was a gigantic publishing venture. For Great Britain as a whole,

[26] C. E. Moss, *Geographical Distribution of Vegetation in Somerset: Bath and Bridgwater District* (London, 1907), p. 2; C. E. Moss, *Vegetation of the Peak District* (Cambridge, 1913), Preface; J. B. Harley, 'The Ordnance Survey and land-use mapping', in Historical Geography Research Group, Research Paper, 2, 1979.

51,000 25-inch sheets had been published by the end of the century, together with 15,000 6-inch sheets that covered every acre of the Kingdom. Although their largeness of scale had fully met the purposes for which they were intended, it also contributed to their becoming rapidly out of date. By the time the initial 25-inch survey was completed in 1893, fresh surveys were already underway, the aim being to avoid any sheet being more than twenty years old. The average age of a sheet was to be ten years. The first revision kept to this objective, and was completed in Yorkshire in 1914. A second national revision was begun in 1904.

As was intended, the maps were used for a variety of purposes as base maps. No longer was it necessary for landowners to commission fresh surveys – improvement schemes, land use and the extent of different tenures could be plotted on the 6- and 25-inch maps. Commercial publishers used the Ordnance Survey 1-inch maps as the basis for their many small-scale maps, adding whatever additional embellishment or detail was required. So long as the published maps were at a different scale, no questions were raised by the Ordnance Survey as to copyright or royalties. The most outstanding examples of such maps were those of John Bartholomew and Son of Edinburgh, whose series of half-inch maps were admired for both their beauty and utility. They covered the whole of the British Isles by 1903.[27]

Bartholomew made full use of the recent advances in colour printing to resolve one of the most difficult challenges in cartography, the depiction of relief. No matter how well executed, hachuring alone could not give a precise impression of steepness. While contour lines went a long way to providing that precision, and gave a sense of steepness where the contours were close together, most map users found it difficult to gain an overall impression of the changes in relief. As an alternative to using some combination of hachures and contours, Bartholomew introduced the device of 'contour layer colouring'. The earliest examples appeared in Baddeley's guide book, *The Thorough Guide to the English Lake District*, published in 1880. A range of green and brown tints indicated 'the contours of altitude at intervals of 500 feet', from sea level up to 3,000 feet, above which the few peaks were shown in white.[28]

iv. The photograph

The revolution in communications, whether in the form of statistics, maps, newspapers or books, had the effect of highlighting how much

---

[27] L. Gardiner, *Bartholomew 150 Years* (Edinburgh, 1976), pp. 25–6.
[28] M. J. B. Baddeley, *The Thorough Guide to the English Lake District* (London, 1880).

there was still to learn about the countryside even closest at hand. After 1870, reappraisal in education sought to make children observant, inquisitive and, at length, more thoughtful. In opening their eyes to the wonder and excitement about them, a powerful part could be played by photography. William Jerome Harrison, as Science Demonstrator to the School Board for Birmingham, was among the first to recognise that potential. Through his method of 'direct learning', Harrison attached great importance to scientific demonstrations and field excursions. The Optical Lantern offered a substitute reality. From the image projected in the classroom, pupils could believe they were 'viewing the original scene' at first hand.[29]

Despite the invention by William Henry Fox Talbot of a practical negative/positive process, it was not until the late 1850s that photography became widely available, and open to commercial development. In 1851, Frederick Scott Archer offered his far quicker and more reliable collodion, or wet-plate process, freely without patent.

Generations of students of the countryside had sought to convey their impressions and discoveries in the form of text, drawings and paintings, frequently focusing on the picturesque and exceptional. Despite the cumbersome equipment and 'colour-blind' plates, photography introduced a new and exciting dimension – one that could be pursued and enjoyed on an individual basis and in the company of others.[30] Of the many whose experiences might be cited, Joseph Gale took up photography in 1859. He had spent many holidays sketching old buildings, but it was so much quicker using photography. A keen naturalist and fisherman, some seven years later he joined the Amateur Photographic Field Club, whose members shared a common interest in 'mead and stream' photography. In Huddersfield in 1893, an amalgamation took place between the local photographic society and the Naturalists' Society, which had been founded in 1850 and prided itself on being the oldest provincial natural history society in the country. As the president of the unified body remarked a few years later, photography had become, over as short a period as twenty years, the second most popular pastime to cycling.[31]

The vision conveyed by the mid-Victorian photographer was one of an ideal, attractive and unchanging countryside. Picturesque ruins of abbeys and castles were obvious subjects, whether presented in Russell

[29] L. C. Miall, *Thirty years of Teaching* (London, 1897), pp. 209–25; P. James, 'William Jerome Harrison, Sir Benjamin Stone and the photographic record and survey movement', MA dissertation, Birmingham Polytechnic, 1989.

[30] H. Gernsham, *The Rise of Photography, 1850–1880* (London, 1988).

[31] B. Coe, *A Victorian Country Album. The Photographs of Joseph Gale* (Oxford, 1988); G. T. Porritt, *Our Society and its Work* (Huddersfield, 1898).

Sedgfield's ambitious part-series, *Photographic Delineations of the Scenery, Architecture and Antiquities of Great Britain and Ireland*, or the cheaper volume by William and Mary Howitt, which included twenty-five prints of abbeys and castles, taken by leading photographers. The reader was no longer left to the mercy of imagination, or the caprices and deficiencies of artists. As Sedgfield wrote in his prospectus, the occasional sacrifice of pictorial effect was more than compensated for by the truthfulness of every detail in the prints.[32]

In that context, the introduction of the more convenient and still faster gelatine dry-plate in the 1880s held out exciting possibilities. There appeared in 1895 a book hailed by contemporary reviewers as the beginning of a new era in natural-history publishing. The book, on British birds' nests, was illustrated by photographs taken by Cherry Kearton. However faithful the woodcuts and coloured pictures of earlier texts, Kearton's photographs were the first to show the nests, eggs and birds *in situ*. It was as if the reader was actually taking part in a birdnesting expedition. A further book, *With Nature and Camera*, contained 180 photographs, and recounted how they were obtained. Intended as a guide and incentive for others to follow, his brother, Richard Kearton, wrote of 'the gratification of having sent hosts of amateur photographers into the fields to study wild life for themselves'.[33]

The more manageable dry-plate process also made it easier for photographers, such as Frank Meadow Sutcliffe and Peter Henry Emerson, to adopt a fresher and more natural approach to photography as an art-form. Sutcliffe was of the first generation to use the camera, rather than a sketching-book, to capture the natural and transitory visual effects on town and countryside, as they changed seasonally or minute by minute. In order to convey those atmospheric effects, none of his prints was 'straight' – most involved some enhancement or holding back of a shadow area, and control of image contrast and tonal depth. If the forceful style of Sutcliffe's photographs had the effect of turning Whitby, on the north Yorkshire coast, into 'the Photographers' Mecca', Emerson focused attention on the Norfolk countryside.[34]

Only some five years after taking up the camera, and in close collaboration with Thomas F. Goodall, a little known but highly articulate

[32] G. Seiberling, *Amateurs, Photography and the Mid-Victorian Imagination* (Chicago, 1986); Russell Sedgfield, *Photographic Delineation of the Scenery, Architecture and Antiquities of Great Britain and Ireland* (London, 1854); W. and M. Howitt, *Ruined Abbeys and Castles of Great Britain* (London, 1862).

[33] R. Kearton, *British Birds' Nests. How, Where, and When to Find and Identify Them* (London, 1895); R. Kearton, *With Nature and a Camera* (London, 1898), pp. vii–viii.

[34] B. E. C. Howarth-Loomes, *Victorian Photography* (London, 1974); B. E. Shaw, *Frank Sutcliffe. A Second Selection* (Whitby, 1978); M. Hiley, *Frank Sutcliffe – Photographer of Whitby* (London, 1974).

painter of the 'naturalistic school', a portfolio was published on the *Life and Landscape on the Norfolk Broads*. The honest and straightforward approach broke free from convention as to what constituted an art-form in photography. The forty plates were printed from untouched negatives. Much less consciously, Emerson introduced a documentary element. Both in this and in later works, each plate was accompanied by a descriptive text, expressing genuine sympathy for the lifestyle of the Broadland dwellers and their desolate landscape. In that sense, Emerson may be regarded as the first 'concerned photographer' of rural life and the countryside.[35]

There was no shortage of examples of what might be recorded by the potentially vast army of information gatherers. Philip Henry Delamotte, the Professor of Drawing at King's College London, was commissioned to record each week the progress made in re-erecting the Crystal Palace on the new site at Sydenham. Between 1875 and 1886, the Society for Photographing the Relics of Old London issued to subscribers 120 photographs of picturesque or historic buildings threatened with demolition. In 1889, a committee was appointed by the Geological Section of the British Association for the Advancement of Science for the 'collection, preservation and systematic registration of photographs of geological interest in the United Kingdom'. Within five years, a collection of some 1,200 photographs had been deposited in the library of the Geological Survey, the number rising to 5,500 prints by 1919.[36]

It was not long before photographs of towns, villages, and the countryside and coast were being sold commercially in large numbers for a few pence. A large proportion was taken by Francis Frith, whose ambition was to build up a complete photographic record of the British Isles. Sold in over 2,000 shops, 'Frith's postcards' became almost a household name. They could be purchased in local post offices, whether to be sent to friends and relatives, or to join the topographical prints and engravings found in many homes. Their success reflected not only the increased use of the post, but a sense of nostalgia among the now predominantly urban population for the peaceful life of green fields and leafy lanes – examples of which abounded in Frith's collection.[37]

---

[35] P. H. Emerson and T. F. Goodall, *Life and Landscape on the Norfolk Broad* (London, 1886); P. Turner and R. Wood, *P. H. Emerson. Photographer of Norfolk* (London, 1974).

[36] G. Winter, *A Country Camera, 1844–1914* (Newton Abbot, 1966); J. Brown and S. B. Ward, *Village Life in England, 1860–1940. A Photographic Record* (London, 1985); P. James, 'A century of survey photography', *Local Historian*, 20, 1990, pp. 166–72; G. Bush, *Old London. Photographs by Henry Dixon and Alfred and John Bool for the Society for Photographing Relics of Old London* (London, 1975); G. McKenna, 'The geological photographs of the BAAS', *Geology Today*, 6, 1990, pp. 157–9.

[37] D. Wilson, *Francis Frith's Travels. A Photographic Journey through Victorian Britain* (London, 1985); S. F. Kelly, *Victorian Lakeland Photographers* (Shrewsbury, 1991).

B. THE EXPERIENCE OF CHANGE

The combination of a national coverage of large-scale maps, an annual agricultural census, and decennial population censuses of a much more detailed kind, was enough to set the late nineteenth century apart from any period that had gone before. It represented an explosion of knowledge on the world in which everyone lived and worked. But that was not all. Through the revolution in transport technology that was taking place – first in the form of the railway, and later the bicycle and the motor car and motorcycle – it was becoming easier to see for oneself what was happening.

i. The railway

The railway was the great connecter, turning what might once have been a major expedition into a comparatively simple journey.[38] The first maps to cater explicitly for the railway traveller were those published by Cruchley from 1855 onwards. The half-inch maps, with the railways standing out boldly, were described as 'a complete guide to the student of history, the lover of antiquarian research, the amateur pedestrian, or the commercialist'. Between 1858 and 1862, the *Weekly Dispatch* newspaper first gave away and then sold travellers' maps by mail order. Their circulation increased greatly when the publishers, Cassell, Petter and Galpin, acquired the plates and stock in 1863, and sold the county maps for as little as 4d. each. Their cover design showed railway passengers watching the passing scene from the windows, and walkers resting by the wayside.[39]

The heroic period of railway building ended about 1852. Route mileage had risen from 97 miles to 1,497 miles during the 1830s. The first tentative steps gave way to a 'more confident routine' and, between 1844 and 1848, 720 railway Bills were authorised. Although not all projects were implemented, over 7,500 miles were in use by 1852, forming the basis of what was to become the future system of trunk routes and their more important branch-lines. In comparison, the 1850s were a comparatively lean period. Another great burst of activity was to follow in the 1860s, which ended equally dramatically in the financial crisis of 1866.[40]

The engineer, F. R. Conder, wrote in 1868 of how 'the profession of the Civil Engineer is under a cloud', paradoxically because so much had

---

[38] J. Simmons, *The Victorian Railways* (London, 1991); H. Perkin, *The Age of the Railway* (London, 1970); A. and E. Jordan, *Away for the Day. The Railway Excursion in Britain* (Kettering, 1991).

[39] T. R. Nicholson, *Wheels on the Road. Maps of Britain for the Cyclist and Motorist, 1870–1940* (Norwich, 1983).

[40] H. J. Dyos and D. H. Aldcroft, *British Transport* (Leicester, 1969), pp. 125–9.

been achieved. He had 'covered the islands with noble viaducts and lofty embankments, furrowed it with excavations, and pierced it with tunnels', much of it ahead of 'the wants of the day' since the traffic was still largely underdeveloped.[41] The lines were not as immutable as expected. Many cut-offs or short-cuts were made, the most famous examples being on the Great Western Railway (dubbed by some the Great Way Round). Third and fourth tracks were added to existing lines, thereby transforming them into essentially new lines where the original contractors 'would probably fail altogether to recognise their own work'. The decision to lay four tracks on the Great Western Railway as far as Didcot was probably the main reason for the eventual abolition of the broad gauge on that system in 1892.[42]

The schemes, which continued into the twentieth century, provided 'the new engineers with an opportunity to improve on the work of the old by learning from their mistakes'. The steam navvy was first put to work on the Midland line between Tilton and Market Harborough in 1876. It removed the blue boulder clay so quickly and cheaply that a further three machines were purchased. Eight machines were employed in the building of the Hull and Barnsley Railway in 1884–5, and the Great Central Railway, completed in 1899, employed forty steam navigators. With a ruling gradient of 1 in 75 and easy curves, the Great Central Railway was probably the finest line ever built from the engineering point of view.[43]

By means of physical adaptations and, more especially, agreements and amalgamations between companies, by 1910 a reasonably unified network had evolved of just over 16,000 miles, of which 9,430 miles were double track, 5,330 miles single track and 1,120 miles quadruple. In the London area, in particular, there was a maze of interconnecting lines, with virtually no dead-end branches. In the densely populated parts of industrial lowland England, no place was more than 2 miles from a railway; in rougher country, with a sparse population, the rail net was much more open, and some parts might be over 20 miles from a railway.[44]

Communications of a different kind became even more pervasive in town and countryside. A feature of many estate collections of the period

[41] F. R. Conder, *A Civil Engineer* (London, 1868), pp. 412–14.

[42] Simmons, *Victorian Railways*, pp. 52–6; W. M. Acworth, *The Railways of England* (1900, repr. London, n.d.), pp. 232–3; C. H. Grinling, 'Fifty years of railway engineering', *Railway Mag.*, 1, 1897, pp. 10–19; E. F. Carter, *An Historical Geography of the Railways of the British Isles* (London, 1959).

[43] Nottinghamshire RO, DD 505/1; J. Simmons, *The Railway in England and Wales 1830–1914* (Leicester, 1978), p. 155; B. Morgan, *Railways: Civil Engineering* (London, 1973), p. 148.

[44] M. Freeman and D. Aldcroft, *The Atlas of British Railway History* (London, 1985) p. 43; A. C. O'Dell and P. S. Richards, *Railways and Geography* (London, 1971), pp. 102–7.

are the wayleave agreements drawn up between landowners and the Postmaster-General, under the Telegraph Acts of 1863 to 1892. In 1873, permissions were sought for the placing of supports for telegraph posts in fields along lengths of the Great North Road in Huntingdonshire, and the cutting of trees from time to time. The poles and wires hardly added to the beauty of the surroundings, and in any case not every settlement was enthusiastic to have the telephone. There were expected to be too few subscribers to warrant the cost of £5 a year to bring a trunk line from Oxford to Thame, and it was not until seven years later, in 1909, that the town was joined with the Aylesbury trunk at Butler's Cross.[45]

By the turn of the century, it was a matter of comment should anyone travel long distances by road as opposed to rail. Whilst most canals had become 'a pathetic and melancholy part of the landscape', the question was how soon the railways would decay into a similar state. The author of a book on *Vanishing England* forecast that in only a few years the air would be conquered by balloons, flying-machines and airships.[46] In 1910, the town of Huntingdon turned out to see a pioneer flight from the flood-meadow of Portholme: a meeting called by the mayor resolved to encourage the establishment of a factory in the town. The launching of the first sea-plane into the sky at Windermere in 1911 prompted the children's writer, Beatrix Potter, to draw up a petition of protest against the prospect of this beautiful lake being turned into another Brooklands or Hendon. A more inappropriate place for 'experimenting with flying machines could scarcely be imagined'.[47]

ii. The bicycle and motor car

Although the railways were the most strikingly visual expression of the transition from an essentially pedestrian society to one of greater mobility, that freedom was subject to many constraints. In a sense, the railways involved as great a degree of regimentation as the stagecoach.

Even in the towns and cities well served by the railways, there was little improvement in personal mobility for the bulk of the population – railways were never intended primarily for cheap, casual, short-distance usage. In that respect, the decision to install an electric tramway system had at least as great an impact on the lives and perceptions of urban families. In Huddersfield, the first trams ran in 1901, and a route of over ten miles had been constructed to Marsden by 1914, making it possible to

---

[45] Huntingdonshire RO, Manchester MSS 12/12; J. H. Brown and W. Guest, *A History of Thame* (Thame, 1935), p. 253.

[46] J. J. Hissey, *Across England in a Dog-cart from London to St Davids and Back* (London, 1891), pp. 400–1; P. H. Ditchfield, *Vanishing England* (London, 1910), pp. 2, 389–91.

[47] J. Taylor, *Beatrix Potter* (London, 1986), pp. 125–7.

spend an afternoon or day tramping the moors of Wissenden Valley. So popular was this amenity that, despite the trams running every seven minutes and many people being left behind at each stopping point, every car was crowded on fine Sunday nights in summer with passengers returning to their homes in the centre of Huddersfield.[48]

However great the potential of the tram and light railway, no form of public transport could provide the independence of movement available in the form of the bicycle – as the Edwardian Mr Polly discovered when he 'cast a strategical eye' over Chertsey or Weybridge in the novel by H. G. Wells.[49] Although the first practical models had been imported from France in the late 1860s, it was not until the introduction of the modern 'safety' bicycle in the early 1890s, with its readily detachable pneumatic tyres, that cycling became relatively safe and comfortable. Not only was it possible for a fit cyclist to cover more ground in a day than a carriage without change of horses, but the fact that bicycles became progressively cheaper, costing £7 or so in 1905, meant that almost everyone had one. Every town and large village had its bicycle and tricycle agents and repairers.[50]

In a survey of cottage accommodation, carried out by the Board of Agriculture in 1919, there were numerous reports of colliers, artisans and mechanics living in houses previously occupied by farm labourers, using a bicycle to travel the three to five miles to work in the colliery or town. In Staffordshire, miners were prepared to bicycle five or six miles from Hamstead Colliery in order to 'obtain cheaper cottages and more rural surroundings'.[51] Thousands of cyclists poured out of the cities, especially London, at weekends. For the first time, perhaps, there was pleasure to be derived from travelling by road.

Popular motoring dated from the first decade of the century, the number of licensed motor cars and motorcycles (for which sidecars soon became available) rising from fewer than 8,500 in 1904 (the first year of licensing) to 90,000 in 1910 and nearly 230,000 in 1914. The botanist, G. C. Druce, had acquired a car by 1909, and even a poorish country vicar like E. S. Marshall had obtained one by 1914, writing of how 'it comes in handy for botanical outings, beyond carriage-drive range'.[52] With speeds and endurance far greater than those of a horse or cyclist, the internal combustion engine conferred both independence and long-distance mobility on its owner. A witness giving evidence to the Royal Commission on Motor Cars in 1905 spoke of its being a favourite run to

[48] R. Brook, *The Story of Huddersfield* (London, 1968), pp. 222–5.

[49] H. G. Wells, *The History of Mr Polly* (Pan edn, London, 1971), p. 85.

[50] W. Plowden, *The Motor Car and Politics, 1896–1970*, (London, 1971), pp. 23–5.

[51] Board of Agriculture, *Wages and Conditions of Employment in Agriculture*, BPP, 1919, IX, vol. I. *General Report*, pp. 139, 144.   [52] D. E. Allen, *The Botanists* (Winchester, 1986), p. 106.

motor the 100 miles from London to Bath for luncheon, and to return in time for dinner. Map publishers were quick to relabel their 'cycling road maps' so as to read 'cycling and motoring' or 'cycling and automobile' maps. In practice, however, motorists, in making their longer journeys, wanted less detail and fewer sheets to handle – the number of quarter-inch and smaller-scale maps published multiplied.[53]

As more and more roads were used for both short-distance journeys (between home and work or market) and by longer-distance traffic, the highway authorities pressed ever more vigorously for the costs of road improvements and maintenance to fall on the motorist, rather than the rural ratepayer. The motoring organisations recognised that the most certain way of securing improvements as cheaply and quickly as possible was to end the anachronistic system whereby responsibility was shared between nearly 2,000 authorities. The local government reforms of the late nineteenth century had begun to prepare the way by conferring complete financial responsibility for main roads and bridges on county and county borough councils. Urban and rural district councils were made responsible for minor roads and bridges. Acting on the recommendations of the Royal Commission of 1906, Lloyd George included provision in his Budget of 1909 for the raising of duties on motor spirits and motor car licences, so as to establish a central road-improvement account. The various local authorities would act as agents to the central Road Board, appointed to disburse the monies.[54]

---

[53] *RC on Motor Cars, Report*, Cd 3080, QQ 11875–84, BPP, 1906, XLVIII; Nicholson, *Wheels on the Road*, pp. 45–8.

[54] Plowden, 'Motorcar and politics', pp. 89–95; Finance (1909–10) Act, 1910, 10 Edward VII, c. 8; Development and Road Improvement Funds Act, 1909, 9 Edward VII, c. 47.

CHAPTER 28

TOWN AND COUNTRY:
AN OVERVIEW

BY GORDON E. CHERRY AND JOHN SHEAIL

INTRODUCTION

During the second half of the nineteenth century and up to the First
World War, the countryside of England and Wales experienced considerable change. The general context was provided by an urbanising and
industrialising nation, and the next chapter considers some of the
specifically urban issues which contributed to rural change. But to begin
with, it is necessary to establish the major elements of transformation
which directly affected the face of the countryside. A brief overview is
all that will be possible (all the aspects are examined elsewhere in an
extensive literature), but it is important to set the scene in terms of a
dynamic historical geography. The emphasis here, and throughout the
volume, is on physical change and visual appearance, but these cannot be
described without establishing their relationship to a range of contribu-
tory factors – social, economic, political and institutional. The threads of
the canvas constitute agriculture; industry and employment; housing;
urban growth, and urban demands on the countryside.

A. AGRICULTURE

Our period of review – sixty-four years – falls conveniently into three
parts. The first extended to the mid-1870s and represented a prosperous
period of high farming; the second covered the last quarter of the century
and featured acute depression accompanied by a sharp shift in farming
practices from arable to pasture; and the third suggested a return to some
stability, the pendulum of prosperity having swung back somewhat. To
pursue the chronological perspective, each period can be summarised in
turn, a much fuller review being given in earlier sections of this volume.

At the outset the elements of the general picture were as follows. The
western counties were for the most part given over to grazing, and in the
east, to 'corn' (wheat and barley). In more detail, Harley suggests that five
types of agricultural region covered England: the sandlands which had

already emerged as districts of mature high farming; fenland and marsh-land, on their way to becoming the new granaries of nineteenth-century England; the south-east chalk and limestone areas, part of arable England though variably with a grain–sheep economy; the claylands dominated by grassland; and fifthly the uplands, with pastoral farming and patches of arable.[1] Large farms of 500 acres and over were most numerous in eastern and southern England and in Northumberland. Small farms of less than 100 acres were more represented in the north Midland counties, and in the north-west and south-west.

Rural Wales was distinctive by reason of its social conditions and rela-tive isolation. The gentry was largely anglicised and Anglican, whilst the remainder of the rural population was poor, increasingly non-conformist and Welsh-speaking.[2] This contributed to a bitter legacy of what at times threatened to become insurrection. The unrest associated with the Rebecca riots of south-west Wales in the early 1840s had been mainly directed at the tollgates of the turnpike roads. But as the province entered the second half of the nineteenth century, there were signs that the lot of the small farmer and the landless poor was improving; the railway system penetrated isolated areas, and gave access to wider markets for the farmers' pastoral produce. Land was, nevertheless, still concentrated in few hands. In the 1870s, six landlords owned half the county of Caernarvonshire. More than 60 per cent of Welsh land was owned by 571 landowners, all with estates of over 1,000 acres. In stark contrast, over 35,000 cottagers owned only 7,000 acres between them.[3]

There was an absence of agricultural statistics, but an enquiry by James Caird into the state of farming in the principal counties of England in 1850–1, originating in *The Times* and published in 1852, suggested a total of 27 million acres of cultivated land (including meadow and arable pasture grounds), 2 million acres of uncultivated land and 3,160,000 acres of moor and mountain. Caird was optimistic about the years ahead: 'we see no reasons to despond, but many to encourage hope, in the future prospects of British agriculture'.[4] And well he might: a Golden Age was dawning.

Sixty years later, R. E. Prothero (later Lord Ernle) described the first thirty-seven years of Victoria's reign as 'an era of advancing prosperity and progress'. To meet the burgeoning urban demand, food had to be produced on land prepared, improved and fertilised for that purpose by a substantial outlay of capital and sustained enterprise. To that end, a

---

[1] J. B. Harley, England *circa* 1850', in *A New Historical Geography of England* ed. H. C. Darby (Cambridge, 1973), pp. 527–94.

[2] P. Morgan and D. Thomas, *Wales: the Shaping of a Nation*, (Newton Abbot, 1984), pp. 84–7.

[3] G. E. Jones, *Modern Wales: a Concise History, c. 1485–1979*, (Cambridge, 1984).

[4] J. Caird, *English Agriculture in 1850–51* (London, 1852), p. 526.

self-sufficing domestic industry had to become a profit-earning manu-factory of bread, beef and mutton. In that enterprise, the large tenant-farmer proved dominant; the rapidly increasing demand for food could never have been met by small yeoman, undercapitalised farmers and commoners.[5]

Whilst the interpretation placed on the period by Prothero and others has been challenged, there was no denying at any rate not over the first two decades, the striking increase in agricultural output, at a time when 80 per cent of the people were fed from home-produced food, from land cultivated by only 14 per cent of the labour force.[6] Vastly improved farming methods had been applied to a considerably increased farm acreage. In some cases, the improvements brought land into farming for the first time. More usually, it was a case of upgrading the downland, heath, moor and other rough land already grazed by livestock. The changes were facilitated by the extension of sub-soil drainage (at least 2 million acres were drained by the early 1870s), the enlargement of fields, straightening of boundaries, adoption of wire fencing, and erection of farm buildings. The outcome in some parts was a radical change in the physical appearance of the countryside.

Farming benefited from a new alliance with science, particularly in regard to fertilisers and breeding. The German chemist, Justus von Liebig, had traced the relations between plant nutrition and soil compo-sition in his book, *Organic Chemistry in its Applications to Agriculture and Physiology*, published in 1840. Artificial manures became more plentiful: nitrate of soda and Peruvian guano, for example. Although mechanisa-tion came late to English agriculture, steam-driven machinery, including steam ploughs, but particularly portable steam threshing-engines, was widely introduced from mid-century.[7] Innovation abounded: new implements, corn and seed drills, reapers, binders and mowing machines; lighter carts; and the cylindrical clay pipe which facilitated the drainage of the clay lands.

The period was therefore one of agricultural prosperity at a time when trade and manufacturing at home were buoyant. While the Continent and America were at war in the 1860s, England enjoyed peace. It was marked by a system of high investment in search of high returns, the justification for which was evidenced by high prices. The most conspic-uous feature was a dominant cereal cultivation, in association with live-stock production, made possible through investment by landlord and tenant alike: in buildings, in drainage and pedigree livestock, and in feeds

[5] R. E. Prothero, *English Farming Past and Present* (London, 1912), p. 346.

[6] G. E. Mingay, 'The Agricultural Revolution in English history: a reconsideration', *AHR* 26, 1963, pp. 123–33.

[7] E. J. T. Collins 'The age of machinery', in Mingay, *Victorian Countryside*, vol. I, pp. 200–13.

and fertilisers. Output was also increased as mixed farming practices were extended into districts of light soil.

The situation was soon to change, with dramatic consequences for the appearance of the countryside and the economic and social conditions of its workforce. Symptoms can be discerned as early as 1862. Sharp depression hit farming in two periods, 1875–84 and 1891–9. Prothero, writing in 1912, considered the last twenty-six years of Victoria's reign 'a period of agricultural adversity – of falling rents, dwindling profits, contracting areas of arable cultivation, diminishing stock, decreasing expenditure on land improvement'.[8]

Contemporary observers, and later historians, have sought to distinguish the effects of what Prothero called a modern type of financial crisis, which affected all sectors of the national economy, from those forms of adversity peculiar to agriculture. Adverse weather conditions were at first held to be the principal factor in the decline of agricultural fortunes. This was the prevailing view of witnesses examined by the Royal Commission on the Depressed Condition of the Agricultural Interest (the Richmond Commission), which reported in 1882. But a deeper and more lasting cause was the growing strength of overseas competition. Some fifteen years later, in 1897, a further Royal Commission concluded,

among all classes of agriculturalists there is a consensus of opinion that the chief cause of the existing depression is the progressive and serious decline in the prices of farm produce.

The financial difficulties were the manifestation of a secular shift in national patterns of food consumption and international trade in food products. The first sector of farming to be affected was cereals: imports of wheat doubled between 1850 and 1872.[9]

Each year's returns to the Board of Agriculture seemed to give substance to the general belief that remarkable changes were taking place in the ratio of arable to grassland. Much publicity was accorded to the fact that whereas arable made up two-thirds of the cultivated area of Great Britain in 1871, the proportions were roughly equal by 1891. The Board's statisticians attributed half the increase in permanent grass to the substitution of grassland for arable, and the remainder to the conversion of mountain and heath land to what was regarded as permanent grass.[10] In England and Wales, permanent grassland made up 43.8 per cent of the total area of crops and grassland between 1866 and 1875, 48.7 per cent (1876–85), 53.4 per cent (1886–95), 55.9 per cent (1896–1905), and 58.6 per cent (1906–15).

---

[8] Prothero, *English Farming Past and Present*, p. 346.

[9] *RC on Agriculture, Report and Minutes of Evidence*, BPP, 1882 XIV, C 3309; *RC on the Agricultural Depression, Final Report*, BPP, 1897. xv, C 8540, p. 43.

[10] *Agricultural Returns of Great Britain*, BPP, 1892, LXXXVIII, C. 6743, pp. ix–xii.

What was generally spoken of as a decline in arable land was in practice a diminution in the land growing corn crops – the combined area
under wheat, barley and oats in England and Wales fell by 1,154,000 acres
(16 per cent) between 1868 and 1891. It was the unanimous view among
collectors of the statistics that low grain prices and the pressure of
American competition, at a time of increasing labour costs and relatively
high prices for meat and dairy produce,[11] were largely responsible for
bringing about the decline. As a collector in 1886 recorded, with experience in the south, west and midland parts of England, a large proportion of the increase in permanent pasture was made up of

land seeded down in ordinary rotation, and which, had wheat been a paying
crop, would have been ploughed up the second or third year and sown with this
cereal, whereas the land has simply drifted into permanent pasture owing to
circumstances over which cultivators have now no control.[12]

Nowhere in the agricultural literature were the 'base-maps' of the
Ordnance Survey used to greater effect than the report of the Royal
Commission on Agriculture, published in 1894, which included, in the
words of Clapham, 'a terrible map, dotted thick with black patches
showing the land gone to "coarse, weedy pastures"'. Information as to
which fields had reverted to grass over the previous fifteen years in an area
of 222,720 acres of south Essex was recorded on 6-inch Ordnance Survey
sheets by an Assistant Commissioner, and the composite map, marking
the areas in black, was reduced by photography to a scale of one-and-a-
half inches to the mile for reproduction in the report. Staff of the Survey
Branch of the Board of Agriculture calculated that 28,222 acres, or 12.67
per cent of the total area, had passed out of cultivation.[13]

It was in the Midland region that the proportionate loss of cereal
farming was greatest. Taken as a whole, the region was one of convert-
ible husbandry, where one type of farming was most easily adjusted to
another. Adaptation usually took one of two forms, either an expansion
of permanent pasture or the development of intensive fruit culture and
market gardening. The first response was most obvious in the higher parts
of Leicestershire and Rutland, and the second in the Vale of Evesham,
where physical and economic conditions were especially favourable.[14]
There were also striking changes on the silt fens of the Wisbech area of
Lincolnshire where almost continuous plantations were laid out of small

[11] BPP, 1881, XCIII, C. 3078, pp. 7–8; BPP, 1893–4, CI, C. 7167, p. ix.
[12] *Agricultural Returns of Great Britain*, BPP, 1886, LXX, C. 4847, p. 14.
[13] J. H. Clapham, *An Economic History of Modern Britain* (Cambridge, 1926–38) vol. III, p. 79; *Royal Commission on Agriculture*, BPP, 1894, XVI, C. 7374. Report by Mr R. Hunter Pringle, pp. 48, 133.
[14] P. Roxby, 'Rural depopulation in England during the nineteenth century', *Nineteenth Century & After*, 71, 1912, p. 185.

fruit, usually overplanted with apples and plums. The industry was still so new at the turn of the century that there were no large trees or old orchards to be seen. Where the silt was coarser, particularly around Spalding, great breadths of land were devoted to daffodils, tulips, crocuses and snowdrops, grown for their bulbs rather than blooms.[15]

Within lowland England, Kent was historically the least arable of the corn-growing counties.[16] Its land-use pattern became even more diverse. The proportion of land under arable declined from three-fifths in 1867 to two-fifths in 1907, the downward trend being most marked in respect of wheat, beans, peas, turnips and clover. An even greater emphasis was placed on animal husbandry, with large-scale poultry farming becoming established in the Kentish Weald, where there were more sheep per acre than in any other English county. About two-thirds of the country's hop acreage was found in Kent and, during the last three decades of the century, the area devoted to orchards and small fruit doubled. Some fruit growers between Woolwich and Orpington devoted as much as 500 acres of their holdings to strawberries.[17]

Changes in land use were particularly striking in Wales, where concurrently the size of the sheep flock rose by a third and the area under permanent grass increased from 57.3 per cent of the total agricultural area to 75.2 per cent, whilst the area under arable fell from 42.7 per cent to 24.8 per cent. Between 1890 and 1902, increasing importance was attached to temporary grass on the remaining arable land. A key factor in enabling these changes to come about was the improved crop yields, brought about by better seed varieties, improvements in the manuring of pastures and better hay-making and, from the 1890s, increasing purchases of concentrated feedstuffs.[18]

H. Rider Haggard, in an impressionistic account of agriculture and social matters in rural England in 1901–2, felt that 'English agriculture seems to be fighting against the mills of God. Many circumstances combine to threaten it with ruin, although as yet it is not actually ruined.' He further commented that, 'some parts of England are becoming almost as lonesome as the veld of Africa'. While Free Trade had filled the towns and emptied the countryside, it had 'gorged the Banks but left our rick-yards bare'. What had been a major benefit to the townsman as cheap food presented a seemingly intractable problem for those whose livelihood depended on the land.[19] Prothero put the matter succinctly: 'How were [English farmers] to hold their own in a treacherous climate on

[15] A. D Hall, *A Pilgrimage of British Farming, 1910–1912*, (London, 1913), pp. 86–7.
[16] A. M. Everitt, 'The making of the agrarian landscape of Kent', *Arch. Cantiana*, 92, 1977, pp. 1–31.
[17] J. Whyman, 'Agriculture in Kent', *Bygone Kent*, 1, 1980, pp. 169–76.
[18] D. Howell, 'Welsh agriculture, 1815–1914', PhD thesis, Univ. of London, 1970, pp. 288–9.
[19] H. Rider Haggard, *Rural England* (London, 1906), vol. 1, pp. 540, 564, 586.

highly rented land, whose fertility required constant renewal, against produce raised under more genial skies on cheaply rented soils, whose virgin richness needed no fertilisers?'[20]

However, by the end of the century the worst was over, and a gradual though uneven recovery was underway. The gross output of English agriculture, which had fallen by 13 per cent between 1871 and 1891, had returned to its 1871 level by 1911. Wheat prices levelled out after 1896. Though agriculture remained relatively unprofitable in the years approaching 1914, the general picture was one of stability. Those engaged in milk production continued to be relatively well-protected. Milk was a highly perishable product, and obvious advantage was taken of the railway. The amount of milk brought to London, for instance, by rail rose from 9 million gallons in 1870 to 40 millions in 1890, and to 53 millions in 1900. The Great Western Railway (dubbed the 'Milky Way') doubled its milk traffic between 1892 and 1910; and that of the London and North Western Railway grew by a half between 1892 and 1904. New uses were found for the milk. In one development, Cadbury's 'Dairy Milk' chocolate became a best-seller. Milk-condensing plant were established among the dairy farms of Shropshire and Gloucestershire, so as to prevent milk going sour before reaching the factory at Bournville.[21]

By the outbreak of the First World War, it was clear that the country's agricultural industry had experienced a major period of change: no passing phenomenon, but a permanent alteration. Farming was of much less importance in the national economy, home agriculture supplying only 42 per cent of the food consumed in the United Kingdom in 1914. Agriculture accounted for 20 per cent of the gross national product in the late 1850s, but for only 6 per cent, in the late 1890s.[22] For many, change was associated with decay. Rural England was delapidated compared with the image it presented in the early 1870s. The nature of the agrarian transformations during the second half of the nineteenth century and into the twentieth century, and the legacy experienced in the concluding years, proved a major feature of the period. The changes should be seen as a result of national and international shifts in the way in which an industrialising, urbanising world fed its population. The consequences only served to provide more opportunities for urban encroachments to gather momentum.

[20] Prothero, *English Farming Past and Present*, p. 377.

[21] J. Brown, *Agriculture in England: a Survey of Farming 1870–1947* (Manchester, 1887); W. Stranz, *George Cadbury* (Aylesbury, 1973), p. 19.

[22] P. J. Perry, *British Farming in the Great Depression 1870–1914* (Newton Abbot, 1974).

B. RURAL INDUSTRY

We can be fairly sure of the geographical pattern of industry at mid-century because the population census of 1851 attempted the first scientific classification of occupations. A good deal of loosely distributed rural industry was in evidence, but the boundaries of industrial England showed a particular concentration, and it was here that the major landscape changes were taking place. The coalfields provided the key to that location; in the Black Country, in particular, an urban countryside was ravaged by a coke-iron industry that surpassed for a time all other parts of the country. Here was a landscape

interspersed with blazing furnaces, heaps of burning coal in process of coking, piles of ironstone calcining, forges, pit-banks, and engine chimneys; the country being besides intersected with canals, crossing each other at various levels; and the small remaining patches of the surface soil occupied with irregular fields of grass or corn, intermingled with heaps of refuse of mines or of slag from the blast furnaces.[23]

In his textbook on metallurgy, John Percy described how a dense cloud of white smoke hung perpetually over the copperworks of the district around Swansea. It could be seen clearly in favourable conditions from the south side of the Bristol Channel, some 27 miles away. 'Copperopolis', as Swansea was called, had become the largest producer of refined copper in the world by the middle of the century. The sulphurous and sulphuric acids, and arsenical fumes, emitted from the low chimneys, destroyed vegetation for a considerable distance. As Percy recounted, 'the inhabitants of Swansea generally seem to be habituated to the inhalation of smoke', submitting to the evil, 'if evil it be regarded, with murmuring resignation'.[24]

But the situation changed. The copper industry of Swansea went into relative decline in the 1860s, and absolute decline by the turn of the century, as other nations developed their own smelting and refining capacity. It was an early example of a successful British industry being outcompeted in the expanding international commodity markets.[25] Between 1885 and 1909–13, the production of copper ore in Britain fell

---

[23] Midland Mining Commission, *First Report, South Staffordshire*, BPP, 1843, XIII, p. iv.

[24] J. Percy, *Metallurgy* (London, 1861), pp. 299–300, 335–8; K. J. Hilton, 'A case history of derelict land in Swansea, 1900–1966', in *The Lower Swansea Valley Project*, ed. K. J. Hilton (London, 1967), pp. 38–46.

[25] E. Newell, '"Copperopolis": the rise and fall of the copper industry in the Swansea district, 1826–1931', *Business History*, 32, 1990. pp. 75–97; E. J. T. Collins, *The Economy of Upland Britain, 1750–1950: an Illustrated Overview* (Reading, 1978), p. 22.

by 86 per cent, tin ore by 45 per cent, and zinc and lead ore each by 47 per cent. A legacy of derelict mines and a discarded landscape of industrial buildings, engine houses, tall chimneys and waste heaps remained. In north Wales, during the second half of the century, Penrhyn and Dinorwic became the world's largest slate quarries. Their years of expansion came to an end in 1879, brought about first by slumps in demand in the building industry and then, more significantly in the longer term, by the closure of overseas markets and innovations in the lowland tile industry which, from 1906 onwards, led to the manufacture of cheaper tiles made of asbestos fibre and cement. A devastating lock-out at Penrhyn between 1900 and 1903 hastened the demise of the industry and community. A workforce of 2,800 in 1900 never again topped 1,800.[26]

In the developing industrial areas too, manufacturing, mining and urban land mingled in a rural setting, and features of the earlier countryside remained. The mining industry was densely concentrated in both the Black Country, where prosperity was based on the highly productive thirty-foot seam, and south Wales, where some 41,000 miners were employed in the Rhondda valley alone in 1913. The development of other industrial areas was more open. The cotton-dominated valleys of Lancashire, particularly in the more remote areas, retained an agricultural character, which even the mills, bleach and dye works, printworks and 'tenter grounds' could not erase. A hybrid landscape of agriculture and industry was encountered in the Yorkshire woollen area, where smallholders produced butter and milk for the local market, with factories and collieries never far from view.

With major advances in prospecting and extractive technologies, new resources could be exploited. The concealed coalfields of Durham and Northumberland, with their distinctive pit-head settlements, were progressively developed. From south-east of Sheffield, through Mansfield, and towards Nottingham, new pits were sunk throughout the second half of the century (the first in Nottinghamshire in 1859 and another sixteen between 1870 and 1880). The deepest borehole ever made was sunk in the purely agricultural district of the Isle of Axholme in 1893. It revealed reserves of 35,000 million tons of workable coal. To contemporary observers, there seemed every prospect of shafts being sunk even in the midst of the Thorne Moors. In Kent, a newly discovered coalfield was in production by the outbreak of the First World War.[27]

The focus of the iron-ore mining industry shifted, as the output of the Coal Measure ores declined. Whereas 95 per cent of output came from

[26] J. Lindsay, *A History of the North Wales Slate Industry* (Newton Abbot, 1974), p. 257.
[27] G. Dunston, *The Rivers of Axholme with a History of the Navigable Rivers and Canals of the District* (London, 1909), pp. xi–xiii.

the ores in 1850, only 10 per cent of an admittedly much larger domestic total was derived from them in 1913. The working of the haematite of Cumberland and north Lancashire, especially around the Barrow area between 1850 and 1870, and the development of the low-grade ores of the Jurassic, accounted for the change. The Jurassic ores of the Cleveland Hills provided the basis for the new Teesside iron industry; the Skinningrove and Eston seams were opened in 1850, and modern furnaces were soon in blast. Peak production was achieved in 1883. Production around Northampton, Wellingborough and Kettering in Northamptonshire, and towards Banbury in Oxfordshire, rose rapidly between the mid-1860s and mid-1870s.

At a local scale, agriculture itself was responsible for some quarrying activity. The excavation of fertiliser for use on the land was far from new, but the working of coprolite deposits at the base of the chalk, and above the gault clay, was entirely without precedent. Whilst visually striking (attaining in a few cases 30 feet in depth), the excavations were ephemeral in the sense that the land was almost always restored to agriculture, albeit at a lower level. The first pits in Cambridgeshire were dug in the 1850s, and later in Buckinghamshire, beneath the chalk scarp. With the exhaustion of deposits and increased foreign competition, the last pits near Cambridge closed in 1898.[28]

Whilst some rural industries survived outside the main mining and industrial areas, others moved or disappeared altogether. Northamptonshire continued to be important for the boot and shoe trade. The decision to rebuild Bliss Mill at Chipping Norton, of some 100,000 square feet, after the disastrous fire of 1873, was another example of confidence in old locations. The hosiery industry was still largely carried out in the workers' own homes, or in small workshops attached to them. The framework knitters' houses, with their elongated windows running the length of the buildings, were a distinctive feature of the landscape around the 'putting out' centres of Belper, Nottingham, Leicester and Loughborough.[29] But industries concerned with the making or processing of foodstuffs experienced major change as new port locations were developed; new methods of milling (using metal rollers), and increasing reliance on imported materials, resulted in large factories being built in London, Liverpool, Hull and Bristol. Windmills were progressively abandoned as the milling industry was mechanised.

For a time at least, some parts of the country practised a dual economy, whereby some form of out-work or cottage industry supplemented the

[28] R. Grove, 'Coprolite mining in Cambridgeshire', *AHR* 24, 1976, pp. 36–43; Buckinghamshire RO, Ashridge MSS, P15/49.

[29] D. A. Smith, 'The British hosiery industry at the middle of the nineteenth century; an historical study in economic geography', *TIBG*, 32, 1963, pp. 125–42.

livelihood to be gained from agriculture.[30] Handloom weaving in Lancashire and Yorkshire, pottery in Staffordshire, and small metal-working in the west Midlands and south Yorkshire, were pursued in conjunction with farming. The lead miners of the northern Pennines were amongst the more independent out-workers in the mining industry. More generally, each community in 1850 contained rural craftsmen whose role it was to serve the needs of agriculture – typically the farrier, blacksmith, wheelwright, saddler and carpenter. In an essentially horse-drawn society, the horse continued to be the basis for the most important of country crafts.[31]

During the middle years of Victoria's reign, the impact of industrialisation, which had earlier transformed basic manufacture, began to have its effect on traditional crafts. A set of reminiscences for the Dover area of Kent, in 1900, recorded how the shoemaker and wheelwright had largely disappeared from villages. The glazier no longer used locally spun window panes; the plumber now bought sheets of lead from the factory.[32] Machine joinery and new technology affected the woodworker's market; machine tools and galvanised iron products made inroads on the trade of the cooper in the brewing industry. Those engaged in the basket-making trades left for urban workshops.

It would be misleading to portray these shifts as simply the outcome of market forces. Other, less direct factors were at work. The author of a history of Congleton, the market town in south-east Cheshire, was in no doubt as to the consequences of the withdrawal of protection for the silk industry under a treaty with France in 1860. Despite the keenest competition from centres, including Manchester, the town's industry had held its own, with population rising from 10,517 to 12,338 in the 1850s. The number of operatives fell from 5,186 in 1861 to some 1,500 in the mid-1880s.[33] Labour legislation and compulsory education were in time to have a profound effect. The Factory and Workshops Regulation Act of 1867 prescribed the minimum age of employment for children to be eight years (raised to ten years in 1878). Between the ages of eight and thirteen, a child could only be employed in accordance with the half-time system already in use in factories and workshops.[34]

Economic rationalisation led to centralisation of production. In the boot and shoe industry, union power demanded that work be concen-

[30] J. Tann, 'Country outworkers: the men's trades', in Mingay, *Victorian Countryside*, vol. 1, pp. 329–40.

[31] J. A. Chartres and G. L. Turnbull, 'Country craftsmen', in Mingay, *Victorian Countryside*, vol. 1, pp. 314–28.   [32] Canterbury Public Library, Kent, U 801.803.

[33] R. Head, *Congleton. Past and Present* (Congleton, 1887, repr. 1987), pp. 154–7; Royal Commission on Technical Instruction, *Seventh Report, Report on the English Silk Industry by Thomas Wardle*, BPP, 1884, XXXI, p. xl.   [34] Factory and Workshops Regulation Act, 1867.

trated in factories, drawing to an end the era of the out-worker. Lacemaking suffered from competition with machine-made lace. Chairmaking became factory-based in the Chilterns. Cheaper imports hastened the demise of straw plaiting.

Meanwhile, a back-to-the-land movement was seeking the revival of handicrafts and peasant arts.[35] With an emphasis on individual skill and artistry, the intention was to revive pre-industrial production processes. The Guild and School of Handicraft, inaugurated in 1882, settled in Chipping Campden in 1902, where it flourished for a few years, and C. R. Ashbee's followers were attracted to other parts of the Cotswolds. Only in a few localities did such philanthropic ventures have any tangible consequences in landscape terms. The real impact was on social attitudes and perceptions.

By the eve of the First World War, rural industry had undergone a massive change. Whilst industrial employment in the countryside had diminished, there had been a significant increase in the amount of rural land devoted to industrial, and particularly mining, purposes, as part of an increasingly secular domination of the countryside by urban interests and demands. As the period of our study came to an end, the distinction between rural and urban, and between agrarian and industrial, was much sharper than in the pre-industrial era. Households became less able to participate in a combination of agricultural production and manufacturing and mining, in a setting which remained rural and agrarian.[36] Rural industrialisation, once the norm, had given way to specialisation in farming or industry. Whole regions, such as East Anglia, Essex, south Devon, Wiltshire and Westmorland, had abandoned their dual economies; old industrial towns, such as Norwich, Ipswich and Colchester, had to learn to take second place to new manufacturing upstarts in the north.

C. RURAL HOUSING

Every guidebook included an account of the local building tradition – the limestone of the Cotswolds, slate of the Lake District, and flint of Sussex and Norfolk. The clay or cob cottage was by far the most commonplace, with its cruckform construction and wattle and daub walls. The timber frame was studded outside with laths, and daubed with plaster or a mixture of clay and chopped straw. But what of these structures as homes to live in?

The focus of attention on nineteenth-century housing has rested, perhaps inescapably, on urban conditions, where the consistent failure of

[35] J. Marsh, *Back to the Land: the Pastoral Impulse in Victorian England from 1880 to 1914* (London, 1982).
[36] B. A. Holderness, 'The Victorian farmer', in Mingay, *Victorian Countryside*, vol. 1, pp. 227–44.

the house-building 'industry' to provide dwellings of either adequate number or adequate quality contributed much to the nature of the late Victorian urban crisis. But rural housing conditions for the labourer were equally bad, and in many cases worse; the similarity, or indeed the adverse distinction, was concealed only by the difference in scale and concentration of the problem.

In a sense the poverty of both rural and urban housing conditions, for farm labourers as much as for artisans in towns, rested on the same set of economic and social conditions which made it well nigh impossible for good-quality housing to be provided at rents that could be afforded, given the low wages that were paid and the irregularity of employment itself. But the rural worker was particularly disadvantaged, the living standards of agricultural labourers being the lowest of any section in the community. As Burnett puts it: 'they were ill-fed, ill-clothed and ill-housed, and until late in the century, uneducated, unenfranchised, unorganized and unrepresented'.[37]

Harriet Martineau protested at how Man himself was largely responsible for the unhealthiness of so many rural settlements. Nature had provided the purest air and water, and ideal building foundations, but the people lived 'in stench, huddled together in cabins, and almost without water', in a condition that was no better than those 'we are apt to pity' in the metropolis.[38] Generally speaking, cottages were newer, larger and better built in the south and east of the country. The longhouse (a single-storey living room and a shelter for livestock placed side by side under a single extended roof line) persisted in the poorer parts of the west, Wales and the north well into the nineteenth century. Whilst water drainage was installed in some of the larger villages in the 1870s and 1880s, concern as to rural water supplies, drainage and sanitation persisted well into the twentieth century.[39]

By the beginning of Victoria's reign, there were, however, some indications of how a change in attitude might bring about change. There was renewed interest on the part of landowners and speculative builders in providing houses for letting to rural workers. As agricultural profits increased after mid-century, this practice gained ground. Housing reformers attacked both the health hazards and moral dangers implicit in overcrowded conditions and lack of privacy, and the publication of books dealing with cottage design offered models for enlightened landowners to follow. A notable example was J. C. Loudon's *Encyclopaedia of Cottage, Farm and Village Architecture*, which first appeared in 1833.

[37] J. Burnett, *A Social History of Housing 1815–1976*, (Newton Abbot, 1978), p. 119.

[38] N. Nicholson, *The Lakers. The Adventures of the First Tourists* (London, 1955), p. 201.

[39] G. Mingay, 'The rural slum', in *Slums*, ed. M. Gaskell (Leicester, 1990), pp. 92–143.

The first national enquiry into rural labourers' dwellings was commissioned in 1864 by Dr John Simon, the Medical Officer of the Privy Council. A survey of 821 country parishes was carried out by Dr H. J. Hunter, and published the following year.[40] During the inter-censal decade 1851–61, the population of the parishes increased and the number of cottages decreased, the occupancy rate rising from 4.41 to 4.57 per dwelling. A detailed investigation of 5,375 typical cottages revealed that 40.8 per cent had only one bedroom, 54.5 per cent had two, and 4.7 per cent had more than two. The average air-space was 156 cubic feet, which contrasted with the legally required minimum of 250 cubic feet in common lodging houses and 500 cubic feet in Poor Law workhouses.

Further surveys amply confirmed these findings. Whilst the housing of the rural labourer may have varied considerably, it was overall very bad, whether measured in terms of poor structural condition, small size, poor ventilation, dinginess or desolation. The decline in housing stock was exacerbated by the fact that, until 1865, the parish poor-rate was levied on house property. Where estates were being consolidated into larger holdings through enclosure, landowners in 'close' parishes had good reason to demolish accommodation not required for farming purposes. Labour could be drawn from neighbouring 'open' parishes.

Under the Pubic Health Act of 1872, the country was divided into urban and rural sanitary districts, the latter corresponding with the Poor Law unions.[41] The performance of these authorities varied greatly. The Guardians of the Poor in Haverfordwest, for example, immediately appointed a Medical Officer of Health and Inspector, their immediate concern being the abatement of such nuisances as the lack of privies and proximity of pigsties to dwelling-houses – it was common to build the sty next to the outside wall of the kitchen fire. One hundred and eighty statutory notices were issued over a two-year period. In 1894–5, the sanitary authority became a rural district council, but there was scarcely any marked change in the attitude of members – meetings of the new Council were held in the board room of the workhouse, immediately after the Guardians' business was concluded.[42]

An obvious course was for the districts to combine into larger groupings. A combined district was formed in Oxfordshire of the six rural and seven smaller urban sanitary authorities. As the district's Medical Officer of Health commented in his second annual report, there was little to encourage the unpaid officials and salaried health officers and inspectors to take any positive action. Not only was a great deal of time and labour

---

[40] Medical Officer of the Privy Council, *Seventh Report*, BPP, 1865, XXVI, *Appendix 6, Inquiry on the State of Dwellings of Rural Labourers by Dr H. J. Hunter.*

[41] Public Health Act, 1872, 35 and 36 Victoria, c. 79.

[42] H. J. Dickman, *History of the Council, 1894–1974* (Haverfordwest, 1976).

required to understand the problems that needed to be tackled, but there were bound to be sharp differences of opinion over drainage and water-supply issues, and no certainty of success. As the Medical Officer continued, attempts to improve the sanitary state of cottages involved a degree of interference in the rights of property that was as dangerous as it was unpopular. Such constraints might have been overcome if parliament had provided the necessary motivation. The statutory powers were, however, permissive rather than mandatory. There was a most elaborate system of checks to ensure that nothing was done too quickly or strongly. The Act could do little more than give the impression that something practical was being attempted.[43]

At least initially, the main contribution of such sanitary authorities was to draw attention to prevailing conditions, whether in the form of the summaries of statistical evidence required by the Local Government Board, or as very detailed appraisals of the conditions prevailing in individual towns and villages. In Oxfordshire, conditions ranged from the Witney rural sanitary district, where there had been striking improvements, to the Woodstock district, and such instances as Wolvercott, where drainage was abominable, the shallow wells were polluted, and the proximity of Port Meadow encouraged 'the people to keep multitudes of pigs and geese in confined situations, and too close to their houses'. As for the small town of Brill in Buckinghamshire, from 'its magnificent position on the top of a hill, its beautiful scenery, fine air, and possession of one of the strongest iron-springs in England', it might have been expected to be 'the sanitorium of the midland counties'. Instead, a combination of 'neglect, meanness, carelessness, blundering, and resistance to all improvement' had turned the town into an almost perpetual 'fever-nest'. The overriding aim had been to resist any increase in expenditure on sanitary improvement.[44]

Whilst a Medical Officer of Health was likely to focus attention on the very worst cottages, instances of bad housing were to be found in almost every settlement. Many cottages were badly constructed, too small, and poorly lit and ventilated. The Oxford Board of Health complained of having no power to prevent houses being built on unhealthy sites. It was reluctant to use its only weapon in regulating existing housing stock, namely to shut cottages as unfit for human habitation. This would only increase the degree of overcrowding. Major improvements had been carried out in some instances, for the most part by the owners of large estates. As the Medical Officer for the combined district in Oxfordshire

[43] G. W. Child, *A Second Report for the Chairman and Members of the Combined Sanitary Authorities of Oxfordshire* (London, 1875).

[44] G. W. Child, *Report upon the Sanitary Condition of the Districts of the Combined Sanitary Authorities of Oxfordshire* (London, 1874).

emphasised, it was the proprietor of narrow means who rented out the worst cottages. It was against this type of person that the legislation designed to improve the balance between supply and demand for accommodation should be directed.[45]

Change was slow to effect. The franchise of the rural worker was extended in 1884, and the question of both the quality and tenure of cottage accommodation took on a wider political dimension. The Royal Commission on Labour of 1892–4 noted some improvement, but its *General Report* found an irregular distribution of cottages, inconvenient locations, poor standards of accommodation, deficient sanitary provision, and an underlying rent problem.[46] With hindsight, the period of housing improvement had been a short one. No sooner had the productive years of high farming suggested the possibility of sustained housing improvements than agricultural depression set in. To compound the problem, the countryside had been hardly affected by successive legislation designed to improve the housing of the poor. Rural sanitary authorities built but a handful of cottages under the Housing of the Working Classes Act of 1890.[47] By 1905, just thirty-two loans had been sanctioned. In the absence of any radical departures, the contrast between good and bad became even more obvious – between the one-bedroomed cottages and decaying shacks at one end of the spectrum, the intermediate brick-built houses with tiled roofs and a modicum of facilities, and, at the other extreme, the new model cottages, with perhaps three bedrooms, water supply, sanitation and garden.

As with the urban situation, an impasse had been reached. Speculative building for the labourer was scarcely being provided; building by philanthropic landowners had been retarded by the depression, and little more could be expected from that source. Local authorities seemed unwilling to take over the mantle of provider. Whilst the Housing, Town Planning, &c., Act of 1909 made systematic surveys of rural housing obligatory, and the Local Government Board acquired some measure of control over local authorities which themselves were given additional powers, a building programme was scarcely put in hand.[48] It was ironic that during the First World War, which brought 'homes for heroes' to the fore as a political issue, the rural cottage was held up by propaganda as an idyll worth fighting for.

At the other end of the scale, there was the Victorian country house. Perhaps 2,000 were built,[49] and they made their unique contribution to

[45] Child, *Sanitary Condition of Oxfordshire*, 1874; Child, *Sanitary Condition of Oxfordshire*, 1875.

[46] *RC on Labour. Fifth and Final Report*, BPP, 1894, xxv, Cd 7421, pp. 110–12.

[47] Housing of the Working Classes Act, 1890, 53 & 54 Victoria, c.70.

[48] Housing and Town Planning, &c. Act, 1909, 9 Edward VII, c.44.

[49] J. Franklin, 'The Victorian country house', in Mingay, *Victorian Countryside*, vol. II, pp. 399–413.

the appearance of the nineteenth-century countryside. By the 1860s, they were usually built in a Gothic style. Architecturally the emphasis was on height and a consequent narrowing of proportions – very different from the country houses following the eighteenth-century norms of 'four square and spreading'. Roofs were higher and steeply pitched, affording a romantic skyline. By the 1870s, under the influence of the architect Norman Shaw, a vernacular tradition was implanted, which exploited tile hanging, half-timbering, casement windows and tall chimneys. This suited the small to medium-sized houses, to which the architects C. A. Voysey and Baillie Scott were attracted. The larger houses, as designed by Shaw and Edwin Lutyens, still reflected a great symmetry and classical tradition.

Victorian country houses emphasised privacy, as opposed to the openness of the Georgian mansion in its landscape. Typically, the drive from the lodge curved through the grounds before the house came into view. Among the many species planted, conifers, laurels and rhododendrons were particularly popular. But the main landscape difference was in the immediate surroundings of the house, where detailed gardening now became a feature incorporating terraces, shrubberies, rose gardens and specimen gardens.

The demand for country houses reached its peak in the 1870s, and by the end of the century a glut of such properties had appeared on the market. Only those who earned their money from non-agricultural sources could still commission or reside in a large house. The successful middle classes continued to move into the country, but to much more modest estates. Grand houses, such as Cliveden, Mentmore, Eaton Hall and Cragside, gave way to lesser ostentation. One might cite the solid comfort of Highbury, the new home of Joseph Chamberlain at Moor Green, on the fringes of Birmingham. A particularly flamboyant period in countryside residential development had run its course.

D. URBAN GROWTH

Urban expansion implied the loss of rural land for housing and related purposes. The countryside was in fact the recipient for much of the additional population of the growing towns and cities; the new housing areas were annexed, not only functionally but, where necessary, by extending the boundaries of the units of local administration. It is a complex and far from uniform scene to review. The developments may be considered in the following ascending order of importance as far as the rolling back of the countryside is concerned: country and market towns; railway and industrial growth, seaside towns, suburbs generally, and the case of London.

E. COUNTRY TOWNS

The relatively static size of many country or market towns constituted one of the more stable features of the population geography of the period. The population of the Oxfordshire town of Banbury increased by almost exactly the national rate of growth, namely by 16.4 per cent in the 1850s, and 14.9 per cent in the 1860s. Between 1851 and 1881, the town's population rose to 12,126 (an increase of 38 per cent), 800 houses were erected, and the town centre was almost entirely rebuilt. A town hall, two corn exchanges, seven new dissenting chapels and arrays of sub-urban villas bore witness to the town's prosperity. As in the case of almost every other market town of consequence, Banbury had an iron foundry.[50] The Britannia works were founded in 1849, and produced an ever-widening range of agricultural implements and steam engines. Wantage in Berkshire was famous for the White Horse Ironworks and its steam threshing machines. In Wales, foundries at Cardigan, Aberystwyth, Llanidloes, Newtown, Welshpool and Machynlleth achieved local fame as manufacturers of waterwheels and other types of plant required for the mines and other purposes.[51]

The immediate impact of changes in employment opportunities on the appearance and layout of settlements should not be exaggerated. Whilst more families were employed in shoe- and lace-making than in farm work, the village of Rushden in Northamptonshire retained its rural appearance. The first factories of the 1850s were little more than barns, unobtrusively set in yards, and in some instances were merely a single room in a shoemaker's house. It was not until the 1870s that dramatic changes took place, as larger factories and extensions were needed to accommodate the increasingly mechanised industry, and the labour force became more centralised. Population rose from 1,460 in 1851 to 12,459 in 1901. Factories and rows of houses sprang up wherever sites could be purchased within the old quarters of the village, and on former open fields. The red-brick factories, chapels, banks and shops stood incongru-ously among the remaining stone and thatched cottages of the High Street. The uniform colour of the bricks reflected the fact that nearly all had come from the nearby 116-acre brickyard of the Rushden Brick and Tile Company.[52]

Guidebooks and topographies often highlighted the changes taking

[50] D. Alderton and J. Booker, *Batsford Guide to Industrial Archaeology of East Anglia* (London, 1980), pp. 184–5.

[51] B. Trinder, 'Victorian Banbury', *Banbury Historical Society*, 19, 1982; D. M. Rees, *The Industrial Archaeology of Wales* (Newton Abbot, 1975), p. 218.

[52] D. Hall and R. Harding, *Rushden. A Duchy of Lancaster Village* (Rushden, 1985), pp. 188, 214, 239–66.

place in the use of building materials and the advances made in the provision of utilities. In the town of Buckingham, it was noted how the houses were for the most part built of brick, with slates or tiled roofs. Only a few thatched cottages survived. A second gas holder was added in 1859, and the original one replaced by one of the telescoping kind in 1907 so as to increase storage capacity further. Electric lighting was provided 'in a moderate style' at the Town Mills in 1888; the streets were first illuminated with electric light in 1905.[53]

Gas and electricity were largely confined to towns and the larger villages. South of Huddersfield, the gas mains extended no further than the village of Honley. Everywhere beyond continued to rely on the 'lanthorn'.[54] The Gas Light and Coke Company in Bourne, Lincolnshire, which was founded in 1840 with the vicar as one of its trustees, flourished to such an extent that the works, costing £2,000, had to be expanded in 1868, by which time the town of 3,800 had 56 public gas lamps. The advances made in the gas and electricity industries were not, however, always taken up so readily. Whilst the borough council of Boston agreed to the dock area becoming one of the first to be lit by electricity in 1897, there continued to be a marked reluctance to proceed with a borough scheme. By the time a provisional order was sought in 1915 for the supply of both the borough and rural district, Boston had become one of the largest places in England to be still dependent almost entirely on gas.[55]

Each part of the country furnished examples of towns that grew in response to some special attribute of location, availability of resources, or perhaps simply civic enterprise, and of other centres where population growth ceased at some point before 1861. For the most part smaller and without any significant industrial base, the latter category of towns was in many respects caught up in the general movement of population away from the countryside.[56]

Examples of such differences in fortune may be cited from west Sussex. In 1841, there were four small inland market towns: Arundel (population 1,624), Petworth (3,364), Midhurst (1,536) and Steyning (1,495). To the north-west, Horsham was the major trading centre of the western Weald with a population of 5,765. Chichester (8,512) was the regional centre in the south-west. By the turn of the century, Horsham's population had more than doubled to 12,994, helped by favourable rail communications. It had an enhanced marketing role, and its industries had expanded. By contrast, Chichester's development was modest; Midhurst, Arundel and Steyning put on a few hundred population between them, and Petworth

[53] J. T. Harrison, *Leisure-hour Notes on Historical Buckingham* (London, 1909), n.p.

[54] M. A. Jagger, *The History of Honley* (Huddersfield, 1914), p. 68.

[55] G. S. Bagley, *Boston. Its Story and People* (Boston, 1986), pp. 293–5.

[56] J. Brown, *The English Market Town. A Social and Economic History 1750–1914* (Ramsbury, 1986).

decreased in size. Meanwhile, the important rail links established in mid-Sussex had stimulated the growth of a new settlement, Hayward's Heath.[57]

New hierarchies were established. The cathedral cities of Chichester, Durham, Winchester, Hereford, Canterbury and Salisbury might show little capacity for growth, whereas the primacy of others increased as they added new roles to their traditional marketing and administrative functions. A combination of railway development and manufacturing industry brought prosperity to Carlisle, York, Gloucester, Peterborough and Rochester.

Growth required space. As towns acquired stature, so their centres needed new buildings, whether for commercial purposes, corn markets, market halls, hotels, town halls or public offices (for example, for the police). If the displaced residents were to be accommodated, the town had to grow outwards. Dorchester expanded into the neighbouring parish of Fordington, where open fields were enclosed in the 1870s. The form of the extensions was often influenced by railway development. The station might be well outside the town centre, as at Hitchin, where the intervening land was mostly taken up with housing. At Newark, on the other hand, malt houses, breweries and iron foundries provided the infill, whilst at Grantham an engineering works dominated the new land use. Parks might be laid out, as with the People's Park at Boston in 1871 and the Albert Park at Abingdon in 1864, which provided a setting for the new residential districts and their churches.

An extreme example of how the burgeoning population of cities might bring a demand for more building land was Nottingham. It was one of 'the most closely built and densely inhabited towns in this kingdom', where great numbers of houses had been built back to back, without any courtyard or means of through ventilation. It came as no surprise that the General Board of Health reported a high rate of 'premature and preventible mortality'. As the Corporation pointed out, such mortality had been the main reason for promoting an Act of 1845 to enclose the commonable lands that had previously abutted on three sides of the city. Promoted as an instrument of 'social, moral, sanitary, commercial and agricultural reform', the Act opened the way for attacking the appalling overcrowding of the city.[58]

The implementation of the Act provided outstanding insights into the opportunities and constraints of civic direction in building development. The city corporation could take heart from the way in which the town

[57] C. W. Chalkin, 'Country towns', in Mingay, *Victorian Countryside*, vol. I, pp. 275–87.
[58] Act for Inclosing Lands in the Parish of St Mary in the Town of Nottingham, 1845, 8 and 9 Victoria, c. vii; Anon., *Records of the Borough of Nottingham*, Nottingham, vol. IX, 1956, pp. 89–92.

expanded (2,101 houses, 74 factories and 41 warehouses were built between 1851 and 1856). As required by the Act, levels were taken and a plan made by the Ordnance Survey. Under the direction of the commissioners, roads were widened and built, and drains and sewers constructed. Principal thoroughfares were to be at least 30 feet wide, and cess-pools at least 10 feet from dwelling-houses. Within this framework, land was, however, allocated according to the claims of the 400 owners of common – each allotment was treated as a self-contained unit, with no attempt to co-ordinate road building within or between each, or with the neighbouring countryside. The result was 'a labyrinth of perfunctory streets and a jumble of buildings which quickly began to swallow up the available land'.[59]

F. RAILWAY AND INDUSTRIAL TOWNS

Urban growth could also take the form of new settlements, where the scale and extent of expansion quite dwarfed anything there before. Typically, the origin of such growth was the railway or industrial development, and increasingly both.[60] In the second half of the nineteenth century, the ancient towns of Doncaster and Darlington were virtually taken over by railway companies; the railways created their own seaside settlements, such as Skegness, Rhyl and Colwyn Bay, and materially affected the timing of the growth of others. Railway-junction towns became new creations in their own right.

Some of the new railway towns dated from the 1840s, but the full extent of their contribution to built urban form constitutes a feature mainly of the second half of the century. As Richard Jefferies remarked in 1867, it was wholly appropriate to describe the railway workshops and housing of Swindon, on Brunel's London–Bristol line, as a New Town – they had sprung into existence within 40 years, and had a population of 8,000, were illuminated by gas and distinguished by many public buildings. The engine works were probably the largest in the west of England, employing a labour force of 1,700 men, drawn originally and for the most part from the surrounding villages. The workshops were open to visitors every Wednesday afternoon.[61] The carriage works of the Midland Railway were established at Wolverton in Buckinghamshire in 1838. The 'new railway colony' continued not only to grow, but to replace what had already been built – a man who had served his apprenticeship there in the 1840s and returned some fifty years later would have recognised only the

---

[59] R. A. Church, *Economic and Social Change in a Midland Town* (London, 1966), pp. 162–92.

[60] C. and R. Bell, *City Fathers: the Early History of Town Planning in Britain* (London, 1969).

[61] G. Toplis (ed.), *Jefferies' Land. A History of Swindon and its Environs* (London, 1896), pp. 52, 66–7.

great engine shed, the *Royal Engineer*, and a couple of little streets. Following the decision of the London and North-West Railway Company in the 1860s to concentrate its scattered carriage and wagon depot into one huge works at Wolverton, most of the early houses, with their dozen shops, were destroyed as the works doubled in size. Nothing like it had ever happened on so large a scale before in Buckinghamshire.[62]

Eastleigh, outside Southampton, was a complete foundation. Bishopstoke Junction was built in 1839 in open fields, and a permanent colony, taking its name from a neighbouring farm (Eastleigh) grew by a process of accretion as new lines (to Gosport and Salisbury) were opened. Speedier expansion came with the concentration of the company's carriage works, repair sheds and locomotive workshops in the town at the turn of the century. Likewise, Crewe began from virtually nothing. The railway station, built in 1837, was located in the old township of Crewe, but the new town, called after the station, was developed around the railway works to the west, in the township of Monks Coppenhall. The Grand Junction line (eventually part of the London and North Western) selected the site as the junction of the Birmingham–Liverpool and Birmingham–Manchester lines, together with the line to Chester and Birkenhead. The first houses were erected in 1842, and a new town laid out on company land. Its population grew from 4,500 in 1851 to 42,000 in 1901, serviced by a range of company-provided facilities.

The railway was associated with the spectacular growth of new manufacturing towns in the second half of the century. Two examples stand out: Barrow and Middlesbrough. Barrow-in-Furness had attracted some industrial development since the beginning of the eighteenth century because of its local iron ore and charcoal, but the town only came into prominence with the arrival of the railway in 1845 and the subsequent activities of the Furness Railway Company, first of all in building company cottages. After the discovery of rich haematite iron-ore in 1850, the Company engineer, James Ramsden, proceeded to lay out a new town, the population of which rose from 3,000 in 1861 to 19,000 in 1871 and 47,000 in 1881.

The growth of Middlesbrough was even more dramatic. Joseph Pease had been involved with his father in the opening of the Stockton to Darlington Railway. No deepwater staithes were available at Stockton for coal exports, and it was therefore decided to build a new dockside terminus 6 miles downstream, in the Chapelry of Middlesbrough, which then consisted of just a few houses in a saltmarsh setting. It was to be called Port Darlington. Although work began on Pease's plan, drawn up

---

[62] F. Markham, *History of Milton Keynes and District*. Volume II, *From 1800 to about 1950* (Luton, 1975), pp. 186–9.

in 1830, the mining of Cleveland ore from 1850 onwards exerted a much greater influence. The old railway town was boxed in by new lines built to bring the ore from the quarries. Under the name of Middlesbrough, it came to be known first and foremost as a steel town, with Henry Bölckow, the ironmaster, a dominant figure in the community. A population of 7,400 in 1851 rose to 91,300 by the turn of the century. The flat marshland of the river Tees was transformed.

Not all new industrial settlements, which mushroomed in the second half of the century, grew so large. Nevertheless, the effect on the neighbouring countryside could be striking. Nowhere is this better demonstrated than around colliery settlements, with their combinations of mining plant and housing development. The coalfields of Durham, Yorkshire and Nottinghamshire produced typically compact settlements, whereas development in south Wales was characteristically more linear, as the railways penetrated the valley heads to tap both coal and iron for export.[63]

No less dramatic in their local impact were the industrial developments taking place on the fringes of some major towns and cities. Whilst the exact pattern varied, according to local topography and other features, Birmingham might be cited as an example of what happened. From mid-century onwards, industrial growth occurred in four directions: to the north-west, along the banks of the Birmingham Canal beyond Ladywood, taking advantage of the barren waste of Birmingham Heath; to the south-west and south following canal and railway, where the green-field site of Bournville was selected; to the south, again following canal and railway to stimulate the growth of Small Heath, where the massive Birmingham Small Arms factory was located in 1861; and to the north-east, where undeveloped land in the Rea valley attracted large railway waggon works, and further to the north in the Tame valley, where the General Electric Company development at Witton began in 1901. Perhaps the most outstanding example of a new industry being located on the outer fringes of the built-up area occurred in 1906, when Herbert Austin chose the derelict premises of a printing works at Longbridge for car manufacture, 7 miles from the city centre.

Extensive tracts of land might be taken for the construction of docks and quays.[64] Much larger vessels, and the greater amounts of tonnage handled, required both accommodation and equipment on a totally new scale. On the Thames, the completion of St Katharine's Dock in 1828 was followed by a long pause until the Victoria Dock (1855), Millwall Dock (1870), the Royal Albert Dock (1880) and the Tilbury Docks in

[63] P. N. Jones, *Colliery Settlement in the South Wales Coalfield 1850–1926*, University of Hull, Occasional Paper in Geography, 14, 1969.

[64] H. J. Dyos and D. H. Aldcroft, *British Transport: an Economic survey from the Seventeenth Century to the Present* (Leicester, 1969).

1886, 26 miles downstream of London Bridge, with an extension built in 1912. In Liverpool, port improvements followed the establishment of the Mersey Dock and Harbour Board in 1857, with docks constructed on both the Liverpool and Birkenhead sides. During the period 1858–1908, the dock estate increased from 880 acres to 1,171 acres, and the water area doubled from 207 to 418 acres. Avonmouth and Portishead were developed in the late 1870s to serve Bristol. Manchester became an inland seaport, following the opening of the Ship Canal in 1894.

Landscapes were transformed. The Royal Family travelled to Holyhead to watch the blasting of some 20,000 tons of rock from the side of Holyhead Mountain for the construction of the breakwater required for the new 'asylum' and packet port on the Isle of Anglesey. Started in 1848, the breakwater took twenty-eight years to complete. A visitors' guide of 1853 recounted how 'the quarries exhibited one of the most active pictures of industry, from the width of the workings and the number of labourers employed'. By obtaining a pass from the Engineer's office, and subject to certain regulations, visitors could inspect the workings.[65]

By the turn of the century, a third of the world's coal exports came from the south Wales field, trans-shipped through Barry, Cardiff, Newport and Swansea. The Lord Windsor had cut the first sod of the integrated scheme for a new dock and railway, only three months after the Royal Assent was given to the Barry Dock and Railways Bill in August 1884.[66] The nucleus of the new town of Barry had been established by the time the first ship entered the dock in 1889. By the outbreak of the First World War, a thriving community of 38,000 people was well provided with essential services, roads, parks and amenities.

The original temporary 'timber town' was built to accommodate the thousands of migrant labourers and craftsmen employed on the new dock, as well as the railways needed to link the mining valleys with Barry. Most of the more permanent housing was built on land purchased by syndicates of local investors. The town grew organically, field by field, extending beyond the original pathways and cart tracks, until some physical or ownership boundary was reached. The temporary limits were often 'fossilised' by abrupt changes in road and housing layout and design. Plans had to be submitted to the Urban District Surveyor to ensure that they complied with the local version of the model by-laws, enacted under the Public Health Act of 1875. Whereas 'by-law' housing could be monotonous in appearance, the naturally hilly site of Barry meant the terraced rows snaked over the contours, with distant views across the rooftops.[67]

[65] D. Lloyd Hughes and D. M. Williams, *Holyhead: the Study of a Port* (Denbigh, 1967), pp. 94–7.
[66] Barry Dock and Railways Act, 1884, 47 & 48 Victoria, c. cclvii.
[67] D. Moore, *Barry. The Centenary Book* (Barry, 1984), pp. 333–65.

G. COASTAL RESORTS

A group of towns to experience significant growth during the second half of the century were the seaside resorts – indeed the period saw the emergence of a large number of such settlements. At mid-century, four inland spa towns (Bath, Leamington, Cheltenham and Tunbridge Wells) were still important, with a population growth rising above the national average. Harrogate and Buxton received a stimulus from the arrival of the railway. With these exceptions, the future belonged to the seaside resort.

Coastal resorts expanded more rapidly than any other group. In 1851, just 2 per cent of the population of England and Wales lived in them.[68] There was already a concentration in Sussex and Kent. Brighton was pre-eminent, with a population of 65,000, four times that of Hastings (17,000), its closest competitor as a specialised seaside watering-place. Kent's place in the hierarchy was well established, with Ramsgate (nearly 15,000) and Margate (10,000) figuring prominently. The popularity of the south coast was confirmed with the growth of Weymouth and various centres on the Isle of Wight. With railway penetration of the West Country, the Somerset coast and the southwest peninsula showed signs of early growth. Torquay was already well established, with a population of nearly 14,000. Elsewhere, resort towns were less prominent, except in Yorkshire, where Scarborough already had nearly 13,000 inhabitants.

The impact of these forces on regional growth was amply demonstrated by the Chester to Holyhead line, 'bounded by the sea on one hand and by almost inaccessible hills on the other'. The effect was to attract such large numbers of summer visitors and a residential population dependent upon them that half the population of Snowdonia came to be concentrated along the narrow coastal belt of north Wales that occupied less than one-eighth of the total area. The poor and thinly populated parish of Llanfairfechan was transformed. Because the bulk of the common lands up to the shoreline at Llandudno had been allocated under an Award of 1843 to the Mostyn family, it was possible to plan systematically the layout and development of the resort. Building plots were leased under stringent conditions from 1849 onwards. Within a decade of the branch-line being built, the town was 'rising into favour as a salubrious and picturesque bathing-place, with accommodation for some 8,000 visitors'.[69] At Bangor, hopes were expressed in the *North Wales Chronicle* in March 1851 of its becoming the 'Brighton of Wales' – hotels were enlarged and the Railway Hotel and British Hotel built. Its nodal point

[68] J. K. Walton, *The English Seaside Resort: a Social History 1750–1914* (Leicester, 1983).

[69] A. H. Dodd, *A History of Caernarvonshire 1284–1900* (Caernarvon, 1968), pp. 271–3.

on the railway network was a key factor in the establishment of a teachers' training college in 1858 and, at the turn of the century, the University College of North Wales.[70]

Brighton, with a population of over 100,000, figured among the forty provincial towns with a population of over 5,000.[71] Seventeen seaside resorts had more than trebled their population since 1851, and that of a further fourteen had doubled. Twelve resorts could claim to have more than 20,000 residents. Between 1881 and 1911, the aggregate population of coastal resorts in England and Wales grew by more than 600,000 to 1.6 millions, representing 4.5 per cent of the national total. There were now thirty-nine resorts with populations of more than 10,000 (taking account of some with non-resort elements, such as port or commercial functions). A further twenty-seven ranged from 5,000 to 10,000.

Whilst over 40 per cent of the seaside population still lived in Sussex, Kent and Hampshire, the surge of working-class holiday demand caused a spectacular growth of resorts on the Lancashire coast. Blackpool reached a population of 58,000, and Southport 51,000. By offering low fares, the Midland Railway encouraged the development of Morecambe as a resort for the West Riding of Yorkshire. The fortunes of more remote resorts varied. Those in Cornwall and north-east Norfolk recorded some of the fastest growth (admittedly from a small base), but Aberystwyth stagnated, and Tenby fell back.

By the 1880s, seaside resorts had reached the peak of their popularity with holiday-makers, and many were developing into important residential towns. Whitley Bay, to the north-east of Newcastle upon Tyne, and Penarth and Porthcawl in south Wales, became almost suburbs of larger centres. On the Essex coast, the population of Southend rose fourfold between 1881 and 1901 thanks to the combined effect of railway building, rising living standards, and the cheapness with which speculators could purchase farmland. The Great Eastern Railway Company pared fares to the bone and provided a high-frequency service to London. It was not, however, until the early 1900s, and the availability of second-hand bicycles for as little as £2 to £5, that workmen could live beyond a short walking distance of the stations. Streets crammed with two-storey, two-bedroom houses in blocks of four soon appeared. The population of Southend doubled between 1901 and 1911 to 70,700.[72]

A new urban geography was successfully implanted; additional building transformed formerly largely rural settlements and, occasionally, took root in quite empty areas. The precise factors behind the growth varied.

[70] P. E. Jones, *Bangor 1883–1983. A Study in Municipal Government* (Cardiff, 1986), pp. 16–18.

[71] S. Farrant, 'London by the sea: resort development on the south coast of England 1880–1939', *Journ. Contemp. Hist.*, 22, 1987, pp. 137–62.

[72] P. J. Hugill, 'The commuters who got on their bikes', *Geogr. Mag.*, 55, 1983, pp. 371–4.

The expansion of Torquay was inspired by one large landowner, Sir Lawrence Palk, from the 1820s, and various developers to whom he let land for building. Southport fell under the control of the Lords of the Manor, one descendant, Fleetwood Hesketh, greatly influencing the spacious layout of the town. Skegness was developed by the agent of the Earl of Scarborough, and Folkestone by the Earls of Radnor. The character of Eastbourne was largely determined by the Duke of Devonshire, who not only planned the Grand Parade along the seafront and the disposition of residential and commercial development, but concerned himself with such matters as mains drainage and street lighting. Such patronage was not, however, to be taken for granted and, as in numerous other cases, local government had to assume much greater responsibility towards the end of the century. In Eastbourne, expenditure on infrastructure was severely curtailed from 1891 onwards, on the succession of the ninth Duke of Devonshire. The borough council (incorporated in 1883) became increasingly involved in providing resort and general facilities, including improvements to the seafront, parades and parks. The electric lighting company was purchased and, in 1903, the borough became the first to establish a municipal bus service.[73]

The singular role of local circumstances was perhaps most graphically illustrated by Bournemouth. Whilst built by the sea, the making of Bournemouth owed at least as much to the affinity felt by its residents for an orderly environment of streets and building development, within a heathland setting. The population of the resort rose from 695 in 1851 to 1,840 in 1861, 5,896 in 1871, 16,859 in 1881, and 59,762 in 1901. Decimus Burton had been engaged to lay out and promote the resort in the 1840s, and a Bournemouth Improvement Act of 1856 provided for the appointment of thirteen commissioners to supervise development. The jurisdiction of the commissioners, initially limited to 1,140 acres, was extended by a further 503 acres in 1876. The Westbourne area and tracts to the east, towards Pokesdown, were added in 1884.[74] Accorded the status of a municipal borough in 1890, and county borough in 1900, Bournemouth was, in the words of C. J. Cornish, one of the economic puzzles of the century, at least as remarkable as Middlesbrough or Barrow-in-Furness. Communications had followed rather than preceded settlement. It was not until the 1870s that Bournemouth acquired two stations. Its population had gathered not to make money, but to spend it. The attraction was not so much the sea, but rather the sand-cliffs and pine trees. Leases and local regulations sought to prevent any wanton burning or injury to the heath.[75]

[73] H. W. Fovargue, *Municipal Eastbourne, 1883–1933* (Eastbourne, 1933).

[74] Bournemouth Improvement Act, 1856, 19 & 20 Victoria, c. xc.

[75] Cornish, *Wild England of Today*, pp. 76–82.

H. THE SUBURBS

The bulk of the urban growth to affect the countryside occurred on the periphery of the larger towns and, above all, London. By 1901, the number of towns in England and Wales which contained populations of more than 50,000 was seventy-seven – an increase of thirty over the figure of 1881. A third of these contained between 100,000 and 250,000 people. It was in this group that the suburban trend, noticeable since the 1860s, became marked from the 1880s onwards.

Suburban growth, albeit one of the great features of the century, was not in itself new. Urban form had long been fashioned by the practice of rich merchants building their villas on the main roads out of a city, so encouraging radial fingers to develop into the countryside. But the suburb in the second half of the century became something rather different; the spilling out of housing and different types of estate development became the norm, fuelled by a rising population and increased ability, as well as incentive, to escape from over-congested, insanitary areas of earlier construction. The new population comprised shop workers, commercial travellers, the rising class of professionals and, above all, clerks (rising from 2.5 per cent of all occupied males in 1851 to 7 per cent in 1911, implying an increase in number from less than 150,000 to more than 900,000). Formerly fashionable suburbs were to be overwhelmed by an invasion of those lower down the social scale. In Birmingham, the village of Acock's Green was absorbed by meaner forms of suburban housing, with the more affluent seeking locations further afield, as in Solihull. It was a sequence of population invasion and succession repeated in many other cities.

The railways were a crucial factor in the development of 'the new ring suburbs' around Nottingham. According to a sales catalogue, the plots offered for sale at Gedling, just outside the city, were ideally suited for village development, 'being very handy for persons having business in Nottingham'. The station, with 'its excellent service of Trains', was only three minutes' walk from the village centre, and a few minutes' travelling time from Nottingham.[76] A location just beyond the city boundary 'had the great advantage of non-liability to the heavy rates of the Town'. In his magazine for July 1883, the rector of Gedling remarked on how a new town had sprung up in the parish of Netherfield, where, over the previous decade, the Great Northern Railway and the London and North-East Railway Company had 'taken up positions of great importance'. The establishment of a large cotton doubling mill, and a lithographic and printing works, had introduced new trades. There was 'not only a swarm

[76] Nottinghamshire RO, DD 296/9.

from the old stock, but a cast as well, to be hived'. As soon as a church was built, Netherfield became a separate ecclesiastical parish in 1883. Growth continued, the sales particulars of a plot of 32 acres near the new church and parsonage drew attention to the demand for 'workmen's houses near the sidings of the Great Northern Railway'. By the turn of the century, the sidings had become the largest in the Company's system, covering 50 acres and employing a thousand men.[77]

The suburban movement was in part associated with shifts in house-tenure choice, as the trend from private renting to owner occupation became discernible. Tenancies remained dominant. The character of development in Cardiff, for example, was set by the controlled use of short-term leases on the three large estates of Bute, Tredegar and Windsor. The number of building societies nevertheless increased from 1,500 in mid-Victorian times to 2,600 in 1895, with a commensurate increase in membership from 300,000 to 600,000. Speculative building was still the principal way in which the suburbs were created, and in this regard we should note the imperfections and inadequacies of the building 'industry'. At one level the needs of the lower end of the working classes were not met at all, there being a failure to provide houses at rents that could be afforded. At another level, the relative inefficiency, whereby general market demands were met, caused occasional over-supply. Vacancy rates were at times high. Empty houses during the last quarter of the century across London reached 4 per cent of the total stock. The wastefulness of such unco-ordinated speculative land-take by large numbers of developers attracted much critical attention by the end of the century.

By 1851, the new patterns of social geography to affect the larger provincial towns were becoming apparent, with spatial and social segregation clearly in evidence. Urban form reflected this. The Ordnance Survey maps of 1848 for Manchester, for example, show the coagulation of the principal settlements:[78] already, there was continuous building along the 7 miles of road from Manchester to Oldham. The towns of Stalybridge, Dukinfield and Ashton were fused by the spread of houses and factories. The villages of Cheadle and Gatley to the south, the small market town of Wilmslow and larger market centre of Altrincham, were engulfed by the spread of suburbs. Didsbury and other nearby villages were colonised by businessmen. In 1851, 92,000 people had lived in central Manchester; the decennial census of 1901 recorded only a third of that number.

Fashionable residential suburbs rapidly became a marked feature of the

[77] C. Gerring, *A History of the Parish of Gedling* (Nottingham, 1908), pp. 214–16; Nottinghamshire RO, DD 296/9.

[78] T. W. Freeman, 'The Manchester conurbation', in *Manchester and its Region*, ed. C. F. Carter (Manchester, 1962), pp. 47–60.

morphology of the larger cities as they spread into the surrounding countryside. In Birmingham, the Calthorpe estate of Edgbaston was systematically developed on long building leases, with covenants to ensure the
highest standards of residence for the gentry of the fast-growing city.
Leeds spilled over into Chapel Allerton and Headingley; business men in
Sheffield sought out the villages of Dore and Totley, and built on the high
ground between the valleys. Liverpool spread inland to places like West
Derby. Clifton was selected by the wealthy of Bristol. The high-status
area of Cardiff was Roath Park and, in Newcastle, Jesmond became the
most favoured district.

A distinctive, intimate landscape came into being from the 1880s
onwards. In the words of J. D. Sedding, 'Everyone who can, now lives in
the country, where he is bound to have a garden.' No longer was gardening 'a merely princely diversion requiring thirty wide acres for its display'.
Science, technology and the fruits of geographical discovery could be
drawn upon, in terms of growing methods, the use of the greenhouse
and conservatory, and access to exotic plantlife from the Empire.
Gardening was taken up with zeal by the villa owner and larger house-
holder. The book *Gardens for Small Country Houses*, by Gertrude Jekyll
and Lawrence Weaver, published in 1912, was an immediate and phe-
nomenal success.[79]

Whereas the designers of house and garden had tended to work independently, the 'Edwardian garden' brought them together. As Jekyll and
Weaver wrote,

It is upon the right relation of the garden to the house that its value and the
enjoyment that is derived from it will largely depend. The connection must be
intimate, and the access not only convenient but inviting.

The challenge was to secure that linkage. The revival of the architectural layout was one solution, as exemplified by the works of Reginald
Blomfield and Francis Inigo Thomas, and the advocacy of the strict formality of the Italianate garden, with shrubbery, gravel paths (in preference to lawns), urns and statuary regularly placed, bedding plants and
ornamental edgings. The garden became an architectural design feature.[80]
At the opposite end of the spectrum, a more 'natural' school of gardening evolved, inspired by the landscape gardener, William Robinson,
whose book, *The English Flower Garden*, published in 1883, had passed
through nine editions by 1905. The trimming and pruning of plants into
a manicured design was rejected in favour of flowers, plants and shrubs

[79] J. D. Sedding, *Garden-craft Old and New* (London, 1894), pp. 176–7; H. Barrett and J. Phillips,
Suburban Style: the British Home, 1840–1960 (London, 1987); G. Jekyll and L. Weaver, *Gardens for
Small Country Houses* (London, 1912).
[80] R. Blomfield and F. I. Thomas, *The Formal Garden* (London, 1892).

displayed in their natural state. The move towards more natural effects was echoed in the design of garden furniture. In sympathy with the Arts and Crafts movement, plain oak or elm benches replaced ornate iron-work. Trelliswork and pergolas came into fashion. Lawns became bold and given over to games, such as croquet and tennis; lawn mowers and heavy iron-rollers kept the lawns firm.[81]

Whilst most Edwardian gardens were formal near the house, the further parts were given over to Nature's wilder self. Here, as Sedding wrote, 'Art should only give things a good start', and help the propagation of those not indigenous to the locality. For her book *Wood and Garden*, which appeared in 1899, Gertrude Jekyll took for her inspiration the country-side of her native Surrey, and the practical lessons learnt in her own house and garden at Munstead Wood. It was in the gardens created with Edwin Lutyens that formal layout and exuberant, informal plantings found great-est expression. A genius in architectural form and geometric invention was combined with her 'mature understanding of the crafts and an enthu-siasm for vernacular and old-fashioned plants'. The outcome was, to quote Ottewill, 'a succession of enchanting designs, original in concept, perfect in scale and exquisite in colour and material'.[82]

The upper middle class might have been the 'pioneers' in selecting new areas for living and according so much attention to the house and garden, but the suburbs quickly assumed an even wider heterogeneity as other income groups moved in. Speculative development embraced estates ranging from those of the highest reputation, where larger villas were set back in their own landscape grounds, to those comprising terraces for clerks and the more modest professional classes. Sometimes their charac-ter might be formed by the nature of the transport network, typically the railway. At other times, the aspirations of local landowners might be the main determinant, using the device of minimum housing values to secure the desired social occupancy and appearance of the estate. Whatever the process, it was a mosaic of suburban building types that ate into the sur-rounding fields.

I. LONDON

Nowhere were the suburbs more a feature of the urban–rural juxtaposi-tion than in London. In 1901, the population of Greater London stood at more than 6.5 million. It is significant that the Census of 1881 intro-duced the term 'Greater London' for the first time, acknowledging the

[81] W. Robinson, *The English Flower Garden* (London, 1883).

[82] G. Jekyll, *Wood and Garden* (London, 1899); D. Ottewill, *The Edwardian Garden* (New Haven, 1989).

outer spread of the metropolis. The outer ring of its suburbs grew by about 50 per cent in each of the three intercensal periods between 1861 and 1891, and by 45 per cent in the decade between 1891 and 1901. The four places which experienced most rapid population growth in the whole country between 1881 and 1891 were London suburbs. It was in this context that Dyos traced the growth of Camberwell, distinguishing the periods before and after the coming of the railway in the 1860s and of the tramway in the 1890s.[83] Throughout London, there was a rich variety of suburbia, each form depending greatly on the nature and aspirations of local landowners, developers and builders.[84]

Middle-class suburbs were to be found in Surbiton, Ealing and Sidcup. Most of the houses were within easy walking distance of the station, but the larger villas were some distance away, maintaining privacy and seclusion. As Jackson has noted, 'woods and fields were never very far away, and *urbs* mixed most harmoniously with *rus*'.[85] The less expensive middle-class suburbs were in such places as Palmers Green, Wood Green and Hornsey, which themselves were more attractive alternatives to Hackney, Islington and Holloway. Suburbs for artisans and clerks were much less attractive, those of Tottenham, Edmonton, Walthamstow and Leyton comprising long, standardised terraces, with minimal garden. Even in these areas, after the turn of the century, densities fell and layouts became more open, thereby increasing the annual land-take *pro rata* to dwellings built. Suburbs matured as building cycles ended, and new areas were selected for development. Edwardian growth in east London came to centre on Ilford, Chingford, Wanstead and Woodford. In the north, Southgate and Enfield expanded; so too did Hendon, a little further west. In the north-west, it was Harrow, Wembley and Ruislip; to the west, Hanwell, Southall and Twickenham. South of the river, building activity favoured Wimbledon, Raynes Park and Streatham, and particularly around Merton and Morden. The notion of London as a collection of villages was maintained.

The importance of the railway in opening up the rural frontiers for an expanding London was crucial, though Kellett has stressed the subordination of railway building to the pre-eminent importance of property ownership and patterns of land values.[86] London's first railway, the London and Greenwich, was an intensive passenger-carrying line, but those built later were generally slow to develop their suburban traffic.[87]

[83] H. J. Dyos, *Victorian Suburb: a Study of the Growth of Camberwell* (Leicester, 1961).

[84] F. M. L. Thompson (ed.), *The Rise of Suburbia* (Leicester, 1982); P. Brandon and B. Short, *The South East from AD 1000* (London, 1990), pp. 279–90.

[85] A. A. Jackson, *Semi-detached London: Suburban Development, Life and Transport, 1900–39* (London, 1973).   [86] J. R. Kellett, *Railways and Victorian Cities* (London, 1969).

[87] P. Hall, 'The development of communications', in *Greater London*, ed. J. T. Coppock and H. C. Prince (London, 1964), pp. 52–79.

However, from the 1850s onwards, new stations were successively opened on existing main lines. Branches were constructed in order to tap the profitable areas in between, thereby opening up any tracts of countryside that remained. The construction of the first long-distance commuter line in 1856, the London, Tilbury and Southend line, dramatically extended the urban reach.

The intimate relationship between the development of a suburban railway network and suburban building is well seen in the changes that took place in north-west London.[88] By 1853, the North London Railway was making determined efforts to transform a set of scattered lines belonging to hostile companies into a coherent network. An immediate result was the expansion of building between Kentish Town and Hampstead. Later, the completion of the Inner Circle, and the extension of the Metropolitan Railway to the north-west, stimulated development between Swiss Cottage and Kilburn. Later still, beyond Hampstead, the arrival of the tube in 1907 was followed by the rapid growth of Golders Green.

The policies of the railway companies also had a bearing on the social character of the suburbs. In the 1850s, the London and North Western Railway Company encouraged what amounted to low-density development by the offer of a free first-class ticket for twenty-one years to those building houses of an annual rented value of £50 and over at places as far out as King's Langley, Boxmoor and Tring.[89] On the other hand, the Great Eastern Railway encouraged third-class travel. It came about in the following way. In 1861, in order to obtain parliamentary approval for the demolition of working-class houses as part of the Broad Street extension, the North London Railway was obliged to provide special workmen's trains at low fares. In 1864, the Great Eastern had to follow suit in order to secure access to Liverpool Street. By the 1880s, more trains were provided than strictly required, and the result was the development of working-class suburbs in Tottenham, Edmonton, Leyton and Walthamstow.

Thus London's suburbs continued to grow, heavily reliant on effective forms of transport; sequentially the horse omnibus, suburban train, tram, tube and latterly the petrol bus. By 1901, over two million people lived in this belt, between 6 and 15 miles from the centre (effectively beyond the range of the horse tram). Surrounding villages became dormitory settlements and, in the process, London's urban fringe became, in the words of Olsen, 'islands of traditional village environments'.[90] Remains of

[88] H. C. Prince, 'North west London 1864–1914', in Coppock and Prince, *Greater London*, pp. 120–41.

[89] J. T. Coppock, 'Dormitory settlement around London', in Coppock and Prince, *Greater London*, pp. 265–91.   [90] D. J. Olsen, *The Growth of Victorian London* (London, 1979), p. 227.

earlier dwellings gave architectural variety and a certain richness; a suburb and each district of a suburb had its own variety in which the marks of a rural past (footpaths and field boundaries, to name but two) were never quite obliterated.

J. OTHER URBAN DEMANDS

By the end of the century, genuine alarm was expressed as to the remorseless outward march of suburban London, where the land-take was most sustained. But housing and industry were not the only demand to be made on either the urban fringe or further afield in the country. There were other components to urban land-use, and their requirements should not be overlooked. A number will be identified here. Others will be reserved for later chapters.

If the countryside was a provider of the raw materials needed to sustain urban life, so too did it provide space for those uses which could no longer be accommodated in town and city. The expanding 'forest of practically imperishable gravestones' was eloquent testimony to the pressure on urban space. The large number of burials and 'enormous loads of monumental marble' made a large extension to Highgate cemetery necessary only a few years after its opening.[91] Unless England was to be divided 'like a chess-board into towns and burial-places, ways had to be found for disposing of the dead without taking so much valuable space from the living'. In the metropolitan area of London, twenty-four new cemeteries, the equivalent of 600 acres, were almost filled, and in some cases, overfilled, during Queen Victoria's reign. Of the 362 grounds identified in a survey of 1895, only forty-one could still be used. Ninety had already been converted to public gardens and playgrounds. Unless full use was made of the Disused Burial Grounds Act of 1884 and Open Spaces Act of 1887, many of the remainder would, if past experience was any guide, be built over, and their value as open space lost.[92]

Nothing came of proposals by the Board of Health for two large cemeteries for London at Kensal Green and Erith in Kent, and, in 1851, a group of businessmen and landowners formed the London Necropolis and National Mausoleum Company in order to exploit another proposal made, namely for a national cemetery or necropolis (city of the dead) at Woking, one of the few remaining unenclosed areas near London. Not only were the light sandy soils of the Bagshot Sands well suited for interment, but the railway from Waterloo bisected the parish. The lordship of

[91] W. Howitt, *The Northern Heights of London* (London, 1869), pp. 94–5.
[92] B. Holmes, *The London Burial Grounds* (London, 1896); C. Brooks, *Mortal Remains. The History and Present State of the Victorian and Edwardian Cemetery* (Exeter, 1989).

the manor was purchased and, under a Private Act of 1852, the Company was empowered to acquire by compulsion, if necessary, almost all the commonland of the parish.[93] In the event, only 450 acres of the 2,118 acres of former common were used as a cemetery. It was, nevertheless, the largest cemetery to be established. Located over 3 miles from Woking station, in the part of the parish least suitable for speculative building development, it became known as the Brookwood Necropolis, and acquired a station of its own on the main railway line to London. The expanse of gorse, bracken and a few pine trees was transformed by the company architect into a grid of paths and sequence of curving concentric avenues. Plantings of rhododendrons and azaleas were interspersed with lawns and groves of conifers. The peak of use was in 1866, when 3,842 burials took place. The opening of further cemeteries nearer to London caused usage to fall significantly in the 1890s.[94]

Urban fringe-development might also include the building, somewhat belatedly, of mental and isolation hospitals. Whether private, county or city asylums, they were situated well beyond the built-up limits. That for Birmingham was erected at Rubery in Worcestershire. The Infectious Disease (Notification) Act of 1889 made it compulsory for doctors to notify their local authorities of the incidence of such diseases in London, and optional throughout the rest of the country.[95] Despite the cost, the building of 'fever' hospitals for the larger towns began in earnest.

Clues as to the availability of open spaces on, or near, the urban fringe at any one time may be provided by the opening of public recreational grounds. One may cite the municipal parks set out in Manchester and Salford in 1846, Norfolk Park in Sheffield (1847), the People's Park in Hull, and Peel Park, Bradford, both in 1863, and Alexandra Park, Manchester, in 1868. In London, the Victoria Park was opened in 1844, Battersea Park in the 1860s and Finsbury Park and Southwark Park in 1869. In Birmingham, the first public park, Adderley Park (provided by Lord Norton in 1856) was followed by Calthorpe Park in 1857, Aston Park (opened by Queen Victoria in 1857), and Highgate Park and Summerfield Park in 1876. Cannon Hill Park, the gift of Louisa Ann Ryland, was opened in 1873, and Small Heath Park, from the same donor, in 1887. A landscape of trimmed walkways, flower gardens, informal open spaces and the ubiquitous bandstand became a distinctive urban feature.[96]

By the end of the century, the golf course had taken its place as a completely new use of land in England and Wales. Those who had enjoyed

---

[93] London Necropolis and National Mausoleum Act, 1852, 15 and 16 Victoria, c. cxlix.

[94] A. Crosby, *A History of Woking* (Chichester, 1982), pp. 54–63.

[95] Infectious Disease (Notification) Act, 1889, 52 & 53 Victoria, c.72.

[96] G. E. Cherry, *Cities and Plans* (London, 1988).

golfing holidays in Scotland, the traditional home of the game, sought similar facilities nearer home. An obvious location were those parts of the London commons near a railway station – the attractions of Mitcham Common were exploited by the establishment of the Prince's Golf Club in 1890, the prime mover being the chairman of the Mitcham Common Conservators. A lack of privacy and the imposition of restrictions (golf was prohibited after 9.30 in the morning on Clapham Common) encouraged co-operative ventures for the purchase or renting of hundred-acre sites. The Woking Golf Club was founded in 1893 by a group of London barristers and judges as the first of many on the heathlands of Surrey and neighbouring counties. By 1900, there were forty-nine clubs within 15 miles of Charing Cross, and a steady increase in the provinces. In the Birmingham area, there were the Robin Hood, Moseley, Harborne, Handsworth and Edgbaston Clubs.[97] A wartime survey identified 100 courses in Lancashire, occupying some 5,900 acres. Most occupied 'open wastes', and would have required large amounts of fertiliser and fencing for any agricultural improvement to be possible.[98]

Proximity to London or large urban centres also counted for much in the development of the race-course. Four of the five top enclosed grounds, as well as Epsom, were in Surrey. Sandown Park, the first enclosed ground, opened its turnstiles in 1875, and Redcar and Stockton, serving the rapidly growing population of Teesside, soon followed. In the Midlands, courses were created at Derby in 1880, Leicester 1884, and Colwick Park, Nottingham, 1892.[99] Racing of a different kind was established at Brooklands, near Weybridge, in 1907 – again on the light sandy soils of Surrey. A labour force of up to 2,000 men, using ten steam grabs and a steam navvy, constructed the first motor course in the world within a year. The 100-feet wide concrete track was roughly oval in layout, and almost 3 miles long. As with horse-racing, the railways ensured the success of motor-racing as a mass-spectator sport, the crowds travelling on the same line, the London and Southwestern Railway, as served Sandown Park and the Brookwood necropolis.[100]

The amounts of refuse generated by cities and towns varied. In London, the Strand area produced an average of 9 hundredweight (cwt) per head per annum, St Pancras 6.66 cwt, and Shoreditch 3.5 cwt. The average for London as a whole (5 cwt) was similar to that estimated for the remainder of the south of England. The Midlands produced an

[97] G. Cousins, Golf in Britain. A Social History from the Beginnings to the Present Day (London, 1975), pp. 48–54.

[98] Lancashire RO, County War Agricultural Executive Committee, Minute Books.

[99] W. Vamplew, The Turf; a Social and Economic History of Horse Racing (London, 1976), pp. 38–48; M. Huggins, 'Horse-racing on Teesside in the nineteenth century', Northern Hist., 23, 1987, pp. 98–118.   [100] W. C. Boddy, The History of Brooklands Motor Course (London, 1948–50), pp. 1–4.

average of 7 cwt per person, and the north 10 cwt. There was a decline in the proportion of refuse sorted and reused. Not only did a recycling system have to be elaborate to be effective, but it became increasingly difficult to find men willing to engage in such 'loathsome and degrading employment' for so small a wage.[101]

Refuse tips became larger and more commonplace, despite the refusal, on the part of the Local Government Board, to sanction loans for the purchase of such sites on the premise that the flies and dust were likely to spread disease, and the smells arising from decomposing matter were so objectionable. In many parts of the country, heaths and commons, which would otherwise have been 'natural health resorts', were defiled by the shooting of rubbish into old gravel, clay or stone pits, and by the litter of glass, crockery and metal, as well as dust and decomposing matter, left around them. As H. B. Woodward pointed out, in a paper to the Royal Sanitary Institution in 1906, the number of abandoned pits had greatly increased over the previous twenty years, reflecting the decline in demand for the poorer kinds of road-material and the concentration of the brick industry on the larger works in better situations.[102] At Kimbolton in Huntingdonshire, the dust cart of the St Neots District Council made a round once a quarter, disposing of its contents into the Stonely clay pits.[103]

The most effective and economical system of refuse disposal was considered to be the modern refuse destructor, whereby all the refuse was burnt to such high temperatures as to leave only a small residue of fixed and harmless products. The earliest destructors were erected in Manchester in 1876, and in Leeds a year later.[104] The question was how far the smaller authorities might be able to adopt the practice. At Bangor in north Wales, household refuse was dumped by contractors on the foreshore between Hirael and the mouth of the Adda, where in time the land was extended across the bay. Mounting concern as to the health risks to nearby built-up areas, and opposition from groups of ratepayers to the use of alternative sites, caused the council to erect a destructor in 1899 on a site adjacent to the electricity works. Whilst the steam which drove an electric generator made the venture economic, there were complaints as to the large quantities of smoke emitted and the accumulation of rubbish when the plant needed servicing and repair.[105]

A harmony between urban and farming interests was most obviously

[101] E. C. S. Moore and E. J. Silcock, *Sanitary Engineering* (London, 1906), vol. II, p. 784.
[102] H. B. Woodward, 'The utilization of old pits and quarries and of cliffs for the reception of rubbish', *Journ. Roy. Sanit. Instn*, 27, 1906, pp. 467–9.
[103] Huntingdon RO, Manchester MSS, 57a.
[104] Moore and Silcock, *Sanitary Engineering*, vol. II, p. 787.
[105] Jones, *Bangor 1883–1983*, pp. 60–1.

achieved at Carrington Moss in Cheshire. An estimated 215,000 tons of refuse were produced annually in nearby Manchester. As a partial solution, the Health Committee of the Corporation suggested transporting the night soil and street sweepings to a locality remote from residential property, where it might be used to manure farmland. Of five sites considered, Carrington was the best in terms of size and access by rail and canal. Loan sanction was obtained for the purchase of the estate of 1,093 acres, containing 600 acres of uncultivated mosslands, which were subdivided into 8-acre rectangular plots and further divided into 2-acre blocks by drains 4 feet deep. The night soil was distributed by three small locomotives pulling over a hundred trucks on a 12-mile light railway constructed beside new roads. In the first year of reclamation as much as 300 tons per acre were applied; as the peat dried and became more consolidated, the amount was reduced to 30 tons.[106]

The Corporation succeeded in its primary aim of disposing of large quantities of rubbish (an average of 54,000 tons were transported annually between 1889 and 1899, with each acre receiving 730 tons), and in its secondary object of converting the mossland to productive farmland. In 1895, the Corporation purchased 2,500 acres of Chat Moss, to the north of the Manchester Ship Canal, where a similar scheme was introduced. By the early 1900s, however, the manure had become so rubbishy as to be hardly worth carting to the fields. The introduction of 'modern methods of sanitation' had caused the proportion of night soil and cinders to become so low as to require supplementing with artificial manures.[107]

Both urban housing and industry looked to the country for raw materials. In south Wales, the monotonously dull, greyish colour of the Pennant sandstone of the terraced housing was a dominating feature of the coalfield valleys. Almost every township had its quarry.[108] There might be both dereliction and pollution. As Spencer's *Illustrated Leicester Almanack* of 1880 remarked, what traveller on the Midland Railway had not encountered in the Soar valley, between Leicester and Derby, the thick clouds of lime dust and smoke that arose from 'the numerous and extensive works carried close to the line in the manufacture of hydraulic cement'?[109] Within a few years in the 1850s, seven cement works were established at Findsbury, and several more on both banks of the Medway above Rochester, in Kent. Entire hills were demolished for the chalk, and enormous quantities of 'blue clay' were excavated from beneath the

[106] A. D. M. Phillips, 'Mossland reclamation in nineteenth-century Cheshire', *Trans Hist. Soc. Lancs. & Cheshire*, 129, 1979–80, pp. 93–107.

[107] E. Price Evans, 'Carrington Moss, with special reference to the weeds of arable ground', *Journ. Ecology*, 11, 1923, pp. 64–77.

[108] E. D. Lewis, *The Rhondda Valleys* (London, 1959), p. 140.

[109] J. Croker, *Charnwood Forest: a Changing Landscape* (Loughborough, 1981), p. 140.

Medway saltings (one firm alone extracted over 2 million tons between 1881 and 1911).[110]

It was, however, in brick-making that the most spectacular increases in land-take were to be seen. Brick was no longer the accepted material only for fashionable buildings of the aristocracy, gentry and clergy, but became in the second half of the nineteenth century the most common building material. Traditionally, bricks and tiles were produced as near to the point of demand as possible. Most brickyards served the needs of their respective estate or parish. Professional brickmakers were few, and labourers engaged only on a seasonal basis. However, even before the half-century, important changes were afoot, particularly following the adoption of brick- and tile-making machines, which made possible the mass production of unprecedented quantities of bricks, when required. A further technological breakthrough came in 1858, when continuous firing in a kiln became a practical possibility. The new kiln saved fuel, and because the same chambers could be filled or emptied as others were being fired, bricks could be made on a continuous basis in the same kiln.

Bedfordshire emerged as a major brickmaking centre, consequent upon the opening of the Great Northern Railway in 1850.[111] Robert Beart built a huge works at Arlesey on the gault clay. By 1858, about 8 million bricks and 1 million agricultural drainage-pipes were produced annually, the bricks being sent to many parts of the country, especially to the northern Home Counties and London. Huge amounts of clay were extracted, the working face of the pit being some 50 to 60 feet deep. In addition to a further four brickworks opened at Arlesey, beside the railway, others were established at Meppershall, linked by tramway to the Bedford–Hitchin line, Elstow and Westoning, on the Midland Railway to London, and Wooton Broadmead on the Bedford to Bletchley line.

The deep and uniform deposits of Lower Oxford Clay in the Fletton area, south of Peterborough, were found around 1880 to be exceptional in having so high a carbonaceous content as to mean each brick, once heated, would virtually heat itself. Coal, or rather coal dust, was needed only to control and help maintain the temperature for the period of firing. This property, together with the ease with which the clay could be ground to powder, and its low moisture content and plasticity, enabled the bricks to be produced more cheaply than anywhere else for the London market. Large quantities of smoke had always arisen from the open clamp kilns used by brickmakers, but the 'oily' fumes from the

---

[110] R. Marsh, *Rochester. The Evolution of the City and its Government* (Rochester, 1974), p. 59; J. M. Preston, *Industrial Medway: an Historical Survey* (Rochester, 1977), pp. 68–90.

[111] A Cox, *Brickmaking: a History and Gazetteer* (London, 1979).

Fletton kilns were particularly obnoxious. A High Court injunction was obtained by a local resident in 1881. The adoption of the more efficient transverse-arched Hoffman kiln brought some relief.[112]

Whether in town or country, air pollution became increasingly obtrusive. In his guide to Epping Forest in 1884, E. N. Buxton warned of how, if distant views were sought, it was best to choose a day when the wind was in the east. Winds from the south and west came laden with London's smoke. A Flora of the Manchester area, published in 1859, noted how the quantity of lichens had declined of late years, as a result of the felling of old woods and 'the influx of factory spoke', which appeared to be 'singularly prejudicial to the lovers of pure atmosphere'.[113]

There were descriptions of how 'the great smoke drift from south and east Lancashire could be seen crossing the Pennine Range of moorlands, and then mingling with the West Riding smoke'. A clear expanse of white snow on the moors might often be 'palpably blackened' within two or three hours. The distance to which soot could be carried was perhaps most graphically indicated in the Lake District, where rain, falling from clouds that had traversed an industrial area, sometimes deposited so much soot on the lakes as to give them a chequered appearance as the tarry matter spread out into black, greasy patches. In time, as a published photograph of Lake Coniston showed, a fringe of black scum appeared round the lake edge.[114]

In the north of England at least, it could be safely inferred that nowhere was free from 'atmospheric contamination from smoke'. Systematic studies carried out on the university farm and other sites within, and around, Leeds suggested that, of the 35,000 tons of soot produced each year, six-sevenths came from the domestic grate. Ten times as much was blown away from the city as was deposited locally. A paper published in the *Journal of Agricultural Sciences* in 1911, drawing on the Leeds observations, described the deleterious effects on crops and soil organisms. Analyses showed that the rain became noticeably rich in suspended matter, chlorides, sulphates (and often sulphur dioxide and other sulphur compounds), nitrogenous compounds (notably ammonia) and free acid. An experiment with perennial meadow grass, timothy, found that the soil, after continued application of 'acid rain', became distinctly poorer in protein and richer in crude fibre, and consequently less

[112] R. Hillier, *Clay that Burns. A History of the Fletton Brick Industry* (London, 1984).

[113] E. N. Buxton, *Epping Forest* (London, 1884), p. 29; P. Brimblecombe, *The Big Smoke. A History of Air Pollution in London* (London, 1987), pp. 63–160; L. H. Grindon, *The Manchester Flora* (London, 1859), p. 513.

[114] C. E. Moss, *Vegetation of the Peak District* (Cambridge, 1913), pp. 25–6; J. B. Cohen and A. G. Ruston, *Smoke: a Study of Town Air* (London, 1912).

nutritious. As the authors of the study remarked, if this held good for meadow grasses in general, the findings had great practical importance, not least in respect of the nutritive value of meadow hay bought from smoke-infested areas.[115]

---

[115] C. Crowther and A. G. Ruston, 'The nature, distribution and effects upon vegetation of atmospheric impurities in and around industrial towns', *Journ. of Agricultural Science*, 4, 1911, pp. 25–66.

CHAPTER 29

THE DYNAMICS OF CHANGE

BY GORDON E. CHERRY AND JOHN SHEAIL

INTRODUCTION

The parameters of countryside change have been established. In agriculture, the unprecedented prosperity of the third quarter of the century was followed by the 'Great Depression'. For many, the countryside seemed to have little to offer, whether in terms of employment or a place to live. Economic fragility was emphasised by the continuing pressure on rural craft industries. There was, however, another side to the picture. Towns of many types were quite literally advancing. Large-scale suburbanisation led to major intakes around London from the 1860s onwards, and around many provincial centres from the 1880s. An increasing range of urban demands had to be met by the countryside. The urban 'hand' became ever more apparent.

It would, however, be misleading to suggest that the late-Victorian countryside was perceived entirely in terms of potential urban-space. There is evidence, both on the ground and in contemporary writings, of a longing on the part of some to live and work in the country. The slight increase in rural population between 1901 and 1911 reflected both suburbanisation and, much less obviously, the scatter of housing that was beginning to appear in remote and hitherto sparsely populated areas. In his *Highways and Byways in Hampshire*, Moutray Read remarked on how, at Medstead, there was 'a small tin town' that might have been picked up in the Wild West and dropped by the roadside.[1] In parts of the Weald, too, there were bungalows and raw-bricked cottages and villas looking as if they had been 'transported bodily from a London suburb', and erected and maintained out of savings earned somewhere else.[2]

Such developments reached their apogee in the 'plotlands', comprising shacks and shanties, first seen around 1900, along coasts and rivers, and in the countryside generally. They were the rural, informal retreats

[1] D. H. Moutray Read, *Highways and Byways in Hampshire* (London, 1908), pp. 409–10.
[2] Hall, *Pilgrimage of Farming*, pp. 47–8.

of freedom-loving town-dwellers.[3] The land for the unregulated and unwanted additions was bought cheaply, although there was also a good deal of squatting. The banks of the rivers Thames and Lea, and the North Downs, were popular among Londoners. Mancunians fled to the Pennine fringes; east Midlanders turned to Charnwood Forest. The larger colonies of the south and east coasts attracted most attention. Shoreham Beach, Peacehaven and Camber Sands, and Jaywick Sands and Canvey Islands all became notorious, particularly after the War. The self-built cottages of the Laindon–Pitsea area of south Essex became to all intents and purposes, an extensive, rural slum.

Whilst some might have been driven to the countryside as a refuge, many more appear to have consciously sought a rural setting in which to live, work or retire. For those who could exercise a choice, what kind of countryside was sought? The geographer, Percy M. Roxby, noted how 'the rural El Dorado' was not so much the flat fenlands of south Lincolnshire, where almost every available acre was cultivated, but rather a region like the Surrey hills, where much remained as heath and forest.[4] The new villas followed 'the line of the sand as closely as collieries follow the line of the coal'. An area of 120,000 acres of the Surrey Hills and Hampshire commons was parcelled out. The roofs of red houses could be seen, thick among the pines, birches and heather, threatening to become 'one immense residential suburb, composed of houses graded to suit all incomes from £500 a year upwards'.[5]

The realisation of this threat was still some way off – not every speculative venture succeeded. An estate of 940 acres, comprising the former common at Fleet, was auctioned in 1878, and acquired by a local land agent, who straightaway laid out roads and carved the land, half of which was still unbroken heath, into 1,500 plots. A few were soon resold, but most were offered for auction in 1880, the sales particulars describing the 'beautiful and rising neighbourhood'. In the event, too much land was put on the market, and many vacant plots remained when the Ordnance Survey came to revise its large-scale maps of the area in 1910.[6]

If these were some of the complex, physical expressions of a move, both from and into, the country, what provisions were made of an institutional kind? What kind of impression was conveyed by those who set out to awaken and shape public consciousness – journalists, men of letters, social theorists and sociologists, historians, clergymen, medical doctors, sanitary experts, architects and municipal administrators of the

---

[3] D. Hardy and C. Ward, *Arcadia for All; the Legacy of a Makeshift Landscape* (London, 1984).

[4] Roxby, 'Rural depopulation in England', pp. 174–5.

[5] Cornish, *Wild England of Today*, pp. 76–82.

[6] British Museum, MSS 136a, 10(1); M. H. Ferguson, 'Land use, settlement and society in the Bagshot Sands region, 1840–1940', PhD thesis, University of Reading, 1979, p. 642.

period? How far did rural nostalgia and a yearning for outdoor life figure among their aspirations?

It is clearly time for us to focus on these individual forces of change, which sought to highlight the economic or social opportunities opening up in the late nineteenth century. The themes to pursue include: institutional reform, the urban demand for rurality, housing reform, industrial villages, colonial plantations, and the garden city.

A. INSTITUTIONAL REFORM

Rural England and Wales were characterised by entrenched social and political power structures. The critical feature was land; its ownership was a principal source of authority, and lack of it often the reason for gross inequality. Both social standing and political power stemmed from wealth, and landed property, itself the source of wealth, attracted special esteem. The hierarchy of landed proprietors had long attracted reformist attention and, at the time of the repeal of the Corn Laws, for example, there had been demands for the aristocratic monopoly of the ownership of land to be broken, and for free trade in corn to be accompanied by free trade in land. In the final quarter of the century, the political stakes were raised in an effort to shift the balance and bring about a redistribution of social and political privileges.[7]

We might begin by considering the rural power structures, and how shifts in property rights in the last quarter of the century, and before the First World War, were part of the process of social, economic and political change. In the transition from a narrow plutocracy to bureaucratic, liberal welfare capitalism, the inherited hierarchy of class and other structures was subject to increasing stresses and strains. In tracing the form these took, it should be noted that the situation was a little different in Wales. Rural grievances were channelled into a radical, Welsh political programme. A series of evictions in 1858 and 1869 fuelled anti-landlord resentment, exacerbated by a long-standing resistance to tithes, which flared up in the depression years of the 1880s. The threat of a mass tenant movement for land reform did not, however, materialise. Tenurial relations were not that imperfect. There was in many cases a much closer relationship between farmers and labourers than in England, arising in part from the ties of non-conformity and language.

The Victorian period was 'the last in which the possession of vast acres

[7] R. Douglas, *Land, People and Politics; a History of the Land Question in the United Kingdom, 1878–1952* (London, 1976); P. Horn, *Labouring Life in the Late Victorian Countryside* (Dublin, 1976); P. Horn, *Rural Life in England in the First World War* (Dublin, 1984); A. Offer, *Property and Politics, 1870–1914* (Cambridge, 1981); H. Newby, *Country Life: A Social History of Rural England* (London, 1987); M. Shoard, *This Land is Our Land* (London, 1987).

was a passport to political power'.[8] The transformation began in the counties with the third Reform Act and the redistribution of seats in 1884–5. The franchise was extended to all (male) householders, and the old two- or three-member county divisions were replaced by smaller, single-member constituencies. Change was, however, slow. Whilst the rural electorate was enlarged, and the farming element was proportionately reduced, the landed interest was not extinguished. The great estates did not break up, and the labouring classes were not emancipated quickly.

For a long period, the aristocracy and gentry had provided together an impressive pinnacle of power. The national survey of land ownership, in 1872, indicated the respective positions. About 400 peers and peeresses owned estates in England and Wales which averaged over 14,000 acres; roughly 1,300 'great landowners' (but not peers) owned estates of less than 7,000 acres; 2,500 squires and lesser gentry owned estates averaging more than 1,700 acres. The smaller estates tended to be nearer London, and the rate of turnover was increasing as suburban pressure rose. The larger estates were rarely sold.

It was against such a background of land tenure, and its attendant social tensions and political antagonisms, that the 'land question' became the focus for attacks on rural privilege and disparities. A number of related issues came to the fore: economic recession in British farming, the condition of the agricultural labourer, migration from rural areas, landlord–tenant relations, and agricultural unrest. The period was conducive for political change and institutional reform.

The Game Laws were a major area of contention. Whatever the intention of the Game Act of 1831, which codified previous legislation and sought to make the taking of game less exclusive, the overall effect was, according to George Shaw Lefevre, the author of a treatise on the Game Laws in 1874, to make the reservation of game by landlords almost universal. Now that game could be sold, it had the sanction of property. Subletting became common to shooting tenants, who had even less of a common interest with the farm tenant. Whereas a hundred or at the most two hundred head of game had been regarded as a good day's sport for a party of gentlemen, battues toward the end of the century 'degenerated' into the slaughter of numbers in excess of a thousand. To achieve such numbers, hare and rabbit, as well as pheasant, partridge and grouse, had to be reared. The number of gamekeepers increased from 9,000 in 1851 to 23,000 by 1911.[9]

The question of land taxation figured prominently, demarcating the creeds of Liberalism and Conservatism. Liberals argued that local taxation fell increasingly on urban dwellings, not rural land, and that the working class, not the landowners, bore the heavier burden. Conservatives argued that tax burdens bore excessively on direct taxpayers, rate-payers, agriculture and landowners. It was in this context that Henry George's *Progress and Poverty*, published in 1880, had so much appeal. His argument was that land hoarding and speculation were the root causes of economic depression, unemployment, urban overcrowding and poverty. He advocated a tax on the entire value of rent, which, he argued, would liberate both capital and labour from taxation, make land freely available to labour, and usher in an era of unfettered free trade.[10] The teaching of George was championed: the Land Reform Union was founded in 1883, changing its name the next year to the English Land Restoration League.

Radical political programmes fed on the George doctrine. In the late 1880s, land agitation became an urban phenomenon too. The advocacy of annual tax on the site value of land proved beguiling, and site-value rating was taken up in political circles. Liberalism's collectivist view of the countryside became clear. Campbell-Bannerman's first speech as Prime Minister, in December 1905, affirmed that

we desire to develop our undeveloped estates in this country (cheers) – to give the farmer greater freedom and greater security in the exercise of his business; to secure a home and career for the labourer who is now in many cases cut off from the soil. We wish to make the land less a pleasure ground for the rich (loud cheers) and more of a treasure house for the nation (renewed cheers).

As Campbell-Bannerman concluded, the health and stamina of the nation were bound up with the maintenance of a large class of workers on the land.[11]

In the 1909 Budget, Lloyd George introduced a central, government-financed Development Fund to encourage scientific and practical agricultural research, to promote rural co-operation, marketing and transport, and to initiate afforestation. His proposals for increased death duties represented a throwing down of the gauntlet to tenurial interests. The Liberal land campaign was launched by a Land Enquiry of June 1912, under the direction of B. S. Rowntree, into economic and social conditions. A political strategy unfolded that included a minimum wage for agricultural labourers and the possibility of rent remissions from landowners. Policy was to be effected by Wages Boards and Rent Courts, with arbitration powers vested in a Land Commission.

Paradoxically, this welter of political activity coincided with a period of reasonable agricultural prosperity, compared with what had gone

[10] H. George, *Progress and Poverty* (1880, repr. London, 1976).

[11] *The Times*, 22 December 1905; Offer, *Property and Politics*, p. 356.

before. However, confidence was shaken in the security of land as a long-term investment.[12] Land sales increased particularly sharply in the five years up to 1914, with perhaps 800,000 acres changing hands. Political attacks on the landowning class by the Liberal Government, and its threats to increase the tax burden upon agricultural estates, were demoralising. A flurry of surveys and books confirmed the lowly status of the agricultural labourer. *The Village Labourer* by J. L. and Barbara Hammond, and Rowntree's *How the Labourer Lives*, were notable examples.[13] The first volume of the Land Enquiry pointed directly to the cause of distress:

over 60 per cent of the ordinary adult agricultural labourers receive less than 18s a week when all their earnings from all sources have been taken into consideration, while there are some twenty to thirty thousand labourers whose total earnings are less than 16s a week.

The report went on to conclude,

we cannot treat the problem of the labourer apart from that of the farmer and the landowner, and we believe that a readjustment of rents is an essential condition for securing a speedy and sufficient rise in wages.[14]

One obvious way in which economic power might be redistributed was through peasant proprietorship in land. Jesse Collings, the protégé of Joseph Chamberlain, advocated the creation of smallholdings by local authorities, founding the Allotments Extension Association in Birmingham in 1882. In the General Election of 1885, the doctrine of 'three acres and a cow' became a national political issue. In 1887, an Allotment Act, pushed through by the incoming Conservative Government, empowered local authorities to purchase land for allotments, but something more was needed. In 1891, county councils were permitted to create smallholdings of up to 50 acres, but the creation of a prosperous land-based yeomanry remained a distant dream.

The Small Holdings and Allotments Acts of 1907 and 1908 greatly extended the possibilities, by obliging county councils to establish smallholdings, through compulsory powers of land acquisition if necessary. The 1907 Act required county councils to prepare schemes for consideration by the Board of Agriculture of the number, nature and size of smallholdings to be supplied, and the land to be acquired for that purpose. Special commissioners were appointed to help the councils. The 1908 (consolidating) Act enabled county councils to compel a landowner to hire out land and sell it (through the local authority) to smallholding tenants.

---

[12] P. Horn, *The Changing Countryside in Victorian and Edwardian England and Wales* (London, 1984).

[13] J. L. and B. Hammond, *The Village Labourer, 1760–1832*, (London, 1911); B. S. Rowntree and M. Kendall *How the Labourer Lives* (London, 1913).

[14] Land Enquiry Committee, *The Land*, 1, *Rural*, 1913, pp. 65–6.

Over the period 1908 to 1914, more than 14,000 holdings were created, and some 200,000 acres acquired by councils for that purpose. Geographically, the response varied: Norfolk was the most energetic county, owning 1,375 smallholdings by 1914, whereas Oxfordshire had created only 200. Moreover, the record has to be seen in perspective. During the same period, there was a net loss of agricultural holdings of under 50 acres, mainly through urban encroachment. Whilst it is hardly possible to speak of a great rural or social transformation taking place, a distinctive landscape feature was established.[15]

An attack on monopolies of rural power might also be mounted through trade union action. A firmer basis for agricultural trade unionism began to emerge in the mid-1880s, with the establishment of a number of local organisations. In 1872, the labourers of south Warwickshire, under the leadership of Joseph Arch, provided the most dramatic example of what could be achieved through a strike and formation of the Warwickshire Agricultural Labourers' Union at Leamington Spa, which brought together all existing union bodies in the country. As the National Agricultural Labourers' Union, it pressed ahead with claims for higher wages.

The county councils and county borough councils, formed in 1888, were part of a much broader reform of local government. The Public Health Act of 1872, consolidated three years later, established for the first time a complete system of health administration for the country. In rural areas, Poor Law Guardians became the rural sanitary authorities, and, under the Local Government Act of 1894, they in turn gave way to rural district councils, based on direct elections. The property qualification was abolished. Control of the Poor Law was no longer in the hands of the well-to-do.

An effort was made in 1894 to revive the smallest unit of government, the parish. A parish meeting, or assembly of all householders, was made an annual statutory requirement and, where the population was over 300, a parish council had to be elected for the discharge of business on a representative basis. The new councils assumed whatever functions remained of the old vestries, together with new powers concerning such matters as allotments, recreation grounds and village greens. The new village parliaments seemed particularly suited to a new yeomanry of England, a counterweight to the vicar and squire. However, when the first elections were held, 60 per cent of the parish councillors were found to be from the upper and middle classes, and only 9 per cent were labourers and other unskilled workers.[16]

---

[15] D. Crouch and C. Ward, *The Allotment: its Landscape and Culture* (London, 1988).

[16] B. Keith-Lucas and P. G. Richards, *A History of Local Government in the Twentieth Century* (Manchester, 1978).

How could a system of counties and county boroughs, districts and parishes cope with urban expansion into the countryside? There were soon signs of the counties losing out to the county boroughs. The Local Government Bill of 1888 had originally proposed ten major towns to be excluded from the jurisdiction of the new county councils, but, by the time of enactment, the figure had risen to sixty-one, and, by 1922, had reached eighty-two. The counties became increasingly concerned. Applications by Cambridge and Luton for county-borough status in 1913 were defeated (in parliament) only with difficulty. Meanwhile, new urban district councils were being carved out of rural districts (170 in the period 1889–1927).

The boundaries of many more county boroughs and urban districts were extended. Birmingham's massive boundary extension in 1911 may have been exceptional, but it makes the point. At the time of the Local Government Act, Birmingham consisted of Birmingham township, Deritend, Bordesley, Duddleston-cum-Nechells and Edgbaston, with a population of 470,000. Within three years, it had annexed Balsall Heath, Saltley, Little Bromwich and Harborne. Quinton fell in 1909, and, in 1911, sweeping additions took in Aston, Yardley, and a huge suburban and rural fringe containing Handsworth, Erdington, King's Norton and Northfield, thus providing 'lebensraum' for the inter-war expansion that was to follow.

B. RURAL NOSTALGIA AND THE OUTDOOR LIFE

During the nineteenth century, 'big cities became primary objects of concern and stimuli to reflection and debate for large numbers of articulate men and women'; this is the theme of Andrew Lees's study of the perception of cities by European and American society in the period 1820–1940. The industrialised city provided many inviting targets.[17] The various material and physical hardships, manifest in such matters as sanitary deficiencies, poor housing conditions, the prevalence of disease and the incidence of high mortality rates, all encouraged the view that big cities undermined the bases of religion and morality, and that they promoted vice and crime. Moreover, it was believed that the ugliness of cities militated against an urban population developing any sensitivity to beauty.

The capacity of cities to spread seemed insatiable, and the scientific utopianism of H. G. Wells's *Anticipations*, published in 1902, could not have reassured many. He envisaged a London area of up to 70 miles from

---

[17] A. Lees, *Cities Perceived: Urban Society in European and American Thought, 1820–1940* (Manchester, 1985).

the centre.[18] The influential Henry George had already expressed his doubts (in 1884) as to the deleterious consequences:

This life of great cities is not the natural life of man. He must, under such conditions, deteriorate, physically, mentally, morally. Yet the evil does not end there. This is only one side of it. This unnatural life of the great cities means an equally unnatural life in the country. Just as a tumour, drawing the wholesome juices of the body into its poisonous vortex, impoverishes all other parts of the frame, so does the crowding of human beings into great cities impoverish human life in the country.[19]

It was argued that big cities provided a thoroughly unnatural setting for man's existence. British cities were raising a puny and ill-developed urban 'race'. Much was made of the fact that army recruiting stations had to reject as medically unfit large proportions of those who enlisted (35 per cent for the period 1893–1902).[20]

The enquiries into poverty, the strident condemnation of drink and vice, and revelations of investigative journalism put into context the perceived virtues of rural England.[21] The corrupting and polluting influences of cities, and the erosion of past vestiges of rural life, prompted a rush of nostalgia. Jan Marsh has outlined the many-sided aspects of 'the pastoral impulse'.[22] One was disenchantment with nineteenth-century materialism. The journalist Robert Blatchford, of the *Clarion*, sold over a million copies of his book, *Merrie England*. Its philosophy was beguilingly simple:[23]

the relative beauty and pleasantness of the factory and country districts do not need demonstration. The ugliness of Widnes and Sheffield and the beauty of Dorking and Monsal Dale are not matters of sentiment nor of argument – they are matters of fact. The value of beauty is not a matter of sentiment: it is fact. You would rather see a squirrel than a sewer rat. You would rather bathe in the Avon than in the Irwell. You would prefer the fragrance of a rose garden to the stench of a sewage works. You would prefer Bolton Woods to Ancoats slums.

A return to nature and a simpler, plainer way of life was urged. Within, say, the counties of Surrey, Suffolk and Hampshire,

you will get pure air, bright skies, clear rivers, clean streets, and beautiful fields, woods and gardens; you will get cattle and streams, and birds and flowers, and you will know that all these things are well worth having, and that none of them can exist side by side with the factory system.

---

[18] H. G. Wells, *Anticipations* (London, 1902), pp. 33–65.

[19] H. George, *Social Problems* (London, 1884, repr. 1931), p. 203.

[20] A. S. Wohl, *Endangered Lives: Public Health in Victorian Britain* (London, 1984).

[21] G. E. Cherry, *Cities and Plans* (London, 1988), pp. 1–77.   [22] J. Marsh, *Back to the Land*.

[23] R. Blatchford, *Merrie England* (London, 1894), pp. 21, 23.

There was no shortage of advocacy. John Ruskin's vision of a quasi-feudal agrarian society found expression in a series of monthly letters in 1871, addressed to 'the workmen and labourers of Great Britain', under the title of *Fors Clavigera*. There followed the founding of the St George's Guild and the revival of handspinning and weaving.[24] As an influential poet of Arcadia, William Morris gave encouragement to craft design. Roused by proposals for the reconstruction of Tewkesbury Abbey, he was instrumental in the founding of the Society for the Protection of Ancient Buildings in 1877. His book, *News from Nowhere*, of 1891, was a stirring, pastoral romance of London as it might be, with no hint of ugliness or manufacture.[25] Edward Carpenter was an apostle of the simple-life movement, living from 1882 on a smallholding at Millthorpe, south of Sheffield. His book, *Civilisation: its Causes and Cure*, developed his ideas on simplicity and a lifestyle in harmony with natural surroundings.[26]

Country locations were consciously sought out. William Morris's firm moved from central London to a weather-board mill on the Wandle at Merton, on the southern edge of London. C. R. Ashbee's Guild of Handicraft went to Chipping Campden in 1902, taking over a silk mill and, later, an old malthouse. Another group of craftsmen settled at Sapperton, west of Cirencester, where furniture-making in the workshops of Ernest Gimson and the brothers Ernest and Sidney Barnsley was already established. Linen-making flourished in Langdale in the Lake District, inspired by the St George's Guild.

A love of the countryside was reflected in the work of artists and writers. Painters descended on the unspoilt country around London, particularly the Surrey Hills. Hardy's Wessex novels drew on nostalgia. Housman's verse gave glimpses of rural Shropshire. Kenneth Grahame's children's story, *The Wind in the Willows*, was steeped in the countryside.[27] In music, the English folksong was popularised and folkdancing rediscovered. Ralph Vaughan Williams's music set out to capture the spirit of the English countryside. The need felt by so many patrons of Anglican churches to have them restored and enlarged provided an opportunity to introduce the conventions of Gothic architecture, which drew so powerfully on a nostalgia for the past.

A yearning for rurality strongly influenced the architectural profession. The country cottage and the old manor house suggested models for a

[24] J. Ruskin, *Fors Clavigera* (London, 1871–7); J. Dixon Hunt, *The Wider Sea; a Life of John Ruskin* (London, 1982).   [25] W. Morris, *News from Nowhere* (London, 1891).

[26] E. Carpenter, *Civilisation; its Cause and Cure and Other Essays* (London, 1889), pp. 1–50.

[27] H. C. Darby, 'The regional geography of Thomas Hardy's Wessex', *Geogr. Rev.*, 38, 1948, pp. 426–43; R. Williams, *The Country and the City* (London, 1973), pp. 197–218; J. R. Short, *Imagined Country. Environment, Culture and Society* (London, 1991), p. 177; K. Grahame, *The Wind in the Willows* (1908).

preferred style of building. Victorian ornateness gave way to plainness and simplicity. Philip Webb's Red House, designed for William Morris at Bexley in 1860, set the style. Voysey and Baillie-Scott were given many commissions for domestic architecture that made use of old crafts and traditional forms of building. Internal design rejected Victorian clutter and ornament in favour of clear lines. The inglenook became a necessary feature; a hall and gallery were incorporated, if possible. Hand-made furniture in a country idiom became fashionable.

As a writer, horticulturalist and garden planter, Gertrude Jekyll drew much of her inspiration from the woods, heaths and cottage gardens of the Surrey Weald. Writing in 1904, she recalled how, 'when I was a child all this country was undiscovered; now, alas it is overrun'. There was no hope of racapturing 'its old charm of peace and retirement'. It was, however, regrettable that no really adequate record or collection had been made. Her book, *Old West Surrey*, was an attempt to record the artefacts of a way of living, and perceptions of life, in the years before steam machinery, when articles were made singly and by one man, rather than being turned out by the hundreds and thousands.[28]

Virtually no aspects of life remained unaffected by rural nostalgia. In matters of dress, the Rational Dress Society founded in 1881 sought the adoption of simpler apparel, based on considerations of health, comfort and beauty. Wool came into fashion with the theories of Dr Gustav Jaeger of Stuttgart, who stressed the health-giving qualities of woollen clothes. Sandals became a popular feature of dress reform. The health and hygienic properties of vegetarianism were advocated. The Vegetarian Society was founded in 1850 and developed strongly with other groups in the 1880s. Interest was taken in the vegetarian sanatorium founded in Michigan by Dr J. H. Kellogg, subsequently of cornflakes fame.

The value of sun and open air was recognised by the families of hop pickers from east London, who spent a working holiday in Kent, accommodated in sheds, barns and tents. The objects of the Yorkshire Ramblers' Club, founded in 1892, were to organise walking and mountaineering excursions, and 'to gather and promote knowledge concerning natural history, archaeology, folklore and kindred subjects'.[29] Camping under canvas became popular. Summer camps were provided by charitable societies for working-class children. The Boys' Brigade, the Church Lads' Brigade and the Scouting movement all embraced the camping tradition. The first venture in camping as a commercial enterprise was made by

[28] G. Jekyll, *Old West Surrey. Some Notes and Memories* (London, 1904, repr. Dorking, 1978); G. Jekyll, *Old English Household Life* (London, 1925); F. Jekyll, *Gertrude Jekyll. A Memoir* (London, 1934); S. Festing, *Gertrude Jekyll* (London, 1991).

[29] H. H. Bellhouse, 'The formation of the Yorkshire Ramblers' Club', *Yorkshire Ramblers' Journ.*, 1, 1899, pp. 3–12.

Joseph Cunningham, who selected a site at Llandudno before moving to the Isle of Man.[30]

An important outlet for the nostalgia and zest for outdoor life was the founding of societies for the protection of old landscapes. One of the earliest and most outstanding examples was the Commons Preservation Society, which developed out of attempts to protect Putney Heath, Wimbledon Common, Hampstead Heath, Epping Forest and other examples of common land around London (see Chapter 31). To succeed, there had to be a shift in public perception from one where the commons were simply regarded as fit only for building and agricultural development to one that placed higher priority on their being maintained as open spaces, serving as lungs for the metropolis. In the twenty years following the General Inclosure Act of 1845, more than 614,000 acres of commons were enclosed, and only a total of 4,000 acres allotted for recreational purposes. The shift in attitude was made tangible in the formation of the Society in 1865 and, much more significantly, in a series of legal victories in which the honorary secretary and solicitor of the Society, Robert Hunter, played a conspicuous part.

Much has been written on the networks of common membership that led to the foundation of new societies and amalgamation of others. In 1905, a number of walking groups combined to form the Federation of Rambling Clubs, out of which sprang the Ramblers' Association. Together with the Cyclists' Touring Club, this lobby became a strong pressure group in countryside affairs, as too did the Society for the Preservation of Ancient Buildings. The National Footpaths Preservation Society was founded by Henry Allnult, a former editor of the *Estate Gazette*, in 1884. In 1899, it merged to become the Commons and Footpaths Preservation Society, the words 'Open Spaces' being added to the title in 1910. Whilst such mergers brought greater influence and scope for action, they did not guarantee success. In 1905, a public inquiry found that the public right of way to Stonehenge was not proven. Attempts in parliament to secure greater public access to the upland moors and mountains made little headway (see Chapter 33).

In many cases, the most certain way of protecting a feature was to purchase the land. As Octavia Hill discovered in 1884, when 2 acres of land and a 'curious old building' were offered to the public, there was no provision in law by which 'the public could assume ownership'. The incident inspired Robert Hunter (in a paper read to the Social Science Congress in Birmingham) to advocate 'the formation of a corporate company' to acquire properties to which common rights were attached, manors, the gardens of squares, and disused churchyards, 'with a view to

---

[30] C. Ward and D. Hardy, *Goodnight Campers: the History of the British Holiday Camp* (London, 1986).

the protection of the public interest in the open spaces of the country'. Copies of the proposal were widely circulated, the name *National Trust* was agreed, and a share list opened.[31]

The scheme languished until 1893 when Canon Hardwicke Rawnsley, the leading figure in campaigns to protect the Lake District, discovered that there was no suitable legal device whereby the Falls of Ladore and the island of Grasmere could be taken into public ownership. On the initiative of Hunter, Rawnsley and Octavia Hill, a formal meeting of a provisional council of the National Trust was convened. In applauding such a move, *The Times* hoped the time had come when 'the lovers of natural beauty' would be as well organised as 'the people who want to build a mile or two of nine-inch brick walls'. It was important, however, for the new body to be directed by persons of taste, who would keep 'artificial treatment' to the minimum. There was already 'a surfeit of statues, fountains, allegorical memorials, and things of that kind'.[32]

The National Trust for Places of Historic Interest or Natural Beauty was registered at the Board of Trade in 1895 as a charitable body under the provisions of the Companies Act. Its objectives were

to promote the permanent preservation, for the benefit of the Nation, of land and tenements (including buildings) of beauty or historic interest: and as regards lands, to preserve (so far as practicable) their natural aspect, features, and animal and plant life.

By 1906, the Trust owned twenty-four properties, and the need for some kind of additional protection for these sites became urgent. This was achieved in 1907, when an Act of Parliament enabled the National Trust to declare its land and buildings inalienable. They could not be sold or given away without the express consent of parliament. Passed without opposition, the Act became a valuable safeguard and stimulus to those who were considering giving further properties to the Trust for safe-keeping in the national interest.[33]

An alternative to land purchase on the part of voluntary bodies was the intervention of Government, whether central or local. Reginald Brabazon, Earl of Meath, was an early advocate of a green girdle,[34] the forerunner of the Green Belt. His reforming zeal in the 1880s and 1890s led him to espouse many causes, including the provision of open spaces and playgrounds in London; his idea of a green girdle or ring for the city had the further objective of improving sanitary and health conditions. In

---

[31] G. Murphy, *Founders of the National Trust* (London, 1987); J. Gaze, *Figures in a Landscape* (London, 1988), pp. 12–55.   [32] *The Times*, 17 November 1893, p. 9.

[33] National Trust Act, 1907, 7 Edward VII, c. cxxxvi.

[34] F. H. A. Aalen, 'Lord Meath, city improvement and social imperialism', *Planning Perspectives*, 14, 1989, pp. 127–52.

February 1889, he was elected an Alderman of the first London county council, and became the first chairman of the LCC Parks and Open Spaces Committee. In that capacity, he visited America to study city public gardens and open spaces. He returned, impressed by the interest shown in circular boulevards which united parks and open spaces, and recommended such a system for London.

In 1901, William Bull, the Member of Parliament for Hammersmith, proposed that 'a continuous chain of verdure' should be created around London as a public memorial to Queen Victoria. A 35-mile green strip, about half-a-mile wide, would be bought to form a 'health girdle', linking the existing parks on the outskirts of London as an outer lung. Meath endorsed the general concept, but suggested including a larger number of parks so as to create a girdle in the form of an outer ring following the outer rim of London, eight or nine miles from east to west, and five or six miles from north to south. The continuous belt of irregular width, as described in *The Sphere* in 1901, would link Epping Forest (working clockwise) with Wanstead Park, Wanstead Flats, East Ham Marshes, Bostall Heath and Woods, Plumstead Common, Woolwich Common, Blackheath, Streatham Common, Tooting Common, Wimbledon Common, Richmond Park, Kew, Gunnersbury Park, Wembley Park, Hampstead Heath, Alexandra Park and Tottenham Marshes.

In the flurry of speculation before the First World War, the idea of 'a green rim' took root. In 1911, George Pepler proposed a parkway around London in the form of a strip of land a quarter of a mile wide, and further from the centre than previous models. It would link existing open spaces and have, in its centre, a network of roads, railways and tramways. In tracing the course of the debate, an important distinction should be discerned. For Meath and Bull, the emphasis had been on inserting 'a green ring' into the development of London. For others, and increasingly after the War, the emphasis was more on the regulation of that growth, with public access to 'green' space or spaces a secondary factor.[35]

C. MODEL HOUSING

Over the previous hundred years, the pressing problem of providing adequate housing for the country labourer had produced opportunities to experiment with a range of models so that, by the mid-nineteenth century, the term 'model housing' had come to be associated with improvement, in terms of either quality or appearance.[36] Enlightened estate owners lost no opportunity to display their new creations to dis-

[35] D. Thomas, *London's Green Belt* (London, 1970), pp. 41–52.
[36] G. Darley, *Villages of Vision* (London, 1975).

tinguished visitors. As in urban situations, housing reformers seized on
the notion to demonstrate desirable advances in both practical knowledge
and aesthetic considerations. In this and the following sections, we consider the part played by model housing, and the important impetus provided by alternative forms of domestic architecture and spatial layout,
which were to become so significant for new building in rural areas at
the turn of the century.

The wealth and taste of the nobility dictated the living conditions of
the lower orders. By mid-century the picturesque style was dominant, as
was reflected at Somerleyton, near Lowestoft, a village built for the estate
employees of Sir Morton Peto in the early 1850s. Twenty-eight cottages
were built around an open green at the gates of Somerleyton Hall. A
school house featured black-and-white East Anglian timbering, a
thatched roof and ornate clustered chimneys. In the Peak District, G. G.
Scott built a village at Ilam in Dovedale for the industrialist Jesse Watts
Russell in 1857. Cottages with steep, tiled roofs and dormer windows,
and a turretted black-and-white school provided the composition. Holly
Village, Highgate, commissioned by Baroness Burdett-Coutts, followed
the same picturesque pattern; a small cluster of cottages was grouped
around a central green in the mid-1860s.

Model housing threw up a range of examples, which attracted the
interest of a combination of philanthropic interests, health reformers, the
eccentric and practically minded. It received a great fillip from the personal patronage of the Prince Consort, as highlighted by his part in providing a display of model houses at the Great Exhibition of 1851. In his
book, *Health in the Village*, Sir Henry Acland, the Regius Professor of
Medicine at Oxford and trustee of a charity in the village of Marsh
Gibbon, in Buckinghamshire, set out the essential requirements for a
healthy environment – dwellings were at the top of his list.[37]

Many of the great landowners, notably the Duke of Bedford, Duke of
Northumberland and Earl Spencer, worked to improve their workers'
accommodation, and the Prince of Wales attended to the improvements at
Sandringham. John Walter, father and son, chief proprietors of *The Times*,
built houses outside the gates of their Bearwood mansion at Sindlesham
Green, Berkshire. The wealthy sisters, Georgina Charlotte and Mary Anne
Talbot, built Talbot village, inland from the town of Bournemouth, in the
1860s, the picturesque cottages set amongst the pine trees. Another personal fortune, this time belonging to R. S. Holford, permitted the rebuilding of the villages of Westonbirt and Beverston in Gloucestershire. Many
of the improved dwellings were of brick or stone, but one scheme used

[37] S. M. Gaskell, *Model Housing from the Great Exhibition to the Festival of Britain* (London, 1986); H. W. Acland, *Health of the Village* (London, 1884).

the new material, concrete, namely the estate cottages of the Hon. Henry Hanbury-Tracy at Grgynog in Montgomeryshire (with long-lasting constructional problems caused by damp).

Rural housing improvement was also implicit in many of the back-to-the-land communities.[38] The Chartist Land Company set out to build a series of such settlements. Before it failed for economic and legal reasons, five communities had been established: O'Connorville, in Hertfordshire, and Lowbands (Worcestershire) in 1846; Charterville, in Oxfordshire, and Snigs End (Gloucestershire) in 1847, and Great Dodford in Worcestershire in 1848. From 1851 onwards, they were first administered by Chancery, and then subsequently dispersed in private ownership.

One important consequence of model housing was its emphasis on vernacular traditions in national architecture. The cottage became a common unit of design, not only for rural but also for urban areas, as it replaced the by-law terrace. Amongst a host of individual architects, Raymond Unwin lent greatest weight to the new trends in design and spatial layout.[39] He and his cousin-in-law, Barry Parker, came together in an architectural partnership in 1896. Parker had been trained as an architect; Unwin moved into architecture from the basic building skills he had acquired with the Staveley Company in Derbyshire, a firm which covered iron, steel and mining interests. Together, they had a common ground in the Arts and Crafts movement. By 1903, Unwin and Parker were well-known provincial architects, with a flourishing practice in small country houses. Their views had already been set out in a book of essays, drawings and photographs. The book, *The Art of Building a Home*, was significant for its analysis of architectural design in relation to social and individual use; it demonstrated the extent to which design principles could be used in site planning and community layout.[40] The concept of a model hamlet was put forward, as a series of irregular, stepped and linked cottages composed around a large open village green.

The twin concept of the cottage and the community, linked by social purpose, also appeared in Unwin's Fabian Society pamphlet, *Cottage Plans and Common Sense*, of 1902.[41] A coherent approach to residential architecture and estate layout was in the making and, for virtually the next thirty to forty years, its principles were to dominate building develop-

---

[38] D. Hardy, *Alternative Communities in Nineteenth Century England* (London, 1979).

[39] M. G. Day, 'The contribution of Sir Raymond Unwin and R. Barry Parker to the development of site planning theory and practice, c. 1890–1918', in *British Town Planning: the Formative Years* ed. A. Sutcliffe (Leicester, 1981), pp.156–99; F. Jackson, *Sir Raymond Unwin: Architect, Planner and Visionary* (London, 1985); M. Miller, 'Raymond Unwin, 1863–1940', in *Pioneers in British Planning* ed. G. E. Cherry (London, 1981), pp. 72–102.

[40] R. Unwin and B. Parker, *The Art of Building a Home* (London, 1901).

[41] R. Unwin, *Cottage Plans and Common Sense*, Fabian Society Tract, 109 (London, 1902).

ment, particularly in forms of garden suburbs, garden cities and, between the wars, private and local-authority estates. It was a form of development typically suited to British circumstances, feeding richly on nineteenth-century trends in which the cottage, the estate village and recurrent notions of housing improvement loomed large.

Unwin adapted many of the innovations already established by Webb and Voysey. Gaskell summarises the new departures in the following way:

In place of the narrow fronted terraced house with a front and back room, the lower streetage costs enabled the evolution of houses with greater widths. This meant that, in the first place, it was possible to provide more variety in accommodation. Secondly, all the accommodation required was brought under the main roof, and long back projections or detached out-buildings were dispensed with, which effected a reduction of gloom and shade. Thirdly, the increased wall space admitted more windows and allowed staircases, landings and larders to be placed on outer walls with direct light and ventilation. Lastly, the proportions of the buildings lent themselves to a treatment more pleasing to the eye.[42]

Unwin's fifteen-page penny tract deserves to be regarded as a seminal statement which marked the introduction of a consistent development motif in the residential landscape.

Cottage Plans and Common Sense listed systematically the major requirements of a dwelling: provision of air and sunlight, abolition of backyards, a spacious setting (perhaps in a quadrangle form), self-containment of the house, internal design to fit the needs of the occupant, and sympathetic furniture design. In addition, there would be a communal centre for a laundry, baths and common rooms. Whilst acknowledging its debt to the nineteenth century, it was none the less a twentieth-century statement. In Unwin's words,

it is along these lines that we must look for any solution of the housing question in town suburbs which shall be satisfactory from the point of view of health and economy, and at the same time afford some opportunity for the gradual development of a simple dignity and beauty in the cottage, which assuredly is necessary, not only to the proper growth of the gentler and finer instincts of men, but to the producing of that undefinable something which makes the difference between a mere shelter and a home.[43]

Unwin's influence on architectural and (shortly) planning practice spread widely. His book, *Town Planning in Practice: an Introduction to the Art of Designing Cities and Suburbs*, of 1909, established his professional pre-eminence.[44] His pamphlet, *Nothing Gained by Overcrowding*, three years later, presented the economic case for low-density layout, though it is

---

[42] Gaskell, *Model Housing*, p. 65.   [43] Unwin, *Cottage Plans*, p. 15.

[44] R. Unwin, *Town Planning in Practice; an Introduction to the Art of Designing Cities and Suburbs* (London, 1909).

memorable today more for its title than for the statistical analysis of the argument.[45] The Cheap Cottage Exhibition at Letchworth in 1905 had sought to discover whether it was possible to build a satisfactory cottage for £150, predicated on the maximum rent of 3 s. per week (nearly £8 per year) from a rural labourer. In fact, it proved not possible, the winning design (by Percy Houfton) costing £250. But the Exhibition produced some useful features in both planning and design, stressing the importance of combining sound materials, simplicity of design and construction, and honest workmanship.

The solution to the difficulties of providing affordable housing for the working classes was likely to be found in the open land on the fringes of the built-up areas. In most cities (basically all except London) there was political objection to local councils engaging to any significant degree in providing local-authority accommodation – the costs were in any case enormous. Private builders were also finding land and development costs prohibitive. Cheap dwellings in spacious surroundings were, however, possible if the cheap land on the fringes could be used, and transport links provided. In Birmingham, Councillor J. S. Nettleford, Chairman of the City Housing Committee, believed so, and vigorously campaigned for planned garden suburb development.[46]

This strategy gained ground elsewhere, and led directly to the Housing, Town Planning, &c. Act of 1909, legislation that gave powers to local councils to prepare schemes for the orderly development (the 'proper sanitary conditions, amenity and convenience') of land scheduled for, or in the course of development.[47] The first application received by the Local Government Board to prepare a Town Planning Scheme was from Birmingham, in respect of 2,320 acres comprising the whole of the parish of Quinton and parts of Harborne, Edgbaston and Northfield. Authority to prepare the scheme was given in 1911. Other schemes followed, covering Birmingham's outer limits. In north-west London, the Ruislip–Northwood scheme covered over 5,900 acres of the urban district, and the parish of Rickmansworth in the neighbouring rural district.

The logic of the model housing movement led to results far removed from the original concerns of house improvement and village development, but while the new profession of town planning clearly advocated artistic beauty and grace of design, the quality of individual dwellings remained a crucial element. Herein lies the link between nineteenth-century model housing and twentieth-century town planning. In a practical manual for those preparing town-planning schemes, the Secretary

[45] R. Unwin, *Nothing to be Gained by Overcrowding* (London, 1912).

[46] G. E. Cherry, *The Politics of Town Planning* (Harlow, 1982).

[47] G. E. Cherry, *The Evolution of British Town Planning* (Leighton Buzzard, 1974).

of the National Housing and Town Planning Council, Henry Aldridge, set down fundamental principles for the new activity. The top priority was the provision of healthy sites for homes; the second was that they should be well planned and grouped.[48] He wrote,

The limitation of the number of the houses per acre and the fixing of a definite and proportionate relation between the whole site and that part of it devoted to buildings, will alone not be sufficient to secure all that is needed. Building regulations specially applicable to each area should be framed to keep within severe limits the tendency to erect rear buildings jutting out from the main building. It is important that the building of narrow-fronted houses should be checked and builders and others encouraged to build houses with rooms broad and deep.

Housing improvement remained the objective of the exercise.

D. FACTORY VILLAGES

The category of factory villages is large enough to warrant separate treatment. The eighteenth and early nineteenth centuries furnished many examples of enlightened factory owners building improved houses and, in some cases, small settlements, laid out and planned to accompany their industrial or commercial enterprise. They set the scene for a number of much more ambitious schemes after 1850.

The notion of the 'new town' received a fillip with the publication of 1849 of James Silk Buckingham's book, *National Evils and Practical Remedies*.[49] A member of parliament and mid-century reformer, he proposed a Model Town Association

for the purpose of building an entirely new Town, to combine within itself every advantage of beauty, security, healthfulness, and convenience, that the latest discoveries in architecture and science can confer upon it.

His Model Town, Victoria, for a population of 10,000 people, was both a carefully detailed physical plan, as well as a social plan. As a social Utopia, it was not so much a scheme for a single factory village but rather a model to which any sober, peaceful, healthy society might aspire. Although never implemented, the idealism behind the proposal had been richly nourished and, over the next half-century, a number of more successful projects were achieved, which aimed at a combination of housing improvement, environmental enhancement and social betterment.

One of the earliest was on the Wirral at Bromborough Pool by the Mersey, where the Wilson brothers, James and George, proprietors of

[48] H. Aldridge, *The Case for Town Planning* (London, 1915), p. 335.
[49] J. S. Buckingham, *National Evils and Practical Remedies*, (London, 1849), p. 141.

Price's candle factory, laid out a new village in an area then largely undeveloped.[50] The Patent Candle Company had been London-based, in Battersea; expansion was impossible, and the Liverpool area was identified for a new location. Sixty-one acres of riverine marshland were bought in 1853, and building of the factory and associated houses began immediately. The houses, designed by Julian Hill of London, were plain and architecturally undistinguished, but they had gardens front and rear, and were significantly in advance of contemporary standards. They were larger than the average artisan house, and more solidly constructed, with water-borne sanitation. By 1858, there was a population of 460. The village developed in fits and starts, reflecting the commercial difficulties of the works itself, and it was only in the last phase that the main public buildings were erected. The last houses were built in 1900, by which time there were 142 dwellings and a population of 728. Earnest reformers, driven by religious impulse towards social amelioration, the Wilsons created a community well provided with gardens, allotments and outdoor recreational facilities, features to which so much importance was attached in later imitations.

Perhaps a more typical industrial village was Street, near Glastonbury in Somerset.[51] Founded in 1825, Clark's, the Quaker family firm of shoe manufacturers, expanded from a cottage industry to the point where a new factory was built in 1857. It produced 208,000 pairs of shoes a year by 1861, and output had quadrupled by 1901 – the population of the village more than doubled to 4,000. A small-community atmosphere was preserved; the company houses boasted gardens, and a careful layout ensured availability of recreational space.

At about the same time, a cluster of model villages was established by factory owners in the West Riding.[52] Colonel Edward Ackroyd, a local member of parliament and textile manufacturer, built Copley, near Halifax, beginning in 1847. Although the housing was back-to-back, it was at least set in the countryside, with the mill a little way off. Shops were provided, and later a church. His second scheme, begun in 1861, was Akroyden, nearer to Halifax. An overall plan was prepared, based on a large square with a central cross; housing was much improved with water and gas laid on. Social integration was attempted, through a mixture of housing types. Two years later, John Crossley inaugurated his West Hill Park Estate in Halifax.

The largest scheme in the area was Saltaire, the creation of another benefactor, Titus Salt, a worsted-stuff manufacturer who, by 1850, was the

[50] A Watson, *Price's Village: a Study of a Victorian Industrial and Social Experiment* (Bromborough Pool, 1966).

[51] M. Havinden, 'The model village', in Mingay, *Victorian Countryside*, vol. II, pp. 414–27.

[52] Darley, *Villages of Vision*.

biggest employer in Bradford, with five mills in operation.[53] During the 1840s machinery had been sufficiently perfected for all types of wool-combing to be concentrated in factory conditions, with the consequence that Salt's many hundreds of woolcomber outworkers could be reorganised. There were substantial economies to be made, as well as opportunities for better supervision, by investing in a single plant. He could also leave Bradford. The cholera epidemic of August 1849 had confirmed Salt's own knowledge of the insanitary conditions – a senior alderman, he was mayor in 1848–9. A new site was selected in an undeveloped area, three miles to the north, on the river Aire, with canal and railway facilities.

Saltaire was built between 1851 and 1871, as a comprehensive economic unit of factory and community. The mill covered 6.75 acres and was the largest in Europe. Expanded, it eventually employed 3,000 people. Designed to resemble an Italian renaissance palace, it did indeed look like a palace of industry. Two Bradford architects, Lockwood and Mawson, laid out the estate in accordance with a strictly geometrical plan of parallel streets. By 1871, there were 775 houses, 45 almshouses, and a population of almost 4,400. There was a proper system of drainage, water and gas were supplied, and separate outside lavatories occupied individual backyards. Most dwellings were two-storeyed terraced cottages, but there were some three-bedroom houses, each with a front garden. Intermixed were a small number of large houses for managers and some bigger three-storeyed houses suitable for lodgers. To complete the community, churches, chapels, schools, hospital, an Institute, public baths, wash houses and a park were provided.

Saltaire may be best seen as an attempt to marry nineteenth-century technology to the intimacy of pre-industrial rural society, away from the squalor of the industrial city. Whilst distinctive, the township had no obvious focus, other than the huge manufactory that dominated it. There was no village green. In sharp contrast to the picturesque tradition of rural estate villages it was rather impersonal, conveying the impression of regimented space, where order and social discipline could prevail. Salt died in 1876, by which time his vision of an ideal industrial community in a rural setting was being lost as a flood of speculative building surrounded the site, but a powerful example had been provided for others to follow.

A very different creation was founded on Merseyside, in this case instigated by a soap magnate, William Hesketh Lever, the son of a successful wholesale grocer.[54] Entering the family firm, he specialised in the marketing of soap, and a factory was leased for the manufacture of Sunlight soap in Warrington. Soon new premises were needed, and 56 acres of

[53] J. Reynolds, *Saltaire: an Introduction to the Village of Sir Titus Salt* (Bradford, 1976).
[54] E. Hubbard and M. Shippobottom, *A Guide to Port Sunlight Village* (Liverpool, 1988).

marshy land were purchased adjoining Bromborough Pool. Given the name Port Sunlight, the factory and original village settlement were completed in 1889 and 1897 respectively. More land was acquired, and gradually the village took shape, with a layout based on 'super blocks' with sets of cottages enclosing allotments. The tidal channels were gradually filled in, and the final plan, prepared by Ernest Prestwich in 1910, emphasised a strong formality with two axes, one, the 'Diamond', parallel to the Birkenhead–Chester railway, and the other, the 'Causeway', directed on the church.

Port Sunlight emerged as one of the earliest and most successful of the factory villages that were to exercise so much influence on English housing and planning in the twentieth century. By 1914, the target of 900 dwellings had almost been reached, with a rich variety of designs provided by a number of architects from Warrington, Chester and Liverpool. A picturesque, vernacular style of cottage architecture prevailed, with generous social and educational provision based on a collection of public buildings erected for that purpose.

But if Saltaire was a rather bleak, technological solution, and Port Sunlight adopted a *beaux-arts* layout, Bournville was the true progenitor of the garden suburb, built around a modern factory. From wool and soap, we turn to chocolate. John Cadbury, a tea-dealer and coffee-roaster, opened a shop in Bull Street, central Birmingham, in 1824. The business had expanded sufficiently by 1847 to justify taking a lease for a large factory in Bridge Street for cocoa and chocolate manufacture. In 1879, his sons, George and Richard, built a new factory four miles to the south-west of Birmingham, adjoining a small stream, the Bourn Brook, in open country. In deference to the advertising appeal of French chocolate, they called their site Bournville. Canal and railway facilities were close at hand.

After Richard's death, George took charge of the enterprise.[55] As a Quaker and social reformer, he set out to show what could be achieved by way of good-quality housing, attractive landscape and social provision, in startling contrast to the overcrowded, unhealthy conditions then prevailing in Birmingham. Having erected sixteen cottages for the factory's key workers, 120 acres of land adjoining the factory were bought in 1893, and work on the estate began in earnest. More acquisitions were made and, in 1900, 330 acres and 313 houses were handed over to the Trustees of a newly founded Bournville Village Trust. By 1912, Bournville had over 900 houses and a population approaching 4,500. A garden suburb not tied to factory employees was in the making. The southern part of the Estate was gradually developed by a number of societies, including the Bournville Tenants Ltd, a housing co-operative, and Weoley Hill, a

---

[55] P. Henslowe, *Ninety Years On: an Account of the Bournville Village Trust* (Bournville, 1984).

society created in 1914 for the leasehold development of houses for office workers.

Tree-lined curving streets, houses with gardens front and rear, generous open space, vernacular architecture with the early houses (by Alexander Harvey, a Birmingham architect) showing rough-cast and half-timbering, and sites for schools, halls and churches formed the main elements. It was a decisive break from the by-law streets of neighbouring Selly Oak and Stirchley. The Green was the functional heart. But the uniqueness of Bournville was not in housing and landscape; it was also a social experiment in terms of a desire to create a mixed community and to foster community-democracy through a village council. Deftly advertised as a garden suburb and a factory in a garden, Bournville soon attracted the reformers' attention. In 1901, the newly founded Garden City Association met at Bournville for its annual conference.

Another garden village sprang from a Quaker chocolate manufacturer: New Earswick on the outskirts of York. Joseph Rowntree was the original mover behind a well-laid-out group of thirty houses built for his employees in 1902–3. A Trust was set up for further development, and the initiative passed to his son, Seebohm, whose work on poverty in York, published in 1901, had led him to see the advantages of a model village environment. Raymond Unwin and Barry Parker were appointed to supervise the development of the extension. Informal, low-density layout principles, earlier seen at Bournville, were developed at New Earswick; tree-lined streets, variety of house type, open areas, children's play space and the use of culs-de-sac for intimacy and human scale.

By now the trend had been set. Architectural fashion was breaking away from the terraced houses of bylaw streets to emphasise the alternative model of a return to a 'cottagey' vernacular style, set in a pattern of informality which stressed greenery, open space and a village atmosphere. The degree to which the new industrial villages conformed to the emergent pattern, even in fairly remote districts, was exemplified by Woodlands colliery village, north of Doncaster, where over 650 houses, laid out by Percy Houfton, were built for the Brodsworth Main Colliery between 1907 and 1912. Located on the 'concealed' coalfield, the new settlement was very different from the straggling pit-villages of earlier years towards Barnsley in the west, where depressing terraces contrasted sharply with Houfton's simple, whitewashed, plaster cottages in a landscaped setting of mature trees.

E. SUBURBAN STYLES

While cities in continental Europe had to defend themselves over many centuries by an elaborate system of fortifications beyond which building was discouraged, British cities could spread. So at least runs the argument:

The continental city grew like a crab, casting its shell at intervals and forming a new one. The English city grew like lichen, in an irregular formless manner, spreading along roads, engulfing villages and leaving, here and there, pockets of countryside to form parks.[56]

The comparison may be over-simplistic, but the territorially spread city, characterised by houses rather than tenements, was an incontrovertible feature of the English and Welsh scene. In this section, we consider the various styles of that development, noting how a very dominant form of layout and style of building came to prevail.

By the end of the Edwardian era, four architectural styles could be identified in domestic building. The new type of house was cheaper and more functional than the extraordinary medley of styles that had gone before, though still gothic in inspiration. George Truefitt, who built largely in Tufnell Park, was the chief exponent in the London area. The curiously named Queen Anne style followed in the mid-1870s. Tudor style half-timber was also popular, as exemplified by Shaw's Cragside in Northumberland and, at the other extreme, some of the buildings at Port Sunlight. In the late 1880s, Shaw changed to the classical traditions of the late seventeenth century and symmetrical neo-georgian. In neat red brick, it was taken up by a host of imitators. Finally, and under the influence of the Arts and Crafts movement, middle-class houses were built in the manner of sixteenth-century farm workers' cottages. Voysey, for example, brought the roofs of his houses down to first-floor level, walls were stone or rough cast, and there were huge chimneys and small-paned windows.

The 'specialised' dwelling form of the bungalow also made its appearance, introduced from India as a middle-class vacation house within commuting distance of London.[57] The first, very few, examples were to be found on the Kent coast, but, between 1890 and 1910, they became more widespread in outer suburban and especially country locations. As a building form, the bungalow was to become very popular after the War. 'Bungaloid growth' was coined as a term of derision in 1927.

Most suburban housing, however, took the form of speculative building at the lower end of the market. There was an infinite number of small terraced or semi-detached houses, red brick with slate roofs, bay windows, stone dressing and cement ornaments. Avenue planting seems to have started around 1851 and became increasingly popular. Another innovation was the front garden, with its small wall and boundary hedge. By and large, architects were not involved and, as the century emerged,

---

[56] A. M. Edwards, *The Design of Suburbia* (London, 1981).

[57] A. D. King, *The Bungalow. The Production of a Global Culture* (London, 1984), pp. 65–90; A. D. King, *Urbanism, Colonialism and the World Economy* (London, 1990).

a uniformity was brought about by a combination of standard drawings, applied decoration and building by-laws. Some measure of individuality might come from the singular nature of the developers. Land societies flourished in some cities in the 1850s and 1860s, stimulated perhaps by the fact that membership, under the Reform Act of 1832, conferred on their members entitlement to a parliamentary vote. The Birmingham Freehold Land Society was founded in 1848 and proceeded to acquire land and then divide it into building plots with large gardens.[58]

Bedford Park was a speculative enterprise which established not only a design prototype for middle-market estates during the last quarter of the century, but also a model for community building as a social organisation.[59] It was begun in 1876 by a speculative builder, Jonathan Thomas Carr, on land adjoining Turnham Green station on the Metropolitan Railway in west London. By 1883, an estate of 113 acres had been laid out, and 490 red-brick houses, with bay windows and gables to the road, had been built. Bedford Park was the first occasion when the Queen Anne style was used for a whole estate. Carr's design consolidated the semi-detached as the norm for middle-class housing. It offered an acceptable solution to the problem of access to the rear; terraces had demanded basements or cellars (for coal for example). Carr employed the best architects of his day, including Norman Shaw. The motif of two large gables to the front elevation became extremely fashionable and, half a century later, standard for the inter-war years.

Bedford Park was important on several counts. It established a perspective from which to assess other forms of estate development in London. 'Camden Town gothic' was particularly reviled. As an exclusive estate, it was intended for people with incomes of between £300 and £1,500 per year, if the annual rents of between £45 and £90 were anything to go by. But it was in its novelty as an escape from regularity and vulgarity, its early maturity by virtue of the intermixture of old trees, and claim for community integration, that marked it out for others to follow and develop.

New heights of environmental quality were set by another estate development, the Hampstead Garden Suburb. For some time, the future of Hampstead had seemed uncertain.[60] Parliament had sanctioned the extension of the Charing Cross, Euston and Hampstead Railway (now the Northern Line of London Transport) to Golders Green. Large-scale housing development could be anticipated around the stations on the line. Enter Mrs (later Dame) Henrietta Barnett, the wife of the Vicar of

[58] S. M. Gaskell, '"The suburb salubrious": town planning in practice', in Sutcliffe, *British Town Planning*, pp. 16–61.

[59] M. Jones Bolsterli, *The Early Community of Bedford Park* (London, 1977).

[60] B. Grafton Green, *Hampstead Garden Suburb 1907–1977: a History* (Hampstead, 1977).

St Jude's, Whitechapel, and Warden of Toynbee Hall. For two days a week, the family used Hampstead as a country retreat. Determined to preserve their pleasant surroundings, Mrs Barnett secured influential support. A 'syndicate of eight' (two earls, two lawyers, two Free Churchmen, a bishop and a woman) prepared a scheme to save a portion of the Heath in return for housing development for the working classes on the remainder. In the event, 80 acres of the Heath Extension were conveyed to the London County Council, and 243 acres were earmarked for building. In August 1906, the Hampstead Garden Suburb Bill, a private measure, was passed, appointing a Trust to administer the proposed housing development as a Garden Suburb. Golders Green station was opened in June 1907 and, below ground, the platforms of an intermediate station were built in the vicinity of Wyldes Farmhouse.

From its very inception, Hampstead was lauded as the very pinnacle of Garden Suburb achievement. It was an exercise in site planning of the highest order. Raymond Unwin had prepared the first plan as early as February 1905, in which quadrangles, greens and culs-de-sac were prominent. Through legislation, it had been possible to remove the constraints imposed by by-laws on road widths, thereby making it possible to group houses around greens, served by a large number of short roads. A greater measure of formality was imposed with the appointment of Edwin Lutyens as consulting architect to the Trust in 1906, particularly through his influence on the Central Square, but Unwin and his group of young architects were able to continue developing the Suburb as 'a flowing relationship between house and garden, building groups and commercial open spaces, residential areas and the Heath extension'.[61] Different parcels of land were developed by a number of co-partnership societies; more land was purchased from Finchley, and leased from the Ecclesiastical Commissioners, in 1911.

By the middle years of the Edwardian period, the advantages to be gained from the new designs, layouts and architectural styles had been demonstrated beyond doubt. The examples of Port Sunlight, Bournville and New Earswick, and of Bedford Park and Hampstead, all pointed in one direction. Unwin captured in words the aesthetic spirit behind the artistic revolution that was transforming suburban land-take:[62]

The truth is that we have neglected the amenities of life. We have forgotten that endless rows of brick boxes, looking out upon dreary streets and squalid backyards, are not really homes for people, and can never become such, however complete may be the drainage system, however pure the water supply, or however detailed the bye-laws under which they are built. Important as all these

---

61 Miller in Cherry, *Pioneers in British Planning*, pp. 72–102.
62 Unwin, *Town Planning in Practice*, p. 4.

provisions for man's material needs and sanitary existence are, they do not suffice. There is needed the vivifying touch of art which would give completeness and increase their value tenfold; there is needed just that imaginative treatment which could transform the whole.

The 'imaginative treatment' seemed at first to depend on particular persons being sympathetic to new design solutions – a paternalist factory owner, or a determined activist as a housing reformer. Clearly other agencies were needed. The building-society movement seemed interested only in the construction of individual houses, and co-operative societies were unwilling to venture capital in forms of estate development other than those of proven character. The challenge passed instead to co-partnerships. Whilst the idea of a housing society sharing profits with its tenants had been successfully pursued since the 1880s, it was not until the example set by the Ealing Tenants Ltd in 1901 that co-partnership in housing began to make rapid advances. Its model plan for an estate in west London demonstrated what might be achieved. Unwin became consultant architect to Co-Partnership Tenants Ltd, a body registered in 1907, which raised capital from all federated societies. By 1913, some 60 estates, containing nearly 11,500 houses, had been developed, largely on new garden-suburb or garden-city lines (the terms were used with little precision). A number were developed in the north-west, on the outskirts of Manchester for example, and at Childwall in Liverpool, and at Bristol (Shirehampton), Cardiff (Clyn Cory), Merthyr, Coventry, Birmingham (Harborne), Wolverhampton (Fallings Park), Hertfordshire (Cuffley Garden), Southampton, Romford and Ilford.

The other obvious agency was local authorities, but only the London County Council (LCC) played any significant part. In 1893, a new group was formed within the Architect's Department of the Council, called the Housing of the Working Classes Branch. In the years up to 1914, its work came to be regarded as among the highest achievements of the Arts and Crafts movement in English architecture.[63] During its early years, the LCC had been principally concerned with rehousing under Parts I and II of the Housing Act, 1890: the provision of dwellings for people made homeless by slum clearance and road improvement schemes.[64] In 1898, the Council resolved to take full advantage of Part III powers: to buy and build on vacant land to create additional accommodation for Londoners. In 1900, a new Housing Act enabled local councils intending to build houses under Part III to do so outside their own boundaries, thus opening up the opportunities to build garden cities.

The first two schemes were within the LCC area. The Totterdown

---

[63] S. Beattie, *A Revolution in London Housing: LCC Architects and Their Work, 1893–1914* (London, 1980). [64] Housing of the Working Classes Act, 1890, 53 & 54 Victoria, c. 70.

Fields Estate of 39 acres at Tooting was laid out for development on a grid system; the two-storey, brick-built cottages had gardens front and rear. Despite the plan, routine standardisation was avoided: the terraces were cut into short lengths, and gables interrupted a uniform roof line, while porches introduced important detail. The scheme was carried between 1903 and 1911, with over 1,200 houses built at a density of more than 30 to the acre. The second was the much larger, and more successful, White Hart Lane estate, covering 177 acres in Tottenham, begun in 1904. Densities were lower, and architectural detailing more subtle as the Housing Branch moved closer to the ideals set out in Parker and Unwin's *Cottage Plans and Common Sense*.

With the third scheme, the LCC went beyond its boundaries to a 30-acre hillside site at Norbury, near Croydon, developed after 1906. The topography enhanced the picturesque grouping, and the inflexibility of the standardised terrace was broken. The fourth scheme was the Old Oak estate, a 54-acre site in Hammersmith, where the influence of Hampstead Garden Suburb was much in evidence. The quadrangles and closes, the angle blocks and the terraces unfolding along the line of a curving street combined to make the estate the most successful of the LCC's contribution to the revival of English domestic architecture.

The example of Manchester makes clear the variety of sources from which the stimulus for new ideas in suburban design might come.[65] The local housing reform movement was particularly strong, and its involvement in the politics of housing was clear to see in the number of different schemes around the city. Thomas Horsfall and T. R. Marr were key figures in the Citizens' Association for the Improvement of the Unwholesome Dwellings and Surroundings of the People, established in 1902. Six hundred people were living in the spacious Blackley Estate of the Manchester Corporation by 1913. Inspired by the Cooperative Wholesale Society, Burnage Garden Village was developed by Manchester Tenants Ltd as a co-partnership scheme for 136 houses, at 12 to the acre. Fairfield Tenants Ltd developed a second co-partnership suburban estate, to the east of the city. A 40-acre private estate was built at Chorltonville on the south side of Chorlton-cum-Hardy; the architects had visited Bournville and Harborne, and the low-density layout emphasised wide frontages to the houses, gardens, and maximum exposure to light and air. The ameliorative effect of these schemes may have been very small in relation to Manchester's overall housing problem, but they proved compelling prototypes for the new low-density estates that followed the War, Wythenshawe being the prime example.

[65] M. Harrison, 'Housing and town planning in Manchester before 1914', in Sutcliffe, *British Town Planning*, pp. 106–53.

By the outbreak of war, the new forms of suburban development had become well established. Indeed, the housing provided for munitions workers was an integral part of that trend. The Well Hall Estate at Eltham, London, by Frank Baines was the first English example of a large working-class housing estate in the picturesque manner. A combination of initiatives, under the influence of housing reformers, and the new town planning movement, had given a new qualitative dimension to English suburban schemes.

F. COLONIES IN THE COUNTRYSIDE

The role of the factory village and garden suburb in providing accommodation for an urban population in a rural setting has been described; it is now time to look at the housing-reform movement from another perspective. The transfer of town dwellers to more salubrious areas was a recurrent theme of those who attacked the unhealthy living-conditions prevalent in towns. General William Booth of the Salvation Army was just one advocate of a conscious resettlement-movement programme. In his book, *In Darkest England and the Way Out*, he proposed the establishment of

self-helping and self-sustaining communities, each being a kind of co-operative society, or patriarchal family, governed and disciplined on the principles which have already proved so effective in the Salvation Army.

He envisaged a three-tier system for these 'colonies': city colonies, as refuges to collect recruits before being redirected to permanent employment; farm colonies 'on an open estate in the provinces' for retreating and reformation of character; and overseas colonies for eventual resettlement.[66]

The farm colonies were relevant to the prevailing back-to-the-land movement, and merit some attention.[67] The setting up of agrarian communes had been a feature throughout the century, and mention has already been made of the Chartist colonies and Ruskin's Guild of St George. The anarchist influence of Kropotkin and Tolstoy bolstered the notion that the industrial system might be renounced in favour of the adoption of self-supportive manual labour on the land, in the form of spade husbandry. Peter Kropotkin settled in England in 1886, and his influential book, *Fields, Factories and Workshops*, was published in 1890, having first appeared in a series of articles over the previous two years. Dispersal held out the possibility of securing the kind of society he wanted.

[66] W. Booth, *In Darkest England and the Way Out* (London, 1890), p. 91.
[67] Hardy, *Alternative Communities*; Marsh, *Back to the Land*.

Kropotkin was dismissive of the abandonment of land and the low productivity of English farming. He recalled how, whether walking north or south from London, he had been struck by the absence of men in the fields, and of how, around Harrow, the meadows would scarcely keep alive 'one milch cow on each 2 acres':

In the Weald I could walk for twenty miles without crossing anything but heath or woodlands, rented as pheasant-shooting grounds to 'London gentlemen', as the labourers said.

Close by was a city of 5 million inhabitants, supplied with 'Flemish and Jersey potatoes, French salads and Canadian apples'.[68]

Kropotkin urged a new economic and social system, centred on small units of production, a revivified agriculture, intensive market gardening, and decentralisation of industry. His advocacy was, 'have the factory and the workshop at the gates of your fields and gardens, and work in them'. In accordance with these principles, a group of communally minded anarchists in Newcastle founded the Clousden Hill Communist and Co-operative colony in 1895. Although a success in terms of horticulture, personality disagreements led to its being disbanded after five years. Other short-lived communes were set up in other parts of the country, based on market gardening.

Cottage farming for ex-urbanites was also extolled, not so much for its communitarian implications, but simply as an articulation of a rustic dream. An early example was the Methwold Fruit Farm colony in Norfolk, founded in 1889–90 by R. K. Goodrich, who had formerly run a small business in London. Within ten years, he was joined by fifty others on two- or three-acre plots similar to his own. There was a steady income from fruit, vegetables and new-laid eggs, and a small jam factory was built. Thomas Smith left his employment as a printer in Manchester in 1895, and invested his savings in eleven acres at Mayland in south Essex; other settlers followed, tempted by the blend of individual ownership and voluntary co-operation. At Cudworth in Surrey, a 400-acre estate was subdivided into smallholdings for forty to fifty families, who had previously been city dwellers.

How far was it practical to consider relocating surplus labour on the land so that agricultural productivity might increase, and labourers' families could be rehoused in conditions of decency, comfort and health? The notion had been around for some time. In 1887, the Mansion House Inquiry into the Condition of the Unemployed set up a sub-committee on agricultural colonies. In 1892, the Rev. Herbert Mills launched the Home Colonisation Society, but his farm colony at Starnthwaite in the

---

[68] P. Kropotkin, *Fields, Factories and Workshops*, 1890, ed. C. Ward (London, 1974), pp. 53–4, 197.

Lake District failed due to rising militancy among the unemployed. More successful was the Salvation Army colony at Hadleigh, near Southend, which opened in 1891 on 3,000 acres of land, comprising three empty farms. Within a year, there were 300 colonists working on the land and in workshops and brickworks. Rider Haggard was commissioned by the government in 1905 to carry out a survey of settlement schemes. He found Hadleigh in a flourishing condition.

A farm colony was established at Laindon, in Essex, by the Poor Law Guardians of Poplar in east London. There had been many examples of workhouse farms, but George Lansbury, the moving force behind the initiative, had much more in mind than the usual forms of workhouse relief. Problems of expenditure (and the local poor rate) were overcome by Joseph Fels, a wealthy American soap manufacturer who came to Britain in 1901. A follower of Henry George and an advocate of small-holdings, he had already been involved in the Mayland colony. Through his financial support, Laindon was opened in 1904, and a much larger venture followed at Hollesley Bay in Suffolk.

Under the Unemployed Workmen Act of 1905, a Central Unemployed Body (CUB) was set up, which *inter alia* could assist farm colonies. However, the incoming President of the Local Government Board, John Burns, had no love for farm colonies, and prevented the CUB from working with permanent resettlement schemes. Both Laindon and Hollesley Bay were relegated to the function of workhouse out-stations, and the farm colony movement had withered away by the outbreak of war. Structural unemployment, such as that experienced in London from the 1880s onwards, could not be remedied by that device.

G. GARDEN CITIES

By the turn of the century, conditions in both town and country were conducive to new attitudes towards housing design, the nature and location of residential development, and aspects of community living. A late Victorian urban crisis led to anxiety about the quality and expectations of city living, while, in the countryside, there was a worrying combination of economic stagnation and inferior living conditions. Reformist movements urged new approaches to urban problems, some of which had far-reaching implications for rural life. One such movement went further than any other in bringing together urban and rural issues. This was the garden city movement.

Context is all important and we can usefully reassemble the main points. During the nineteenth century various measures of public sector control had been established over the built urban environment. Important steps had been taken to establish the proper regulation of street widths, new highway construction, public health and sanitation, fire control, building construction and standards of space around buildings. By the end of the century, an effective local-government system was in place. Overcrowding rates were declining. Terraced, by-law suburbs became the homes for increasing numbers of skilled artisans, clerks and shopkeepers, who had access to open countryside in the outer suburbs. In some very favoured instances, factory villages and garden suburbs brought together the benefits of town and country living.[69]

But the processes of improvement were slow and fearsome residual problems were encountered, suggesting to some reformist circles that completely new forms of city life were needed. The fact that the problems were at their most intractable in the innermost parts of cities, and in London especially, caused many to look to the urban fringes and beyond for a solution. The evidence from the LCC schemes was telling. Whereas the cost of acquiring 12.25 acres of Boundary Street, a central area of slums in Bethnal Green, was £500,000, some 39 acres of Tooting in south London could be bought for £44,000. Alfred Marshall, Professor of Economics at Cambridge, advocated urban colonies 'well beyond the range of London smoke'.[70] The arguments about health were increasingly persuasive: the new residential areas could be built at lower density, they would provide and have access to more open space, and housing design would give more spacious accommodation, which in turn would admit more sunlight. The technique had been proven at a number of factory villages and in the new garden suburbs.

These were essentially rural solutions to an urban problem, with land available for development at the urban fringe. Agricultural rents were low and some land had actually fallen out of cultivation. For decades, the depopulation of the countryside had been in marked contrast to urban overcrowding. Reversing the trend, living in the countryside now became an exciting prospect for increasing numbers of people. As we have observed, the pastoral impulse heightened ideas for a return to the land.

It was against this background that Ebenezer Howard developed his scheme for a garden city, in which town and country might be married in a new urban community.[71] Howard, the son of a baker and confectioner, was born in 1850 in the City of London. He started work in a City stockbroker's office, and occupied successive office jobs, learning

[69] Cherry, *Cities and Plans*.

[70] A. Marshall, 'The housing of the London poor. 1. Where to house them', *Contemporary Review*, 45, 1884, pp. 224–31.

[71] R. Beevers, *The Garden City Utopia: a Critical Biography of Ebenezer Howard* (London, 1988); D. Hardy, *From Garden Cities to New Towns: the History of an Environmental Campaign* (London, 1991); D. Macfadyen, *Sir Ebenezer Howard and the Town Planning Movement* (Manchester, 1933).

shorthand on the way. He went to America and took up farming on a 160-acre holding in Nebraska. Failing in agriculture, he became an official shorthand writer in a Chicago office, before returning to England in 1876, as a shorthand writer in the House of Commons and elsewhere. Howard read voraciously, and moved in reformist, liberal and rather earnest circles. He joined the Zetetical (debating) Society, toyed with spiritualism and wrestled with problems of religion and society.

In a biography, Beevers has recounted how, from the late 1880s, a concern for social and urban questions led Howard to the notion of a garden-city utopia. One trigger was his reading of Edward Bellamy's *Looking Backward*, an American novel in which a sleeper awakes in the year 2000 to find his city of Boston transformed from the past.[72] That vision was of a socialist state which owned property and imposed forms of community living that made life easier for all. Rather different was William Morris's view of the future, as seen in *News from Nowhere*, with its delight in medieval landscapes and a society of co-operation and harmony.[73] Howard came to prefer the Morris vision, but, whatever viewpoint was taken, it was the comparison of London in Howard's day with what might be created under a new social and economic system that spurred him on to crystalising his own imagination.

Howard drew up a plan for a new city which combined industrial, residential and agricultural land uses, built by 'private enterprise pervaded by public spirit'.[74] He tested it by means of public lectures, and meetings with the regular chapel-goers of progressive London sects. The notion of an imaginary city of 30,000 people, the capital cost of which would be £300,000, took shape. By 1893, he and his supporters were confident enough to hire the Memorial Hall of the Congregational Union for an open meeting. The outcome was a resolution that 'a society be formed to be called the Co-operative Land Society, having for its object the promotion of a colony or colonies on the basis of common ownership of land'. A committee was formed, but broke up under competing interests.

Howard resolved to concentrate on the draft of a book. His philosophy and approach remained remarkably constant throughout the middle 1890s. All the major social problems of the day could be resolved by a package of reforms through the means of building a new city on new land – that was his 'Master Key'. He first called this Unionville; he dallied with the option of Rurisville, and only quite late did the preference for Garden City emerge. A paper submitted to the *Contemporary Review* was rejected on the grounds of length; it was in fact a summary of the book. Howard approached possible publishers, but they were unwilling to risk

[72] E. Bellamy, *Looking Backwards* (London, 1888).   [73] Morris, *News from Nowhere*.
[74] Beevers, *Garden City Utopia*, pp. 30, 37.

loss. Fortunately, a private benefactor put up a loan in late 1897 of £50, and his book finally emerged in October 1898, *Tomorrow: a Peaceful Path to Real Reform*.[75]

In his Introduction, he asserted that it was 'well-nigh universally agreed by men of all parties' that 'it is deeply to be deplored that the people should continue to stream into the already over-crowded cities and should thus further deplete the country districts'. The problem then was to restore people to the land. Howard argued that each city might be regarded as a magnet, and each person as a needle. Existing cities had the force of old attractions; these could be overcome by the force of newly created attractions. The challenge was to make the country more attractive than the town,

> to make wages, or at least the standard of physical comfort, higher in the country than in the town; to secure in the country equal possibilities of social intercourse, and to make the prospects of advancement for the average man and woman equal, not to say superior, to those enjoyed in our large cities.

Instead of the two alternatives of town and country life, there should be a third, 'in which all the advantages of the most energetic and active town life, with all the beauty and delight of the country, may be secured in perfect combination'. Howard returned to the analogy of the magnets: town and country were two magnets, each striving to attract people, but there was the possibility of a third Town–Country magnet, where human society and the beauty of nature could be enjoyed together.[76]

Howard went on to call this third magnet Garden City, and to describe it. It was idealised as an agricultural estate of 6,000 acres, purchased at £40 per acre, the money raised on mortgage debentures of 4 per cent annual interest. Trustees would be 'four gentlemen of responsible position and of undoubted probity and honour'.[77] The development would be paid for out of rents; land would be owned by the 'municipality'. The objective of the exercise would be to assist the working class to find better paid and regular work, and to live in healthier surroundings. Enterprising manufacturers and professions would secure new and better employment, and agriculturalists would find new markets for their produce. About one-sixth of the Estate would be the town of Garden City, for 30,000 people. Howard offered an idealised description built up of circular rings: outwards from a core of public buildings in a garden setting, to a central park, via avenues of various dimensions and functions, through the residential districts to the industrial quarters on the outskirts.

Never before had a scheme for a model city been set out so compre-

---

[75] E. Howard, *Tomorrow: a Peaceful Path to Real Reform* (London, 1898).   [76] *Ibid.*, pp. 2–7.
[77] *Ibid.*, p. 13.

hensively. Moreover, it was clear that Howard was a practical reformer, devoted to consensus means. His book did not simply urge reform, but showed how it could be achieved.[78] In Howard's words, it was offered:

in a manner which need cause no ill-will, strife or bitterness; is constitutional; requires no revolutionary legislation; and involves no direct attack on vested interests.

Further, Howard was an incrementalist, recognising that it was not possible to create a set of perfect conditions overnight. Garden City was an experiment and a stepping stone; a carefully planned town which in time might become one of a cluster of such towns. Such a cluster would be called Social Cities, where, in an area of 66,000 acres (a little less than that administered by the LCC) would reside a total population of a quarter of a million, each of six smaller Garden Cities having an area of 9,000 acres and a population of 32,000. The Central City would have an area of 12,000 acres and a population of 58,000. The whole would be united by a system of common waterways and railways.

What is may hinder What Might Be for a little while, but cannot stay the tide of progress.

So Howard.[79] Crowded cities had had their day. Each generation should build to suit its own needs.

The simple issue to be faced and faced resolutely is, can better results be obtained by starting on a bold plan on comparatively virgin soil than by attempting to adapt our old cities to our newer and higher needs?

Howard believed that the effect on London of such Social Cities would be dramatic: rents would fall, slumland would be reduced and the ill-housed would migrate. Howard's theme was essentially that of co-operative self-help: co-operation was the 'true path to real reform', not socialism and trade unionism based on class struggle. It was a message of social reform taken up more readily by liberals and conservatives, than by socialists and trade unionists.

With the publication of the book, and from a small nucleus of supporters, the Garden City Association (GCA) was launched in June 1899 at a public meeting, chaired by Sir John Leng, Liberal MP for Dundee. It may never have become more than an obscure propagandist body had Ralph Neville KC not been persuaded to become Chairman early in 1901. Thomas Adams, a dairy farmer from the Lothians and a political agent for the Liberal Party, became Secretary in April of the same year. With George Cadbury's help, a first conference of the Association was held at Bournville. By the time Howard's book was reissued, with small

---

[78] *Ibid.*, p. 119.   [79] *Ibid.*, pp. 134–5.

modifications, in 1902, under the title, *Garden Cities of Tomorrow*, the membership of the Association had risen to 1,300.[80]

A major step forward was taken in July 1902 with the flotation of a public company with a share capital of £200,000 to be called The Garden City Pioneer Company Ltd. A vital clause in the Memorandum of Association enabled the company to carry out any experiment in the common ownership of land related to the objectives of the GCA. Now at the age of fifty-three, Howard was appointed by the directors as a salaried Managing Director. Neville's first co-directors were Aneurin Williams, Liberal MP and Middlesbrough ironmaster, George Cadbury's son, Edward, Thomas Rizema, manager of the *Daily News*, Thomas Idris, London mineral water manufacturer, Franklin Thomasson, cotton manufacturer, and Howard Pearsall, civil engineer. Thomas Adams left the GCA to become secretary of the Pioneer Company.

Howard visited a number of sites in Warwickshire, Staffordshire, Essex and Nottinghamshire, and initially preferred the Chartley Castle estate in Staffordshire. But a 3,800-acre estate at Letchworth in Hertfordshire was more attractive to the London-based directors. It was purchased, and the Pioneer Company disbanded itself to become, in September 1903, the First Garden City Ltd for the purpose of building the Garden City. From this point onwards, Howard concentrated on publicity and promotion. He had been overtaken in influence by Adams and, later, by Charles Purdom, who joined the Pioneer Company as account clerk in November 1902. Vital features of Howard's concept of the garden city were successively discarded, crucially on the question of creating a Trust. Neville refused any form of self-government to future citizens in order not to alienate potential shareholders.

No one in recent history had been presented with the opportunity to design a town on a virgin site. A competition was held, though its arrangements seemed irregular and assessment unorthodox.[81] A plan prepared by Parker and Unwin was accepted in 1904. Though low-density housing remained, Howard's famous diagrams were quite overtaken by their pragmatic adjustment of layout to site. Whilst there was the formality of the grand axis, earlier ideas were also retained, including 'urban' quadrangles, village greens, sites for detached houses, and neighbourhood clusters.

For some years, Letchworth gave little impression of a city, or even a township, with its scatter of neighbourhoods, stunted civic centre and scattered factories, though it did have a fairly consistent domestic architecture (despite the multiplicity of architects). Population was less than

[80] E. Howard, *Garden Cities of Tomorrow* (London, 1902).
[81] Miller in Cherry, *Pioneers in British Planning*.

2,000 in late 1905, and under 9,000 even by 1914, but the town acquired a distinctive spirit and social character.[82] Growth was helped by the successful establishment of industry, the biggest single employer being W. H. Smith, whose mechanised bookbinding department moved from London in 1907. The typical garden-city worker proved to be a skilled or semi-skilled migrant (with his firm) from London. Rents were still too high for the unskilled labourer who lived in neighbouring towns and villages.

Meanwhile, the purity of the garden-city movement was being diluted. In 1906, the GCA formally added to its objectives the creation of garden suburbs, and changed its name in the next year to the Garden Cities and Town Planning Association. By 1909, the original first object of building garden cities had been replaced by a more general, advisory role, embracing garden suburbs and even garden villages. Howard still persisted, proposing in 1910 the building of a King Edward Memorial Garden City, later called King Edwardstown. He met with no support.

Towards the end of the Great War, Purdon and Frederick Osborn, whose association with Letchworth began in 1912, formed a group called the New Townsmen, and within a year had captured control of the GCA, then almost dormant.[83] As the centre of a national campaign for the building of a hundred new towns, it took on the character of a housing programme, rather than a scheme of social and economic reform. The message became one of town and country planning on a massive scale, rather than a redistribution of wealth – a reversal of Howard's original intention. But that is a post-War story.

The garden city movement had already made its mark with a distinctive impact on the countryside, contributing massively to the perceived need to decentralise city populations and to revitalise rural areas. Letchworth may not have become a city owned and governed by its community. A new social and industrial order may not have been achievable. But Howard's Association did address the problems of the great cities and their relationships to the surrounding countryside; it left an indelible impression on urban/rural change.

[82] M. Miller, *Letchworth: the First Garden City* (Chichester, 1989).
[83] Beevers, *Garden City Utopia*.

CHAPTER 30

EXPERT GUIDANCE TO THE COUNTRYSIDE

BY GORDON E. CHERRY AND JOHN SHEAIL

INTRODUCTION

Never before had it been so easy to move about the country. As early as 1841, Thomas Cook organised his first venture, a philanthropic temperance excursion from Leicester to Loughborough, using 'the newly developed power and facilities of the railways'.[1] In a preface to a new handbook on 'the botanical localities of the Metropolitan Districts', the author wrote of how

the facilities of locomotion afforded by railway communication are now so good that students can readily extend their researches to the more distant sandy heathlands and chalk ranges of Kent and Surrey, to the woodlands and cornfields of Essex and Herts, as well as to the banks of the Thames above Richmond and below Erith.[2]

As early as 1848, the Midland Railway Company ran Saturday-afternoon excursion trains from Derby to the picturesque parts of the county.[3] The construction of branch-lines of the Great Eastern Railway to Loughton in 1856, Epping and Ongar in 1865, and Chingford in 1873 made it easier for individuals, families and organised parties to visit Epping Forest at weekends. There were between 100,000 and 200,000 visitors on a Bank Holiday in the 1890s. Most stayed near the tea gardens or purpose-built retreats, positioned near beauty spots, where hundreds of people could enjoy inexpensive meals at one sitting and take part in the various 'fairground' amusements.[4]

The prospect of ever-increasing numbers of visitors filled some with horror. There were soon complaints as to how the hills of Ambleside in the Lake District were darkened in summer by 'swarms of tourists', and of how

---

[1] P. Brendon, *Thomas Cook: 150 Years of Popular Tourism* (London, 1991).
[2] E. C. De Crespigny, *A New London Flora; or, Handbook to the Botanical Localities of the Metropolitan Districts* (London, 1877), Preface.
[3] H. Spencer, *An Autobiography* (London, 1904), vol. I, p. 332.
[4] B. Ward, *The Retreats of Epping Forest* (Epping, 1978).

a great steam monster ploughs up our lake and disgorges multitudes upon the pier; the excursion trains bring thousands of anxious, vulgar people.[5]

The romance of the wild scenery of Lands End had been more or less spoilt by the numerous and boisterous excursion parties, that left abundant traces of their day's picnicking, 'in the shape of broken bottles and sundry other fragments'.[6]

If the prospect filled some with despair, others saw it as their duty to meet the challenge through guidance and education. Of the many examples that might be cited, we might instance the considerable increase of tourism to the Cheddar Gorge, in the limestone scenery of the Mendips, following the discovery of Gough's Cave in 1893. A number of cheap guides sought to meet the immediate needs of the visitor, and, in 1907, there appeared the first detailed account of 'the modern sport of cave exploring (speleology)' in *The Netherworld of Mendip*, and the scientific results already obtained in unravelling the archaeology of the human and animal remains found.[7]

In promoting such 'background' reading to a region or subject, publishers could take advantage of higher levels of literacy; books were becoming much more familiar objects about the home. Real wages rose by a third between 1875 and 1900, and the working week fell from an average of 63 to 54 hours. The Bank Holidays Act of 1871 introduced paid holidays and, by 1890, the traditional day of rest on Sunday had been extended for an increasing number of families to Saturday afternoon as well.

Whatever the type of visitor, there could be no doubt of the need for guidebooks. For E. N. Buxton, the inspiration for writing a guide to Epping Forest in 1884 was seeing how few summer visitors actually ventured far from where the trains set them down. The reluctance sprang not from indifference but from a dread of losing the way. In addition to the detailed instructions given in his guidebook, illustrated by annotated extracts from Ordnance Survey maps at the scale of 3 inches to the mile, Buxton (in his capacity as a Forest Verderer) cut the distinguishing letter of each of his routes in the bark of conspicuous trees. As he remarked, it was to be hoped that this would not be taken as justification for the 'practice of engraving initials and love emblems in similar positions'. The timetables of the Chingford and Loughton Railways were set out in an appendix to the book. On Bank Holidays, the trains ran every quarter of an hour.[8]

If Buxton wrote for the general excursionist, an increasing number of

---

[5] S. Margetson, *Leisure and Pleasure in the Nineteenth Century* (London, 1969), p. 82.

[6] Hissey, *Across England*, p. 230.

[7] P. Johnson, *The History of Mendip Caving* (Newton Abbot, 1967); E. A. Baker and H. E. Balch, *The Netherworld of Mendip. Explorations in the Great Caverns of Somerset, Yorkshire, Derbyshire and Elsewhere* (Clifton, 1907).   [8] Buxton, *Epping Forest*.

guides catered for the specialist, whether in the works of man or nature. During the 1850s and 1860s, the natural history of the coasts attracted armies of eager naturalists, who turned the results of their labours into elegant sketchbooks, collections of shells and rock samples, albums and pictures made of pressed seaweeds, and even small aquaria.[9] Margaret Gatty's guide for the amateur explorer gave accounts of the individual genera and species of seaweed, according to a 'simple' system of Mrs Gatty's own, using words 'in the vulgar tongue' rather than scientific terms. As well as the 384 figures reduced and reproduced in colour from Harvey's *Phycologia Britannia*, there was guidance on such practical points as clothing and apparatus, and rules for preserving and laying out the sea-weeds collected.[10]

How well equipped was the world of science to respond to the need for accurate, comprehensive guides? What significance might such an activity have in a scientific age that came to be so closely associated with the name of Darwin? What relevance did the distinction between 'informant' and 'tutee', or that between 'professional' and 'amateur', have in Victorian and Edwardian England and Wales? There was so much still to discover and explain that all were dependent on the findings of others. Everyman still had a role to play as 'Explorer in his own Backyard'.

A. THE SCIENTIFIC APPROACH

In publishing his book, *On the Origin of Species*, in 1859, Charles Darwin focused even greater attention on what was already one of the most burning questions of his day. Whilst the theory of 'evolution' was by no means new, no one had previously made so great an impact in putting it forward. As well as explaining how all organisms (including man) were descended through many generations, Darwin suggested a mechanism whereby this could have occurred, which he called 'natural selection'.[11] The implications were of immense importance, not only for biologists in their understanding of the plant and animal kingdom, but for all concerned with the history of man and his environment.[12]

To dwell on the remarkable advances made in the philosophical and theoretical aspects of the sciences is to run the risk of overlooking what, for the most part, really engaged the time and energies of scientists and their respective societies in the second half of the century. It would be to neglect much of the work of even such figures as Darwin, whose own

[9] E. Holden, *The Country Diary of an Edwardian Lady* (London, 1977).

[10] A. Gatty, *British Sea-weeds* (London, 1872), 2 vols.

[11] C. Darwin, *On the Origin of Species* (1859, repr. London, 1950); H. B. Woodward, *The History of the Geological Society of London* (London, 1907), pp. 212–13.

[12] M. Ruse, *The Darwinian Revolution* (Chicago, 1979), pp. ix–xiv.

observations on earthworms, kept in earth pots in his study over many months, led eventually to the publication of a pioneer study on *The Formation of Vegetable Mould through the Action of Worms with Observations on their Habits*.[13] The great mass of books, maps and memoirs of the period were self-consciously empiricist, setting out to explain what had been seen at first hand in the course of fieldwork. Rather than seeking to enunciate some cosmological theory, Victorian scientific periodicals were much more concerned with practical intellectual activity, the aim of which was to erect some kind of classification into which the rocks, plants and animals, and early man and his artefacts, might be fitted.[14]

Whilst the small number of 'professionals', occupying for the most part academic posts, might provide leadership and rigour in meeting these various goals, much of the exploratory work was carried out by 'amateurs'. The country parson continued to make a significant contribution. By the time he was fifty, the Reverend Octavius Pickard-Cambridge had published almost eighty scientific papers on spiders alone, in addition to others on antiquarian subjects, meteorology and general entomology. From his rectory at Bloxworth, he depended for his specimens on what could be collected by bicycle or on walks over the local heathlands. Although a member of numerous societies, he rarely left Dorset, maintaining instead an unflagging correspondence with scientists all over the world, willingly identifying and classifying other people's specimens. He was elected to the Royal Society in 1887.[15]

The Marshioness of Huntly was 'the epitome of the Victorian collector, for whom the very size of a collection came to be regarded as worthy in itself, implying the strong-minded devotion of many hours of loving toil and effort'.[16] Her journals record visits made by carriage to fifty-nine localities near Orton Longueville, near Peterborough, mostly within an afternoon. A typical entry of 20 April 1874 reads, 'to the hills and holes of Barnack and got roots of *Anemone pulsatilla* and coming home found *Pinguicula vulgaris*, *Valeriana dioica* and *Pedicularis* in wet ground east side of Wittering Brook'. Once a source of specimens had been found, repeated visits were made to secure more specimens for the herbarium and rockery. Increasingly, too, as her eldest son recalled later, collecting became the object of family expeditions, imbuing her children with 'a love of Nature'.[17] The Marchioness's 'engaging eccentricity' was shared

[13] C. Darwin, *The Formation of Vegetable Mould* (London, 1881, repr. London, 1958).

[14] J. A. Secord, *Controversy in Victorian Geology. The Cambrian–Silurian Dispute* (Princeton, 1986), pp. 3–13.   [15] B. Colloms, *Victorian Country Parsons* (London, 1977), pp. 56–78.

[16] D. E. Allen, *The Naturalist in Britain. A Social History* (London, 1976).

[17] Marquis of Huntly, *Milestones* (London, 1926), p. 28; J. Sheail and T. C. E. Wells, 'The Marchioness of Huntly: the written record and the herbarium', *Biolog. Journ. Linnean Soc.*, 13, 1980, pp. 315–30.

by many, in both fact and fiction. In her novel, *Magnum Bonum, or Mother Carey's Brood*, Charlotte M. Yonge portrayed Mrs Joe Brownloe as always tramping her children over the country, collecting specimens for their museum.[18]

Every community was likely to have its naturalists, mathematicians, musicians and poets – self-educated persons whose social influence was out of all proportion to their number.[19] In her novel, *Mary Barton*, Elizabeth Gaskell referred to the warm and devoted followers of botany among the common handloom weavers of Lancashire, who knew

the name and habitat of every plant within a day's walk from their dwellings; who steal the holiday of a day or two when any particular plant should be in flower.

There were entomologists too, who used a rude-looking net to take winged insects, or a dredge to rake 'green and slimy pools; practical, shrewd, hard-working men, who pore over every new specimen with real scientific delight'.[20] For the most part, these were sociable people, eager to share and display their information, whether at meetings or in the published proceedings of their respective society. Some ten county societies were founded in the 1840s, eight in the 1850s, and eleven in the next two decades. In 1913, the London Natural History Society was formed by the union of two earlier, Victorian societies.

Not only did the meetings and publications of the various societies provide an outlet for the findings of individual members, but it was soon realised that the only way to obtain an overall picture was to share and disseminate information. Nowhere was the need for collaboration more obvious than in the recording of meteorological phenomena. In order to make comparisons, both spatial and temporal, there had to be a willingness on the part of observers to submit copies of their records to some kind of central point for synthesis and publication. In 1860, George J. Symons published a volume, summarising the statistics for rainfall recorded at 168 stations throughout England and Wales. It became the first of a series of annual volumes, that of 1898 being based on 2,782 stations, and a further 622 stations in Ireland and Scotland. Unmatched by any other country, the mass of data, compiled by volunteers according to standard criteria, proved invaluable for such purposes as the planning of reservoir schemes.[21]

[18] C. M. Yonge, *Magnum Bonum, or Mother Carey's Brood* (London, 1879), vol. I, p. 142.

[19] J. F. C. Harrison, *Learning and Living, 1790–1960. A Study in the History of English Adult Education Movement* (London, 1961), pp. 44–5, 49; J. Sheail, 'T. W. Woodhead and the study of vegetation and man in the Huddersfield district', *The Naturalist*, 113, 1988, pp. 125–39.

[20] E. Gaskell, *Mary Barton. A Tale of Manchester Life* (1848, repr. Harmondsworth, 1970), p. 75.

[21] S. Lee (ed.), *Dictionary of National Biography. Supplement, Volume III* (London, 1901), pp. 374–5; AB, 'George James Symons', *Proc. Roy. Soc.*, 75, 1905, pp. 104–5.

Most society members were interested in more than one aspect of the countryside. A man of extraordinarily diverse interests was John Lubbock, the future Lord Avebury, head of a banking family, sometime member of parliament, friend of Charles Darwin and, throughout his life, president of countless mercantile associations and nearly every scientific society.[22] It was through his search for prehistoric stone implements in the river deposits of south-east England that Lubbock first became interested in geology. A close association was struck up with Lyell, Prestwich and others, who were trying at that time to determine the relative ages of the various terrace deposits. The fact that deposits were known to date back to a 'glacial period' gave added zest to Lubbock's frequent holidays in Switzerland, almost always in the company of geologists and others making a special study of the country. The outcome was a handbook, *The Scenery of Switzerland*.[23] While on holiday in Snowdonia in March 1899, he began work on 'a book on the Physical Geography of England'. Visits were made to numerous parts of the country, including the headwaters of the Thames to study 'the early story of the river'. The book was finished in late 1901, and published as *The Scenery of England and the Causes to which it is due*.[24]

B. FIELD GEOLOGY

Victorian geologists were deeply conscious of the fact that no other part of the world of equal area had a rocky crust that was so varied in its structure and scenery, or so rich in mineral resources.[25] In the words of one prize essayist, 'we have the geological alphabet'.[26] The immediate priority was to establish a stratigraphical order and to map the countryside. Many of the type localities of different rocks found in Britain became the basis for classifications adopted in other parts of the world.

Far from being a non-controversial fact-gathering exercise, there was long and acrimonious debate over almost every major division of the geological column – the most famous being the controversy as to where the boundary should be placed between the Cambrian and Silurian divisions in the oldest fossil-bearing rocks. Previous experience, early training, institutional loyalties, personal temperament and theoretical outlook

[22] AES, 'John Lubbock, Baron Avebury', *Proc. Roy. Soc.*, 87, 1914, pp. i–iii.

[23] J. Lubbock, *The Scenery of Switzerland and the Causes to Which It is Due* (London, 1896).

[24] Lord Avebury, *The Scenery of England and the Causes to Which It is Due* (London, 1902); H. G. Hutchinson, *Life of Sir John Lubbock*. vol. II (London, 1914), pp. 103, 130; A. Smith Woodward, 'Geology', in A. Grant Duff (ed.), *The Life-work of Lord Avebury* (London, 1934), pp. 105–14.

[25] D. MacKintosh, *The Scenery of England and Wales. Its Character and Origin* (London, 1869), p. 22.

[26] J. E. Thomas, *A Geographical and Geological Description of Caernarvonshire, with Special Reference to its Quarries, Mines and Mineral Products* (London, 1874), p. 1.

were all brought to bear in deciding on the names of the different rocks, where their boundaries should be drawn, and how far the fossil record or physical characteristics of the rocks should determine the stratigraphical classification.[27]

If one of the ultimate objects of geology was to discover the physical conditions which had prevailed over different areas at successive periods, the other great object, equally scientific, was to develop the practical applications of geological knowledge.[28] Mining and industry would gain from an inventory of coal, iron and other mineral reserves, and agriculture from an improved knowledge of soils and sub-soils. The geological maps, cross-sections and memoirs published by the Geological Survey of Great Britain, an official body appointed in 1835, were at least the equal to anything produced on the Continent or in America. Demand for its products outstripped supply. A see-sawing of priorities began. They were essentially threefold. Whilst striving for excellence in executing each map and memoir, there was a pressing need to complete the primary survey and to publish the maps at the one-inch scale. A third priority was to update the six-inch maps of such important areas as the coalfields.

In 1866, a Royal Commission was appointed to take stock of available coal resources and to estimate probable dates of exhaustion.[29] The presence of so many Survey personnel on the Commission provided an outstanding opportunity to draw attention to the work done – the first map and memoir of a coalfield (South Staffordshire) had appeared in 1853. In order to expedite the survey of coal and mineral resources, and the mapping of drift deposits (which were particularly important to farming, civil engineering and sanitary bodies), government approval was obtained for doubling the number of field geologists (bringing the number to fifty-seven by 1871). By the mid-1880s, maps of the solid geology had been published at the one-inch scale for the whole of England and Wales. Such work was greatly facilitated by the increasing availability of the Ordnance Survey six-inch maps.[30]

It was clear that a separate series of drift maps would be required, especially for those parts where the surface deposits exerted a powerful influence on scenery, agriculture and water supplies. They were 'as truly geological formations as the more regularly stratified rocks upon which they rested'. If local agricultural colleges and research institutions used the maps to carry out soil surveys and analyses, the agronomist, A. D.

[27] Secord, *Controversy in Geology*.

[28] H. B. Woodward, *The Geology of Soils and Substrata, with Special Reference to Agriculture, Estates and Sanitation* (London, 1912).

[29] *RC on Several Matters relating to Coal in the United Kingdom, Report*, BPP, 1871, XVIII.

[30] E. Bailey, *Geological Survey of Great Britain* (London, 1952), pp. 75–6.

Hall, wrote, every farmer would be in possession of 'that exact knowledge of the soil which is fundamental to all farming operations'.[31]

The growth of urban populations, rising standards of public health, and the insistent demand for more abundant supplies of water were causing local authorities, engineers and private persons to look increasingly to the Geological Survey for relevant data upon which schemes could be formulated. Memoirs were published from 1899 onwards, describing the water supplies of individual counties, based upon well-records.[32] Large-scale construction works presented outstanding opportunities to investigate the geology of extensive tracts of countryside. A detailed paper on the Ordovician and Silurian complex of central Wales was given to the Geological Society in 1900, based on observations made during the construction of the westernmost length of aqueduct carrying water from the reservoirs of the Claerwen valley to the waterworks of Birmingham Corporation – the author had acted as assistant engineer to that part of the impounding project. Descriptions of the site of the reservoirs, and of the rest of the aqueduct, were published in the Geological Survey's annual survey of progress, published for 1901.[33]

There was increasing interest in the landforms encountered. In the memoir on *The Geology of Rutland*, the effects of the position and inclination of the rocks, and their varying resistance to 'denudation agencies' on local scenery, were illustrated by a series of lithographic plates, made from pencil drawings.[34] Because the rocks of the Quaternary period were the most recent to be formed, their importance was out of all proportion to their thickness. They occupied much of the land surface and, therefore, exerted a direct influence on agriculture. They afforded the earliest undoubted evidence of the presence of man.[35]

Whilst there was consensus that the existing configuration of the land depended on three main factors – the original configuration, the nature of the rocks, and their inclination – there was fierce controversy in the late 1860s as to whether the sea or rain and frost were primarily responsible for 'composing the more abrupt inequalities of the surface of the earth'. As Daniel MacKintosh wrote, in his study of the origin of escarpments, it would be difficult to exaggerate 'the battering, disentangling, undermining, excavating, upthrowing and transporting power of the sea'. The forces that were creating Beachy Head had once assuredly

[31] Woodward, *Geology of Soils*, pp. 19–20, 35–6.

[32] W. Whitaker and C. Reid, *The Water Supply of Sussex* (London, 1899); J. S. Flett, *The First Hundred Years of the Geological Survey* (London, 1937), p. 129.

[33] H. Lapworth, 'The Silurian sequence of Rhayador', *Qtly Journ. Geol. Soc.*, 56, 1900, pp. 67–137; Geological Survey, *Summary of Progress of the Geological Survey of the United Kingdom for 1901* (London, 1902), pp. 53–64.   [34] J. W. Judd, *The Geology of Rutland* (London, 1875).

[35] Avebury, *Scenery of England*, p. 45.

formed the succession of headlands, bays and combes that comprised the escarpment of the South Downs. Whereas a hollow excavated by rain-torrents would assume a more or less V-form, and would necessarily shallow out towards the summit, such combes as those between Brighton and Lewes, and at Kingley Vale near Chichester, were 'as much a sea-cove as if it were still washed by the spray'. Their geometrical regularity, combined with other unmistakeable features, indicated that they had been literally scooped out by a sweeping, undermining and gyratory action.[36]

For MacKintosh and those of similar outlook, it seemed scarcely credible that rain and frost could be responsible by themselves for the major landforms of the countryside. In both chalk and oolitic limestone districts, rain was a solvent rather than a transporting agent. Proof that rain could not change the general contour of the chalk was afforded by the thousands of raised beaches, many of only a few feet in height, that might be seen 'undulating along the side of the valley to the east of Mere in Wiltshire'. The terrain was so intimately associated with the valleys, combes and escarpments as to make abundantly evident their common origin. There soon appeared in the *Geological Magazine* an article pointing out that the 'raised sea beaches' were in fact lynchets, or cultivation strips. Far from providing evidence of the impotence of rain in moulding the earth's surface, the speed with which terraces could be formed where soil was broken down by the plough, and washed downslope, provided some of the most convincing proofs of the role of subaerial denudation in altering the configuration of hill-slopes.[37]

It was William Whitaker who finally disposed of the view that the chalk and Lower Tertiary escarpments of south-east England had been sculptured by marine action. In a paper to the Geological Society of London in 1867, he recounted how, in the course of detailed fieldwork, no evidence had been found to suggest that any of the features were sea-cliffs. In assessing how far water, wind and the sun might have been responsible, it was essential to bear in mind the lengths of time over which such forces could have operated. In the same way as astronomy revealed 'our liveliest picture of infinity', so geology provided 'our best idea of eternity'. There were times when the climate would have been more severe, and the agents more powerful. Even in historic times, the 'wearing action of surface water' had been appreciably reduced, as a result

[36] D. MacKintosh, 'The sea against rain and frost', *Geological Mag.*, 3, 1866, pp. 63–70; R. J. Chorley, A. J. Dunn and R. P. Beckinsale, *The History of the Study of Landforms or the Development of Geomorphology* (London, vol. 1, 1964), pp. 325–9.

[37] G. Poulett Scrope, 'The terraces of the chalk downs', *Geological Mag.*, 3, 1866, pp. 293–6; Anonymous, 'A terraced hill-side', *St James' Gazette*, 2, 1881, p. 1084.

of the draining of the land, the embankment and canalisation of rivers, and 'like human handiworks'.[38]

Studies on the impact of glaciation had been held back by the fact that so much of the relevant material was scattered among foreign publications, and that it required 'an elastic and original mind' to bridge the gap between the diminutive Alpine glaciers and what was conceived to have taken place in the Quaternary period. With the rapid strides being made in polar exploration, however, all that was changing. The scientific observations made by many of the explorers had removed all grounds for scepticism as to the importance of ice as a geological agent, and the extent to which it could once have overridden the continents. Rather than transcending all experience, the phenomena of the Ice Age had provided 'the last and most striking example' of the principle that the agencies which had fashioned the earth throughout the geological ages were those which still could be seen in action in various parts of the world at the present day.[39]

C. THE FIELD NATURALIST

The effort invested in the recording of the plant and animal life of the countryside was far from even. Botanists continued to list and count the species found. In his *Flora of Cumberland*, William Hodgson classified 946 of the 1,196 species as natives, 60 as colonists (well established as weeds on cultivated ground), and 43 as denizens, now quite wild but possibly introduced by human agency. 'Aliens' was the name given to those plants which were the least thoroughly established. There were 144 in Cumberland, excluding those found in docks and harbours.[40]

Whilst the enumeration of animal life was in a much more primitive state, there had never been a more favourable time for studying the history and habits of the different species.[41] During the 1880s, there appeared four major ornithological works. They included the last volume of William Yarrell's *A History of British Birds*, the first systematic textbook of its kind, and Henry Seebholm's *History of British Birds*.[42] Not only did each provide essential guidance in recognising the different species, but also a challenge as to what could be achieved by the dedicated amateur,

---

[38] J. S. Flett, 'William Whitaker', *Qtly Journ. Geol. Soc.*, 81, 1925, pp. lxi–lxii; W. Whitaker, 'On sub-aerial denudation, and on cliffs and escarpments of the Chalk and Lower Tertiary beds', *Geological Mag*, 4, pp. 447–54, 483–93. [39] W. B. Wright, *The Quaternary Ice Age* (London, 1914).

[40] W. Hodgson (ed.), *The Flora of Cumberland* (Carlisle, 1898), p. x.

[41] J. C. Mansel-Pleydell, *The Birds of Dorsetshire* (London, 1888), pp. vii–xvi.

[42] W. Yarrell, *A History of British Birds* (London, 4th edn, 1871–4, revised and enlarged by A. Newton, 1883–5).

in terms of both original bird observation and the formulation of ideas as to bird distribution and behaviour.

Observers along the south coast were particularly well placed to study bird migration. In his *Ornithological Rambles in Sussex*, A. E. Knox put forward evidence to suggest that it was not only summer and winter visitors that performed a double migration each year. Numerous flocks of birds, normally regarded as permanent residents, also passed in rapid succession along the coast in an easterly direction during the early autumn, so as to cross the Channel by the Straits of Dover.[43] The Irish naturalist, Alexander Gordon More, was the first to make a systematic bird census of the British Isles, based on the evidence collected during the nesting season by correspondents in many parts of the country.[44]

Whatever their particular interests, increasing numbers of naturalists ventured into the field. Not only did the railways save a great deal of time and leg-weariness, but improved postal communications made it easier to consult friends and have specimens identified by the best authorities.[45] In the papers of the Duke of Manchester's estate in Huntingdonshire, there survives a letter from a local solicitor, asking the bailiff whether the Reverend Augustus Ley might

botanise for one day this month (August) in West Wood and Calpher Wood. He is coming to stay with me for the purpose of studying the brambles and wild roses of our county.[46]

The appearance of a species list often acted as a stimulus to further searches. In the six years following the circulation of copies of Hamilton Davey's 'A Tentative List of the Flowering Plants, Ferns, etc. of Cornwall' in 1902, some forty-two species and a large number of varieties were found, of which two species and one variety were new to science, and two species had never figured in a British list before.[47]

However great the deficiencies in recording the flora and fauna, it was obvious that striking changes were taking place. *The Flora of the Isle of Wight*, published in 1856, gave the date of each species record. For some of the rarer or more local species, it often proved to be the last date, as 'waste' lands were enclosed, and localities, not so long free, became dotted with tenements, their sites fenced 'with the jealous exclusiveness

---

[43] A. E. Knox, *Ornithological Rambles in Sussex, with a Systematic Catalogue of the Birds of the County* (London, 1849).

[44] A. G. More, 'On the distribution of birds in Great Britain during the nesting-season', *Ibis*, new ser., I, 1865, pp. 1–27, 119–42, 425–58.

[45] J. W. White, 'Remarks on the preparation of a local flora', *Proc. Bristol Naturalists' Soc.*, 3, 1881, pp. 97–106.

[46] Huntingdonshire RO, Manchester MSS, 57a, letter from E. W. Hunnybun, 9 August 1908.

[47] F. Hamilton Davey, *Flora of Cornwall* (Penryn, 1909), pp. lxi–lxii.

of individual appropriation'.[48] Until 1864, the autumn crocus (*Colchicum autumnale*) was fairly abundant in a large meadow near Blennerhasset, on the banks of the river Eden in Cumberland. At that date, the property changed hands, and 'the comparatively rare species was extirpated' as the new owner 'threw down the old hedges, drained the meadows, and reduced them to arable'.[49]

Whole regions could be affected. Babington's *Flora of Cambridgeshire* was one of the first to draw explicit comparisons between the past and present status of plants. By doing so, it was able to record 'the great alterations caused by modern enclosures and drainage'. With a very few slight exceptions, all the open 'field' of the clayey district was now enclosed, and the long, narrow 'balks' or strips of 'ancient turf', with the various plants which grew upon them, had been destroyed by the plough. Until the last sixty years, most of the chalk district had been covered with 'a beautiful coating of turf, profusely decorated with *Anemone pulsatilla*, *Astragalus hypoglottis*, and other interesting plants'. It was now converted to arable land, and 'its peculiar plants mostly confined to small waste spots by road-sides, pits, and the very few banks which are too steep for the plough'. Even the tumuli, entrenchments and 'other interesting works of the ancient inhabitants' had seldom escaped 'the rapacity of the modern agriculturist, who too frequently looks upon the native plants of the country as weeds, and its antiquities as deformities'.[50]

As the Reverend E. A. Woodruffe-Peacock remarked, it was only complicating matters, and causing endless unnecessary difficulties, to exclude man and his works from any assessment as to how Nature had evolved. Man was 'as natural and as much a circumstance of environment, as soil, climate or elevation'.[51] By far the greatest source of alien species in the late nineteenth century was the grain imported for flour-milling and distilling. Every sack contained countless seeds of cornfield weeds, which had to be sifted out and thrown onto waste ground, or sold for feeding to domestic fowls and game. Whilst the wheat imported annually from America was comparatively clean, that from Russia was more than enough to account for the constant occurrence of Eastern weeds around large towns. Likewise, whilst the barley imported from France and Germany was usually clean, Persian, Danubian and Turkish barley normally contained very large quantities of other seeds.[52]

Vegetation surveys were an obvious way of sustaining the interest of local naturalists at a time when 'the registration of new species was pretty

[48] W. A. Bromfield, *Flora Vectis* (London, 1856), pp. xvi–xvii.

[49] W. Hodgson, 'Disappearance of plants in Cumberland', *Naturalist, Hull*, 1891, pp. 7–12.

[50] C. C. Babington, *Flora of Cambridgeshire* (London, 1860).

[51] E. A. Woodruffe-Peacock, 'The Flora of Lincolnshire', *Trans Lincs. Naturalists' Union*, 4, 1916, pp. 22–40.   [52] S. T. Dunn, *Alien Flora of Britain* (London, 1905).

well exhausted'.[53] A Committee for the Survey and Study of British Vegetation (later shortened to the British Vegetation Committee) was formed in 1904 to co-ordinate further work, and to secure greater uniformity of method in so far as it was desirable.[54] One of its leading members was Charles Moss. Drawing upon his considerable field experience and grasp of the literature in the incipient science of ecology, Moss put forward a hierarchy, within which vegetation might be classified. The term 'plant formation' described the whole of the vegetation occurring in a definite and essentially uniform habitat. A 'plant association' characterised minor differences within the otherwise uniform habitat, and the third term, 'plant society', corresponded with even less fundamental differences in the habitat. Some idea of the areal extent of each of the units was conveyed by the fact that plant formations could be recorded on quarter-inch (1:253,440) maps, whereas 1-inch (1:63,660) maps were needed to show plant associations. Most societies and the smaller examples of associations could only be plotted on 6-inch (1:10,560) maps.[55]

The value of this conceptual framework was demonstrated by Moss's monograph, published by the Royal Geographical Society, on the vegetation of the Bath and Bridgwater district of Somerset. Based on mapping carried out (at the 1- and 6-inch scales), the aim was to record and explain the interplay of climate and soil on vegetation. To do this, Moss illustrated how a plant formation began its history as an open or unstable plant association, passed through intermediate associations, and eventually became a closed or stable association. In the case of the coastal sand dunes, the plant formation had not progressed beyond intermediate associations of pasture or marsh plants. In the lowland peat moors, the native associations had been largely destroyed, but most of the plant species survived in what Moss called 'the substituted plant associations' that were evolving in the wet hollows left by the peat-cutters, or in the artificially straightened courses of rivers, canals and rhynes.[56]

The opportunity to apply what was already coming to be known as an ecological approach to the survey of the British Isles arose in 1911, when the first International Phytogeographical Excursion took place. It was organised on behalf of the British Vegetation Committee by A. G. Tansley, who also acted as editor of what became the first attempt at 'a scientific description of British vegetation'. From an analysis of the Annual Returns to the Board of Agriculture, Tansley estimated that up to 20 per cent of England and 40 per cent of Wales might be covered by natural or semi-natural vegetation, namely mountain and heath land, and

[53] C. E. Moss, 'Norland Clough', *Halifax Naturalist*, 5, 1900, pp. 40–5.

[54] J. Sheail, *Seventy-five Years in Ecology. The British Ecological Society* (Oxford, 1987), pp. 22–35.

[55] C. E. Moss, 'The fundamental units of vegetation', *New Phytologist*, 9, 1910, pp. 18–53.

[56] Moss, *Geographical Distribution of Vegetation in Somerset*, pp. 32–3.

lands so barren and remote as to be used for grazing only. There were in addition extensive areas of woodland and permanent grassland, where management was minimal. If these areas were included, as much as a quarter of England, and half of Wales, might be regarded as covered by natural vegetation 'more or less modified by the activities of man'.[57]

Seven members of the British Vegetation Committee contributed to Tansley's Excursion volume. Fourteen plant formations, and their associations and societies, were identified. They included those occurring on clays and loams, sandy soils, older siliceous soils, and calcareous soils. Freshwater, marsh and fen formations, moorland, arctic–alpine vegetation, and the salt-marsh and sand-dune formations of the coast were identified. The compilation of the volume, and attempt to illustrate diagramatically the relationships between the formations (Figure 30.1) highlighted the many gaps in knowledge that remained. There had, for example, been scarcely any comprehensive studies of aquatic vegetation, or of any of the plant formations and associations that characterised Wales. It was to help repair these glaring deficiencies that the British Ecological Society was founded in 1913 – the first national ecological society to be established anywhere in the world.

D. THE FIELD ARCHAEOLOGIST

The Victorians were fascinated by the past, reading historical novels and even the works of academic historians in large numbers. The extension of history into new areas excited them, especially where it cast further light on the origins of their own society, and its development to the present pinnacle of human achievement.[58] The discoveries made in the course of attending meetings and excursions, or reading the journals or accounts published in local newspapers, provided an assurance that it was possible to maintain a link with the past without impeding the march of improvement.

There was obvious pride in recording what could be learned of the history of individual localities and parts of the country. The authors of an inventory of the old cottages of Snowdonia recounted how the buildings could be studied from both an artistic and a historical viewpoint. Like the many-branched Welsh oaks, peculiar to the Principality, the buildings were a natural product of the country, revealing 'as clearly as any written history the development of the life of the people'. There was no time to lose. Most were in the last stages of decay. Within a few years,

---

[57] A. G. Tansley (ed.), *Types of British Vegetation* (Cambridge, 1911).

[58] P. J. Bowler, 'The invention of the past', in *The Making of Britain*, ed. L. M. Smith (London, 1987), pp. 159–71.

FORMATION OF MOUNTAIN TOP DETRITUS

Arctic–alpine formations

CHOMOPHYTE FORMATION

Association of exposed ledges

Association of sheltered ledges

Association of Hydrophilous Chomophytes

Moss–lichen open association

Shade Chomophyte association

Rhacomitrium heath

Upland Rhacomitrium moor

Heather–bilberry moor

Bilberry moors

ARCTIC–ALPINE GRASSLAND FORMATION

Cotton-grass moor — Scirpur moor

Grass moor

FORMATION OF SILICEOUS SOILS

Siliceous grassland

Scrub

Birchwood (Highland type)

Sessile oakwood

Damp sub-association

Pedunculate oakwood

Dry sub-association

Sandy oakwood

Scrub

Grass heath

FORMATION OF SANDY SOILS

FORMATION OF CALCAREOUS SOILS

Limestone grassland

Limestone scrub

Ash–oakwood

Ashwood

Scrub

Marly grassland

Chalk grassland

Chalk scrub

Beechwood

(Lowland moor)

FEN FORMATION

Fen carr (Alder wood)

Fen association

Swamp carr (Alder wood)

Alder–willow wood

Marsh associations

FORMATION OF CLAYS AND LOAMS

Neutral grassland

MARSH FORMATION

Sub-formation richer in salts

Sub-formation poorer in salts

FRESHWATER AQUATIC FORMATION

Maritime Formations { SALT MARSH FORMATION / SAND DUNE FORMATION

Heathy sub-association

Oak birch heath — Heath

Pinewood (sub-spont)

Birchwood

Beechwood

HEATH FORMATION

Upland heath

Pinewood (Highland type)

Heather moor

Birchwood

Heather moor

Cotton-grass moor

Sphagnum bog

Birchwood (Fen)

MOOR FORMATION

Lowland

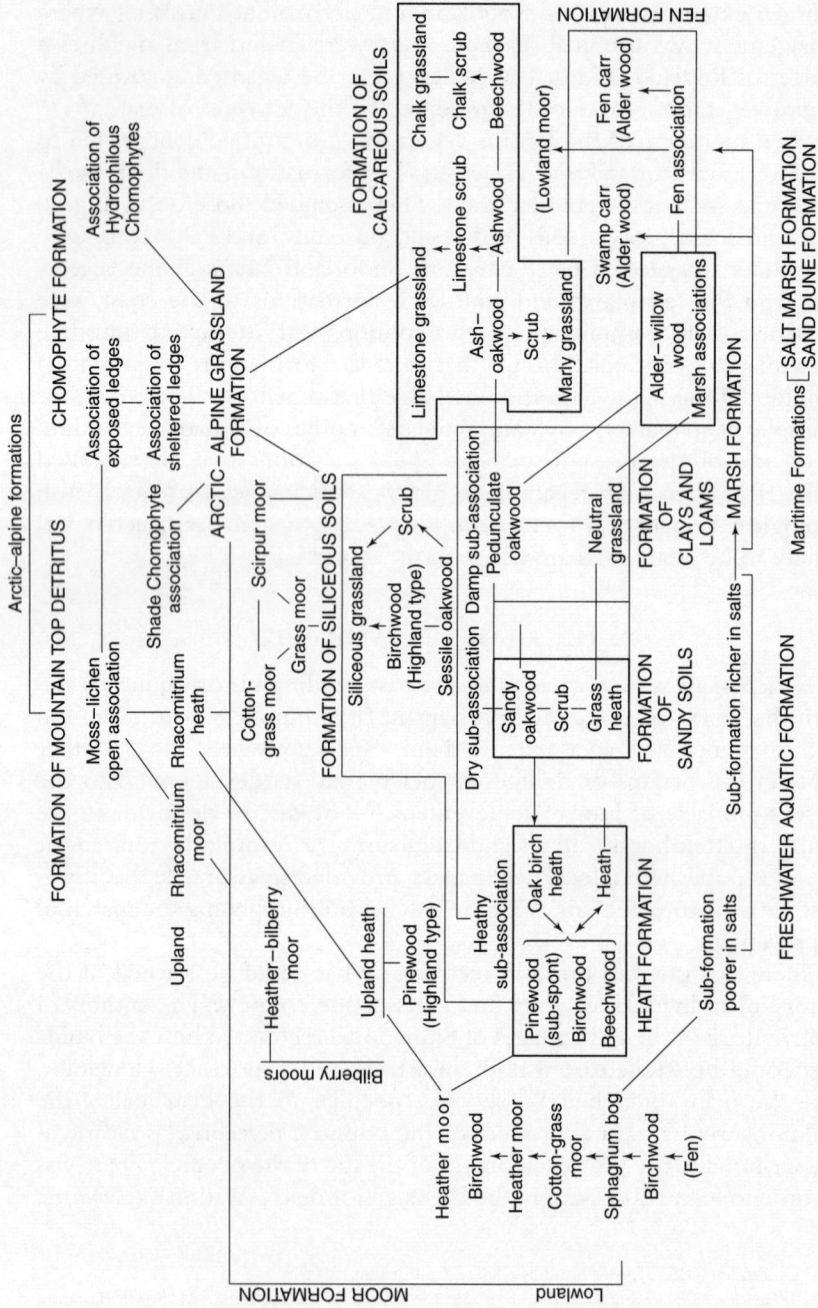

30.1 The relationships of the fourteen British plant formations, as described in *Types of British Vegetation*, edited by A. G. Tansley, and published in 1911.

'these valuable pages of national history and native building' would be lost for ever.[59]

The county archaeological and historical societies, professional journals, and the innumerable local histories and guidebooks, all had their part to play. The aims of the Hampshire Field Club, founded in 1885, were serious and scientific. According to the rules, each member must have an interest in some aspect of natural history or antiquities. Each year, about eight excursions were organised. The report on the first included criticism of the 'improvements' to the church at Compton, which marred the Norman doorway. The second meeting deplored the ruinous state of West Worldham church. Right from the start, the Field Club's *Proceedings* included papers, reports and records of observations.[60] By far the most ambitious publishing venture was the commissioning of a history to commemorate the Queen's Diamond Jubilee. Planned from the outset as a whole, the 'Victoria History of the Counties of England' was to comprise 160 volumes. By 1907, twenty-seven volumes had appeared.[61]

Whilst the discoveries of local antiquaries had long aroused interest, and Stonehenge and other 'memorials of antiquity' were admired as examples of ancient skill and perseverence, little progress could be made in interpreting these 'pages of human history' until biblical catastrophes were replaced by what many contemporaries were pleased to call 'scientific geology'.[62] Sir John Lubbock's book, *Prehistoric Times*, which passed through seven editions between 1865 and 1913, did much to ensure the adoption of a fourfold division of prehistory, namely the palaeolithic and neolithic periods, and the bronze and iron ages. From such a classificatory system, based on the presence and craftsmanship of the stone, metal and other artefacts, the archaeologist could join the ethnographer and anthropologist in tracing advances in the human economy and society from geological times to the historical period.[63]

The 1860s have been identified as the decade when archaeology ceased to be an essentially dilettantist avocation, and became an organised discipline, and ethnology was resuscitated with a decidedly archaeological orientation. As in other fields of science, a crucial role was played by personal connections and the institutions or societies that

[59] H. Hughes and H. L. North, *The Old Cottages of Snowdonia* (1908, repr. Capel Curig, 1979).

[60] C. Dellheim, *The Face of the Past. The Preservation of the Medieval Inheritance in Victorian England* (Cambridge, 1982); B. Taylor, 'One hundred years of the Hampshire Field Club', *Proc. Hants. Field Club & Archaeol. Soc.*, 41, 1985, pp. 5–20.

[61] R. B. Pugh, *The Victoria History of the Counties of England. General Introduction* (Oxford, 1970), pp. 1–10.

[62] A. S. Goudie, 'Geography and archaeology: the growth of a relationship', in J. M. Wagstaff (ed.), *Landscape and Culture* (Oxford, 1987), pp. 11–15.

[63] J. Lubbock, *Pre-historic Times as Illustrated by Ancient Remains, and the Manner and Custom of Modern Savages* (London, 1865); G. E. Daniel, *A Hundred Years of Archaeology* (London, 1950), pp. 77–86.

made them possible. Typically, Augustus Henry Lane Fox Pitt Rivers began as an amateur collector, in his case of antique arms. Membership first of the Geographical Society and then, in 1861, of the Ethnological Society, helped to widen Pitt Rivers's field of contacts. As for many others, evidence of man's existence extending back perhaps millions of years struck him with the force of revelation and, from 1860 onwards, he was a committed convert to the 'scientific method'. As a member of the rejuvenated Society of Antiquaries and the Archaeological Institute, Pitt Rivers joined with Lubbock and John Evans, a leading paper manufacturer, collector and archaeologist, in using such bodies as a vehicle for securing greater recognition for the scientific approach.[64]

The implications of the shift from the 'mere relic grubbing' to 'scientific excavations' gradually became clear. As Pitt Rivers insisted, it was only through the closest study of common objects thrown away by the inhabitants of the site, that the date, nature and significance of earthworks could be determined.[65] If the excavations by Pitt Rivers at Cranborne Chase in Wiltshire emphasised the need for excavators to think in terms of the archaeology of areas, as distinct from isolated sites, so too did they provide confidence for such major initiatives as that taken on the Roman town-site of Silchester in Hampshire. Over the period 1864–78, the rector of Strathfield Saye, under the patronage of the site's owner, the Duke of Wellington, had drawn attention to the types of building present, but there had been no settled plan of excavation. The general character and relation of one part of the town to another remained unclear. With the active encouragement of John Evans, who, as president of the Society of Antiquaries of London, had established a research fund for such purposes, a proposal was made in 1890 to carry out a complete and systematic excavation of the entire 100-acre site. The first summer season of six months began that year, and, for twenty years, work continued under the supervision of members of an executive committee of experts. Monies for the Silchester Excavation Fund were provided by the Society and local and national subscriptions.[66]

The term 'field archaeology' was coined by J. P. Williams-Freeman in 1915, who had been struck by the use of 'field naturalist' to distinguish the study of plants and animals in the field from that in the laboratory or

---

[64] W. Chapman, 'The organizational context in the history of archaeology: Pitt Rivers and Other archaeologists in the 1860s', *Antiquaries Journ.* 69, 1989, pp. 23–42.

[65] Lieut. General Pitt Rivers, *Excavations in Cranborne Chase* (privately printed, 1892), vol. III, pp. xi–xii.

[66] G. E. Fox and W. H. St John Hope, 'On the desirability of the complete and systematic excavation of the site of Silchester', *Proc. Soc. Antiquaries Lond.*, 2nd ser., 13, 1890, pp. 85–97; G. E. Fox, 'The Romano-British town of Calleva Atrebatum', in *VCH Hampshire and the Isle of Wight*, ed. H. A. Doubleday, vol. I, pp. 350–72; G. C. Boon, *Roman Silchester* (London, 1957), pp. 37–9.

museum. As he wrote in his introduction to *Field Archaeology as Illustrated by Hampshire*, it was a branch of science quite in its infancy, where every student could record new facts and make his own discoveries.[67] In his *Illustrated Map of Dorsetshire, its Vestiges, Celtic, Roman, Saxon and Danish*, Charles Warne had represented the various sites and earthworks by different colours. In Wiltshire, the Reverend A. C. Smith, rector of Yatesbury, laboured for thirty years to produce his map of 100 square miles around Avebury, the area that he could reach on horseback. The enlarged Ordnance Survey one-inch map, published in 1884, was annotated with field names and details of earthworks. James Fergusson's near-global study of a single class of field monument, the rude stone monuments, in 1872, was based on a combination of fieldwork and excavation.[68]

An attempt on the part of the Society of Antiquaries to secure a better co-ordination of research effort led in 1889 to the institution of the Congress of Archaeological Societies, and the publication of annual reports.[69] An Earthworks Committee was appointed in 1901 to provide a basis for systematic field research. The value of such an approach was soon demonstrated in Hadrian Allcroft's major study of the earthworks of England, which appeared in 1908.[70] In the same year, a standing Royal Commission was appointed to make 'an inventory of the Ancient and Historical Monuments and Constructions connected with or illustrative of the contemporary culture, civilization and conditions of life' from earliest times to 1700. The first inventory to be published was that of Hertfordshire in 1911.[71] Ostensibly in response to the Earthworks Committee's appeal for regional surveys, Heywood Sumner published his earthwork surveys of Cranborne Chase (1913) and the New Forest (1917). The distinctive artistic style of his maps and plans – he was a product of the pre-Raphaelite movement and a friend of William Morris – set new standards in archaeological draughtsmanship.[72]

As Heywood Sumner remarked, the farmer could not be expected to pay rent on behalf of Archaeology. The burden of preserving ancient earthworks had to be placed on broader shoulders than those that

---

[67] J. P. Williams-Freeman, *An Introduction to Field Archaeology as Illustrated by Hampshire* (London, 1915), pp. xix–xxii.

[68] O. G. S. Crawford, *Archaeology in the Field* (London, 1953), p. 36; P. Ashbee, 'Field archaeology; its origins and development', in *Archaeology and the Landscape*, ed. P. J. Fowler (London, 1972), pp. 38–69.

[69] B. H. St J. O'Neil, 'The Congress of archaeological societies', *Antiquaries Journ.*, 26, 1946, pp. 61–6.   [70] A. H. Allcroft, *Earthworks in England* (London, 1908).

[71] Royal Commission on Historical Monuments, *An Inventory of the Historical Monuments in Hertfordshire* (London, 1911), pp. ix–xvi.

[72] H. Sumner, *The Ancient Earthworks of Cranborne Chase* (London, 1913); *The Ancient Earthworks of the New Forest* (London, 1917).

chanced to bear them. That situation would only come about when the national imagination was fired, and the meaning of earthworks explained. Whatever the objects to be preserved, natural or man-made, there was a crucial educative role to be fulfilled by those in the relevant disciplines and professions.

E. PRESERVATION

The late nineteenth century was noteworthy for the number of expert bodies founded for the protection of whatever was worthy of preservation in town and country. The fragmented nature of the movement reflected the extent to which each voluntary society focused on its own discrete concerns and was at first oblivious to the way in which such issues as the promotion of amenity, wild plant and animal life, and public access to the countryside, were intertwined. Once the societies had been founded, the personalities who had provided so much of the driving force were reluctant to see the uniqueness of their achievement merged into something larger. There was in any case some merit in numbers. Whatever its merits in terms of efficiency, a single comprehensive organisation could never convey the same sense of 'mass' feeling as many smaller groups.[73]

The overlapping membership of the various societies ensured that there was a considerable fund of expertise and experience to draw on. An obvious model were those bodies combatting bad housing conditions and such social vices as cruelty. By the 1870s, the Royal Society for the Prevention of Cruelty to Animals, had become embroiled in defending all types of animal from deliberate acts of cruelty. Whether raising funds, strengthening liaison between its own inspectors and the police, conducting legal cases, or promoting legislation, the Society brought considerable professionalism to its various activities. Although it did not flinch from appealing to emotion and sentiment when attacking cruelty, the Society also became a leading authority on animal questions, and was closely consulted by government, various organisations and individual persons.[74]

At a time of unprecedented change in land use, it was not surprising that some observers sought a more active role. As the Leicestershire botanist, Arthur Horwood, asked,

What will be the effect in another hundred years, when many towns now only springing up into existence, will probably have expanded into veritable New Yorks and Chicagos?

[73] J. Ranlett, '"Checking Nature's Desecration": late-Victorian environmental organization', *Victorian Studies*, 26, 1983, pp. 197–222.

[74] B. Harrison, 'Animals and the state in nineteenth-century England', *EHR*, 88, 1973, pp. 786–820.

In a review of the reasons for decline among cryptogamic plants, Horwood identified eight main causes, and twenty-eight minor reasons.[75]

The form any action could take depended on the nature of the threat. A contributor to the *South-eastern Naturalist* drew a distinction between those causes which were the inevitable result of the 'progress of civilisation', and the less necessary forms of destruction.[76] Particularly near great centres of population, changes in the abundance of native plants, and even their extinction were probably inevitable, whereas other 'modern tendencies' could be combatted. Cyclists were frequently encountered, returning from an outing laden with the roots of ferns and primroses. The fact that blue violets (*Viola odorata*) were much less common than the variety *alba* in the Bristol area could be attributed to so many being dug up, and transplanted into gardens or hawked for sale.[77]

Most commentators reserved their greatest wrath for the collector, who 'is ubiquitous, and all but omnivorous'.[78] Any publicity given to the presence of rare species was likely to lead to their extinction. Lord de Tabley might have published his *Flora of Cheshire* sooner had he not 'scrupled to lay bare the secret of the whereabouts of some rarities'.[79] Hamilton-Davey omitted from his *Flora of Cornwall* all the localities of the Royal fern (*Osmuda regalis*) because of the continued depradations of 'local and itinerant fern-vendors' – a collector had once boasted to him of despatching a truck-load of roots, weighing over 5 tons, from a local railway station.[80]

Natural history societies had an especially important educative role. In a paper to the first meeting of the Wiltshire Archaeological and Natural History Society in 1853, the Reverend A. C. Smith expressed the hope that the diffusion of more correct information regarding the economy, habits and usefulness of 'the furred and feathered tribes' would end 'the system of wanton persecution', especially among the uneducated classes.[81] The president's Prize of the Northamptonshire Natural History Society was awarded in 1898 for an analysis of many hundreds of owls' pellets, taken from estates where game was reared, showing the extent to which rats, voles, mice, shrews and small birds, chiefly sparrows, made up the staple diet of the owl.[82]

[75] A. R. Horwood, 'The extinction of cryptogamic plants', *South-eastern Naturalist* 1910, pp. 56–86.

[76] J. J. Scargill, 'What can be done to save our fauna and flora from unnecessary destruction?', *South-eastern Naturalist*, 1897, pp. 8–11. [77] J. W. White, *The Flora of Bristol* (Bristol, 1912), pp. 30–1.

[78] G. S. Boulger, 'The preservation of our indigenous flora, its necessity and the means of accomplishing it', *South-eastern Naturalist*, 1902, pp. 28–35.

[79] Lord de Tabley, *The Flora of Cheshire* (London, 1899), p. xxii.

[80] F. Hamilton Davey, *Flora of Cornwall* (Penrhyn, 1909), p. 543.

[81] A. C. Smith, 'A plea for the moles', *Wilts. Arch. & Nat. Hist. Mag.*, 15, 1987, p. 308; G. Abbey, *The Balance of Nature and Modern Conditions of Cultivation. A Practical Manual of Animals' Foes and Friends for the Country Gentleman, the Farmer, the Forester, the Gardener, and the Sportsman* (London, 1909).

[82] L. E. Adams, 'A plea for owls and kestrels', *Journ. Northants. Natural Hist. Soc.*, 10, 1898, pp. 45–55.

The first national society to be explicitly concerned with the preservation of wildlife was the Selborne League, which included among its objects the preservation of wild flowers, rare birds and those with beautiful plumage. It took its name from the eighteenth-century naturalist, Gilbert White of Selborne, whose writings had been among the first to bring warmth and sympathy to the study of plants and animals in the ordinary countryside. The Plumage League was established in the same year, 1885. The two leagues merged a year later as the Selborne Society for the Preservation of Birds, Plants and Pleasant Places. By 1893, the Society could boast of forty-five branches and, at the turn of the century, of a membership of over 1,500. Its popular nature study journal, *Nature Notes*, was supplied to school libraries and sold by booksellers. Hundreds of countryside outings were organised.[83]

Whatever the name and avowed intentions of a society, a great deal continued to depend on individual and local initiative. The first of a series of wild bird protection measures arose from concern over the cruel and indiscriminate slaughter of seabirds in such localities as the Bempton cliffs on the Yorkshire coast. By the 1860s, 'scores of excursionists poured from the neighbouring towns' to shoot the birds as they fed their young. Largely at the instigation of local clergymen, an East Riding Association for the Protection of Seabirds was formed. Circulars were printed and the support of the Archbishop of York was obtained. The local member of parliament, Christopher Sykes, and the Duke of Northumberland promoted a Sea Birds Protection Bill, which they justified on humanitarian grounds and as a means of protecting those birds of value to farmers and fishermen. The Bill, which introduced a close season from 1 April until 1 August, received the Royal Assent in 1869. Further Bills were soon promoted to extend the precedents established to a much wider range of birdlife.[84]

Important legislative precedents were also established in respect of the preservation of ancient earthworks. Whilst still plentiful, numbers were lost each year through such activities as ploughing and gravel working. In response to mounting anxiety, the First Commissioner of Works was empowered by the Ancient Monuments Act of 1882 to take into guardianship any of the sixty-eight prehistoric and famous sites specified in a schedule to the Act. The schedule included Avebury Circle, Maes Howe, Silbury Hill and Stonehenge.[85] Pitt Rivers held the post of 'Inspector of Ancient

---

[83] J. Sheail, *Nature in Trust: the History of Nature Conservation in Britain* (Glasgow, 1976), pp. 10–12; Ranlett, '"Checking Nature's Desecration".'

[84] Sea Birds Preservation Act, 1869, 32 & 33 Victoria, c. 17; E. W. Wade, 'The birds of Bempton Cliffs', *Trans Hull Scientific & Field Naturalists' Club*, 3, pp. 1–26; Sheail, *Nature in Trust*, pp. 22–36.

[85] Ancient Monuments Protection Act, 1882, 45 and 46 Victoria, c. 73; C. P. Kains-Jackson, *Our Ancient Monuments and the Land Around them* (London, 1880); M. W. Thompson, 'The first inspector of ancient monuments', *Journ. Brit. Archaeol. Soc.*, 3rd ser., 23, 1960, pp. 103–24.

Monuments' from 1883 until his death in 1900. There was, however, no power to force landowners to hand over the ancient monuments. The Office had to stand impotently by, even when the warnings of its Chief Inspector were ignored at Stonehenge, and 'one of the finest of the upright stones fell in a storm, breaking a smaller stone nearby and itself'. As a frustrated member of the Office of Works commented, 'one cannot but regret that in this country, unlike several Foreign States, there is no legal power to take possession of Stonehenge as a National Monument; paying the owner whatever may be the fair value of his interest in it'.[86]

The limited value of legislation was also plain to see in the case of curbs imposed on air pollution. Lord Palmerston had sponsored a series of Acts intended to regulate smoke pollution in London. In a further attempt to encourage the enforcement and extension of regulations, a Coal Smoke Abatement Society was founded in 1898. Despairing of strong government action, the Society came to occupy a kind of middle ground. As well as exerting pressure on local authorities to take action, the Society followed the example of the Leeds Sanitary Aid Society of appointing its own smoke control inspectors. From the 2,000 observations made in 1903, 1,460 cases of nuisance were recorded and 1,298 official complaints lodged. The frustrations were many, but the Society could point to some successes, in terms of both influencing the public conscience and the breadth of its approach to campaigning. Local authorities were encouraged to appoint smoke inspectors, and a training programme was inaugurated at the Borough Polytechnic Institute. A research and publication programme was launched with the *Lancet*, the joint editors of which were Society committee members. All this was in addition to a voluminous correspondence with manufacturers and local authorities, in Britain and Europe, America and the Empire.[87]

Legislative initiatives were required not only to preserve and maintain the natural and man-made heritage, but also to prevent new and offensive intrusions into the countryside. In 1893, the Society for Checking the Abuses of Public Advertising (SCAPA) was founded by Richardson Evans and Octavia Hill. The 'hideous creations, usually of tin, with vile glazed inscriptions' had 'entirely ruined' the view along almost the whole length of the railway between London and Oxford. At its first annual meeting, the Society discussed a draft Bill to regulate such advertising.[88]

The problem may have been comparatively new, but the vested interests were already in place. The United Billposters' Association had

---

[86] PRO, WORK 14, 213; C. Chippendale, *Stonehenge Complete* (London, 1983); Ancient Monuments Consolidation and Amendment Act, 1913, 3 and 4 George V, c. 32.

[87] E. Ashby and M. Anderson, *The Politics of Clean Air* (Oxford, 1981), pp. 1–90.

[88] J. J. Hissey, *Across England in a Dog-cart from London to St Davids and Back* (London, 1891), p. 65: Ranlett, '"Checking Nature's Desecration".'

existed since the 1860s and, for many years, its parliamentary committee blocked any move towards introducing regulatory legislation. Despite the support of a large number of public figures and a membership of over 750, SCAPA was unable to make any headway until 1900, and the public outcry that followed the erection by an American company of two enormous boards advertising a breakfast cereal on the cliffs of Dover. A number of local authorities, including the Dover Corporation, took powers under Local Acts to regulate advertising and, from 1903, discussions were held between SCAPA and the industry with a view to drafting legislation acceptable to both sides. The outcome was the Advertisements Regulation Act of 1907, permitting local authorities to make by-laws regulating the kinds of public advertising that SCAPA had found so obnoxious.[89]

The year 1907 also saw the enactment, without opposition, of the Bill granting the National Trust powers to declare its properties inalienable. By 1910, the Trust owned thirteen sites of special interest to the naturalist, but still there was impatience with what was being achieved. Largely on the initiative of the Hon. N. C. Rothschild, who had assisted in the purchase of other properties, a new body was formed in 1912, the Society for the Promotion of Nature Reserves (SPNR), explicitly to stimulate the National Trust and other bodies and individuals to create reserves. Help was offered in persuading landowners to give or sell land on favourable terms to the National Trust. By 1916, lists had been compiled of 251 areas 'worthy of protection' in England, Wales and Scotland. There was no intention of the Society itself owning or managing the land.[90]

The founding of the Society and its wartime surveys might be taken as an important stage in the development of the preservation movement. The advances made in an individual society had always been heavily dependent on personal initiative and diligence, and much of the history of the period might be written in terms of personal networks and overlapping membership. Rothschild's initiative in founding the 'pressure group', the SPNR, might be taken as an early intimation of a maturity, whereby networks were beginning to evolve between the respective institutions. Their relevant skills and experience were being deployed in promoting the aspirations and stature of others. It was a kind of working alliance that was to be developed to such great effect some years later in the formation of the Council for the Preservation of Rural England in 1926.

[89] C. Sheldon, *A History of Poster Advertising* (London, 1937); G. E. Cherry, *Urban Change and Planning* (Henley on Thames, 1972), pp. 116–18; Dover Corporation Act, 1901, 1 Edward VII, c. ccxllll; 1907, 7 Edward VI, c. 27.

[90] Sheail, *Nature and Trust*, pp. 58–67, 127–30.

CHAPTER 31

WOODLANDS

BY GORDON E. CHERRY AND JOHN SHEAIL

INTRODUCTION

Paradoxically, there is comparatively little statistical information on one of the most conspicuous features in the countryside, namely woodlands. Unlike farmland, there were no *annual* crops of any significance to be gathered, and never any risk of the nation running short of wood products before the next harvest. Steel replaced wood in shipbuilding and for many other purposes, and there was an abundance of imported timber to meet any deficit in home-grown wood. There was accordingly no attempt to compile the same *annual* inventory of resources as in agriculture. Rather, it was left to the Ordnance Survey to record the extent of woodlands, and the proportions of land under conifers and deciduous species, as part of its normal surveying programme. Many years might elapse before the woodlands of a particular county or region might be surveyed, and statistical abstracts of what had been found were never published. For the most part, commentators had to rely on the special Returns of woodlands made, as part of the Annual Returns to the Board of Agriculture, on seven occasions between 1871 and 1913.

The first three woodland Returns, made in 1871, 1880 and 1888, succeeded in highlighting how little land was covered by woodland. Only about 4 per cent of England and 3 per cent of Wales were recorded as being 'woods – coppices or plantations, excepting gorseland and garden shrubberies'. Despite 'great pains' being taken to ensure that the Return for each district was accurate and complete, there were many omissions – the county totals computed in 1871 and 1880 were so erroneous that revised totals had to be published a year later. A further Return of 1891 recorded 1,788,816 acres of woodland in England and Wales, an increase of 159,992 acres over the total for 1880. It was also the first Return to distinguish plantations, namely woodlands planted during the previous ten years.[1]

---

[1] *Agricultural Returns of Great Britain*, BPP, 1871, LXIX, C.460, p. 7; 1872, LXIII, C.675, p. 10; 1880, LXXVI, C.2727, p. 3; 1881, XCIII, C.3078, pp. 6–7; 1888, CVI, C.5493, pp. 11–12; 1890–1, XCI, C.6524, pp. viii–ix.

As the Board of Agriculture quickly discovered, it was not enough to add a further column to the forms sent to the occupiers of farmland – only a small proportion of woodland was owned or managed by farmers. Right from the start, the offices of the Inland Revenue were required to consult parish rate-books and valuations lists, but these were soon found to be highly misleading. By far the best method was to make direct contact with proprietors, and their stewards and agents, using the Ordnance Survey maps that were becoming increasingly available as a guide and check on the measurements given. Even then, it was difficult to ensure that woodlands were defined in the same way throughout the country, and between the Returns of different years. As the Return for 1891 remarked, 'areas whereof the surface is grazed, although studded more or less closely with trees, may conceivably be dealt with either as woodland, or as permanent grass or mountain land'.

Further Returns were made in 1895, 1905 and 1913 – the latter two surveys as a direct response to growing concern over the nation's dependence on imports of foreign timber. Over three-quarters of all woodland were situated in the southern counties and west Midlands – almost a quarter of the English acreage was found on the light sands of the Weald, and the Lower Greensand and Eocene series of the four south-eastern counties, Hampshire, Kent, Surrey and Sussex, where woods and plantations made up over 11 per cent of the surface area (Figure 31.1).[2]

The Returns highlighted the deceptive appearance of many parts of lowland England. Their apparently well-wooded character, as seen for instance from the train, was due entirely to the large number of trees in the hedgerows separating the comparatively small fields into which the prevailing pasture lands were divided. Seen from a distance, the trees and hedges blended, giving an entirely false impression of woodland. Not even this degree of deception was perpetrated on the higher ground of the north and west, where the wholesale destruction of woodland in earlier times had encouraged surface water to collect and form bogs and deep peat and where, even on the drier slopes, exposure to wind and grazing minimised any chances of woodland growth.[3]

A. SEMI-NATURAL WOODLANDS

The country had been cultivated and thickly populated for so long that it seemed legitimate to ask whether any natural woodland could have survived. Despite the backwardness of present-day forestry, had not planting and other forms of interference destroyed long ago any native communities, and left 'mere congeries of indigenous or introduced

---

[2] *Ibid.*, BPP, 1896, XCII, pp. viii–ix; 1906; CXXXIII, pp. xiii–xiv; 1914, XCVIII, pp. 6–7.
[3] Tansley, *Types of British Vegetation*, pp. 70–1.

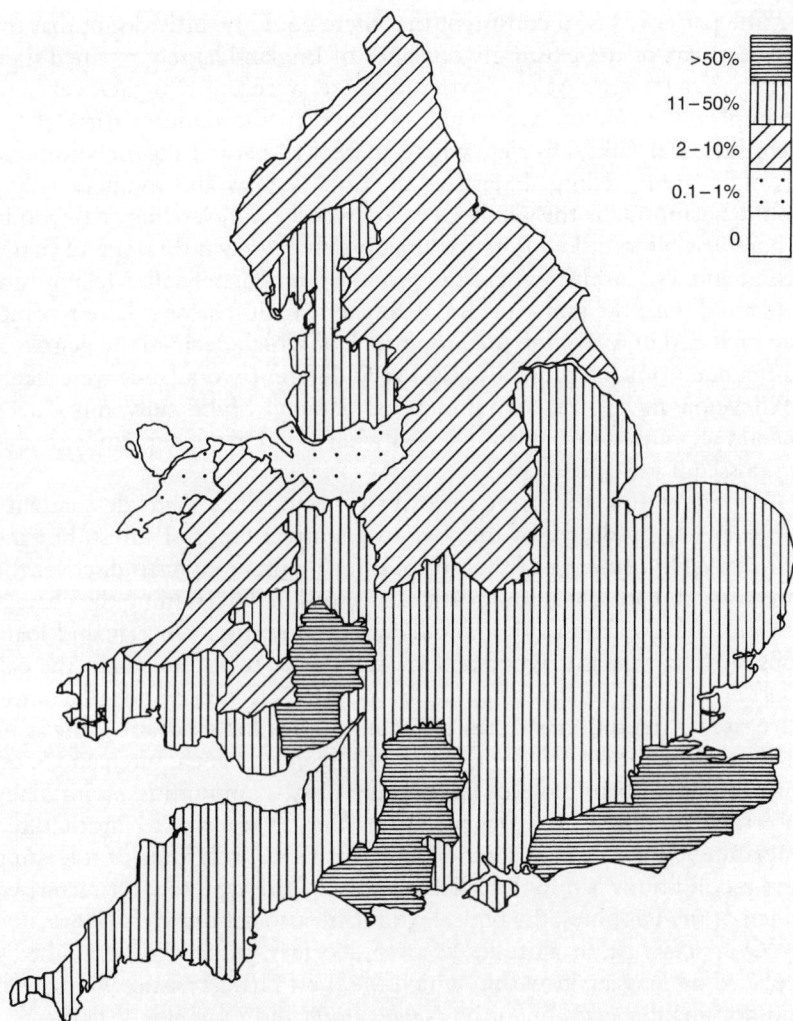

31.1 Distribution and importance of coppice management
(as a percentage of total woodland), as recorded in the Agricultural Returns of
1905 (BPP, 1906, CXXXIII).

species'? Attempts on the part of ecologists to classify woodlands
revealed, however, a very different situation. Whilst there were clearly
many examples of plantations, the most striking discovery was 'the con-
stancy in the general type of woodland on a given type of soil over wide,
and often widely separated, stretches of country'.[4]

---

[4] C. E. Moss, W. M. Rankin and A. G. Tansley, 'The woodlands of England', *New Phytologist*, 9,
1910, pp. 113–49.

The pattern was so consistent that there could be little doubt that the 'greater part of the existing woodlands of England largely retained their original character'. At one extreme, there were the comparatively few examples of primitive woodland, as found in the remoter parts of the Pennines and Lake District, where there was natural regeneration and sporadic or no felling. Examples of more regular and frequent felling could be found in the Chiltern beechwoods and Wealden oakwoods. There was often little difference ecologically between this type of native woodland and a third, where the ground, immediately after felling, was replanted with the same species. A fourth situation arose where replanting included new species, the proportion of which decided the degree of difference from the previous class. Very different woodlands were likely to develop in the last two instances, namely where only introduced species were used for replanting, and where land previously under arable or grassland was afforested.[5]

Confident that the great majority of woodlands were descendents, more or less modified, of the various primitive types of forest, bearing unmistakable evidence of their origins, ecologists sought to discover the effects of climate and, more particularly, soils in determining the character and composition of the woodland communities. On clay and loam soils, the characteristic woodland associations were dominated by the oak *Quercus robur* (*Q. pendunculata*). Such woods on clearance or degeneration gave rise to 'neutral' grasslands. On coarser sands and sandstones, the associations were more varied – *Q. robur* and *sessiflora*, beech, birch and Scots pine might all form more or less definite plant-communities, with many intermediate stages. On degeneration, these gave rise to heath often intermingled with grassy patches containing species capable of tolerating acidic conditions. On older siliceous soils of the type that characterised much of the Pennines, the typical woodland associations were dominated by *Q. sessiflora* or, at altitudes of over 800 feet, birch. These might be replaced by grasslands, within which the most characteristic, widespread and dominant species might be *Nardus stricta* and *Deschampsia flexuosa*.[6]

Soils containing a comparatively large proportion of lime were marked by an abundance of certain species – calcicoles – and absence of others – the calcifuges. The most extensively developed limestones were the Carboniferous or Mountain Limestones of the Pennines, Wales and Somerset, the Magnesian Limestones of Permian Age in northern England, the Oolitic Limestones of the Jurassic series in the west Midlands, and the chalk. Ashwoods were characteristic of the oldest limestones, the ash being most frequently associated with the wych elm and hawthorn, both of which were much more abundant than in typical

⁵ *Ibid.*   ⁶ Tansley, *Types of British Vegetation*, pp. 76–94.

oak and birch woods. A particularly noticeable feature was the large number of associated species of trees and shrubs, including the two conifers, the juniper and yew – the latter being particularly abundant in north Lancashire. Whereas sheets of wood garlic (*Allium ursinum*) and the lesser celandine (*Ranunculus ficaria*) were common in the woods of the moister soils, dog's mercury (*Mercurialis perennis*) often occurred where the soils became very dry in summer.

The younger limestone areas were characterised by two types of woodland. The purest stands of the beechwood association occurred on the steeper slopes, often called hangers in chalkland areas, where the rock might be covered by only a few inches of mild humus and the roots largely embedded in the chalk. So dense was the shade that shrubs were practically absent. Sheets of dog's mercury might occur in those parts where the canopy was thinnest. The association was largely confined to the escarpment and valley sides of the North and South Downs, the Chiltern hills centred on Buckinghamshire and south Oxfordshire, and on the Oolitic Limestones of the Cotswolds. The other type of woodland, an ashwood association similar in most respects to that of the older limestones, occurred on the south-western extremities of the chalk outcrop, lying in the Isle of Wight, Dorset and east Devon – the line of contact being near Butser Hill, where the South Downs reached their highest elevation (889 feet). The factors determining the distribution of the beech and ashwood were far from clear.

About a third of the woodland in England, and a tenth in Wales, was managed as coppice (Figure 31.1). As Tansley explained, the origins of such coppices could probably be traced back to the felling of trees in the once natural woodlands, and the cutting of the shrub growth as small wood.[7] Such woodlands were composed of species which, when cut close to the ground, sent out fresh shoots from their 'stools'. They required little attention beyond the occasional renewal of plants, produced an annual return in a short time, and afforded good covert for game. Oak was usually grown on a fifteen- to twenty-five-year cycle; where the object was to extract tannin, the shoots had to be cut before the bark fissured. Ash was usually cut on a fifteen-year rotation in England, and twenty-five year in Wales, and sweet chestnut, that grew best on siliceous soils, a shorter cycle of eight to fifteen years.

In establishing new coppices, it was the practice on the Cowdray estate in Sussex to clear the ground of existing growth, cut trenches two spits deep, and plant three-year-old saplings in November, 3 feet apart. In the following summer, briars would be cut and rubbish removed so as to allow plenty of light and air. By cutting down the chestnut saplings at the

---

[7] *Ibid.*, pp. 144–6.

end of the third year, more vigorous growth was promoted, which soon smothered the weeds. They would be ready to cut after eight years, when any dead stock could be replaced. The underwood grew best in warm, dry seasons. Growth rates of 6 to 9 feet were recorded in the first season after cutting in the hot, dry summers of 1868–70.[8]

The woodland industries could, in some instances, assume a regional and even national importance. The richly wooded countryside of ash, birch, alder and willow coppice about Langdale and the Tilberthwaite side of the Coniston range in Cumbria provided the basis for a large business in bobbins and in charcoal for the gunpowder works at Elterwater. Millions of bobbins were sent from the sixty or seventy small mills and workshops in High Furness to the textile industries of Lancashire and Yorkshire.[9] The fact that three-quarters of the woodland in Kent was coppiced, and that this far exceeded the area of any other county, reflected the importance of hop-farming, which reached its maximum extent of 45,000 acres in the mid-nineteenth century. For as long as it was the practice to train every hop-bine up its own pole, coppice was essential. Where grown on a ten-year rotation, an acre of underwood was required to supply a sufficient number of poles for replenishing the needs of 10 acres of hops.

With hindsight, the decades up to 1850 were something of a golden age for woodlands. Sales of oak coppice declined sharply following the adoption of chemically prepared substitutes for tannin – the Chepstow bark-export trade ended in the 1880s. Bark, which had once sold for £8 a ton, was offered for as little as 47s. 6d. by the turn of the century.[10] The bobbin-mills of Cumbria encountered increasingly severe competition from Scandinavian supplies.[11] Far-reaching changes were taking place in the layout of hop-gardens, involving the intricate use of string and wire, the use of creosote for protecting the life of the poles and, in some cases, the use of larch thinnings as hop-poles led to a significant decline in demand for chestnut and ash coppice. Poles, which had realised between 25s. and 30s. per hundred in 1878, were worth only 7s. 6d. by the early 1900s. Throughout the country, wood account-books highlight the growing dependence on sales of pit and log wood to the coalfields.

[8] H. Evershed, 'Report of the farming of Kent, Sussex and Surrey', *Journ. Bath & West of England Soc.*, 3rd ser., 3, 1871, pp. 42–4.

[9] J. Somervell, 'Water power and industries in Westmorland', *Trans. Newcomen Soc.*, 18, 1937–8, pp. 235–44; J. D. Marshall and M. Davies-Shiel, *The Industrial Archaeology of the Lake Counties* (Berkemet, Cumbria, 1977), pp. 62–3.

[10] W. Linnard, *Welsh Woods and Forests: History and Utilization* (Cardiff, 1982), pp. 93–4; F. A. Osmond Smith, 'The conversion of underwood and coppice-with standards into highwood', *Qtly. Journ. Forestry*, 2, 1908, pp. 154–65.   [11] Somervell, 'Water power'.

The geological structure of the south Wales coalfield gave rise to particularly heavy demands; many millions of pit-props were also exported to the northern coalfields from Kent.[12]

B. THE CLEARANCE OF WOODLAND

Woodlands might be closely managed, cleared, or simply neglected. Decisions as to which course to follow were likely to take account not only of the profitability of wood sales, but also of the rapidity with which the costs of clearing the land and preparing it for agriculture or some other use could be recouped and an annual profit made. Where incomes from both wood sales and farmland were depressed, the wood might survive, but in a neglected state. Profitability was not, however, the only consideration. As Clutterbuck remarked of Hertfordshire, numerous woods had been grubbed up, especially where 'their frequent interlacing with the arable land made improved cultivation impossible', but the steam plough was unlikely to destroy many game coverts in a county with so many resident proprietors.[13]

Estate papers provide insights into the range of opportunities that had earlier been available. A valuation of Calpher Wood in Easton, Huntingdonshire, estimated in 1863 that the clearance and sale of the timber and underwood were likely to yield an income of about £3,000. If reclaimed, the land would be worth £25 per acre, or £1 per acre if let annually. Allowing £15 per acre for the costs of reclamation, the overall value to the purchaser of the land was likely to be £4,260, namely £10 per acre.[14]

The Worcestershire botanist, Edwin Lees, recalled in *The Phytologist* of 1851 how, until three years previously, there had survived near Henwick a small wood, a 'preserve', to which constant reference could be made in studying 'the most remarkable and rarer *Rubi*', and those among 'the Cryptogamic tribes'. The estate had changed hands, however, and the new owner ruthlessly levelled both coppices and all sheltering hedges. A visit the next year revealed the loss of 'one of my cherished localities'.[15] As Lees remarked,

so the country changes year after year, and progression effaces the haunts and footsteps of our fathers, with their older plants, and we are compelled to observe what new arrangements and altered cultivation bring to light.

[12] A. M. Everitt, 'The making of the agrarian landscape of Kent', *Archaeologia Cantiana*, 92, 1977, pp. 1–31.   [13] J. Clutterbuck, 'Agricultural notes on Hertfordshire', *JRASE*, 25, 1864, p. 314.

[14] Huntingdonshire RO, Manchester MSS, Box 10B, 19 & Box 39B, 2.

[15] E. Lees, 'Records of observations on plants appearing upon newly-broken ground', *Phytologist*, 4, 1851, pp. 136–7.

By far the most famous instances of woodland clearance occurred in respect of royal forests. They were the subject of legislation and in some instances involved the wholesale destruction of extensive areas in a short space of time. By an Act of 1851, commissioners were appointed to partition and then disafforest that part of Hainault Forest in Essex owned by the Crown and subject to rights of common grazing and fuel.[16] An area of 1,917 acres was awarded to the Crown, and 970 acres to the commoners. The deer were either removed or killed and, over a period of six weeks, the woodland of the Crown allotment was converted to arable land, a manufacturer of steam ploughs being engaged to drag the roots of the old oaks from the soil.[17]

The Royal Forest of Wychwood, to the east of Burford in Oxfordshire, provided another instance of large-scale clearance of oak, ash and beech wood. Under the terms of a Disafforesting Act of 1853,[18] the Crown allotment of 2,543 acres (subsequently enlarged to 2,937 acres by purchase) was transformed in both use and appearance, in as little as sixteen months for a cost of £5,815. Amid great publicity, tenders were invited for the leases of thirty-one years. Seventy were received. The seven holdings, with their farmsteads built of local stone, presented anything but a smooth inviting appearance in their early years.

Wide ditches, and long, irregular high banks, that had formed the boundaries of the different coppices; deep pits and hollows, where stone had been dug for the use of byegone generations; small straggling briars that had escaped the notice of the woodgrubbers; roots of trees and underwood, left a few inches below the surface, by oversight or intentional neglect on the part of dishonest woodsmen; large patches of rough brown fern-stems, that had afforded covert to the fawns: all of these and many other impediments stood in the way of the 'forest farmers'.

By 1863, the Crown's income from rents had nevertheless risen to £5,104 7s. 6d., compared with annual receipts of as little as £1,813 7s. 1d. prior to the Act.[19]

As the author of a prize essay on the destruction of the Forest recounted in the *Journal of the Royal Agricultural Society of England*, 'poets and painters may sigh because some fine woodland scenery has been swept away', but magnificent views could be of little consequence when they were compared with 'that plenty which has taken the place of poverty, or those habits of industry now firmly established, where dissipation and crime once abounded'. A visit made by members of the Land Surveyors' Club, nine months after the farms were tenanted, found that,

[16] 1851, 14 and 15 Victoria, cap. 43.

[17] Buxton, *Epping Forest*, pp. 17–18; J. Nisbet, 'Forestry', in *VCH* Essex, II (London, 1907), pp. 622–3. [18] Wichwood Disafforesting Act, 1853, 16 and 17 Victoria, c. 35.

[19] C. Belcher, 'On the reclaiming of waste lands as instanced in Wichwood Forest', *JRASE*, 24, 1863, pp. 271–85; V. J. Watney, *Cornbury and the Forest of Wychwood* (London, 1910), pp. 202–10.

for promptitude of execution and quality of work, nothing like it had been seen before. The first grain crops were at least equal to the average of the older cultivated lands of the district. A hamlet of farm labourers' cottages was built at Fordwells.[20]

Not only did the decline in the agricultural value of land soon make such ventures less attractive, but experience indicated that the soils resulting from the conversion of woodlands were far less fertile than had been supposed. The clearance of 30 acres of chestnut and ash coppice with oak standards in Brenchley, Kent, between 1911 and 1913 provided the agronomist, E. J. Russell, with an opportunity to draw comparisons. The decomposition of organic material and nutrient losses appeared to be so rapid in woodland soils that the neighbouring old arable fields seemed to be much richer in everything except potash. Far from being able to draw on a reserve of fertility, the farmer had to build up the status of the land, using clover and farmyard manure to raise nitrogen levels, and apply phosphates to the equivalent of 4 tons of 26 per cent superphosphate, or 2 tons of bone meal, per acre. Rather than risk capital in clearing the wood-covered clay areas, which were still dotted all over the country, Russell concluded that it was much more prudent to invest in 'proper afforestation methods'.[21]

C. PLANTATIONS

By the turn of the century, the United Kingdom imported over 10 million tons of timber each year. The fact that this represented 84 per cent of the nation's needs, and that for many purposes foreign timber was preferred to home-grown timber, reflected not so much the unsuitability of the soils and climate for growing timber, but the 'extremely backward condition' of forestry practice. A Departmental Committee appointed by the President of the Board of Agriculture in 1902 recalled how, until the previous ten years, woodland owners had perceived the eventual shape, size and quality of their trees to be very largely the outcome of accident, over which they had virtually no control.[22]

In so far as expert guidance was sought, it came initially from Scotland. A widely used textbook had been published in 1847, written by James Brown, a forester of Arniston in Mid Lothian. The book outlined the benefits of establishing mixed plantations, with large-scale thinning over the first twenty years.[23] In practice, the main outcome was the adoption

[20] Belcher, 'On the reclaiming of waste lands'.
[21] Anonymous, '"Grubbing" as a "Paying Proposition"', Country Life, 38, 1915, pp. 50–1: E. J. Russell, 'The reclamation of waste land', JRASE, 80, 1919, pp. 112–33.
[22] Report of Departmental Committee on British Forestry, BPP, 1902, xx, Cd 1319.
[23] J. Brown, The Forester (Edinburgh, 1847).

of a routine approach which led to increased tree growth, but took little account of the type and quality of timber eventually harvested. A Royal Commission report of 1909 complained of how owners tended to be much more concerned with arboriculture than sylviculture – with the tree rather than the wood. The influence of the oak and larch had been little short of calamitous – it was assumed that practices which suited those species would be appropriate for all types of tree. Far too little account had been taken of 'the sylvicultural requirements of different species'.[24]

By far the most important coniferous species was the larch. Large areas had been planted in northern England since the late eighteenth century. In his book, James Brown had emphasised the need to avoid straight lines when laying out plantations. Straight boundaries were 'disagreeable to the eye of taste', and never occurred in nature, not least because they were 'without strength to resist outward pressure'.[25] Experience indicated that larch plantations were particularly vulnerable to windthrow. A gale of January 1884 uprooted and snapped in half many of the trees of a second-generation plantation near Castle Madoc, Breconshire, where trees of sixty to seventy years in the first generation had also suffered similarly extensive damage during a summer storm of July 1853.[26] The need to pay even greater attention to management practices was emphasised by the damage inflicted by the larch cancer. The fungus, *Dessycypha calyzina*, had spread to almost every young plantation. The only protective measure that seemed to be at all effective was to promote as vigorous a growth of the tree as possible in situations that resembled in their climatic rigour the natural home of the larch in the Alps, Carpathians and Siberia.[27]

The introduction of exotic species from beyond Europe and the eastern seaboard of North America did little to raise standards. The first to achieve some commercial success was the Douglas Fir. Introduced in 1827, the earliest known plantings in Wales took place at Penrhyn Castle in 1840, from where fertile seeds were soon obtained, and saplings planted out at various localities in Caernarvonshire.[28] Among other species to be introduced successfully to England and Wales were the Sitka spruce and great silver fir (1831), western hemlock (1851), western red cedar (1853), lodgepole pine (1853–4) and Lawson cypress (1854). By the end of the century, there were so many species to choose from, and so little guidance as to which would flourish in the different conditions

[24] A. C. Forbes, *English Estate Forestry* (London, 1904), pp. 17, 20–5; *RC on Coast Erosion, Second Report (on Afforestation)*, BPP. 1909, XIV, Cd.4460, p. 4.   [25] Brown, *The Forester*.

[26] Linnard, *Welsh Woods*, p. 142.

[27] C. O. Hanson, *Forestry for Woodmen* (Oxford, 1911), pp. 133–5.

[28] Linnard, *Welsh Woods*, pp. 124–6.

encountered, that many woodlands took on the appearance of an arboretum or pinetum, rather than a timber-producing plantation.[29]

It was through a concern for the forests of India that many foresters first encountered the techniques and approaches developed over a long period on the continent of Europe. Founded in 1855, the Indian Forest Service owed much of its success to three outstanding German foresters, who served in turn as Inspector-General of Forests. Because of the great difficulties in securing properly qualified assistants, arrangements were made from 1866 onwards for newly appointed staff to spend an initial two years eight months studying in France or Germany. It was only a matter of time before these officers sought to apply what they had learned to English and Welsh woodlands. A working-plan for the Forest of Dean and Highmeadows was prepared by H. C. Hill, Conservator of Forests in India, during his furlough of 1897. The plan set out explicitly to introduce 'a more scientific and systematic approach to forest cultivation', providing 'a good practical object lesson' of the approaches adopted in France and Germany. Instead of trying to impose some kind of uniform scheme, each area was to be stocked and managed according to what suited local conditions best.[30]

By far the most influential figure in forestry was William Schlich, who had left Germany to take up an appointment in the Indian Forest Service in 1866, where he eventually became Inspector-General, before being deputed, in 1885, to leave for England in order to found a forestry training school at the Royal Indian Engineering College at Cooper's Hill. It was transferred to Oxford in 1905, where Schlich became professor of forestry in the university. Whilst still at Cooper's Hill, Schlich had written the first three volumes of his *Manual of Forestry*, dealing in turn with the value of forestry, its fundamental principles, and a wide range of practical issues related to forest management. It was reprinted many times.[31]

Invitations to draw up working plans for large estates provided Schlich with outstanding opportunities to demonstrate the approaches and techniques set out in his *Manual*. One such instance was the Boughton estate of the Duke of Buccleuch in Northamptonshire, where there had been a succession of neglect and reckless cutting. On Schlich's advice, it was decided in 1904 to replace over a period of time the old coppice woods with high forest. As the coppice of each section reached its most profitable age, it was to be cut and the stools interplanted with suitable timber trees.[32]

[29] Forbes, *English Estate Forestry*, pp. 20–5.

[30] *Seventy-fifth Report of Commissioners of Her Majesty's Woods, Forests and Land Revenues*, BPP 1897, XXIII, p. 4; C. E. Hart, *Royal Forest. A History of Dean's Woods as Producers of Timber* (Oxford, 1966), pp. 223–6.   [31] W. Schlich, *A Manual of Forestry* (London, 1899), vol. I.

[32] B. Bellamby, *Geddington Chase. The History of a Wood* (privately printed, 1986), pp. 60–1.

In later years, Schlich came to regard such advisory work as the least successful aspect of his varied career. So often, he was beaten by the frequent changes in estate ownership and lack of any forestry tradition.[33]

Although there were some signs of improvement, forestry practice remained in 'an extremely backward condition', the majority of woodlands, particularly in the south of England, being 'a scandal and a disgrace' from the forester's point of view. Not only were owners deterred by the long time taken to recoup investment, but, in many cases, the primary concern of owners was to produce not timber but as large a head of game as possible.[34] The overall effects of the introduction of the breach-loading gun that used cartridges, a central-fire system and choked barrel, had been to achieve more accurate shooting at a faster rate over a longer range. In such cases, it was the business of the forester to help find ways of raising as large and healthy a stock of pheasants in the coverts as possible, without reducing the timber yield more than absolutely necessary. The key to achieving this was to maintain a full and dense underwood through periodic cutting and the protection of new shoots from ground game; the overwood should be thin enough to allow sufficient light to reach the underwood. By carrying out such operations in the early part of the year, disturbance of game should be kept to the minimum.[35] Such attempts at reconciling the interests of the forester and gamekeeper were often jeopardised by the depredations of the rabbit.[36]

If there was to be any substantial increase in home timber production, far greater emphasis had to be placed on afforestation. The terms of reference and membership of a Royal Commission on Coastal Erosion were extended in 1908 to encompass an investigation of the merits of large-scale tree planting. Enquiries made by the Board of Agriculture indicated that, of the 2,826,000 acres of land classified in the Annual Returns as 'mountain and heath', located below 1,500 feet, it could be assumed that 1,500,000 acres were afforestable. To this area, a further million acres might be added as representing poor tillage land that would be more profitable under forests.[37]

In his Budget Statement of 1909, Lloyd George announced his intention of bringing all the modest grants made towards such schemes as light railways and harbours together in the form of a Development Grant, which would be substantially enhanced and extended to cover such activities as the purchase and preparation of land for afforestation. The

---

[33] W. E. Hiley, 'Sir William Schlich's work in Britain', *Qtly. Journ. Forestry*, 20, 1926, pp. 12–14.

[34] Board of Agriculture and Fisheries, *Wages and Conditions of Employment in Agriculture*, BPP, 1919, IX, Cmd 24, *Volume I*, pp. 46–7, & 99, and Cmd 25, *Reports of Investigators, Volume II*.

[35] W. Schlich, *Forestry in the United Kingdom* (London, 1904), pp. 51–61.

[36] Forbes, *English Estate Forestry*, pp. 36–7.   [37] *RC on Coast Erosion*, BPP 1909, XIV, p. 32.

requisite powers were obtained under the Development and Road Improvement Fund Act of 1909.[38]

An Advisory Committee was appointed by the President of the Board of Agriculture in 1912 to consider and advise upon proposals for a forest survey, draw up plans for experimental plantings, and review the provision of instruction for woodsmen. It was suggested that a preliminary, or flying, survey should be followed by mapping at a 6-inch scale, distinguishing the several classes of land according to their possible productivity for afforestation.[39] Meanwhile, the Return for 1913 indicated that whilst there appeared to be an overall increase of 8,420 acres, or 12 per cent, in the area of such plantations since the last Return of 1905, clearance was still taking place from eastern England and south Wales.[40]

D. WOOD PASTURES

A distinctive type of woodland had evolved where considerable importance was attached not only to the tree crop but also to the value of the ground cover for pasturing animals. The wood pastures, usually found on commonland and in the royal forests, formed the most complex type of multiple land use encountered in England and Wales.[41] Perhaps not surprisingly, the long-term maintenance of both trees and pastures on the same ground presented chronic problems, so many interested parties being involved. Too many trees reduced the herbage available for deer or farm animals, or perhaps a combination of both, whilst unrestricted grazing might eliminate the trees.

The difficulties of securing a balance were highlighted by the far-reaching changes initiated by the New Forest Deer Removal Act of 1851.[42] Although the Crown, through the Office of Woods, owned over two-thirds of the Forest area (to which the public had access), much of this area was subject to common rights of pasture, pannage and estover. By the mid-century, the Crown had lost interest in the Forest as a hunting ground, and sought to promote timber growth wherever possible. Under the terms of the Act of 1851, the Crown was compensated for relinquishing its interest in the deer by being granted the right to enclose 10,000 acres, in addition to the 6,000 acres already enclosed, within which common rights would cease to operate. By means of a rolling programme of some 16,000 acres at any one time, the Crown

[38] Development and Road Improvement Fund Act, 1909, 9 Edward VII, c. 47.
[39] Reports of Advisory Committee on Forestry, July to October 1912, BPP, 1913, xxv, Cd 6713.
[40] Agricultural Returns of Great Britain, BPP, 1914, xcviii, Cd 7487, pp. 6–7.
[41] O. Rackham, Trees and Woodland in the British Landscape (London, 1976), pp. 152–65; G. F. Peterken, Woodland Management and Conservation (London, 1981), pp. 12–17.
[42] New Forest Deer Removal Act, 1851, 14 and 15 Victoria, c.76.

planned in time to cover every part of the Forest suitable for planting with trees.[43]

The commoners, who had clamoured for the removal of the deer, expected the losses in grazing area to be amply compensated by the increased herbage for their cattle on the areas still open. In fact, the reverse happened. The numbers of deer had already fallen dramatically, and only a few remained by the mid-1850s, after which numbers slowly rose to about 200 in the early 1900s.[44] There were soon complaints that the removal of the deer had been injurious in as much as the deer had consumed, during the winter months, 'the coarse herbage, heather and other rough undergrowth which now overspreads the pasture'.[45] Scrub and self-sown trees were overrunning the lawns at an alarming rate. The longer-term significance of the relaxation of grazing pressures was highlighted by growth-ring counts of 530 trees in the unenclosed parts some hundred years later. Of the three generations of trees distinguished, the older dated from the periods 1660–1760, and from 1840 to 1870.[46]

Whatever the precise implications of the removal of deer, a Select Committee of the House of Lords in 1868 concluded that it was inevitable that the greater part of the Forest should become thickly covered with timber to the detriment of the pasture. Whilst the Crown was required to throw open existing enclosures in order to form new ones, once the total area enclosed reached 16,000 acres, the herbage under the thickly planted trees would be far inferior to that of the areas about to be enclosed. In order to avoid what would inevitably be 'a perpetual struggle of conflicting interests', the Committee recommended the appointment of a Commissioner to allot certain portions to the Crown, 'leaving the residue to the commoners to deal with in such manner as they may think best'.[47]

Moves to disafforest and partition the Forest had, in any case not taken account of what was later called 'the great and growing craving for open spaces free to the public'. In the words of Gerald Lascelles, the Deputy Surveyor of the New Forest from 1880 onwards, they had overlooked the force of the aesthetic movement, and the view of a growing body of public opinion that it was worth waiving some thousands of pounds of additional income from timber in order to preserve a 'magnificent park to take their pleasure in'. The London and Dorchester railway had been built through

[43] G. Lascelles, *Twenty-five Years in the New Forest* (London, 1915), pp. 11–13.

[44] C. R. Tubbs, *The New Forest. An Ecological History* (Newton Abbot, 1969), pp. 137–9; C. R. Tubbs, *The New Forest* (London, 1986), pp. 76–9.

[45] *Report and Proceedings of Lords Select Committee on the New Forest Deer Removal Act 1851*, BPP, 1867–8, VIII.

[46] G. F. Peterken and C. R. Tubbs 'Woodland regeneration in the New Forest, Hampshire, since 1650', *Journ. Applied Ecology*, 2, 1965, pp. 159–70; Tubbs, *The New Forest*, 1986, p. 155.

[47] *Lords Select Committee on New Forest Deer Removal Act*, BPP, 1867–8, VIII.

the Forest, 'bringing thousands of people . . . to discover what it was worth'.[48] In 1871, the House of Commons passed a resolution to the effect that, pending further legislation, there should be no further felling of ornamental trees or fresh enclosures, except for the purposes of thinning the young plantations and satisfying the fuel rights of the commoners.[49]

The eventual outcome was the New Forest Act of 1877.[50] No further areas were to be enclosed and planted beyond those already 'enclosed, planted, laid open, replanted or re-inclosed'. The New Forest was otherwise to remain open and unenclosed. The traditional Verderers' Court was reconstituted, becoming in effect a legislature, judiciary and executive in the defence of the commoners and common rights of the Forest.[51] The Act had done nothing, however, to eliminate the potential for conflict between the Crown's agent, the Office of Woods, and the Verderers as to their respective rights. No positive steps had been taken to protect the distinctive character of the Forest. As Lascelles remarked, a better course might have been to have widened the responsibilities of the Office of Woods so as to take explicit account of the public's desire for 'a great public park'.[52]

The value of wood pastures as public open spaces was even more obvious in the vicinity of cities and large towns. Reference has already been made to the circumstances in which the Commons Preservation Society was founded in 1865. In his *Year-book of the Country*, William Howitt wrote of how in London, on Easter Mondays,

out of all the alleys and close courts of the huge metropolis, men, women and children pour to catch a breath of fresh air, for once, on the heights of Greenwich Park; to partake in all the fun of a country fair; and to see the youngsters take their roll down the hill.[53]

In 1863, a House of Commons' address called upon the Crown to prevent further enclosure on Crown lands within fifteen miles of the metropolis.[54] Although the Metropolitan Commons Act of 1866 made the enclosure of London's commons practically impossible, the continued sale of Crown rights had the effect of further reducing the area of Epping Forest. During the mid-1860s, a further 1,300 acres of the manor of Loughton was enclosed by the lord of the manor, who was also rector of the parish. There was a public outcry when a father and two sons, found lopping wood, were sentenced to prison on a charge of malicious trespass.[55]

[48] Lascelles, *Twenty-Five Years in the New Forest*, pp. 15–24.   [49] PD, 3rd series, CCVII, 328–43.

[50] New Forest Act, 1877, 40 and 41 Victoria, c. cxxi.

[51] *Report and Proceedings of Commons Select Committee on the New Forest*, BPP, 1875, XIII.

[52] Lascelles, *Twenty-Five Years in the New Forest*.

[53] W. Howitt, *The Year-book of the Country* (London, 1850), p. 108.   [54] PD, 3rd ser., CLXVIII, 309.

[55] Lord Eversley, *Commons, Forests and Footpaths* (London, 1910), pp. 82–90.

It was one thing to protest at such enclosures, but another to find a corporation or person with the sufficient status, determination and resources to stand up to the manorial lords. At the prompting of leading figures in the Commons Preservation Society the House of Commons in 1870 passed a motion, calling upon the Crown to take such steps as would preserve Epping Forest 'as an open space for the recreation and enjoyment of the public'.[56] An Act was passed in 1871, appointing Commissioners to prepare a scheme for disafforestation, and for preserving and managing the waste lands. No more than 600 acres were to be set aside as a public open space.[57]

It was at this point that the City Corporation of London intervened. Through an earlier acquisition of land in Ilford for use as a cemetery, the Corporation had acquired common rights in Epping Forest. It was accordingly in a position to challenge the legality of recent enclosures in the Forest, and therefore likely to achieve great popularity at a time when the Corporation felt threatened by general demands for 'a single Municipal Government of London'.[58] Whatever the motives, the Corporation succeeded in its law suits, and set about purchasing the rights of the lords of the manor over 3,000 acres of the Forest. The costs of £240,000 were met by the commutation of the ancient metage dues on grain in the City.[59] The Forest was thrown open to the public by Queen Victoria in person, in the company of a great assembly of people, in 1882.

A formula for protecting the Forest had been found. Under the Epping Forest Act of 1878, the Corporation became Conservators of the Forest, responsible at all times for keeping the Forest 'uninclosed and unbuilt on, as an open space for the recreation and enjoyment of the public'.[60] The Corporation would, 'as far as possible preserve the natural aspect of the Forest', including the protection of 'the timber and other trees, parklands, shrubs, underwood, heather, gorse, turf, and herbage'.[61] These were duties that the Corporation was unlikely to overlook. There were, for example, in the Essex Field Club,

some hundreds of members more or less interested in natural history, and who bring to bear a vigorous public opinion in favour of the preservation of all the wild life which finds a home in the 'waste'.

In the words of E. N. Buxton, one of the four verderers elected by the commoners, 'the general opinion, so unmistakably evinced, that the

[56] PD, 3rd ser., cxix, 246–66.   [57] 34 and 35 Victoria, c. 93.

[58] Eversley, *Commons*, pp. 91–100.

[59] D. Owen, *The Government of Victorian London 1855–1889* (Cambridge, Mass., 1982), pp. 247–59, 251.   [60] Epping Forest Act, 1878, 41 and 42 Victoria, c. ccxiii.

[61] W. R. Fisher, *The Forest of Essex* (London, 1887), pp. 347–72.

Forest shall remain a forest and not be civilised into a park, is but the expression of a true instinct'.[62]

The outcome was a vegetation pattern that was neither that of an urban park, nor what was held to be the original wood pasture. The absence of 'rugged and venerable giants' was attributed to 'the destructive custom of *pollarding*, or cutting back the stem and branches every fifteen years at ten feet from the ground, for purposes of fuel'. The wood-cutting rights were accordingly abolished under the Act so as to remove 'the mop-head growth', and enable the trees to assume their 'natural shape'. The pollards were, however, so closely spaced (up to 3,000 per acre in some parts) that, if left alone, they would soon spoil one another, and prevent light from reaching the undergrowth. Without thinning, the unrestrained growth would bring about far-reaching changes to the appearance and wild plant and animal life of the Forest: views and vistas would be blotted out.[63]

Whilst Nature could not be left to take its course, there were bound to be serious misgivings in using the axe in a Forest where the declared aim had been to protect the existing trees at all costs. The *Pall Mall Gazette* warned in April 1894 that, unless operations were stopped, 'in a few weeks Epping Forest will be a thing of the past'. A field meeting was convened which enabled members of the Essex Field Club to see the thinnings for themselves. As the Club's founder-president emphasised, it was important not to base opinions on the *present* appearance of the Forest – the Conservators held the Forest 'in trust not only for the present, but for the future'. In tabling a motion supporting the actions of the Conservators, George E. S. Boulger, the professor of natural history at Cirencester, spoke of how the Act had sought the impossible in stipulating that the Forest should be maintained in its natural condition. There was no natural condition to maintain. The Conservators had been confronted with

the unspeakably difficult task of regenerating a forest all but destroyed by the vandalism of generations, and they had no precedents to guide them in their action. Such action must, therefore, be very largely experimental, and it was also a process demanding a considerable term of years.

It was hardly surprising that the Conservators had failed to satisfy everyone. The motion of support was carried by forty-one to eight votes.[64]

---

[62] Buxton, *Epping Forest*, pp. v–vii.   [63] *Ibid.*, pp. 105–7.
[64] Anonymous, 'The Management of Epping Forest', *Essex Naturalist*, 8, 1894, pp. 52–71.

CHAPTER 32

THE LOWLANDS

BY GORDON E. CHERRY AND JOHN SHEAIL

INTRODUCTION

Frederick Law Olmsted, the pioneer American landscape architect, made a thirteen-week tour of England in 1850. In his published account, *Walks and Talks of an American Farmer*, written in 1858, Olmsted confessed to being puzzled as to why he was so fascinated by 'the common-place scenery' of midland, rural England. There was nothing striking about the scenery, no particular part was especially notable or memorable, and yet, although he had since visited scores of places celebrated for the grandeur of their scenery and had spent months in the most beautiful, natural scenery of west Texas, he had never been so charmed as when walking through those parts of England. The scenery was beautiful without intention or artifice – picturesque, without being ungentle or shabby.[1]

Geography and history had combined to create an extraordinary variety of lowland countryside. In Shropshire, for example, the author of the county flora described plateaux and plains, hills and vales, boggy flats and heathery moors, cornlands and pastures, wooded slopes and barren crags, meres and ponds, streams and, most important of all, the river Severn. The wide variety of bird and mammal life brought about by such a combination of physical features was further enhanced by the mobility of individual species. A surprising number of sea- and shore-birds came as visitors, not so much driven by storms as attracted by such sheets of water as Ellesmere and by certain reaches of the Severn. Some, such as the sandpiper, followed the course of the Severn right from the sea.[2]

Some observers believed they could discern a progression as urban expansion cut rudely across the countryside. The meadows and pastures were first ploughed, and used as urban allotments or market gardens, before being taken over for housing. The ample hedges and broad, roadside wastes disappeared, and the odd corners and 'headlands' of fields

---

[1] F. L. Olmsted, *Walks and Talks of an American Farmer in England*, 1859 (repr. Ann Arbor, 1967), p. 229.   [2] H. E. Forrest, *The Fauna of Shropshire* (Shrewsbury, 1899), pp. 11–12.

were more closely cultivated. Even the herbage of the hedgebanks in the lanes was chopped and trimmed for tidiness, 'lest rude uncultured wild flowers should offend the eye'.[3] Far-reaching change also seemed to be taking place deeper in the countryside. At the inaugural meeting of the Northamptonshire Natural History Society in 1876, the Reverend M. J. Berkeley complained of how one could no longer expect to make any great discoveries. Over the previous ten years, 'the most prolific portion' of the fenland had been drained, woodlands disfigured, and many of the best localities destroyed by general progress in farming.[4]

In practice, refugia could still be found, reflecting the transient and complex patterns of land husbandry. Before the introduction of foreign and artificial manures, the practice of marling or 'chalking' the fields had resulted in the excavation of large numbers of pits in the corners of fields. In Cheshire, thousands occurred where the deposits of Keuper marl and mildly calcareous drift of the Dee valley and south Wirral were at or near the surface. Often circular, and ranging from 15 to 60 yards in diameter, they were seldom very deep. It was not long before they became water-filled and 'strongholds of retreat' for the genus *Potamogeton*, which was otherwise displaced by 'its New World exterminator', *Elodea*, which had, by the late nineteenth century, already reached the Cheshire meres.[5]

Estate-management practices might also help to sustain continuity. Covenants in a lease for Tyneham Farm in Dorset in 1848 prescribed a penalty of £5 for every acre of grassland where the sward was 'destroyed or injured by stagnant water or by unnecessary poaching of livestock or other negligent or improper treatment'.[6] A tenancy agreement for 54 acres of land in Carew, Pembrokeshire, in 1915, stipulated that no portion of grass could be mown in successive years unless manured 'not later than is usual in the neighbourhood' at a rate of at least 1 lb per acre of artificial manure (nitrate of soda excepted), or its equivalent manurial value of farmyard manure. The grass was to be kept free of nettles, rushes, docks, thistles, fern and tussocks.[7]

It would be wrong to suggest, however, that owners and occupiers had full control over what was happening on their land. At the turn of the century, there was considerable anxiety as to the damage caused to the beechwoods of the Chilterns by the felted beech coccus, which lived on the sap of the tree. It was spreading with alarming speed and, short of felling the trees, no remedy was known.[8] If the sucking insect was an

---

[3] White, *Flora of Bristol*, p. 2.

[4] M. J. Berkeley, 'Northamptonshire as a field of study for naturalists', *Journ. Northants Natural Hist. Soc.*, 1, 1876, pp. 347–51.

[5] Lord de Tabley, *The Flora of Cheshire* (London, 1899), pp. lxvi–lxvii.

[6] Dorset County Museum, MSS 6109.   [7] Pembroke RO, D/CAR/81.

[8] G. Eland, *The Chilterns and the Vale* (London, 1911), pp. 132–4.

obvious pest, the position of some mammals and birds was much more equivocable. In a sense, the term 'wild' animal was misleading because a species, regarded by one person as a pest might be highly prized by another. Although never a game animal in a legal sense, the rabbit was both preserved and introduced by sporting landowners and tenants to further parts of the country so that, by the late nineteenth century, numbers became so high that rabbit shooting took on the dimensions of a battue. In his manual on game and game coverts, published by the Country Gentlemen's Association in 1907, John Simpson wrote of how

rabbits undoubtedly contribute very largely to the head of game on some estates, afford good sport for the gun, require the least care, and probably pay better than any on the list.[9]

The amount of damage caused by the animals was a source of much argument. Some commentators claimed that almost all farmland was affected by losses in corn and grass, and damage to hedgerows and plantations, as a result of the animals browsing the young shoots, ring-barking, and burrowing among the root systems. Simpson contended that many of these claims for damage were fictitious. The alleged damage was caused by bad cultivations; the crops were so miserable that 'there was next to nothing for the rabbits to damage'.

In an attempt to restore harmony between farm tenant and the landlord and his sporting tenant, the Ground Game Act of 1880 extended the right to take and kill hares and rabbits to the farm tenant, subject to limitations. The principal restrictions were that only one person beside the occupier could use firearms, and only members of his resident household or regular employees, or one other person specifically employed, could remove the animals. Existing tenancy agreements had to be honoured until their expiry.[10] An analysis of the petty sessional reports for Devon between 1880 and 1899 indicates that, in that county at least, every use was made of the restrictions to constrain as severely as possible the tenant's scope for reducing the numbers of ground game.[11]

Whatever its intentions, the Act did little to arrest the increase in rabbit population, which spread to most parts of the country in the nineteenth century and locally attained densities of well in excess of twenty per acre. In those instances where tenants were able to exploit its potential, sport and the sale of carcases often took precedence over extermination. As A. D. Hall remarked, it was wonderful what damage the farmer would tolerate for the sake of a little shooting.[12] In other instances, the right to take

[9] J. Simpson, *Game and Game Coverts* (London, 1907), pp. 51–4.

[10] Ground Game Act, 1880, 43 & 44 Victoria, c. 47.

[11] J. H. Porter, 'Tenant right: Devonshire and the 1880 Ground Game Act', *AHR*, 34, 1986, pp. 188–97.   [12] Hall, *Pilgrimage of Farming*, pp. 185–6.

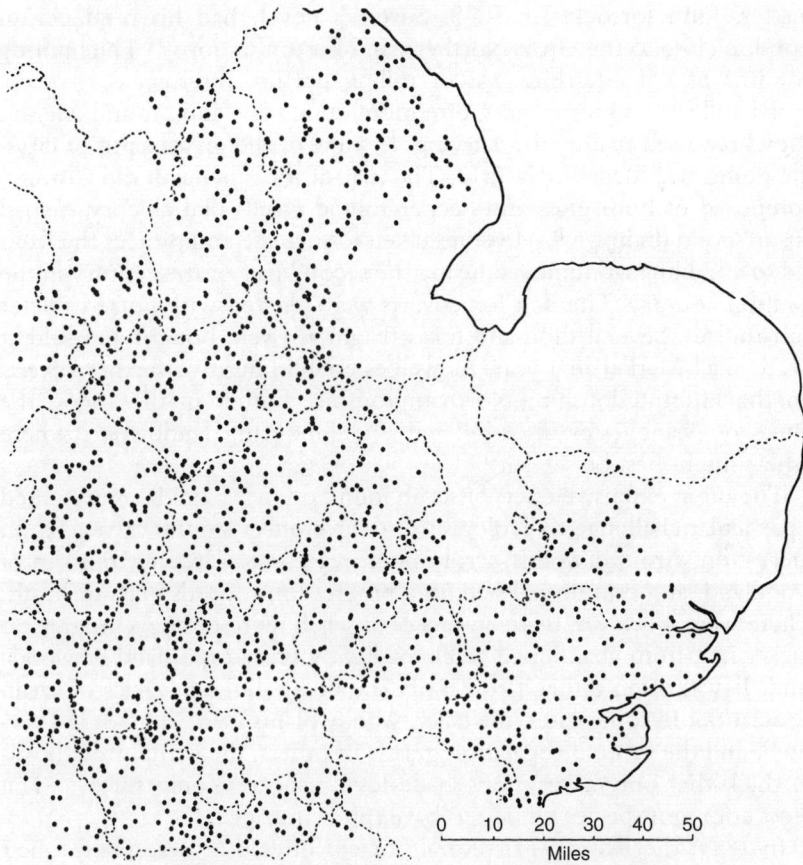

32.1 Where the foxhounds met in the mid-nineteenth century
(after *Hobson's Fox Hunting Atlas*, published in 1850).

rabbits might be 'let' to a professional trapper, who moved on once the value of the catch fell below what he could obtain from other rabbit-infested areas. The scope for large-scale trapping was greatly enhanced by the mass production of the 4-inch steel trap, called the gin or spring trap.[13]

With hindsight, the mid- to late-nineteenth century was the golden age of foxhunting, despite fears as to the 'scarifying' effect of the railways on the countryside, and the hazards posed by barbed-wire fencing. The principal effect of the railways was to make it easier for foxhunters and those who followed the hounds to venture further afield. The Cockney

[13] J. Sheail, *Rabbits and their History* (Newton Abbot, 1971).

master, John Jorrocks, in R. S. Surtees's novel, had his residence in London close to the Great Northern and Euston stations.[14] The hunting country of Leicestershire was of rolling upland, intersected by deep, broad valleys – Quorn and Cottesmore country – and around Melton Mowbray, used mainly for dairying. In view of the great depth of clays, the going was surprisingly firm. The mat of roots beneath old pastures, composed of both grass and deeper-rooted weeds like chicory, helped ensure good drainage.[15] Measures taken to increase and protect the foxes led to a trebling of numbers during the second half of the century in the Grafton country. The best fox coverts were planted with gorse to deter cub-thieves. Several thousand foxes (bagmen) were bought and sold at Leadenhall Market in a year. As well as establishing two further coverts on the Huntingdonshire–Northamptonshire border in the 1860s, the Duke of Manchester discussed with neighbouring landowners where others might best be set out.[16]

The often extensive tracts of scrub found on many holdings presented a particular challenge to ecologists in their pioneer studies of vegetation succession. Around woods, scrub might represent a degenerate form of woodland, at a transitional state in a progression towards grass or heath. There were, however, many instances of scrub representing a progressive succession, from abandoned arable or lightly grazed grassland to woodland. It was often difficult to classify the many different facies of scrub association. In some cases, the taller, woody plants were more or less isolated, and the ground vegetation grassy. In others, the shrub growth was so thick that only a very few shade-loving species could survive. The most common species tended to have spiney foliage.[17]

In 1845, the Royal Agricultural Society offered a prize for the best essay on hedges. The winning essay was simply entitled, 'On Fences'. Another focused 'On the Advantages of Reducing the Size and Number of Hedges', and a third more dogmatically 'On the Necessity for the Reduction or Abolition of Hedges'. All were united in their condemnation of hedges, which took up so much land, made the use of machinery difficult, acted as weed magazines and asylums for pests, impoverished the soil, and prevented the free circulation of air. The prize winner, James Grigor, estimated that, on the basis of a sample survey of four arable districts of Norfolk, there were 25 miles of hedgerow per square mile, covering over 10 per cent of the surface area. Applying the formula to 'the

[14] R. S. Surtees, *Handley Cross; or Mr Jorrocks's Hunt* (London, 1854), pp. 57–8; W. C. Hobson, *Hobson's Fox-hunting Atlas* (London, 1850).

[15] G. Thompson, *History of the Fernie hunt* (Leicester, 1987), p. 11.

[16] J. M. K. Elliott, *Fifty Years' Fox-hunting with the Grafton and Other Packs of Hounds* (London, 1900), p. 205; J. N. P. Watson, *The Book of Foxhunting* (London, 1977), pp. 26–7; Huntingdon RO, Manchester MSS, 10b, 24.   [17] Tansley, *Types of British Vegetation*, pp. 83–4.

forty divisions of England', the total area occupied was equivalent to 'two of the largest counties'.[18]

To others, it seemed extraordinary that a feature, so widely admired on the Continent should be threatened with destruction. The landowners of Norfolk, for example, seemed to have forgotten the ameliorating effects which the hedgerows and plantations had brought to the 'open Commons or Wire uninclosed Districts in Cultivation'. Claims that the hedges had to be removed so as to allow English farmers to compete more successfully with foreign producers reflected the wisdom of city counting-houses, and scientific lecture-rooms, which looked upon the land as little more than 'a manufactory of agricultural produce'. Fortunately, as William Johnston wrote, the countryside was held in such affection by the great body of people that it was likely to prevail over the mercantile spirit and devotion to profit among the middle classes.[19]

Whatever the determining factor, the loss of hedgerows was modest. Some were destroyed when fields were enlarged to accommodate machinery, particularly following the introduction of the steam plough. Richard Jefferies protested in the 1870s of how, in arable districts, 'modern agriculture endeavours to cut down trees and grub up hedges', not only on account of the shade cast and injury done by their roots, but because of the shelter afforded to sparrows and other birds. But in his *Pilgrimage of British Farming*, written in 1910–12, A. D. Hall continued to be astonished at the extent to which hedgerows survived as obstacles to farming. A green sheltered country of little fields might make a charming property but, to the farming eye, such a spectacle denoted 'the same retail way of business as the endless tiny shops in the suburbs of a manufacturing town'.[20]

Hedges were still needed in pastoral areas, and the shift from arable to pasture may have led to their extension. In his paper, 'Hedges and Hedge-making', published in the *Journal of the Royal Agricultural Society of England* in 1899, W. J. Malden made considerable use of the experience of the Midland Railway Company. The fact that whitethorn (*Crataegus oxycantha*) could be planted alongside the entire length of track said much for its wide adaptability.[21] Much depended on how well the hedges were managed. One of the more obvious signs of agricultural depression was

---

[18] J. Grigor, 'On fences', *JRASE*, 6, 1845, pp. 194–228; W. Cambridge, 'On the advantages of reducing the size and number of hedges', *JRASE*, 6, 1845, pp. 333–42; J. H. Turner, 'On the necessity of the reduction or abolition of hedges', *JRASE*, 6, 1845, pp. 479–88.

[19] Huntingdon RO, Manchester MSS, Box 12/7; W. Johnston, *England as it is, Political, Social, and Industrial, in the Middle of the Nineteenth Century* (London, 1851), vol. I, pp. 1–9.

[20] R. Jefferies, *Wild Life in a Southern County* (1879; repr. London, 1949), p. 312; Hall, *Pilgrimage of Farming*, pp. 206–7.

[21] W. J. Malden, 'Hedges and hedge-making', *JRASE*, 3rd ser., 10, 1899, pp. 87–115.

the skimping of such work. Between Oxford and Thame, the hedgerows, once kept 'so painfully low and well-trimmed', had been allowed by the 1890s to grow high. In parts of Essex and Suffolk, the hedges took on the appearance of 'shaws' or lines of woodland, growing up to 25 feet in height, and encroaching onto fields and roadside wastes.[22]

From the 1840s onwards, factory-made iron railings and posts, and wire for strengthening fences, became increasingly available. The latter might have been adopted more widely and quickly if it had not been for foxhunting. At first, the wire was of a stout, heavy type, about a quarter to half an inch thick, which at least meant it was visible to the horse. Lodged on the top, or stapled to the sides of the posts, it offered little resistance. Barbed wire had no redeeming features. Arrangements were made by most hunts to remove the strands during the hunting season. The Fernie Hunt set up a special Wire Sub-committee in 1895, and contemplated the appointment of a wire inspector in 1904.[23] Whilst there was no possibility of a general curb being placed on the use of barbed wire, a Bill was introduced by a group of members of parliament in 1893, making it easier and cheaper to obtain legal redress where injury was caused by 'any wire with jagged projections' on a public highway. The preamble to the draft Bill recalled how there had been many accidents, as well as 'danger, injury and cruelty to animals'. The Bill, as amended by the Local Government Board, enabled local authorities to seek the removal of such wire, where it constituted a nuisance to users of the highway.[24]

Whilst city architecture in England could not compare in terms of grandeur and beauty with that on the Continent, no other country could boast of so much 'ornamental cultivation' in the countryside. For the most part, 'the delightfulness of natural beauty had been enhanced by the association of shelter, comfort, and not inappropriately skill'. Chronologically, the countryside of particularly south-east England had been remodelled by successive styles of landscape design, each overlapping in time and evolving incessantly according to changing visual preference and economic and social conditions.[25]

Whilst there seemed little scope for further originality in the laying out of ornamental parklands, there were exciting challenges to be met in cre-

---

[22] Hissey, *Across England*, p. 383; E. J. T. Collins, 'Agriculture and conservation in England: an overview, 1880–1939', *JRASE*, 146, 1985, p. 41.

[23] Elliot, *Fifty years' Fox-hunting*, p. 205; F. P. de Costobadie, *Annals of the Billesdon Hunt* (London, 1914), pp. 112–13.

[24] PRO HLG 29, 41; PD, 4th ser., XII, 302–7; Barbed Wire Act, 1893, 56 and 57 Victoria, c.32.

[25] Johnston, *England as it is*; P. F. Brandon, 'The diffusion of designed landscapes in South-England', in *Change in the Countryside. Essays on Rural England, 1500–1900*, ed. H. S. A. Fox and R. A. Butlin, Institute of British Geographers, Special Publication, X, London, pp. 165–87.

ating wider landscapes, encompassing whole estates. At its simplest, hedgerows and the 'tail-ends' of fields might be planted up, so softening their outline and helping to create an illusion of continuous woodland from afar. By opening up prospects through thick woods, planting out unsightly views, enhancing the effects of physical features with trees of contrasting shape and colour, the various components of an estate – the house and park, home farm and tenants' holdings, woods and village – might be brought into a single, all-embracing composition which visually, as well as economically, was conceived to be indivisible. Whether a 'toy estate' or 'pleasure farm', or something more extensive, each estate had an individuality 'hardly less real than a kingdom', and yet the aggregate effect over large parts of south-east England was to bring about a 'vast, created landscape, natural enough to our eyes, but in reality arranged as much for picturesque appearances as for economic returns'.

A. RUDERAL HABITATS

The term 'ruderal' was adopted by botanists to describe the plant communities of roadsides, rubbish tips and other 'artificial' habitats. The word itself was derived from the plural form of the Latin word 'rudus' meaning rubble, ruins or heaps of mortar. In his *Types of British Vegetation*, Tansley wrote of how instructive studies might be of this 'battleground for aliens and casuals'. Owing to the peculiar nature of the substratum and supply of seed for colonisation, the successions did not often lead to the establishment of communities belonging to the normal plant-formations of the country.[26] This section will focus on two of the most conspicuous forms of ruderal habitat, the railway formations and roadside verges.

Without doubt, the most eye-catching feature to be seen in the countryside was the railway. From the 1860s onwards, the railway companies increasingly used standardised designs for their country stations, partly out of economy and partly to promote a corporate message. Those on the London, Brighton and South Coast Railway, along the Sussex coast, were sedate two-storey Italianate villas. Those of the Furness Railway were Swiss-chalet-type stone-and-timber buildings.[27] It was, however, the construction and initial appearance of the embankments and cuttings that aroused the greatest wonder and, in some cases, abhorrence.

The first intimation of what was to come might be the engineer traversing the intended route of a railway, with 'Ordnance Map in his hand, and a mountain barometer to take "flying levels"'. The gradients and

---

[26] Tansley, *Types of British Vegetation*, p. 9.
[27] J. Richards and J. M. Mackenzie, *The Railway Station. A Social History* (New York & Oxford, 1986), pp. 27–8.

curves of the railway had to be kept to the minimum. Engineers took advantage of the Thames's floodplain and terraces in laying out the London to Windsor line, and the railway to the north and south of Barmouth on the Welsh coast was built as close to the base of the sea-cliffs as possible. A cutting of 2 miles in length and up to 60 feet in depth was required to traverse the high ground between Twyford and Reading on the Great Western Railway.[28]

One way of reducing the time and costs of construction was to select a route whereby the amount of material excavated from cuttings corresponded with that required for the embankments. As the author of *The Practical Railway Engineer* pointed out, great skills were required in making such estimates. The fragments of rock used in the embankments occupied a larger space than the rock in its original state in the cuttings. Spoil heaps were often found where a contractor finished his length of line, and some material had in any case to be left for repairing any slips in the cuttings or subsidence on the embankments.[29] A prominent feature of the excavation of the Tring cutting, on the London to Birmingham line, was the famous series of forty horse-runs of up to 50 feet in height, used to lift the surplus material up to the spoil banks aligning the route.[30]

Experience indicated that thorough and systematic drainage was essential if landslips were to be avoided, and maintenance costs kept to a minimum. In making new banks, careful allowance had to be made for settlement in both depth and width. In the words of one contractor, it was better 'to trim a bit of surplus muck off than to try and add a bit later on, particularly as regards the top of the slopes'.[31] An obvious way of improving stability and reducing erosion was to clothe the slopes in vegetation. In many cases, meadows and pastures on the route of a railway were cut into turves of 6 inches in thickness, lifted, and quickly relaid on the slopes of the new cuttings and embankments as a 'preservative'.[32] Among the detailed stipulations included in the contracts for the building of the Wrexham, Mold and Connah's Quay Railway in 1893, the contractor was required

to remove and carefully put aside sufficient top soil from the site of the cuttings and embankments to soil the slopes when trimmed, and also the whole of the

[28] F. S. Williams, *Our Iron Roads* (1883, repr. London, 1968), pp. 64–5; J. H. Appleton, *The Geography of Communications in Great Britain* (Oxford, 1962), p.68; H. Greenleafe and G. Tyers, *The Permanent Way* (London, 1948), p. 110; G. Biddle, *The Railway Surveyor* (London, 1990).

[29] G. D. Dempsey, *The Practical Railway Engineer* (London, 1855), p. 63.

[30] Williams, *Our Iron Roads*, pp. 115–16.

[31] Anonymous, 'Minutes', *Proc. Instn. Civil Engineers*, 4, 1844, pp. 135–73, and 62, 1879–80, pp. 272–87; Nottinghamshire RO, DD 505/1.

[32] J. Day, *A Practical Treatise on the Construction and Formation of Railways* (London, 1848), pp. 98–100; Williams, *Our Iron Roads*, pp. 128–30.

formation at the site of excavation, to a thickness of at least 6 inches, measured at right angles to the line of the slope; and where it may be in grassland, the turf is to be carefully cut into sods for the same purpose.

To facilitate the colonisation and maintenance of vegetation on the slopes, the contract specified that

in all cuttings a row of sods, not less than 1 feet 6 inches wide and 6 inches thick, is to be laid where directed by the Engineer at the foot of slopes to support the soiling or sodding above.

The slopes, when soiled, were to be sown at the proper season with a mixture of 3 bushels of ryegrass seeds and 1½ lbs of white clover seeds per acre, with further sowings until a good turf was established.[33]

The earthworks may have been 'raw and brutish' in the short term, but they were soon concealed by grass, bushes and hedgerow-trees. There was, however, an ever-increasing risk of fires as railway companies introduced more powerful engines to draw the heavier loads. A survey conducted by the Central Chamber of Agriculture in 1901 recorded seventy-five separate incidents in eleven counties. By 1905, almost every agricultural society and farmers' club of importance had made representations to the Board of Agriculture. Agreement was reached with the railway companies in 1905 for a Bill making it considerably easier for aggrieved persons to claim compensation for loss of crops, hedgerow or wood as a result of a fire started by a railway engine. The hope was that such payments, which did not become operable until 1908, would encourage the railway companies to install more effective fire-boxes.[34]

The railways were the first to bring the visual experience of the industrial revolution to many parts of the country.[35] In his *Geological Treatise on the Cleveland District*, Joseph Bewick commented on how, in the remote moorland pass between Commondale and Kildale in north-east Yorkshire, with its slopes of scant vegetation and high ridges, thickly studded with purple heather, 'you might well suppose you were "out of humanity's reach"', were it not for 'the railway twisting its serpent-like form round the projecting portions of the hills by the side of the stream'. The hissing, puffing and rapid motion of the locomotive at once 'tells you that industry and enterprise have reached this bleak and desolate region'.[36]

Geologists were the first to recognise the unprecedented insights which railway construction could bring to their discipline. The name of the famous engineer, Brunel, was given to a rock formation (the Brunel

[33] PRO, RAIL 535/3.
[34] PD, XC, 737–63; CXLII, 348–74; CXLIX, 1300–4; Railway Fires Act, 1905, 5 Edward VII, c. 11.
[35] C. Barman, *Early British Railways* (Harmondsworth, 1950).
[36] J. Bewick, *Geological Treatise on District of Cleveland* (London, 1861).

Beds) at the base of the Ordovician, through which his railway to Fishguard in Pembrokeshire was constructed.[37] Among the numerous articles to appear was one in the *Geological Magazine*, with the title, 'Railway Geology of Devon', describing the geological phenomena encountered along lengths of railway in the county. The geology of the country traversed by the Hull to Barnsley railway (extending from the most recent formations to the Coal Measures) was outlined, based entirely on observations made during the construction of the line.[38]

For the antiquarian and naturalist, the opportunities presented by the railways were less obvious. Although steps were taken to avoid any damage to Maiden Castle, outside Dorchester, railways were driven through the priory at Lewes, and castles at Northampton and Berwick upon Tweed. In Lewes, however, the public interest aroused by the unearthing of the remains of William de Warren and Gundrada, the Norman founders of the priory, soon led to the establishment of the Sussex Archaeological Society, one of the first in the country.[39] The biography of the railway contractor, Joseph Firbank, described how the construction of the Bedford to Cambridge line led to the discovery of many vases and other relics of the Roman site of Salenoe at Sandy, but generally only a small proportion of the artefacts found by navvies was ever recorded or preserved.[40] The Annual Report of the institution of Civil Engineers urged contractors to impress on their men the need to preserve and surrender the relics, so that they could be forwarded to the Society of Antiquities for incorporation in national collections.[41]

Railways constituted yet another threat to the most accessible sites for botanising. One London botanist bemoaned the destruction of 'the wild Botanic gardens' of Battersea Fields in Surrey by the proposed railway to Nine Elms.[42] He called on fellow botanists to co-operate in recording and mapping the area so that,

at a future period when railroads, and such like public undertakings have demolished our richest locality in the immediate neighbourhood of the metropolis, we shall have at least a plan to show our descendants that a place existed which abounded in so profuse a supply of plants at a stone's throw from London.

[37] T. R. Owen, *Geology Explained in South Wales* (Newton Abbot, 1973), pp. 35–6.

[38] D. Mackintosh, 'Railway geology in Devon', *Geol. Mag.*, 4, 1867, pp. 390–401; E. M. Cole, *Notes on the Geology of the Hull, Barnsley and West Riding Junction Railway and Dock* (Hull, 1886).

[39] Anonymous, 'Report', *Sussex Archaeolog. Coll.*, 1, 1848, p. vii; W. Figg, 'On two relics found at Lewes', *Sussex Archaeolog. Coll.*, 1, pp. 43–5.

[40] F. McDermott, *The Life and Work of Joseph Firbank, Railway Contractor* (London, 1887).

[41] Anonymous, 'Annual Report: Session 1854–55', *Minutes of Proceedings of the Institution of Civil Engineers*, 14, 1855, pp. 103–4.

[42] D. Cooper, 'On the distribution of plants in Battersea Fields, Surrey', *Proc. Bot. Soc. of London*, 1, 1839, pp. 22–5.

There were, however, compensations. The earthworks of railways provided outstanding opportunities for studying the patterns of change in plant populations. Seeds and roots were shaken from trucks, or taken long distances within the soil and rubbish of ballast. Except for periodic burning and scything, the embankments and cuttings were seldom disturbed, and, from their high banks, seeds were blown onto neighbouring land.

As the author of the *Flora of Bristol* observed, native species often did unusually well on the embankments and cuttings of railways, which offered good drainage and a sunny exposure. The rush of the trains would carry seeds and spores long distances. *Lathyrus nissolia* was remarkably plentiful along the Great Western Railway, east of Brislington. The Midland line, near Staple Hill and Mangotsfield, was gay with the blossoms of *Hieracia* at midsummer, and the cuttings through the Cheddar valley abounded in sage and red valerian. A few species, of which *Arenaria tennuifolia* and *Linaria viscida* were the more conspicuous, were rarely found except on railway ballast.[43]

Perhaps the most detailed study of the colonisation of railway formations appeared in the *Flora of Halifax*, with its examples of how a railway bank

not only furnishes an interesting object lesson in the succession of plants that occupy it for some years after it has been made, but also offers a permanent home of a peculiar character, exactly adapted to the requirements of those that eventually win the day.

The *Flora* compared the flora of an embankment near Wyke station (made a number of years previously) with a new embankment at Hipperholme station. Thirty species were recorded at Wyke, compared with seventy at the other site. The number, the character of the dominant ones, and the subordinate position of the grasses highlighted the 'immaturity of the flora' at Hipperholme.[44]

If the railway formations won grudging respect, there seemed to be no mitigating features about the state of the roads, following the introduction of the motor car. Whilst it took some time for the implications of the internal combustion engine to be recognised, it was clear by the turn of the century that something radical was required, whether in terms of limiting the speed of motor traffic, or in providing the roads with a more durable type of surface.

At first, there was little attempt to adapt the roads to the bicycle or motor car. For example, the practice in the Oxfordshire town of Thame had been to dump occasional loads of flints from the fields of the

[43] White, *Flora of Bristol*, pp. 30–1.
[44] W. H. Crump and C. Crossland, *The Flora of the Parish of Halifax* (Halifax, 1904), pp. xxx–xxxii.

Chilterns, and to depend on road users to wear them into the ruts. Granite chippings were used from about 1890 onwards, spread loosely over the surface; a steam roller was used for the first time in 1894.[45] The Cyclists' Touring Club succeeded in 1897 in drawing the attention of the Fenny Stratford Urban District Council to 'the damage done to pneumatic tyres of cycles by thorns left upon the Highways by parties responsible for the upkeep and trimming of adjacent hedges', but in general the cycling organisations protested in vain over the state of the roads. It took the dust of the summers of 1904 and 1905, raised by the growing number of motor cars, to bring about some improvement.[46]

The iron studs of the wheels of early motor cars (designed to prevent skidding) soon tore up the loosely consolidated surfaces, raising such clouds of dust that an estate agent from Windsor and Ascot described to the Royal Commission in 1905 how the herbage on both sides of the Bath Road, within 50 yards of the hedge, was absolutely useless, whether for feeding cattle or for harvesting. It was impossible to sell or let some of the houses near the road because of the dust. The Newport Pagnell Rural District Council was one of many to urge a lower speed limit on motor cars, 'for they create such great clouds of dust as render the sight of other vehicles almost impossible and they have no time to get out of the way'.[47]

The Royal Commission on Motor Cars regarded the possibility of designing 'dustless' cars as a chimera: the solution lay in dustless roads. Every market town and large village had long been forced to employ a water-cart; adequate supplies of tar did not become available until 1905. Although early trials were not very successful, tarring gradually became commonplace. The whole of the High Street in Stony Stratford in Buckinghamshire was tarred in 1907, and the Stratford Road in Wolverton in the following year. In 1909, the county council agreed to the cost of tarring the central 10 feet of all main roads through towns. The practice of sanding the roads after tarring was adopted a year later.[48]

The dust problem was largely conquered when, by the 1920s, most main roads had been sealed with a tarred or bituminous surface. The new surfaces were not, however, without their problems. They were regarded as being particularly dangerous for horse-traffic. In 1910, much to the consternation of the County Surveyors' Society, a Kent farmer won a court action for compensation for the loss of a cow, which had allegedly died from drinking streamwater contaminated by tar washed off a nearby road. Much more serious was the threat posed to inland fisheries. Studies

---

[45] J. H. Brown and W. Guest, *A History of Thame* (Thame, 1935), pp. 240–1.

[46] Markham, *History of Milton Keynes*, Vol. II, pp. 201–2; W. R. Jefferies, in *Motors and Motor-driving*, ed. Lord Northcliffe, 1906.

[47] *Report of Royal Commission on Motor Cars*, BPP, 1906, XLVIII, Cd 3080, QQ 13500–10.

[48] Markham, *History of Milton Keynes*.

conducted under the aegis of the Road Board suggested that the danger was greatest during the first rains to fall on the tar surface and much later, when it began to break up as a result of traffic use and frost damage. Highway authorities were urged to use bitumen as a road dressing, wherever the washwater was likely to flow directly into a watercourse.[49]

B. THE HEATHLANDS

In many parts of Britain, a loose distinction was drawn between moorlands, which generally occurred on the wetter and more upland soils, and heathlands, which were found on acidic, nutrient-poor, mineral soils below about 800 feet. Lowland heaths occurred at one time in most southern counties from Cornwall to East Anglia, and from Sussex to Yorkshire. There was considerable interest among British botanists as to their affinities with the heathlands of north-west Germany and Jutland. The heathlands of the south-west of England had developed in an oceanic climate, where rainfall exceeded 40 inches per annum and there was less seasonal variation than further east, whereas, in the East Anglian heathlands, rainfall might be less than 25 inches and the climate was much more continental in character.[50]

The heathlands were thought to occupy the sites of former dry oak-woods. An oak–birch heath association represented a transitional stage. The oaks had often been felled, and the birch, which in any case cast only a light shade, was usually separated by open spaces, in which occurred such species as bilberry (*Vaccinium myrtillus*), heath grass (*Deschampsia flexuosa*) and, in the well-lighted spots, the ling (*Calluna vulgaris*). In its purest form, this dwarf shrubby Ericaceae was the dominant, and frequently the only, plant present. In the drier areas, *Erica cinerea* might be freely mixed with the dominant ling, *Erica tetralix*, taking its place in moister parts. In many soil profiles, the junction between the leached sand and unaltered underlying sand was marked by a hard continuous layer, or moorpan, of some inches in thickness, made up of humous compounds of iron oxide. The consequent shallowing of the soil acted as a further impediment to tree growth.[51]

The most important heathland area in the south-west peninsula was the extensive outcrop of serpentine rock on the Lizard in Cornwall. By 1848 and the appearance of C. A. John's *A Week at the Lizard*, the area was already famous among botanists for its Cornish or Goonhilly heath (*Erica vagans*).[52] The Dorset heathlands, on sandy, acid soils derived from

[49] A. Smith, *A History of the County Surveyors' Society* (Shrewsbury, 1985), p. 29; J. Sheail, 'Road surfacing and the threat to inland fisheries', *Journ. Transport History*, 12, 1991, pp. 135–47.

[50] N. Webb, *Heathlands* (London, 1986), p. 14.   [51] Tansley, *Types of British Vegetation*, pp. 98–106.

[52] C. A. Johns, *A Week at the Lizard* (London, 1848), pp. 269–70.

Tertiary deposits, formed an important transitional area between the more oceanic heathland and the more continental heathlands of East Anglia. The dry heath was dominated by *Calluna*, and the damper areas by cross-leaved heath (*Erica tetralix*) and purple moor grass (*Molinia caerulea*). The wetter areas to the south of Poole Harbour were the British stronghold of the Dorset heath (*Erica ciliaris*).[53]

The grass-heaths of Breckland, covering some 400 square miles around Thetford on the Norfolk–Suffolk border, remained the wildest and least populated part of East Anglia. Naturalists were fascinated to discover what they regarded as 'relics of a seaside flora, together with several species of coast-loving insects, especially beetles'. A strait was thought to have once divided Norfolk from the rest of East Anglia. Two or three plants of sand-dune habitats still grew plentifully, including the sand sedge (*Carex arenaria*), the underground runners of which helped to bind the wind-heaped sand. Eight species of Lepidoptera, usually confined to the sea coast, were more or less common. The stone curlew, or Norfolk plover, was more numerous than anywhere else in the British Isles.[54]

The first detailed ecological study of a lowland heath was carried out by E. Pickworth Farrow on Cavenham Heath in Breckland. His first paper in the *Journal of Ecology* identified the different associations encountered. Breckland wildlife represented not so much a relict of a former sandy coastline as an outstanding example of the adaptation of species to soils, climate and biotic pressures. As Farrow remarked, the long occupation of the district by man and his domestic stock had probably exerted a considerable effect on the vegetation, causing the once primitive woodland to be replaced by what were often practically pure heath associations.[55]

Although frequently dismissed as barren and worthless, the heathlands nevertheless provided a great amount of 'hard but healthy herbage' for the large flocks that were 'the sheet anchors of the occupiers of these lands'.[56] During the 1870s, they sustained an estimated 110 sheep per 100 acres of farmland, compared with an average 71 sheep for the remainder of Norfolk.[57] As many as 20,000 silver-grey rabbits were sent to London from Thetford Warren in the late 1860s.[58] Among his field notes for the Brandon and Icklingham district in 1877, the geologist, S. B. J. Skertchley, described the processes by which the rabbit wool and skins

[53] J. C. Mansell-Pleydell, *The Flora of Dorsetshire* (Dorchester, 1895), p. ii.; Webb, *Heathlands*.

[54] W. A. Dutt, *Wild Life in East Anglia* (London, 1906), pp. 58, 84–5.

[55] E. P. Farrow, 'On the ecology of the vegetation of Breckland', *Journ. of Ecology*, 3, 1915, pp. 211–28.

[56] C. S. Read, 'Recent improvements in Norfolk farming', *JRASE*, 19, 1858, p. 271.

[57] M. R. Postgate, 'Historical geography of Breckland, 1600–1850', MA thesis, Univ. of London, 1961.   [58] W. G. Clarke, 'Thetford Warren', *Zoologist*, 4th ser., 7, 1903, pp. 100–4.

were removed and processed in an increasingly mechanised industry employing between 300 and 400 persons.[59] By the end of the century, foreign competition had caused prices to fall so low that it was no longer worth conserving the rabbits through winter and protecting them from poaching. Numbers were nevertheless still very high.[60]

Tracts of heathland were often cultivated for a number of years as a form of shifting cultivation. As many writers pointed out, 'brek' was the name given to heathland that had, at some time, been ploughed, and then allowed to revert. Ridge and furrow could still be traced across many of the flint-strewn tracts of dry, sandy soil around Weeting. The dependence of some species on these periodic disturbances was highlighted by the common Breckland plant, the cut-leaved mignonette (*Reseda lutea*), which was particularly conspicuous on Rushford Heath in the early autumn of 1905, after many acres were broken up by means of a scarifier. The alien species, the Canadian fleabane (*Erigeron canadense*), first recorded in the early 1860s, soon occupied hundreds of acres of the larger brecks, in some cases becoming a pest.[61]

The evidence of the Tithe Commutation Surveys of the late 1830s and 1840s suggests that the area of grass-heath may have shrunk to only a third of the total area, and that, by the Annual Returns of the early 1870s, may have fallen still further in some parishes at least.[62] On the Merton estate, the fifth Baron Walsingham (1839–60) established four farms and converted extensive areas of sheepwalk and rabbit warren into arable land. Much of the estate's success was attributed to the strict insistence on covenants to regulate the size and disposition of sheep flocks on the arable land.[63] The first large-scale survey of the Ordnance Survey in 1883 may have recorded the category of rough pasture at its minimum extent in the nineteenth century.

The increasing interest in game preservation, and the fall in cereal prices, brought about a marked change in land use and management. The Breckland became almost 'one vast game preserve'; it was so jealously guarded by game-preservers as to be practically a forbidden land to the public at large.[64] Investment in game management on the Merton estate rose fivefold between the 1830s and 1870s.[65] During a four-day shoot in 1876, 4,842 head of game were shot. The most famous battues took place

[59] British Geological Survey, manuscript notebooks of S. B. J. Skertchley, volume 6.

[60] J. Sheail, 'Documentary evidence of the changes in the use, management and appreciation of the grass-heaths of Breckland', *Journ. Biogeography*, 6, 1979, pp. 277–92.

[61] Dutt, *Wild Life in E. Anglia*, pp. 85–6.

[62] E. H. Griffith, 'Changes in land utilization in the East Anglian Breckland during the past one hundred years', MA thesis, Univ. of Wales, 1955, pp. 34, 46.

[63] G. Crabbe, *Some Materials for a History of the Parish of Thompson* (Norwich, 1892).

[64] Dutt, *Wild Life in East Anglia*, p. 58.   [65] Nortfolk RO, 21 Walsingham MSS, 18/7.

at Sandringham, to the north of Breckland. The Prince of Wales purchased the estate in 1861 and, by renting neighbouring properties, soon amassed a shooting estate of 12,000 acres. The best game bags were recorded in 1896–7, when 13,958 pheasants, 3,965 partridges, 836 hares, 6,185 rabbits, 77 woodcock, 8 snipe, 52 teal, 271 wild duck, 18 pigeons and 27 other bird species were taken.[66]

At the time of the Tithe Commutation Surveys, a special inquiry was necessary to assess the extent to which farm output was adversely affected,[67] the Surveyor concluding that

game appears to be the great object of the proprietors in this neighbourhood and from the immense quantity of it and the light weak nature of the soil they do much damage to the crops.

It was through a book by a widowed tenant-farmer that conditions on the Sandringham estate attained some notoriety – the royal agent tried to buy up and destroy as many copies as possible.[68] As Mrs Cresswell recalled, it was soon clear that 'game, game, nothing but game, was to be the order of the day'. The newly imported head keeper, with 'an organised staff of officials, took possession of the place in military style'. The aim was to rear as much game as the land would carry. The tenant farms were parcelled out like policeman's beats, strips cut across the fields like a gridiron and planted for game shelters, and,

until the trees were sufficiently grown to smother them, a mass of noxious weeds grew, seeded, and blew all over my crops and the newly cleaned land, no one being allowed to cut them down for fear of disturbing the game.

The proportion of Breckland under plantations and belts of fir increased from as little as 7 per cent in the 1840s to over 10 per cent by 1891, and 13 per cent by 1905.[69]

Whereas initially the numbers of game might be attributed to the eccentricities and wealth of individual landowners – the sixth Lord Walsingham spent much of his time touring the world as a naturalist and organising battues on his Merton estate – the decline in agricultural prices from the 1870s onwards caused many to look increasingly to game as their principal source of income. The rental for Waterloo Farm, of 740 acres on the Merton estate, fell from £600 in 1863 to £500 in 1870, and to £127 by 1900.[70] The shooting of the Merton estate was leased in the

[66] W. A. Dutt, *The King's Homeland. Sandringham and North-west Norfolk* (London, 1904), p. 109.

[67] PRO IR 18, 6333.

[68] G. Cresswell, *Eighteen Years on Sandringham Estate* (London, 1887); C. Chevenix Trench, *The Poacher and the Squire. A History of Poaching and Game Preservation* (London, 1967), pp. 171–80.

[69] Griffith, 'Changes in land utilization', p. 70.

[70] H. Rider Haggard, *Rural England* (London, 1902), vol. II, pp. 489–96.

1890s. As the Assistant Commissioner for Norfolk reported to the Royal Commission on Agriculture in 1895,

it seemed to be generally believed that the game were, on the whole, a blessing to the county, as bringing into it a number of wealthy men, who, for the sake of shooting, take over a considerable amount of land which otherwise would be practically abandoned.[71]

Most commentators regarded the 1890s as the worst point in the depression. Between 1870 and 1905, the arable area of Breckland fell from 70,640 acres to 45,297 acres. The decline might have been greater had it not been for the belief that pheasants and partridges flourished best where there was some cultivation. The report of the Royal Commission stressed how it was very difficult to establish any kind of turf on the abandoned area.[72] The prediction that it would all revert to heath seemed to be borne out by the fact that the area of permanent grass, as recorded in the Annual Census, rose by only 3 per cent between 1870 and 1905. A slight expansion of the arable occurred after 1905, assisted by the use of steam ploughs for breaking up the derelict brecks and heath.[73]

Elsewhere, the heathlands had become much more fragmented and, in some places, disappeared. Large tracts of heathland had been converted to arable on the Suffolk 'sand-lands', the belt of Pliocene and Pleistocene deposits that extended up to 10 miles inland from the coast, following the discovery of the valuable fertilising qualities of the coprolites at the base of the Red Crag.[74] Charles Babington found that the heaths on the Lower Greensand outcrop on the Bedfordshire–Cambridgeshire border, at Sandy and Gamlingay, had been enclosed and, for the most part, cultivated. As he remarked, in 1862, 'every spot is so thoroughly under cultivation now that the botanizing is poor'. The last bog to have been drained, in 1853, now supported a crop of potatoes.[75]

Even where the heathlands survived, their species composition might change. On most Hertfordshire commons, furze (*Ulex* spp.) and bracken (*Pteridium aquilinum*) grew 'comparatively amicably together', with gains in one spot being met by losses in another. This was the case on Harpenden Common until the turn of the century, when bracken began to gain the upper hand. In a paper to the Hertfordshire Natural History Society, E. J. Salisbury attributed the changes to two man-induced factors. Over the previous two decades, the abstraction of underground water for supply purposes had caused the soils of first the higher, and then the lower parts of the Common, to become progressively drier and,

[71] *Report of Royal Commission on Agriculture*, BPP, 1895, XVII, C. 7915, p. 58.   [72] *Ibid.*, p. 20.
[73] Griffith, 'Changes in land utilization', p. 76.
[74] W. A. Dutt, *Wild Life in East Anglia* (London, 1906), pp. 290–1.
[75] C. C. Babington, *Memorials, Journal and Botanical Correspondence* (Cambridge, 1897).

therefore, more favourable to bracken. The spread of bracken had been further accelerated by furze fires so that, in some parts, only the tallest bushes were found in an otherwise continuous cover of bracken. Experience indicated that the only effective method of stopping the spread of bracken was to cut it constantly.[76]

The French topographer, de Lavergne, referred to the heaths of the Berkshire–Hampshire–Surrey border as 'the Sologne of England'.[77] There were few plantations of Scots pine until the early nineteenth century, when enclosure encouraged their establishment. Despite the heavy losses caused by drought and browsing, the region had taken on a much more wooded appearance by the third quarter of the century. The spread of self-sown Scots pine across the heathland would have been even more rapid had it not been for the frequent fires in summer and autumn. One in 1854 was caused by sparks from a railway engine at Byfleet, and others on the Bramshill estate in 1891 were the result of picnic litter and out-of-hand burning. The intensity of such fires reflected the length of time over which the forest litter had been allowed to accumulate.[78]

Just over a third of the Bagshot Sands region was affected by parliamentary enclosure, the aim being not so much to raise the productivity of the soils but rather to suit the administrative convenience of the larger landowners and, in some cases, to secure land for institutional use.[79] Exercises involving over 16,000 troops in the summer of 1853 on Chobham Common in Surrey had highlighted the value of acquiring training grounds that could be used exclusively for combined exercises of the cavalry, artillery and infantry. The obvious location was Reigate in Surrey, at the junction of several railways, but there were no wastelands available. The next best position for collecting troops for covering the capital, and affording speedy reinforcements to the southern counties in the event of an invasion, was the extensive heathlands of Aldershot, Farnham and Ash. Unencumbered by crops, hedges and trees, there was little risk of claims for compensation from farmers for the damage caused by field exercises and 'sham' fights. It was, however, necessary to act quickly – parliament had just approved a measure to enclose the commons extending over twenty-seven parishes in the Aldershot area.[80] Within a year, the Principal Officers of Her Majesty's Ordnance were empowered to purchase commons and some enclosed lands in the par-

[76] E. J. Salisbury, 'The competition of furze and bracken, particularly on Harpenden Common', *Trans Hertford. Natural Hist. Soc.*, 15, 1912, pp. 71–2.

[77] L. de Lavergne, *The Rural Economy of England, Scotland and Ireland* (Edinburgh, 1855), p. 210.

[78] PRO RAIL 411, 162; R. G. Kingsley, 'The flora of Eversley and Bramshill fifty years ago', *Proc. Hants. Field Club*, 8, 1918, pp. 129–38.

[79] M. H. Ferguson, 'Land use, settlement and society in the Bagshot Sands Region, 1840–1940', PhD thesis, Univ. of Reading, 1979.   [80] PRO WO 33, 1/3.

ishes of Aldershot, Yateley, Farnborough, Crondall and Farnham for 'the purposes of a Military Camp, and for the instruction and training of troops in the Science of War'. By a further measure of 1856, many rights of way were extinguished, and others created.[81]

Within four years, Aldershot had ceased to be 'one of the most pleasant and picturesque hamlets in Hampshire'. Its population rose from 875 in 1851 to about 3,500. As a contributor to Charles Dickens's weekly journal, *All the Year Round*, wrote in 1859, 'the mushroom village' could not grow fast enough. Scaffold-poles and unfinished brickwork were seen sprouting at both ends of the straggling mile of shops and houses. By 1859, a bank, military tailor and outfitter, and printing works, had appeared. But there was still 'a somewhat disjointed and unconnected appearance'.[82]

Never before had so large an area been acquired explicitly for military training purposes. Whilst the immediate effect was a great flurry of building activity within the camps, the longer-term effect was to ensure the survival of large tracts of heathland so near to London. Further substantial purchases were made in the years up to the First World War.

C. THE CHALK AND LIMESTONE COUNTRY

The typical grasslands of limestone country were a short turf, composed mainly of sheep's fescue-grass (*Festuca ovina*). The incidence of ancient trackways, camps and other earthworks was taken to be evidence of the great age of the grasslands and of how they might never have been wooded. Sheep were the main grazing animal and, particularly where also nibbled by rabbits, scarcely any herbaceous plants grew to a height of over an inch or so. A multitude of species could be found in the close mat of springy, wiry herbage. Besides the grasses and less abundant sedges, there were many dicotyledonous species – their association varying from place to place, owing to minor variations in the depth and nature of the soil and to differences in the incidence and exposure to grazing. Wild thyme and ants were often found together, the conical-shaped hills resembling thousands of miniature volcanoes. As Richard Jefferies remarked, they were often so close to one another that one could walk some 20 to 30 yards without touching the proper ground.[83] A particularly characteristic feature was the numerous species of Orchidaceae, some of which were otherwise extremely rare.

---

[81] Commons, and Rights (Ordnance) Act, 1854, 17 and 18 Victoria, cap. 67; Aldershot Camp Act, 1856, 19 and 20 Victoria, c. 66.

[82] Anonymous, 'Aldershot town and camp', *All the Year Round*, 1, 1859, pp. 401–8.

[83] Tansley, *Types of British Vegetation*, pp. 173–5; R. Jefferies, *Wild Life in a Southern County* (1879; repr. London, 1949), pp. 292–4.

Salisbury Plain formed the heart of the English chalklands. Turning outwards, say, from Battlebury Camp behind Warminister, farms ran in long strips down the steep scarp face to the water-meadows, encompassing the wide variety of soils across the terrace of chalk marl and narrow valley of gault clay.[84] Turning inwards to the Plain, the chalk turf might be covered by a scatter of yew trees (*Taxus baccata*) and bushes of box (*Buxus sempervirens*) and juniper (*Juniperus communis*), or by a few gnarled old thorn bushes. On his tour of England, the American, F. L. Olmsted, wrote of 'a strange weary waste of elevated land, undulating like a prairie, sparsely greened over its gray surface with short grass; uninhabited and treeless'. The trails ran crookedly, divided and crossed frequently. There were few guide-posts, and 'twice or thrice we were as completely lost as Oregon emigrants might be in the wilderness'.[85]

Throughout the nineteenth century, more and more of the higher downland acquired the name of 'bakeland'. The turf was broken up, burned, and prepared as a seed-bed, but, whereas previously many of the intakes reverted back to pasture after only a few years because of the farmer's inability to provide sufficient farmyard dung, the increasing availability of crushed bones and guano made it possible to adopt 'a more regular system of culture', namely a five-year rotation of wheat, turnips or swedes, oats or barley, and ley grass for two years. Sheep grazed the herbage at the end of the rotation, helping to make the soil more compact. The more permanent nature of the cultivation was reflected in the appearance of numerous barns and yards on the downlands during the 1840s and 1850s.[86]

In his *Flora of Oxfordshire*, G. C. Druce recalled how the tract of country above Burford, once grassy down, 'fragrant with thyme, and adorned with *Pulsatilla*, *Herminium*, *Cineraria*, and *Ophrys aranifera*', was now bare and bleak, with traces of its 'former glories' to be found only alongside roads or in abandoned quarries too barren to be utilised.[87] Whereas shepherds had once trapped in summer hundreds of dozens of wheatears on the South Downs for sale as a delicacy in the coastal resorts, only a few dozen were caught by the 1860s. The reason was not so much over-trapping, but 'the breaking-up and bringing under tillage of thousands of acres of sheep-walk, down, heath, common and warren, which were the ancient nurseries in this country of this prolific species'.[88] As the 'elderly gentleman' remarked, in Thomas Hughes's *The Scouring of the White Horse*, there seemed no limit to the endeavours of farmers. In an

---

[84] Hall, *Pilgrimage of Farming*, pp. 1–7.   [85] Olmsted, *Walks and Talks*, p. 272.
[86] E. Little, 'Farming of Wiltshire', *JRASE*, 5, 1845, pp. 161–4.
[87] G. C. Druce, *The Flora of Oxfordshire* (Oxford, 1886), p. xxxii.
[88] Yarrell, *History of British Birds*, pp. 349–51.

age that venerated nothing, 'they would plough and grow mangold-wurzels on their father's grave'.[89]

Farmers had become too sanguine in expecting to be able to continue growing corn on such light soils in direct competition with farms on more naturally suitable land.[90] Whilst these immense tracts of wheatland might seem profitable at first, the French topographer, de Lavergne, travelling through Salisbury Plain in the 1850s, warned of how the transformation might not be judicious in every case.[91] One commentator went so far as to assert that 'England is not strictly a good wheat-growing country'.[92] In travelling across

the wide plain of Salisbury, the eye meets hundreds of acres, over which small broken flints lie scattered. These testify that the down was once arable; but not being found profitable it has been suffered to go out of cultivation.

Ploughing too deeply might bring to the surface a vein of rubble chalk, or small broken flints – the dross of the land. Particularly after hard frosts, the ploughed, light soils around the base of the young wheat plants might be blown away, leaving the stems 'hung up by one leg'.

Whilst soil erosion was rarely mentioned in the abundant literature of the period, its incidence should not be overlooked. No matter how rarely surface water flowed across the exposed, light, thin soils, the longer-term impact on soil structure could be significant. A cloudburst over Towthorpe in the Yorkshire Wolds in June 1888 caused innumerable streams to form on the sides of the sloping fields, converging into larger courses that swept away the soil and hollowed out deep trenches in the broken chalk subsoil. Some four years later, a water-spout at Huggate caused a river to flow a distance of four miles down the dale, destroying several houses and damaging seventy more in the village of Langtoft.[93] Such events might represent only the most catastrophic phases of soil wash which, as A. D. Hall pointed out, underlined the need for the most careful soil husbandry, and the establishment of more hedgebanks to break the force of the stormwater and reduce its excavating power.[94]

Whatever the precise reasons, the growth of 'Wild England' proceeded by leaps and bounds. In the words of one commentator, writing in the 1890s, it would be an interesting exercise for the newly founded County

[89] T. Hughes, *The Scouring of the White Horse* (London, 1859), p. 29.

[90] J. Caird, *English Agriculture in 1850–1* (London, 1852), pp. 79–87.

[91] Lavergne, *Rural Economy of England*, p. 248.

[92] J. Stevens, *The Farm Labourer, at Home, and in the Field* (n.p., 1874), p. 13.

[93] E. Maule Cole, 'Water-spouts on the Yorkshire Wolds', *Trans Hull Scientific & Field Naturalists' Club*, I, 1900, pp. 225–8; E. Hatfield, 'The way of a water-spout', *Pearson's Mag.*, 10, 1900, pp. 531–5.   [94] Hall, *Pilgrimage of Farming*, pp. 118–19.

Councils to issue a map, on which the land withdrawn from cultivation and running wild was coloured in a bold tint.[95] Large areas of 'bakeland' were laid down to grass. In his report on the Andover district of Hampshire, W. Fream, an Assistant Commissioner, described to the Royal Commission on Agriculture in 1894 how, in every direction from Andover, there was corn stubble, or old 'seeds' or sainfoin layers, left long beyond their usual time.[96]

The grass sward that had grown up since 'the great depression' looked very different from the 'generally old sheep-walk strewn with typical chalk flowers'. Nature took a long time to recover. Whereas the plough-lands of Roman times might be covered by a thick and rich turf, those abandoned during the nineteenth century remained, for the most part, barren, thirsty and weedy wastes. The exact species composition varied greatly. By chance, two or three species might have fallen on a suitable, unoccupied spot, whereas a completely different selection of plants might have colonised or come to dominate the next spot. As W. H. Hudson pointed out, the different ages of the grasslands became especially apparent during times of drought. Except in the very driest seasons, the 'true native flora' remained green and blossoming throughout the months of March to October, whereas the 'intruders which have become natives' were past their best by May and June.[97]

It was expected that many tracts would be planted as game covert, and developed as 'a residential and sporting locality, resorted to by those who had been fortunate in making money in other pursuits than agriculture'.[98] Many of the landowners on the downs already had considerable experience of the large-scale rearing of game. Rough sheds were used, with wire-netting on the top and sides, to rear pheasants at Godmersham Park, in the North Downs. As many as 1,700 head of pheasants, partridges, hare, rabbits and woodcock might be laid out for carting to the game larder after a big shoot. On the Berkshire downs, 'grass-burning' was 'an exciting minor branch of husbandry'. In March of each year, long trails of straw were set alight and the flames driven by the wind across the down, removing the dry haulm and leaving 'the good green undergrowth sprinkled with invigorating ashes'.[99]

Some of the higher and more remote downs became little more than rabbit warrens – a use which a land agent for Rawlence and Squarey pos-

[95] C. J. Cornish, *Wild England of To-day and the Wild Life in it* (London, 1895), p. 187.

[96] *RC on Agriculture*, BPP, 1894, XVI, C. 7365. Report by Dr W. Fream upon the Andover District of Hampshire, p. 7.

[97] Hall, *Pilgrimage of Farming*, pp. 1–7; W. H. Hudson, *Nature in Downland* (London, 1900), pp. 39–42.   [98] *RC on Agriculture*, BPP, 1894, XVI, p. 15.

[99] Canterbury Public Library, Kent, Wells MSS; Cornish, *Wild England*.

itively recommended in his evidence to the Royal Commission of 1881. In his book, *A Shepherd's Life*, W. H. Hudson described how the former arable lands were enclosed with big wire fences and rabbit netting, within which 'little but wiry weeds, moss, and lichen' grew.[100] On Fyfield Down in north Wiltshire, the keepers set aside 10 acres of swedes each year, which were spread among the sarsen stones for the rabbits whenever snow covered the ground. About 5,000 rabbits were killed each year by visiting sportsmen, and the carcases sold to poulterers and game salesmen in the Central Market in London. On the night before a shoot, sticks dipped in a mixture of brimstone were thrust down the burrows. The rabbits found refuge in 6 acres of low gorse and elder, set aside for that purpose, from which they were flushed by beaters as the shoot was about to begin.[101]

It was during the 1890s that the War Department began to build up an estate on Salisbury Plain. The first purchase of 364 acres was made at Market Lavington. By 1900, over 41,000 acres had been acquired in the eastern part of the Plain, including the Tidworth Park estate, with its mansion, 101 cottages, 12 residences, and the Ram Inn. Rows of green-painted barracks rose in Bulford Fields, and the number of military personnel visiting the Plain rose substantially each year. Purchasing was resumed in 1909, and this time took in the lands to the west and south, the aim being all the time to build up a compact ring-fence holding. There were criticisms of the way sheep were excluded from large areas, and fears were expressed for the trout-streams – the Avon, Porton Water, Wylye and Ebble which flowed through the scenes of 'mimic warfare'.[102] But the impact was less than anticipated. Military activity was largely confined to the summer months; the flocks of sheep were readmitted once firing ceased. As Ella Noyes wrote in 1913,[103]

the landscape is so large and open that the camps scattered here and there from April to September – and even a permanent settlement, such as Bulford Camp – are soon lost and forgotten in its immensity. A fold in the downs, which looks from a distance a mere wrinkle in the surface, can hide a whole army corps from sight.

As on the heathlands of Aldershot, military occupation was a force for preservation, rather than of destruction, of the wild scenery associated with the chalklands of central lowland England.

---

[100] W. H. Hudson, *A Shepherd's Life* (London, 1910), p. 16.

[101] N. E. King and J. Sheail, 'The old rabbit warren on Fyfield Down, near Marlborough', *Wilts, Archaeol. & Nat. Hist. Mag.*, 65, 1970, pp. 1–6.

[102] A. G. Bradley, *Round about Wiltshire* (London, 1907), p. 263; G. A. B. Dewer, *Wild Life in Hampshire Highlands* (London, 1899), p. 23.

[103] E. Noyes, *Salisbury Plain* (London, 1913), p. vii.

D. THE ALLUVIAL GRASSLANDS

In his *Types of British Vegetation*, Tansley used the term 'neutral grasslands' to distinguish those frequently found on alluvial soils from the grasslands more commonly associated with acidic or downland soils.[104] The hinterland of seaside resorts tended to be well botanised. There was a chapter on 'bog-botany' in P. H. Gosse's *Tenby: a Sea-side Holiday*, published in 1856,[105] and it was from Tenby that C. C. Babington compiled his species list of south Pembrokeshire, published in the first volume of the *Journal of Botany*. Although most of the area was under cultivation, there were still two marshy districts of considerable size. One was the valley of the Ritec, described as rough pasture with seasonal characteristic wetland species. The other was the Castlemartin Corse, which had been drained some sixty years previously, but had since 'fallen back into the state of coarse wet pastures'. It was 'nearly deprived of all its peculiar plants, a few only of them remaining in the ditches'.[106]

An important distinction was drawn between flood-meadows, which were inundated whenever the river or stream was in spate, and water-meadows, where flooding was more carefully regulated, in terms of both its volume and timing. Weirs and hatches were constructed so as to enable the water to flow more easily into the meadows, directing the water along a series of carriers cut into the crown or centre of ridges, from where it overflowed into drains made along the bottom of the furrows. Sods of earth and turf might be used to ensure that the entire surface of the 'corrugated' pasture was covered, and grass growth was therefore uniform. In order to prevent the water from stagnating, it was introduced at a 'trot', and removed at a 'gallop'. The length and width of the ridges varied according to the gradient and free-drainage of the meadows.[107]

As Olmsted recounted, 'the effect of irrigation' could often be seen at a considerable distance in the deeper green and greater density of the grass. The quality and quantity of the water were of critical importance. Dark, muddy water was not necessarily the richest in nutrients.[108] Spring-water was about 50° Fahrenheit when it emerged from the ground. In his *Wild England of To-day and the Wild Life in it*, Cornish recalled how the river Itchen in Hampshire flowed 'like a vein of warm life through the cold body of the hills', even in the coldest of weather. Although 'ribbed across with multitudinous channels of white and

---

[104] Tansley, *Types of British Vegetation*, p. 84.

[105] P. H. Gosse, *Tenby: a Seaside Holiday* (London, 1856).

[106] C. C. Babington, 'On the botany of south Pembrokeshire', *Journ. of Botany* 1, 1863, pp. 258–70.

[107] H. P. Moon and F. H. W. Green, 'Water-meadows in southern England', in *The Land of Britain, Part 89. Hampshire*, ed. F. H. W. Green (London, 1940), pp. 373–90.

[108] Olmsted, *Walks and Talks*, pp. 258–60.

Table 32.1. *Comparison of the species composition of water-meadows and dry meadows, 1888 (after W. Fream, 'On the Flora of Water-meadows', Journ. Linnean Soc. – Botany, 24, 1888, pp. 454–64)*

	Gramineae	Leguminosae	Miscellaneous	Total
Total on water-meadows	26	7	52	85
Total on dry meadows	19	10	55	84
Exclusively on water-meadows	10	1	32	43
Exclusively on dry meadows	3	4	35	42

crackling ice', the water-meadows remained green, and were crowded with 'plovers and redwings, snipe and water-hens, sea-gulls, field-fares and missel thrushes, pipits and larks, and all the soft-billed birds in search of food'.[109]

In his account of Hampshire farming, written in 1861, Wilkinson described how irrigation began in earnest in early autumn, the ground being thoroughly soaked for about a week, making sure that the top of the grass was always well above the water. Thereafter, it occurred on perhaps six days a week, falling to five and four days a week in January and February respectively. Irrigation ended in March, when the meadows were thoroughly drained and used to provide an early bite for ewes and their lambs. The animals were withdrawn five or six weeks later, irrigation was resumed, and the depth of flooding increased as the hay crop grew taller. The hatches were closed whenever there was a risk of the water stagnating, and closed permanently after six weeks, when the hay crop was ready. Dairy and store cattle grazed the aftermath.[110]

The uniform pattern of annual management provided outstanding opportunities for research on meadow herbage. In a paper read to the Linnean Society in 1888, William Fream, the professor of natural history at the College of Agriculture at Downton, near Salisbury, compared the flora of the water-meadows in Nether Charford, on the Hampshire Avon, with that of the old grassland plots at Rothamsted in Hertfordshire (Table 32.1). As Fream pointed out, the cycle of flooding and drainage minimised the effects of rainfall. The list of eighty-five flowering plants was nevertheless surprisingly high in the water-meadows, where grazing pressure would have been intense. The inclusion of some plants growing within the watercourses may have inflated the list.[111]

---

[109] Cornish, *Wild England*, p. 132.
[110] J. Wilkinson, 'The Farming of Hampshire', *JRASE*, 22, 1861, pp. 288–90.
[111] W. Fream, 'On the flora of water-meadows', *Journ. Linnean Soc., Botany*, 24, 1888, pp. 454–64.

The management of water-meadows had always had to take close account of other river interests, namely the mill and navigation interests as well as the competing claims of other meadow-owners. The increasing popularity of sport-fishing in the late nineteenth century added a further complication. It was far from easy to assess how far fish stocks might be affected. Migrating parr and smolt were thought to lose their way in the rafts of floating weed, cut adrift by the meadsmen clearing the channels. In 1909, there were complaints of so much water being diverted through a side-hatch at Itchen Abbas that the fish had foresaken the mainstream.[112] Herons and other predators could be seen taking fish stranded by sudden falls in water level on the meadows. Much harder to assess was the extent to which the stranding of insects, entomostraca and plankton in the vegetation after irrigation affected the fish of the mainstream.[113]

By the 1850s, there were signs that some water-meadows were not receiving as much attention as previously. Labour costs were rising, and the greater availability of artificial feedstuffs meant that farmers were no longer so dependent on an 'early bite' in the meadows. In some instances, it became more profitable to let the watercourses for the artificial culture of watercress, the demand for which rose dramatically in the mid-nineteenth century. In his 'Agricultural notes on Hertfordshire', Clutterbuck recounted how the ditches were levelled and in many cases widened so as to form a series of shallow lakes, either permanently or temporarily impounded. Borings might be made in the chalk to facilitate the issue of spring-water, and the height of the water regulated so as to bring out 'a proper succession in the ripening of the crop' within the different parts of the beds. Constant vigilance was required in protecting the plants from birds, especially blackbirds.[114]

A wide range of conditions was encountered, both within and between the low-lying parts of the country. In the Somerset Moors, West Sedgemoor was famed for the grazing of young stock, bullocks and horses, whereas a form of shifting cultivation was practised over large parts of nearby King's Sedgemoor, whereby the surface vegetation was pared and burned, and a few crops of wheat, oats and potatoes or beans taken.[115] Everywhere, the character and productivity of the Moors closely reflected the degree to which husbandry practices made use of

---

[112] Hampshire RO, 18M 51, 5036.

[113] J. Sheail, 'The formation and maintenance of water-meadows in Hampshire', *Biological Conservation*, 3, 1970, pp. 101–6.

[114] Wilkinson, '*Farming of Hampshire*', pp. 320–1; J. Clutterbuck, 'Agricultural notes on Hertfordshire', *JRASE*, 15, 1864, p. 312.

[115] T. D. Acland and W. Sturge, *The Farming of Somerset* (London, 1851), pp. 56–8; J. A. Clark, 'On the Bridgwater and other Levels of Somersetshire', *Journ. Bath and West of England Soc.*, 2, 1854, pp. 99–128.

local conditions. Whilst the soils of the higher levels contained the most 'proof', many of the grasslands on the thinner soils along the rivers produced a higher-quality cheese for as long as winter flooding continued to sustain fertility. So as to ensure that sufficient sediment was deposited, a system of artificial warping was adopted between the higher level of the Brue and lower level at Glastonbury. By means of hatches, or small sluices, formed in the bank of the higher river, the winter floodwaters, 'thick' with sediment, were directed onto the meadows where they were retained for as long as it took for the sediment to be deposited.[116]

Such forms of husbandry called for a considerable understanding and control of the watercourses. In regulating water levels, account had to be taken of the need to minimise the risk of flooding when livestock were present, and conversely the importance of conserving drinking water, perhaps through the installation of temporary dams, or clyses, in the event of prolonged drought. Even if the manifold needs of local farmers could be met through a degree of unified water management, much depended on the course of events elsewhere in the river catchment. The volume of water flowing through the inefficient channel of the river Parrett to the sea was thought to have been altered in two ways during the mid-nineteenth century. First, a fresh cut was made from Middleney Lock to just below Langport, which had the effect of sending down 'freshes' in about a third of the time previously taken. Secondly, there was mounting concern as to how far the field-drainage and ditching schemes in the catchment above the Moors and Levels would bring about changes in the scale and timing of flooding in the lower reaches.[117]

One could never be sure that a species had become extinct – the distribution of some populations had always been extremely localised. In his *Flora of Bristol*, J. W. White recorded *Ranunculus lingua*, *Utricularia neglecta*, *U. minor*, *Polygonum minus*, *Sparganium minimum*, *Cladium mariscus* and *Carex teretriuscula* as each occurring in only a single enclosure.[118] There was, nevertheless, unmistakable evidence of the plants becoming much rarer as deeper drains were cut across the once 'prolific botanizing grounds'. For example, the drained pastures of the Gordano valley, between Portishead and Cleveden, had become much coarser and, at Walton in Gordano, the site of the rare sedge, *Cyperus longus*, had been ploughed up and planted with potatoes. Although many plants were to be found on the sides of the ditches and in the crop in the following few years, the fact that growth was so retarded, and flowering delayed by two months, suggested that the species would soon become extinct.[119]

---

[116] Clark, 'Bridgwater and other Levels'.
[117] Acland and Sturge, *Farming of Somerset*; Clark, 'Bridgwater and other Levels'.
[118] White, *Flora of Bristol*.   [119] J. W. White, *Flora of the Bristol Coal-field* (Bristol, 1887), pp. 213–14.

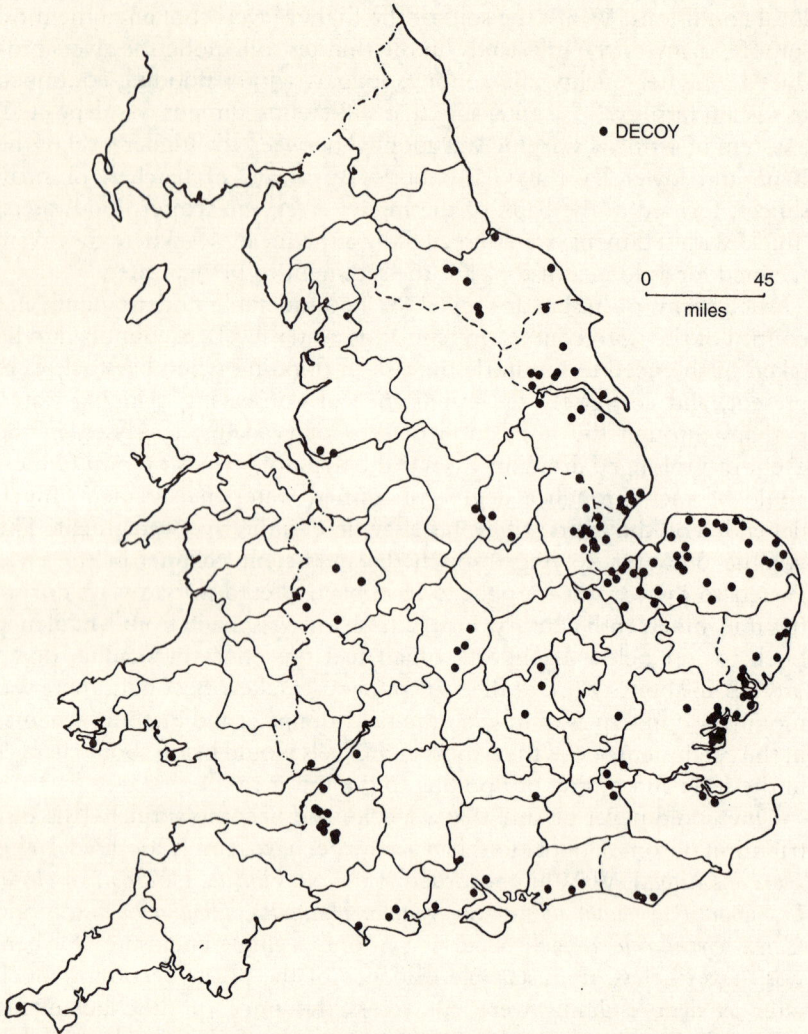

32.2 Distribution of duck decoys (after *The Book of Duck Decoys* by
R. Payne-Gallwey, published in 1886)

A distinctive feature of many of these wetlands had been the duck
decoy. 'A cunning and clever combination of water, nets and screens', it
was the only way to catch large numbers of duck in a short space of time.
Three decoys were constructed a quarter of a mile apart, in a straight line
across King's Sedgemoor. An average of 1,200 duck, teal and mallard was
netted each year between 1868 and 1882. One of the best-known decoys
was constructed in the 1830s at the foot of the Lias escarpment at Ashby

in Lincolnshire. Like many others, it had a roughly square lake, from each corner of which ran curving, narrowing channels into the encircling woodland. The channels or pipes, enclosed by netting, were in effect blind tunnels at the end of which, behind an arrangement of screens, the birds could be quietly slaughtered. The number of birds taken each year varied from 936 in 1865–6 to 6,357 in 1834–5, with an average of 2,800 birds in its first 35 seasons. Mallard and teal were the easiest to catch – widgeon had become very shy by the 1850s.[120]

The decades after 1850 saw the virtual demise of the decoy as a commercial venture. Not only did local conditions often become unsuitable, but competition from the Netherlands and other parts of the Continent, following the introduction of steamer services, caused prices to fall so low as to remove any incentive to carry out maintenance work or employ a decoyman. The two decoys, 5 miles to the east of Newport in Monmouth, were abandoned following the construction of the railway; the Great Western Railway paid £500 in compensation for the disturbance caused by the trains. By 1900, there were only six in Somerset. The one at Nayland, in the Axe valley, had been drained and converted to pasture. Even the decoy at Ashby in Lincolnshire, the 'most successful in the kingdom', became 'grown up and delapidated'. The only decoys to survive were those maintained as part of a game preserve – they made 'a far more agreeable feature in a park than a rookery'.[121]

E. THE EASTERN FENS

History is full of turning points, but the nineteenth century may have been particularly significant in the development of the fenlands of eastern England. Prior to that, the main effect of human activity had been to contribute to the variety of landscapes and wildlife. By the 1850s, however, changes in land use and management were taking place so quickly and over so large an area as to have much more of an 'impoverishing' effect.

One area to undergo such a degree of change was the Thorne Moors, Hatfield Chase and Isle of Axholme on the borders of Lincolnshire, Nottinghamshire and Yorkshire. It was because of the long and complex

---

[120] R. Payne-Gallwey, *The Book of Duck Decoys* (London, 1886), pp. 17–18; E. Harting, 'Wildfowl decoys in Somersetshire', *Devon. Rev.*, 5, 1886, pp. 218–24; D. A. E. Spalding, 'Notes on Ashby decoy', *Journ. Scunthorpe Museum Soc.*, 2, 1965, pp. 7–10; H. C. Folkard, *The Wild-fowler* (London, 1859), p. 68.

[121] T. Audas, 'Old wild duck decoys in Lincolnshire and the East Riding of Yorkshire', *Trans. Hull Scientific & Field Naturalists' Club*, 1, 1900, pp. 91–7; Payne-Gallwey, *Book of Duck Decoys*; M. Williams, *The Draining of the Somerset Levels* (Cambridge, 1970), p. 173; Folkard, *The Wild-Fowler*, pp. 40–1.

history of small-scale human activity that such species as those normally found on the wet sandy-heaths (of such localities as Scotton Common) had been able to gain a footing in the tracts of *Sphagnum* bog – the 'scurf' would otherwise have 'been far too overpowering'. Such incidents as the cutting of drains, dykes, turbaries, and 'wells' for trapping or shooting duck had created many open-water bodies, ideal for colonisation by plants introduced by duck, wader and other waterfowl.[122]

Whilst extensive areas continued to be covered by ling and cross-leaved heather and bracken, many species were undergoing a decline. The principal threat came from excavation and the practice of warping. A system of peat digging was developed, known locally as 'honey combing'. First, deep drains, with shallow tributary dykes, were excavated, and the turves were then cut to a depth of 3 to 5 feet – as far as the peat was brown and fibrous. Once the turves had begun to dry out, they were transferred from small to large stacks, whence they could be transported by small trucks, running along tramways, to a mill, where the peat was passed between toothed rollers and torn up. The coarser fragments were baled and wired for sale as stable or moss litter. By the turn of the century, several companies were engaged in the industry, affecting some 2,000 acres of moor.[123]

It was in the 1840s that the tremendous opportunities were recognised in the Crowle district for extending the natural warping that took place along the rivers Ouse and Trent to wherever the excavation of peat had reduced the surface of the raised bog. Woodruffe-Peacock anticipated that there would come a time when the entire surface of Thorne Waste would have been cut away and covered with alluvium suitable for cultivation.[124] Areas of some 200 acres were embanked at a time, and a connection was made with a warping drain, a straight channel leading to the estuary with powerful sluice gates at the entrance. At high tide, the sluice was opened, and the thick muddy estuarine water poured up the channel and flooded over the embanked area, where it was allowed to stand for three to four hours, until the tide had fallen and the water was able to flow back into the open river. It was important for the water to be introduced as rapidly as possible, and for the warp to dry out between tides. A paper-thin layer of fine sand and mud was left. Invariably, the strong spring tides deposited much more silt than the slower-moving neaps.[125]

[122] E. A. Woodruffe-Peacock, 'The ecology of Thorne Waste', *Naturalist, Hull*, 1921–2, pp. 301–4, 353–6, 381–4, 421–5.

[123] T. Bunker, 'The natural history of Goole Moor and the immediate vicinity', *Trans. Hull Scientific & Field Naturalists' Club*, 1, pp. 9–16.   [124] Woodruffe-Peacock, 'Ecology of Thorne Waste'.

[125] R. Creyke, 'Some account of the process of warping', *JRASE*, 5, 1844, pp. 398–405; T. J. Herepath, 'The improvement of land by warping, chemically considered', *JRASE*, 11, 1850, pp. 93–113.

During the process of warping, the 'sloppy' surface, which was drowned twice a day, became a favourite haunt for wildfowl, and carried a strange vegetation, in which plants of the 'slub land', such as the sea aster and Samphire, grew luxuriantly alongside heather and fern, still flourishing on little floating islands of peat that had broken away and kept rising and falling with the waters. Finally, when sufficient deposit had accumulated and the rain had washed out the salt, a crop of clover and rye-grass was sown and left down for two years while the ground became consolidated. When firm enough, and any irregularities in deposition removed, the newly formed field passed into general cultivation.[126]

In the East Anglian Fens, even those parts furthest from the sea in Cambridgeshire, Huntingdonshire and Lincolnshire were affected by large-scale drainage and reclamation schemes in the nineteenth century. Because of the vital role played by the numerous ditches in dividing the new fields and collecting the water, the truly aquatic species did not suffer as severely as others, but even these declined in number as steam pumping became so efficient as to leave the ditches dry in summer.

By the 1860s, Wicken Sedge Fen had become the only locality in the Cambridgeshire fenland where 'the ancient vegetation' survived, and the land was sufficiently wet 'to allow of its coming to perfection'.[127] The timing and frequency of the sedge (*Cladium mariscus*) harvest had a significant effect on the vigour and abundance of the plant. Cutting seems to have taken place between May and August on a three- to five-year cycle. The practice of feeding the litter, the poorer-quality material, to livestock may have led to a mixed sedge, or grass-dominated cover, in some parts. The mosaic of vegetation supported a flora and fauna of national repute.[128]

The drainage of Burwell Fen in the 1840s, and improvements to the drainage of Swaffham and Bottisham Fens in the 1850s, may have led directly to the 'discovery' of Wicken Sedge Fen by naturalists. G. S. Gibson noted in 1848 how, whilst Burwell Fen had been drained, there was 'a fen beyond still undrained', where *Liparis loeselii*, once so plentiful in Burwell Fen, might be found. An entomologist from Cambridge, F. Bond, may have been the first to collect systematically from the site which he called 'Wicken Fen' in 1850. The first ornithological records were made in 1851–2. The earliest botanical reference was that of Babington, who noted the presence of *Senecio paludosus*. Records began to accumulate quickly, following the publication of a list of 130 species, collected by Babington at Wicken Fen, in his *Flora of Cambridgeshire*.[129]

In time, the whole of the Sedge Fen was bought and donated to the

---

[126] Hall, *Pilgrimage of Farming*, pp. 109–14. [127] Babington, *Flora of Cambridgeshire*, pp. xvii–xviii.

[128] T. A. Rowell, 'Sedge in Cambridgeshire: its use and production since the seventeenth century', 34, 1986, pp. 140–8.

[129] T. A. Rowell, 'History and management of Wicken Fen', PhD thesis, Univ. of Cambridge, 1983.

newly founded National Trust as a means of preventing destructive forms of management and over-collecting. The series of sedge meadows, on which the wildlife depended, was not a natural swamp but 'the product of a certain kind of cultivation'. A regular cutting regime was essential in order to arrest the spread of alder, buckthorn and sallow, which were already beginning to establish themselves in some parts, strangling all other growth. Sedge was no longer required for thatching, and sales for forage did not cover the costs of clearance. It was at Wicken Fen that naturalist-owners gained their first insights into the practical difficulties of managing vegetation for the purposes of nature preservation.[130]

It was perhaps the drainage and reclamation of Whittlesea Mere in Huntingdonshire that most caught the Victorian imagination. By the mid-nineteenth century, it was possible to contemplate the drainage and reclamation of the last of the Fenland meres, including the largest, Whittlesea Mere. Once the new Marshland Cut was finished across the Middle Level in 1850, Whittlesea Mere was drained, leaving 'a vast mud bed'.[131]

Whittlesea Mere had been 'an awkward place to go to see', behind its 'phalanx of tall reeds near the margin of the water'. Nevertheless, enough was recorded to make it clear that the lake had been diminishing in size for some decades, and was little more than a thousand acres in size by 1850. Writers commented on how 'cultivation had been gradually creeping on for many years past', destroying the habitat of such plants as the great water dock (*Rumex hydrolapathum*), the foodplant of the large copper butterfly (*Lycaena dispar dispar*), the last specimens of which were taken in the 1840s. In an article in the *Phytologist* for 1851, the Reverend W. T. Bree described how the sites, from which he had gathered bog and aquatic plants some ten years previously, were now under crops of oats and barley.[132]

Although steam pumps were obviously far more reliable than wind-pumps in keeping the fenlands drained, there was considerable controversy as to the best kind of steam pump. The prime mover in the drainage of Whittlesea Mere, William Wells, had chosen to invest in a centrifugal pump, which had previously been used only in the Netherlands. A model was exhibited at the Great Exhibition in 1851. Amid great publicity, a

---

[130] W. G. Sheldon, 'Wicken Fen: its past, its present condition, and its future', *The Entomologist*, 49, 1916, pp. 1–4.

[131] W. Wells, 'The drainage of Whittlesea Mere', *JRASE*, 21, 1860, pp. 134–53; J. Sheail and T. C. E. Wells, 'The Fenlands of Huntingdonshire, England', in *Mires, Swamp, Bog, Fen and Moor. Regional Studies*, ed. A. J. P. Gore (Amsterdam, 1983), pp. 375–93.

[132] W. T. Bree, 'Recollections of a morning's ramble in the Whittlesea Fens', *Phytologist*, 4, 1851, pp. 98–103; E. Duffey, 'Economical studies on the large copper butterfly at Woodwalton Fen National Nature Reserve', *Journ. Applied Ecology* 5, 1968, pp. 69–96.

large party was assembled on the east side of Whittlesea Mere in October 1851 to see the formal opening of the pumphouse, with its 25-horsepower engine. The centrifugal wheel was 4 feet 6 inches in diameter and, after a few revolutions, 'the troubled water rose to the top of the sluice, and was hurled over the gauge-boards in a roaring torrent that fell with a discharge of 16,521 gallons of water per minute'. The advantage of the centrifugal pump over the scoop wheel was the comparative ease with which it could be adapted to combat the falling level of the land. Nevertheless, the pump had to be lowered in 1871, and an engine twice as powerful had to be installed in 1877 to lift the water over a height of 9 feet.[133]

A general lowering of the land surface had been expected as the water-filled peat dried and shrank, making it more vulnerable to bacterial and frost action. In high winds, the powdery mass was carried 'in great clouds like miniature copies of the sandstorms of the deserts'. So much loose and friable peat was washed or blown into the field drains that each had to be cleared at frequent intervals.[134] A cast-iron post, graduated in feet and inches, was somehow driven vertically into the peat and underlying clay at Holme Fen in 1848 as a datum from which to measure the lowering of the surface. By 1860, as much as 6 feet was exposed, and 8 feet 2 inches by 1875. The first farmhouses, built on piles driven into the spongy soil, had to be frequently repaired and rebuilt.[135]

The installation of steam pumping improved the drainage of about 5,000 acres, making possible a tenfold increase in the value of the land. About 200 acres were already under cultivation in 1850, and a further 1,500 acres, comprising the old mere bed and reed shoals, were under a regular system of cultivation by 1860. When S. B. J. Skertchley of the Geological Survey visited the area in 1870, he described in his notebooks how the white shell marl of the former Ugg Mere varied from 'a mere trace' to a metre thick, but was 'rapidly disappearing under cultivation'. A slight elevation marked the former shorelines of the meres, with alluvium extending beyond them marking the former limits of flooding.[136]

[133] C. Bede, 'Whittlesea Mere', *Illustrated London News*, 19, 1851, p. 613; R. L. Hills, *Machines, Mills and Uncountable Costly Necessities* (Norwich, 1967), pp. 75–126; E. B. Watts, 'Experience of the working of Appold's centrifugal pump in the drainage of Whittlesea Mere', *Journ. Bath and West of England Soc.*, 2, 1854, pp. 221–3; J. M. Heathcote, *Scoop Wheel and Centrifugal Pump* (London, 1877); Cambridgeshire RO, R60/19/1.

[134] Huntingdon RO, Conington MSS, Box 10, and Fielden MSS; S. B. J. Skertchley, *Geology of the Fenland* (London, 1878).

[135] J. N. Hutchinson, 'The record of peat wastage in the East Anglian Fenlands at Holme Post, 1848–1978 AD', *Journ. Ecol.*, 68, 1980, pp. 229–49; R. F. Grantham, 'The drainage of the fens', *Proc. Inst. Mechanical Engineers*, 11, 1913, pp. 777–85; C. Bede, 'Fen and mere', *Leisure Hour*, 26, 1877, pp. 296–300.

[136] British Geological Survey, Geology Museum, MS Surveyors' Notebooks, Skertchley, vol. 2, pp. 74, 80–1.

The far greater challenge of reclaiming the acid peatlands to the south and west of Whittlesea Mere was highlighted by such contracts as that drawn up between William Wells and the landowners, the West of England and South Wales Land Drainage and Inclosure Company, in 1856, for the drainage, warping and improvement of the newly drained land. The mortgages raised to finance such work, and the 'heavy tax' of the drainage rates, made it imperative that wheat, by far the most remunerative crop, was grown. The future fields were marked out, and the rank vegetation was pared from the ground. Very deep ploughing brought lumps of the 'sterile seam of red "moory" peat' to the surface. These were also raked into heaps and burned, until at length the band was broken up and the roots of future crops could reach 'the softer and richer soil beneath'.[137]

Wheat grown on pure peat was both thick skinned and easily blown down. The best remedy was to 'dry warp' or 'clay' the land. Not only did this overcome the deficiency in 'argillaceous and siliceous matter', but it reduced the likelihood of the soil being blown away. The Kimmeridge or Oxford Clay which underlies the peat was usually excavated from narrow trenches, cut at regular intervals in the fields to be treated; men were employed to dig the clay from depths of up to 10 feet. Because the clay deposits to the south of Whittlesea Mere were at an average depth of 15 feet, a portable railway of 2½ miles was used to import clay from further afield. Using 50 trucks and 10 horses, it was possible to spread a layer of clay of up to 4 inches over 1 acre of land in a week. The work was completed in 1866.[138]

In his account written for the *Illustrated London News*, the curate of the nearby parish of Glatton-with-Holme hailed the reclamation of the meres and peatland as a fine example of contemporary enterprise. The loss of a distinctive and irreplaceable scenery and wildlife would be small compared with the expected profits from cultivation.[139] By the end of the century, not only had the expected profits failed to materialise, but the enormity of the losses in distinctive plant and animal life had become clear. The 'once rich entomological hunting grounds' had been destroyed. The favourite locality for the large copper butterfly was 'a field of stinking cole-seed, with a flock of sheep eating it off'. Because earlier naturalists had ignored the commonplace, and recorded only what they perceived to be rare, it was impossible to make any exact inventory of the changes that had taken place, but clearly most of the characteristic insects of the district had disappeared, to be replaced by those of the higher

---

[137] W. Wells, 'On the treatment of the reclaimed bog-land of Whittlesea Mere', *JRASE*, 2nd ser., 6, 1870, pp. 203–8. [138] *Ibid.*, pp. 142–4.

[139] C. Bede, 'The drainage of Whittlesea Mere', *Illustrated London News*, 18 supplement, 1851, pp. 324, 326.

grounds to the north and west of the fenlands, as the foodplants of these species became more plentiful in the newly reclaimed fields.[140]

There is evidence that not all the fenland was transformed. A lepidopterist described Holme Fen in 1878 as 'to some extent still unaltered'. Herbarium specimens and references in the day-journals of the Marchioness of Huntly testify to the continued presence of a few refugia. On her visit to the 'fens below Yaxley' in May 1880, she wrote,

I went on . . . to the uncultivated ground in the vicinity of what was Whittlesea mere − explored for plants, of these many of the old ones remain although the land is gradually becoming dry from the continued drainage.

In 1886, the Marchioness found grass of Parnassus (*Parnassia palustris*) by the side of one of the ditches, 'a remnant of the old marsh times'. A visit of June 1888 resulted in the discovery of *Carduus pratensis* (*Cirsium dissectum*), which, she noted, 'I was not aware lingered still in the fens. It must be 40 years since I found it on the borders of Whittlesea Mere.'[141]

The fall in wheat prices in the 1870s and growing interest in game preservation may have saved a few tracts of fenland from the full impact of drainage and reclamation. Plantations and tracts of rough pasture can be distinguished on the large-scale Ordnance Survey maps of Holme Fen, surveyed in 1886. They covered an even greater area when the maps were revised in 1902. Managed as a game covert, large numbers of silver birch, bracken and bramble were planted. In 1905, the sum of £819 had to be paid to neighbouring farmers in compensation for damage caused by the pheasants and hares.[142]

The other traces of fenland to escape the full impact of reclamation was in Woodwalton Fen. The first large-scale Ordnance Survey Map of 1887 showed a dense and regular network of water-courses, five of which had been excavated as boat dykes for the purpose of transporting the large quantities of turf excavated as fuel for the brickyards and villages of the district, including the town of Ramsey.[143] In 1910, the site of 300 acres was purchased by a leading international banker and entomologist, N. C. Rothschild, ostensibly as a game covert, but primarily as a means of protecting its wildlife from overcollecting.[144]

Woodwalton Fen was the only British station of *Viola canina* spp. *montana* and *Luzula pallescens*. These and the other distinctive species

[140] J. R. Charnley, 'An extinct butterfly', *Field Naturalists' Qtly*, 1, pp. 294–9; S. H. Miller and S. B. J. Skertchley, *The Fenland* (London, 1878).

[141] Sheail and Wells, 'Fenlands of Huntingdonshire', pp. 325–6.

[142] Huntingdon RO, Valuation of Holme Wood Estate, 1905.

[143] S. Marshall, *Fenland Chronicle* (Cambridge, 1967).

[144] J. Sheail, *Nature in Trust: the History of Nature Conservation in Britain* (Glasgow, 1976), pp. 54–5, 173–4.

flourished on the most recently abandoned peat workings. As in the case of Wicken Sedge Fen, this site had also been previously overlooked, being several miles from the nearest railway station and a boat being needed to actually enter the fen. It was its 'discovery' by a local solicitor (and talented naturalist) about 1905 that forced Rothschild to intervene to protect the wildlife.[145] As the *Manchester Guardian* commented[146]

unfortunately nothing but absolute seclusion can save many birds, plants, insects from molestation. Given, however, a few of these reserves or 'nurseries', our animals and plants will increase and spread into the surrounding country, where the public will have the benefit of observing or even collecting them: so long as the reserves or preserves exist the supply will be kept up.

Such optimism was likely to be misplaced unless ways could be found of sustaining the drainage regime and management practices previously associated with the peat-workings.

[145] Sheail and Wells, 'Fenlands of Huntingdonshire', pp. 387–8.
[146] *Manchester Guardian*, 18 June 1910.

CHAPTER 33

THE UPLANDS

BY GORDON E. CHERRY AND JOHN SHEAIL

INTRODUCTION

The Annual Returns made to the Board of Agriculture excluded gardens and allotments and, at the other end of the scale, mountain, hill and heath land, despite the fact that the livestock which might graze such areas were included. It was reluctantly decided in 1871 that it would be impracticable for collecting officers 'to obtain satisfactory particulars' of the commons and unenclosed wastes.[1]

In default of such estimates, the only data available on a national basis were those derived from a return made by the Copyhold, Inclosure and Tithe Commissioners in response to an address by the House of Commons in 1873 for a survey of 'The waste land subject to rights of common'. Reference to the earlier Tithe Commutation Surveys, individual tithe agreements and awards, and subsequent enclosure awards indicated that the extent of common land in England and Wales was likely to exceed 2,368,000 acres (with a further 264,000 acres classified as common field land). Taking into account the height of the common lands, as indicated by the relevant one-inch Ordnance Survey maps, the Commissioners estimated that 43 per cent of the 1.7 million acres classified as common land in England, and 23 per cent of the 668,000 acres in Wales, were capable of cultivation. Such commons made up a quarter of the land surface of Westmorland and four counties in Wales – Breconshire, Merionethshire, Radnorshire and Montgomeryshire (Figure 33.1).[2]

A first attempt was made by the Board of Agriculture in 1891 to estimate 'that no inconsiderable surface of unenclosed mountain and heath land, which has not hitherto formed the subject of investigation'.[3] From 1892 onwards, statistics for 'mountain and heath land used for grazing'

[1] *Agricultural Returns of Great Britain*, BPP, 1871, LXXIX, C.460, p. 7.
[2] *Return of the Acreage of Waste Lands*, BPP, 1874, LII; M. Williams, 'The enclosure and reclamation of waste land in England and Wales in the eighteenth and nineteenth centuries', *TIBG*, 51, 1970, pp. 55–69.   [3] *Agricultural Returns of Great Britain*, BPP, 1890–1, XCI, C.6524, vii–viii.

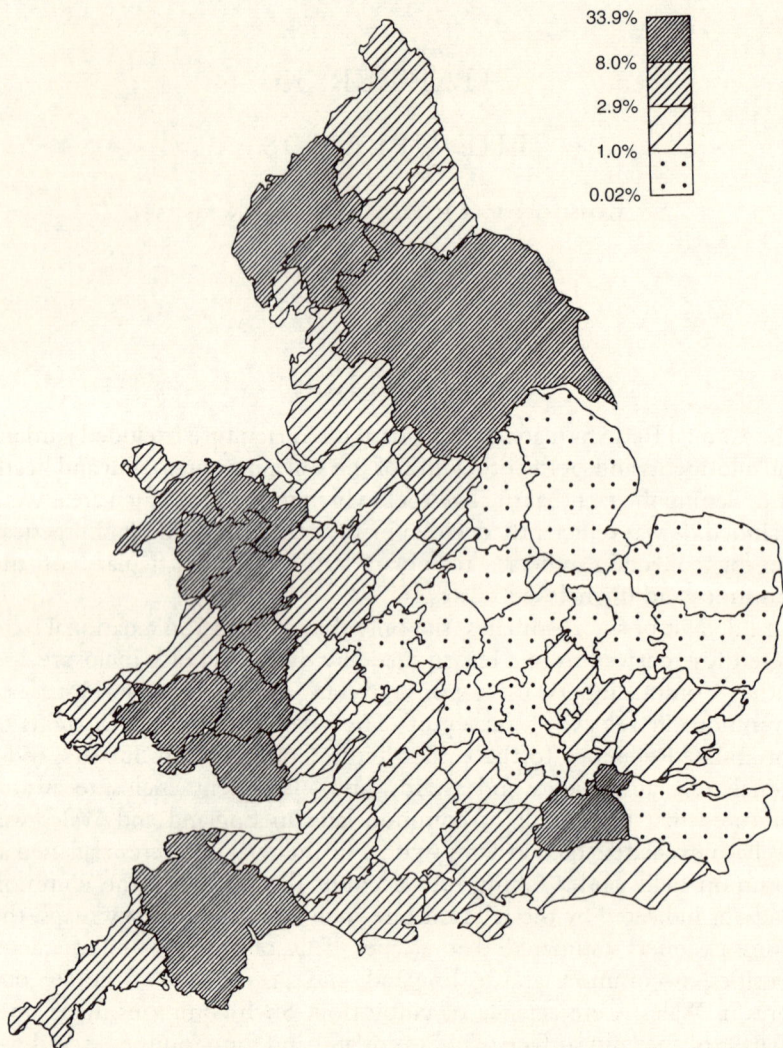

33.1 Extent of waste land subject to rights of common, as a proportion of the county area (after M. Williams, *Trans. Institute of British Geographers*, 51, 1970, p. 60).

were published annually. Of the 5 per cent of England classified in that way, two-thirds occurred in the four northern counties of Westmorland (37 per cent of the county's land area), Northumberland (35 per cent), the North Riding of Yorkshire (21 per cent), and Cumberland (20 per cent). In Wales, such land comprised one-fifth of the total area, making up 41 per cent of Breconshire, about one-third of Merionethshire and

Table 33.1. *The relationship of plant species to the temperature regions in the north of England*

Regions	Britain	North Yorkshire	Northumberland and Durham	Lake District
Agrarian				
infer	1,225			
mid	1,070	948	920	859
super	760	413	418	301
Arctic		126	108	125
infer	293			28
mid	244			
super	111			
Total of known species	1,425	992	935	893

Radnorshire, and a quarter of Cardiganshire and Caernarvonshire. By 1911, the proportion of land recorded as rough mountain and heath land had risen to about 8 per cent of the total area of England, and 28 per cent of Wales.[4]

A. THE UPLAND MOORS

For many naturalists, it was not enough to describe what they saw; they also wanted to explain the various patterns of distribution and abundance. John G. Baker gave his *Flora of North Yorkshire* the subtitle, 'Studies in its botany, geology, climate and physical geography'. He confessed to finding it extremely difficult to relate at all precisely the relationship of plants to different levels of heat and moisture, light and shade, and soil conditions, particularly over large tracts of land. Temperature seemed to be the most important factor, when expressed in terms of 'the sums of summer heat and the extreme minima of the colder parts of the year'. A plant was not, however, 'a mere machine, like a thermometer, but a living organism'. The way it reacted to temperature and other physical phenomena opened 'out a wide field for research and consideration'.[5]

A framework of reference for authors of county floras was provided by H. C. Watson's subdivision of the British Isles into two temperature regions, as determined by latitude and altitude. Despite its obvious limita-

[4] BPP, 1892, LXXXVIII, C.6597, pp. viii–ix & 1912–3, CVI, p. 6.
[5] J. G. Baker, *North Yorkshire. Studies of its Botany, Geology, Climate and Physical Geography* (London, 1863), pp. 179–93.

tions, an upper, or arctic, region was distinguished from a lower, or agrarian, region. The boundary between the two was taken to be the 1,800-feet contour, the upper limit of possible cultivation of grain. Within each region, three zones were distinguished (infer, mid and super), corresponding with a range of about 3° of mean annual temperature (Table 33.1). The number of species decreased rapidly from zone to zone, reflecting climate and the smaller extent of surface area available – whereas there were 920 species in the mid-agrarian zone of up to 900 feet in Northumberland and Durham, and 418 in the super-agrarian zone of up to 1,800 feet, there were only 108 on the two Cheviot peaks that rose distinctly into the infer-arctic zone, namely Cheviot itself and Hedgehope. The only parts of England within Watson's mid-arctic zone were the hilltops above 2,700 feet in the Lake District. Of the twenty-eight species found in this zone, the only characteristic plants were the least willow (*Salix herbacea*) and stiff sedge (*Carex rigida*).[6]

The total population of the towns and villages near the edge of the upland moors might number millions, but the moorlands themselves remained a *terra incognita*. An occasional drive along a moorland road, or a visit by tramcar to some moor-edge, was as far as the bulk of people ever ventured. The local naturalist was hardly more adventurous – organising a hurried visit to a clough-head, or a sharp walk in the fading light of the evening across a small heather-clad piece of ground.[7] Such a partial knowledge encouraged the misleading notion that the moors were practically overgrown with heather. As the pioneer studies of Charles Moss and other members of the British Vegetation Committee were to reveal at the turn of the century, this was far from being the case.

In one of his earliest papers, Charles Moss drew attention to the importance of the cotton-grass moorland, and the extent to which the distribution of this and other types of vegetation reflected physiographical factors and differences in the history of land use and management. In the course of a systematic survey of his home area, the moors of south-west Yorkshire, Moss distinguished seven type areas (Figure 33.2). The bilberry (*Vaccinium myrtillus*) was dominant in the plant associations of Boulsworth Hill and other parts where the sandstone rose to a distinct peak.[8] The great bulk of the moorland was, however, covered by cotton grass (*Eriophorum vaginatum* and *E. angustifolium*) (Rishworth type). Forming a more or less complete ring round the cotton-grass moor was a belt dominated by heather or, where the moor edge was steep and shaley, rough grass (Rumbles types). Midgley Moor was an

---

[6] J. G. Baker and G. R. Tate, *A New Flora of Northumberland and Durham* (Newcastle, 1868), pp. 48–9, 59–60; J. G. Baker, *A Flora of the English Lake District* (London, 1885), pp. 3–6.

[7] C. E. Moss, 'Moors of south-west Yorkshire', *Halifax Naturalist*, 7, 1902–3, p. 88.

[8] *Ibid.*, pp. 88–94.

MOORLAND ASSOCIATIONS

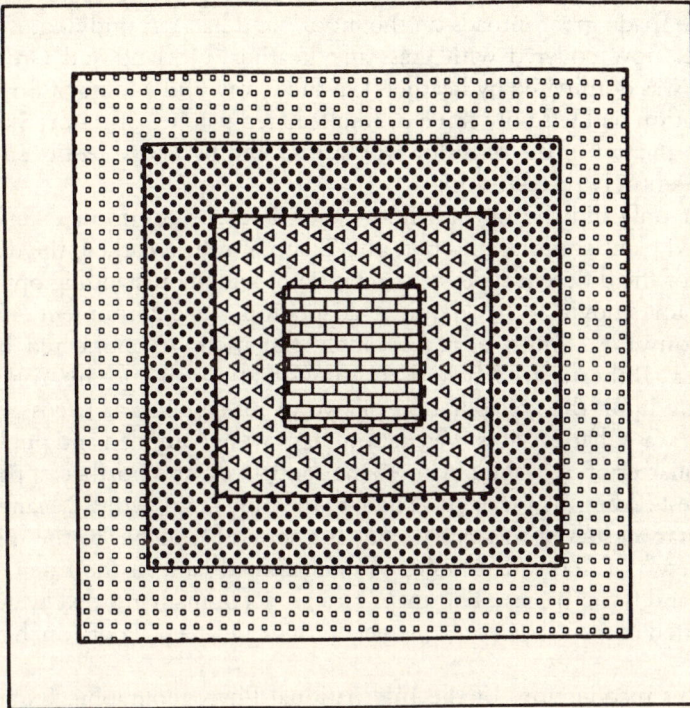

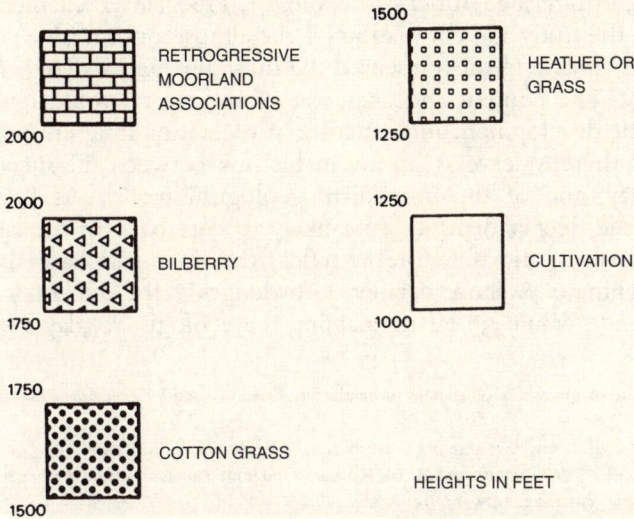

33.2 Diagram of the moorland associations encountered on a flat-topped
eminence reaching an altitude of 2,000 feet (610 metres) in the Peak District
(after C. E. Moss, *Vegetation of the Peak District*, 1913, p. 195.)

example of a moor that extended 'long strips or tongues far into the main valleys'. Moss explained how 'in times past the spade and the plough made great inroads on the moor', and had left outlying, isolated patches now covered with grass and heather. Norland and Greetland Moor was dominated by heather. On the Crow Hill Moors of Sowerby, cultivation had left only three or four isolated patches of moor: Skircoat Moor, the last in the series of type-areas cited by Moss, had been converted entirely to cultivation.

Not only did the upland moors soon become the most extensively mapped plant communities but, as William Smith remarked, the distribution of the different types of vegetation provided outstanding opportunities for studying 'the exquisite adaptations to environment'.[9] From 1898 onwards, Smith began to apply the methods pioneered by his brother, Robert, in Scotland, to the situations found in Yorkshire, drawing upon the collaboration of two of his students at Leeds, one of whom was Charles Moss. Vegetation maps were compiled and published for a quarter of Yorkshire, namely the south-west and north-east parts of the West Riding, at a scale of ½ inch to the mile (1:26,720).[10] Meanwhile, a vegetation survey was carried out at a scale of 6 inches to the mile by F. J. Lewis of 560 square miles of uncultivated land in the Eden, Tees, Wear and Tyne basins. The maps and an accompanying text were also published in the *Geographical Journal*, in this case at a scale of 1 inch to the mile.[11]

In his recollections of the International Phytogeographic Excursion of 1911, the distinguished American ecologist, Frederic E. Clements, wrote of how the study and correlation of 'the almost innumerable peat sections' in the upland moors presented the most alluring challenge for British botany. The content and character of the layers would reveal much about the development and structure of vegetation long since disappeared, and thereby serve as 'an invaluable link between the successions of to-day, and of the immediate geological past'.[12] As Lewis pointed out, the deeper deposits were likely to date back to the later phases of glaciation, and therefore to reflect the shifts that had taken place in the climate. A more detailed knowledge of the dynamics of moorland change would go far in enabling the ecologist to play a key

[9] W. G. Smith, 'The origin and development of heather moorland', *Scottish Geogr. Mag.*, 18, 1902, pp. 587–97.

[10] W. G. Smith and C. E. Moss, 'Geographical Distribution of Vegetation in Yorkshire', *Geog. Journ.*, 21, 1903, pp. 375–401; W. G. Smith and W. M. Rankin, 'Geographical distribution of vegetation in Yorkshire', *Geog. Journ.*, 22, 1903, pp. 149–78.

[11] F. J. Lewis, 'Geographical distribution of vegetation of the basins of the Rivers Eden, Tees, Wear and Tyne', *Geog. Journ.*, 23, 1904, pp. 313–31, and 24, 1904, pp. 267–85.

[12] F. E. Clements, 'Some impressions and reflections', *New Phytol.*, 11, 1912, pp. 177–9.

role in not only the scientific, but also the economic and practical, aspects of moorland management.[13]

Perhaps the most exciting dimension of these early studies was the discovery of the remains of fully grown trees in some of the peat deposits. As Moss concluded, in a paper read to the meeting of the British Association in 1903 on the age, origin and utilisation of the peat moors of the Pennines, the moorland vegetation seemed to have originated 'in morasses formed probably by the destruction of primitive Pennine woods, which in their prime were not only much more extensive, but ascended to a much higher elevation than their meagre remains'. Knowledge that such areas were once forested might be of relevance in identifying how the grassy and heathy moor-edges might again be planted with trees or turned over to farmland. Even where the moors were to remain shooting grounds, their value for grouse might be enhanced through the better cultivation of the heather and bilberry.[14]

The most comprehensive, yet detailed, survey was carried out by Moss in the Peak District from 1903 onwards. Publication of the maps and accompanying monograph was long delayed because of the refusal of the Treasury to sanction the Board of Agriculture meeting the costs of printing the coloured maps.[15] Publication was eventually made possible by a grant from the Royal Geographical Society. Six broad categories of vegetation were identified on the sandstones and shales – the land under cultivation, siliceous grasslands, heather moor, bilberry moor, cotton-grass, and retrogressive moor (Figure 33.2). There was in general a marked zonation of the different associations, reflecting for the most part the effects of altitude and, at a more local scale, the influence of physiographical factors. The heather moors tended to be found in the drier, more sandy, shallower and less elevated regions, whereas the cotton-grass moors dominated the wetter, purer and deeper peats at higher elevations. Often the only vascular plants present, the dead green hue of the shoots of the cotton-grass in late summer and early autumn turned to a dull red in late autumn and winter. Only in June was the moor attractive to the eye, when the pure white fruits appeared like suspended snowflakes.[16]

Whilst there was evidence of a continued build-up of the peat deposits under the closed associations of cotton-grass, and to a lesser extent the heather and bilberry associations, the peat of the most elevated portions was in a retrogressive phase. The process appeared to begin with the

[13] F. J. Lewis, 'The sequence of plant remains in the British peat mosses', *Science Progress*, 2, 1907, pp. 307–25.

[14] C. E. Moss, 'Peat moors of the Pennines; the age, origin and utilization', *Geog. Journ.*, 23, 1904, pp. 660–71.

[15] J. Sheail, *Seventy-five Years in Ecology: the British Ecological Society* (Oxford, 1987), pp. 25–6.

[16] C. E. Moss, *Vegetation of the Peak District* (Cambridge, 1913).

cutting back of streams at their sources. Moss recounted how those streams shown on the revised Ordnance Survey maps of 1870–80 seemed to be three-quarters of a mile longer than those recorded when the Peak was originally surveyed in 1830, and were by the early 1900s a quarter of a mile longer than depicted on the revised maps of the 1870s. Every storm removed further quantities of peat from the bare banks. Over time, both the streams and their tributaries became broader and deeper, and further watercourses evolved, so that eventually even the detached masses of peat, known locally as 'peat hags', shrank in size and disappeared.

B. THE HEATHER MOORS

The most detailed study of upland heather communities was carried out in the moors of north-east Yorkshire. Having first made a study of their insect life and then geology, it was in 1905–6 that Frank Elgee, an assistant curator at Middlesbrough Museum, began to investigate the 'moors' in their totality, taking account of the flora, fauna, geology, and even works of man. If sound conclusions were to be reached on any aspect, the moors had to be regarded as 'a unique assemblage of factors of intense interest, which owe their present status to innumerable causes that have been operating for ages'. From the course of the roads, position of entrenchments, absence of moorland villages, and the word 'moor' itself, Elgee believed the high moors had been clear of trees from early times, perhaps 2,000 to 3,000 years ago.[17]

In a sense, the identification and classification of the different kinds of moor on the basis of their interdependence with all the other aspects of the moorland did little more than to extend and formalise the way in which the native dalesmen perceived the moors, using such expressive terms as the 'Fat Moors' and 'Thin Moors', the 'Mosses' and 'Swangs'. The Fat Moors embraced several types of vegetation flourishing on the blackish-brown peats from 1 to 4 feet in thickness, and more or less damp even in drier weather. Four distinct associations were identified by Elgee, namely purple heather moor; heather and bilberry moor; heather, flying bent, cotton-grass and common rush moor; and, fourthly, heather, flying bent, common rush and sweet gale moor. The Thin Moor was more localised and found on the edge of the moors. Seven types of vegetation could be distinguished. Whilst the extremes of each were easily recognised, the combinations of different heathers, grasses, club rush and gorse merged into one another 'as slight changes of conditions first favour one species and then another'.[18]

[17] F. Elgee, *The Moorlands of North-eastern Yorkshire. Their Natural History and Origin* (London, 1912), pp. 10–11; F. Elgee, 'The vegetation of the eastern moorlands of Yorkshire', *Journ. Ecology*, 2, 1914, pp. 1–18.   [18] Elgee, *Moorlands of N. E. Yorks.*, p. 38.

The extent and character of the heather reflected both the intensity of grazing and more direct forms of management. Shepherds had tradition-ally burned wide tracts of up to a tenth of the moorland each year as a means of encouraging young, more palatable growth. The practice con-tinued; Elgee remarked on how, in spring, small fires could be observed sending their columns of smoke high above the moorland of north-east Yorkshire.[19] A game lease for part of upper Teesdale in 1898 required the tenant to 'judiciously and in a workmanlike manner burn such quantities of heather and ling as may be considered necessary'. Another lease, for an estate south of Alwinton in Northumberland, limited the area of 'heath, bent or grass' to be burnt in any one year to one sixth part of the whole, with the landlord's agent being given every opportunity 'to attend and superintend such burning and fix upon what parts shall be burnt'.[20]

From the 1850s onwards, grouse-shooting became popular, encour-aged by the extension of the railways, improvements to the sporting gun and, above all, fashion. Moors were taken in hand, or leased to shooting tenants. Lodges, approach roads and other conveniences were provided. It was not long, however, before there were complaints of an increase in disease among the grouse, and a major epidemic in 1872–3. In a paper to the British Association in 1867, the ornithologist, H. B. Tristram, attrib-uted much of the blame to 'the indiscriminate slaughter of predatory animals'. Birds of prey were 'the sanitary police of nature, and that if they had existed in their original strength they would have stamped out the grouse disease'.[21]

A Committee of Inquiry was appointed by the President of the Board of Agriculture in 1905. Despite the considerable interest and investment in grouse over the previous half-century, a systematic study had never previously been made of the life-history, food, physiology and diseases of the species. Having examined nearly 2,000 cases of death from other than natural causes and 200 separate outbreaks of the Grouse Disease, and carried out experiments whereby healthy birds were artificially infected, the Committee became convinced that adult deaths were caused by the threadworm, *Trichostrongulus pergracilis*. The two factors common to all outbreaks were the large numbers of adult strongyle found in the caecum of the grouse, and the low resistance of the birds.[22]

---

[19] *Ibid.*, pp. 33–4.

[20] J. M. Britton, 'Farm, field and fell in Upper Teesdale, 1600–1900: a study in historical geogra-phy', MA thesis, Univ. of Durham, 1974, p. 128; Northumberland RO, ZSA 26/92. 3; Elgee, *Moorlands of N. E. Yorks.*, pp. 33–4.

[21] British Association for the Advancement of Science, *Report, Dundee Meeting*, 1867, p. 97, and *Report, Exeter Meeting*, 1869, pp. 91–6.

[22] Committee of Inquiry into Grouse Disease, *Report to the President of the Board of Agriculture and Fisheries*, BPP, 1911, XXVI, Cd 5871; Committee of Inquiry on Grouse Disease, *The Grouse in Health and in Disease* (London, 1911), 2 vols.

The incidence of Grouse Disease depended on the relationship between the degree of infection and the birds' powers of resistance, which in turn depended to a large extent on the level of food supplies in early spring – a point usually overlooked by proprietors who often only visited their moors in summer. To secure as high a carrying capacity as possible, it was essential that burning took place on a rotational basis, so that all the heather was less than twenty years old. Where this occurred, the new heather would spring from the roots in the same year as burning took place.[23] By arresting the vegetation at the same point in the succession, a form of monoculture would be introduced of greater extent and purity than that seen in even the most intensely arable areas of lowland England.

C. ENCLOSURE

As the Agricultural Census for 1887 remarked, it seemed at first glance anomalous that the area under cultivation should rise each year at a time of 'undoubtedly intensified depression in agriculture'. The area of England and Wales under crops and grass rose by 1,127,000 acres between 1871 and 1881, and by a further 552,000 acres to reach a peak of 28,001,000 acres in 1891. In Cumberland, Westmorland and Northumberland, and in some Welsh counties, significant amounts of land, previously regarded as unenclosed mountain and heath land, were classified as permanent pasture.[24]

Whilst the increase might in part be 'due to more correct and exhaustive enumeration', it was also clear that 'a real and material extension of the margin of cultivation' was taking place, 'by the reclamation and fencing in of land previously more or less completely waste'. Many landlords in the early 1880s, believing that the depression in prices was only temporary, had continued to reclaim tracts of the upland edge.[25]

Many upland tracts were neither natural in their appearance, nor used and managed in the most efficient way possible. For example Holystone Common in Alwinton, Northumberland, as described in a report of 1877, may have been representative of many in as much as it bore the evidence all too clearly of how 'every man has done what has seemed right in his own eyes'. The common of 860 acres had a north-easterly exposure and attained a considerable elevation at the west end. It had in practice been appropriated by five or six cottagers, who had no 'infield' of their own. They purchased whatever stock they could afford, the animals

---

[23] Elgee, *Moorlands of N. E. Yorks.*, pp. 46, 104.
[24] *Agricultural Returns of Great Britain*, BPP, 1887, LXXXVIII, C.5187, p. 3; BPP 1880–1, XCI, pp. vii–viii. [25] BPP, 1890, LXXIX, C.6143, p. viii and 1887, LXXXVIII, C.5187, pp. 3–4.

remaining on the moor the whole year round and often starving in winter. The ground was largely covered by heather and stones. Freestone had been quarried at different times from all parts of the common. Although turf was no longer cut for fuel, the remains of the diggings could still be traced. Everyone pulled heather and cut bushes as they pleased.[26]

Rights of common might be abolished in one of three ways – by their coming into the hands of the owner of the soil and so ceasing to exist; by agreement between the owner and the commoners; and, thirdly, by special act of parliament. Many years might elapse before fences were erected between the various allotments of land, and, even when erected, there was often little, if any, intention to embark on large-scale reclamation.[27] In north Wales, for example, the acts were perceived mainly as an effective device for creating and redistributing proprietory claims. The Royal Commission on Land in Wales, which sat between 1893 and 1896,[28] remarked on how it would be

idle to suppose that the main motive of the Welsh landowners who eagerly used the facilities given by Parliament was to extend the margin of cultivation. They saw clearly enough that the movement gave them the opportunity of acquiring the sheep-walks and pasture lands till then unenclosed as their own in severalty.

Even so, the new owners were bound to recognise the practical benefits to be derived from the new walls, not least in providing additional shelter for livestock and making it easier to regulate numbers, and encouraging the improvement of the breed and general health of the stock.[29]

Whatever the motives and resources available to the new owners, there was often little immediate change in the appearance of the country. When the Royal Forest of Exmoor was disafforested and enclosed in 1815, an area of 12 acres was set aside for a church and churchyard, should they be needed.[30] The remaining 10,250 acres of the Crown allotment were purchased in 1820, and became part of the estate of the Knight family, extending over four-fifths of the old Forest. Despite the vast sums invested in draining, fencing, building and cultivation, little change was effected on the wildest parts of the moor, other than the erection of fences. Having rejected an earlier approach, the Treasury agreed in 1853 that the increase of inhabitants on the various holdings established since

[26] Northumberland RO, ZAN Bell, 59/8.

[27] A. H. Dodd, *The Industrial Revolution in North Wales* (Cardiff, 1951), pp. 84–8.

[28] *Report of the Royal Commission on Land in Wales and Monmouthshire*, BPP, 1896, xxxiv, C.8221, p. 214.

[29] D. W. Howell, 'Welsh agriculture, 1815–1914', PhD thesis, Univ. of London, 1970, pp. 57–8.

[30] An Act for vesting in His Majesty certain parts of the Forest of Exmoor, 1815, 55 George III, c. 138.

1846, and the mining works in progress, warranted the erection of a church accommodating 109 persons. The new parish came into being three years later, some forty years after the original act of inclosure.[31]

It would be wrong to suggest that proposals to extend and improve the cultivation of the uplands arose from ignorance as to the conditions likely to be encountered. In his prize essay on the cultivation of Dartmoor, the estate agent, Henry Tanner, was highly conscious of how Nature held an unrelenting sway over the moor, unchecked by the hand of man. Everything bore the impress of grandeur, solitude and duration – the granite tors, capping the hills, had changed little since light first shone on them, and a search was likely to reveal the remains of the very first people to inhabit the land. But equally, a close knowledge of the moor revealed the potential for extending the limits of farming. The rain falling on the central parts of the moor was much less than that on the south and west sides. Wet enough to support abundant vegetation, there was sufficient dry weather for agricultural purposes. On the premise that the average temperature fell by 1° Fahrenheit for every 270 feet of ascent, Tanner reasoned that the temperatures at 1,300 feet on the moor were comparable with those of land at 120 feet above sea level in County Durham, some 4° of latitude north of Dartmoor – but with the vital difference that Dartmoor was on a promontary of land over which 'the warm sea breezes of Southern latitudes pass with freedom'.[32]

There seemed at least some justification for believing that the very acts of good land management would help to ameliorate conditions. The climate of Dartmoor was thought to be influenced not only by elevation but also by defective drainage. The atmosphere resting upon the large quantities of surface water became charged with vapour and, through evaporation, large quantities of heat were abstracted from the air. If the land was drained, Tanner speculated, the percolation of rainwater would carry air and warmth into the soil – the former preparing food for vegetable growth, and the latter providing the energy by which plants assimilated the nutrients thus made available. Drainage was likely to be improved both through making surface cuts and by breaking up the 'moor-pan', an impervious, compact layer of 2 to 5 inches in thickness, formed where iron in the surface soils had been dissolved and redeposited. Through a study of the 'natural habits' of the crops planted, and the extent to which they would flourish under the local pecularities of soil and climate, about 10,000 acres of Dartmoor might be used as arable land, and a further 50,000 acres adapted for 'the luxuriant growth of grass'. The optimal size of the enclosures was from 6 to 12 acres, being

[31] MacDermot, *History of Forest of Exmoor*, pp. 438–40; C. S. Orwin, *The Reclamation of Exmoor Forest* (Oxford, 1929).   [32] H. Tanner, *The Cultivation of Dartmoor. A Prize Essay* (London, 1854).

smaller as the land became more elevated or had a northerly or easterly aspect.

Whilst the success of convicts in turning the heath and furze around the prison at Princetown on Dartmoor into luxuriant crops proved that such schemes were technically feasible, it was far from clear whether they would on balance bring commensurate benefits to individual landowners or indeed the nation at large. What was praiseworthy patriotism in the extreme case of saving communities confronted with starvation might become reckless speculation in other circumstances. There was never any doubt that the only possible use of the Crown allotment on Exmoor Forest was stock farming – the main question was how to produce sufficient keep for the higher numbers of sheep and cattle throughout the year at an economic cost. The 1850s and 1860s were the most active and important in the history of Exmoor cultivation, the gross return was considerable, but the net income was practically nil.[33]

In his study of the reclamation of Exmoor Forest, Orwin wrote of how commentators, in the earlier period, tended to focus on promise rather than fulfilment, being perhaps unduly optimistic in their outlook.[34] More modern writers had gone to the other extreme, failing to see the real achievement of what had been accomplished through the 'haze of costly experiment and misdirected effort'. For his part, the agricultural journalist, Samuel Sidney, highlighted both the setbacks of the earlier attempts of the Knight family to reclaim the moor, and the extremely encouraging results of an experiment conducted by the new agent in the late 1860s to substitute rape for the more expensive turnip crops, with their preliminary cultivations and aftercare. Drawing on the experience of previous agents and his own as a tenant reclaiming parts of Challacombe Common, the agent arranged for the wet peat land to be pared, burned, ploughed and sown with rape seed, with 3 tons of lime applied to each acre. By the time a further three to four crops had been sown and each eaten down by sheep, the combined effects of sheep-treading and penetration by tap-roots were judged to have 'decomposed the peat almost or quite down to the pan, which was broken up by the subsoil plough'. After a crop of rape and grass seeds, the land was laid down to permanent pasture. Timothy grass (*Phleum pratense*), Yorkshire fog (*Holcus lanatus*) and cock's foot (*Dactylis glomerata*), with rye grass and perennial clovers, were thought best suited for the improved peat.[35]

Even those who conceded the urgent need for more water for the towns and cities were often concerned at the impact of the 'sweeping powers' sought by the water undertakers on the catchment areas. In order

[33] Orwin, *Exmoor Forest*, pp. 85–6.   [34] Orwin, *Exmoor Forest*, p. x.
[35] S. Sidney, 'Exmoor Reclamation', *JRASE*, 2nd ser., 14, 1878, pp. 72–97.

to safeguard the purity of the water from its new gathering-grounds on the upper Wye, the Birmingham Corporation insisted on purchasing, by compulsion if necessary, the freehold and manorial rights of the catchment of 45,560 acres. Critics attacked the Bill to secure these powers as 'a gigantic Enclosure Bill' that provided no safeguards for the commoners or general public. About thirty small farmers depended for their livelihood on renting the common rights that extended over 32,000 acres. A spokesman for the Corporation argued that any local distress was a small price to pay for protecting the health of half a million people in the Birmingham area. During the Parliamentary Committees' hearings on the Bill in 1892, however, legal counsel for the Commons Preservation Society challenged the need to interfere with the traditional forms of upland management. Under considerable pressure, the Corporation agreed reluctantly to substitute a series of by-laws to protect the purity of the water, which would be subject to the approval of the Board of Agriculture. The existing freeholders would be granted 999-year leases, and existing tenants would be given 21-year leases on the same terms as hitherto.[36]

On most gathering grounds, it was the practice to withdraw all livestock so as to eliminate any chance of the surface run-off being polluted, or purification plant being needed. Unless prohibited by altitude or soils, the only alternative to the land becoming derelict was afforestation. Samuel Margerison, a leading timber merchant and well-known Yorkshire botanist, acted as expert adviser to the Leeds Corporation Waterworks Committee in pioneering the planting of trees around the Fewston and Swinsty reservoirs in the Washburn valley, near Otley.[37] Margerison argued that such undertakers, through their permanency as 'non-dying Corporations', were extremely well placed to reap the profits from the timber famine which experts confidently expected to occur within 'almost measurable distance'. The principal difficulty was the concern expressed by most water engineers as to whether the coniferous forests might reduce the runoff from the catchments. In his evidence to the Royal Commission on Afforestation in 1902, and articles in scientific and agricultural journals, Margerison sought to allay such fears, arguing to the contrary that evaporation would be reduced and the tree roots would make it easier for rainwater to penetrate and filter through the ground.[38]

In many parts, mining and agriculture were interdependent, the

---

[36] House of Lords RO, Commons Select Committee, 1892, minutes of evidence; Birmingham Corporation Act, 1892, 55 and 56 Victoria, c. clxxviii.

[37] HEW, 'In memoriam: Samuel Margerison', *Naturalist, Hull*, 1917, pp. 235–6.

[38] S. Margerison, 'The afforestation of waterworks catchment areas', *Trans. Royal English Aboriculturalists' Soc.*, 6, 1906, pp. 276–84.

expansion of one often acting as the stimulus to the other.[39] The same large-scale Ordnance Survey maps that help to identify the incidence of industrial activity, also show, through the use of rough grazing and shrub symbols, the expansion and perhaps later retreat of improved farming. The thin seams of coal that occurred under the North York Moors and Yorkshire Dales were exploited as fuel, primarily for burning the lime upon which so many schemes for land improvement depended. The fact that the largest group of pits, at Rudland in the North York Moors, was set out in rows of twenty-five to twenty-eight pits suggests that some kind of overall control was exerted.[40] Lead mining in upper Teesdale gave rise not so much to separate smallholdings, but rather to a close and informal interaction between the mines and the existing well-established farms which, by the late nineteenth century and the demise of the industry, had enclosed the greater part of the dale from Middleton right up to Harwood, and onto the fells of Middleton and Newbiggin Common.[41]

Not only were many of the so-called agricultural improvement schemes and industrial ventures visually striking, but their promoters were often accomplished publicists. Accordingly, it was easy for opponents to portray a picture of conflict between two opposing forces in which, as usual, the aggressor was fast becoming the conqueror – threatening to 'civilise' such areas as Dartmoor off the face of the earth. It was little consolation to know that the costs of enclosure frequently outstripped any profits that accrued. On Dartmoor, it was melancholy to witness the irreparable mischief caused by abortive attempts to reclaim the bleak hill-sides,[42] where

after a Rock-pillar has been demolished for a gate-post, and a Cromlech overthrown for a foot-bridge, or a Kistvaen destroyed for a New-take wall, the injudicious effort has been abandoned as hopeless.

In bringing such a 'serious and alarming question' before members of the Devonshire Association, a speaker in 1876 emphasised how the object was not to discuss the mysteries of land tenure, or to alarm the worshippers of vested interests, but rather to indicate how glorious an inheritance it was to possess any portion of wild and beautiful Dartmoor, and how great a desecration (and dead loss of money) it would be to enclose an acre of it.[43] A Dartmoor Preservation Association was founded in 1883,

---

[39] E. J. T. Collins, *The Economy of Upland Britain, 1750–1950: an Illustrated Overview*, Centre for Agricultural Strategy, paper 4 (Reading, 1978), p. 19.

[40] J. McDonnell, *A History of Helmsley, Rievaulx and District* (York, 1963), pp. 459–60.

[41] Britton, 'Farm, field and fell', pp. 133–45; B. K. Roberts, 'Man and land in Upper Teesdale', in *Upper Teesdale: the Area and its Natural Beauty*, ed. A. R. Clapham (London, 1978), pp. 141–59.

[42] S. Rowe, *A Perambulation of the Antient and Royal Forest of Dartmoor* (Plymouth, London, edn, 1856), pp. 11–12.   [43] W. F. Collier, 'Dartmoor', *Rep. & Trans. Devon. Assoc.*, 8, 1876, pp. 370–9.

the objects of which were to prevent and remove recent enclosures of open ground made by local landowners in contravention of the rights of Venville owners and commoners.

Such changes as did occur had to be seen in perspective. The granite quarries may have spoiled the appearance of the area around Princetown and Walkhampton Common, but, as William Crossing pointed out in his compilation, *Amid Devonia's Alps*, Dartmoor was 'a big place, and we must not grudge to enterprise such a comparatively small portion of surface as such an undertaking needs'. Beyond the intakes, there was nothing but 'wild moor, untouched by any attempts at cultivation, extending for miles in every direction'.[44] Livestock raising remained the predominant enterprise – the newtakes acting as enclosures for the cattle and sheep in winter and exclosures perhaps for growing hay in summer. Even at the high-water mark of farming on Exmoor Forest, there remained some 8,000 acres of open land utilised as a pony stud, and for the summer grazing of farmstock.

D. ACCESS TO THE MOUNTAINS

It was because the uplands were perceived to be so remote and untouched by human endeavour that they came to attract large numbers of visitors. The qualities were considered so precious that the future development of the uplands became a matter of more than local concern. Not surprisingly, much of the debate was focused on the impact of the revolution of communications. By the turn of the century, the case had been argued for building a road over every major pass in the Lake District. The longest-running debate centred on the Sty Head Pass, which would have reduced the travelling distance between Keswick and Wastwater from 56 to 17 miles, via Borrowdale. On Dartmoor, the 'pure invigorating air and solemn grandeur' were expected to remain, whatever happened, but a guidebook warned the tourist to hurry if he wanted to study 'the primitive manners and customs of the people' before the Highway Boards and School Boards had completed their work of assimilation with the rest of the country. Although still for the most part steep and narrow, the roads had nevertheless been vastly improved to accommodate wheeled transport.[45]

Even in the remotest parts, the original stimulus to road improvements can often be traced to the opening of a railway line. Both railway companies and local communities were alive to the potential benefits of the lines for agriculture, quarrying and mineral development, and tourism.

[44] B. L. Messurier (ed.), *Crossing' Amid Devonia's Alps* (Newton Abbot, 1974), p. 24.
[45] R. Dymond (ed.), *Widecombe in the Moor* (Torquay, 1876), p. 3.

Established industries might take on a new lease of life. Following the opening of the Settle to Carlisle railway, two major limeworks were opened at Langcliffe, to the north of Settle, in 1872–3, one with a traditionally designed set of three draw kilns, and the other with an even more massive Hoffman-type kiln. Large quantities of quicklime were sent to Bradford and Sheffield.[46] In anticipation of the completion of the railway through Wensleydale, linking the Settle to Carlisle line in the west with the North-eastern Railway beyond Leyburn in the east, a market for dairy produce and pens for 10,000 sheep were constructed at Hawes in time for the opening of the line in 1878. Large quantities of both liquid milk and building stone were exported to the expanding towns and cities of northern England. During the 1880s, when demand for good-quality stone, flags and slate reached its peak, 80 tons of stone were carted to the station each day from the Burtersett quarries.[47]

Whilst the first trunk-line through north Wales, the Chester to Holyhead railway, was opened in 1848, and the Cambrian and Central Wales line reached the more remote counties of Merioneth, Montgomery and Radnor in the 1860s, it was the narrow-gauge railway, with its steeper gradients and sharper curves, that opened up the prospects of every valley having its iron road. A key factor in the rapid development of the world's greatest complex of slate quarries, in the Cambrian rocks of Snowdonia, had been the installation of tramways and then, in the 1840s, of iron roads designed for transporting the product to local harbours. By the 1870s, 250,000 tons of slate were produced annually from the Penrhyn quarries, which were worked by open galleries in the form of a horseshoe penetrating the mountain sides. Each of the sixteen galleries was up to 70 feet high, with a terrace of similar width between each gallery upon which tramways were laid to remove both slate and the disproportionate quantity of debris, which was deposited beyond the terraces up to a distance of half a mile over the hillsides. The fast-expanding town of Blaenau Ffestiniog was described by a journalist in 1873 as the 'City of Slate' – the walls, parapets and kerbstones were of slate, and the mud of the roads had a 'blue slaty colour'.[48]

Railways were a major stimulus to a very different type of development in the Lake District. The pull of fashion had long drawn individuals from the moneyed, leisured and educated classes to the uplands, and the reputation of the Lake Poets helped to ensure that Cumbria received

[46] M. Trueman, 'A lime burning revolution', *Bull. Assoc. Industrial Archaeologists*, 17, 1990, pp. 1–2.

[47] C. S. Hallas, 'The social and economic impact of a rural railway: the Wensleydale line', *AHR*, 34, 1986, pp. 29–44.

[48] J. E. Thomas, *A Geographical and Geological Description of Caernarvonshire, with Special Reference to its Quarries, Mines and Mineral Products* (London, 1874); J. Lindsay, *A History of the North Wales Slate Industry* (Newton Abbot, 1974), pp. 117–51.

a disproportionate number. They appeared from Easter onwards, reaching a flood in August, and disappeared 'with amazing abruptness' in late September. The opening of the branch railway from Oxenholme junction on the London to Carlisle railway, through Kendal to Windermere, in 1847, had reaffirmed the importance of Bowness and Ambleside as centres from which those with limited time could follow carefully prepared itineries drawn from any of several popular guidebooks. On Whit Monday 1883, an estimated 8,000 trippers came via the Kendal railway, and a further 2,000 by the Furness line and the lake steamers.[49]

Commuters, summer residents and the semi-retired sought places to live. The author of a prize essay on the farming of Westmorland[50] wrote of how

in the neighbourhood of the Lakes a new class of competitors for the ownership of the soil has arisen in the merchant princes of the manufacturing districts, who eagerly buy up any nook where they may escape from their own smoke, and enjoy pure air and bracing breezes, with shooting and fishing.

Both the speed and manner of development were remarkable. Harriet Martineau, who moved to a new house near Ambleside in 1846, had to revise her guidebook almost as soon as it appeared, there was so much new to describe. Another writer recounted how the inhabitants of this 'modern, wealthy and well-adapted' new village had made the most of the natural advantages of the area. Becoming sites were chosen for mansions fitted for people with deep purses and liberal education. A fine old tree might be left standing, perhaps enclosed with rustic palings, if it accorded well with the newer buildings. A tough ledge of rock might be incorporated in a garden wall. It was as if nature had been brought under 'the tuition of a landscape gardener, smoothed and combed and daintily trimmed – Wordsworth's mountain child with a perpetual Sunday frock on, and curls newly taken out of paper'.[51]

Hoteliers, with increasing support from lesser tradesmen and the encouragement of Cooks, the travel agents, formed in 1876 an English Lake District Association, whose objectives included the promotion of railways, maintenance and improvement of roads, and provision of outdoor amenities, without impairing the natural beauty of the area. The difficulties of meeting all these objectives soon became clear. Vastly expanded quarrying activities were essential if there was to be an ade-

[49] C. E. Benson, *Crag and Hound in Lakeland* (London, 1902), Introduction; J. D. Marshall and J. K. Walton, *The Lake Counties from 1830 to the Mid-twentieth Century* (Manchester, 1981), pp. 180–2.
[50] C. Webster, 'On the farming of Westmorland', *JRASE*, 2nd ser., 4, 1868, p. 8.
[51] H. Martineau, *Guide to Windermere* (Windermere, 2nd edn, 1854), p. 3; F. Fenwick Miller, *Harriet Martineau*, 1884 (repr. Port Washington, 1972), pp. 169–79; E. Lynn Linton, *The Lake Country* (London, 1864), pp. 4–5.

quate return for the railway companies on the expensive engineering works and heavy operating costs. The immediate pretext for the forming of a Lake District Defence Society in the early 1880s was a proposal to build a railway from near Keswick to the slate quarries on the Buttermere side of the Honister Pass. A marked shift in the perception of such ventures could also be discerned. Whereas in the late 1860s, it was possible for the author of a guidebook to speak of the Glenridding lead mines as an added attraction to the ascent of Helvellyn, Baddeley's guidebook of 1891 criticised the mines for being 'the one blot on Patterdale's otherwise perfect loveliness'. Such comments highlighted the extent to which Wordsworth's earlier strictures on intrusive housebuilding, tree-planting and such speculative ventures as railway building, were being increasingly deployed to provide 'a ready-made battery of arguments, sanctified by the literary fame of their author'.[52]

The threat of ill-considered speculative development was by no means confined to the Lake District. The opening of the railway through Wensleydale enabled large numbers of tourists to admire such attractions as the waterfalls of Aysgarth, described by Ruskin as 'out and out the finest thing in water I've seen in these islands'. There was accordingly great consternation when, in the 1880s, proposals were put forward for a further railway from Skipton through Wharfedale and Bishopdale to Aysgarth, which would have required a bridge to have been built across the Ure. A professor in the University of London compared its impact on the view of the waterfalls to that of inserting a page of Bradshaw in the midst of Spenser's *Fairie Queene*. An Aysgarth Defence Association was formed, and meetings and letters of protest organised.[53] Not all schemes were defeated. In north Wales, a rack-and-pinion railway was built from Llanberis to just below the summit of Snowdon in the 1890s, giving rise to demands for a parliamentary commission to safeguard the national interest in scenery, and to see that recreation grounds and the wider countryside were not 'jeopardised by the will or whim of private owners, or the selfishness of speculators'. It seemed inevitable that the railway would degrade 'one of the chief glories of our country' to 'the level of a tea garden'.[54]

The dilemma posed by the need to embark on further development and, at the same time, to protect amenity found most explicit acknowledgment in the Light Railways Bill, first promoted by James Bryce, as President of the Board of Trade, in 1895, and then by his successor in the Conservative government of the following year. The

[52] Marshall and Walton, *The Lake Counties*, p. 194.
[53] C. S. Hallas, *The Wensleydale Railway* (Clapham, 1984), pp. 15–17.
[54] Ranlett, '"Checking Nature's Desecration",' p. 203.

construction of light railways was seen as one of the few practical ways of assisting agriculture in its depressed state. By encouraging the establishment of industries in small towns and rural districts, it might also help to stem the influx of people into the larger towns and cities. Three Light Railway Commissioners were to be appointed by the Board of Trade, whose responsibility it would be to find the quickest and cheapest way of considering and expediting proposals. On the premise that the rolling stock would travel at low speeds, the normally rigorous standards of construction were to be relaxed.[55] By the time the Bill reached the Commons Standing Committee on Trade, the wider implications for the countryside of a marked increase in railway development had become clearer. Bryce, who, from the Opposition benches had continued to give the Bill the greatest support, introduced a series of amendments, one of which was to require the Commissioners and Board of Trade to give explicit consideration to any objection made on the grounds of the damage likely to be inflicted on natural scenery, buildings or other objects of historical interest. The clause, together with another seeking to minimise the impact on commons and commoners' rights, was approved.[56]

It was one thing to secure the protection of amenity, but quite another to ensure that access was granted to everyone who wanted to enjoy it. The increased interest in game preservation and the extension of grouse moors had led to the closure of many footpaths, and denial of accustomed access to the open hillsides. A Keswick and District Footpaths Association was founded as early as 1856, and a National Footpaths Preservation Society in the 1880s. In parliament, James Bryce, whose rambling and botanical ventures had included the climbing of peaks in the eastern United States and a solo ascent of Mount Ararat, introduced an Access to Mountains Bill, designed to prevent any 'owner or occupier of uncultivated mountains or moorland' from excluding 'any person from walking or being on such land for the purpose of recreation or artistic study'. Whilst the Bill made no progress, the issue was kept alive by such incidents as the demonstrations and removal of barriers erected by landowners across footpaths on Skiddaw, above Keswick. The value of rambling was extolled, both as a form of physical recreation and for its contribution to spiritual health. Further Bills of 1892 and 1908 achieved nothing beyond a resolution favourable to the principle of legislation.[57]

The 'guardians of amenity' became such accomplished publicists that it would be easy to gain an exaggerated impression of the threat posed to the uplands. The fact that the Lake District Defence Society suffered no

---

[55] PD, Commons, 3rd ser., XXXII, 1700–26, and XXXVII, 736–71 and 1524–74.
[56] Standing Committee on Trade, *Report on the Light Railway Bill*, BPP, 1896, x; Light Railways Act, 1896, 59 and 60 Victoria, c. 48.   [57] Ranlett, '"Checking Nature's Desecration"'.

major defeats in respect of railway construction had more to do with the estimated costs of building such lines and discouraging traffic forecasts, rather than the strength of opposition on aesthetic grounds. In practice, most parts of the uplands remained open to those prepared to brave their physical rigours. It was during the latter years of the nineteenth century that the highest and most inaccessible ground was explored, by rock-climbers or cragsmen.

The higher parts of the Lake District and Snowdonia offered the greatest freedom to walkers, scramblers and climbers. Fell walking and climbing the peaks had long been popular. Ponies could be hired for 7s. 6d. a day, 11s. being charged for a pony and guide to the top of Coniston Old Man, and 18s. for the ascent of Scafell from Dungeon Gill and 10s. from Wasdale Head. Often parties of students from Oxford and Cambridge would stay at a hotel or farmhouse, combining days on the fells with evenings of intellectual discussion. As a member of such a reading party, Walter Parry Haskett Smith began to reconnoitre for the first time in a systematic way the gullies, crevices and chimneys, and finally the open faces, scaling the Nape's Head above Wasdale Head in 1886. Until the mid-1890s, the great cliffs of Snowdonia had been virtually unknown. Largely through the initiative of James Merriman Archer Thompson, over thirty climbers' routes had been pioneered on Lliwedd alone by the autumn of 1914, and over a hundred on the other crags. Sometimes as a direct extension of rock climbing, the sport of cave exploration (or pot-holing) was also established, first in the High Peak and then beneath the Mendip Hills in Somerset.[58]

An important distinction has to be drawn. Whereas, in the Alps, there was a close affinity between the mountaineer and the adventurer-explorer who opened up the unknown parts of Africa and Arabia, the Lake District was treated much more as a gymnasium, where muscle and nerve might be tested on the steepest crags, often only a hundred feet in height and all within a few miles of one another. It was a challenge taken up predominantly by those in their twenties and thirties, prosperously engaged in business or the professions, living in London and the larger cities. It was a form of escapism from the smoke, bustle and conventions of city life. In that sense, rock climbing was, as Hankinson has remarked, 'a maverick by-product of the Industrial Revolution'.[59]

[58] A. Hankinson, *The First Tigers. The Early History of Rock Climbing in the Lake District* (London, 1972); H. and M. Jackson (eds.), *Lakeland's Pioneer Rock-Climbers* (Clapham, 1980); W. P. Haskett-Smith, *Climbing in the British Isles* (London, 1894); O. Glynne Jones, *Rock-climbing in the English Lake District* (Keswick, 1897); E. A. Baker, *Moors, Crags and Caves of the High Peak and Neighbourhood* (London, 1903).

[59] A. Hankinson, *Camera on the Crags* (London, 1975); A. Hankinson, *A Century on the Crags* (London, 1988).

CHAPTER 34

COASTS AND RIVERS

BY GORDON E. CHERRY AND JOHN SHEAIL

INTRODUCTION

The coast was most likely to be in retreat where the seacliffs were composed of clays or sands. On a tour of inspection in 1889, the Director-General of the Ordnance Survey gave instructions for a 'diagram' to be prepared of the recent and rapid encroachment by the sea of the east Yorkshire coast. The striking changes since the time of the 6-inch survey would be of public interest. It was estimated that, over the thirty-seven-year period, the cliffs had retreated by an average of 215 feet. Further measurements were made along the Lancashire and Yorkshire coasts, following the appointment of a committee of the British Association for the Advancement of Science to investigate coastal erosion in 1892. It was noted that whilst there were significant losses in some parts, the coastline in others might be greatly extended. Two obvious examples from Lancashire were the extensive reclamation of the Ribble estuary, and the construction of the Furness railway.[1]

At first, attempts to secure Government support for schemes to combat coastal erosion met with no success. In 1899, the Fisheries and Harbours Department of the Board of Trade rejected proposals made in the correspondence columns of *The Times* for an official department to be set up to monitor the situation and carry out protection works. It would lead to very great expense and probably involve the revision of many acts of parliament. There was a more positive response in 1906, when the members of parliament for Holderness in east Yorkshire and Thanet in Kent moved an amendment to the King's Speech, drawing attention to 'the serious encroachments of the sea on our coast'. In his reply, David Lloyd George, as President of the Board of Trade in the new Liberal Government, acknowledged that the question was one of national importance. Local authorities, with small populations and low rateable income, could not be expected to take adequate precautions. A Royal

[1] PRO OS 1, 9/4.

Commission was appointed to assess the extent to which further statutory powers and resources were needed. As the Commissioners remarked in their third and final report of 1911, it was the first time that the subject had been investigated in any detail.[2]

It was easy to exaggerate the extent and danger of coastal erosion. Whereas the growth of land in sheltered estuaries and bays was slow and gradual, erosion tended to be dramatic, removing large slices of land at one time. Furthermore, as the Parliamentary Secretary of the Board of Trade conceded in response to a Parliamentary Question in 1905, no records were kept of even the approximate losses of land to the sea.[3] In response to a questionnaire sent by the Royal Commission to local authorities and landowners, the Rural District Council of Blything, responsible for one of the most severely eroded lengths of coastline in the country, wrote of how there were no measurements nor 'any maps showing the extent of the erosion and the rate at which it has taken place'. At the request of the Royal Commission, the Ordnance Survey made a comparison of maps compiled thirty-five years previously with those of the present day. The most seriously affected length of coastline was that between Bridlington and Spurn in Yorkshire. An area of 774 acres had been lost in Yorkshire, followed by Lancashire (545 acres) and Suffolk (518 acres).[4]

Whilst the Royal Commission endorsed the need for more effective measures where erosion was particularly severe, it did not regard the overall situation as alarming. A comparison of Ordnance Survey maps suggested that in England and Wales, over the previous thirty-five years, only 4,692 acres had been lost, compared with 35,444 acres gained through natural accretion and reclamation. The principal need was for greater expertise and co-ordination in implementing protective schemes and for quicker and cheaper ways of reclaiming land from the sea. As the Commission pointed out, some schemes had been so badly devised as to make matters worse.[5] To the east of Newhaven on the Sussex coast, the erection of a harbour wall as a breakwater, under the provisions of the Newhaven and Seaford Sea Defence Act of 1898, had halted the eastward movement of material, and thereby exposed the otherwise unprotected cliffs of Seaford to rapid erosion.[6]

[2] *The Times*, 4 October 1899; PRO BT 13, 42; PD, 4th ser., CLII, 862–76; *First Report of Royal Commission on Coast Erosion and the Reclamation of Tidal Lands in the United Kingdom*, BPP, 1907, XXXIV, Cmd 3683. [3] PD, 4th ser., CXLIII, 1529.

[4] *Minutes of Evidence of Royal Commission on Coast Erosion and the Reclamation of Tidal Lands in the United Kingdom*, BPP, 1907, XXXIV, Cd 3684, appendix xxiii; *Third (and Final) Report of Royal Commission on Coast Erosion and Afforestation*, BPP, 1911, XIV, Cd 5708, pp. 43–4, QQ 15656–837.

[5] BPP, 1907, XXXIV, Cmd 3684, appendix xxiii.

[6] Newhaven and Seaford Sea Defence Act, 1898, 61 & 62 Victoria, cap. clxxi.

One of the many witnesses to the Royal Commission was Francis W. Oliver, the professor of botany at University College London. A distinguished palaeobotanist, Oliver's interest in plant geography developed through the annual field parties he led to the Bouche d'Erquy, to the west of St Malo in Britanny, with the object of studying the relationship of plants to physiographic conditions and to measure the speed with which changes took place. Drawing on the data, photographs and charts acquired during the fortnightly expeditions, Oliver emphasised to the Royal Commission how the special qualities of plants could be used to further the aims of the maritime engineer, using their rooting systems and the shelter afforded by their projecting parts to induce stability in sand-dune systems and to accelerate natural accretion in estuaries.[7] In a book, the *Tidal Lands*, Oliver illustrated how

Nature is conquered by obeying her; man would remain a puppet until he had learned the lesson of obedience.

Written in collaboration with an engineer, the book described in detail how the various branches of horticulture might be 'enlisted' in the cause of conservation.[8]

The coast provided outstanding opportunities for the naturalist to study the zonation and dynamics of vegetation. In her early guide to the collecting of seaweed, Mrs Gatty described how the large tangle or oar weed (*Lamnaria digitata*) was always found in the same situation, namely at extreme low-water mark. Even those plants found anywhere provided some clues to algological geography through their colour and character of growth. *Ceramium rubrum* was a dirty stone-colour in upper pools, but a fine red in deeper ones.[9] In a paper of 1912, Frank Oliver put forward a tentative classification of shingle beaches, of which there were an estimated 300 miles in England and Wales, drawing particular attention to the effects of differences in exposure and mobility, and the nature of soils, on plant life. In order to discover more about these fundamental relationships, a study was made of how *Suaeda fruticosa* established itself on shingle banks and how, over the course of time, its rooting system might slow down or arrest the landward movement of banks.[10]

Whilst sand dunes were not confined to the coast, it was here that their presence attained crucial importance. As the sand accumulated on the

[7] J. Sheail, *Seventy-five Years in Ecology: the British Ecological Society* (Oxford, 1988), pp. 49–51; *Minutes of Evidence of Royal Commission on Coast Erosion and Afforestation*, BPP, xiv, 1909, Cd 4461.

[8] A. E. Carey and F. W. Oliver, *Tidal Lands: a Study of Shore Problems* (London, 1918).

[9] Gatty, *British Sea-weeds*, p. xiv.

[10] F. W. Oliver, 'The shingle beach as a plant habitat', *New Phytol.*, 11, 1912, pp. 73–99; F. W. Oliver and E. J. Salisbury, 'Vegetation and mobile ground as illustrated by *Suaeda fruticosa* on shingle', *Journal of Ecology*, 1, 1913, pp. 249–72.

seaward side of the dune, it was to some extent held in place by the binding root system of marram grass (*Ammophila arenaria*). So valuable was this 'upward-growing' plant that it was often deliberately planted to hold dunes in check. Without it, extensive tracts of cultivated land behind the sandhills towards Berrow and Burnham on the north Somerset coast would have been covered by drifting sand and rendered useless.[11] Where a dune system became fixed, and the supply of fresh sand ceased, a dense scrub of sea buckthorn (*Hippophae rhamnoides*), elder (*Sambucus nigra*) and bramble might develop. Such a progression might be interrupted by a change in tidal currents, or more abruptly by a storm surge − large numbers of rabbits were drowned when the sea broke through the Norfolk sandhills in several places in 1897.[12] The other major form of disturbance was building activity and the laying out of golf courses.

In his *Types of British Vegetation*, Tansley recounted how the first or pioneer association of the salt marsh was an open one made up of herbaceous species of the glasswort, *Salicornia*. On the mud flats of estuaries, this association extended as far as the upper limit of neap tides. On the outer fringes, the plants were few and far between, whereas, further inshore, they were closer set, but never formed a close association. At higher levels, covered only by medium tides, a more general salt-marsh association developed, within which the species varied according to local conditions and 'the accidents of colonisation'. A closed turf of *Glyceria maritima* might develop on those parts covered only by the higher tides, studded by such species as sea lavender (*Limonium vulgare*) and thrift (*Statice maritima*). The highest zone of the salt marsh, reached only by the highest spring tides, was frequently occupied by an association dominated by the sea-rush (*Juncus maritimus*).[13]

During the early years of the century, considerable botanical interest was focused on the cord-grass association, which largely replaced the glasswort association along lengths of the south and east coasts. *Spartina stricta* occurred in occasional patches. *S. alterniflora* was confined to Southampton Water, where *S. townsendii* was first noticed in 1870. Generally regarded as a hybrid, it quickly spread from Southampton Water across the Solent, and into Poole Harbour and the channels and inlets between Fareham and Chichester. By 1911, remarkably pure associations covered up to 8,000 acres of previously bare mud, the thick forest of stems and leaves playing a very important part in fixing large areas of mud, raising the levels of the flats and so preparing them for eventual reclamation from the sea. Whether the plant was regarded 'as

[11] White, *Flora of Bristol*, p. 231.   [12] Dutt, *Wild Life in East Anglia*, p. 138.
[13] Tansley, *Types of British Vegetation*, pp. 330−5.

a botanical phenomenon, a weed which seriously threatens navigation, or a gift of providence capable of being put to a variety of uses', there was an urgent need to discover more about its life-history and invasive qualities.[14]

There was a long and complex history of marsh reclamation. On the north Lincolnshire coast, schemes of the previous two centuries were extended in the 1840s, when the Donnanook Newmarsh, a strip of 300 acres, was taken in, and a bank and sea sluice were constructed to enclose 309 acres (Figure 34.1). North of Saltfleet Haven, 250 acres of salt marsh were reclaimed in 1853–4 by fixing a line of fascines to arrest the sand blown from the shore. The North Coates Fitties of about 450 acres were taken in to form a new foreland in 1856. The Gramthorpe and Marsh Chapel Outmarsh Enclosure Award of 1858 facilitated the reclamation of 446 acres of the 'outmarsh' and, finally, a 125-acre strip of the New East Marsh was enclosed in 1863. The indirect effect of the straightening and embanking of Saltfleet Haven was to encourage the formation of further foredunes and saltings to the south, leading ultimately to the establishment of a brackish and then freshwater fen, encountered nowhere else on the east coast.[15]

The ague or marsh fever had always been closely associated with freshwater and salt marshes. An official inquiry of 1864, largely based on interviews with surgeons in charge of fifty-three poor law unions, drew attention to a marked decline over the previous two decades in the frequency and severity of this form of malaria spread by the mosquito, as a result of improved drainage. Great significance was attached to the fact that cases continued to be most numerous in the Huntspill district of Somerset, and the marshes of the river Swale in Kent, where there was still need for large-scale drainage schemes (Figure 34.2).[16] The most dramatic changes were likely to have occurred where drainage was accompanied by the conversion of pasture to arable, and by improvements to housing and the general well-being of the human population. Acting together, they may have been enough to break what could have always been a precarious cycle of infection.[17]

---

[14] F. W. Oliver, 'Spartina problems', *Annals of Applied Biology*, 1920, pp. 25–39.

[15] D. N. Robinson, 'Coastal evolution in north-east Lincolnshire', in *Geographical Essays in Honour of K.C. Edwards*, ed. R. H. Osborne (Nottingham, 1970), pp. 62–9; D. N. Robinson, 'The Saltfleetby–Theddlethorpe coastline', *Trans. Lincs. Naturalists' Union*, 21, 1984, pp. 1–12.

[16] *Sixth Report of the Medical Officer of the Privy Council*, BPP, 1864, XXVIII, pp. 31–7; S. P. James, 'The disappearance of malaria from England', *Proc. Roy. Soc. Medicine*, 23, 1929, pp. 71–87.

[17] P. MacDougall, 'Malaria: its influence on a north Kent community', *Archaeologia Cantiana*, 95, 1979, pp. 255–64; M. J. Dobson, '"Marsh fever" – the geography of malaria in England', *JHG*, 6, 1980, pp. 357–89.

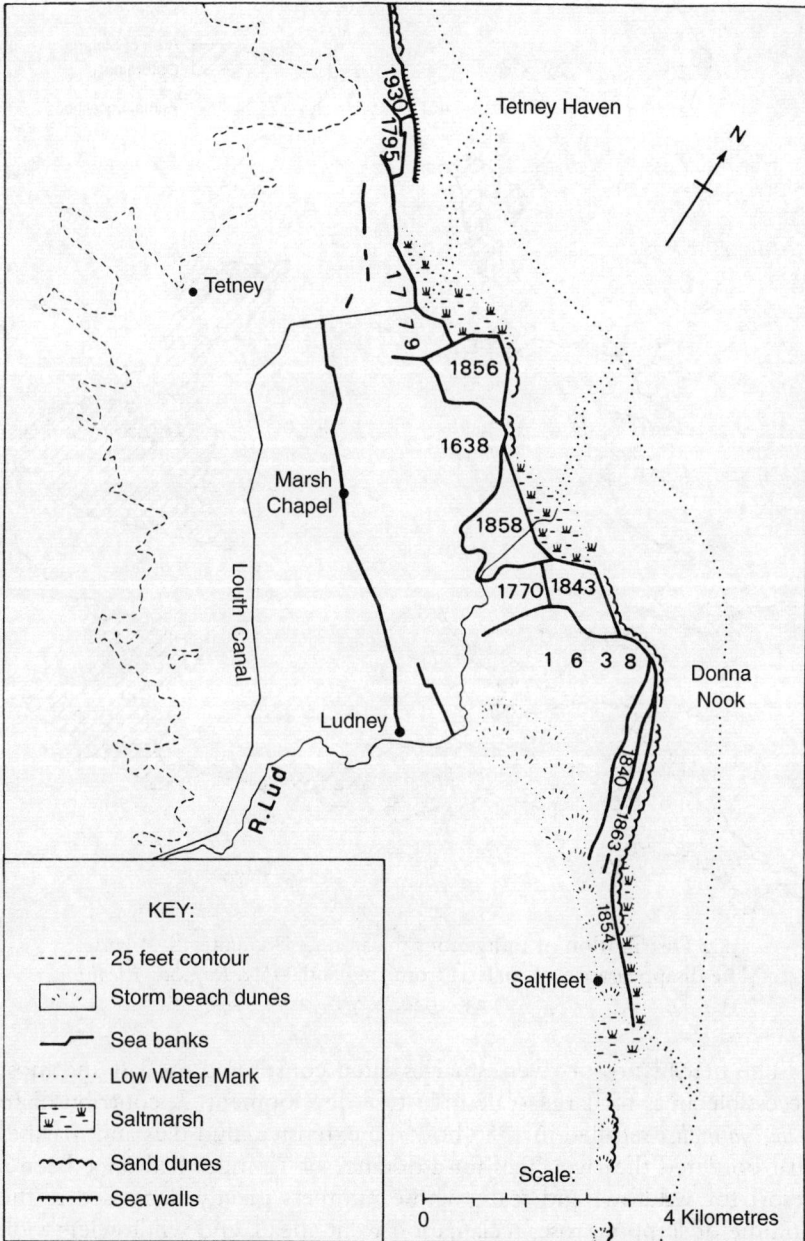

34.1 Coastal evolution in north-east Lincolnshire (after D. Robinson, in *Geographical Essays*, edited by R. H. Osborne *et al.*, 1978, p. 65)

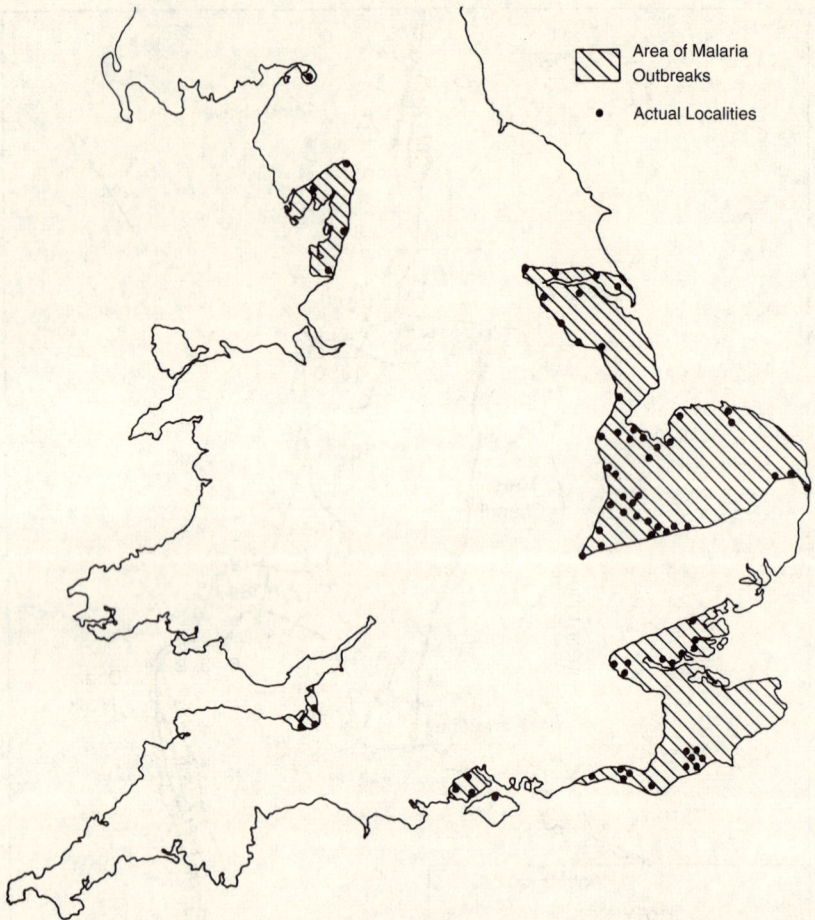

34.2 Distribution of indigenous malaria, *c.* 1860 (after S. P. James,
'The disappearance of malaria from England', *Proc. Roy. Soc. Medicine*,
23, 1929, p. 73).

The marshlands of river estuaries often constituted the last and most
accessible areas for large-scale industrial development. A contributor to
The Naturalist recalled in 1887 how the extensive mud flats and marshes
that bordered the river Tees for a distance of 12 miles had once been a
resort for wildfowl and seals, whose numbers greatly decreased as the
volume of shipping rose, reclamation went ahead, and wildfowlers took
advantage of the new embankments – chiefly working men from
Middlesbrough, using guns of every conceivable kind. The marshes on
the south side of the river, and much of the foreshore between Stockton
and Eston, had been reclaimed and filled with slag, and were now occu-

pied by ironworks, wharves and shipyards. An ironworks had been established on the Coatham marsh, opposite the extreme mouth of the river, where a duck decoy had survived until the 1870s. The open and unoccupied north side of the river was broken at Port Clarence, where ironworks had been erected opposite Middlesbrough.[18]

The authors of almost every County Flora and Fauna with a seaboard referred to the development taking place along river estuaries and the coast. From Greenwich to Gravesend in Kent, the river banks were lit by gas lamps, and from Herne Bay to Hythe, with the exception of the Sandwich Flats, the coast was increasingly given over to 'the tripper and the jerry builder'. Until about 1875, the coast from Blackpool to Lytham had been 'truly a botanist's paradise', but, by the turn of the century, those parts not already built over between St Annes and Lytham were '"improved" and drained, and given over to crowds of excursionists or converted into golf links'.[19]

In the longer term, it was not only visual intrusion and loss of wildlife habitat that threatened the coast. As in the Thames estuary, sewage and industrial pollution, and the dumping of refuse, severely affected the oyster and other shell-fish fisheries. In 1907, a seven-masted sailing vessel, the *Thomas W. Lawson*, bound for London with a cargo of oil, sank off the Scilly Isles. As the tanks burst open, the oil was washed onto the shore, killing many of the rabbits and birds on Annet. The smell persisted for months. By the First World War, no part of the coast was free from the dangers of such pollution, whether as a result of an accident or the deliberate cleaning of the ship's tanks at sea.[20]

A. WATER MANAGEMENT

There were traditionally three main parties to any debate over the management of rivers. There were the riparian owners and those who owned or rented the mills and fisheries, secondly the navigation interests, and thirdly (and far less well organised) the riverside population, for whom the risk of flooding was the principal preoccupation. Any improvements brought about by one interest were bound to affect others. There were, for example, complaints of damage to fishing and amenity on the river Rothay in Westmorland, following a shortening of the river course in order to reduce flooding. The river had become one continuous run,

---

[18] R. Lofthouse, 'The River Tees: its marshes and their fauna', *Naturalist*, Hull, 1887, pp. 1–16.

[19] R. J. Babton, C. W. Shepherd and E. Bartlett, *Notes on the Birds of Kent* (London, 1907), p. xiii; J. A. Wheldon and A. Wilson, *The Flora of West Lancashire* (Eastbourne, 1907), p. 34.

[20] J. Sheail, 'An historical perspective on the development of a marine resource: the Whitstable oyster fishery', *Marine Environmental Research*, 19, 1986, pp. 279–93; J. Mothersole, *The Isles of Scilly* (London, 1910), pp. 128–30.

without variety of current or shaded pool. There were no easy waters for the trout to lie up in, in times of flood.[21]

There were frequent complaints as to how the regulation of the river took no account of the changing circumstances of the individual river interests. Although sufficient money had been subscribed in the late 1840s to improve the outfall of Windermere at Fell Foot, little was likely to be achieved until a much larger operation was carried out, which would include the removal of a weir that served the corn mill at Newby Bridge. Now that steam power was universally available, Harriet Martineau reasoned, it was intolerable that hundreds of acres should be turned into swamps, and hundreds of lives lost by fever, ague and rheumatism brought about by flooding, for the sake of operating a water mill that made a profit of perhaps £30–£40 a year.[22]

The most dramatic changes in the use of watercourses were likely to arise from the abstraction of potable water and the disposal of sewage and industrial effluent. Although it was not until the 1880s, and the work of Robert Koch on microbes, that bacteriology became an established science, there was plenty of circumstantial evidence to suggest a causal relationship between epidemics and polluted drinking-water.[23] Henry W. Acland was well placed to study the cholera epidemic that broke out in Oxford in 1854 – he was a physician at the Radcliffe Infirmary and Aldrechian professor of clinical medicine. Whilst the outbreak of such epidemics focused public attention on the many instances of bad drainage and sewerage in the cities and towns affected, he realised that any significant and longer-term improvements to hygiene had to extend far beyond the built-up area. There had to be two priorities. One was to prevent surface flooding, and the other to find ways of removing the surface water and sewage of a settlement without contaminating the river. For as long as private interests and local convenience prevailed, and no attempt was made to manage the waters of each valley as a whole, such towns as Oxford would continue to be flooded for part of the year and to suffer from 'offensive exhalations of decaying substances left by receding waters' at other times. The Thames valley would remain a national disgrace.[24]

A more integrated approach to river management could be discerned. The preamble to an Act of 1857 cited the need to place the entire length of the river Thames below Staines under one body so as to secure the necessary authority and resources to dredge the river, remove unauthor-

---

[21] G. F. Braithwaite, *The Salmonidae of Westmorland. Angling Reminiscences* (Kendal, 1884), pp. 71–2.

[22] N. Nicholson, *The Lakers. The Adventures of the First Tourists* (London, 1955), p. 201.

[23] W. M. Stern, 'Water supply in Britain: the development of a public service', *Journ. Roy. Sanit. Inst.*, 74, 1954, pp. 998–1004.

[24] H. W. Acland, 'Town drainage', *Journ. Bath and West of England*, 5, 1857, pp. 188–200.

ised encroachments, and regulate steamer traffic on an adequate scale. By a further measure of 1866, the jurisdiction of the Thames Conservancy was extended to the whole of the navigable river. Whilst the rights of millowners were to be respected, the weirs were to be transferred to the Conservancy on payment of compensation so that 'private interests should no longer interfere with the navigation of one of the most important highways of the kingdom'.[25] Annual reports chronicled the incidence and types of improvement carried out. In 1913, 85,499 cubic yards of material were raised by steamer dredgers, and 22,062 cubic yards by hand punts. Shoals were removed in twelve places – a steam grab dredge being used to remove 'accumulations above Oxford' and a rock cutter 'in breaking up the limestone formation below Sandford Lock'.[26]

Natural sedimentation and weed growth were likely to pose particular problems. P. H. Emerson, in his *Life and Landscape on the Norfolk Broads*, recounted how tidal currents gradually silted up and narrowed the mouth of an estuary, whilst the river brought down detritus from the higher ground. Water, thick with mud, was pumped from the drains of adjacent arable and pasture land.[27] The rivers might be kept clean by 'dydleing', but comparisons of Ordnance Survey maps suggested that many of the broads in east Norfolk had decreased in size. Surlingham and Chapman's Broad had virtually disappeared. Womack Broad had been a 'nice sheet of water', but, by the early 1900s, it was already growing up very rapidly, 'each year seeing an accretion to the growth of spongy marsh, and an additional layer of mud on the bottom'. Although still extensive, Hickling Broad was increasingly covered by 'bunches of weed'. Promontaries of reed were pushing themselves further and further into the lake, and the bays between became shallower.[28]

However successful the attempts to manage watercourses in a more integrated manner, they could not take account of what was happening throughout the entire catchment. There were numerous instances of ditching and under-drainage schemes, carried out by landowners and farmers, being blamed for increases in the frequency and extent of flooding downstream. The effects of under-drainage were almost limitless. It was estimated that 22.8 million acres, out of a total of 56.352 million acres in Great Britain, were capable of marked improvement through drainage, and that only 1.5 million acres had so far benefited from such schemes. In a paper, first published in the *Journal of the Royal Agricultural Society* and then in the *Proceedings of the Institution of Civil Engineers*, an account was given of trials carried out over several hundred

[25] F. S. Thacker, *The Thames Highway*. Vol. 1. *General History* (1914, repr. Newton Abbot, 1968).

[26] Thames Conservancy, *General Report*, BPP, 1914, XLVIII.

[27] Emerson & Goodall, *Life and Landscape*, pp. 33–5.

[28] Tansley, *Types of British Vegetation*, pp. 214–29.

acres of farmland at Hinxworth in Hertfordshire, in an attempt to gain a more precise understanding of the relationship between the volume and duration of rainfall and the discharge of water from the under-drains.[29]

It was in this wider context that attempts were made to conserve fish stocks. The overriding purpose of the large body of statute and common law, historically, had been to protect the property rights of those owning the adjacent banks. Late nineteenth-century attempts to give more explicit consideration to the fish stocks themselves had to take account of five different kinds of interest, namely: the sportsmen, anglers and rod fishermen, whose main concern was the protection of the spawning salmon; those who used nets to gather the largest share of river salmon, but showed less interest in replenishing stocks; tidal fishermen who enjoyed common-law rights to take the salmon on a commercial scale before they entered freshwater; and the mill and factory owners.[30]

Sustained agitation from fisheries' interests led to the appointment of a Royal Commission and a series of broad recommendations, based on three fundamental principles – a closed season, the free ascent of salmon, and prevention of pollution. Whilst the eventual Salmon Act of 1861 was shorn of thirty-six of its original fifty-seven clauses, it did include provisions for the appointment of inspectors, who were to play a key role in stimulating moves to introduce a further Act of 1865, which set up locally elected boards of conservators over fishery districts that comprised complete watersheds.[31] The principal effect was to highlight the otherwise unprotected state of fish stocks. A contributor to the recently founded *Fishing Gazette* noted how perhaps as much as two-thirds of the 5,000 miles of rivers and canals in England had been rendered almost barren by netting and the destruction of fish.[32]

Fishing was becoming an increasingly popular pastime. Norwich was only three to four hours by rail from London, and every guidebook to the Norfolk Broads included details of 'railway access to the fishing stations', where boats could be hired. The question was whether there would be sufficient fish to meet the demand. Such practices as the use of long, small-meshed nets, and even trawls, to collect bait for sea-fishing were thought to have reduced stocks considerably.[33] The Norfolk and

[29] J. Bailey Denton, 'On the discharge from under-drainage, and its effect on the artificial channels and outfalls of the country', *Proc. Inst. Civil Engin.*, 21, 1862, p. 48; A. D. M. Phillips, *The Underdraining of Farmland in England During the Nineteenth Century* (Cambridge, 1989).

[30] R. M. MacLeod, 'Government and resource conservation: the Salmon Acts Administration, 1860–1886', *Journ. British Studies*, 7, 1967, pp. 114–50; P. Bartrip, 'Food for the body and food for the mind: the regulation of freshwater fisheries in the 1870s', *Victorian Studies*, 28, 1985, pp. 285–304.   [31] Salmon Act, 1861, 24 and 25 Victoria, c. 109.

[32] Anonymous, 'The preservation of coarse fish', *Fishing Gazette*, 2, 1878, pp. 49–50.

[33] G. C. Davies, *The Handbook to the Rivers and Broads of Norfolk and Suffolk* (London, 1883).

Suffolk Fisheries Act of 1877 sought 'to preserve the fisheries in the navigable rivers and broads' of these counties through the establishment of a board of conservators, appointed by the respective justices of the peace and mayors, aldermen and burgesses. It would have powers to impose a close season for all kinds of freshwater fish and types of fishing (except rod and line), and regulate the 'nets, engines, trimmers, liggers, or instruments of any kind' used for taking fish.[34] A close season, 1 March to 30 June, was enacted for most species, and several preservation societies were set up to combat poaching and encourage 'fair angling' generally.

Sheffield, with its eighty clubs and combined membership of some 8,000, could plausibly claim to be the angling capital of England. Its member of parliament, A. J. Mundella, promoted a Freshwater Fisheries Bill in 1878, making it illegal to capture, purchase or sell coarse fish during close seasons. Fisheries boards could be set up for the protection of trout and char in any part of the country.[35] The measure was hailed by the *Fishing Gazette* as 'another step in the general conservation of our indigenous fauna'.[36] It did not pose any threat to significant commercial interests. On the contrary, the Select Committee on the Bill commented that it would hold out the promise of 'food for the mind and food for the body', and perhaps even greater business for tackle-makers and dealers, boat-builders and hirers, and railway operatives.[37]

B. POSITIVE AND NEGATIVE POLLUTION

Water assumed an ever-widening role as Britain became 'the workshop of the world' and over half the population lived in towns. As a direct consequence, the rivers of particularly the more industrialised parts became notorious for two forms of pollution, namely positive pollution, where sewage and trade effluent were directed into streams, and negative pollution, where naturally clean water was abstracted.[38]

Paradoxically, the situation was made worse by attempts to improve the health of towns and cities by exploiting the fact that water, passing through a small-diameter pipe, could act as the basis of a self-cleansing system of sewage disposal. The innovation enabled every urban property to become part of a unified drainage and water-supply system. At first, no one foresaw the consequences of allowing a greater proportion of sewage to be discharged into streams, particularly where water had already been abstracted for domestic and industrial purposes. As Wohl

---

[34] Norfolk and Suffolk Fisheries Act, 1877, 40 and 41 Victoria, c. xcviii.

[35] Freshwater Fisheries Act, 1878, 41 & 42 Victoria, c. 39.

[36] Anonymous, 'The Bill for the preservation of fresh-water fish', *Fishing Gazette*, 2, 1878, pp. 172–3.

[37] *Report of the Select Committee on Freshwater Fishery Protection*, BPP, 1878, XIII, pp. 28, 30.

[38] M. M. Paterson, *Compensation Discharge in the Rivers and Streams of the West Riding* (London, 1896).

commented, in his book, *Endangered Lives: Public Health in Victorian Britain*, rivers and streams were so conveniently located, running down to the sea, involving no construction costs, constantly moving and, theoretically at least, self-cleansing.[39]

i. Positive pollution

Stream life even in the remotest parts of the country had always been liable to such disruptions as sheep washing. There was evidence, however, to suggest a marked deterioration in water quality during the mid-nineteenth century. In Westmorland, the sawdust from the bobbin and other mills was thought to be especially harmful to fishlife – the small particles impeding the action of their gills. The pollution caused by the waste liquor from paper works, or simply from emptying the dye-pan into the water, was considered to be particularly reprehensible – the damage could have been avoided so easily. Rarely did all stream life die at once – as on the river Kent, a pattern could be discerned. The first to be destroyed were the lower forms of life, which constituted the food of fish over a large part of the year. Then the trout fry and yearly fish would become scarce, and finally the mature trout.[40]

The development of extractive industries could have a dramatic impact on water purity and stream life. Long after the traveller left the china-clay works between Bodmin and St Austel in Cornwall, the river water used for washing the clay had 'the hue and exact appearance of milk'. The river Webber on Dartmoor was coloured by the workings of the copper mines. Leases for the Glasdir Copper Mines in Llanfachreth, Merionethshire, in the early 1900s, stipulated that the mining company had to construct 'filter beds or slime pits' so as to prevent pollution. A substantial wall was to be built at the lower end 'as shall effectively retain the tailings deposited and prevent their refuse from percolating into the Mawddach River'.[41]

Important legal precedents were set in the 1860s, when the owner of a paper mill at Wookey Hole, at the foot of the Mendips, brought an action against the St Cuthbert's Lead Works at Priddy. It was alleged that, since the ore had been puddled with stream water which flowed through the minery and then disappeared down swallets in the limestone rock, the water emerging at Wookey Hole had become polluted with lead, and less

---

[39] A. S. Wohl, *Endangered Lives: Public Health in Victorian Britain* (London, 1983), pp. 233–56.

[40] Braithwaite, *The Salmonidae of Westmorland*, pp. 40–5; J. Watson, *The English Lake District Fisheries* (London, 1899), pp. 11–14.

[41] C. A. Johns, *A Week at the Lizard* (London, 1848), p. 3; J. G. Croker, *A Guide to the Eastern Escarpment of Dartmoor* (London, 1852), p. 5; University College, North Wales, Nannau MSS, 3822.

suitable for paper manufacture. An enquiry established that it took thirty-six hours for the water to emerge. Before the Queen's Bench, the plaintiff's counsel argued that the case was analogous to that of an ancient mill, which was entitled to the use of a watercourse. Counsel for the defendant insisted that account had to be taken of the peculiarities of sub-terranean water. The Lord Chief Justice, the other judges concurring, gave judgment in favour of the millowner on the grounds that pollution was caused by percolation.[42]

Few watercourses escaped the effects of pollution. Soon after the establishment of a paper works in the Lancashire town of Clitheroe in 1860, a sediment (the colour of pea-soup) could be seen rolling into the clear waters of Mearley Brook and Pendleton Brook, keeping to its own side for a distance and then spreading over the whole watercourse until 'one side was no better than the other'.[43] In the case of the river Thames, not only were large amounts of sewage discharged at low water, but much of it was carried upstream on the rising tide and back to London on the ebb. The Royal Commission on the State of the City of London in 1853, which led to the establishment of the Metropolitan Board of Works, rec-ommended the construction of interceptor sewers to take the sewage further downstream, where it was discharged through balancing reser-voirs at Barking Creek and Crossness. The works, completed in 1865, had an average discharge of 200 million gallons a day. Within ten years, they had been outstripped by the volume of sewage, an increasing pro-portion of which again flowed directly into the Thames.[44]

Some intimation of the range and effects of pollutants may be gained from the river Erewash, which rose in the Derbyshire parish of Kirkby-in-Ashfield and, for the first part of its course, consisted mainly of sewage in dry weather. For much of its length, the river formed a county bound-ary. From the Nottinghamshire side, the collieries at Eastwood dis-charged large quantities of turbid and much discoloured effluent. The waters of the tributary, the Giltbrook, consisted mainly of effluent, much of it from a chemical works and brewery. The most serious pollution on the lower course of the Erewash came from the sewage works of the Derbyshire borough of Ilkeston. A survey of 1902 revealed that, whereas the water above the works was relatively clear and odourless, with a cover of green vegetation, the stream bed below the works was covered with black mud, the waters were turbid and foul-smelling, and the banks covered with sewage fungus.[45]

It was not difficult for the satirist to bring out the irony of what was

---

[42] J. W. Gough, *The Mines of the Mendips* (Oxford, 1930, repr. 1967), pp. 14–16.

[43] Lancashire RO, DDX 54, 88–9.

[44] J. M. Sidwick, 'A brief history of sewage treatment', *Effluent Water Treatment Journ.*, 16, 1976, pp. 65–71.   [45] Nottinghamshire RO, CC/HE/1/4/1–2.

happening. The scientific processes that had destroyed domestic manufacture in the countryside were now bringing about the physical degradation of the crowded towns. The condition of the river Thames was so
bad as it flowed past parliament in the summer of 1858 that the year
became known as the 'Year of the Great Stink'.[46] With a little further
development, it seemed as if the air would become poisoned and the
dwellers on the bank killed. It was extraordinary to Dr Opimian, in
Thomas Love Peacock's novel, *Gryll Grange*, that parliament should turn
to science for solutions. Science would no doubt oblige by spending large
sums of money, the main outcome of which would be to demonstrate
the need for even greater expenditure.[47]

There was no obvious 'mode of ultimate disposal' for the various forms
of effluent. In choosing between the earth (privy), precipitation and
water (irrigation) methods, there were limits to what even a spa town, for
example, jealous of its reputation, could afford. A Local Board of Health
was established for the Midlands resort of Leamington Spa, the population of which had risen from 6,250 in 1831 to nearly 17,000 by the late
1850s. A scheme was implemented to pump all the town's sewage to a
series of 'deodorising tanks', from where, having been treated with
common lime, it passed through a series of depositing canals and sieves
before being discharged into the river Leam. Property-owners downstream contended that the river was still polluted. The owner of Myton
Grange described how the waters were covered in 'a foul scum', and solid
sewage clung to the banks and bushes. Fish died, and cattle refused to
drink the water. Under the Act confirming the appointment of the Local
Board, downstream interests had secured a special clause, prohibiting the
discharge of any water containing noxious and offensive matter.[48] The
Earl of Warwick had been the leading figure in securing the safeguard –
the river flowed through his park. The Board contended that the foetid
mud had been deposited before the present works were opened, and that,
in any case, parliament had only sought to minimise pollution. The Board
was using 'the best available means that science could afford'. To leave the
sewage in cess-pools, or irrigate it over land, would encourage disease.[49]

The Court of Chancery was unconvinced, and imposed an injunction
in 1866, prohibiting the discharge of any water containing sewage into
the Leam, on pain of a penalty of £5,000. The Board extended the
depository canals, and installed further screens. When these had little
effect, work began on embanking and cleaning the river. The patience

[46] L. B. Wood, *The Restoration of the Tidal Thames* (Bristol, 1982), pp. 20–6.
[47] T. L. Peacock, *Gryll Grange* (London, 1861), pp. 2–3; A. E. Dyson, *The Crazy Fabric* (London, 1965), pp. 68–9.
[48] Public Health Supplementary Act (No. 2), 1852, 15 and 16 Victoria, c. lxix.
[49] Warwickshire RO, CR 1538, 187; *Leamington Spa Courier*, 1 and 8 July 1865.

of the owner of Myton Grange finally ran out in 1868. The Board's property was sequestrated, leaving the Clerk to meet out of his own pocket the costs of policing, lighting, and even the wages of labourers attending the deodorising tank. Newspaper advertisements had already begun to invite local landowners and occupiers to join the Board in establishing a 'sewage farm'.[50] A thirty-year agreement was concluded with the Earl of Warwick a few months after the Order, whereby the Earl was to contribute £450 per annum towards the cost of pumping and conducting raw sewage to the various parts of his Model Farm at Heathcote. The untreated sewage would be pumped by means of a high-pressure, condensing beam-type engine a distance of $2\frac{1}{4}$ miles, communication being maintained by a telegraph of 'very simple and easily understood construction'. The daily dry-weather flow was to be 600,000 gallons, rising to 1.5 million gallons in wet weather.[51]

In 1857, a Royal Commission was appointed 'to inquire into the best mode of distributing the sewage of towns'. From experiments conducted at Rugby, it recommended some eight years later the continuous application of sewage to the land.[52] Open carriers, or grips, distributed the sewage to the different fields of the farm, where it overflowed into smaller carriers excavated along the centre of raised beds or ridges, resembling in many respects the appearance of a water-meadow. The method was of obvious interest to farming, combining as it did 'perfect cleanliness with considerable economy'. In 1879, the Royal Agricultural Society offered prizes for the best-managed farms. At the garrison town of Aldershot, a rye-grass crop of up to 50 tons per acre was produced on what had previously been sterile land. The Leamington sewage farm was awarded first prize among those entries taking the sewage of over 20,000 people. It comprised 83 acres of grassland and 284 acres of arable, managed as part of a larger holding of 764 acres.[53] By the time the sewage water found its way into the brook, it was 'as clear and bright as the most pellucid stream water ever recorded'.[54]

Moves to arrest and reverse the degradation of rivers were hampered by difficulties in defining polluting substances and enforcing whatever standards were agreed. A Royal Commission, appointed in 1865, and

[50] Warwickshire RO, CR 1563, 178 & 187–8.

[51] Warwickshire RO, 1563, 179; R. Davidson, 'Description of the Leamington sewage pumping station', *Proceedings of the Association of Municipal and Sanitary Engineers and Surveyors*, 1, 1874, pp. 140–3.

[52] *Third Report of the Royal Commission on the Best Mode of Distributing the Sewage of Towns*, BPP, 1865, XXVII, pp. 3–4.

[53] Anonymous, 'Report of the Judges appointed by the Royal Agricultural Society of England to adjudicate the prizes in the Sewage Farm Competition, 1879', *JRASE*, 2nd ser., 16, 1880, pp. 1–80.   [54] *Leamington Spa Courier*, 4 May 1872.

another in 1868, issued a series of reports, focusing on different aspects of river management and the setting of standards, in both physical and chemical terms.[55] Whilst the resulting Pollution of Rivers Act of 1876 placed restrictions on the discharge of human and trade waste into streams, other than tidal waters, little of practical value was achieved.[56] The enforcement of the Act was entrusted to the same local authorities that used the rivers for sewage disposal. If the discharge of trade effluent had been closely regulated, industrial costs might have risen, leading perhaps to greater unemployment and the loss of rateable income. Even if a town council had decided to impose controls, its jurisdiction was so limited geographically that any isolated action would have had little general effect on water quality.[57]

It was in this context that the establishment of county councils under the Local Government Act of 1888 had far-reaching implications. Among their responsibilities as sanitary authorities was the enforcement of the Pollution of Rivers Act. Because of their geographical extent, strength in terms of rateable income, and the fact that they had no direct responsibility for water supplies and sewage, the new councils were uniquely placed to take a more detached view of river management. However, even in the West Riding of Yorkshire, where the greatest use was made of the powers, there were severe constraints as to what the Council could achieve as 'the single authority' for all the rivers of the Riding. The Local Government Act of 1888 had conferred the status of County Borough on the towns of Bradford, Halifax, Huddersfield, Leeds and Sheffield. Each was equal in status to, and separate from, the administrative county. The Act envisaged that a joint committee would be formed where a river flowed through two or more administrative areas. In practice, the committees offered no guarantee of effective action.[58]

It was a further Royal Commission, appointed in 1898, that marked the transition from folklore to a more scientific approach. In its nine reports, issued over a seventeen-year period, the Commission sought to investigate the treatment and disposal of sewage (including any liquid from a factory) with due regard to the protection of public health and the economical and efficient discharge of the duties of local authorities. On the basis of investigations made of a large number of rivers, it was found that tests for the amount of putrescible material in the water (the dis-

[55] *First Report of the Royal Commission into the Best Means of Preventing the Pollution of Rivers. River Thames*, BPP, 1866, XXXIII, pp. 3–4; *Report of the Royal Commission into the Best Means of Preventing the Pollution of Rivers. Mersey and Ribble Basins*, BPP, 1870, XL, C37.

[56] Pollution of Rivers Act, 1976, 39 and 40 Victoria, c. 75.

[57] Wohl, *Endangered Lives*, pp. 233–56.

[58] J. Sheail, 'Government and the perception of reservoir development in Britain', *Planning Perspectives*, 1, 1986, 53–5.

solved oxygen absorption test), whilst arbitrary, provided a fair measure of cleanliness. A watercourse containing 0.1 part per 100,000 of dissolved oxygen, absorbed over 5 days, could be rated as clean, whereas those with 1.0 part per 100,000 rated as bad.[59]

The Commission's findings also helped resolve the long-running controversy between those local authorities which wanted to build 'artificial' treatment works, and the Local Government Board which insisted, as a condition of sanctioning loans for any schemes, that the sewage had to be purified by passing it through land before being discharged into streams. Whilst evidence indicated that land treatment was perhaps the best form of treatment, it required immense areas of land because of the way in which the ground soon became 'sewage sick'. The need for artificial methods was highlighted by the Birmingham, Tame and Rea District Drainage Board, where reliance on land irrigation meant that an additional 50 acres of land per year were needed to meet the needs of the developing city. The Commission's reports not only put 'artificial' methods on a respectable footing, but gave guidance as to which were the more effective. In the case of the Birmingham works, 35 acres of biological filter beds were commissioned at Minworth between 1904 and 1914, thereby reducing the demand for more land to an average of 1 acre per annum. Since even this rate could not be sustained indefinitely, there was incentive to investigate the potential of the activated sludge process, not so much as a substitute but as a supplement to the filters.[60]

ii. Water abstraction

The water requirements of the larger cities rose appreciably during the century. In Birmingham, for example, demand rose from 8.30 million gallons a day in 1876 to a daily total of 16.82 million gallons in 1891. The twofold increase stemmed from the rising population of the supply area and the more extensive use of water, as in fire fighting, industrial processes and domestic hygiene. Integrated systems of water supply and water-borne removal made essential a complete overhaul of the water-supply system.[61]

From the mid-century onwards, the responsibilities of the local joint-stock companies were increasingly taken over by local government in the larger industrial towns. One such was Halifax, which feared its textile industries might be threatened if demand for water outstripped supply.

---

[59] *Report of the Royal Commission on the Treating and Disposing of Sewage*, BPP, 1901, xxxiv, Cd 685; H. B. N. Hynes, *The Ecology of Running Waters* (Liverpool, 1972), p. 443.

[60] Sidwick, *History of Sewage Treatment*, pp. 193–9.

[61] J. A. Hassan, 'The growth and impact of the British water industry in the nineteenth century', *EcHR*, 2nd ser., 38, 1985, pp. 531–47.

An Act of 1853 granted the Corporation powers of compulsory purchase for three upland reservoirs, and another for a further two schemes. The Pennine gathering grounds, available to Yorkshire and Lancashire, soon became prize possessions, and corporations had to look further afield for supplies. As early as 1848, Manchester had gone to the Longdendale valley, 15 miles from the city, and, by the 1880s, it was forced to seek powers to impound Thirlmere in the Lake District.[62]

The impact of water abstraction has been so considerable that major schemes have always been subject to an exceptional degree of public scrutiny. They were submitted to parliament in the form of Local and Private Bills, seeking powers not otherwise available under public law. Whilst considerable importance was attached to precedents, account also had to be taken of the peculiar circumstances of each Bill, in respect of both the demand for water and the manner in which the further regulation of the watercourses might affect other users of the catchment.

Parliamentary committees spent more time considering the question of compensation than any other aspect of reservoir development. How could recompense be made to other users for the amounts of water abstracted? Undertakers argued that it was far more convenient for all parties if compensation was given in the form of a guaranteed minimum flow of water below the dam rather than as monetary payments. Not only would it be impracticable to negotiate and reach a monetary settlement with every individual owner on a highly industrialised stream, but some millowners had themselves built reservoirs so as to take advantage of a more predictable flow of water.[63]

The principal question was how to decide what proportion should be set aside as compensation water. Before the Royal Commission on Water Supply in 1869, the engineer, Thomas Hawkesley, recalled how it was only economical to impound the average runoff of a catchment, as represented by the three driest consecutive years on record. The amount was estimated from rainfall data on the assumption that only 80 per cent of the average rainfall fell on such occasions. From this figure, the equivalent of up to 15 inches was deducted to allow for evaporation and percolation. The resulting figure, called the available rainfall or reliable yield, could then be expressed in terms of gallons per day for the drainage area in question. Experience showed that in the catchment areas of the Pennine Chain, the equivalent of 8 to 10 inches of rainfall was sufficient for industrial purposes, if sent down during the working hours of the mills. This represented a third of the reliable yield, as calculated on the

[62] Sheail, 'Historical perspective on a marine resource'.
[63] N. Smith, *A History of Dams* (London, 1971), pp. 170–81; K. Smith, *Water in Britain: a Study in Applied Hydrology and Resource Geography* (London, 1972), pp. 12–20; J. Sheail, 'Constraints on water-resource development in England and Wales: the concept and management of compensation flows', *Journ. Environmental Management*, 19, 1984, pp. 351–61.

basis of Hawkesley's formula. Hawkesley himself insisted that the rule of a third was only 'a centre point on which you might turn rather widely'. Because most early schemes were promoted in the Pennines, the ratio of two-thirds for supply purposes, and one-third for compensation (one-quarter where no industry was present), came to be widely applied. It seemed to hold out 'the prospect of abundant supply to the towns without injuring existing interests'.[64]

Much of the controversy stemmed from confusion as to the precise purpose of the water which, in turn, reflected an even more basic lack of appreciation of the different uses to which the water could be put and the importance attached by each user to changing economic and social circumstances within and beyond the catchment. By the time canal companies, and then water undertakers, began to abstract water in the more industrialised areas, 'probably every available foot of fall in the larger rivers had then been utilised'. Masters of the situation, the millowners could drive hard bargains as to the amount of water to be left in the watercourses, and establish the precedent by which the compensation water was to be discharged from the reservoirs in an intermittent, rather than a continuous flow, so as to ensure that there was a greater volume available during the working hours of the mills. Such a timetable took no account of the role of the watercourse for the removal of sewage and trade effluent. 'The continuity of flow, regularity of volume and coolness of temperature' were important factors in keeping down 'the undue production of offensive gases'.[65]

The increasing scarcity of potable water led to the promotion of schemes envisaging the transfer of water from one river basin to another. Some intimation of the outcry which this would cause was given by the opposition to a Bill for the transfer of 1 million gallons a day of spring-water from the headwaters of the river Thames to the Severn valley in 1855. Whereas a town on the Thames would use and return the water, albeit polluted, not a drop of the water abstracted under the Bill would find its way back to the Thames. Opponents warned of how shoals and flats would appear, and the proportion of sewage to pure water would rise. The Bill was rejected on its second reading by 118 votes to 88 votes.[66] In a report of 1869, the Royal Commission on Water Supply to the Metropolis affirmed that it would be wrong in principle that any one town or district should take possession of the gathering grounds belonging geographically to another.[67]

[64] House of Lords RO, Commons Select Committee, 1854, vols. 2 and 3, Minutes of Evidence, and Lords Select Committee, 1854, vol. 5, Minutes of Evidence; B. Law, 'River conservation in England with special reference to the Yorkshire Calder', MSc thesis, Univ. of London, 1956.

[65] Paterson, *Compensation Discharge*, pp. 21–2, 27–9.   [66] PD, Commons, 3rd ser., 177, 490–5.

[67] BPP, 1868–69, XXXIII, *Report of the Royal Commission on Water Supply*, BPP, 1868–9, pp. cxxiv–v, cxxviii.

It was a sentiment frequently voiced during the promotion of Bills by the Corporation of Manchester to impound Thirlmere in the Lake District (1878–9), and of Liverpool (1880) and Birmingham (1892) respectively to construct reservoirs in the Vyrnwy, and in the Elan and Claerwen valleys, of Wales. The Bills raised such large questions of public importance that each was debated on the second reading, and exceptional steps were taken to ensure that the Select Committees made a close scrutiny of such questions as compensation. Together, the schemes represented a new phase in the scale of water undertakings, in terms of both the distances involved and the quantities of land and water exploited. Not only were the Welsh gathering grounds, for example, over 70 miles from Birmingham, but their area of 70 square miles far exceeded that of the city.[68]

There was considerable resentment at the further 'export' of Welsh water to England, as envisaged by the Birmingham Bill of 1892. Legal counsel for the opponents compared the gathering grounds of the river Wye to a carcase, around which the 'water-works eagles' were gathered. During the second reading of the Bill, the Birmingham member, Joseph Chamberlain, felt obliged to emphasise how the desired water 'does not spring from Wales. It comes from Heaven and goes to the sea, and no more belongs to Welshmen than anybody else who stands in need of it.' A Welsh member of parliament, Sir Hussey Vivian, retorted that the English had to pay for Welsh coal and iron, and by the same token Birmingham should make 'some compensation for these valuable and immemorial rights'. No other part of the United Kingdom had grown so fast in terms of population as the colliery and coastal districts of south Wales. Vivian claimed it was in the interests of both suppliers and consumers that the apportionment of gathering grounds should be treated as a national question.[69]

In the case of the Vyrnwy scheme, parliament found the downstream interests divided in how they believed the scheme might affect the river. Navigation interests stressed the dangers posed to the river Severn – this 'great artery of the trade and commerce of the Midland Counties'. The riparian interests between Tewkesbury and Gloucester petitioned in favour of the Bill, arguing that it 'would tend to regulate the flow of water down the river Severn and diminish the flooding by which much damage was done to their property'. Between sittings of the Select Committee, Liverpool Corporation agreed to raise the amount of compensation water to about a quarter of the available yield, and created the precedent of offering an additional amount 'for flushing purposes'. The Severn

---

[68] A. Briggs, *History of Birmingham: Borough and City. 1865–1938* (Oxford, 1952), pp. 89–91.
[69] PD, Commons, 4th ser., 2, 265–307.

Navigation Commissioners could ask for 40 million gallons a day from a total of 1,280 million gallons to be discharged over a period of 4 successive days in each month between 28 February and 1 November each year, with at least 14 days between each flush.[70]

The Birmingham Bill was the first occasion when considerable attention was paid to the needs of fisheries, as opposed to industry, in assessing compensation needs. The Elan and Claerwen valleys included some of the finest spawning grounds for salmon on the river Wye. Even if practical, it would have been too expensive to build fish passes for the six proposed dams of over 95 feet in height. In a report on the Bill, the Board of Trade recommended, on behalf of the fisheries interests, that the Corporation should set up a fund for such purposes as the removal of obstacles to the ascent of fish on other tributaries of the Wye. Under the Act, the sum of £7,500 was paid to the Board of Trade, to be held in trust for the improvement of the fisheries by the Conservators of the Wye Fisheries District.[71]

The first major clash between water undertakers and the protagonists of amenity occurred in 1878, when Manchester Corporation turned to the Lake District for water supplies. The earlier water schemes of Glasgow and Whitehaven to abstract water from Lock Katrine and Ennerdale respectively had made no difference to the appearance of these lakes. Manchester wanted to impound Thirlmere and increase its size from 335 to 800 acres. Opposition was so considerable that the Chairman of the Committee of Ways and Means, with Government support, agreed to the appointment of a Hybrid or Special Select Committee, with an Instruction to consider not only the aspects raised by petitioners against the Bill, but all aspects of the scheme.[72] The evidence taken on the Bill highlighted differences in the way amenity was perceived. There were those who wanted to keep Thirlmere very much like a preserve, where, in the words of the member of parliament for Cockermouth, these 'aesthetic gentlemen' could come 'down surveying the district and building Gothic villas' on every vantage point. These gentlemen would be displaced if the Corporation was able to purchase all 'the surrounding hills so as to preserve the gathering ground either from houses or mines'.

[70] PD, Commons, 3rd ser., 250, 1278–95; *Special Report of the Select Committee on the Liverpool Corporation Water Bill*, BPP, 1880, x; Liverpool Corporation Waterworks Act, 1880, 43 and 44 Victoria, c. cxliii.

[71] House of Lords RO, Select Committee, 1892, Birmingham Corporation Water Bill, vols. 2–4, Minutes of Evidence; Board of Trade, *Report on Birmingham Corporation Water Bill, 1892*, BPP, 1892, LXXII; Birmingham Corporation Water Act, 1892, 55 and 56 Victoria, c. clxxviii; E. L. Mansergh and W. L. Mansergh, 'The works for the supply of water to the City of Birmingham, from mid-Wales', *Proc. Inst. Civil Engineers* 190, 1912, pp. 3–88.

[72] PD, Commons, 3rd ser., 237, 1503–33.

By doing so, the Corporation would preserve 'the natural state' of the hills for future generations. One of the best-known and most visited lakes would be transformed into 'an extensive people's park, with a serpentine containing two artificial islands in the middle, and a great broad path all round'.[73]

The Select Committee upheld the principle of safeguarding the amenity of the catchment, but dismissed the case against the Manchester Bill on points of detail. Thirlmere possessed 'many advantages as a source of water supply for a large district (such as Manchester), if a portion of water could be removed without destroying the beauty of a scenery which is a valued possession of the whole nation'. The proposed engineering works were of 'great simplicity'. The highest point of the embankment would be only 50 feet, one well 'in proportion to the lofty hills around'. By raising the level of the lake, the Corporation might not only enhance the appearance but would in effect be restoring the water to 'its ancient condition'. The difference between high and low water in the natural lake was 8 feet; it would take a drought of exceptional severity to promote this difference in the reservoir. The temporary fluctuations would hardly be conspicuous, especially as the margins consisted of shingle and not mud.[74]

There was no doubt in the mind of George Bolam, the author of *Wild Life in Wales*, that the formation of the large reservoir on the Vyrnwy had added to the beauty and attractiveness of the Montgomeryshire mountains. Whilst the site of the ancient village was covered by the reservoir, the houses and churchyard had been moved to a new site near the base of the dam. On an eminence at the other end of the embankment was a palatial hotel for the convenience of anglers and visitors to the district. The lake itself was well stocked with common and rainbow trout. Whilst they appeared to be increasing in number and size, the dearth of food in early spring, a common feature of 'most waters at a highish altitude', acted as a limiting factor. As the trees grew up, and other vegetation increased around the margin, the lake was likely to improve as a hatching ground for all kinds of insect and crustacean life. Meanwhile, as Bolam pointed out, a great deal was being done by the authorities to assist Nature.[75]

Not every water undertaker could look to the hills. For some, the only uncommitted and potable supplies were underground. The tripling of London's population in the first half of the nineteenth century made the

---

[73] House of Lords RO, Commons Select Committee, 1879, Manchester Corporation Bill, Minutes of Evidence.

[74] *Report of the Select Committee on the Manchester Corporation Water Bill*, BPP, 1878, XVI; Manchester Corporation Act, 1879, 42 Victoria, c. xxxvi.

[75] G. Bolam, *Wild Life in Wales* (London, 1913), p. 116.

search for new sources especially urgent. The main task of the Royal Commission on Metropolitan Water Supply, appointed in 1891, was to identify how these additional supplies might be obtained, bearing in mind the claims of others to the same underground reserves. The report and proceedings of the Commission marked an important stage in a debate that had been developing for over half a century.[76]

The dichotomy of urban and rural interests was most marked in Hertfordshire, where, with minor exceptions, all the rivers originated in the county and flowed southwards into the metropolitan area.[77] Although the New River and East London Companies obtained nearly five-sixths of their needs from these sources, and supplied 2.6 million people living in the London area, hardly any water was supplied to consumers in Hertfordshire, who had to look to more local water-undertakings. Even if there had been ample supplies for both metropolitan and Hertfordshire needs, guidelines were still needed as to how they might be allocated equitably. If inadequate, as Hertfordshire maintained, ways had to be found for deciding who should be obliged to take the more expensive course of securing additional supplies.[78]

Particularly following the establishment of the Metropolitan Water Board in 1902, it became even more important that the supply areas in Hertfordshire should be able to mount a powerful and co-ordinated response. It was in that context that the creation of the Hertfordshire County Council was so important. Right from the start, the Council petitioned against every scheme for the construction of wells in the Lea and other valleys for the supply of water to 'outside areas'.[79]

Everything hinged on the Council being able to prove, on the basis of detailed observation and hydrological theory, that further abstraction would harm the interests of those living in the supply areas. The Royal Commission rejected both the empirical and theoretical evidence put forward to demonstrate that pumping had led to falling water-levels. It recommended that a further 56 million gallons a day might be safely abstracted. For the County Council, the most urgent priority was to begin monitoring well-levels systematically, both in areas where water was being abstracted and where exploitation was expected to start.[80] In a letter to *The Times* in 1898, Sir John Evans (a leading millowner and chairman of

[76] *Report of the Royal Commission on Metropolitan Water Supply*, BPP, 1893–4, XL, C7172.

[77] W. Whitaker, *The Water Supply of Kent* (London, 1908); W. Whitaker, *The Water Supply of Buckinghamshire and Hertfordshire from Underground Sources* (London, 1921); J. Sheail, 'Underground water abstraction: indirect effects of urbanization on the countryside', *JHG*, 8, 1982, pp. 395–408.

[78] J. Mansergh, 'Presidential Address', *Proc. Inst. Civil Engineers*, 142, 1900, pp. 22–6.

[79] Metropolitan Water Board, *The Water Supply of London* (London, 1961).

[80] Hertfordshire RO, County Council, Minute Books and Highways Department, Water Supplies Papers; Royal Commission on Metropolitan Water Supply, BPP, 1893–4, XL.

the Council's parliamentary committee) warned of how the increased demands of the water companies were turning springs into swallow-holes – the direction of flow underground was being reversed. A few weeks later, even the famous Chadwell Spring became a swallow-hole, despite the wet season. A decline of 40 per cent in its annual flow reflected an almost threefold rise in abstraction rates. For all practical purposes, the flow was irreversible. There were estimates that it might take up to forty years for the levels recorded in the chalk beneath Hertfordshire in the 1860s to be restored, even if all pumping stopped.[81]

Evidence of a widening concern as to the consequences of unregulated underground water abstraction was evinced during the consideration of a Bill promoted by the Metropolitan Water Board in 1907, when the Lord Chairman of a Joint Committee of both Houses of Parliament decided that it was no longer enough for parliament to grant powers of abstraction, and to leave it to technology and the resources of the individual undertaker to decide the location and capacity of the pumps. Limits had to be set on the amounts of water to be abstracted.[82] A further important precedent was set in 1909 when, following expressions of concern in the House of Commons, an Instruction was made requiring Select Committees on Water Bills to make provision for recompense to those adversely affected by the abstraction of underground water.[83]

In 1909, a Private Members' Bill, drafted by the County Solicitor of the Hertfordshire County Council, sought to prohibit the sinking of any further wells without the explicit approval of parliament, and to make it obligatory for undertakers to meet legitimate claims for compensation. A Joint Committee of both Houses heard evidence of how contour maps of water levels were being compiled for an area of 900 square miles, based on readings taken at 800 wells in Hertfordshire. According to the Council's expert witness, the general and considerable depletion of underground reserves could be correlated to 'a very extraordinary degree' with the increase in abstraction.[84]

In its report, the Joint Committee found the need for legislation proven in such instances as Hertfordshire, but feared that the Bill, as drafted, might substitute 'other grievances scarcely less serious than those' it sought to remedy. It would be difficult to cast a general measure sufficiently widely, and yet rigidly, to take account of the manifold peculiarities of water schemes and in 'the absence of trustworthy and indeed often any information as to the subsoil water available in any par-

---

[81] *The Times*, 31 August 1898; J. Hopkinson, 'Water and water supply, with special reference to the supply of London from the chalk of Hertfordshire', *Trans. Hertford. Nat. Hist. Soc.*, 6, 1891, pp. 129–61; J. Hopkinson, 'The Chadwell Spring and the Hertfordshire Bourne', *Trans. Hertford. Nat. Hist. Soc.*, 10, 1898, pp. 69–83.   [82] Hertfordshire RO, County Council, Minute Books.
[83] PD, Commons, 4th ser., 2, 255–87.   [84] PD, Lords, 2, 926–42, and 5, 452–3.

ticular area'. With reluctance, the Joint Committee concluded that control was best secured by inserting any safeguards required, as and when the individual Water Bills came before parliament. Such individual scrutiny should be conducted in the light of knowledge gained from a comprehensive and detailed survey of water supplies and needs throughout England and Wales.[85]

[85] *Report of the Joint Select Committee on the Water Supplies (Protection) Bill*, BPP, 1910, VI.

CHAPTER 35

SUMMARY AND REFLECTIONS

BY GORDON E. CHERRY AND JOHN SHEAILS

In the period between the middle of the nineteenth century and the outbreak of the Great War, Britain changed from a nation which was still essentially agrarian and rural to one unmistakably industrial and urban. England was at the forefront of this transition. The growth of the population of London and other major built-up areas was so striking as to attract a new name. The sociologist, biologist and planner, Patrick Geddes, called these centres, 'conurbations'.[1] There was a considerable expansion of manufacturing in cities and those regions whose economies were dominated by the availability of raw materials. Coal mining and iron manufacture enabled Wales to share the drama of this transition.

A new balance was being struck between town and country; the extent of rural land contracted, while urban requirements and influences impacted more and more on rural life. How far the period was pivotal, in the sense of marking the end of one set of conditions and the beginning of another, remains for discussion. This chapter will seek to address that question.

Of the reality of change, there could be no doubt, given all the contemporary evidence of observation. George Bourne (alias Sturt) took his pen-name from the Surrey heathland where he lived (The Bourne, south of Farnham in Surrey). He described some of these transformations in his own community in *Change in the Village*, published in 1912.[2] The most major change was the enclosure of the common in 1861, whereby 'a few adjacent landowners obtained the lion's share, while the cottagers came in for small allotments'. The latter were soon sold and became the sites of the first new cottages for a newer population. Then, after the turn of the century, greater change was effected when the valley was 'discovered' for residential development. A water company provided a good supply and speculators proceeded to build. A population of 500 increased to over 2,000;

---

[1] P. Geddes, *Cities in Evolution: an Introduction to the Town Planning Movement and to the Study of Civics* (London, 1915).   [2] G. Bourne, *Change in the Village*, 1912 (repr. London, 1966), pp. 3, 11.

the final shabby patches of old heath are disappearing; on all hands glimpses of new building and raw new roads defy you to persuade yourself that you are in a country place.

Bourne concluded, perhaps sadly but finally without regret, that

the old life is being swiftly obliterated. The valley is passing out of the hands of its former inhabitants. They are being crowded into corners, and are becoming as aliens in their own home; they are receding before newcomers with new ideas, and, greatest change of all, they are yielding to the dominion of new ideas themselves.

We need a broad canvas, geographically as well as in time, and we can well begin with a resumé of earlier chapters. We might put ourselves in the position of a young man of twenty, in 1850, and consider what he would have seen as the essential features of the countryside in his early manhood; and then in his mid-eighties, in 1914, noticing both the continuities and the upheavals of his personal experience. We shall see that some things remained remarkably constant, in spite of outward appearances, whilst in other respects a virtual revolution had occurred. In order to give a more focused impression, we select one area, Hardy's Dorset, for a more detailed account. No one case study can be typical, and certainly Wessex was not, but the novelist captured many of the rural changes of his day, and the example is as good as any.

We seek to place the period in context, in the sense of taking both a backward and forward look. Retrospectively, we shall see that some of the changes experienced, at least in their essentials, were not novel to the period of the 1850s; rather, they represented the continued working-out of secular trends of previous centuries. Prospectively, we shall appreciate that other changes, first recognised in the second half of the century and particularly around 1900, continued their full thrust into the twentieth century. The continuities are certainly striking: a report on the countryside, published in 1990, portrayed England's rural areas as an 'arena' – 'an arena in which different concerns and aspirations stimulated by social and economic change, meet and must somehow be reconciled'.[3] There was the same arena in 1890 too.

We then turn to a key feature of the late nineteenth century, the 'back-to-the-land' movement and the concerns of pastoralism and the preservation of rural traditions. This issue is selected for further observation because of the insight it offers into certain perceptions of town and country held in the twenty years or so before the Great War. In many ways, we shall find that our own twentieth-century countryside agenda was set in this formative period, in that 'modern' issues, such as countryside and

[3] Archbishop's Commission on Rural Areas, *Report. Faith in the Countryside* (Worthing, 1990), p. 4.

coastal protection, rural leisure and recreation, and notions of a planned countryside for public enjoyment, are derived from the circumstances of this time.

Last, we indulge in a little speculation about the medium-term future, into the next century. The theme here is to argue that the period between 1850 and 1914 saw a divorce between town and country in economic and social affairs, a formerly significant inter-relationship progressively fractured. At the same time, towns spilled over into the countryside, and urban influences dominated. The twenty-first century may see a return to a much less distinct demarcation between urban and rural in economic, social and land-use terms. If the thesis has any merits, it might suggest that one of the cornerstones of twentieth-century land-use planning, namely the separation of town from country, may have run its course, the time for a rather more flexible strategy having arrived.

A. THE EXTENT AND NATURE OF CHANGE

Changes in agriculture were particularly pronounced, the dominant rural economy experiencing three distinctly different fortunes, all within the space of little more than half a century: the prosperity of high farming, followed by twenty years of depression, and, last, a return to some stability but little overall profitability. During this time, secular shifts in patterns of food consumption and international trade in food products led to pronounced swings in farm output, with the result that pasture took on a new significance, even in the drier eastern counties.

By the early 1900s, farming was much less important in the national economy, measured in terms of gross national product and numbers of working population. In the early nineteenth century, it provided one-third of the national product, compared with one-fifth by the 1850s and less than one-tenth by the end. Its share of national employment declined in approximately the same proportions, although the numbers actually employed changed little overall (a figure of 1,700,000 being estimated) for both 1801 and 1881). Depending upon exactly where he lived, our twenty-year-old in 1850 might well have concluded by the end of his life that agricultural prosperity was a thing of the past; lower food prices for the town-dweller, made possible by overseas competition, had seen to that.

Yet during a time of relative decline, important advances were recorded. Except in the worst years, between the 1870s and mid-1890s, and in the hardest-hit areas, farming improvements were sustained by a capitalist agriculture. Land was better and more systematically drained, new farm buildings were erected, older ones improved, better farming methods were adopted, mechanisation was introduced and scientific innovations advanced. Overall increases in land and especially labour productivity were recorded. Furthermore, certain sections of national agriculture were growing in importance; dairy farming and market gardening made significant advances.

The signals were, therefore, conflicting, but for many on the land the overall impression may well have been one of continuity. The improvement schemes involving the drainage of wetlands or reclamation of heathland extended and, in some cases, completed work that had been under way for centuries. The further disafforestation of England maintained a trend of long standing. The farming profile remained remarkably the same: it was a land of small and medium-size farms, where agriculture was practised according to a delicate balance struck between landlord and tenant. In that balance, the unfortunate lot of the landless farm labourer proved remarkably enduring.

Perhaps our young observer in 1850 might have concluded, by the end of his life, that industry had undergone greater change. Manufacturing had once been well represented in rural areas. Mills and factories in the early years of the Industrial Revolution were as likely to be found in the countryside as in towns. By the early 1900s, manufacturing was unambiguously concentrated on the coalfields, in the larger cities, and at tidewater locations. Developments had been set in train that, during the second half, virtually completed the shift in balance. Opportunities for practising a dual economy of farming and manufacture all but disappeared. The economics of urban production simply eroded the productive capacity of rural crafts; machine technology swept aside handicraft industries. Farm engineering, brewing, food preparation and the like were given urban locations. The 'back to the land' imperative, encouraged by the Arts-and-Crafts movement, saw a very modest revival of some rural trades, but it was too little to reverse the century-long trends.

When we turn to housing, the scene again becomes complex. Far-reaching initiatives were taken, but the underlying dichotomy persisted, namely the poverty of housing conditions for the labourers and the opulence of the country house for the gentry and larger landowner. Model housing schemes suggested sensible improvements, but the gap between what the labourer could afford, and what the housing reformer thought he should have, could rarely be bridged. Model villages were even more impressive portents of what could be achieved in the way of improved living-conditions, but enlightened estate developers found it easier to rehouse better-paid employees than their labourers. The housing reform movement focused on urban situations rather than rural. Whilst the questions of health and the moral dangers of overcrowding had equal applicability to town and country, housing legislation brought little respite to country towns and villages.

The improvement in rural housing that did take place could hardly be called rural, but rather peri-urban or suburban. A succession of factory villages demonstrated what could be achieved in a rural setting. At least one (Bournville) was called 'a factory in a garden'. Bromborough Pool, Street, Saltaire and the Bradford–Halifax group of settlements, Port Sunlight, Earswick and Woodlands Colliery Village, provided an impressive lineage of housing improvements for workers in countryside locations. Many living in the countryside must have been conscious of the striking differences between their own homes and the sturdy, well-designed and relatively commodious dwellings being erected on farmland on the edge of larger towns, and particularly London, ranging from the villa residences of the well-to-do to the regular terraced houses of clerks, shopkeepers and city-workers.

The irony is that around the turn of the century a new generation of architects and house-builders found inspiration in traditional, rural housing-stock for an urban population. Cottage design came into its own, Unwin-inspired, as at Letchworth, the numerous garden suburbs, or in some of the housing estates of the London County Council. The essential ingredients of twentieth-century suburban development came to be based on a nineteenth-century idyll that was in part a myth and in part a parody of an underprivileged, ill-housed section of the rural community.

Our aged rural observer in 1914 would have been most impressed, perhaps, by population changes that had taken place. There was no mistaking the growth of towns; men left for construction work and the wide variety of other trades available in the larger country towns or cities, women for domestic service, and an unknown number of individuals and families departed for America or the colonies. Whilst new forms of employment appeared in the countryside, as in the police, public services and railways, the overall imperative, particularly for the landless, was to leave. The unmistakable attractions were the range of job opportunities, higher pay, better housing and the other perceived facilities of the towns, and the stimulus of education. Absolute numbers may have changed little (rural populations in some parts actually increased), but relatively a decisive shift had taken place in the balance of urban and rural populations.

A new settlement geography took shape. In the more remote rural areas, where farming was marginal, there was depopulation. Elsewhere, the broad picture was one of stagnation or gentle decline after 1861 in the countryside, interspersed with instances of dramatic growth: a country market town favoured, perhaps by railway development; seaside resorts particularly in the south and east of England; new towns created by the railways or industry, and of course suburban development, which reached its greatest manifestation around London.

The countryside was being used unrestrainedly for urban purposes. As the informal 'plotlands' so clearly indicated, it was at the urbanite's disposal. Parts of Bedfordshire were systematically excavated for the clays to make the bricks needed for the building of suburban London. Pennine watercourses were impounded for the urban populations and industries of Lancashire and Yorkshire. Industrial land, railway sidings, docks, sewage works, cemeteries, hospitals and golf courses were all typically located on the urban periphery. Urban land-hunger, increasing exponentially, could only be met at the expense of the rural land reserve.

Changes in the governance of rural England and Wales would have been harder to detect. In spite of determined efforts to alter the traditional power structure, the observer in 1914 might have concluded that not all that much had changed. The privilege of landownership was still paramount. Yet there had been a redistribution of a kind in social and political advantage. The electorate had been enlarged. National, urban politics impinged on rural concerns. Trade union power was recruited. Radical political programmes and the appeal of land taxation promised to tackle the 'land question'; the creation of allotments and small holdings suggested the re-establishment of a prosperous, land-based yeomanry; and local-government reform gave renewed status to the parish, partly in an effort to break down the influence of the 'squirearchy'. Such changes were too subtle for ready observation and, in any case, needed time to have their effect. Rural England changed slowly and, with its different cultural emphases, Wales even more slowly.

B. CONTINUITY AND CHANGE: HARDY'S DORSET

The changes summarised above have all been described in greater detail in earlier chapters. Even so, the treatment is synoptic, and there is a danger of distorting the reality experienced in particular locations. The rural areas were (and are) heterogeneous, and important differences obtained between different tracts. For this reason, distinctions have been drawn between woodland areas, the lowlands and uplands, and the coasts and river valleys, and more detailed accounts given of these. Many more important subdivisions could be identified. To give some flavour of the period in a well-defined district, we may use the example of Wessex, or more narrowly the Dorset of Thomas Hardy, helped as we are by his acute observations of rural life, as it had been and become.[4]

Hardy was born in the hamlet of Upper Bockhampton in an uncultivated area of wood and heath, east of Dorchester. With the nearby village of Puddletown, it formed an isolated and undeveloped countryside, with

---

[4] J.Fowles and J. Draper, *Thomas Hardy's England* (London, 1984).

its own dialect, customs and traditions. The wretched conditions of the Dorset labourers won national notoriety in the 1830s, through the agricultural riots, rick burning and fate of the Tolpuddle Martyrs from the nearby village. Emigration was one answer; the population of Puddletown fell from 1,334 in 1854 to its lowest level of 934 in 1901. Domestic life was arduous, no cottage had piped water, and overcrowding was severe. In the Report of the Royal Commission on the Employment of Women and Children in Agriculture, of 1869, it was reported that, in the forty-two parishes that comprised the unions of Wimborne and Cerne, 16 per cent of the cottages had only one bedroom, and a third of these contained families with more than three children.[5]

Half the labour of Puddletown was engaged in farming – a high level, but broadly representative of Dorset as a whole. As early as 1860, the markets of the small cottage industries, such as button-making and lace-making, had been taken over by factories elsewhere. The impact of the Ashton patent machine button on the hand-made button industry of east Dorset in the 1850s was plain for all to see. Hundreds of wire-makers, paperers and button makers, including 350 in Sherborne, were thrown out of employment. Many were 'sent off' to Perth, Moreton Bay and Quebec by 'the noblemen of the county'.[6] Only a few specialised crafts, such as rope-making at Bridport and flax-spinning at Burton Bradstock, survived. Many villages still had at least one mill for grinding corn. There were sufficient blacksmiths, wheelwrights, shoe makers and other crafts-men for the larger settlements to preserve a degree of self-sufficiency.

About a third of the county was owned by seventeen families, mostly nobility, each estate being of over 7,000 acres. Those who actually worked the land owned little of it – even their rights of common had, for the most part, been lost. The Royal Commission of 1868 took a typical farm of 660 acres in Dorset. It had 500 acres of arable, 120 acres of pasture, and 40 acres of downland. Twenty-nine men and boys were employed throughout the year, with an extra four men and eleven women for harvest. The high but typical figure of three men and two boys per 100 acres reflected how little farmwork was mechanised. The principal exception was threshing, although new machinery was begin-ning to appear in other sectors. Eddison's Steam Plough Works was set up in Dorchester in 1870. Milking was still done by hand and, as milk production rose in response to urban demand and the general improve-ment in prosperity, more dairymaids and men were required. Whilst parliamentary enclosure might lead to displacement, reclamation might

[5] *Reports from the Assistant Commissioners to the Royal Commission on the Employment of Children, Young Persons, and Women in Agriculture*, BPP, 1868–9, XIII, p. 6.

[6] J. E. Acland, 'Dorset "Buttony"', *Proc. Dorset Nat. Hist. & Antiquarian Field Club*, 35, 1914, pp. 71–4.

open up fresh opportunities for employment. Fordington Great Field, adjoining Dorchester, contained 1,500 acres (and no fences) when the parish was enclosed in 1874. Three farms were established where forty tenants had once held the land. On the heathlands of the county, to the east of Bournemouth on what Thomas Hardy called Edgon Heath, after the Iron Age hillfort of Eggardon, more labour was required to establish and maintain the fields and farmsteads created out of the heath.

During Hardy's lifetime, the train completely replaced road and sea transport for long-distance travel. The railway reached Dorchester from Southampton in 1847, and another from Yeovil ten years later. Whilst horse-drawn transport was further challenged by the cycle and motor car, the horse-drawn carrier continued to play a crucial role as the principal means of contact, plying between villages and particularly those towns with a railway station, two or three times a week. Especially for the poor, walking remained the common means of movement between one place and another.

Dorchester (Casterbridge) was the centre of Hardy's life. The focal point of a region, its growth reflected the modest changes taking place. The hiring fair continued to be held on Old Candlemas Day (14 February). The railway had arrived, brewing was prosperous, and mechanical engineering works had been established. As a garrison town, it had two barracks. There were banks, a market, shops, museum, county court, hospital, prison and workhouse. The modest population of 6,500 in 1851 had risen by a third by the turn of the century, at a time when that of so many surrounding settlements was in decline.

C. THE WIDER CONTEXT

Events are by definition time specific and, therefore, different from anything that has gone before – but they may share the same, underlying causes. If we follow the argument of Raymond Williams, the root causes of change in the countryside had for centuries been the use of land for the benefit of different forms of agrarian capitalism, and the consequent demand for labour.[7] The simmering discontent that marked the late nineteenth century in England and Wales falls into place: the dislocation wrought by falling food prices as home production contracted in the face of foreign imports; the dispute over rents and leases which caused a rift to develop between landowners and tenants; the struggle between employers and workers, culminating in the spread of unionism. Within a seemingly stable structure, an important dynamic was provided by agrarian capitalism, the process by which the countryside was transformed.

[7] R. Williams, *The Countryside and the City* (London, 1973).

For centuries, this dominant interest had led to the enclosure of commonland, clearance of woodland, and foundation of settlement. Williams argues that it is a history of disturbance, rather than of natural evolution. For the majority of people, it was a case of substituting one form of domination for another. By the mid-nineteenth century, parliamentary enclosure had secured the appropriation of more than six million acres of land under nearly 4,000 Acts. Enclosure, by whatever method, had led to the concentration of ownership, the stratification of owners and tenants, and the increase of landless families. The system of landlord–tenant–labourer had become even more explicit. The attendant class struggle, highlighted by the ill-feeling aroused over game preservation, was the essential background to the political agitation associated, at the end of the century, with the Land Question.

There had emerged, however, another, crucial form of disturbance. The national economy was now dominated by an increasingly organised industrial and urban market, the effects of which fell heavily on the countryside. Rural structures were already under strain through foreign competition on agriculture; the industrial and urban pressures compounded the difficulties. Rural manufacture was badly hit, and some rural trades disappeared. The contrast between town and country life became increasingly stark. Cities were seen to be draining away the life-blood of the countryside, and the rural economy and society was in no position to withstand the loss. In little more than a decade after the Great War, a quarter of the land of England and Wales was transferred from landlords to owner-occupiers. In that sense, what happened between 1850 and 1914 was part of an unfinished story. The social and economic consequences were still being worked out between the wars and later.

If we now look forward from 1914, the urban influence on the countryside, already apparent, is strengthened considerably. The beginnings of a pronounced take-up of rural land for urban purposes are compounded by the argument that the needs of the urban dweller can best be met on cheaper land in rural locations. Positive forms of dispersal should be adopted so as to tackle the problems of overcrowding in London and other major centres of population. Whilst the garden city might be the preferred strategic model, low-density layouts made up of cottage architecture in informal settings became the new, dominant expression. Letchworth Garden City was accompanied by a range of garden suburbs before 1914. They were the modest forerunners of a veritable tidal wave of suburban building in the 1920s and 1930s, when the development reached a peak of 25,000 hectares per annum. Letchworth was followed by Welwyn Garden City; early garden suburbs burgeoned into private and local-authority estates of huge dimensions. In London, 'metroland' of the private speculator vied with Becontree, then the largest local-authority planned community in the world. What had been presaged, came to full flood after the Great War.

The period 1850–1914 was therefore no isolated one. Its economic and social structures had been prepared. The challenge of industrial and urban influences gave rise to further change and tension during the Victorian and Edwardian periods. The urban expansion that had already begun to affect rural England, particularly around London, was greatly extended after the War. Remedial measures, such as the Green Belt, conceived before the turn of the century, were taken up with greater urgency.

D. CONSERVATION AND PRESERVATION

The release of rural nostalgia, that became so conspicuous a feature at the turn of the century, continued to show great intellectual vigour through-out the century. Indeed, it can be argued that the agenda still used for the maintenance of rural traditions and protection of scenic quality was first set by the late Victorians and Edwardians through the importance they attached to pastoralism.

In spite of a prevalent Victorian belief in progress associated with the power of science and technology, there was, in intellectual circles at least, an increasing fear of the squalor and disease generated by the city, and of industrialism becoming a wholly philistine force. It was argued that the industrial city provided no suitable setting for the natural life of man. A 'back to the land' philosophy fed on these sentiments. For those prepared to consider the evidence, nor was all well in the countryside. A combina-tion over the previous few decades of intensive farming, a near-maximum area of cultivation, dense rural population, game preservation and recreational shooting had impoverished wildlife to a degree never previously experienced.[8]

A landscape, plant or animal, or outdoor recreation, is never so cher-ished as when it is threatened with extinction. By the turn of the century, a love of the countryside was explicitly expressed by painters and writers; campaigns were fought to prevent the destruction of ancient woodlands and the surviving commonlands. Societies were formed for the preservation and enhancement of rural amenities. The Arts-and-Crafts movement flourished, its search for past traditions affecting modes of dress and fashions in architecture, furniture and food. The conferment of royal patronage on the Royal Society for the Protection of Birds in 1904 gave notice of how man's domination of the natural world was being questioned. Strong medical approval was given

[8] E. L. Jones, 'Reconstructing former bird communities', *Forth Naturalist and Historian* 6, 1981, pp. 101–8.

to the burgeoning popularity of outdoor pursuits, whether as cycling, walking or camping.

All this has a distinctly modern ring. There are many aspects of the 1890s that still seem relevant a century later. The concepts may be expressed differently, but a concern for 'green' issues, and the assertion that man has to live in balance with nature, can be discerned a hundred years or more ago. The terms of reference of those early bodies varied from the bland sentiments of the Kyrle Society, dedicated to the promotion of beauty in all its forms, to the more tangible objectives of such groups as the Society for the Protection of Ancient Buildings, but the underlying motives of these pioneers was the same, namely a desire to combat ignorance and greed. In preserving tracts of countryside, a building, or a rare plant or animal, each society believed it was helping to preserve and enhance the essential fabric of human society.[9]

Individuals might lead and shame others into action, but they could not take on the organisation alone. The pressures for rural preservation had to be institutionalised. The founding of societies to protect the commons (1865), ancient buildings (1877), footpaths (1884) and birds (1889) represented a first step, but at least as critical was the founding of the National Trust for Places of Historical Interest or Natural Beauty in 1894. This, together with the conferment of inalienability on its properties under the Act of 1907, represented a major reconciliation between the public interest and rights of private property-ownership – at first in a largely symbolic sense and, more recently, in the size of the Trust's landholding. By the early 1990s, the Trust had come to own over one-hundreth of the land surface of England and Wales, over 200 historic buildings, and about a quarter of undeveloped coastline. A public agency in the true sense of the word, the estate and its management have been built up through specific acts of generosity and the abundance of support and goodwill implied by a membership of almost 2 million.

What then of the role of Government? Again, important precedents can be found at the turn of the century. The Liberal Land Enquiry and Lloyd George's concern for aspects of rural reform may be perceived as a forerunner of later examples of State intervention, first in the agricultural industry through education and research, pricing and marketing mechanisms, and then in forms of land-use planning, including control over development and provision of facilities. Through extensions to legislation concerned with bird preservation, ancient monuments and town planning, and then through more comprehensive statutory devices, more than one-third of the land surface has come to be covered by some

---

[9] Sheail, *Nature in Trust*; P. Lowe, 'The rural idyll defended: from preservation to conservation', in Mingay, *The Rural Idyll* (London, 1989), pp. 113–31.

form of protective designation, in the form of National Parks, Areas of Outstanding Natural Beauty, Green Belts, Environmentally Sensitive Areas, National Nature Reserves and Sites of Special Scientific Interest. The measures taken by both central and local government have both supplemented and complemented the efforts of individual enterprise.

E. THE NEXT CENTURY

We can demonstrate no singular role for the period 1850–1914. Its character owed much to earlier centuries, and it passed on to the twentieth century issues for further evolution. Yet the full force of the impact of industrialisation and the growth of cities did bring new elements to bear on the rural scene, which were to have novel and profound consequences. How might the twenty-first century relate to this scheme of things?

The message of the nineteenth century was that the countryside was for farming, and the towns for industry. The dual economy of rural areas had been shattered. The competitiveness of small-scale industries in the countryside had been undermined. Manufacturing industry had come to be concentrated in cities and major towns. Progress in science and technology seemed to favour large units of production; growing towns commanded immediate markets; higher education and training in new skills were more accessible in towns.

This new separation of town from countryside came at a time when the urban influence on rural areas had never been so striking. Towns and cities were physically breaking out from their historically compact forms. Whereas the built-up area had once abutted hard against the surrounding fields, there was now a more extended and ragged edge. Centripetal forces were replaced by centrifugal. Urban overlapped with rural, as low-density estates spread and, at times, 'jumped' in a formless manner into the countryside, as in the Fleet area of north-east Hampshire. As urban and rural grew apart economically, they physically began to inter-penetrate. Howard's strategic model of the Garden City was followed by countless examples of urban housing which sought out a rural setting and often adopted pseudo-rural forms of architecture.

The twentieth-century response has been to endorse the one trend and curb the other. The official Scott Report, *Land Utilisation in Rural Areas*, of 1942, believed the key to preserving the countryside and providing the necessary level of support to rural communities lay in a 'healthy agriculture'.[10] Rural trades and crafts had their place, but the report's antipathy

---

[10] Ministry of Works and Planning, *Report of the Committee on Land Utilisation in Rural Areas*, Cmd 6378 (London, 1942).

towards industry in the countryside has been generally upheld. Meanwhile, the rash of residential development in the countryside in the first thirty years of the century encouraged a determined effort to regulate and, where possible, prevent its happening. The use of the Green Belt to protect the countryside from such large-scale suburban development has been one of the more enduring aspects of planning since the Second World War.

Yet circumstances may be changing. In June 1989, the architect Leon Krier unveiled his proposals for development on 200 acres of Duchy of Cornwall land at Dorchester; four separate quarters focusing on a new town centre will comprise a new settlement which will increase the size of Dorchester by about one-third over a fifteen-year period.[11] The countryside preservation lobbies are as strong as ever, and no one argues for a general relaxation of planning controls. But there is now a widespread recognition that an increasing proportion of future housebuildings, previously in the form of suburban accretions, might now be accommodated as new communities in 'green' settings. A new type of co-existence is contemplated, whereby small communities are planned and protected in their own 'green' environments, recreating village life, maintaining population dispersal from the cities, yet, through good communications, permitting access to city facilities. In south-east England and around major cities elsewhere, this model of population distribution can already be seen at work, turning its back on fifty years of planning policies. Through the further development of electronic methods of communication, such dispersed forms of residential and occupational location open up opportunities for renewed integration of industrial development in the countryside.

The dispersed city of the twenty-first century would imply, through the return of industry and population, a reutilised, reinhabited countryside. It would also suggest more, rather than less, planning. In recompense, certain forms of rural land-use would return to cities. In the early 1990s, there are already plans for urban forests and special forms of farming in urban environments. At a time when there is a surplus of productive farmland, a return to a more flexible interpenetration of town and country could well have advantages. The challenges in land-use change, met by the Victorians and Edwardian periods, with such profound implications for later generations, may well have to be tackled differently this time round.

[11] G. E. Cherry, 'Town and country planning', in *Contemporary Britain: an Annual Review*, ed. Peter Catterall (London, 1990), pp. 405–11.

PART VII

THE STATISTICAL BASE OF AGRICULTURAL PERFORMANCE IN ENGLAND AND WALES, 1850–1914[1]

[1] This 'Statistical Digest' was completed with the aid of an ESRC Research Grant which was awarded during the academic year 1988–9, for the project *The Statistical Base of Agricultural Performance in England and Wales, 1850–1914*, for which we are grateful.

INTRODUCTION

BY BETHANIE AFTON AND MICHAEL TURNER

A feature in the construction of volumes in the *AHEW* series covering the years from 1500 onwards has been an appendix of 'Statistical sources'.[2] These have been dominated by agricultural prices, though other subjects have also been covered. Some have been compendia constructed from ready-made extracts of data from printed sources, others have come entirely from manuscripts, and the raw data have been transformed into a comparable index. But there has been no extended commentary to accompany what amounts to a plethora of data. Volume VI ventured further and supplied data on many more broad topics than prices, giving appropriate commentaries to elaborate on three of its six main headings. There were commentaries for the sections on Output, Labour and Wages combined, and Land.

It was decided very early on in the planning of the statistics for Volume VII to depart in three ways from the earlier volumes. First, it was felt that it would be useful to cover a larger number of the most important themes in some detail and, secondly, to provide a commentary for each. These take the form of critical 'sketches' of the problems involved in the nature and derivation of the original sources and in their subsequent use. Third, it was deemed necessary not to rely on printed sources alone, but to reconstruct tables that included manuscript sources, and, where necessary, to place them in a contextual framework of their own and also in relation to the surrounding events of which they formed a part. A number of graphs and maps have also been produced from the raw data. All of this treatment is variable in character.

The chapters which follow are arranged in terms of the traditional and broad input factors of production, and their associated returns – Land, Labour, Capital – though normally there would be some overlap. For

---

[2] This was the case in volumes IV and V which covered the periods 1500–1640 and 1640–1750, but in volumes covering earlier periods the 'Statistical Appendix' for volume II (1042–1350) was an appendix of prices and wages, and for volume III (1348–1500) was an appendix on prices, each appended to an appropriate chapter in the main body of the volume.

example, land and livestock might also be considered as part of the capital stock, and rent is traditionally a return to the land factor, but it is also a reward more broadly for capital employed on the land.

Under Land we begin with the broad statistics of agricultural land use, but then also include data on livestock and crop yields, as well as statistics on the farming units themselves, the farms or agricultural holdings. The Land chapter is concluded with a section on the broad return to land, agricultural output and income, followed by a section on the traditional return to the land factor, rent. Under Labour we produce sections on the agrarian population and on agrarian occupations, and then a section on wages as the return to the labour factor. The chapter on Capital is a composite commentary on landlord and tenant capital. Finally, two chapters deal with other essential aspects of the statistical base. One concerns product prices, and the other foreign trade, thus reflecting the growing influence of this aspect of the agricultural economy in relatively modern times.

Finally, a note of caution is offered. By the middle and late nineteenth century, it is no longer useful to talk about the *Agrarian History* simply of *England and Wales*. The agriculture of Britain, or the United Kingdom, is at times a more useful entity for statistical analysis. Indeed, it would have been impossible to disentangle England and Wales from the broader picture of Great Britain or the United Kingdom for every schedule which is produced below. Conversely, it would have been very difficult always to replicate for the other parts of the United Kingdom, for Scotland and Ireland, all those items which are only specified for England and Wales. The best service which the Statistical Base can provide is to make it perfectly clear to which country the separate items refer.

CHAPTER 36

BASIC STATISTICAL DATA

BY BETHANIE AFTON AND MICHAEL TURNER

A. INTRODUCTION

A statistical starting point for the study of modern English and Welsh, or British and UK agriculture, has traditionally been the annual census known as the June Returns. For basic data on acreages under crops and animal numbers this has been the primary data source, and indeed conveniently we already have relatively modern compilations which summarise the main features of those returns in B. R. Mitchell's compendium of *British Historical Statistics*, and also in the centenary history of the returns. The returns were first collected in 1866.[1] But 1866 is not exactly remote from the present. The neglect of successive governments to create the machinery to find out, tabulate, let alone analyse such basic data in what was the premier economy of the mid- and late nineteenth century was bemoaned by contemporaries. For example, in 1851 James Caird remarked that 'There are statistical returns on almost every other subject connected with the business or welfare of the country, but that which may be well regarded as the most important of all – the annual supply of food – is still left to conjecture.' He was to become a constant advocate in the House of Commons for the systematic collection of agricultural returns.[2] The subsequent contemporary analysis of the usefulness of the annual returns is a reminder that the Civil Service as a collecting

---

[1] B. R. Mitchell, *British Historical Statistics* (Cambridge, 1988), chapter III, pp. 180–235. Collection problems in 1866 suggest that 1867 is the first reliable year. In general see J. T. Coppock, 'The statistical assessment of British agriculture', *AHR*, 4, 1956, pp. 17–20. See also, Ministry of Agriculture, Fisheries and Food, *A Century of Agricultural Statistics: Great Britain 1866–1966* (London, 1968), pp. 5–7.

[2] J. Caird, *English Agriculture in 1850–51* (London, 1852), p. 520. On the background in general see Coppock, 'A statistical assessment', pp. 7–16; J. T. Coppock and R. H. Best, *The Changing Use of Land in Britain* (London, 1962), chapter 1; G. E. Fussell, 'The collection of agricultural statistics in Great Britain: its origins and evolution', *Agricultural History*, 18, 1944, pp. 161–7; MAFF, *A Century*, pp. 1–4; L. Napolitan, 'The centenary of the agricultural census', *JRASE*, 127, 1966, pp. 81–96; E. Thomas, 'The June returns one hundred years old', *Agriculture*, 73, 1966, pp. 245–9; J. A. Venn, *The Foundations of Agricultural Economics* (Cambridge, 2nd edn, 1933), chapter 20, pp. 424–40.

agency did not always listen to the advice offered by interested parties. In particular there was worrying interest over the state of the nation's food supply.[3] For example, official estimates (not returns) of crop yields were not made until 1884, before which data we must rely on the independent calculations made by J. B. Lawes and J. H. Gilbert (but for wheat only) and others.[4]

The background to the systematic collection of agricultural data could be a study in its own right, but when it did begin it was provoked by a crisis, the cattle plague (rinderpest) of 1865. In due course livestock returns were collected on 5 March 1866, followed by acreage returns on 25 June. Thereafter the collection for both was regularised as 24 or 25 June until 1877, from which year it was conducted on 4 June, avoiding weekends by adopting the Friday preceding that date, when necessary.[5] Early June is not an inappropriate time in the agricultural calendar for surveying crops at least. It is early enough to capture early Cornish potatoes, but late enough to capture the sowing of spring corn and turnips in the north.[6] It should be noted that the subsequent annual collection of agricultural returns was not without its problems, including the reluctance of land occupiers to furnish details, and the overburden of work placed on the collection agencies. But as far as Volume VII is concerned it is sufficient to note that the returns were voluntarily submitted by holders of agricultural land, of whom perhaps fewer than 3 per cent ever

[3] Caird was an early critic, see particularly, James Caird, 'On the agricultural statistics of the United Kingdom', *JRSS*, 31, 1868, pp. 127–45. See, in particular, a series of articles by R. H. Rew presented to the *Committee Appointed to Inquire into the Statistics Available as a Basis for Estimating the Production and Consumption of Meat and Milk in the United Kingdom*, and presented as, 'Production and consumption of meat and milk: second report', *JRSS*, 67, 1904, pp. 368–84; 'Production and consumption of meat and milk: third report', in *ibid.*, pp. 385–412; 'Observations on the production and consumption of meat and dairy products', in *ibid.*, pp. 413–27. This series of articles was preceded by his 'An inquiry into the statistics of the production and consumption of milk and milk products in Great Britain', *JRSS*, 55, 1892, pp. 244–86. See also his, 'The nation's food supply', *JRSS*, 76, 1912, pp. 100–1. See also P. G. Craigie, 'Statistics of agricultural production', *JRSS*, 46, 1883, pp. 1–58; Craigie, 'On the production and consumption of meat in the United Kingdom', *Report of the British Association for the Advancement of Science* (London, 1884), Section F, pp. 841–7; J. B. Lawes and J. H. Gilbert, 'Home produce, imports, consumption, and price of wheat, over forty harvest-years, 1852–53 to 1891–92', *JRASE*, 3rd ser., 4, 1893, pp. 77–133; R. C. Turnbull, 'The household food supply of the United Kingdom', *Transactions of the Highland and Agricultural Society*, 15, 1903, pp. 197–211.

[4] Lawes and Gilbert, 'Home produce'. See also M. J. R. Healy and E. L. Jones, 'Wheat yields in England, 1815–59', *JRSS*, 125, 1962, pp. 574–9, for relatively modern estimates of mid-nineteenth-century wheat yields based on contemporary sources. On the general background see Venn, *Agricultural Economics*, pp. 447–54.

[5] This basic story is told in all of the sources in note 2 above. For a general narrative on the history of collecting the wider agricultural statistical base see D.K. Britton and K. E. Hunt, 'Agriculture', in M. G. Kendall and A. Bradford Hill (eds.), *The Sources and Nature of the Statistics in the United Kingdom* (London, 1952), vol. I, pp. 35–74.   [6] Venn, *Agricultural Economics*, p. 432.

refused to comply.[7] It should be emphasised that the annual returns were enumerations of holdings, and therefore of units of occupancy, and not of ownership, though of course owners do themselves occupy land. In addition the method of enumeration was in terms of physical quantities, not by valuations.[8]

The tardiness in collecting data on a regular basis is surprising in view of the fact that the Irish had collected equivalent data on an annual basis since 1847, and even included useful material on landholding sizes and animal numbers in the 1841 Irish Population Census. Scotland too gave serious attention to the problem in the period from 1853 to 1857.[9] To the extent that there had been prior investigations or pilot projects we highlight the sample survey of 1854 which looked at nine English and two Welsh counties for that year, and derived details of crop acreages and animal numbers. From this partial survey a national estimate was formed. Scotland had been fully covered in 1854.[10]

In this section of the statistical appendix we present, mainly, a digest of official agricultural statistics incorporating land use, livestock, and crop yields. Numbers of other statistical digests are available, though they do not all distinguish between the constituent countries of the United Kingdom.[11] As far as possible we present data separately for England and Wales.

B. LAND USE

How much land was under crops and grass at any one time between 1850 and 1914? Even with the availability of the annual census returns this is

---

[7] MAFF, *A Century*, pp. 3–4; Napolitan, 'The centenary', pp. 83–7; Thomas, 'The June returns', p. 247; Venn, *Agricultural Economics*, p. 434.   [8] MAFF, *A Century*, p. 1.

[9] *Census of Ireland for the Year 1841*, BPP, 1843, p. 24; Saorstat Eireann, *Agricultural Statistics 1847–1926: Reports and Tables* (Department of Industry and Commerce, Dublin, 1930); Napolitan, 'The centenary', p. 82.

[10] The 1854 survey was preceded by a trial exercise on Norfolk and Hampshire in England, and Haddingham, Roxburgh and Sutherland in Scotland. The history of this 1854 data-gathering experiment is given in a number of critical essays by J. P. Dodd, but see in particular his 'The agricultural statistics for 1854: an assessment of their value', *AHR*, 35, 1987, pp. 159–70. See also S. W. Martins, *A Great Estate at Work: The Holkham Estate and its Inhabitants in the Nineteenth Century* (Cambridge, 1980), pp. 260–2, for a detailed breakdown of the 1854 survey for Norfolk at the level of the Poor Law Unions. The 1854 survey can be found in *Reports by Poor Law Inspectors on Agricultural Statistics (England), 1854*, BPP, 1854–5, LIII, p. 495. See also G. E. Mingay, ed., *AHEW*, vol. VI, *1750–1850* (Cambridge, 1989), pp. 1042–4, for more accessible full details at the county level. See also Fussell, 'Collection of agricultural statistics', p. 164 for notes on Scotland.

[11] See in particular Coppock and Best, *The Changing Use of Land*, Appendix, pp. 234–5; B. R. Mitchell, *Abstract of British Historical Statistics* (Cambridge, 1962), pp. 78–91; Napolitan, 'The centenary', pp. 94–5; L. D. Stamp, *The Land of Britain: its Use and Misuse* (London, 1948), Appendix; Venn, *Agricultural Economics*, Appendix, pp. 551–8.

not always a question capable of an easy answer. Definitions of agricultural land were sometimes adjusted, and in particular this was the case in the definition of permanent and rough grazing, and permanent and temporary grass. The Welsh hill farmers' permanent pasture might be the Leicester graziers' rough pasture.[12] In addition there were changes in definition over what was and was not regarded as bare fallow. For example, in 1869 the definition changed from 'Bare Fallow or Uncropped Arable Land' to 'Bare Fallow or Ploughed Land from which a Crop will not be taken this year'. During the agricultural depression of the last quarter of the century when some cultivated land was 'temporarily' abandoned, there was uncertainty whether this should be regarded as bare fallow. And after how long out of cultivation should land be classified as permanent grass? The introduction of a new category, 'Market Gardens', in 1872 caused much confusion to landholders who could not decide whether some of their acres of peas and potatoes were market-garden crops or field crops. The separate category of market gardens was abandoned in 1897.[13]

The data generated by the Tithe Commutation Act of 1836 might make an appropriate base from which to construct a long-run series of land-use statistics. The resulting survey did not cover the whole country because over large swathes of the land the tithes had been commuted already. One estimate has put the extent of the arable land in England and Wales *c.* 1836 at 15.1 million acres (but excluding seed crops for western counties), with an additional 16.4 million acres of grass (including rough grazing), giving a grand total for the cultivated area of 31.5 million acres. The total area of land was 37.3 million acres, and the residual 5.8 million acres must be assumed to have been woodland, common and unfarmed land.[14]

James Caird's mid-nineteenth-century datum gives an arable estimate of 13.7 million acres, in addition to 13.3 million acres of meadow and pasture, with a final cultivated area of 27 million acres.[15] This compares

[12] MAFF, *A Century*, pp. 7, 10, 12, 90; Coppock and Best, *The Changing Use of Land*, p. 56, and in particular pp. 63–7 for full details and the rather bleak conclusion 'that detailed comparisons of total acreages of land in these categories over long periods of time cannot be made'. Their solution to the problem, which they suggest will also help the comparison of other crops and animals over time, 'is to consider changes, not in total acreages, but in the proportion of land under the various crops. Densities and ratios of livestock should similarly be compared, rather than total numbers', p. 67.

[13] See in particular Coppock and Best, *The Changing Use of Land*, pp. 60–1, and for problems over the returns for small fruits, see pp. 61–2; for difficulties over the definition of temporary/permanent grass see MAFF, *A Century*, p. 12.

[14] R. J. P. Kain, *An Atlas and Index of the Tithe Files of Mid-Nineteenth-Century England and Wales* (Cambridge, 1986), pp. 458–9; R. J. P. Kain and H. C. Prince, *The Tithe Surveys of England and Wales* (Cambridge, 1985), p. 174. [15] Caird, *English Agriculture*, p. 521.

with an estimate for 1854 of 15.3 million acres of arable (including clover and seeds) and 12.4 million acres of meadow and pasture, and a final cultivated area of 27.7 million acres. This estimate is based on the 1854 survey of eleven counties (summarised in Table 36.1).

Apart from these early public and private attempts at estimating the extent of agricultural land, it is only from 1866–7 and the advent of the June Returns that the systematic collection of agricultural data began in earnest. The accuracy of these returns has been questioned, but the degree of inaccuracy is unknown. And as long as there are no, or few, comparable sources against which to compare them, the precise nature of inaccuracy is necessarily blurred. Nevertheless, it is readily accepted that they contain some accidental, and some wilful errors.[16] These arose from three main sources: simply not knowing the true facts; inability to supply the facts; and not putting the facts which had been gathered into the correct broad categories of different crops and animals. Not least of the problems was not knowing the true extent (acreage) of the land.[17] In addition, the definition of an agricultural holding was not properly settled until 1892. In 1866 acreage details were collected only from occupiers with at least five acres; in 1867 the collection was made from all occupiers; in 1869 a limit was drawn at one quarter of an acre, which was raised to one acre in 1892.[18]

Nevertheless, to the extent that we can identify the essentials of the annual returns they form part of this statistical appendix. Table 36.2 gives separately for England and Wales the acreage sown to the corn crops, the principal root and green crops, the rotation grasses, the permanent pastures not normally broken up in rotation, the flax acreage, the acreage under small fruit plantations, and the bare fallow or arable land which was not otherwise cultivated. The practice of measuring the land under small fruits only began in 1888. It was also recognised, and often recorded separately, that fruit could be grown in what was otherwise other cultivated land (orchards in meadows for example). Such fruit land was recorded by the administrators of the annual returns along with the principal crops. By twists and turns there was often, but not always, systematic recording of other crops. In particular, the acreage under carrots was usually recorded separately, or at least it was until the 1880s, though thereafter it was usually classified as an 'other green crop'. In the tables below it is included with vetches, tares, lucerne and 'other green crops'. The acreage under carrots in England declined from nearly 16,000 acres in 1866 to 9,000 by 1914, and in Wales it never exceeded 500 acres. Flax and

---

[16] In general see Coppock and Best, *The Changing Use of Land*, chapter 2.

[17] *Ibid.*, especially pp. 54–5, including the sometime local use of customary acres.

[18] Napolitan, 'The centenary', p. 85.

hops were also marginal crops in Wales and in the tables which refer to Wales their precise area is hidden in the total acreage. It is also possible from the annual returns to gauge the size of the acreage under the different fruits separately, and also some of the other crops such as onions, mustard, buckwheat and so on. All such marginal crops have been collected as 'other' green crops, and also counted in the total acreage. The annual returns also separately recast the rotation grasses into those which were used for hay and those which were not.

From the private and official estimates or returns from the mid-nineteenth century onwards it is possible to construct Table 36.3, the extent of cultivated land in England and Wales. Part of the increase in the cultivated acreage from the 1860s to the 1890s was due to land improvement and reclamation, but an untold amount must have been due to the increasing accuracy of the June Returns.[19]

C. LIVESTOCK NUMBERS

We are less well placed to know the size of the animal herds before the advent of the June returns, though the 1854 survey included animals. The estimated totals for England and Wales for c. 1854 are included in Table 36.1. Table 36.4 records the numbers of different types of animals of different ages from 1866 to 1914, separately for England and Wales. The amount of detail varied slightly over time, especially from 1893 when the authorities systematically separated the ewes and boars from the sheep and pig totals, and they also distinguished cattle specifically below one year of age from all cattle below two years of age. Until 1919 the bulls were included with all cattle over two years of age. In 1866 dairy cattle meant cows, but thereafter cows and heifers in milk and in calf were brought together, though very often it is possible to separate them by a careful inspection of the enumerated annual returns. Horses were first enumerated in 1869. The distinction between horses used solely for agricultural purposes and other horses is a little blurred since at times this definition included breeding mares and at other times it did not.[20] In addition, non-agricultural horses were often recorded as if they were agricultural horses until 1911, at which time a category called 'other' horses was introduced, though it was not closely defined until 1913. Thus the apparent decline in agricultural horses between 1910 and 1913 is due 'largely, if not entirely to the removal of horses from one category to another'.[21]

---

[19] Coppock and Best, *The Changing Use of Land*, p. 58.

[20] See also F. M. L. Thompson, 'Nineteenth-century horse sense', *EcHR*, 29, 1976, pp. 60–81.

[21] E. J. T. Collins, 'The farm horse economy of England and Wales in the early tractor age 1900–40', in F. M. L. Thompson, ed., *Horses in European Economic History* (British Agricultural History Society, Reading, 1983), p. 75.

The main problems involved in comparing livestock numbers are related to definitions, and the significance of different age categories. This is particularly the case when defining the dairy herd since some heifers in some places were reared in order to replace spent cows in other places. This might lead to an exaggeration of the importance of dairying in one place and underscoring it in another. In addition, some cows were genuinely used to produce milk and therefore contributed to the dairy industry, while others were used to suckle beef calves and therefore contributed to the meat industry. And what of the definition of the beef industry? Beef cattle were usually equated with other cattle aged two years or over, but fat cattle could be slaughtered at younger ages, and spent cows also contributed to the meat industry. A reliance on the number of cattle of two years or over, therefore, will not necessarily capture the trends of the beef industry.[22] On odd occasions the returns also recorded poultry, but these were not systematically enumerated.[23]

Therefore we should not be fooled into thinking that these tables are perfectly reliable. In 1885, for example, a request for the number of livestock born on holdings within the year of the return resulted in 220,980 replies from 509,186 known occupiers. Many occupiers simply did not know when all of the animals in their possession were born.[24]

Acreages are pretty well fixed in any one year, but animals at any one time represent some of this year's new acquisitions (and these are not always new births), some of last year's, and some older stock. And in 1867 some livestock which were grazing on common land or moorland were omitted in error because apparently such land was not returnable. When it was felt that there were clear errors of omission, the enumerators adjusted the returns by estimation before they were published in their aggregate form.[25] The seasonality, and in particular the regional seasonality, of lambing and slaughter can also produce problems in interpreting the annual census.[26]

D. CROP YIELDS

Crop yields, and hence essential information towards an estimate of agricultural, or correctly, crop, production, were not systematically collected

---

[22] Coppock and Best, *The Changing Use of Land*, pp. 62–3, though they come to the useful conclusion that most of these definitional and interpretive issues 'are of relatively minor importance. Provided that the steady improvement in the accuracy and completeness of the returns is taken into account, acreages of crops and numbers of livestock may, in general, be safely compared.'

[23] MAFF, *A Century*, p. 54, records that their numbers were collected in 1884, 1885, 1908 and 1913 in the period covered by volume VII, and in 1921, and annually from 1926.

[24] Coppock and Best, *The Changing Use of Land*, p. 55.

[25] *Ibid.*, p. 57; Venn, *Agricultural Economics*, p. 434.　　[26] MAFF, *A Century*, p. 50.

or estimated until the June Returns were close to two decades old, in 1885. Even then, in fact, they were 'estimates' of yield taken before and after the harvest, but checked by testing weights during threshing. Nevertheless, as estimates they were then combined with 'facts' about acreages to produce what must in turn be 'estimates' of production.[27] A number of complementary estimates are available which not only complete data for the years between 1866 and 1884, but also for the period from the early 1850s to 1866. James Caird produced an estimate of wheat yields on a county basis in 1851. This is reproduced in Table 36.5 along with other county-based estimates of corn yields for various dates up to the 1880s.[28] For the longer period 1815–59, M. J. R. Healy and E. L. Jones produced annual estimates of English wheat yields from records held by the Royal Statistical Society. Most famous of all are the estimates of wheat yields produced by Lawes and Gilbert. These last two estimates are reproduced in Table 36.6, along with the officially recorded wheat yield estimates for the years from 1885 to 1914.[29] Long-term yields for England and Wales separately, for selected corn crops, root and green crops, and for hay, are given in Table 36.7, along with estimates of crop production.[30]

E. COUNTY DISTRIBUTIONS

The statistics presented here are a mixture of long-run national annual data, and cross-sectional data at the county level. To a large degree they can be presented more systematically at the regional level. Initially, the self-defined region was, and remains, the county. These may once have had important regional meanings, and indeed were seen administratively as units of collection and analysis, but in terms of agricultural distributions or agricultural regions the individual county does not always make much sense. Nevertheless, the county remains the most convenient unit for constructing regional comparisons. From the point of view of subsequent analysis the main problem was the practice of some occupiers who made their returns as from their county of residence, rather than from the county in which their land was situated. In addition, land-

---

[27] See Venn, *Agricultural Economics*, p. 435, on the work of the Crop Reporters whose job it was to furnish these estimates of crop yields. Each reporter was responsible for about 80,000 acres covering perhaps 1,200 holdings over 40 parishes. The national estimates which were produced were then derived by a system of weighting the many sample estimates.

[28] Caird for 1850 and other estimates for 1861, 1870, 1879 and 1882, and averages for pre-1878, for 1876–82 and for 1863–82, in Craigie, 'Statistics', pp. 40–2.

[29] Healy and Jones, 'Wheat yields', p. 578; Lawes and Gilbert, 'Home produce', p. 133; *Annual Agricultural Returns*, BPP, various. See also chapter 2 above where E. J. T. Collins reviews contemporary yield estimates for the period 1840–80, pp. 86–93.

[30] Reconstructed from *Annual Agricultural Returns*, BPP, various.

holdings were not always co-extensive with county boundaries. Some occupiers clearly returned details as if their consolidated single holdings were in fact a series of smaller holdings. This might have arisen after a period of consolidation where some occupiers continued to make returns separately for the once-separate holdings. Such problems were certainly encountered early on, but largely they may have been rectified by the 1890s.[31]

The final tables in this section therefore reproduce agricultural distributions at the regional (county) level for the years 1875, 1895 and 1915, for crops (Table 36.8), animals (Table 36.9) and landholding distributions (Table 36.10). The first date represents the heyday of 'High Farming', the second the depths of the late-nineteenth-century depression, and the third the end of the period of recovery but before the major changes provoked by the First World War. Some of the most important aspects of the changing agricultural geography of the period are picked out in a selection of maps.[32]

---

[31] Coppock and Best, *The Changing Use of Land*, pp. 59, 68.
[32] *Annual Agricultural Returns*, BPP, various, and the crop yields also from Table 36.5.

APPENDIX

Table 36.1. *Estimates of land use distributions (crops, grass, other) and livestock numbers in England and Wales, c. 1854*

	Sample of eleven counties[a]	Estimated totals for England and Wales[b]
(A) LAND USE (in acres)		
Gross total acreage	7,743,850	37,324,915
Crops		
Wheat	790,019	3,807,846
Barley	553,487	2,667,776
Oats	270,290	1,302,782
Rye	15,297	73,731
Beans and peas	144,854	698,188
Vetches	45,343	218,551
Turnips	470,379	2,267,200
Mangolds	36,777	177,263
Carrots	2,622	12,638
Potatoes	39,894	192,287
Flax	2,107	10,156
Hops	3,937	18,976
Osiers	224	1,079
Other crops, e.g. cabbages	20,194	97,334
Bare fallow	185,888	895,969
Total crops	2,581,312	12,441,776
Grass		
Clover, lucerne and other artificial grasses	585,083	2,820,066
Permanent pasture	1,841,297	8,874,946
Irrigated meadow	268,121	1,292,329
Sheep walks and downs	461,595	2,224,862
Total Grass	3,156,096	15,212,203
Other		
Houses, gardens, roads, fences etc.	202,533	976,197
Waste	163,209	786,658
Woods and plantations	352,154	1,697,362
Commons	401,906	1,937,164
Holdings less than 2 acres	95,322	459,447
Acres unaccounted for	791,318	3,814,108
Total other	2,006,442	9,670,936

(continued)

Table 36.1 *(cont.)*

	Sample of eleven counties[a]	Estimated totals for England and Wales[b]
(B) LIVESTOCK NUMBERS		
Horses		
Horses	218,038	1,050,931
Colts	53,544	258,079
Cattle		
Milch cows	285,629	1,376,708
Calves	146,722	707,192
Other cattle including working oxen	277,860	1,339,270
Sheep		
Tups	50,645	244,106
Ewes	1,514,523	7,299,915
Lambs	1,449,806	6,987,982
Other sheep	862,891	4,159,085
Pigs		
Pigs	490,405	2,363,724

Notes:
[a] These counties were, in the order in which they appear in the original, Hampshire, Wiltshire, Leicestershire, Norfolk, Suffolk, Berkshire, Worcestershire, Brecknock, Salop, Denbigh and Yorkshire West Riding. Full details are available for all these counties.
[b] The method of estimation employed at the time was to increase each land use and animal number in the ratio 7,743,850 to 37,324,915, the ratio of the combined acreage of the eleven counties to the acreage of England and Wales.

Sources: Reports by Poor Law Inspectors on Agricultural Statistics (England), 1854.
BPP, 1854–5, LIII, p. 495. See also G. E. Mingay, ed., *AHEW* vol. VI, *1750–1850* (Cambridge, 1989), pp. 1042–4, for more accessible full details at the county level.

Table 36.2a. *England: land use: crops annually (thousands of acres to the nearest thousand), 1866–1914*

These tables are summaries of the original annual returns and are expressed in 000s of acres. They are subject to rounding errors which accounts for any apparent differences when the totals are summed across the page.

Date	Wheat	Barley and bere	Oats	Rye	Beans	Peas	Total corn
1866	3,126	1,877	1,504	51	493	314	7,365
1867	3,140	1,892	1,506	43	506	312	7,399
1868	3,397	1,780	1,488	38	504	292	7,499
1869	3,417	1,864	1,512	53	548	391	7,785
1870	3,248	1,964	1,491	53	504	312	7,570
1871	3,313	1,964	1,454	58	513	382	7,684
1872	3,337	1,896	1,442	52	496	353	7,577
1873	3,253	1,926	1,419	40	553	311	7,502
1874	3,391	1,890	1,357	37	527	304	7,505
1875	3,129	2,090	1,422	44	533	311	7,529
1876	2,823	2,109	1,534	46	488	288	7,288
1877	2,987	2,001	1,490	49	470	306	7,303
1878	3,041	2,062	1,430	50	412	278	7,275
1879	2,719	2,236	1,425	40	420	274	7,113
1880	2,746	2,061	1,520	32	404	231	6,994
1881	2,641	2,029	1,627	32	418	213	6,961
1882	2,829	1,858	1,533	47	409	243	6,919
1883	2,467	1,912	1,675	41	422	236	6,752
1884	2,531	1,808	1,620	38	422	226	6,645
1885	2,349	1,894	1,648	42	409	227	6,569
1886	2,161	1,899	1,772	47	360	211	6,450
1887	2,198	1,760	1,768	46	350	226	6,347
1888	2,419	1,742	1,616	64	321	238	6,401
1889	2,322	1,776	1,624	59	305	222	6,307
1890	2,256	1,776	1,648	45	340	216	6,281
1891	2,192	1,772	1,673	38	337	202	6,215
1892	2,103	1,710	1,765	39	295	192	6,103
1893	1,799	1,752	1,914	47	229	208	5,950
1894	1,827	1,766	1,978	81	229	241	6,122
1895	1,340	1,838	2,045	60	229	207	5,719
1896	1,609	1,779	1,846	65	237	194	5,729
1887	1,786	1,698	1,829	67	214	187	5,781
1898	1,987	1,563	1,731	60	217	173	5,731
1899	1,900	1,636	1,782	44	235	160	5,755
1900	1,745	1,645	1,861	46	249	154	5,699
1901	1,618	1,635	1,832	50	237	152	5,524
1902	1,631	1,579	1,893	61	229	177	5,569

(continued)

Table 36.2a (*cont.*)

Date	Wheat	Barley and bere	Oats	Rye	Beans	Peas	Total corn
1903	1,497	1,545	1,954	52	227	179	5,455
1904	1,302	1,544	2,060	49	241	174	5,370
1905	1,704	1,410	1,880	55	243	173	5,467
1906	1,661	1,440	1,881	57	276	152	5,467
1907	1,537	1,411	1,968	54	296	164	5,430
1908	1,549	1,383	1,959	46	284	162	5,382
1909	1,734	1,379	1,840	49	303	182	5,487
1910	1,717	1,449	1,858	42	258	167	5,491
1911	1,804	1,338	1,841	40	300	166	5,489
1912	1,822	1,365	1,866	53	276	200	5,582
1913	1,663	1,470	1,772	51	267	163	5,387
1914	1,770	1,420	1,730	52	293	168	5,434

Date	Potatoes	Turnips and swedes	Mangold	Cabbage, kohl-rabi and rape	Vetches lucerne & other green	Total green
1866	311	1,611	254	160	425	2,760
1867	290	1,621	254	129	398	2,692
1868	327	1,606	244	111	296	2,585
1869	357	1,615	287	141	359	2,759
1870	359	1,642	300	140	319	2,760
1871	392	1,593	352	175	387	2,898
1872	339	1,512	321	172	435	2,779
1873	309	1,540	317	169	414	2,749
1874	315	1,561	315	163	411	2,764
1875	320	1,569	352	183	423	2,848
1876	305	1,561	339	173	375	2,752
1877	304	1,496	348	176	435	2,759
1878	302	1,467	334	164	414	2,681
1879	324	1,458	353	162	440	2,736
1880	325	1,473	334	155	373	2,659
1881	348	1,479	339	138	378	2,665
1882	332	1,463	326	143	401	2,685
1883	335	1,469	322	139	382	2,647
1884	360	1,472	319	140	396	2,688
1885	359	1,461	346	146	426	2,738
1886	364	1,448	341	145	399	2,697
1887	369	1,419	352	146	395	2,681
1888	391	1,391	352	151	396	2,681
1889	385	1,370	318	137	308	2,518

(*continued*)

Table 36.2a (*cont.*)

Date	Potatoes	Turnips and swedes	Mangold	Cabbage, kohl-rabi and rape	Vetches lucerne & other green	Total green
1890	348	1,393	323	147	311	2,523
1891	355	1,368	345	146	317	2,530
1892	350	1,390	352	142	277	2,510
1893	356	1,424	339	146	263	2,527
1894	341	1,400	344	165	292	2,542
1895	373	1,362	326	142	264	2,467
1896	400	1,337	329	148	296	2,511
1897	352	1,288	345	151	314	2,450
1898	365	1,237	343	150	303	2,399
1899	388	1,204	363	156	299	2,410
1900	397	1,160	402	178	305	2,442
1901	415	1,144	386	162	294	2,401
1902	413	1,092	426	176	317	2,424
1903	403	1,085	388	163	299	2,338
1904	403	1,091	386	159	271	2,310
1905	435	1,084	392	160	282	2,353
1906	397	1,083	418	161	298	2,357
1907	382	1,058	436	166	326	2,368
1908	391	1,052	415	152	294	2,305
1909	406	1,057	443	154	312	2,371
1910	377	1,064	429	142	276	2,288
1911	403	1,067	439	140	283	2,331
1912	437	1,016	473	155	316	2,397
1913	417	997	409	131	289	2,243
1914	436	990	421	132	324	2,303

Date	Clover, sainfoin or grass in rotation	Permanent grass not broken in rotation	Flax	Hops	Small fruit	Bare fallow	Total acreage
1866	2,296	8,998		57		761	22,237
1867	2,478	9,546		64		753	22,932
1868	2,371	9,704	16	64		800	23,039
1869	2,005	10,096	19	62		644	23,371
1870	2,767	9,680	22	61		549	23,409
1871	2,694	9,882	16	60		484	23,718
1872	2,822	9,991	14	62		585	23,830
1873	2,678	10,238	14	63		649	23,894
1874	2,619	10,438	9	66		607	24,008
1875	2,608	10,536	7	69		515	24,112

(*continued*)

Table 36.2a (*cont.*)

Date	Clover, sainfoin or grass in rotation	Permanent grass not broken in rotation	Flax	Hops	Small fruit	Bare fallow	Total acreage
1876	2,787	10,689	7	70		608	24,202
1877	2,737	10,858	7	71		576	24,312
1878	2,785	11,010	7	72		588	24,418
1879	2,675	11,234	7	68		671	24,504
1880	2,646	11,462	9	67		760	24,596
1881	2,549	11,656	6	65		745	24,664
1882	2,546	11,801	5	66		735	24,736
1883	2,585	12,009	4	68		731	24,795
1884	2,545	12,198	2	69		698	24,844
1885	2,750	12,230	2	71		520	24,880
1886	2,763	12,411	3	70		521	24,915
1887	2,825	12,547	3	64		456	24,922
1888	2,747	12,616	2	58	33	427	24,964
1889	2,887	12,701	2	58	37	480	24,991
1890	2,791	12,836	2	55	41	479	25,008
1891	2,762	13,085	2	56	53	410	25,113
1892	2,720	13,037	1	56	57	439	24,924
1893	2,675	13,128	1	58	60	498	24,898
1894	2,605	13,128	2	60	62	360	24,881
1895	2,826	13,245	2	59	68	459	24,845
1896	2,696	13,354	2	54	70	416	24,831
1897	2,886	13,192	1	51	64	369	24,794
1898	2,923	13,254	1	50	63	336	24,757
1899	2,807	13,324		52	65	323	24,736
1900	2,768	13,392		51	67	294	24,714
1901	2,863	13,458	1	51	68	329	24,694
1902	2,825	13,463	1	48	68	282	24,680
1903	2,822	13,581	1	48	69	337	24,651
1904	2,718	13,693	1	48	71	420	24,630
1905	2,574	13,761		49	71	336	24,611
1906	2,539	13,817		47	72	302	24,601
1907	2,612	13,808		45	73	249	24,585
1908	2,557	13,901		39	76	301	24,560
1909	2,383	13,912		33	78	277	24,541
1910	2,360	13,923		33	76	343	24,515
1911	2,327	13,903		33	76	319	24,478
1912	2,237	13,818		35	77	269	24,414
1913	2,240	14,013		36	76	381	24,375
1914	2,122	14,061		37	76	335	24,368

Notes: Small fruit not separately distinguished before 1888.

Table 36.2b. *Wales: land use: crops annually (thousands of acres to the nearest thousand), 1866–1914*

Date	Wheat	Barley and bere	Oats	Rye	Beans	Peas	Total corn
1866	114	146	252	2	4	3	521
1867	117	148	247	3	3	3	521
1868	131	152	257	2	4	3	548
1869	136	158	253	3	4	3	556
1870	127	164	253	2	4	3	554
1871	126	170	254	2	4	5	561
1872	126	168	256	2	3	6	562
1873	117	164	245	2	4	5	537
1874	118	152	236	2	4	5	516
1875	112	154	237	2	3	4	512
1876	94	154	242	1	4	3	499
1877	100	147	239	1	3	4	495
1878	102	148	235	2	3	3	492
1879	95	152	227	1	3	3	482
1880	90	143	240	2	3	2	478
1881	90	142	244	2	3	2	482
1882	95	135	251	2	3	2	489
1883	78	134	255	2	3	2	475
1884	78	130	249	2	3	2	464
1885	74	126	247	1	3	2	452
1886	69	125	250	1	2	2	449
1887	69	119	255	1	2	2	449
1888	77	118	251	1	2	2	450
1889	68	122	249	1	2	1	444
1890	69	120	241	1	2	2	434
1891	62	117	234	1	2	2	417
1892	55	115	233	1	2	1	407
1893	55	112	241	1	2	1	411
1894	56	112	251	2	2	1	423
1895	44	112	242	1	1	1	402
1896	47	108	242	2	1	2	401
1897	54	104	239	2	1	2	402
1898	59	103	231	2	1	2	397
1899	54	106	220	2	1	2	385
1900	52	105	216	1	1	2	377
1901	47	102	209	1	1	2	362
1902	48	101	210	1	1	1	364
1903	43	99	213	1	1	1	359
1904	35	96	212	1	1	1	347
1905	44	91	208	1	1	1	347

(continued)

Table 36.2b (*cont.*)

Date	Wheat	Barley and bere	Oats	Rye	Beans	Peas	Total corn
1906	44	93	205	1	1	1	346
1907	40	91	204	1	2	1	338
1908	35	87	202	1	1	1	326
1909	40	85	199	1	1	1	326
1910	39	88	205	1	1	1	335
1911	38	87	206		2	1	334
1912	41	91	207	1	1	1	343
1913	38	89	202		1	1	332
1914	37	84	200	2	1	1	325

Date	Potatoes	Turnips and swedes	Mangold	Cabbage, kohl-rabi and rape	Vetches lucerne & other green	Total green
1866	44	62	4	1	27	140
1867	45	68	3	1	21	138
1868	47	70	4	1	6	128
1869	49	67	5	1	6	127
1870	49	70	5	1	5	130
1871	52	70	7	1	6	137
1872	48	69	7	1	10	136
1873	45	71	7	1	9	133
1874	45	71	6	1	8	132
1875	45	70	7	1	8	131
1876	43	72	7	1	6	129
1877	43	71	8	1	7	130
1878	41	68	7	1	6	123
1879	43	67	8	1	7	127
1880	39	65	8	1	7	120
1881	42	66	7	1	7	125
1882	42	68	7	1	7	124
1883	40	70	7	1	6	124
1884	41	70	7	1	6	125
1885	41	69	7	1	5	124
1886	40	69	7	1	5	122
1887	41	70	8	1	5	125
1888	42	72	8	1	5	128
1889	40	72	7	2	3	124
1890	40	73	7	2	3	125
1891	38	71	8	2	4	122
1892	37	71	8	2	3	121

(continued)

Table 36.2b (cont.)

Date	Potatoes	Turnips and swedes	Mangold	Cabbage, kohl-rabi and rape	Vetches lucerne & other green	Total green
1893	35	71	7	2	3	119
1894	34	73	8	2	3	121
1895	34	72	8	2	3	119
1896	34	71	7	3	3	118
1897	33	70	8	3	3	117
1898	33	68	8	3	3	115
1899	33	67	9	4	3	115
1900	33	63	10	4	3	112
1901	32	62	10	4	3	110
1902	31	61	11	4	2	110
1903	30	61	10	4	2	108
1904	30	61	10	4	2	107
1905	29	60	10	5	2	107
1906	29	59	11	5	2	105
1907	28	58	11	5	2	104
1908	27	57	10	5	2	102
1909	27	58	11	5	2	103
1910	26	58	11	5	2	102
1911	27	58	11	5	2	102
1912	26	57	12	6	2	102
1913	25	56	10	6	2	99
1914	25	56	11	7	2	99

Date	Clover, sainfoin or grass in rotation	Permanent grass not broken in rotation	Small fruit	Bare fallow	Total acreage
1866	257	1,258		110	2,285
1867	301	1,472		86	2,519
1868	328	1,415		84	2,504
1869	261	1,528		59	2,531
1870	398	1,428		38	2,548
1871	375	1,494		38	2,605
1872	371	1,532		35	2,636
1873	361	1,582		35	2,647
1874	365	1,634		32	2,679
1875	361	1,666		26	2,696
1876	360	1,698		26	2,712
1877	352	1,732		23	2,731

(continued)

Table 36.2b (*cont.*)

Date	Clover, sainfoin or grass in rotation	Permanent grass not broken in rotation	Small fruit	Bare fallow	Total acreage
1878	356	1,748		27	2,747
1879	347	1,774		29	2,759
1880	332	1,806		31	2,768
1881	331	1,815		31	2,785
1882	314	1,837		29	2,793
1883	309	1,865		27	2,800
1884	310	1,886		24	2,810
1885	332	1,893		18	2,818
1886	319	1,914		17	2,822
1887	312	1,929		15	2,831
1888	309	1,939	1	15	2,841
1889	318	1,949	1	18	2,854
1890	332	1,956	1	16	2,864
1891	325	2,012	1	10	2,888
1892	338	1,983	1	9	2,860
1893	318	1,998	1	8	2,856
1894	321	1,983	1	8	2,857
1895	329	1,978	1	9	2,838
1896	328	1,977	1	9	2,834
1897	374	1,930	1	8	2,833
1898	381	1,924	1	9	2,827
1899	395	1,920	1	8	2,823
1900	397	1,929	1	8	2,824
1901	400	1,941	1	8	2,823
1902	406	1,924	1	6	2,811
1903	386	1,939	1	7	2,800
1904	362	1,974	1	6	2,799
1905	345	1,989	1	6	2,795
1906	347	1,987	1	6	2,793
1907	320	2,022	1	5	2,792
1908	311	2,041	1	6	2,788
1909	293	2,053	1	5	2,782
1910	285	2,049	1	5	2,777
1911	282	2,046	1	5	2,770
1912	286	2,022	1	5	2,760
1913	256	2,058	1	7	2,755
1914	260	2,055	1	6	2,746

Notes: Small fruit not separately distinguished before 1888.
Sources: Annual Agricultural Returns, BPP, various.

Table 36.3. *Extent of cultivated land (England and Wales), 1836–1914*

Date	Millions of acres	Source
c. 1836/40	31.5	Tithe (Kain)
c. 1850	27.0	Caird
1854	27.7	1854 survey
1866	24.5	Annual Returns
1880	27.4	Annual Returns
1890	27.9	Annual Returns
1900	27.5	Annual Returns
1914	27.1	Annual Returns

Sources: R. J. P. Kain, *An Atlas and Index of the Tithe Files of Mid-Nineteenth-Century England and Wales* (Cambridge, 1986), pp. 458–9; J. Caird, *English Agriculture in 1850–51* (London, 1852), p. 521; *Reports by Poor Law Inspectors on Agricultural Statistics (England), 1854*, BPP, 1854–5, LIII, p. 495; *Annual Agricultural Returns*, BPP, various.

Table 36.4a. *England: Animal totals annually, 1866–1914 (in thousands)*

	Horses			Cattle					Sheep				Pigs	
Date	Solely agricultural	Others	Total horses	Cows and heifers in milk or calf	Other cattle >2 years	Cattle <2 years	Of which <1 year	Total cattle	Sheep >1 year	Of which ewes	Sheep <1 year	Total sheep	Total pigs	Of which sows
1866				1,291	947	1,070		3,307	10,620		4,504	15,125	2,066	
1867				1,411	920	1,138		3,469	12,383		7,415	19,798	2,549	
1868				1,505	977	1,298		3,780	13,231		7,700	20,931	1,982	
1869				1,499	987	1,221		3,707	12,512		7,309	19,822	1,630	
1870	756	222	978	1,529	978	1,250		3,757	12,003		6,937	18,940	1,814	
1871	733	230	963	1,461	981	1,229		3,671	11,145		6,386	17,530	2,079	
1872	732	231	963	1,523	1,052	1,327		3,092	11,296		6,617	17,913	2,348	
1873	737	242	979	1,581	1,052	1,541		4,174	11,908		7,261	19,170	2,141	
1874	739	268	1,007	1,614	1,106	1,585		4,306	12,442		7,418	19,860	2,059	
1875	745	286	1,032	1,595	1,163	1,461		4,218	11,973		7,142	19,115	1,875	
1876	759	299	1,058	1,574	1,150	1,353		4,076	11,586		6,734	18,320	1,924	
1877	761	309	1,070	1,558	1,072	1,350		3,980	11,482		6,848	18,330	2,115	
1878	767	322	1,089	1,568	1,085	1,381		4,035	11,410		7,034	18,444	2,125	
1879	770	331	1,101	1,605	1,033	1,491		4,129	11,521		6,925	18,446	1,771	
1880	767	326	1,092	1,593	1,076	1,489		4,158	10,630		6,199	16,829	1,698	
1881	806	289	1,094	1,621	1,103	1,435		4,160	9,819		5,564	15,383	1,733	
1882	805	280	1,084	1,618	1,027	1,437		4,082	9,313		5,635	14,948	2,123	
1883	811	273	1,084	1,651	1,010	1,556		4,217	9,629		5,966	15,595	2,231	
1884	811	276	1,087	1,715	1,026	1,710		4,452	10,010		6,419	16,428	2,207	

(continued)

Table 36.4a (cont.)

| | Horses | | | Cattle | | | | | Sheep | | | | Pigs | |
Date	Solely agricultural	Others	Total horses	Cows and heifers in milk or calf	Other cattle >2 years	Cattle <2 years	Of which <1 year	Total cattle	Sheep >1 year	Of which ewes	Sheep <1 year	Total sheep	Total pigs	Of which sows
1885	798	283	1,081	1,831	1,082	1,800		4,711	10,130		6,680	16,810	2,037	
1886	802	293	1,094	1,837	1,171	1,761		4,769	10,012		6,390	16,402	1,883	
1887	802	295	1,097	1,842	1,161	1,621		4,624	9,931		6,522	16,453	1,941	
1888	792	300	1,092	1,765	1,065	1,523		4,353	9,455		6,334	15,789	2,018	
1889	800	291	1,091	1,752	1,082	1,519		4,353	9,526		6,314	15,840	2,118	
1890	805	295	1,100	1,833	1,061	1,724		4,618	10,055		6,786	16,841	2,356	
1891	842	301	1,143	1,917	1,114	1,839		4,870	10,783		7,092	17,875	2,461	
1892	850	319	1,169	1,915	1,246	1,808		4,969	10,918		7,076	17,994	1,829	
1893	840	334	1,174	1,841	1,162	1,742	828	4,744	10,254	6,152	6,552	16,805	1,793	260
1894	831	345	1,176	1,759	1,114	1,578	775	4,451	9,337	5,728	6,173	15,510	2,014	295
1895	837	348	1,185	1,786	1,063	1,623	846	4,473	9,360	5,692	6,198	15,558	2,471	355
1896	837	353	1,190	1,809	1,007	1,757	894	4,574	9,584	5,822	6,447	16,031	2,476	336
1897	824	345	1,169	1,823	974	1,771	866	4,568	9,383	5,849	6,338	15,721	1,991	282
1898	830	333	1,164	1,873	1,021	1,780	888	4,674	9,487	5,878	6,400	15,887	2,079	306
1899	839	324	1,164	1,946	1,008	1,888	956	4,842	9,585	6,096	6,676	16,261	2,225	317
1900	834	318	1,152	1,900	1,036	1,913	923	4,849	9,466	6,012	6,379	15,845	2,021	280
1901	844	318	1,162	1,887	1,059	1,845	900	4,792	9,266	5,845	6,282	15,548	1,842	269
1902	832	323	1,155	1,841	988	1,782	862	4,612	8,814	5,681	6,221	15,034	1,956	295
1903	857	323	1,179	1,876	1,068	1,802	894	4,746	8,690	5,542	6,211	14,901	2,306	333

1904	870	326	1,196	1,962	1,034	1,921	952	4,917	8,642	5,571	6,107	14,749	2,476	328
1905	871	333	1,204	1,990	1,049	1,982	975	5,021	8,534	5,582	6,164	14,698	2,083	285
1906	866	335	1,200	2,020	1,068	1,973	941	5,061	8,608	5,665	6,232	14,840	1,984	285
1907	864	325	1,189	2,032	1,043	1,912	920	4,988	8,745	5,750	6,354	15,099	2,257	322
1908	867	313	1,180	2,047	1,039	1,913	946	4,998	9,347	5,980	6,612	15,959	2,439	316
1909	879	309	1,188	2,073	990	2,037	1,018	5,100	9,777	6,192	6,718	16,495	2,046	268
1910	884	300	1,184	2,054	1,025	2,047	985	5,126	9,478	6,140	6,795	16,274	2,020	281
1911	844	415	1,258	2,109	1,039	2,027	990	5,174	9,320	5,960	6,420	15,740	2,415	334
1912	817	429	1,245	2,062	1,017	2,008	955	5,087	8,535	5,653	5,970	14,504	2,270	291
1913	727	515	1,242	1,999	1,051	1,942	952	4,991	7,927	5,275	5,809	13,736	1,912	247
1914	713	527	1,240	2,185	879	2,056	1,061	5,119	7,735	5,320	5,917	13,652	2,260	307

Table 36.4b. *Wales: animal totals annually, 1866–1914 (in thousands)*

	Horses			Cattle					Sheep				Pigs	
Date	Solely agricultural	Others	Total horses	Cows and heifers in milk or calf	Other cattle >2 years	Cattle <2 years	Of which <1 year	Total cattle	Sheep >1 year	Of which ewes	Sheep <1 year	Total sheep	Total pigs	Of which sows
1866				223	134	185		541	1,288		381	1,669	192	
1867				245	111	189		545	1,514		713	2,227	230	
1868				255	121	217		593	1,816		853	2,669	187	
1869				256	122	211		589	1,870		851	2,721	172	
1870	71	45	116	256	123	225		605	1,892		815	2,706	199	
1871	69	48	117	251	118	228		597	1,886		821	2,706	225	
1872	69	49	118	251	109	243		603	1,966		901	2,867	238	
1873	70	51	120	260	107	276		643	2,050		917	2,967	211	
1874	69	54	124	264	125	276		665	2,111		954	3,065	214	
1875	70	55	125	261	141	249		651	2,082		870	2,952	203	
1876	71	57	128	259	135	242		637	2,002		871	2,873	215	
1877	71	59	130	254	120	241		616	1,974		888	2,862	231	
1878	72	60	132	253	112	244		608	1,998		928	2,926	218	
1879	73	63	136	262	112	270		644	2,012		861	2,873	193	
1880	73	62	135	261	126	267		655	1,905		813	2,718	182	
1881	84	54	138	260	133	262		655	1,771		696	2,467	192	
1882	85	54	138	260	116	268		645	1,745		773	2,518	234	
1883	85	54	139	260	107	285		652	1,767		814	2,581	230	
1884	85	55	140	267	112	302		681	1,786		871	2,657	217	
1885	84	56	140	280	123	306		709	1,847		920	2,768	216	
1886	84	56	140	283	140	297		720	1,769		746	2,515	205	
1887	85	56	140	283	132	283		697	1,804		936	2,740	223	
1888	84	55	139	275	120	271		666	1,841		897	2,738	231	

Year													
1889	86	55	141	272	114	280		666	1,882		959	2,841	241	
1890	87	57	143	282	106	317		705	1,987		1,082	3,070	258	
1891	93	57	150	295	127	338		759	2,139		1,095	3,234	270	
1892	91	58	149	291	145	319		754	2,157		1,040	3,198	197	
1893	90	57	147	281	139	319	157	739	2,075	1,138	1,027	3,102	201	34
1894	90	57	148	272	120	303	155	695	2,041	1,129	1,038	3,079	228	37
1895	92	62	153	275	104	324	173	704	2,029	1,138	971	3,001	260	41
1896	92	64	156	275	94	344	178	713	2,080	1,217	1,128	3,208	258	40
1897	88	65	153	273	95	342	172	709	2,103	1,241	1,093	3,195	216	36
1898	90	62	152	274	91	336	176	702	2,101	1,273	1,168	3,269	239	39
1899	91	63	154	286	87	364	195	737	2,188	1,333	1,229	3,416	258	41
1900	91	62	153	287	93	379	193	758	2,235	1,360	1,198	3,433	228	36
1901	92	63	155	281	94	368	188	743	2,210	1,364	1,218	3,428	213	35
1902	91	63	154	276	93	353	179	722	2,209	1,372	1,254	3,463	215	37
1903	94	63	157	274	87	350	186	711	2,230	1,400	1,282	3,511	244	39
1904	94	66	160	277	84	367	190	728	2,243	1,415	1,246	3,490	241	37
1905	95	67	162	280	90	368	190	739	2,248	1,435	1,287	3,535	211	33
1906	95	67	162	284	95	369	188	748	2,281	1,461	1,305	3,586	210	34
1907	96	66	162	288	96	355	181	739	2,335	1,513	1,368	3,703	233	40
1908	97	64	161	285	85	362	189	732	2,363	1,546	1,359	3,721	241	38
1909	97	64	161	286	83	376	195	745	2,414	1,582	1,381	3,795	205	33
1910	97	61	158	282	85	372	190	740	2,352	1,538	1,333	3,685	196	33
1911	93	69	163	284	90	366	192	740	2,306	1,510	1,286	3,591	236	42
1912	89	71	161	286	95	373	187	754	2,258	1,495	1,291	3,549	227	43
1913	81	80	160	265	100	360	189	726	2,193	1,424	1,201	3,394	191	34
1914	79	81	160	300	74	385	206	758	2,255	1,518	1,353	3,608	222	34

Notes:

Horses kept for agriculture were variously defined as including those mares kept for breeding purposes. But at other times this definition was relaxed. For the purposes of this table the separation point is 1881. 'Other' horses were mainly unbroken horses. Cattle less than two years of age were separated into those less than one and those between one and two years of age in 1893, and in the same year ewes were distinguished from other sheep greater than one year of age, and sows from other pigs.

Sources: Annual Agricultural Returns, BPP, various.

Table 36.5. *County crop yields, 1850–1882*

(a) Wheat (bushels per acre)

County	1850	1861	1870	Pre-1878	1879	Average (1876–82)	Average (1863–82)	1882
Beds	25	28.5	30.0	30.5	19.6	25.7	28.6	23.8
Berks	30	33.5	31.5	32.0	20.2	24.3	32.0	27.5
Bucks	25	28.5	29.0	30.0	15.2	26.3	29.2	27.2
Cambridge	32	32.3	33.0	33.0	23.2	27.1	31.6	34.2
Chesh	28	29.0	30.0	28.0	20.0	28.0	25.5	26.0
Corn		23.7	25.0	29.7	14.1	24.0	29.6	23.8
Cumb	27		29.0			29.2	22.7	21.3
Derby		29.0	29.0	28.0	20.0	26.0	24.0	24.3
Devon	20	22.2	21.5	24.6	23.2	20.7	21.9	19.0
Dorset	21	29.7	29.0	30.7	18.3	24.0	28.8	25.3
Durham	16	25.2	26.0	28.0	17.8	22.5	29.1	28.1
Essex	28	31.0	33.0	33.6	22.9	26.0	30.2	29.2
Glos	23	27.5	28.0		15.2	22.5	25.7	21.7
Hants	30	27.7	29.5	28.5	19.4	25.8	26.4	25.9
Heref		25.2	29.5	23.0	12.5	21.2	29.7	27.0
Herts	22	28.0	28.5	26.3	16.3	23.0	28.2	28.7
Hunts	32	29.0	32.5	30.5	20.2	26.0	30.5	27.0
Kent		33.0	33.7	34.0	27.4	27.0	32.0	32.1
Lancs	28	34.2	32.0	29.2	23.7	22.2	26.4	28.4
Leics	21	29.7	31.0	33.0	17.3	26.6	29.0	27.4
Lincoln	26	31.0	32.7	31.6	19.6	23.0	29.0	27.0
Middx		30.0	31.0			30.0		
Mon			29.0			22.7	25.4	23.0
Norfolk	32	33.5	31.1	31.6	22.0	28.0	31.5	31.5
N'hants	28	32.5	32.2	32.5	20.4	25.6	31.3	26.4
N'land	23	26.0	27.0	28.6	30.0	21.0	28.1	29.7
Notts	32	29.7	30.0	30.7	22.7	23.0	28.1	25.1
Oxford	25	31.0	31.0	36.0	19.0	25.3	30.1	25.9
Rutland		33.0	31.2			25.0		
Salop		24.2	26.0	21.0	19.1	19.6	24.4	23.2
Som		29.0	29.0	31.6		26.0	27.6	24.2
Staffs	28	28.5	29.5	29.9	18.0	19.4	24.0	22.7
Suffolk	32	28.7	28.7	30.4	21.7	27.0	30.0	26.4
Surrey	22	27.0	28.0	28.0	22.2	28.2	29.0	27.3
Sussex	22	29.7	30.0	31.0	23.0	23.0	32.0	31.1
Warws	30	30.0	30.0	31.7	19.4	22.3	28.0	23.7
Westmor		28.7	28.0	20.0	15.0		20.0	24.5
Wilts	26	28.7	29.0	28.0	19.0	24.5	27.0	23.7
Worcs		29.2	30.0			26.8	29.0	24.8
Yorks, ER	30	29.6	30.0	29.0	17.7		26.4	24.2

(continued)

Table 36.5 (*cont.*)

(a) Wheat (*cont.*)

County	1850	1861	1870	Pre-1878	1879	Average (1876–82)	Average (1863–82)	1882
Yorks, NR	20	29.5	30.0	28.0	19.0	22.0	27.1	21.7
Yorks, WR	30	29.5	30.0	25.0	17.6		25.7	24.7
Unweighted mean	26.3	29.1	29.6	29.4	19.8	24.6	27.9	26.0

(b) Barley (bushels per acre)

County	1861	Pre-1878	1879	Average (1876–82)	Average (1863–82)	1882
Beds	35.5	32.0	21.0	27.7	33.0	31.4
Berks	41.7	36.1	25.0	29.3	37.6	33.0
Bucks	37.7	28.0	24.0	31.3	36.4	32.0
Cambridge	41.0	37.4	26.2	34.7	37.5	41.3
Chesh	31.5			30.0	28.3	30.4
Corn	31.7	34.6	19.8	25.5	39.0	32.6
Cumb		36.0	27.0	25.6	35.0	35.4
Derby	40.5	34.0	26.0	33.2	32.0	29.3
Devon	31.7	30.5	24.7	28.7	30.2	27.4
Dorset	38.0	39.3	27.3	31.0	33.7	29.3
Durham	33.0	30.0	24.7	28.0	34.6	35.0
Essex	40.0	39.9	24.0	33.0	38.0	37.8
Glos	34.5			31.0	32.2	29.2
Hants	36.5	34.0	28.3	31.0	32.8	28.7
Heref	32.0	24.5	16.0	25.5	33.2	33.6
Herts	35.5	33.0	19.2	30.4	37.0	37.3
Hunts	40.2	37.0	19.0	32.4	35.8	32.9
Kent	40.5	43.0	36.6	36.0	40.6	41.0
Lancs	39.0	48.0	47.0	36.5	32.0	31.4
Leics	37.7	38.0	19.0	28.0	36.0	32.8
Lincoln	39.5	37.4	22.4	27.3	34.5	34.7
Middx	37.5			39.0		
Mon				25.5	28.6	24.6
Norfolk	42.7	34.8	25.0	33.8	37.0	37.3
N'hants	44.0	37.7	23.0	32.0	37.2	34.5
N'land	35.7	36.1	40.0	27.0	36.0	40.3
Notts	41.7	36.7	27.6	26.4	34.4	33.0
Oxford	39.5	35.0	21.6	28.0	35.7	28.7
Rutland	43.2			34.0		
Salop	29.0			21.8	26.6	27.3

(continued)

Table 36.5 (cont.)

(b) Barley (cont.)

County	1861	Pre-1878	1879	Average (1876–82)	Average (1863–82)	1882
Som	36.0	37.3		33.5	35.1	32.5
Staffs	35.7	36.0	20.0	29.2	30.7	26.0
Suffolk	35.5	37.0	24.4	35.7	36.3	34.6
Surrey	35.7	34.1	31.0	31.5	34.5	34.3
Sussex	41.7	37.0	26.7	28.0	36.4	39.6
Warws	39.5	34.7	18.4	28.2	31.7	30.0
Westmor	37.0	30.5	27.0	20.0	33.0	29.0
Wilts	36.5	31.0	23.1	31.0	32.5	25.9
Worcs	39.2			34.2	35.0	32.2
Yorks, ER	39.5	35.0	25.0		35.0	30.2
Yorks, NR	39.5	36.0	23.0		34.7	27.0
Yorks, WR	39.5	29.0	22.5		33.0	32.7
Unweighted mean	37.7	35.2	25.2	30.1	34.3	32.4
N. Wales			22.5	34.0	28.0	27.0
S. Wales		30.2	19.1	27.0	27.5	28.0

(c) Oats (bushels per acre)

County	1861	Pre-1878	1879	Average (1876–82)	Average (1863–82)	1882
Beds	47.7	40.8	32.0	37.8	41.5	37.6
Berks	56.0	54.7	54.0	37.7	53.0	56.0
Bucks	48.5	45.3		33.3	47.0	44.7
Cambridge	59.5	58.0	53.6	48.7	56.6	68.6
Chesh	41.7	44.0	34.0	43.7	34.8	38.6
Corn	37.7	45.0	39.2	30.5	46.6	40.4
Cumb		52.0	39.0	27.4	34.5	37.2
Derby	48.0	40.0	24.0	40.6	36.0	36.7
Devon	37.0	37.0	40.0	41.8	34.6	34.9
Dorset	45.0	50.7	44.7	36.7	43.2	40.7
Durham	41.0	36.0	34.7	37.5	39.3	41.9
Essex	51.0	50.8	34.0	44.7	44.7	48.4
Glos	40.7			37.0	41.2	38.7
Hants	47.5	45.0	45.5	43.8	40.4	49.2
Heref	31.2			29.8	37.0	40.0
Herts	44.2	42.3	35.0	39.8	47.0	52.6
Hunts	54.2	54.0	38.6	47.3	45.3	45.6
Kent	53.0	58.3	48.8	45.7	51.0	58.9

(continued)

Table 36.5 (*cont.*)

(c) Oats (*cont.*)

County	1861	Pre-1878	1879	Average (1876–82)	Average (1863–82)	1882
Lancs	40.0	51.0	49.3	46.2	47.2	49.0
Leics	47.0	42.2	25.5	38.8	41.8	40.4
Lincoln	54.5	51.8	43.3	43.8	49.3	50.8
Middx	57.2			50.6		
Mon				34.7		
Norfolk	55.5	56.0	34.0	41.2	51.3	54.9
N'hants	53.7	43.0	34.0	38.2	44.6	43.7
N'land	40.7	52.8	80.0	30.8	37.2	41.0
Notts	52.0	44.8	37.0	37.0	38.4	42.7
Oxford	49.7			38.3	44.8	45.2
Rutland	53.0			39.8		
Salop	31.5			23.6	29.8	33.2
Som	44.0			39.2	41.6	43.5
Staffs	40.0			30.6	37.3	41.7
Suffolk	47.0	46.7	43.0	46.0	55.4	48.0
Surrey	42.5	50.0	32.0	47.4	49.0	49.4
Sussex	51.0	44.6	46.0	37.7	52.0	61.4
Warws	49.0			36.0	38.0	40.4
Westmor	34.7	47.5	33.0	37.0	38.0	39.8
Wilts	47.2	44.5	36.0	41.3	43.3	41.5
Worcs	44.2			37.5		32.0
Yorks, ER	51.0	51.0	48.6		48.3	49.6
Yorks, NR	51.0	53.5	43.6		47.4	44.0
Yorks, WR	51.0	39.0	39.0		36.1	39.2
Unweighted mean	46.8	47.5	40.7	39.0	43.3	44.7
N. Wales			36.3	33.7	30.0	33.4
S. Wales		35.0	28.0	32.0	37.4	36.4

Sources:
P. G. Craigie, 'Statistics of agricultural production', *JRSS*, 46, 1883, pp. 40–2.
Based in turn on:

1850	James Caird
1861	*Mark Lane Express*
1870	*Journal of the Chamber of Agriculture*
Average before 1878	Reports of the Royal Commission on Agriculture
1879	Reports of the Royal Commission on Agriculture
Average 1876–82	*Mark Lane Express*
Average 1863–82	Inquiry of *Farmer and Chamber of Agriculture Journal*
1882	Inquiry of *Farmer and Chamber of Agriculture Journal*

Table 36.6. *English wheat yields (in bushels per acre), 1850–1914*

	(A)	(B)	(C)	(D)		(A)	(B)	(C)	(D)
1850	41.8	29.0			1883			28.5	
1851	48.9	34.0			1884			29.9	
1852	45.0	31.3	23.3		1885			30.8	31.5
1853	37.9	26.3	21.3		1886			29.8	26.9
1854	57.3	39.8	35.4		1887			28.9	32.3
1855	46.3	32.2	27.9		1888			27.4	28.2
1856	52.7	36.6	27.5		1889			30.0	29.9
1857	57.3	39.8	33.7		1890			32.0	30.8
1858	57.9	40.2	32.0		1891			30.0	31.3
1859	55.1	38.3	26.5		1892				26.2
1860			22.5		1893				25.8
1861			25.7		1894				30.7
1862			29.8		1895				26.2
1863			39.4		1896				33.9
1864			35.9		1897				29.0
1865			31.1		1898				34.8
1866			25.5		1899				32.8
1867			21.4		1900				28.4
1868			34.7		1901				30.8
1869			27.5		1902				32.8
1870			30.5		1903				30.1
1871			24.4		1904				26.5
1872			24.4		1905				32.7
1873			22.9		1906				33.6
1874			29.8		1907				34.0
1875			23.3		1908				32.2
1876			25.4		1909				33.6
1877			27.0		1910				30.2
1878			30.5		1911				32.6
1879			15.8		1912				28.7
1880			24.9		1913				31.3
1881			24.4		1914				32.4
1882			26.0						

Sources:

(A) M. J. R. Healy and E. L. Jones, 'Wheat yields in England, 1815–59', *JRSS*, 125, 1962, p. 578.

(B) Derived from *ibid.*, but reduced in the ratio 50:72 on the suggestion of the authors, p. 576, because the original estimates related to a wheat acreage which included paths, headlands, hedges, etc.

(C) J. B. Lawes and J. H. Gilbert, 'Home produce, imports, consumption, and price of wheat, over forty harvest-years, 1852–3 to 1891–2', *JRASE*, 3rd ser., 4, 1893, p. 133, rounded to first decimal place.

(D) *Annual Agricultural Returns*, BPP.

Table 36.7. *Estimated yields of English and Welsh produce, 1885–1914*

36.7a. *Estimated yields of English produce: CORN CROPS AND PULSES (in bushels per acre and 000s bushels)*

	Wheat		Barley		Oats		Beans		Peas	
Date	Yield	Prod	Yield	Prod	Yield	Prod	Yield	Prod	Yield	Prod
1885	31.51	74,021	35.63	67,503	40.63	66,934	19.89	8,135	18.71	4,240
1886	26.87	58,071	32.23	61,201	39.98	70,860	26.84	9,655	27.40	5,779
1887	32.25	70,875	31.32	55,113	36.45	64,441	22.37	7,823	24.50	5,537
1888	28.18	68,159	33.14	57,740	40.11	64,836	28.83	9,268	24.30	5,784
1889	29.87	69,336	31.55	56,036	41.94	68,109	28.55	8,693	26.32	5,840
1890	30.79	69,442	35.06	62,250	43.75	72,104	32.63	11,103	28.76	6,223
1891	31.33	68,694	34.36	60,901	41.72	69,786	29.54	9,966	28.31	5,703
1892	26.20	55,107	34.81	59,511	41.50	73,266	21.70	6,390	25.91	4,966
1893	25.81	46,429	27.99	49,033	35.08	67,164	18.58	4,257	22.64	4,705
1894	30.71	56,088	34.65	61,194	44.63	88,289	29.00	6,633	25.58	6,167
1895	26.21	35,120	31.61	58,092	38.45	78,645	22.58	5,172	22.64	4,672
1896	33.88	54,523	33.64	59,843	37.60	69,402	25.27	5,974	25.40	4,912
1897	28.97	51,724	32.48	55,159	40.26	73,639	28.71	6,123	27.64	5,168
1898	34.76	69,074	35.44	55,378	43.49	75,283	30.83	6,692	27.69	4,782
1899	32.83	62,380	34.34	56,164	41.48	73,905	29.90	7,005	27.31	4,359
1900	28.39	49,528	30.99	50,977	39.56	73,604	27.88	6,928	25.94	3,995
1901	30.84	49,883	30.30	49,558	37.05	67,863	23.63	5,602	26.03	3,947
1902	32.82	53,529	34.80	54,947	46.00	87,065	31.25	7,131	28.59	5,040
1903	30.12	45,102	31.76	49,081	42.37	82,790	31.20	7,064	26.60	4,764
1904	26.52	34,536	30.47	47,028	40.82	84,079	22.61	5,435	25.77	4,406
1905	32.66	55,670	33.53	47,288	39.41	74,117	32.13	7,807	25.73	4,403
1906	33.61	55,824	34.71	49,969	43.34	81,533	34.67	9,526	30.23	4,475
1907	33.96	52,210	35.67	50,340	46.61	91,715	34.45	10,168	29.49	4,702
1908	32.16	49,801	32.48	44,937	40.82	79,950	29.94	8,460	28.24	4,323
1909	33.61	58,284	36.78	50,725	42.45	78,111	28.42	8,562	25.90	4,369
1910	30.19	51,831	32.62	47,287	41.87	77,774	32.20	8,259	26.17	3,973
1911	32.63	58,873	31.44	42,052	39.03	71,854	24.96	7,341	26.39	3,672
1912	28.74	52,354	30.47	41,587	35.56	66,340	27.41	7,400	22.53	3,885
1913	31.32	52,095	32.57	47,876	38.51	68,253	28.30	7,287	26.41	3,364
1914	32.43	57,408	32.90	46,732	40.01	69,226	30.49	8,635	23.00	2,969

36.7b. *Estimated yields of English produce: ROOT CROPS AND HAY*
(in tons per acre and 000s tons)

Date	Potatoes		Turnips & swedes		Mangolds		Hay (from grasses)		Hay (from permanent pasture)	
	Yield	Prod	Yield	Prod	Yield	Prod	Yield	Prod	Yield	Prod
1885	6.08	2,183	9.03	13,190	15.43	5,336				
1886	5.81	2,112	14.55	21,073	20.94	7,145	1.50	2,449	1.35	5,132
1887	6.23	2,300	8.05	11,420	15.05	5,291	1.38	2,364	1.06	4,159
1888	5.38	2,103	12.61	17,549	17.36	6,112	1.37	2,267	1.46	6,006
1889	6.08	2,337	14.03	19,225	18.79	5,978	1.73	3,185	1.50	6,507
1890	5.62	1,959	13.65	19,012	18.29	5,903	1.52	2,564	1.36	5,620
1891	5.78	2,051	12.94	17,704	19.10	6,598	1.49	2,373	1.21	4,659
1892	5.96	2,085	13.76	19,122	18.52	5,620	1.22	1,921	0.95	3,638
1893	6.64	2,362	12.08	17,206	12.74	4,313	0.83	1,248	0.61	2,189
1894	5.80	1,975	13.52	18,932	18.57	6,395	1.62	2,524	1.47	6,121
1895	6.75	2,519	12.12	16,511	16.61	5,414	1.39	2,429	0.97	3,950
1896	6.35	2,539	10.96	14,656	15.10	4,968	1.13	1,806	0.88	3,487
1897	5.38	1,896	13.28	17,106	18.76	6,480	1.44	2,434	1.27	4,963
1898	6.17	2,256	10.58	13,083	17.68	6,063	1.70	3,033	1.50	5,883
1899	5.81	2,254	7.95	9,574	17.56	6,378	1.36	2,208	1.17	4,380
1900	5.00	1,986	13.66	15,855	20.51	8,244	1.41	2,261	1.24	4,670
1901	6.33	2,628	10.88	12,450	19.54	7,543	1.23	2,125	0.83	3,107
1902	5.39	2,226	14.67	16,024	21.29	9,073	1.64	2,856	1.39	5,523
1903	5.07	2,041	11.97	12,997	18.00	6,984	1.58	2,847	1.32	5,450
1904	6.11	2,463	13.05	14,241	18.81	7,252	1.50	2,543	1.25	5,140
1905	6.02	2,619	12.93	13,907	20.43	8,002	1.43	2,244	1.09	4,395
1906	6.15	2,439	13.03	14,104	19.81	8,288	1.40	2,206	1.12	4,619
1907	5.49	2,098	13.86	14,666	19.92	8,691	1.67	2,739	1.39	5,941
1908	6.95	2,717	13.79	14,516	21.08	8,757	1.60	2,597	1.27	5,419
1909	6.52	2,643	15.65	16,543	21.03	9,316	1.44	2,091	1.16	4,731
1910	6.66	2,467	15.53	16,532	21.20	9,105	1.59	2,360	1.27	5,442
1911	6.65	2,675	8.73	9,317	16.51	7,246	1.21	1,774	0.91	3,898
1912	4.84	2,115	11.90	12,085	18.11	8,572	1.31	1,805	1.26	5,531
1913	6.61	2,754	12.03	11,936	18.17	7,434	1.61	2,472	1.27	5,705
1914	6.44	2,807	12.77	12,598	18.37	7,720	1.37	1,906	1.09	4,604

Notes:
At times, for some root crops and hay, the Annual Returns reported variously
in terms of tons and sometimes in cwts. The required conversion has been
conducted in order to standardise the table.
Subject to rounding errors.
Sources: Annual Agricultural Returns, BPP.

36.7c. *Estimated yields of Welsh produce: CORN CROPS*
(*in bushels per acre and 000s bushesl*)

| | Wheat | | Barley | | Oats | | Beans | | Peas | |
Date	Yield	Prod	Yield	Prod	Yield	Prod	Yield	Prod	Yield	Prod
1885	22.65	1,673	27.91	3,503	32.53	8,024	24.00	63	21.33	43
1886	21.86	1,501	26.48	3,303	31.95	8,003	29.94	56	20.84	40
1887	23.37	1,622	26.52	3,154	30.99	7,915	26.69	55	18.55	40
1888	21.36	1,641	26.39	3,111	30.02	7,521	25.79	48	16.75	35
1889	24.43	1,673	29.07	3,548	32.73	8,150	28.92	47	21.16	31
1890	24.94	1,713	30.24	3,622	33.65	8,116	31.35	59	23.62	40
1891	23.73	1,462	29.36	3,439	32.89	7,699	29.11	55	19.98	31
1892	23.86	1,319	29.26	3,351	34.18	7,977	28.34	44	19.73	25
1893	22.09	1,205	25.06	2,803	30.94	7,452	21.11	35	16.16	20
1894	25.15	1,420	30.01	3,348	35.93	9,013	21.44	33	19.09	24
1895	21.61	952	26.78	2,997	31.60	7,654	22.94	33	17.38	24
1896	22.95	1,078	26.21	2,823	29.71	7,180	20.68	31	16.64	26
1897	24.76	1,332	29.86	3,116	32.56	7,766	20.74	30	20.35	35
1898	26.83	1,585	32.82	3,377	36.37	8,390	28.25	36	21.87	34
1899	25.62	1,381	31.41	3,328	34.18	7,528	27.29	37	21.22	35
1900	25.79	1,332	31.81	3,342	33.44	7,238	25.34	33	21.65	33
1901	24.67	1,157	29.60	3,016	31.09	6,490	21.76	26	19.53	31
1902	27.96	1,348	33.66	3,410	36.55	7,681	25.57	31	19.94	27
1903	24.59	1,059	29.17	2,890	31.06	6,623	30.06	38	20.27	21
1904	25.41	891	30.97	2,983	34.99	7,426	24.27	30	21.61	20
1905	26.59	1,167	30.88	2,817	33.87	7,042	25.89	30	20.54	19
1906	28.57	1,268	32.54	3,021	37.11	7,817	29.27	38	27.15	23
1907	27.64	1,103	30.82	2,793	37.22	7,590	28.68	45	20.96	18
1908	27.08	936	29.99	2,600	34.30	6,915	27.43	30	22.46	17
1909	28.09	1,112	31.88	2,719	35.32	7,012	26.91	36	21.86	15
1910	27.59	1,088	32.06	2,808	37.90	7,773	28.74	39	23.59	16
1911	28.15	1,083	30.48	2,645	33.35	6,870	24.38	28	23.40	13
1912	26.31	1,089	30.09	2,752	32.99	6,825	24.83	28	22.75	14
1913	27.34	1,043	30.39	2,707	33.48	6,778	27.77	30	23.42	10
1914	28.32	1,049	31.50	2,660	36.10	7,204	29.45	35	23.38	10

36.7d. Estimated yields of Welsh produce: ROOT CROPS AND HAY
(in tons per acre and 000s tons)

Date	Potatoes		Turnips & swedes		Mangolds		Hay (from grasses)		Hay (from permanent pasture)	
	Yield	Prod	Yield	Prod	Yield	Prod	Yield	Prod	Yield	Prod
1885	5.21	212	11.90	825	15.26	111				
1886	5.30	215	14.37	994	16.79	114	1.26	234	0.92	421
1887	6.96	282	10.37	731	13.96	108	1.08	189	0.74	359
1888	4.50	188	12.57	907	14.05	109	1.23	218	1.00	503
1889	5.95	238	14.99	1,084	17.53	122	1.27	236	1.07	545
1890	4.43	177	15.91	1,155	17.22	126	1.24	224	1.04	513
1891	5.44	208	14.31	1,010	16.55	130	1.16	197	0.93	448
1892	5.66	207	16.04	1,138	17.97	142	1.05	188	0.84	415
1893	6.63	232	15.02	1,072	16.38	123	0.75	126	0.54	268
1894	5.48	186	15.79	1,150	17.25	140	1.37	233	1.11	569
1895	6.74	227	15.08	1,086	14.82	115	1.06	186	0.81	411
1896	6.45	218	13.47	957	13.28	99	0.91	161	0.65	325
1897	5.10	166	15.84	1,114	16.07	126	1.26	248	1.01	479
1898	5.62	184	14.84	1,011	16.39	129	1.43	285	1.16	549
1899	5.24	173	10.99	734	14.64	129	1.21	240	0.90	414
1900	4.61	153	15.34	966	17.37	171	1.24	244	1.00	463
1901	5.70	182	15.28	947	17.98	176	1.00	202	0.70	325
1902	4.95	156	16.50	1,005	18.32	198	1.42	296	1.06	501
1903	4.37	132	14.31	874	15.10	155	1.12	228	0.89	433
1904	4.84	144	16.40	1,001	17.69	181	1.24	253	1.02	511
1905	5.57	164	12.78	771	16.55	166	1.21	233	0.95	483
1906	4.91	143	15.81	935	19.15	203	1.36	258	1.07	546
1907	4.09	115	15.08	882	18.44	204	1.38	253	1.09	562
1908	5.55	152	16.26	933	18.80	196	1.28	228	1.06	560
1909	5.57	150	16.49	960	18.98	211	1.13	193	0.92	485
1910	5.06	132	17.11	1,001	18.58	205	1.38	235	1.07	585
1911	6.58	175	14.30	829	17.15	191	1.08	186	0.81	445
1912	4.85	126	14.09	803	17.31	215	1.29	226	1.03	564
1913	5.53	140	15.19	858	17.14	177	1.41	237	1.13	638
1914	5.73	146	15.34	852	18.05	199	1.29	211	1.00	545

Notes:
At times, for some root crops and hay, the Annual Returns reported variously
in terms of tons and sometimes in cwts. The required conversion has been
conducted in order to standardise the table.
Subject to rounding errors.
Sources: Annual Agricultural Returns, BPP.

Table 36.8. *All crop distributions by counties (in ooos acres), 1875, 1895, 1915*

County	Corn & grain	Root & green	Rotation grasses	Permanent pasture & meadow	Other	Total
Bedford						
1875	117	35	17	76	10	256
1895	90	30	22	100	14	255
1915	89	30	15	111	9	255
Berks						
1875	152	60	39	113	6	369
1895	105	44	42	164	14	368
1915	101	31	30	170	13	344
Bucks						
1875	137	38	27	190	9	401
1895	95	27	32	236	11	401
1915	83	19	20	265	7	394
Cambridgeshire						
1875	261	84	43	76	17	481
1895	222	77	52	117	21	489
1915	233	86	39	114	18	490
Cheshire						
1875	85	32	57	347	2	523
1895	78	39	64	356	2	540
1915	88	39	65	337	1	530
Cornwall						
1875	146	60	136	163	20	525
1895	122	49	178	247	8	604
1915	121	34	149	298	5	607
Cumberland						
1875	97	47	100	298	5	546
1895	86	46	116	330	2	579
1915	72	37	76	367	0	552
Derby						
1875	74	22	34	363	9	502
1895	48	20	25	406	3	503
1915	43	19	16	402	2	483
Devon						
1875	292	157	181	426	28	1083
1895	223	128	219	633	11	1214
1915	209	99	165	729	7	1208

(continued)

Table 36.8 (*cont.*)

County	Corn & grain	Root & green	Rotation grasses	Permanent pasture & meadow	Other	Total
Dorset						
1875	116	63	55	235	6	474
1895	82	51	50	300	5	487
1915	75	40	38	318	5	477
Durham						
1875	95	35	56	202	21	409
1895	65	33	54	277	10	439
1915	68	32	36	288	4	428
Essex						
1875	424	110	73	179	37	822
1895	314	96	103	261	53	827
1915	327	91	64	283	26	790
Glos						
1875	178	68	85	300	11	642
1895	121	50	88	387	10	656
1915	113	36	65	432	7	563
Hants						
1875	262	139	110	167	20	699
1895	184	111	118	266	27	706
1915	188	89	81	287	26	672
Hereford						
1875	112	38	40	232	14	435
1895	74	28	42	289	13	445
1915	69	19	27	325	9	449
Hertford						
1875	152	44	36	94	11	337
1895	120	34	46	118	18	336
1915	126	29	33	124	11	323
Hunts						
1875	103	22	14	58	10	208
1895	71	20	21	86	12	211
1915	80	20	12	88	10	210
Kent						
1875	247	85	53	290	51	726
1895	167	73	51	386	72	748
1915	143	71	28	437	50	729

(*continued*)

Table 36.8 (*cont.*)

County	Corn & grain	Root & green	Rotation grasses	Permanent pasture & meadow	Other	Total
Lancashire						
1875	102	50	73	533	3	761
1895	103	54	76	586	4	824
1915	114	58	63	549	2	786
Leicester						
1875	111	26	28	294	10	470
1895	67	21	26	351	8	472
1915	56	18	18	377	4	473
Lincoln						
1875	632	244	159	417	20	1473
1895	552	241	194	501	30	1518
1915	577	230	157	525	30	1520
Middlesex						
1875	19	13	4	79	1	117
1895	11	15	3	88	6	123
1915	7	13	1	68	5	94
Monmouth						
1875	40	14	23	149	5	232
1895	21	10	15	194	2	243
1915	16	6	10	203	1	235
Norfolk						
1875	463	205	162	231	7	1067
1895	404	191	171	297	14	1077
1915	425	187	149	285	19	1066
Northampton						
1875	188	46	31	274	14	553
1895	128	37	36	344	14	559
1915	120	29	25	377	10	561
Northumberland						
1875	134	60	96	378	15	683
1895	93	48	85	476	6	707
1915	78	39	61	520	2	700
Nottingham						
1875	165	55	53	157	15	444
1895	120	50	59	209	11	448
1915	121	44	42	223	9	440

(*continued*)

Table 36.8 (*cont.*)

County	Corn & grain	Root & green	Rotation grasses	Permanent pasture & meadow	Other	Total
Oxfordshire						
1875	170	61	41	137	6	414
1895	123	46	50	188	9	416
1915	120	35	37	213	6	411
Rutland						
1875	28	9	6	41	2	85
1895	21	8	6	52	1	87
1915	20	6	5	55	1	87
Salop						
1875	177	64	76	369	10	696
1895	132	57	70	456	4	718
1915	116	44	59	494	2	715
Somerset						
1875	151	70	60	544	9	834
1895	95	52	54	653	6	861
1915	84	37	34	688	4	846
Stafford						
1875	121	44	50	367	11	593
1895	83	41	49	427	3	604
1915	75	36	38	440	2	591
Suffolk						
1875	394	128	79	146	20	766
1895	330	104	110	188	36	769
1915	364	103	75	194	22	758
Surrey						
1875	100	47	30	108	13	297
1895	60	36	24	152	13	286
1915	46	29	15	159	6	255
Sussex						
1875	212	81	65	264	30	652
1895	145	62	61	381	29	679
1915	125	51	39	417	13	644
Warwick						
1875	151	34	40	244	15	485
1895	93	30	38	325	9	495
1915	82	24	26	366	6	505

(*continued*)

Table 36.8 (*cont.*)

County	Corn & grain	Root & green	Rotation grasses	Permanent pasture & meadow	Other	Total
Westmorland						
1875	21	10	18	190	0	239
1895	18	10	17	206	0	251
1915	14	7	13	209	0	244
Wilts						
1875	227	108	81	313	14	743
1895	153	88	81	424	13	579
1915	131	57	57	444	13	703
Worcester						
1875	126	29	29	186	15	386
1895	78	31	30	248	16	404
1915	67	24	17	262	14	383
Yorkshire East Riding						
1875	281	113	87	169	21	670
1895	245	109	94	206	15	670
1915	249	109	85	230	13	686
Yorkshire North Riding						
1875	221	81	76	418	29	824
1895	183	76	76	509	20	864
1915	179	68	61	540	12	860
Yorkshire West Riding						
1875	245	103	85	721	18	1172
1895	192	95	78	825	12	1202
1915	188	89	57	826	7	1167
Anglesey						
1875	29	10	29	77	0	146
1895	25	9	28	90	0	152
1915	21	8	28	91	0	148
Brecon						
1875	33	8	25	121	4	191
1895	23	7	20	151	1	202
1915	16	6	14	161	0	196
Cardigan						
1875	64	14	38	151	1	269
1895	53	13	42	161	1	270
1915	48	12	30	167	1	258

(*continued*)

Table 36.8 (*cont.*)

County	Corn & grain	Root & green	Rotation grasses	Permanent pasture & meadow	Other	Total
Carmarthen						
1875	71	11	46	286	2	417
1895	57	10	30	344	1	441
1915	43	9	22	362	1	436
Caernarvon						
1875	23	9	37	113	3	185
1895	19	9	29	135	0	192
1915	17	8	27	119	0	171
Denbigh						
1875	63	15	43	131	2	254
1895	49	14	42	166	1	272
1915	38	12	29	181	1	261
Flint						
1875	34	9	17	62	3	125
1895	24	8	17	76	1	126
1915	17	6	14	88	0	125
Glamorgan						
1875	37	16	28	181	3	265
1895	26	13	21	217	1	279
1915	20	9	15	213	1	258
Merioneth						
1875	18	4	15	113	0	150
1895	15	4	14	121	0	154
1915	13	3	11	122	0	149
Montgomery						
1875	55	13	31	147	3	249
1895	43	11	30	186	2	271
1915	34	8	25	205	1	272
Pembroke						
1875	60	14	36	181	3	294
1895	50	13	42	210	1	315
1915	47	11	31	216	0	306
Radnor						
1875	24	8	17	101	2	152
1895	19	7	14	122	1	163
1915	17	6	14	124	0	161

(*continued*)

Table 36.8 (*cont.*)

County	Corn & grain	Root & green	Rotation grasses	Permanent pasture & meadow	Other	Total
England						
1875	7,529	2,848	2,608	10,536	591	24,112
1895	5,719	2,467	2,826	13,244	588	24,844
1915	5,603	2,147	2,102	14,038	413	24,303
Wales						
1875	512	131	361	1,666	26	2,696
1895	402	119	329	1,978	10	2,838
1915	331	96	261	2,049	5	2,742
England and Wales						
1875	8,041	2,980	2,969	12,203	617	26,808
1895	6,121	2,586	3,155	15,223	598	27,683
1915	5,934	2,266	2,362	16,087	418	27,067

Table 36.9. *Animal distributions by counties (in 000s), 1875, 1895, 1915*

County	Horses kept solely for agriculture	Total horses	Cows & heifers in milk or calf	Other cattle >2 years	Other cattle <2 years	Total cattle	Total sheep	Total pigs
Bedford								
1875	9	11	10	11	13	34	179	28
1895	9	12	11	8	11	30	96	33
1915	7	12	12	8	18	37	65	31
Berks								
1875	12	15	16	7	12	35	301	35
1895	11	16	19	7	13	39	185	33
1915	9	14	25	7	21	54	115	19
Bucks								
1875	12	17	27	21	21	69	292	35
1895	13	18	29	16	21	66	193	40
1915	10	19	33	17	35	84	166	29
Cambridgeshire								
1875	19	27	15	14	18	48	322	44
1895	20	31	16	13	20	48	213	56
1915	19	31	23	11	27	61	109	66
Cheshire								
1875	13	20	96	18	48	162	111	61
1895	17	26	103	13	54	170	92	78
1915	15	27	119	11	70	199	104	80
Cornwall								
1875	22	30	50	45	62	157	439	56
1895	23	34	67	46	81	194	421	100
1915	19	31	81	36	108	224	358	95
Cumberland								
1875	13	19	39	34	52	125	511	25
1895	14	23	46	33	63	142	518	24
1915	13	23	56	30	80	166	628	17
Derby								
1875	13	19	65	29	48	143	262	37
1895	16	27	69	22	45	136	177	41
1915	14	27	79	17	56	152	141	31
Devon								
1875	35	51	75	63	80	218	976	85
1895	36	60	95	62	107	264	874	124
1915	29	52	114	44	153	311	830	107

(continued)

Table 36.9 (*cont.*)

County	Horses kept solely for agriculture	Total horses	Cows & heifers in milk or calf	Other cattle >2 years	Other cattle <2 years	Total cattle	Total sheep	Total pigs
Dorset								
1875	13	15	49	11	16	76	518	41
1895	13	16	52	10	21	83	371	66
1915	10	16	63	7	31	101	292	53
Durham								
1875	11	17	20	18	25	63	208	12
1895	13	20	26	17	28	72	225	13
1915	12	22	30	18	35	83	241	15
Essex								
1875	32	41	23	28	28	79	399	90
1895	31	41	31	20	29	80	283	108
1915	24	38	43	13	41	97	161	75
Glos								
1875	18	24	37	33	46	116	454	55
1895	19	31	38	24	47	110	347	84
1915	16	29	49	25	73	146	316	73
Hants								
1875	22	28	34	11	17	62	596	63
1895	21	29	44	10	22	76	366	82
1915	18	29	55	8	34	96	283	63
Hereford								
1875	13	21	26	22	33	81	348	25
1895	14	24	29	20	40	89	319	34
1915	12	23	36	17	59	112	334	29
Hertford								
1875	11	14	12	9	12	32	188	31
1895	11	16	13	6	11	30	121	34
1915	8	14	18	7	19	43	71	28
Hunts								
1875	7	10	8	9	9	26	156	19
1895	7	11	7	9	11	28	101	22
1915	6	10	7	8	15	30	61	23
Kent								
1875	25	30	28	20	25	73	1008	58
1895	23	28	30	15	23	68	904	68
1915	17	27	37	13	34	85	826	60

(*continued*)

Table 36.9 (*cont.*)

County	Horses kept solely for agriculture	Total horses	Cows & heifers in milk or calf	Other cattle >2 years	Other cattle <2 years	Total cattle	Total sheep	Total pigs
Lancashire								
1875	22	34	121	34	76	231	315	38
1895	29	45	130	27	69	226	307	60
1915	27	43	146	23	82	251	361	82
Leicester								
1875	12	17	34	62	41	137	453	26
1895	13	21	39	54	39	132	305	31
1915	11	23	46	52	54	152	264	22
Lincoln								
1875	48	63	53	74	82	209	1,555	101
1895	52	73	62	76	96	233	1,162	114
1915	45	70	73	60	122	255	822	86
Middlesex								
1875	4	6	18	5	5	28	40	14
1895	5	8	14	4	4	22	26	19
1915	3	6	10	3	5	18	15	19
Monmouth								
1875	6	10	16	12	16	44	212	14
1895	7	14	18	10	18	46	201	19
1915	7	13	20	9	25	54	228	18
Norfolk								
1875	41	59	29	41	39	109	729	83
1895	44	68	31	48	46	125	521	119
1915	37	60	37	31	61	129	352	117
Northampton								
1875	14	20	24	60	34	118	553	30
1895	15	22	27	59	33	120	409	40
1915	12	23	32	57	50	139	346	30
Northumberland								
1875	14	19	20	44	33	97	914	13
1895	13	19	25	44	39	108	807	13
1915	11	18	29	55	49	134	1099	13
Nottingham								
1875	14	20	23	25	31	79	285	27
1895	15	23	27	24	31	81	222	34
1915	12	22	30	22	38	90	160	30

(*continued*)

Table 36.9 (*cont.*)

County	Horses kept solely for agriculture	Total horses	Cows & heifers in milk or calf	Other cattle >2 years	Other cattle <2 years	Total cattle	Total sheep	Total pigs
Oxfordshire								
1875	13	17	18	15	22	55	356	37
1895	13	18	19	13	22	53	242	45
1915	10	17	23	15	35	73	179	28
Rutland								
1875	2	3	3	9	6	18	106	3
1895	2	3	3	9	6	18	83	3
1915	2	3	4	8	9	20	71	3
Salop								
1875	19	29	55	34	54	143	496	57
1895	20	35	59	37	69	165	458	82
1915	17	36	82	31	102	216	495	79
Somerset								
1875	22	32	100	48	57	206	721	84
1895	25	38	104	43	74	221	539	141
1915	21	38	130	33	94	257	429	115
Stafford								
1875	15	22	67	26	46	139	317	51
1895	18	28	75	22	56	153	247	62
1915	15	30	92	19	76	187	212	51
Suffolk								
1875	32	42	20	20	25	65	476	111
1895	32	42	23	16	28	67	388	168
1915	26	40	27	7	40	74	256	164
Surrey								
1875	11	14	22	9	15	45	101	33
1895	9	13	21	8	13	41	79	27
1915	7	13	22	4	16	43	42	25
Sussex								
1875	21	26	36	29	36	101	565	41
1895	20	26	43	26	35	104	446	48
1915	15	23	56	17	51	123	344	40
Warwick								
1875	15	19	31	33	32	97	393	41
1895	14	22	33	29	33	95	273	46
1915	12	24	41	32	56	128	233	36

(*continued*)

Table 36.9 (*cont.*)

County	Horses kept solely for agriculture	Total horses	Cows & heifers in milk or calf	Other cattle >2 years	Other cattle <2 years	Total cattle	Total sheep	Total pigs
Westmorland								
1875	4	7	21	11	26	58	331	5
1895	5	10	24	10	31	64	360	5
1915	5	10	29	9	39	77	436	5
Wilts								
1875	18	22	51	14	24	89	749	59
1895	19	24	60	13	33	106	534	79
1915	15	24	83	9	41	133	376	53
Worcester								
1875	14	18	25	15	20	60	229	39
1895	14	21	25	13	22	60	165	51
1915	11	20	27	13	40	80	144	45
Yorkshire East Riding								
1875	23	37	25	27	33	85	514	49
1895	24	46	25	25	34	83	436	63
1915	22	37	28	24	48	99	454	59
Yorkshire North Riding								
1875	26	39	49	47	62	159	704	53
1895	27	45	52	45	67	164	677	59
1915	24	42	59	42	88	189	719	54
Yorkshire West Riding								
1875	33	50	105	64	79	248	732	67
1895	38	61	123	59	79	260	673	103
1915	33	58	133	51	97	282	688	108
Anglesey								
1875	4	6	14	11	16	41	54	15
1895	5	8	16	11	23	49	59	19
1915	4	7	16	9	31	56	123	12
Brecon								
1875	5	10	14	8	15	36	466	9
1895	5	13	15	6	19	40	456	10
1915	5	12	17	4	20	41	523	7
Cardigan								
1875	8	13	26	13	23	62	205	21
1895	8	16	26	7	33	65	222	24
1915	9	17	27	3	40	70	265	21

(*continued*)

Table 36.9 (*cont.*)

County	Horses kept solely for agriculture	Total horses	Cows & heifers in milk or calf	Other cattle >2 years	Other cattle <2 years	Total cattle	Total sheep	Total pigs
Carmarthen								
1875	10	18	49	19	37	105	226	27
1895	12	22	52	12	52	116	235	40
1915	12	23	55	7	62	124	289	33
Caernarvon								
1875	3	7	21	12	18	51	224	19
1895	5	9	22	8	24	54	236	24
1915	4	8	24	5	28	56	287	16
Denbigh								
1875	6	12	24	11	24	59	260	22
1895	7	15	25	8	32	65	297	30
1915	6	13	29	6	45	80	416	27
Flint								
1875	4	6	14	5	10	29	60	14
1895	4	7	16	4	14	34	69	20
1915	3	7	20	3	22	44	111	19
Glamorgan								
1875	7	12	21	13	17	51	298	15
1895	8	17	24	9	19	52	293	19
1915	8	16	28	9	25	61	325	16
Merioneth								
1875	3	6	15	9	16	39	383	9
1895	3	5	14	5	19	38	377	10
1915	3	5	14	4	21	39	434	7
Montgomery								
1875	7	14	23	16	28	67	357	22
1895	7	16	23	13	34	70	368	27
1915	7	16	26	9	44	79	480	21
Pembroke								
1875	8	13	31	17	33	81	118	24
1895	9	16	32	13	42	88	120	32
1915	8	16	34	9	55	98	146	30
Radnor								
1875	4	8	10	8	13	31	299	6
1895	4	10	10	7	15	33	271	6
1915	4	10	12	4	18	34	300	5

(*continued*)

Table 36.9 (*cont.*)

County	Horses kept solely for agriculture	Total horses	Cows & heifers in milk or calf	Other cattle >2 years	Other cattle <2 years	Total cattle	Total sheep	Total pigs
England								
1875	745	1,032	1,595	1,163	1,461	4,218	19,115	1,875
1895	784	1,188	1,786	1,063	1,623	4,473	15,365	2,471
1915	656	1,138	2,140	922	2,225	5,287	13,824	2,176
Wales								
1875	70	125	261	141	249	651	2,952	203
1895	77	153	275	104	324	704	3,001	260
1915	73	149	301	72	410	783	3,698	213
England and Wales								
1875	815	1,156	1,856	1,304	1,709	4,870	22,066	2,079
1895	861	1,341	2,062	1,167	1,947	5,176	18,365	2,731
1915	729	1,287	2,441	994	2,635	6,070	17,523	2,389

Table 36.10. *Landholdings by counties, 1875, 1895, 1915*

County	<50 acres	50–100 acres	100–300 acres	>300 acres	Total holdings
Bedford					
1875	2,802	286	574	236	3,898
1895	2,695	294	546	239	3,774
1915	3,142	346	590	194	4,272
Berks					
1875	2,652	331	643	405	4,031
1895	2,493	359	682	345	3,879
1915	2,364	385	677	319	3,745
Bucks					
1875	3,433	510	1,126	291	5,360
1895	2,975	580	1,116	278	4,949
1915	3,335	588	1,115	266	5,304
Cambridgeshire					
1875	5,747	608	945	420	7,720
1895	5,238	635	898	430	7,201
1915	6,393	685	939	398	8,415
Cheshire					
1875	10,816	1,655	1,561	69	14,101
1895	9,523	1,718	1,622	86	12,949
1915	8,631	1,706	1,649	69	12,055
Cornwall					
1875	10,710	1,658	1,433	118	13,919
1895	10,224	1,846	1,661	138	13,869
1915	9,752	2,028	1,748	112	13,640
Cumberland					
1875	4,354	1,721	1,593	190	7,858
1895	3,943	1,663	1,785	210	7,601
1915	3,919	1,785	1,737	155	7,596
Derby					
1875	10,698	1,463	1,298	121	13,580
1895	9,296	1,513	1,335	110	12,254
1915	8,018	1,567	1,354	84	11,023
Devon					
1875	11,628	2,962	3,552	339	18,481
1895	9,172	2,876	4,095	419	16,562
1915	9,570	3,205	4,196	310	17,281

(continued)

Table 36.10 (*cont.*)

County	<50 acres	50–100 acres	100–300 acres	>300 acres	Total holdings
Dorset					
1875	3,686	501	772	472	5,431
1895	3,194	531	847	452	5,024
1915	3,404	622	924	405	5,355
Durham					
1875	4,098	849	1,297	164	6,408
1895	4,278	918	1,404	172	6,772
1915	4,314	968	1,409	140	6,831
Essex					
1875	5,254	1,314	2,053	650	9,271
1895	4,919	1,231	2,041	647	8,838
1915	5,275	1,298	2,075	554	9,202
Glos					
1875	8,475	1,007	1,534	499	11,515
1895	6,943	972	1,499	502	9,916
1915	6,658	1,101	1,547	459	9,765
Hants					
1875	6,105	726	1,124	725	8,680
1895	6,382	841	1,168	642	9,033
1915	7,220	945	1,206	554	9,925
Hereford					
1875	5,395	671	1,133	291	7,490
1895	4,117	677	1,253	285	6,332
1915	4,152	817	1,312	250	6,531
Hertford					
1875	2,776	373	768	324	4,241
1895	2,234	362	705	345	3,646
1915	2,240	314	724	321	3,599
Hunts					
1875	2,117	225	409	308	2,959
1895	1,637	214	451	195	2,497
1915	1,627	245	416	208	2,496
Kent					
1875	6,760	1,285	1,814	502	10,361
1895	6,826	1,337	1,828	494	10,485
1915	7,553	1,543	1,899	396	11,391

(*continued*)

Table 36.10 (*cont.*)

County	<50 acres	50–100 acres	100–300 acres	>300 acres	Total holdings
Lancashire					
1875	18,210	2,873	1,468	87	22,638
1895	15,372	3,359	1,747	96	20,574
1915	14,025	3,378	1,700	78	19,181
Leicester					
1875	5,974	966	1,406	193	8,539
1895	5,179	1,005	1,352	206	7,742
1915	4,542	1,095	1,414	191	7,242
Lincoln					
1875	19,706	2,181	2,888	1,215	25,990
1895	16,796	2,269	2,879	1,251	23,195
1915	15,190	2,413	3,092	1,187	21,882
Middx					
1875	2,406	275	298	51	3,030
1895	2,343	264	321	50	2,978
1915	1,694	215	256	35	2,200
Monmouth					
1875	3,242	719	659	61	4,681
1895	3,398	763	687	62	4,910
1915	3,064	779	678	49	4,570
Norfolk					
1875	12,493	1,465	1,875	935	16,768
1895	8,950	1,520	1,813	982	13,265
1915	9,509	1,619	2,024	880	14,032
Northampton					
1875	4,406	695	1,324	493	6,918
1895	3,825	700	1,303	482	6,310
1915	3,298	713	1,345	485	5,841
Northumberland					
1875	3,070	536	1,313	695	5,614
1895	3,146	575	1,357	737	5,815
1915	3,170	600	1,419	709	5,898
Nottingham					
1875	6,194	757	1,079	281	8,311
1895	5,373	813	1,188	248	7,622
1915	4,290	830	1,254	242	6,616

(*continued*)

Table 36.10 (*cont.*)

County	<50 acres	50–100 acres	100–300 acres	>300 acres	Total holdings
Oxfordshire					
1875	2,789	437	939	390	4,555
1895	2,652	474	907	394	4,427
1915	2,596	480	932	378	4,386
Rutland					
1875	950	139	207	66	1,362
1895	665	136	189	68	1,058
1915	500	140	197	72	909
Salop					
1875	8,281	903	1,925	504	11,613
1895	8,121	999	2,000	461	11,581
1915	7,904	1,085	2,056	420	11,465
Somerset					
1875	11,999	1,812	2,354	407	16,572
1895	9,972	1,751	2,429	439	14,591
1915	9,291	1,936	2,515	377	14,119
Stafford					
1875	10,870	1,406	1,566	269	14,111
1895	9,589	1,435	1,596	256	12,876
1915	8,483	1,478	1,628	212	11,801
Suffolk					
1875	5,667	1,436	2,043	568	9,714
1895	4,671	1,313	1,859	582	8,425
1915	4,591	1,407	1,935	523	8,456
Surrey					
1875	4,159	539	771	187	5,656
1895	3,567	552	704	162	4,985
1915	3,321	558	709	104	4,692
Sussex					
1875	5,717	1,140	1,523	471	8,851
1895	5,555	1,251	1,457	481	8,744
1915	5,839	1,325	1,461	414	9,039
Warwick					
1875	5,210	726	1,393	298	7,627
1895	4,991	832	1,377	296	7,496
1915	4,794	898	1,476	284	7,452

(continued)

Table 36.10 (*cont.*)

County	<50 acres	50–100 acres	100–300 acres	>300 acres	Total holdings
Westmorland					
1875	2,134	821	575	94	3,624
1895	1,993	842	656	93	3,584
1915	1,834	816	676	95	3,421
Wilts					
1875	5,291	613	1,156	788	7,848
1895	4,503	675	1,152	710	7,040
1915	4,106	788	1,179	649	6,722
Worcester					
1875	6,210	765	1,121	187	8,283
1895	6,246	797	1,153	182	8,378
1915	6,295	778	1,146	149	8,368
Yorkshire East Riding					
1875	5,573	819	1,503	601	8,496
1895	4,463	767	1,547	587	7,364
1915	4,288	803	1,695	558	7,344
Yorkshire North Riding					
1875	10,260	1,851	2,554	330	14,995
1895	8,411	1,897	2,752	349	13,409
1915	7,704	1,956	2,818	316	12,794
Yorkshire West Riding					
1875	25,152	2,823	2,881	384	31,240
1895	21,776	3,018	3,020	415	28,229
1915	19,286	3,048	3,063	384	25,781
Anglesey					
1875	2,757	417	374	28	3,576
1895	3,180	455	382	23	4,040
1915	3,625	460	348	16	4,449
Brecon					
1875	1,927	735	539	37	3,238
1895	1,731	816	644	29	3,220
1915	1,793	820	608	24	3,245
Cardigan					
1875	3,700	1,013	822	37	5,572
1895	4,371	1,025	805	16	6,217
1915	4,645	1,044	732	10	6,431

(*continued*)

Table 36.10 (*cont.*)

County	<50 acres	50–100 acres	100–300 acres	>300 acres	Total holdings
Carmarthen					
1875	5,071	1,785	1,214	39	8,109
1895	5,195	1,865	1,295	25	8,380
1915	5,503	1,954	1,204	35	8,696
Caernarvon					
1875	4,430	583	390	43	5,446
1895	5,324	630	395	33	6,382
1915	5,481	607	314	17	6,419
Denbigh					
1875	4,016	940	687	42	5,685
1895	3,992	1,051	717	44	5,804
1915	4,047	993	692	30	5,762
Flint					
1875	3,356	427	318	24	4,125
1895	2,938	406	328	23	3,695
1915	2,657	434	321	16	3,428
Glamorgan					
1875	3,483	989	687	62	5,221
1895	3,975	999	752	52	5,778
1915	3,768	999	688	34	5,489
Merioneth					
1875	2,233	580	341	28	3,182
1895	2,052	644	368	22	3,086
1915	1,995	618	375	19	3,007
Montgomery					
1875	3,731	910	663	50	5,354
1895	3,700	1,019	764	44	5,527
1915	3,825	1,043	769	39	5,676
Pembroke					
1875	4,190	857	806	82	5,935
1895	4,094	865	908	84	5,951
1915	3,951	917	900	65	5,833
Radnor					
1875	1,267	420	475	55	2,217
1895	1,180	442	538	48	2,208
1915	1,218	482	544	37	2,281

(*continued*)

Table 36.10 (*cont.*)

County	<50 acres	50–100 acres	100–300 acres	>300 acres	Total holdings
England					
1875	293,469	44,842	58,450	15,579	412,340
1895	257,646	46,574	60,381	15,578	380,179
1915	247,181	49,286	62,185	13,985	372,637
Wales					
1875	40,161	9,656	7,316	527	57,660
1895	41,732	10,217	7,896	443	60,288
1915	42,508	10,371	7,495	342	60,716
England and Wales					
1875	333,630	54,498	65,766	16,106	470,000
1895	299,378	56,791	68,277	16,021	440,467
1915	289,689	59,657	69,680	14,327	433,353

Sources for Tables 36.8, 36.9, 36.10

1875 from *Agricultural Returns of Great Britain, with Abstract Returns for the United Kingdom, British Possessions and Foreign Countries*, BPP, 1875, LXXIX, pp. 26–7, 31, 46–51, 56–61, 66–71, 78–83.

1895 from *Agricultural Returns for Great Britain*, C. 8073, BPP, 1896, XCII, pp. 4–21; *Returns as to the Number and Size of Agricultural Holdings in Great Britain in the year 1895*, BPP, 1896, LXVII, pp. 16–19.

1915 from Board of Agriculture and Fisheries, *Agricultural Statistics. 1915*, BPP, 1916, XXXII, pp. 22–31, 65, 73–5.

Notes:

City of York always included in Yorkshire East Riding.

London distinguished separately in 1895 and 1915, here counted in Middlesex.
Isle of Ely, Isle of Wight, the three divisions of Lincolnshire, the Soke of Peterborough, East and West Suffolk and Sussex distinguished separately in 1915, but here counted in Cambridgeshire, Hampshire, Lincolnshire, Northamptonshire, Suffolk and Sussex, respectively.

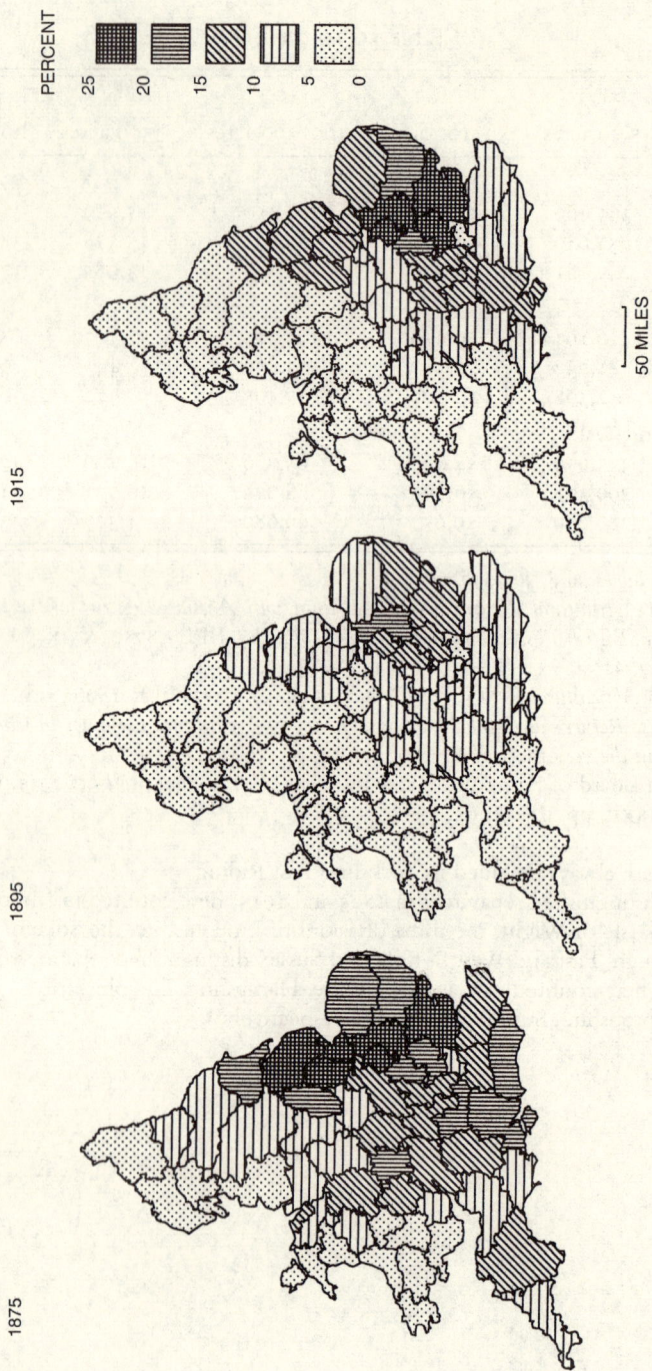

36.1 Wheat acreage as a proportion of the cultivated area, 1875, 1895, 1915

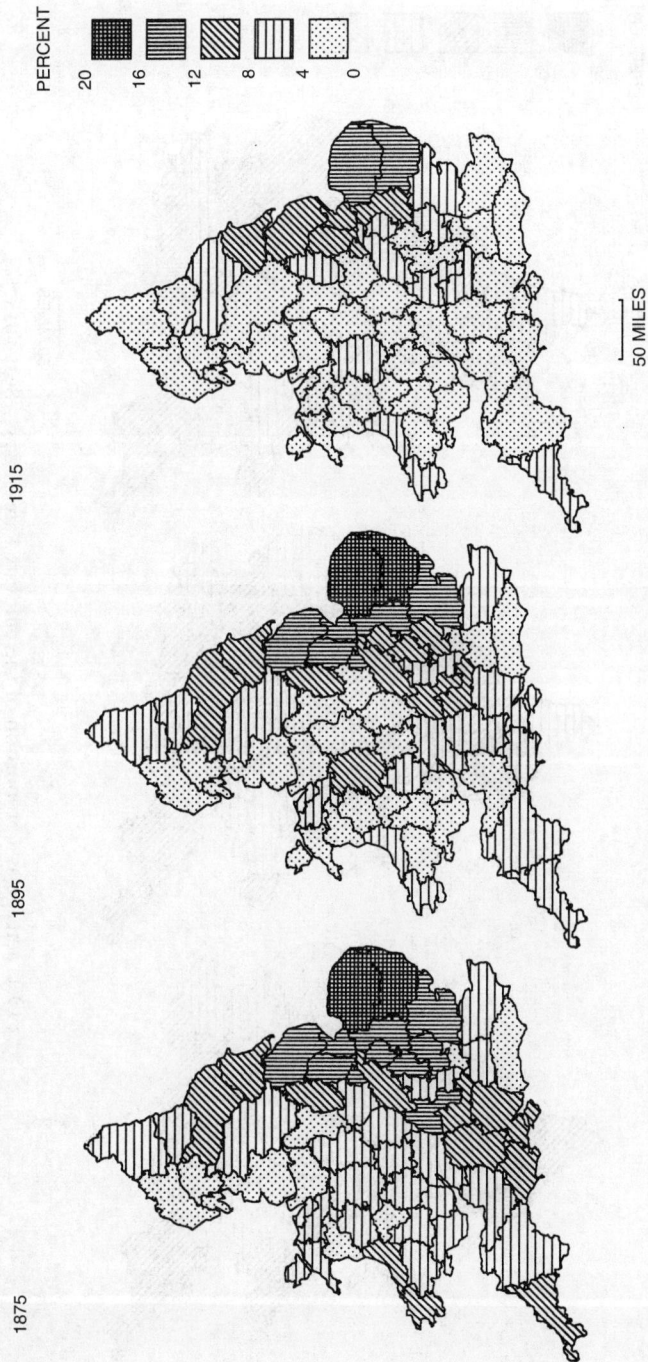

36.2 Barley acreage as a proportion of the cultivated area, 1875, 1895, 1915

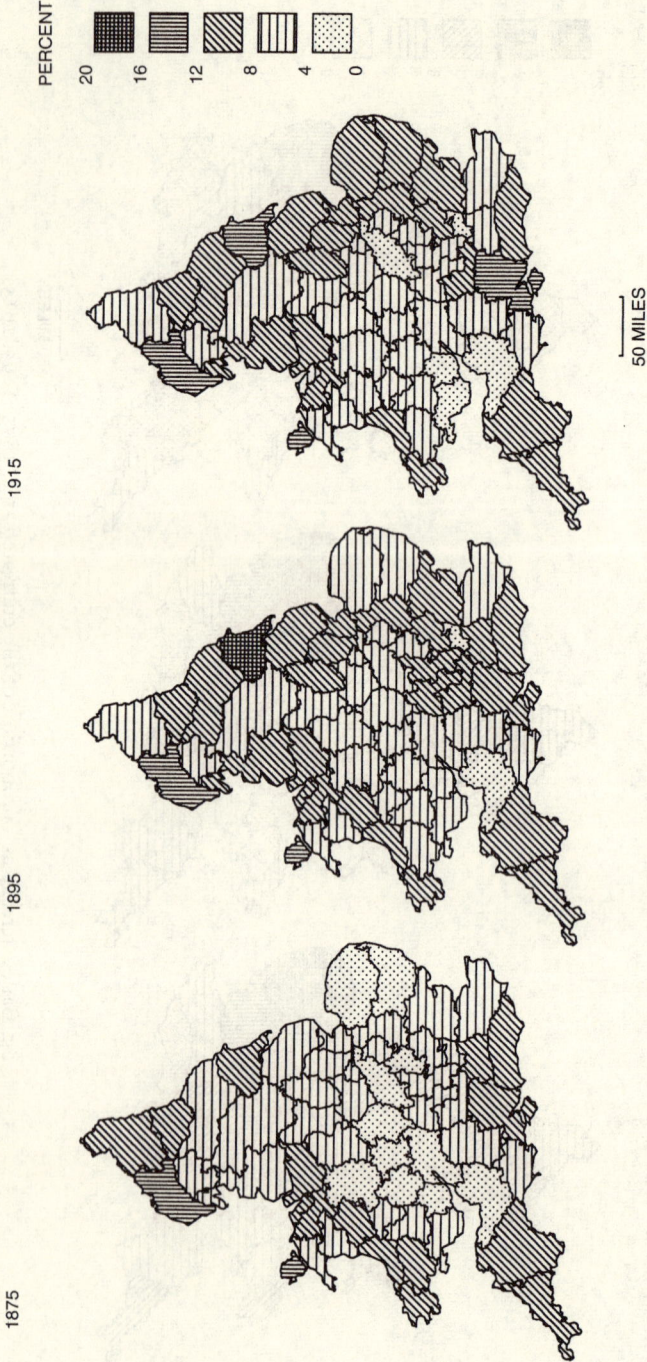

36.3 Oats acreage as a proportion of the cultivated area, 1875, 1895, 1915

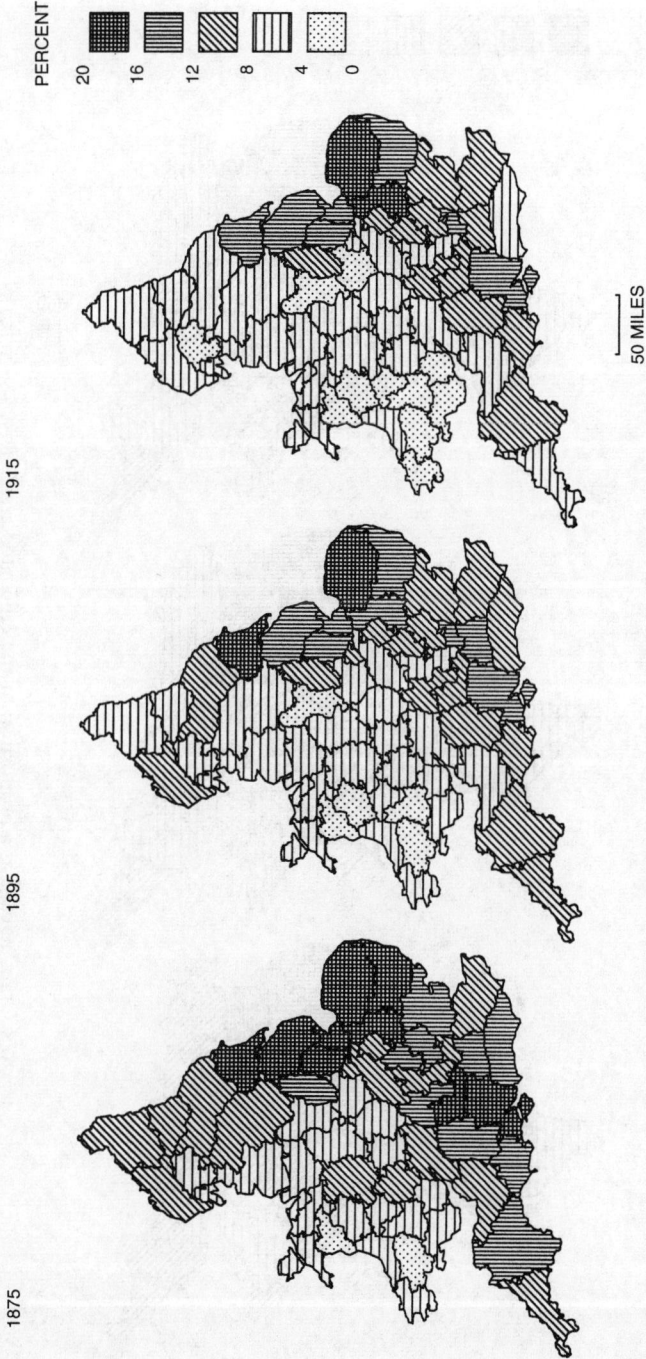

36.4 Root and green crop acreage as a proportion of the cultivated area, 1875, 1895, 1915

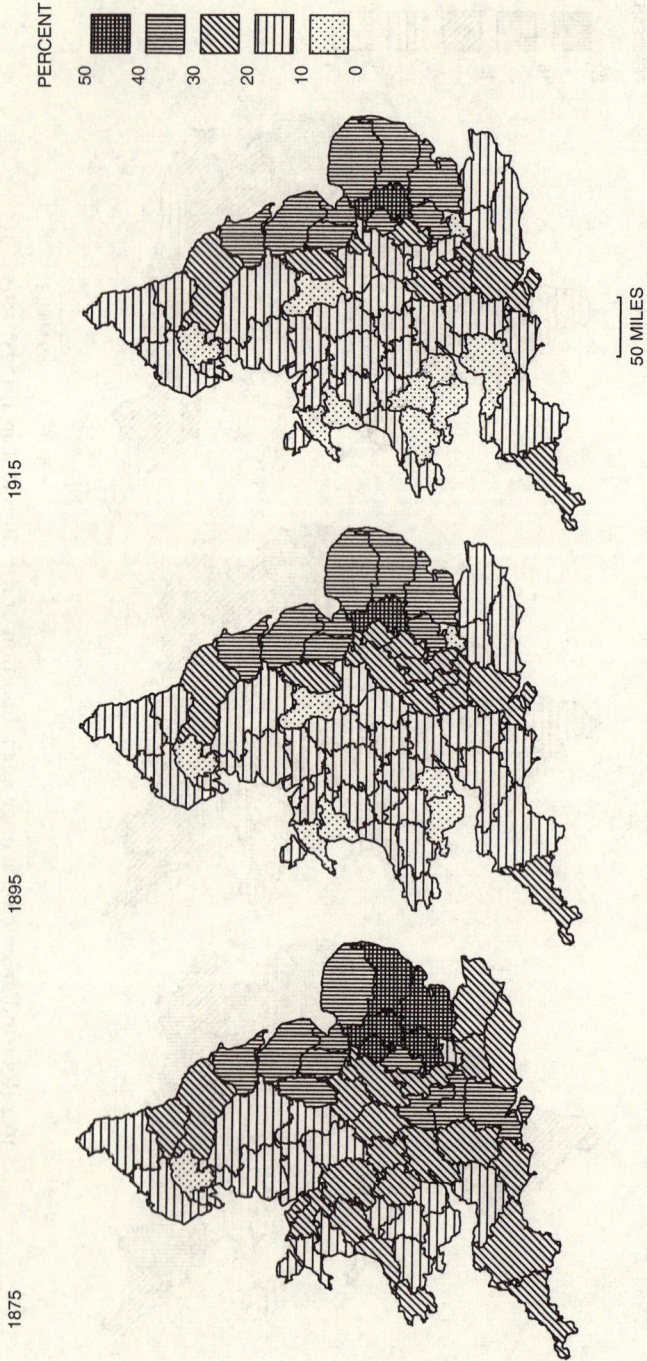

36.5 Grain acreage as a proportion of the cultivated area, 1875, 1895, 1915

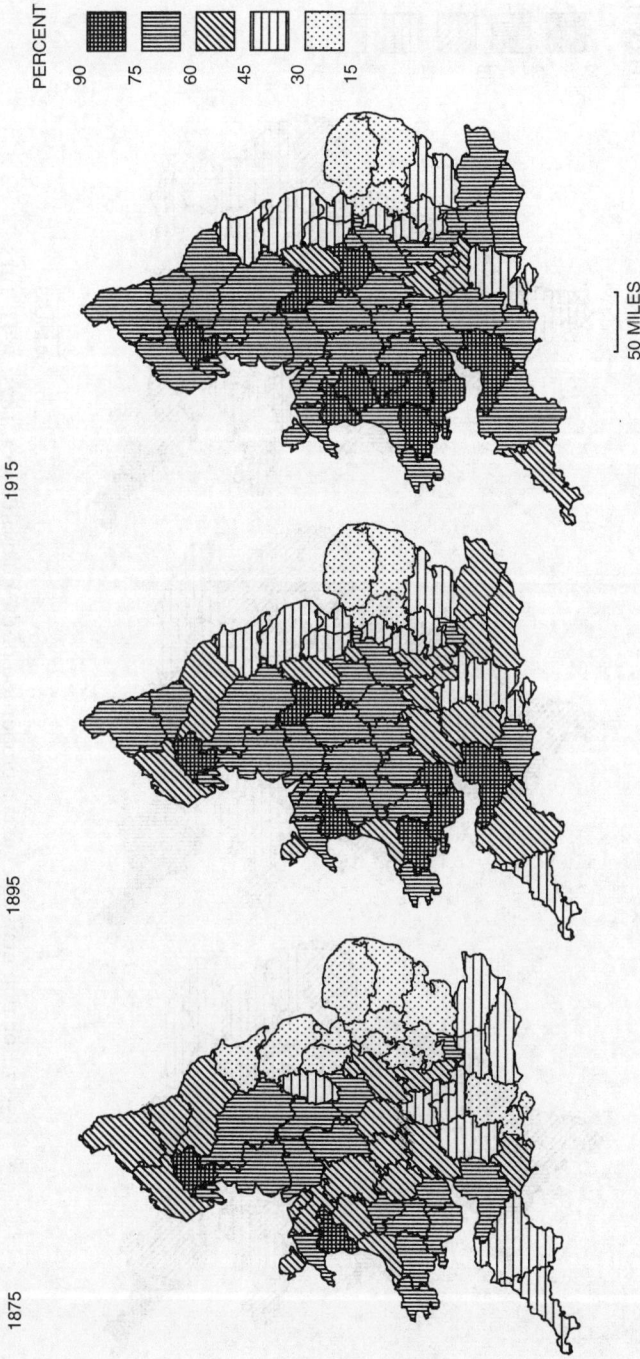

36.6 Permanent pasture acreage as a proportion of the cultivated area, 1875, 1895, 1915

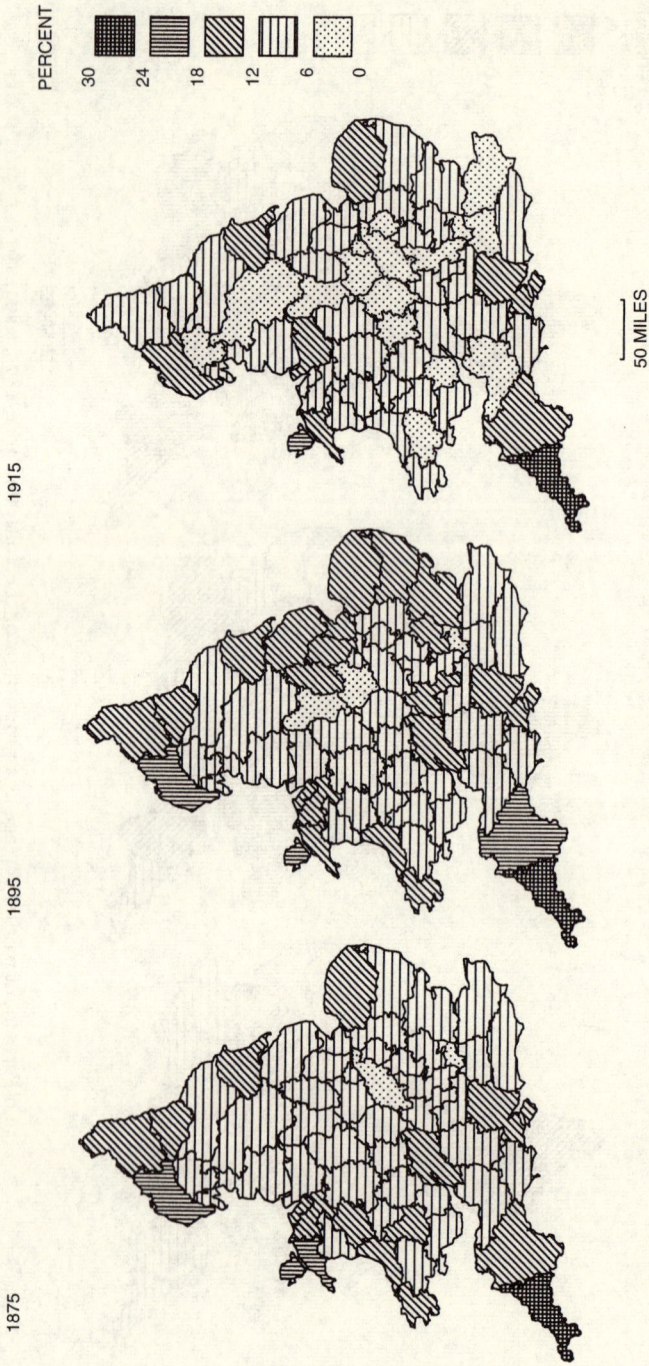

PERCENT

30
24
18
12
6
0

1915

1895

1875

50 MILES

36.7 Rotation grass acreage as a proportion of the cultivated area, 1875, 1895, 1915

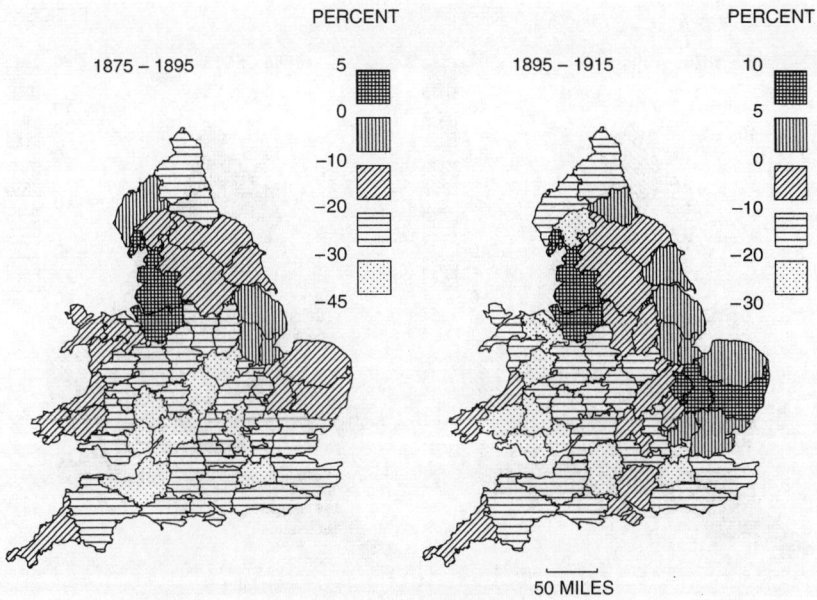

36.8 Changes in tillage acreage, 1875–95, 1895–1915

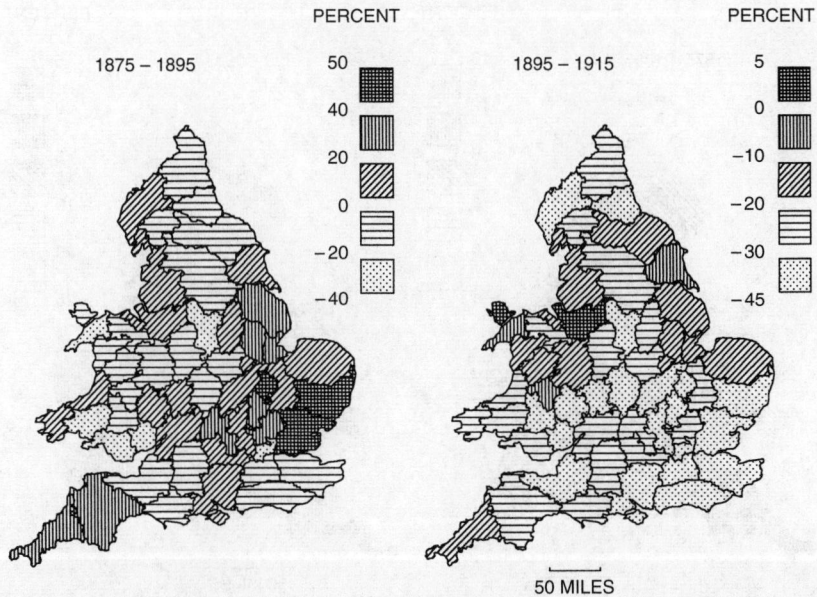

36.9 Changes in hay acreage, 1875–95, 1895–1915

PERCENT

1875 – 1895

60
40
30
20
10
0

PERCENT

1895 – 1915

20
10
0
–10
–20
–30

50 MILES

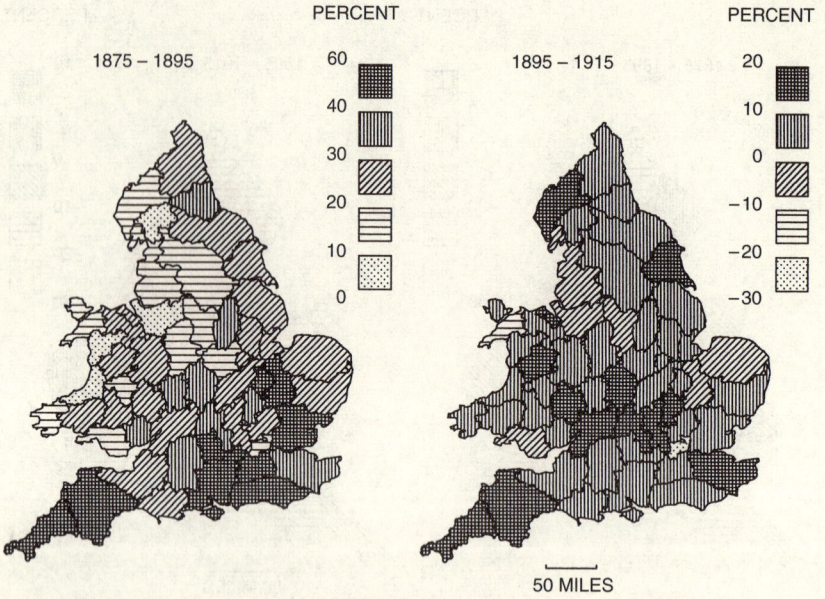

36.10 Changes in permanent pasture acreage, 1875–95, 1895–1915

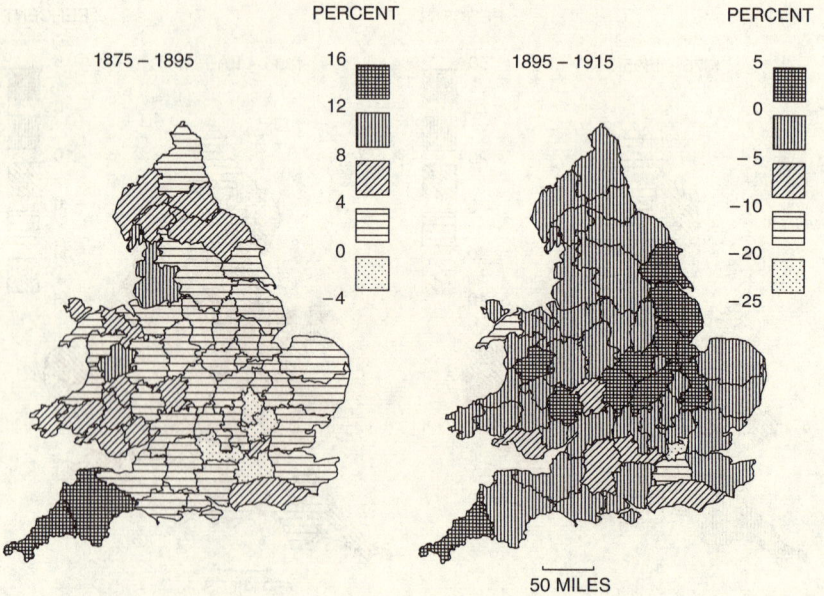

PERCENT

1875 – 1895

16
12
8
4
0
–4

PERCENT

1895 – 1915

5
0
–5
–10
–20
–25

50 MILES

36.11 Changes in total cultivated acreage, 1875–95, 1895–1915

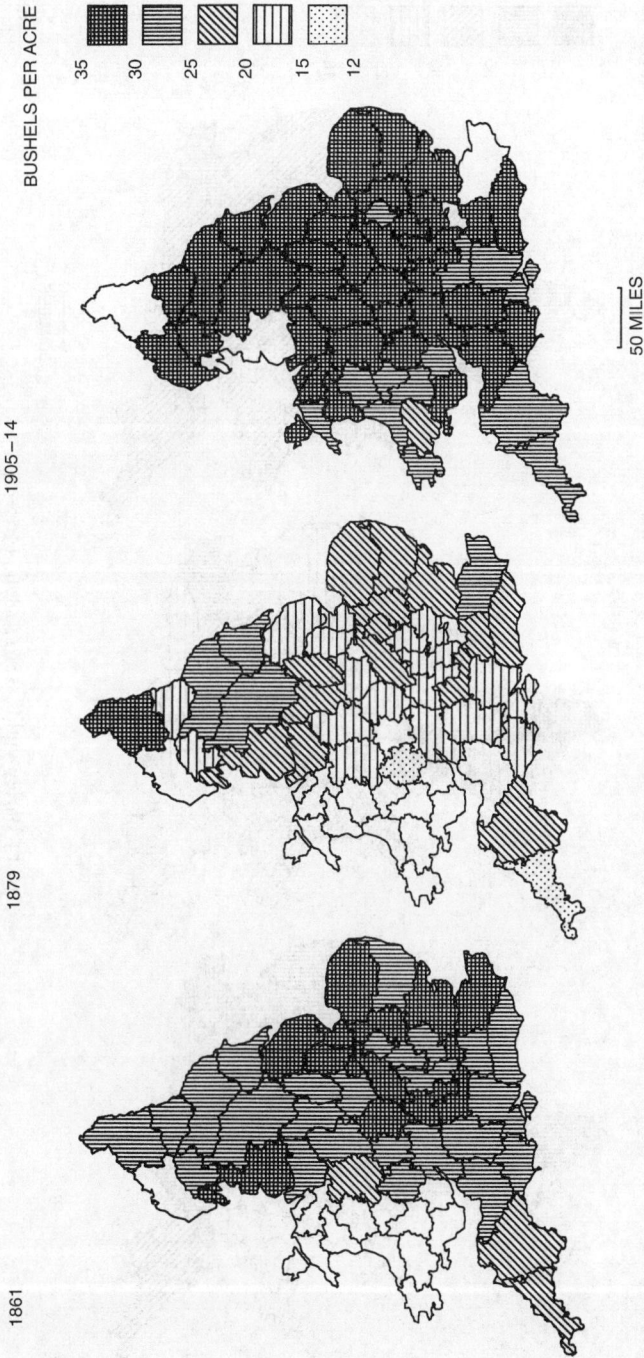

BUSHELS PER ACRE

35
30
25
20
15
12

1905–14

1879

1861

50 MILES

36.12 Wheat yields, 1861, 1879, 1905–14

1824

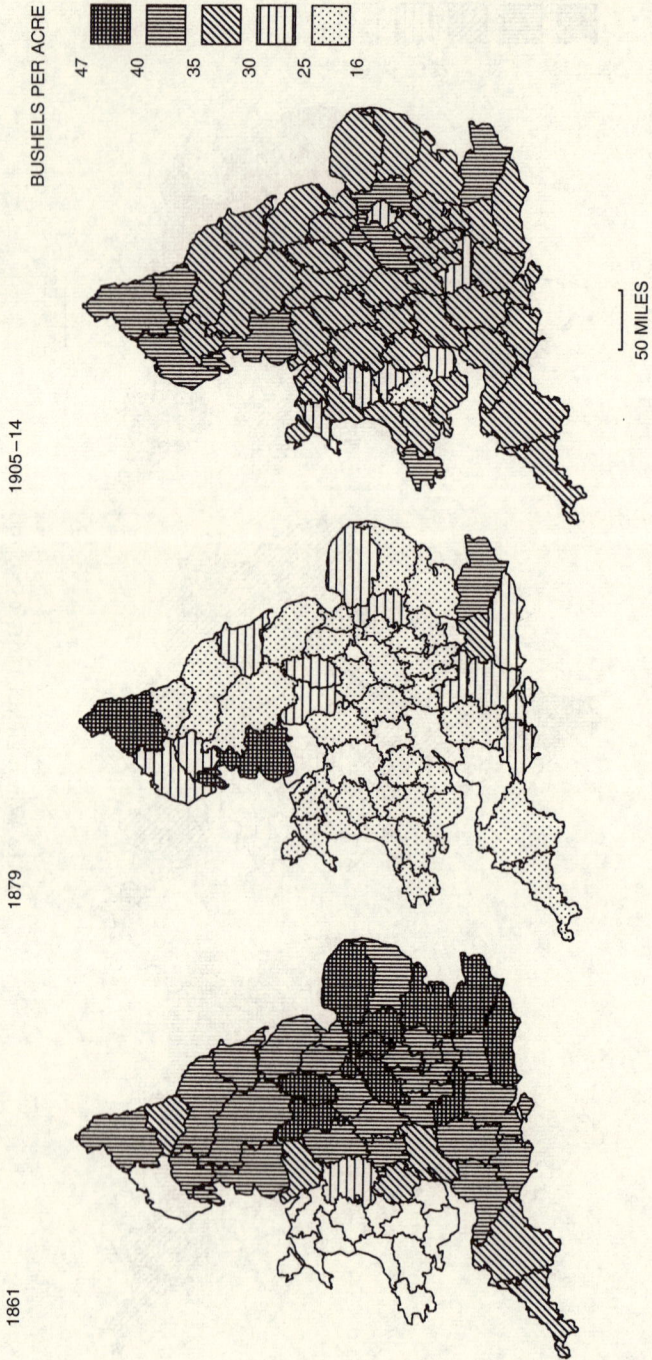

BUSHELS PER ACRE

47
40
35
30
25
16

1905–14

1879

1861

50 MILES

36.13 Barley yields, 1861, 1879, 1905–14

BUSHELS PER ACRE

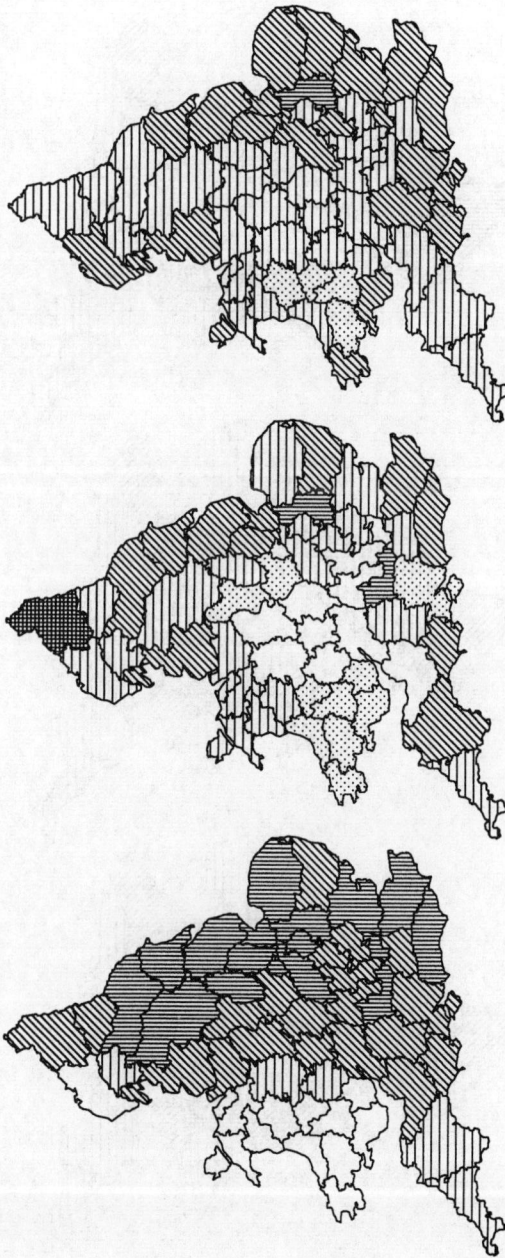

80
60
50
40
30
24

1905-14

50 MILES

1879

1861

36.14 Oats yields, 1861, 1879, 1905-14

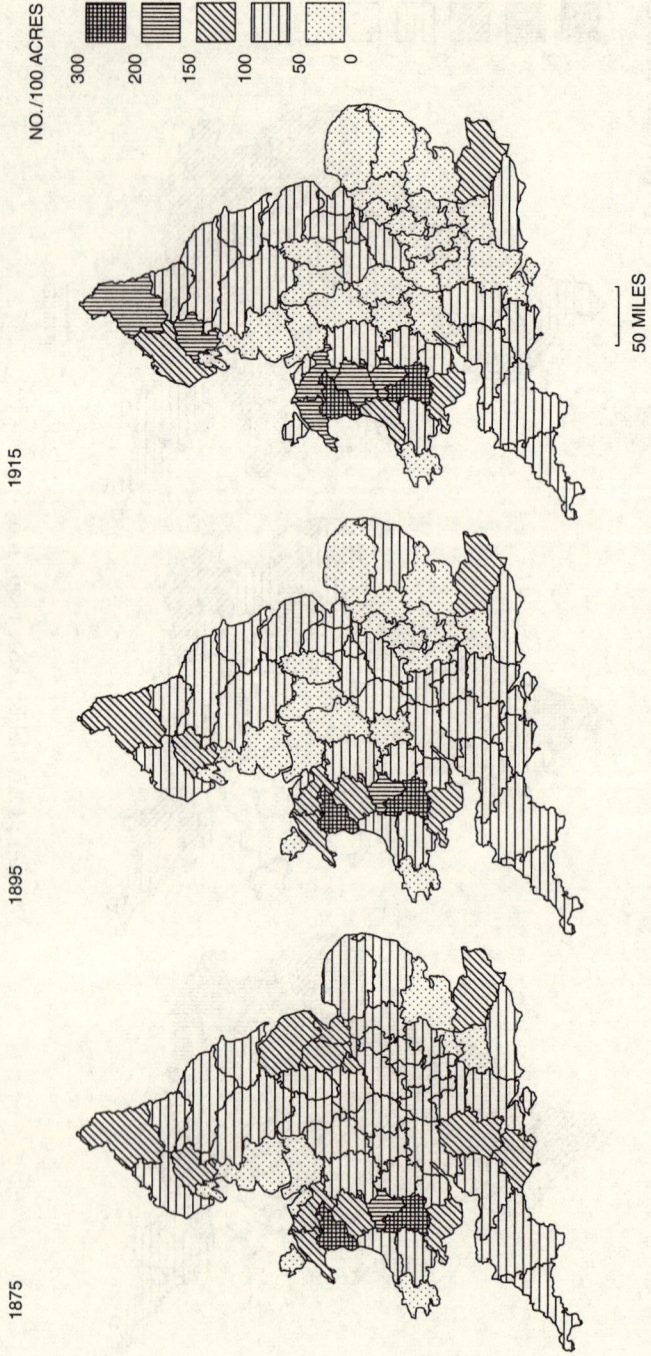

NO./100 ACRES

300
200
150
100
50
0

1875

1895

1915

50 MILES

36.15 Sheep per 100 acres of cultivated land, 1875, 1895, 1915

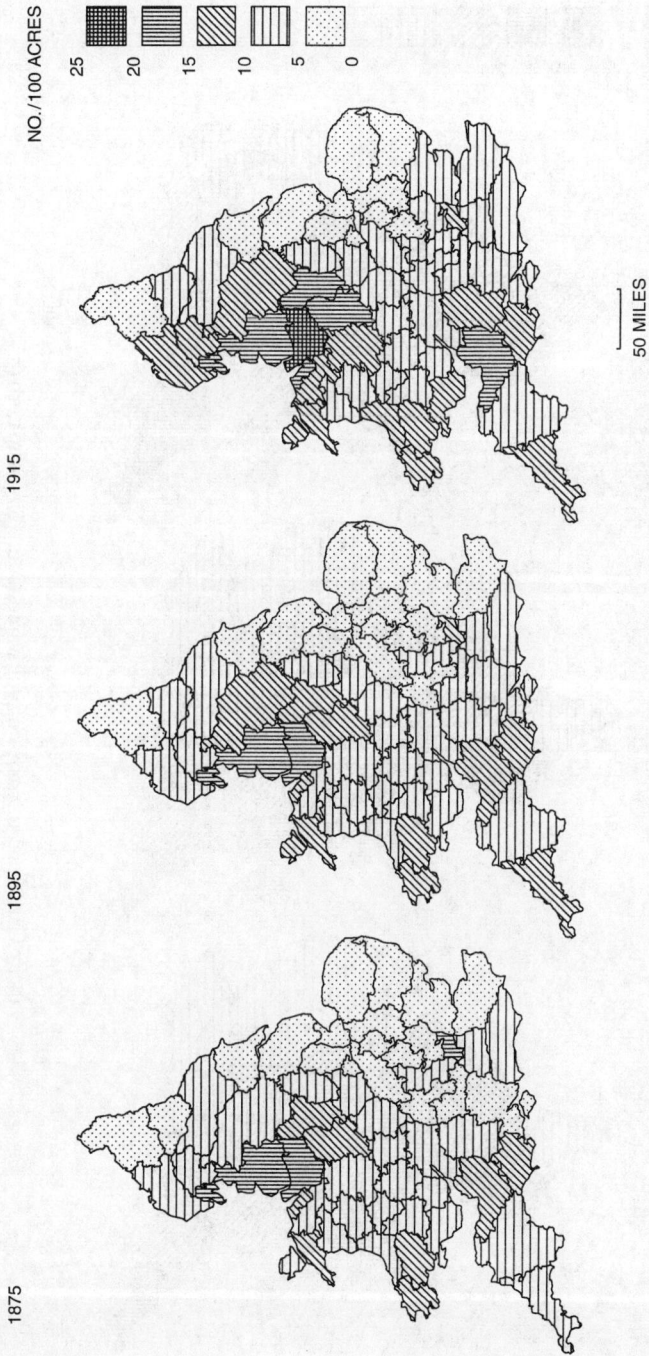

NO./100 ACRES

25
20
15
10
5
0

1915

1895

1875

50 MILES

36.16 Dairy cattle per 100 acres of cultivated land, 1875, 1895, 1915

1828

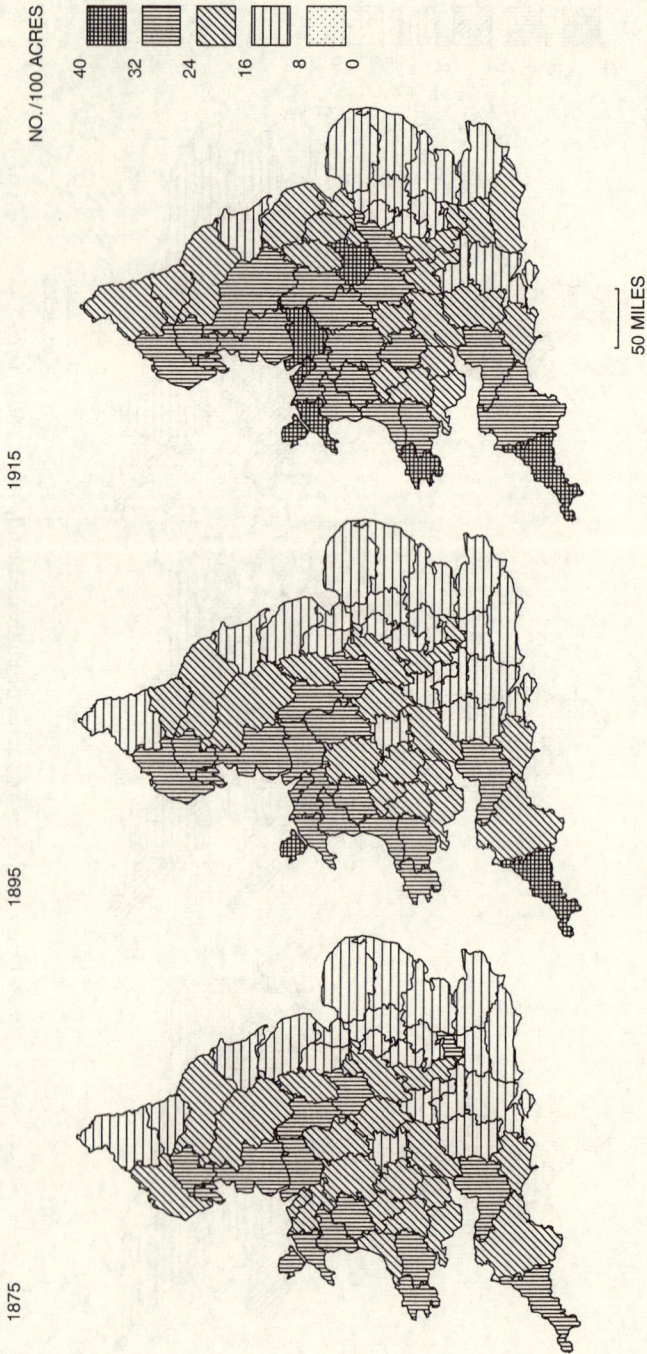

NO./100 ACRES

40
32
24
16
8
0

1915

1895

1875

50 MILES

36.17 Cattle per 100 acres of cultivated land, 1875, 1895, 1915

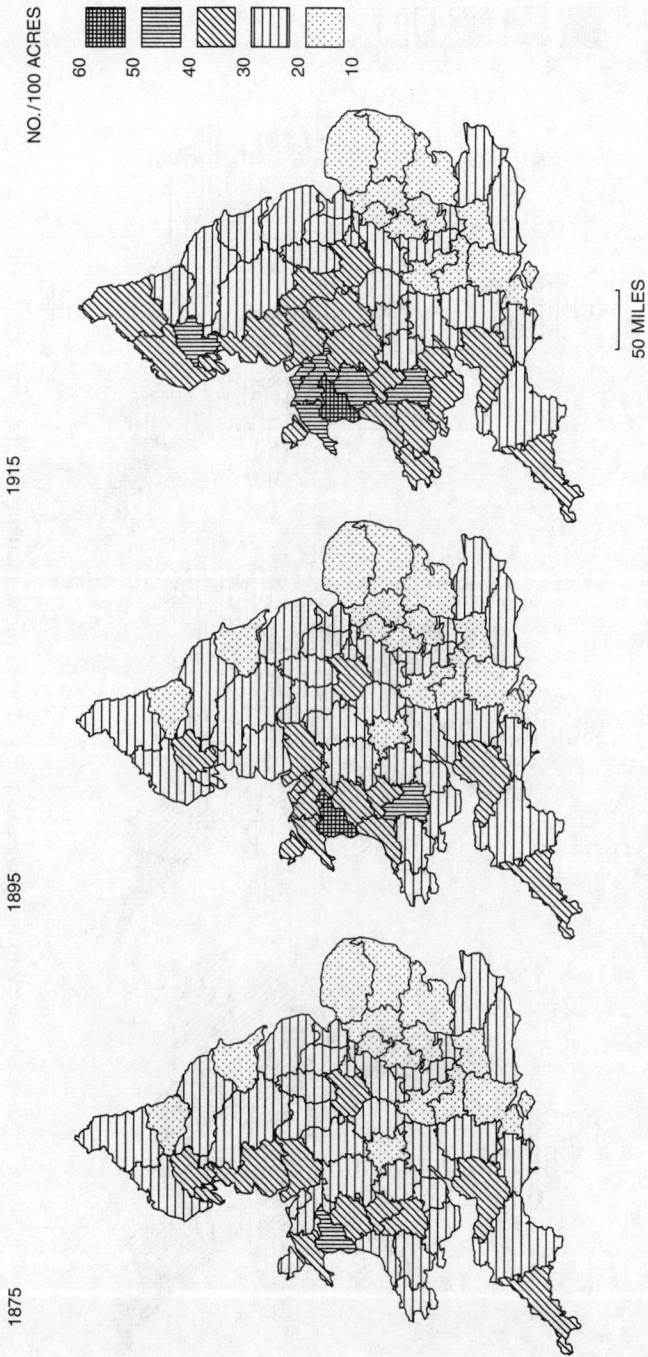

NO./100 ACRES

60
50
40
30
20
10

1875   1895   1915

50 MILES

36.18 Livestock unit equivalents per 100 acres of cultivated land, 1875, 1895, 1915

1830

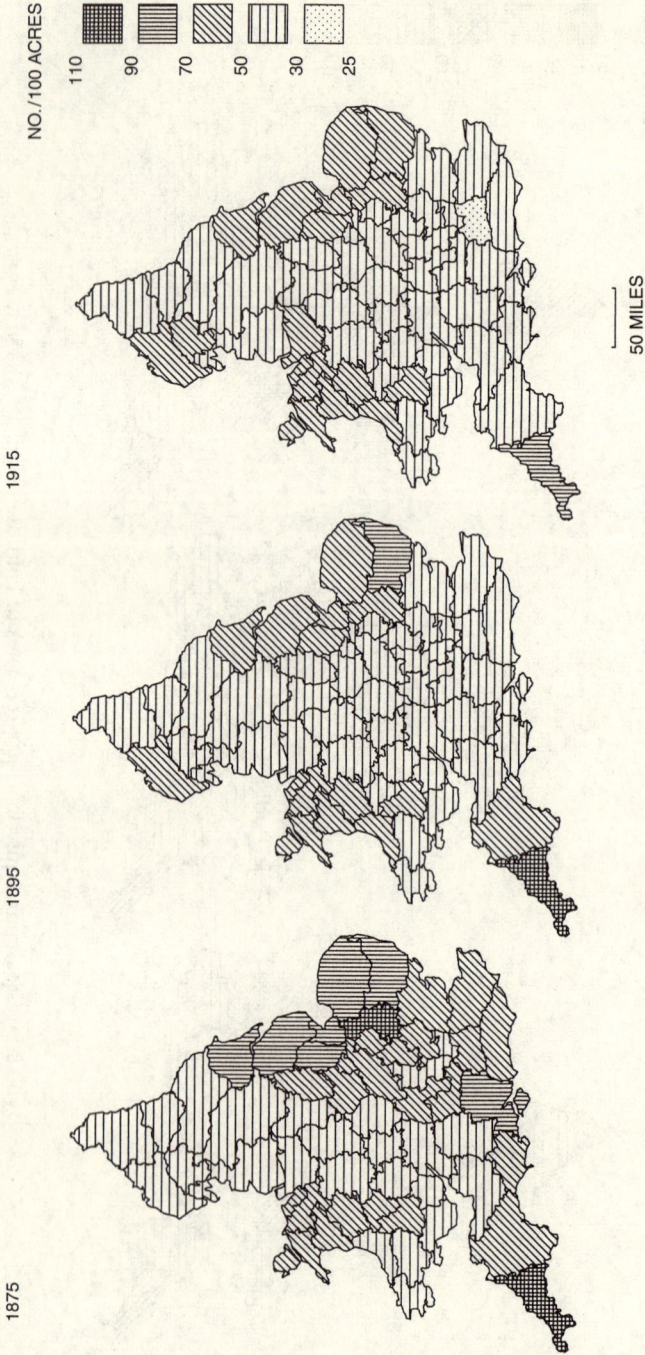

NO./100 ACRES

110
90
70
50
30
25

50 MILES

1915

1895

1875

36.19 Livestock unit equivalents per 100 acres of permanent pasture, 1875, 1895, 1915

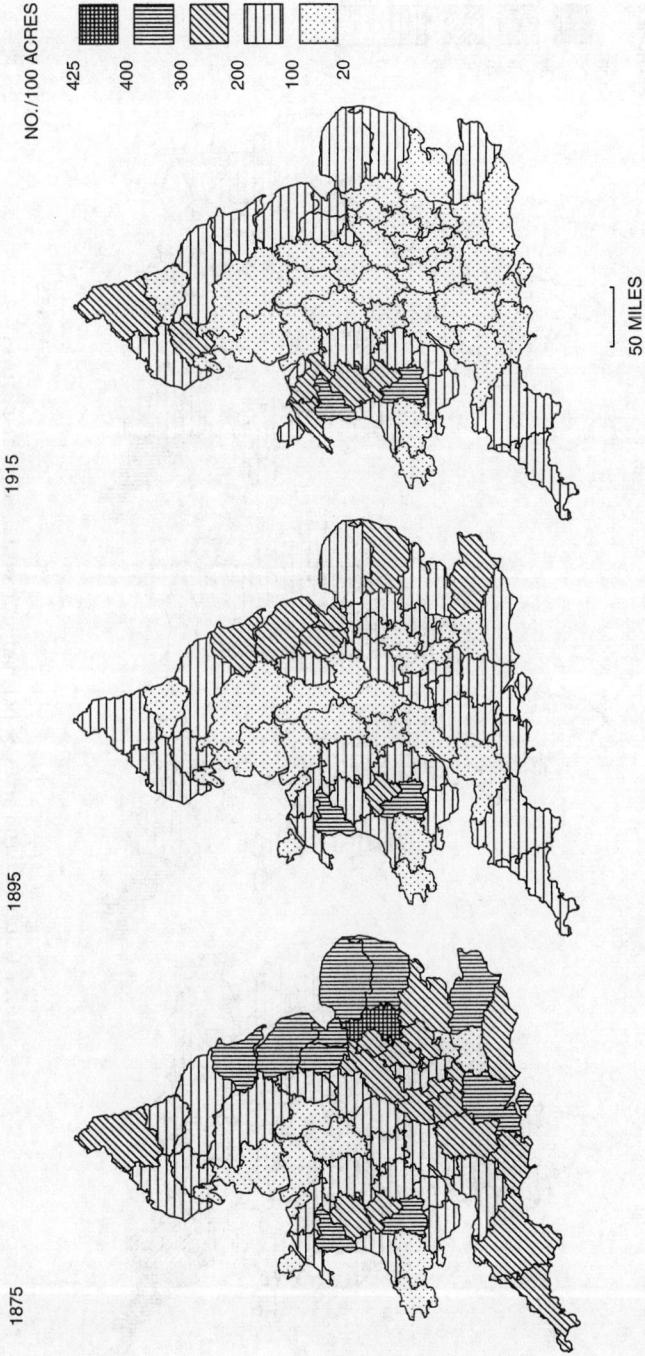

36.20 Sheep per 100 acres of permanent pasture, 1875, 1895, 1915

NO./100 ACRES

425
400
300
200
100
20

50 MILES

1915

1895

1875

NO./100 ACRES

100

90

70

50

30

15

1915

1895

1875

50 MILES

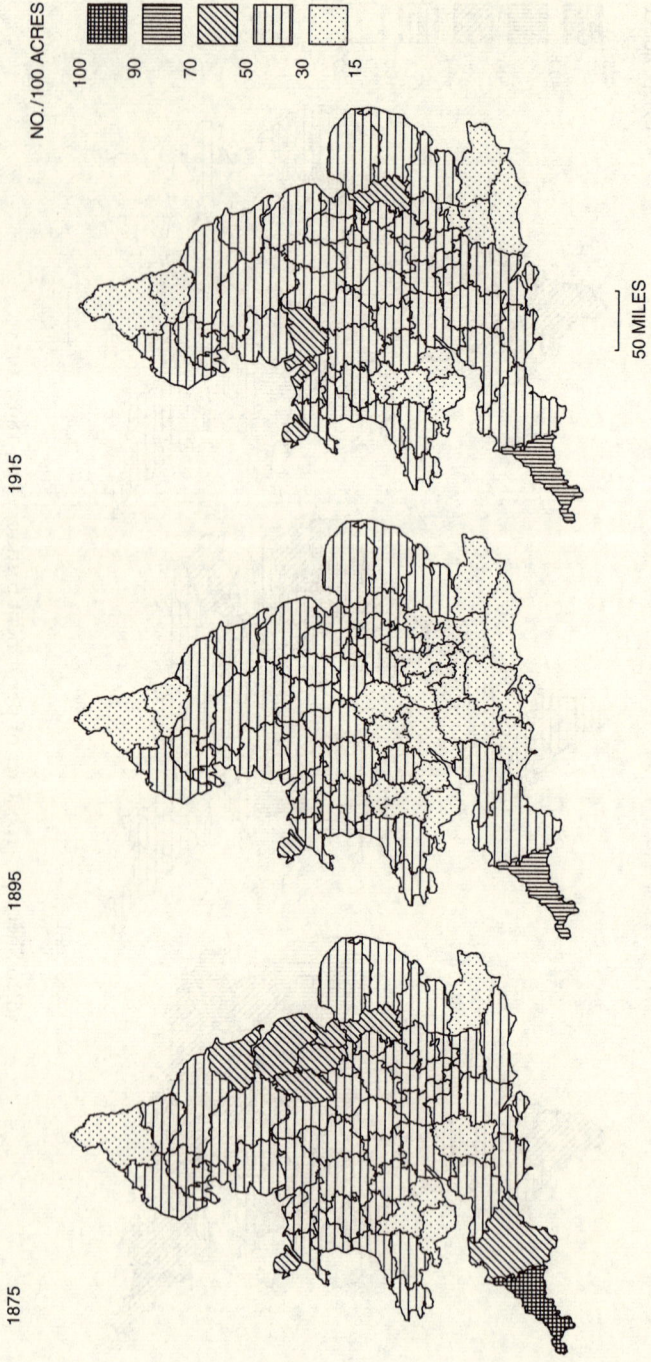

36.21 Cattle per 100 acres of permanent pasture, 1875, 1895, 1915

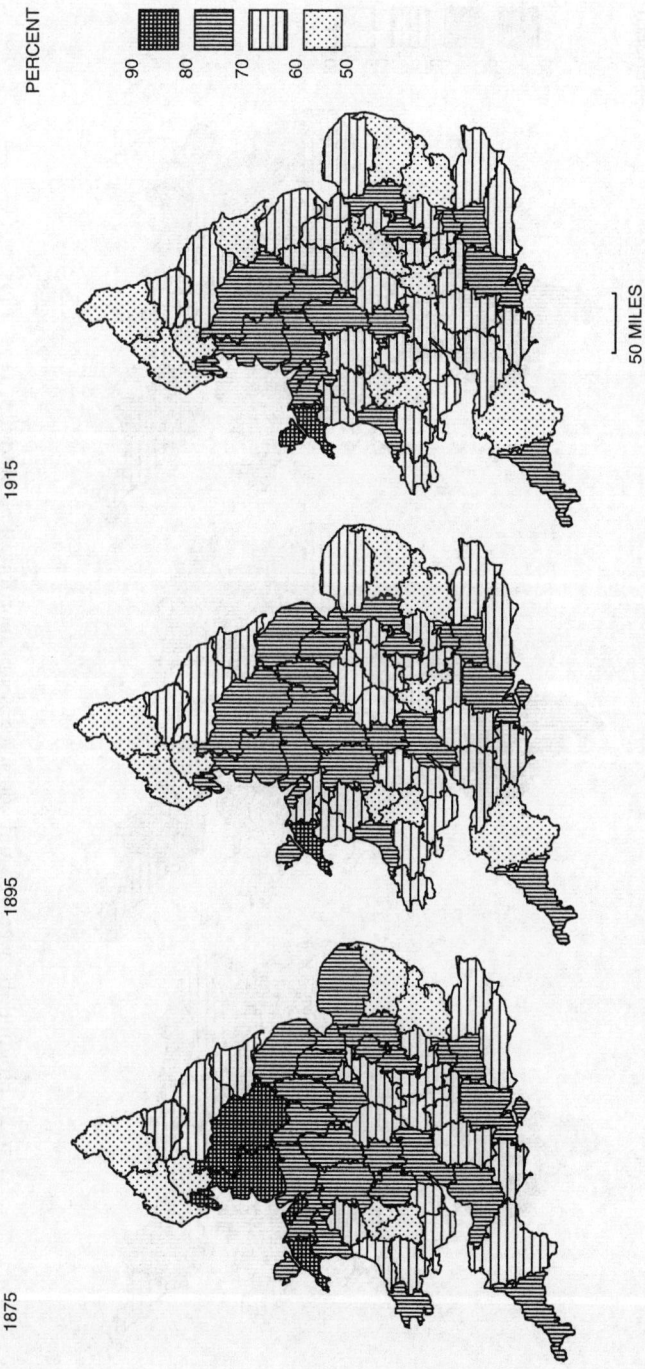

PERCENT

90
80
70
60
50

1915

1895

1875

50 MILES

36.22 Holdings under 50 acres as a proportion of all holdings, 1875, 1895, 1915

1834

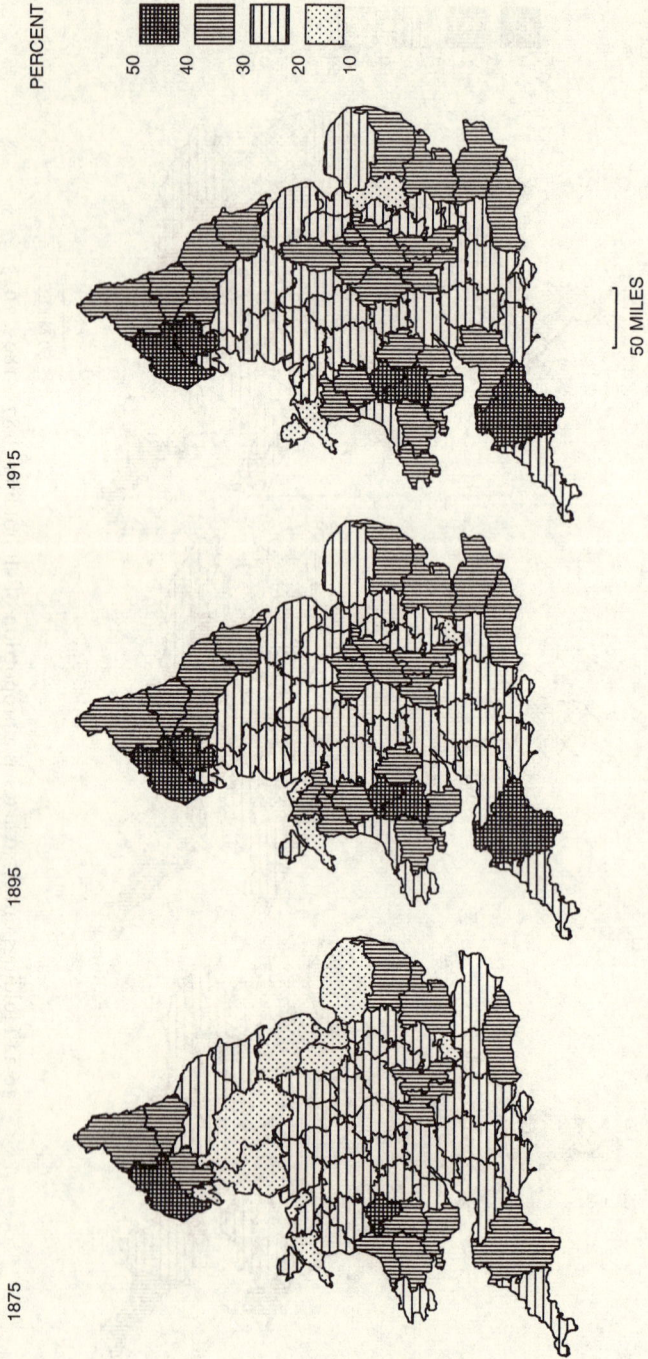

PERCENT

50 40 30 20 10

50 MILES

1915

1895

1875

36.23 Holdings between 50 and 300 acres as a proportion of all holdings, 1875, 1895, 1915

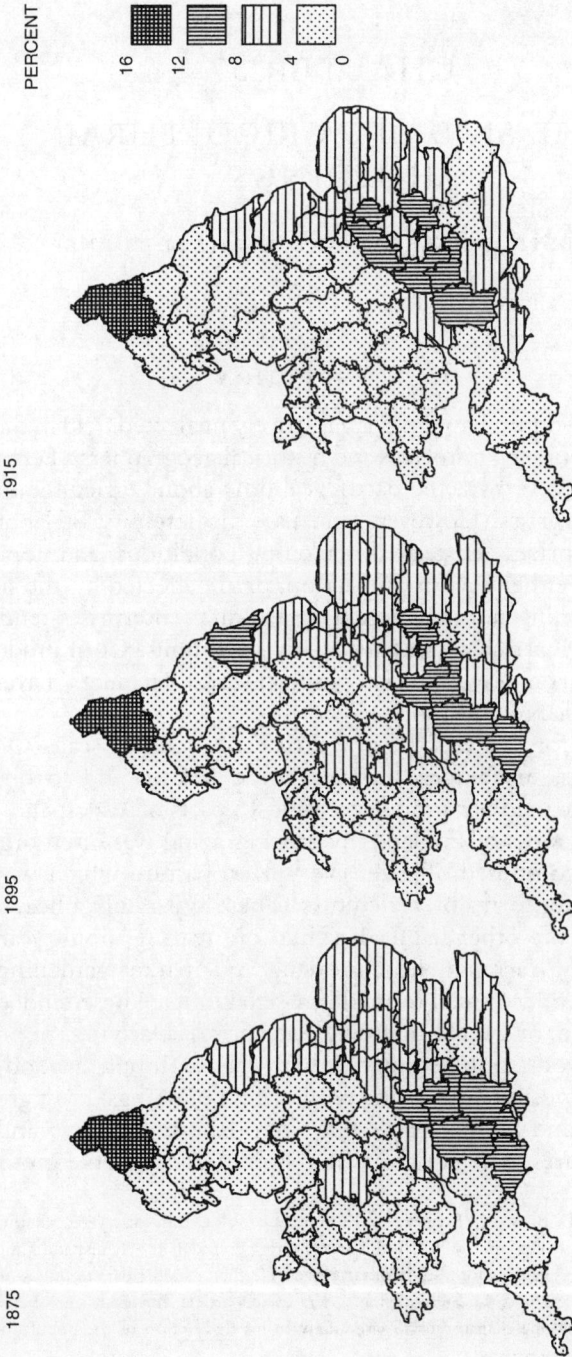

36.24 Holdings over 300 acres as a proportion of all holdings, 1875, 1895, 1915

CHAPTER 37

THE SIZE OF AGRICULTURAL HOLDINGS

BY BETHANIE AFTON AND MICHAEL TURNER

A. INTRODUCTION

The size of agricultural holdings is a relatively neglected field in the work conducted recently by historians and historical geographers.[1] Perhaps this is because, on their own, the data reveal little about agricultural conditions and enterprises. However, land use, the intensity of labour and capital inputs, prices and related marketing conditions, and population distribution all either have an impact on, or are affected by, the size and number of farming units. Different agricultural enterprises tend to be most efficiently carried on in particularly sized units. Corn production, and mixed farming, particularly sheep–corn farming, have been identified with large units.

One reason for this was the potential economies of scale. On large units, buildings, machinery, motive power, labour, and so on, were more efficiently used per productive unit. In parts of Wales, and on the Pennines and moors of England, livestock grazing was often organised in large units composed of extensive upland pasture which would be unsuited to other forms of agriculture. There was rarely a heavy input of labour. On the other hand, dairying, pig rearing, poultry and egg production, hop farming, fruit growing and market gardening were usually carried out in small to medium-sized units. These tended to be high labour and/or capital intensive enterprises. Dairying, pig rearing and poultry production required a high labour input, a good water supply and, in most situations, the provision of buildings. Fruit growing, hops farming and market gardening were labour intensive and were high-risk ventures. The last two also required good, fertile ground. To

---

[1] One exception is D. B. Grigg, 'Farm size in England and Wales, from early Victorian times to the present', *AHR*, 35, 1987, pp. 179–98. In this article Grigg considers the format and limitations of the various sources available for a study of farm sizes. He then uses the data to discuss the trends in size between 1851 and the present time. Users of this article, however, must be wary of a number of inconsistencies in the text, particularly in his description of the population census material. See *infra* notes 5 and 9.

organise any of these in very large units would have been prohibitively expensive.

Because of the close correlation between size and land use, marketing conditions largely determined the dominant unit size in the agricultural sector. Thus, as corn and meat prices fell during the late-nineteenth-century Agricultural Depression, the proportion of large units also tended to fall. Where possible there was a shift into perishable and high-value products which, in turn, increased the proportion of smaller units. There were often clusters of these enterprises near urban areas with good market outlets. As the town spread, many of the small units were subsumed, and larger holdings were fragmented to replace the smaller sub-urban enterprises.[2] The economic conditions favouring the growth of small units during the Agricultural Depression and its aftermath were augmented by social considerations. The size of agricultural units received considerable attention from about 1880 from those politicians who advocated the growth of small holdings, because more labour per acre was used on smaller units. This was seen both as a means of increasing the well-being of the rural lower classes, and as a way to engender the conservative attitudes associated with agrarian society.[3]

There are two main sets of sources from which to assemble a database on the size of agricultural units. The first attempt to establish the precise number of holdings in Great Britain was the 1851 Census of Population. Each farmer was asked to make a return of the size of his farm and the number of men employed on it. The material was then collated and published by county.[4] There are several important limitations associated with it. The first point to remember when using it is that this was a record of *farms*. In itself, this is not a problem and is a logical unit to adopt. However, it is impossible to make direct comparisons with later surveys, such as the Annual Agricultural Returns, which were based on 'holdings' rather than 'farms'. A second problem is one of possible underenumeration. The number of farmers in the census does not correspond to the number of farms. Not all farmers completed the returns. 249,431 farmers were listed in England and Wales but only 225,318 farms were returned. In addition, the sizes of 2,047 farms were not given by farmers returning the forms. Two circumstances may partially account for the discrepancy. Retired farmers and those in institutions were not expected to complete

---

[2] *The Agricultural Output of England and Wales, 1925*, BPP, 1927, XXV, pp. 81–2.

[3] H. Levy, *Large and Small Holdings: a Study of English Agricultural Economics* (Cambridge, 1911), pp. 94–5; P. G. Craigie, 'The size and distribution of agricultural holdings in England and abroad', *JRSS*, 50, 1887, pp. 107–8.

[4] *Census of Great Britain, 1851. Population Tables*, BPP, 1852–3, LXXXVIII, Part I, pp. lxxviii–lxxxi, clxxiv–clxxv, 30, 120–23, 236–40, 312–15, 408, 513–16; *ibid.*, Part II, pp. 596–8, 657–8, 728–9, 800–2, 884–5.

the return on farm size, although they were included in the occupations returns. On the other hand, those whose secondary occupation was farming were not enumerated as farmers, but they sometimes completed farm size returns.[5] A third problem with the census data is the lack of acreage information on a county basis, although the number of holdings for a range of areas was recorded for England and Wales as a whole in 1851, and for a limited group of counties in both 1861 and 1871.[6] This prevents us using the material to calculate the average size of holding by county. Finally, the returns instructed farmers to exclude hill pasture from the acreage.[7] When it is remembered that the cultivated area of Wales in the 1871 Census was only 59 per cent of the total area and for England 75 per cent, the exclusion of this land, much of which would have been used for livestock grazing, seriously affects the comprehensiveness of the data.[8] Similar material was collected with the population censuses of 1861 and 1871.[9] However, in 1861 details were only published for ten English counties and, in 1871, for seventeen.[10]

The second important set of data on agricultural holding sizes is included in the annual agricultural returns collected in Great Britain from 1866.[11] Unlike the census returns, the data from the agricultural returns are for 'holdings' rather than for 'farms'. This had two main consequences, both of which tended to inflate the supposed number of 'farms' at the lower end of the size scale. Some farms consisted of a number of separate holdings which, although they were cultivated as a single unit, were officially returned separately.[12] Additionally, because the returns were not specifically collected from farmers, a number of the smaller holdings were believed to be for holdings of land not used for agricultural purposes. The returns applied only to land in crops and grass, excluding 'all mountain and heath land used for grazing', with the same results as those mentioned for the census material. Another problem with

[5] *Census 1851*, Part I, p. 30n, and all the following county tables itemised in note 4 *supra*; this is contrary to information given in Grigg, 'Farm size', pp. 181, 184.

[6] *Census, 1851*, Part I, p. lxxx; *Census of England and Wales, 1871. General Report*, BPP, 1873, LXXI, Part II, pp. xliv–xlv.   [7] *Census, 1851*, Part I, p. lxxix; Grigg, 'Farm size', p. 181.

[8] Craigie, 'Size and distribution', pp. 88–9.

[9] *Census of England and Wales, 1861. General Report*, BPP, LIII, Part I, pp. 140–3; *Census, 1871*, Part II, pp. 124–9.

[10] For 1861 these counties were Buckinghamshire, Cambridgeshire, Cheshire, Cumberland, Lincolnshire, Norfolk, Shropshire, Sussex, Wiltshire and Yorkshire North Riding; for 1871, Surrey (extra Metropolitan), Kent (extra Metropolitan), Sussex, Hampshire, Berkshire, Essex, Suffolk, Norfolk, Leicestershire, Rutland, Lincolnshire, Nottinghamshire, Derbyshire, Durham, Northumberland, Cumberland and Westmorland. There is a discrepancy between this list and the information given in Grigg, 'Farm size', p. 181, in which he does not mention the ten counties in the 1861 returns, and states that in 1871 the seventeen counties were all in eastern and southern England.   [11] These are located in *Accounts and Papers, Annual Agricultural Statistics*, BPP.

[12] *The Agricultural Output, 1925*, p. 82.

the returns was caused by the system of size classification adopted. The twenty-four categories used for the census returns were abandoned, and a smaller number was substituted. From 1895 a more standardised system of classification was adopted which used holdings of above one acre as a baseline. Prior to this the system varied and, because the divisions used were not consistent over time, it is less easy to make direct comparisons. The use of fewer classes made changes in farm sizes, caused by new marketing conditions, less obvious in the statistics. According to one agricultural economist of the period, this masked significant changes in the structure of British agriculture.[13] Up to and including 1875, the number of holdings was annually published on a county basis for England and Wales. From this information it is possible to calculate the average holding size for each county. Thereafter, the number of holdings organised in various size groups is available for 1880, 1885, 1889, 1895, and annually from 1903. Acreage information for each county was included in 1875, 1880, 1885, 1895, and annually from 1912.

Several other official sources also contain useful details on the sizes of agricultural holdings. The reports of the Inland Revenue Commissioners included the numbers, in addition to the value, of separate properties assessed under Schedule B in respect of the occupation of land.[14] These returns have the advantage of including all types of land, including all upland grazing areas. However, they were collected for all land from which an income was obtained and may include many holdings that were not used specifically for agriculture.[15] In compliance with the Census of Production Act of 1906, the report prepared by the Ministry of Agriculture and Fisheries between 1906 and 1908 contains some data on holdings, but is more important for the use it made of the data to analyse the effect of size on agriculture in general.[16] The material was collected in the same way as the other investigations carried out for the annual returns and, consequently, has the same limitations.

B. THE DATA

Tables 37.1a and 37.1b contain information on the size of agricultural holdings in England and Wales between 1870 and 1914. The data were originally published in the Annual Agricultural Returns and reproduced

---

[13] Levy, *Large and Small Holdings*, pp. 98–9.

[14] These were published in most years in the annual Report of the Commissioners of the Inland Revenue. For example, *The 28th Annual Report of the Commissioners of the Inland Revnue, 1885*, BPP, 1884–5, XXII, has the data from 1869/70 to 1883/4.

[15] Craigie, 'Size and distribution', pp. 87–8; *28th Report*, pp. 80–1.

[16] *The Agricultural Output of Great Britain ... in connection with the Census of Production Act, 1906*, BPP, 1912–13, X.

in *The Agricultural Output of England and Wales, 1925*. One of the most obvious difficulties encountered when using the information arises when trying to compare one year with another. This is particularly so for comparisons between years before and after 1895. In that year both the holdings equal to, or above, one acre, and those above one acre were included; but before that date holdings of one acre were included, and thereafter they were not. Consequently it is impossible to relate much of the pre-1895 data to that from later years. A similar, but less troublesome problem, is the variation in size categories found in the various years. Hence Table 37.1a includes those holdings between 300 and 1,000 acres and above 1,000 acres, while Table 37.1b only lists those above 300 acres.

Tables 37.2a and 37.2b give the acreage of the holdings of various size categories. This information was again published in the Annual Agricultural Returns, but appeared less frequently than measurements simply of the number of holdings. It suffers from the same limitations as the information in Tables 37.1a and 37.1b.

The somewhat vaguer information from the Schedule B Income Tax Returns is contained in Table 37.3. Tax under this schedule was the charge on the profits made by the occupier of land from the exercise of his capital and skill in husbandry. However, not all the properties included were agricultural holdings.

Figures 37.1a–i, 37.2a–b, and 37.3a–d contain details of the average size of farms, rather than holdings, as recorded on a county basis in the *Census of Population* for 1851, 1861 and 1871. In 1851 all counties were included, thereafter only selectively. Three counties, Cumberland, Norfolk and Sussex, were included in the Census Reports from all three years. Perhaps the most interesting point to note was the greater concentration of small farms in north-western England and in Wales.

Figure 4 again considers the average size of agricultural units on a county basis. The three maps illustrate the average size of holdings at the height of the High Farming period in 1870, at the bottom of the agricultural depression in 1895, and in 1914 when the peacetime recovery was complete. As in the case of Figures 37.1 and 37.2, the average for 1870 includes all properties, while in 1895 and 1914 properties of one acre and under are excluded. The impact of the exclusion of holdings of exactly one acre obviously did not affect each county equally.

Figures 37.5–37.8 relate holdings and land use in 1885 and 1908, with 37.5a–c and 36.6a–c illustrating land use according to the proportion of holdings in particular size categories, and 37.7a–c and 38.8a–c the land use in terms of the proportion of the acreage they occupied. While the exact criteria for allocating holdings to each land use category were not stated in the data, the predominance of mixed farms, particularly in holdings over 50 acres, is striking. Between 1885 and 1908 there was a significant reduction in the number and acreage of arable holdings. The shift in profitability in favour of livestock husbandry during and after the agricultural depression at the end of the nineteenth century is apparent on all sizes of holdings.

Figures 37.9a and 37.9b consider the effect of holding size on livestock numbers in 1908. Figure 37.9a considers the actual number of each type of animal. However, in Figure 37.9b the animals have been reduced to common units. Although there is a long tradition of using multiples to equate different types of livestock, a formalised convention of weighting was established in the mid-twentieth century. Using the estimated feeding requirements of farm animals provides a means of comparing one animal with another in terms both of the feeding capacities of grazing land and feedstuffs, and of the capacity of these to maintain or fatten the animals. Each variety of livestock is reduced to a cow equivalent. The conversion can be quite complex, with variations according to age, condition and location of each animal.[17] However, a very general weighting of five sheep and three and one-third pigs to one cow has been used in Figure 37.9b. Dairying and pig keeping were most closely associated with smaller holdings, beef-cattle rearing with medium-sized holdings, and sheep grazing with larger units.

[17] For further guidance on the weighting of livestock see W. B. Morgan and R. J. C. Munton, *Agricultural Geography* (London, 1971), pp. 106–7; J. T. Coppock, *An Agricultural Atlas of England and Wales* (London, 1964), pp. 165, 167, 226; and the various editions of J. Nix, *Farm Management Pocketbook* (e.g. 6th edn, Department of Agricultural Economics, Wye College, Ashford, Kent, 1974).

Table 37.1a. *Number of holdings of various sizes in England and Wales,*
1870–1895

	¼ acre but less than 1 acre	1 acre to 5 acres inclusive	Total not exceeding 5 acres	Above 5 acres to 20 acres	Total not exceeding 20 acres
1870	–		113,050	127,761	240,811
1871	–				237,999
1872	18,659	103,189	121,848		
1875					
1880					
1885	22,162	114,273	136,425	126,674	263,099
1889	27,352	121,826	149,178	129,250	278,428
1895★		133,372	133,372	126,714	260,086

	Above 20 acres to 50 acres	Total not exceeding 50 acres	Above 50 acres to 100 acres	Total not exceeding 100 acres
1870	75,418	316,229	54,569	370,798
1871				
1872				
1875		333,630	54,498	388,128
1880		336,149	54,369	390,518
1885	73,472	339,571	54,937	391,508
1889	74,611	353,039		
1895★	74,846	334,932	56,791	391,723

(continued)

Table 37.1a (*cont.*)

	Above 100 acres to 300 acres	Above 300 acres not exceeding 1000 acres	Above 1000 acres	Total above 100 acres
1870	–			78,749
1871				
1872				
1875	65,766	15,633	473	81,872
1880	66,373	16,241	506	83,120
1885	67,024	16,035	573	83,632
1889				
1895*	68,277	15,494	527	84,298

Notes: ★ Excluding holdings of less than 1 acre, but including 35,554 holdings of exactly 1 acre.
Source: Agricultural Output of England and Wales, 1925, BPP, 1927, p. 143.

Table 37.1b. *Number of holdings of various sizes in England and Wales, 1895–1914*

	Above 1 acre to 5 acres	Above 5 acres to 20 acres	Total above 1 acre, but not exceeding 20 acres	Above 20 acres to 50 acres	Total above 5 acres, but not exceeding 50 acres
1895	97,818	126,714	224,532	74,846	201,560
1903	91,797				198,874
1908	89,958				197,218
1913	92,302	122,117	214,419	78,027	200,144
1914	91,570	121,698	213,268	78,454	200,152

	Above 50 acres to 100 acres	Above 100 acres to 150 acres	Above 150 acres to 300 acres	Total above 50 acres, but not exceeding 300 acres	Above 300 acres
1895	56,791		(68,277)	125,068	16,021
1903				126,980	15,351
1908				127,864	15,041
1913	59,287	31,838	37,593	128,718	14,513
1914	59,514	31,860	37,615	128,989	14,413

Source: Agricultural Output of England and Wales, 1925, p. 144.

Table 37.2a. *Acreage of holdings of various sizes in England and Wales,*
1875–1885

	¼ acre but less than 1 acre	1 acres to 5 acres inclusive	Total not exceeding 5 acres	Above 5 acres to 20 acres	Total not exceeding 20 acres
1875					
1880					
1885	10,518	321,058	331,576	1,419,832	1,751,408

	Above 20 acres to 50 acres	Total not exceeding 50 acres	Above 50 acres to 100 acres	Total not exceeding 100 acres
1875		4,182,346	3,957,979	8,140,325
1880		4,176,427	3,940,796	8,117,223
1885	2,462,852	4,214,260	4,021,021	8,235,281

	Above 100 acres to 300 acres	Above 300 acres not exceeding 1000 acres	Above 1000 acres	Total above 100 acres
1875	11,183,618	6,928,237	584,935	18,696,790
1880	11,400,011	7,068,708	644,487	19,113,206
1885	11,519,362	7,206,932	745,511	19,471,805

Sources: Annual Agricultural Returns, 1875, 1880, 1885.

Table 37.2b. *Acreage of holdings of various sizes in England and Wales, 1895–1914*

	Above 1 acre to 5 acres	Above 5 acres to 20 acres	Total above 1 acre, but not exceeding 20 acres	Above 20 acres to 50 acres	Total above 1 acre, but not exceeding 50 acres
1895	300,901	1,421,983	1,722,884	2,501,710	4,224,594
1912	284,864				4,289,489
1914	282,980	1,366,990	1,649,970	2,636,094	4,286,064

	Above 50 acres to 100 acres	Above 100 acres to 300 acres	Total above 50 acres, but not exceeding 300 acres	Above 300 acres
1895	4,153,226	11,672,707	15,825,933	7,632,520
1912			16,114,691	6,770,510
1914	4,340,952	11,788,767	16,129,719	6,698,221

Sources: Annual Agricultural Returns, 1895, 1912, 1914.

Table 37.3. *Number of non-metropolitan properties in England and Wales assessed under Schedule B, 1872–98*

Year ending	Number of properties
1872	716,362
1873	725,420
1874	724,364
1875	726,527
1876	726,952
1877	738,459
1878	734,032
1879	733,677
1880	781,654
1881	771,075
1882	772,950
1883	798,342
1884	796,292
1885	796,175
1886	794,165
1887	792,946
1888	791,285
1889	–
1890	794,251
1891	796,374
1892	797,851
1893	799,992
1894	796,274
1895	796,227
1896	799,308
1897	803,377
1898	803,608

Source: 'Annual Reports', *Commissioners of the Inland Revenue*, 1884/5–1898.

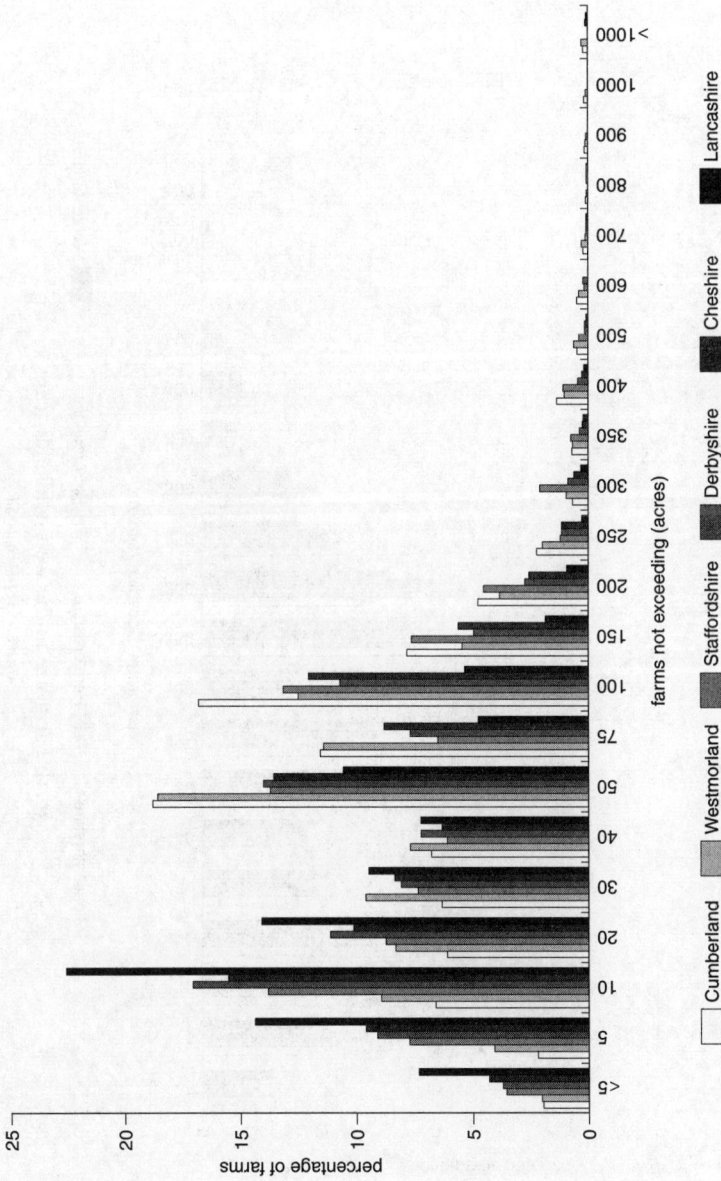

37.1a Farm size in north-western England, 1851

Source: Census of Great Britain, 1851, BPP, 1852–3, LXXXVIII

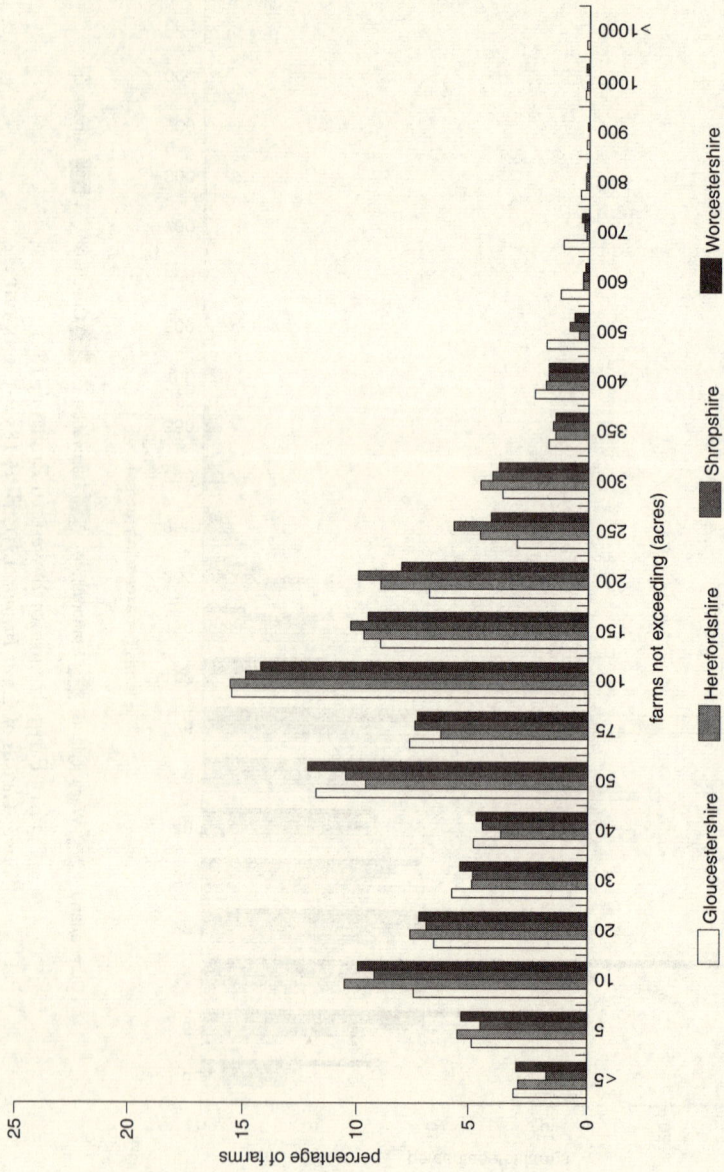

37.1b Farm size in the west Midlands, 1851

Source: Census, 1851

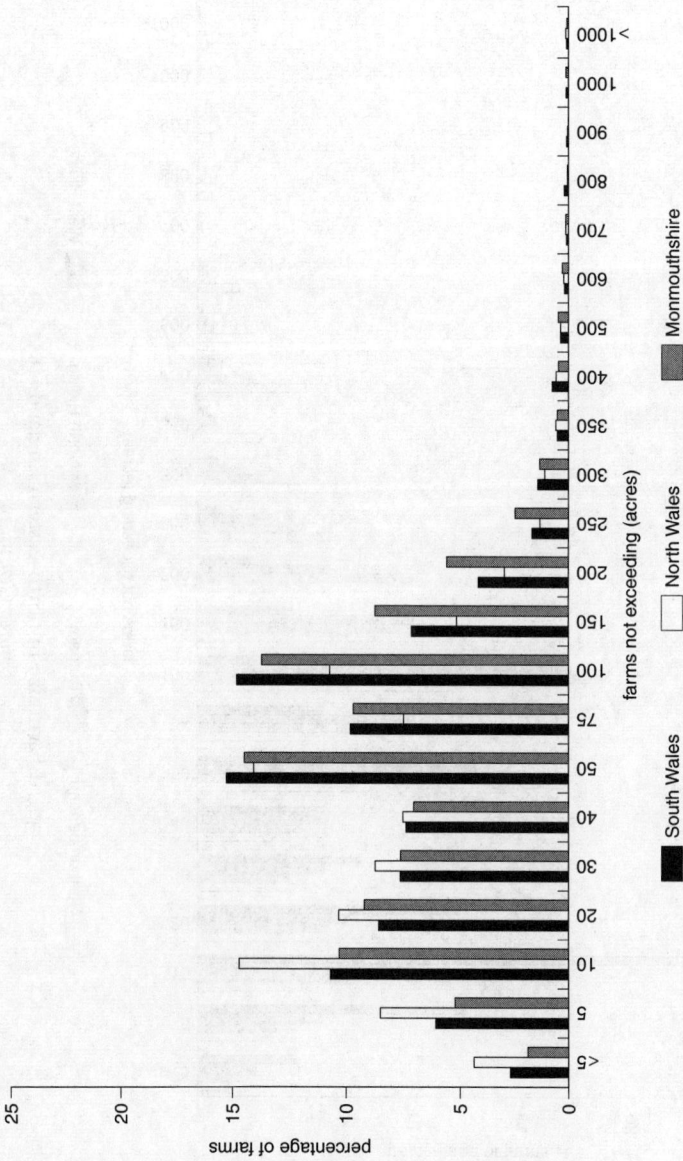

37.1c Farm size in Wales, 1851
Source: Census, 1851

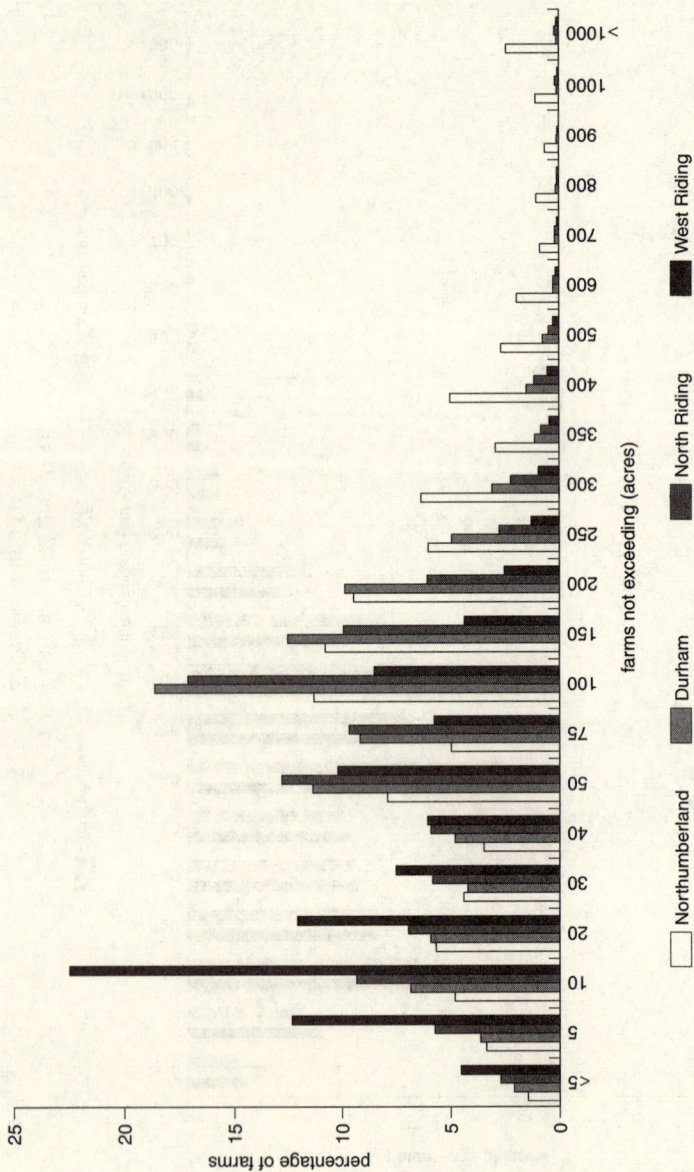

37.1d Farm size in north-eastern England, 1851

Source: Census, 1851

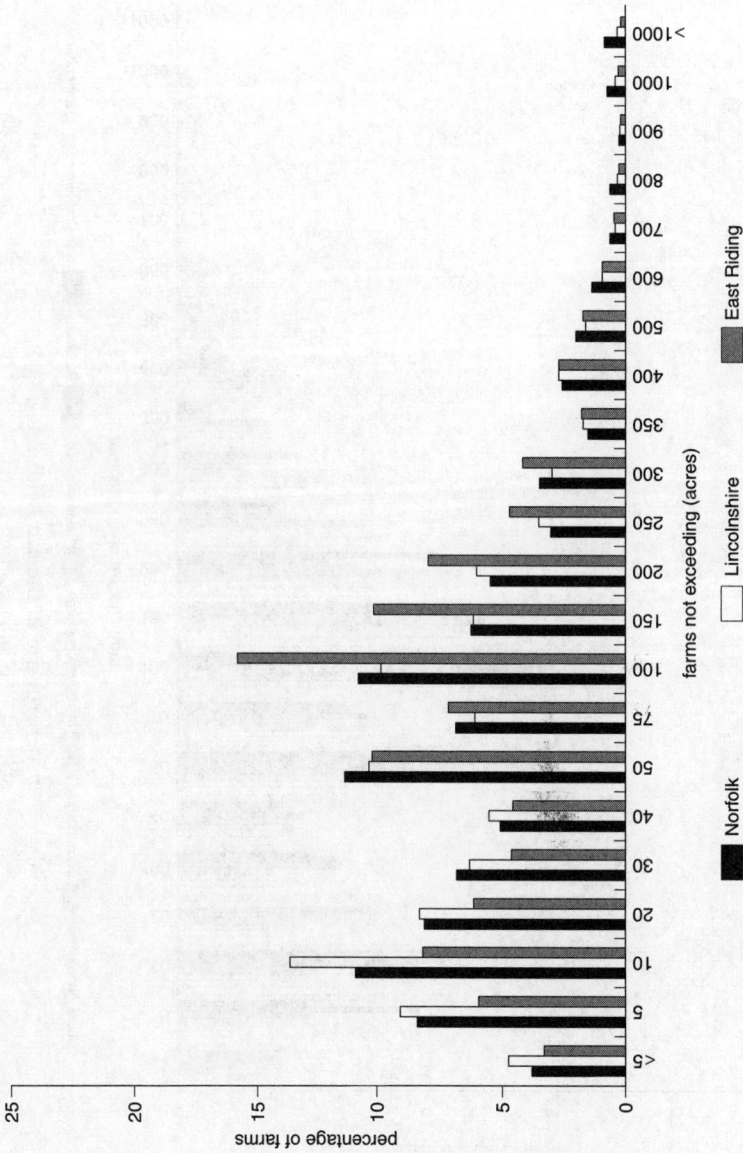

37.1e Farm size in eastern England, 1851
Source: Census, 1851

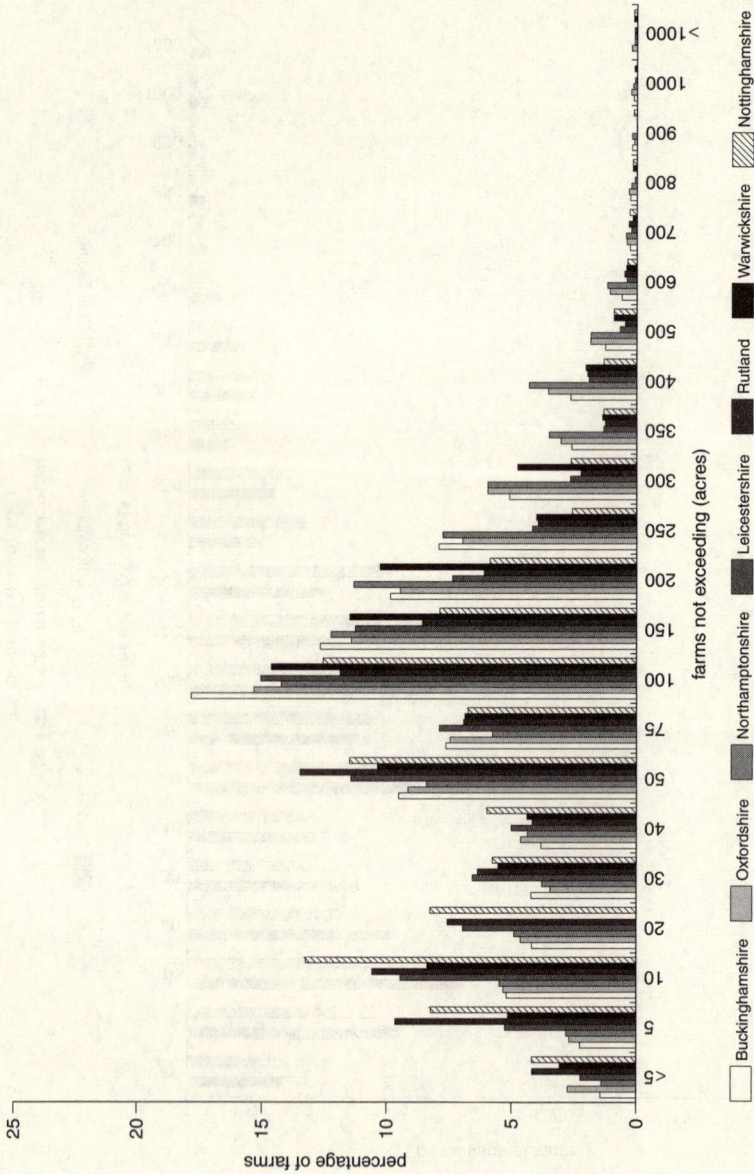

37.1f Farm size in the east Midlands, 1851
Source: Census, 1851

Buckinghamshire   Oxfordshire   Northamptonshire   Leicestershire   Rutland   Warwickshire   Nottinghamshire

farms not exceeding (acres)

percentage of farms

<5 5 10 20 30 40 50 75 100 150 200 250 300 350 400 500 600 700 800 900 1000 >1000

0 5 10 15 20 25

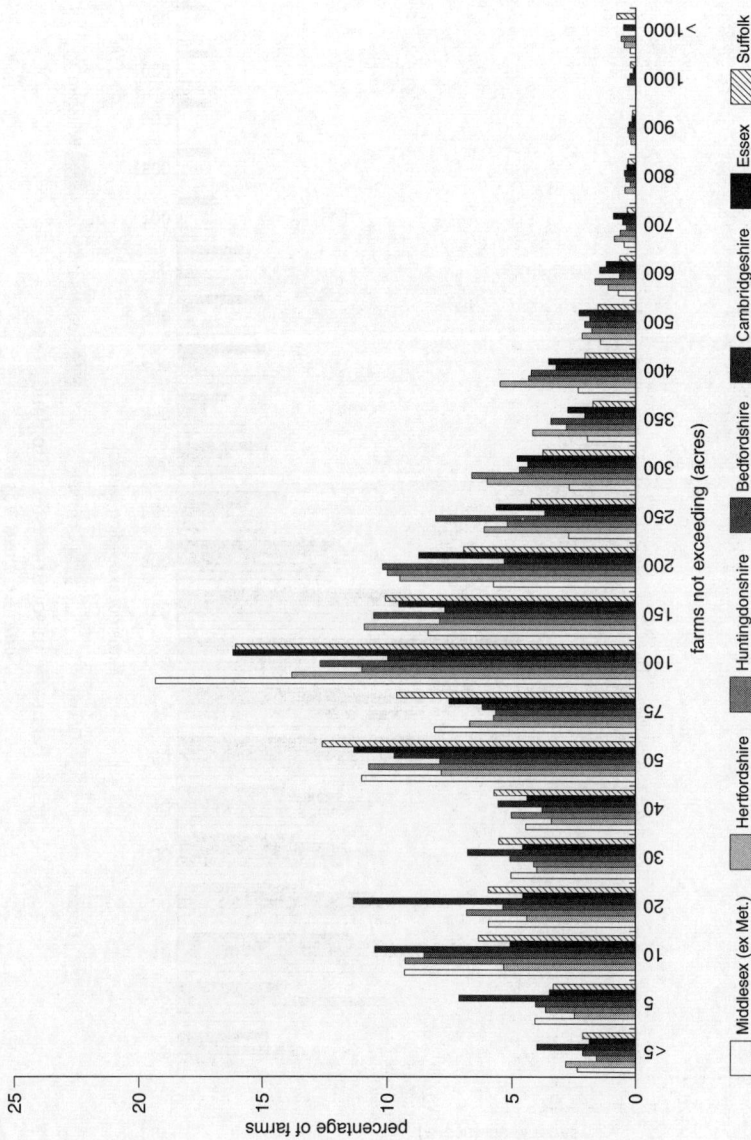

37.1g Farm size in the south Midlands, 1851

Source: Census, 1851

Legend: Middlesex (ex Met.), Hertfordshire, Huntingdonshire, Bedfordshire, Cambridgeshire, Essex, Suffolk

x-axis: farms not exceeding (acres) — >5, 5, 10, 20, 30, 40, 50, 75, 100, 150, 200, 250, 300, 350, 400, 500, 600, 700, 800, 900, 1000, >1000

y-axis: percentage of farms — 0, 5, 10, 15, 20, 25

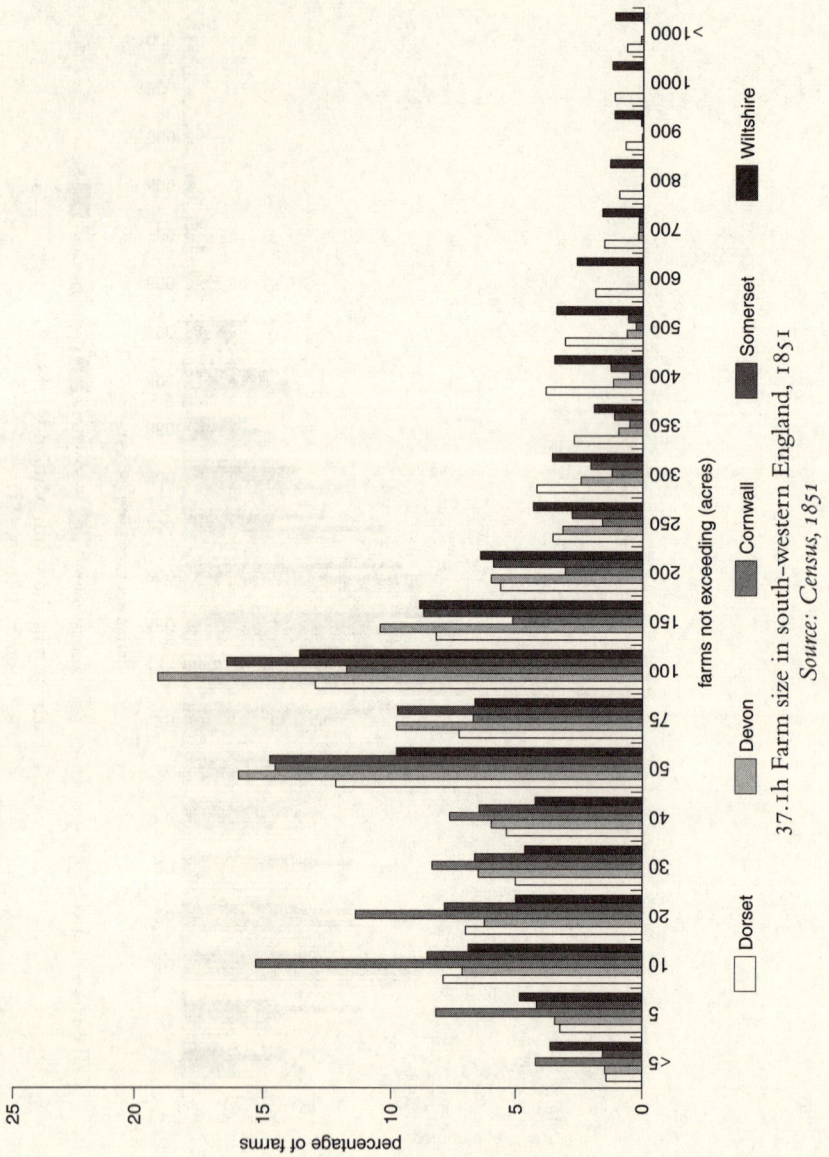

37.1h Farm size in south-western England, 1851
Source: *Census, 1851*

Legend: Dorset, Devon, Cornwall, Somerset, Wiltshire

x-axis: farms not exceeding (acres) — <5, 5, 10, 20, 30, 40, 50, 75, 100, 150, 200, 250, 300, 350, 400, 500, 600, 700, 800, 900, 1000, >1000

y-axis: percentage of farms — 0, 5, 10, 15, 20, 25

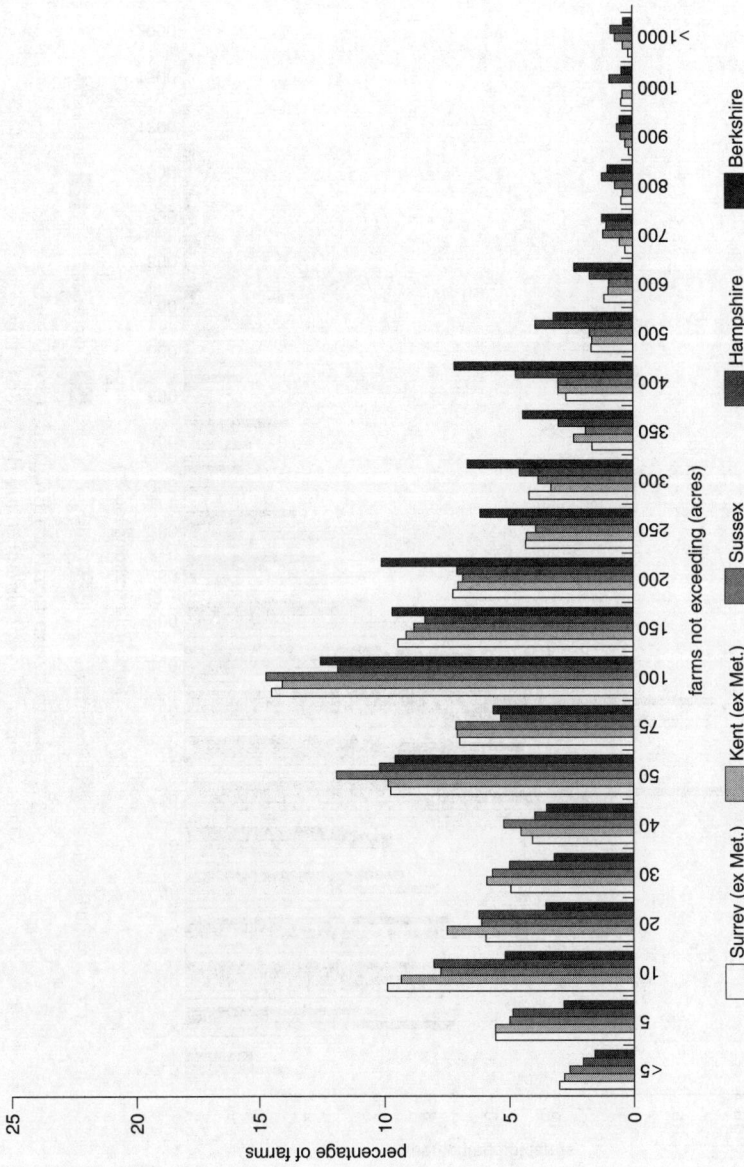

37.ii Farm size in south–eastern England, 1851

Source: Census, 1851

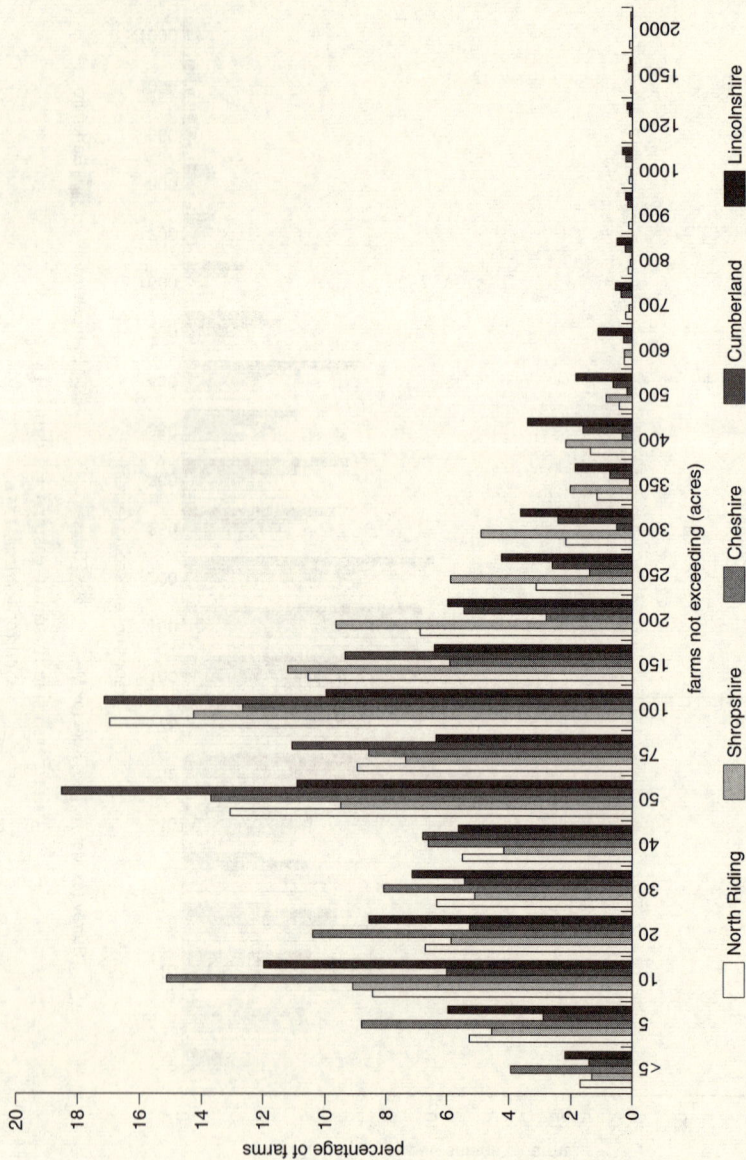

37.2a Farm size in northern England, 1861
Source: *Census of Great Britain, 1861, BPP, 1863, LIII*

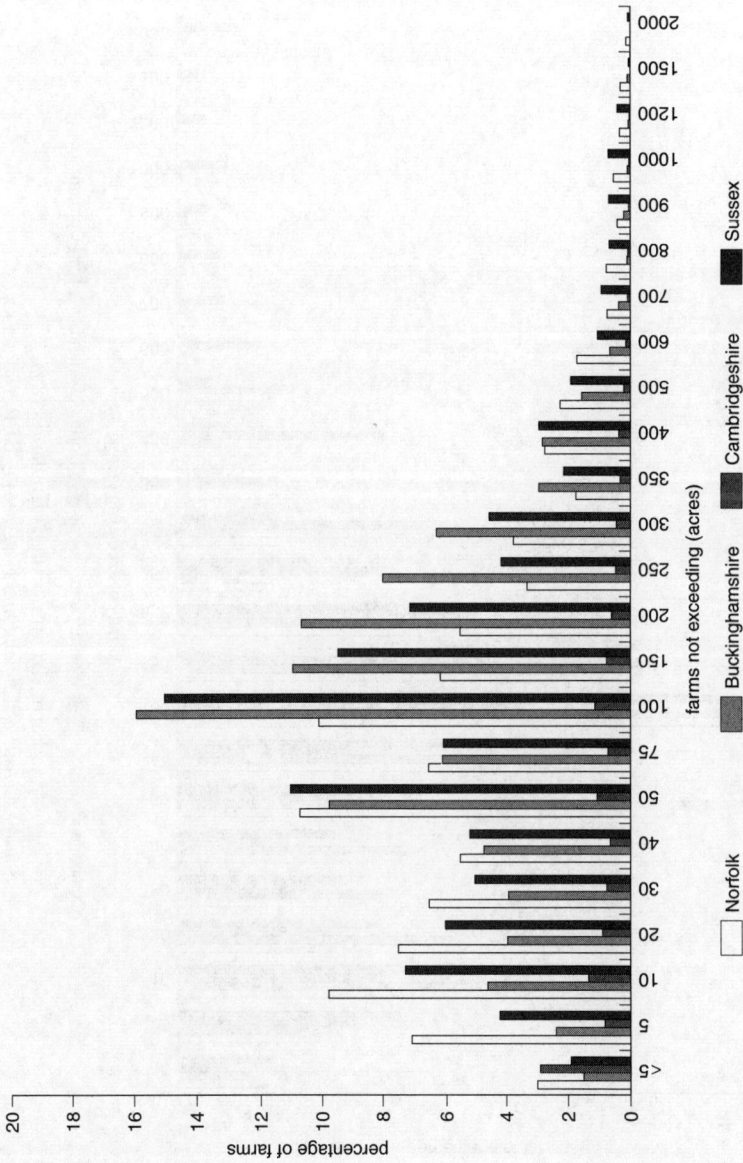

37.2b Farm size in southern England, 1861

Source: Census, 1861

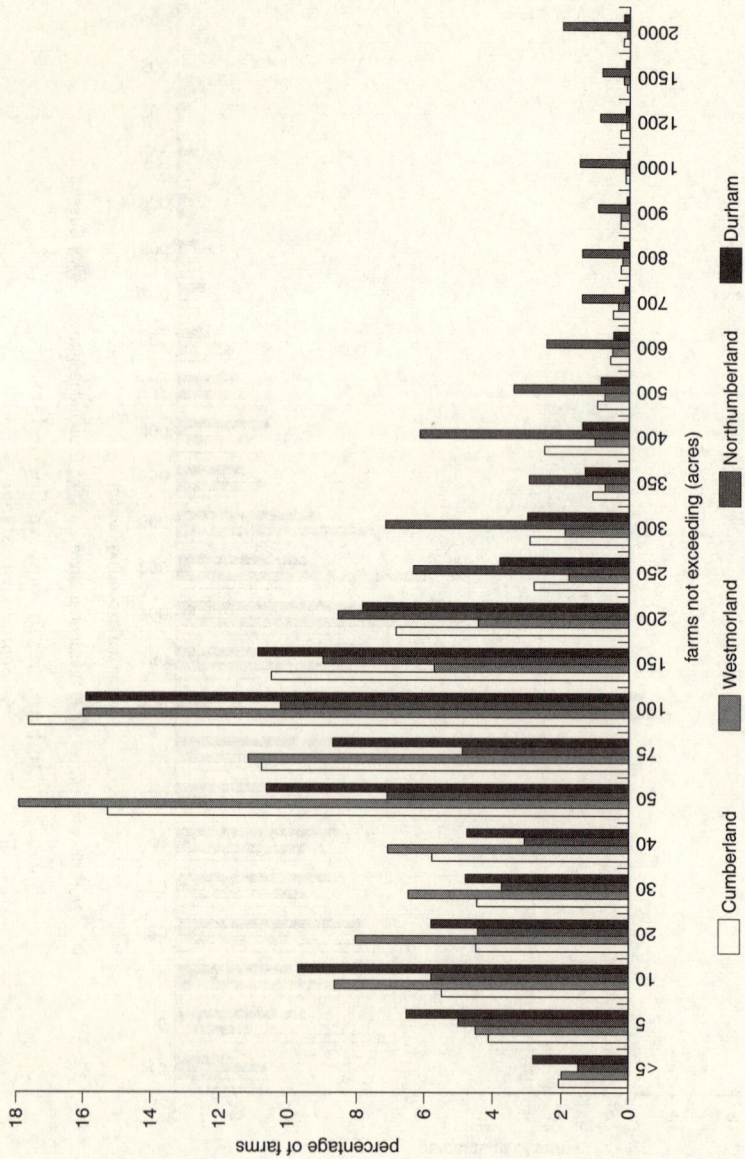

37.3a Farm size in northern England, 1871
Source: *Census of England and Wales, 1871*, BPP, 1873, LXXI

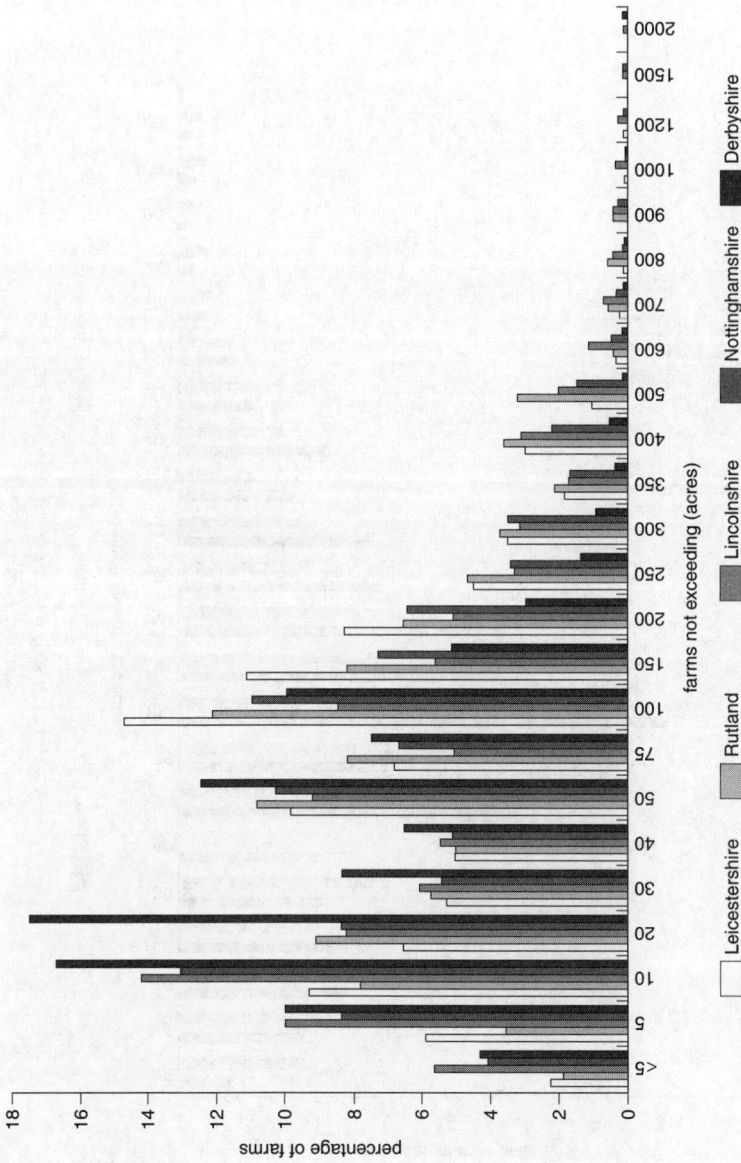

37.3b Farm size in the Midlands, 1871
Source: Census, 1871

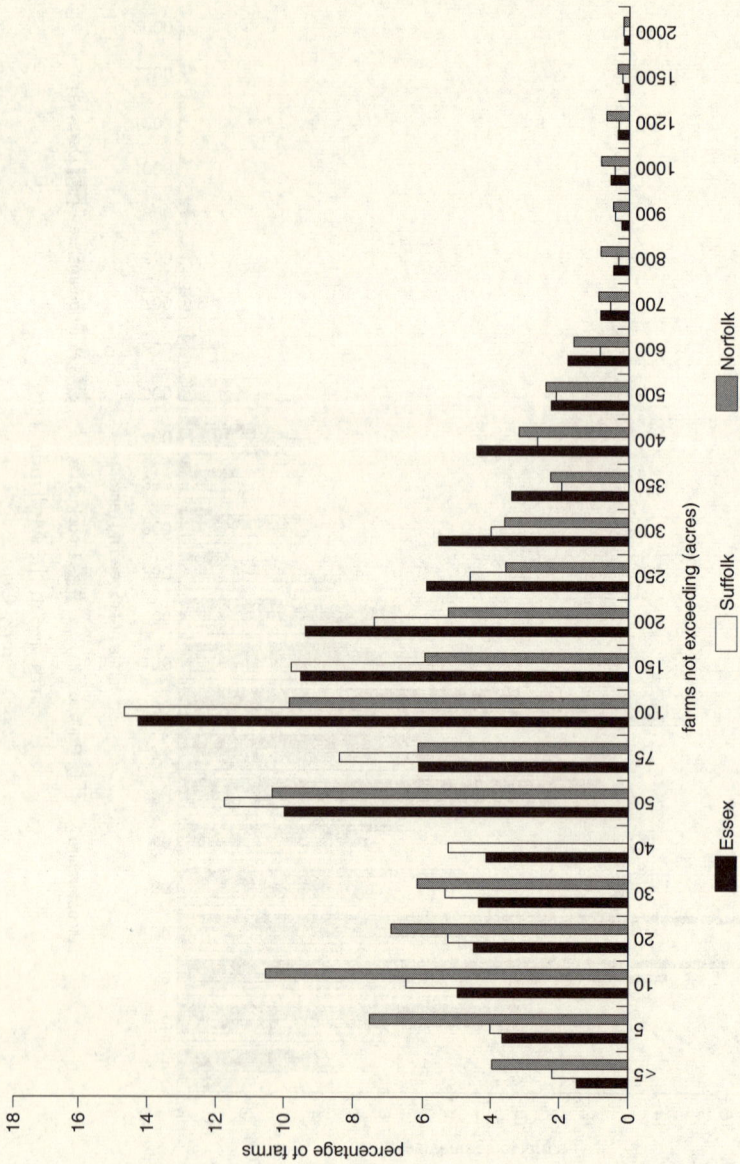

37.3c Farm size in eastern England, 1871
Source: Census, 1871

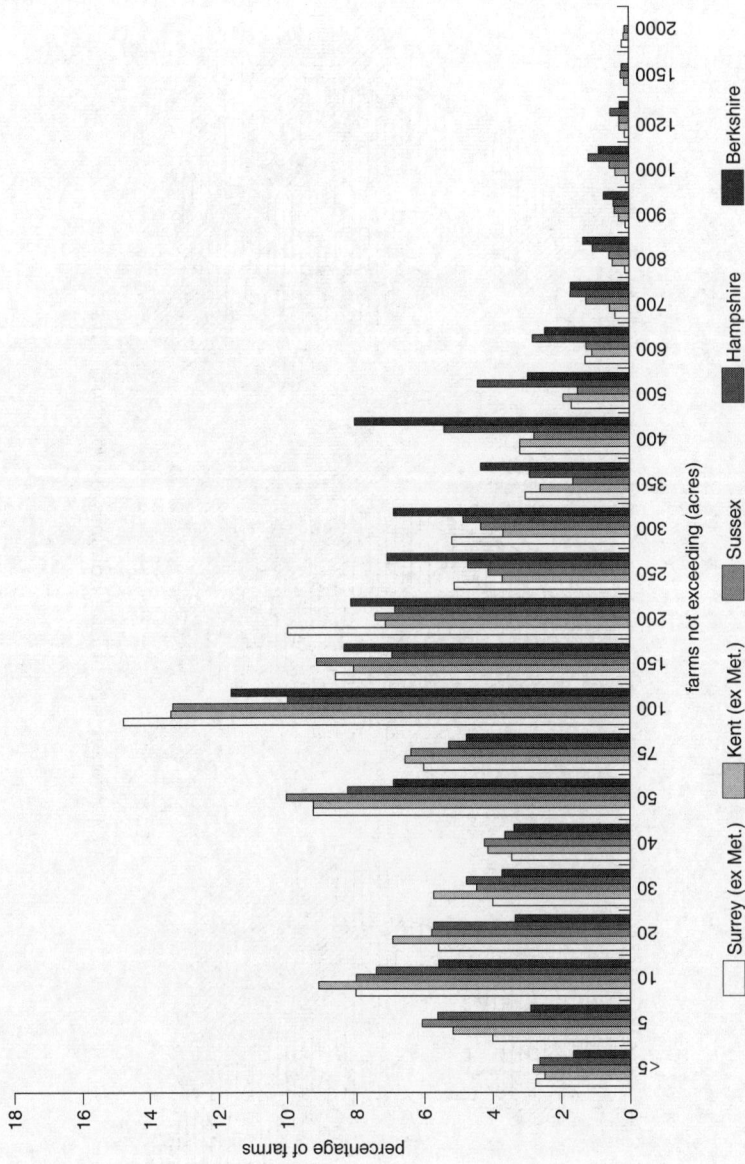

37.3d Farm size in southern England, 1871
Source: Census, 1871

1862

ACRES

130
110
90
70
50
30

1914

1895

1870

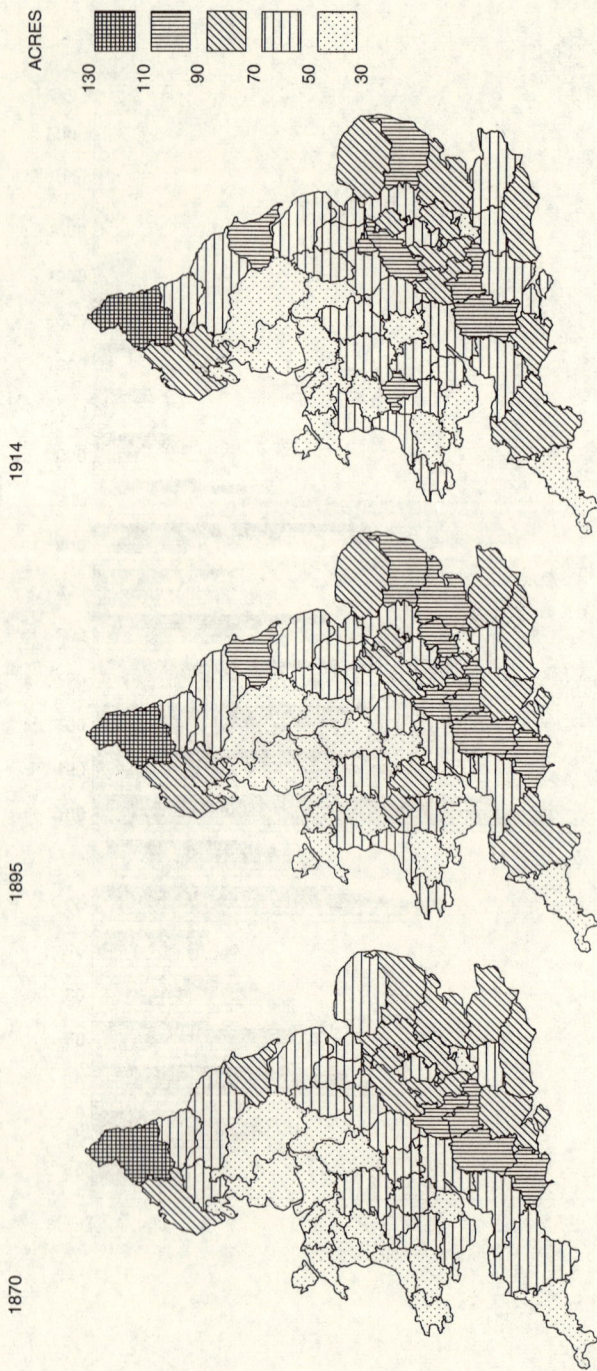

37·4 Average size of land holdings, 1870, 1895, 1914

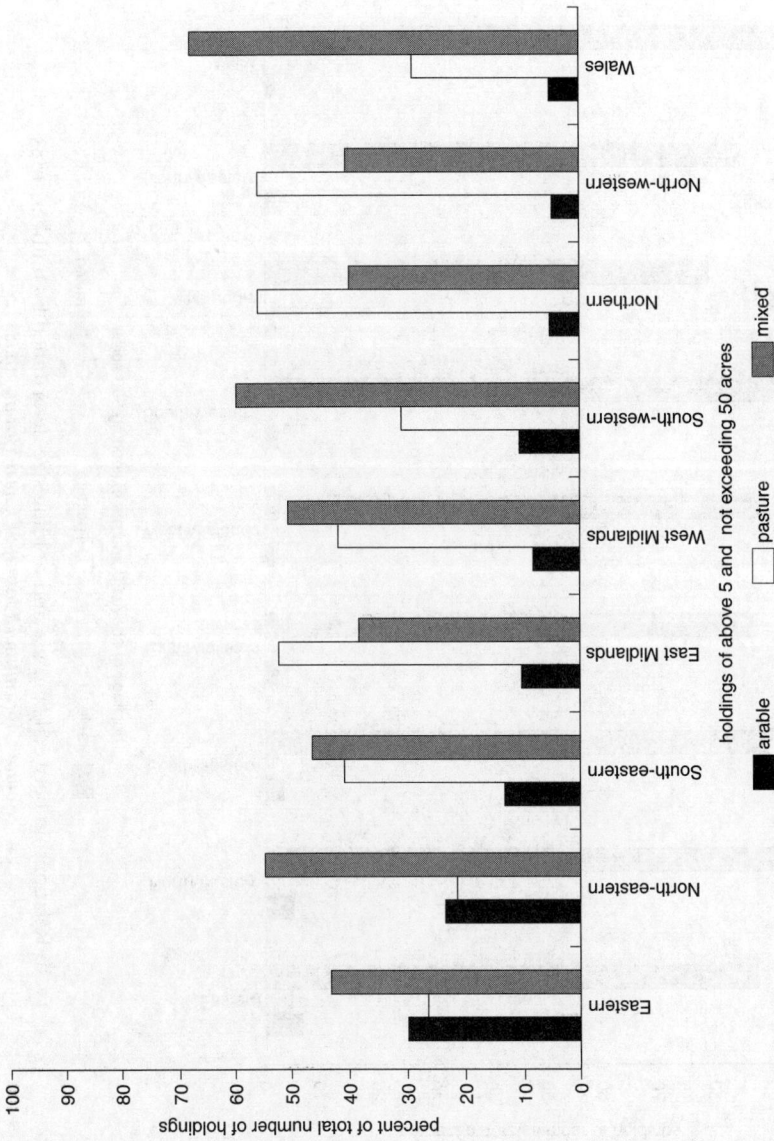

37.5a Relationship between land use and holding size determined by number, 1885

Source: *The Agricultural Output of Great Britain, 1908, in connection with the Census of Production Act, 1906, BPP, 1912–13, X*

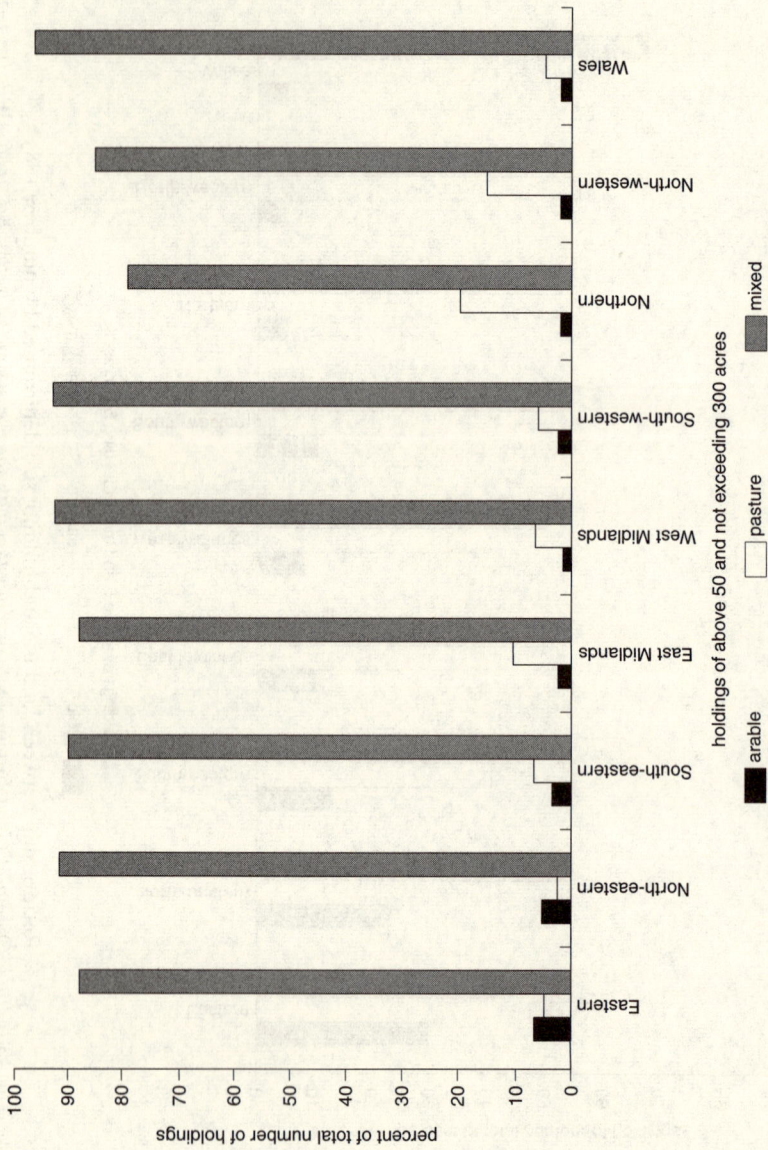

37.5b Relationship between land use and holding size determined by number, 1885

holdings of above 50 and not exceeding 300 acres

Source: Agricultural Output of Great Britain, 1908

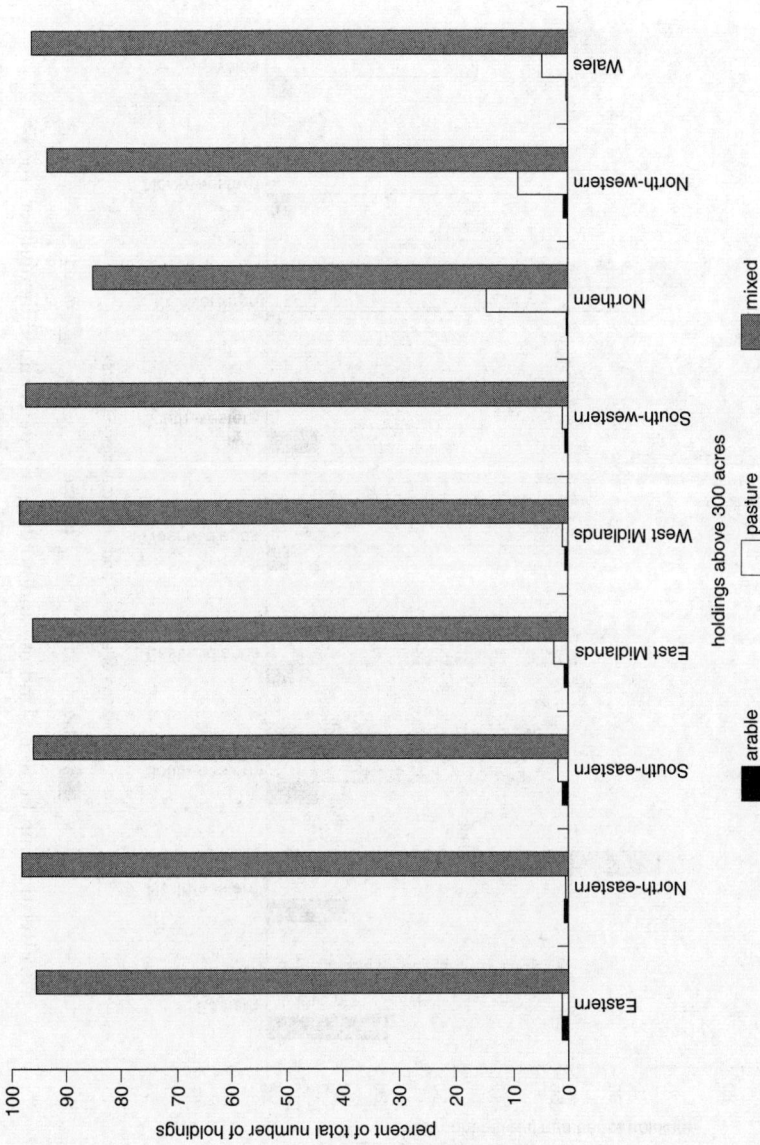

37.5c Relationship between land use and holding size determined by number, 1885

Source: *Agricultural Output of Great Britain, 1908*

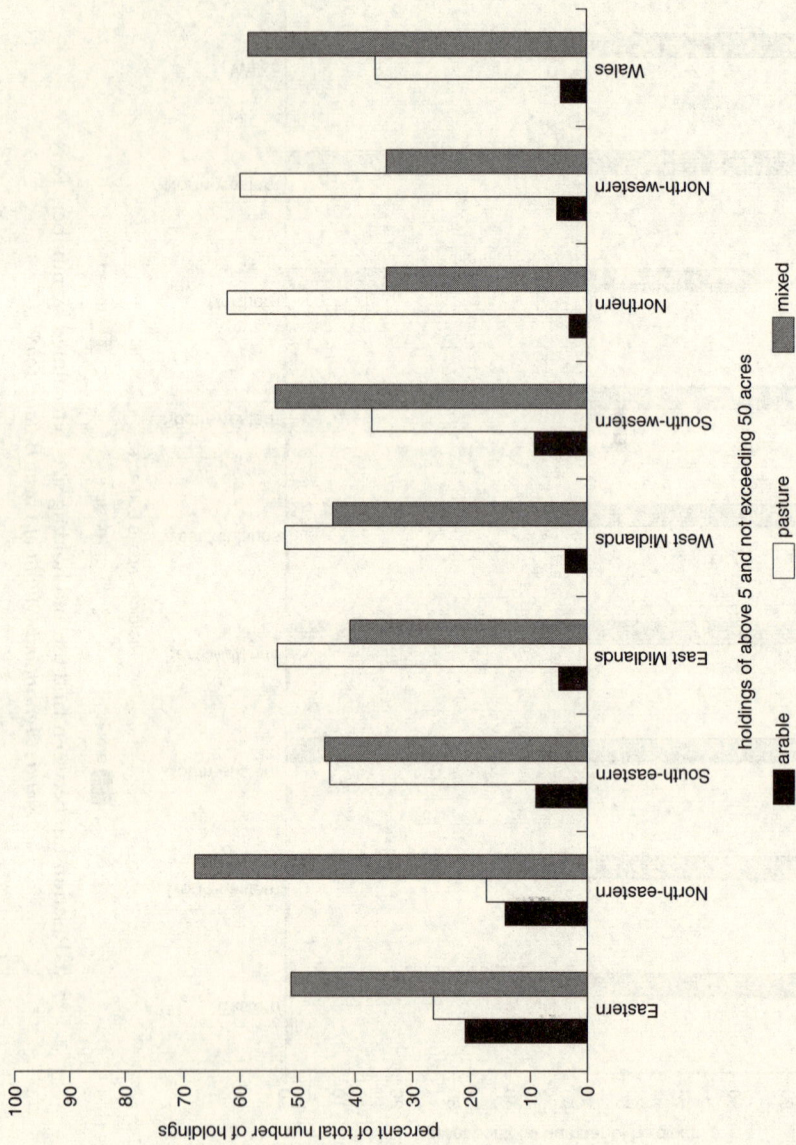

37.6a Relationship between land use and holding size determined by number, 1908

holdings of above 5 and not exceeding 50 acres

arable   pasture   mixed

Source: *Agricultural Output of Great Britain, 1908*

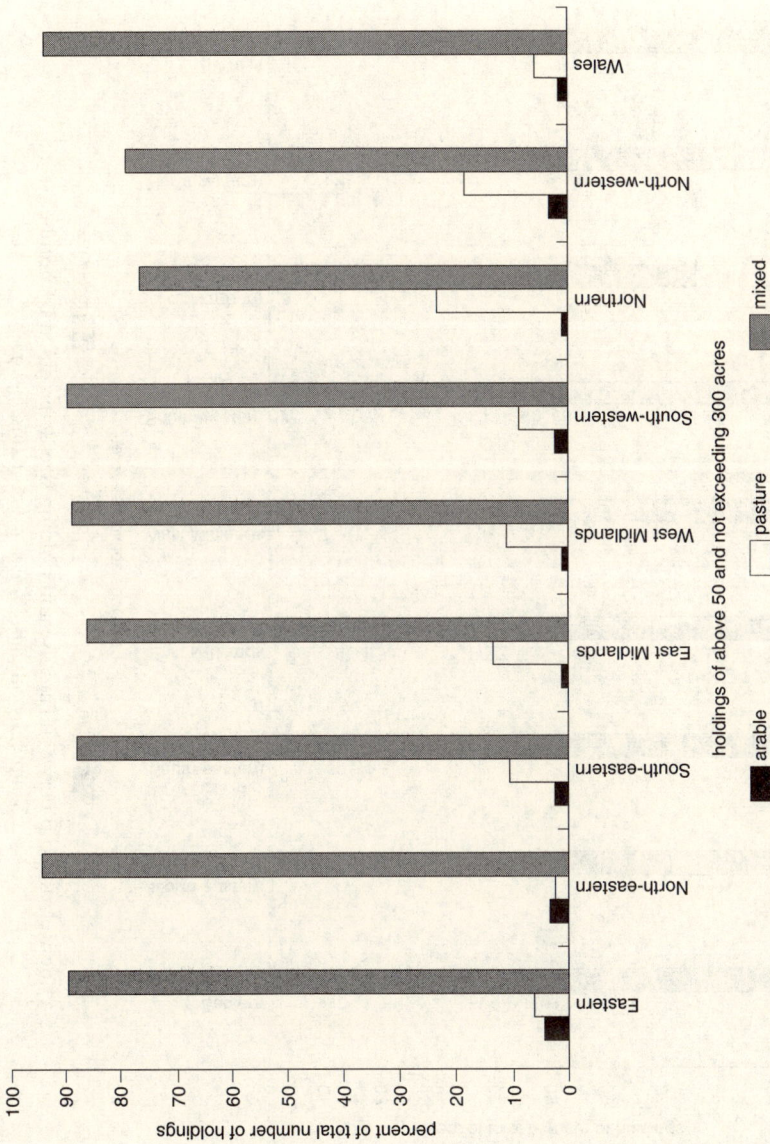

37.6b Relationship between land use and holding size determined by number, 1908

holdings of above 50 and not exceeding 300 acres

mixed
pasture
arable

Source: Agricultural Output of Great Britain, 1908

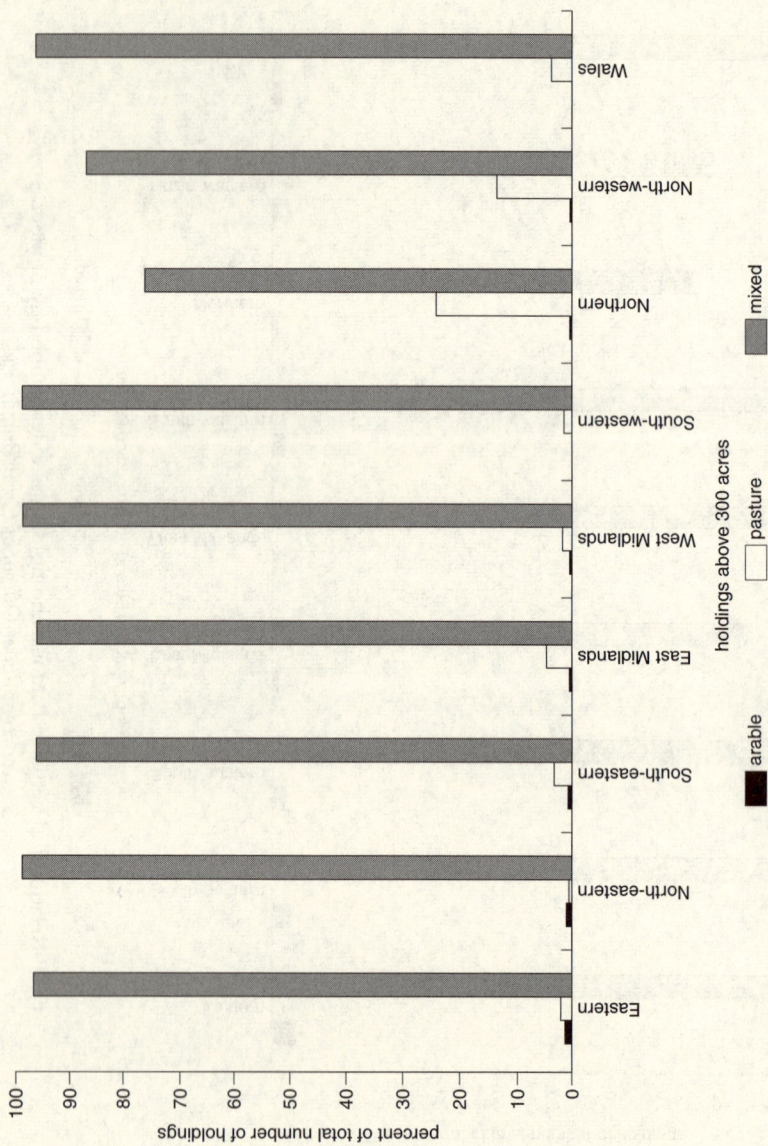

37.6c Relationship between land use and holding size determined by number, 1908
Source: Agricultural Output of Great Britain, 1908

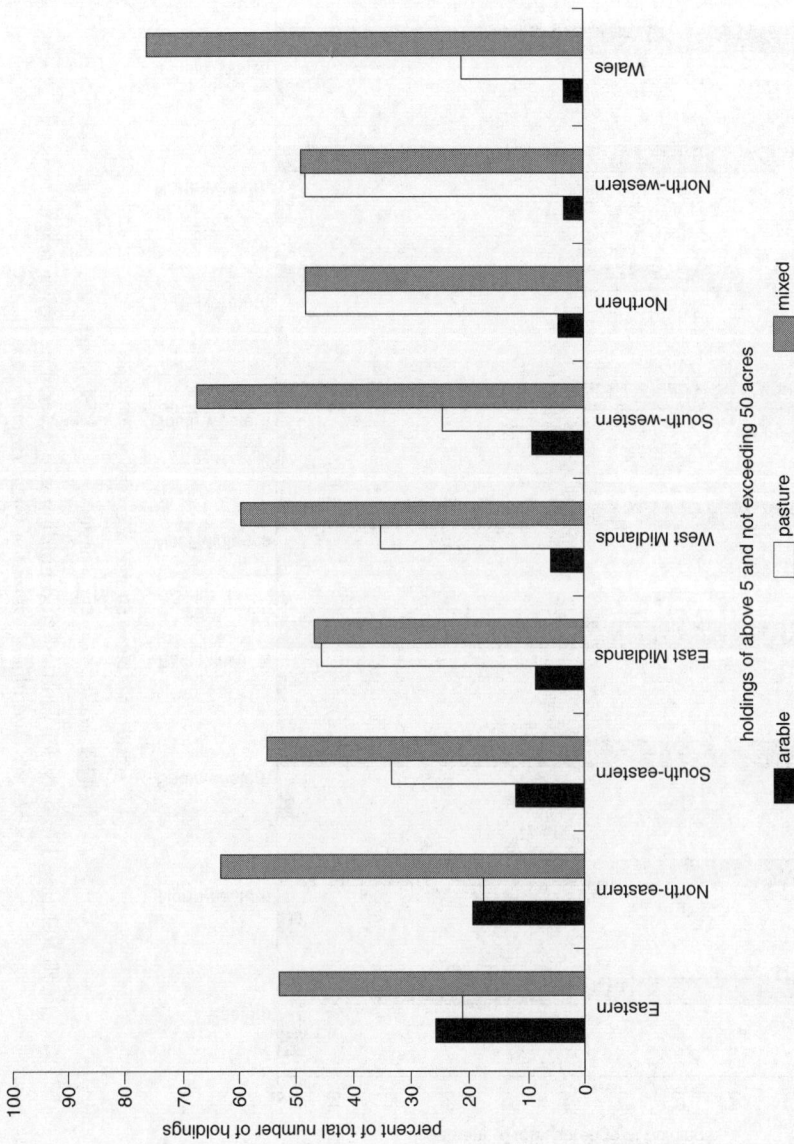

37.7a Relationship between land use and holding size determined by acreage, 1885
Source: *Agricultural Output of Great Britain, 1908*

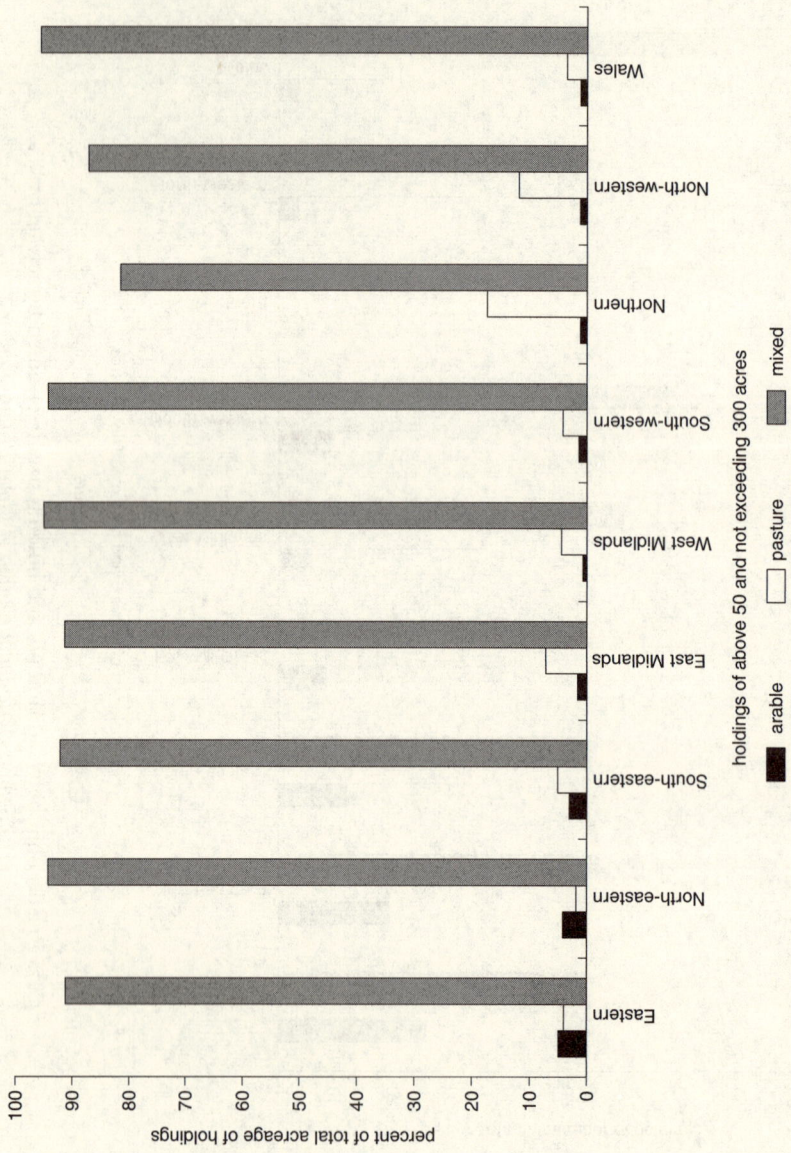

37.7b Relationship between land use and holding size determined by acreage, 1885
Source: Agricultural Output of Great Britain, 1908

holdings of above 50 and not exceeding 300 acres

arable   pasture   mixed

percent of total acreage of holdings

Eastern   North-eastern   South-eastern   East Midlands   West Midlands   South-western   Northern   North-western   Wales

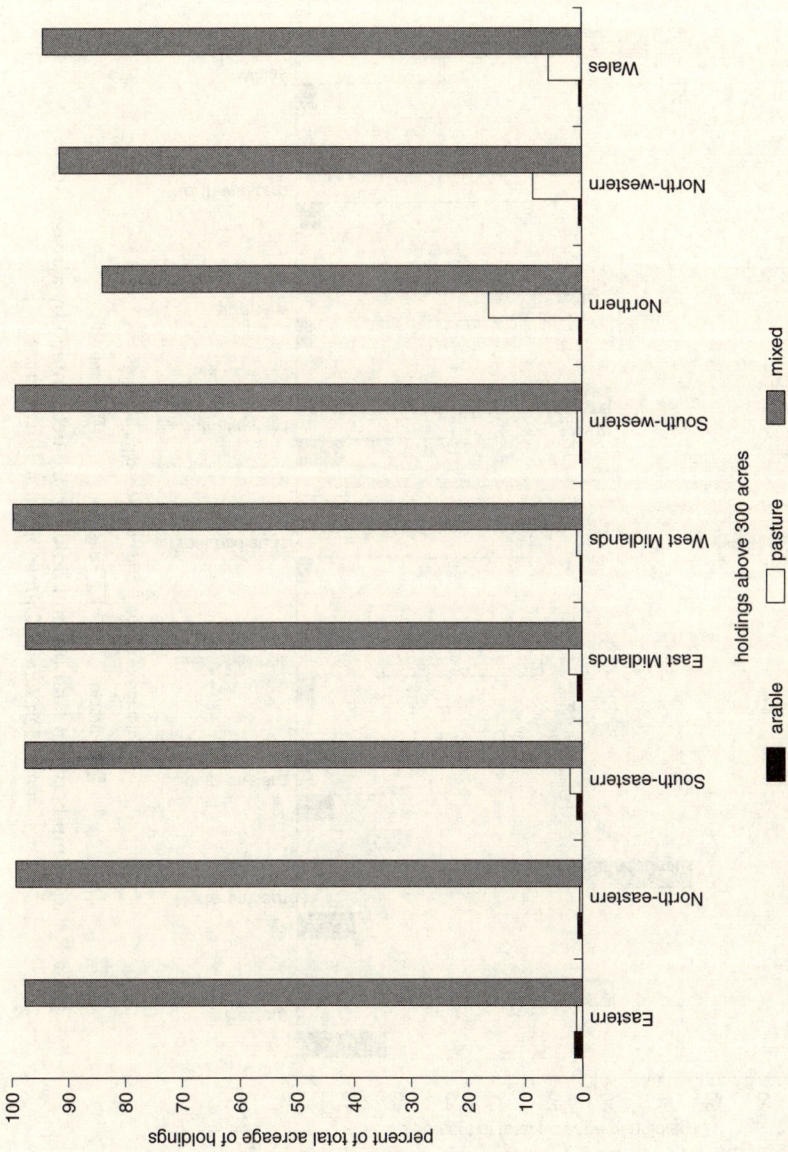

37.7c Relationship between land use and holding size determined by acreage, 1885
Source: *Agricultural Output of Great Britain, 1908*

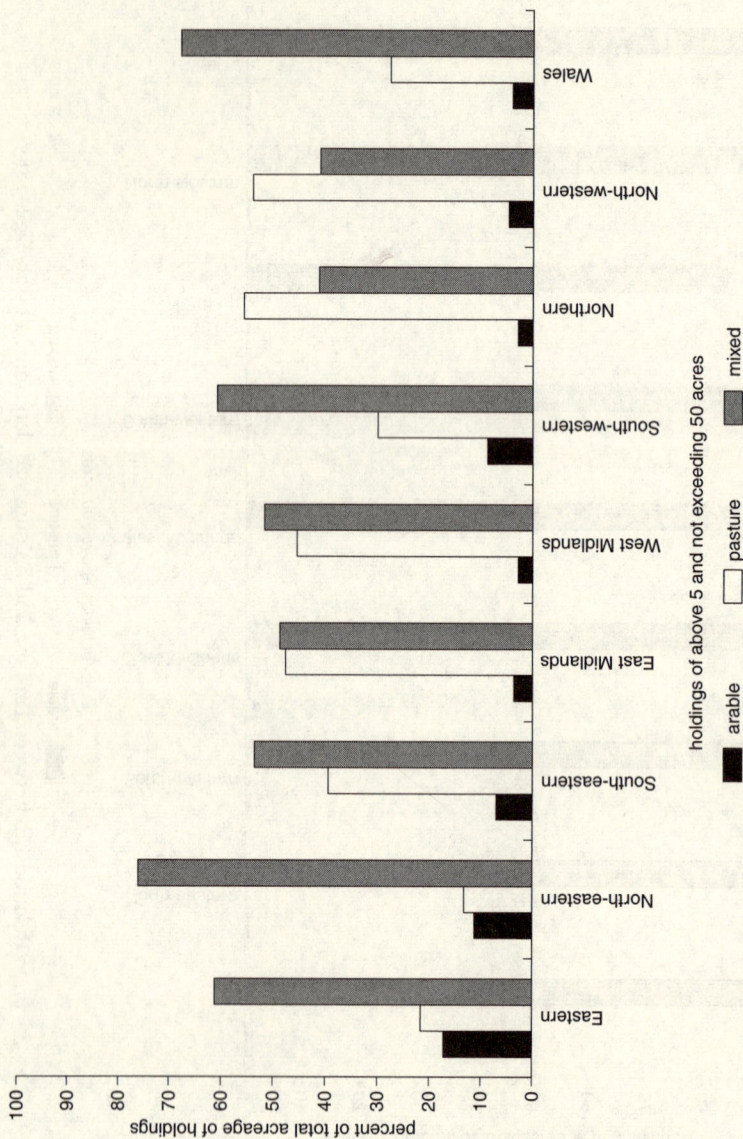

37.8a Relationship between land use and holding size determined by acreage, 1908

Source: *Agricultural Output of Great Britain, 1908*

holdings of above 5 and not exceeding 50 acres

■ arable □ pasture ▨ mixed

percent of total acreage of holdings

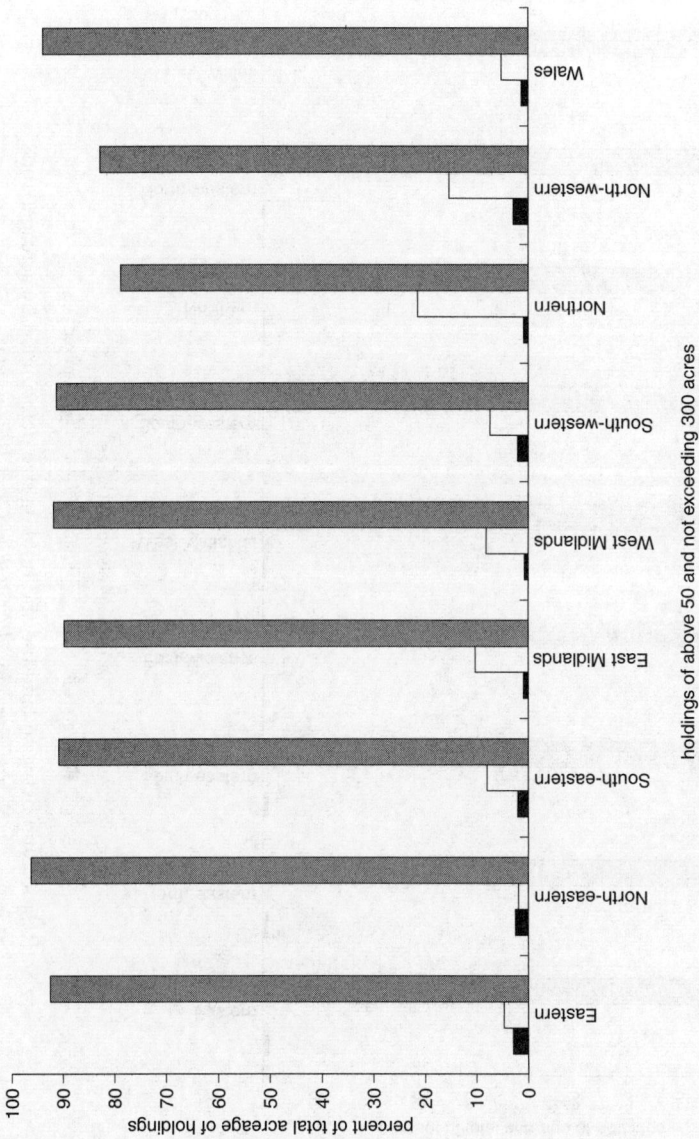

37.8b Relationship between land use and holding size determined by acreage, 1908

Source: *Agricultural Output of Great Britain, 1908*

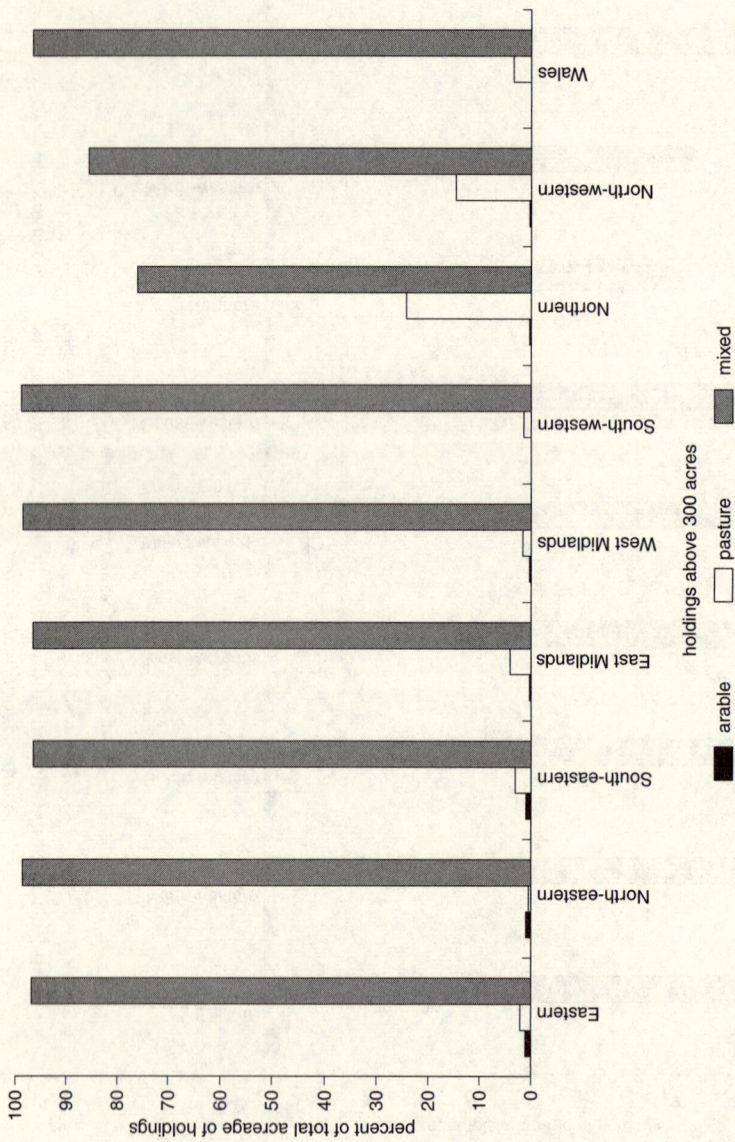

37.8c Relationship between land use and holding size determined by acreage, 1908

Source: Agricultural Output of Great Britain, 1908

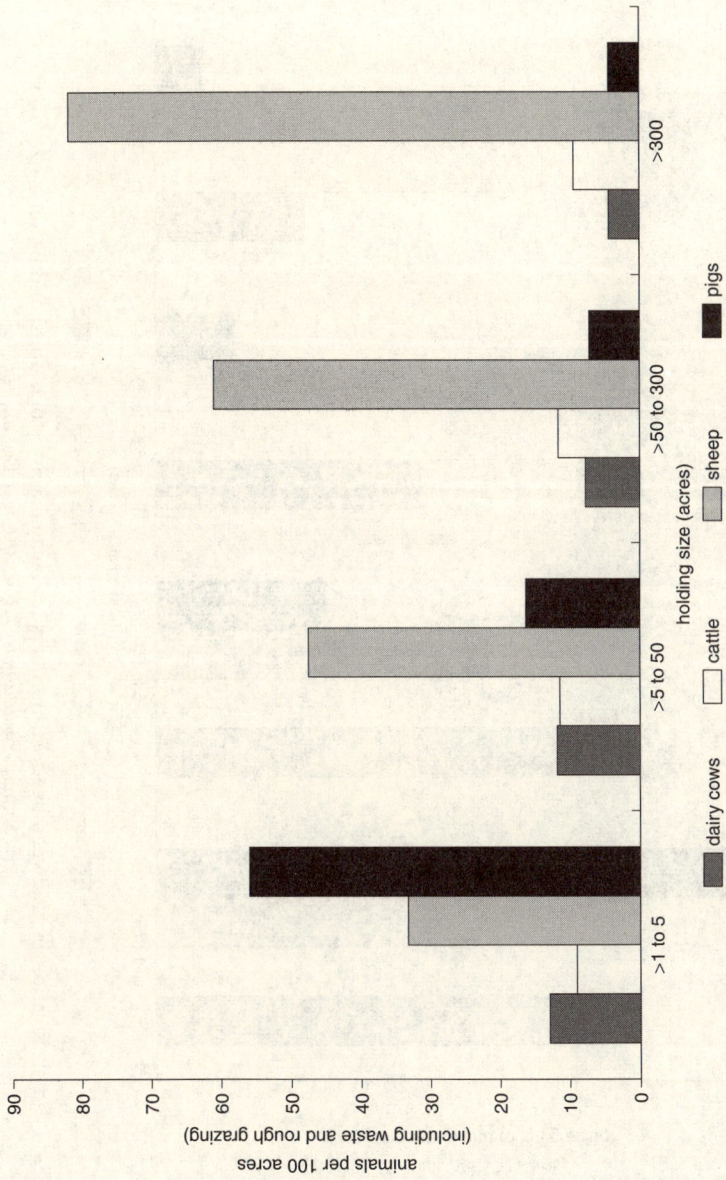

37.9a Relationship between livestock numbers and holding size, 1908

Source: *Agricultural Output of Great Britain, 1908*

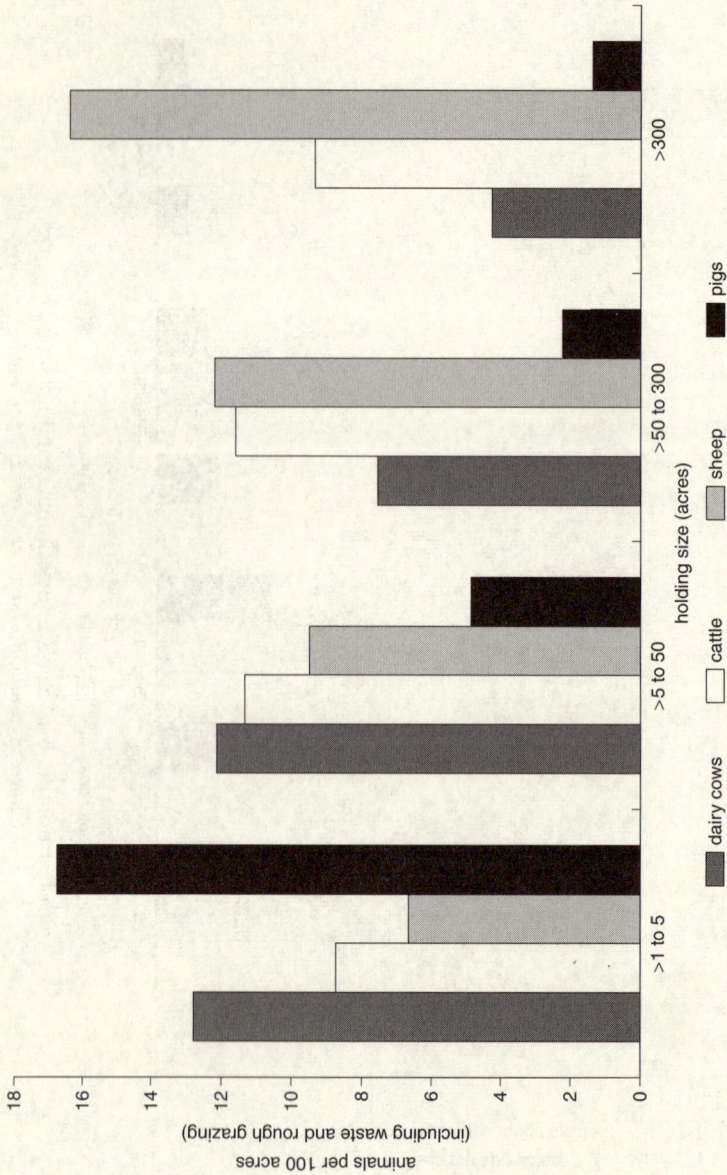

37.9b Relationship between livestock numbers (calculated as livestock units) and holding size, 1908

Source: Agricultural Output of Great Britain, 1908

CHAPTER 38

AGRICULTURAL OUTPUT

BY BETHANIE AFTON AND MICHAEL TURNER

A. INTRODUCTION

In two of the major sections which follow – those on rent and wages – data are presented on the returns to factors of production. Rent is a return to land, but in our terms it appears as an income to the landlords, and wages represent the return to the hired labour force. The prior return which services both of these is the income from farming itself. Rent and wages represent two ways in which that income is divided. But there is a third way: there is the residual which the farmer keeps for himself after paying his landlord the rent and his labourers their wages. And part of that residual he loses in the payment of other costs before he arrives at his net income. In this section the total agricultural income is measured in terms of output, and some of the ways in which that income is divided among its various claimants are indicated.

The assessment of output usually takes the form of a financial measure. This is the one common factor which can link acres of crops, measured in bushels and tons, with numbers of animals of various shapes and sizes. Most of what follows, therefore, is presented in financial terms, though occasionally it is presented in physical quantities.

B. THE DATA

This section, unlike others, is inextricably linked to one of the main chapters in the volume. Therefore, Chapter 3 should be consulted for details of the context in which agricultural output, income and productivity are discussed.

The first large-scale, serious appraisal of food supplies was not made until 1916. In a public sense the UK was demonstrably a laggard in this matter compared with the then Developed World.[1] The first Census of

---

[1] *The Food Supply of the United Kingdom: A Report drawn up by a Committee of the Royal Society at the request of the President of the Board of Trade*, BPP, 1916, IX. See also D. K. Brutton and K. E. Hunt,

Production, relating to 1908 as far as agriculture was concerned, appeared in 1912,[2] but essentially it was a cross-sectional appraisal of British output. Therefore, to the extent that there has ever been a long-term view, it has been constructed mainly by individuals rather than by government agencies. Chapter 3 furnishes details of definitions, and the historical development of our appraisal of agricultural output.

The emphasis is on financial measures, but the first estimate is a physical appraisal of output. Table 38.1 presents O. J. Beilby's assessment of agricultural production in England and Wales in terms of calorific values. This is one of the few studies which is set at the level of England and Wales separately from Britain or the UK.[3] It took a German scholar, Leo Drescher, to make an annual estimate of agricultural output, or in fact indexes of the economic and physical volume of agricultural production. The physical index is given in Table 38.2 (shown for the period 1866–1914, but available up to 1931), and his comparable economic index is given in Table 38.3.[4] The next major contributor was E. M. Ojala whose estimates were constructed in the 1940s and early 1950s and published in 1952. These are reproduced in Tables 38.4, 38.5 and 38.6.[5] They

Footnote 1 (*cont.*)

'Agriculture', in M. G. Kendall and A. Bradford Hill (eds.), *The Sources and Nature of the Statistics of the United Kingdom* (London, 1952), vol. 1, p. 60. And for details of the non-official appraisals see in particular a series of articles by R. H. Rew presented to the Committee 'Appointed to Inquire into the Statistics Available as a Basis for Estimating the Production and Consumption of Meat and Milk in the United Kingdom', as in, 'Production and consumption of meat and milk: second report', *JRSS*, 67, 1904, pp. 368–84; Rew, 'Production and consumption of meat and milk: third report', in *ibid.*, pp. 385–412; Rew, 'Observations on the production and consumption of meat and dairy products', in *ibid.*, pp. 413–27. This series of articles was preceded by his 'An inquiry into the statistics of the production and consumption of milk and milk products in Great Britain', *JRSS*, 55, 1892, pp. 244–86. See also P. G. Craigie, 'Statistics of agricultural production', *JRSS*, 46, 1883, pp. 1–58; Craigie, 'On the production and consumption of meat in the United Kingdom', *Report of the British Association for the Advancement of Science* (London, 1884), Section F, pp. 841–7; J. B. Lawes and J. H. Gilbert, 'Home produce, imports, consumption, and prices of wheat, over forty harvest-years, 1852–53 to 1891–92', *JRASE*, 3rd ser., 4, 1893, pp. 77–133. Finally see R. C. Turnbull, 'The household food supply of the United Kingdom', *Transactions of the Highland and Agricultural Society*, 15, 1903, pp. 197–211; R. H. Rew, 'The nation's food supply', *JRSS*, 76, 1912, pp. 98–105.

[2] Board of Agriculture and Fisheries, *The Agricultural Output of Great Britain [1908] . . . in Connection with the Census of Production Act, 1906*, BPP, 1912–13, x; and for Ireland see Department of Agriculture and Technical Instruction for Ireland, *The Agricultural Output of Ireland, 1908* (London, 1912).

[3] O. J. Beilby, 'Changes in agricultural production in England and Wales', *JRASE*, 100, 1939, pp. 62–73, esp. p. 64. This article also refers to the earlier attempt by the Irish to make the same sort of calculation, for which see Saorstat Eireann, *Agricultural Statistics 1847–1926* (Department of Industry and Commerce, Dublin, 1930), pp. 6–13. See also M. E. Turner, *After the Famine: Irish Agriculture 1850–1914* (Cambridge, 1996), pp. 139–53.

[4] L. Drescher, 'The development of agricultural production in Great Britain and Ireland from the early nineteenth century', *The Manchester School*, 23, 1955, pp. 153–75.

[5] E. M. Ojala, *Agriculture and Economic Progress* (Oxford, 1952), pp. 191–217.

were not annual estimates, but instead were presented in groups of years, those years based on the peaks in the trade cycle identified by Colin Clark.[6] An important difference with respect to Drescher is that Ojala took fully into account the distinction between gross production and gross output. In Ojala's language, gross agricultural output equalled sales off farms plus consumption in farm houses, the latter being an estimate for what the farmers and their families would have had to buy at the market for their own table. Table 38.4 is presented in current prices and Table 38.5 in constant, 1911/13, prices, thus providing an indication of the changing volume of gross output. Table 38.6 is Ojala's estimates of the inputs to agriculture, essentially the non-labour and rent costs.

Apart from Drescher's estimates, until the early 1950s no annual estimate of national agricultural output had been made to cover the period up to 1914. This was rectified by J. R. Bellerby.[7] Working from the University of Oxford Agricultural Economics Research Institute his main objective was the measurement of agricultural income, but he also constructed estimates of output. Table 38.7 presents several of his annual output estimates. The first column defines 'the quantities of basic food supplies which a wartime Minister of Food would take as his starting point in planning the distribution of farm products for both human and animal consumption'.[8] Thus no deductions were made for seed or fodder. The second column, 'gross output 1', shows home food supply after deducting seed and animal fodder. The difference between columns 1 and 2, therefore, is an estimate of the value of UK output which was recycled in one form or another. Column 4 represents Bellerby's estimate of net output, or gross output after allowing for the cost of productive inputs bought from abroad by farmers and from non-farm sources. This is mainly feedstuffs, and it also takes into account depreciation and landowners' extra-estate payments for upkeep. Column 5 is another measure of net income after a further deduction is made for land taxes paid by landowners, and any remnants of tithe which were still in force, with final adjustments made for the valuation of farmers' self-supplies at retail prices.

Columns 6 and 7 show Bellerby's separation of output into grain and livestock components. He called this separation the adjusted gross revenue, or the value of the total crop of wheat, barley, oats and straw, less seed, and the value of gross output of cattle, sheep, pigs (plus and

---

[6] C. Clark, *National Income and Outlay* (London, 1937), p. 246.

[7] J. R. Bellerby, 'The distribution of farm income in the UK, 1867–1938', *Journal of the Proceedings of the Agricultural Economics Society*, 10, 1953, pp. 127–44; Bellerby, revised with annual estimates in W. E. Minchinton (ed.), *Essays in Agrarian History* (Newton Abbott, 1968), vol. II, pp. 261–79; and for the original estimates see Rural History Centre, University of Reading, Bellerby Mss. See also Bellerby, *Agriculture and Industry Relative Income* (London, 1956), chapter IV, especially pp. 55–6.   [8] Bellerby, 'The distribution of farm income', in Minchinton, *Essays*, p. 275.

minus inventory changes for the animals), milk and milk products, and clipwool, less the purchase of manufactured feed and imports of store livestock. This then is a mixture of definitions. By taking effectively the total value of grain, without adjusting for that proportion which went into animal feed, and conversely not deducting the same from the gross output/revenue accruing to livestock products but instead deducting other costs – which in the case of manufactured feed included some allowance for grain which went into feed – the grain cannot be legitimately added to the livestock to represent total output. To do so would include an element of double counting on the one hand and cost deduction on the other. Furthermore, the revenue which supposedly accrued to grain did not include a comparable deduction for fertiliser. Column 9 is Bellerby's construction of an agricultural price index.

The remaining column 3 is the prototype of column 2. It is included here because there still survives in the manuscripts from which this column was taken a breakdown of output into individual crop and livestock components. These unpublished data are included in Table 38.8, and Table 38.9 contains the accompanying adjustment factors and inputs which when combined with gross output produce estimates of net income, which is net of rent but not net of the wage bill. This, therefore, is the net return to the human factor, the sweat and toil of the tenant and his labour force. Bellerby commented that, 'This total incentive income will later be further divided into the wages bill, and the return to farmers.' Unfortunately, that final calculation is missing from the surviving manuscripts.[9]

When C. H. Feinstein came to assess the contribution of the agricultural sector to the UK economy in his mammoth compilation of the National Income Accounts he was bound to use or adapt many of the existing estimates, though with certain adjustments.[10] His version is presented in Table 38.10, including estimates for rent and wages.

Table 38.8 is an estimate of output for different crop and livestock products for the UK. A complicated decomposition of the UK estimates has been made in order to isolate the likely contribution coming from England and Wales alone. The results of this decomposition are given in

---

[9] The quotation is in Bellerby Mss., D/84/8/24, p. 159. For the procedures which Bellerby adopted see *ibid.*, pp. 1–3 for an introduction, pp. 86–95 for valuation of inputs of feedingstuffs, pp. 96–8 for fertilisers, pp. 99–103 for seed and livestock inputs, pp. 104–5 for expenditure on machinery, fuels and repairs, pp. 106–7 for land tax and rates, p. 108 for miscellaneous expenses, p. 109 for a summary of total inputs. See pp. 110–18 for estimates of livestock inventory changes, pp. 119–23 for estimates of revaluation of farmers' own consumption at rural retail prices, pp. 125–31 for estimates of UK rent, and pp. 132–56 for estimates of occupiers' capital expressed as an interest charge on holding stocks and equipment.

[10] C. H. Feinstein, *National Income, Expenditure and Output of the United Kingdom 1855–1965* (Cambridge, 1972), pp. 23–9, 38–43, 212–13, and T60.

Table 38.11. Before readers attempt to use these estimates, they are strongly advised to read the details of construction in Chapter 3.

As already indicated, Table 38.7 contains the annual agricultural price index which Bellerby himself constructed. He claimed it was derived from the Paasche method of construction. This index is repeated in Table 38.12 alongside three versions of a new index derived from the Laspeyres method of construction. These three indexes employ product weights appropriate to the income contribution of the different products in 1870, 1895 and 1910, respectively.[11]

Apart from the estimated decomposition of UK output into its combined English–Welsh component (Table 38.11), there have been few other concerted attempts at estimating output for the countries of the UK.[12] A significant exception is an English estimate made by F. M. L. Thompson which goes further than all others by presenting the estimates at the county level for the two years 1873 and 1911. The results of this are included as Table 38.13.[13]

---

[11] Also printed in M. E. Turner, 'Output and prices in UK agriculture, 1867–1914, and the Great Agricultural Depression reconsidered', *AHR*, 40, 1992, pp. 38–51, esp. pp. 46–7.

[12] T. W. Fletcher, 'The Great Depression of English agriculture', in Minchinton, *Essays*, p. 256. See also P. E. Dewey, *British Agriculture in the First World War* (London, 1989), pp. 244–8 for 'British' estimates for 1909–18.

[13] F. M. L. Thompson, 'An anatomy of English agriculture, 1870–1914', in B. A. Holderness and M. E. Turner, (eds.), *Land, Labour and Agriculture, 1700–1920: Essays for Gordon Mingay* (London, 1991), chapter 11, pp. 211–40, especially 233, 237.

APPENDIX

Table 38.1. *Calorific value of agricultural production in England and Wales, 1885–1914*

| | | Crops | | Livestock | | Imported feed | | 'True' |
	Total crops	For human food	For animal fodder	Gross	From home-produced feed	Total	In terms of livestock	output of food
	1			2	3	4	5	6
1885–9	55,333	9,630	40,357	5,109	3,522	11,015	1,509	13,152
1890–4	51,985	9,054	38,236	5,363	3,613	12,238	1,676	12,667
1895–9	50,253	8,943	36,553	5,352	3,190	15,131	2,073	12,133
1900–4	50,251	8,166	37,481	5,516	3,235	15,914	2,180	11,401
1905–9	52,613	8,634	38,972	5,845	3,634	15,376	2,106	12,268
1910–14	49,114	8,323	36,064	5,969	3,636	16,407	2,248	11,959

Notes:
1 Includes hay and straw sold off farms for non-agricultural purposes and which are not included in other crop totals.
2 Includes estimates of output of poultry and eggs.
3 Gross livestock production less production from imported feed and imported store cattle.
4 Imports for use in England and Wales taken as 80 per cent of the total UK imports.
5 This is the calorie value of livestock output estimated to be obtained from the imported feed.
6 This is the sum of crops for human food, plus net livestock production. It does not include fruit and vegetables.

(continued)

Table 38.1 (*cont.*)

Calorific values

Product	Calories per lb	Product	Calories per lb
Wheat	1,350	Eggs, hen	660
Oats	1,200	Eggs, duck	810
Barley	1,280	Poultry	445[2]
Potatoes	430	Poultry	540[3]
Hay	960–990	Store cattle	500[4]
Straw (oat)	720	Maize	1,500
Turnips	125	Oat products	1,200
Swedes	180	Maize products	1,400
Mangolds	190	Wheat offals	1,200
Hops	560[1]	Cotton	1,380[5]
Sugar beet	340[1]	Ground nuts	1,430[6]
Sugar beet, pulp	1,080[1]	Ground nuts	1,430[7]
Milk	300	Linseed	1,450[6]
Beef/veal	1,100	Soya beans	1,550[1]
Mutton/lamb	1,400	Copra	1,430[1]
Pigmeat	1,700	Palm kernels	1,600[1]

Notes:

1 Calculated or estimated from starch equivalents.

2 Chicken.

3 Average of ducks, turkeys and chickens.

4 Average weight estimated at 750 lb, and 375,000 calories per animal.

5 Seed, cake and meal.

6 Cake and meal.

7 Decorticated.

Source: O. J. Beilby, 'Changes in agricultural production in England and Wales', *JRASE*, 100, 1939, pp. 64, 72.

Table 38.2. *Drescher's Physical Volume Index, 1866–1914 (1909–13=100)*

Date	Arable production			Livestock production		
	GB	Ireland	UK	GB	Ireland	UK
1866	101.1	99.9	100.9	55.4	69.4	59.4
1867	98.0	105.4	99.3	60.4	70.1	63.1
1868	118.9	117.8	118.7	62.0	66.9	63.5
1869	111.1	104.9	110.0	60.7	70.1	63.4
1870	115.6	118.3	116.0	62.2	73.7	65.5
1871	111.2	102.0	109.6	62.1	75.7	66.0
1872	100.5	82.8	97.4	65.4	75.7	68.4
1873	103.4	94.4	101.9	67.5	74.2	69.5
1874	121.7	109.4	119.5	69.1	73.8	70.4
1875	119.6	115.7	118.9	68.0	76.4	70.4
1876	108.5	114.8	109.6	67.6	78.1	70.6
1877	94.4	78.2	91.5	68.2	78.4	71.1
1878	112.2	92.0	109.2	68.9	76.9	71.2
1879	73.1	61.8	71.1	69.1	75.3	70.9
1880	109.5	97.7	107.4	68.7	71.5	69.6
1881	110.7	102.1	109.2	69.3	73.8	70.6
1882	100.4	81.7	97.1	71.1	76.5	72.7
1883	109.2	99.8	107.5	73.3	77.5	74.6
1884	113.0	89.5	108.9	76.3	76.0	76.2
1885	104.6	81.5	100.6	79.8	78.9	79.5
1886	109.7	87.3	105.7	79.7	79.1	79.5
1887	100.3	83.8	97.4	80.2	79.7	80.0
1888	104.2	84.5	100.8	78.6	79.4	78.9
1889	110.0	91.0	106.6	79.5	79.7	79.6
1890	109.9	75.3	103.8	84.4	85.8	84.8
1891	107.6	98.5	106.0	89.1	85.8	88.2
1892	103.5	87.4	100.7	86.9	85.2	86.4
1893	94.1	95.0	94.3	84.3	85.3	84.6
1894	108.0	79.2	102.9	82.9	87.5	84.2
1895	96.9	98.3	96.9	86.0	86.9	86.2
1896	100.8	87.0	98.4	87.8	88.0	87.9
1897	98.8	67.0	93.2	86.5	88.7	87.1
1898	105.1	95.6	103.4	89.6	89.1	89.5
1899	96.9	90.5	95.8	93.3	91.6	92.8
1900	90.3	77.4	88.0	91.0	92.3	91.4
1901	96.4	98.4	96.8	90.1	94.0	91.3
1902	107.1	97.1	105.3	89.5	96.7	91.6
1903	94.0	76.3	90.9	92.7	96.7	93.9
1904	95.0	83.1	92.9	96.2	96.9	96.4

(continued)

Table 38.2 (*cont.*)

Date	Arable production			Livestock production		
	GB	Ireland	UK	GB	Ireland	UK
1905	101.7	95.1	100.5	95.8	95.0	95.7
1906	104.2	88.0	101.4	96.8	96.7	96.8
1907	103.6	83.2	100.0	99.3	101.4	100.0
1908	103.2	99.3	102.5	101.2	100.4	101.1
1909	108.8	103.1	107.8	100.6	98.0	99.8
1910	104.4	94.1	102.6	99.7	98.5	99.4
1911	95.0	104.5	96.7	103.0	100.4	102.3
1912	94.6	90.4	93.9	100.6	101.9	101.0
1913	97.2	107.9	99.0	96.1	101.1	97.6
1914	102.9	101.5	102.7	103.9	104.5	104.1

Source: L. Drescher, 'The development of agricultural production in Great Britain and Ireland from the early nineteenth century', *The Manchester School*, 23, 1955, p. 174.

Table 38.3. *Drescher's Economic Volume Index, 1866–1914 (1909–13=100)*

Date	GB	Ireland	UK
1866	75.1	77.4	75.7
1867	77.5	79.6	78.0
1868	85.3	80.8	84.2
1869	82.4	79.3	81.6
1870	84.2	86.1	84.7
1871	83.6	82.6	83.3
1872	80.6	76.2	79.6
1873	83.9	79.3	82.8
1874	91.4	83.7	89.6
1875	91.9	87.5	90.9
1876	85.4	88.5	86.2
1877	79.5	76.6	78.8
1878	87.8	80.4	86.0
1879	69.8	68.9	69.6
1880	86.8	78.3	84.8
1881	86.4	81.3	85.2
1882	82.4	76.7	81.1
1883	88.5	84.0	87.0
1884	93.0	79.7	89.8
1885	89.6	81.5	87.7
1886	94.4	80.8	91.2
1887	88.3	80.7	86.5
1888	90.0	83.5	88.5
1889	93.8	82.9	91.2
1890	96.2	82.3	92.9
1891	97.8	90.1	96.0
1892	95.8	85.9	93.5
1893	90.7	88.6	90.2
1894	94.8	84.3	92.4
1895	92.4	90.7	92.0
1896	94.2	88.2	92.8
1897	93.6	81.6	90.7
1898	96.1	91.6	95.0
1899	93.8	91.5	93.2
1900	92.6	87.8	91.5
1901	93.5	96.1	94.1
1902	97.9	97.0	97.7
1903	89.8	89.9	89.9
1904	96.9	92.9	95.9
1905	98.4	95.0	97.6

(*continued*)

Table 38.3 (*cont.*)

Date	GB	Ireland	UK
1906	99.8	93.6	98.3
1907	100.7	94.8	99.4
1908	103.0	100.5	102.4
1909	104.8	99.6	103.5
1910	103.0	97.0	101.6
1911	98.4	102.6	99.4
1912	97.8	97.7	97.8
1913	96.0	103.1	97.7
1914	101.5	103.2	101.9

Source: L. Drescher, 'The development of agricultural production in Great Britain and Ireland from the early nineteenth century', *The Manchester School*, 23, 1955, p. 175.

Table 38.4. *United Kingdom gross agricultural output and product distribution at current prices, 1867/9 to 1911/13 (in £s million)*

Product	1867/69	1870/76	1877/85	1886/93	1894/1903	1904/10	1911/13
Crops							
Wheat	35.38	27.56	19.35	11.72	7.72	8.69	9.18
Barley	16.78	17.56	14.01	10.51	9.43	8.80	9.47
Oats	10.54	9.07	7.22	5.28	4.51	5.02	4.31
Potatoes	14.02	13.82	12.06	8.55	7.70	8.70	10.65
Hops	1.95	3.35	2.87	2.59	2.08	1.81	2.65
Beans	1.49	1.36	1.01	0.64	0.48	0.83	0.79
Peas	1.08	1.01	0.69	0.51	0.43	0.59	0.49
Rye	0.12	0.11	0.08	0.10	0.10	0.09	0.10
Flax	2.70	1.75	1.40	0.89	0.55	0.37	0.57
Hay + straw	10.28	9.76	8.33	7.82	8.24	6.48	7.47
Fruit	5.40	4.84	4.80	4.46	4.86	5.07	4.97
Vegs	4.43	4.80	4.17	3.68	3.67	4.23	5.58
Total	104.17	94.99	75.99	56.75	49.77	50.68	56.23
As percent of grand total	45.3	38.4	34.7	30.2	27.2	25.2	25.3
Livestock							
Beef + veal	34.90	44.53	43.14	37.95	39.88	43.39	47.43
Mutton	25.92	29.77	27.34	23.94	23.16	23.87	25.01
Pigmeat	18.60	22.55	19.90	18.05	18.31	19.19	20.63
Horses	0.40	1.60	3.00	3.20	2.55	2.40	2.31
Milk	33.78	38.51	37.99	36.36	36.89	46.06	52.83
Wool	7.49	8.27	4.48	3.80	3.24	3.51	3.71
Eggs	3.43	5.24	5.54	5.84	6.77	8.79	10.51
Poultry	1.14	1.72	1.82	1.91	2.21	2.86	3.46
Total	125.66	152.19	143.21	131.05	133.01	150.07	165.89
As percent of grand total	54.7	61.6	65.3	69.8	72.8	74.8	74.7
Total	229.83	247.18	219.20	187.80	182.78	200.75	222.12

Note: Gross Agricultural Output is sales off farms, plus farmers' consumption.
Source: E. M. Ojala, *Agriculture and Economic Progress* (Oxford, 1952), p 208.

Table 38.5. *United Kingdom gross agricultural output and product distribution at constant, 1911–13, prices (in £s million)*

Product	1867/69	1870/76	1877/85	1886/93	1894/1903	1904/10	1911/13
Crops							
Wheat	19.65	17.22	14.44	12.08	9.40	9.15	9.18
Barley	11.66	13.11	11.88	11.07	10.97	10.23	9.47
Oats	7.75	6.93	6.33	5.62	5.18	5.57	4.31
Potatoes	10.37	9.95	9.34	9.17	8.74	9.87	10.65
Hops	2.67	4.53	3.50	3.70	3.99	3.07	2.65
Beans	1.48	1.40	1.21	0.93	0.71	0.89	0.79
Peas	1.15	1.13	0.87	0.77	0.66	0.60	0.49
Rye	0.08	0.08	0.07	0.10	0.12	0.10	0.10
Flax	1.51	1.12	1.04	0.81	0.51	0.45	0.57
Hay + straw	7.33	7.51	7.59	7.67	7.84	7.70	7.47
Fruit	2.92	2.92	3.29	3.84	4.50	4.92	4.97
Vegs	3.75	3.96	4.17	4.38	4.49	4.99	5.58
Total	70.37	69.86	63.73	60.14	57.21	57.54	56.23
Index	125	124	113	107	102	102	100
Livestock							
Beaf + veal	35.59	39.07	39.90	43.14	44.34	46.42	47.43
Mutton	27.58	26.33	23.98	24.68	24.39	24.11	25.01
Pigment	19.19	20.72	19.90	20.76	21.25	20.88	20.63
Horses	0.35	1.40	2.10	2.80	2.98	2.80	2.31
Milk	34.46	36.68	38.37	42.77	45.54	50.62	52.83
Wool	4.51	4.33	3.93	4.04	4.00	3.86	3.71
Eggs	4.84	5.70	6.60	7.48	9.15	9.76	10.51
Poultry	1.60	1.88	2.17	2.45	2.99	3.18	3.46
Total	128.12	136.11	136.95	148.12	154.64	161.63	165.89
Index	77	82	83	89	93	97	100
Total	198.49	205.97	200.68	208.26	211.85	219.17	222.12
Index	89	93	90	94	95	99	100

Notes:
Gross Agricultural Output is sales off farms, plus farmers' consumption.
Faithfully reproduced from Ojala. Note! The crop total in 1867/9 was £70.37 m, and in 1894/1903 £57.21 m whereas by addition of the respective columns these totals should come to £70.32 m and £57.11 m. Whether the errors lie in the totals or in the products is uncertain.
Source: E. M. Ojala, *Agriculture and Economic Progress* (Oxford, 1952), p. 209.

Table 38.6. *Inputs and expenses in United Kingdom agriculture, 1867/69 to 1911/13*

Product	1867/69	1870/76	1877/85	1886/93	1894/1903	1904/10	1911/13
(i) By value and quantity £s million 000s tons							
Wheat	0.444	0.486	0.588	0.452	0.476	0.764	0.861
imported	35	42	57	60	72	97	104
Wheat	9.854	11.000	10.784	7.625	7.364	6.127	9.351
offals	1,422	1,403	1,534	1,436	1,531	1,761	1,848
Maize	4.797	7.875	9.470	8.154	10.592	11.351	12.502
imported	619	1,090	1,561	1,614	2,437	2,092	2,161
Barley			0.373	0.576	0.794	1.206	1.442
imported			50	100	150	200	200
Brewers'	3.061	4.108	3.575	2.741	2.728	2.655	2.549
offals	396	470	456	463	509	457	452
Pulses	0.470	0.578	0.545	0.534	0.458	0.323	0.473
imported	51	67	73	86	76	41	53
Meals	0.041	0.058	0.341	0.266	0.806	1.524	2.041
imported	3	4	32	34	104	244	301
Molasses	0.153	0.101	0.054	0.050	0.096	0.185	0.270
imported	12	10	7	8	21	45	65
Oilcakes	1.236	1.512	1.747	1.790	2.100	2.150	2.322
imported	146	158	203	276	358	340	376
Oilcakes	2.989	4.554	4.920	4.131	4.108	6.153	6.235
home	305	411	495	550	605	836	872
Fertilisers	5.267	6.357	3.903	2.909	3.371	3.998	4.890
	508	757	650	647	912	1,050	1,281
Totals							
£s million	28.312	36.629	36.3	29.228	32.893	36.436	42.936
000s tons	3,497	4,412	5,118	5,274	6,775	7,163	7,713
(ii) By value only in £s million							
Margin on feeds	2.083	2.898	3.543	3.896	5.341	5.832	6.342
Margin on fertilisers	0.617	1.046	0.898	0.949	1.448	1.746	2.202
Animals	3.673	4.402	4.895	5.176	5.745	5.583	5.387
Seeds	3.060	3.277	3.049	2.907	3.369	3.438	3.796
Machinery	1.530	1.680	1.420	1.210	1.350	1.580	1.890
Misc.	13.303	14.193	11.351	9.874	9.571	11.101	12.475

(continued)

Table 38.6 (*cont.*)

Product	1867/69	1870/76	1877/85	1886/93	1894/1903	1904/10	1911/13
Land Tax	0.880	0.880	0.848	0.832	0.714	0.612	0.612
Rates	(8.500	8.000	6.500	5.000	4.000	4.000	4.000)
Total	33.646	36.376	32.504	29.844	31.538	33.892	36.704
Grand total expenses £s million	61.958	73.005	68.804	59.072	64.431	70.328	79.640

Note: The costs of Rates, in parenthesis, are approximations.
Source: E. M. Ojala, *Agriculture and Economic Progress* (Oxford, 1952), pp. 212–13.

Table 38.7. *United Kingdom Gross Agricultural Output, 1867–1914. J. R. Bellerby's estimates of output*

Year	Domestic supply (1) £m	Gross output 1 (2) £m	Gross output 2 (3) £m	Net output (4) £m	Net income (5) £m	Adjusted gross revenue			Price index
						Grain (incl.straw) (6) £m	Livestock and livestock products (7) £m	Total adjusted revenue (8) (=6+7) £m	Agricultural prices (at farm) (9) 1867–77 = 100
1867	246.8	205.5	196.7	155.0	153.8	73.9	79.4	153.3	97.1
1868	250.5	207.9	234.0	150.4	150.0	79.2	73.3	152.5	93.7
1869	230.7	193.1	194.6	134.2	132.6	63.0	71.0	134.0	93.2
1870	249.9	209.0	206.5	149.8	148.8	71.9	69.0	140.9	94.8
1871	254.2	216.7	205.1	152.5	151.4	68.3	78.0	146.3	99.6
1872	254.1	221.4	212.6	153.8	153.0	66.7	83.3	150.0	103.3
1873	271.8	233.1	214.0	166.0	165.4	75.7	88.2	163.9	108.3
1874	273.4	227.6	230.5	157.1	156.3	77.5	79.4	156.9	100.1
1875	281.0	230.4	213.5	155.8	154.4	71.8	79.1	150.9	103.6
1876	283.5	230.3	208.9	153.9	152.4	76.4	71.4	147.8	102.9
1877	254.9	214.8	216.9	145.7	144.6	67.2	74.9	142.1	100.3
1878	249.3	210.6	224.9	138.2	137.2	62.9	68.8	131.7	94.0
1879	215.8	180.3	166.4	115.9	113.4	52.0	62.5	114.5	89.0
1880	233.4	191.6	192.8	122.4	120.7	59.4	58.2	117.6	91.9
1881	233.8	192.7	184.2	125.0	123.1	59.5	60.0	119.5	91.2
1882	227.3	194.8	190.6	130.6	129.3	55.4	71.1	126.5	92.0
1883	226.3	194.8	189.8	124.6	123.2	52.3	66.6	118.9	88.5
1884	212.4	184.6	178.9	121.4	120.4	50.9	66.0	116.9	83.7

Year									
1885	208.6	178.8	179.7	115.1	114.5	47.8	62.2	110.0	78.6
1886	200.6	171.2	176.8	113.4	113.2	41.7	63.9	105.6	77.4
1887	197.5	166.7	170.6	110.3	110.5	40.5	61.0	101.5	73.8
1888	195.4	165.9	179.3	110.4	109.4	41.6	61.9	103.5	75.1
1889	197.3	172.3	178.4	111.3	109.9	45.1	57.0	102.1	76.2
1890	199.7	178.8	171.6	117.7	115.7	49.0	61.3	110.3	76.2
1891	213.3	185.4	189.4	124.3	121.7	48.4	68.4	116.8	75.7
1892	203.1	173.2	177.0	114.4	112.1	40.4	65.0	105.4	75.8
1893	199.2	164.1	162.2	108.6	106.7	36.8	61.3	98.1	76.5
1894	187.5	158.2	173.0	103.6	101.3	35.2	60.3	95.5	73.0
1895	181.5	156.6	162.1	104.0	101.6	30.6	64.6	95.2	70.9
1896	186.7	161.1	159.3	107.4	105.1	36.8	61.9	98.7	68.6
1897	191.4	165.8	165.6	111.7	108.9	41.3	61.5	102.8	74.2
1898	195.3	169.2	185.5	108.8	105.5	41.3	55.3	96.6	70.9
1899	204.0	178.6	183.0	115.4	112.4	38.9	62.9	101.8	72.3
1900	204.7	179.9	185.9	115.4	112.5	36.8	64.6	101.4	76.5
1901	209.7	179.7	184.1	115.6	112.6	38.7	62.7	101.4	76.7
1902	213.4	187.6	192.6	123.2	120.1	39.2	67.9	107.1	78.3
1903	204.0	178.8	180.1	112.8	110.5	33.1	64.1	97.2	76.5
1904	204.8	177.6	178.5	113.6	111.6	33.3	65.8	99.1	75.2
1905	213.5	185.6	183.2	118.3	115.7	37.1	64.8	101.9	77.8
1906	222.3	194.1	187.3	126.3	123.6	38.5	68.0	106.5	80.9
1907	228.3	201.4	195.0	127.8	125.1	40.2	68.5	108.7	82.7
1908	226.3	198.7	186.2	133.4	130.1	39.7	75.8	115.5	79.6
1909	225.7	196.9	185.7	123.4	120.0	39.8	64.3	104.1	80.2
1910	227.1	198.0	180.0	122.6	120.5	37.7	63.4	101.1	83.1
1911	249.9	212.7	198.7	140.9	137.8	44.1	72.3	116.4	85.1

(continued)

Table 38.7 (cont.)

Year	Domestic supply (1) £m	Gross output 1 (2) £m	Gross output 2 (3) £m	Net output (4) £m	Net income (5) £m	Adjusted gross revenue			Price index
						Grain (incl. straw) (6) £m	Livestock and livestock products (7) £m	Total adjusted revenue (8) (=6+7) £m	Agricultural prices (at farm) (9) 1867–77 = 100
1912	251.8	218.9	210.2	136.9	134.1	40.0	72.7	112.7	89.8
1913	243.5	210.0	206.0	131.3	127.8	41.0	65.5	106.5	88.8
1914	279.3	237.7	210.8	164.7	160.9	57.5	86.9	144.4	94.8

Note:

An explanation of this table can be found in Chapter 3 on Agricultural Output, pp. 249–56. It has not been reproduced faithfully from the original source. It will be seen that net income was always 98 or 99 per cent of net output, indicating that in relation to the value of output the costs of land tax, tithes and farmers' own supplies were fairly small. But in the original source, for the period from 1888–1914, net income apparently was always greater than 100 per cent of net output. This was not a period of government price support. I suggest therefore that the original table was incorrectly printed. If we make this assumption, and reverse the figures for the period 1888–1914, as I have done in this table then indeed net income was always about 98 per cent or so of net output, but declined towards the end of the period to about 96 per cent.

Source: Columns 1, 2, 4, 5, 6, 7 and 8 from, J. R. Bellerby, 'Distribution of farm income in the United Kingdom, 1867–1914', in W. E. Minchinton (ed.), *Essays in Agrarian History* (Newton Abbot, 1968), vol. II, pp. 276–7. Column 3 from, Rural History Centre, University of Reading, Bellerby Mss., D/84/8/24.

Table 38.8. *Gross value of United Kingdom farm output, 1867–1914*

(1) Crops (gross value in £000s)

Year	Wheat	Barley	Oats	Potatoes	Hay	Straw	Vegetables	Fruit
1867	27,353	11,415	5,540	14,516	13,182	8,022	4,125	1,418
1868	37,339	21,217	12,685	20,469	13,853	12,597	4,303	1,461
1869	22,898	15,355	8,547	12,356	13,663	9,729	4,777	1,495
1870	25,541	15,881	8,584	14,712	12,443	9,202	3,293	1,091
1871	26,903	13,922	6,651	11,629	13,263	7,930	4,591	1,418
1872	25,612	14,004	5,873	19,287	10,465	6,647	6,382	1,667
1873	24,221	15,185	5,499	19,415	10,690	6,111	6,589	1,749
1874	26,770	22,951	9,033	15,347	12,544	8,616	6,336	1,801
1875	16,317	15,934	6,116	12,858	12,407	6,239	6,439	1,797
1876	18,949	14,792	6,831	16,822	12,387	5,898	5,549	1,650
1877	21,441	18,365	7,155	16,427	14,685	6,288	5,371	1,782
1878	21,181	21,220	7,883	20,718	13,775	6,744	5,281	1,735
1879	10,171	9,044	1,940	11,097	14,145	3,147	3,725	1,693
1880	14,829	13,286	5,765	14,543	15,879	6,033	4,572	1,825
1881	14,211	12,424	5,624	9,997	15,961	5,002	3,927	1,807
1882	12,576	10,720	6,310	11,537	17,820	5,411	5,009	2,264
1883	10,261	11,864	7,435	13,279	18,236	5,697	3,820	1,736
1884	13,145	10,875	7,732	10,183	19,147	5,352	4,443	2,133
1885	9,281	11,963	7,909	9,793	19,718	6,194	4,794	2,246
1886	10,756	9,328	7,697	9,985	19,978	5,566	4,390	1,991
1887	10,098	7,621	5,815	12,168	17,046	4,695	4,434	2,244
1888	10,292	9,571	6,356	9,807	24,878	5,014	4,517	2,100
1889	10,445	8,857	6,931	10,889	25,931	5,388	4,658	2,133
1890	11,458	10,945	7,335	7,627	19,988	5,565	4,438	2,068
1891	12,218	10,903	7,219	12,055	16,185	5,684	5,477	2,790
1892	8,102	8,775	7,170	8,703	14,740	5,217	4,614	2,572
1893	5,666	6,730	6,941	11,896	12,817	4,764	4,044	2,152
1894	5,594	8,413	7,424	7,775	21,599	5,359	4,880	2,478
1895	3,578	7,114	5,580	11,358	14,766	3,981	4,659	2,663
1896	6,448	7,866	5,152	7,368	13,720	4,194	3,960	2,427
1897	8,307	8,281	5,958	7,267	18,561	4,690	4,231	2,708
1898	12,473	10,579	7,091	10,596	22,079	5,464	5,070	3,169
1899	7,970	8,979	6,350	8,732	16,376	4,500	4,428	3,076
1900	7,379	7,895	6,499	9,309	17,358	4,098	5,418	3,406
1901	6,514	8,834	7,934	9,146	15,078	4,068	6,083	3,896
1902	6,711	8,119	8,927	7,332	23,344	4,983	6,530	4,207
1903	4,629	5,926	7,662	7,761	18,454	3,436	6,013	3,833
1904	4,763	7,050	6,763	9,866	17,623	3,557	5,368	3,624
1905	8,161	7,699	7,396	8,467	15,188	4,039	5,833	3,875
1906	7,296	7,141	9,395	7,242	14,787	3,597	5,705	5,177
1907	8,644	7,301	8,137	8,308	16,690	3,459	6,028	3,971

(continued)

Table 38.8 (cont.)

(1) Crops (cont.)

Year	Wheat	Barley	Oats	Potatoes	Hay	Straw	Vegetables	Fruit
1908	7,665	6,595	7,058	10,598	11,571	2,919	5,953	4,031
1909	10,871	7,023	7,504	7,372	9,778	3,289	7,345	3,267
1910	7,850	6,045	6,466	8,195	13,134	3,414	6,175	5,127
1911	8,497	7,425	6,629	11,851	11,645	3,119	8,382	4,963
1912	7,805	7,439	7,530	8,538	14,464	3,098	7,914	4,682
1913	9,658	7,334	7,228	12,454	13,774	3,241	7,655	5,491
1914	10,376	7,686	7,043	9,688	9,141	2,620	10,103	4,289

(2) Livestock

Year	Cattle	Sheep	Pigs	Milk + milk products	Eggs	Poultry
1867	17,384	10,118	24,347	33,170	3,570	1,210
1868	20,153	11,632	13,774	38,263	3,884	1,307
1869	20,886	12,820	9,631	37,089	3,850	1,303
1870	22,026	12,709	16,359	38,650	3,952	1,493
1871	20,139	11,808	22,147	33,914	5,111	1,725
1872	19,091	11,729	24,126	35,362	5,365	1,785
1873	26,082	13,337	13,737	37,281	5,825	1,968
1874	27,862	13,524	15,753	37,995	5,849	1,933
1875	30,301	15,099	17,846	40,150	6,021	2,035
1876	30,191	14,973	11,964	37,047	6,650	2,300
1877	26,096	14,111	18,077	37,608	6,218	2,062
1878	23,824	14,253	25,285	34,088	6,206	2,089
1879	25,578	14,774	16,219	30,443	5,745	1,926
1880	25,457	14,508	15,701	34,287	5,581	1,888
1881	25,648	12,959	16,225	34,941	5,902	2,074
1882	23,780	12,538	24,814	34,300	5,582	1,935
1883	24,116	13,199	22,584	33,735	5,375	1,868
1884	21,348	12,270	16,298	33,725	5,055	1,742
1885	25,972	11,724	16,108	31,449	5,554	1,925
1886	26,540	12,139	15,644	30,853	5,259	1,854
1887	24,369	10,997	17,218	32,265	5,559	1,933
1888	23,358	11,381	18,393	31,212	5,359	1,856
1889	18,495	10,948	18,425	30,539	5,582	1,914
1890	19,094	10,793	16,481	31,371	5,730	1,998
1891	24,164	11,210	23,443	33,962	5,734	2,011
1892	27,290	12,294	19,359	34,481	5,936	2,070
1893	28,062	12,603	7,599	34,073	6,432	2,285

(continued)

Table 38.8 (*cont.*)

(2) Livestock (*cont.*)

Year	Cattle	Sheep	Pigs	Milk + milk products	Eggs	Poultry
1894	23,585	11,226	19,071	32,251	5,782	2,036
1895	21,667	10,926	21,903	30,605	5,807	2,026
1896	21,842	11,403	20,717	30,961	6,185	2,159
1897	23,313	10,790	17,805	31,316	6,219	2,216
1898	22,641	10,564	21,187	31,198	6,162	2,159
1899	25,437	12,569	27,330	33,882	6,358	2,256
1900	27,120	12,728	24,824	35,396	6,709	2,392
1901	27,178	12,039	24,224	35,202	6,785	2,423
1902	28,665	11,438	22,021	36,069	6,950	2,495
1903	24,068	12,606	24,630	37,310	6,817	2,436
1904	25,346	12,304	20,859	35,557	7,031	2,572
1905	25,895	11,847	19,093	38,761	7,641	2,822
1906	27,248	12,171	17,080	42,312	8,442	3,053
1907	27,504	11,941	20,227	44,771	8,452	3,254
1908	28,818	11,933	18,077	45,249	8,328	3,089
1909	29,589	11,340	14,465	45,808	8,708	3,427
1910	29,641	12,339	7,645	45,940	8,681	3,400
1911	29,163	12,644	14,905	48,275	9,449	3,508
1912	31,846	13,696	21,627	50,334	9,581	3,816
1913	29,172	12,976	14,182	50,140	9,976	4,179
1914	35,111	13,363	15,628	52,755	10,953	4,054

(3) Other

Year	Horses	Wool	Hops	Flax	Rye	Other	Total other
1867	2,940	9,483	2,191	1,908	17	4,798	21,337
1868	2,910	9,203	1,805	1,429	18	5,708	21,073
1869	2,940	9,262	1,553	1,676	26	4,746	20,203
1870	3,008	8,197	2,449	1,845	22	5,036	20,557
1871	3,303	9,990	4,949	690	17	5,003	23,952
1872	3,836	12,341	2,839	978	23	5,185	25,202
1873	4,066	12,505	3,455	1,066	17	5,220	26,329
1874	3,997	11,007	2,611	891	19	5,621	24,146
1875	4,086	9,934	3,464	1,187	10	5,206	23,887
1876	4,088	9,059	3,266	1,424	8	5,096	22,941
1877	4,203	8,447	2,159	1,097	26	5,290	21,222
1878	4,120	7,275	2,747	981	40	5,486	20,649

(*continued*)

Table 38.8 (*cont.*)

(3) Other (*cont.*)

Year	Horses	Wool	Hops	Flax	Rye	Other	Total other
1879	3,993	6,488	1,502	729	0	4,059	16,771
1880	4,106	6,883	1,923	1,053	15	4,703	18,683
1881	3,948	5,149	2,865	1,025	7	4,492	17,486
1882	3,784	4,617	2,201	687	22	4,648	15,959
1883	3,433	4,262	3,594	615	28	4,628	16,560
1884	3,375	4,429	2,795	507	26	4,365	15,497
1885	3,422	4,407	2,048	765	21	4,385	15,048
1886	2,921	4,345	2,329	886	25	4,312	14,818
1887	2,988	4,686	1,766	524	17	4,161	14,142
1888	3,191	4,497	2,317	611	28	4,543	15,187
1889	3,144	4,840	4,164	586	36	4,522	17,292
1890	3,400	5,205	2,986	580	30	4,511	16,712
1891	3,279	4,879	2,778	360	26	4,978	16,300
1892	3,371	4,350	2,010	277	12	5,653	15,673
1893	3,521	4,874	2,795	547	6	4,419	16,162
1894	3,330	4,512	2,130	747	56	4,711	15,486
1895	3,350	5,354	1,816	369	14	4,569	15,472
1896	3,191	5,285	1,609	291	40	4,490	14,906
1897	2,982	4,435	1,648	201	39	4,668	13,973
1898	2,861	4,026	2,549	178	55	5,403	15,072
1899	3,121	3,932	2,196	181	33	5,331	14,794
1900	3,752	3,646	2,002	355	11	5,589	15,355
1901	3,519	3,181	1,911	543	12	5,536	14,702
1902	3,494	2,833	2,211	423	21	5,790	14,772
1903	3,139	3,223	2,293	314	11	5,585	14,565
1904	3,401	4,262	2,676	372	11	5,536	16,258
1905	3,468	5,252	1,553	386	11	5,854	16,524
1906	3,865	4,955	1,372	439	18	5,983	16,632
1907	3,862	4,335	1,524	390	19	6,231	16,361
1908	3,079	3,581	1,310	219	12	6,123	14,324
1909	3,047	4,800	1,689	228	19	6,105	15,888
1910	3,076	4,871	1,561	339	17	6,087	15,951
1911	3,043	4,886	3,331	480	12	6,727	18,479
1912	3,226	4,661	2,302	532	36	7,109	17,866
1913	3,277	5,230	2,374	498	37	7,160	18,576
1914	3,109	5,189	1,962	328	55	7,325	17,968

(*continued*)

Table 38.8 (*cont.*)

(4) Totals

Year	Crops	Livestock	Other	Grand total
1867	85,571	89,799	21,337	196,707
1868	123,924	89,013	21,073	234,010
1869	88,820	85,579	20,203	194,602
1870	90,747	95,189	20,557	206,493
1871	86,307	94,844	23,952	205,103
1872	89,937	97,458	25,202	212,597
1873	89,459	98,230	26,329	214,018
1874	103,398	102,916	24,146	230,460
1875	78,107	111,452	23,887	213,446
1876	82,878	103,125	22,941	208,944
1877	91,514	104,172	21,222	216,908
1878	98,537	105,745	20,649	224,931
1879	54,962	94,685	16,771	166,418
1880	76,732	97,422	18,683	192,837
1881	68,953	97,749	17,486	184,188
1882	71,647	102,949	15,959	190,555
1883	72,328	100,877	16,560	189,765
1884	73,010	90,438	15,497	178,945
1885	71,898	92,732	15,048	179,678
1886	69,691	92,289	14,818	176,798
1887	64,121	92,341	14,142	170,604
1888	72,535	91,559	15,187	179,281
1889	75,232	85,903	17,292	178,427
1890	69,424	85,467	16,712	171,603
1891	72,531	100,524	16,300	189,355
1892	59,893	101,430	15,673	176,996
1893	55,010	91,054	16,162	162,226
1894	63,522	93,951	15,486	172,959
1895	53,699	92,934	15,472	162,105
1896	51,135	93,267	14,906	159,308
1897	60,003	91,659	13,973	165,635
1898	76,521	93,911	15,072	185,504
1899	60,411	107,832	14,794	183,037
1900	61,362	109,169	15,355	185,886
1901	61,553	107,851	14,702	184,106
1902	70,153	107,638	14,772	192,563
1903	57,714	107,867	14,565	180,146
1904	58,614	103,669	16,258	178,541
1905	60,658	106,059	16,524	183,241
1906	60,340	110,306	16,632	187,278

(continued)

Table 38.8 (*cont.*)

(4) Totals (*cont.*)

Year	Crops	Livestock	Other	Grand total
1907	62,538	116,149	16,361	195,048
1908	56,390	115,494	14,324	186,208
1909	56,449	113,337	15,888	185,674
1910	56,406	107,646	15,951	180,003
1911	62,511	117,944	18,479	198,934
1912	61,470	130,900	17,866	210,236
1913	66,835	120,625	18,576	206,036
1914	60,966	131,864	17,968	210,798

Source: Rural History Centre, University of Reading,
Bellerby Mss., D/84/8/17.

Table 38.9. *Outputs, inputs and incentive income in United Kingdom agriculture, 1867–1914 (all in £000s)*

(A)	Adjustments		Inputs				Capital	
	Farmers' consumption	Inventory changes	Feed	Fertiliser	(Machinery, fuel & repairs)	Rates & Land Tax	Rent	Interest charges
1867	11,611	1,975	23,653	5,258	15,900	9,417	38,208	21,610
1868	12,348	1,386	29,211	5,753	17,045	9,215	38,504	25,270
1869	11,171	−1,574	28,720	6,384	16,938	9,151	38,864	22,750
1870	11,497	−110	29,473	6,707	16,894	8,891	38,934	21,590
1871	11,388	2,812	31,891	7,215	18,482	8,922	39,024	21,800
1872	11,619	6,997	34,465	7,569	18,692	8,629	43,175	22,230
1873	11,578	5,686	33,372	7,709	19,509	8,267	43,361	23,100
1974	11,739	−8	36,648	7,547	19,703	8,485	43,241	24,840
1975	11,058	−4,603	39,200	7,311	20,308	8,242	43,430	22,750
1876	10,857	−2,935	41,106	6,658	20,403	8,092	42,778	21,250
1877	11,354	−502	35,774	6,245	19,858	8,194	42,780	21,530
1878	11,533	1,410	38,741	5,780	19,877	8,032	41,676	22,360
1879	9,869	−1,601	33,082	5,393	18,226	7,886	37,470	15,490
1880	10,499	−3,962	36,852	4,993	19,644	7,684	36,792	19,210
1881	10,328	−2,232	36,455	4,711	18,968	7,692	35,373	17,570
1882	10,720	1,349	32,867	4,484	19,676	7,579	36,649	17,360
1883	10,633	4,172	36,878	4,218	20,517	7,546	36,601	18,450
1884	10,510	4,647	32,141	4,096	19,633	7,082	34,809	17,520
1885	10,435	1,475	33,047	4,047	18,895	6,865	30,899	17,370
1886	10,423	−1,470	29,157	3,949	18,226	6,631	30,577	16,910
1887	10,274	−2,205	28,646	3,869	17,436	6,471	27,812	14,820
1888	10,593	−1,210	28,506	3,871	18,340	5,891	28,254	16,290
1889	10,672	5,052	32,778	3,782	18,928	5,676	28,042	16,800
1890	10,574	8,671	33,414	3,796	19,142	5,094	27,462	16,450
1891	10,976	5,057	34,139	3,751	19,204	4,856	27,271	17,410
1892	10,493	−1,673	34,665	3,740	15,606	4,777	25,575	16,120
1893	9,971	−5,316	32,266	3,732	15,056	4,636	25,801	14,360
1894	10,365	−2,926	31,693	3,847	15,779	4,779	24,713	15,290
1895	10,283	2,055	29,556	3,816	16,401	4,568	24,587	13,510
1896	10,123	1,631	30,671	3,999	15,591	4,654	23,907	12,360
1897	10,270	360	31,086	4,209	15,554	4,275	23,901	13,620
1898	10,704	2,226	36,637	4,382	16,480	4,290	23,505	15,230
1899	10,512	1,292	38,702	4,669	16,683	4,432	22,815	14,040
1900	10,611	−189	39,457	4,866	16,938	4,671	23,070	15,930
1901	10,420	443	39,915	4,892	16,797	4,390	23,677	16,280
1902	10,773	2,868	40,023	5,016	16,761	4,678	23,044	18,310

(continued)

Table 38.9 (cont.)

(A)			Inputs				Capital	
	Adjustments				(Machinery,	Rates &		
	Farmers'	Inventory			fuel &	Land		Interest
	consumption	changes	Feed	Fertiliser	repairs)	Tax	Rent	charges
1903	10,257	782	40,620	5,050	16,707	4,885	23,410	15,530
1904	10,210	713	39,022	5,105	16,538	5,141	22,774	15,760
1905	10,242	362	43,024	5,341	16,382	4,620	23,493	15,840
1906	10,355	19	42,653	5,528	16,603	4,705	23,830	15,830
1907	10,421	1,021	47,949	5,624	17,266	4,735	23,929	17,660
1908	10,425	2,395	41,079	5,711	16,595	4,150	24,146	16,620
1909	10,093	117	48,740	5,981	16,626	3,713	24,314	17,450
1910	9,821	−595	49,749	6,205	16,575	4,789	24,548	18,450
1911	10,488	−1,541	46,349	6,415	17,286	4,463	24,714	18,360
1912	10,473	−3,532	53,864	6,931	18,565	4,541	25,102	20,870
1913	10,483	−1,143	52,987	6,867	18,025	3,889	25,393	20,850
1914	10,814	818	47,639	6,962	18,099	3,951	25,642	20,400

(B) Totals

Year	Gross output 1	Total adjustment 2	Total inputs 3	Total capital 4	Incentive Income after adjustment (1+2−3−4)	Incentive Income before adjustment (1−3−4)	Net Income (Gross + adjustment− inputs) (1+2−3)
1867	196,707	13,586	54,228	59,818	96,247	82,661	156,065
1868	234,010	13,734	61,224	63,774	122,746	109,012	186,520
1869	194,602	9,597	61,193	61,614	81,392	71,795	143,006
1870	206,493	11,387	61,965	60,524	95,391	84,004	155,915
1871	205,103	14,200	66,510	60,824	91,969	77,769	152,793
1872	212,597	18,616	69,355	65,405	96,453	77,837	161,858
1873	214,018	17,264	68,857	66,461	95,964	78,700	162,425
1874	230,460	11,731	72,383	68,081	101,727	89,996	169,808
1875	213,446	6,455	75,061	66,180	78,660	72,205	144,840
1876	208,944	7,922	76,259	64,028	76,579	68,657	140,607
1877	216,908	10,852	70,071	64,310	93,379	82,527	157,689
1878	224,931	12,943	72,430	64,036	101,408	88,465	165,444
1879	166,418	8,268	64,587	52,960	57,139	48,871	110,099
1880	192,837	6,537	69,173	56,002	74,199	67,662	130,201
1881	184,188	8,096	67,826	52,943	71,515	63,419	124,458
1882	190,555	12,069	64,606	54,009	84,009	71,940	138,018
1883	189,765	14,805	69,159	55,051	80,360	65,555	135,411

(continued)

Table 38.9 (*cont.*)

(B) Totals (*cont.*)

Year	Gross output 1	Total adjustment 2	Total inputs 3	Total capital 4	Incentive Income after adjustment (1+2−3−4)	Incentive Income before adjustment (1−3−4)	Net Income (Gross + adjustment− inputs) (1+2−3)
1884	178,945	15,157	62,952	52,329	78,821	63,664	131,150
1885	179,678	11,910	62,854	48,269	80,465	68,555	128,734
1886	176,798	8,953	57,963	47,487	80,301	71,348	127,788
1887	170,604	8,069	56,422	42,632	79,619	71,550	122,251
1888	179,281	9,383	56,608	44,544	87,512	78,129	132,056
1889	178,427	15,724	61,164	44,842	88,145	72,421	132,987
1890	171,603	19,245	61,446	43,912	85,490	66,245	129,402
1891	189,355	16,033	61,950	44,681	98,757	82,724	143,438
1892	176,996	8,820	58,788	41,695	85,333	76,513	127,028
1893	162,226	4,655	55,690	40,161	71,030	66,375	111,191
1894	172,959	7,439	56,098	40,003	84,297	76,858	124,300
1895	162,105	12,338	54,341	38,097	82,005	69,667	120,102
1896	159,308	11,754	54,915	36,267	79,880	68,126	116,147
1897	165,635	10,630	55,124	37,521	83,620	72,990	121,141
1898	185,504	12,930	61,789	38,735	97,910	84,980	136,645
1899	183,037	11,804	64,486	36,855	93,500	81,696	130,355
1900	185,886	10,422	65,932	39,000	91,376	80,954	130,376
1901	184,106	10,863	65,994	39,957	89,018	78,155	128,975
1902	192,563	13,641	66,478	41,354	98,372	84,731	139,726
1903	180,146	11,039	67,262	38,940	84,983	73,944	123,923
1904	178,541	10,923	65,806	38,534	85,124	74,201	123,658
1905	183,241	10,604	69,367	39,333	85,145	74,541	124,478
1906	187,278	10,374	69,489	39,660	88,503	78,129	128,163
1907	195,048	11,442	75,574	41,589	89,327	77,885	130,916
1908	186,208	12,820	67,535	40,766	90,727	77,907	131,493
1909	185,674	10,210	75,060	41,764	79,060	68,850	120,824
1910	180,003	9,226	77,318	42,998	68,913	59,687	111,911
1911	198,934	8,947	74,513	43,074	90,294	81,347	133,368
1912	210,236	6,941	83,901	45,972	87,304	80,363	133,276
1913	206,036	9,340	81,768	46,243	87,365	78,025	133,608
1914	210,778	11,632	76,651	46,042	99,717	88,085	145,759

Source: Rural History Centre, University of Reading, Bellerby Mss., D/84/8/17 & 24.

Table 38.10. *Farm income distribution in the United Kingdom, 1855–1914 (all in £s million)*

Year	Net output	Wages	Rent	Farmers' net income
1855	146	61	45	40
1856	149	60	47	42
1857	155	57	49	49
1858	146	56	49	41
1859	143	57	49	37
1860	142	56	49	37
1861	154	56	50	48
1862	155	56	50	49
1863	158	55	50	53
1864	148	55	50	43
1865	151	56	49	46
1866	156	56	50	50
1867	160	55	51	54
1868	156	55	52	49
1869	140	53	53	34
1870	156	57	53	46
1871	159	56	54	49
1872	161	62	55	44
1873	172	61	56	55
1874	164	61	56	47
1875	163	62	57	44
1876	161	61	57	43
1877	154	61	58	35
1878	146	60	56	30
1879	126	58	55	13
1880	132	58	53	21
1881	135	58	53	24
1882	141	57	55	29
1883	136	57	55	24
1884	133	56	53	24
1885	127	55	51	21
1886	127	54	49	24
1887	124	54	48	22
1888	125	54	48	23
1889	126	54	48	24
1890	133	55	48	30
1891	139	54	48	37
1892	130	54	47	29
1893	125	54	46	25

(continued)

Table 38.10 (*cont.*)

Year	Net output	Wages	Rent	Farmers' net income
1894	121	54	45	22
1895	122	54	44	24
1896	124	54	43	27
1897	127	54	42	31
1898	123	54	41	28
1899	130	55	41	34
1900	130	56	41	33
1901	130	56	41	33
1902	138	56	42	40
1903	127	56	41	30
1904	127	56	41	30
1905	131	56	41	34
1906	139	57	42	40
1907	140	57	42	41
1908	145	56	43	46
1909	136	56	43	37
1910	135	57	43	35
1911	152	58	43	51
1912	148	59	43	46
1913	142	60	43	39
1914	173	62	44	67

Source: C. H. Feinstein, *National Income, Expenditure and Output of the United Kingdom 1855–1965* (Cambridge, 1972), Table 23, p. T60.

Table 38.11. *Estimates of agricultural output for England and Wales, 1867–1914 (all in £000s)*

(A) Crops

Year	Wheat	Barley	Oats	Potatoes	Hay	Straw	Fruit & veg.
1867	24,475	9,548	4,391	6,479	9,096	5,385	4,113
1868	33,332	17,458	9,912	9,692	9,559	8,429	4,311
1869	20,431	12,504	6,735	6,136	9,427	6,525	4,692
1870	22,841	12,879	6,766	7,307	8,586	6,162	3,276
1871	24,150	11,353	5,209	6,082	9,151	5,341	4,586
1872	23,097	11,362	4,594	9,545	7,221	4,478	5,927
1873	22,241	12,325	4,359	9,639	7,376	4,169	6,178
1874	24,520	18,694	7,034	7,776	8,655	5,902	6,200
1875	15,045	13,003	4,859	6,555	8,561	4,269	6,164
1876	17,694	12,120	5,647	8,411	8,547	4,029	5,370
1877	19,930	14,875	5,837	8,184	10,133	4,288	5,363
1878	19,684	17,230	6,365	10,412	9,505	4,628	5,199
1879	9,366	7,369	1,603	5,847	9,760	2,160	3,948
1880	13,712	10,942	4,841	7,666	10,957	4,129	4,752
1881	13,081	10,133	4,887	5,404	11,013	3,410	4,347
1882	11,626	8,713	5,304	6,217	12,296	3,678	5,404
1883	9,626	9,764	6,564	7,323	12,583	3,882	4,139
1884	12,471	8,984	6,756	5,944	13,211	3,663	5,051
1885	8,808	9,875	6,995	5,778	13,625	4,228	5,352
1886	10,168	7,756	7,049	5,915	13,887	3,793	4,895
1887	9,586	6,350	5,328	7,299	11,880	3,197	5,073
1888	9,628	7,863	5,682	6,040	17,290	3,425	5,012
1889	8,809	7,258	6,270	6,706	18,438	3,688	5,176
1890	10,720	9,014	6,697	4,480	14,154	3,809	5,054
1891	11,513	8,963	6,670	7,305	11,381	3,900	6,398
1892	7,605	7,210	6,764	5,261	10,221	3,563	5,539
1893	5,370	5,570	6,744	7,366	8,718	3,236	4,697
1894	5,320	6,966	7,316	4,733	15,138	3,659	5,651
1895	3,401	5,913	5,639	7,320	10,372	2,687	5,578
1896	6,158	6,490	4,996	4,993	9,511	2,836	4,993
1897	7,879	6,744	5,831	4,686	12,959	3,196	5,400
1898	11,826	8,513	6,790	7,023	15,509	3,723	6,441
1899	7,578	7,245	6,186	6,012	11,389	3,080	5,885
1900	6,971	6,361	6,512	6,525	11,988	2,804	6,913
1901	6,219	7,191	7,907	6,746	10,437	2,777	7,987
1902	6,366	6,567	9,072	5,412	16,382	3,408	8,531
1903	4,400	4,830	7,836	5,677	12,989	2,335	7,830
1904	4,533	5,784	7,094	7,169	12,295	2,414	7,119
1905	7,774	6,187	7,500	6,414	10,436	2,756	7,825

(continued)

Table 38.11 (*cont.*)

(A) Crops (*cont.*)

Year	Wheat	Barley	Oats	Potatoes	Hay	Straw	Fruit & veg.
1906	6,915	5,678	9,516	5,220	10,182	2,441	8,790
1907	8,197	5,827	8,420	5,976	11,673	2,346	8,077
1908	7,248	5,321	7,314	7,711	8,058	1,985	8,097
1909	10,329	5,628	7,612	5,527	6,716	2,250	8,812
1910	7,423	4,898	6,515	5,830	8,980	2,330	9,015
1911	8,027	6,021	6,699	8,736	7,855	2,136	10,714
1912	7,377	5,978	7,658	6,545	9,814	2,126	10,179
1913	9,183	5,924	7,206	9,386	9,492	2,209	10,627
1914	9,842	6,182	7,010	7,478	6,170	1,794	11,912

(B) Animals

Year	Cattle	Sheep	Pigs	Milk + milk products	Eggs + poultry
1867	7,584	6,582	16,102	15,639	2,151
1868	9,703	7,708	9,398	18,269	2,336
1869	9,884	8,453	5,759	17,560	2,319
1870	10,404	8,404	9,062	18,182	2,450
1871	9,197	7,622	12,382	16,387	3,076
1872	8,847	7,573	14,994	17,060	3,218
1873	12,371	8,706	9,121	18,101	3,507
1874	13,472	8,920	10,167	18,632	3,502
1875	14,520	9,969	10,655	19,478	3,625
1876	14,236	9,560	6,884	18,095	4,028
1877	12,325	9,302	10,683	18,418	3,726
1878	11,332	9,372	15,790	16,870	3,733
1879	12,255	9,789	10,068	15,143	3,452
1880	12,413	9,396	10,354	17,370	3,361
1881	12,468	8,310	9,934	18,222	3,589
1882	11,430	7,995	14,838	17,854	3,383
1883	11,626	8,948	14,014	17,546	3,259
1884	10,513	7,989	10,157	17,753	3,059
1885	12,956	7,647	9,879	16,292	3,366
1886	13,398	7,949	9,376	16,419	3,201
1887	12,187	7,195	10,051	17,280	3,371
1888	11,416	7,300	10,885	16,755	3,247
1889	9,036	6,951	11,168	16,483	3,373
1890	9,420	6,802	9,917	16,771	3,478
1891	11,994	7,073	15,039	18,041	3,485

(*continued*)

Table 38.11 (*cont.*)

(B) Animals (*cont.*)

Year	Cattle	Sheep	Pigs	Milk + milk products	Eggs + poultry
1892	13,559	7,762	12,064	18,276	3,603
1893	13,729	7,914	4,639	18,105	3,923
1894	11,258	6,964	11,309	17,111	3,518
1895	10,429	6,826	14,168	16,290	3,525
1896	10,554	7,126	13,221	16,572	3,755
1897	11,179	6,693	10,710	16,741	3,796
1898	10,916	6,521	13,246	16,694	3,744
1899	12,509	7,826	17,027	18,072	3,876
1900	13,275	7,920	15,298	18,814	4,095
1901	13,106	7,428	14,646	18,753	4,144
1902	13,439	7,057	13,177	19,080	4,250
1903	11,512	7,845	15,428	19,814	4,164
1904	12,362	7,729	13,568	18,869	4,321
1905	12,777	7,447	12,209	20,624	4,708
1906	13,538	7,697	10,501	23,137	5,173
1907	13,544	7,501	12,738	24,114	5,268
1908	16,524	7,516	11,989	24,231	5,138
1909	14,704	7,247	9,224	24,741	5,461
1910	14,781	7,923	4,774	24,761	5,436
1911	14,535	8,040	9,326	26,227	5,831
1912	15,614	8,559	13,568	27,135	6,029
1913	13,971	8,068	9,050	26,994	6,370
1914	16,937	8,271	9,841	28,186	6,735

(C) Totals

Year	Total crops	Total animals	Interim total (91% of true total)	Other (9% of true total)	True total	Allowance for output adjustments	Adjusted Total Output
1867	63,488	48,058	111,546	11,032	122,578	8,466	131,044
1868	92,692	47,413	140,106	13,857	153,962	9,036	162,998
1869	66,451	43,974	110,425	10,921	121,346	5,984	127,330
1870	67,816	48,502	116,319	11,504	127,823	7,049	134,871
1871	65,872	48,664	114,536	11,328	125,864	8,714	134,578
1872	66,224	51,691	117,915	11,662	129,577	11,346	140,923
1873	66,288	51,805	118,093	11,680	129,772	10,468	140,241
1874	78,781	54,693	133,474	13,201	146,675	7,466	154,141
1875	58,456	58,248	116,703	11,542	128,246	3,878	132,124
1876	61,817	52,803	114,620	11,336	125,956	4,776	130,732

(*continued*)

Table 38.11 (cont.)

(C) Totals

Year	Total crops	Total animals	Interim total (91% of true total)	Other (9% of true total)	True total	Allowance for output adjustments	Adjusted Total Output
1877	68,609	54,454	123,063	12,171	135,234	6,766	142,000
1878	73,023	57,096	130,119	12,869	142,988	8,228	151,216
1879	40,054	50,707	90,761	8,976	99,737	4,955	104,692
1880	56,998	52,894	109,892	10,868	120,760	4,094	124,854
1881	52,275	52,523	104,798	10,365	115,163	5,062	120,225
1882	53,238	55,500	108,738	10,754	119,492	7,568	127,060
1883	53,880	55,393	109,273	10,807	120,080	9,368	129,448
1884	56,080	49,471	105,551	10,439	115,990	9,825	125,815
1885	54,662	50,140	104,802	10,365	115,167	7,634	122,801
1886	53,462	50,343	103,805	10,266	114,071	5,777	119,848
1887	48,714	50,084	98,798	9,771	108,569	5,135	113,704
1888	54,941	49,603	104,544	10,340	114,884	6,013	120,896
1889	57,345	47,011	104,356	10,321	114,677	10,106	124,783
1890	53,928	46,389	100,317	9,921	110,238	12,363	122,601
1891	56,131	55,632	111,763	11,054	122,817	10,399	133,216
1892	46,163	55,264	101,427	10,031	111,458	5,554	117,012
1893	41,702	48,310	90,012	8,902	98,914	2,838	101,752
1894	48,784	50,160	98,943	9,786	108,729	4,676	113,406
1895	40,911	51,239	92,149	9,114	101,263	7,707	108,970
1896	39,976	51,228	91,204	9,020	100,224	7,395	107,619
1897	46,695	49,118	95,813	9,476	105,289	6,757	112,046
1898	59,826	51,122	110,948	10,973	121,921	8,498	130,419
1899	47,376	59,311	106,686	10,551	117,237	7,561	124,798
1900	48,074	59,403	107,477	10,630	118,106	6,622	124,728
1901	49,263	58,076	107,340	10,616	117,956	6,960	124,915
1902	55,737	57,004	112,741	11,150	123,891	8,776	132,668
1903	45,896	58,762	104,659	10,351	115,010	7,048	122,057
1904	46,408	56,849	103,257	10,212	113,469	6,942	120,411
1905	48,893	57,765	106,658	10,549	117,206	6,783	123,989
1906	48,742	60,046	108,788	10,759	119,547	6,622	126,169
1907	50,515	63,164	113,679	11,243	124,922	7,328	132,250
1908	45,735	65,397	111,132	10,991	122,123	8,408	130,531
1909	46,875	61,376	108,251	10,706	118,957	6,541	125,499
1910	44,990	57,676	102,665	10,154	112,819	5,783	118,602
1911	50,188	63,958	114,146	11,289	125,435	5,641	131,077
1912	49,677	70,906	120,583	11,926	132,509	4,374	136,883
1913	54,027	64,453	118,480	11,718	130,198	5,902	136,100
1914	50,390	69,988	120,377	11,905	132,283	7,299	139,582

Source: See Chapter 3 for derivation of this Table.

Table 38.12. *United Kingdom agricultural price indexes, 1867–1914 (in all cases 1867–77 = 100)*

Year	1870 Weights	1895 Weights	1910 Weights	Bellerby Index
1867	98.9	95.7	95.9	97.1
1868	99.4	95.6	96.7	93.7
1869	95.7	96.6	96.2	93.2
1870	93.1	94.4	94.0	94.8
1871	97.3	97.1	97.5	99.6
1872	100.9	100.5	100.5	103.3
1873	107.4	107.4	106.9	108.3
1874	104.0	104.1	104.3	100.1
1875	101.8	104.9	104.9	103.6
1876	100.1	103.0	101.5	102.9
1877	101.4	101.1	101.5	100.3
1878	97.5	98.9	98.0	94.0
1879	89.2	91.0	89.6	89.0
1880	94.2	96.9	95.3	91.9
1881	90.4	92.8	92.5	91.2
1882	93.6	96.4	96.3	92.0
1883	89.8	92.6	92.3	88.5
1884	84.3	88.3	88.7	83.7
1885	79.4	83.4	83.2	78.6
1886	77.2	81.3	80.4	77.4
1887	74.7	78.2	78.0	73.8
1888	76.5	79.8	80.0	75.1
1889	76.4	80.6	80.2	76.2
1890	77.2	80.4	80.1	76.2
1891	79.5	82.0	82.6	75.7
1892	77.0	81.8	81.2	75.8
1893	76.4	81.8	80.5	76.5
1894	72.5	78.2	77.7	73.0
1895	70.0	75.1	75.7	70.9
1896	68.0	71.7	73.0	68.6
1897	73.6	78.0	77.4	74.2
1898	76.9	80.6	79.4	70.9
1899	71.7	76.6	77.2	72.3
1900	76.0	81.5	81.6	76.5
1901	75.7	81.9	81.2	76.7
1902	78.8	85.1	85.0	78.3
1903	74.6	80.5	80.8	76.5
1904	75.1	79.6	79.9	75.2

(continued)

Table 38.12 (*cont.*)

Year	1870 Weights	1895 Weights	1910 Weights	Bellerby Index
1905	76.3	81.4	81.7	77.8
1906	78.2	84.7	85.5	80.9
1907	80.1	85.9	86.6	82.7
1908	76.3	81.0	83.1	79.6
1909	79.1	83.4	84.7	80.2
1910	80.2	87.2	87.4	83.1
1911	83.5	89.3	90.4	85.1
1912	89.6	95.3	96.1	89.8
1913	88.1	95.5	95.0	88.8
1914	85.9	92.2	93.6	94.8

Source: See Chapter 3.

Table 38.13. *Agricultural output in England in 1873 and 1911*
(by county in £ooos)

County	1873	1911	County	1873	1911
Bedford	1,600	814	Middlesex	665	425
Berkshire	2,078	1,120	Monmouth	1,113	746
Buckingham	2,120	1,320	Norfolk	6,376	3,573
Cambridge	3,125	2,058	Northampton	3,056	1,757
Cheshire	3,037	3,321	Northumberland	2,792	1,996
Cornwall	3,392	2,997	Nottingham	2,451	1,436
Cumberland	2,548	1,938	Oxford	2,370	1,298
Derby	2,521	2,039	Rutland	463	267
Devon	6,757	4,881	Shropshire	3,659	2,928
Dorset	2,809	2,049	Somerset	5,298	4,402
Durham	1,647	1,298	Stafford	3,163	2,568
Essex	4,836	2,727	Suffolk	4,754	2,786
Gloucester	3,882	2,508	Surrey	1,644	908
Hampshire	3,897	2,424	Sussex	3,838	2,444
Hereford	3,095	1,839	Warwick	2,637	1,574
Hertford	1,885	1,069	Westmorland	1,142	956
Huntingdon	1,268	645	Wiltshire	3,942	2,720
Kent	6,632	4,989	Worcester	2,646	1,863
Lancashire	4,294	4,465	Yorkshire East Riding	3,483	2,272
Leicester	2,540	1,663	Yorkshire North Riding	4,083	2,923
Lincoln	9,205	6,042	Yorkshire West Riding	5,833	4,712

Notes:
Based on the raw data in the *Annual Agricultural Returns* and the following
proportions sold off farms at the prices indicated:

Product	Proportion sold (% ages)		Price		
	1873	1911	1873	1911	
Wheat	90	75	£13.5	£7.4	per ton
Barley	93	68	£10.2	£7.6	per ton
Oats	25	16	£8.8	£6.8	per ton
Hay	40	40	£2.8	£2.1	per ton
Potatoes	70	70	£3.0	£3.7	per ton
Hops	100	100	£6.2	£8.3	per cwt
Beef	128.9 cwt per 100 cattle enumerated		£3.5	£2.0	per cwt
Mutton	20 cwt per 100 sheep enumerated		£4.5	£2.8	per cwt
Pigmeat	212.1 cwt per 100 pigs enumerated		£2.8	£1.8	per cwt
Wool	3.5 lb for every sheep enumerated		18d.	9d.	per lb
Milk	350 & 420 galls per cow in milk or calf		7.35d.	7.0d.	per gallon
Fruit	1.86 tons per acre of orchard fruit		£17.6	£10.6	per ton

(continued)

Notes to Table 38.13 *(cont.)*

Thus omitting the following main products: straw, flax, small (soft) fruit, vegetables, poultry, eggs and horses, which are estimated to account for about 10 per cent of output in 1873 rising to about 15 per cent in 1911.

Source: F. M. L. Thompson, 'An anatomy of English agriculture, 1870–1914, in B. A. Holderness and Michael Turner (eds.), *Land, Labour and Agriculture, 1700–1920* (London, 1991), chapter 11, pp. 233–4, 237.

CHAPTER 39

RENT AND LAND VALUES

BY BETHANIE AFTON AND MICHAEL TURNER

A. INTRODUCTION

It might be imagined that something so central as rent and land values to the study of landownership and occupancy might already have a ready-made database to hand. But the details of land values, and of the landlord and tenant relationship which was embodied in the rent exchange which took place between them, is not so accessible as we might imagine. Data on land values, on any large scale, can only be derived indirectly through the income yield taken by the Inland Revenue through Income Tax, and the collection of official, reliable data on the level of agricultural rents, again on any large scale, did not begin until just at the time of or after the Second World War. The censuses of agricultural output in 1925 and 1931, for example, presented estimates not measurements.[1] This is an inauspicious start, but it does not mean that a database for the period 1850–1914 cannot be constructed, giving a reasonable geographical cover in which we can have some confidence, and a temporal cover of which the general trend is generally recognised and trusted as representative of its time.

At the national level, for individual years, rent estimates are available, such as those for 1851/2 and 1850 of 22s. 0d. and 27s. 2d. per acre for England and Wales proposed by J. R. McCulloch and James Caird.[2] Thereafter, until recently, we have been reliant on a surrogate indication of long-run rent changes based on Income Tax returns and some well-known conflations of estate material. From the latter it is possible to con-

---

[1] H. A. Rhee, *The Rent of Agricultural Land in England and Wales* (Central Landowners Association, London 1946), p. 33. See also D. R. Denman, 'Farm rent surveys 1938–1959', *The Farm Economist*, 9, 8, 1960, pp. 372–7; D. K. Britton, *An Enquiry into Agricultural Rents and the Expenses of Landowners in England and Wales 1938 and 1946* (Central Landowners Association and the Ministry of Agriculture and Fisheries, London, 1949).

[2] Quoted by R. J. Thompson, 'An inquiry into the rent of agricultural land in England and Wales during the nineteenth century', *JRSS*, 70, 1907, pp. 587–625, reprinted in W. E. Minchinton (ed.), *Essays in Agrarian History* (Newton Abbot, 1968), vol. II, pp. 55–86, this reference p. 64.

struct a long-term national rent index for England and Wales, though at the lower regional level this would be sporadic and mainly based on individual estates. This indirect and somewhat outdated database has now been superseded by a serious attempt to construct afresh an agricultural rent index. But the construction of that rent index is best understood within an historiographical and developmental context.

B. THE DATA

Table 39.1 summarises some of the basic printed data which have thus far been available. Column 1 is the Gross Assessments to Schedule A Income Tax as they related to land (including tithes) for England and Wales for the period 1850–1914.[3] Schedule A Income Tax is the income arising from the ownership of landed property and it was introduced in 1842 when the fiscal basis of the United Kingdom accounts was revised, and direct taxation on income steadily superseded indirect taxes, especially those arising from import duties. The era of free trade had arrived, but state income and finance had to be catered for in different ways. It is possible to construct some meaningful temporal and spatial patterns from the Income Tax returns, and though there are problems in using them, they were recognised centrally as an indication of rental trends. *The Royal Commission on Land in Wales and Monmouthshire*, for example, was in no doubt about this relationship: 'After careful consideration, we have ourselves come to the conclusion that the agricultural rental is most accurately represented by the returns of the gross assessments under the heading "Lands" under Schedule A, and it will be these returns that we shall therefore use as our data in the inquiry.'[4] One of the major problems in using Schedule A in an annual index is that the land values on which the tax was assessed, and which we assume more or less reflected rent levels, were, like most valuation systems, from the late seventeenth-century land tax, through the parish and property rates of the early nineteenth century, and into the modern systems of the community charge. That is, the process of revaluation was not regular. In the case of the

[3] In fact 1849–50, 1850–51 . . . and so on to . . . 1913–14, taken from J. C. Stamp, *British Incomes and Property* (London, 1927), p. 49. There is a slightly different rendition of the same material for the period 1850–1904 in Thompson, 'An inquiry', in Minchinton, *Essays*, p. 84. It is not clear why Thompson differs from Stamp but when the two sources are compared statistically they are as near identical as makes little difference. Schedule A was published on a parish basis in 1842 and 1860 according to D. B. Grigg, 'An index of regional change in English farming', *TIBG*, 1965, pp. 55–67, especially p. 58.

[4] See the discussion in M. E. Turner, J. V. Beckett and B. Afton, *Agricultural Rent in England, 1690–1914* (Cambridge, 1997), pp. 64–6. *The Royal Commission on Land in Wales and Monmouthshire*, BPP, 1896, XXXIV, Report, pp. 361–469, quoting p. 362. See also D. W. Howell, 'Welsh agriculture, 1815–1914', PhD thesis, University of London, pp. 343–4. See also Grigg, 'An index', p. 58.

Schedule A values the revaluation initially took place every three or four years, before settling at a revaluation every five years.

The second column of Table 39.1 is J. R. Bellerby's estimates of gross rent for England and Wales for the period from 1867 which purportedly took into account some of the problems arising from using and interpreting Schedule A.[5]

The third and fourth columns of Table 39.1 are composite rent indexes which have been derived from the pioneer work of R. J. Thompson. They arise from splicing together four separate indexes. The first was based on *c.* 72,000 acres from estates in Lincolnshire, Essex, Hereford and north Wales, and available for the period 1800–1900; the second from estates totalling over 120,000 acres in Lincolnshire, Hereford, Buckinghamshire, Bedfordshire, Cambridge, Essex and north Wales, for the period 1816–1900; the third from estates totalling *c.* 390,000–400,000 acres distributed throughout more than twenty-four English counties and one Welsh county, from 1872–1900; and the fourth from a single estate of 290,000 acres for 1900–14, though available up to 1933.[6] This last sample formed a large proportion of the land in the third sample. The differences between the two indexes derived from Thompson's data arise from a suggestion by J. T. Ward. He observed that the rents from the larger, post-1871 samples were higher than those from the smaller samples which pre-dated 1872. He suggested that this arose from the fact that the earlier samples were dominated more by larger farms than the later samples, which therefore produced smaller unit rents, akin to an economies of scale effect. For the period 1872–6 the difference in rents between the third sample and the weighted average of samples one and two was a factor of 1.24. Thus Ward suggested that a composite index should reflect these differences, and should include samples one and two weighted by a factor of 1.24 and then spliced to the two later samples.[7] Thus column three is the unweighted, and column four the weighted index.

---

[5] J. R. Bellerby, 'Gross and net farm rent in the United Kingdom, 1867–1938', *Proceedings of the Journal of the Agricultural Economics Society*, 10, 1954, pp. 356–62, especially pp. 357–8. For some of the problems see *Royal Commission on Agriculture* BPP, 1897, xv, Final Report, p. 26; Rhee, *The Rent of Agricultural Land*, p. 34. The data included land whether in cultivation or not, and also ornamental gardens, gardens over one acre, farm houses and buildings; they refer to gross rent payable and not to actual rents received after any abatements or remissions; the valuation was made every five years, formerly every three; and there are other, minor problems. See also Grigg, 'Index', p. 58.

[6] Full details of the construction, including the full tabulation of the four separate samples, can be found in Turner, Beckett and Afton, *Agricultural Rent*, pp. 125–9. For the originals see Thompson, 'An inquiry', in Minchinton, *Essays*, pp. 82–4, for the period up to 1900, and Thompson's evidence to the *Royal Commission on Tithe Rentcharge, Minutes of Evidence* (London, 1934), p. 44 and reprinted in Rhee, *The Rent of Agricultural Land*, pp. 34 and 42.

[7] Both indexes reproduced from Turner, Beckett and Afton, *Agricultural Rent*, pp. 127–8. See also J. T. Ward, 'A study of capital and rental values of agricultural land in England and Wales between 1858 and 1958', PhD thesis, University of London, 1960, pp. 67–82, esp. pp. 72–8.

Until recently this was the closest approximation we had to an index of national rents, and to all intents and purposes it served its purpose well. There is close harmony, in the trends at least, of the Schedule A returns and this rent index.[8] On the upturn to the 1870s there was a fairly smooth rise in gross values and also in the course of the rents to a degree, with rents mostly ahead of values. On closer inspection the three- or four-year revaluation in land values is apparent. On the downturn there was a smooth fall in values but greater fluctuations in the rent estimate. This may reflect the problems tenants faced in paying their rents: a saw-tooth effect appears in the rent profile, demonstrating the influences of rent arrears. (Figure 39.2 shows the profile of the changing land values against the profile of two new rent indexes, the construction of which is outlined below. A comparison with Thompson's index would look much the same.)

There are reasons to suspect that the printed rental data, given their provenance from mainly large landowners and landed estates, tended to be for farms and holdings which were larger than the average for the country as a whole. The average size of holdings at the time was of the order of 70 to 80 acres, but the estates in Thompson's survey represented holdings closer to 150–200 acres. Thus the average rents derived from Thompson probably understated the true average rents.[9] For 1872–3 it has been estimated that the average rent for agricultural land (without rough grazing) was 28s. (or 25s. with rough grazing) per acre, and this was for medium to large-sized farms. The average rent for all holdings was reckoned at 34.5s. without rough grazing, or 30.75s. with it.[10]

The famous contemporary land sales and rental estimate of the firm of land agents, Norton, Trist and Gilbert, which is based on their own records of land sales, is shown in Table 39.2.[11] Comparisons between all of these estimates, the Income Tax, Bellerby, Thompson, and Norton, Trist and Gilbert, have been made elsewhere.[12] Table 39.3 takes one

---

[8] See a similar illustration in Thompson, 'An inquiry', in Minchinton, *Essays*, pp. 68–70. As Thompson himself indicates (p. 67), these cannot be direct comparisons because although 'The income tax figures have been relied upon by various writers and Royal Commissions as affording an indication of the rise and fall of rents . . . they are subject to serious qualifications.' There is even more harmony in the movements of Schedule A and the weighted Thompson index.

[9] For an illustration of the change in unit rents between large and small farms, see Turner, Beckett and Afton, *Agricultural Rent*, pp. 55–6, 116–22. See also Britton, *An Inquiry into Agricultural Rents*, p. 26, and a table showing gross rents per acre for different sized estates, though for the years 1938, 1946. See also T. W. Beastall, 'The history of the Earl of Scarborough's estate, 1860–1900', MA thesis, University of Manchester, where on p. 99 there is a partial demonstration of the same phenomena.   [10] Rhee, *The Rent of Agricultural Land*, pp. 31–2 in a prefatory note by A. W. Ashby.

[11] Letter from the firm of Norton, Trist and Gilbert to *The Times*, 20 April 1889, p. 11 and reprinted as, 'A century of land values: England and Wales', *JRSS*, 54, 1891, pp. 128–31. See also G. H. Peters, S. T. Parsons and D. M. Patchett, 'A century of land values 1781–1880', *Oxford Agrarian Studies*, 11, 1982, pp. 93–107; A. Offer, 'Farm tenure and land values in England, c. 1750–1950', *EcHR*, 44, 1991, pp. 1–20, especially pp. 13–14.

[12] Turner, Beckett and Afton, *Agricultural Rent*, pp. 166–74.

aspect of the Norton, Trist and Gilbert material and picks out the sale price element. While this material is available from the 1780s, unfortunately it stops in 1880. However, in celebration of the centenary of the journal *The Estates Gazette* in 1958, J. T. Ward reconstructed the average price of land for 100 years from the 1850s on the basis of auctioneers' reports submitted to the Estates Exchange. The average annual schedule of these prices is included in Table 39.3, and Figure 39.1 compares the Ward and Norton sale-price data with the new rent index.[13]

Tables 39.4 and 39.5 have been constructed from the *RC on Agriculture* (1894–7), and show the kind of evidence available at the macro (estate) and micro (farm) level. Table 39.4 summarises the evidence from twenty-nine estates from twenty-two English and one Welsh county. They ranged from the 49,000 acres of the Earl of Ancaster's Lincolnshire and Rutland estate down to a little over 2,000 acres of the single Welsh estate in Flintshire. The average size of estate was 16,700 acres. The Commission reported the acreage of the estate let out, the rents agreed or due, and the rents received. This full range of data was not recorded for every year, thus the acreage for estates where rents were due does not match the acreage for rents received, but from 1872 onwards there are data common to rents due and rents received for up to 300,000, and at times well over 400,000 acres.[14] In contrast, Table 39.5 shows the level and movement of rents on a sample of individual farms.

Yet all of these indexes have now been superseded by a new (England only) rent index. This is based on an amalgamation of the Thompson material, material contained in otherwise unpublished theses, the evidence in the *RC on Agriculture*, individual published estate accounts, but most crucially on a nationwide project to extract otherwise unresearched estate material. All of these data have been drawn together, and a summary is contained in Table 39.6. It shows rent assessment and rent collection, that is, what was meant to be paid by tenants, and what they actually paid. The difference between the two can mainly be accounted for by rent arrears, and sometimes rent abatements. Table 39.6 therefore gives a schedule of rents agreed or due, and rents received. By inspection and implication it also shows that not all estate accounts, or indeed the worthy efforts of historians, always record both kinds of rent. However, in a large number of cases both 'rents' have been recorded, and therefore Table 39.6 also gives the two rent indexes where these are based on a common coincidence of material. This is not the place to elaborate further on the sources and methods of construction, though it should be

---

[13] Norton, Trist and Gilbert, 'A century of land values'; J. T. Ward, 'Farm sale prices over a hundred years', *The Estates Gazette*, Centenary Supplement, May 1958, pp. 47–9.

[14] Table 39.5 reconstructed from *RC on Agriculture*, 'Particulars of Expenditures and Outgoings on Certain Estates in Great Britain', BPP, 1896, XVI, pp. 6–33.

noted that the extent of England covered in these indexes, and demonstrated by the acreages involved, is substantial. Figure 39.2 shows the profiles of the rents due and rents received compared with the profile of land values, and Figure 39.3 is a measure of the difference between the two rent indexes, essentially the rent arrears.

This then is a national picture. To the extent that we can have some idea of the regional variations in rents we can call upon various contemporary estimates. Thus Table 39.7 and Figure 39.4 show Caird's incomplete county estimates of average agricultural rent levels, and Table 39.8 the complete estimates of county rents associated with the 1873 'Returns of Owners of Land'.[15] And we can systematise these regional data through time by looking at a sample of county assessments of Schedule A Income Tax. Table 39.9 shows this for English and Welsh counties for selected years.[16] To gauge the geographical variations in long-term change we have also calculated some intercensal changes. These corroborate the picture we obtain from the general literature of the period. From 1860–95, essentially the adjustment from the High Farming period into the Great Depression, the greatest reductions in land values occurred, generally speaking, in the arable east of England, and the smallest reductions or even increases, occurred in the pastoral west and in Wales, and this geographical variation is shown in Figure 39.5 along with the geographical variation in the subsequent recovery phase leading up to the First World War.[17]

[15] James Caird, *The Times*, 20 December 1851; *Returns of Owners of Land in England and Wales*, BPP, 1876, LXXX, pp. 3–21. And for the problems this produced in terms of rent reductions see, amongst a wider literature, J. H. Brown, 'Agriculture in Lincolnshire during the Great Depression, *c.* 1873–1896', PhD thesis, University of Manchester, 1978, where on p. 80 there is a list of nineteen, mainly large estates, showing rent reductions for the period 1871–98 of 20–50 per cent. See also D. B. Grigg, *The Agricultural Revolution in South Lincolnshire* (Cambridge, 1966), pp. 123–5; D. W. Howell, *Land and People in Nineteenth Century Wales* (London, 1977), p. 13.

[16] 1859, 1903 and 1912 taken from Stamp, *British Incomes*, pp. 54–5; 1860 from *Gross Annual Value of Property Assessed to the Property and Income Tax in England and Wales, for the Year Ended 5 April 1860*, BPP, 1860, XXXIX, part II, pp. 386–7; 1870 from Miscellanea, 'The Domesday Book of 1873', *JRSS*, 39, 1876, pp. 409–10; 1895 and 1896 from *Income Tax – Schedule A*, BPP, 1897, XXIV. In addition, for Wales and Monmouth for 1842, 1852, 1862, 1872, 1882 and 1892 see Howell, 'Welsh agriculture', p. 343. See also Grigg, 'Index', pp. 60–2 for maps of England and Wales based on county units of 'rent per acre' based on Schedule A Income Tax for the years 1815, 1860, 1878 and 1912, and pp. 58–9 for a discussion about the method of isolating strictly agricultural land from other lands and farm buildings for this purpose. As with Schedule A in general the rents in question are akin to rent assessed rather than rent paid, and therefore in the late nineteenth century do not take into account temporary rent reductions or abatements. For a county-based study see D. B. Grigg, 'Changing regional values during the agricultural revolution in south Lincolnshire', *TIBG*, 1962, pp. 91–103, and Grigg, *The Agricultural Revolution*, especially pp. 175–7.

[17] See also F. M. L. Thompson, 'An anatomy of English agriculture, 1870–1914', in B. A. Holderness and M. E. Turner (eds.), *Land, Labour and Agriculture, 1700–1920: Essays for Gordon Mingay* (London, 1991), pp. 211–40, especially pp. 225–6.

Table 39.1. *Schedule A Income Tax and rental estimates, England and Wales,*
1850–1914

Date	Gross assessed value to Schedule A Income Tax £000	Gross rent £million	Composite rent indexes shillings/acre	
1850	42,829		18.9	23.4
1851	42,790		18.4	22.8
1852	41,490		18.8	23.4
1853	41,449		18.8	23.3
1854	41,430		19.0	23.5
1855	41,597		19.3	23.9
1856	41,485		20.1	24.9
1857	41,545		20.9	25.9
1858	42,895		21.2	26.3
1859	42,912		21.1	26.2
1860	42,995		21.1	26.2
1861	43,036		21.5	26.6
1862	44,686		21.4	26.5
1863	44,663		21.5	26.7
1864	44,724		21.4	26.6
1865	46,462		21.4	26.5
1866	46,482		21.5	26.7
1867	46,556	42.7	21.8	27.1
1868	47,767	43.9	22.4	27.8
1869	47,799	44.6	22.6	28.1
1870	47,857	44.6	22.0	27.3
1871	49,011	45.1	22.4	27.8
1872	49,027	46.1	26.7	28.4
1873	49,035	46.9	27.0	28.8
1874	49,956	47.4	27.2	29.0
1875	50,272	47.8	27.3	29.1
1876	50,408	48.3	27.4	29.2
1877	52,016	48.6	27.9	29.7
1878	51,934	47.4	27.8	29.7
1879	51,870	45.6	24.5	26.0
1880	52,041	43.9	24.4	25.9
1881	51,847	44.3	24.0	25.5
1882	51,419	45.1	25.3	27.0
1883	48,659	45.5	25.4	27.0

(continued)

Table 39.1 (*cont.*)

Date	Gross assessed value to Schedule A Income Tax £000	Gross rent £million	Composite rent indexes shillings/acre	
1884	48,211	43.8	24.5	26.1
1885	47,864	42.1	22.0	23.3
1886	46,255	39.9	23.4	24.8
1887	45,635	39.5	20.6	21.8
1888	44,732	39.0	21.4	22.8
1889	42,534	39.4	21.7	23.1
1890	42,053	39.4	21.5	22.8
1891	41,635	39.1	21.4	22.7
1892	41,385	38.2	20.8	22.1
1893	41,059	37.2	20.0	21.3
1894	40,335	36.5	19.5	20.7
1895	39,942	35.8	19.4	20.6
1896	39,624	35.3	19.3	20.5
1897	39,045	34.8	19.0	20.2
1898	38,378	34.2	19.0	20.3
1899	37,526	33.9	18.8	20.0
1900	37,332	34.1	18.8	20.1
1901	37,160	34.2	20.7	20.7
1902	37,017	34.5	20.3	20.3
1903	36,836	34.0	20.5	20.5
1904	37,149	34.1	19.9	19.9
1905	36,896	34.0	20.4	20.4
1906	36,810	34.4	20.6	20.6
1907	36,715	34.5	20.6	20.6
1908	36,629	34.7	20.8	20.8
1909	36,584	34.8	20.8	20.8
1910	36,567	34.7	20.8	20.8
1911	37,044	34.8	20.8	20.8
1912	36,990	35.1	21.1	21.1
1913	37,013	35.3	21.3	21.3
1914	37,071	35.5	21.3	21.3

Note: For difference between the two composite rent indexes see the text.
Sources: J. C. Stamp, *British Incomes and Property* (London, 1927), p. 49.
J. R. Bellerby, 'Gross and net farm rent in the United Kingdom, 1867–1938',
Journal of the Proceedings of the Agricultural Economics Society, 10, 1954, pp. 357–8.
M. E. Turner, J. V. Beckett and B. Afton, *Agricultural Rent in England,
1690–1914* (Cambridge, 1997), pp. 127–8, and based on R. J. Thompson, 'An
inquiry into the rent of agricultural land in England and Wales during the
nineteenth century', *JRSS*, 60, 1907, pp. 587–625, reprinted in W. E.
Minchinton (ed.), *Essays in Agrarian History* (Newton Abbot, 1968), pp. 57–86.
Royal Commission on Tithe Rentcharge. Minutes of Evidence (London, 1934).

Table 39.2. *Land sales conducted by Norton, Trist and Gilbert, 1850–1880*

Date	Acres sold	Rental thereof (£s)	Rent per acre (shill)	Rent per acre (Index)	Price sold for (£s)	Price per acre (£s)	Years' purchase
1850	3,556	3,806	21.4	100.0	101,160	28.4	27
1851	6,172	7,204	23.3	109.1	232,300	37.6	32
1852	4,066	6,392	31.4	146.9	166,650	41.0	26
1853	13,225	17,542	26.5	123.9	485,000	36.7	28
1854	10,245	16,680	32.6	152.1	332,340	32.4	20
1855	12,753	19,285	30.2	141.3	388,850	30.5	20
1856	13,594	18,424	27.1	126.6	484,820	35.7	26
1857	15,105	20,339	26.9	125.8	580,740	38.4	29
1858	24,658	36,829	29.9	139.5	1,055,422	42.8	29
1859	7,890	13,365	33.9	158.3	357,860	45.4	27
1860	8,764	13,333	30.4	142.1	329,930	37.6	25
1861	11,768	11,098	18.9	88.1	337,655	28.7	30
1862	17,564	17,853	20.3	95.0	618,340	35.2	35
1863	16,291	18,958	23.3	108.7	609,620	37.4	32
1864	15,274	24,500	32.1	149.9	990,150	64.8	40
1865	10,946	17,320	31.6	147.8	431,035	39.4	25
1866	12,021	16,900	28.1	131.4	621,840	51.7	37
1867	14,333	17,526	24.5	114.2	474,000	33.1	27
1868	6,862	8,909	26.0	121.3	360,600	52.6	40
1869	8,314	9,000	21.7	101.1	330,945	39.8	37
1870	9,082	11,254	24.8	115.8	462,813	51.0	41
1871	7,599	11,982	31.5	147.3	480,480	63.2	40
1872	6,511	13,534	41.6	194.2	399,000	61.3	29
1873	16,509	24,335	29.5	137.7	922,940	55.9	38
1874	15,143	21,420	28.3	132.2	937,150	61.9	44
1875	12,412	18,130	29.2	136.5	553,660	44.6	31
1876	9,355	13,846	29.6	138.3	436,000	46.6	31
1877	11,755	19,110	32.5	151.9	676,650	57.6	35
1878	8,603	12,545	29.2	136.2	407,000	47.3	32
1879	9,623	13,710	28.5	133.1	352,500	36.6	26
1880	12,519	14,230	22.7	106.2	426,500	34.1	30

Note:
Rent per acre, price per acre and years' purchase calculated rather than
transcribed from the original. There are arithmetical errors in the original.
Source: 'A century of land values: England and Wales', *JRSS*, 54, 1891, pp.
128–31, reprinted from *The Times*, 20 April, 1889, p. 11.

Table 39.3. *Land prices, 1850–1914*
(£s *per acre, based on five-year averages*)

	Norton Series	Ward Series			Ward Series
1850	33.6		1881	38.0
1851	34.3		1882	38.0
1852	35.2		1883	35.0
1853	35.6		1884	33.0
1854	35.3		1885	31.0
1855	34.7		1886	31.0
1856	36.0		1887	27.0
1857	38.6		1888	26.0
1858	40.0		1889	27.0
1859	38.6	39.0	1890	25.0
1860	37.9	40.0	1891	24.0
1861	36.9	39.0	1892	21.0
1862	40.8	39.0	1893	20.0
1863	41.1	38.0	1894	19.0
1864	45.7	37.0	1895	19.0
1865	45.3	38.0	1896	19.0
1866	48.3	39.0	1897	20.0
1867	43.3	41.0	1898	20.0
1868	45.6	41.0	1899	20.0
1869	47.9	44.0	1900	20.0
1870	53.6	44.0	1901	20.0
1871	54.2	48.0	1902	20.0
1872	58.7	49.0	1903	20.0
1873	57.4	52.0	1904	20.0
1874	54.1	53.0	1905	21.0
1875	53.3	54.0	1906	20.0
1876	51.6	52.0	1907	21.0
1877	46.5	51.0	1908	21.0
1878	44.4	49.0	1909	22.0
1879	43.9	45.0	1910	22.0
1880	39.3	43.0	1911	24.0
			1912	24.0
			1913	23.0
			1914	23.0

Sources: 'A century of land values'; J. T. Ward, 'Farm sale prices over a hundred years', *The Estates Gazette*, Centenary Supplement, May 1958, pp. 47–9.

Table 39.4. *Royal Commission on Agriculture – Rent due and rent received,*
1842–92

Date	Rent and acres due			Rent and acres received		
	Acres 000s	Rent £000s	Rent/acre shillings/acre	Acres 000s	Rent £000s	Rent/acre shillings/acre
1842	85.5	110.5	25.9	128.7	148.6	23.1
1852	113.1	135.3	23.9	156.3	197.8	25.3
1863	118.1	186.5	31.6	161.7	240.0	29.7
1872	273.1	420.0	30.8	301.6	456.8	30.3
1873	289.1	446.1	30.9	317.6	484.7	30.5
1874	291.3	451.2	31.0	319.8	487.4	30.5
1875	291.7	454.0	31.1	320.2	489.3	30.6
1876	315.8	477.0	30.2	344.4	512.0	29.7
1877	316.3	485.0	30.7	345.3	523.6	30.3
1878	347.4	533.1	30.7	370.9	554.7	29.9
1879	341.6	524.8	30.7	370.7	477.1	25.7
1880	377.7	559.8	29.6	414.1	532.7	25.7
1881	437.6	612.5	28.0	479.6	584.6	24.4
1882	432.0	602.3	27.9	473.9	615.5	26.0
1883	431.7	598.8	27.7	488.2	649.0	26.6
1884	429.2	585.9	27.3	485.6	623.8	25.7
1885	425.8	577.4	27.1	482.0	565.5	23.5
1886	428.4	560.5	26.2	484.5	573.8	23.7
1887	426.8	551.1	25.8	483.0	535.9	22.2
1888	426.3	541.4	25.4	482.4	553.4	22.9
1889	432.2	544.3	25.2	488.1	579.7	23.8
1890	433.4	542.3	25.0	489.7	561.3	22.9
1891	433.4	545.4	25.2	489.7	569.1	23.2
1892	434.7	548.7	25.2	490.9	548.9	22.4

Source: Reconstructed from *Royal Commission on Agriculture: Particulars of Expenditures and Outgoings on Certain Estates in Great Britain and Farm Accounts,* BPP, 1896, XVI of 1896, pp. 6–33.

Table 39.5. *Rents paid on a sample of individual farms, 1868–95*
(all in shillings per acre)

County Acres	Camb 565	Camb 401	Camb 372/ 952[1]	Dors 840	Essex 950	Suff 260[2]	Suff 327/ 590[3]	Wilts 806/ 840[4]	Wilts 760	Yorks 837
1868								24.8		
1869								24.9		
1870								23.1		
1871								23.2		
1872								22.7		
1873								22.7		
1874			41.1					22.7		
1875	41.6	45.0	41.1					22.6		
1876	41.7	45.0	41.1	17.1				23.6		
1877	41.8	45.0	33.0	17.1		23.3		23.6		
1878	42.0	49.9	33.0	17.1		25.1		23.6		
1879	42.1	49.9	31.6	17.1		25.0		21.3		27.5
1880	42.1	49.9	35.3	17.1	24.6	22.9		19.0		
1881	41.9	44.9	28.2	16.7	24.9	23.6		21.1		
1882	42.0	37.4	35.5	16.1	24.1	19.4		19.8		
1883	38.1	40.5	31.0	15.6		21.4		19.8		
1884	38.1	40.5	29.1	16.5		20.2	28.4	20.1		
1885	38.1	40.5	38.8	15.9		18.7	28.1	19.9		
1886	38.1	40.5	38.8	12.0		18.8	29.5	18.6		
1887	38.1	20.2	29.3	12.1		18.5	22.2	17.0		
1888	37.9	44.7	27.8	12.1		17.0	21.9	17.3		
1889	38.0	46.9	27.9	10.3		14.8	24.1	13.3		
1890	38.1	40.5	27.9	9.8		14.2	23.3	13.0		
1891	38.1	40.5	27.9	9.7		14.4	22.9	12.9		
1892	38.1		28.0	9.8		14.2	21.1	12.9		
1893	38.1		28.0	8.5		14.8	17.5	12.6	15.2	
1894	38.1					13.0	17.8			
1895	37.9									

Notes:

1 372 acres in 1874, 445 acres 1875–82, 754 acres 1884–6, and 952 acres 1887–94.
2 The rent includes tithes.
3 327 acres in 1884–6, 590 acres 1887–94.
4 836 acres in 1868–9, 806 acres 1870–6, 840 acres 1877–88, and 827 acres 1889–93.

Source: RC on Agriculture: Particulars of Expenditure, pp. 74–9, 86, 91, 174–5, 186–95.

Table 39.6. *New indexes of rents due and rents received, 1850–1914*

A.	Total rent assessed £s	Total no. of acres acres	Rent/ acre shillings	Total rent received £s	Total no. of acres acres	Rent/ acre shillings
1850	414,032	415,319	19.9	461,345	443,588	20.8
1851	468,316	452,750	20.7	437,650	421,574	20.8
1852	496,809	465,247	21.4	467,731	436,619	21.4
1853	443,330	424,929	20.9	406,489	391,733	20.8
1854	452,880	426,469	21.2	439,087	404,772	21.7
1855	506,387	456,802	22.2	488,908	437,737	22.3
1856	516,366	460,281	22.4	561,867	487,673	23.0
1857	503,653	445,334	22.6	567,226	483,284	23.5
1858	510,438	447,418	22.8	567,521	477,484	23.8
1859	517,572	451,470	22.9	512,789	432,666	23.7
1860	545,436	459,174	23.8	550,503	455,744	24.2
1861	505,983	432,316	23.4	513,404	429,795	23.9
1862	582,581	471,022	24.7	651,600	516,687	25.2
1863	555,644	453,758	24.5	632,929	502,309	25.2
1864	528,965	419,561	25.2	536,550	434,511	24.7
1865	537,813	424,399	25.3	607,853	486,174	25.0
1866	545,190	421,102	25.9	635,629	495,049	25.7
1867	552,257	421,574	26.2	598,517	474,373	25.2
1868	595,458	446,073	26.7	657,001	502,655	26.1
1869	582,200	432,618	26.9	644,997	491,669	26.2
1870	548,914	391,318	28.1	596,306	452,125	26.4
1871	620,512	439,068	28.3	803,928	602,104	26.7
1872	818,919	572,555	28.6	978,257	709,404	27.6
1873	785,582	543,763	28.9	936,096	677,831	27.6
1874	781,253	550,393	28.4	846,121	621,887	27.2
1875	811,019	562,218	28.9	981,559	702,880	27.9
1876	812,684	565,093	28.8	1,006,580	722,211	27.9
1877	868,320	602,172	28.8	1,038,059	745,530	27.8
1878	912,055	630,083	29.0	1,063,966	750,490	28.4
1879	1,030,624	714,083	28.9	944,063	743,557	25.4
1880	1,224,851	869,752	28.2	1,161,415	898,350	25.9
1881	1,319,006	973,149	27.1	1,319,339	1,065,226	24.8
1882	1,335,397	971,462	27.5	1,419,264	1,110,230	25.6
1883	1,303,746	972,796	26.8	1,515,084	1,199,251	25.3
1884	1,267,233	984,509	25.7	1,448,014	1,187,433	24.4
1885	1,268,444	1,006,403	25.2	1,398,631	1,204,520	23.2
1886	1,235,458	1,005,130	24.6	1,365,882	1,212,744	22.5
1887	1,223,966	1,012,268	24.2	1,365,508	1,237,355	22.1
1888	1,222,397	1,050,118	23.3	1,401,268	1,279,531	21.9

(continued)

Table 39.6 (*cont.*)

	Total rent assessed £s	Total no. of acres acres	Rent/ acre shillings	Total rent received £s	Total no. of acres acres	Rent/ acre shillings
1889	1,192,429	1,012,391	23.6	1,399,194	1,252,895	22.3
1890	1,190,088	1,015,733	23.4	1,387,767	1,258,545	22.1
1891	1,200,070	1,031,882	23.3	1,412,658	1,274,966	22.2
1892	1,238,264	1,058,753	23.4	1,387,350	1,277,346	21.7
1893	556,069	489,934	22.7	677,796	672,904	20.1
1894	491,467	424,720	23.1	627,130	620,529	20.2
1895	483,247	422,718	22.9	613,072	608,865	20.1
1896	472,292	420,031	22.5	563,366	558,535	20.2
1897	435,693	403,590	21.6	576,452	571,246	20.2
1898	429,927	401,594	21.4	575,797	579,950	19.9
1899	422,249	395,679	21.3	575,895	573,603	20.1
1900	415,245	385,952	21.5	561,554	563,876	19.9
1901	302,070	288,618	20.9	701,305	710,528	19.7
1902	329,766	314,937	20.9	739,332	741,297	19.9
1903	315,742	304,581	20.7	713,753	726,441	19.7
1904	328,170	312,931	21.0	592,307	600,614	19.7
1905	327,565	312,738	20.9	590,395	587,501	20.1
1906	286,952	272,833	21.0	574,457	557,018	20.6
1907	314,442	300,179	21.0	597,634	587,862	20.3
1908	280,111	270,008	20.7	568,821	557,691	20.4
1909	286,010	273,029	21.0	566,497	552,672	20.5
1910	288,254	276,656	20.8	577,362	556,299	20.8
1911	292,257	276,905	21.1	717,198	697,148	20.6
1912	240,162	233,368	20.6	535,564	514,269	20.8
1913	238,491	230,706	20.7	536,588	513,207	20.9
1914	242,233	230,821	21.0	538,157	513,322	21.0

B. Where for each year the two rents come from the same acres

	Rents assessed £s	Rents received £s	Common acres acres	Assessed rent/acre shillings	Received rent/acre shillings	Putative arrears £s	Arrears as a % of assessed rents
1850	367,187	354,155	361,393	20.3	19.6	−13,032	3.5
1851	356,513	352,375	350,082	20.4	20.1	−4,137	1.2
1852	384,523	378,099	362,579	21.2	20.9	−6,424	1.7
1853	331,820	322,197	322,261	20.6	20.0	−9,623	2.9
1854	362,514	353,680	334,869	21.7	21.1	−8,834	2.4
1855	409,738	395,569	365,202	22.4	21.7	−14,169	3.5

(*continued*)

Table 39.6 (*cont.*)

B (*cont.*)

	Rents assessed £s	Rents received £s	Common acres acres	Assessed rent/acre shillings	Received rent/acre shillings	Putative arrears £s	Arrears as a % of assessed rents
1856	489,145	467,155	418,958	23.4	22.3	−21,990	4.5
1857	473,532	449,948	404,011	23.4	22.3	−23,584	5.0
1858	479,479	463,645	406,095	23.6	22.8	−15,834	3.3
1859	417,898	405,981	359,870	23.2	22.6	−11,917	2.9
1860	448,855	434,068	368,953	24.3	23.5	−14,787	3.3
1861	414,217	398,436	345,944	23.9	23.0	−15,781	3.8
1862	551,541	536,055	432,730	25.5	24.8	−15,486	2.8
1863	524,457	510,305	415,466	25.2	24.6	−14,153	2.7
1864	458,140	442,450	366,744	25.0	24.1	−15,690	3.4
1865	535,998	420,400	421,859	25.4	24.7	−15,598	2.9
1866	544,036	529,825	418,562	26.0	25.3	−14,211	2.6
1867	523,131	499,742	402,372	26.0	24.8	−23,389	4.5
1868	572,897	553,878	430,613	26.6	25.7	−19,020	3.3
1869	522,477	506,205	387,016	27.0	26.2	−16,272	3.1
1870	432,850	421,418	308,359	28.1	27.3	−11,433	2.6
1871	619,050	601,908	436,528	28.4	27.6	−17,142	2.8
1872	817,389	806,055	570,015	28.7	28.3	−11,334	1.4
1873	763,135	750,910	528,303	28.9	28.4	−12,225	1.6
1874	677,141	667,254	480,914	28.2	27.7	−9,887	1.5
1875	809,433	798,232	559,678	28.9	28.5	−11,201	1.4
1876	810,980	803,452	562,553	28.8	28.6	−7,528	0.9
1877	835,144	816,124	578,342	28.9	28.2	−19,021	2.3
1878	867,384	853,750	597,981	29.0	28.6	−13,634	1.6
1879	915,929	837,627	645,356	28.4	26.0	−78,302	8.5
1880	1,129,803	1,039,859	798,532	28.3	26.0	−89,944	8.0
1881	1,317,190	1,214,564	970,609	27.1	25.0	−102,626	7.8
1882	1,333,674	1,224,918	968,922	27.5	25.3	−108,756	8.2
1883	1,302,078	1,233,325	970,256	26.8	25.4	−68,753	5.3
1884	1,254,135	1,177,382	969,832	25.9	24.3	−76,753	6.1
1885	1,252,507	1,167,873	988,863	25.3	23.6	−84,634	6.8
1886	1,233,623	1,126,778	1,002,590	24.6	22.5	−106,846	8.7
1887	1,221,890	1,126,137	1,009,728	24.2	22.3	−95,753	7.8
1888	1,220,414	1,155,851	1,047,578	23.3	22.1	−64,563	5.3
1889	1,190,776	1,130,583	1,009,851	23.6	22.4	−60,193	5.1
1890	1,188,638	1,130,197	1,013,193	23.5	22.3	−58,441	4.9
1891	1,198,614	1,153,379	1,029,342	23.3	22.4	−45,235	3.8
1892	1,190,983	1,124,012	1,018,654	23.4	22.1	−66,971	5.6

(*continued*)

Table 39.6 (*cont.*)

B (*cont.*)

	Rents assessed £s	Rents received £s	Common acres acres	Assessed rent/acre shillings	Received rent/acre shillings	Putative arrears £s	Arrears as a % of assessed rents
1893	506,439	465,333	440,500	23.0	21.1	−41,107	8.1
1894	490,089	452,666	422,180	23.2	21.4	−37,423	7.6
1895	483,247	451,082	422,718	22.9	21.3	−32,166	6.7
1896	472,292	446,110	420,031	22.5	21.2	−26,182	5.5
1897	435,693	428,280	403,590	21.6	21.2	−7,414	1.7
1898	429,927	420,972	401,594	21.4	21.0	−8,955	2.1
1899	422,249	416,785	395,679	21.3	21.1	−5,464	1.3
1900	415,245	400,693	385,952	21.5	20.8	−14,552	3.5
1901	254,280	244,565	243,725	20.9	20.1	−9,715	3.8
1902	281,833	281,839	270,044	20.9	20.9	6	−0.0
1903	267,970	258,953	259,688	20.6	19.9	−9,017	3.4
1904	280,349	270,935	268,038	20.9	20.2	−9,415	3.4
1905	264,903	261,899	254,925	20.8	20.5	−3,004	1.1
1906	239,383	237,844	227,940	21.0	20.9	−1,540	0.6
1907	267,050	262,612	255,286	20.9	20.6	−4,438	1.7
1908	232,673	233,117	225,115	20.7	20.7	443	−0.2
1909	238,517	232,357	228,136	20.9	20.4	−6,159	2.6
1910	240,846	242,750	231,763	20.8	20.9	1,903	−0.8
1911	244,700	246,291	232,012	21.1	21.2	1,591	−0.7
1912	192,505	193,384	188,475	20.4	20.5	879	−0.5
1913	190,654	189,614	185,812	20.5	20.4	−1,040	0.5
1914	193,962	191,505	185,929	20.9	20.6	−2,457	1.3

Source: Turner, Beckett and Afton, *Agricultural Rent*, pp. 312–13, 317–18, 322–23.

Table 39.7. *James Caird's rent estimates, 1851*
(Average rent per cultivated acre in shillings)

County	Rent	County type	County	Rent	County type
Beds	25.0	c	Norf	25.5	c
Berks	30.0	c	N'ton	30.0	c/g
Bucks	26.0	c/g	Northumb	20.0	c
Camb	28.0	c	Notts	32.5	c/g
Chesh	30.0	c/g	Oxon	30.0	c/g
Corn	n.a		Rutland	n.a	
Cumb	25.0	c/g	Salop	n.a	
Derby	26.0	c/g	Som	n.a	
Devon	30.0	c/g	Staff	30.0	c/g
Dorset	20.0	c	Suff	24.0	c
Durham	17.0	c	Surr	18.5	c
Essex	26.0	c	Suss	19.0	c
Glos	28.0	c/g	Warw	32.5	c/g
Hants	25.0	c	Westmor	n.a	
Heref	n.a		Wilts N	35.0	c/g
Herts	22.5	c	Wilts S	17.5	c
Hunts	26.5	c	Worcs	n.a	
Kent	n.a		Yorks E	22.5	c
Lancs	42.0	c/g	Yorks N	29.0	c
Leics	35.0	c/g	Yorks W	40.0	c/g
Lincs	30.0	c			
Middx	n.a		Wales	n.a.	
Mon	n.a				

Notes:
n.a. Data not available
c/g Defined by Caird as mixed corn and grass
c   Defined by Caird as chiefly corn producing
Average of rent of cultivated land in all counties   27.2
Average of rent of cultivated land in c/g counties   31.4
Average of rent of cultivated land in c counties   23.7
Source: James Caird, *The Times*, 20 December 1851.

Table 39.8. *Summary of 'Return of owners of land, 1873' in England and Wales (by County exclusive of the Metropolis and excluding land <1 acre)*

County	Acres 000s	Estimated rental £000	Rent/ acre shillings	County	Acres 000s	Estimated rental 000	Rent/ acre shillings
Beds	278	583	42.0	Staffs	634	2,656	83.8
Berks	430	900	41.9	Suff	908	1,429	31.5
Bucks	455	932	41.0	Surr	396	1,310	66.2
Camb	521	1,058	40.6	Suss	865	1,331	30.8
Chesh	597	2,011	67.4	Warw	535	1,509	56.4
Corn	758	1,072	28.3	Westmor	335	399	23.8
Cumb	729	902	24.7	Wilts	827	1,404	34.0
Derby	620	1,389	44.8	Worcs	436	1,241	56.9
Devon	1,514	2,267	29.9	Yorks E	702	1,279	36.4
Dorset	562	837	29.8	Yorks N	1,015	1,543	30.4
Durham	513	1,910	74.5	Yorks W	1,519	5,027	66.2
Essex	946	1,849	39.1				
Glos	728	1,511	41.5	ENGLAND	29,015	65,844	45.4
Hants	879	1,535	34.9				
Heref	506	811	32.1	Anglesey	162	151	18.6
Herts	383	914	47.7	Brecknock	302	198	13.1
Hunts	225	414	36.8	Cardigan	391	174	8.9
Kent	943	2,320	49.2	Carmarthen	509	376	14.8
Lancs	932	7,340	157.5	Caernarvon	301	297	19.7
Leics	517	1,079	41.7	Denbigh	348	400	23.0
Lincs	1,603	2,880	35.9	Flintshire	142	338	47.6
Middx	136	991	145.7	Glamorgan	428	1,417	66.2
Mon	288	710	49.3	Merioneth	302	165	10.9
Norf	1,233	1,976	32.1	Montgomery	380	352	18.5
N'ton	590	1,349	45.7	Pembroke	356	367	20.6
Northumb	1,189	1,585	26.7	Radnor	207	145	14.0
Notts	506	1,106	43.7				
Oxon	448	879	39.2	WALES	3,828	4,380	22.9
Rutland	93	174	37.4				
Salop	787	1,342	34.1	ENGLAND+			
Som	935	2,090	44.7	WALES	32,843	70,224	42.8

(continued)

Table 39.8 (*cont.*)

Summary by size groups

Acreage size groups	No. of owners	Extent of land 000s acres	Gross rent (estimated) £000s	Rent/acre shillings
<1	703,289	151.2	29,127.7	3,852.9
1–<10	121,983	478.7	6,438.3	269.0
10–<50	72,640	1,750.1	6,509.3	74.4
50–<100	25,839	1,791.6	4,302.0	48.0
100–<500	32,317	6,827.3	13,680.8	40.1
500–<1,000	4,799	3,317.7	6,427.6	38.7
1,000–<2,000	2,719	3,799.3	7,914.4	41.7
2,000–<5,000	1,815	5,529.2	9,579.3	34.6
5,000–<10,000	581	3,974.7	5,522.6	27.8
10,000–<20,000	223	3,098.7	4,337.0	28.0
20,000–<50,000	66	1,917.1	2,331.3	24.3
50,000–<100,000	3	194.9	188.7	19.4
>100,000	1	181.6	161.9	17.8
No. areas	6,448		2,831.5	
No. rentals	113	1.4		
Total	972,836	33,013.5	99,352.4	60.2
Total (less land of <1 acre and the last 2 categories)				
	262,986	32,860.9	67,393.2	41.0

Source: Collated from *Summary of the Return of Owners of Land in England and Wales*, BPP, 1876, LXXX, pp. 3–21.

Table 39.9A. *Gross assessments to Schedule A Income Tax by county,*
1859–1912 (in £000s)

County	1859	1860	1870	1895	1896	1903	1912
Beds	416.0	415.7	389.2	380.5	381.1	335.0	332.0
Berks	612.0	612.0	677.5	484.2	477.8	429.0	436.0
Bucks	656.0	655.4	535.3	605.5	602.4	553.0	547.0
Camb	874.0	873.8	792.5	739.9	733.1	669.0	721.0
Chesh	1,043.0	1,043.2	960.6	1,123.1	1,120.8	1,130.0	1,113.0
Cornwall	721.0	720.7	766.5	889.1	887.5	843.0	853.0
Cumb	684.0	683.9	636.5	740.5	731.8	681.0	680.0
Derby	842.0	842.1	655.6	836.1	829.1	793.0	761.0
Devon	1,581.0	1,580.9	1,485.0	1,719.1	1,712.1	1,597.0	1.625.0
Dorset	703.0	703.4	612.1	643.5	637.3	566.0	570.0
Durham	578.0	577.8	651.8	568.6	566.1	563.0	564.0
Essex	1,441.0	1,440.7	1,251.4	1,054.4	1,045.3	889.0	898.0
Glocs	1,171.0	1,170.9	964.7	975.9	963.5	876.0	853.0
Hants	972.0	972.0	919.2	844.3	832.1	765.0	763.0
Heref	679.0	678.6	535.0	637.5	635.5	594.0	600.0
Hertford	552.0	551.5	540.1	479.0	476.7	428.0	427.0
Hunts	327.0	326.7	281.5	255.3	252.8	234.0	244.0
Kent	1,477.0	1,476.0	1,330.7	1,371.3	1,343.4	1,180.0	1,112.0
Lancs	1,606.0	1,604.3	1,589.1	1,649.4	1,644.7	1,567.0	1,514.0
Leics	933.0	933.1	856.5	793.3	785.6	730.0	735.0
Lincs	2,641.0	2,640.1	2,207.6	2,212.7	2,186.8	2,104.0	2,141.0
Middx	386.0	342.8	326.0	355.6	355.7	307.0	286.0
Mon	332.0	331.7	316.7	310.2	307.4	303.0	302.0
Norf	1,832.0	1,832.0	1,591.9	1,431.9	1,402.8	1,229.0	1,248.0
N'ton	1,019.0	1,018.0	920.0	819.2	810.5	727.0	730.0
Northumb	907.0	906.5	922.0	879.1	874.0	873.0	870.0
Notts	779.0	778.2	807.1	643.8	640.2	598.0	598.0
Oxford	653.3	653.3	654.7	537.4	530.5	477.0	482.0
Rutland	142.0	141.7	159.7	121.6	119.9	112.0	106.0
Salop	1,087.0	1,087.4	1,128.5	1,046.1	1,043.1	1,009.0	1,010.0
Som	1,755.0	1,755.1	1,663.4	1,664.8	1,658.3	1,514.0	1,549.0
Staff	1,087.0	1,086.6	1,084.1	1,052.2	1,044.3	1,034.0	1,013.0
Suff	1,265.0	1,264.7	1,142.2	898.9	879.4	744.0	743.0
Surr	491.0	489.3	395.6	504.7	503.6	464.0	452.0
Suss	918.0	917.6	808.3	914.9	906.2	826.0	810.0
Warw	946.0	945.2	869.7	755.1	745.7	712.0	707.0
Westmor	266.0	266.4	271.7	315.8	313.7	296.0	293.0
Wilts	1,098.0	1,097.8	975.4	820.7	805.9	739.0	760.0
Worc	756.0	756.2	691.3	678.0	670.5	639.0	641.0
Yorks	4,128.0	4,126.2	3,836.3	3,799.7	3,787.0	3,697.0	3,620.0
Metropolitan	0.0	0.0	118.8	84.1	83.3	76.0	68.0

(continued)

Table 39.9A *(cont.)*

County	1859	1860	1870	1895	1896	1903	1912
Anglesey	151.0	150.7	166.5	183.8	183.5	191.0	193.0
Brecknock	160.0	160.1	187.4	190.4	190.1	190.0	189.0
Cardigan	173.0	173.0	198.9	234.3	234.3	220.0	224.0
Carmarthen	334.0	333.7	371.1	426.1	426.1	426.0	427.0
Caernarvon	175.0	175.2	187.7	210.3	209.8	215.0	213.0
Denbigh	327.0	326.9	330.7	323.4	322.5	318.0	318.0
Flint	206.0	205.6	209.9	203.4	202.9	205.0	209.0
Glamorgan	272.0	272.2	293.9	317.5	316.6	316.0	330.0
Merioneth	120.0	119.6	126.7	134.6	134.4	132.0	130.0
Montgomery	287.0	287.2	310.2	315.6	314.4	306.0	302.0
Pembroke	301.0	300.8	330.9	344.9	344.5	347.0	352.0
Radnor	136.0	135.6	158.4	159.1	159.1	153.0	153.0

Table 39.9B. *England and Wales*

	1859	1860	1870	1895	1896	1903	1912
ENGLAND	40,356.0	40,299.6	37,322.2	36,636.9	36,327.7	33,902.0	33,777.0
WALES	2,642.0	2,640.6	2,872.0	3,043.4	3,038.1	3,019.0	3,040.0
ENGLAND and WALES	42,998.0	42,940.2	40,194.5	39,680.3	39,365.8	36,921.0	36,817.0
England % of total	93.9	93.9	92.9	92.3	92.3	91.8	91.7
Wales % of total	6.1	6.1	7.1	7.7	7.7	8.2	8.3

Table 39.9C. *Percentage changes*

	1860–95	1895–1912	1860–1912		1860–95	1895–1912	1860–1912
Beds	−8.5	−12.7	−20.1	Salop	−3.8	−3.5	−7.1
Berks	−20.9	−9.9	−28.8	Som	−5.1	−7.0	−11.7
Bucks	−7.6	−9.7	−16.5	Staff	−3.2	−3.7	−6.8
Camb	−15.3	−2.6	−17.5	Suff	−28.9	−17.3	−41.3
Ches	7.7	−0.9	6.7	Surr	3.1	−10.5	−7.6
Cornwall	23.4	−4.1	18.4	Suss	−0.3	−11.5	−11.7
Cumb	8.3	−8.2	−0.6	Warw	−20.1	−6.4	−25.2
Derby	−0.7	−9.0	−9.6	Westmor	18.5	−7.2	10.0
Devon	8.7	−5.5	2.8	Wilts	−25.2	−7.4	−30.8
Dorset	−8.5	−11.4	−19.0	Worc	−10.3	−5.5	−15.2
Durham	−1.6	−0.8	−2.4	Yorks	−7.9	−4.7	−12.3
Essex	−26.8	−14.8	−37.7	Metropolitan		−19.1	
Glos	−16.7	−12.6	−27.2				
Hants	−13.1	−9.6	−21.5				
Heref	−6.1	−5.9	−11.6				
Hertford	−13.1	−10.9	−22.6	Anglesey	22.0	5.0	28.1
Hunts	−21.9	−4.4	−25.3	Brecknock	18.9	−0.7	18.1
Kent	−7.1	−18.9	−24.7	Cardigan	35.5	−4.4	29.5
Lancs	2.8	−8.2	−5.6	Carmarthen	27.7	0.2	28.0
Leics	−15.0	−7.3	−21.2	Caernarvon	20.0	1.3	21.6
Lincs	−16.2	−3.2	−18.9	Denbigh	−1.1	−1.7	−2.7
Middx	3.7	−19.6	−16.6	Flint	−1.0	2.7	1.7
Mon	−6.5	−2.6	−8.9	Glamorgan	16.6	3.9	21.2
Norf	−21.8	−12.8	−31.9	Merioneth	12.6	−3.4	8.7
N'ton	−19.5	−10.9	−28.3	Montgomery	9.9	−4.3	5.2
Northum	−3.0	−1.0	−4.0	Pembroke	14.6	2.1	17.0
Notts	−17.3	−7.1	−23.2	Radnor	17.3	−3.9	12.8
Oxford	−17.7	−10.3	−26.2				
Rutland	−14.2	−12.8	−25.2				
ENGLAND	−9.1	−7.8	−16.2	WALES	15.3	−0.1	15.1
ENGLAND and WALES	−7.6	−7.2	−14.3				

Notes:
Metropolitan London not separately distinguished in 1859/60 from other lands in Middlesex, Surrey and Kent.
1870 for Wales is in fact 1873
Source: See Note 16.

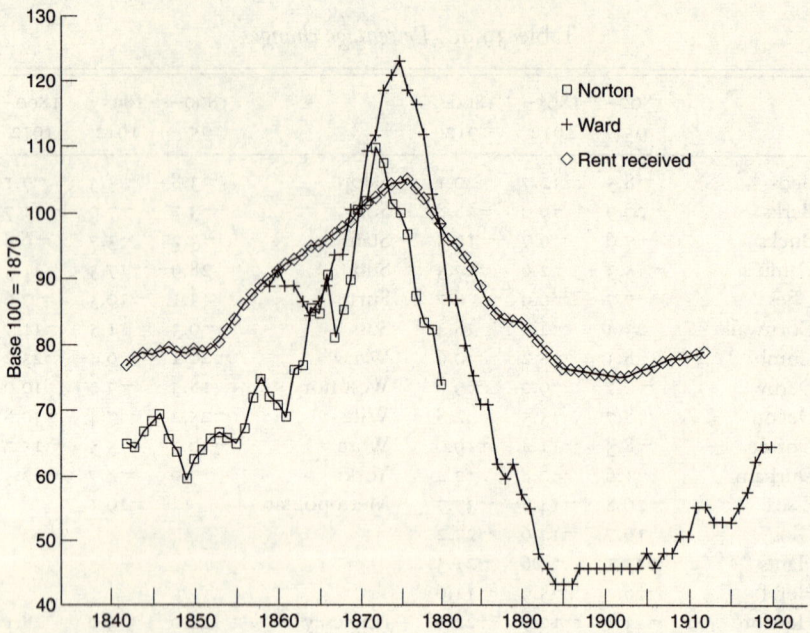

39.1 Land prices index and a rent index, 1840–1920

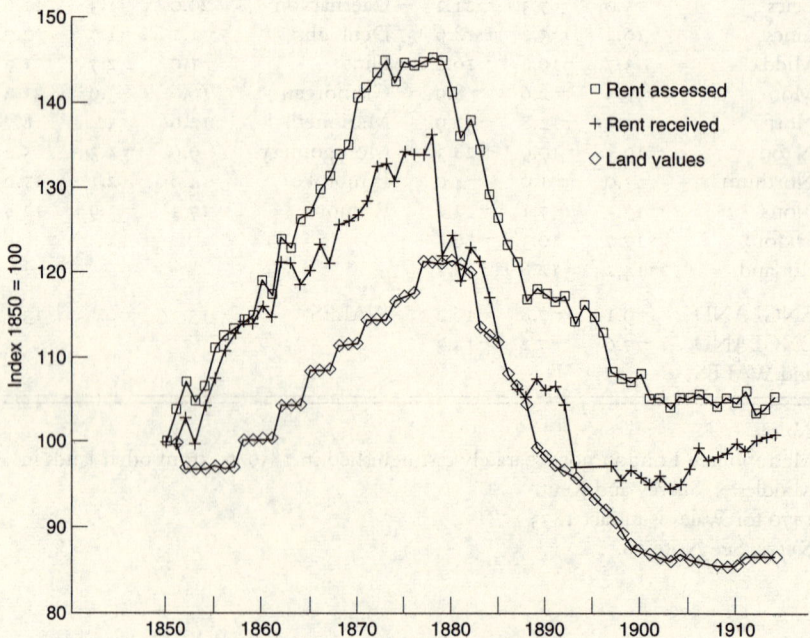

39.2 Index of English rent and land values, 1850–1914

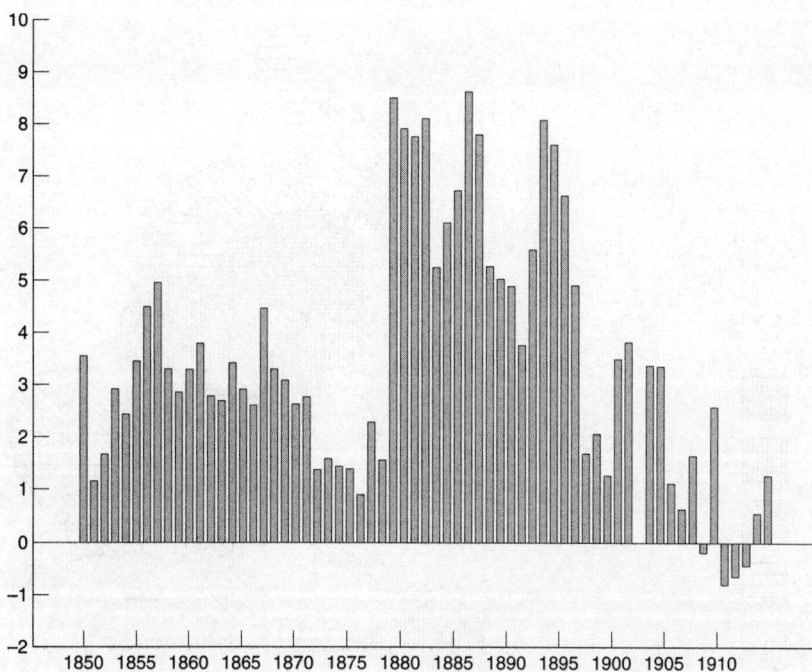

39.3 Rent arrears, 1850–1914, as a percentage of rent assessed

SHILLINGS

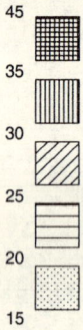

45

35

30

25

20

15

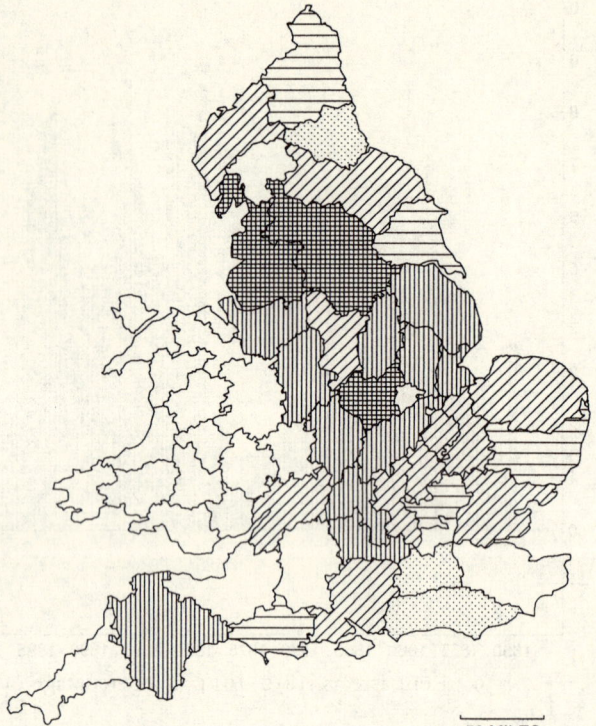

50 MILES

39.4 Rent per acre, 1850–1

PERCENTAGE CHANGES BASED ON SCHEDULE A INCOME TAX

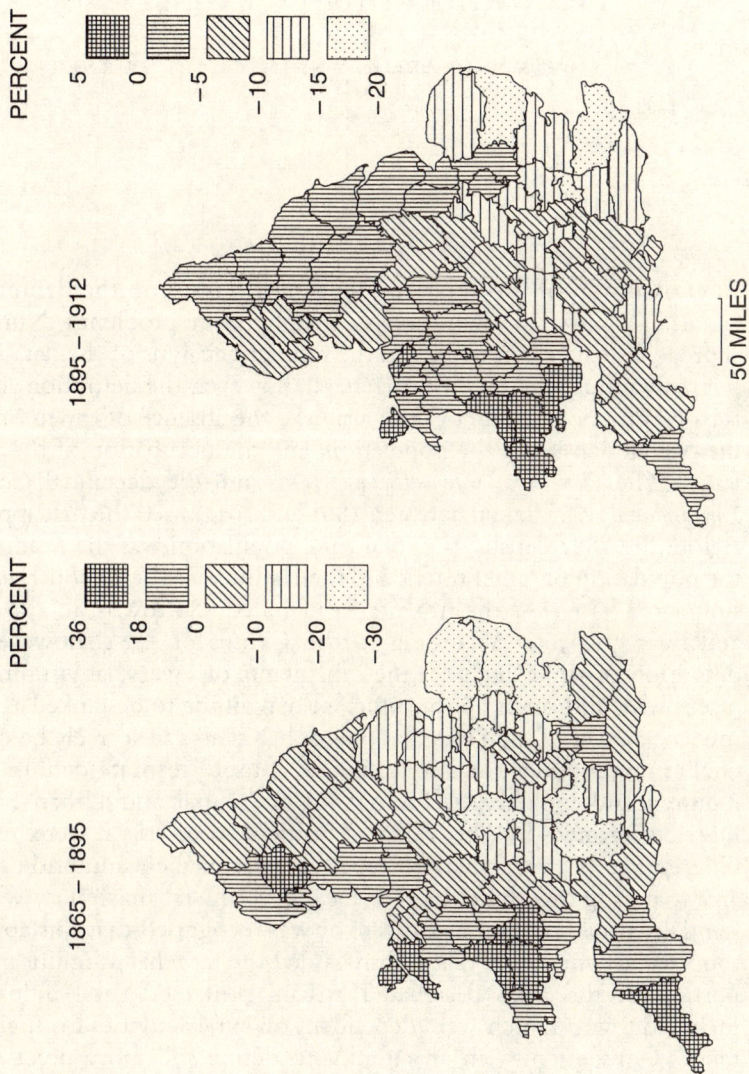

1865 – 1895

PERCENT

36
18
0
−10
−20
−30

1895 – 1912

PERCENT

5
0
−5
−10
−15
−20

50 MILES

39.5 Changes in land values: percentage changes based on Schedule A income tax, 1865–95, 1895–1912

CHAPTER 40

THE AGRARIAN POPULATION

BY BETHANIE AFTON AND MICHAEL TURNER

A. INTRODUCTION

Any discussion of the agrarian population centres on the definition of 'agrarian', and that definition is fraught with problems. Numerous approaches have been used, and the resultant analysis of the data is considerably influenced by this. In general, however, the definition adopted is based on one or more of three criteria: the absence of towns or cities; the overall density of the population; and the occupation of the inhabitants. The *Census of Great Britain, 1851*, and the decennial *Census of England and Wales*, taken between 1861 and 1911, used the first approach. Before the 1881 census, the total rural population was the residue after the population of 'chief towns and their immediate neighbourhood' was subtracted.[1] In 1851 the urban area included 580 towns, in 1861, 781, and by 1871, 966. According to the Census of 1881, however, this definition of rural 'includes the inhabitants of a very large number of places which, though not of sufficient magnitude to be ranked as "chief towns", are yet of such size that their inhabitants can scarcely be considered as living under rural conditions'.[2] A more accurate definition was required and looked for. From 1881 the Rural and Urban Sanitary Districts created by the Public Health Act of 1872 were used to differentiate between the two types of area on the assumption that an area requiring the higher level of local government machinery was substantially urban in nature. By 1891 it was recognised that this approach too was inadequate, in that it disregarded the fact that a significant proportion of the areas designated 'urban' contained rural industry or included towns which were dependent on rural activities for their existence. Consequently, in an attempt to define still more precisely the remoteness from rural influence, urban districts, though still classified as 'urban', were further divided into three groups: those above 10,000 in population; those with between 5,000 and 10,000; and those under

---

[1] *Census of England and Wales, 1881. General Report*, BPP, 1883, LXXX, p. 8.   [2] *Ibid.*, p. 9.

5,000. In 1911 out of a total of 1,137 urban districts, 661, or 58 per cent were below 10,000.[3]

Others dealing with demographic questions sought a more accurate division and modifed the definition.[4] T. W. Welton excluded populous areas of rural districts where manufacturing, mining and other industrial pursuits occupied a large number of people. G. B. Longstaff included only those districts which were declining in population so as not to include rural districts which were becoming residential. J. Saville, when investigating rural decline, reasoned that, because at the end of the nineteenth century well over half (61 per cent) of the parishes of England and Wales had a population of under 500, it should be at that level that such a study should begin. While each author made choices which were useful for their own studies, none of the definitions gave sufficient information to formulate a comprehensive differentiation between agrarian and urban.[5]

Using the overall population density of an area – a second means of distinguishing agrarian from urban population – has the advantage of adding the dimension of space to that of simple numbers. The censuses did include information on density, but this was not used to help define 'rural'. However, A. L. Bowley made a detailed study of agrarian population in England and Wales which did consider density. His definition included rural districts which in 1891 had a maximum density of 0.3 persons to the acre. He then added to these districts any area, investigated at parish level, with a density of between 0.3 and 0.5 which did not include concentrations of industry, manufacturing, mining, suburban residential areas or military camps. Because it did not redefine the area considered rural over time, it is likely that the agrarian population in the earlier periods is underestimated. However, this could be calculated using the detailed printed information left by Bowley. This work by Bowley was one of the most useful and complete contemporary attempts to measure the population of rural England and Wales.[6]

An approach integrating density with other variables is often used by modern researchers. C. M. Law integrated size, density and the degree of nucleation in his study which, though concerned with urban growth,

[3] *Census of England and Wales, 1911. General Report*, BPP, 1917–18, xxxv, p. 36.

[4] For a discussion of nineteenth-century writings on population and migration, see J. Saville, 'Internal migration in England and Wales during the past hundred years', in J. Sutter (ed.), *Human Displacements: Measurement, Methodology, Aspects* (Monaco, 1962), pp. 1–21.

[5] T. A. Welton, 'On the distribution of population in England and Wales, and its progress in the period of ninety years from 1801 to 1891', *JRSS*, 63, 1900, pp. 527–28; G. B. Longstaff, 'Rural depopulation', *JRSS*, 56, 1893, pp. 383–4; Saville, 'Internal migration', p. 5.

[6] A. L. Bowley, 'Rural population in England and Wales: a study of the changes in density, occupation, and ages', *JRSS*, 77, 1913–14, pp. 597–8.

also defined rural population.[7] R. Lawton has mapped the density in each registration district from the 1851 and 1911 censuses without deducting urban concentrations.[8] He also used changes in population density to map migration during the nineteenth century.

The third criterion sometimes used to define rural population is occupation. W. Ogle selected counties of England in which 10 per cent or more of the population was employed in agriculture. He then subtracted towns of 10,000 or more, and considered the remaining numbers as a fair representation of the rural population.[9] The problem with such an approach is that 'rural' and 'agricultural' are not synonymous terms or concepts. Using Bowley's corrected workings of the 1911 census, 20 per cent of those employed in agriculture, excluding domestic gardeners and gamekeepers, were enumerated as residing in urban districts.[10] The divergence grew as specialist agricultural enterprises, most notably milk dairies and market gardens, were attracted by marketing conditions to the peripheries of towns and cities. These were often labour-intensive enterprises, and, as such, their inclusion inflated the numbers employed in agriculture who were located around urban areas. Again in 1911, 60 per cent of those returned as gardeners, seedsmen and florists, but excluding domestic gardeners, were located in urban districts.[11]

Other occupations were common to both town and country. The change in numbers of those employed in rural crafts and industries, as well as those who provided services for rural society, was masked in the census by urban and suburban employment which was identically described even when the tasks were performed by fundamentally different techniques in different types of establishment. Thus, any attempt to measure the agrarian nature of an area based on employment would necessitate constant decisions, often subjective, concerning the degree to which an occupation was associated with rural life. While Ogle's calculations, so heavily based on employment in agriculture, are problematical, those of Bowley and Welton which also depend somewhat on occupation are not beyond criticism. Their definitions eliminate groups as 'urban' which today might well be controversial. Mining, for example, was regarded as urban employment, part of the force behind industrialisation, whereas a modern demographer might make a different choice. It becomes clear that to define the population which should be classified as rural requires careful consideration. Differing

---

[7] C. M. Law, 'The growth of urban populations in England and Wales, 1801–1911', *TIBG*, 41, 1967, p. 129.

[8] R. Lawton, 'Population and society 1730–1900', in R. A. Dodgshon and R. A. Butlin (eds.), *An Historical Geography of England and Wales* (London, 1978), pp. 313–66.

[9] W. Ogle, 'The alleged depopulation of the rural districts of England', *JRSS*, 52, 1889, pp. 208–9.

[10] Bowley, 'Rural populations', p. 60.   [11] *Ibid.*, p. 60.

definitions may be useful when asking different questions, but this must be recognised by anyone using the information. Saville, for example, made use of both a definition based on towns with under 500 people and of one based on the much broader distinction adopted by the decennial censuses. He warned that any definition adopted is critical, and 'only approximate'.[12]

The *Census of Great Britain, 1851*, and the decennial *Census of England and Wales*, 1861–1911 are the main data sources, and are considered reasonably accurate for most ordinary purposes regarding population.[13] The user must be aware, however, of alterations, particularly in the districts, which, though limited, did occur, and which would particularly affect local studies. The report which accompanies each census provides good information about such changes, as does the work of Saville. M. Drake provides a glossary of the terms involved and a useful summary of the contents of each census.[14] The registration of births and deaths, required by the Civil Registration Act of 1837, supplements the census. These registers are used in migration studies. Their limitations and accuracy are discussed by R. Lawton.[15]

B. THE DATA

The data presented in this section are reproduced in order to give the user a selection of statistics with which to work rather than to present a definitive answer as to the size of the agrarian population. Table 40.1 illustrates the wide variation between the different calculations of rural population which were discussed above. Figure 40.1 shows these figures as a percentage of the total population.

Table 40.2 reproduces the calculations of rural population by Bowley constructed in 1913–14, and Figure 40.2 illustrates his rural population as a percentage of the total in 1861 and 1911. One of the greatest values of Bowley's work is the information which he provided in several appendices. He listed, and defined, all parishes in rural districts which he excluded from his calculations. He then gave, by district, his revised rural population for 1861, 1901 and 1911. His calculations must, however, be used with some caution. For example, according to Bowley, Cardigan

---

[12] J. Saville, *Rural Depopulation in England and Wales, 1851–1951* (London, 1957), pp. 59–61.

[13] P. M. Tillott, 'Sources of inaccuracy in the 1851 and 1861 censuses', in E. A. Wrigley (ed.), *Nineteenth-century Society: Essays in the Use of Quantitative Methods for the Study of Social Data* (Cambridge, 1972), p. 83.

[14] Saville, *Rural Depopulation*, pp. 60–1; M. Drake, 'The census, 1801–1891', in Wrigley (ed.), *Nineteenth-century Society*, pp. 32–46.

[15] R. Lawton, 'Population changes in England and Wales in the later nineteenth century: an analysis of trends by registration districts', *TIBG*, 44, 1968, pp. 55–6.

contained a greater rural population than the entire population recorded in the official census.

Because density of population has been one of the important criteria on which to base calculations of agrarian population, Table 40.3 has been included. This reproduces the densities by county found in the official censuses. Figure 40.3 shows this information for the years 1851 and 1911.[16]

The use of occupational structure as a measure of agrarian population is illustrated in Table 40.4. In this table the population engaged in agriculture, arboriculture and horticulture (excluding wives of farmers and landed proprietors in 1851, and landed proprietors in 1871) is used as a surrogate for 'agrarian' occupations. Figure 40.4 maps this information. This proportion of agrarian occupations to total population was the starting point of Ogle's work. He then deleted counties with less than 10 per cent engaged in these occupations, and subtracted towns of 10,000 or more in each county. This he defined as the agrarian population. An important obstacle encountered with this method of calculating the agrarian population is the presence of agricultural and horticultural workers in essentially urban areas. Table 40.5 shows the estimates of these made by Bowley from the 1911 census.

[16] Lawton, 'Population and society', pp. 338–9.

Table 40.1. *Calculation of the rural population of England and Wales based on differing definitions (in thousands), 1851–1911*

	Census (1)	Welton (2)	Longstaff (3)	Bowley (4)	Law (5)	Ogle (6)
1851	8,937	5,919			8,240	2,381
1861	9,105	5,845		4,939	8,282	
1871	8,671	5,856	5,033	4,936	7,910	
1881	8,338	5,668	4,874	4,764	7,794	2,358
1891	8,107	5,534		4,625	7,402	
1901	7,469			4,454	7,156	
1911	7,908			4,581	7,603	

Definitions:

1 Residue of total population minus chief towns or Urban Districts
2 Rural areas minus mining, manufacturing and other industries
3 Rural Districts with declining populations
4 Based on population density, subtracting mining, manufacturing, industry, residential areas and military camps
5 Based on population size, density and degree of nucleation
6 Counties in which >10% of the population is agricultural minus large towns

Sources: 'General report', *Census of Great Britain, 1851*, BPP, 1852–3, LXXXVIII; 'General Report', *Census of Great Britain, 1861*, BPP, 1863, LIII; 'General report', *Census of England and Wales, 1871*, BPP, 1873, LXXI; 'General report', *Census of England and Wales, 1881*, BPP, 1883, LXXX; 'General report', *Census of England and Wales, 1891*, BPP, 1893, LV; 'General report', *Census of England and Wales, 1901*, BPP, 1904, CVIII; 'General report', *Census of England and Wales, 1911*, BPP, 1912–13, CXI; T. A. Welton, 'On the distribution of population in England and Wales', *JRSS*, 63, 1900, p. 560; G. B. Longstaff, 'Rural depopulation', *JRSS*, 56, 1893, p. 419; A. L. Bowley, 'Rural population in England and Wales', *JRSS*, 77, 1913–14, p. 606; C. M. Law, The growth of urban populations in England and Wales', *TIBG*, 41, 1967, p. 129; W. Ogle, 'The alleged depopulation of rural districts of England', *JRSS*, 52, 1889, p. 210.

Table 40.2. *The population of the registration counties of England and Wales after subtracting urban and industrial regions (thousands), 1861–1911*

	1861	1871	1881	1891	1901	1911
Bedfordshire	89	92	85	80	75	76
Berkshire	109	110	111	110	103	108
Buckinghamshire	80	81	74	75	72	75
Cambridgeshire	111	114	106	103	100	105
Cheshire	98	103	104	105	102	108
Cornwall	165	160	139	128	122	120
Cumberland	99	99	101	100	95	95
Derbyshire	56	55	56	55	54	56
Devon	257	249	233	226	212	214
Dorset	116	116	108	103	94	95
Durham	40	42	37	36	35	35
Essex	171	171	159	157	152	162
E. Riding	94	93	92	87	83	86
Gloucestershire	136	137	127	123	117	120
Hampshire	141	144	142	145	46	159
Herefordshire	85	85	79	74	71	69
Hertfordshire	80	81	77	76	72	75
Huntingdonshire	41	39	35	32	29	30
Kent	167	177	180	182	178	184
Lancashire	108	105	106	103	101	103
Leicestershire	81	78	78	75	73	76
Lincolnshire	268	168	258	240	227	233
Monmouthshire	48	50	44	43	41	43
Norfolk	273	265	260	256	245	249
Northamptonshire	129	124	121	114	108	109
Northumberland	99	94	91	85	84	84
Nottinghamshire	77	72	70	67	65	67
N. Riding	135	137	133	124	118	120
Oxfordshire	111	109	101	97	87	88
Rutland	21	20	20	19	17	16
Shropshire	152	154	151	145	142	145
Somerset	233	230	215	204	193	194
Staffordshire	102	107	110	111	114	118
Suffolk	219	214	203	196	183	185
Surrey	41	44	46	50	55	61
Sussex	168	178	183	185	183	197
Warwickshire	103	106	106	101	97	101
Westmorland	38	38	38	38	35	35
Wiltshire	156	151	141	136	128	129
Worcestershire	76	80	78	76	76	78
W. Riding	165	164	167	163	171	178

(continued)

Table 40.2 (*cont.*)

	1861	1871	1881	1891	1901	1911
Anglesey	25	22	21	20	19	19
Brecknock	37	35	34	32	30	31
Caernarvon	64	69	75	71	73	72
Cardigan	82	82	78	71	66	64
Carmarthen	46	44	45	45	40	44
Denbigh	39	37	36	33	32	31
Flint	29	32	33	29	31	33
Merioneth	36	38	39	37	36	35
Montgomery	53	53	50	44	40	39
Pembroke	55	52	49	47	45	44
Radnor	13	13	12	11	11	10
England and Wales English and Welsh rural regions	4,939	4,936	4,764	4,625	4,454	4,581
English and Welsh industrial regions★	15,127	17,776	21,210	24,378	28,074	31,489
Total	20,066	22,712	25,974	29,003	32,528	36,070

Note: ★ All Middlesex and Glamorgan included.
Source: Bowley, 'Rural population', pp. 605–6.

Table 40.3. *Population density of English and Welsh counties (persons per square mile), 1851, 1911*

1851		1911	
London	19,375	London	38,680
Lancashire	1,003	Middlesex	4,848
Middlesex (ex-met.)	546	Lancashire	2,554
Staffordshire	534	Glamorgan	1,383
Yorks, West Riding	508	Durham	1,350
Warwickshire	501	Surrey	1,172
Cheshire	391	Staffordshire	1,158
Worcestershire	381	Warwickshire	1,147
Gloucestershire	375	Yorks, West Riding	1,099
Durham	349	Cheshire	931
Nottinghamshire	314	Essex	883
Kent (ex-met.)	306	Monmouthshire	725
Derbyshire	299	Nottinghamshire	716
Somerset	289	Worcestershire	703
Leicestershire	283	Kent	686
Surrey (ex-met.)	272	Derbyshire	673
Bedfordshire	272	Isle of Wight	599
Glamorgan	268	Sussex, East	588
Monmouthshire	262	Gloucestershire	585
Hertfordshire	260	Hampshire	576
Cornwall	259	Leicestershire	572
Hampshire	243	Soke of Peterborough	535
Flint	235	Hertfordshire	492
Suffolk	231	Bedfordshire	411
Sussex	229	Berkshire	375
Buckinghamshire	228	Yorks, East Riding	369
Oxfordshire	227	Flintshire	364
Berkshire	226	Northumberland	345
Essex	224	Northamptonshire	332
Yorks, East Riding	223	Suffolk, East	318
Northamptonshire	216	Buckinghamshire	293
Cambridgeshire	215	Somerset	283
Devon	214	Sussex, West	281
Norfolk	213	Devon	268
Wiltshire	198	Oxfordshire	265
Anglesey	188	Cambridgeshire	261
Dorset	184	Lincolnshire, Lindsey	244
Huntingdonshire	178	Norfolk	243
Northumberland	154	Cornwall	242
Denbigh	153	Dorset	228
Caernarvon	151	Caernarvon	219

(continued)

Table 40.3 (*cont.*)

1851		1911	
Pembroke	149	Denbigh	217
Herefordshire	149	Wiltshire	212
Lincolnshire	147	Lincolnshire, Holland	197
Cumberland	125	Yorks, North Riding	197
Carmarthen	117	Suffolk, West	191
Cardigan	102	Isle of Ely	188
Yorks, North Riding	101	Anglesey	185
Montgomery	89	Shropshire	183
Salop	88	Cumberland	175
Brecknock	86	Carmarthen	174
Westmorland	77	Lincolnshire, Kesteven	153
Merioneth	65	Huntingdonshire	152
Radnor	58	Pembroke	146
Rutland	54	Herefordshire	136
		Rutland	134
		Cardigan	86
		Brecknock	81
		Westmorland	81
		Merioneth	69
		Montgomery	67
		Radnor	48

Note: For 1851, registration counties are used for England, and ancient counties for Wales; in 1911 administrative counties and County Boroughs are used.

Source: Census, 1851, Census, 1911.

Table 40.4. *Agrarian occupations of England and
Wales by county as a percentage of the total population
(percent), 1851–1911*

	1851	1871	1891	1911
Bedfordshire	19	14	11	8
Berkshire	21	15	10	6
Buckinghamshire	16	13	9	8
Cambridgeshire	20	18	16	8
Cheshire	9	6	4	3
Cornwall	12	10	8	9
Cumberland	15	10	7	7
Derbyshire	10	5	3	3
Devonshire	15	10	7	7
Dorset	16	12	10	9
Durham	6	3	1	1
Essex	16	12	6	3
Gloucestershire	9	6	4	3
Hampshire	12	8	5	3
Herefordshire	18	16	14	15
Hertfordshire	18	13	10	6
Huntingdonshire	19	18	16	16
Kent	13	9	7	5
Lancashire	4	2	1	1
Leicestershire	11	8	5	3
Lincolnshire	10	16	13	11
Middlesex (ex-met.)	10	5	3	1
Monmouth	10	6	4	2
Norfolk	9	14	12	10
Northamptonshire	16	12	8	5
Northumberland	9	6	4	2
Nottingham	10	8	5	3
Oxfordshire	18	15	11	9
Rutland	19	18	15	13
Shropshire	17	14	12	10
Somerset	14	11	9	8
Staffordshire	7	4	3	2
Suffolk	18	15	12	10
Surrey	14	7	5	2
Sussex	15	11	7	5
Warwickshire	7	5	3	2
Westmorland	19	14	11	12
Wiltshire	19	15	11	9
Worcestershire	9	7	7	6

(*continued*)

Table 40.4 (*cont.*)

	1851	1871	1891	1911
E. Riding	14	10	6	5
N. Riding	17	11	7	7
W. Riding	6	4	2	1
Anglesey	15	12	10	12
Brecknock	18	14	12	10
Caernarvon	14	10	8	6
Cardiganshire	16	15	11	9
Carmarthen	27	17	16	26
Denbighshire	20	13	8	8
Flintshire	6	4	3	6
Glamorgan	8	3	2	1
Merioneth	30	18	15	12
Montgomery	28	21	19	19
Pembroke	15	12	10	12
Radnorshire	33	19	15	19

Sources: Census, 1851, 1871, 1891, 1911.

Table 40.5. *The distribution of agrarian occupations in rural and urban districts of England and Wales, 1911 ('000 males over 20 years)*

	Rural	Urban
Farmers	179.3	28.8
Farmers' relatives	55.5	7.7
Bailiffs	18.8	3.2
Shepherds and labourers	424.4	73.8
Gardeners, seedsmen, and florists	48.3	73.3
Others, including woodmen	22.8	3.7
Total	749.1	190.5

Source: Bowley, 'Rural population', p. 610.

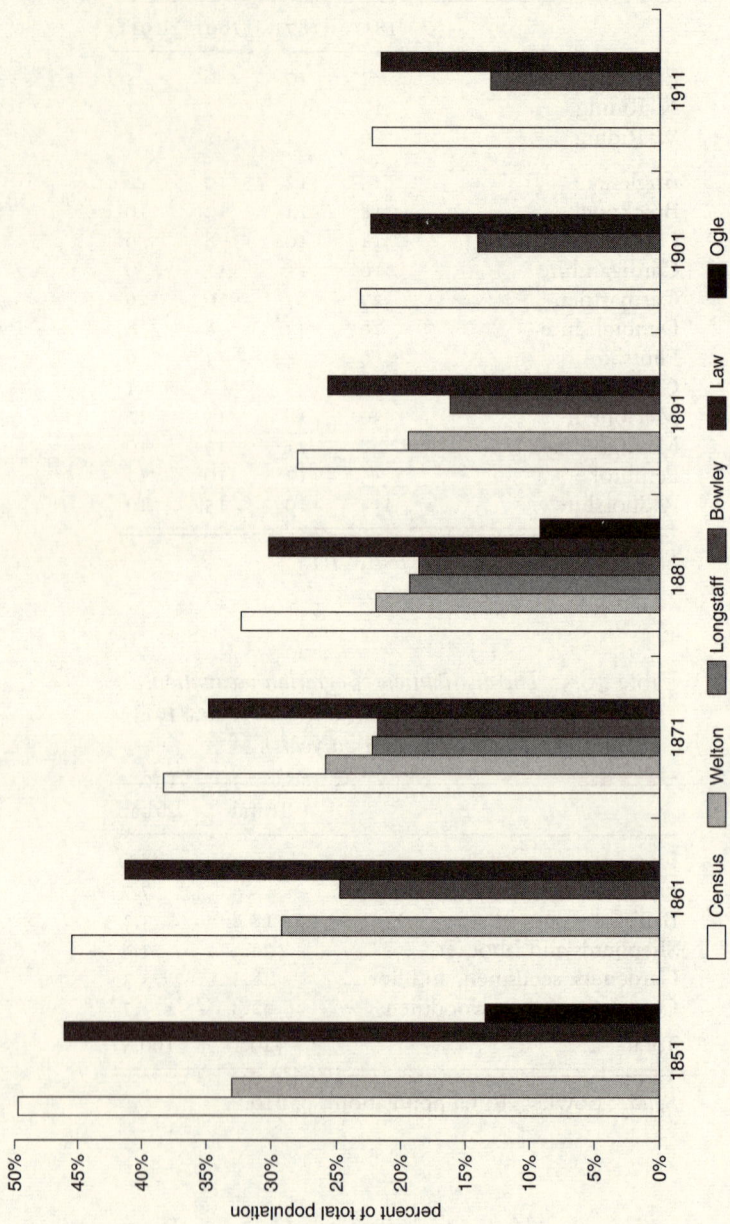

40.1 Agrarian population of England and Wales using differing definitions, 1851–1911

Sources: as for Table 40.1

1861   1911   PERCENT

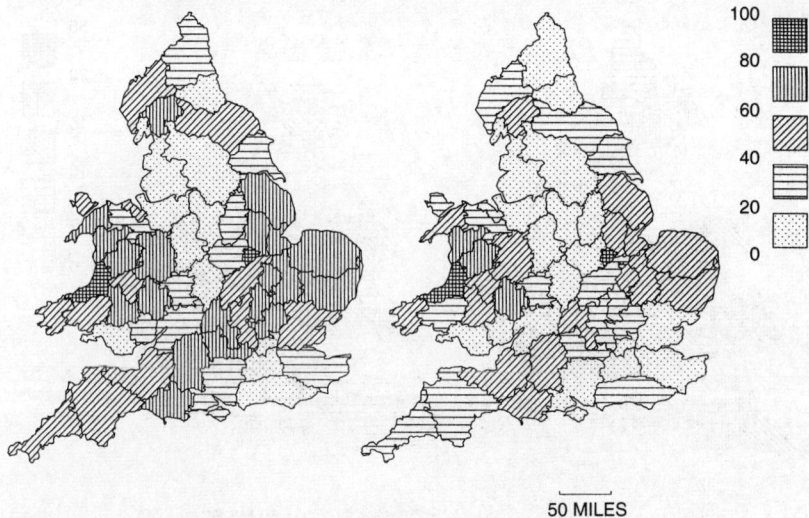

50 MILES

40.2 Percentage of total population not engaged in industrial or
urban occupations, 1861, 1911

1851   1911   PERSONS PER SQ. MILE

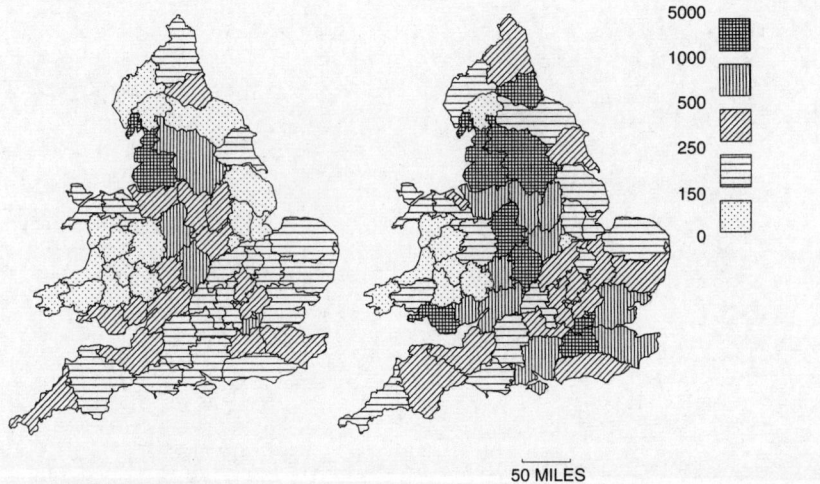

50 MILES

40.3 Density of population, excluding London in both years and excluding
metropolitan parts of Middlesex, Surrey and Kent in 1851, 1911

1851   1911

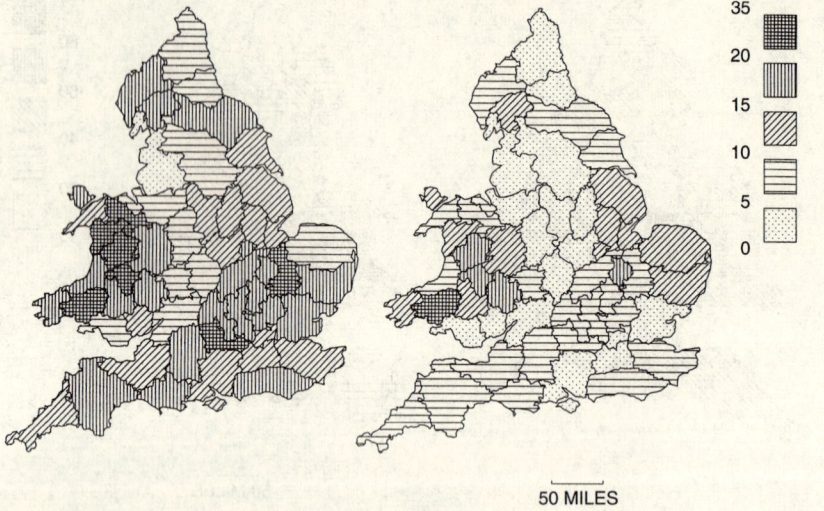

PERCENT

35
20
15
10
5
0

50 MILES

40.4 Proportion of total population engaged in agrarian occupations, 1851, 1911

CHAPTER 41

AGRARIAN OCCUPATIONS

BY BETHANIE AFTON AND MICHAEL TURNER

A. INTRODUCTION

It is easy to list a number of agrarian occupations: there were those who worked directly on the land, that is, those employed in agriculture, horticulture and forestry; and there were those, such as hurdlemakers, bodgers and millers, involved in processing the product of the land within the agrarian community, who would likewise have been included. The product from the land, in either its raw or its processed state, required transportation, adding carters, drovers and the like to the list. Furthermore, the community needed the craftsmen, tradesmen and the professional sector to provide various goods and services. The agrarian community was composed not only of those in occupations closely associated with the land, but also the many people who could have easily been found in any community, rural or urban. Consequently, quantifying these occupations is problematical. In part, this is because by 1850 the division between town and country had become increasingly nebulous. Towns grew, and areas once perceived as rural became engulfed by industry and housing. Communication and movement between country and town increased, and the draw of the large population centres disrupted traditional marketing patterns and servicing networks. There was a gravitation towards the cities both of those who provided goods and services for the agrarian sector and of those in certain types of agrarian enterprise who traded directly with large urban markets. During the period it became possible to live in the country and commute to jobs in the cities, so remaining outside of the rural occupational sector entirely, but living within it. Thus, even careful sampling of rural communities to establish employment structure would mask the true occupational composition of agrarian England and Wales.

The problem of enumeration is further complicated by the many occupations which superficially might be thought rural, but which were located in the cities, or which had taken on a new description when transposed into an urban setting. Even those with a direct link to the land, that is those working in agriculture, forestry and horticulture, were sometimes, as in the cases of market gardeners and town dairymen, essentially in urban occupations. According to the 1901 census, almost 20 per cent of those classified as agriculturalists were located in urban districts.[1] This makes it somewhat arbitrary when trying to distinguish them from rural farmers, graziers or agricultural labourers, and when quantifying the agrarian workforce. An even greater problem is found when measuring the occupational composition of the larger agrarian communities. This requires eliminating all those whose employment was not directly related to the agrarian sector. Thus, for example, farriers who dealt with farm horses would be classified as agrarian while those doing essentially the same job in a city would be urban. The miller grinding flour and animal provender for the farmers in the immediate vicinity could hardly be classified in the same way as the miller working a large dockside roller mill.

In spite of the problems encountered when quantifying occupational structure, it is important to be aware of the sources available for such a study and of the limitations of these sources. The most important national source for occupational information is the decennial census of population. The first four censuses collected in 1801, 1811, 1821 and 1831 enumerated occupations according to three categories: those chiefly engaged in agriculture; those in trade, manufacture and handicrafts; and those otherwise employed.[2] By 1841 it was felt that more detailed information was needed. The census for that year listed people in 877 alphabetically listed occupations.[3] From 1851 a system was initiated which remained essentially unchanged, even to recent times. This organised employment into a number of 'orders' and 'suborders' containing a total of between 350 and 400 different professions and occupations.[4] Agriculture, arboriculture and horticulture were classified together, along with other occupations relating to animals in two of the censuses.

Because it is the only comprehensive national source for occupational information, any survey of the occupational structure of England and Wales is likely to utilise the census. However, it has been criticised for its inherent inaccuracies and, consequently, must be used with care.[5] A majority of the problems with the source fall into two somewhat overlapping categories, those inaccuracies introduced as the material was collected, edited and published, and those which relate to the measurement of the 'actively engaged' population.

[1] *Census of England and Wales, 1901. General Report*, BPP, CVIII, p. 101.

[2] C. H. Lee, *British Regional Employment Statistics 1841–1971* (Cambridge, 1979), p. 3.

[3] *Ibid.*, p. 4.

[4] *Census of Great Britain, 1851. Population Tables*, BPP, 1852–3, LXXXVIII, Part I, pp. lxxxi–c.

[5] For example, *Census of England and Wales, 1881. General Report*, BPP, 1883, LXXX, p. 25; *Census of England and Wales, 1911. General Report*, BPP, 1917–18, XXXV, pp. 11–14.

When a census was taken, those drawing up the instructions for the data collection and enumeration attempted to clarify at the same time questions relating to occupations. However, the ability of those selected to be enumerators could greatly affect the quality of the census. In 1861 they were required to be literate, respectable and acquainted with the district.[6] However, in the 1881 census report the quality of the 35,000 enumerators required was unlikely to have been even, and 'the requisite precision of statement as to occupations was far from being universally observed'.[7] As the occupations were categorised, the difficulty experienced in determining the exact nature of a listing caused problems. The scope of this can be understood when it is remembered that the number of occupations listed was reduced from near 900 in the 1841 census to approximately 400 in subsequent years. Many entries could not easily be put into a particular class. In agriculture, serious underenumeration occurred when agricultural labourers simply entered themselves as labourers. In 1881 collectors were directed to pay special attention to the specific jobs of a labourer.[8] In the 1911 Census the task was simplified when a question was included which asked the nature of a person's employer's business. In that year there was a substantial reduction in the numbers listed as general labourers, a large number of whom were listed as agricultural labourers.[9] Over time, there were also considerable alterations in the types of job included under each occupation. A reasonable degree of conformity was maintained in the 1851, 1861 and 1871 censuses. Thereafter, considerable changes were made at each census.[10] While the reports which accompanied each census clarified the changes implemented in that year, the material would have been more directly comparable had no such reorganisation been made.

Another problem encountered is that of multiple occupation. Many members of the rural community worked in numerous jobs. The list in Table 41.2 of the twenty-five most common additional occupations of farmers in ten counties in 1861 helps to demonstrate the scope of the problem. Up until 1881 householders were asked to list their occupations

---

[6] *Census of England and Wales, 1861. General Report*, BPP, 1863, LIII, Part I, p. 1.

[7] *Census, 1881*, p. 27.

[8] *Census of England and Wales, 1871. General Report*, BPP, 1873, LXXI, Part II, p. xlv; *Census of England and Wales, 1891. General Report*, BPP, 1893–4, CVI, p. 35; *Census, 1881*, pp. 36–7.

[9] *Census, 1911*, pp. 112–13; W. E. Bear, 'The agricultural population and the census of 1901', *JRASE*, 64, 1903, p. 123.

[10] *Census, 1891*, pp. 35–7; J. Dent, 'Agricultural jottings from the General Report of the Census of England and Wales for the year 1871', *JRASE*, 2nd ser., 10, 1874, pp. 390–401; Bear, 'The agricultural population', p. 123; F. D. W. Taylor, 'Food and agriculture: notes and statistics', *The Farm Economist*, 8, pt 4, 1955, pp. 36–8; Lee, *Regional Employment*, p. 610; Edward Higgs, *Making Sense of the Census: The Manuscript Returns for England and Wales, 1801–1901* (Public Record Office Handbooks No. 23, London, 1989), pp. 10–17, 78–95.

in order of importance. Generally the first on the list was then used by the enumerator. In 1881, however, enumerators were given the responsibility for assessing the principal occupation. In that year they were asked to select, 'Firstly, that a mechanical handicraft or constructive occupation should invariably be preferred to a mere shopkeeping occupation; secondly, that, if one of the diverse occupations seemed of more importance than the others, it should be selected; and, thirdly, that in default of such apparent difference the occupation first mentioned should be taken on the ground that a person would be likely to mention his main business first.' Thus an innkeeper and farmer would have been classified as an innkeeper until 1881 and a farmer thereafter.[11]

Further inaccuracies were introduced as the material was processed. The statistics collected were entered, checked and often altered, at three different stages of preparation.[12] An important example of a serious cause of confusion was the unsystematic transfer, during recessions, of boys who gave no occupation on the original form to those listed by their fathers. This and similar alterations could seriously distort the returns.[13] Another difficulty arises when the census material is used for comparative purposes. The geographical areas on which enumeration was based did not remain constant throughout the period. Initially the data were published according to Poor Law Union boundaries, but in 1841 the basic unit was the registration county. In 1894 local government was reorganised and from 1901 the new administrative county was used. In that same census the data were included for both registration and administrative counties, making it possible to calculate the effect of the change in geographical unit.

Calculation of the 'actively employed' population was made difficult both by the decisions made by those responsible for the collection and publication of the censuses, and by the attitudes prevalent in society at the time of each census. The question of those underemployed or working only part-time was particularly relevant in the agricultural sector. Far greater numbers worked during certain parts of the year, particularly during the haymaking, harvest, hop and fruit picking seasons. The timing of the census in late March or early April undoubtedly resulted in under-enumeration of part-time workers, particularly women and migrant labourers. These were an essential part of the agricultural workforce. Edward Higgs has cautioned that 'analysis based on these data [i.e. the Victorian census] tends to proceed as if the figures in the occupational tables represent units of labour of an equal duration and inten-

---

[11] *Census, 1851*, Part I, p. lxxxii; *Census, 1861*, Part I, p. 30; *Census, 1881*, p. 28; *Census, 1891*, p. 37.

[12] P. M. Tillott, 'Sources of inaccuracy in the 1851 and 1861 censuses', in E. A. Wrigley (ed.), *Nineteenth-century Society: Essays in the Use of Quantitative Methods for the Study of Social Data* (Cambridge, 1972), pp. 83–4.   [13] *Ibid.*, pp. 121–7.

sity, but this is a highly dubious assumption . . . There is evidence that casual and part-time work was often not considered worthy of an occupational designation, and that the recording of the work of women and children suffered as a result.'[14] On the basis of a reworking of Charles Booth's population figures for 1851–71, Higgs estimates that, although working only part-time, female relatives of farmers, all relatives of farm labourers, those women who did part-time work on the farm as well as their tasks in the home, and both male and female migrant workers added very significantly to the total labour workforce. This labour, when converted to full-time equivalents, along with female labourers not recorded in the census and a proportion of labourers who did not specify their precise occupation, could easily increase the size of the agricultural workforce as calculated by Booth by 32 per cent in 1851, by 33 per cent in 1861 and by 41 per cent in 1871.[15]

C. H. Lee in his comparative analysis of the census warns that, even after 1881 when an effort was made to calculate the economically active proportion of the population more accurately, it should be assumed that the figures are not adequately adjusted for unemployment or casual employment.[16] Until 1881 both institutionalised and retired people were included under their former occupations. From that census the retired were omitted from the lists of actively occupied persons. The report accompanying the 1881 census calculated that the inclusion of the retired in any particular heading would have increased that category by approximately 2 per cent.[17] People in various institutions were listed according to former occupations except in the case of the mentally insane, who were considered to be unoccupied. In the same year the age at which children were considered to be employable changed from five years of age and over to ten years and above. The change roughly coincided with the rise in the age for leaving full-time education. However, this again masked the numbers employed on a part-time basis, particularly during the harvest. The number of actively employed children in the census as a whole was decreasing. Between 1851 and 1871, before alterations in the age were made, the number of actively employed children decreased from 3.4 per cent of the work force to 1.5 per cent.[18]

Throughout the period there were doubts about the correct method by which to enumerate the occupations of women. Many were believed to have returned themselves according to the occupations of their husbands. This was the suggested practice when they worked with or alongside them. However, often no indication of the capacity in which the

[14] Edward Higgs, 'Occupational censuses and the agricultural workforce in Victorian England and Wales', *EcHR*, 48, 1995, p. 704. [15] *Ibid.*, p. 709. [16] Lee, *Regional Employment*, pp. 7–8. [17] *Census, 1881*, p. 28; Lee, *Regional Employment*, p. 8. [18] Lee, *Regional Employment*, p. 8.

women were actually employed was given. In 1891, women doing domestic work in the home were enumerated along with domestic servants.[19] The possibility of an accurate calculation of the degree and structure of female employment was made particularly difficult by the change in social attitudes to working women which occurred during the period. In agriculture, for example, the tendency to omit women from the census categories of the 'usefully employed' coincided with a period in which contemporaries believed that women were actually doing less field work. While evidence from parliamentary commissions seems to confirm the decline in female employment, it can be argued that this was prejudiced by the same social bias which may have led to a possible understatement of employed women in the census enumerations. If this is true, it is difficult to suggest a printed source which would not be similarly affected.[20]

Thus quantifying the active agrarian workforce with confidence is problematical. C. H. Lee has calculated a fall in the indicated numbers of those actively employed of approximately 5 per cent for males and 8 per cent for females when those no longer judged to be actively employed were removed from the 1881 census data. However, he warns that this could mask an overall increase in the total activity rate as more people switched from casual to full-time employment.[21]

The Board of Agriculture and Fisheries' *Report on the Agricultural Output of Great Britain, 1908* includes data which are particularly relevant to the question of the size and structure of the agricultural labour force.[22] This report is based on the result of an additional set of questions which were to be answered by farmers in their annual agricultural return completed in June 1908. It is particularly useful to compare the report data with those in the population censuses. The basis and scope of the two surveys are entirely different. Information about occupation included in the decennial census of population is based on the perception of the head of the household concerning the status – social as well as economic – of the members of that household. In contrast, the 1908 returns were completed by occupiers of holdings of one acre and above. Each farmer was instructed to indicate the number of members of his or her family assisting on the farm, the number of other persons permanently employed

[19] *Census, 1891*, pp. 57–8; Lee, *Regional Employment*, p. 9.

[20] For a discussion of the usefulness of the census as a source for occupational data and the problems associated with its use see W. A. Armstrong, 'The use of information about occupation', in Wrigley, *Nineteenth-Century Society*, pp. 191–210; Tillott, 'Sources of inaccuracy', in *ibid*, pp. 82–128; L. L. Price, 'The census return of 1891 and rural depopulation', *JRASE*, 5, 1894, pp. 44–6.

[21] *The Agricultural output of Great Britain, 1908, in connection with the Census of Production Act, 1906*, BPP, 1912–13, X.   [22] *Ibid.*, pp. 16–17.

during the previous twelve months, and the number of persons temporarily employed on 4 June 1908. Thus the information, though less
detailed than the population census material, provides an important, and
otherwise unavailable, insight into the structure of the agricultural workforce.

Two important points should be noted about the data. First, there is a
considerable discrepancy between the number of returns completed by
'occupiers of land' in 1908 (430,081) and the number of 'farmers and graziers' in the 1901 census (224,299).[23] A number of holdings were not
'farmed as a business'. In 1907 this was 6 per cent of the annual returns.[24]
Although a small proportion occupying more than one holding may have
made more than one return, this does not account for the difference in
the two numbers. It does suggest that many farmers described themselves
in another way in the census. The second point concerns the date of the
return. Because this was made at the beginning of June before the
numbers had been swollen by haymaking, harvesting, fruit and hop
picking and the like, but after the winter lull, the size of the temporary
workforce would have been neither a maximum nor a minimum.

B. THE DATA

Table 41.3a gives the number of people in England and Wales employed
in agriculture as recorded in the decennial censuses. The data of those
employed in agriculture for the years 1851, 1871, 1891 and 1911 are
shown in Figure 41.1a–h. These have been calculated on a county basis
and then arranged by region on each graph. To improve comparability
the category 'wives of farmers' used in 1851 has been excluded from those
data. In contrast, Table 41.3b shows the complete workforce on 4 June
1908.

As discussed above, the census data are best used comparatively after a
number of alterations, often of a complex nature, have been made.
Several contemporary and modern attempts to adjust the data exist. Table
41.4a is from the 'Report' to the 1911 census. In an effort to facilitate
comparison, categories have been added or removed. For example,
domestic gardeners were added to the figures for 1901 and 1911 while
farmers' male relatives under fifteen years of age and all female relatives
were excluded. However, the figures for 1851–71 include the 'retired'.
Numerous unofficial approximations of the corrected data were also
made. In 1886 Charles Booth considered the 'Occupations of the people
of the United Kingdom'. Table 41.4b records his findings. One of the
more significant, and now controversial, adjustments he included was the

---

[23] *Annual Agricultural Returns, 1907.*   [24] Lee, *Regional Employment*, p. 10.

exclusion of males under twenty years of age and females of any age who were not self-supporting. Therefore, he excluded partially employed farmers' wives. Again, after the 1881 census, S. B. L. Druce attempted to adjust the returns in order to compare the very different censuses of 1871 and 1881. W. E. Bear made a similar attempt after the 1901 census. However, Bear did not carry his work back beyond 1881. In 1907 Lord Eversley calculated the numbers employed in agriculture from 1851 through to 1901. He was particularly aware of the problematical nature of the data for women and for males under twenty years of age. In the case of women, the 1851 and 1861 censuses included 'wives of farmers' and 'daughters and other female relatives living with farmers' as employed in agriculture. Because these were not similarly recorded in later censuses, he removed them from his calculations. He was also aware that both women and youths were often 'only employed at certain periods of the year, in such seasonal work as the picking of fruit and hops, in haymaking and harvesting'.[25] Such persons were not included in the 1881 and 1901 censuses. Table 41.4c shows the extension of the work of Eversley by A. L. Bowley, who excluded women and men under the age of twenty in an effort to arrive at figures which could justifiably be compared over the period. The totals were modified to take into account those retired but included before 1881, and to adjust for the numbers who were absent during the Boer War.[26]

Three more modern calculations of the census data for agricultural occupations have also been included. In 1955, F. D. W. Taylor of the Agricultural Economics Research Institute, Oxford, assessed the numbers employed in agriculture as part of a broader project on agricultural output. Woodmen, foresters, dealers, landed proprietors, and gardeners working in private gardens were excluded, as were retired people and female relatives, particularly the wives of farmers. The results are included as Table 41.4d. Table 41.4e is the work of W. A. Armstrong taking Charles Booth as the base. Thus, it included all males over twenty years of age, but none under, nor any females unless they were specifically included in a particular category in the census.[27] Table 41.4f is based on the estimates of Edward Higgs. The 'orthodox total' is as calculated by Charles Booth. The figure for women in the agricultural workforce

---

[25] Lord Eversley, 'The decline in number of agricultural labourers in Great Britain', *JRSS*, 70, 1907, p. 270.

[26] Eversley, 'Decline in numbers', pp. 267–303; C. Booth, 'Occupations of the people of the United Kingdom', *JRSS*, 49, 1886, pp. 314–35; A. L. Bowley, 'Rural population in England and Wales', *JRSS*, 77, 1914, pp. 597–645; S. B. L. Druce, 'The alteration in the distribution of the agricultural population of England and Wales between the returns of the census of 1871 and 1881', *JRASE*, 2nd ser., 21, 1885, pp. 96–126; Bear, 'The agricultural population, pp. 123–37.

[27] Armstrong, 'Information about occupation', in Wrigley, *Nineteenth-Century Society*, pp. 255–6.

includes the females originally included in the census but removed by Booth, the female relatives of both the farmers and the labourers, a number of women not included in the census but extrapolated from the data in the 1908 Agricultural Output Report, female servants who are assumed to have worked part-time on agricultural tasks, and female migrant labour. Additional males include male relatives of agricultural labourers, a proportion of general labourers, and migrant labourers. All have been converted to full-time labour equivalents.

'Landed proprietor' was an agrarian occupational category in the censuses up to, and including, 1871. In that decade, a survey of all landowners occupying one acre and over was undertaken.[28] Data from this survey, corrected by John Bateman in *The Great Landowners of Great Britain and Ireland* (1883), are included in Table 41.5. The table gives the owners of the three largest estates in each county along with the value of the estate in that county and the ranking, by size of total holding in England and Wales, of the landowner.

[28] *Return of Owners of Land 1872–3.* BPP, 1874, LXXII.

APPENDIX

Table 41.1. *Classification of agricultural orders in different population censuses, 1851–1911*

1851	1861	1871	1881	1891	1901	1911
Land proprietor	Land proprietor	Land proprietor				
Farmer	Farmer, grazier	Farmer, grazier	Farmer, grazier	Farmer, grazier	Farmer, grazier	Farmer, grazier
Grazier						
Farmer, grazier wife	Farmer, grazier wife					
Farmer's son, brother, grandson, nephew, daughter, sister, niece, granddaughter	Farmer's son, brother, grandson, nephew, daughter, sister, niece, granddaughter	Farmer's son, brother, grandson, nephew, daughter, sister, niece, granddaughter	Farmer's son, brother, grandson, nephew	Farmer's son, brother, grandson, nephew (males only, females referred to as 'Others' in the unoccupied class)	Farmers', graziers' sons, daughters, or other relatives assisting in the work of the farm	Farmers', graziers' sons, daughters, or other relatives assisting in the work of the farm
Farm bailiff	Farm bailiff	Farm bailiff	Farm bailiff	Farm bailiff	Farm bailiff, foreman	Farm bailiff, foreman
Agricultural labourer (out-doors)	Agricultural labourer	Agricultural labourer	Agricultural labourer, farm servant, cottager	Agricultural labourer, farm servant	Agricultural labourer, farm servant not otherwise distinguished	Agricultural labourer, farm servant not otherwise distinguished

Farm servant (indoor)	Farm servant (indoor)	Farm servant (indoor)	Horsekeeper, horseman, teamster, carter	Horsekeeper, horseman, teamster, carter	Agricultural labourer, farm servant in charge of horses / Agricultural labourer, farm servant in charge of cattle	Agricultural labourer, farm servant in charge of horses / Agricultural labourer, farm servant in charge of cattle
Shepherd	Shepherd	Shepherd	Shepherd	Shepherd	Shepherd	Shepherd
	Agricultural implement proprietor	Agricultural machine proprietor, attendant	Agricultural machine proprietor, attendant	Agricultural machine proprietor, attendant	Agricultural machine proprietor, attendant	Agricultural machine proprietor, attendant
	Agricultural engine and machine worker					
	Agricultural student, pupil	Agricultural student, pupil				
	Land surveyor, estate agent	Land surveyor, estate agent				
	Land drainage service	Land drainage service				
	Hop grower					
	Willow, rod grower, dealer	Willow, rod grower, dealer				
	Teazle grower, merchant	Teazle grower, merchant				
	Colonial planter	Colonial planter				

(continued)

Table 41.1 (*cont.*)

1851	1861	1871	1881	1891	1901	1911
Others connected with agriculture	Others engaged in agriculture	Others engaged in agriculture	Others engaged in, or connected with agriculture	Others engaged in, or connected with agriculture	Others engaged in, or connected with agriculture	Others engaged in, or connected with agriculture
Woodman	Woodman, wood gather	Woodman	Woodman	Woodman	Woodman	Woodman
Others connected with arboriculture	Others connected with arboriculture					
Gardener	Gardener (not domestic)	Gardener, nurseryman, seedsman -woman florist	Gardener, nurseryman, seedsman	Gardener, nurseryman, seedsman	Gardener (not domestic), nurseryman, seedsman, florist	Nurseryman, seedsman, florist
						Market gardeners (including labourers)
						Other gardeners (not domestic)
	Watercress grower					
Others connected with horticulture	Others connected with horticulture	Others connected with horticulture				
		Cattle-, sheep-, pig-dealer, salesman		Cattle-, sheep-, pig-dealer, salesman		

Drover
Gamekeeper
Dog-, bird-, animal-keeper, dealer
Knacker, catsmeat dealer, vermin destroyer

Drover
Gamekeeper
Dog-, bird-, animal-keeper, dealer
Knacker, catsmeat dealer, vermin destroyer

Sources: 'General report', *Census of Great Britain, 1851*, BPP, 1852–3, LXXXVIII; 'General Report', *Census of Great Britain, 1861*, BPP, 1863, LIII; 'General report', *Census of Great Britain, 1871*, BPP, 1873, LXXI; 'General report', *Census of England and Wales, 1881*, BPP, 1883, LXXX; 'General report', *Census of England and Wales, 1891*, BPP, 1893, LV; 'General report', *Census of England and Wales, 1901*, BPP, 1904, CVIII; 'General report', *Census of England and Wales, 1911*, BPP, 1912–13, CXI.

Table 41.2. *The fifty most common pursuits engaged in by occupiers of land returning one or more pursuits besides farming in ten sample counties, 1861*

Trade, profession, etc. besides farming	Order in Census	Number of persons
Innkeeper, hotelkeeper	V	525
Miller	XII	518
Publican	V	456
Butcher, meat salesman	XII	295
Grocer, tea dealer	XII	220
Carpenter, joiner	X	141
Farm bailiff	VIII	135
Carman, carrier, carter, drayman	VII	127
Shoemaker, bootmaker	XI	120
Maltster	XII	111
Blacksmith	XV	108
Brick-maker, dealer	XV	97
Cattle-, Sheep-dealer	IX	85
Beerseller	V	70
Lead-miner	XV	66
Brewer, etc.	XII	59
Baker	XII	58
Land surveyor, estate agent	VIII	54
Gardener	VIII	51
Coal merchant, dealer	XV	47
Protestant minister	III	44
Shopkeeper	VI	44
Wheelwright	X	40
Builder	X	38
Auctioneer, appraiser, valuer	VI	36
Farrier, veterinary surgeon	IX	33
Corn merchant, dealer	XII	32
Tailor	XI	31
Gamekeeper	IX	29
Coal-miner	XV	29
Agricultural labourer	VIII	28
Poulterer, game dealer	XII	28
Mason, pavior	X	27
Magistrate	I	25
Other, general dealer	VI	25
Bricklayer	X	25
Engine and machine maker	X	22
Timber merchant, dealer	XIV	22
Union & Parish Officer	I	20
Solicitor	III	19
Clergyman	III	19

(continued)

Table 41.2 (*cont.*)

Trade, profession, etc. besides farming	Order in Census	Number of persons
Provision curer, dealer	XII	19
Limestone quarrier, burner	XV	18
Horse dealer	IX	17
Parish clerk	III	16
Pig merchant, dealer	IX	16
Merchant	VI	16
Cheesemonger	XII	16
Tanner	XIII	16
Iron manufacturer	XV	16

Notes:
1. Sample taken from the counties of Buckinghamshire, Cambridgeshire, Cheshire, Cumberland, Lincolnshire, Norfolk, Shropshire, Sussex, Wiltshire and the North Riding of Yorkshire
2. A total of 4,671 people were included as having two or more occupations
Source: Census, 1861.

Table 41.3a. *Number of persons in England and Wales engaged in agriculture, arboriculture and horticulture, 1851–1911*

	Agriculture	Arboriculture	Horticulture	Total
Males				
1851	1,479,673	7,992	72,097	1,559,762
1861	1,457,075	8,917	79,675	1,545,667
1871	1,264,031	7,855	101,056	1,372,942
1881	1,135,763	8,151	70,539	1,214,453
1891	1,050,363	9,448	174,290	1,234,101
1901	935,881	12,034	123,125	1,071,040
1911	988,111	12,301	140,103	1,140,515
Females				
1851	449,742	16	2,260	452,018
1861	376,577	9	1,857	378,443
1871	183,450	6	2,639	186,095
1881	61,073		3,098	64,171
1891	45,999		5,046	51,045
1901	52,459	1	5,104	57,564
1911	90,518	2	4,202	94,722
Total number of persons				
1851	1,929,415	8,008	74,357	2,011,780
1861	1,833,652	8,926	81,532	1,924,110
1871	1,447,481	7,861	103,695	1,559,037
1881	1,196,836	8,151	73,637	1,278,624
1891	1,096,362	9,448	179,336	1,285,146
1901	988,340	12,035	128,229	1,128,604
1911	1,078,629	12,303	144,305	1,235,237

Source: Census, 1851–1911.

Table 41.3b. *Number of persons in the agricultural workforce, 1908*

	Male	Female	Total
Occupiers of holdings of 1 acre and over			430,081
Members of farmers' families working on the holding	262,000	144,000	406,000
Other persons permanently employed	508,000	68,000	576,000
Temporary labourers working on 4 June 1908	114,000	32,000	146,000
Total	884,000	244,000	1,128,000

Source: Agricultural Output of Great Britain, 1908.

Table 41.4a. *Decennial census calculations of numbers employed in agriculture, 1851–1911*

	Males	Proportion of total males %	Females	Proportion of total females %
1851	1,544,087	23.50	168,652	2.40
1861	1,539,965	21.20	115,213	1.50
1871	1,371,304	16.80	85,667	1.00
1881	1,288,173	13.80	64,216	0.60
1891	1,233,936	11.60	51,045	0.40
1901	1,153,185	9.50	38,782	0.30
1911	1,253,859	9.20	37,969	0.30

Notes:
1. The figures for 1851–71 include the 'Retired'.
2. The 1901 Census was taken during the Boer War.
Source: Census, 1911.

Table 41.4b. *Calculations by Charles Booth of persons employed in agriculture (thousands), 1851–81*

	1851	1861	1871	1881
Males				
Farmers and relatives	348.8	334.9	318.6	297.9
Agricultural labourers and shepherds	1,110.3	1,098.3	922.0	830.5
Nurserymen and gardeners	83.6	89.3	110.0	80.2
Drainage and machinery attendants		3.2	3.5	5.9
Breeding and dealing	47.6	59.0	64.4	61.9
Females				
Farmers and relatives	22.9	22.8	24.3	20.6
Agricultural labourers and shepherds	143.5	90.6	58.1	40.3
Nurserymen and gardeners	2.3	2.0	2.7	3.2
Drainage and machinery attendants	–	–	–	0.1
Breeding and dealing	0.6	0.1	0.3	0.4
Total				
Farmers and relatives	371.7	357.7	342.9	318.5
Agricultural labourers and shepherds	1,253.8	1,188.9	980.1	870.8
Nurserymen and gardeners	85.9	91.3	112.7	83.4
Drainage and machinery attendants		3.2	3.5	6.0
Breeding and dealing	48.2	59.1	64.7	62.3
Total in agriculture	1,759.6	1,700.2	1,503.9	1,341.0

Source: C. Booth, 'Occupations of the people of the United Kingdom', *JRSS*, 49, 1886, pp. 351–2.

Table 41.4c. *Calculations by A. L. Bowley of land occupations of males over the age of twenty (thousands), 1861–1911*

	1861	1871	1881	1891	1901	1911
Farmers	226.00	224.50	202.40	200.50	202.00	208.10
Farmers' relatives	60.00	47.90	47.10	43.40	55.70	63.30
Bailiffs	15.60	16.30	19.40	18.00	22.40	22.00
Shepherds & labourers	809.40	657.80	606.00	545.60	458.40	498.20
Gardeners, seedsmen & florists	28.20	46.70	58.90	78.90	105.40	121.50
Agricultural machinists	1.40	2.00	3.90	4.30	6.10	6.90
Others in agriculture	8.40	7.80	9.70	10.30	16.40	19.70
Domestic gardeners	50.20	61.00	62.10	68.80	75.20	100.60
Gamekeepers	9.40	11.80	11.30	12.70	15.10	15.50
Total	1,208.60	1,075.80	1,020.80	982.50	956.70	1,055.80
Total as modified by Bowley	1,180.00	1,049.00	1,020.00	983.00	977.00	1,056.00

Source: A. L. Bowley, 'Rural population in England and Wales', *JRSS*, 77, 1914, p. 610.

Table 41.4d. *Calculations by F. D. W. Taylor of the numbers employed in agriculture, 1851–1911*

	1851	1861	1871	1881	1891	1901	1911
				Farmers			
adult males	225,747	226,019	224,506	202,392	200,527	202,015	208,065
youths	768	938	1,063	937	1,391	736	696
females	22,916	22,448	24,338	20,614	21,692	21,548	20,027
Total	249,431	249,735	249,907	223,943	223,610	224,299	228,788
				Relatives			
adult males	73,634	60,044	47,854	47,121	43,409	55,685	63,250
youths	38,070	32,277	28,612	28,076	23,878	33,480	34,439
females						18,618	17,000
Total	111,704	92,321	76,466	75,197	67,287	107,783	114,689
			Contract workers on farms				
adult males	836,452	825,070	692,710	625,951	565,329	486,008	528,270
youths	287,955	288,995	246,881	227,641	211,221	151,477	146,105
females	143,493	90,569	58,152	40,421	24,240	12,228	13,575
Total	1,267,900	1,204,634	997,743	891,013	800,790	649,713	687,950
				Others			
adult males	69,107	83,267	101,819	71,074	90,451	114,096	131,188
youths	7,693	9,957	11,953	9,915	13,813	18,347	19,083
females	2,227	1,857	2,639	3,140	4,170	5,469	4,263
Total	79,027	95,081	116,411	84,129	108,434	137,612	154,534
				Total			
adult males	1,204,940	1,197,700	1,066,889	946,538	899,716	857,804	930,773
youths	334,486	332,167	288,509	263,569	250,303	204,040	200,323
females	168,636	11,204	85,129	64,175	50,102	57,563	54,865
Total	1,708,062	1,641,771	1,440,527	1,274,282	1,200,121	1,119,407	1,185,961

Source: F. D. W. Taylor, 'United Kingdom: numbers in agriculture', *The Farm Economist*, 8, 1955, pp. 38–9.

Table 41.4e. *Calculations by W. A. Armstrong of persons over the age of twenty engaged in agrarian occupations (thousands), 1851–91*

	1851	1861	1871	1881	1891
	Males				
Farmers, graziers	226.5	226.9	225.6	203.3	201.9
Farm bailiffs, stewards	10.6	15.7	16.5	19.4	18.2
Farmers' male relatives	111.7	92.3	76.5	75.2	67.3
Agricultural labourers	1,097.8	1,072.7	898.7	807.7	735.0
Shepherds	12.5	25.6	23.3	22.8	21.6
Woodmen	7.8	8.9	7.9	8.1	9.4
Nurserymen, seedsmen, florists	2.4	2.8	5.1	7.0	
Gardeners	69.6	76.7	95.8	63.6	84.7
Others in agriculture	3.8	0.9	1.2	1.5	2.6
	Females				
Farmers, graziers	22.9	22.8	24.3	20.6	21.7
Farm bailiffs, stewards	–	–	–	–	–
Farmers' male relatives	–	–	–	–	–
Agricultural labourers	143.5	90.6	58.1	40.3	24.2
Shepherds	–	–	–	–	–
Woodmen	–	–	–	–	–
Nurserymen, seedsmen, florists	–	0.1	0.4	0.8	
Gardeners	2.2	1.8	2.2	2.3	5.0
Others in agriculture	0.1	0.1	0.1	0.1	0.1
	Persons				
Farmers, graziers	249.4	249.7	249.9	223.9	223.6
Farm bailiffs, stewards	10.6	15.7	16.5	19.4	18.2
Farmers' male relatives	111.7	92.3	76.5	75.2	67.3
Agricultural labourers	1,241.3	1,163.3	956.8	848.0	759.2
Shepherds	12.5	25.6	23.3	22.8	21.6
Woodmen	7.8	8.9	7.9	8.1	9.4
Nurserymen, gardeners	74.2	81.4	103.5	73.7	89.7
Others in agriculture	3.9	1.0	1.3	1.6	2.7
Total	1,711.4	1,637.9	1,435.7	1,272.7	1,191.7

Source: W. A. Armstrong, 'The use of information about occupation', in E. A. Wrigley (ed.), *Nineteenth-century Society: Essays in the Use of Quantitative Methods for the Study of Social Data* (Cambridge, 1972), pp. 255–6.

Table 41.4f. *Calculations by E. Higgs of the numbers of agricultural workers (thousands), 1851–71*

	1851	1861	1871
'Orthodox total' of the agricultural workforce in England and Wales based on Booth's reworking of the census data	1,625.60	1,546.70	1,323.10
Additional female worker full-time equivalents	405.90	401.90	414.40
Additional male worker full-time equivalents	120.30	107.00	122.90
Total number of agricultural workers	2,151.80	2,055.60	1,860.40

Source: E. Higgs, 'Occupational censuses and the agricultural workforce in Victorian England and Wales', *EcHR*, 48, 1995, pp. 709–10.

Table 41.5. *The three largest landowners in each of the English and Welsh counties, c. 1875*

County	Surname	Forenames	Title	Ranking of landowner by area of English & Welsh estates	Area owned in county	Gross Annual Value
Bedfordshire	Bedford		Duke	5	32,269	45,687
	Whitbread	Samuel		247	13,257	20,399
	Cowper		Earl	33	9,105	13,394
Berkshire	Loyd-Lindsay	Robert	Sir	139	20,528	26,492
	Craven		Earl	59	19,225	21,767
	Benyon	Richard		195	10,129	13,303
Buckinghamshire	Carington		Lord	89	16,128	26,805
	Brownlow		Earl	13	11,785	15,450
	Rothschild	N.	Sir	211	9,959	17,216
Cambridge	Hardwicke		Earl	155	18,978	26,349
	Bedford		Duke	5	18,800	34,325
	Childers	John Walbanke		265	7,402	12,587
Cheshire	Tollemache		Lord	38	28,651	33,614
	Cholmondeley		Marquis	44	16,992	29,213
	Westminster		Duke	149	15,138	32,387
Cornwall	Rashleigh	Jonathan		63	30,156	9,000
	Falmouth		Viscount	60	25,910	35,953
	Robartes		Lord	121	22,234	30,730
Cumberland	Carlisle		Earl	6	47,730	16,850
	Lonsdale		Earl	9	28,228	42,818
	Graham	Frederick	Sir	93	25,270	26,696

(*continued*)

Table 41.5 (cont.)

County	Surname	Forenames	Title	Ranking of landowner by area of English & Welsh estates	Area owned in county	Gross Annual Value
Derbyshire	Devonshire		Duke	3	89,462	89,557
	Rutland		Duke	7	27,069	31,710
	Crewe	John	Sir	76	12,923	24,204
Devon	Rolle	M.	Hon.	15	55,592	47,170
	Bedford		Duke	5	22,607	45,907
	Fortescue		Earl	83	20,171	17,245
Dorset	Fox-Pitt-Rivers	George	Maj.-Gen.	77	24,942	33,682
	Wingfield-Digby	Walter Ralph		85	21,230	36,106
	Bankes			156	19,228	14,985
Durham	Cleveland		Duke	4	55,837	29,219
	Boyne		Viscount	78	18,023	76,855
	Durham		Earl	62	14,664	63,929
Essex	Petre		Lord	160	19,085	22,595
	Braybrooke		Lord	259	9,820	13,160
	Rayleigh		Lord	469	8,632	12,800
Gloucestershire	Fitz-Hardinge		Lord	141	18,264	31,836
	Beaufort		Duke	20	16,610	21,220
	Sherborne		Lord	197	15,773	21,345
Hampshire	Portsmouth		Earl	41	17,460	14,731
	Wellington		Duke	158	15,847	16,873
	Ashburton		Lord	35	15,330	13,289

County	Name		Title			
Herefordshire	Arkwright	John Hungerford		352	10,559	14,972
	Rouse–Boughton–Knight	Andrew		354	10,348	11,786
	Harley	Robert		362	9,901	11,203
Hertfordshire	Salisbury		Marquis	143	13,389	18,372
	Smith	Abel		325	11,212	14,617
	Cowper		Earl	33	10,122	13,540
Huntingdonshire	Fellowes	Edward		145	15,629	22,128
	Manchester		Duke	219	13,835	20,589
	Heathcote	John		615	7,144	11,386
Kent	Holmesdale		Viscount	170	16,209	28,731
	Sondes		Earl	159	14,157	23,000
	Hothfield		Lord	30	10,144	16,108
Lancashire	Derby		Earl	8	57,000	156,735
	Sefton		Earl	142	20,250	43,000
	Clifton	John Talbot		199	15,802	41,965
Leicestershire	Rutland		Duke	7	30,188	46,241
	Loudoun		Earl	235	10,174	17,722
	Howe		Earl	49	9,755	18,024
Lincolnshire	Yarborough		Earl	14	56,795	84,000
	Willoughby d'Eresby		Baroness	16	24,696	36,530
	Chaplin	Henry		112	23,370	30,517
Middlesex	Strafford		Earl	591	4,993	9,676
	Jersey		Earl	154	1,993	7,117
	Newdegate	Charles		661	1,491	2,524
Norfolk	Leicester		Earl	23	44,090	59,578
	Townshend		Marquis	146	18,343	20,717
	Cholmondeley		Marquis	44	16,995	11,960

(continued)

Table 41.5 (cont.)

County	Surname	Forenames	Title	Ranking of landowner by area of English & Welsh estates	Area owned in county	Gross Annual Value
Northamptonshire	Fitzwilliam	George		104	18,116	32,492
	Buccleuch and Queensferry		Duke	82	17,965	26,531
	Spencer		Earl	81	16,800	29,060
Northumberland	Northumberland	W.	Duke	1	181,616	161,874
	Selby			64	30,000	10,550
	Tankerville		Earl	55	28,930	31,416
Nottinghamshire	Portland		Duke	11	43,036	35,752
	Newcastle		Duke	40	34,467	73,098
	Manvers		Earl	32	26,771	36,788
Oxfordshire	Marlborough		Duke	110	21,944	34,341
	Ducie		Earl	241	8,798	13,430
	Abingdon		Earl	130	8,173	12,944
Rutland	Gainsborough		Earl	163	15,076	23,716
	Aveland		Lord	56	13,633	19,797
	Finch	George		178	9,183	15,098
Shropshire	Powis		Earl	12	26,986	29,000
	Cleveland		Duke	4	25,604	32,605
	Brownlow		Earl	13	20,233	29,717
Somerset	Portman		Viscount	46	24,339	35,557
	Poulett		Earl	123	22,123	21,885
	Acland	Thomas Dyke	Sir	28	20,300	10,680

County	Name		Title			
Staffordshire	Lichfield		Earl	129	21,433	41,560
	Shrewsbury and Talbot		Earl	37	18,954	29,898
	Anglesey		Marquis	66	17,441	91,304
Suffolk	Rendlesham		Lord	106	19,869	19,275
	Tomline	G.	Col.	84	18,473	24,005
	Duleep-Singh	H.		179	17,210	4,755
Surrey	Onslow		Earl	266	11,761	8,761
	Lovelace		Earl	164	10,214	12,384
	Leveson-Gower	Granville		617	6,930	5,234
Sussex	Leconfield		Lord	10	30,221	29,688
	Norfolk		Duke	21	21,446	27,557
	De La Warr		Earl	113	17,185	10,827
Warwickshire	Leigh		Lord	135	14,891	23,043
	Willoughby de Broke		Lord	168	12,621	16,540
	Aylesford		Earl	151	12,453	19,653
Westmorland	Lonsdale		Earl	9	39,229	27,141
	Hothfield		Lord	30	17,093	13,830
	Headfort		Marquis	136	12,851	13,686
Wiltshire	Pembroke		Earl	26	42,244	40,500
	Ailesbury		Marquis	17	37,993	40,334
	Bath		Marquis	51	19,984	29,325
Worcestershire	Dudley		Earl	99	14,698	48,545
	Coventry		Earl	232	13,021	22,367
	Beauchamp		Earl	172	10,624	17,789
Yorkshire	Londesborough		Lord	18	52,655	67,876
	Feversham		Earl	29	39,312	34,328
	Bowes	John		24	34,887	5,283

(continued)

Table 41.5 (cont.)

County	Surname	Forenames	Title	Ranking of landowner by area of English & Welsh estates	Area owned in county	Gross Annual Value
Anglesey	Tapps-Jervis-Meyrick	George	Sir	131	16,918	13,283
	Williams–Bulkeley	Richard Mostyn Lewis	Sir	65	16,516	17,997
	Anglesey		Marquis	66	9,620	9,784
Brecon	Bailey	Joseph Russell	Sir	74	21,979	19,367
	Powell	George		48	11,704	556
	Thomas		Miss	236	9,235	3,417
Caernarvon	Penrhyn		Lord	22	41,348	62,622
	Duff–Assheton-Smith	G.		42	33,752	42,255
	Willoughby d'Eresby		Baroness	16	30,391	7,966
Cardigan	Lisburne		Earl	25	42,720	13,616
	Pryse	Pryse	Sir	52	28,684	10,634
	Powell	George		48	21,933	9,024
Carmarthen	Cawdor		Earl	19	33,782	20,780
	Jones	Morgan		292	11,031	5,867
	Cowell–Stepney	A.	Sir	396	9,841	7,047
Denbigh	Williams–Wynn	Watkin	Sir	2	28,721	24,368
	Bagot		Lord	61	18,044	7,200
	Hughes	Hugh		192	13,287	19,229
Flint	Hanmer	Wydham	Sir	302	7,318	10,970
	Gladstone	William	Right Hon.	644	6,908	17,565
	Rowley–Conwy	Conwy Grenville Hercules		849	5,526	6,995

County	Surname	First name	Title	No.		
Glamorgan	Talbot	C.		45	33,920	44,057
	Dunraven		Earl	100	23,751	23,974
	Bute		Marquis	111	21,402	100,000
Merioneth	Williams-Wynn	Watkin	Sir	2	42,044	7,438
	Price	Richard		27	40,500	10,600
	Vaughan	J.		186	16,443	4,300
Monmouth	Beaufort		Duke	20	27,229	24,582
	Tredegar	John	Lord	31	25,500	35,000
	Hanbury			337	10,210	20,660
Montgomery	Williams-Wynn	Watkin	Sir	2	70,559	18,139
	Powis		Earl	12	33,545	28,000
	Sudeley		Lord	105	17,158	13,539
Pembroke	Philipps	Charles		115	19,745	21,151
	Cawdor		Earl	19	17,735	14,207
	Scourfield	Owen	Sir	262	11,243	8,722
Radnor	Ormathwaite		Lord	217	12,428	8,126
	Lewis	Gilbert	Rev.	382	10,000	7,000
	Green-Price	Richard	Sir	458	8,774	7,638

Source: J. Bateman, *The Great Landowners of Great Britain and Ireland* (Leicester, 1971). The work involved in entering the data was funded by a grant provided by Professor J. V. Beckett and the Department of History, University of Nottingham.

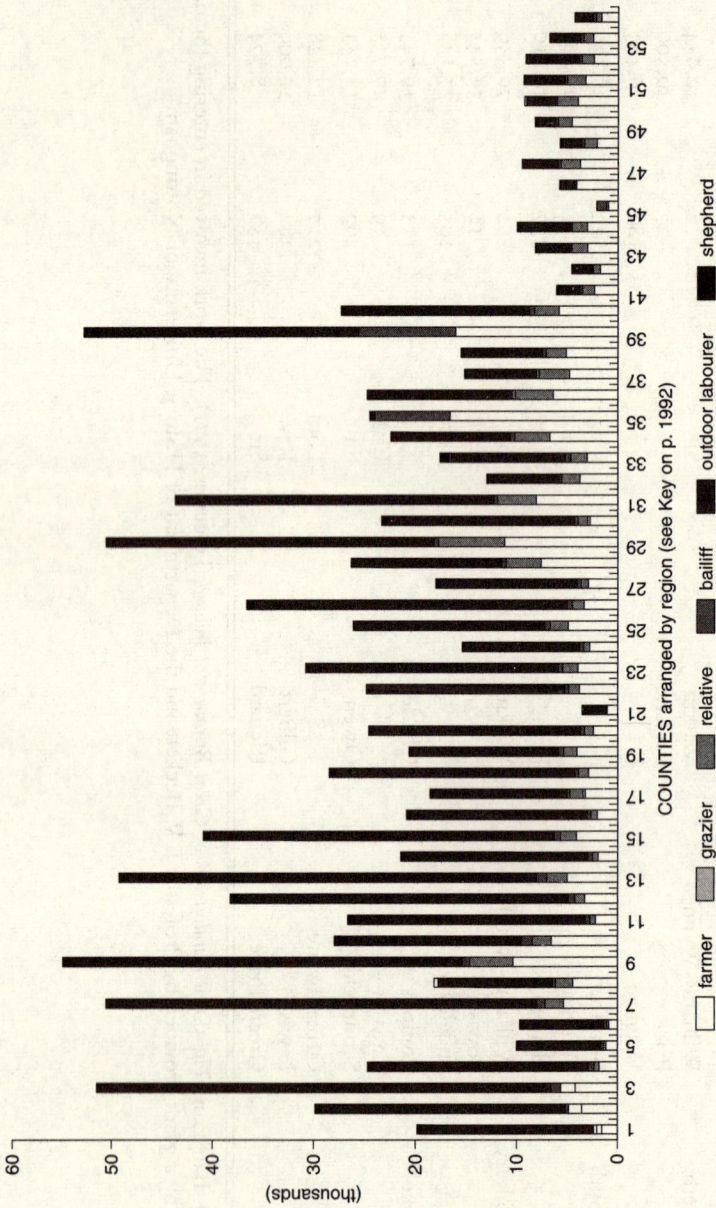

41.1a Agricultural occupations of males aged ten years and over, 1851

Source: Census, 1851

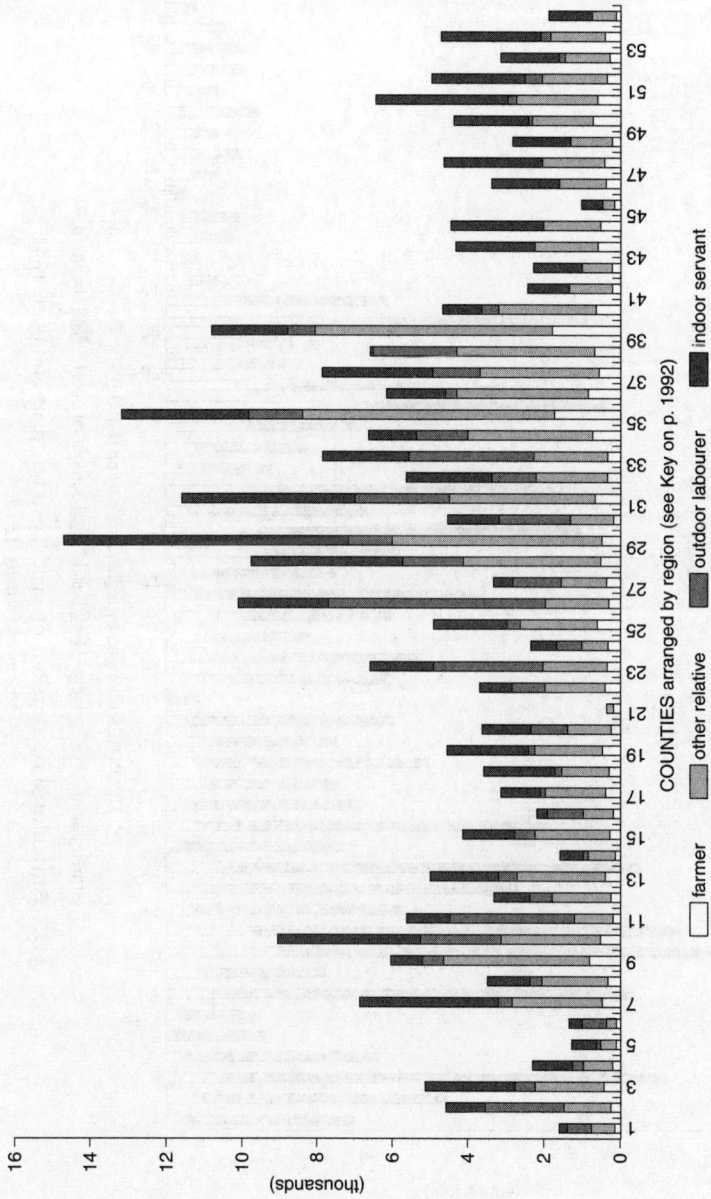

41.1b Agricultural occupations of females aged ten years and over, 1851

Source: *Census, 1851*

COUNTIES arranged by region (see Key on p. 1992)

Key: farmer | other relative | outdoor labourer | indoor servant

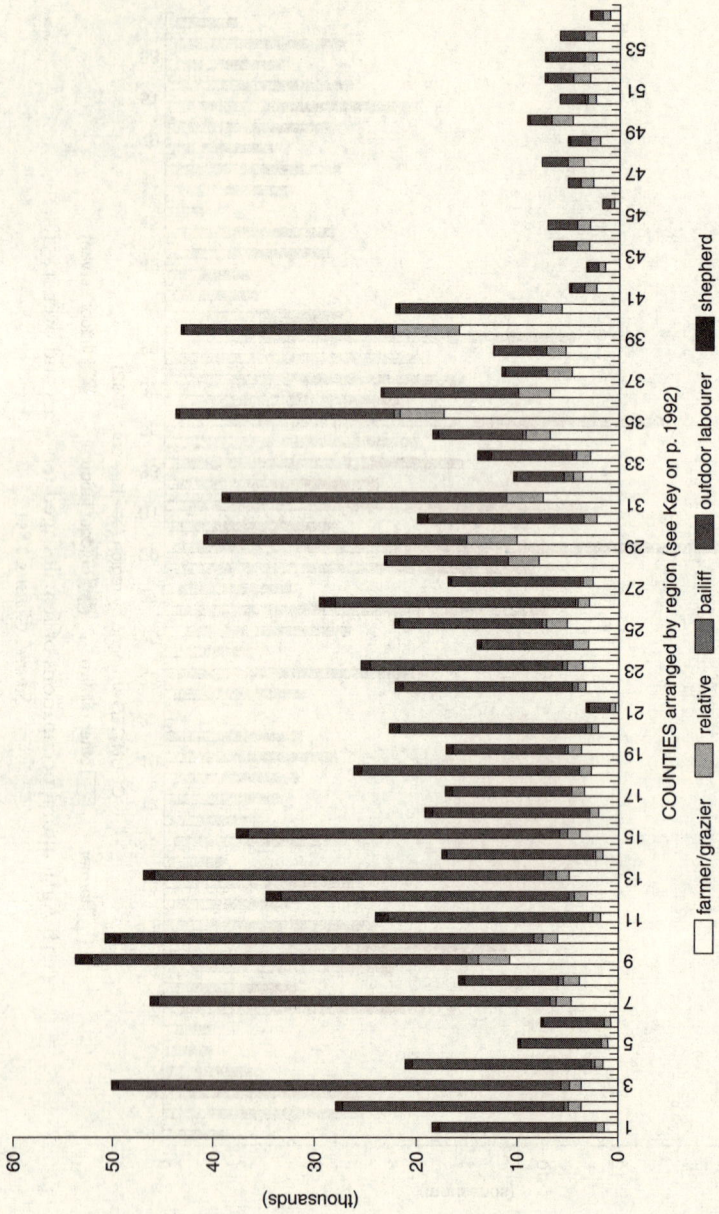

COUNTIES arranged by region (see Key on p. 1992)

□ farmer/grazier ▨ relative ▥ bailiff ■ outdoor labourer ■ shepherd

41.1c Agricultural occupations of males aged ten years and over, 1871

Source: Census, 1871

(thousands)

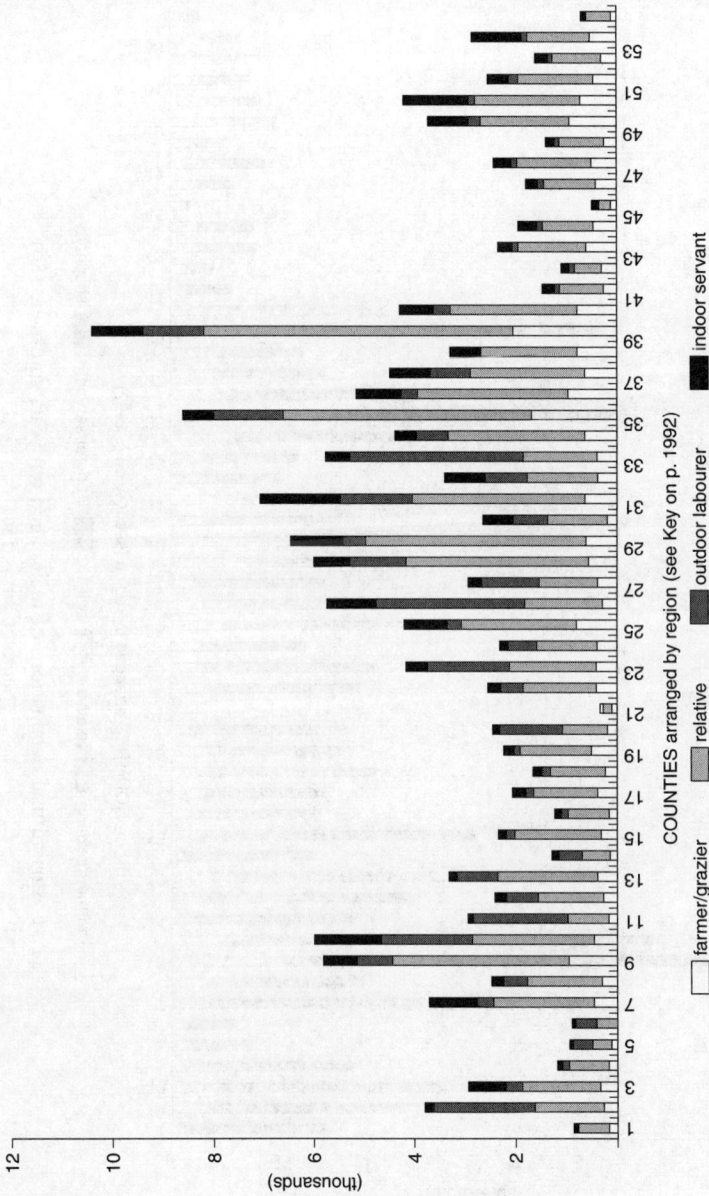

41.1d Agricultural occupations of females aged ten years and over, 1871

Source: Census, 1871

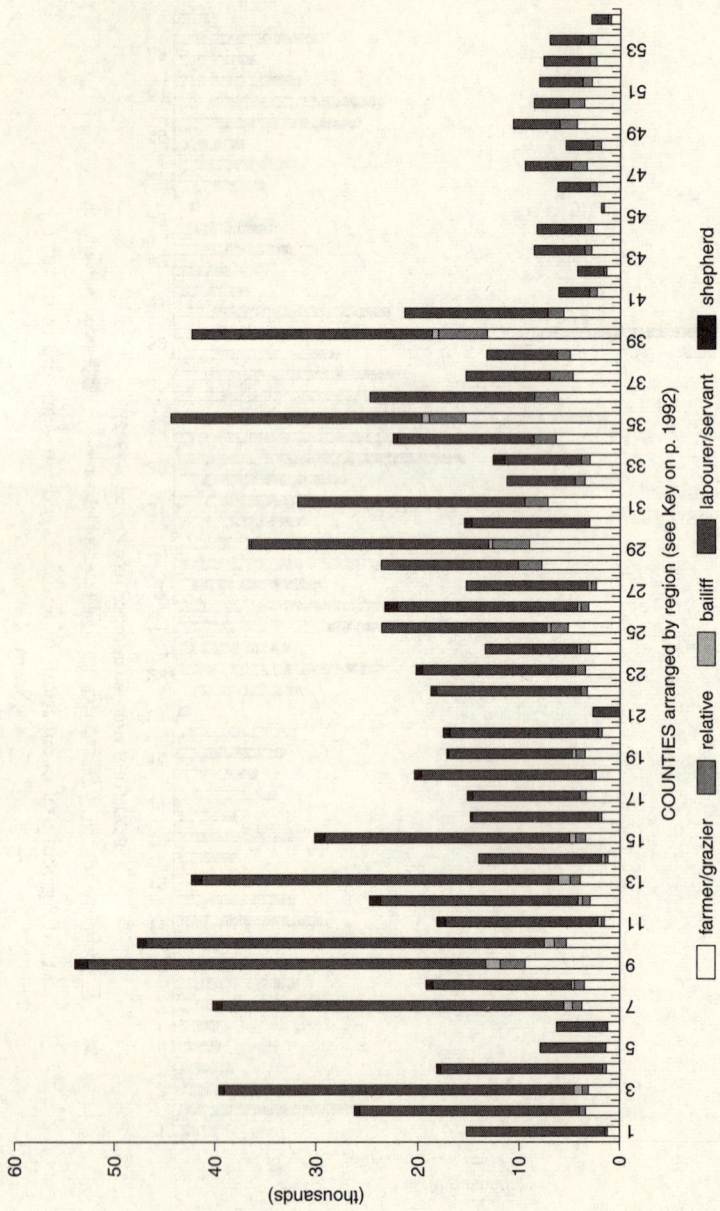

41.1e Agricultural occupations of males aged ten years and over, 1891

Source: Census, 1891

COUNTIES arranged by region (see Key on p. 1992)

farmer/grazier   relative   bailiff   labourer/servant   shepherd

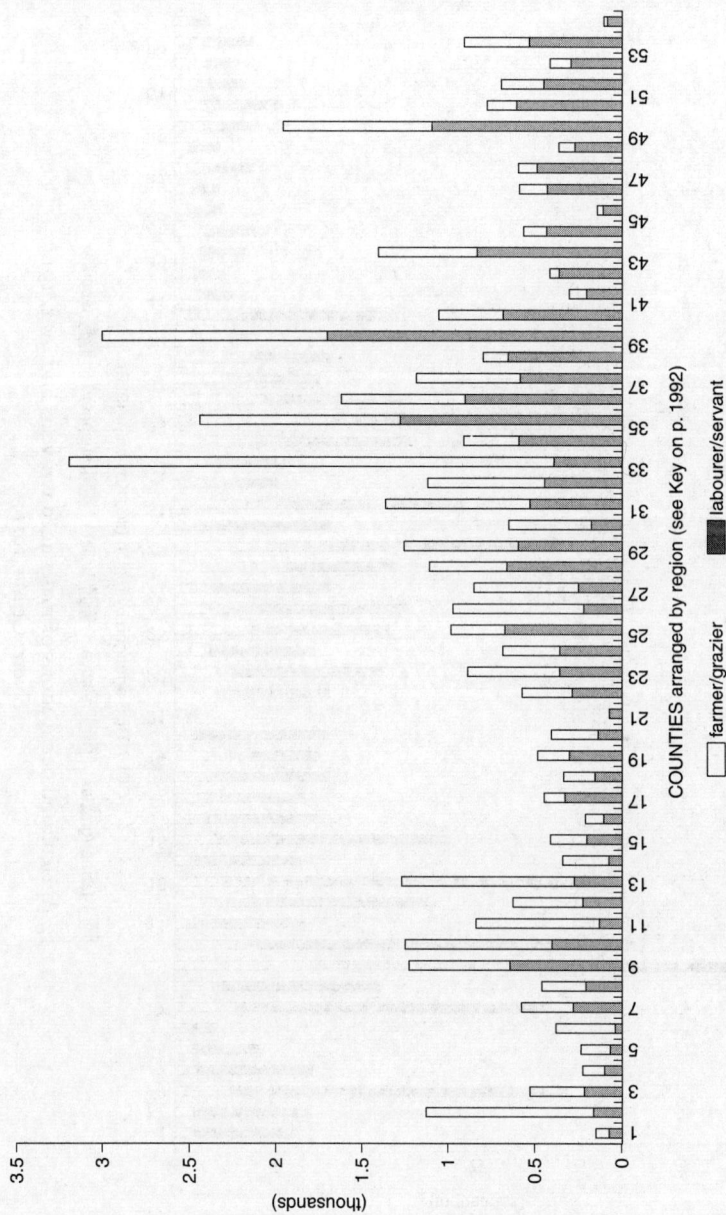

COUNTIES arranged by region (see Key on p. 1992)

■ labourer/servant

□ farmer/grazier

41.1f Agricultural occupations of females aged ten years and over, 1891

Source: Census, 1891

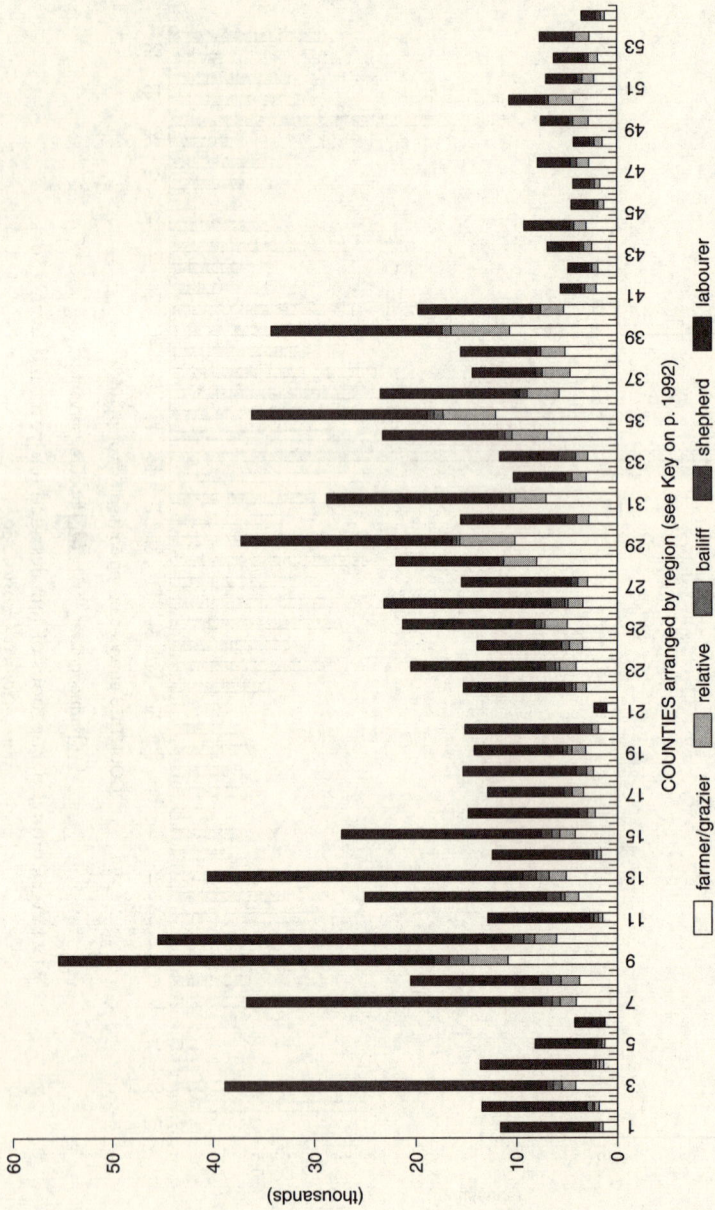

41.1g Agricultural occupations of males aged ten years and over, 1911

COUNTIES arranged by region (see Key on p. 1992)

farmer/grazier   relative   bailiff   shepherd   labourer

Source: Census, 1911

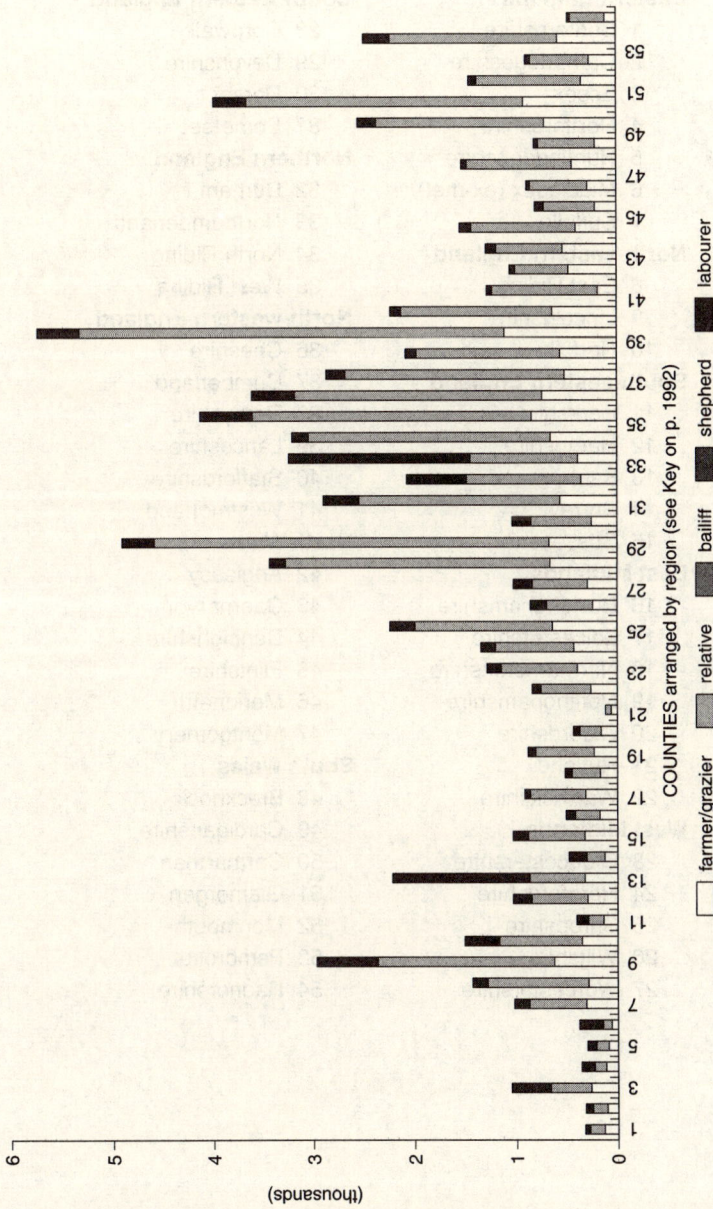

41.1h Agricultural occupations of females aged ten years and over, 1911

Source: Census, 1911

Key of counties for Figures 41.1a–41.1h

Eastern England
1 Bedfordshire
2 Cambridgeshire
3 Essex
4 Hertfordshire
5 Huntingdonshire
6 Middlesex (ex-met)
7 Suffolk

North-eastern England
8 East Riding
9 Lincolnshire
10 Norfolk

South-eastern England
11 Berkshire
12 Hampshire
13 Kent
14 Surrey
15 Sussex

East Midlands
16 Buckinghamshire
17 Leicestershire
18 Northamptonshire
19 Nottinghamshire
20 Oxfordshire
21 Rutland
22 Warwickshire

West Midlands
23 Gloucestershire
24 Herefordshire
25 Shropshire
26 Wiltshire
27 Worcestershire

South-western England
28 Cornwall
29 Devonshire
30 Dorset
31 Somerset

Northern England
32 Durham
33 Northumberland
34 North Riding
35 West Riding

North-western England
36 Cheshire
37 Cumberland
38 Derbyshire
39 Lancashire
40 Staffordshire
41 Westmorland

North Wales
42 Anglesey
43 Caernarvon
44 Denbighshire
45 Flintshire
46 Merioneth
47 Montgomery

South Wales
48 Brecknock
49 Cardiganshire
50 Carmarthen
51 Glamorgan
52 Monmouth
53 Pembroke
54 Radnorshire

CHAPTER 42

WAGES

BY BETHANIE AFTON AND MICHAEL TURNER

A. INTRODUCTION

A major problem which historians of all periods encounter when discussing wages is precisely what we mean by the term. Very often, at best, we are presented with information on wage rates (per hour, per day, sometimes per week or year). At times this may be only part of the story because there are supplements to wages through subsidiary employments or by-employments which it is necessary to know about to gain a full impression of earnings. At other times it may be because the wage rates are often those of the breadwinner alone, and ignore the employment of others who also contributed to family earnings. But at times the quoted rates, or indeed estimates of income, may be more optimistic than was, in fact, the case because full-employment is silently assumed. What of periods of slack employment, or no employment at all because of seasonal rhythms, and what of longer-term cycles of activity and, in our case, the repercussions arising from agricultural depression? In addition, wages in all trades in the eighteenth and nineteenth centuries might or might not have included allowances for food and other perquisites. Finally, the quoted wages or wage rates may hide a multitude of skills under the single banner 'agricultural wages'. The term might subsume the skills of shepherds, foremen, horse-keepers, ploughmen and general labourers.

The system for engaging agricultural labourers could, and did, vary regionally. In 1905, reviewing the period up to c. 1903, A. Wilson Fox observed that in Scotland and the north of England (Durham and Northumberland) farm servants were mainly hired on yearly or half-yearly terms, they were paid a regular wage, and they might receive board and lodging perquisites according to marital status. Elsewhere in the north and in Wales unmarried men were engaged by the year or half-year, and married men were engaged weekly. Elsewhere in England (except along the Welsh border) agricultural labourers were usually hired on weekly engagements, and men who were in charge of animals were engaged for longer periods. But by the early twentieth century he reports, 'Throughout the greater part of England the custom of lodging and boarding men in the farmhouses has practically ceased to exist', and whilst the hiring fair remained in place in the north of England and parts of Wales, elsewhere it was extinct, and even where it still remained it was in decline.[1]

If the terms of engagement varied across space and through time, so also did the method of remuneration: there could be time-payments with or without extra cash for piecework, and with or without allowances in kind. To this extent Wilson Fox summarises neatly the problem we face, where 'A comparison between the rates of wages of agricultural labourers in different parts of the UK is, therefore, of not much practical advantage without an accompanying statement of their actual annual earnings, including the extra amounts earned in cash from all sources, and the value of allowances in kind.'[2] To this effect he gives details from twenty particular farms in England and seven in Wales for the period 1902–3, evidence which is not capable of satisfactory tabulation.[3] We could not do better than repeat the warning issued by C. S. Orwin and B. I. Felton that 'while it is necessary for the purposes of an historical review to base estimates on the assumption of a full week's work, it must be remembered that this was not always forthcoming, and that by a certain proportion of the agricultural labourers the full weekly wage was not invariably attainable'.[4]

B. THE DATA

In this section we simply signal some of the problems associated with the data without offering a solution. Usefully we reiterate what E. H. Hunt says. Quoting many authorities, he said that agricultural wages are important in their own right because agriculture was a major employer of labour, but also important because 'farm wages were a reference point by which others fixed their position'.[5]

Table 42.1 is an estimate of average weekly wage 'rates' for agricultural labourers for 1850–1, produced by James Caird and based on coun-

---

[1] *Second Report on the Wages, Earnings, and Conditions of Employment of Agricultural Labourers in the United Kingdom*, by A. Wilson Fox, BPP, 1905, XCVII (but reprinted London, 1913), p. 2.

[2] *Ibid.*, p. 2, and in detail on both the terms of engagement and the methods of remuneration see pp. 9–26.   [3] *Ibid.*, pp. 46–64, 79–83.

[4] C. S. Orwin and B. I. Felton, 'A century of wages and earnings in agriculture', *JRASE*, 92, 1931, p. 249. We might simply add to this near-contemporary bibliography on wages the following, all of which with the foregoing is instructive in sounding out warnings about basing too much security on the available evidence: on wages for all trades see A. L. Bowley, *Wages in the United Kingdom in the Nineteenth Century* (Cambridge, 1900); and specifically on agriculture see A. L. Bowley, 'The statistics of wages in the United Kingdom during the last hundred years', Part I, 'Agricultural wages', *JRSS*, 61, 1898; and A. Wilson Fox, 'Agricultural wages in England and Wales during the last fifty years', *JRSS*, 66, 1903, pp. 273–348, reprinted in W. E. Minchinton (ed.), *Essays in Agrarian History* (Newton Abbot, 1968), vol. II, pp. 121–97.

[5] E. H. Hunt, *Regional Wage Variations in Britain, 1850–1914* (Oxford, 1973), p. 4.

ties. We parenthesise the word 'rates' because Caird is not consistent in his descriptions. By twists and turns he records his information as at times a rate, and at others a wage, and even as wages and rates in consecutive sentences.[6] We think that these are not yearly earnings divided into weekly averages, but rather they are weekly rates which make no allowances for winter and summer differences and off-season unemployment.[7] Caird omitted a number of counties, but he did signal differences in both broad regional and broad agricultural terms, which we also include. Finally we include in Table 42.1 an assessment of pauperism based on poor-relief expenditure, as calculated by Caird from contemporary parliamentary papers. These demonstrate the inverted relationship between high wages and low pauperism.

Table 42.2 (see also Figure 42.3) based on counties for various dates, shows nominal average weekly wages (i.e. wage rates), and Table 42.3 shows average weekly earnings which take into account allowances in kind (see also Figure 42.4). The main source of reference for wages is A. L. Bowley's collation of material for the nineteenth century, and for earnings we derive the data from E. H. Hunt's reworking of, and additions to, Bowley.[8] Table 42.3 contains a regional summary of earnings based on regions devised by Hunt.[9] Table 42.4 demonstrates a different regional summary (for England only), this time of weekly cash wages derived from Orwin and Felton, but based on a regional classification devised by Bowley. For comparison we include in this table Bowley's own regional summary for a selection of dates.[10]

---

[6] James Caird, *English Agriculture in 1850–51* (London, 1852), pp. 510–13; see also pp. 474–5 for a comparison with 1770.

[7] And yet we cannot be sure because examples are quoted which state boldly that the hiring of labourers was done by the year and the quoted weekly rates have been calculated accordingly. *Ibid.*, p. 420.

[8] Bowley, *Wages*, end page, and see also Bowley, 'The statistics of wages', pp. 704–7; *Second Report on the Wages*, by Wilson Fox, pp. 28–9, 72, 76, and 148–9, and see pp. 172–216 for a detailed breakdown over time of the summer and winter weekly cash wages for 142 specific farms; Hunt, *Regional Wage Variations*, pp. 61–3; see also E. H. Hunt, 'Industrialization and regional inequality: wages in Britain, 1760–1914', *JEH*, 46, 1986, pp. 935–66. We should note here for reference the important official report of W. C. Little, Senior Assistant Commissioner to the Royal Commission on Labour, especially the section *The Agricultural Labourer*, part of the *Memorandum on the Reports and Proceedings of the Royal Commission on Agricultural Interests, 1879–1882* (otherwise known as the Richmond Commission), BPP, 1893–4, XXXV–VI; and F. Purdy, 'On the earnings of agricultural labourers in England and Wales, 1860', *JRSS*, 24, 1861, pp. 328–73.

[9] Hunt, *Regional Wage Variations*, p. 64. Hunt reports that 'the regional average is the unweighted average of the counties in the region . . . Bowley found that weighting gave results practically identical to the arithmetic average'.

[10] Orwin and Felton, 'A century of wages', p. 233; Bowley, 'The statistics of wages', pp. 704–7. See also J. J. Macgregor, 'Labour costs in English forestry since 1824', *Forestry*, 20, 1946, pp. 30–43, who uses the Orwin and Felton data as a proxy for labour costs in forestry on the assumption, based on private estate records, that 'there is a close relationship between agricultural wages and the rates paid to forestry workers', p. 31.

At this stage we should point out that the Bowley material, and the use of that material by Orwin and others, depends very much on the dates which are chosen as reference points. For example, in Table 42.2 we note that at times around 1870 we have up to three observations from which to choose, 1867/70, 1869/70 and 1870/1, and we could add a fourth, 1872. For 1860 we have in fact 1860 or 1860/1, and for 1892 we have 1892 or 1892/3. Sometimes the data are incomplete, though Bowley in his original presentation provided the reader with interpolations. Readers should bear these points in mind when using sources which in turn are partly derived from Bowley, such as Orwin and Felton. With this in mind we produce Table 42.5, which is a regional assessment of weekly earnings for labourers derived from Orwin and Felton.[11]

Tables 42.6 (see also Figure 42.5) and 42.7 are compiled from A. Wilson Fox's report of 1905 but based on a farm survey for the year 1902. For English and Welsh counties they show average weekly cash wages, distinguishing summer and winter rates, and average weekly earnings, including the value of allowances in kind. In the case of Wales they show average yearly cash wages and estimated earnings, and for the English counties alone they show average weekly wages and earnings for horsemen, cattlemen and shepherds.

Thus far we have shown mainly county or regional estimates of wages and/or earnings but only for spot years. We continue this section with Tables 42.8 and 42.9 showing annual estimates of average weekly cash wages paid to 'Ordinary agricultural labourers' for the period 1850–1905. These are taken from an unpublished study by G. H. Wood, commissioned by the Board of Trade, Labour Department, and two studies by A. Wilson Fox. Table 42.8 shows estimates from 69 farms in England and Wales, from 12 farms from the Eastern Counties of England, and finally from 128 farms from England and Wales but for the truncated period 1874–1905. Table 42.9 shows estimates from 67 farms from England and Wales with a regional breakdown as indicated in the table. The counties within the regions are the same as those in Tables 42.6 and 42.7.[12] Figures 42.1 and 42.2 are the diagrammatic representation of Table 42.9 in both its nominal and real wage form, where the latter is the nominal wage deflated by the standard Sauerbeck-Statist wholesale price index. Finally, Table 42.10 presents index numbers of weekly rates of cash wages of ordinary labourers for England and Wales derived from returns to the Labour

[11] Orwin and Felton, 'A century of wages', pp. 233, 247.

[12] G. H. Wood, 'Rates of wages and hours of labour in various industries in the UK for a series of years', (London, 1907, Board of Trade, Labour Department, unpublished, copy consulted in Royal Statistical Society Library, University College London; *Second Report on the Wages*, by Wilson Fox, p. 68; Wilson Fox, 'Agricultural wages', esp. pp. 331–2, which in Minchinton (ed.), *Essays*, is pp. 181–2.

Department from 156 farms. Similar information is available for smaller numbers of farms for Scotland and Ireland, and for comparative purposes we include these figures and also those for the United Kingdom.[13]

---

[13] *Seventeenth Abstract of Labour Statistics*, BPP, 1914–16, LXI, p. 67. Whether it is methodologically correct to impute money wages using this table in conjunction with earlier tables is unclear. In general terms we do not think that this would be a too inaccurate procedure. See also C. H. Feinstein, 'New estimates of average earnings in the United Kingdom, 1880–1913', *EcHR*, 43, 1990, pp. 608–9, 614–16, for the same index, but rebased on 1911.

Table 42.1. *Agricultural labourers' wage rates and pauperism in 1850/51, by county*

County	Type of county	Wages (shillings per week)	Poor relief (shillings per £ on property)	Poor relief (shillings per head of population)	Paupers as a % of population (per cent)
Bedford	c s	9.0	2.04	7.65	11.6
Berkshire	c s	7.5	2.23	9.94	12.8
Buckingham	c/g s	8.5	2.37	10.23	14.6
Cambridge	c s	7.5	1.77	9.10	10.7
Cheshire	c/g n	12.0	1.02	3.67	5.5
Cornwall	n/a				
Cumberland	c/g n	13.0	1.08	4.25	6.2
Derbyshire	c/g n	11.0	1.04	3.71	4.2
Devon	c/g s	8.5	1.94	7.04	10.6
Dorset	c s	7.5	2.23	9.63	15.7
Durham	c n	11.0	1.29	3.63	5.2
Essex	c s	8.0	2.15	9.77	14.2
Gloucester	c/g s	7.0	1.69	6.88	9.9
Hampshire	c s	9.0	2.23	8.19	11.9
Hereford	n/a				
Hertford	c s	9.0	1.71	7.77	11.3
Huntingdon	c s	8.5	2.00	9.54	11.6
Kent	n/a				
Lancashire	c/g n	13.5	1.06	3.60	7.2
Leicester	n/a				
Lincoln	c n	10.0	1.19	6.65	7.5
Middlesex	n/a				
Monmouth	n/a				
Norfolk	c s	8.5	2.17	9.71	12.8
Northampton	c/g s	9.0	1.96	9.29	11.3
Northumberland	c n	11.0	1.21	5.63	6.7
Nottingham	c/g n	10.0	1.42	5.25	7.9
Oxford	c/g s	9.0	2.42	10.37	15.1
Rutland	n/a				
Shropshire	n/a				
Somerset	n/a				

(continued)

Table 42.1 (*cont.*)

County	Type of county	Wages (shillings per week)	Poor relief (shillings per £ on property)	Poor relief (shillings per head of population)	Paupers as a % of population (per cent)
Stafford	c/g n	9.5	1.08	3.69	4.3
Suffolk	c s	7.0	2.17	9.29	13.6
Surrey	c s	9.5	1.92	6.75	8.8
Sussex	c s	10.5	2.15	9.13	12.7
Warwick	c/g s	8.5	1.29	5.13	5.0
Westmorland	n/a				
Wiltshire N	c/g s	7.5	2.27	10.42	16.1
Wiltshire S	c s	7.0	2.27	10.42	16.1
Worcester	n/a				
Yorkshire E	c n	12.0	1.23	4.90	7.9
Yorkshire N	c n	11.0	1.08	5.90	6.5
Yorkshire W	c/g n	14.0	1.46	4.10	6.0
Wales	n/a				

Notes:

n/a   Data not available
c n   Predominantly corn growing – northern counties
c s   Predominantly corn growing – southern counties
c/g n   Mixed corn/grazing – northern counties
c/g s   Mixed corn/grazing – southern counties
Average wages per week   9.5 shillings
Average wages predominantly corn growing   9.08 shillings
Average wages predominantly mixed corn/grazing   10.0 shillings
Average wages of all northern counties   11.5 shillings
Average wages of all southern counties   8.42 shillings
Source: J. Caird, *English Agriculture in 1850–51* (London, 1852), pp. 512, 514.

Table 42.2. *Nominal weekly wages for agricultural labourers, by county (shillings and pence per week converted to shillings),*
c. 1850–1907

	Region	c. 1850	c. 1860	c. 1870	c. 1880	c. 1892	1903[a]	1907[b]
Bedford	5	9.0	10.25	12.0	12.5	12.5	14.0/13.0	13.75
Berkshire	6	7.5	11.0	9.5	12.25	11.0	13.0/12.0	12.75
Buckingham	5	8.5	10.667	11.417	12.75	12.333	14.0	14.833
Cambridge	5	7.5	10.0	12.083	13.5	12.0	12.0	13.25
Cheshire	1	12.0	11.5	13.5	15.667	15.0	17.8/17.0	16.583
Cornwall	4	8.667	10.5	11.0	13.75	14.0	15.0	15.0
Cumberland	1	13.0	13.5	15.0	18.0	18.0	18.0	19.25
Derbyshire	2	11.0	12.0	14.0	16.0	16.0	18.0	17.417
Devon	4	8.5	9.75	10.25	13.0	13.5	13.0	14.5
Dorset	4	7.5	10.0	10.25	10.75	10.0	12.0	12.083
Durham	1	11.0	14.25	16.0	17.75		20.0	
Essex	7	8.0	11.5	11.0	12.5	11.5	13.0	13.583
Gloucester	3	7.0	9.5	10.25	13.25	10.5	14.0/13.0	13.917
Hampshire	6	9.0	12.0	10.5	12.0	11.5	13.0	15.083
Hereford	3	8.417	9.083	10.0	11.75	11.0	13.0	13.917
Hertford	5	9.0	9.917	12.25	13.5	11.5	15.0	14.667
Huntingdon	5	8.5	10.75	11.0	12.5	13.0	13.0	13.417
Kent	6	11.5	11.917	14.0	15.75	14.5	16.0/15.0	16.333
Lancashire	1	13.5	14.0	15.0	18.0	18.0	20.0	16.167
Leicester	2	9.5	13.25	13.0	13.0	15.0	16.0	16.583
Lincoln	2	10.0	13.083	13.5	14.5	14.667	16.5/15.0	15.25
Middlesex	6	11.0	12.75	13.0	15.0	14.083	16.0	18.583
Monmouth	6	9.667	11.75	13.75	12.0	12.5	16.0/15.0	14.75

County								
Norfolk	7	8.5	10.5	11.5	12.5	12.0	13.0/12.0	12.583
Northampton	5	9.0	11.0	11.5	13.5	14.0	14.0	14.5
Northumberland	1	11.0	14.5	16.5	17.0	17.0	20.0	17.167
Nottingham	2	10.0	12.75	13.0	14.0	15.0	16.5	12.917
Oxford	5	9.0	10.667	11.417	12.75	12.0	12.0	
Rutland	2	11.5	12.0	12.0		15.0/14.0	15.0/14.0	
Shropshire	3	7.25	10.083	12.0	13.25	14.0	15.0	14.583
Somerset	4	8.583	9.833	10.0	12.5	11.0	14.0/13.5	14.0
Stafford	3	9.5	12.5	13.0	14.5	16.0	16.0	15.833
Suffolk	7	7.0	10.5	11.5	12.5	12.0	13.0	12.417
Surrey	6	9.5	12.75	13.5	15.5	15.0	16.0	16.667
Sussex	6	10.5	11.5	12.0	13.5	12.0	15.0	15.083
Warwick	3	8.5	10.75	12.0	14.25	11.5	15.0	15.333
Westmorland	1	12.0	14.25	15.5	18.0		18.0	19.083
Wiltshire	4	7.25	9.5	10.25	11.75	10.0	13.0/12.5	12.0
Worcester	3	7.667	10.0	11.0	13.0	12.0	15.0/14.0	14.5
Yorkshire E	1	12.0	10.917	13.167	15.0	15.5	16.5	16.417
Yorkshire N	1	11.0	13.833	13.5	17.0	15.5	18.0	16.75
Yorkshire W	1	14.0	13.583	15.25	17.0	16.0	18.0	17.5

Note:
For Bedfordshire, Berkshire, Cheshire, Gloucester, Kent, Lincoln, Monmouth, Norfolk, Rutland, Wiltshire and Worcester in 1903, the wages for the median summer and median winter rates are shown. For all other counties in that year those two rates were the same.

Table 42.2 (cont.)

	c. 1850	c. 1860	c. 1870	c. 1880	c. 1892	1902[a]
Wales	6.917	11.0	11.833	18.0	14.75	
Anglesey						16.0
Brecknock						17.0
Cardigan						14.5
Carmarthen						17.0
Caernarvon						18.0
Denbigh						17.0
Flint						17.0
Glamorgan						20.0
Merioneth						16.583
Montgomery						15.0
Pembroke						15.0
Radnor						14.0

Notes:

For completeness we could have included 1861, 1867/9, 1867/70, 1869, 1869/70, 1870/1, 1875, 1892/3.

For an explanation of the regions see Table 42.3.

[a] Derived from, *Second Report on the Wages, Earnings, and Conditions of Employment of the Agricultural Labourers in the United Kingdom*, by A. Wilson Fox, BPP, 1905, XCVII, pp. 148–9, 72. The data are also available at the rural district level. See also the Summer and Winter ranges and median rates for 1903 in Table 42.6 below.

[b] A. L. Bowley, 'Rural population in England and Wales: a study of the changes of density, occupations, and ages', *JRSS*, 77, 1914, pp. 597–645, esp. p. 645, where there are also revised versions for England only for 1861, 1870, 1880, 1892 and 1902.

Sources: a combination of A. L. Bowley, 'The statistics of wages in the United Kingdom during the last hundred years', 'Part 1, Agricultural wages', *JRSS*, 61, 1898, pp. 704–6. A. L. Bowley, *Wages in the United Kingdom in the Nineteenth Century* (Cambridge, 1900), end page.

Table 42.3. *Agricultural labourers' weekly earnings, by county, 1867/70–1907 (shillings and pence per week converted to shillings)*

County	Region	1867–70	1892	1898	1907
Bedford	3	14.250	14.833	16.167	16.250
Berkshire	3	13.500	14.750	15.083	16.667
Buckingham	3	14.250	15.000	15.167	16.917
Cambridge	3	14.250	14,833	16.417	16.250
Cheshire	8	16.000		18.000	19.000
Cornwall	2	12.500	16.250	16.583	17.583
Cumberland	9	18.500		18.750	19.250
Derbyshire	6	15.500	18.500	19.917	20.417
Devon	2	12.500	15.667	16.333	17.750
Dorset	2	11.500	14.500	14.750	16.083
Durham	10	20.000		20.750	21.750
Essex	1	14.250	15.500	15.500	16.333
Gloucester	2	12.750	15.000	15.083	16.250
Hampshire	3	14.000	15.000	16.583	17.417
Hereford	5	12.750	14.750	15.833	17.083
Hertford	3	13.500	14.333	16.083	16.833
Huntingdon	3			15.333	16.167
Kent	1	17.000	16.333	19.833	18.833
Lancashire	8	17.750	19.667	19.333	19.833
Leicester	6	13.500	16.250	17.167	18.750
Lincoln	7	16.250	15.750	16.254	19.417
Middlesex	1	17.250		19.417	20.250
Monmouth	4	13.500	15.333	16.667	18.083
Norfolk	3	14.750	15.000	14.750	15.333
Northampton	3	15.250	15.750	16.667	16.750
Northumberland	10	17.500	20.750	20.167	21.167
Nottingham	6	15.000		19.167	19.417
Oxford	3	13.500	15.000	14.667	14.917
Rutland	7				17.000
Shropshire	6	12.250	17.500	17.417	18.000
Somerset	2	12.250	12.500	15.833	17.250
Stafford	6	14.000	17.000	17.917	18.667
Suffolk	3			14.417	15.750
Surrey	1	17.500	16.000	19.000	18.750
Sussex	3	16.500	15.000	17.833	17.570
Warwick	6	15.000	14.583	16.167	17.167
Westmorland	9	18.500		18.750	19.083
Wiltshire	2	13.000	14.750	15.000	16.000
Worcester	6	13.500	13.500	17.083	16.250
Yorkshire E	7	17.500	17.000	18.500	19.250
Yorkshire N	7	17.500	17.000	18.667	19.583
Yorkshire W	8	17.500	17.000	18.583	20.000

(continued)

Table 42.3 (*cont.*)

County	Region	1867–70	1892	1898	1907
Anglesey	5	11.500		15.500	17.500
Brecknock	5	13.500		16.667	18.750
Cardigan	5	11.500		14.750	16.500
Carmarthen	4	11.500		16.583	18.083
Caernarvon	5	13.500		17.167	18.583
Denbigh	5	13.500		16.750	18.083
Flint	5	13.500		17.250	18.833
Glamorgan	4	14.000		19.083	19.250
Merioneth	5	13.500		16.417	18.167
Montgomery	5	13.500		15.417	16.583
Pembroke	4	11.500		15.833	17.250
Radnor	5	13.500		15.500	16.667

	Regions	1867–70	1898	1907
1	London and Home Counties	16.500	18.417	18.542
2	South-west	12.417	15.853	16.833
3	Rural south-east	14.372	15.750	16.417
4	South Wales	12.625	17.042	18.167
5	Rural Wales and Hereford	13.000	16.125	17.667
6	Midlands	14.083	17.833	18.372
7	Lincs, Rutland, E & N Riding	17.083	18.000	18.833
8	Lancs, Chesh., W Riding	17.083	18.667	19.583
9	Cumb. and Westmorland	18.500	18.750	19.167
10	Northumb. and Durham	18.750	20.459	21.459

Sources: 1892 from: A. L. Bowley, *Wages in the United Kingdom in the Nineteenth Century* (Cambridge, 1900), end page.
Other years from: E. H. Hunt, *Regional Wage Variations in Britain 1850–1914* (Oxford, 1973), pp. 61–3.

Table 42.4. *Weekly cash wages of agricultural labourers in the nineteenth century, by region (shillings and pence per week converted to shillings)*

1 According to Orwin and Felton

Region	1850/1	1860/1	1867/71	1879/81	1892/3
1 North	12.167	12.25	15.083	16.167	16.417
2 North Midlands	10.083	12.5	13.333	14.417	15.167
3 West Midlands	8.333	10.333	11.333	13.333	12.5
4 South-west	7.75	9.583	10.5	12.333	11.667
5 E & S Midlands	8.667	10.583	11.583	13.0	12.333
6 South and SE	9.083	11.833	11.667	13.833	12.833
7 East Anglia	7.833	11.083	11.0	12.5	11.833
National average	9.583	11.583	12.417	13.75	13.333

Region	1898	1902	1907	1914
1 North	16.833	18.0	17.25	22.25
2 North Midlands	16.167	16.75	16.25	16.583
3 West Midlands	13.833	13.833	14.667	15.25
4 South-west	12.583	13.25	13.667	14.25
5 E & S Midlands	13.0	13.583	13.917	15.583
6 South and SE	14.833	15.333	15.0	17.333
7 East Anglia	11.917	12.917	12.833	15.833
National average	14.417	14.667	14.917	16.75

2 According to Bowley

Region	1850	1860	1870	1880	1892
1 North	12.167	13.667	14.833	16.583	16.417
2 North Midlands	10.083	12.5	13.083	14.167	15.167
3 West Midlands	8.083	10.333	11.417	13.333	12.5
4 South-west	8.083	9.917	10.333	12.333	11.667
5 E & S Midlands	8.667	10.417	11.667	13.0	12.5
6 South and SE	9.083	11.833	12.083	14.0	13.0
7 East Anglia	7.083	11.417	11.167	12.5	11.833
English Average	12.167	13.667	14.833	16.583	16.417
Monmouth	9.667	11.75	13.75	12.0	12.5
Wales	6.917	11.0	11.833		
National Average	9.5	11.583	12.417	13.583	13.417

Notes: The regions are those defined by Bowley, as follows:
1 North   Cumberland, Westmorland, Northumberland, Durham, Yorkshire, Lancashire, Cheshire.
2 North Midlands Derby, Nottingham, Lincolnshire, Rutland, Leicester.

(continued)

Notes to Table 42.4 (*cont.*)

3 West Midlands	Warwick, Worcester, Stafford, Shropshire, Hereford, Gloucester.
4 South-west	Somerset, Cornwall, Devon, Dorset, Wiltshire.
5 E & S Midlands	Cambridge, Bedford, Huntingdon, Northamptonshire, Hertford, Buckingham, Oxford.
6 South and SE	Hampshire, Sussex, Kent, Surrey, Middlesex, Berkshire.
7 East Anglia	Essex, Suffolk, Norfolk.

Sources: C. S. Orwin and B. I. Felton, 'A century of wages and earnings in agriculture', *JRASE*, 92, 1931, pp. 233, 247, 255.
A. L. Bowley, 'The statistics of wages in the United Kingdom during the last hundred years. (Part 1) Agricultural Wages', *JRSS*, 61, 1898, pp. 704–7.

Table 42.5. *Weekly earnings of agricultural labourers, by region (shillings and pence per week converted to shillings), 1867/70–1912/13*

Region	1867–70	1892	1898	1902	1907	1912–13	1912–13
1 North	17.583	19.167	19.083	20.083	19.833	21.583	21.917
2 North Midlands	15.083	16.833	18.5	19.083	19.0	20.417	20.667
3 West Midlands	13.333	15.25	16.583	16.75	17.25	19.083	18.083
4 South-west	12.333	14.75	15.667	16.5	16.917	17.833	17.583
5 E & S Midlands	14.167	15.0	15.75	16.167	16.25	18.333	16.833
6 South and SE	15.667	15.417	17.917	18.5	18.25	19.917	18.333
7 East Anglia	14.5	15.25	14.833	15.75	15.833	18.167	16.333

Note: There are two estimates for 1912–13, derived from two different sources.
Sources: C. S. Orwin and B. I. Felton, 'A century of wages and earnings in agriculture', *JRASE*, 92, 1931, pp. 233, 247.

Table 42.6. *Average weekly wages/earnings of agricultural labourers, by county (shillings and pence per head converted to shillings), 1902–3*

A England

County	Region	a 1902	b 1902	Summer Wages (Range) 1903	Winter Wages (Range) 1903	Summer Wages (Median) 1903	Winter Wages (Median) 1903
Bedford	M	13.5	16.5	13 to 15	12 to 15	14	13
Berkshire	S	13.167	15.917	11 to 15	11 to 15	13	12
Buckingham	M	14.666	16.333	12 to 16	12 to 16	14	14
Cambridge	E	12.666	16.083	12 to 15	12 to 15	12	12
Cheshire	M	17.0	18.75	15 to 20	15 to 20	18	17
Cornwall	S	14.5	17.333	13 to 16	13 to 16	15	15
Cumb/Westmor	N	18.333	20.0	16 to 21	15 to 20	18	18
Derbyshire	M	18.666	20.583	18 to 21	17 to 21	18	18
Devon	S	13.750	17.083	11 to 16	11 to 16	13	13
Dorset	S	11.917	15.5	10 to 15	10 to 15	12	12
Durham	N	20.0	22.167	17 to 24	17 to 24	20	20
Essex	E	13.75	16.917	11 to 18	11 to 17	13	13
Gloucester	S	12.917	15.417	11 to 18	11 to 17	14	13
Hampshire	S	13.75	17.75	12 to 15	12 to 15	13	13
Hereford	S	13.25	16.25	12 to 16	11 to 15	13	13
Hertford	M	14.666	17.167	12 to 18	12 to 18	15	15
Huntingdon	E	13.666	16.167	13 to 15	13 to 15	13	13
Kent	S	16.333	19.583	13.5 to 20	13.5 to 19	16	15
Lancashire	N	18.83	20.583	18 to 22	18 to 22	20	20
Leics/Rutland	M	15.75	17.333	15 to 20	14 to 20	16	16
Lincoln	E	15.5	18.666	15 to 18	13.5/18	17	15
Middlesex	M	17.83	20.333	16 to 21	16 to 21	16	16
Monmouth	S	16.5	18.83	15 to 18	15 to 18	16	15
Norfolk	E	12.333	15.25	12 to 15	12 to 14	13	12
Northampton	M	14.083	16.167	12 to 16	12 to 15	14	14
Northumberland	N	19.333	21.583	17 to 22	17 to 22	20	20
Nottingham	M	17.25	19.75	16 to 20	15 to 20	17	17
Oxford	M	12.0	14.50	11 to 15	11 to 14	12	12
Rutland	M			14 to 15	14 to 15	15	14
Shropshire	M	14.666	18.0	14 to 16	14 to 16	15	15
Somerset	S	13.5	16.917	13 to 18	11 to 18	14	14
Stafford	M	15.917	18.333	14 to 18	14 to 18	16	16
Suffolk	E	12.75	15.5	12 to 14	12 to 14	13	13
Surrey	S	16.333	20.0	15 to 20	15 to 20	16	16
Sussex	S	14.83	17.583	12 to 18	12 to 17	15	15
Warwick	M	14.333	16.333	12 to 20	11 to 18	15	15
Wiltshire	S	12.75	15.666	12 to 16	11 to 16	13	13

(continued)

Table 42.6 (*cont.*)

A England (*cont.*)

County	Region	1902[a]	1902[b]	Summer Wages (Range) 1903	Winter Wages (Range) 1903	Summer Wages (Median) 1903	Winter Wages (Median) 1903
Worcester	M	14.167	16.417	12 to 19	12 to 19	15	14
Yorkshire E	N	16.75	19.167	15 to 19	15 to 19	17	17
Yorkshire N	N	16.75	18.83	16 to 21	15.5 to 20	18	18
Yorkshire W	N	17.25	19.83	15.5 to 22	15 to 22	18	18

Note:

[a] Average (median) weekly rates of cash wages paid through the year.

[b] Average (median) weekly earnings including the value of allowances in kind.

B Wales (In this Welsh case, this is specifically 'Married Labourers finding their own Food', and based on the modal rather than median averages.) (shillings and pence per head converted to shillings)

	1902[a]	1902[b]	Summer Wages (Range) 1902	Winter Wages (Range) 1902	Summer Wages (Modal) 1902	Winter Wages (Modal) 1902
Anglesey	16.0	16.583	14 to 17	14 to 17	16	16
Brecknock	17.0	18.5	13 to 20	13 to 20	17	17
Cardigan	14.5	15.666	12 to 18	12 to 18	14/15	14/15
Carmarthen	17.0	17.75	14 to 20	14 to 20	17	17
Caernarvon	18.0	18.666	17 to 21	17 to 21	18	18
Denbigh	17.0	17.666	15 to 20	15 to 18	17	17
Flint	17.0	18.417	16 to 19	16 to 19	17	17
Glamorgan	20.0	21.25	15 to 22	15 to 22	20	20
Merioneth	16.583	17.666	16.5 to 20	15 to 17	17	16
Montgomery	15.0	16.0	14 to 17	14 to 17	15	15
Pembroke	15.0	16.583	12 to 18	12 to 18	15	15
Radnor	14.0	16.83	12 to 15	12 to 15	14	14

Note:

[a] Average (modal) weekly rates of cash wages paid through the year.

[b] Average (modal) weekly earnings including the value of allowances in kind.

(continued)

Table 42.6 (*cont.*)

C Wales (Ranges in £s)

	Yearly cash wages of hired men 1902	Yearly total earnings of hired men including value of board and lodging 1902
Anglesey	22 to 32	39 to 49
Brecknock	20 to 30	37 to 47
Cardigan	20 to 28	37 to 45
Carmarthen	22 to 32	39 to 49
Caernarvon	24 to 34	41 to 51
Denbigh	18 to 28	35 to 45
Flint*		
Glamorgan	25 to 34	42 to 51
Merioneth	22 to 30	39 to 47
Montgomery	18 to 25	35 to 42
Pembroke	20 to 28	37 to 45
Radnor	20 to 27	37 to 44

Notes:
* According to Wilson Fox few men were hired in Flintshire, and in the counties of Anglesey, Brecknock and Caernarvon many of the hired men were on half-yearly engagements.
The regions, as defined by Wilson Fox, are:
E   Eastern Counties   M Midland Counties
N Northern Counties   S   Southern and South-Western Counties
Source: Second Report on the Wages, Earnings, and Conditions of Employment of Agricultural Labourers in the United Kingdom, by A. Wilson Fox, BPP, 1905, xcvii (and reprinted London, 1913), pp. 28–9, 72, 76, 148–9.

Table 42.7. *Average weekly wages/earnings of specialist labour in 1902, by county (shillings & pence per head converted to shillings)*

County	Region	Horsemen		Cattlemen		Shepherds	
		1	2	1	2	1	2
Bedford	M	15.917	18.25	16.0	18.333	15.833	18.167
Berkshire	S	14.833	17.917	14.667	17.667	14.5	18.333
Buckingham	M	15.917	17.917	16.25	17.25	16.0	18.083
Cambridge	E	14.583	17.917	13.917	17.25	14.667	18.917
Cheshire	M	18.667	20.667	17.5	19.833	18.0	20.417
Cornwall	S	14.583	18.417	14.833	19.083	14.917	19.333
Cumb/Westmor	N	18.0	20.25	18.667	21.167	17.833	20.417
Derbyshire	M	19.833	21.5	19.333	21.167	19.917	22.833
Devon	S	14.083	17.583	14.583	18.583	14.5	18.75
Dorset	S	13.0	16.25	13.833	16.583	13.167	17.833
Durham	N	20.0	22.167	20.417	22.333	21.25	24.667
Essex	E	15.167	18.917	16.167	19.25	15.417	19.583
Gloucester	S	14.0	17.167	14.083	17.083	14.25	17.667
Hampshire	S	14.25	17.917	14.917	18.083	14.167	18.5
Hereford	S	14.417	18.083	14.333	17.833	14.75	18.833
Hertford	M	16.167	19.0	16.75	19.167	16.333	19.75
Huntingdon	E	15.25	18.25	14.75	18.25	15.667	18.833
Kent	S	17.917	21.0	18.25	21.0	17.833	22.0
Lancashire	N	20.5	21.583	19.75	20.917	23.333	25.417
Leics/Rutland	M	17.5	19.833	17.667	19.917	16.917	19.583
Lincoln	E	15.25	19.667	14.917	19.583	15.083	20.583
Middlesex	M	18.75	22.917	18.25	22.667	20.0	22.083
Monmouth	S	16.833	19.667	15.667	19.0	16.25	20.417
Norfolk	E	14.083	17.167	14.0	17.5	14.417	18.167
Northampton	M	15.833	18.667	15.75	18.167	15.833	19.0
Northumberland	N	19.333	21.583	20.083	22.583	19.5	23.083
Nottingham	M	19.25	22.333	18.833	22.333	18.25	21.75
Oxford	M	14.25	16.917	14.5	17.5	14.333	17.667
Shropshire	M	15.25	19.083	15.583	19.333	16.083	20.333
Somerset	S	14.417	18.167	15.0	18.0	14.5	19.25
Stafford	M	16.417	19.5	16.667	20.0	16.5	20.25
Suffolk	E	14.5	17.583	14.5	17.417	14.75	18.833
Surrey	S	17.417	20.417	17.417	20.083	17.333	20.5
Sussex	S	16.417	19.417	17.083	19.917	17.083	19.917
Warwick	M	16.333	19.083	16.083	18.917	15.833	18.917
Wiltshire	S	13.917	17.333	14.167	17.167	13.917	17.583
Worcester	M	14.83	17.917	15.0	18.5	15.083	18.75

(continued)

Table 42.7 (*cont.*)

County	Region	Horsemen		Cattlemen		Shepherds	
		1	2	1	2	1	2
Yorkshire E	N	17.5	20.083	17.25	20.083	17.75	20.25
Yorkshire N	N	17.417	20.417	17.667	20.667	17.0	20.583
Yorkshire W	N	18.583	21.417	17.833	21.083	18.0	21.75

Notes:
1 Average weekly rates of cash paid throughout the year
2 Average weekly earnings (including the Value of Allowances in Kind)
E Eastern Counties   M Midland Counties
N Northern Counties   S Southern and South-Western Counties
In the cases of horsemen in Northumberland and Durham, these are in fact ploughmen or hinds.
Source: Second Report on the Wages, Earnings, and Conditions of Employment of Agricultural Labourers in the United Kingdom, by A. Wilson Fox, BPP, 1905, XCVII (and reprinted London, 1913), pp. 37–8, and, at the rural district level for 1903, see pp. 160–71.

Table 42.8. *Average weekly cash wages of ordinary agricultural labourers, by years, 1850–1905 (shillings and pence per week converted to shillings)*

Date	England and Wales 69 farms	Eastern counties 12 farms	England and Wales 128 farms	Date	England and Wales 69 farms	Eastern counties 12 farms	England and Wales 128 farms
1850	9.292	8.667		1878	13.667	13.042	14.042
1851	9.209	8.25		1879	13.292	12.459	13.708
1852	2.25	8.542		1880	13.209	12.083	13.625
1853	9.917	9.959		1881	13.167	12.0	13.625
1854	10.667	11.167		1882	13.209	12.083	13.625
1855	10.959	11.417		1883	13.25	12.042	13.667
1856	11.042	11.417		1884	13.209	11.959	13.625
1857	10.959	10.917		1885	13.083	11.417	13.459
1858	10.792	10.459		1886	12.917	11.209	13.333
1859	10.708	10.209		1887	12.792	10.959	13.209
1860	10.917	10.667		1888	12.792	10.625	13.209
1861	11.083	10.833		1889	12.875	10.917	13.333
1862	11.083	10.583		1890	13.042	11.042	13.5
1863	11.0	10.125		1891	13.333	11.792	13.792
1864	11.042	10.25		1892	13.417	11.667	13.833
1865	11.25	10.417		1893	13.292	11.375	13.75
1866	11.5	10.125		1894	13.25	11.083	13.666
1867	11.917	11.542		1895	13.209	11.0	13.708
1868	12.0	11.75		1896	13.333	11.125	13.75
1869	11.708	11.25		1897	13.417	11.5	13.875
1870	11.875	11.125		1898	13.708	12.25	14.125
1871	12.083	11.625		1899	13.875	12.542	14.333
1872	12.708	12.375		1900	14.459	13.125	14.833
1873	13.333	13.042		1901	14.542	13.209	14.917
1874	13.583	13.209	13.959	1902	14.583	13.209	14.959
1875	13.583	12.959	14.0	1903	14.583	13.209	14.959
1876	13.667	13.083	14.125	1904	14.583		14.959
1877	13.667	12.875	14.125	1905	14.625		15.0

Note: Cash wages are exclusive of extra payments for piecework, hay and corn harvests, overtime, etc., and also of the value of allowances in kind.
Sources: Second Report on Wages, Earnings and Conditions of Employment of Agricultural Labourers in the United Kingdom, by A. Wilson Fox, BPP, 1905, XCVII (and reprinted London, 1913), p. 68.
G. H. Wood, 'Rates of wages and hours of labour in various industries in the UK for a series of years' (unpublished, Board of Trade, Labour Department, 1907), p. 292. Copy in Royal Statistical Society Library, University College, London. The Wilson Fox series ends in 1903 and was extended to 1905 in Wood.

Table 42.9. *Average weekly cash wages of ordinary agricultural labourers, by years, 1850–1902 (in shillings and pence per week converted to shillings)*

Date	England and Wales	Northern counties	Midland counties	Eastern counties	Southern and SW counties
1850	9.292	11.417	9.708	8.625	8.458
1851	9.208	11.375	9.688	8.292	8.479
1852	9.25	11.354	9.646	8.521	8.5
1853	9.979	11.875	10.354	9.917	9.042
1854	10.75	12.604	11.208	11.146	9.438
1855	11.042	12.75	11.479	11.417	9.833
1856	11.104	13.354	11.5	11.375	9.813
1857	10.958	13.583	11.271	10.833	9.833
1858	10.75	13.458	10.854	10.354	9.833
1859	10.688	13.542	10.813	10.104	9.792
1860	10.917	13.646	11.021	10.604	9.958
1861	11.063	13.708	11.188	10.711	10.083
1862	11.021	13.625	11.313	10.458	10.083
1863	10.979	13.604	11.292	10.042	10.229
1864	11.021	13.604	11.269	10.167	10.313
1865	11.188	13.813	11.375	10.292	10.5
1866	11.5	14.208	11.563	11.042	10.667
1867	11.938	14.271	12.188	11.521	10.854
1868	12.021	14.646	12.188	11.688	10.896
1869	11.833	14.75	12.0	11.146	10.813
1870	11.854	14.875	12.042	11.063	10.854
1871	12.063	15.188	12.208	11.521	10.938
1872	12.688	15.854	13.021	12.333	11.333
1873	13.313	17.229	13.771	13.021	11.542
1874	13.521	17.708	14.042	13.208	11.667
1875	13.542	17.833	14.083	12.958	11.71
1876	13.667	17.708	14.104	13.063	11.958
1877	13.667	17.771	14.146	12.875	12.042
1878	13.625	17.479	14.042	13.0	12.063
1879	13.271	16.604	13.708	12.438	12.0
1880	13.188	16.729	13.667	12.063	12.042
1881	13.167	16.708	13.625	11.979	12.063
1882	13.208	16.792	13.583	12.063	12.104
1883	13.25	16.875	13.646	12.021	12.125
1884	13.188	16.979	13.521	11.917	12.104
1885	13.0	16.938	13.313	11.375	12.083
1886	12.875	16.688	13.229	11.167	11.938
1887	12.729	16.563	12.917	10.958	11.979
1888	12.688	16.667	12.875	10.667	12.083
1889	12.792	16.667	12.917	10.938	12.167

(continued)

Table 42.9 (*cont.*)

Date	England and Wales	Northern counties	Midland counties	Eastern counties	Southern and SW counties
1890	13.042	16.938	13.271	11.021	12.396
1891	13.333	17.25	13.479	11.813	12.458
1892	13.417	17.375	13.625	11.667	12.583
1893	13.292	17.375	13.5	11.333	12.563
1894	13.229	17.375	13.396	11.063	12.563
1895	13.229	17.375	13.396	11.0	12.583
1896	13.313	17.375	13.479	11.104	12.729
1897	13.417	17.375	13.521	11.458	12.792
1898	13.708	17.438	13.708	12.25	13.0
1899	13.917	17.479	14.021	12.5	13.146
1900	14.479	18.125	14.417	13.021	13.708
1901	14.583	18.229	14.417	13.125	13.833
1902	14.583	18.229	14.417	13.125	13.854

Notes:
England and Wales based on 64 English plus 3 Welsh farms.
Northern counties based on 8 farms.
Midland counties based on 22 farms.
Eastern counties based on 13 farms.
Southern and south-western counties based on 21 farms.
Sources: A. Wilson Fox, 'Agricultural wages in England and Wales during the last fifty years', *JRSS*, 66, 1903, pp. 331–2, reprinted in W. E. Minchinton (ed.), *Essays in Agrarian History* (Newton Abbot, 1968), vol. II, pp. 121–98, esp. pp. 181–2.

Table 42.10. *Wages of agricultural labourers in the UK, by years, 1880–1913*
(Index of weekly cash wages of ordinary labourers 1900 = 100)

Date	England and Wales 156 farms	Scotland 98 farms (Wages)	Scotland 98 farms (Earnings)	Ireland 27 farms	United Kingdom
1880	92.6	83.8	89.7	85.8	90.7
1881	92.5	84.2	89.2	85.5	90.6
1882	92.8	84.5	91.0	87.3	91.3
1883	93.0	85.1	91.4	88.3	91.7
1884	92.5	86.0	89.9	89.2	91.5
1885	91.2	86.3	90.3	89.4	90.7
1886	90.4	86.9	90.3	90.3	90.4
1887	89.5	87.6	89.5	91.3	90.0
1888	89.4	87.4	90.6	91.8	90.1
1889	90.1	89.0	90.2	91.9	90.6
1890	91.4	89.3	93.4	93.0	92.6
1891	93.3	91.0	94.1	93.4	93.4
1892	93.8	92.8	94.0	94.0	93.9
1893	93.1	94.2	93.3	93.7	93.3
1894	92.6	94.8	94.4	95.2	93.4
1895	92.8	96.5	93.7	95.8	93.6
1896	93.1	97.1	94.4	97.0	94.2
1897	93.9	97.3	98.2	97.3	95.1
1898	95.7	97.6	95.6	97.7	96.2
1899	97.1	98.6	98.4	99.4	97.8
1900	100.0	100.0	100.0	100.0	100.0
1901	100.7	101.7	100.3	101.0	100.7
1902	101.1	103.0	100.6	102.2	101.3
1903	101.2	104.4	101.8	103.4	101.8
1904	101.4	105.2	100.9	103.8	102.0
1905	101.7	105.9	101.2	105.4	102.6
1906	102.0	106.9	101.4	105.6	102.9
1907	102.0	107.3	104.3	106.2	103.2
1908	102.4	107.9	102.0	107.2	103.6
1909	102.6	108.2	102.8	108.1	104.0
1910	103.1	109.4	103.1	109.6	104.7
1911	103.2	110.9	104.1	112.1	105.5
1912	104.9	112.8	106.4	114.2	107.4
1913	109.0	115.8	108.5	117.8	111.2

Notes:
In 1913, England and Wales, and Ireland based on 135 and 26 farms only.

For Scotland the estimate is based on yearly, not weekly cash wages, and for married horsemen, not ordinary labourers. In addition, because allowances formed a larger proportion of the total earnings of farm servants in Scotland than elsewhere, the original source included an estimate of total earnings.
Source: Seventeenth Abstract of Labour Statistics, BPP, 1914–16, LXI, p. 67.

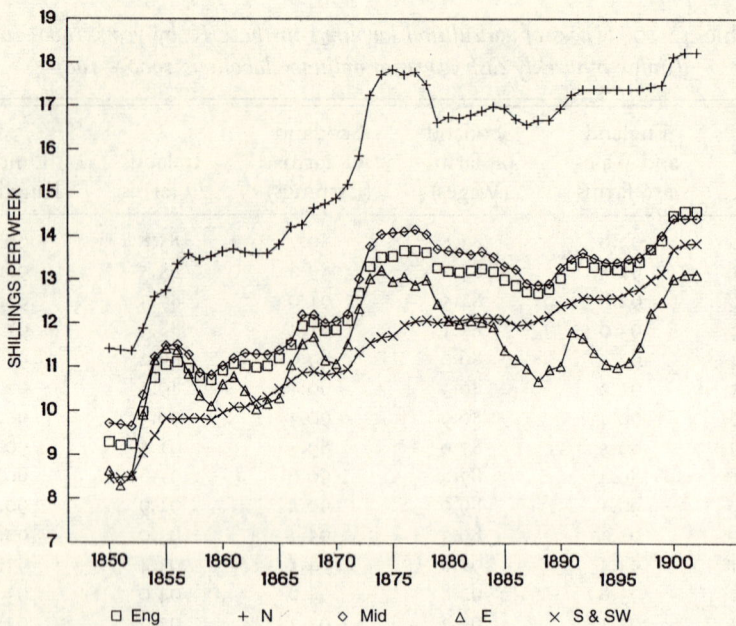

42.1 Agricultural labourers' wages, 1850–1900, by region

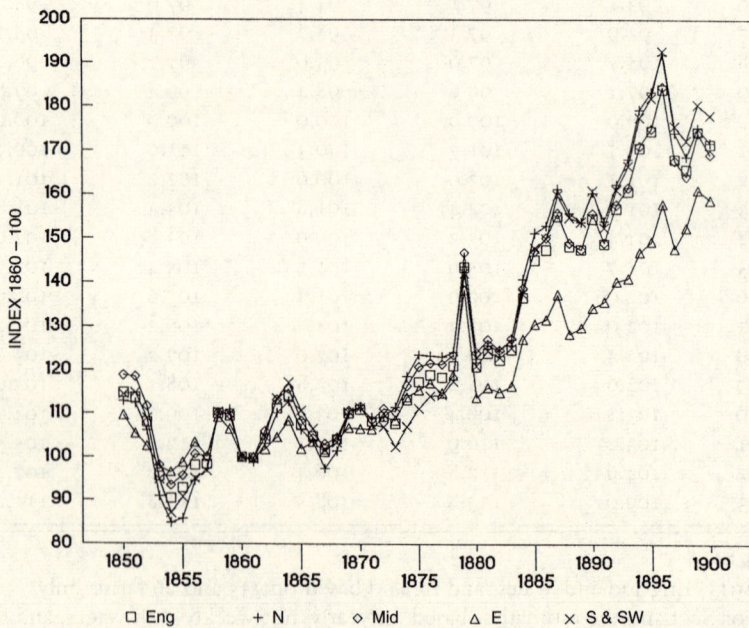

42.2 Agricultural labourers' real wages, 1850–1900, by region

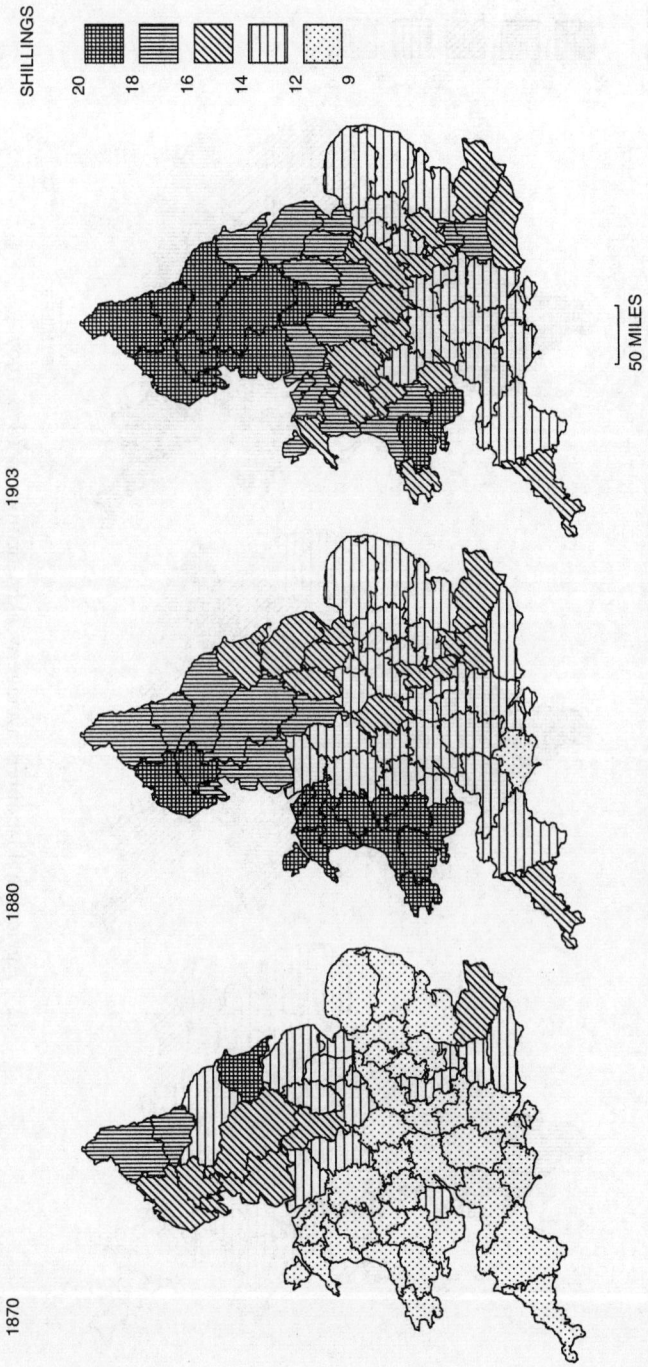

42.3 Agricultural labourers' average weekly wages, 1870–1903

SHILLINGS

20
18
16
14
12
9

50 MILES

1870   1880   1903

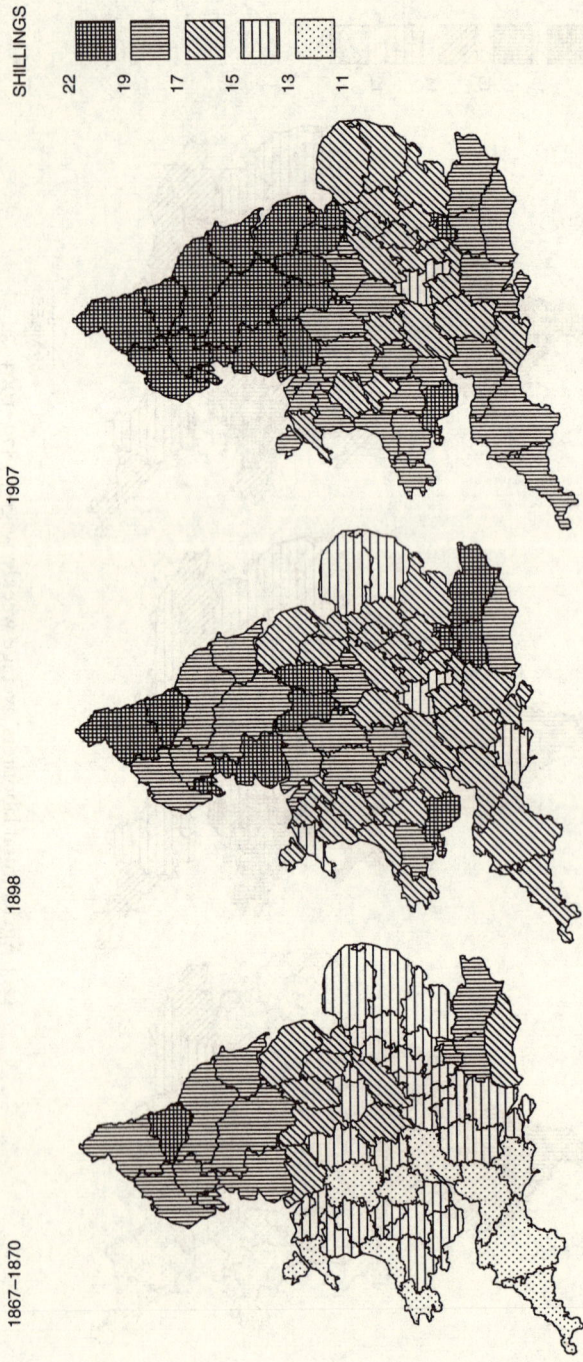

42.4 Agricultural labourers' average weekly earnings, c. 1870–1907

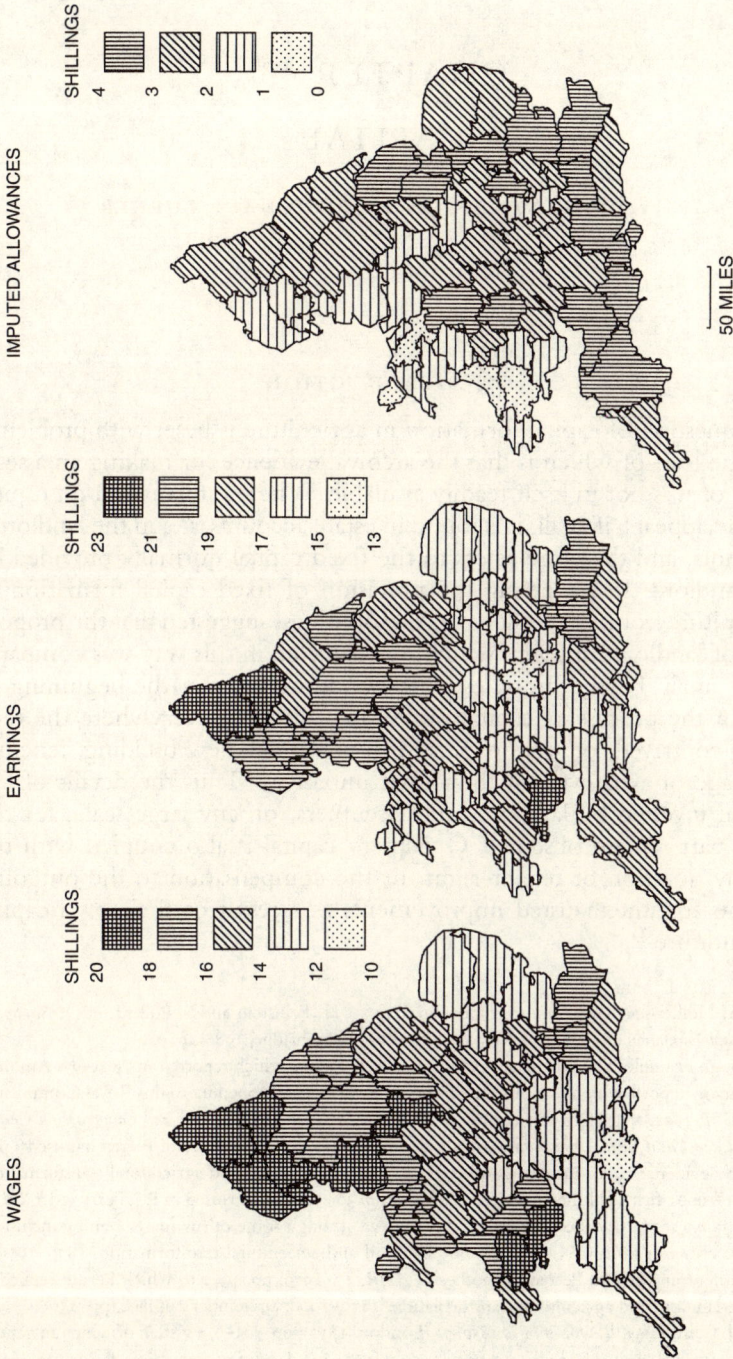

42.5 Agricultural labourers' wages, earnings and allowances in 1902: average weekly estimates

CHAPTER 43

CAPITAL

BY BETHANIE AFTON AND MICHAEL TURNER

A. INTRODUCTION

The question of capital formation in agriculture is beset with problems, not the least of which is that the archival evidence for making an assessment of it is not in itself readily available. Where information on capital usually appears, if at all, it is found in estate accounts, i.e. in the landlords' accounts, and therefore refers to the fixed capital normally provided by the landlord. In reviewing the provision of fixed capital formation in agriculture from 1770–1860, B. A. Holderness suggested that the proportion of landlords' income which was recycled in this way was comparatively small. Furthermore, estate expenditure was not the beginning or end of the process of capital provision, 'Virtually everywhere the tenantry contrived, or were expected to undertake new building, fencing, drainage or whatever on their own initiative.'[1] Thus the details of the capital invested by landlords and occupiers, on any large scale, for the most part remain obscure.[2] Occupiers' capital is also coupled with the thorny question of tenant-right, or the compensation to the outgoing tenants for unexhausted improvements as a result of their own capital expenditure.[3]

---

[1] B. A. Holderness, 'Agriculture, 1770–1860', in C. H. Feinstein and S. Pollard (eds.), *Studies in Capital Formation in the United Kingdom 1750–1920* (Cambridge, 1988), p. 9.

[2] Though examples litter the pages of the *RC on Agriculture* which reported in 1894–97. Amongst the many reports which comprise the minutes, evidence and appendix to this Royal Commission see BPP, 1894, XVI; 1897, XV; and especially *Particulars of the Expenditures and Outgoings on Certain Estates in Great Britain and Farm Accounts*, C. 8125, BPP, 1896, XVI. Richard Perren has used a part of the evidence in C. 8125 to good effect in his 'The landlord and agricultural transformation, 1870–1900', first published in *AHR*, 18, 1970, pp. 36–51, and reprinted in P. J. Perry (ed.), *British Agriculture 1875–1914* (London, 1973), pp. 109–28; though some of his analyses and conclusions drew criticism from C. Ó Gráda, 'The landlord and agricultural transformation, 1870–1900: a comment on Richard Perren's hypothesis', *AHR*, 27, 1979, pp. 40–2 to which Perren replied in 'The landlord and agricultural transformation, 1870–1900: a rejoinder', in *ibid.*, pp. 43–6.

[3] See J. Caird, *English Agriculture in 1850–51* (London, 1852), pp. 503–9, a system of which in general Caird was very critical. He felt it was subject to fraud and deception. See also E. P. Squarey, 'Farm

A very detailed estimate of 1861 by a Warwickshire tenant, Mr Charles Wratislaw, gives an appreciation of the thinking which must have engaged a tenant in the decision whether or not to take on a farm. On a notional 200 acres, equally shared between arable and pasture, Wratislaw estimated that the cost of supplying a variety of livestock, including 51 assorted cattle, 149 sheep, 4 pigs and a team of 6 horses, would come to £1,013; the cost in preparing the arable ground, seeding it, and all the other attendant costs in bringing it to the first year's harvest he estimated at a little under £264 for cultivation and £163 for labour; and the costs of implements might amount to a little under £207. Thus the input of tenant capital would total £1,647, or £8.23 per acre.[4] The capital input on an alternative 200 acres, where the arable portion allowed the cultivation of turnips and barley as part of a Norfolk four-course rotation which also included wheat, and seeds or pulse, amounted to £272 for cultivation, close to £128 for labour, £1,217 for stocking with animals and again nearly £207 for implements, or a little over £9 per acre. Indeed, four separate calculations for an equal quantity of 200 acres which varied from light soil to clay land and of four different qualities amounted to between £8 and £9.5 per acre.[5] This is very close to other late-nineteenth-century estimates. The rate of capital input by the tenant was related to the value of the land; the higher the value the higher the capital input per unit acre. Thus E. P. Squarey in 1878 reckoned that an annual value of 50s. per acre might be met with a tenant input of £12 per acre. In the examples he quotes, at 30s. per acre the tenant contribution was also £12, but at 20s.

---

capital', *JRASE*, 2nd ser., 14, 1878, especially pp. 435, 443–4; C. S. Orwin and E. H. Whetham, *History of British Agriculture 1846–1914* (Newton Abbot, 1964), pp. 247–8, 298–301. The tenants' right to the unexhausted improvements became recognised in law through the *Agricultural Holdings Act* of 38 & 39 Victoria, cap. 92, 1875. Initially this recognition remained a permissive not a compulsory contract between landlord and tenant, though it was revised more in the tenant's favour in 1883 when it became compulsory, and was revised again in 1906. For comment on the 1875 act see F. Clifford and J. A. Foote, 'English land law', *JRASE*, 2nd ser., 14, 1878, pp. 365–84. But for much comment on the failures of the 1883 act, see *RC on Agriculture, 'Garstang and Glendale': Reports by Assistant Commissioner Mr Wilson-Fox*, BPP, 1894, XVI, p. 21 for specific reference to the working of the act in Lancashire, and more generally for the country at large in *Final Report of her Majesty's Commissioners appointed to inquire into the Subject of Agricultural Depression*, BPP, 1897, XV, p. 301.

[4] Charles Wratislaw, 'The amount of capital required for the profitable occupation of a mixed arable and pasture farm in a midland county', *JRASE*, 1st ser., 22, 1861, pp. 167–89, especially pp. 169–73.

[5] *Ibid.*, pp. 174–8. It should be added that this article was published after the death of its author. It is clear from the appended remarks on pp. 182–9 that the editor would normally have required some revisions to the estimates to answer some practical, and hence financially important agricultural matters which he thought the article needed to address. Had these questions been addressed it seems unlikely that the final estimates would have strayed much beyond a capital input measure of £8–£9 per acre.

it would amount to £10 per acre, and at the highest quoted example of 63s. per acre it would rise to £15 per acre. By a reverse rule, however, the distribution of landlord and tenant capital would operate in the other direction. Thus the higher value land, even though it would attract a higher input per unit acre from the tenant, would nonetheless result in a lower proportionate contribution from the tenant. At a value of 20s. per acre per annum and a tenant input of £10 per acre, the tenant would provide 25 per cent of the capital, but at 63s. per acre per annum and a tenant input of £15 per acre the tenants' share of capital would fall to 13.7 per cent.[6]

B. THE DATA

Farm capital of course is not simple to define. Unlike rent and wages, which in their own respects are regular periodic payments from tenants to landlords in one direction, and tenants to labourers in another, farm capital payments can work in a number of different directions and invariably on an irregular basis. Some of that capital was fixed in the form of one-off payments, and some of it was what we call circulating, that is, it took the form of regular injections for the more efficient running of a farm. In general, the fixed capital was the responsibility of the landlord, and the circulating was the responsibility of the tenant, though there are grey areas regarding this division of responsibilities.[7] Quite a number of late-nineteenth-century estimators turned their hands to deriving national estimates of farmers' capital, which in most cases was an inventory of moveable capital, including crops, livestock and associated products, implements, machinery and tenants' fixtures such as harnesses. These national estimates are listed in Table 43.1[8]

When it is understood that these are little more than informed guesses, and not necessarily based on the same procedures, there is an evident bunching on or around the range of £360 million to £400 million. In addition, three further estimates by *The Economist* for 1895, 1905 and 1909 put UK farmers' capital at £368 million, £340 million and £348 million, respectively.[9]

---

[6] For these estimates and more general contemporary ideas on the distribution of capital, see, Squarey, 'Farm capital', pp. 429–44, especially pp. 431–2.

[7] *Ibid.*, pp. 431–2. On that blurred distinction see also B. A. Holderness, 'The Victorian farmer', in G. E. Mingay (ed.), *The Victorian Countryside* (London, 1981), vol. 1, p. 233.

[8] From R. H. Rew, 'Farm revenue and capital', *JRASE*, 3rd ser., 6, 1895, pp. 36, 45, where these individual estimates are collated. See also J. Stamp, *British Incomes and Property* (London, 1927), pp. 386–8.

[9] Quoted in J. R. Bellerby, 'Farm occupiers' capital in the United Kingdom before 1939', *The Farm Economist*, 7, no. 6, 1953, p. 259; also in Stamp, *British Incomes*, p. 387.

Even to gain an appreciation of one item of capital improvement, such as underdrainage, is not simple. As the most recent historian of nineteenth-century underdrainage explains, there is not even agreement on the amount of land which was underdrained, let alone other considerations.[10] Nevertheless, some statistics can be compiled to show the extent of drainage in England as a capital improvement. Table 43.2 summarises some of the main findings.[11]

As will be apparent from the main chapter on output and production, and from the section on output in this statistical appendix, the modern pioneer of annual estimates of output, input and capital was J. R. Bellerby and his team of researchers working at Oxford in the first decade after the Second World War. Table 43.3 has been constructed from the surviving manuscripts and working papers of that team.[12] The estimation of the elements in the table is explained in the manuscripts, and these are simplified and more readily accessible in an article written by A. J. Boreham, the principal researcher within the Bellerby team.[13] Boreham's annual averages are reproduced as Table 43.4.

There is a certain incompatability between the annual figures in Table 43.3 and the annual averages in Table 43.4. The main difference involves a revision of the estimate for crops. It is quite clear in all of the Bellerby manuscripts, and subsequently in the published work, that minor amendments of the original estimates were made, usually of an unspecified nature. In addition, the estimate for machinery and implements took the form of a quoted value for 1937/8 which was extrapolated back to the 1860s. This has drawn the criticism from Feinstein that the result 'does not look plausible'.[14]

The most modern estimators have been B. A. Holderness and C. H. Feinstein. For 1850 for Great Britain Holderness has estimated that fixed capital formation amounted to £6.1 million, or 13 per cent of gross rents and tithes, which was an estimated rise from £1.5 million or 7 per cent from 1770. By 1860 he suggested that the rising trend had continued to £7.0 million or 14 per cent of gross rents and tithes.[15] We have already

---

[10] For this and other debates about underdraining see A. D. M. Phillips, *The Underdraining of Farmland in England during the Nineteenth Century* (Cambridge, 1989), especially chapter 1.

[11] *Ibid.*, pp. 74–6, 124–5, 178–83, all of which data are also available on a county basis.

[12] University of Reading, Rural History Centre, Bellerby Manuscripts, D/84/8/17, pp. 152 and 154, and between pages 133–9 and 142–3.

[13] A. J. Boreham, 'A series of estimates of occupiers' capital, 1867–38', *The Farm Economist*, 7, no. 6, 1953, pp. 260–3. See also J. R. Bellerby, 'Farm and non-farm capital, 1867–1938', *The Farm Economist*, 8, no. 3, 1955, pp. 17–20.

[14] C. H. Feinstein, 'Agriculture', in Feinstein and Pollard (eds.), *Studies in Capital Formation*, pp. 274–5n.

[15] Holderness, 'Agriculture, 1770–1860', p. 10. And for the late nineteenth century, see Rew, 'Farm revenue', pp. 33–4. And for our period, Holderness, *supra*, Chapter 13.

quoted estimates of tenants' capital amounting to £8–£9 per acre, and Holderness quotes another for 1850 suggesting that the tenant of a mixed farm, farmed intensively, might have required about £10 per acre of farming capital. Late-nineteenth-century estimates fluctuated from anything near to £8 per acre, rising to £14 in one inflated estimate.[16]

The most substantial capital items of stock in farming were the buildings, but such provision did not happen as a single event at the same time throughout the farming world. Nevertheless, it is possible to take a snapshot of the current inventory of buildings and to give it a value. Holderness has estimated that the undepreciated aggregate capital value of farm buildings in Britain in 1860 was £134 million without dwellings, and £253.8 million with dwellings. When depreciation is taken into consideration his estimates reduce to £82 million and £156 million respectively.[17] Table 43.5 summarises estimates for various other items of capital formation which Holderness has pieced together for the years 1850 or 1860. It is important to understand that a good deal of intuitive guesswork has gone into these estimates.[18]

For the period from the mid-nineteenth century the actual archival references to capital are either not available, or yet await a substantial research project to extract them. We are guided therefore by indirect evidence. The Bellerby estimates fall into this category, but they may now have been superseded by those of C. H. Feinstein. He has estimated the level of fixed capital formation on farm buildings and works indirectly by assuming a relatively fixed proportion of rent received by landlords as an estimate of their outlay on these items. This is based on the informed research of a number of nineteenth-century specialists and the voluminous evidence of the *Royal Commission on Agriculture*, which presented its evidence from 1894–7. According to this approach the level of capital expenditure as a proportion of rent received varied from a low of 10 per cent at the bottom of the late-nineteenth-century depression in the 1890s, to a high of 13.5 per cent during the height of 'High farming'. These estimates are summarised in Table 43.6.[19] Table 43.7 is the current best-guess estimates of fixed capital formation in British agriculture. They measure the capital value of the gross and net stock of farm buildings and works at constant (1900) prices, also that of

[16] Orwin and Whetham, *British Agriculture*, p. 35. And even as low as £6, as summarised in Stamp, *British Incomes*, p. 388.   [17] Holderness, 'Agriculture, 1770–1860', p. 18.

[18] Frankly stated throughout Holderness, 'Agriculture, 1770–1860', pp. 9–34. See also B. A. Holderness, 'Prices, productivity, and output', chapter 2 in G. E. Mingay (ed.), *AHEW*, vol. VI 1750–1850 (Cambridge, 1989), p. 189.

[19] For which, see Feinstein, 'Agriculture', pp. 268–9 and 272, and the intervening scripted explanation. See also the scepticism regarding the value to the agricultural historian of these and the estimates in Table 43.6, in F. M. L. Thompson's review in *AHR*, 37, 1989, pp. 104–5.

the equivalent capital formation with respect to farm machinery, implements and vehicles, and finally contain an estimate of the total fixed assets within agriculture.[20] Table 43.8 gives a current prices estimate of farm stocks for benchmark years in Great Britain and the UK, and Table 43.9 gives the best estimates available for annual gross domestic fixed capital formation, and gross and net stock of domestic reproducible fixed assets in UK agriculture (these figures are available also in constant-price terms).[21]

The Royal Commission on Agriculture (1894–97) provides a wealth of information on capital formation, both at the tenant level throughout its reports, and more systematically for landlords' provision in its in depth review of twenty-nine, mainly substantial, estates in England and one in Wales.[22] Table 43.10 summarises the expenditure and other outgoings for these estates. Most of them report appropriate data for the period 1872 to 1892. The data are not consistently reported or summarised for the years before the 1870s. In addition, it is not possible in most cases to extract the precise details of landlords' capital formation. The best we can do is present a composite assessment of landlords' expenditure on items of fixed capital, such as new buildings and other unspecified permanent improvements, combined with reported expenditure on what might be taken to be allied items, the so-called 'Repairs, Fences and Insurance'. Nevertheless, as a proportion of rent received it appears as though the landlords relaxed their direct involvement in the expenditure on their estates during the period of 'High Farming', but increased it slightly with the onset of the Great Depression. Indeed, it was reported directly that the evidence 'shows generally that the expenditure by many landlords, and especially the owners of large estates has been heavy and continuous for many years'.[23] If on further investigation this turns out to be more generally the case, then it seems to overturn much orthodox thinking. Certainly this characterisation seems to contradict Feinstein's quantitative appraisal, though the interpretation of the available evidence and documentation is highly problematical. Not the least of the problems is the biased nature of the estate sample, the average size of which was 16,000 acres.[24] Perren used a subset of the

---

[20] Feinstein, 'Agriculture', pp. 272, 278, 279, and intervening script.

[21] Feinstein, 'Agriculture', pp. 394, 429–30, 433–4, 437–8, 444–5, 448–9, 452–3.

[22] They cover some Scottish estates as well. See *Particulars of the Expenditures and Outgoings*, BPP, 1896, XVI. See also the references to Perren in note 2 above.

[23] *Final Report of her Majesty's Commissioners*, BPP, 1897, XV, p. 287 para. 155, with examples in subsequent paragraphs, pp. 287–90.

[24] A point made by Feinstein who recognised the possible bias but suggested that in his calculations this had been allowed for. The nature of the allowance is not specified. Feinstein, 'Agriculture', p. 268, n7. Average size of estate taken from Table 43.7.

same evidence to make a distinction between arable and livestock areas, and he also indicated that some estate owners returned answers to questions which blurred the distinction between capital expenditure on new items from the similar expenditure on maintaining and repairing old items.[25]

[25] Perren, 'The landlord', in Perry (ed.), *British Agriculture*, p. 114.

APPENDIX

Table 43.1. *Estimates of farmers' or farm occupiers' capital in the UK, 1878–94*

Authority	Date	Capital (£million)
Caird	1878	400.000
Craigie	1878	376.000
Giffen	1878	667.520
HM Treasury	1885	300.000
Giffen	1890	521.864
Turnbull	1893	366.744
Rew	1891/3	319.014
Harris	1894	352.180

Source: R. H. Rew, 'Farm revenue and capital', *JRASE*, 3rd ser., 6, 1895, pp. 36, 45.

Table 43.2. *Loan capital in English underdrainage, 1847–99*

A. Source of loan capital 1847–99
    Public Money Draining Acts   £2,008,803
    General Land Drainage Company   £1,108,962
    Land Improvement Company   £1,962,981
    Land Loan Company   £301,344
    Improvement of Land Act   £117,558

Total   £5,499,648

B. Chronological distribution of loans, 1847–99 (percentages)

1847–9	1.7	1870–5	9.6
1850–4	14.8	1876–9	5.2
1855–9	20.7	1880–4	9.3
1860–4	18.9	1885–9	3.8
1865–9	14.2	1890–4	1.5
		1895–9	0.3

C. Number of landowners using draining-loan capital, average amount of
    loan, and intensity of use, by estate size, 1847–99

Under 1,000 acres	1668 owners	at £626 per estate
	467,336 acres	at £2.23 per acre
	which was 4.6 per cent of all owners	
1,000–2,999 acres	451	at £2,522
	821,956 acres	at £1.38 per acre
	which was 21.1 per cent of all owners	
3,000–9,999 acres	342	at £5,547
	1,836,728 acres	at £1.03 per acre
	which was 29.8 per cent of all owners	
10,000+ acres	124	at £9,780
	2,296,587 acres	at £0.53 per acre
	which was 44.4 per cent of all owners	
Untraceable	118	at £1,766

Source: A. D. M. Phillips, *The Underdraining of Farmland in England during the
Nineteenth Century* (Cambridge, 1989), pp. 74–6, 124–5, 178–83.

Table 43.3. *Farm occupiers' capital in the UK, 1867–1914 (in £s million)*

	Crops	Livestock	Machinery and implements	Stock of feedstuff	Other	Total
1867	238	164.9	5.3	3.9	21.7	433.9
1868	329	157.3	5.3	4.9	26.1	522.6
1869	263	170.1	5.3	4.8	23.3	466.5
1870	243	170.9	5.4	4.9	22.3	446.5
1871	239	177.4	5.4	5.3	22.5	449.6
1872	234	193.7	5.4	5.7	23.1	461.9
1873	234	212.7	5.5	5.6	24.1	481.9
1874	280	197.5	5.5	6.1	25.7	514.8
1875	237	205.5	5.4	6.5	23.9	478.4
1876	219	198.9	5.4	6.9	22.6	452.8
1877	232	192.6	5.3	6.0	22.9	458.8
1878	252	189.4	5.3	6.5	23.9	477.1
1879	139	174.4	5.2	5.5	17.1	341.1
1880	213	180.2	5.0	6.1	21.3	425.6
1881	187	174.0	4.9	6.1	19.6	391.5
1882	201	181.7	4.8	5.5	20.7	413.6
1883	200	186.2	4.7	6.1	20.9	417.9
1884	188	177.5	4.7	5.4	19.8	395.3
1885	192	164.6	4.6	5.5	19.3	386.0
1886	190	161.3	4.4	4.9	19.0	379.5
1887	164	143.8	4.3	4.8	16.7	333.6
1888	181	153.7	4.2	4.8	18.1	361.7
1889	199	158.5	4.1	5.5	19.3	386.4
1890	187	163.4	4.1	5.6	19.0	379.0
1891	205	163.1	4.1	5.7	19.9	397.8
1892	181	163.7	4.2	5.8	18.7	373.3
1893	154	161.2	4.1	5.4	17.1	341.7
1894	184	154.5	4.1	5.3	18.3	366.2
1895	156	156.3	4.0	4.9	16.9	338.2
1896	144	150.4	4.1	5.1	16.0	319.6
1897	169	155.9	4.2	5.2	17.6	351.9
1898	202	154.8	4.1	6.1	19.3	386.3
1899	157	165.8	4.1	6.5	17.5	350.9
1900	172	178.3	4.1	6.6	19.0	380.0
1901	175	168.2	4.1	6.7	18.6	372.6
1902	211	175.9	4.2	6.7	20.9	418.7
1903	167	167.1	4.3	6.8	18.2	363.3
1904	172	169.3	4.4	6.5	18.5	370.8
1905	175	170.2	4.4	7.2	18.8	375.6
1906	162	178.1	4.6	7.1	18.5	370.3
1907	181	184.2	4.7	8.0	19.9	397.8

(continued)

Table 43.3 (cont.)

	Crops	Livestock	Machinery and implements	Stock of feedstuff	Other	Total
1908	175	178.1	4.9	6.8	19.2	384.1
1909	186	174.7	5.4	8.1	19.7	393.9
1910	185	185.3	5.9	8.3	20.2	404.8
1911	183	178.8	6.1	7.7	19.8	395.3
1912	206	192.0	6.5	9.0	21.8	435.2
1913	194	194.8	7.2	8.8	21.3	426.2
1914	191	197.8	8.2	7.9	21.3	426.3

Note:
Category 'Other' calculated as 5 per cent of Total.
Source: Reading University, Rural History Centre, Bellerby Manuscripts, D/84/8/17: Crops p. 152; Livestock between pp. 133–9, Machinery and Implements between pp. 142–3; Stock of Purchased Feeds, p. 155.

Table 43.4. *Farm occupiers' capital in the UK, 1867–1914, Boreham's estimates (annual averages in £s million)*

Period	Livestock	Crops	Machinery and implements	Other items	Total
1867–73	178.155	261.652	5.500	27.529	472.7
1874–8	196.791	248.323	5.400	29.226	479.7
1879–83	179.278	190.669	4.900	24.908	399.8
1884–96	159.386	180.009	4.200	22.710	366.3
1897–1910	171.849	179.751	4.500	25.040	381.1
1911–14	190.849	194.908	7.000	28.424	421.2

Note:
In which other capital items cover the value of occupiers' fixtures, purchased feeds, miscellaneous stores such as tools and harnesses and stocks of livestock products. In point of fact in the main they were not estimated individually. An annual estimate for the stock of purchased feeds was constructed and the remainder of 'other' capital was reckoned to be the equivalent of about 5 per cent of the total capital involved.
Source: A. J. Boreham, 'A series of estimates of occupiers' capital, 1867–1938', *The Farm Economist*, 7, no. 6, 1953, pp. 260–3.

Table 43.5. *Capital formation in mid-nineteenth-century Great Britain (in £s million), 1850, 1860*

Category of investment	1850	1860
A. *Fixed capital*		
1. Fixed capital formation in agriculture	6.100	7.000
2. Undepreciated current aggregate capital value of farm buildings without dwellings		134.000
3. Ditto, with dwellings		243.800
4. Estimated gross new value of farm buildings, without houses		0.145
5. Depreciated value of British farm buildings		82.000
6. Ditto, with dwellings		156.000
7. Total investment in parliamentary enclosure of common field, 1850–9 and 1860–9*	0.228	0.182
8. Investment in underdraining	0.540	0.585
9. Investment in fencing and ditching	0.600	
10. Value of standing timber		70.000
B. *Tenants' or working capital*		
11. Implements		20.000
12. Capital value of livestock		128.800
of which sheep		40.600
of which cattle		50.000
of which horses		35.200
of which swine		3.000
13. Net gross capital value of crops		74.700
of which wheat		32.090
of which barley		14.080
of which oats		14.160
of which rye		1.050
of which non-cereals		13.320

Note: * Estimated investment in the reclamation and enclosure of waste, 1830–69, was £1.350 million.
Source: B. A. Holderness, 'Agriculture, 1770–1860', in C. H. Feinstein and S. Pollard (eds.), *Studies in Capital Formation in the United Kingdom 1750–1920* (Oxford, 1988), pp. 9–34.

Table 43.6. *Fixed capital formation in farm buildings and works, Great Britain, 1851–1920 (decade averages)*

	Rent received £s million	Capital expenditure as a percentage of rent %	Fixed capital formation at current prices £s million p.a.	Fixed capital formation at 1900 prices £s million p.a.
1851–60	44.3	13.2	5.8	6.7
1861–70	48.7	13.5	6.6	7.5
1871–80	53.2	13.5	7.2	7.1
1881–90	47.9	12.0	5.8	6.8
1891–1900	41.2	10.0	4.1	4.8
1901–10	39.5	11.0	4.3	4.8
1911–20	41.8	11.5	4.8	3.5

Source: C. H. Feinstein, 'Agriculture', in C. H. Feinstein and S. Pollard (eds.), *Studies in Capital Formation in the United Kingdom 1750–1920* (Oxford, 1988), p. 269.

Table 43.7. *Capital formation in agriculture in Great Britain at 1900 prices, 1850–1911/20*

A Farm buildings and works

	Gross fixed capital formation £ million p.a.	Retirements (or allowances for demolition and obsolescence) £ million p.a.	End-of-period stocks	
			Gross £ million	Net £ million
1850			321.8	170.3
1851–60	6.70	3.23	356.5	187.6
1861–70	7.52	4.26	389.1	206.8
1871–80	7.09	5.25	407.5	215.1
1881–90	6.84	5.96	416.3	217.1
1891–1900	4.80	6.42	400.1	199.5
1901–10	4.80	7.09	377.2	184.1
1911–20	3.54	8.19	330.7	160.1

(continued)

Table 43.7 (*cont.*)

B Farm machinery, implements, and vehicles

	Gross fixed capital formation £ million p.a.	End-of-period stocks	
		Gross £ million	Net £ million
1850		31.0	17.1
1851–60	1.60	37.0	19.3
1861–70	3.50	59.0	32.9
1871–80	3.30	68.0	32.6
1881–90	3.40	67.0	33.3
1891–1900	4.00	74.0	38.0
1901–10	5.00	90.0	47.3
1911–20	5.50	105.0	52.9

C Total fixed assets in agriculture

	Gross fixed capital formation £ million p.a.	End-of-period stocks	
		Gross £ million	Net £ million
1850		385.0	204.4
1851–60	8.98	429.8	225.7
1861–70	11.79	487.7	260.4
1871–80	11.11	516.8	269.2
1881–90	10.93	525.1	272.1
1891–1900	9.28	513.9	257.4
1901–10	10.30	504.9	249.7
1911–20	9.40	468.9	229.1

Note: For details of construction, refer to original source.
Source: C. H. Feinstein, 'Agriculture', in C. H. Feinstein and S. Pollard, (eds.), *Studies in Capital Formation in the United Kingdom 1750–1920* (Oxford, 1988), pp. 272, 278. 279.

Table 43.8. *Estimated farm stocks in Great Britain and the UK, 1850–1920*
(£s million at current prices)

	Livestock	Harvested and growing crops	Total
GB			
1850	106	94	200
1860	143	120	263
UK			
1850	150	116	266
1860	203	149	352
1870	192	171	363
1880	202	160	362
1890	176	113	289
1900	188	91	279
1910	216	112	328
1920	546	384	930

Source: C. H. Feinstein, 'Agriculture', in C. H. Feinstein and S. Pollard (eds.),
Studies in Capital Formation in the United Kingdom 1750–1920 (Oxford, 1988),
p. 394.

Table 43.9. *Gross Domestic Fixed Capital Formation (GDFCF), and Gross and Net Stock of Domestic Reproducible Fixed Assets (GDRFA and NDRFA) in UK agriculture, 1850–1914 (In £s million in current prices with UK economy totals for comparison)*

	GDFCF Agric	GDFCF UK	GDRFA Agric	GDRFA UK	NDRFA Agric	NDRFA UK
1850			304	1,583	162	997
1851	7.1	46.3	303	1,591	161	1,001
1852	7.3	52.6	315	1,671	167	1,053
1853	7.6	59.3	369	1,925	195	1,213
1854	8.0	63.5	387	2,061	205	1,302
1855	8.2	62.2	389	2,103	205	1,328
1856	8.3	55.8	386	2,083	203	1,312
1857	8.4	53.0	382	2,113	201	1,326
1858	8.6	51.3	364	2,075	191	1,299
1859	8.7	54.2	367	2,089	193	1,301
1860	8.9	58.6	379	2,136	199	1,328
1861	9.6	63.2	381	2,162	200	1,345
1862	9.9	67.5	381	2,188	201	1,362
1863	10.5	77.9	392	2,291	207	1,430
1864	10.6	88.2	409	2,421	217	1,516
1865	10.6	91.1	411	2,478	218	1,558
1866	11.1	89.3	421	2,610	223	1,644
1867	11.6	79.9	418	2,620	222	1,648
1868	11.8	76.2	423	2,632	225	1,652
1869	11.6	77.0	432	2,687	231	1,678
1870	11.8	86.9	445	2,795	238	1,743
1871	11.9	98.9	469	2,929	251	1,829
1872	12.4	118.4	535	3,322	285	2,074
1873	13.2	125.2	595	3,740	316	2,335
1874	13.4	140.7	594	3,817	314	2,385
1875	13.0	136.9	549	3,666	290	2,301
1876	12.1	138.9	534	3,639	281	2,297
1877	11.4	134.7	525	3,638	275	2,305
1878	10.8	119.7	506	3,557	264	2,253
1879	10.3	106.3	476	3,485	248	2,202
1880	10.4	107.2	482	3,706	250	2,338
1881	10.4	109.4	466	3,665	242	2,311
1882	10.7	110.5	471	3,797	245	2,391
1883	10.7	112.9	463	3,791	241	2,385
1884	10.3	105.8	448	3,740	233	2,353
1885	9.8	96.2	440	3,721	229	2,336
1886	9.1	85.1	424	3,657	220	2,289
1887	9.0	86.5	426	3,659	222	2,283

(continued)

Table 43.9 (*cont.*)

	GDFCF Agric	GDFCF UK	GDRFA Agric	GDRFA UK	NDRFA Agric	NDRFA UK
1888	8.6	90.0	430	3,720	224	2,315
1889	8.8	100.1	451	3,911	234	2,426
1890	8.7	105.9	471	4,146	244	2,568
1891	8.5	106.9	457	4,117	236	2,540
1892	8.4	108.3	452	4,146	233	2,551
1893	8.1	108.6	439	4,122	225	2,532
1894	8.0	110.7	434	4,153	222	2,548
1895	7.8	114.8	425	4,178	216	2,564
1896	7.8	126.8	428	4,323	218	2,657
1897	7.9	144.1	436	4,512	221	2,782
1898	8.0	172.1	451	4,796	228	2,974
1899	8.3	191.6	473	5,160	238	3,216
1900	8.7	204.9	514	5,619	258	3,515
1901	8.7	210.3	490	5,621	245	3,537
1902	9.4	213.0	468	5,563	233	3,518
1903	9.1	208.0	466	5,606	231	3,561
1904	9.0	203.2	465	5,678	231	3,615
1905	9.2	198.4	452	5,767	224	3,676
1906	9.9	191.9	457	6,029	226	3,836
1907	10.3	175.5	466	6,309	231	3,995
1908	10.5	144.7	464	6,240	230	3,922
1909	10.3	153.8	460	6,275	228	3,917
1910	10.4	158.4	463	6,477	229	4,018
1911	10.5	163.4	470	6,733	233	4,150
1912	10.8	171.3	486	7,137	241	4,368
1913	11.2	192.2	518	7,502	257	4,565
1914	11.5	192.8	525	7,671	260	4,642

Sources: C. H. Feinstein, 'Agriculture', in C. H. Feinstein and S. Pollard (eds.), *Studies in Capital Formation in the United Kingdom 1750–1920* (Oxford, 1988), pp. 429–30, 433–4, 437–8.

Table 43.10. *Distribution of landlords' expenditure as a percentage of rent received (Columns 4–9 all in percentages of rent received), 1842–92*

Date	1	2	3	4	5	6	7	8	9
A Weighted by rents									
1842	5	149	129	27.2	0.2	14.4	4.6	46.4	53.6
1852	6	198	156	31.4	0.4	17.0	4.1	52.9	47.1
1862	6	240	162	26.1	0.5	17.6	3.3	47.5	52.5
1872	14	457	301	26.0	0.6	16.4	4.2	47.2	52.8
1882	26	603	462	30.3	0.7	16.6	5.7	53.3	46.7
1892	29	538	479	31.1	1.0	18.6	6.4	57.0	43.0
B Weighted by acreage									
1842				25.8	0.3	13.9	4.6	44.6	55.4
1852				30.1	0.4	16.6	4.0	51.1	48.9
1862				26.2	0.5	17.8	3.4	47.9	52.1
1872				26.6	0.6	15.9	4.2	47.3	52.7
1882				31.0	0.8	16.0	5.9	53.6	46.4
1892				31.3	1.0	17.9	6.9	57.1	42.9

Notes:
1 The number of estates on which the calculation is based in each year.
2 Aggregate of rent received – in £000s
3 Acreage from which rent received – in 000s acres
4 Percentage of rent the landlord expends on repairs, fences, insurance, new buildings and permanent improvements
5 Percentage of rent the landlord expends on local rates
6 Percentage of rent the landlord expends on tithe, land tax, drainage rates and miscellaneous outgoings
7 Percentage of rent the landlord expends in managerial charges
8 Total expenditure
9 Net income to the landlords
Source: Reconstructed from *RC on the Agricultural Depression*, 'Particulars of the Expenditure and Outgoings on Certain Estates in Great Britain and Farm Accounts', BPP, 1896, XVI, pp. 54–8. Reproduced in M. E. Turner, J. V. Beckett and B. Afton, *Agricultural Rent in England 1690–1914* (Cambridge, 1997), p. 23.

CHAPTER 44

PRICES

BY BETHANIE AFTON AND MICHAEL TURNER

A. INTRODUCTION

Statistical information on prices forms an integral part of any attempt to evaluate the health of the agrarian economy. It is therefore important not only to collect price series, but also to be fully aware of the problems associated with their collection and subsequent use. None is ideal and some can only be used because nothing better exists. The most accurate use of the available data is achieved only when each series of prices is evaluated in order to determine their consistency over both time and location.

The published price series available to the historian tend to derive from four main types of source.[1] One of the most easily used is the series of official and quasi-official prices published, at least in part, for comparative reasons. These, by definition, represent an effort to maintain consistency of both quality and quantity over time and place. The Corn Market Returns for England and Wales are one such price series. These were published in the *London Gazette* from 1772. From the time of the Tithe Commutation Act of 1836 they were officially used to provide price data from which to calculate tithe rent payments. As such, they are the most complete and probably the most accurate of all market series available. However, even these cannot be used uncritically.[2] One problem is that the volumetric unit of measurement varied even amongst the reporting markets. As late as 1893 there were 46 measures for wheat, 26 for barley, and 36 for oats, making direct regional comparisons difficult.[3] The number of markets included also altered. In 1850 there

---

[1] The authors are indebted to Dr Richard Perren for his assistance with parts of this section.

[2] For references to the use of the Corn Returns and the associated problems see particularly J. A. Venn, *The Foundations of Agricultural Economics* (Cambridge, 2nd edn, 1933); W. Vamplew, 'A grain of truth: the nineteenth century corn average', *AHR*, 28, 1980, pp. 1–17; L. Adrian, 'The nineteenth century Gazette corn returns from East Anglian markets', *JHG*, 3, 1977, pp. 217–36; S. Fairlie, 'The nineteenth century corn law reconsidered', *EcHR*, 18, 1965, pp. 562–75; S. Fairlie, 'The corn laws and British wheat production, 1829–76', *EcHR*, 22, 1969, pp. 88–116.

[3] Venn, *Foundations*, p. 279.

were 290, subsequently reduced to 150 in 1865, but increased again to 187 in 1883, and to 196 in 1890.[4] In addition, some contemporaries believed that the market quotations were often distorted by the inaccurate reporting of sales on the part of both inspectors and farmers. At Winchester, for example, there were many weeks when no official prices were recorded for one or more types of corn even though the market itself was active. Farmers were expected to report sales to the inspector. When corn was sold by sample, many failed to do this.[5] More important was the quantity of corn which never passed through an official market. In 1880, for example, a sown area of 2,835,462 acres of wheat, 2,203,321 acres of barley, and 1,759,651 acres of oats was returned for England and Wales. However, the quantities of grain recorded in the returning markets were 1,607,908 quarters of wheat, 1,591,925 quarters of barley, and only 164,791 quarters of oats.[6] Large quantities of grain which were diverted as animal feed, had they been recorded, might have lowered the overall price. In spite of this, a Parliamentary Committee considering the problem in 1881 concluded that although the quantities returned were not representative of the total national supply, the prices were sufficiently reliable to be of use.[7] The Gazetted corn returns do provide both a national and a regional weekly guide to the movement of prices.

Another set of prices which were collected for comparative purposes are the meat prices from several major national markets published in *The Mark Lane Express*. These gave the prices of various meats according to quality, thus allowing a degree of comparability. However, it is difficult to make comparisons from one place to another because of the differences in the way a carcass was marketed. London and much of the south sold fat livestock by the 8 lb stone 'sinking the offal'. Hence, the skin, tallow, bone and so forth were excluded from the price quoted. In the north the livestock was sold by the 14 lb stone which included the weight of the entire animal. To an extent this made the measurements more directly comparable than they would at first appear because the calculated ratio of meat to the total weight of a fat beast was 8 to 14.[8] However, with the improvement in livestock which took place during the period this ratio became less reliable. After the 1891 *Market and Fairs [Weighing of Cattle] Act*, easily usable data on livestock, which had been gathered in accordance with the Act, were published in the annual Agricultural Returns. These returns included information as to the numbers and values returned

---

[4] Vamplew, 'A grain of truth'.

[5] Letter written by J. T. Twynham, *Hampshire Chronicle*, 10 February 1868.

[6] *Return showing the Total Quantity of Wheat, Barley, and Oats . . . in England and Wales in the year 1880*, BPP, 1881, LXXXIII, p. 7.   [7] *Report from the Select Committee on Corn Averages*, BPP, 1888, x.

[8] P. McConnell, *The Agricultural Note-Book* (London, 6th edn, 1897), pp. 369, 388.

from reporting markets as well as details of numerous breeds and conditions of both fat and store livestock.[9]

The second group of sources available to the historian includes the market and fair reports found in many local newspapers as well as some national papers including *The Times*, *Bell's Weekly Messenger* and *The Mark Lane Express*. Because these tended simply to give the reader an impression of the trends in prices, they often require considerable interpretation by the historian. A farmer was not disconcerted to be told that the price was not much changed from the previous week, but for the historian such vagueness is frustrating, particularly when a similar report was made the week before. Such reports were made to supply information to an informed, participating audience. Unlike the corn market returns or *The Mark Lane Express* series of meat market prices, little conscious effort was made to provide long-run comparable data. This is particularly apparent with store stock prices. It was unusual for an average price to be quoted consistently. Thus, most runs either record highest and lowest prices achieved at the market or fair, or they represent some sort of compromise, striving for an acceptable average. Further difficulties are added by the variations in the size, age, potential use, condition and quality of the stock being sold. Store, breeding and ready-to-fatten stock were marketed by the head. Value was added to the animal as it moved from one area of the country to another in the course of the fattening process. Furthermore, the stock could vary in quality from top pedigree breeding stock to barren, aged animals being culled at the end of their useful lives. Consequently, any attempt to construct a national store-stock price series is extremely difficult. The data are functional, however, if used to demonstrate relative trends, rather than absolute values of store-stock prices from year to year or location to location. Similar problems occur with fat-stock sale prices. However, these series do not demonstrate the degree of variation in use and condition associated with store stock.

Contract prices form a third source. The official prices series of this sort were largely based on data from London institutions such as St Thomas's, Greenwich and Bethlem Royal Hospitals, and the London County Council asylums. Other series are also available, from other cities, from the Naval victualling department, amongst other sources. Although these prices are wholesale, rather than market or farm-gate, they provide a consistent series, often over a long number of years, which is representative of a largely unchanging class of consumers. Contract price series give some of the best available information covering the period for dairy

---

[9] *Report on Wholesale and Retail Prices in the United Kingdom in 1902 with Comparative Statistical Tables for a series of years*, BPP, 1903, LXVIII; R. H. Rew, 'English markets and fairs', *JRASE*, 3rd ser., III, 1892, pp. 112–15.

products, particularly milk, and for some vegetables used for human consumption such as beans, peas and potatoes.[10] Many of the contracts were negotiated for a period of months or even a year. The Asylum Committee of the London County Council, for example, regularly made contracts in March and October which ran either for the seven months from April through to October, or for five months from November to March. It is essential therefore to be aware that the prices given are not a monthly or weekly average for the year. It is also important to remember that the quality of the produce in the contract was likely to have been lower than the national average.[11]

The value of imported foodstuffs and agricultural raw materials provides a fourth source of prices. These were published by the Board of Trade in the 'Statistical Abstract of the United Kingdom' and in the 'Annual Statement of Trade of the United Kingdom with Foreign Countries and British Possessions'.[12] If the product was consigned for sale, the import value was the most recent sale value of that product. Otherwise it represented the cost, insurance and freight of the imported goods. It did not include customs duty, details of which can be found in 'Customs Tariffs of the United Kingdom from 1800 to 1897'.[13] These prices do not suffer from the fluctuations associated with the market prices of seasonal goods. They also provide a series going back to 1854 when the value of imports was first recorded. For some articles no comparable run of market prices is known to exist. However, while some of these values can be used to reproduce home market prices, a number of problems are associated with them when they are used as price series. First, it is difficult to determine whether they reflect the ruling prices for the area around the port of entry, which would give them a bias towards English prices, or whether they represent an average United Kingdom sales value. Secondly, imported products do not necessarily correspond exactly to the home product. Much of the imported wheat was a hard variety with a higher gluten and protein content. This was preferred for bread making and attracted a higher price than English wheat. Imported barley, on the other hand, was used mainly for animal feed while a large portion of English barley was of sufficiently high quality to be used for malting. Meat products had to be shipped live or preserved by salting, canning or freezing. The method used greatly influenced the value of the product. Fertilisers were often imported in a raw state to be processed in the United Kingdom. Hence, the price for phosphate of lime or rock

[10] On milk see E. H. Hunt and S. J. Pam, 'Prices and structural response in English agriculture, 1873–1896', *EcHR*, 50, 1997, pp. 477–505, esp. pp. 491–7.

[11] *Report on Wholesale and Retail Prices*, App. I.

[12] These are located in *Accounts and Papers* of the British Parliamentary Papers.

[13] *Customs Tariffs of the United Kingdom from 1800 to 1897*, BPP, 1898, LXXXV.

phosphate, for example, can be expected to show a similar price trend to that of the manufactured product, provided no significant alterations in processing occurred during the period. Were they available, the *export* values would be more representative of the United Kingdom superphosphate price.

Printed sources of information fall basically into two categories: official publications; and national and local newspapers and society journals. Prices for an increasing range of agricultural products were included in the annual Agricultural Returns from 1879. By the end of the period there were data for most agricultural products including corn, meat, wool, fruit and vegetables, and dairy products. Two other useful official sources are the *Miscellaneous Statistics of the United Kingdom* and the *Statistical Abstract for the United Kingdom*.[14] A government publication particularly useful as a collection based on diverse sources of price information is the 1903 *Report on Wholesale and Retail Prices*.[15] This Parliamentary report, published by the Department of Labour, contains tables of wholesale prices for coal, iron, steel and other metals, cotton and wool and other textiles, corn, flour, cattle and meat, dairy produce and eggs, fish, sugar and other articles of food and drink, oils, seeds, building materials and numerous miscellaneous articles. The sources used by the compilers were listed and agreed to have represented the most accurate source for each price series. These continued as far back into the nineteenth century as the source was felt to have been reliable. Thus, for example, the report included market prices for corn, fat cattle and live meat, contract prices from London institutions for dairy products, eggs and some vegetables, and import values for numerous other products. The retail price information is, unfortunately, less informative for the agrarian historian because it is taken exclusively from urban sources.[16] The 1880 Richmond Commission and the 1890s Royal Commission on Agriculture also contain price series for agricultural products.[17] These are often more parochial, thus providing local and regional data. The *London Gazette*, a quasi-official source, published corn prices from reporting markets weekly. Other national publications provided weekly price data. Among the best known of these, especially for agricultural prices, were *The Mark Lane Express* which gave prices of meat from markets and fairs around the country, and *The Grocer* which gave, amongst others, the prices of dairy products. Agricultural journals also often contained price data, as did journals such as *The Journal of the Royal Statistical Society* and *The Economist*, both of which also published

[14] These are located in *Accounts and Papers* of BPP.   [15] See note 9 *supra*.   [16] *Ibid.*, App I.

[17] *Royal Commission on the Distressed Condition of Agricultural Interests 1880–2*, BPP, 1880, XVIII; 1881, XV, XVI, XVII; 1882, XIV, XV, *passim*; *Royal Commission on Agriculture 1894–7*, BPP, 1894, XVI, Parts I–III; 1894–5, XVI; 1895, XVII; 1896, XVI, XVII; 1897, XV, *passim*.

price indexes. Local newspapers are a good source for prices from local and regional fairs and markets.

B. THE DATA

Tables 44.1 to 44.36 form a collection of official and unofficial price series for the major agricultural products. Some were widely disseminated at the time, others were less well known. Some have been gleaned from the work of modern researchers who have gathered prices from local newspapers and other, often unpublished but local sources. Some, however, are more reliable than others and all need to be carefully evaluated prior to use.

The series derived from the Corn Market Returns (Table 44.1) are undoubtedly the best known. Table 44.2 provides a regional comparison of grain prices for the year 1880. Regional variations were generally caused by unusual growing or harvest conditions. However, although there could be considerable variation, for example the price of oats in Pembroke was 25 per cent below the national average and that of wheat in Cumberland was 14 per cent above, the market for grain was considered to have been essentially a national market. Consequently, annual variations, available from the Returns, have not been included for each year. Table 44.3 gives the annual prices for a number of vegetable products.

Livestock price series represent a more complex range of products. Each animal may have been sold a number of times in various markets around the country as breeding stock, store stock and, ultimately, as fat stock. Consequently, more emphasis is placed on livestock than on other products in this study of price information. Tables 44.4 to 44.20 relate to live cattle, sheep and pig prices for store stock, breeding stock and live fat animals. Irish store-stock prices have been included because these animals were an important source of supply of English store stock. Tables 44.21 to 44.24 contain prices of beef, mutton and pig carcasses, and Tables 44.25 to 44.32 the prices of animal commodities, both those grown as an end product and those which were essentially by-products of meat production. Again, Irish butter prices have been included because Irish butter was commonly imported, particularly when the consumption of raw milk increased in England and Wales.

Finally, Tables 44.33 to 44.36 are for some of the more commonly purchased agricultural inputs. From the latter half of the nineteenth century agriculture increasingly relied on off-farm sources to provide seeds for a wide variety of crops, to maintain land fertility, to feed livestock, and to replace increasingly expensive labour. Such inputs, along with rent and wages were the major expenses of the farmer.

Table 44.1. *Gazette corn prices (shillings per quarter),*
1850–1914

	Wheat	Barley	Oats
1850	40.25	23.42	16.42
1851	38.50	24.75	18.58
1852	40.75	28.50	19.08
1853	53.25	33.17	21.00
1854	72.42	36.00	27.92
1855	74.67	34.75	27.42
1856	69.17	41.08	25.17
1857	56.33	42.08	25.00
1858	44.17	34.67	24.50
1859	43.75	33.50	23.17
1860	53.25	36.58	24.42
1861	55.33	36.08	23.75
1862	55.42	35.08	22.58
1863	44.75	33.92	21.17
1864	40.17	29.92	20.08
1865	41.83	29.75	21.83
1866	49.92	37.42	24.58
1867	64.42	40.00	26.00
1868	63.75	43.00	28.08
1869	48.17	39.42	26.75
1870	46.92	34.58	22.83
1871	56.67	36.17	25.17
1872	57.00	37.33	23.17
1873	58.67	40.42	25.42
1874	55.75	44.92	28.83
1875	45.17	38.42	28.67
1876	46.17	35.17	26.25
1877	56.75	39.67	25.92
1878	46.42	40.17	24.33
1879	43.83	34.00	21.75
1880	44.33	33.33	23.08
1881	45.33	31.92	21.75
1882	45.08	31.17	21.83
1883	41.58	31.83	21.42
1884	35.67	30.67	20.25

(continued)

Table 44.1 (*cont.*)

	Wheat	Barley	Oats
1885	32.83	30.08	20.58
1886	31.00	26.58	19.00
1887	32.50	25.33	16.25
1888	31.83	27.83	16.75
1889	29.75	25.83	17.75
1890	31.92	28.67	18.58
1891	37.00	28.17	20.00
1892	30.25	26.17	19.83
1893	26.33	25.58	18.75
1894	22.83	24.50	17.08
1895	23.33	21.92	14.50
1896	26.17	22.92	14.75
1897	30.17	23.50	16.92
1898	34.00	27.17	18.42
1899	25.67	25.58	17.00
1900	26.92	24.92	17.58
1901	26.75	25.17	18.42
1902	28.08	25.67	20.17
1903	26.75	22.67	17.17
1904	28.33	22.33	16.33
1905	29.67	24.33	17.33
1906	28.25	24.17	18.33
1907	30.58	25.08	18.83
1908	32.00	25.83	17.83
1909	36.92	26.83	18.92
1910	31.67	23.08	17.33
1911	31.67	27.25	18.83
1912	34.75	30.67	21.50
1913	31.67	27.25	19.08
1914	34.92	27.17	20.92

Source: Report on Wholesale and Retail Prices in the United Kingdom in 1902, BPP, 1903, LXVIII; Annual Agricultural Returns, 1902–16.

Table 44.2. *Regional grain prices with variation from the national average, 1880 (shillings per quarter)*

	Wheat		Barley		Oats	
	price	variation	price	variation	price	variation
Bedfordshire	45.75	1.42	38.83	5.75	21.25	−1.83
Berkshire	47.33	3.00	38.58	5.50	25.67	2.58
Buckinghamshire	45.58	1.25	29.08	−4.00	25.25	2.17
Cambridgeshire	42.25	−2.08	32.92	−0.17	20.08	−3.00
Cheshire	44.42	0.08	27.25	−5.83	21.17	−1.92
Cornwall	42.92	−1.42	27.50	−5.58	19.83	−3.25
Cumberland	50.67	6.33	33.00	−0.08	26.00	2.92
Derbyshire	48.50	4.17	36.17	3.08	25.25	2.17
Devon	46.33	2.00	32.33	−0.75	22.50	−0.58
Dorset	40.50	−3.83	33.67	0.58	21.25	−1.38
Durham	47.42	3.08	32.33	−0.75	25.42	2.33
Essex	43.08	−1.25	35.67	2.58	22.83	−0.25
Gloucestershire	44.50	0.17	34.67	1.58	22.75	−0.33
Hampshire	41.83	−2.50	33.00	−0.08	21.25	−1.83
Herefordshire	44.33	0.00	35.25	2.17	19.25	−3.83
Huntingdonshire	42.17	−2.17	37.58	4.50	18.58	−4.50
Kent	43.33	−1.00	35.92	2.83	21.08	−2.00
Lancashire	41.50	−2.83	29.67	−3.42	22.83	−0.25
Leicestershire	45.58	1.25	35.92	2.83	25.67	2.58
Lincolnshire	45.67	1.33	34.08	1.00	22.83	−0.25
Middlesex	47.17	2.83	33.75	0.67	24.42	1.33
Monmouth	41.08	−3.25	40.67	7.58		

Norfolk	43.50	−0.83	31.67	−1.42	22.33	−0.75
Northamptonshire	44.33	0.00	36.00	2.92	24.00	0.92
Northumberland	47.08	2.75	33.83	0.75	26.08	3.00
Nottinghamshire	48.00	3.67	36.58	3.50	22.50	−0.58
Oxfordshire	44.17	−0.17	35.67	2.58	20.08	−3.00
Somerset	44.58	0.25	32.67	−0.42	20.58	−2.50
Suffolk	43.83	−0.50	34.58	1.50	23.00	−0.08
Surrey	45.67	1.33	37.50	4.42	21.92	−1.17
Sussex	44.58	0.25	33.00	−0.08	22.33	−0.75
Warwickshire	46.33	2.00	36.08	3.00	26.92	3.83
Westmorland	42.58	−1.75			25.33	2.25
Wiltshire	39.83	−4.50	32.50	−0.58	22.00	−1.08
Worcestershire	45.42	1.08	32.75	−0.33		
Yorkshire	45.25	0.92	32.17	−0.92	21.08	−2.00
Caernarvonshire			29.58	−3.50	17.92	−5.17
Carmarthenshire					18.67	−4.42
Denbighshire	44.83	0.50	33.25	0.17	24.17	1.08
Glamorganshire	42.50	−1.83	38.92	5.83	21.00	−2.08
Pembrokeshire			31.08	−2.00	17.33	−5.75
National Average	44.33		33.08		23.08	

Source: Returns Relating to Wheat, Barley, and Oats, BPP, 1881, LXXXIII.

Table 44.3. *Green crop prices, 1850–1914*

	Contract potatoes (s./cwt.)	Market returns		Wholesale import value		
		beans (s/qr.)	peas (s./qr.)	potatoes (s./cwt.)	beans (s./cwt.)	peas (s./cwt.)
1850		26.83	27.33			
1851		28.58	27.17			
1852		32.25	30.58			
1853		40.08	38.50			
1854		47.25	45.58	3.00	10.10	10.43
1855	6.25	46.50	43.33	3.50	9.25	10.67
1856	5.33	43.92	41.58	3.50	8.17	9.44
1857	7.67	43.00	41.33	3.67	8.61	8.52
1858	7.00	41.92	42.92	3.92	7.95	8.60
1859	6.50	42.25	39.75	3.43	8.17	8.18
1860	7.83			4.87	8.42	8.74
1861	7.33			5.83	8.24	7.95
1862	7.33			4.93	7.22	8.07
1863	7.00			3.85	7.21	7.59
1864	5.17			3.84	7.59	7.68
1865	5.58			4.00	8.26	8.14
1866	5.83			4.84	9.02	8.33
1867	7.58			5.78	8.78	9.10
1868	7.75			4.74	9.56	9.84
1869	6.92			4.73	8.78	8.63
1870	6.67			6.35	8.62	8.35
1871	6.67			5.31	8.53	8.77
1872	7.42			5.48	8.07	8.70
1873	7.83			5.65	8.51	8.70
1874	7.25			5.19	9.46	9.28
1875	7.00			4.56	9.06	9.25
1876	7.75			5.78	8.07	8.72
1877	8.00			5.90	7.31	8.58
1878	7.50			5.46	8.00	7.88
1879	8.25			5.76	7.74	7.72
1880	7.92			5.84	8.15	8.14
1881	7.25			5.44	7.92	8.08
1882	6.92			6.67	7.78	7.90
1883	7.33			6.16	7.60	7.88
1884	6.00			6.74	6.54	6.92
1885	5.92			6.33	6.15	6.73
1886	5.75			5.90	6.23	6.25
1887	6.00			7.06	6.36	5.85

(continued)

Table 44.3 *(cont.)*

	Contract potatoes (s./cwt.)	Market returns		Wholesale import value		
		beans (s/qr.)	peas (s./qr.)	potatoes (s./cwt.)	beans (s./cwt.)	peas (s./cwt.)
1888	5.50			6.73	6.11	5.83
1889	5.50			7.90	6.27	6.56
1890	5.08			7.36	5.94	6.57
1891	6.50			7.55	6.57	7.13
1892	5.42			6.32	6.16	6.90
1893	5.33			6.41	5.71	6.33
1894	5.50			7.62	5.12	5.70
1895	5.92			6.23	5.23	5.73
1896	4.25			8.09	5.40	5.65
1897	4.33			6.12	5.37	5.47
1898	5.50			5.07	5.84	6.33
1899	4.50			6.12	6.11	6.53
1900	4.08			5.02	6.25	6.94
1901	4.42			5.23	6.74	7.32
1902	3.83			5.58	6.81	7.27
1903				5.69	6.74	7.55
1904				4.88	6.20	7.04
1905				7.67	6.76	7.19
1906				6.97	7.31	8.46
1907				5.75	5.25	9.68
1908				5.59	7.15	10.15
1909				6.58	6.98	9.18
1910				7.07	7.34	9.03
1911				8.01	7.29	9.22
1912				5.99	7.49	10.03
1913				5.49	7.38	10.18
1914				9.22	6.98	11.11

Source: Wholesale and Retail Prices, 1902.

Table 44.4. *Welsh store-cattle prices (£/head), 1850–94*

	Hereford bullocks	Store cattle
1850		4.20
1851		4.63
1852		5.90
1853		
1854		
1855		
1856		6.91
1857		8.48
1858		8.22
1859		
1860		
1861		
1862		6.88
1863		7.90
1864		8.17
1865		8.80
1866		8.34
1867		8.59
1868		
1869		
1870		
1871	14.00	
1872	15.50	
1873	17.00	
1874	15.70	
1875	14.75	
1876		11.50
1877	12.00	11.60
1878	15.25	12.69
1879	12.15	11.61
1880	11.00	11.42
1881	10.25	10.71
1882	14.00	12.48
1883		
1884		
1885	9.50	
1886	9.25	
1187		
1888	9.50	
1889	10.50	
1890		
1891		
1892	12.50	
1893	7.50	
1894	13.00	

Sources: RC on Agriculture, BPP, 1896, IV; D. W. Howell, *Welsh Agriculture 1815–1914*, London, 1977.

Table 44.5. *Hawick store-cattle prices (£/head),
1859–1902*

	Shorthorn stirks	2-year-old shorthorns
1859	7.13	10.38
1860	6.63	10.75
1861	7.63	11.50
1862	8.13	12.00
1863	9.15	13.15
1864	9.00	13.50
1865	9.25	13.50
1866	9.25	12.40
1867	8.88	12.38
1868	8.50	12.00
1869	8.50	12.75
1870	9.15	13.65
1871	11.63	16.00
1872	11.25	14.50
1873	11.10	16.00
1874	10.40	14.25
1875	10.25	14.88
1876	9.00	15.00
1877	9.50	14.25
1878	11.00	16.00
1879	8.38	12.75
1880	9.13	14.50
1881	8.75	13.65
1882	11.63	16.50
1883	12.25	16.25
1884	9.75	14.00
1885	6.25	11.50
1886	6.75	11.00
1887	6.88	10.50
1888	8.88	14.13
1889	11.00	14.55
1890	10.25	13.03
1891	8.38	12.38
1892	6.50	10.50
1893	6.75	11.25
1894	8.25	11.88
1895	8.63	12.75
1896	8.25	11.63
1897	8.00	11.50
1898	7.38	11.13
1899	8.00	11.75
1900	9.00	13.25
1901	8.75	11.50
1902	8.75	12.50

Source: Wholesale and Retail Prices, 1902.

Table 44.6. *Lincolnshire store-cattle prices (£/head), 1874–96*

	Lincoln Beast Fair						Boston May Fair					
	Yearling		2 year old		Drape		Yearling		2 year old		Drape	
	min.	max.	min	max.	min.	max.	min.	max.	min.	max.	min.	max.
1874								10.00	15.00	17.00		
1875												
1876												
1877												
1878												
1879												
1880												
1881												
1882	8.00	12.00	15.00	20.00	16.00	24.00						
1883	11.00	15.00	15.00	21.00	16.00	25.00	12.00	13.00	20.00	21.00	16.00	21.00
1884							9.00	13.00	13.00	18.00	15.00	20.00
1885												
1886												
1887												
1888							7.00	11.00			11.50	16.50
1889	10.00	12.00	15.00	18.00	14.00	18.00						
1890	10.00	12.00	16.00	18.00			8.50	10.00	16.00	17.00	16.00	17.00
1891	9.00	11.00	12.00	16.00	13.00	16.00						
1892							5.00	7.00	10.00	13.00	12.00	16.00
1893	6.00	8.00	10.00	12.00		13.00	7.00	10.00	11.50	14.50		20.00
1894	6.00	8.00	12.00	17.00	10.00	16.00						
1895												
1896							7.00	8.00	9.00	14.00	15.50	16.00

Source: J. H. Brown 'Agriculture in Lincolnshire during the Great Depression', PhD thesis, University of Manchester, 1978.

Table 44.7. *Irish store-cattle prices, Dublin Market (£s per/head), 1850–1914*

	2-year-old cattle			1-year-old cattle		
	min.	max.	avg.	min.	max.	avg.
1850	4.00	12.00		1.50	5.00	
1851	6.00	12.00		2.00	5.00	
1852						
1853	4.00	13.00		2.00	5.00	
1854	6.00	15.00		2.50	5.00	
1855	6.00	16.00		4.00	7.00	
1856	7.00	17.00		4.00	8.00	
1857	7.00	20.00		3.00	4.00	
1858	8.00	20.00		5.00	7.00	
1859	7.00	18.00		3.50	7.00	
1860	8.00	20.00		4.00	7.00	
1861	7.00	20.00		3.50	7.00	
1862	8.00	18.00		5.50	7.50	
1863	9.00	20.00		3.75	7.00	
1864	9.00	18.00		5.50	7.00	
1865	8.00	20.00		4.00	9.00	
1866	8.00	18.00		4.50	8.00	
1867	6.00	22.50		3.50	5.00	
1868	9.00	20.00		4.00	6.00	
1869	8.00	24.00		4.00	6.50	
1870	9.00	23.00		3.50	7.50	
1871	10.00	21.00		6.00	9.00	
1872	13.00	21.00		7.00	8.40	
1873	9.00	25.00		5.00	11.00	
1874	12.00	28.00		6.00	9.00	
1875	10.00	24.00		7.00	9.00	
1876	10.00	24.00		5.00	12.00	
1877	10.00	26.00		5.00	10.00	
1878	10.00	24.00		6.00	11.00	
1879	9.00	23.00		5.00	10.00	
1880	9.50	22.00		5.00	11.75	
1881	9.00	25.00		5.50	10.00	
1882	10.00	25.50		5.00	11.00	
1883	11.00	25.00		5.00	12.60	
1884	8.00	25.00		5.00	10.00	
1885	7.00	20.00		4.00	8.50	
1886	5.50	20.00		3.50	7.85	
1887						
1888			9.70			6.40
1889			10.30			7.15

(continued)

Table 44.7 (cont.)

	2-year-old cattle			1-year-old cattle		
	min.	max.	avg.	min.	max.	avg.
1890			10.25			7.25
1891			9.55			6.30
1892			8.35			5.05
1893			8.10			4.95
1894			8.25			5.30
1895			8.70			9.05
1896			8.65			8.90
1897			9.05			6.15
1898			8.75			6.45
1899			9.10			6.70
1900			9.65			6.85
1901			9.65			6.85
1902			9.50			9.60
1903			10.00			7.20
1904			9.75			7.10
1905			9.60			6.85
1906			9.40			6.75
1907			9.60			7.20
1908			10.10			7.35
1909			10.35			7.55
1910			10.90			8.00
1911			11.05			8.45
1912			11.05			8.30
1913			11.60			8.75
1914			11.85			8.85

Source: Annual Agricultural Returns, Ireland, 1908 & 1915.

Table 44.8. *England and Wales Market Returns – store cattle (£/head),*
1905–14

	Yearlings 1st class	Yearlings 2nd class	2 year olds 1st class	2 year olds 2nd class	3 year olds 1st class	3 year olds 2nd class
			Shorthorns			
1905	9.05	7.80	12.75	11.20	15.75	14.10
1906	8.80	7.50	12.55	11.00	15.65	13.90
1907	9.70	8.25	13.90	12.10	16.65	14.95
1908	10.10	8.75	14.25	12.65	16.75	15.00
1909	10.10	8.50	14.10	12.35	17.15	15.15
1910	10.40	8.90	14.55	12.70	17.70	15.40
1911	9.90	8.40	13.75	12.05	17.20	15.20
1912	10.35	8.70	14.65	12.65	18.30	16.20
1913	11.10	9.65	15.45	13.45	19.05	16.60
1914	11.90	10.20	16.00	14.05	19.60	17.20
			Devons			
1905	8.85	7.65	12.95	11.20	16.65	15.00
1906	10.55	9.50	12.00	10.70	16.00	14.45
1907	10.45	9.00	13.60	11.85	17.40	15.75
1908	10.70	9.05	14.70	13.10	18.15	16.45
1909	10.05	8.35	14.50	12.65	17.60	15.95
1910	10.45	8.70	14.30	12.65	17.70	15.75
1911	10.45	8.60	13.85	11.80	17.25	15.45
1912	10.20	8.25	14.35	12.50	17.65	15.70
1913	11.85	9.95	15.75	14.00	18.90	16.75
1914	12.45	10.75	16.30	14.40	19.40	17.35
			Herefords			
1905	11.05	9.45	15.25	13.15	17.10	14.25
1906	11.25	9.90	14.85	13.30	16.65	15.30
1907	11.15	9.65	14.75	13.20	17.05	15.45
1908	11.65	10.05	14.95	13.60	17.30	15.90
1909	11.90	10.40	15.35	14.00	17.65	16.10
1910	11.95	10.35	15.40	13.95	18.00	16.30
1911	11.45	9.65	15.55	13.60	18.10	16.05
1912	11.90	10.25	15.70	13.90	18.45	16.80
1913	12.95	11.00	17.45	15.05	18.90	17.10
1914	12.75	11.35	17.30	15.40	20.15	18.25

(continued)

Table 44.8 (*cont.*)

	Yearlings 1st class	Yearlings 2nd class	2 year olds 1st class	2 year olds 2nd class	3 year olds 1st class	3 year olds 2nd class
			Welsh Runts			
1905	6.75	5.90	12.55	10.00	15.00	13.35
1906	9.00	7.40	12.45	10.55	15.30	13.75
1907	9.85	8.70	13.55	11.65	16.35	14.15
1908	9.65	8.15	13.95	12.30	16.85	14.90
1909	9.50	7.85	14.25	11.60	16.95	14.55
1910	9.75	8.55	15.20	13.35	17.05	15.00
1911	9.50	7.75	17.10	12.50	19.10	16.55
1912	9.65	8.45	15.05	13.30	17.30	15.25
1913	11.35	9.60	15.70	13.85	18.50	16.30
1914	11.50	10.15	16.55	14.55	18.70	16.75

Source: Annual Agricultural Returns, 1905–15.

Table 44.9. *Metropolitan Cattle Market – cattle*
live meat (d/lb sinking the offal), 1850–90

	Inferior	Prime	Min.	Max.
1850	4.03	5.16		
1851	3.72	4.78		
1852	3.66	4.97		
1853	4.59	5.75		
1854	5.06	6.41		
1855	5.31	6.75		
1856	5.09	6.69		
1857	5.09	6.72		
1858	4.88	6.34		
1859	5.00	6.56		
1860	5.16	6.97		
1861	4.94	6.75		
1862	4.69	6.41		
1863	5.47	7.00		
1864	5.63	7.16		
1865	5.66	7.22		
1866	5.72	7.34	5.50	8.00
1867	5.13	6.78	5.25	7.75
1868	4.91	6.56	4.75	7.75
1869	5.19	7.38	5.25	8.50
1870	5.38	7.44	5.25	8.25
1871	5.91	8.06	5.75	8.75
1872	5.44	7.94	6.25	8.75
1873	7.13	8.78	7.50	9.50
1874	6.75	8.56	6.75	9.25
1875	6.19	8.53	5.75	9.25
1876	6.88	8.63	6.50	9.00
1877	7.03	8.50	6.50	8.75
1878	7.09	8.47	6.75	9.00
1879	6.44	7.75	6.00	8.25
1880	6.72	8.44	6.25	8.75
1881			6.00	8.25
1882			6.00	9.00
1883			6.25	9.00
1884			6.00	8.50
1885			5.75	8.00
1886			4.88	7.25
1887			4.50	6.63
1888			3.50	7.38
1889			4.13	7.25

(continued)

Table 44.9 (*cont.*)

	Inferior	Prime	Min.	Max.
1890			3.50	7.25
1891			4.13	7.38
1892			4.38	7.13
1893			4.25	7.13
1894			3.63	6.75
1895			4.00	6.75
1896			3.50	6.63
1897			3.63	6.75
1898			3.50	6.38
1899			3.75	6.88
1900			4.63	7.25
1901			3.50	6.88
1902			4.38	7.38
1903			4.25	7.00
1904			4.13	6.88
1905			4.25	6.88
1906			4.13	6.88
1907			4.13	7.00
1908			4.00	7.13
1909			4.25	7.38
1910			4.38	7.63
1911			3.88	7.38
1912			4.75	8.25
1913			4.75	8.00
1914			5.25	8.25

Note: The two runs of prices represent different averages; 1850–80 is the average for two distinct grades of meat, i.e. 'inferior' and '3rd class large prime', while 1866–1914 is the minimum and maximum average price of all grades.

Source: Returns Showing the Quantities of Various Kinds of Grains and Flours Imported into the United Kingdom; and the Average Annual Prices of Butcher's Meat etc., BPP, 1881, LXXXIII; *Report of Wholesale and Retail Prices in the UK 1902, with Comparative Statistical Tables for a Series of Yews*, BPP, 1903, LXVIII; *Agricultural Returns*, 1878, 1896, 1915.

Table 44.10. *London Christmas Beast Market*
live meat (d/lb sinking the offal), 1850–88

	Min.	Max.
1850	4.50	5.75
1851	4.00	6.25
1852	4.00	6.00
1853	4.75	7.25
1854	5.25	8.00
1855	5.50	6.25
1856	5.00	7.50
1857	5.00	7.00
1858	5.00	7.50
1859	5.25	8.00
1860	5.00	8.25
1861	5.00	7.50
1862	5.00	7.50
1863	5.25	7.75
1864	5.50	8.50
1865	5.00	8.00
1866	5.50	8.25
1867	5.00	7.50
1868	5.00	8.50
1869	5.25	9.25
1870	5.25	9.25
1871	5.75	9.25
1872	5.50	9.00
1873	6.50	9.75
1874	6.50	10.00
1875	6.75	9.75
1876	6.50	9.50
1877	6.75	9.00
1878	6.75	9.00
1879	6.00	9.50
1880	6.00	9.00
1881	6.00	9.25
1882	6.75	9.50
1883	6.00	9.50
1884	6.00	9.25
1885	5.25	8.00
1886	5.25	7.50
1887	3.75	8.00
1888	3.50	7.75

Source: 'Statistics affecting British agricultural
interests', *JRASE*, 2nd ser., 25, 1889, p. xviii.

Table 44.11. *England and Wales Market Returns – fat cattle*
(shilling per stone of 8 lb), 1905–14

	Polled Scots		Shorthorn	
	1st class	2nd class	1st class	2nd class
1905	7.75	7.33	7.58	7.00
1906	7.67	7.33	7.50	6.92
1907	7.92	7.58	7.83	7.17
1908	8.17	7.75	7.92	7.25
1909	8.42	7.92	8.17	7.42
1910	8.75	8.25	8.58	7.75
1911	8.42	7.92	8.17	7.42
1912	9.25	8.67	9.00	8.08
1913	9.25	8.75	9.00	8.25
1914	9.33	8.92	9.17	8.42

	Herefords		Devons	
	1st class	2nd class	1st class	2nd class
1905	7.75	7.08	7.75	7.17
1906	7.67	7.17	7.83	7.17
1907	8.00	7.42	8.17	7.50
1908	8.08	7.58	8.25	7.50
1909	8.42	7.67	8.42	7.75
1910	8.75	8.08	8.75	7.92
1911	8.42	7.67	8.33	7.58
1912	9.17	8.42	9.00	8.08
1913	9.25	8.58	9.17	8.25
1914	9.25	8.67	9.17	8.42

Source: Annual Agricultural Returns, 1905–15.

Table 44.12. *Welsh and Hawick store-sheep prices*
(shillings per head), 1850–1902

	Welsh	Cheviot lamb	Cheviot ewe
1850	7.83		
1851	8.50		
1852	9.17		
1853			
1854			
1855			
1856	10.08		
1857	10.75		
1858	10.00		
1859		10.25	20.50
1860		13.75	21.50
1861		10.75	22.00
1862	9.75	10.50	23.00
1863	11.00	13.25	24.25
1864	12.00	16.75	27.00
1865	12.67	18.75	33.25
1866	17.33	17.00	28.50
1867	11.58	8.50	19.00
1868	8.17	8.75	17.50
1869	9.17	11.25	25.75
1870		11.67	26.50
1871		16.50	36.50
1872		19.33	36.00
1873		17.00	32.50
1874		10.50	26.00
1875		13.75	36.00
1876		13.50	33.25
1877		14.75	31.50
1878		15.50	37.00
1879		13.50	21.50
1880		13.75	31.00
1881		13.50	30.50
1882		19.50	42.00
1883		20.00	38.75
1884		13.75	29.50
1885		12.00	18.75
1886		12.00	27.00
1887		12.67	23.75
1888		16.25	28.00
1889		18.92	34.08

(continued)

Table 44.12 (*cont.*)

	Welsh	Cheviot lamb	Cheviot ewe
1890		17.25	30.50
1891		11.75	23.00
1892		8.75	15.00
1893		11.75	21.50
1894		14.00	24.50
1895		15.75	27.75
1896		15.75	23.00
1897		15.50	24.00
1898		14.25	24.25
1899		11.00	21.00
1900		15.25	24.00
1901		12.50	22.50
1902		14.50	23.00

Sources: RC on Agriculture; Howell, *Welsh Agriculture*.

Table 44.13. *Lincolnshire store-sheep prices*
(shillings per head), 1874–97

| | Lincoln | Boston hogget | |
		(min.)	(max.)
1874		54.00	78.00
1875			
1876			
1877	68.31		
1878	70.42		
1879	56.38	50.00	60.00
1880	64.00	48.00	74.00
1881	59.00	50.00	65.00
1882	64.31	53.00	72.00
1883	67.46	50.00	60.00
1884	59.21	50.00	65.00
1885	48.25	32.00	46.00
1886	55.50		50.00
1887	49.31	44.00	52.00
1888	50.98		
1889	56.83		
1890	62.10	50.00	70.00
1891	49.42	42.00	55.00
1892	50.92		
1893	45.58		
1894	53.25	50.00	60.00
1895	53.17	45.00	58.00
1896		48.00	60.00
1897		46.00	51.00

Source: Brown, 'Agriculture in Lincolnshire'.

Table 44.14. *Hampshire Down breeding and store-sheep (shillings per head),*
1850–1914

	Draft ewe		9-month-old lamb (breeding)		6-month-old lamb (store)		Ram lamb
	min.	max.	min.	max.	min.	max.	avg.
1850	26	35	18	36	21	37	
1851	28	35	18	25		31	
1852			25	35			
1853	35	44	25	35	24	39	
1854	32	42	24	35			
1855	32	40	24	36	21	29	
1856	35	45	30	42	23	35	
1857	30	35	28	38	21	38	
1858	30	40	25	40	21	35	
1859	33	45	18	30	20	36	
1860	35	43			25	38	
1861			23	35	20	35	
1862	32	37	24	40	20	42	
1863	37	46			30	35	
1864	35	48	22	30	27	38	
1865	36	47	35	42	25	42	
1866			47	48	30	51	
1867			32	43	24	42	
1868	20	38	16	38	28	38	
1869	35	42	26	36	23	30	
1870	30	45	17	35	24	36	
1871	40	65	32	52	28	55	
1872	55	62			35	50	
1873	50	78	35	55	40	60	
1874			32	48	20	36	
1875	48	56	30	48	20	45	
1876	60	65	25	50	40	46	
1877	45	66	38	50	30	45	
1878	55	70		51	35	57	
1879	46	62	25	38	30	43	149
1880	50	63	32	54	30	45	147
1881	50	60	32	46	32	45	165
1882		60	45	55	44	52	283
1883	69	74			53	58	204
1884	59	69	26	46	44	51	228
1885	42	57		29	33	38	166
1886							126

(continued)

Table 44.14 (cont.)

	Draft ewe		9-month-old lamb (breeding)		6-month-old lamb (store)		Ram lamb
	min.	max.	min.	max.	min.	max.	avg.
1887	38	52	23	30	29	35	106
1888	39	47	35	45	31	48	162
1889	57	62	40	44	36	46	137
1890	48	61	35	45	35	38	124
1891	38	58			32	40	149
1892	29	50			20	36	126
1893	24	54	24	38	16	37	107
1894	42	52		38	29	47	110
1895	41	49	36	40	29	39	155
1896	24	47			21	34	147
1897	28	53	21	54	23	47	165
1898	29	53	26	31	27	44	204
1899	34	50	24	40		36	148
1900	30	47	31	44	25	49	169
1901	33	49	29	44	25	48	170
1902	43	52			25	49	171
1903	30	49			29	46	201
1904	26	52	29	44	23	42	163
1905	21	50	30	47	32	55	149
1906	27	55	30	46	34	52	181
1907	33	55			36	46	191
1908	39	54		35	35	46	222
1909	22	49	22	38	24	40	283
1910	21	47	21	48	27	43	232
1911	16	40	16	37	24	45	
1912	19	47	34	44	23	42	148
1913	26	55	31	50	34	48	172
1914	42	65	39	47	34	56	

Source: B. Afton, 'Mixed farming on the Hampshire Downs 1837–1914', PhD thesis, University of Reading, 1993.

Table 44.15. *Irish store-sheep prices, Dublin market*
(shillings per head), 1850–1914

| | Lamb | | |
	max.	min.	avg.
1850	14.00	23.00	
1851	16.00	24.00	
1852			
1853	19.00	22.00	
1854	21.00	27.00	
1855	18.00	25.00	
1856	18.00	25.00	
1857	20.00	31.00	
1858	20.00	32.00	
1859	20.00	28.00	
1860	23.00	35.00	
1861	24.00	32.00	
1862	22.00	33.00	
1863	20.00	33.00	
1864	16.00	38.00	
1865	20.00	40.00	
1866	30.00	38.00	
1867	36.00	40.00	
1868	25.00	32.00	
1869	22.00	37.00	
1870	35.00	38.00	
1871	26.00	36.00	
1872	25.00	40.00	
1873	30.00	50.00	
1874	30.00	45.00	
1875	25.00	42.00	
1876	26.00	50.00	
1877	30.00	45.00	
1878	30.00	48.00	
1879	22.00	50.00	
1880	37.00	47.00	
1881	20.00	48.00	
1882	30.00	52.00	
1883	24.00	50.00	
1884	20.00	48.00	
1885	18.00	50.00	
1886	16.00	42.00	
1887			
1888			26.33

(*continued*)

Table 44.15 (*cont.*)

	Lamb		
	max.	min.	avg.
1889			28.25
1890			26.08
1891			27.08
1892			21.67
1893			23.25
1894			25.50
1895			24.92
1896			25.08
1897			24.33
1898			23.83
1899			24.33
1900			24.25
1901			22.58
1902			23.42
1903			25.92
1904			27.25
1905			27.17
1906			30.92
1907			30.50
1908			25.50
1909			22.75
1910			24.17
1911			24.58
1912			24.17
1913			28.50
1914			30.00

Source: Annual Agricultural Returns, Ireland, 1908 & 1915.

Table 44.16. *England and Wales Market Returns –
store-sheep (shillings per head), 1905–14*

	Teg & lamb	
	1st class	2nd class
1905	40.67	36.33
1906	42.25	37.25
1907	44.50	39.00
1908	41.43	36.00
1909	34.58	29.25
1910	37.17	31.50
1911	35.00	28.75
1912	37.58	31.50
1913	44.25	38.67
1914	46.08	40.25

Source: Annual Agricultural Returns, 1905–15.

Table 44.17. *Metropolitan Cattle Market – Sheep*
live meat (d/lb sinking the offal), 1850–1914

	Inferior	Prime	Min.	Max.
1850	4.53	6.00		
1851	4.53	6.09		
1852	4.47	6.22		
1853	5.38	7.41		
1854	5.25	7.31		
1855	5.38	7.50		
1856	5.69	7.75		
1857	5.63	8.00		
1858	5.06	7.44		
1859	5.53	7.81		
1860	5.75	8.38		
1861	5.28	5.22		
1862	5.50	8.06		
1863	6.00	8.19		
1864	6.31	8.50		
1865	7.00	9.56		
1866	6.16	9.34	5.50	9.25
1867	5.34	7.91	5.25	8.00
1868	5.22	7.53	4.75	7.75
1869	5.50	8.63	5.25	8.75
1870	5.25	8.41	5.25	8.50
1871	6.13	9.63	6.25	9.75
1872	6.69	10.00	7.25	10.00
1873	7.78	10.19	8.50	10.25
1874	7.19	8.72	7.25	8.75
1875	7.31	10.22	6.75	10.50
1876	7.56	10.34	7.00	10.50
1877	8.50	10.31	8.25	10.50
1878	8.31	10.06	8.00	10.25
1879	7.34	9.72	6.75	10.00
1880	7.53	10.19	7.25	10.25
1881			7.25	10.00
1882			8.00	10.75
1883			8.25	10.75
1884			7.50	9.50
1885			6.00	8.50
1886			6.25	9.00
1887			5.38	7.88
1888			4.88	8.75
1889			5.25	9.50

(continued)

Table 44.17 (cont.)

	Inferior	Prime	Min.	Max.
1890			6.75	9.38
1891			5.63	8.75
1892			5.63	8.38
1893			5.50	8.13
1894			5.38	8.75
1895			5.88	8.88
1896			4.88	8.13
1897			5.50	8.50
1898			4.75	8.25
1899			5.00	8.50
1900			5.00	9.00
1901			4.88	8.50
1902			5.38	8.38
1903			5.50	8.75
1904			5.75	8.88
1905			5.88	8.88
1906			6.63	9.13
1907			6.50	9.25
1908			5.63	8.75
1909			4.63	7.88
1910			5.25	8.75
1911			5.00	8.38
1912			5.63	9.38
1913			6.25	9.63
1914			6.75	10.38

Note: The two runs of prices represent different averages: 1850–80 is the average for two distinct grades of meat, 'inferior' and '4th class Southdown', while 1866–1914 is the minimum and maximum average prices of all grades. *Source:* see Table 44.9, p. 2058, *supra.*

Table 44.18. *England and Wales market returns –*
fat sheep (pence per pound), 1905–13

| | Longwool | |
	1st class	2nd class
1905	8.00	7.50
1906	8.50	7.75
1907	8.50	7.75
1908	7.75	7.00
1909	6.75	6.00
1910	7.75	6.75
1911	7.25	6.50
1912	8.25	7.50
1913	9.00	8.00

| | Downs | |
	1st class	2nd class
1905	8.75	8.00
1906	8.75	8.00
1907	9.00	8.25
1908	8.50	7.50
1909	7.50	6.50
1910	8.25	7.25
1911	7.75	7.00
1912	8.75	8.00
1913	9.25	8.50

| | Crossbred | |
	1st class	2nd class
1905	8.50	7.75
1906	8.75	8.25
1907	9.00	8.25
1908	8.25	7.50
1909	7.25	6.50
1910	8.25	7.50
1911	7.75	7.00
1912	8.75	7.75
1913	9.25	8.50

Source: Annual Agricultural Returns, 1905–15.

Table 44.19. *Hampshire fat sheep prices – Basingstoke Repository (shilling per head), 1878–1914*

| | Prime young | | 13 mo. teg | | Old ewe | | Ram | | 3 mo. lamb | |
	min.	max.	min.	max.	min.	max.	min.	max.	min.	max.
1878	69.00	81.00	55.50	64.00	62.00	78.00	59.00	80.00	39.00	50.50
1879	35.00	75.00	52.50	55.50	61.50	61.50	50.00	80.00	39.00	52.00
1880	48.00	69.00	45.50	67.00	48.00	70.50			33.00	52.00
1881	52.00	81.00	42.50	65.00	53.00	78.00	55.00	88.00	41.00	54.00
1882	50.00	75.00	50.00	74.00	57.00	78.00			42.00	54.00
1883	77.00	100.00	73.50	91.50	81.00	100.00			43.00	57.00
1884	71.00	89.00	68.00	80.50	64.00	75.00			47.00	56.00
1885	55.50	66.00			55.00	65.00			39.00	43.00
1886	52.50	64.00	46.00	56.50	48.00	54.00			39.00	48.00
1887	58.00	69.00	42.00	58.00	49.50	57.00			40.00	51.00
1888	43.00	52.00	40.50	55.00	40.50	55.00			35.50	39.00
1889	59.50	71.00	55.00	68.00	55.00	66.00			39.00	50.00
1890	61.00	70.00	55.00	62.50	50.50	63.50			43.00	48.00
1891	53.50	68.00	44.00	60.00	43.50	50.50			28.50	45.50
1892	43.50	63.50			43.50	63.50			30.00	40.50
1893	40.00	58.00			36.50	46.50			30.50	50.00
1894	42.00	63.00	39.00	50.00	35.00	50.00			28.00	41.50
1895	48.00	62.00	34.00	62.00	37.00	53.00			28.00	42.50
1896	36.00	50.00			35.00	57.00			27.00	42.00
1897	35.00	57.00			34.00	48.00			28.00	42.00
1898	44.00	60.50			37.00	53.00			27.00	42.00
1899	46.00	62.00	38.00	52.50	36.00	52.00			25.00	45.50

Year										
1900	40.00	50.00	33.00	48.00	32.00	40.00			33.00	43.00
1901	46.00	58.00	39.00	56.00	36.00	45.00	42.00	50.00	34.00	42.00
1902	44.00	56.00	36.00	50.00	35.00	42.00	30.00	44.00	32.00	40.00
1903	48.00	57.50	42.00	56.00	38.00	42.00	45.00	59.00	34.00	42.00
1904	48.00	53.50	45.00	60.00	34.00	38.00	42.00	50.00	40.00	45.50
1905	52.00	58.00	53.00	57.00	38.00	42.00	42.00	52.00	36.00	41.00
1906			54.00	60.00	36.00	42.00	46.00	52.00	38.00	42.00
1907	57.00	66.00	46.00	62.50	36.00	56.50	50.00	61.00	38.00	44.00
1908			46.00	60.00	36.00	42.00	42.00	52.00	38.00	45.00
1909	40.00	51.00	48.00	56.50	30.00	36.00	46.50	50.00	38.00	46.00
1910	40.00	47.00	42.00	49.50	25.00	36.00	40.00	48.00	33.00	43.00
1911		48.00	51.00	54.50	30.50	34.00			37.50	44.00
1912	40.00	53.50	47.00	58.50	33.50	42.50	29.50		33.00	40.00
1913	43.00	57.00	52.00	63.50	40.00	54.00	51.00	60.00	37.00	47.00
1914			45.00	67.50	41.00	55.00	31.00	52.00	33.50	44.00

Source: Afton, 'Mixed farming'.

Table 44.20. *Fat pork and bacon prices, 1850–1914*

					England & Wales Market Returns – Fat Pigs			
	Metropolitan Cattle Market							
			Pigs	Pigs	Bacon		Porkers	
	Porkers	Hogs	1st class	inferior	1st class	2nd class	1st class	2nd class
	(live meat, d/lb sinking the offal)					(d/lb)		
1850	5.88	4.75						
1851	5.53	4.56						
1852	5.53	4.56						
1853	6.09	5.06						
1854	6.84	5.44						
1855	6.75	5.50						
1856	7.25	5.75						
1857	7.50	6.06						
1858	6.59	5.56						
1859	6.75	5.38						
1860	7.50	6.09						
1861	7.47	6.13						
1862	7.25	6.00						
1863	6.94	5.75						
1864	7.00	5.91						
1865	7.50	6.50						
1866	7.19	6.28						
1867	6.06	5.22						
1868	6.13	5.28						
1869	7.78	6.44						
1870	8.53	7.25						
1871	7.03	5.75						
1872	7.19	6.06						
1873	7.34	6.19						
1874	7.25	6.19						
1875	7.78	6.91						
1876	8.00	7.13						
1877	7.06	6.22						
1878	6.91	6.16						
1879	6.66	5.94						
1880	7.41	6.50						
1881	7.69	7.06						
1882	7.31	6.75						
1883	7.00	6.44						
1884	6.53	5.84						
1885	6.16	5.34						
1886	6.09	4.53						

(continued)

Table 44.20 (*cont.*)

				England & Wales Market Returns – Fat Pigs			
	Metropolitan Cattle Market						
		Pigs	Pigs	Bacon		Porkers	
	Porkers Hogs	1st class	inferior	1st class	2nd class	1st class	2nd class
	(live meat, d/lb	sinking the offal)			(d/lb)		
1887	5.81 4.25						
1888	5.50 4.22						
1889		6.66	3.84				
1890		6.09	3.59				
1891		5.59	3.84				
1892		6.88	4.31				
1893		7.38	4.75				
1894		6.72	4.44				
1895		5.41	3.59				
1896		5.16	3.16				
1897		6.28	4.13				
1898		6.59	4.34				
1899							
1900		6.50	4.00				
1901		7.03	3.91				
1902		6.75	4.00				
1903							
1904							
1905				9.63	9.00	10.50	9.87
1906				10.38	9.75	11.25	10.50
1907				10.01	9.38	10.76	10.13
1908				9.26	8.51	9.87	9.26
1909				10.62	9.87	11.25	10.50
1910				11.75	11.13	12.50	11.75
1911				10.01	9.26	10.88	10.13
1912				11.00	10.25	11.51	10.76
1913				12.63	11.88	13.38	12.50
1914				11.75	11.00	12.50	11.88

Source: Wholesale and Retail Prices 1902; Annual Agricultural Returns, 1902–16.

Table 44.21. *Beef prices (d/lb), 1850–1914*

	London	Newcastle
1850	5.13	4.56
1851	4.85	4.48
1852	4.59	4.63
1853	5.72	5.97
1854	6.46	6.02
1855	6.60	6.40
1856	6.57	6.31
1857	6.62	6.47
1858	5.97	6.12
1859	6.54	6.47
1860	7.01	6.89
1861	6.47	6.36
1862	6.65	6.58
1863	6.98	6.87
1864	7.03	6.79
1865	7.78	7.54

Source: R. Edwards and R. Perren, 'A note on regional differences in British meat prices, 1825–1865', *Economy and History*, 22, 1979, p. 134.

Table 44.22. *Beef carcass prices, 1850–1914*

	London Central Meat Market d/lb		Liverpool Meat Market d/lb		Exeter Market	Reading Market	Bangor Market d/lb
	min.	max.	min.	max.	shillings per 8 lb	shillings per 8 lb	avg.
1850					3.20	3.68	
1851					3.04	3.46	
1852					3.18	3.54	
1853					3.78	4.03	
1854					4.11		
1855					4.16	4.63	
1856					4.17	4.72	
1857					4.32	4.73	
1858					3.92	4.40	
1859					4.03	4.67	
1860					4.29		
1861					4.40		
1862					4.03	4.68	
1863					4.31	4.80	
1864					4.48		
1865					4.73	5.10	
1866	4.75	7.00	6.00	8.00	4.70		8.00
1867	4.75	6.75	4.75	7.75	4.77	5.14	8.00
1868	4.50	6.75			4.63	4.82	7.00
1869	4.75	7.50	4.75	6.75	5.06	5.23	7.75
1870	4.75	7.25	4.75	6.75	5.00	5.19	7.75

(continued)

Table 44.22. (cont.)

| | London Central Meat Market | | Liverpool Meat Market | | Exeter Market | Reading Market | Bangor Market |
| | d/lb | | d/lb | | shillings per 8 lb | | d/lb |
	min.	max.	min.	max.			avg.
1871	5.25	7.75	5.25	7.00	5.32		
1872	5.25	7.75	5.25	7.25	5.23		9.75
1873	5.50	8.50	6.25	8.25	5.82		10.25
1874	5.00	8.00	4.75	8.25	5.68	6.14	
1875	5.25	8.25	5.00	8.00	5.56	6.19	
1876	5.00	8.25	5.75	7.50	5.67	5.91	
1877	4.25	8.00	5.25	7.00	5.58		
1878	4.25	8.25	5.50	7.50	5.52		
1879	4.00	7.25	5.25	7.25	4.96		
1880	4.25	7.75	4.75	6.75	5.41	5.99	
1881	4.50	7.25	4.75	6.75			
1882	4.75	8.00	4.75	7.00			10.75
1883	5.00	8.00	5.25	7.25			9.75
1884	4.25	7.75	4.25	6.50			9.00
1885	3.75	6.75	3.75	6.50			9.75
1886	3.36	6.36	3.25	6.00			8.25
1887	2.86	5.64	3.36	6.50			7.50
1888	3.25	6.36	3.64	6.00			7.50
1889	3.50	6.36	3.64	5.88			8.00
1890	3.36	6.36	3.86	6.00			8.25

Year					
1891	3.25	6.50	2.86	5.75	8.25
1892	3.00	6.25	2.75	5.00	7.75
1893	3.13	6.38	3.00	5.25	7.75
1894	2.75	6.13	2.88	5.13	7.75
1895	2.88	6.25	4.25	5.50	
1896	2.38	5.88	3.25	5.00	
1897	3.25	6.13	3.75	5.25	
1898	3.13	5.88	3.38	5.13	
1899	3.38	6.00	3.63	5.63	8.75
1900	3.88	6.25	4.00	6.00	8.00
1901	3.50	6.00	4.00	5.75	
1902	4.25	6.75	4.13	6.63	8.25
1903	3.50	6.00	4.13	6.00	8.50
1904	3.38	6.13	3.88	5.75	8.00
1905	3.25	6.00	3.63	5.88	7.50
1906	3.13	5.88	3.75	5.63	7.50
1907	3.38	6.13	3.75	5.75	7.50
1908	3.38	6.50	3.75	6.13	
1909	3.25	6.50	4.38	6.13	
1910	3.50	6.63	4.63	6.25	
1911	3.13	6.38	4.75	6.13	
1912	3.25	7.00	4.50	6.50	
1913	3.13	6.75	4.38	6.63	
1914	3.38	6.88	4.88	6.88	

Sources: Annual Agricultural Returns, 1886, 1915; Howell, *Welsh Agriculture*; A. K. Copus, 'Changing markets and the response of agriculture in south-west England 1750–1900', PhD thesis, University of Aberystwyth, 1986.

Table 44.23. *Mutton prices, 1850–65 (d/lb)*

	London	Newcastle
1850	5.50	4.62
1851	5.29	5.16
1852	5.04	5.05
1853	6.63	6.48
1854	6.61	6.33
1855	6.72	6.29
1856	6.93	6.60
1857	7.21	6.43
1858	6.60	6.42
1859	6.88	6.80
1860	7.46	7.37
1861	7.12	7.23
1862	7.31	6.96
1863	7.53	7.21
1864	7.52	7.46
1865	8.61	8.03

Source: Edwards and Perren, 'A note on regional differences'.

Table 44.24. *Mutton carcass prices, 1850–1914*

	London Central Meat Market (d/lb)		Liverpool Meat Market (d/lb)		Exeter Market s/8lb	Reading Market s/8lb	Bangor Market d/lb avg.
	min.	max.	min.	max.			
1850					3.35	3.50	
1851					3.28	3.88	
1852					3.36	3.90	
1853					4.17		
1854					4.19		
1855					4.11	4.95	6.75
1856					4.25	4.86	6.50
1857					4.61	4.94	6.25
1858					4.05	4.65	6.50
1859					4.23	4.99	7.25
1860					4.53		7.25
1861					4.59		7.75
1862					4.40	4.97	8.50
1863					4.66	4.89	7.75
1864					4.89		7.75
1865					5.23	5.65	8.75
1866	5.50	8.00			5.24	5.63	9.25
1867	5.00	7.00			4.51	5.25	9.50
1868	4.50	6.75			4.05	4.63	7.25
1869	5.00	7.50	5.00	7.25	4.87	4.87	8.25
1870	5.25	7.75	5.50	7.50	5.29	5.25	8.50

(continued)

Table 44.24 *(cont.)*

	London Central Meat Market d/lb		Liverpool Meat Market d/lb		Exeter Market s/8lb	Reading Market s/8lb	Bangor Market d/lb avg.
	min.	max.	min.	max.			
1871	5.50	8.25	6.25	8.00	5.45		9.00
1872	6.00	8.75	7.00	8.50	6.01		9.50
1873	6.25	9.00	6.75	9.00	6.07		
1874	5.00	8.00	6.00	8.25	5.20	5.23	
1875	5.50	9.00	7.00	9.25	5.46	6.05	
1876	5.25	9.50	6.75	8.75	6.13	6.32	
1877	4.75	9.25	7.50	9.50	6.28		
1878	4.75	9.25	7.00	9.25	5.79		
1879	4.50	8.50	6.75	9.00	5.50		
1880	4.75	8.75	6.75	8.75	5.38	6.03	
1881	5.00	9.00	6.50	8.50			
1882	5.50	9.50	7.50	9.50			10.75
1883	5.75	9.75	7.25	9.25			10.75
1884	5.00	8.75	6.50	8.00			9.50
1885	4.25	7.50	4.75	7.75			10.50
1886	4.13	8.13	5.00	7.63			9.25
1887	3.38	7.13	5.88	7.13			9.25
1888	3.63	7.75	5.00	7.38			9.25
1889	4.00	8.38	6.38	8.25			9.00
1890	3.50	8.00	5.38	8.13			8.75

1891	3.00	7.38	5.25	7.25	8.75
1892	3.13	7.13	4.50	7.00	9.00
1893	3.00	6.88	4.25	6.88	8.50
1894	2.75	7.38	4.50	7.13	8.25
1895	2.88	7.63	5.13	7.50	
1896	2.38	7.00	4.50	7.13	
1897	2.50	7.25	4.50	7.50	
1898	2.75	7.13	5.00	7.00	
1899	3.00	7.25	5.25	7.38	8.75
1900	3.25	7.63	5.38	7.88	8.75
1901	3.00	7.25	5.75	7.75	
1902	3.63	7.13	5.00	7.88	8.75
1903	3.63	7.75	5.38	8.25	8.50
1904	3.75	7.88	5.13	8.25	8.50
1905	3.25	7.63	5.38	8.13	8.25
1906	3.25	7.88	5.50	8.50	8.25
1907	3.38	7.75	5.63	8.38	8.75
1908	3.13	7.63	4.75	8.25	
1909	2.63	6.63	3.75	7.50	
1910	3.25	7.38	4.50	8.38	
1911	3.13	6.75	4.50	8.50	
1912	3.38	7.75	4.38	7.88	
1913	3.75	8.25	5.50	8.88	
1914	4.00	8.25	5.13	8.88	

Sources: Annual Agricultural Returns, 1886, 1915; Howell, *Welsh Agriculture*; Copus 'Changing markets and the response of agriculture'.

Table 44.25. *Wool prices (d/lb), 1850–1914*

	Southdown	Hampshire Down	Leicester	Lincoln	Cheviot	Welsh
1850	11.75	11.00		11.00		
1851	12.50	12.00		12.50		
1852	12.25	13.25		13.38		
1853	15.25	17.00		16.00		
1854	17.25	18.75		15.50		
1855	12.75	14.00		13.00		
1856	15.00	15.75		16.00		13.00
1857	18.50	17.75		20.50		
1858	14.50	14.00		15.38		14.00
1859	18.25	21.33		18.63		
1860	18.50	19.00		20.13		
1861	18.25	24.08		19.50		
1862	17.00	28.00		20.50		
1863	20.00	31.55		22.38		
1864	21.75	31.02		27.38		
1865	23.25	30.00		25.75		
1866	21.50	16.50		23.50		20.00
1867	17.50	16.75		18.88		
1868	14.00	31.38		17.50		
1869	14.25	30.58		18.13		10.00
1870	13.50	12.25		16.75		8.50
1871	13.00	16.50		21.38		15.00
1872		21.00		25.63		
1873	23.00	18.75		24.50	20.00	
1874		17.00		20.75	17.00	
1875	18.25	17.75		19.75	18.00	
1876	17.50	14.25		17.75	17.25	
1877	16.50	15.75		16.25	15.00	12.00
1878	15.50	15.00		15.00	16.00	
1879	13.00	11.50		12.50		
1880	14.50	15.75		15.13	16.00	
1881	15.50	13.00		12.38	15.00	
1882	14.00	13.00		11.25	14.00	
1883	14.00	13.00	9.50	10.00	14.00	9.00
1884	13.50	11.50	9.25	10.00	14.00	
1885	12.25	9.75	9.00	9.88	13.00	
1886	12.50	11.00	9.75	10.00	13.75	
1887	12.75	11.00	10.25	10.50	14.00	
1888	11.75	9.75	10.00	10.75	14.00	
1889	12.50	11.25	10.50	11.00	14.00	8.50

(continued)

Table 44.25 (*cont.*)

	Southdown	Hampshire Down	Leicester	Lincoln	Cheviot	Welsh
1890	13.00	10.50	10.50	11.00	14.00	
1891	13.00	10.75	10.00	9.75	14.00	7.50
1892	12.50	9.75	9.00	8.75	14.00	8.00
1893	12.00	9.50	9.25	10.25	13.50	7.50
1894	12.00	9.75	10.00	10.13	13.00	7.00
1895	11.50	9.75	10.50	12.00	12.50	
1896	11.25	10.00	11.00	11.50	12.00	
1897	10.50	9.25	10.00	9.38	11.50	
1898	9.75	8.75	8.75	8.75	10.00	
1899	11.00	9.50	8.00	8.25	9.25	
1900	12.00	10.00	7.50	7.88	9.25	
1901	9.25	7.25	6.00	6.88	8.25	
1902	9.13	7.50	5.63	6.25	8.50	
1903	11.50	9.50	6.75	7.25	9.00	
1904	11.75	10.75	9.63	10.50	10.50	
1905	13.25	13.25	12.00	12.50	13.00	
1906	15.13	14.25	13.00	14.13	14.25	
1907	15.00	11.75	12.88	12.25	12.00	
1908	12.50	10.25	8.88	8.25	9.00	
1909	13.38	13.00	8.63	8.75	13.00	
1910	15.00	13.00	9.75	9.75	15.00	
1911	14.63	12.50	10.13	9.88	15.00	
1912	14.50	13.00	10.38	10.50	12.43	
1913	15.88	14.75	12.13	12.50	12.86	
1914	16.25	13.75	12.75	12.75	12.43	

Sources: Annual Agricultural Returns, 1886, 1895, 1912–13, 1914; *RC on Agriculture; Wholesale and Retail Trade 1902*; Howell, *Welsh Agriculture*; 'Statistics affecting British agricultural interests', *JRASE*, 77, 1916; *Transactions of the Highland and Agricultural Society of Scotland*, 5th ser., 27, 1915.

Table 44.26. *Animal by-products*
(import/wholesale value), 1854–1914

	Hides (£/cwt.)	Tallow (s/cwt)
1854	2.52	
1855	2.90	55.68
1856	3.73	52.19
1857	4.06	54.54
1858	2.97	49.42
1859	3.44	55.01
1860	3.44	56.36
1861	3.04	50.74
1862	2.93	45.80
1863	2.72	42.29
1864	2.79	41.07
1865	2.62	46.10
1866	2.57	45.64
1867	2.75	44.05
1868	3.07	48.05
1869	3.03	45.33
1870	3.21	43.35
1871	2.99	42.02
1872	3.42	42.88
1873	3.53	41.28
1874	3.63	40.36
1875	3.48	42.30
1876	3.13	42.77
1877	3.09	41.96
1878	2.93	39.39
1879	2.88	35.87
1880	3.12	35.12
1881	3.18	35.24
1882	3.15	40.35
1883	3.18	40.48
1884	3.19	37.78
1885	3.15	31.32
1886	2.96	25.68
1887	2.72	23.99
1888	2.58	25.00
1889	2.51	26.46
1890	2.42	24.94
1891	2.42	25.85
1892	2.29	25.41

(continued)

Table 44.26 (*cont.*)

	Hides (£/cwt.)	Tallow (s/cwt)
1893	2.31	27.72
1894	2.17	25.52
1895	2.22	23.67
1896	2.28	21.26
1897	2.30	19.17
1898	2.35	20.44
1899	2.30	23.09
1900	2.47	26.04
1901	2.49	26.14
1902	2.58	30.40
1903	2.68	28.50
1904	2.63	25.59
1905	2.79	26.00
1906	3.04	28.91
1907	3.26	33.38
1908	2.94	30.26
1909	3.14	30.59
1910	3.34	34.06
1911	3.34	33.32
1912	3.50	33.51
1913	4.02	34.01
1914	4.25	32.95

Source: Wholesale and Retail Prices 1902; Annual Agricultural Returns, 1902–16.

Table 44.27. *Regional butter prices, 1850–1914*

	Kirkby Lonsdale Market (d/lb)	Carmarthen Market (d/lb)	Dorset butter/ Leadenhall (d/lb)
1850		7.50	8.13
1851		7.75	9.69
1852		8.25	9.88
1853		9.00	11.01
1854		10.50	11.55
1855		10.50	11.37
1856		11.50	11.92
1857		11.50	12.37
1858		10.75	12.28
1859		10.75	12.40
1860		11.75	12.41
1861		10.75	12.40
1862		10.50	12.14
1863		9.50	11.66
1864		10.75	12.70
1865		12.25	13.46
1866		12.75	13.90
1867		11.50	12.38
1868	14.75	10.25	13.46
1869	15.25	14.00	14.07
1870	15.25		13.43
1871	15.50	13.50	15.47
1872	15.25	12.50	15.02
1873	16.75	13.50	14.41
1874	17.75	14.00	16.05
1875	17.00	14.25	15.90
1876	17.25	14.00	16.34
1877	16.75	14.50	16.33
1878	15.75	13.00	15.42
1879	14.25	11.00	13.24
1880	16.25	13.25	
1881	15.75	13.00	
1882	15.25	13.25	
1883	15.00	13.00	
1884	14.25	12.75	
1885	13.25	12.00	
1886	12.25	11.00	
1887	13.00	11.00	
1888	12.50	11.50	

(continued)

Table 44.27 (*cont.*)

	Kirkby Lonsdale Market (d/lb)	Carmarthen Market (d/lb)	Dorset butter/ Leadenhall (d/lb)
1889	12.75	12.00	
1890	12.00	11.00	
1891	13.25	12.00	
1892	13.50	12.75	
1893	13.00	12.50	
1894	12.00	11.00	
1895		10.75	
1896		10.50	
1897		11.00	
1898		9.75	
1899		11.25	
1900		12.00	
1901		12.25	
1902		11.50	
1903		11.50	
1904		12.25	
1905		11.75	
1906			
1907			
1908		12.75	
1909		12.50	
1910		12.50	
1911		12.00	
1912		13.25	
1913		13.50	
1914			

Sources: RC on Agriculture; Howell, *Welsh Agriculture*; Copus, 'Changing markets and the response of agriculture'.

Table 44.28. *Irish butter prices (1st quality), 1851–1902*

	Mild cured		Salted	
	June (s/cwt)	December (s/cwt)	June (s/cwt)	December (s/cwt)
1851			74	79
1852			71	90
1853			84	100
1854			93	98
1855			103	115
1856			105	122
1857			100	105
1858			100	104
1859			102	120
1860			110	114
1861			94	107
1862	98	110	94	104
1863	93	113	86	109
1864	98	123	89	114
1865	98	130	91	126
1866	107	119	98	107
1867	100	118	96	100
1868	105	138	101	129
1869	103	132	100	127
1870	107	140	104	132
1871	115	135	114	129
1872	114	137	112	121
1873	116	149	113	142
1874	120	159	116	150
1875	120	150	118	142
1876	126	153	123	147
1877	119	145	117	127
1878	115	149	103	121
1879	100		87	133
1880	112	150	106	138
1881	109	160	98	127
1882	113	157	102	135
1883	111	136	100	131
1884	103	142	94	129
1885	90	143	83	116
1886	89	123	70	114
1887	77	110	70	117
1888	88	122	78	113
1889	83	123	80	115

(continued)

Table 44.28 (*cont.*)

	Mild cured		Salted	
	June (s/cwt)	December (s/cwt)	June (s/cwt)	December (s/cwt)
1890	85	128	81	118
1891	84	127	78	117
1892	90	126	82	116
1893	89	127	84	121
1894	84	116	80	99
1895	72	104	71	109
1896	78	110	79	102
1897	75	106	74	93
1898	76	101	72	89
1899	79	99	74	90
1900	84	103	78	92
1901	89	109	83	100
1902	97	103	87	95

Source: Wholesale and Retail Prices, 1902.

Table 44.29. *English cheese prices, 1868–89*

	'Fine' Cheddar		'Good' Wiltshire		'New' Cheshire		'English'	
	(January market prices)							
	(s/cwt) min.	(s/cwt) max.	(s/cwt) min.	(s/cwt) max.	(s/cwt) min.	(s/cwt) max.	(s/cwt) min.	(s/cwt) max.
1868	78	88	52	62	76	84		
1869	86	94	62	64	80	90		
1870	90	94	62	68	84	90		
1871					78	90		
1872	66	84	50	60	70	84		
1873	70	90	56	60	70	84		
1874	76	92	60	66	78	85		
1875	74	94	66	68	84	88		
1876	74	92			76	86		
1877	60	94			78	90		
1878	78	90	64	70	78	84		
1879			50	66	74	82		
1880	72	86			64	86	66	86
1881	76	90			74	88	70	90
1882	76	82			72	82	60	82
1883	68	82			68	80	62	82
1884	72	86			74	85	64	86
1885							64	85
1886							54	78
1887							38	78
1888							46	78
1889							36	86

Sources: 'Statistics of dairy produce', *JRASE*, 6, 1870; 15, 1879; 20, 1884; 25, 1889.

Table 44.30. *English cheese prices, 1905–14*

	Cheddar 1st class annual average (s/cwt)	Cheshire 1st class average (London) (s/cwt)
1905	71.50	
1906	75.50	
1907	78.50	
1908	74.50	
1909	75.50	78.50
1910	74.50	76.50
1911	79.00	83.00
1912	82.00	81.00
1913	76.50	78.00
1914	82.00	83.50

Source: Annual Agricultural Returns, 1905–15.

Table 44.31. *London contract milk prices*
(d per gallon), 1850–1914

	St Thomas	Bethlem	London (average)
1850		10.00	
1851		8.00	
1852		7.00	
1853		8.00	
1854		7.50	
1855	9.19	10.00	
1856	9.82	10.00	
1857	9.74	10.00	
1858	9.75	9.00	
1859	9.25	9.00	
1860	9.00	9.00	
1861	9.75	10.00	
1862	10.00	10.00	
1863	10.00	9.00	
1864	10.00	9.50	
1865	11.00	13.00	
1866	13.00	12.00	
1867	11.00	12.00	
1868	10.00	12.00	
1869	9.50	12.00	
1870	8.00	12.00	
1871	8.25	12.00	
1872	9.20	12.00	
1873	10.50	12.00	
1874	10.50	15.00	
1875	10.50	15.00	
1876	10.50	15.00	
1877	11.00	15.00	
1878	11.00	15.00	
1879	10.40	15.00	
1880	10.00	12.00	
1881	10.00	12.00	
1882	10.00	12.00	
1883	9.84	13.00	
1884	9.75	11.00	
1885	9.40	11.00	
1886	8.76	11.00	
1887	8.20	11.50	
1888	8.16	11.50	

(continued)

Table 44.31 *(cont.)*

	St Thomas	Bethlem	London (average)
1889	8.10	10.50	
1890	8.00	10.50	
1891	7.90	10.50	
1892	9.00	11.00	
1893	9.70	10.50	
1894	10.50	9.25	
1895	8.80	9.00	
1896	8.00	8.00	
1897	8.20	8.00	
1898	7.92	8.00	
1899	8.00	8.25	
1900	8.28	8.50	
1901	9.10	8.75	
1902	9.30	9.00	
1903			
1904			
1905			
1906		8.06	7.88
1907		8.31	8.25
1908		8.44	8.38
1909		8.38	8.50
1910		8.38	8.25
1911		8.81	8.50
1912		9.06	9.13
1913		9.16	9.00
1914			9.31

Source: Wholesale and Retail Prices 1902;
R. Cohen, *The History of Milk Prices,*
(Oxford, 1936).

Table 44.32. *Eggs, 1854–1914*

	Import value (d/doz.)	Aberystwyth Market (d/doz.)	English markets	
			min. (d/doz.)	max. (d/doz.)
1854	5.40			
1855	6.84			
1856	6.84			
1857	7.20			
1858	6.49			
1859	6.52			
1860	8.22			
1861	7.80			
1862	7.36			
1863	7.27			
1864	7.17			
1865	7.37			
1866	7.26			
1867	7.16			
1868	7.57			
1869	7.34			
1870	7.37			
1871	9.09			
1872	9.55			
1873	10.29			
1874	10.30			
1875	9.95			
1876	10.02			
1877	9.48			
1878	9.23	8.25		
1879	8.62	7.00		
1880	8.61			
1881	8.84	7.25		
1882	8.44	8.75		
1883	8.37	11.75		
1884	8.44	9.75		
1885	8.42	8.50		
1886	8.02	8.75		
1887	8.15	8.75		
1888	7.88	8.25		
1889	7.96	8.50		
1890	8.00	8.25		
1891	7.92	10.25		
1892	8.18	9.25		

(continued)

Table 44.32 (*cont.*)

	Import value (d/doz.)	Aberystwyth Market (d/doz.)	English markets	
			min. (d/doz.)	max. (d/doz.)
1893	8.42	9.75		
1894	7.65	10.25		
1895	7.55	9.25		
1896	7.58	9.75		
1897	7.45			
1898	7.42			
1899	7.48	8.75		
1900	7.69	10.50		
1901	7.73	10.00		
1902	7.98	9.50		
1903	8.00	10.00		
1904	8.10	10.75		
1905	8.69		12.00	13.80
1906	9.03	12.75	12.70	13.90
1907	9.22	12.00	13.00	14.20
1908	9.47		12.80	14.10
1909	9.80		13.30	14.30
1910	9.55		13.20	14.50
1911	10.03		13.60	14.80
1912	10.56		13.80	14.80
1913	10.67		14.10	15.20
1914	11.60		14.80	16.20

Sources: Annual Agricultural Returns, 1902–16; *RC on Agriculture*; Howell, *Welsh Agriculture*.

Table 44.33. *Livestock feed prices, 1850–1914*

	Hay			Linseed cake	
	min. (£/load)	max.	Maize (s/cwt)	min. (£/ton)	max.
1850	2.40	3.90		7.25	7.50
1851	2.40	4.25		7.25	7.50
1852	2.75	4.50		8.00	8.25
1853	3.25	5.75		9.25	10.00
1854	2.50	5.50	9.50	11.00	
1855	2.50	6.25	10.14	12.50	13.00
1856	3.40	3.75	7.43	10.50	
1857	2.50	4.50	8.26	11.00	
1858	2.50	4.50	7.04	10.50	11.00
1859	2.50	4.75	6.69	9.35	9.50
1860	2.50	5.75	7.98	10.50	10.75
1861	1.80	5.50	7.37	11.00	11.50
1862	1.80	5.25	6.53	10.50	11.00
1863	1.80	4.75	6.35	9.75	10.25
1864	3.00	6.00	6.29	10.75	11.25
1865	2.50	6.00	6.30	10.00	10.25
1866	4.00	6.00	6.33	11.00	11.25
1867	3.00	4.50	8.98	11.00	11.25
1868	2.75	5.25	8.43	12.00	12.75
1869	3.50	6.00	6.72	11.35	11.50
1870	3.00	6.30	6.91	11.75	12.50
1871	3.75	7.75	7.69	11.25	11.50
1872	2.50	4.50	7.09	11.25	11.50
1873	1.80	4.40	7.06	11.75	12.50
1874			8.46	12.50	12.75
1875			7.95	12.00	
1876			6.39	11.00	11.50
1877			6.47	10.75	
1878			6.04	9.75	10.00
1879	2.00	5.25	5.43	9.00	9.25
1880	1.40	5.50	6.00	10.00	
1881	1.50	6.70	6.22	9.50	10.50
1882	2.75	6.20	7.15	8.00	
1883	1.80	4.60	6.53	8.75	9.00
1884	2.00	4.75	5.89	9.00	
1885	1.25	4.90	5.39	8.00	8.25
1886	2.25	4.60	4.91	7.75	
1887	2.50	4.63	4.84	6.63	6.75
1888	2.25	6.50	5.43	7.38	7.50

(continued)

Table 44.33 (*cont.*)

	Hay			Linseed cake	
	min.	max.	Maize	min.	max.
	(£/load)		(s/cwt)	(£/ton)	
1889	1.25	6.50	4.74	7.75	8.00
1890	1.30	4.60	4.54	7.25	7.50
1891	1.30	4.50	6.27	7.88	8.00
1892	1.75	5.40	5.33	7.63	8.25
1893	3.50	8.50	4.80	8.00	9.15
1894	3.00	8.50	4.50	6.25	9.50
1895	1.80	5.00	4.60	5.25	8.50
1896	1.80	4.50	3.64	5.50	5.75
1897	2.30	4.60	3.42	7.00	7.25
1898	2.25	4.20	3.95	7.00	7.38
1899	2.00	4.20	4.14	7.25	8.00
1900	2.50	4.38	4.55	8.50	8.63
1901	2.50	5.50	4.82	7.88	8.13
1902	4.00	5.63	5.27	7.38	7.50
1903	3.00	4.50	4.98	6.50	6.75
1904	3.00	4.00	4.78	7.38	7.63
1905	3.00	3.85	5.24	7.75	7.88
1906		4.13	4.92	8.00	8.13
1907		4.25	5.47	7.63	7.75
1908		3.90	6.14	8.13	8.25
1909		4.00	6.16	8.38	8.50
1910		4.00	5.56	9.00	9.13
1911		5.00	5.55	9.50	9.63
1912		5.75	6.20	9.00	9.25
1913		5.38	5.60	7.75	7.88
1914		3.88	6.02	8.63	8.88

Notes: Prices for hay are for London, extracted from *The Times* and *The Annual Register*; cake prices are from the first week in October, maize prices are import values.
Sources: P. J. Atkins, 'The milk trade of London *c.* 1790–1914', PhD thesis, University of Cambridge, 1977; *The Economist*, 1850–1914.

Table 44.34. *Machinery prices, 1850–1914*

	Ransomes' plough (shillings)			Ransomes' threshing drum (£)	Ransomes' steam engine (£)
	2 wheel	1 wheel	swing		
1850					
1851					
1852					
1853	90.00	80.00	74.00		
1854					
1855					
1856	92.50	82.50	76.50		
1857	92.50			130.00	215.00
1858	92.50	82.50		130.00	215.00
1859	92.50			130.00	215.00
1860	92.50	82.50	76.50	130.00	215.00
1861	92.50	82.50	76.50	130.00	215.00
1862	92.50	82.50	76.50	130.00	215.00
1863	92.50	82.50	76.50	130.00	
1864					
1865					
1866					
1867					
1868	95.00	85.00	76.50	125.00	210.00
1869	95.00		76.50	125.00	210.00
1870	95.00	85.00	76.50	125.00	210.00
1871	100.00	90.00	82.50	125.00	210.00
1872	122.50	112.50	112.50	150.00	210.00
1873					
1874					
1875					
1876					
1877					
1878					
1879	105.00	95.00	95.00	150.00	210.00
1880					
1881	102.50	95.00	95.00	150.00	210.00
1882					
1883					
1884	102.50	95.00	92.50	150.00	210.00
1885	102.50	95.00	92.50	140.00	195.00
1886	100.00	92.50	90.00	140.00	190.00
1887	100.00	92.50	90.00	140.00	190.00
1888	100.00	92.50	90.00		

(continued)

Table 44.34 (*cont.*)

	Ransomes' plough (shillings)			Ransomes' threshing drum (£)	Ransomes' steam engine (£)
	2 wheel	1 wheel	swing		
1889	100.00	92.50	90.00	140.00	190.00
1890					
1891	110.00	102.50	100.00	150.00	210.00
1892	110.00	102.50	100.00	150.00	210.00
1893					
1894	110.00	102.50	100.00	150.00	210.00
1895					
1896					
1897	110.00	102.50	100.00	150.00	210.00
1898				150.00	
1899					
1900					
1901	122.00	114.50	110.00	155.00	240.00
1902	115.00	107.50	105.00	150.00	225.00
1903					
1904	115.00	107.50	105.00	150.00	225.00
1905	115.00	107.50	105.00	150.00	225.00
1906	115.00	107.50	105.00	150.00	225.00
1907	120.75	112.88	110.25	160.00	235.00
1908					
1909	115.00	107.50	105.00	160.00	235.00
1910					
1911					
1912					
1913					
1914	117.50	110.00	105.00		

Note: From 1856 to 1863 the Ransomes' plough is a YL class; from 1868 an RNE class; the thresher is an A1 class; the engine is 8 HP.
Source: Ransomes, Sims, & Jeffries Catalogues (RHC TR-RAN, Rural History Centre, University of Reading).

Table 44.35. *Fertiliser prices, 1850–1914*

| | Nitrate of soda | | Guano ($£$/ton)[b] | Phosphate of lime/ rock phosphate ($£$/ton)[b] |
	min. (s/cwt)[a]	max.		
1850	13.00	13.25		
1851	14.50	15.00		
1852	13.75	14.00		
1853	20.50	21.00		
1854	20.00	20.50		
1855	17.25	17.50		
1856	18.00			
1857	19.00	19.50		
1858	17.00	18.00		
1859	16.50	17.00		
1860	15.00	16.50		
1861	13.00	14.00	11.33	
1862	13.50	14.50	11.55	
1863	13.50	14.50	11.38	
1864	14.00	16.00	11.09	
1865	14.00	15.00	11.27	
1866	13.00	13.50	10.61	
1867	12.00	13.00	10.97	
1868	12.50	13.50	11.18	
1869	16.00	16.50	12.58	
1870	17.50	18.00	12.40	
1871	16.00	17.50	11.43	
1872	16.50	17.00	10.12	
1873	16.00	16.50	11.41	
1874	11.00	15.50	12.00	
1875	12.50	13.00	11.30	
1876	11.25	11.50	11.52	
1877	12.50		10.88	
1878	16.00		10.16	
1879	13.25	13.50	9.15	
1880	18.50	18.75	10.06	
1881	15.00		9.73	
1882	14.25	14.50	8.64	1.95
1883	12.25	12.50	9.76	2.93
1884	9.25	9.60	9.10	2.94
1885	9.75		9.70	2.63
1886	11.25		7.81	2.42
1887	11.00	11.25	8.22	2.17
1888	10.25	10.75	8.04	2.11

(continued)

Table 44.35 (cont.)

| | Nitrate of soda | | | Phosphate of lime/ |
| | min. | max. | Guano | rock phosphate |
	(s/cwt)[a]		(£/ton)[b]	(£/ton)[b]
1889	9.75	10.00	6.98	2.31
1890	8.75	8.63	6.07	2.47
1891	9.00	9.25	5.87	2.45
1892	9.00	9.25	6.80	2.12
1893	9.88	10.00	5.17	1.84
1894	9.50	9.75	5.12	1.90
1895	8.75	9.13	5.16	1.76
1896	8.00	8.25	5.37	1.60
1897	8.13	8.25	4.99	1.51
1898	7.50	7.63	5.21	1.52
1899	7.75	8.00	5.21	1.63
1900	7.75	8.00	5.27	1.66
1901			4.60	1.55
1902			3.92	1.52
1903			5.51	1.43
1904			4.85	1.48
1905			4.75	1.46
1906			5.13	1.53
1907			4.75	1.64
1908			4.62	1.73
1909			4.39	1.65
1910			5.29	1.59
1911			5.66	1.58
1912			5.78	1.62
1913	12.25	12.75	5.84	1.62
1914	11.00	11.50	5.91	1.73

Notes:
[a] Prices for nitrate of soda are from the first week of April each year.
[b] Guano and phosphate of lime/rock phosphate were calculated by dividing import values by quantities.
Sources: The Economist, 1850–1914; Annual Agricultural Returns, 1878–1915.

Table 44.36. *Seed prices, 1850–1914*

	Red clover		White clover		Clover and grass
	min.	max.	min.	max.	import value
	(s/cwt)		(s/cwt)		(s/cwt)
1850	30.00	44.00	32.00	46.00	
1851	38.00	52.00	40.00	54.00	
1852	42.00	55.00	48.00	58.00	
1853	45.00	55.00	52.00	60.00	
1854	44.00	56.00	72.00	80.00	
1855	52.00	66.00	60.00	75.00	
1856	68.00	80.00	70.00	90.00	
1857	60.00	78.00	62.00	75.00	
1858	48.00	65.00	55.00	62.00	
1859	80.00	100.00	65.00	80.00	
1860	40.00	50.00	70.00	90.00	
1861	50.00	60.00	70.00	90.00	47.03
1862	42.00	60.00	50.00	76.00	45.89
1863	56.00	68.00	70.00	90.00	47.50
1864	42.00	50.00	40.00	70.00	46.82
1865	70.00	100.00	50.00	80.00	56.10
1866	42.00	56.00	46.00	76.00	53.92
1867	40.00	56.00	66.00	90.00	54.95
1868	44.00	56.00	50.00	70.00	52.79
1869	60.00	66.00	45.00	65.00	54.64
1870	52.00	84.00	60.00	90.00	56.44
1871	70.00	90.00	74.00	90.00	54.45
1872	50.00	56.00	70.00	86.00	51.46
1873	60.00	86.00	70.00	86.00	47.79
1874	40.00	60.00	56.00	70.00	46.01
1875	40.00	60.00	56.00	70.00	47.00
1876	80.00	95.00	76.00	90.00	51.62
1877	75.00	110.00	90.00	115.00	51.86
1878					47.56
1879					43.81
1880	45.00	80.00	56.00	100.00	43.26
1881					44.28
1882					42.66
1883					47.59
1884					45.27
1885					46.67
1886					40.82
1887					41.19
1888					41.26
1889					41.02

(continued)

Table 44.36 (*cont.*)

	Red clover		White clover		Clover and grass import value (s/cwt)
	min. (s/cwt)	max.	min. (s/cwt)	max.	
1890					39.95
1891					43.05
1892					42.72
1893					47.51
1894					47.82
1895					43.18
1896					38.88
1897					38.62
1898					38.23
1899					36.74
1900					38.85
1901					43.51
1902					43.84
1903					44.05
1904					40.79
1905					41.23
1906					40.92
1907					40.38
1908					44.42
1909					45.62
1910					46.29
1911					47.09
1912					48.26
1913					47.84
1914					46.70

Note: Prices for red and white clover are from the first week of April each year, import values are calculated by dividing quantity imported by value.

Sources: The Economist, 1850–1914; *Annual Agricultural Returns*, 1878–1915.

CHAPTER 45

THE IMPACT OF FOREIGN TRADE

BY BETHANIE AFTON AND MICHAEL TURNER

A. INTRODUCTION

One of the most significant external factors affecting the agrarian community in England and Wales during the period 1850 to 1914 was the rapidly expanding supply of imported agricultural goods, particularly those which were in direct competition with home production. This was a result of considered government policy which promoted free trade, ensuring worldwide markets for the nation's manufactured products and a cheap supply of raw materials for the industrial sector and food for consumption. During the 1840s the major acts protecting British agriculture, including restrictions on imported livestock, the Corn Laws and the Navigation Acts, were repealed or scheduled to be phased out. Between 1850 and 1885 free trade flourished and, at its peak in 1860, embraced most of the trading nations of the world. By the mid-1880s many countries had re-imposed restrictions on trade. However, the United Kingdom, in spite of considerable agitation in the closing years of the nineteenth century and the beginning of the twentieth to introduce legislation to protect agriculture and establish a preferential trading network with the empire, essentially maintained a policy of free trade throughout most of the period.[1]

Initially English and Welsh agriculture was relatively unaffected by

---

[1] Amongst a huge literature see, W. Newmarch, 'On the progress of the foreign trade of the United Kingdom since 1856, with especial reference to the effects produced upon it by the protectionist tariffs of other countries', *JRSS*, 41, 1878, pp. 193–7, 267–79; E. A. Benians, 'Finance, trade, and communications 1870–1895', and G. S. Graham, 'Imperial finance, trade, and communications, 1895–1914', in E. A. Benians, J. Butler and C. E. Carrington (eds.), *Cambridge History of the British Empire: The Empire and Commonwealth 1870–1918* (Cambridge, 1959), vol. III, pp. 181–229, 438–52, 865–71; P. Bairoch, 'European trade policy 1815–1914', in P. Mathias and S. Pollard (eds.), *Cambridge Economic History of Europe: The Industrial Economies: the Development of Economic and Social Policies* (Cambridge, 1989), vol. VIII, pp. 25, 86; *British and Foreign Trade and Industry (Memoranda)*, BPP, 1905, LXXXIV.

foreign competition. Continental surplus production was not sufficiently cheap to threaten seriously home production.[2] However, supplies of agricultural imports increasingly came from further afield. The cost of transporting grain from its place of origin to Britain was said to have been reduced by half between 1869 and 1893, and trans-Atlantic freight costs in general were 53 per cent of their 1850 price in 1910.[3] Once cheap transport was available to carry agricultural products to ports for shipping to Britain, the export of grain from India and the rich virgin lands of the Americas and Australia was facilitated. An increasingly international market for agricultural products was developing.[4]

Opening British ports to unrestricted trade in livestock had a largely unanticipated, pernicious effect on agriculture. Diseases, endemic to the Continent, were imported into the United Kingdom. Foot and mouth disease, pleuropneumonia, sheep pox, swine fever, rinderpest and anthrax periodically spread across the English Channel. Estimates suggest that the annual losses from disease rose from between 1.5 per cent and 2.5 per cent in 1840 to between 5 per cent and 11 per cent in the 1860s.[5] Efforts to prevent this were the only major pieces of restrictive legislation imposed on agricultural imports during the period. In 1848 an act was passed which allowed for the prevention of importing diseased animals or for their quarantine once imported. The Cattle Diseases Prevention Act of 1866 provided for the landing of foreign animals to be limited to certain parts of ports, 'Foreign Animal Wharves', where they could be inspected and slaughtered if necessary. Restrictions on movements inland were also imposed. This was extended in 1869 to limit ports to which livestock could be sent. From 1878 foreign livestock could be imported only for slaughter unless, in exceptional circumstances, this was waived. An act of 1884 enabled livestock to be banned altogether from Britain if they came from areas in which listed diseases existed. In 1896 it became compulsory for all imported animals to be slaughtered at Foreign Animal Wharves.[6] The 'sanitary restrictions' altered the international trading patterns in livestock. When the restrictions were seriously enforced, the number of

---

[2] Newmarch, 'On the progress', pp. 191–97; S. Fairlie, 'The nineteenth-century corn laws reconsidered', *EcHR*, 18, 1965, pp. 562–75; S. Fairlie, 'The corn laws and British wheat production, 1829–76' *EcHR*, 22, 1969, esp. pp. 88–90; R. Perren, *The Meat Trade in Britain 1840–1914* (London, 1978), pp. 74–9.

[3] Bairoch, 'European trade', p. 56. For a detailed discussion of the reduction in costs of transport between 1869 and 1893, see R. F. Crawford, 'An inquiry into wheat prices and wheat supplies', *JRSS*, 58, 1895, pp. 84–99.

[4] Bairoch, 'European trade' pp. 24–5; Perren, *Meat Trade*, pp. 114–32.

[5] Perren, *Meat Trade*, p. 69.

[6] *Committee on Combinations in the Meat Trade*, BPP, 1909, xv, appendix IV; G. T. Brown, 'The progress of legislation against contagious diseases of live stock, pt. II', *JRASE*, 3rd ser., 4, 1893, pp. 281–5; Perren, *Meat Trade*, pp. 106–14.

live imports decreased. Trade in pigs became negligible by 1890, and that of sheep diminished considerably. Exporters generally responded by shipping increased quantities of dead meat.[7] Thus, even when the powers to restrict imports were fully implemented, home producers were only partially sheltered from foreign competition.[8]

The returns of the Board of Trade are the main data source for import and export statistics. Although their format and reliability have altered over the centuries, records of overseas trade are the oldest set of regularly collected statistics in Britain.[9] The statistics for the period were published monthly in *Accounts relating to the Trade and Navigation of the United Kingdom*, and annually in the *Annual Statement of the Trade of the United Kingdom*. There are problems when using these statistics. First, without totalling the entries from each individual port, it is not possible to disaggregate the statistics for England and Wales from the whole. Furthermore, the integrated nature of the agriculture of the United Kingdom in the late nineteenth century makes such calculations problematic. A more useful division would place Wales with Scotland and Ireland, as areas of upland grazing which exported to the lowland areas of England.[10] There are also problems in determining the country of origin of imports. Until 1904 the recorded country of origin of United Kingdom imports was taken from information provided from Customs entries. Where possible this was recorded as the country from which the goods were originally dispatched. However, more often, the country from which the goods were finally shipped was listed. In 1904 both the 'country of shipment' and the 'country of consignment' were listed, and from 1909 the 'country of consignment' became the substantive system of classification.[11] A third problem with the statistics is in the measurement of value and quantity. Quantitative measurement did not present particular difficulties except in situations when the method for measuring changed between volume and weight or when fabricated items like agricultural machinery were simply recorded by the ton. By using value measurements, goods of completely different natures can be compared. However, when using these values it must be remembered that historically fixed unit values were often used, such that when the value *per unit* of a product changed this is not always

[7] *Agricultural Returns for 1893*, BPP, 1893/4, CI, p. xxxi.

[8] A significant exception to all of these restrictions was the animal trade with Ireland, a discussion of which can be found in Chapter 3 *supra*. See, for example, *Agricultural Statistics, 1907 . . . Prices and Supplies of Corn, Live Stock and other Agricultural Produce*, BPP, 1908, CXXI, in which pp. 272–5 is 'Trade in Livestock with Ireland'.

[9] A. Maizels, 'Overseas Trade', in M. G. Kendall (ed.), *The Sources and Nature of the Statistics of the United Kingdom* (London, 1952), vol. I, pp. 17–33.

[10] F. Capie and R. Perren, 'The British market for meat 1850–1914', *Agricultural History*, 54, 1980, p. 505.   [11] Maizels, 'Overseas trade', p. 30.

apparent if total values alone are used.[12] In addition the method of calculating values changed over time. From 1798 the recorded values of exports were based on the actual value of the goods they were exporting. However, imports continued to be valued according to a 'price list' established in 1696 and only occasionally revised. In 1854 a system of 'computed real values' based on current value was adopted. In 1876 the system was again altered so that the value of all goods was based on the open-market value at the port at the time of import or export.[13]

The data extracted from the Board of Trade returns were included in other governmental publications. Particularly useful are the 'Memoranda, Statistical Tables and Charts' which were produced for a series of reports on *British and Foreign Trade and Industry*.[14] Other useful publications include *Returns showing the Quantities of Various Kinds of Grains and Flour Imported into the United Kingdom*, and the specific inquiries which were made into various aspects of trade, including those on *Combinations in the Meat Trade* and *British Export Trade in Livestock with the Colonies and other Countries*.[15] Periodically, particularly as the effect of foreign trade on agriculture increased, data on imports and, to a lesser extent, exports were included in the *Annual Agricultural Returns*. For the agrarian historian this is especially useful because trade statistics relating to agriculture were brought together into one source along with a wide range of other information. From the annual return in 1885, for example, in addition to information on the agriculture of Great Britain it is also possible to find data on the quantities and values of livestock, meat, grain, fruit, vegetables, manures, oilcake and seeds imported between 1865 and 1884, the country of origin of meat, grain, cotton and wool imported between 1869 and 1884, as well as the acreages and produce of crops and numbers of livestock in both British Possessions and foreign countries over a number of years.[16] The *Reports* accompanying the Annual Returns and, from 1895, those published in the *Journal of the Board of Agriculture and Fisheries* provide detailed analysis of foreign trade in agricultural products. Numerous other contemporary agricultural and statistical journals, including the *Journal of the Royal Agricultural Society* and the *Journal of the Royal Statistical Society*, contain useful analytical articles on agricultural import and export statistics.

---

[12] Problems in using trade statistics for comparative purposes are discussed in S. Bourne, 'On variations in the volume and value of exports and imports of the United Kingdom in recent years', *JRSS*, 52, 1889, pp. 401–3.   [13] Maizels, 'Overseas trade', pp. 21–2.

[14] As in, *British and Foreign Trade and Industry (Memoranda)*, BPP, 1903, LXVII; 1909, CII.

[15] *Returns showing the Total Quantities of the Various Kinds of Grain and Flour Imported into the United Kingdom in 1879 and 1880*, BPP, 1881, LXXXIII; *Combinations in the Meat Trade; Report of the Departmental Committee appointed to inquire into the Report as to the British Export Trade in Livestock with the Colonies and other Countries*, BPP, 1911, XXII, and Evidence in 1912–13, XXV.

[16] *Agricultural Returns of Great Britain for 1885*, BPP, 1884–5, LXXXIV.

B. THE DATA

Because the statistical data of imports and exports for the United Kingdom (separate enumeration for those products imported into or exported from England and Wales does not exist) are readily available, Tables 45.1a–g contain a wide range of products which in some way affected the agrarian, and more specifically the agricultural, community in the years 1854, 1874, 1894 and 1914. From 1854 both the quantities and the value of imports and exports were regularly collected. The year 1874 was at or close to the height of relative prosperity before the onset of the Agricultural Depression. Imports and exports of a wide range of agricultural products were considered to have been in part responsible for the depth of the depression in agriculture. Hence, the information for 1894 has been included, a year which was at or close to the bottom of the depression. The year 1914 represents the final year of what may be considered Britain's nineteenth-century world, before the dramatic changes arising from the First World War took effect, not only in agriculture and the agricultural community, but more widely. While the imports may have been in direct competition with home agriculture, a number of the products actually helped to reduce the costs of the home producer, through purchased inputs including feeds and fertilisers. Grains were easily and often profitably imported.

Figures 45.1a–d plot the annual imports of wheat, wheat flour, barley and oats. Wheat, particularly, was extensively imported during the period. Figures 45.2a and 45.2b detail the country of origin of wheat and wheat flour/meal, and illustrate the shift from Europe to North America in the importance of sources of imports. Table 45.2 compares the price of wheat in the United Kingdom, France, Prussia and the United States, along with the duties imposed by each nation on imports. Table 45.3 considers wheat production in the United States and the cost of shipping it to England.

The pattern of livestock and meat imports into the United Kingdom is illustrated in Figures 45.3a–d. Live imports declined at the end of the period while the import of meat, and especially beef, increased rapidly. The country of origin of imports of sheep, cattle, beef and mutton in Figures 45.4a–d help to demonstrate the impact on trade of improving technology, including the introduction of refrigeration and bulk-cargo holds on ships, and the opening of new farming lands in the New World. The importance of Europe declined as that of North and South America, Australia and New Zealand increased.

Table 45.4 has been included to demonstrate the general fall in the cost of freight from various countries to the United Kingdom for a variety of major agricultural products.

Table 45.1a. *United Kingdom imports and exports of grains and grain products, 1854–1914*

	Quantities (cwt)				Values (£s)			
	1854	1874	1894	1914	1854	1874	1894	1914
Imports								
Wheat	14,868,650	41,527,638	70,126,232	103,926,743	11,693,737	25,236,932	18,760,505	44,734,079
Barley	1,974,900	11,335,396	31,241,384	16,044,422	836,798	5,291,287	7,090,579	5,660,312
Oats	2,791,110	11,387,768	14,979,214	14,156,715	1,377,226	5,116,732	3,900,096	4,674,417
Maize	5,784,420	17,693,625	35,365,043	39,040,747	2,748,606	7,482,720	7,952,238	11,760,912
Peas	491,296		22,272,623	98,694			647,194	546,470
Beans	1,652,854		5,259,895	1,441,559			1,346,096	502,926
Peas & beans		4,186,698				1,964,164		
Rye	24,517	470,018	1,009,226	953,230		197,736	215,075	349,681
Buckwheat	1,025	67,519	134,893	40,065		27,483	38,891	15,058
Bere & bigg								
Other grains[a]					1,102,499			
Wheat meal & flour	3,646,515	6,236,044	19,134,605	10,060,223	3,970,549	5,685,076	7,994,673	5,549,048
Barley meal								
Oatmeal		239,782	239,782	609,992			135,989	502,938
Maize meal		87,120	87,120	232,469			40,968	78,895
Other meals[b]	47,464	39,994	447,051	1,899,057	30,868	18,651	97,921	1,092,615

(continued)

Table 45.1a. (cont.)

	Quantities (cwt)				Values (£s)			
	1854	1874	1894	1914	1854	1874	1894	1914
Malt[c]		96	10,737	42,972		60		31,386
Hops	119,040	145,944	189,155	97,306	1,133,649	929,041	774,378	558,741
Exports								
Wheat	165,468	344,666	15,477	119,827	122,405	230,066	5,374	43,653
Barley			21,556	36,571			10,721	18,163
Oats			251,364	343,912			77,020	88,702
Rye			819	4,256			164	1,624
Beans			574	178,635			251	68,783
Peas			24,130	70,724			11,323	62,949
Other grains					366,474	340,888		
Wheat meal & flour	38,190	95,213	271,553	1,908,478	39,452	86,484	106,420	1,141,180
Oatmeal			49,103	75,144			43,784	77,018
Maize meal				52,267				37,989
Other grains, meals, and farinaceous substances			11,543	870,448			3,517	595,781
Hops	5,238		13,317	9,972	57,740		65,388	65,485
Malt[c]			80,776	272,649	16,875		158,164	193,113

Notes:

[a] Excluding wheat, barley, oats, & maize.

[b] Excluding all but wheat in 1854 and 1874 and wheat, oat and maize meal in 1894 and 1914.

[c] Initially, malt was listed in quarters. A factor ×3 was used to convert quarters to cwt. This has been employed here.

Sources: Annual Statement of the Trade of the United Kingdom, 1855, 1875, 1895, 1915; Agricultural Returns of Great Britain, 1855, 1875, 1895, 1915.

Table 45.1b. *United Kingdom imports and exports of live animals, meats, fish, and dairy products, 1854–1914*

	Quantities (cwt unless otherwise indicated)				Values (£s)			
	1854	1874	1894	1914	1854	1874	1894	1914
Imports								
Live cattle for food (nos.)	114,338	193,862	475,440	2,234	1,163,016	3,296,460	8,285,044	46,295
Live sheep for food (nos.)	183,436	758,915	484,597	1,707	271,605	1,610,355	804,823	3,000
Live swine for food (nos.)		115,389	8	—		358,296	16	
Beef, salted		226,928	242,311	29,841		434,817	342,814	65,262
Beef, fresh or slightly salted		34,793	2,104,104	8,844,567		88,509	4,213,688	19,060,371
Beef, salted or otherwise preserved	192,274				377,809			
Beef, preserved not by salting			291,056	803,402			813,698	4,239,570
Mutton, fresh			2,295,066	5,199,731			4,341,227	11,410,310
Mutton, preserved			112,928	61,274			198,166	183,370
Unenumerated, salted or fresh		119,403		946,265		335,846		25,636,664
Unenumerated, otherwise preserved		265,223				757,001		
Unenumerated, fresh, salted, preserved			340,139	1,083,826			892,762	
Pork, salted and fresh	160,898	322,574	405,042	1,122,344	379,135	704,435	772,102	2,663,199
Bacon and hams	423,510	2,542,095	4,819,388	5,936,910	892,462	5,902,429	10,855,715	21,288,646
Fish	167,788	661,406	16,286,537	3,573,203	146,065	981,950	19,466,780	6,324,940
Eggs (thousands)	121,947	680,552	1,425,236	2,148,577	228,650	2,433,134	3,786,329	8,652,800
Lard	274,595	374,328	1,400,516	1,765,107	707,082	884,596	2,758,416	4,750,943
Butter	482,514	1,619,808	2,574,835	3,984,204	2,171,194	9,050,025	13,456,699	24,014,276
Cheese	388,714	1,485,265	2,266,145	2,433,864	906,078	4,483,927	5,474,940	7,966,162

(continued)

Table 45.1b (cont.)

	Quantities (cwt unless otherwise indicated)				Values (£s)			
	1854	1874	1894	1914	1854	1874	1894	1914
Milk, condensed			529,465	1,225,316			1,079,235	2,154,169
Milk, fresh or preserved other than condensed			161,633	30,601			21,371	91,346
Milk powder, unsweetened				37,396				74,552
Exports of British and Irish produce								
Live cattle (nos.)			8,280	3,597			135,125	191,361
Live sheep (nos.)			4,638	3,040			39,522	36,029
Live pigs (nos.)			1,368	414			7,921	5,399
Live poultry & game (nos.)				24,864				22,403
Live animals, unenumerated (nos.)			28,402	32,336			33,960	20,723
Beef and pork	52,745				171,522			
Bacon and hams	23,935				96,534			
Meat, beef, pork, bacon, hams			99,410	262,552			360,493	1,059,444
Poultry & game				1,886				36,792
Fish, herrings (barrels)	285,453	852,630			297,999	1,216,782		
Fish, herrings			1,386,510	4,051,940			1,456,246	2,340,777
Fish, other kinds					88,612	224,435	596,933	1,417,076
Cheese	16,987	18,689	10,205	7,186	64,588	81,553	39,760	34,872
Butter	92,269	42,688	17,013	10,913	424,192	259,331	97,773	64,934
Milk, condensed			70,883	420,955			147,605	990,652

Sources: Annual Statement of the Trade of the United Kingdom, 1855, 1875, 1895, 1915; Agricultural Returns of Great Britain, 1855, 1875, 1895, 1915.

Table 45.1C. *United Kingdom imports and exports of fruits and vegetables, 1854–1914*

	Quantities (cwt unless otherwise indicated)				Values (£s)			
	1854	1874	1894	1914	1854	1874	1894	1914
Imports								
Potatoes	16,446	3,986,662	2,703,803	3,332,164	17,467	1,034,835	1,030,091	1,535,515
Onions (bus.)		1,370,449	5,288,512	7,513,513		279,926	765,040	1,480,773
Currants	120,253	972,455	1,307,403	1,239,840	130,672	1,290,574	860,510	1,592,296
Raisins	286,857	505,361	721,306	289,062	452,532	915,973	895,460	1,021,546
Oranges & lemons (bus.)	814,065	2,405,054	7,882,994	5,938,401				
Oranges & lemons					367,074	1,163,296	2,206,217	2,817,726
Apples (bus.)	399,465		4,968,669	2,929,649				
Apples					83,047		1,389,421	2,046,824
Pears (bus.)			1,310,074	409,871				
Pears							411,316	364,894
Cherries (bus.)			311,215	167,966				
Cherries							166,899	239,468
Sugar beet & cane raw/unrefined	9,112,364	14,130,041	14,306,004	21,983,003	9,615,802	15,837,617	8,347,711	16,505,147
Exports								
Potatoes			1,073,336	1,014,240			207,333	172,662

Sources: *Annual Statement of the Trade of the United Kingdom*, 1855, 1875, 1895, 1915; *Agricultural Returns of Great Britain*, 1855, 1875, 1895, 1915.

Table 45.1d. *United Kingdom imports of beverages, 1854–1914*

	Quantities (lb)				Values (£s)			
	1854	1874	1894	1914	1854	1874	1894	1914
Imports								
Cocoa	5,258,367	17,909,478	39,115,963	93,511,294	73,124	531,892	1,255,201	2,487,937
Coffee	66,500,358	157,433,381			1,575,185	7,064,788		
Coffee (cwt)			730,746	1,037,300			3,521,731	3,549,038
Tea	85,792,032	162,782,810	224,310,500	371,932,596	5,540,735	11,532,896	9,764,652	14,221,496

Sources: Annual Statement of the Trade of the United Kingdom, 1855, 1875, 1895, 1915; Agricultural Returns of Great Britain, 1855, 1875, 1895, 1915.

Table 45.1e. United Kingdom imports of fibres, hides and skins, 1854–1914

	Quantities (cwt unless otherwise indicated)				Values (£s)			
	1854	1874	1894	1914	1854	1874	1894	1914
Imports								
Wool (lb)	106,121,955	344,470,897	707,337,713	721,459,946	6,499,004	21,116,184	25,135,337	31,640,621
Cotton, raw	7,922,617	13,989,861	15,965,326	16,644,047	20,175,395	50,696,496	32,944,341	55,359,626
Flax	1,303,235	2,026,113	1,238,540	1,458,860	3,384,216	4,939,706	2,296,365	3,663,604
Hemp	729,564	1,236,475	1,741,960	2,532,240	1,817,905	2,190,124	1,955,404	3,353,143
Jute	481,733	4,270,164	6,777,680	4,753,620	553,993	3,553,179	4,622,137	6,412,520
Silk, raw (lb)	7,535,407	5,911,831	1,436,631	1,030,502	5,321,432	4,966,325	938,296	667,034
Sheep & lamb skins, undressed (nos.)	2,705,521	6,559,311	13,831,508		136,711	894,784	1,394,827	2,701,048
Hides, tanned & untanned	638,523	1,538,079			1,899,130	6,831,432		
Hides, raw, wet & dry		1,264,274	1,027,789	1,392,495		4,592,579	2,231,080	5,911,739
Exports								
Wool (lb)	12,901,249	10,077,619	12,984,900	38,458,000	775,092	920,415	491,014	2,294,638
Flax and hemp (tons)	2,358		1,822	5,263			192,793	334,331
Sheep & lamb skins, wooled (lb)				389,663				17,530
Sheep & lamb skins, pickled (nos.)			3,497,273	7,027,602			207,108	487,360
Other skins, undressed (nos.)				55,646,664				433,448
Other skins, dressed (nos.)				3,664,126				310,570
Hides, raw (cwt)			124,108	162,252			153,602	544,770
Hides (nos.)	72,189							

Sources: Annual Statement of the Trade of the United Kingdom, 1855, 1875, 1895, 1915; Agricultural Returns of Great Britain, 1855, 1875, 1895, 1915.

Table 45.1f. United Kingdom imports and exports of timber and wood, 1854–1914

	Quantities (load)				Values (£s)			
	1854	1874	1894	1914	1854	1874	1894	1914
Imports								
Hewn	1,216,920	2,446,525	2,338,062	3,125,298	4,985,342	7,884,209	4,187,073	6,529,111
Sawn, split	934,579	3,841,011	5,446,487	4,841,570	5,013,138	12,508,653	11,899,533	15,611,576
Exports								
Rough hewn, sawn,								
Split and staves			1,296	39,806			11,488	277,328

Sources: *Annual Statement of the Trade of the United Kingdom*, 1855, 1875, 1895, 1915; *Agricultural Returns of Great Britain*, 1855, 1875, 1895, 1915.

Table 45.1g. *United Kingdom imports and exports of off-farm inputs, 1854–1914*

	Quantities				Values (£s)			
	1854	1874	1894	1914	1854	1874	1894	1914
Imports								
Manure & Fertilisers								
bones, exc. whalefins (tons)	56,422	92,158	96,590		296,215	633,535	468,614	232,086
bones, agricultural only (tons)		83,443	88,664	34,404		549,125	412,529	186,001
guano (tons)	235,111	112,429	28,582	39,285	2,530,272	1,348,849	146,361	232,086
nitrate of soda (tons)			125,300	171,910			1,166,800	1,721,138
phosphate of lime (tons)			380,269	562,242			723,205	970,337
others (tons)			77,404	119,065			145,406	358,696
Seeds								
clover & grass (cwt)		256,159	345,118	175,905		589,268	825,155	410,737
clover (qrs.)	141,114				365,893			
Feeds								
oil seed cake (tons)	76,230	157,718	274,351	329,431	697,504	1,579,254	1,707,358	1,988,839
hay (tons)			254,214	10,558			1,174,619	42,763
molasses (cwt)	1,023,172	338,552	853,478	3,377,046			226,568	905,352
Other								
horses (nos.)	6,063	12,083	22,866	8,662	133,386		548,058	315,887
agricultural machinery								612,775
Exports								
Manures								
manures (tons)			354,303	639,115		657,449	2,329,454	4,886,474
Feeds								
oil-seed cake & other animal food (tons)								580,908
molasses, treacle, syrup & glucose (cwt)								281,064

(continued)

Table 45.1g (cont.)

	Quantities				Values (£s)			
	1854	1874	1894	1914	1854	1874	1894	1914
Seeds unenumerated (cwt)	2,346	3,050	16,457	490,680				745,472
Other horses (nos.)				37,706	117,719	205,483	449,804	1,065,022
agricultural machinery & implements							1,898,517	2,313,671

Sources: Annual Statement of the Trade of the United Kingdom, 1855, 1875, 1895, 1915; Agricultural Returns of Great Britain, 1855, 1875, 1895, 1915.

Table 45.2. *Comparative annual price of wheat per imperial quarter (shillings per quarter), 1850–1902*

	United Kingdom		France		Prussia		United States	
	average price[a]	import duty	average price[b]	import duty	average price[c]	import duty	average price[d]	import duty
1850	40	1	33	sliding scale	34	2.92	43	20% ad val.
1851	39	1	34	sliding scale	37	2.92	36	20% ad val.
1852	41	1	40	sliding scale	42	2.92	37	20% ad val.
1853	53	1	52	free	50	2.92	45	20% ad val.
1854	72	1	67	free	63	2.92	67	20% ad val.
1855	75	1	68	free	70	2.92	77	20% ad val.
1856	69	1	71	free	66	2.92	59	20% ad val.
1857	56	1	57	free	50	1.16	53	15% ad val.
1858	44	1	39	free	44	1.16	39	15% ad val.
1859	44	1	39	free	44	1.16	49	15% ad val.
1860	53	1	47	0.58	51	1.16	46	15% ad val.
1861	55	1	57	0.88	53	1.16	43	6.88
1862	55	1	54	1.08	52	1.16	44	6.88
1863	45	1	46	1.08	44	1.16	51	6.88
1864	40	1.06	41	1.08	39	1.16	66	6.88
1865	42	1.06	38	1.08	39	free	68	6.88
1866	50	1.06	46	1.08	43	free	77	6.88
1867	64	1.06	61	1.08	56	free	95	6.88
1868	64	1.06	62	1.08	54	free	83	6.88
1869	48	1.06	47	1.08	42	free	52	6.88
1870	47	free		1.08	44	free	44	6.88
1871	57	free	60	1.08	51	free	53	6.88
1872	57	free	54	1.08	53	free	54	6.88

(continued)

Table 45.2 (cont.)

	United Kingdom		France		Prussia		United States	
	average price[a]	import duty	average price[b]	import duty	average price[c]	import duty	average price[d]	import duty
1873	59	free	60	1.08	58	free	59	6.88
1874	56	free	58	1.08	52	free	49	6.88
1875	45	free	45	1.08	43	free	44	6.88
1876	46	free	48	1.08	46	free	45	6.88
1877	57	free	55	1.08	50	free	55	6.88
1878	46	free	49	1.08	44	free	42	6.88
1879	44	free	52	1.08	43	2.16	39	6.88
1880	44	free	52	1.08	48	2.16	43	6.88
1881	45	free	52	1.08	48	2.16	35	6.88
1882	45	free	45	1.08	45	2.16	44	6.88
1883	42	free	44	1.08	40	2.16	39	6.88
1884	36	free	41	1.08	38	2.16	33	6.88
1885	33	free	38	5.23	35	6.54	31	6.88
1886	31	free	39	5.23	34	6.54	30	6.88
1887	33	free	41	8.79	36	6.54	30	6.88
1888	32	free	44	8.79	38	10.88	32	6.88
1889	30	free	42	8.79	40	10.88	31	6.88
1890	32	free	44	8.79	42	10.88	31	6.88
1891	37	free	48	5.23	48	10.88	35	8.58
1892	30	free	42	8.79	41	7.63	30	8.58
1893	26	free	38	8.79	33	7.63	25	8.85
1894	23	free	35	12.21	29	7.63	30	20% ad val.
1895	23	free	33	12.21	31	7.63	22	20% ad val.

							20% ad val.
1896	26	free	33	12.21	33	7.63	26	8.58
1897	30	free	44	12.21	36	7.63	32	8.58
1898	34	free	46	12.21	41	7.63	32	8.58
1899	26	free	35	12.21	34	7.63	27	8.58
1900	27	free	34	12.21	33	7.63	27	8.58
1901	27	free	35	12.21	35	7.63	27	8.58
1902	28	1	39	12.21	36	7.63	28	8.58

Notes:

[a] Gazette price of British wheat

[b] Official average

[c] Official average from Kingdom of Prussia

[d] Average price of winter wheat at New York Market

Source: *British and Foreign Trade and Industrial Conditions*, BPP, 1903, LXVII.

Table 45.3. *Wheat production in the United States, 1869–93*

	Wheat acreage 000s acres	Wheat production 000s bus.	Chicago spring wheat prices (d/bus.)	Ocean freight rates NY to UK (d/bus.)	Rail rates Chicago to NY (d/bus.)
1869	19,389	235,017	41	6	17
1870			43	6	
1871			52	8	
1872	22,666	271,391	54	8	16
1873			50	11	
1874			47	9	
1875	26,762	305,966	43	8	10
1876			49	8	
1877			59	7	
1878	34,214	441,935	46	8	9
1879			48	6	
1880			51	6	
1881	37,077	423,098	56	4	8
1882			56	4	
1883			49	5	
1884	36,824	429,097	40	3	7
1885			41	4	
1886			37	3	
1887	37,701	440,591	37	3	8
1888			44	3	
1889			41	4	
1890	38,186	493,724	43	3	8
1891			47	3	
1892			38	3	
1893	34,629	384,248	33	2	8

Note: Average cost per acre of wheat in the US in 1893 including rent, manure, seed, labour and marketing was £2 8s. 8½d.
Source: R. F. Crawford, 'An enquiry into the price of wheat', *JRSS*, 58, 1895, pp. 92–8.

Table 45.4. *Mean annual ocean freight rates – inwards, 1884–1903*

	New York to Liverpool				New York to London			
	grain d/bus.	flour s/ton	beef s/tierce	pork s/barrel	grain d/bus.	flour s/ton	beef s/tierce	pork s/barrel
1884	4	13	4	3	4	14	5	3
1885	3	11	3	2	4	13	4	3
1886	3	12	3	3	4	13	4	3
1887	3	9	3	3	3	11	3	4
1888	3	10	3	4	3	13	4	3
1889	4	14	4	3	5	17	5	4
1890	2	10	3	3	3	12	4	3
1891	3	22	4	3	4	14	4	3
1892	3	10	3	2	3	11	3	2
1893	2	9	3	2	3	10	3	2
1894	2	8	2	2	2	9	3	2
1895	2	8	2	1	2	8	2	2
1896	3	11	3	2	3	12	3	2
1897	3	11	3	2	3	13	3	2
1898	3	14	4	3	4	14	4	3
1899	2	11	3	2	3	12	3	3
1900	3	13	4	3	4	17	5	4
1901	1	6	2	1	2	9	3	3
1902	1	7	2	2	2	8	3	2
1903	1	8	2	2	2	8	3	2

	Sulina grain	Azoff grain s/unit, new charter[a]	Odessa grain	Australia frozen mutton d/lb	New Zealand frozen mutton d/lb	New Zealand greasy wool d/lb	Alexandria cotton seed s/ton
1884				2	2	.75 & .63	
1885	12	16	12	2	2 & 1.75	.75 & .63	10
1886	10	13	10	1	1.75	.63	7
1887	12	15	12	1	1.75 & 1.5	.75 & .63	10
1888	15	20	15	1	1.5 & 1.25	.75 & .63	13
1889	15	19	15	1	1.5 & 1.25	.75 & .63	13
1890	13	16	13	1	1.5, 1.25, & 1	.75 & .63	11
1891	13	16	14	1	1.5, 1.25, & 1	.75 & .63	11
1892	10	14	10	1	1.13, 1, & .88	.63 & .5	8
1893	11	14	10	1	1.13, 1, &. .88	.63 & .5	10
1894	11	13	10	1	1	.63 & .5	10
1895	10	13	10	1	1, .81, & .75	.63 & .5	9

(continued)

Table 45.4 (*cont.*)

	Sulina grain	Azoff grain s/unit, new charter*a*	Odessa grain	Australia frozen mutton	New Zealand frozen mutton d/lb	New Zealand greasy wool d/lb	Alexandria cotton seed s/ton
1896	11	13	10	1	1 & .75	.63 & .5	10
1897	10	11	9	1	1, .75, & .5	.63 & .5	8
1898	11	13	10	1	.75 & .5	.5	10
1899	10	12	9	1	.75 & .5	.63, .56, & .5	9
1900	12	15	11	1	.75 & .5	.44	9
1901	10	11	9	1	.75 & .5	.38	10
1902	9	11	9	1	.75 & .5	.44	8
1903	9	10	8	1	.75 & .5	.38	8

Note: ★ The rates given in the original source were per ton tallow for Azoff and Odessa and per quarter of wheat in Sulina, then converted to 'per unit new charter'.
Source: British and Foreign Trade and Industrial Conditions, BPP, 1904, LXXXIV.

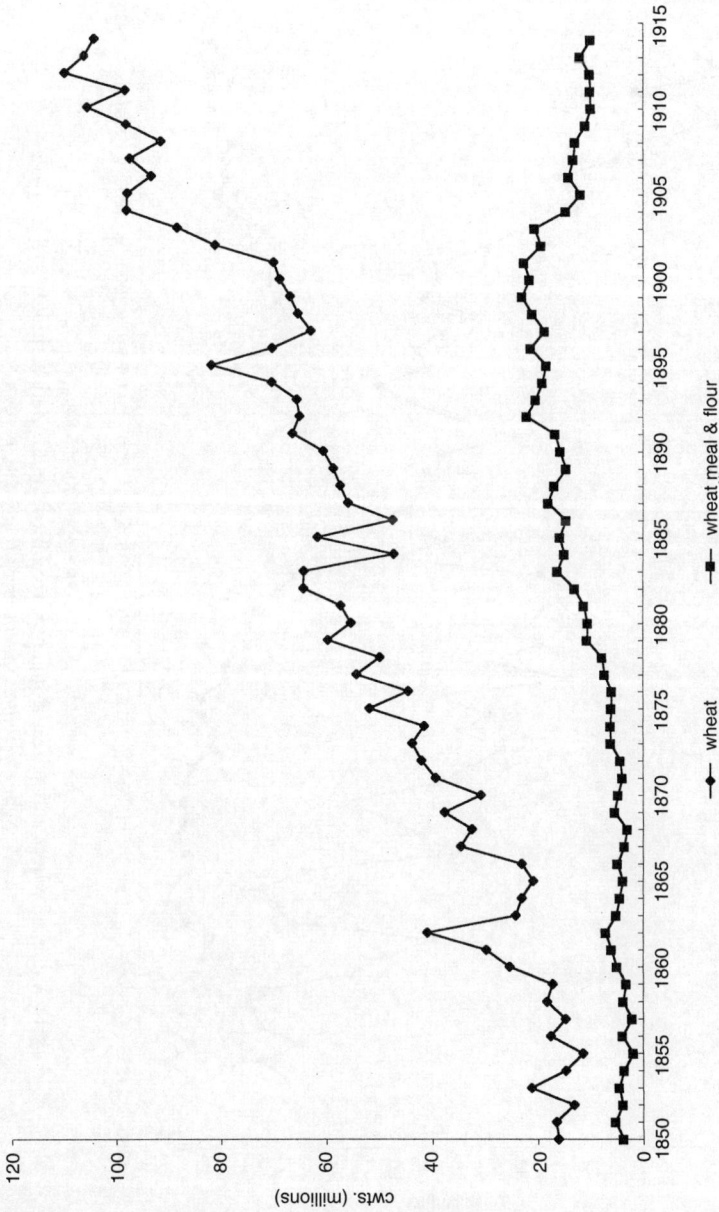

45.1a Quantity of wheat and wheat flour imported into the United Kingdom, 1850–1914

Source: Annual Statement of Trade, 1850–1915

wheat   wheat meal & flour

cwts. (millions)

120 100 80 60 40 20 0

1850 1855 1860 1865 1870 1875 1880 1885 1890 1895 1900 1905 1910 1915

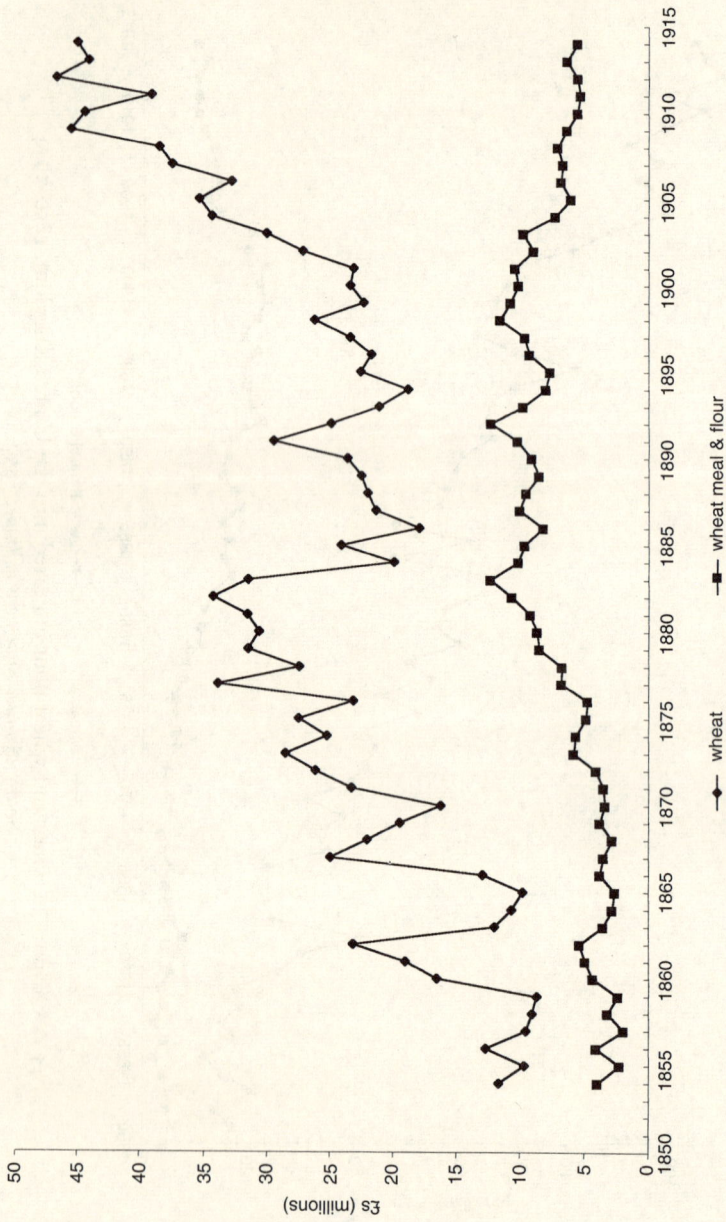

45.1b Value of wheat and wheat flour imported into the United Kingdom, 1850–1914

Source: *Annual Statement of Trade, 1850–1915*

◆ wheat ■ wheat meal & flour

£s (millions)

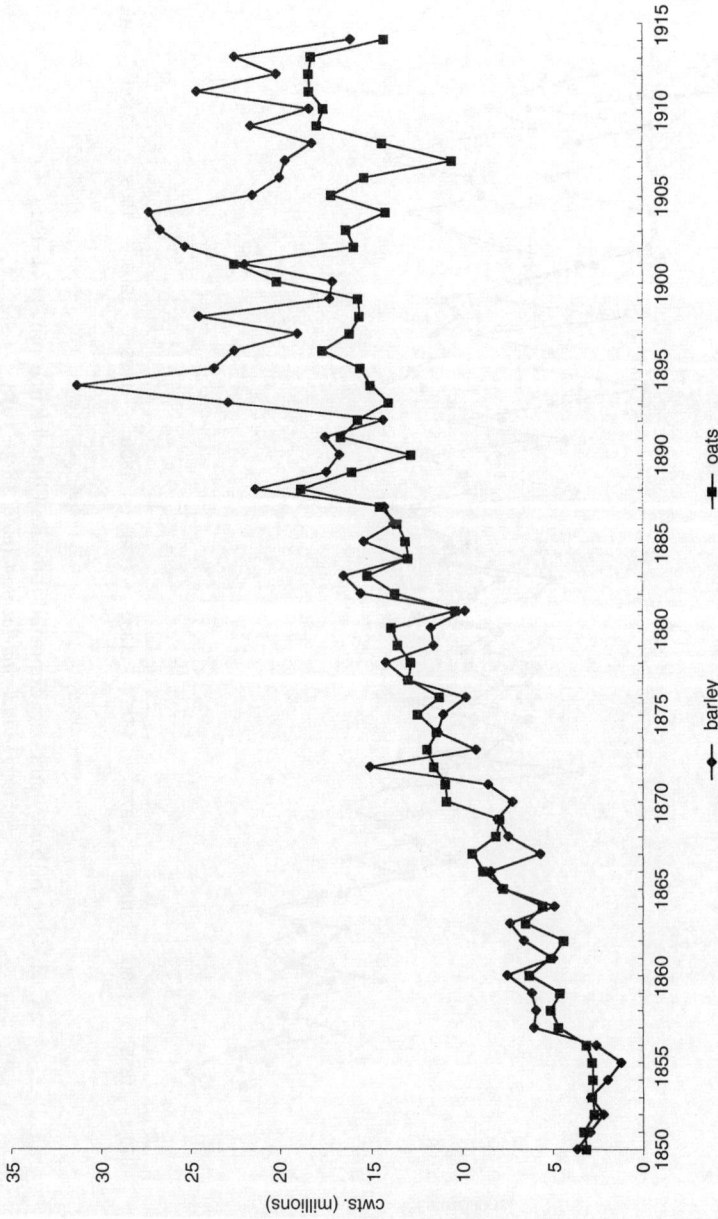

45.1c Quantity of barley and oats imported into the United Kingdom, 1850–1914
Source: *Annual Statement of Trade*, 1850–1915

barley ◆
oats ■

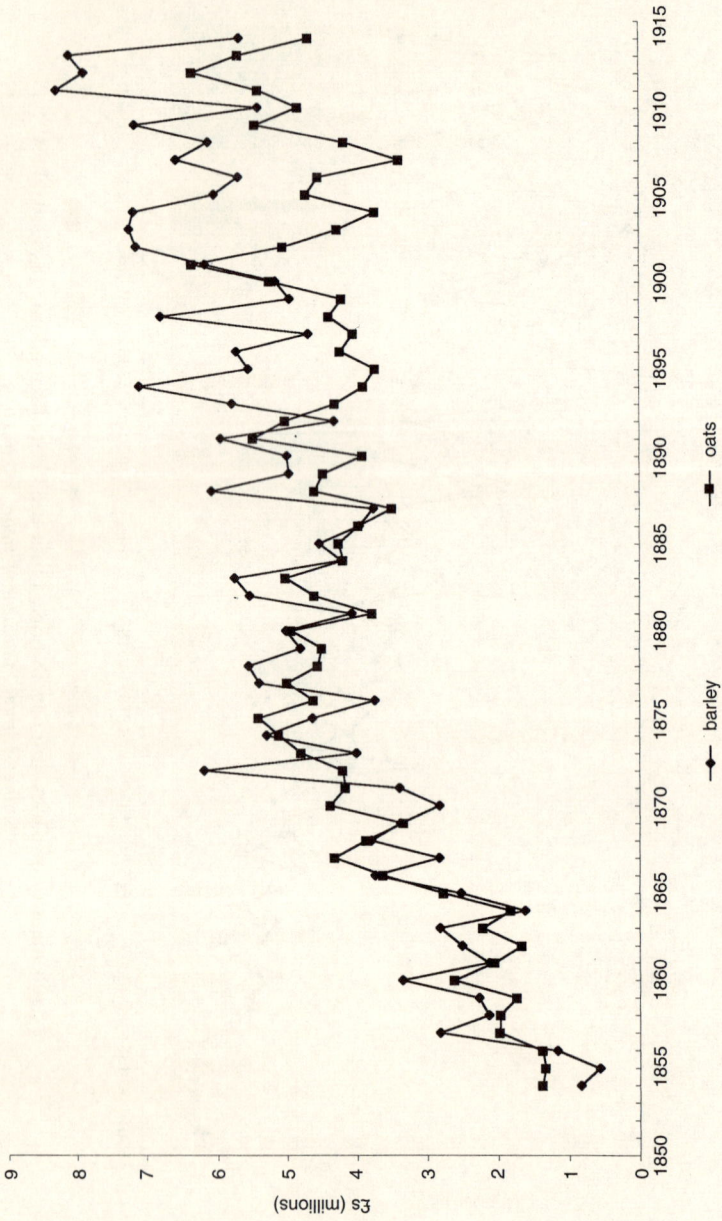

45.1d Value of barley and oats imported into the United Kingdom, 1854–1914
Source: *Annual Statement of Trade, 1850–1915*

barley

oats

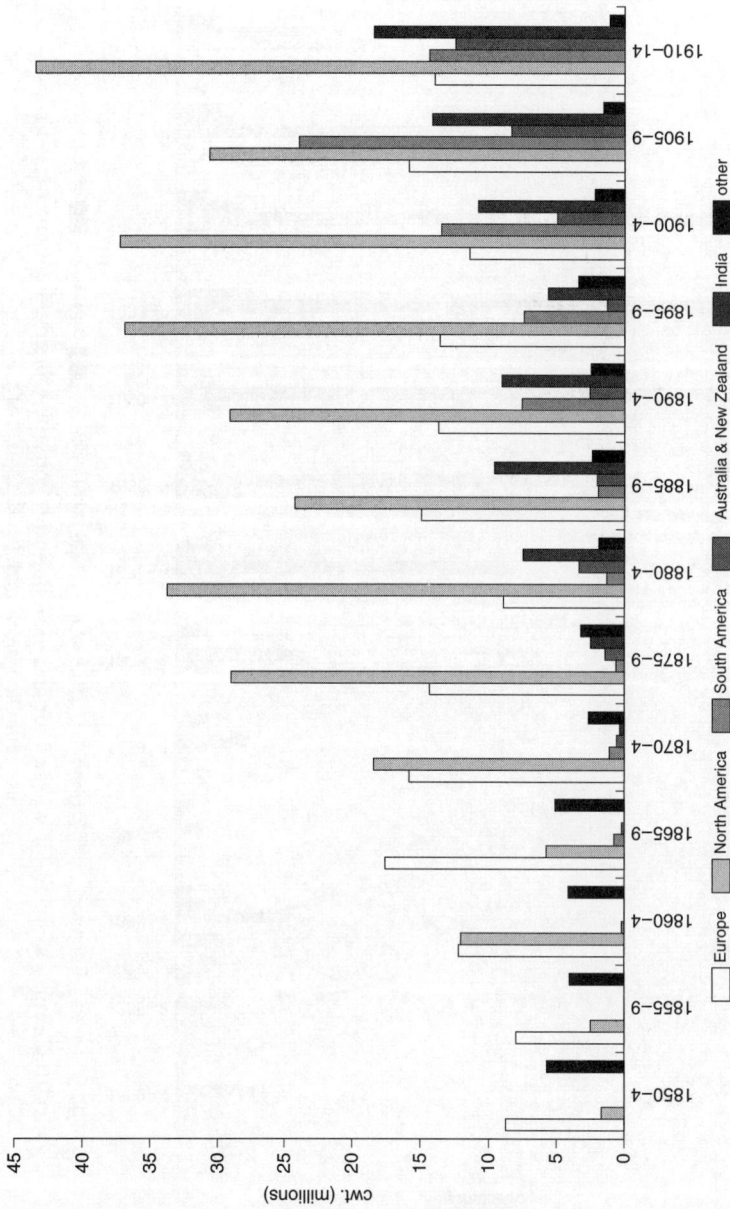

45.2a Country of origin of wheat imported into the United Kingdom, 1850–1914
Source: Annual Statement of Trade, 1850–1915

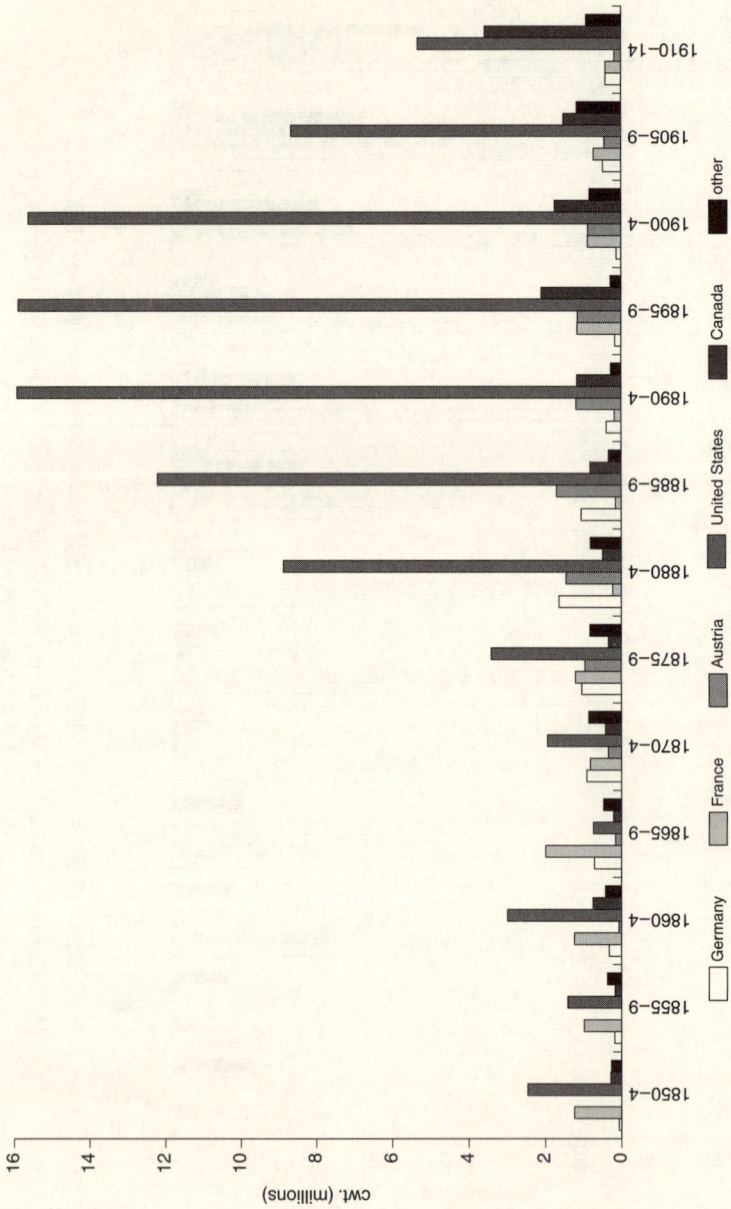

45.2b Country of origin of wheat flour and meal imported into the United Kingdom, 1850–1914
Source: *Annual Statement of Trade, 1850–1915*

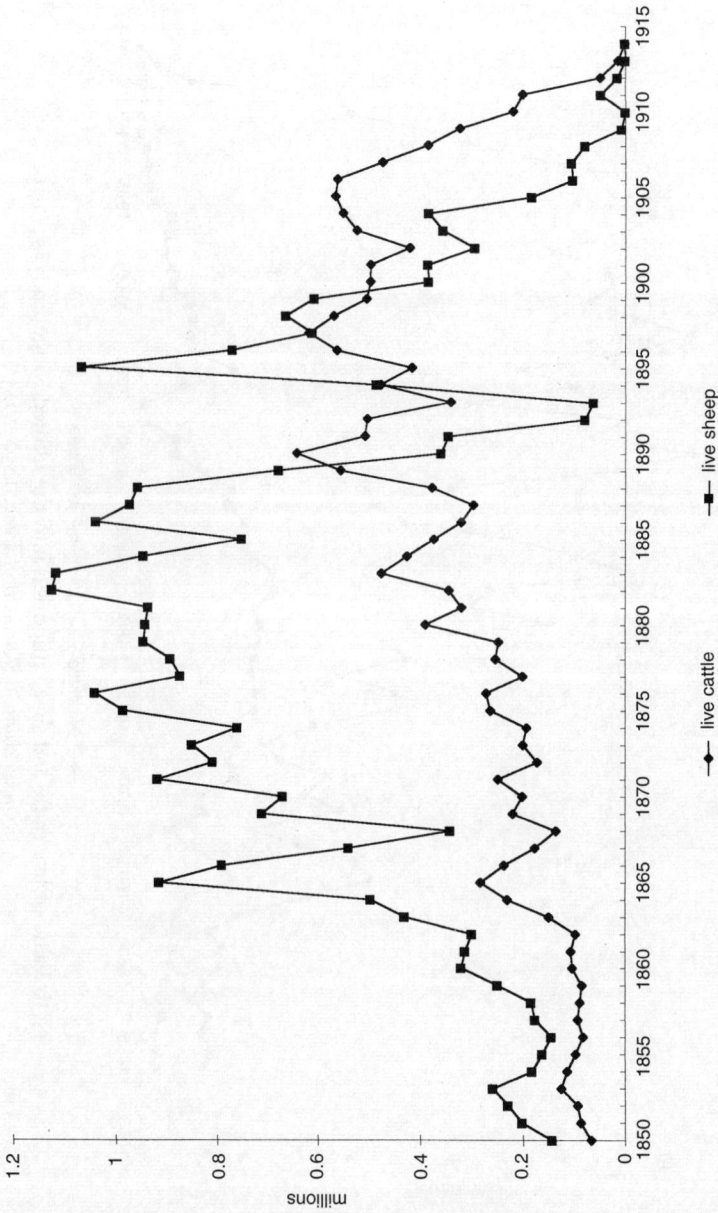

45.3a Quantity of live cattle and sheep imported into the United Kingdom, 1850–1914
Source: *Annual Statement of Trade*, 1850–1915

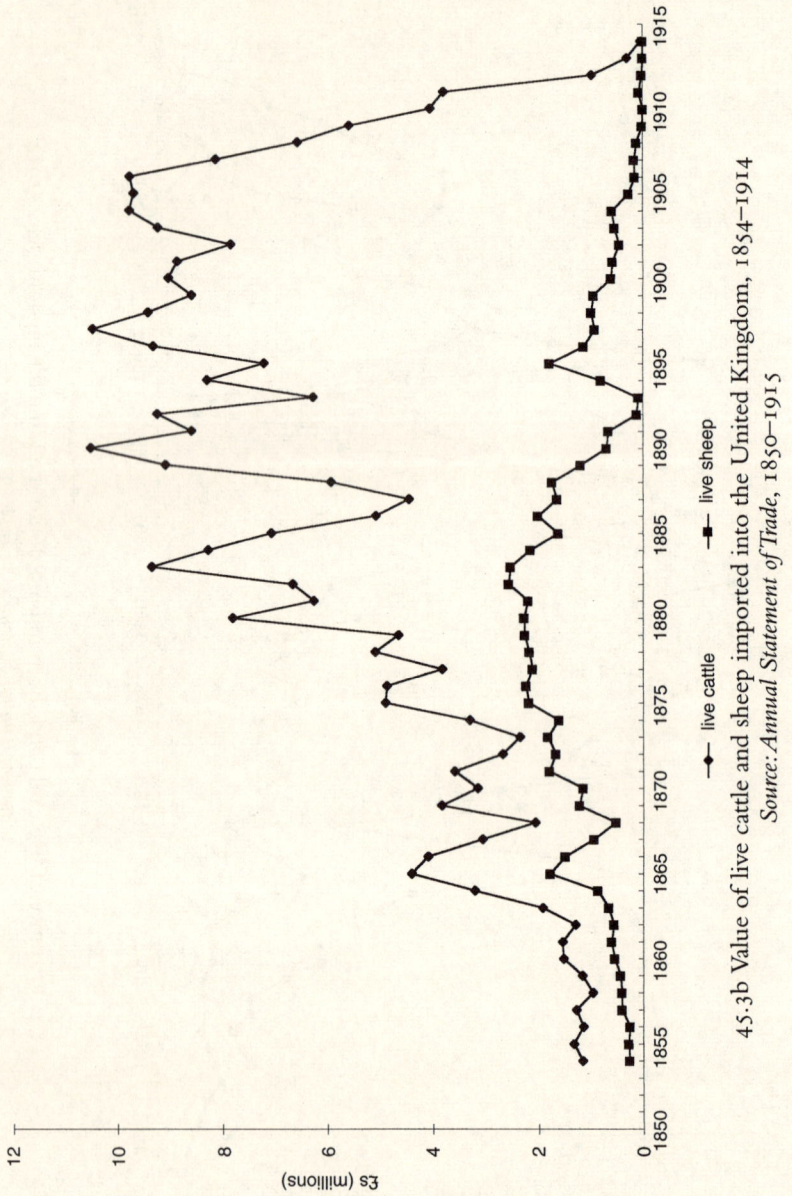

45.3b Value of live cattle and sheep imported into the United Kingdom, 1854–1914
Source: *Annual Statement of Trade*, 1850–1915

live cattle ◆   live sheep ■

£s (millions)

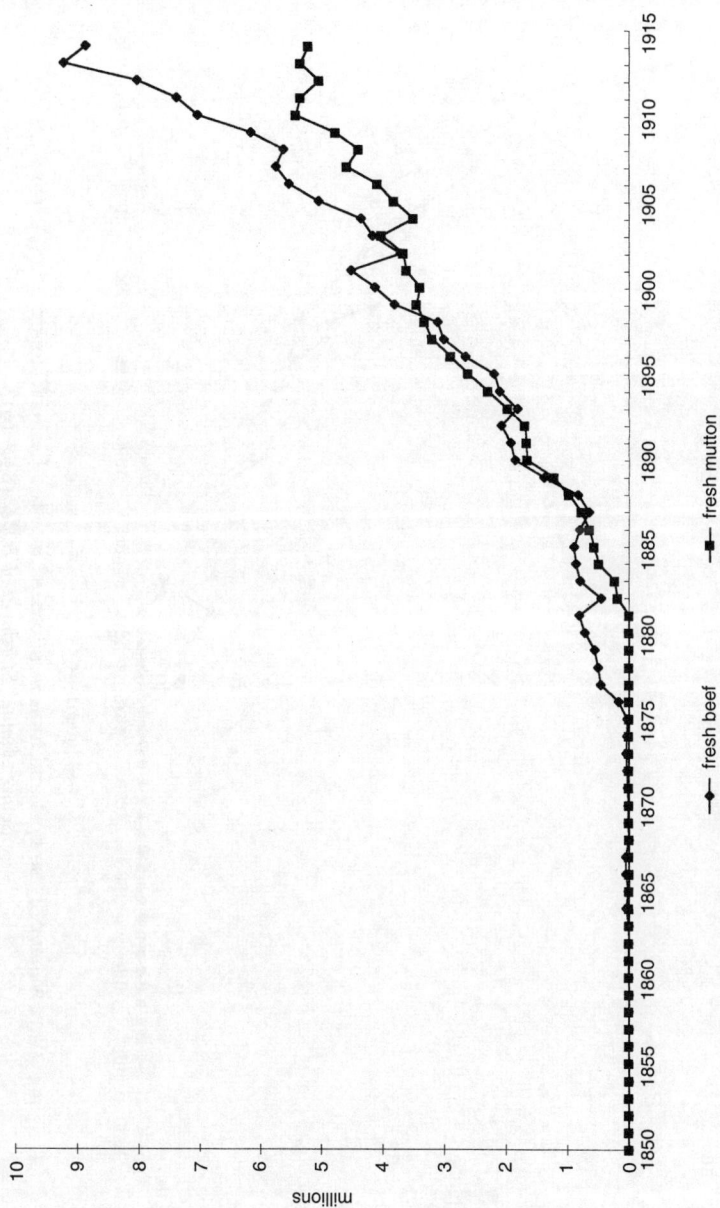

45.3c Quantities of fresh beef and mutton imported into the United Kingdom, 1850–1914
Source: *Annual Statement of Trade*, 1850–1915

fresh beef   fresh mutton

millions

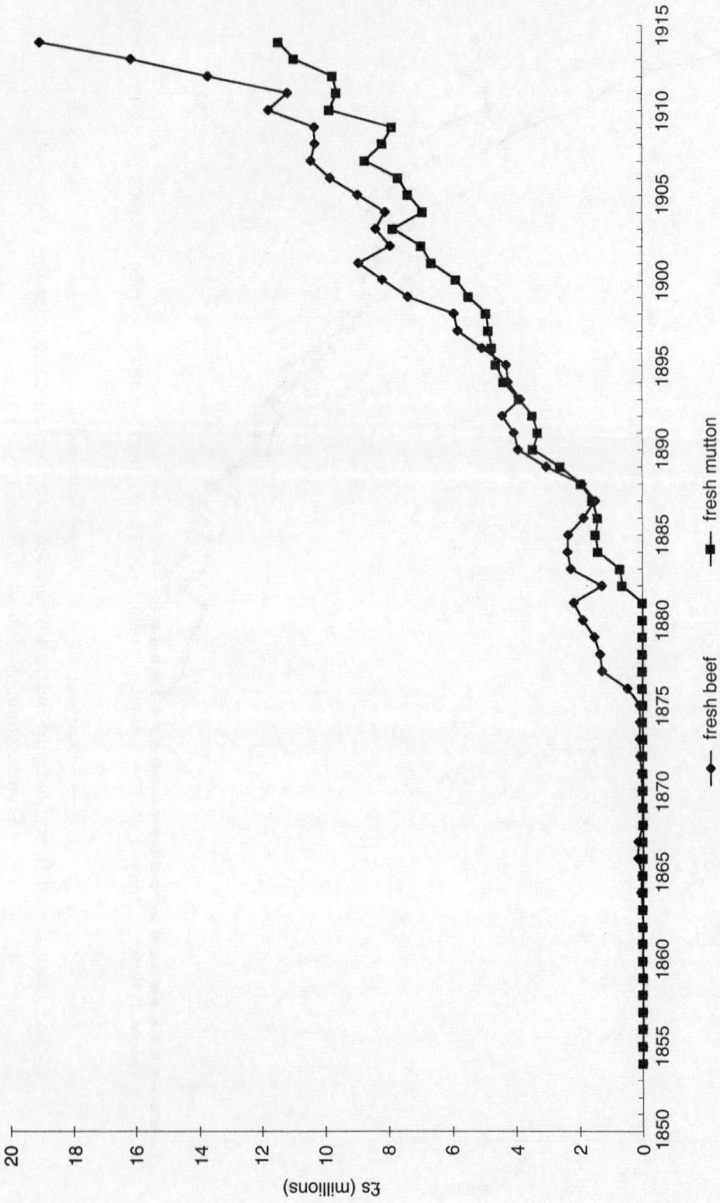

45.3d Value of fresh beef and mutton imported into the United Kingdom, 1854–1914
Source: *Annual Statement of Trade*, 1850–1915

— fresh beef   — fresh mutton

£s (millions)

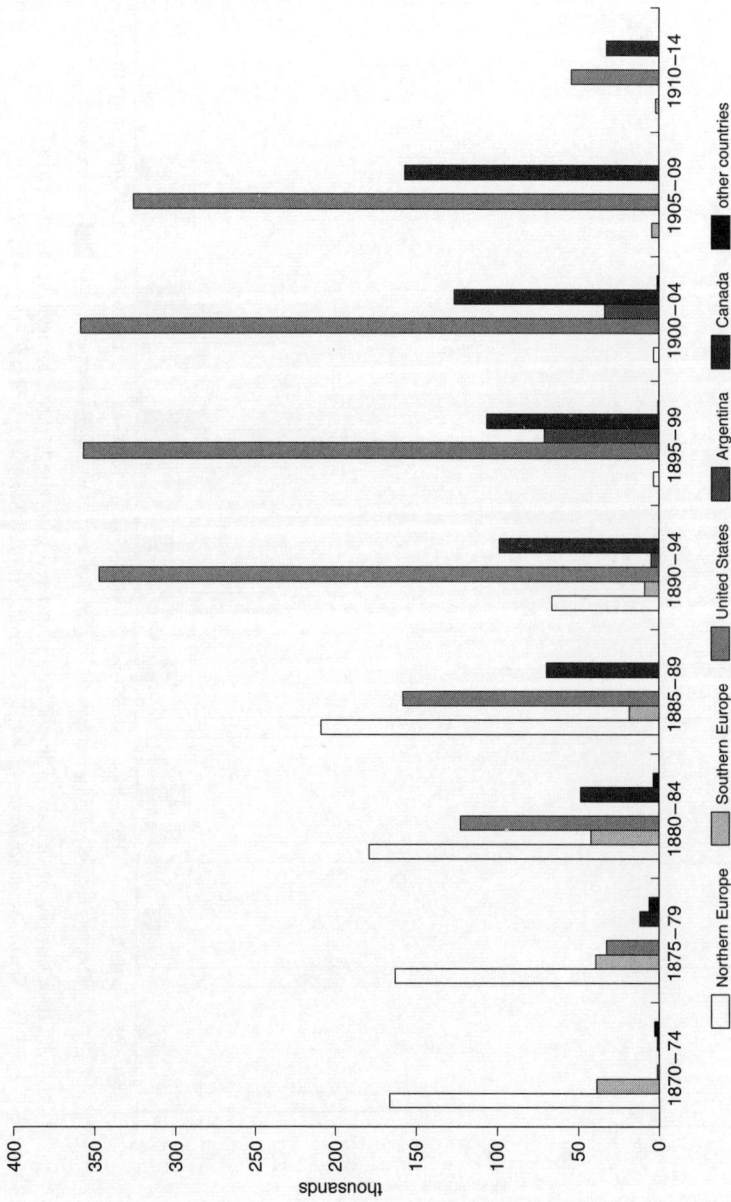

45.4a Country of origin of live cattle imported into the United Kingdom, 1870–1914
Source: Combinations in the Meat Trade, BPP, 1909, xv; *Annual Statement of Trade*, 1870–1915

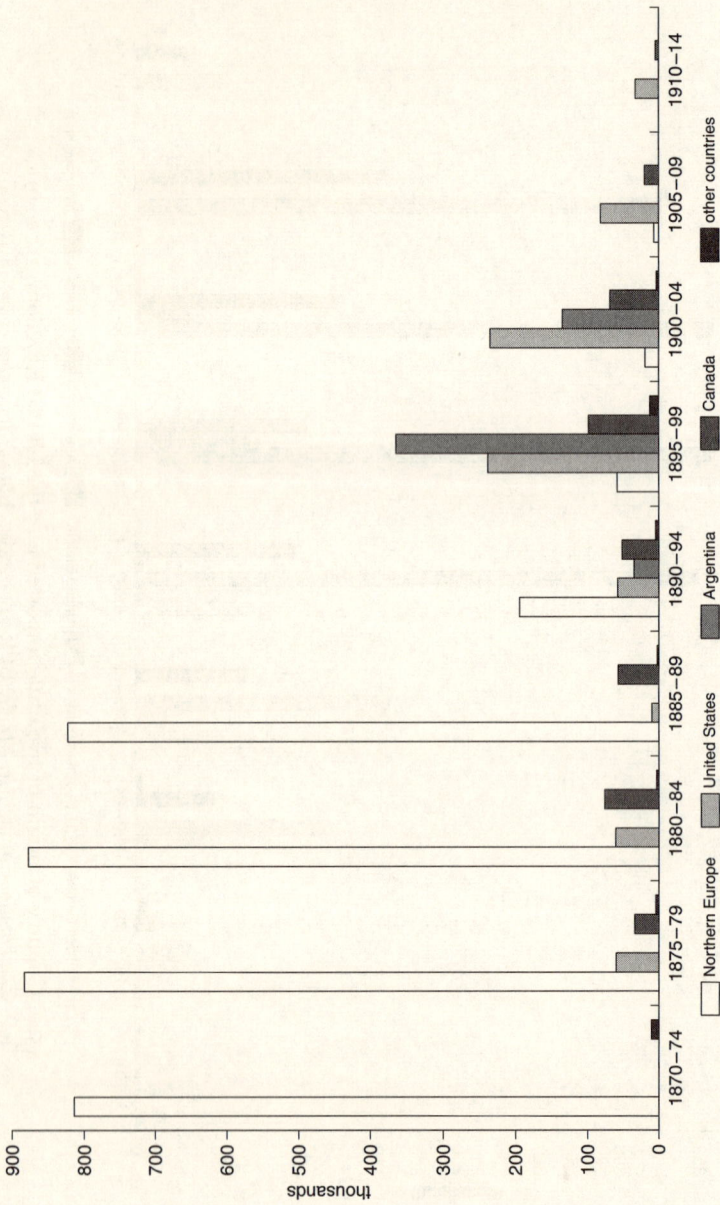

45.4b Country of origin of live sheep imported into the United Kingdom, 1870–1914
Source: Combinations in the Meat Trade; Annual Statement of Trade, 1870–1915

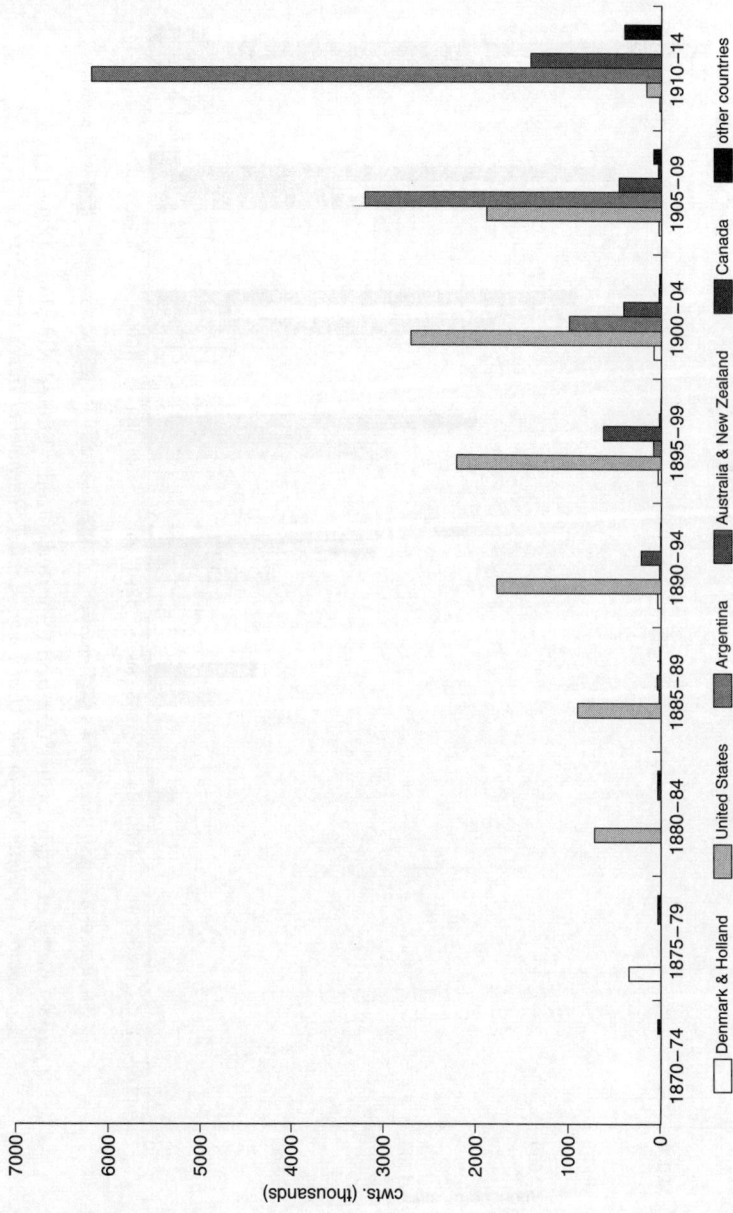

45.4c Country of origin of fresh beef imported into the United Kingdom, 1870–1914
Source: Combinations in the Meat Trade; Annual Statement of Trade, 1870–1915

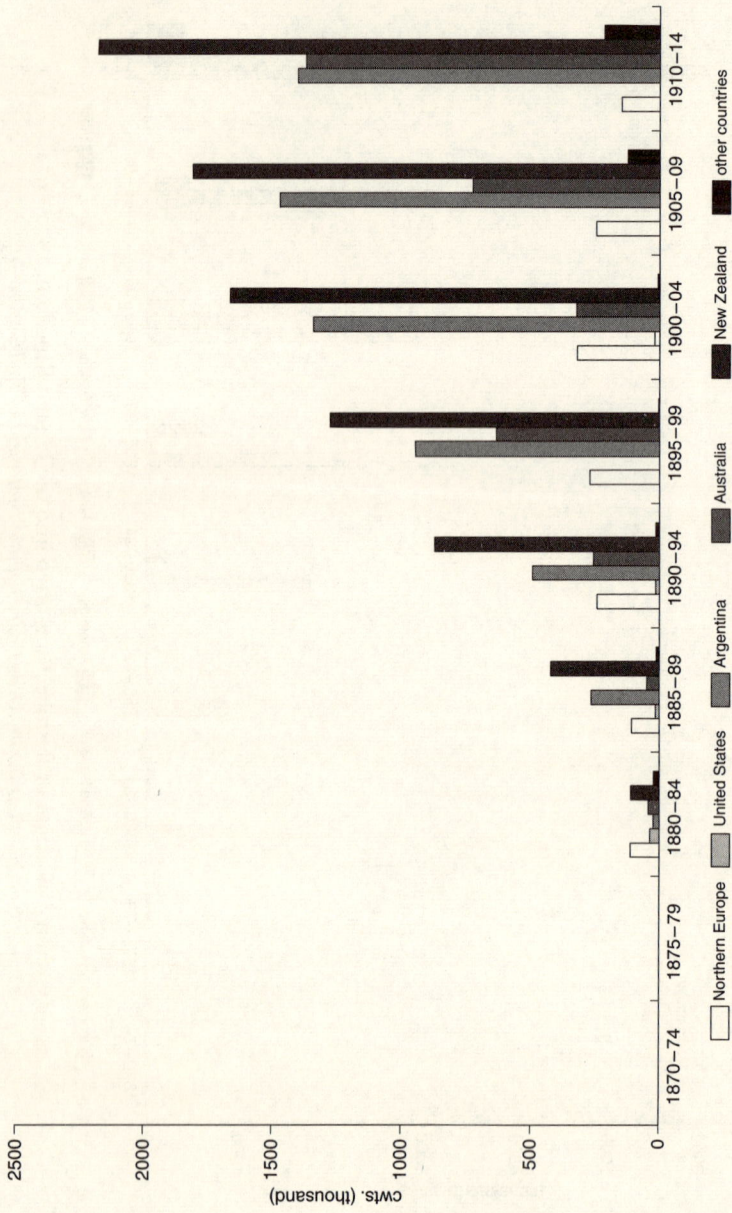

45.4d Country of origin of fresh mutton imported into the United Kingdom, 1880–1914
Source: Combinations in the Meat Trade; Annual Statement of Trade, 1870–1915

CONCLUSION

BY E. J. T. COLLINS

A. FINAL OVERVIEW

The period from the Repeal of the Corn Laws up to the Great War marked the final stage in Britain's transition from an agrarian to a mature industrial economy. By its close, agriculture was contributing a mere 7 per cent of the national product and supporting only 8 per cent of its workforce; agricultural production was about one-fifth greater, but was now supplying a little over 40 per cent of the food requirement, the greater part of which was now obtained from overseas.

The agrarian structure proved remarkably resilient in the face of these developments. At the end of the period, still nearly 90 per cent of the land in England and Wales was used for agriculture and forestry, and just over 5 per cent for urban and industrial purposes. Of that, over half was in the hands of large landowners of more than 1,000 acres, while the number of agricultural holdings in excess of 5 acres totalled about 340,000, averaging 80 acres apiece, nearly 90 per cent of which were tenant-occupied, all very little different from the situation in the mid-nineteenth century.[1] Nor had the face of the countryside much altered, the most drastic changes occurring on the marginal soils where the least profitable land was abandoned, reverting to scrub or rough pasture, or planted to woodland. The increase in grassland between 1875 and 1914 represented only one-eighth of the cultivated area of England and Wales, and an even smaller proportion of that of the arable counties, as shown in Table 1. The reduction in the arable, too, was limited, being most pronounced on the heavy clays of eastern and midland England and on the thin chalk soils of the southern downlands, large areas of which degenerated into ranch-land.

Whereas output in the depression may have declined in the arable

---

[1] For the following related statistics, D. Grigg, *English Agriculture: An Historical Perspective* (Oxford, 1989); and for land use, R. H. Best and J. T. Coppock, *The Changing Use of Land in Britain* (London, 1962), p. 229.

Table 1. *Principal agricultural land-use changes in England and Wales, 1875–1914 (per 1,000 acres)*

	English grazing districts		English arable districts		Wales		% change England and Wales
	1875	1914	1875	1914	1875	1914	1875–1914
Wheat	1,135	512	1,193	1,259	112	37	−24.7
Total corn	2,870	1,919	4,658	3,516	512	324	−31.2
Roots and green crops	1,003	712	1,526	1,154	80	75	−25.6
Rotation grass	1,440	1,143	1,167	979	361	260	−18.0
Bare Fallow	244	79	272	256	26	5	−37.3
Total arable	5,740	4,044	9,760	6,150	1,030	689	−24.7
Permanent grass	7,028	9,050	3,508	5,012	1,666	2,055	+32.1

Sources: C. S. Orwin and E. H. Whetham, *History of British Agriculture, 1846–1914* (London 1964), pp. 251–2, 350–1; HMSO, *A Century of Agricultural Statistics* (London, 1966).

counties, in the pastoral counties and the upland zone generally, it held up or even increased. Many of the farming practices observed by A. D. Hall in 1910–12 would have been recognisable to James Caird sixty years previously. Apart from a larger breadth of green crops and modest use of purchased feedingstuffs, production methods in the pastoral zone – on the hill farms of Wales and the Pennines and in the Midland Dairy Belt where, except in parts of Leicestershire, railway milk had now largely superseded butter and cheese – had changed little in the interim. Hall, an arable enthusiast, could still observe with undisguised pleasure the pursuit of the four-course rotation, with its alternating courses of white straw and green crops, yarded bullocks and folded sheep, over a wide area of England from Sussex to Northumberland. On the Yorkshire Wolds he noted the 'strictest and most conservative adherence to traditional practices, the land scrupulously clean'. On the Wiltshire Downs, up to the First World War, force of tradition ensured the perpetuation of the extended Wiltshire four-course, a system 'as unalterable as the law of the Medes and Persians'. A very few clayland districts, such as the Essex Rodings, still followed the traditional course of wheat, beans and fallow.[2] Physical productivity, too, had changed little; only the hill-farming

[2] A. D. Hall, *Pilgrimage of British Agriculture 1910–12* (London, 1913), chapters I, V, VII, XIII, XVI, XVII; A. G. Street, *Farmer's Glory* (London, new edn, 1959), chapters I–V.

Table 2. *Estimated value of agricultural output in England and Wales,*
1868/70–1911/13 (at current prices and %)

	1868–70		1911–13		% change in value £
	£ millions	%	£ millions	%	
Cereals	47.6	38.9	21.4	18.2	−55.0
Hay and straw	16.1	13.2	11.2	9.5	−30.0
Fruit and vegetables	4.1	3.4	10.5	8.9	+156.1
Total crops (inc. potatoes)	75.6	61.8	51.3	43.6	−32.2
Cattle	10.0	8.2	14.7	12.4	+47.0
Sheep	8.2	6.7	8.2	7.0	0
Pigs	8.1	6.6	10.6	9.0	+29.3
Total livestock	26.3	21.5	33.5	28.4	+27.4
Milk & milk products	18.0	14.7	26.8	22.8	+48.9
Poultry & eggs	2.4	2.0	6.1	5.2	+154.2
Total animal products	46.6	38.1	66.5	56.4	+21.0
Total crops & animals	122.3		117.7		−3.8

Source: Part VII, Chapter 38, Table 38.11. These are 'unadjusted' totals. For derivation of this Table, see Turner, Part I, Chapter 3 above.

regions and the more backward districts of Wales and south-west England having registered a significant improvement in corn and milk yields over those obtaining in the 1850s.

The most significant changes were more those of substance than of outward form. The product composition of farm revenues underwent a more radical transformation than the pattern of land use. The principal changes in the structure of agricultural output by value in England and Wales between 1868–70 and 1911–13 are analysed in Table 2. This reveals, first and foremost, a major shift in the balance between crops and live-stock; and, within it, a decrease in corn, hay and straw; increase in milk, fruit, vegetables and poultry; and little change in meat. By 1914, milk – principally whole milk for direct consumption – had overtaken cereals as the most valuable commodity. These movements were broadly reflective of changes in land use and the number and composition of farm live-stock. Between 1870 and 1914, total livestock units increased by about one-tenth and cattle by about one-third while sheep declined by about one-fifth. In 1914 the dairy cow was the predominant farm animal, com-prising about 40 per cent of the national herd.[3]

[3] HMSO, *A Century of Agricultural Statistics* (London, 1966), pp. 122–9.

Down to 1914 the landowning classes remained the dominant force in rural affairs. True, probably half the gentry families recorded in Debrett in 1863 had by then disappeared, but their places were taken either by representatives of the cadet branches of established families, or by the new rich, who shared the same aristocratic values.[4] The upper-middle classes had in large measure modelled their taste and manners, and in part their values, on those of landed society. 'The entrepreneurial ideal', it seemed to one historian, 'had triumphed only to throw in its lot with the seemingly defeated aristocrat'.[5] Indeed, the *belle époque*, the decade or so just prior to the Great War, saw plutocratic influence on country house life at its zenith, business millionaires making a spectacular assault on fashionable society, setting a pace which only the very richest of the established aristocracy could match. Why it was that Britain, the world's first industrial nation, should also have been among the most traditional and aristocratic, remains one of the unresolved issues of nineteenth-century social history?

Its seemingly effortless dominance concealed the fact that the old order had become unsteady in its foundations. The 'estate system' emerged from depression bruised but to outward appearances remarkably intact, though with a now much wider gap between those landowners with substantial urban or industrial sources of income, and those primarily dependent on agricultural rents. The period from 1880 saw the gradual relinquishing of that hitherto essential adjunct to the country estate, the London town-house. Whereas in 1880, an estimated two-thirds of landed aristocrats had fixed London addresses, by the turn of the century they were fewer than one-half.[6] The old nobility now appeared in the great sale rooms as sellers rather than buyers. The sale of the contents of their houses, and of national treasures such as the four fifteenth-century Italian manuscripts sold by the Earl of Leicester to the American banker, James Pierpoint Morgan, in 1912 for £100,000, can be analysed as 'a concealed form of payment for cheap American wheat'.[7] The way ahead was signalled by the sale of Camden Place, at Chislehurst, Kent, as a golf club in 1894, and of Wycombe Abbey, Buckinghamshire, as a girls' boarding school a decade later.[8]

The trimming back of property portfolios culminating in the much

---

[4] F. M. L. Thompson, 'English landed society in the twentieth century: II, New poor and new rich,' *TRHS*, 6th ser., I, 1991, pp. 15–16.

[5] H. Perkin, *The Origins of Modern Industrial Society 1780–1880* (London, 1969), pp. 436–7.

[6] F. M. L. Thompson, 'Moving frontier and the fortunes of the aristocratic town-house, 1830–1930,' *London Journal*, 20, 1995, pp. 67–75.

[7] G. Reitlinger, *The Economics of Taste*, vol. I (London, 1961), p. 176; vol. II (London, 1963), pp. 84–5, 231 and *passim*.

[8] P. Mandler, *The Fall and Rise of the Stately Home* (New Haven, Conn., 1997), p. 120.

larger disposals from 1908 onwards signified the decline in the positional as well as the economic value of agricultural land. In 1865 at least 326 Members of Parliament had aristocratic connections whereas, following the Liberal Landslide of 1906, only about 40 per cent could be classed as 'territorialists'.[9] Their privileged economic and constitutional status was directly challenged by the 1909 Budget and the Parliament Act of 1911, while aristocratic influence at the local level was weakened by electoral reform and the democratisation of local government. This, together with the surrender of agricultural leadership to the scientists and educators, convinced growing numbers of the aristocracy that the economic opportunity costs of landownership outweighed its social and political advantages. The last decade of our period was in many respects a milestone in 'the liquidation of the landed interest', which was to reach full momentum in the inter-war and modern post-war periods.[10]

The period 1850–1914 was a crucial watershed in rural demography. The 1851 Census recorded a decline in population in Wiltshire and Montgomeryshire, a trend which by 1901 had affected nine English and six Welsh counties. Whilst the *rural* population of England and Wales declined by a modest one-tenth between 1851 and 1911, from 46 per cent to 21 per cent of the *total* population, this concealed much more sizeable reductions in many of the purely agricultural districts, in individual parishes by upwards of 50 per cent.[11] Growth began to falter too in the market towns, many of which were in decline by the end of the century. As Rider Haggard so succinctly put it: 'the plethoric population-bogey of 1830' had been replaced by 'the lean exodus-skeleton'.[12] Over this period, the active farm population contracted by about 40 per cent, at the same time undergoing a restructuring, as the proportion of casual and seasonal workers, and females generally, steadily decreased. By 1914, these classes of worker were to be found mainly in the hop and fruit harvests or in market gardening. As the workforce became more specialised, so wages and living standards improved, and the income gap between the industrial north and agricultural south and east gradually narrowed.

For the majority of rural people, not least the farm worker, the quality of life, as measured by the standard indicators – health, diet, housing,

[9] G. Best, *Mid-Victorian Britain 1851–70* (Fontana edn, London, 1979), p. 263; J. P. Cornford, 'The parliamentary foundations of the Hotel Cecil', in R. Robson (ed.), *Ideas and Institutions of Victorian Britain* (London, 1967), p. 310; W. L. Burn, *The Age of Equipose* (London, 1954), p. 313; D. Thompson, *England in the Nineteenth Century* (London, 1950), pp. 122–3.

[10] See Chapter 10 above, and J. V. Beckett, *The Aristocracy in England 1660–1914* (Oxford, 1986), chapter 14; F. M. L. Thompson, *English Landed Society in the Nineteenth Century* (London, 1963), chapters 10, 11.

[11] B. R. Mitchell and P. A. Deane, *Abstract of British Historical Statistics* (Cambridge, 1961), pp. 20–2; F. Crouzet, *The Victorian Economy*, trans. by A. S. Forster (London, 1982), pp. 89–92.

[12] H. Rider Haggard, *Rural England* (London, 1902), vol. II, p. 566.

education – unquestionably improved after 1870, albeit from a low base point. The labourer, his horizons broadened by newspapers and the oratory of popular politics, was more questioning of his condition. Great underlying forces were at work: a traditional rural culture was being modified and absorbed by the mass-culture of the industrial towns; the authority of the local state – parish, vestry and Justices – was being usurped by the institutions of the national state. The social and ecological pattern of what has tended to be regarded as the traditional English village may have become established only as recently as the third quarter of the nineteenth century.[13] The apparent harmony of English rural life in the relative calm following the upheavals of the Swing Riots and the Repeal, was rudely shattered by agricultural depression, agricultural trades unionism, conflicts of interest between landowner and tenant, and urban development.[14] The variety of village types and social structures – ranging from the stereotype estate villages and close-knit woodland communities of the southern counties, the agro-mining communities of the Yorkshire Dales, and smallholder settlements in the Fenlands, to the villages of the London fringe – renders generalisation about the human experience difficult and, in specific cases, meaningless.

The later Victorian period stands out as the final stage in the de-industrialisation of the countryside. Many villages had by then assumed a predominantly agricultural function where their industrial and occupational structures had once been much more diversified. A combination of factors – urban and overseas competition, the exhaustion of mining deposits, agricultural depression, depopulation – took a heavy toll of rural industry. Thus, while the total population of the agricultural county of Rutland was just 11 per cent smaller in 1911 than in 1851, blacksmiths declined by 28 per cent, wheelwrights by 43 per cent, millers by 70 per cent, and coopers and turners by 87 per cent.[15] In the majority of agricultural counties, employment in manufacturing as a percentage of total employment changed little or may even have risen over the period, but this was centred on the larger towns or in districts adjoining the urban conurbations, and was in no sense rural industry.[16]

Rural depopulation concealed, however, a countervailing trend of equal importance. As the working population on farms decreased, so the agricultural linkage industries, servicing agriculture and processing and distributing its products, increased. Between 1851 and 1911, employment in the 'food, drink and tobacco' sector in Britain rose by one-third to 1.112 millions, while that in agriculture, horticulture and forestry fell by

[13] H. Newby, *Country Life: A Social History of Rural England* (London, 1987), p. 76.

[14] A. Howkins, *Reshaping Rural England: A Short History 1850–1925* (London, 1991).

[15] J. Saville, *Rural Depopulation in England and Wales, 1851–1951* (London, 1957), p. 74.

[16] M. Overton *et al.*, (eds.), *Atlas of Industrialising Britain, 1780–1914* (London, 1986), pp. 32–3.

nearly one-quarter to 1.553 millions.[17] The effect of the growth of the 'agricultural substructure' is estimated to have increased the ratio of supporting and servicing workers to farmers and farm workers in the English and southern Scottish counties from c. 4:1 in 1841 to c. 1.3:1 in 1914.[18] The Victorian and Edwardian periods thus witnessed a transformation in the structure of the food chain, in the form of a shift of added value and employment away from the farm, backwards into the servicing sector and forwards into processing and distribution.

Though the countryside may have suffered a material decline, its importance to the townsman as a place in which to live or to visit, as a source of intellectual or aesthetic satisfaction, or for sport, perceptibly increased. Part VI of this History has analysed the changing interface between town and country, and growing popular interest in the natural environment – in wildlife and scenery, historic towns and villages, and rural crafts and customs. In the late Victorian period village communities came to occupy a unique place in the imagination of the educated middle-classes. Country people and country values were deemed morally and spiritually superior to those of the towns, and an antidote to the squalor of industrialism. 'Green radicalism' had nineteenth-century roots, although neither rural conservation nor agricultural sustainability were then regarded as major issues. Karl Marx's prediction, that capitalist agriculture would in the long run ruin soil fertility, commanded little attention.[19] In the High Farming period, agricultural progress was equated with the Conquest of Nature, while in the Great Depression, the overriding concerns were declining soil fertility and deteriorating infrastructure.[20] Compared with most other European countries Britain did very little to develop a national heritage policy, the unassailable principle of laissez-faire extending to the preservation of historic buildings, landscape and works of art. Voluntary societies helped make good this deficiency: the period between 1870 and the Great War saw the forming of the majority of the currently existing county archaeological and natural history societies, as well as the beginnings of a serious academic interest in rural and agricultural history.

Though its importance in the national economy may have been declining, agriculture was nevertheless the predominant economic activity in

---

[17] Mitchell and Deane, *Abstract of British Historical Statistics*, pp. 60–1.

[18] F. M. L. Thompson, 'Rural society and agricultural change in nineteenth century Britain', in G. Grantham and C. S. Leonard (eds.), *Agrarian Organisation in the Century of Industrialisation: Europe, Russia, and North America, Research in Economic History*, Supplement 5, 1989, pp. 197–202.

[19] K. Marx, *Capital*, edited by F. Engels (London, 17th edn, 1920), p. 513.

[20] J. Sheail, 'Elements of sustainable agriculture: the UK experience, 1840–1940', *AHR*, 43, 1995, pp. 178–92; E. J. T. Collins 'Agriculture and conservation in England: an historical overview', *JRASE*, 146, 1985, pp. 38–46.

the countryside, on whose fate that of the entire rural sector very largely hung. A principal objective of this *History* has been to record and assess that performance in the light of recent research. It poses the question: how 'golden' was the 'Golden Age', and how far was its prosperity due to higher output and rising standards of technical efficiency, and how much to high prices, due to favourable market conditions and slow growth in output? In the Great Depression: did agriculture fail, pulled down by the weight of tradition and an obsolescent agrarian structure; or was it pushed, by the pressure of falling prices and foreign competition into a downward spiral of static or declining output, and low returns? And what were the impact and limitations of nineteenth-century agricultural science and technology; how successful was the 'Second Agricultural Revolution'? In short, did Victorian farming fail?

Agricultural growth rates in the High Farming period are shown to have been very substantially lower than in the preceding quarter century, the de-celeration reflected by the levelling off from about 1860 in the yields of cereals and roots, stagnation in traditional dairying, and a modest rise only in meat production. This less than impressive performance to some extent exonerates the arable sector for its poor showing in the Great Depression. The blame lay more with the livestock sector where market conditions and the relationship between costs and prices were far more favourable to growth. Depression was not to be met by heroic measures, such as peasant proprietorship or headlong retreat from arable to pasture and market gardening. Though a desirable objective, the opportunities for diversification tended in practice to be far more limited than its proponents had farmers believe. Is J. L. van Zanden correct in his analysis that the English problem was one of an inappropriate farm-size structure at a stage when the rising price of livestock products and falling costs of feedingstuffs and fertilisers favoured the small family farmer?[21] Or, as A. Offer has postulated, were the root causes of arable depression sociological and institutional – the structural and social inefficiencies of the estate system, excessive rents in support of a rural governing class and, on the part of the farmer, holdings too small for him to use his head, too large for him to use his hands, and aversion to physical labour?[22]

The discussion on farming techniques in Part II of this volume emphasises the increasingly important role played by industrial and scientific inputs in farming production. Physical volumes of fertilisers and feedingstuffs expanded rapidly over the High Farming period, fell back in the 1880s, and moved ahead sharply from the turn of the century, but

---

[21] J. L. van Zanden, 'The first green revolution: the growth of production and productivity in European agriculture, 1870–1914', *EcHR*, 44, 1991, pp. 215–39.

[22] A. Offer, *The First World War: An Agrarian Interpretation* (Oxford, 1989), chapters 6–8.

had a limited impact on crop yields, which stagnated from the 1860s. Purchased feeds and manures accounted for only a small proportion of the total supply of nutrients, however, while these may often have been purchased not as supplements but as substitutes for home-grown supplies of feed and fertility. Agricultural science may not yet have been sufficiently advanced to provide practical guidelines that could be applied to plant and animal nutrition or plant breeding. Disillusionment with science had set in by the 1860s, to re-awaken after the turn of the century. Low investment in research and education was at least partly responsible for Britain's poor international performance after 1880.

Mechanisation afforded more certain returns, and savings in labour and labour costs had already been achieved on the larger arable holdings by the end of the High Farming period. Between c. 1840 and 1910 farm power supplies in England and Wales increased by an estimated 45 per cent and power per worker by an estimated 97 per cent.[23] The mid-nineteenth century marked the onset of the steam age, but by 1910 steam was itself challenged by the oil engine and motor tractor, though with the horse still the predominant power force, supplying over 90 per cent of draught power and 60 per cent of stationary power. Mechanisation rather than increased output was responsible for the improvement in labour and total factor productivity over this period. In 1850, the average farm worker had fed fewer than six other people, in 1914 approximately ten, at a higher standard of diet.

Agriculture emerged from the depression 'leaner and fitter', producing the same output as at the end of the High Farming period with less labour and capital. High farming could still be justified but only for high-value products, such as fat lambs, best malting barley, vegetables and potatoes, or winter milk. By this stage, reported A. D. Hall, those who had not altered their systems in the low-price years had been forced out of the industry, to be replaced by a new generation, more energetic, more businessmanlike, and more open-minded to the possibilities of science and education.[24] The situation was nicely summarised by the historian and writer on agricultural affairs, W. H. R. Curtler: 'more brains are put into it than formerly, and though there is less "polish" in farming and holdings do not look so "smart", the essential business is being done'.[25] The new race of farmers was of a different social complexion, with a larger proportion of working farmers, many of whom had come into the

[23] E. J. T. Collins, 'Power availability and agricultural productivity in England and Wales, 1840–1939', in E. Buyst et al. (eds.), Historical Benchmark Comparisons in Output and Productivity, 1750–1900 (Centrum voor Economische Studien, Leuven, 1996), pp. 78–98.

[24] Hall, Pilgrimage, pp. 150–1, chapter XVI.

[25] W. H. R. Curtler, 'Enquiry into the rate of wages per acre in England, 1913–1914', International Review of Agric. Economics, 7, 1916, nos. 8–10, p. 19.

industry with little capital and by hard work and strict economy risen up the farming ladder. Much depended upon the standards to which farmers were expected by their class to conform, many of the complaints regarding income were, it was claimed, in reality founded on the difficulty which some of them experienced in maintaining a scale of expenditure approaching that of the well-to-do tradesman, or solicitor, in the towns.[26]

The clear impression is that from about 1906 the average working farmer was more than just making a living. In terms of purchasing power his average income was 30–40 per cent higher than at the end of the Golden Age. The ratio of farm to non-farm income, the yardstick of his relative prosperity, though it never recovered to the dizzy heights of the early 1870s when it briefly exceeded 75 per cent, had progressed from a low-point of 36 per cent in 1892–6 to 47 per cent in 1906–10 and 55 per cent in 1911–14.[27] By comparison the landowners, or at any rate those whose incomes were derived mainly from agricultural rents, had fared badly. The majority, mainly those in the arable counties where per-acre rent had declined by upwards of 25 per cent, were little better off, or worse off even, in 1914 than in the mid-1870s.[28] Somewhere between the two sat the farm worker, his real income one quarter higher, but the improvement that had commenced with the fall in grain prices in the 1870s was checked by a further round of price inflation from early in the new century.[29]

By 1914, Cobbett's ideal physiocratic society had long ceased to exist other than in the imagination. D. W. Brogan's famous statement that 'there is no country in the world in which the feeling for the soil as a *factor of production* is as rare as in England, or where the knowledge of farming as a way of making a living is so much a specialised knowledge', was true up to a point.[30] A feature of the later nineteenth century was the deep and sentimental attachment of the townsman to the countryside, and the extraordinary degree to which it had become rooted in the national consciousness. Agriculture, too, held a special fascination. Charles Dickens's son, for example, brought up in London, studied at the Royal Agricultural College, while the novelist himself wrote an article entitled 'Farm and College'.[31] In industrial Birmingham, hobby farms had been fashionable among wealthy manufacturers since Boulton and Watt; in Edwardian times, Joseph Chamberlain at Highbury, George Cadbury at The Manor, Northfield, and the screw manufacturer, J. S. Nettlefold, at Winterbourne, Edgbaston, had their own little 'farmeries',

[26] E. N. Bennett, *Problems of Village Life* (London, n.d.), pp. 51–2.

[27] J. R. Bellerby, *Agriculture and Industry Relative Income* (London, 1956), p. 56.

[28] Part VII, chapter 39.   [29] Part VII, chapter 42.

[30] D. W. Brogan, *The English People* (London, 1963), quoted in R. S. R. Fitter, *London's Natural History* (London, 1945), p. 208.   [31] *Agricultural Gazette*, 17 October 1868, p. 1099.

all within the present city limits.[32] The bond was epitomised by Gladstone no less: the son of a Liverpool merchant and arch-proponent of free trade, he took a keen interest in practical agriculture, and wrote about fruit-growing. In 1896, in one of his last speeches, at Hawarden near his country house, he declared: 'I have been a townsman most of my life; but I am a rural man, one of the country folk now . . .'[33]

Agriculture had emerged from a prolonged depression, having lost her headship among industries and Britain her position as the world's leading agricultural nation. The depression had not been a universal disaster, as the decline in corn had been to a large measure offset by the growth in dairying, horticulture and poultry. In the words of F. M. L. Thompson, agriculture had 'coped with its decline with more flexibility and less long-term anguish and distress than other staple industries, such as coal, cotton or shipbuilding, which went down the same road later'.[34] The final few years of the period witnessed changes of far-reaching consequence. The establishment of the Development Commission in 1909 signalled the intention of government to play a more positive role in promoting the welfare of rural people. A national farmers' union had been established with a view to pursuing farmers' interests independently of their landlords. Farmers were now more willing to concede that the future of the industry lay with science and education and, though few were yet prepared to admit it, a closer and more binding relationship between agriculture and the state. The 'intellectual revival', as Hall termed it, began about 1905 with an increase in enrolments in educational classes and, between then and 1914, the sale of some 23,000 copies of Fream's *Elements of Agriculture*.[35]

The period had begun sourly, with a large and unruly gathering of farmers in Drury Lane, London, demanding the reinstatement of the Corn Laws, and farmers in the eastern counties declaring they would rather march on Manchester, the capital of free trade, than on Paris.[36] It ended on a similar note. The period 1912–14 saw a rash of strikes from Lancashire to Kent, and at Helions Bumpstead in Essex, in July 1914, the spectacle of 50 young farmers in motor cars confronting some 150 strikers in the roadway and, having routed them, proceeding to bring in the

---

[32] Phillada Ballard, 'A commercial and industrial elite: A study of Birmingham's upper middle class, 1780–1914', PhD thesis, University of Reading, 1983, pp. 906–24.

[33] F. W. Hirst, *Gladstone as Financier and Economist* (London, 1931), pp. 260–1.

[34] F. M. L. Thompson, 'Agriculture and economic growth in Britain, 1870–1914', in P. Mathias and J. A. Davis (eds.), *International Trade and British Economic Growth from the Eighteenth Century to the Present Day* (Oxford, 1996), p. 59.

[35] Hall, *Pilgrimage*, p. 219; N. Goddard, *Harvests of Change: The Royal Agricultural Society of England, 1838–1988* (London, 1988), p. 126.

[36] A. Briggs, *Victorian People* (Pelican edn, London, 1965), p. 29.

hay harvest.[37] In Parliament, the Diehards were mounting a last bitter campaign in defence of aristocratic privileges;[38] in the country, protectionism was not yet dead, many in the corn counties were still smarting over the defeat of Tariff Reform. The First World War was a major turning point in the relationship between agriculture and the state and, in its gruesome way, the slaughter on the battlefields of 'half the great families of England, heirs of large estates and wealth' marked the death of the aristocratic era.[39]

B. PATHWAYS FOR FUTURE RESEARCH

Building upon the steady accumulation of published research since the 'great rewriting' of the 1960s, this volume has, it is hoped, made a useful, and at a number of points original, contribution to our knowledge and understanding of the rural history of a crucial stage in Britain's social and economic development, her transformation into a modern industrial state. The achievement, though, is uneven, and much remains to be done. In this closing section, the part editors attempt to identify the leading areas for future research.

Almost fifty years ago, Asa Briggs described the mid-Victorian period as 'one of the least understood in English history'.[40] His comment is in many respects still true for agricultural history, large areas of which, in the Great Depression as well as the High Farming period, are seriously lacking detailed local and regional studies. This is particularly true for Wales and northern England. The large- versus small-farm controversy, and the debates over the alleged inefficiency of the 'estate system', are as yet unresolved. The contribution of the small farm to agricultural output, and its role in the local economy, demand further investigation; as do hill-farming in Wales and northern England, farming on the urban fringe, and horticultural production outside the London area and Vale of Evesham. A central issue raised by this volume is the extent to which English agriculture, in the Golden Age as well as the Great Depression, can be seen to have failed. Why did the arable sector appear to lose momentum from the late 1850s? Why did the dairying sector seemingly not share in the general prosperity of the High Farming period? The picture is especially fuzzy at the chronological extremes. Building upon the work of Williamson and others, comparisons should be made between agricultural growth and efficiency before and after Repeal.[41] Impressions of the Recovery,

---

[37] A. Armstrong, *Farmworkers* (London, 1988), p. 151; A. J. F. Brown, *Meagre Harvest: The Essex Farm Workers' Struggle Against Poverty, 1750–1914* (Chelmsford, 1990), p. 234.

[38] D. Spring, 'Land and politics in Edwardian England', *Agric. Hist.*, 58, 1984, pp. 17–42; G. D. Phillips, *The Diehards: Aristocratic Society and Politics in Edwardian England* (Cambridge, Mass., 1979).

[39] C. F. G. Masterman, *England after War* (London, 1922), pp. 31–3.

[40] A. Briggs, *Victorian People* (Pelican edn, London, 1965), p. 9.

1897–1914, are largely derived from A. D. Hall's *Pilgrimage of British Farming* (London, 1913), which may be an unrepresentative view, favouring the arable at the expense of the pastoral sector. A comprehensive history to fill the gap between the *Royal Commission on Agricultural Depression* (1894–7) and P. E. Dewey's admirable account of *British Agriculture in the First World War* (London, 1989), is long overdue.

The Editor of Part II questions whether technically 1850–1914 is a coherent period, or whether 1835–75 (the most active phase of the 'Second Agricultural Revolution'), or 1900–39 (the motor and new scientific age), or indeed 1835–1939, is more appropriate for analytical purposes. How far did most farmers abandon 'sustainability', in the sense of relying on the resources of their farms as opposed to purchased inputs? Was mechanisation a more significant innovation than fertilisers and feedingstuffs, or scientific knowledge, and how much were improvements in output and productivity due to increases in existing inputs (movements along the same supply curve) as against genuine innovations (shifting the supply curve to the right). And outside the farm gate, what was the impact of changes in transport technology – railways, motor vehicles and bicycles?

These and related questions require the more rigorous application of production economics. Probably enough work has been done for the moment on agricultural output, although some of the assumptions on which the present estimates are based, especially those concerning meat and horticultural production, need refinement. Econometric historians should see as their goal the construction of output and profits prior to the official statistics, back in the 1830s, in place of the time-hardened indexes, by Deane and Cole and C. H. Feinstein. More reliable measures of productivity await more work on agricultural inputs, especially capital, of 'occupiers' as well as 'landlords'.

Since the chapter on labour was completed, several lines of research have been pursued to amplify its findings and those of Alan Armstrong's definitive study of farm workers.[42] These have included useful work on the gender and social composition of the farm workforce, agricultural trades unionism, and the allotment movement.[43] Suggested areas for future

---

[41] J. G. Williamson, 'The impact of the Corn Laws just prior to Repeal', *Explorations in Entrepreneurial Hist.*, 27, 1990, pp. 123–56.

[42] A. Armstrong, *Farmworkers: a Social and Economic History* (London, 1988).

[43] G. R. Boyer and T. Hatton, 'Did Joseph Arch raise agricultural wages? Rural trade unions and the labour market in late nineteenth-century England', *EcHR*, 47, 1994; D. A. Pretty, *The Rural Revolt that Failed: Farm Workers' Trade Unions in Wales, 1889–1950* (Cardiff, 1989); E. Higgs, 'Occupational censuses and the agricultural workforce in Victorian England and Wales', *EcHR*, 48 1995; S. Horrell and J. Humphries, 'Women's labour participation and the transition to the male-breadwinner family, 1790–1865', *EcHR*, 48, 1995; J. Burchardt, 'Rural social relations, 1830–50: opposition to allotments for labourers', *AHR*, 45, 1997; A. Howkins, 'Peasants, servants and labourers: the marginal workforce in British agriculture, c. 1870–1914', *AHR*, 42, 1994; S. Hussey, '"The last survivor of an ancient race": the changing face of Essex gleaning', *AHR*, 45, 1997.

enquiry include the workforce on the small farm, of which 'farmers' wives' and 'relatives' formed the greatest proportion, and the effect of the decline of rural trades and industries on the occupational structure.

'De-industrialisation' should rank as a central theme in any study of the rural economy in the nineteenth and early twentieth centuries, and a major effort is needed to amplify and extend back into the nineteenth century the seminal work carried out by H. E. FitzRandolph, M. D. Hay, A. M. Jones, and K. S. Woods at Oxford in the early 1920s.[44] In the history of the agricultural supply and processing industries, as much has been added by business historians as by agricultural historians, the output of the former driven by the availability of business records and by firms prosperous and enterprising enough to commission their company histories. The emphasis is invariably on successful large-scale enterprises in industries where the typical firm was small and vulnerable and ceased trading or was absorbed, leaving little or no record of its activities. There so far exists some four major histories of agricultural engineering firms and one each for animal feedingstuffs and agro-chemicals. In the processing sector, attention had largely centred on milling and malting. Whilst the coverage of large firms is patchy, still less is known about small and medium-sized companies. The plea is for more broad-based studies of individual firms and industries, beginning perhaps with a systematic analysis of entries in the trade sections of the local directories and specialist trade publications, linked where possible to the official occupational censuses. The persistence of the small firm in sectors of the industry dominated at the national and international level by large firms requires further exploration, as does the contribution of these industries to the local economy, and their agricultural linkages.

'More is known', remarks the editor of Part v, 'about cows and ploughs than those who used them or deployed them'. Our knowledge of farms, especially small farms, is still very deficient, particularly in the rural counties of upland Wales. The role of women in the social, cultural and economic life of the countryside – not just women farm and industrial workers, but also women farmers, farmers' wives, farmers' female relatives living in the farmhouse, and 'women of the gentry' – remains seriously under-researched. The social and occupational composition of the village community, where by 1900 often far fewer than half the adult population was involved directly with the land, is a problem area. On the urban fringe especially, the 'newcomers' comprised a distinctive social class. But who they were, where they came from, why they came and their impact on

---

[44] H. E. FitzRandolph and M. D. Hay, *The Rural Industries of England and Wales* (Oxford, 1926); A. M. Jones, *Rural Industry in Wales* (Oxford, 1926); K. S. Wood, *Rural Industries Round Oxford* (Oxford, 1921).

the countryside they moved into, cry out for investigation. The market towns are *terra incognita*. Victoria de Bunsen saw them as the 'last stronghold of the vested interest' in the early twentieth century.[45] Having escaped the attentions of the social reformers, they fell between the villages which had been transformed by mass migration, and the industrial towns transformed by trades unions, political meetings, co-operative societies and social clubs. Here too is a challenging area for research, as are the middle classes and their place in rural, social and cultural life, church and chapel, and the welfare services, notably health and poor relief.

Since the completion of Part VI of this volume in 1990 important new and exciting work, much of it interdisciplinary, has been done in the fields of rural environment and the urban impact. The great historical significance of the period 1850–1914 is confirmed by the posthumously published book by the late Gordon Cherry, the co-editor of Part VI, written with Alan Rodgers, *Rural Change and Planning: England and Wales in the Twentieth Century* (London, 1996). Urbanisation required adjustment in both a physical and less tangible sense. The countryside was expendable in terms of providing seemingly unlimited space for urban growth. It was a repository for what could not be accommodated within the expanding physical limits of town and city. The pace and character of that development continue to excite considerable interdisciplinary interest in such journals as *Planning Perspectives*. Many studies have come to focus on what may be characterised as 'the armchair view'. As Michael Bunce wrote in his volume *The Countryside Ideal* (London, 1994), one of the contradictions of modern urban civilisation has been a nostalgia for rural life and landscape. In both Britain and North America, comparative studies have been attempted of how an 'idealised' form of the countryside was consciously sought from the nineteenth century onwards both as a recreational pursuit and, for an increasingly wide spectrum of society, as a permanent residence.

Questions arise as to how farming perceived itself, and was perceived by others. Was it just another industry, or form of investment and employment? How far did it remain uniquely distinctive, to be cherished or abandoned? Does this period mark a turning point when farming ceased to be regarded as the indispensable industry? Michael Winter found the early 1900s highly pertinent, in his volume *Rural Politics* (London, 1996), in locating the first rudimentary steps in both agricultural corporatism and what might be described with hindsight as an environmental movement.

The co-editor of Part VI reminds us of the difficulty which contributors to this and previous volumes of the *Agrarian History* have had in

---

[45] V. de Bunsen, *Old and New in the Countryside* (London, 1920).

defining 'farming', or indeed 'rural'. Beyond the stimulus provided by the reprintings and reworkings of the themes of Raymond Williams's seminal book, *The Country and the City* (London, 1973), there remains considerable scope for closer understanding of the self-consciousness that developed within and between the range of situations that the Registrar-General and local government sought, with little success, to distinguish as urban and rural. How far was there continuity with what went before? How far did the First World War mark a new beginning, or simply a temporary dislocation of anxieties already firmly established that were to find expression in the formation of the Council for the Preservation of Rural England in 1926?

The late nineteenth century witnessed an explosion of interest in, and knowledge about, the countryside. There are fascinating research agendas to explore in terms of discovering how those living in those times coped in comprehending and making use of the deluge of descriptive topographies and such innovations as the photograph. How far did they open up 'new worlds' through travel by train or the bicycle? Peter Bowler's *The Environmental Sciences* (London, 1992) provides valuable context for exploring the regional and local expression of what had become almost an obsessive quest for knowledge, if not always for understanding, of one's rural 'backyard'. As Part VI illustrates, and the commemorative publications of the National Trust and numerous county archaeological and natural history societies have subsequently highlighted, the members of such Victorian institutions were a model of interdisciplinary and common endeavour. Highly complex networks of membership and collaboration by diligent correspondents evolved. Beyond the intrinsic fascination for the particular collections of specimens, whether of shells of coastal waters or perhaps axeheads and other artefacts of the prehistoric period, there was the knowledge that such activities accorded well with the social preoccupations of the age and none more so than that of self-improvement.

As in previous volumes of the *Agrarian History*, this present volume has explored the geography of farming, seeking out the salient attributes of each region. But for a fuller appreciation of farming and its unique contribution to the countryside, it becomes ever more important to discern how the countryside appeared to residents and visitors alike, and more particularly those of towns and cities who now constituted the majority of the population. Their broader perceptions of uplands and lowlands, the coast, woodlands, river valleys and fens help in measuring the impact of farming on the landscape, wildlife and recreational interests.

It is no surprise therefore that those writing of the agricultural history of the period 1850–1914 find themselves, consciously or otherwise, addressing a broader theme, *The Rural History of England and Wales*.

Beyond its life-giving role as a place for food production, the same land-space had come to assume a much wider role in terms of enhancing what, for most people, had become an urban way of life.

Rurally and agriculturally, Britain is believed to have ploughed a lonely furrow. Isolated historiographically from the political and social issues dominating European agriculture in the later nineteenth century – large versus small farms, peasant politics, syndicalism – as well as trade policy – research into British agricultural history has tended to focus on the narrower topics of agricultural output and productivity, farming techniques, landownership and estate management, and the condition of the agricultural labourer. New frameworks may be required to set the British experience in a broader perspective, and to determine to what extent it was unique or merely a regional variation of a European theme. This volume has tried to demonstrate the importance of the wider approach, setting agriculture within the framework of an evolving urban and industrial society. A case can be made also for comparative international studies, to determine the differences and similarities, comparing perhaps the experience of East Anglian arable farming with the 'grande culture' of Beauce and the Paris Basin, or small-scale vegetable growing in the Vale of Evesham and Belgian Flanders. To what extent can small farmers, in Wales and the west of England especially, be considered as part of the 'global' peasantry?

SELECT BIBLIOGRAPHY

Note: British Parliamentary Papers and Publications of the Ministry of Agriculture and Fisheries (MAF) and of the Ministry of Agriculture, Fisheries and Food (MAFF) appear at the end of this list.

Acland, A. H. D. *The Land*. London, 1913.

Acland, T. D. 'On the farming of Somersetshire', *JRASE*, 11, 1850.

Acland, T. D. and Sturge, W. *The Farming of Somerset*. London, 1851.

Adams, M. G. 'Agricultural change in the East Riding of Yorkshire, 1850–80: an economic and social history'. PhD thesis, University of Hull, 1977.

Addison, W. *English Markets and Fairs*. London, 1953.

Adonis, A. 'Aristocracy, agriculture, and liberalism; the politics, finances and estates of the 3rd Lord Carrington', *Historical Journal*, 31, 1988.

  Making Aristocracy Work: the Peerage and the Political System in Britain, 1884–1914. Oxford, 1993.

Adrian, L. 'The nineteenth-century Gazette corn returns from East Anglian markets', *JHG*, 3, 1977.

Alderton, D. and Booker, J. *Batsford Guide to Industrial Archaeology of East Anglia*. London, 1980.

Aldridge, H . *The Case for Town Planning*. London, 1915.

Allcroft, A. H. *Earthworks in England*. London, 1908.

Allen, D. A. *The Naturalist in Britain. A Social History*. London, 1976.

Amos, P. A. *Processes of Flour Manufacture*. London, 1912.

Anderson, M. 'Marriage patterns in Victorian Britain: an analysis based on registration district data for England and Wales, 1861', *Journal of Family History*, 1, 1976.

Arch, J. *The Story of His Life, Told by Himself*, ed. by Countess of Warwick. London, 1898.

Archer, John. *'By a flash and a scare'. Incendiarism, animal maiming and poaching in East Anglia, 1815–1870*. Oxford, 1990.

Armstrong, A. *Farmworkers: A Social and Economic History, 1770–1980*. London, 1988.

Armstrong, W. A. 'The use of information about occupation', in E. A. Wrigley, ed., *Nineteenth-century Society: Essays in the Use of Quantitative Methods for the Study of Social Data*. Cambridge, 1972.

'The influence of demographic factors on the position of the agricultural labourer in England and Wales, *c.* 1750–1914', *AHR*, 29, 1981.

Arnold, Arthur. *Free Land*. London, 1880.

'The indebtedness of the country gentry', *Contemporary Review*, 47, 1885.

Ashbee, P. 'Field archaeology: its origins and development', in P. J. Fowler, ed., *Archaeology and the Landscape*. London, 1972.

Ashby, A. W. and Evans, I. L. *The Agriculture of Wales and Monmouthshire*. Cardiff, 1944.

Ashby, M. K. *Joseph Ashby of Tysoe, 1859–1919*. Cambridge, 1961, London, 1974.

Aslet, Clive. *The Last Country Houses*. New Haven and London, 1982.

Atkins, P. J. 'The milk trade of London, *c.* 1790–1914', PhD thesis, University of Cambridge, 1977.

'London's intra-urban milk supply', *TIBG*, new ser., 2, 1977.

'The growth of London's railway milk trade, *c.* 1845–1914', *JTH*, new ser., 4, 1977–8.

'The production and marketing of fruit and vegetables, 1850–1950', in D. J. Oddy and D. S. Miller, eds., *Diet and Health in Modern Britain*. London, 1985.

Avebury, Lord. *The Scenery of England and the Causes to which it is due*. London, 1902.

Axe, Wortley. *The Horse: its Treatment in Health and Disease*. London, 1900.

Babington, C. C. *Flora of Cambridgeshire*. London, 1860.

Baddeley, M. J. B. *The Thorough Guide to the English Lake District*. London, 1880.

Bagwell, Philip S. *The Transport Revolution from 1770*. London, 1974.

Bagwell, P. S. and Armstrong, J. 'Coastal shipping', in M. J. Freeman and D. H. Aldcroft, eds., *Transport in Victorian Britain*. Manchester, 1988.

Baillie, E. J. 'Willows and their cultivation', *JRASE*, 3rd ser., 5, 1894.

Baines, D. E. *Migration in a Mature Economy: Emigration and Internal Migration in England and Wales, 1861–1900*. Cambridge, 1985.

Bairoch, P. 'European trade policy, 1815–1914', in P. Mathias and S. Pollard, eds., *Economic History of Europe*, vol. VIII, Cambridge, 1989.

Baker, J. G. *North Yorkshire. Studies of its Botany, Geology, Climate and Physical Geography*. London, 1863.

Ballard, P. D. 'A commercial and industrial élite: a study of Birmingham's upper middle class, 1780–1914', PhD thesis, University of Reading, 1983.

Banks, Sarah. 'Nineteenth-century scandal or twentieth-century model? A new look at "open" and "close" parishes', *EcHR*, 2nd ser., 42, 1988.

Barker, T. C. 'Urbanization and rising earnings', in T. C. Barker, J. C. Mackenzie and J. Yudkin, eds., *Our Changing Fare*. London, 1966.

Barnard, A. *Breweries of Great Britain and Ireland*. 4 vols. London, 1889–91.

Barnes, P. B. 'The economic history of landed estates in Norfolk since 1880', PhD thesis, University of East Anglia, 1984.

Barrett, H. and Phillips, J. *Suburban Style: the British Home, 1840–1960*. London, 1987.

Barrow, W. 'The agriculture of Pembrokeshire', *JRASE*, 2nd ser., 23, 1887.

Barty-King, H. *Food for Man and Beast*. London, 1978.

Bateman, John. *The Great Landowners of Great Britain and Ireland*. London, 1883.

Baugh, D. 'The myth of the Old Poor Law and the making of the new', *Journal of Economic History*, 23, 1963.

Baugh, D. C., ed. *V.C.H. Shropshire, IV. Agriculture*. Oxford. 1989.

Bear, W. E. 'The food supply of Manchester', *JRASE*, 3rd ser., 8, 1897.

'The agricultural population and the census of 1901', *JRASE*, 64, 1903.

Bearn, W. 'On the farming of Northamptonshire', *JRASE*, 13, 1852.

Beastall, T. W. 'A South Yorkshire estate in the late nineteenth century', *AHR*, 14, 1966

A North Country estate: the Lumleys and Saundersons as landowners, 1600–1900. London, 1975.

'Landlords and tenants', in G. E. Mingay, *The Victorian Countryside*, II. London, 1981.

Beaven, E. S. *Barley*. London, 1947.

Beckett, J. V. *A History of Laxton: England's Last Open Field Village*. Oxford, 1989.

'The pattern of landownership in England and Wales, 1660–1880', *EcHR*, 37 1984.

'The decline of the small landowner in England and Wales, 1660–1900', in F. M. L. Thompson, ed., *Landowners, Capitalists and Entrepreneurs*. Oxford, 1994.

The Aristocracy in England, 1660–1914. Oxford, 1986.

The Rise and Fall of the Grenvilles: Dukes of Buckingham and Chandos, 1710–1921. Manchester, 1994.

Beckett, J. V. and Wood, B. A. 'Land agency in the agricultural depression: R. W. Wordsworth at Thoresby, 1883–1914', *Trent Geographer*, 11, 1987.

Bedford, Duke of. *A Great Agricultural Estate*. 3rd edn, London, 1897.

Beevers, R. *The Garden City Utopia: a Critical Biography of Ebenezer Howard*. London, 1988.

Beilby, O. J. 'Changes in agricultural production in England and Wales', *JRASE*, 100, 1939.

'Agricultural output and income: discussion with special reference to the expanded sample of financial accounts', *Farm Economist*, 7 (5), 1953.

Belcher, C. 'On the reclaiming of waste lands as instanced in Wichwood Forest', *JRASE*, 24, 1863.

Bellerby, J. R. 'Agricultural output and income: farm incentive income', *Farm Economist*, 7 (5), 1953.

'Farm occupiers' capital in the United Kingdom before 1939', *Farm Economist*, 7 (6), 1953.

'The distribution of farm income in the UK, 1867–1938', *Journal of the Proceedings of the Agricultural Economics Society*, 10, 1953. Reprinted in W. E. Minchinton, ed., *Essays in Agrarian History*, vol. II. Newton Abbot, 1968.

'Gross and net farm rent in the United Kingdom, 1867–1938', *Journal of the Proceedings of the Agricultural Economics Society*, 10, 1954.

'Farm and non-farm capital, 1867–1938', *Farm Economist*, 8 (3), 1955.

Agriculture and Industry Relative Income, London, 1956.

'National and agricultural income in 1851', *Economic Journal*, 69, 1959.

Bellerby, J. R. and Boreham, A. J. 'Farm occupiers' capital in the United Kingdom before 1939', *Farm Economist*, 7, 1953.

Bell's Weekly Messenger, 30 March, 13 & 20 April, 23 & 30 May, 1864.

Bell, T. G. 'A report upon the agriculture of the county of Durham', *JRASE*, 17, 1856.

Benedetta, Mary. *The Street Markets of London*. London, 1936.

Bennett, R. and Elton, J. *History of Corn Milling*, IV. London, 1904.

Berridge, V. 'Health and medicine', in F. M. L. Thompson, *The Cambridge Social History of Britain, 1750–1950*. Cambridge, 1990.

Bettey, J. H. *Rural Life in Wessex, 1500–1900*. Bradford-on-Avon, 1977.

Bibby, J. and C. L. *A Miller's Tale: a History of J. Bibby & Sons Ltd.* Liverpool, 1978.

Blackman, Janet. 'The food supply of an industrial town; a study of Sheffield's public markets, 1780–1900', *Business History*, 5, 2, 1963.

Blatchford, R. *Merrie England*. London, 1894.

Blaug, M. 'The Poor Law report re-examined', *Journal of Economic History*, 14, 1964.

Bone, Q. D. 'Legislation to revive small farming in England, 1887–1914', *Agricultural History*, 49, 1975.

Booth, Charles. 'Occupations of the people of the United Kingdom, 1801–81', *JRSS*, 49, 1886.

Booth, J. B. *Bits of a Character. A Life of Henry Hall Dixon, 'The Druid'*. London, 1936.

Booth, W. *In Darkest England and the Way Out*. London, 1890.

Boreham, A. J. 'A series of estimates of occupiers' capital, 1867–1938', *Farm Economist*, 7 (6), 1953.

Bourke, J. 'Dairywomen and affectionate wives: women in the Irish dairy industry, 1890–1914', *AHR*, 38, 1990.

Bourne, G. *Change in the Village*. 1912, repr. London, 1966, Harmondsworth, 1984.

Bourne, S. 'On variations in the volume and value of exports and imports of the United Kingdom in recent years', *JRSS*, 52, 1889.

Bowley, A. L. 'The statistics of wages in the United Kingdom during the last hundred years, Part I. Agricultural Wages', *JRSS* 61, 1898.

Wages in the United Kingdom in the Nineteenth Century. Cambridge, 1900.

'Rural population in England and Wales: a study of the changes of density, occupations and ages', *JRSS*, 77, 1914.

Boyer, G. R. *An Economic History of the English Poor Law, 1750–1850*. Cambridge, 1990.

Boyer, R. and Hatton, T. J. 'Did Joseph Arch raise agricultural wages? Rural trade unions and the labour market in late nineteenth-century England', Centre for Economic Policy Research, Discussion Paper series, no. 484, December 1990.

Brace, H. W. *History of Seedcrushing in Great Britain*. London, 1960.

Brandon, P. F. 'The diffusion of designed landscapes in south England', in P. Brendon, *Thomas Cook: 150 Years of Popular Tourism*. London, 1991.

Brandon, Peter and Short, Brian. *The South-East from A.D. 1000*. London, 1990.

Brassley, Paul. 'Silage in Britain, 1880–1990: the delayed adoption of an innovation', *AHR*, 44, 1996.

Breckles, R. W. 'The social history of Ashfield, 1880–1930', MPhil thesis, University of Nottingham, 1993.

Brigden, R. *Victorian Farms*. Ramsbury, 1986.

British Yearbook of Agriculture, London, 1913.

Britton, D. K. *An Enquiry into Agricultural Rents and the Expenses of Landowners in England and Wales, 1838 and 1946*. Central Landowners' Association and the Ministry of Agriculture and Fisheries, London, 1949.

Britton, D. K. and Hunt, K. E. 'Agriculture', in M. G. Kendall and A. Bradford Hill, eds., *The Sources and Nature of the Statistics of the United Kingdom*. London, 1952.

Brodrick, G. C. *English Land and English Landlords*. London, 1881.

Brooks, C. *Mortal Remains. The History and Present State of the Victorian and Edwardian Cemetery*. Exeter, 1989.

Brown, B. H. *The Tariff Reform Movement in Great Britain*, New York, 1996.

Brown, Edward. *Poultry Husbandry*. London, 1915.

   British Poultry Husbandry: its Evolution and History. London, 1930.

Brown, G. T. 'The progress of legislation against contagious diseases of livestock, Part II', *JRASE*, 3rd ser., 4, 1893.

Brown, J. 'Scottish and English land legislation, 1905–1911', *Scottish Historical Review*, 47, 1968.

Brown, Jonathan. *The English Market Town. A Social and Economic History, 1750–1914*. Marlborough, 1986.

   Agriculture in England: a Survey of Farming, 1870–1947. Manchester, 1987.

   'The malting industry', in G. E. Mingay, ed., *AHEW* VI, *1750–1850*. Cambridge, 1989.

Brown, Jonathan and Ward, S. B. *Village Life in England, 1860–1940. A Photographic Record*. London, 1985.

Brown, J. H. 'Agriculture in Lincolnshire during the Great Depression, *c*. 1873–1896', PhD thesis, University of Manchester, 1978.

Brown, K. D. *A Social History of the Nonconformist Ministry in England and Wales, 1800–1930*. Oxford, 1988.

Buckingham, J. S. *National Evils and Practical Remedies*. London, 1849.

Buckland, G. 'On the farming of Kent', *JRASE*, 6, 1845.

Burke, J. F. *British Husbandry, vol. I*. Society for the Diffusion of Useful Knowledge, London, 1834.

Burnett, J. *Plenty and Want. A Social History of Diet in England from 1815 to the Present Day*. London, 1968.

   A Social History of Housing, 1815–1976. Newton Abbot, 1978.

   'Trends in bread consumption', in T. C. Barker, J. C. Mackenzie, and J. Yudkin, *Our Changing Fare*, London, 1966.

Burnett, R. G. *Through the Mill: the Life of Joseph Rank*. London, 1945.

Bushaway, Bob. *By Rite, Custom, Ceremony and Community in England, 1700–1880*. London, 1982.

Butlin, R. A., see H. S. A. Fox and R. A. Butlin, eds., *Change in the Countryside*.

Essays on Rural England, 1500–1900. Institute of British Geographers, Special Publication, x, 1979.

Buxton, E. N. *Epping Forest.* London, 1884.

Bythell, Duncan. *The Sweated Trades: Outwork in Nineteenth-Century Britain.* London, 1978.

Caird, J. *English Agriculture in 1850–51.* London, 2nd edn, 1852, repr. 1968.

'On the agricultural statistics of the United Kingdom', *JRSS*, 31, 1868.

The Landed Interest and the Supply of Food. London, 1878.

'A general view of British agriculture', *JRASE*, 14, 1878.

'Fifty years of progress of British agriculture', *JRASE*, 3rd ser., 1, part 1, 1897.

Cairncross, A. K. 'Internal migration in Victorian England', *Manchester School*, 17, 1949.

Calton, H. A. 'Hard wheats winning their way', US Department of Agriculture Yearbook, 1914.

Calvertt, John Simpson. *Rain and Ruin. The Diary of an Oxfordshire Farmer, 1875–1900.* Edited Celia Miller. Gloucester, 1983.

Cambridge, W. 'On the advantages of reducing the size and number of hedges', *JRASE*, 6, 1845.

Campbell, B. M. S. and Overton, M., eds. *Land, Labour and Livestock: Historical Studies in European Agricultural Productivity.* Manchester, 1991.

Cannadine, D. 'The landowner as millionaire: the finances of the Dukes of Devonshire', *AHR*, 25, 1977.

'Aristocratic indebtedness in the nineteenth century: the case re-opened,' *EcHR*, 30, 1977.

Lords and Landlords: the Aristocracy and the Towns, 1774–1967. Leicester, 1980.

The Decline and Fall of the British Aristocracy. New Haven, 1990.

Aspects of Aristocracy: Grandeur and Decline in Modern Britain. New Haven, 1994.

Capie, F. and Perren, R. 'The British market for meat, 1850–1914', *Agricultural History*, 54, 1980.

Carey, A. E. and Oliver, F. W. *Tidal Lands: a Study of Shore Problems.* London, 1918.

Carpenter, E. *Civilisation; its Cause and Cure, and Other Essays.* London, 1889.

Carrington, W. T. 'Pastoral husbandry', *JRASE*, 2nd ser., 14, 1878.

Cartwright, J., ed. *The Journals of Lady Knightley of Fawsley, 1856–84.* London, 1915.

Caunce, S. *Amongst Farm Horses; the Horselads of East Yorkshire.* Stroud, 1991.

Cawood, C. L. 'The history and development of farm tractors', *Industrial Archaeology*, 7, 1970.

Chadwick, O. *The Victorian Church, Part II, 1860–1901.* London, 1970.

Victorian Miniature. Cambridge, 1960.

Chambers, D. J. and Mingay, G. E. *The Agricultural Revolution, 1750–1880.* London, 1966.

Channing, F. A. *The Truth about Agricultural Depression.* London, 1897.

Chapman, S. D. 'Sudeley in the City: the financial problems of the fourth Lord Sudeley', in The Manorial Society, *The Sudeleys of Toddington.* London, 1987.

Charnock, J. H. 'On the farming of the West Riding of Yorkshire', *JRASE*, 9, 1848.

Chartres, J. A. 'Country trades, crafts, and professions', in G. E. Mingay, ed., *AHEW*, VI, *1750–1850*. Cambridge, 1989.

Chase, Malcolm. *'The People's Farm'. English Radical Agrarianism, 1775–1840*. Oxford, 1988.

Cherry, G. *The Politics of Town Planning*. Harlow, 1982.
  Cities and Plans. London, 1988.
  The Evolution of British Town Planning. Leighton Buzzard, 1974.

Church, R. A., *The History of the British Coal Industry*, III, *1830–1913: Victorian Pre-eminence*. Oxford, 1986.

Churchill, R. S. *Lord Derby, King of Lancashire*. London, 1959.

Clapham, J. H. *An Economic History of Modern Britain, II and III*. 2nd edn, Cambridge, 1932, 1938, 1964.

Clark, Colin. *National Income and Outlay*. London, 1937.
  'Capital in agriculture', *Farm Economist*, 4, 1959.

Clarke, J. A. 'Practical agriculture', *JRASE*, 2nd ser., 14, 1878.

Clarke, R. H. *Chronicle of a Country Works*. Thetford, 1952.

Clarkson, L. A. 'The manufacture of leather', in G. E. Mingay, *AHEW*, VI, *1750–1850*, Cambridge, 1989.

Clemenson, H. *English Country Houses and Landed Estates*. London, 1982.

Clifford, F. *The Agricultural Lock-Out of 1874*. Edinburgh, 1875.

Clifford, F. and Foote, J. A. 'English land law', *JRASE*, 2nd ser., 14, 1878.

Clutton, J. 'The cost of conversion of forest and woodland into cultivated land', *Trans. Institution of Surveyors*, 4, 1871–2.

Coale, A. J. and Watkins, S. C. *The Decline of Fertility in Europe*. Princeton, NJ, 1986.

Cobbold, R. and R. Fletcher, eds., *Features of Wortham, 1860, The Biography of a Victorian Village*. London, 1977.

Cocks, R. 'The Great Ashdown Forest case', in T. G. Watkin, ed., *Legal Record and Historical Reality*. London, 1989.

Coe, B. *A Victorian Country Album. The Photographs of Joseph Gale*. Oxford, 1988.

Cohen, Ruth L. *The History of Milk Prices*. Oxford, 1936.

Colbeck, T. L. 'On the agriculture of Northumberland', *JRASE*, 8, 1847.

Collings, J. and Green, F. L. *The Life of the Right Honourable Jesse Collings*. New York, 1920.

Collins, E. J. T. 'Harvest technology, and labour supply in Britain, 1790–1870'. PhD thesis, University of Nottingham, 1970.
  'Harvest technology and labour supply in Britain, 1790–1870', *EcHR*, 2nd ser., 22, 1969.
  A History of the Orsett Estate. Thurrock, 1978.
  The Economy of Upland Britain, 1750–1950: an Illustrated Review. Reading, 1978.
  'The age of machinery', in G. E. Mingay, ed., *The Victorian Countryside*, vol. I. London, 1981.
  'The farm horse economy of England and Wales in the early tractor age,

1900–40', in F. M. L. Thompson, ed., *Horses in European Economic History*. British Agricultural History Society, Reading, 1983.

'The "consumer revolution" and the growth of factory foods: changing patterns of bread and cereal eating in Britain in the nineteenth century', in D. J. Oddy and D. Miller, eds., *The Making of the Modern British Diet*. London, 1976.

'The rationality of surplus agricultural labour: mechanization in English agriculture in the nineteenth century', *AHR*, 35, 1987.

'The agricultural servicing and processing industries, 1, Introduction', in G. E. Mingay, ed., *AHEW*, VI. Cambridge, 1989.

'Did mid-Victorian agriculture fail? Output, productivity and technological change in nineteenth-century farming', *ReFresh*, 21, 1995.

'Industrialisation and the demand for food in Europe', in M. A. Havinden and E. J. T. Collins, eds., *Agriculture in the Industrial State*. Reading, 1995.

'Power Availability and Agricultural Productivity in England and Wales, 1840–1939', in B. J. P. van Bavel and E. Thoen, eds., *Land Productivity and Agro-systems in the North Sea Area*, Corn Publication Series 2, Turnhout, Belg., 1999.

'Agricultural hand tools and the industrial revolution', in N. Harte and R. Quinault, eds., *Land and Society in Britain, 1700–1914. Essays in Honour of F. M. L. Thompson*. Manchester, 1996.

Collins, E. J. T. and Jones, E. L. 'Sectoral advance in English agriculture, 1850–80', *AHR*, 15, 1967.

Colls, Robert and Dodd, Phillip. *Englishness, Politics and Culture, 1880–1920*. London, 1986.

Colyer, R. J. 'Welsh cattle drovers in the nineteenth century – 1', *Journ. National Library of Wales*, 17, 1971–2.

'Nanteos: a landed estate in decline, 1800–1930', *Ceredigion*, 9, 1980.

'The gentry and the country in nineteenth-century Cardiganshire', *Welsh History Review*, 10 (4), 1981.

See also Moore-Colyer, R. J.

Conder, F. R. *A Civil Engineer*. London, 1868.

Cooper, A. F. *British Agricultural Policy, 1912–36*. Manchester, 1989.

Coppock, J. T. 'The statistical assessment of British agriculture', *AHR*, 4, 1956.

'Agricultural changes in the Chilterns, 1875–1900', *AHR*, 9, 1961.

An Agricultural Atlas of England and Wales. London, 1964.

An Agricultural Geography of Britain. London, 1967.

'The changing face of England: 1850-circa 1900', in H. C. Darby, ed., *A New Historical Geography of England*. Cambridge, 1976 & 1986.

Coppock, J. T. and Best, R. H. *The Changing Use of Land in Britain*. London, 1962.

Copus, A. K. 'Changing markets and the response of agriculture in south west England, 1750–1900', PhD thesis, University of Wales, 1986.

'Changing markets and the development of sheep breeds in southern England, 1750–1900', *AHR*, 37, 1989.

Corley, T. A. B. *Quaker Enterprise in Biscuits: Huntley and Palmers of Reading, 1872–1922*. London, 1972.

'Nutrition, technology and the growth of the British biscuit industry, 1820–1900', in D. J. Oddy and D. S. Miller, *The Making of the Modern British Diet*. London, 1976.

Cornish, C. J. *Wild England of Today and the Wild Life in it*. London, 1895.

Cottingham, R. W. 'The agriculture of Nottinghamshire', *JRASE*, 6, 1845.

Cox, G., Lowe, P. and Winter, M. 'The origins and early development of the National Farmers' Union', *AHR*, 39, 1991.

Crafts, N. F. R. *British Economic Growth during the Industrial Revolution*. Oxford, 1985.

Cragoe, Matthew. *An Anglican Aristocracy: the Moral Economy of the Landed Estate in Carmarthenshire, 1832–1895*. Oxford, 1996.

Craigie, P. G. 'Statistics of agricultural production', *JRSS*, 46, 1883.

'On the production and consumption of meat in the United Kingdom', *Report of the British Association for the Advancement of Science, Section F*, 1884.

'The size and distribution of agricultural holdings in England and abroad', *JRSS*, 50, 1887.

Crawford, R. F. 'An inquiry into wheat prices and wheat supplies', *JRSS*, 58, 1895.

'The food supply of the United Kingdom', *JRASE*, 3rd ser., 11, 1900.

Crosby, A. *A History of Woking*. Chichester, 1982.

Crosby, Travis L. *English Farmers and the Politics of Protection, 1815–52*. Hassocks, 1977.

Crossley, A., ed. *A Victoria History of the County of Oxford*, XII. Oxford, 1990.

Crowther, M. A. *The Workhouse System, 1834–1929*. London, 1981.

Currie, R., Gilbert, A., and Horsley, L. *Churches and Churchgoers. Patterns of Church Growth in the British Isles since 1700*. Oxford, 1977.

Dale, H. E. *Daniel Hall: Pioneer in Scientific Agriculture*. London, 1956.

Dalton, R. 'Agricultural change in southern Derbyshire, 1770 to 1870, with special reference to the dairy industry,' PhD thesis, University of Nottingham, 1995.

Daunton, M. J. 'Firm and family in the City of London in the nineteenth century: the case of F. G. Dalgety', *Historical Research*, 62, 1989.

'The political economy of death duties: Harcourt's budget of 1894', in N. Harte and R. Quinault, eds., *Land and Society in Britain, 1700–1914*, Manchester, 1996.

David, P. A. *Technical Choice, Innovation and Economic Growth: Essays on American and British Experience in the Nineteenth Century*. Cambridge, 1975.

Davies, G. C. *The Handbook to the Rivers and Broads of Norfolk and Suffolk*. London, 1883.

Davies, J. 'The end of the great estates and the rise of freehold farming in Wales', *Welsh History Review*, 7, 1974.

Davies, J. H. *Cardiff and the Marquesses of Bute*. Cardiff, 1981.

Day, Clive. 'The distribution of industrial occupations in England, 1841–1861', *Trans. Connecticut Academy of Arts and Sciences*, 28, 1927.

Denman, D. R. *Estate Capital*. London, 1945.

'Farm rent surveys, 1938–1959', *Farm Economist*, 9 (8), 1960.

Dent, J. 'Agricultural jottings from the General Report of the Census of England and Wales for the year 1871', *JRASE*, 2nd ser., 10, 1874.

Denton, J. B. *Agricultural Drainage*. London, 1830.

The Farm Homesteads of England. London, 1863.

Devine, T. M., ed. *Farm Servants and Labour in Lowland Scotland, 1770–1914*. Edinburgh, 1984.

De Vries, Jan. *European Urbanization, 1500–1800*. London, 1984.

Dewey, P. E. *British Agriculture in the First World War*. London, 1989.

Dickinson, W. 'On the farming of Cumberland', *JRASE*, 13, 1852.

Digby, Anne. 'The labour market and the continuity of social policy after 1834: the case of the eastern counties', *EcHR*, 2nd ser., 28, 1975.

'The rural poor law', in D. Fraser, ed., *The New Poor Law in the Nineteenth Century*. London, 1976.

Pauper Palaces. London, 1978.

'The rural poor', in G. E. Mingay, ed., *The Victorian Countryside*. London, 1981.

Making a Medical Living. Doctors and Patients in the English Market for Medicine, 1720–1971. Cambridge, 1994.

Dobson, M. J. '"Marsh fever" – the geography of malaria in England', *JHG*, 6, 1980.

Dodd, G. *The Food of London*. London, 1856.

Dodd, J. P. 'The agricultural statistics for 1854: an assessment of their value', *AHR*, 35, 1987.

Douglas, Roy. *Land, People and Politics: The Land Question in the United Kingdom, 1878–1952*. London, 1976.

Dowell, S. *A History of Taxation and Taxes in England*, III. London, 1884.

Drake, M. 'The census, 1801–1891', in E. A. Wrigley, ed., *Nineteenth-century Society: Essays in the Use of Quantitative Methods for the Study of Social Data*. Cambridge, 1972.

Drescher, L. 'The development of agricultural production in Great Britain and Ireland from the early nineteenth century', *The Manchester School*, 23, 1955.

Druce, G. C. *The Flora of Oxfordshire*. Oxford, 1886.

Druce, S. B. L. 'The alteration in the distribution of the agricultural population of England and Wales between the returns of the census of 1871 and 1881', *JRASE*, 2nd ser., 21, 1885.

Drummond, J. C. and Wilbraham, A. *The Englishman's Food*. 2nd edn, London, 1957.

Dunbabin, J. P. D. *Rural Discontent in Nineteenth-Century Britain*. London, 1974.

Durant, H. 'The development of landownership, 1873–1925, with special reference to Bedfordshire', *Sociological Review*, 28, 1936.

Dutt, W. A. *The King's Homeland. Sandringham and North-west Norfolk*. London, 1904.

Wild Life in East Anglia. London, 1906.

Dyke, G. V. *John Bennet Lawes: the Record of his Genius*. Taunton, 1991.

Dyos, H. J. *Victorian Suburb: a Study of the Growth of Camberwell*. Leicester, 1961.

Edwards, B. *The Burston School Strike*, London, 1974.

Edwards, G. *From Crow Scaring to Westminster*. London, 1922.

Edwards, R. and Perren, R. 'A note on regional differences in British meat prices, 1825–1865', *Economy and History*, 22 (2), 1979.

Elgee, F. *The Moorlands of North-eastern Yorkshire. Their Natural History and Origin.* London, 1912.

Ellmore, W. P. and Okey, T. 'Planting, cleaning and cutting willows', *Journ. of Board of Agriculture*, 18, 1911–12.

Emerson, P. H. and Goodall, T. F. *Life and Landscape on the Norfolk Broad.* London, 1886.

Emy, H. V. 'The land campaign: Lloyd George as a social reformer', in A. J. P. Taylor, ed., *Lloyd George: Twelve Essays.* Aldershot, 1971.

Engels, F. *The Condition of the Working-Class in England from Personal Observation and Authentic Sources, 1845*, ed. E. J. Hobsbawm. London, 1989.

English Labourers' Chronicle, 1 August 1885. London.

English, Barbara. 'On the eve of the great depression: the economy of the Sledmere estate, 1869–1878', *Business History*, 24, 1982.

　 'Patterns of estate management in East Yorkshire, *c.* 1840–*c.* 1880', *AHR*, 32, 1984.

　 The Great Landowners of East Yorkshire, 1530–1910. Hemel Hempstead, 1990.

Ernle, Lord, *English Farming Past and Present.* 6th edn, London, 1961.

Evans, D. Gareth. *A History of Wales, 1815–1906.* Cardiff, 1989.

Evans, Eric J. *The Contentious Tithe: the Tithe Problem and English Agriculture, 1750–1850.* London, 1976.

Evans, George Ewart. *Where Beards Wag All. The Relevance of the Oral Tradition.* London, 1970.

Evans, L. T. *Crop Evolution, Adaptation and Yield.* Cambridge, 1993.

Everitt, A. *The Pattern of Rural Dissent in the Nineteenth Century.* Occasional Paper, no. 4, 2nd ser., University of Leicester, Dept of English Local History, 1971.

　 'Town and country in Victorian Leicestershire: the role of the village carrier', in Alan Everitt, ed., *Perspectives in English Urban History.* London, 1973.

　 'Country carriers in the nineteenth century', *Journ. Transport History*, 2nd ser., 3, 3, 1976.

　 'The making of the agrarian landscape of Kent', *Archaeologia Cantiana*, 92, 1977.

　 'Country, county and town: patterns of regional evolution', *TRHS*, 5th ser., 29, 1979.

Evershed, H. 'On the farming of Surrey', *JRASE*, 14, 1853.

　 'Improvements of the plants of the farm', *JRASE*, 2nd ser., 20, 1884.

Eversley, Lord, 'The decline in the number of agricultural labourers in Great Britain', *JRSS*, 70, 1907.

　 Commons, Forests, and Footpaths. London, 1910.

Fairlie, S. 'The nineteenth century Corn Laws reconsidered', *EcHR*, 18, 1965.

　 'The Corn Laws and British wheat production, 1829–76', *EcHR*, 2nd ser., 22, 1969.

Farmer, 2 Feb. & 23 March 1889.

Farmer and Stockbreeder, 20 April and 4 June 1896.

Farmer's Magazine, 3rd ser., 9 & 10, 1856. London.

Farrall, T. 'Report on the agriculture of Cumberland', *JRASE*, 2nd ser., 10, 1874.

Farrant, S. 'The management of four estates in the lower Ouse valley, Sussex, and agricultural change, 1840–1920', *Southern History*, 1, 1979.

'London by the sea: resort development on the south coast of England, 1880–1939', *Journal of Contemporary History*, 22, 1987.

Feinstein, C. H. *National Income, Expenditure, and Output of the UK, 1855–1965*. Cambridge, 1972.

Statistical Tables of National Income, Expenditure and Output in the UK, 1855–1965. Cambridge, 1976.

'Agriculture', in C. H. Feinstein and S. Pollard, eds., *Studies in Capital Formation in the United Kingdom, 1750–1920*. Oxford, 1988.

'New estimates of average earnings in the United Kingdom, 1880–1913', *EcHR*, 42, 1990.

Finch, G. 'The experience of peripheral regions in an age of industrialisation: the case of Devon, 1840–1914', PhD thesis, University of Oxford, 1984.

Fisher, J. R. 'Public opinion and agriculture, 1875–1900', PhD thesis, University of Hull, 1973.

Clare Sewell Read: an Agricultural Spokesman of the Late Nineteenth Century. Hull, 1975.

'The Farmers' Alliance: an agricultural protest movement of the 1880s', *AHR*, 26, 1978.

'The economic effects of cattle disease in Britain and its containment, 1850–1900', *Agricultural History*, 54, no. 2, 1980.

'The limits of deference: agricultural communities in a mid-nineteenth century election campaign', *Journal of British Studies*, 21, 1981.

'Animal health and the Royal Agricultural Society in its early years', *JRASE*, 143, 1982.

'Landowners and English tenant right, 1845–52,' *AHR*, 31, 1983.

FitzRandolph, Helen E. and Hay, M. Dorel. *Rural Industries of England and Wales*, I. *Timber and Underwood Industries and Some Village Workshops*. Oxford, 1926; II. *Osier-growing and Basketry and Some Rural Factories*. Oxford, 1926, repr. East Ardsley, Wakefield, 1977; III. *Decorative Crafts and Rural Potteries*. Oxford, 1927.

Fletcher, T. W. 'The economic development of agriculture in east Lancashire, 1870–1939.' MSc thesis, University of Leeds, 1954.

'Drescher's index: a comment', appended to L. Drescher, 'The development of agricultural production in Great Britain and Ireland from the early nineteenth century', *Manchester School*, 23, 1955.

'The agrarian revolution in arable Lancashire', *Trans. Lancs. & Cheshire Antiq. Soc.*, 72, 1965.

'The great depression of English agriculture, 1873–1896', *EcHR*, 2nd ser., 13, 1961, reprinted in W. E. Minchinton, ed., *Essays in Agrarian History*, vol. I. Newton Abbot, 1960; also in P. J. Perry, *British Agriculture, 1875–1914*. London, 1973.

'Lancashire livestock farming during the Great Depression', *AHR*, 9, 1961, reprinted in P. J. Perry, ed., *British Agriculture, 1875–1914*. London, 1973.

Flett, J. S. *The First Hundred Years of the Geological Survey*. London, 1937.

Ford, A. G. 'The trade cycle in Britain, 1860–1914' in R. Floud and D.

McCloskey, eds., *The Economic History of Britain since 1870*, vol. II, *1860–1970s*. Cambridge, 1981.

Foreman, Susan, *Loaves and Fishes. An Illustrated History of the Ministry of Agriculture, Fisheries and Food, 1889–1989*. London, 1989.

Fox, A. Wilson. 'Agricultural wages in England and Wales during the last fifty years', *JRSS*, 66, 1903. Reprinted in W. E. Minchinton, ed., *Essays in Agrarian History*, vol. II. Newton Abbot, 1968.

Fox, H. S. A. 'Local farmers' associations and the circulation of agricultural information in nineteenth-century England', in H. S. A. Fox and R. A. Butlin, eds., *Change in the Countryside. Essays on Rural England, 1500–1900*. Institute of British Geographers, Special Publication 10. London, 1979.

Fream, W. *Elements of Agriculture*. 3rd edn, London, 1892.

Freeman, M. D. 'A history of corn milling *c.* 1750–1914, with special reference to south central and south eastern England', PhD thesis, University of Reading, 1967.

Fussell, G. E. 'The collection of agricultural statistics in Great Britain, its origins and evolution', *Agricultural History*, 19, 1944.

The Farmer's Tools, A.D. 1500–1900. London, 1952.

The English Dairy Farmer, 1500–1900. London, 1966.

Garnett, W. J. 'Farming of Lancashire', *JRASE*, 10, 1849.

Gaskell, S. M. *Model Housing from the Great Exhibition to the Festival of Britain*. London, 1986.

Gatrell, V. A. C. and Hadden, T. B. 'Criminal statistics and their interpretation', in E. A. Wrigley, ed., *Nineteenth-Century Society*. Cambridge, 1972.

'The decline of theft and violence in Victorian and Edwardian England', in V. A. C. Gatrell, B. Lenman and G. Parker, eds., *Crime and the Law. The Social History of Crime in Western Europe since 1500*. London, 1980.

'Crime, authority, and the policeman state', in F. M. L. Thompson, ed., *Cambridge Social History of Britain, 1750–1950*, vol. III. Cambridge, 1990.

Gavin, Sir William. *Ninety Years of Family Farming: the Story of Lord Rayleigh's and Strutt and Parker Farms*. London, 1967.

Genovese, E. Fox. 'The many faces of moral economy', *PP*, 58, 1973.

George, Henry. *Progress and Poverty*. 1st English edn, London, 1879.

Gerard, Jessica. 'Lady Bountiful: women of the landed classes and rural philanthropy', *Victorian Studies*, 30, 1987.

Gielgud, J. 'Nineteenth-century farm women in Northumberland and Cumbria; the neglected workforce', DPhil thesis, University of Sussex, 1992.

Giffen, R. *The Growth of Capital*. London, 1889.

Gilbert, A. D. 'The land and the church', in G. E. Mingay, ed., *The Victorian Countryside*. 2 vols., London, 1981.

Gilbert, B. B. *David Lloyd George: a Political Life*. Columbus, 1987.

Girouard, Mark. *The Victorian Country House*. Revised edn, New Haven and London, 1979.

Glendinning, D. R. 'Potato introduction and breeding up to the early twentieth century', *New Phytology*, 94, 1983.

Goddard, Nicholas, 'Agricultural societies', in G. E. Mingay, ed., *The Victorian Countryside*, vol. I. London, 1981.

'The development and influence of agricultural periodicals and newspapers, 1780–1880', *AHR*, 31, 1983.

Harvests of Change:The Royal Agricultural Society of England, 1838–1988. London, 1988.

'Agricultural literature and societies', in G. E. Mingay, ed., *AHEW*, VI, *1750–1850*, Cambridge, 1989.

'Information and innovation in early Victorian farming systems', in B. A. Holderness and M. E. Turner, eds., *Land, Labour and Agriculture, 1700–1920. Essays for Gordon Mingay*. London, 1991.

Gosden, P. H. J. H. *The Friendly Societies in England, 1815–1875*. Manchester, 1961.

Gough, J. W. *The Mines of Mendip*. Oxford, 1930, repr. 1967.

Gould, P. C. 'The back to the land experiment at Starnthwaite, Westmorland (1892–1900)', *Journal of Regional and Local Studies*, 6, 1986.

Gourvish, T. *Norfolk Beers from English Barley*. Norwich, 1987.

Gourvish, T. R. and Wilson, R. G. *The British Brewing Industry, 1830–1980*. Cambridge, 1994.

Grace, D. 'The agricultural engineering industry', in G. E. Mingay, ed., *AHEW*, VI, *1750–1850*. Cambridge, 1989.

Grace, D. R. and Phillips, D. C. *Ransomes of Ipswich: a History of the Firm and Guide to its Records*. University of Reading, 1975.

Graham, P. Anderson. *The Rural Exodus*. London, 1892.

Grantham, G. and Leonard, C. S., eds. *Agrarian Organisation in the Century of Industrialization: Europe, Russia, and North America*. Research in Economic History, Supplement 5. Greenwich, Conn., 1989.

Green, J. L. *The Rural Industries of England*. London, 1895.

Grigg, D. B. 'Changing regional values during the agricultural revolution in south Lincolnshire', *TIBG*, 30, 1962.

'An index of regional change in English farming', *TIBG*, 36, 1965.

The Agricultural Revolution in South Lincolnshire. Cambridge, 1966.

'Farm size in England and Wales, from early Victorian times to the present', *AHR*, 35, 1987.

English Agriculture: an Historical Perspective. Oxford, 1989.

Grigor, J. 'On Fences', *JRASE*, 6, 1845.

Gunstone, D. P. 'Stewardship and landed society: a study of the stewards of the Longleat Estate, 1779–1895', MA thesis, University of Exeter, 1972.

Gurnham, R. 'The creation of Skegness as a resort town by the 9th Earl of Scarborough', Lincolnshire History and Archaeology, 7, 1972.

Gutzke, D. W. 'The social status of landed brewers in Britain since 1840', *Histoire Sociale – Social History*, 17, 1984.

Habakkuk, Sir John. *Marriage, Debt and the Estate System: English Landownership, 1650–1950*. Oxford, 1994.

Haggard, H. Rider, *A Farmer's Year, being his Commonplace Book for 1898*. London, 1899, 1906.

Rural England: being an Account of Agricultural and Social Researches carried out in the Years 1901 and 1902. 2 vols., 1902, 2nd edn, London, 1906.

Haig, A. *The Victorian Clergy.* London, 1984.

Haining, J. and Tyler, C. *Ploughing by Steam.* Hemel Hempstead, 1970.

Hall, Adrian, *Fenland Worker-Peasants. The Economy of Smallholders at Rippingale, Lincolnshire, 1791–1871, AHR*, Supplement Series, 1, 1992.

Hall, A. D. *The Book of the Rothamsted Experiments.* New York, 1905.

A Pilgrimage of British Farming, 1910–12. London, 1914.

'The development of agricultural education in England and Wales', *JRASE*, 83, 1922.

Hall, A. D. and Russell, E. J. *A Report on the Agriculture and Soils of Kent, Surrey, and Sussex.* London, 1911.

Hall, Sherwin, 'The great cattle plague of 1865', *British Veterinary Journal*, 122, 1966.

Hall, S. J. G. and Clutton-Brock, J. *Two Hundred Years of British Farm Livestock.* London, 1989.

Hallas, C. S. 'The social and economic impact of a rural railway: the Wensleydale line', *AHR*, 34, 1986.

'Economic and social change in Wensleydale and Swaledale in the nineteenth century', PhD thesis, Open University, 1987.

'Supply responsiveness in dairy farming: some regional considerations', *AHR*, 39, 1991.

Hardy, D. *Alternative Communities in Nineteenth-Century England.* London, 1979.

Hardy, D. and Ward, C. *Arcadia for All; the Legacy of a Makeshift Landscape.* London, 1984.

Haresign, S. R. 'Small farms and allotments as a cure for rural depopulation in the Lincolnshire fenland, 1870–1914', *Lincolnshire History and Archaeology*, 18, 1983.

Harley, J. B. *The Historian's Guide to Ordnance Survey Maps.* London, 1964.

'The Ordnance Survey and land-use mapping', Research Paper, 2, Historical Geography Research Group, 1979.

Harrison, B. 'Animals and the state in nineteenth-century England', *EHR*, 88, 1973.

Harte, N. and Quinault, R., eds. *Land and Society in Britain, 1700–1914: Essays in Honour of F. M. L. Thompson.* Manchester, 1996.

Hasbach, W. *A History of the English Agricultural Labourer.* 2nd edn, London, 1908.

Hassan, J. A. 'The growth and impact of the British water industry in the nineteenth century', *EcHR*, 38, 1985.

Havinden, M. A. *Estate Villages. A Study of the Berkshire Villages of Ardington and Lockinge.* Reading, 1966.

Hawke, G. *Railways and Economic Growth in England and Wales, 1840–1870.* Oxford, 1970.

Hay, D. 'Crime and justice in eighteenth and nineteenth-century England', *Crime and Justice*, 2, 1980.

Healy, M. J. R. and Jones, E. L. 'Wheat yields in England, 1815–59', *JRSS*, 125, 1962.

Heath, Richard. *The English Peasant.* new edn, London, 1978.

Hendrick, J. 'The growth of international trade in manures and foods', *Trans. Highland and Agric. Soc. of Scotland*, 5th ser., 29, 1917.

Higgs, Edward. 'Women, occupations and work in the nineteenth-century censuses', *History Workshop Journal*, 23, 1987.

Making Sense of the Census: The Manuscript Returns for England and Wales, 1801–1901. Public Record Office Handbooks, no. 23. London, 1989.

'Occupational censuses and the agricultural workforce in Victorian England and Wales', *EcHR*, 2nd ser., 48, 1995.

Hill, 'Bridget. 'Women, work and the census: a problem for historians of women', *History Workshop Journal*, 35, 1993.

'Occupational censuses and the agricultural workforce in Victorian England and Wales', *EcHR*, 2nd ser., 48, 1995.

Hinde, P. R. A. 'The fertility transition in rural England', PhD thesis, University of Sheffield, 1985.

Hobson, W. C. *Hobson's Fox-hunting Atlas.* London, 1850.

Hoelscher, L. 'Improvements in fencing and drainage in mid-nineteenth century England', *Agricultural History*, 38, 1963.

Holderness, B. A. 'Capital formation in agriculture', in S. Pollard and J. P. P. Higgins, eds., *Aspects of Capital Investment in Great Britain, 1750–1850.* London, 1971.

'"Open" and "close" parishes in England in the eighteenth and nineteenth centuries', *AHR*, 20, 1972.

'The English land market in the eighteenth century: the case of Lincolnshire', *EcHR*, 27, 1974.

'Agriculture and industrialization in the Victorian economy', in G. E. Mingay, ed., *The Victorian Countryside*, vol. 1. London, 1981.

'Agriculture, 1770–1860', in C. H. Feinstein and S. Pollard, eds., *Studies in Capital Formation in the United Kingdom, 1750–1920*, Oxford, 1988.

'Prices, productivity, and output', in G. E. Mingay, ed., *AHEW*, VI. Cambridge, 1989.

'The origins of high farming', in B. A. Holderness and M. Turner, eds., *Land, Labour and Agriculture, 1700–1920: Essays for Gordon Mingay.* London, 1991.

Holderness, B. A. and Turner, M. E., eds. *Land, Labour and Agriculture, 1700–1920: Essays for Gordon Mingay.* London, 1991.

Hollis, Patricia. *Ladies Elect. Women in Local Government, 1865–1914.* Oxford, 1987.

Hooker, R. H. 'The meat supply of the United Kingdom', *JRSS*, 72, 1909.

Hooson, D. J. M. 'The straw industry of the Chilterns in the nineteenth century', *East Midland Geographer*, 4, 1966–9.

Horn, P. *The Victorian Country Child.* Kineton, 1974.

Labouring Life in the Late Victorian Countryside. Dublin, 1976.

Education in Rural England, 1800–1914. London, 1978.

The Changing Countryside in Victorian and Edwardian England and Wales. London, 1984.

Hoskins, W. G. *The Midland Peasant.* London, 1951.

Houghton, C. T. 'A new index number of agricultural prices', *JRSS*, 101, 1938.

Howe, A. *The Cotton Masters, 1830–1860*. Oxford, 1984.

Howell, D. W. 'The impact of railways on agricultural development in nineteenth-century Wales', *Welsh Hist. Rev.*, 7, 1974–5.

Land and People in Nineteenth-Century Wales. London, 1977.

Howkins, Alun. *Whitsun in Nineteenth-Century Oxfordshire*. Oxford, 1973.

Poor Labouring Men. London, 1985.

'The discovery of England', in R. Colls and P. Dodd, eds., *Englishness: Politics and Culture, 1880–1920*. London, 1986.

Reshaping Rural England. London, 1991.

'Peasants, servants and labourers: the marginal workforce in British agriculture, c. 1870–1914', *AHR*, 42, 1994.

Howkins, Alun and Merricks, Linda. 'The ploughboy and the plough play', *Folk Music Journal*, 1991.

'"We be blacke as hell"; ritual disguise and rebellion', *Rural History*, 4 (1), 1993.

Hudson, Pat. *The Genesis of Industrial Capital. A Study of the West Riding Wool Textile Industry, c. 1750–1850*. Cambridge, 1986.

Hudson, W. H. *A Shepherd's Life*. London, 1910.

Humphreys, M. *The Crisis of Community: Montgomeryshire, 1680–1815*. Cardiff, 1996.

Humphries, A. E. 'Modern developments in flour milling', *Journ. of the Society of Arts*, 55, 1906.

Hunt, E. H. 'Labour productivity in English agriculture, 1850–1914', *EcHR*, 20, 1967.

Regional Wage Variations in Britain, 1850–1914. Oxford, 1973.

'Industrialization and regional inequality: wages in Britain, 1760–1914', *Journal of Economic History*, 46, 1986.

Hunt, E. H. and Pam, S. J. 'Essex agriculture in the "Golden Age", 1850–73', *AHR*, 43, 1995.

'Prices and structural response in English agriculture, 1873–1896', *EcHR*, 50, 1997.

Hurt, J. S. *Elementary Schooling for the Working Class, 1860–1918*. London, 1979.

Hutchinson, J. N. 'The record of peat wastage in the East Anglian fenlands at Holme Post, 1848–1978 AD', *Journal of Ecology*, 68, 1980.

Inglis, K. S. *Churches and the Working Classes in Victorian England*. London, 1963.

Innes, J. W. *Class Fertility Trends in England and Wales, 1876–1934*. Princeton, 1938.

Jackson, A. A. *Semi-detached London: Suburban Development, Life and Transport, 1900–39*. London, 1973.

Jago, W. and Jago, W. C. *The Technology of Bread Making*. London, 1911.

James, B. L. 'The "great" landowners of Wales in 1873', *National Library of Wales Journal*, 14, 1966.

Jebb, L. *The Small Holdings of England*. London, 1907.

Jefferies, R. *Hodge and his Masters*. London, 1880 edn; also 1966, 1979.

Wild Life in a Southern County. 1879, repr. London, 1949.

Jeffreys, J. B. *Retail Trading in Great Britain, 1850–1914*. Cambridge, 1954.

Jekyll, F. *Gertrude Jekyll. A Memoir*. London, 1934.

Jekyll, G. *Wood and Garden*. London, 1899.
  Old West Surrey. Some Notes and Memories. Dorking, 1904, repr. 1978.
  Old English Household Life. London, 1934.
Jekyll, G. and Weaver, L. *Gardens for Small Country Houses*. London, 1912.
Jenkins, David. *The Agricultural Community in South-West Wales at the Turn of the Twentieth Century*. Cardiff, 1971.
  The West Riding Wool Textile Industry, 1770–1835. Edington, 1975.
Jenkins, D. T. and Ponting, K. G. *The British Wool Textile Industry, 1770–1914*. London, 1982.
Jenkins, H. M. 'Report on the cheese factory system and its adaptability to English dairy districts', *JRASE*, 2nd ser., 7, 1870.
  'Report on the practice of ensilage', *JRASE*, 20, 1884.
Jenkins, J. G. *Agricultural Transport in Wales*. Cardiff, 1962.
  Traditional Country Craftsmen. London, 1969.
Jewell, C. A., ed. *A Sourcebook of Victorian Farming*. Winchester, 1975.
Johnson, J. H. 'Harvest migration from nineteenth-century Ireland', *TIBG*, 41, 1967.
Johnson, P. *Savings Behaviour, Fertility and Economic Development in Nineteenth-Century Britain and America*. Centre for Policy Research Paper, 203, 1987.
Johnson's Official Handbook of Cattle Fairs, 1894.
Jones, Anna M. *The Rural Industries of England and Wales, IV, Wales*. Oxford, 1927, repr. East Ardsley, Wakefield, 1978.
Jones, A. W. 'Agriculture and the rural community of Glamorgan, *c.* 1830–1896', PhD thesis, University of Wales, 1980.
Jones, D. *Crime, Protest, Community and Police in Nineteenth-Century Britain*. London, 1982.
  'Rural crime and protest', in G. E. Mingay, ed., *The Victorian Countryside*. London, 1981.
  'The new police, crime, and people in England and Wales, 1829–1888, *Trans. Royal Historical Society*, 5th ser., 33, 1983.
Jones, E. L. *Seasons and Prices. The Role of the Weather in English Agricultural History*. London, 1964.
  'The changing basis of English agricultural prosperity, 1853–73', *AHR*, 10, 1962.
Jones, Gwyn E. 'William Fream: agriculturist and educator', *JRASE*, 144, 1983.
Jones, G. E. and Tattersfield, B. K. 'John Wrightson and the Downton College of Agriculture', *Agricultural Progress*, 55, 1980.
Jones, L. E. *A Victorian Boyhood*. London, 1955.
Jones, M. 'Y chwarelwyr: the slate quarrymen of North Wales', in R. Samuel, ed., *Miners, Quarrymen and Saltworkers*. London, 1977.
Kain, R. J. P. 'Tithe surveys and landownership', *JHG*, 1, 1975.
  An Atlas and Index of the Tithe Files of Mid-Nineteenth-Century England and Wales. Cambridge, 1986.
Kain, R. J. P. and Prince, H. C. *The Tithe Surveys of England and Wales*. Cambridge, 1985.
Kearton, R. *With Nature and a Camera*. London, 1898.

Kellett, J. R. *Railways and Victorian Cities*. London, 1969.

Kendall, M. G. 'Farm occupiers' capital in the United Kingdom', *JRSS*, 104, 1941.

Kent, W. *The Agricultural Implement Manufacturers' Directory of England*. London, 1867.

Kerr, B. *Bound to the Soil: a Social History of Dorset*. London, 1968.

Kightly, Charles. *Country Voices. Life and Lore in Farm and Village*. London, 1984.

Kilby, Kenneth. *The Cooper and his Trade*. London, 1971.

King, A. D. *The Bungalow. The Production of a Global Culture*. London, 1984. *Urbanism, Colonialism and the World Economy*. London, 1990.

King, P. 'Gleaners, farmers, and the failure of legal sanctions in England, 1750–1850', *PP*, 125, 1989.

Kitchen, Fred. *Brother to the Ox*. 1943, new edn, West Firle, 1981.

Kitteringham, Jennie. 'Country work girls in nineteenth-century England', in Raphael Samuel, ed., *Village Life and Labour*. London, 1975.

Kropotkin, P. *Fields, Factories and Workshops Tomorrow*. London, 1890, repr. London, 1974.

Lascelles, G. *Twenty-five Years in the New Forest*. London, 1915.

Lavergne, Leonce de, *The Rural Economy of England, Scotland, and Ireland*. Edinburgh and London, 1855.

Law, C. M. 'Luton and the hat industry', *East Midland Geographer*, 4, 1966–9. 'The growth of urban populations in England and Wales, 1801–1911', *TIBG*, 41, 1967.

Lawes, J. B. and Gilbert, J. H. 'Allotments and smallholdings', *JRASE*, 3rd ser., 3, 1892. 'Home produce, imports, consumption, and price of wheat over forty harvest-years, 1852–53 to 1891–2', *JRASE*, 3rd ser., 4, 1893.

Lawley, F. *Life and Times of 'The Druid'*. London, 1895.

Lawton, R. 'Rural depopulation in nineteenth century England', in R. W. Steel and R. Lawton, eds., *Liverpool Essays in Geography: a Jubilee Collection*. Liverpool, 1967. 'Population changes in England and Wales in the later nineteenth century: an analysis of trends by registration districts', *TIBG*, 44, 1968. 'Population and society, 1730–1900', in R. A. Dodgshon and R. A. Butlin, eds., *An Historical Geography of England and Wales*. London, 1978.

Layton, W. T. *An Introduction to the Study of Prices*. London, 1920.

Lee, C. H. *British Regional Employment Statistics, 1841–1971*. Cambridge, 1979. 'Regional inequalities in infant mortality in Britain, 1861–71: patterns and hypotheses', *Population Studies*, 45, 1991.

Legard, G. 'On the farming of the East Riding of Yorkshire', *JRASE*, 9, 1848.

Levy, H. *Large and Small Holdings: a Study of English Agricultural Economics*. Cambridge, 1911.

Lindert, P. H. 'Lucrens Angliae: the distribution of English private wealth since 1670', Working Papers, 18 and 19, Agricultural History Center, University of California, Davis, 1985. 'Who owned Victorian England?: the debate over landed wealth and inequality', *Agricultural History*, 61, 1987.

Lindsay, J. *A History of the North Wales Slate Industry*. Newton Abbot, 1974.

Long, J. *The Book of the Pig*. London, 1886.

Longstaff, G. B. 'Rural depopulation', *JRSS*, 56, 1893.

Macdonald, J. *Book of the Farm*, vol. 1. 4th edn, Edinburgh, 1891.

Macgregor, J. J. 'Labour costs in English forestry since 1824', *Forestry*, 20, 1946.

MacMaster, Neil. 'The battle for Mousehold Heath, 1857–1884: "Popular Politics" and the Victorian public park', *PP*, 127, 1990.

Macrosty, H. W. 'The grain milling industry: a study in organisation', *Economic Journal*, 13, no. 51, 1903.

The Trust Movement in British Industry. London, 1907.

Madden, M. 'The National Union of Agricultural Workers, 1906–1956', BLitt thesis, Oxford University, 1956.

Malden, W. J. 'Hedges and hedge-making', *JRASE*, 3rd ser., 10, 1899.

Mandler, Peter. *The Fall and Rise of the Stately Home*. New Haven, 1997.

Margetson, S. *Leisure and Pleasure in the Nineteenth Century*. London, 1969.

Mark Lane Express, 14, 21 and 28 July, 20 and 27 October 1862, and 70th Birthday Number, 31 March 1902.

Marrison, A. J. 'The Tariff Commission, agricultural protection and food taxes, 1903–13', *AHR*, 34, 1986.

Marsh, J. *Back to the Land: the Pastoral Impulse in Victorian England from 1880 to 1914*. London, 1982.

Marshall, J. D. and Davies-Shiel, M. *The Industrial Archaeology of the Lake Counties*. Newton Abbot, 1969.

Martineau, H. *Guide to Windermere*. 2nd edn, Windermere, 1854.

Martins, Susannah Wade. *A Great Estate at Work: the Holkham Estate and its Inhabitants in the Nineteenth Century*. Cambridge, 1980.

Mathew, W. M. 'Peru and the British guano market, 1840–70', *EcHR*, 2nd ser., 23, 1970.

Mathias, P. *The Retailing Revolution*. London, 1967.

Matthews, A. H. H. *Fifty Years of Agricultural Politics: A History of the Central Chamber of Agriculture*. 1915.

McClatchey, D. *Oxfordshire Clergy, 1777–1869*. Oxford, 1960.

McConnell, P. 'Experiences of a Scotsman on the Essex clays', *JRASE*, 3rd ser., 2, 1891.

The Diary of a Working Farmer. London, 1906.

The Agricultural Notebook. 9th edn, London, 1919.

McCulloch, J. R. *A Dictionary, Practical, Theoretical, and Historical of Commerce and Commercial Navigation*. 1871 edn, London.

McGregor, O. R. 'Introduction: after 1815', in Lord Ernle, *English Farming Past and Present*. 6th edn, London, 1961.

McInnis, R. M. 'Output and productivity in Canadian agriculture, 1870–71 to 1926–27', in S. L. Engerman and R. E. Gallman, eds., *Long-Term Factors in American Economic Growth*. Chicago, 1986.

McKinnon, M. 'English poor law policy and the crusade against outrelief' *JEH*, 47, 1987.

McQuiston, J. R., 'Tenant right: farmer against landlord in Victorian England, 1847–1883', *Agricultural History*, 47, 1973.

Menzies, George 'Report on the transit of stock', *Trans. Highland and Agric. Soc. of Scotland*, 4th ser., 2, 1868–9.

Michie, R. C. 'Income, expenditure and investment of a Victorian millionaire: Lord Overstone, 1823–83', *BIHR*, 58, 1985.

'The international trade in food and the City of London since 1850', *Journ. of European Economic History*, 25, 1996.

Middleton, T. H. *Food Production in War*. Oxford, 1923.

Milburn, M. M. 'On the farming of the North Riding of Yorkshire', *JRASE*, 9, 1848.

Miller, C. 'The hidden workforce: female field workers in Gloucestershire, 1870–1901', *Southern History*, 6, 1984.

Miller, C., ed. *Rain and Ruin: the Diary of an Oxfordshire Farmer*. Gloucester, 1983.

The Account Books of Thomas Smith, Ireley Farm, Hailes, Gloucestershire, 1865–71. Gloucester, 1985.

Miller, M. *Letchworth: the First Garden City*. Chichester, 1989.

Miller, S. H. and Skertchley, S. B. J. *The Fenland*. London, 1878.

Mills, Dennis R. *Lord and Peasant in Nineteenth-Century Britain*. London, 1980.

'The nineteenth-century peasantry of Melbourn, Cambridgeshire', in R. M. Smith, ed., *Land, Kinship and Life Cycle*. Cambridge, 1984.

Milward, R. 'The emergence of wage labour in early modern England', *Explorations in Economic History*, 18, 1981.

Mingay, G. E. *Rural Life in Victorian England*. London, 1976.

ed. *The Victorian Countryside*, London, 1981.

British Friesians: an Epic of Progress. Rickmansworth, 1982.

ed. *AHEW*, VI, *1750–1850*. Cambridge, 1989.

A Social History of the English Countryside. London, 1990.

Mitchell, B. R. and Deane, P. *Abstract of British Historical Statistics*. Cambridge, 1962.

Moore-Colyer, R. J. *The Welsh Cattle Drovers*. Cardiff, 1976.

A Land of Pure Delight: Selections from the Correspondence of Thomas Johnes of Hafod, 1748–1816. Gwasg Gomer, 1992.

Morgan, D. H. *Harvesters and Harvesting, 1840–1900*. London, 1982.

Morgan, Kenneth O. *Wales. Rebirth of a Nation, 1880–1980*. Oxford, 1981.

Morgan, Raine. 'The root crop in English agriculture, 1650–1870', PhD thesis, University of Reading, 1978.

Dissertations on British Agrarian History. Reading, 1981.

Morris, W. *News from Nowhere*. London, 1891.

Morton, J. C. *The Farmer's Calendar*. London, 1884.

A Cyclopedia of Agriculture, Practical and Scientific. Glasgow, 1855.

'Agricultural progress: its helps and hindrances', *Journal of Society of Arts*, 12, 1863.

'Dairy farming', *JRASE*, 2nd ser., 14, 1878.

Moscrop, W. J. 'A report on the farming of Leicestershire', *JRASE*, 2nd ser., 2, 1866.

Moss, C. E. *Geographical Distribution of Vegetation in Somerset: Bath and Bridgwater District*. London, 1907.

Vegetation of the Peak District. Cambridge, 1913.

Moss, M. S. 'History of Fisons, Ltd'. Unpublished typescript by author, Archives and Business Record Centre, University of Glasgow.

Mounfield, P. R. 'The shoe industry in Staffordshire, 1767 to 1951', *N. Staffs. Journ. of Field Studies*, 5, 1965.

   'The footwear industry of the East Midlands, I. Present locational pattern and the problem of its origins; III, Northamptonshire, 1700–1911; IV. Leicestershire to 1911', *East Midland Geographer*, 3, 1964–5; 4 (1), 1966.

Munting, R. 'Ransomes in Russia', *EcHR*, 2nd ser., 31, 1978.

Murphy, G. *Founders of the National Trust.* London, 1987.

Murray, K. A. H. *Factors affecting the Prices of Livestock in Great Britain: a Preliminary Study.* Oxford, 1931.

Mutch, Alistair, 'Rural society in Lancashire, 1840–1914,' PhD thesis, University of Manchester, 1980.

   'Farmers' organisations and agricultural depression in Lancashire, 1890–1900', *AHR*, 31, 1983.

   Rural Life in South-West Lancashire, 1840–1914. Lancaster, 1988.

Napier, C. J. 'Aristocratic accounting: the Bute estate in Glamorgan, 1814–1890' *Accounting and Business Research*, 21, 1991.

Napolitan, L. 'Agricultural output and income: national output and income accounting in agriculture', *Farm Economist*, 7 (5), 1953.

   'The centenary of the agricultural census', *JRASE*, 127, 1966.

Neave, David. *Mutual Aid in the Victorian Countryside: Friendly Societies in the Rural East Riding, 1830–1914.* Hull, 1991.

Neeson, J. M. *Commoners: Common Right, Enclosure and Social Change in England, 1700–1820.* Cambridge, 1993.

Newell, E. '"Copperopolis": the rise and fall of the copper industry in the Swansea district, 1826–1931', *Business History*, 32, 1990.

Newmarch, W. 'On the progress of the foreign trade of the United Kingdom since 1856, with especial reference to the effects produced upon it by the protectionist tariffs of other countries', *JRSS*, 41, 1878.

Nicholson, N. *The Lakers. The Adventures of the First Tourists.* London, 1955.

Nicholson, T. R. *Wheels on the Road. Maps of Britain for the Cyclist and Motorist, 1870–1940.* Norwich, 1983.

Norman, E. R. *Church and Society in England, 1770–1970: A Historical Study.* Oxford, 1976.

Norton, Trist and Gilbert. 'A century of land values: England and Wales', *The Times*, 20 April 1889, reprinted in *JRSS*, 54, 1891.

Obelkevich, J. *Religion and Rural Society. South Lindsey, 1825–1875.* Oxford, 1976.

O'Brien, P. K. 'Agriculture and the home market for English industry, 1660–1820', *English Historical Review*, 100, no. 397, 1985.

O'Brien, P. K. and Guiomard, C. 'Agricultural output in the Irish Free State area before and after independence', *Irish Economic and Social History*, 12, 1985.

O'Dowd, A. *Spalpeens and Tattie Hokers: History and Folklore of the Irish Migratory Worker in Ireland and Britain.* Dublin, 1991.

Offer, A. *Property and Politics, 1870–1914.* Cambridge, 1981.

    The First World War: An Agrarian Interpretation. Oxford, 1989.

    'Farm tenure and land values in England, *c.* 1750–1950', *EcHR*, 2nd ser., 44, 1991.

Ogle, W. 'The alleged depopulation of the rural districts of England', *JRSS*, 52, 1889.

Ó Gráda, C. 'The landlord and agricultural transformation, 1870–1900: a comment on Richard Perren's hypothesis', *AHR*, 27, 1979.

    'Irish agricultural output before and after the famine', *Journal of European Economic History*, 13, 1984.

    Ireland before and after the Famine: Explorations in Economic History, 1800–1925. Manchester, 1988.

    'British agriculture, 1860–1914', in R. Floud and D. McCloskey, eds., *The Economic History of Britain since 1700*, vol. II, *1860–1939.* 2nd edn, Cambridge, 1994.

Ojala, E. M. *Agriculture and Economic Progress.* Oxford, 1952.

Olmsted, F. L. *Walks and Talks of an American Farmer in England.* 1859, repr. Ann Arbor, 1967.

Olney, R. J. *Lincolnshire Politics, 1832–1885.* Oxford, 1973.

    Rural Society and County Government in Nineteenth-Century Lincolnshire. Lincoln, 1979.

Olsen, D. J. *The Growth of Victorian London.* London, 1979.

Ordish, G. *The Constant Pest.* London, 1976.

Orwin, C. S. 'Land tenure in England', *Proc. of First International Conference of Agricultural Economists*, 1929.

    The Reclamation of Exmoor Forest. Oxford, 1929.

Orwin, C. S. and Felton, B. I., 'A century of wages and earnings in agriculture', *JRASE*, 92, 1931.

Orwin, C. S. and Whetham, E. H. *A History of British Agriculture, 1846–1914.* London, 1964.

Ottewill, D. *The Edwardian Garden.* New Haven, 1989.

Overton, M. *Agricultural Revolution in England.* Cambridge, 1996.

Owen, Trefor M. *Welsh Folk Customs.* Cardiff, 1959.

Pahl, Ray. *Urbs in Rure. The Metropolitan Fringe in Hertfordshire.* London, 1964.

Parker, J. Oxley, *The Oxley Parker Papers from the Letters and Diaries of an Essex Family of Land Agents in the Nineteenth Century.* Colchester, 1964.

Parton, A. G. 'Parliamentary enclosure in nineteenth-century Surrey – some perspectives on the evaluation of land potential', *AHR*, 33 1985.

Paterson, M. M. *Compensation Discharge in the Rivers and Streams of the West Riding.* London, 1896.

Patisson, Iain, *The British Veterinary Profession, 1791–1948.* London, 1983.

Pawson, H. C. *Cockle Park Farm.* London, 1960.

Payne-Gallwey, R. W. F. *The Book of Duck Decoys, their Construction, Management and History.* London, 1886.

Peacock, A. J. 'Land reform, 1880–1919', MA thesis, University of Southampton, 1963.

Pearson, J. W. 'The seed crushing industry', *Journ. of the Society of Arts*, 12, Dec. 1919.

Pell, A. 'The making of the land in England', *JRASE*, 2nd ser., 23, 1887.

Pelling, H. M. *Social Geography of British Elections, 1885–1914*. Aldershot, 1967.

Perkin, H. *The Age of the Railway*. London, 1970.

Perren, R. 'The effects of agricultural depression on the English estates of the Dukes of Sutherland, 1870–1900', PhD thesis, University of Nottingham, 1967.

'The landlord and agricultural transformation, 1870–1900', *AHR*, 18, 1970.

'The meat and livestock trade in Britain, 1850–1870', *EcHR*, 2nd ser., 28, 1975.

'The North American beef and cattle trade with Great Britain, 1870–1914', *EcHR*, 2nd ser., 24, 1971.

The Meat Trade in Britain, 1840–1914. London, 1978.

'The landlord and agricultural transformation, 1870–1900: a rejoinder', *AHR*, 27, 1979.

'Markets and marketing', in G. E. Mingay, ed., *AHEW*, VI, *1750–1850*. Cambridge, 1989.

'The manufacture and marketing of veterinary products from 1850 to 1914', *Veterinary History*, new ser., 6, no. 2, Winter 1989/90.

'Structural change and market growth in the food industry: flour milling in Britain, Europe and America, 1850–1914', *EcHR*, 2nd ser., 43, 1990.

Agriculture in Depression, 1870–1940. Cambridge, 1995.

Perry, P. J. 'Where was the "Great Agricultural Depression"?', *AHR*, 20, 1972, reprinted in P. J. Perry, ed., *British Agriculture, 1875–1914*. London, 1973.

British Farming in the Great Depression, 1870–1914: An Historical Geography. Newton Abbot, 1974.

'High farming in Victorian Britain: the financial foundations', *Agricultural History*, 52, 1978.

Peters, G. H., Parsons, S. T. and Patchett, D. M., 'A century of land values, 1781–1880', *Oxford Agrarian Studies*, 11, 1982.

Philips, D. 'Good men to associate and bad men to conspire. Associations for prosecution of felons in England, 1760–1860', in D. Hay and F. Snyder, eds., *Policing and Prosecution in Britain, 1750–1850*. Oxford, 1989.

Phillips, A. D. M. 'Underdraining and agricultural investment in the Midlands in the mid-nineteenth century', in A. D. M. Phillips and B. J. Turton, eds. *Environment, Man and Economic Changes*. London, 1975.

'Mossland reclamation in nineteenth-century Cheshire', *Trans. Hist. Soc. Lancs. and Cheshire*, 129, 1979–80.

The Underdraining of Farm Land in England during the Nineteenth Century. Cambridge, 1989.

'Landlord investment in farm buildings in the English Midlands in the mid-nineteenth century', in B. A. Holderness and M. E. Turner, eds., *Land, Labour, and Agriculture, 1700–1920: Essays for Gordon Mingay*. London, 1991.

Phythian-Adams, Charles. 'Local history and national history: the quest for the peoples of England', *Rural History, Economy, Society, Culture*, 2(1), 1991.

Pidgeon, Dan. 'The development of agricultural machinery', *JRASE*, 3rd ser., 1, 1890.

Plowman, Thomas F. 'Agricultural societies and their uses', *J. Bath and West*, 3rd ser., 17, 1885–6.

Porter, J. H. 'Tenant right: Devonshire and the 1880 Ground Game Act', *AHR*, 34, 1986.

Pratt, E. A. *The Transition in Agriculture*. London, 1906.

Pretty, David A. *The Rural Revolt that Failed. Farm Workers' Trade Unions in Wales, 1889–1950*. Cardiff, 1989.

Price, L. L. 'The census return of 1891 and rural depopulation', *JRASE*, 3rd ser., 5, 1894.

Prize Reports on County Agriculture, listed in Lord Ernle, *English Farming Past and Present*, London, new sixth edn, London, 1961; Introduction, Part 2, pp. cii–ciii.

Prochaska, F. K. *Women and Philanthropy in Nineteenth-Century England*. Oxford, 1980.

The Voluntary Impulse. Philanthropy in Modern Britain. London, 1988.

Prothero, R. E. 'English land, law and labour', *Edinburgh Review*, 165, 1887.

'English agriculture in the reign of Queen Victoria', *JRASE*, 62, 1901.

English Farming Past and Present, 1912, new sixth edn, 1961.

Pugh, R. B. *The Crown Estate*. London, 1960.

Punchard, F. 'Farming in Devon and Cornwall', *JRASE*, 3rd ser., 1, 1890.

Purdy, F. 'On the earnings of agricultural labourers in England and Wales, 1860', *JRSS*, 24, 1861.

Pusey, P. 'Some introductory remarks on the present state of agriculture as a science in England', *JRASE*, 1, 1840.

Quinault, R. see under Harte, N.

Radzinowicz, L. and Hood, R. *A History of English Criminal Law and its Administration from 1750*. Oxford, 1986.

Rae, J. 'Why have the yeomanry perished?', *Contemporary Review*, 44, 1883.

Raper, A. C. *Weyhill Fair '. . . the greatest fair in the kingdom'*. Buckingham, 1988.

Ravenstein, E. G. 'The laws of migration', *Journal Statistical Society of London*, 48, 1885.

'The laws of migration. Second paper', *JRSS*, 42, 1889.

Raybould, T. J. *The Economic Emergence of the Black Country*. Newton Abbot, 1973.

Raynbird, H. 'On the farming of Suffolk', *JRASE*, 8, 1847.

Read, C. S. 'On the farming of South Wales', *JRASE*, 10, 1849.

Reay, Barry. *The Last Rising of the Agricultural Labourers. Rural Life and Protest in Nineteenth-Century England*. Oxford, 1990.

Redford, A. *Labour Migration in England, 1800–1850*. 1926, 3rd edn, London, 1976.

Reed, Michael. 'The peasantry of nineteenth-century England: a neglected class?', *History Workshop Journal*, 18, 1984.

'"Gnawing it out". A new look at economic relations in nineteenth-century rural England', *Rural History*, 1, 1990.

Reid, D. B. *Elements of Chemistry*. 3rd edn, London, 1838.

Rew, R. H. 'An inquiry into the statistics of the production and consumption of milk and milk products in Great Britain', *JRSS*, 55, 1892.

'English markets and fairs', *JRASE*, 3rd ser., 3, 1892.

'Production and consumption of milk', *JRASE*, 3rd ser., 3, 1892.

'Farm revenue and capital', *JRASE*, 3rd ser., 6, 1895.

'Report of the Committee to enquire into the statistics available as a basis for estimating the production and consumption of meat and milk in the UK', *JRSS*, 65, 1902.

'The food production of British farms', *JRASE*, 3rd ser., 14, 1903.

'The production and consumption of meat and milk in the UK: second report', *JRSS*, 67, 1904.

'Production and consumption of meat and milk: third report', *JRSS*, 67, 1904.

'Observations on the production and consumption of meat and dairy products', *JRSS*, 67, 1904.

'The nation's food supply', *JRSS*, 76, 1912.

An Agricultural Faggot. A Collection of Papers on Agricultural Subjects. Westminster, 1913.

Rhee, H. A. *The Rent of Agricultural Land in England and Wales*. Central Landowners' Association, London, 1946.

Richards, E. 'An anatomy of the Sutherland fortune: income, consumption, investments and returns, 1780–1880', *Business History*, 21, 1979.

'The land agent', in G. E. Mingay, ed., *The Victorian Countryside*, vol. II. London, 1981.

Richards, S. 'Agricultural science in British higher education, 1790–1914', MSc thesis, University of Kent, 1982.

'"Masters of Arts and Batchelors of Barley": the struggle for agricultural education in mid-nineteenth-century Britain', *History of Education*, 12, no. 3, 1983.

Rimmer, W. G. 'Leeds leather industry in the nineteenth century', *Thoresby Miscellany*, Thoresby Soc. Publications, XLVI, 1963.

Ritvo, H. *The Animal Estate*. Cambridge, Mass., 1987.

Robin, J. 'Prenuptial pregnancy in a rural area of Devonshire in the mid-nineteenth century: Colyton, 1851–1881', *Continuity and Change*, 1, 1986.

Robinson, G. M. *West Midlands Farming, 1840s to 1970s*. Cambridge, 1983.

Agricultural Change: Geographical Studies of British Agriculture. Edinburgh, 1988.

Rogers, G. 'Social and economic changes on the Lancashire landed estates during the nineteenth century with special reference to the Clifton estate', PhD thesis, University of Lancaster, 1981.

'Lancashire landowners and the great agricultural depression', *Northern History*, 22, 1986.

'The nineteenth-century landowner as urban developer: the Clifton estate and the development of Lytham St Annes', *Trans. Lancs. and Cheshire Historic Society*, 145, 1995.

Rose, Walter. *The Village Carpenter*. Cambridge, 1937.

Rossiter, M. W. *The Emergence of Agricultural Science: Justus Liebig and the Americans, 1840–80*. New Haven and London, 1975.

Rostow, W. W. *How it all Began: Origins of the Modern Economy*. London, 1975.

Rothenberg, W. B. 'A price index for rural Massachusetts, 1750–1855', *JEH*, 39, 1979.

Rothstein, M. 'American wheat and the British market, 1860–1905', PhD thesis, University of Cornell, 1960.

Rowell, T. A. 'Sedge in Cambridgeshire: its use and production since the seventeenth century', *AHR*, 34, 1986.

Rowlandson, T. 'On the agriculture of North Wales', *JRASE*, 7, 1946.

Rowntree, B. S. and Kendall, M. *How the Labourer Lives. A Study of the Rural Labour Problem*. London, 1913.

Roxby, P. 'Rural depopulation in England during the nineteenth century', *Nineteenth Century and After*, 71, 1912.

Rubinstein, W. D. 'British millionaires, 1809–1949', *BIHR*, 48, 1974.

'New men of wealth and the purchase of land in nineteenth-century Britain,' *PP*, 92, 1981.

'Men of Property'. The Very Wealthy in Britain since the Industrial Revolution. New Brunswick, NJ, 1981.

'Cutting up rich: a reply to F. M. L. Thompson', *EcHR*, 2nd ser., 45, 1992.

'Businessmen into landowners: the question revisited', in N. Harte and R. Quinault, eds., *Land and Society in Britain, 1700–1914*. Manchester, 1996.

Rural Industries Intelligence Bureau, 'The country wheelwright and his outlook', *Journ. of Board of Agriculture*, 31, August 1924.

Ruskin, J. *Fors Clavigera*. London, 1871–7.

Russell, E. J. *A History of Agricultural Science in Great Britain, 1620–1954*. London, 1966.

Russell, R. *The 'Revolt of the Field' in Lincolnshire*. Lincoln, 1956.

Ryan, R. J. 'A history of the Norwich Union Fire and Life Insurance Societies from 1797 to 1914', PhD thesis, University of East Anglia, 1983.

Ryle, G. *Forest Service: the First 45 Years of the Forestry Commission of Great Britain*. Newton Abbot, 1969.

Salaman, R. N. *The History and Social Influence of the Potato*. Cambridge, rev. edn, 1985.

Samuel, R. 'Comers and goers', in H. J. Dyos and M. Wolff, eds., *The Victorian City*. 2 vols., London, 1973.

ed. *Village Life and Labour*. London, 1975.

'"Quarry roughs": life and labour in Headington Quarry, 1860–1920. An essay in oral history', in Raphael Samuel, ed., *Village Life and Labour*. London, 1975.

'Mineral workers' in R. Samuel, ed., *Miners, Quarrymen and Saltworkers*. London, 1977.

Saul, S. B. 'The market and the development of mechanical engineering industries in Britain, 1860–1914', *EcHR*, 2nd ser., 20, 1967.

The Myth of the Great Depression, 1873–1896. London, 1969.

Saville, J. *Rural Depopulation in England and Wales, 1851–1951*. London, 1957.

'Internal migration in England and Wales during the past hundred years', in J. Sutter, ed., *Human Displacements: Measurement, Methodology, Aspects*. Monaco, 1962.

Schlich, W. *Forestry in the United Kingdom*. London, 1904.

Scola, Roger, ed. W. A. Armstrong and Pauline Scola, *Feeding the Victorian City. The Food Supply of Manchester, 1770–1870*. Manchester, 1992.

Searby, Peter. 'Great Dodford and the later history of the Chartist land scheme', *AHR*, 16, 1968.

Searle, C. E. '"The odd corner of England": a study of a rural social formation in transition, Cumbria, *c.* 1700–1914', PhD thesis, University of Essex, 1983.

Secord, J. A. *Controversy in Victorian Geology. The Cambrian–Silurian Dispute*. Princeton, 1986.

Self, P. and Storing, H. J. *The State and the Farmer*. London, 1962.

Seymour, W. A. *A History of the Ordnance Survey*. London, 1980.

Shannon, F. A. *The Farmer's Last Frontier: Agriculture, 1860–97*. New York, 1945.

Sharples, M. 'The Fawkes–Turner connection and the art collection at Farnley Hall, Otley, 1792–1937: a great estate enhanced and supported', *Northern History*, 26, 1990.

Sheail, J. *Rabbits and their History*. Newton Abbot, 1971.

Nature in Trust: the History of Nature Conservation in Britain. Glasgow, 1976.

'Underground water abstraction: indirect effects of urbanization on the countryside', *JHG*, 8, 1982.

'Government and the perception of reservoir development in Britain', *Planning Perspectives*, 1, 1986.

Seventy-five Years in Ecology. The British Ecological Society. Oxford, 1987.

Short, B. *The Geography of England and Wales in 1910: an Evaluation of Lloyd George's 'Domesday' of Landownership*. Historical Geography Research Series, 22, 1989.

ed. *The English Rural Community: Image and Analysis*. Cambridge, 1992.

Simmons, J. *The Railway in Town and Country, 1830–1914*. Newton Abbot, 1986.

The Victorian Railways. London, 1991.

Sinclair, Upton, *The Jungle*. London, 1906.

Smith, D. *Victorian Maps of the British Isles*. London, 1985.

Snell, K. D. *Annals of the Labouring Poor. Social Change and Agrarian England, 1600–1900*. Cambridge, 1985.

Somerville, A. *The Whistler at the Plough* (1852), ed. with intro. by K. D. M. Snell. London, 1989.

Spencer, S. *Pigs*. London, 1897.

Spring, D. *The English Landed Estate in the Nineteenth Century: its Administration*. Baltimore, 1963.

'Land and politics in Edwardian England', *Agricultural History*, 58, 1984.

'Willoughby de Broke and Walter Long: English landed society and political extremism, 1912–24', in N. Harte and R. Quinault, eds., *Land and Society in Britain, 1700–1914*. Manchester, 1996.

Spring, David and Eileen. 'Debt and the English aristocracy', *Canadian Journal of History*, 31, 1996.

Spring, E. 'Landowners, lawyers, and land law reform in nineteenth-century England', *American Journal of Legal History*, 21, 1977.

Springall, L. M. *Labouring Life in Norfolk Villages, 1834–1914*. London, 1936.

Squarey, E. P. 'Farm capital', *JRASE*, 2nd ser., 14, 1878.

Stamp, J. C. *British Incomes and Property*. London, 1916.

Stamp, L. D. *The Land of Britain: its Use and Misuse*. London, 1948.

Steedman, C. *Policing the Victorian Community. The Formation of the English Provincial Police Forces, 1856–1880*. London, 1984.

Stephens, H. *Book of the Farm*. 2nd edn, Edinburgh and London, 1851.

Stern, R. M. 'A century of food exports', *Kyklos*, 13, 1960.

Stewart, R. *The Politics of Protection*. Cambridge, 1971.

Stone, L. and J. F. C. *An Open Elite? England, 1540–1880*. Oxford, 1983.

Stopes, H. *Malt and Malting*. London, 1885.

Storch, R. D. *Popular Culture and Custom in Nineteenth-Century England*. London, 1982.

'Policing rural southern England before the police: opinion and practice, 1830–1856', in D. Hay and F. Snyder, eds., *Policing and Prosecution in Britain, 1750–1850*. Oxford, 1989.

Storck, J. and Teague, W. D. *Flour for Man's Bread: a History of Milling*. Minneapolis, 1952.

Stovin, Jean, ed. *Journals of a Methodist Farmer*. London, 1982.

Street, A. G. *Farmer's Glory*. London, 1932, new edn, 1959.

Sturgess, R. W. 'The agricultural revolution on the English clays', *AHR*, 14, 1966.

'The agricultural revolution on the English clays: a rejoinder', *AHR*, 15, 1967.

Sturmey, S. G. 'Owner-farming in England and Wales, 1900–50', *Manchester School*, 23, 1955.

Sturt, George. *The Wheelwright's Shop, 1923*. Cambridge, 1993.

The Bettesworth Book, 1901. Facsimile of 2nd edn of 1902, Firle, 1978.

Sumner, H. *The Ancient Earthworks of Cranborne Chase*. London, 1913.

The Ancient Earthworks of the New Forest. London, 1917.

Supple, B. *The Royal Exchange Assurance: a History of British Insurance, 1720–1970*. Cambridge, 1970.

Surtees, R. S. *Handley Cross: or Mr Jorrocks's Hunt*. London, 1854.

Sutherland, D. *The Landowners*. 2nd edn, London, 1988.

Swan, E. E., ed. *The Diary of a Farm Apprentice: William Carter Swan, 1909–10*. Gloucester, 1984.

Sykes, A. *Tariff Reform in British Politics, 1903–13*. Oxford, 1979.

Sykes, J. D. 'Agriculture and science', in G. E. Mingay, ed., *The Victorian Countryside*, vol. I. London, 1981.

Tann, J. 'Corn milling', in G. E. Mingay, ed., *AHEW*, VI, *1750–1850*. Cambridge, 1989.

Tann, J. and Jones, R. G. 'Technology and transformation: the diffusion of the roller mill in the British flour milling industry, 1870–1907', *Technology and Culture*, 37, 1996.

Tanner, H. 'The farming of Devonshire', *JRASE*, 9, 1848.

Tansley, A. G., ed. *Types of British Vegetation*. Cambridge, 1911.

Tariff Commission, vol. IV, 1909, *Engineering Industries*.

Taylor, A. M. *Gilletts: Bankers at Banbury and Oxford*. Oxford, 1964.

Taylor, Christopher. *Village and Farmstead. A History of Rural Settlement in England.* London, 1983.

Taylor, David. 'The English dairy industry, 1860–1930', *EcHR*, 2nd ser., 29, 1976.

'Growth and structural change in the English dairy industry, *c.* 1860–1930', *AHR*, 35, 1987.

'London's milk supply, 1850–1900: a reinterpretation', *Agricultural History*, 45, 1971.

'London's milk trade, 1850–1900', *AHR*, 45, 1971.

Taylor, F. D. W. 'Food and agriculture: notes and statistics', *Farm Economist*, 8 (4), 1955.

Taylor, J. C. 'Townmilk', *JRASE*, 2nd ser., 4, 1868.

Teitelbaum, M. S. *The British Fertility Decline: Demographic Transition in the Crucible of the Industrial Revolution.* Princeton, NJ, 1984.

The Times, 20 December 1892.

The Times Food Number, 1915.

Thirsk, Joan. *Agricultural Regions and Agrarian History in England, 1500–1750.* London, 1987.

ed. *AHEW*, v, 1, 1640–1750. Cambridge, 1984.

English Peasant Farming. The Agrarian History of Lincolnshire from Tudor to Recent Times. London, 1957, repr. 1981.

Alternative Agriculture: A History From the Black Death to the Present Day. Oxford, 1997.

Thirsk, Joan and Imray, Jean. *Suffolk Farming in the Nineteenth Century.* Suffolk Records Society, 1, 1958.

Thomas, E. 'The June Returns one hundred years old', *Agriculture*, 73, 1966.

Thompson, D. M. *Nonconformity in the Nineteenth Century.* London, 1972.

Thompson, E. P. 'The moral economy of the English crowd in the eighteenth century', *PP*, 50, 1971.

Thompson, Flora. *Lark Rise to Candleford.* Oxford, 1939, repr. 1945, Harmondsworth, 1973.

Thompson, F. M. L. 'The land market in the nineteenth century', *Oxford Economic Papers*, 1957, reprinted in W. E. Minchinton, ed., *Essays in Agrarian History*, vol. II. Newton Abbot, 1968.

'Agriculture since 1870', *VCH Wilts.*, vol. IV. London, 1959.

English Landed Society in the Nineteenth Century. London, 1963.

'The second agricultural revolution, 1815–1880', *EcHR*, 2nd ser., 21, 1968.

Chartered Surveyors. The Growth of a Profession. London, 1968.

'Nineteenth-century horse sense', *EcHR*, 2nd ser., 29, 1976.

'Britain', in D. Spring, ed., *European Landed Elites in the Nineteenth Century.* Baltimore, 1977.

'Free trade in land', in G. E. Mingay, ed., *The Victorian Countryside*, vol. I. London, 1981.

'Landowners and the rural community', in G. E. Mingay, ed., *The Victorian Countryside*, vol. II. London, 1981.

'Social control in Victorian Britain', *EcHR*, 2nd ser., 33, 1981.

ed. *The Rise of Suburbia*. Leicester, 1982.

'Horses and hay in Britain, 1830–1918', in F. M. L. Thompson, ed., *Horses in European Economic History*. Reading, 1983.

'Life after death; how successful nineteenth-century businessmen disposed of their fortunes', *EcHR*, 2nd ser., 43, 1990.

ed. *The Cambridge Social History of Britain, 1750–1950*, vol. 1. *Regions and Communities*. Cambridge, 1990.

'An anatomy of English agriculture, 1870–1914', in B. A. Holderness and M. Turner, eds., *Land, Labour and Agriculture, 1700–1920. Essays for Gordon Mingay*. London, 1991.

'Desirable properties: the town and country connection in British society since the later eighteenth century', *Historical Research*, 64, 1991.

'English landed society in the twentieth century, I, Property: collapse and survival'; II, 'New poor and new rich'; III, 'Self help and outdoor relief'; IV, 'Prestige without power?', *TRHS*, 5th ser., 40, 1990, 6th ser., 1, 1991; 2, 1992; 3, 1993.

'Stitching it together again', *EcHR*, 2nd ser., 45, 1992.

'Business and landed elites in the nineteenth century' in F. M. L. Thompson, ed., *Landowners, Capitalists, and Entrepreneurs*. Oxford, 1994.

ed. *Landowners, Capitalists, and Entrepreneurs. Essays for Sir John Habakkuk*. Oxford, 1994.

'The development of the agricultural sub-structure', in M. A. Havinden and E. J. T. Collins, eds., *Agriculture in the Industrial State*. Reading, 1995.

'Agriculture and economic growth in Britain', in P. Mathias and J. Davies, eds., *Agriculture and Industrialization: from the Eighteenth Century to the Present*. Oxford, 1996.

Thompson, H. S. 'On the management of grassland with especial reference to the production of meat', *JRASE*, 2nd ser., 7, 1872.

Thompson, M. W. 'The first inspector of ancient monuments', *Journ. British Archaeolog. Soc.*, 3rd ser., 23, 1960.

Thompson, R. J. 'An inquiry into the rent of agricultural land in England and Wales during the nineteenth century', *JRSS*, 70, 1907, reprinted in W. E. Minchinton, ed., *Essays in Agrarian History*, II, Newton Abbot, 1968.

Thompson, Ruth D'Arcy, *The Remarkable Gamgees: a Story of Achievement*. Edinburgh, 1974.

Thompson, R. N. 'The working of the Poor Law Amendment Act in Cumbria, 1836–1871', *Northern History*, 15, 1979.

Thomson, D. 'The decline of social welfare: falling state support for the elderly since early Victorian times', *Ageing and Society*, 4, 1984.

Tillott, P. M. 'Sources of inaccuracy in the 1851 and 1861 censuses', in E. A. Wrigley, ed., *Nineteenth-Century Society. Essays in the Use of Quantitative Methods for the Study of Social Data*. London, 1972.

Tostlebe, A. S. *Capital in Agriculture, its Formation and Financing since 1870*. New Haven, 1957.

Tracy, M. *Agriculture in Western Europe*. 1964, 2nd edn, London, 1982.

Trainor, R. 'The gentrification of Victorian and Edwardian industrialists', in

A. L. Beier, D. Cannadine and J. M. Rosenheim, eds., *The First Modern Society*. Cambridge, 1989.

Trebilcock, C. *Phoenix Assurance and the Development of British Insurance*. Cambridge, 1985.

Trow-Smith, R. *A History of British Livestock Husbandry: II, 1700–1900*. London, 1959.

Tubbs, C. R. *The New Forest. An Ecological History*. Newton Abbot, 1969.

Turnbull, R. C. 'The household food supply of the United Kingdom', *Trans. Highland and Agricultural Society*, 15, 1903.

Turnbull, R. F. 'Capital and revenue of agriculture', *Journal of the Farmers' Club*, 1994.

Turner, J. H. 'On the necessity of the reduction or abolition of hedges', *JRASE*, 6, 1845.

Turner, J. W. 'The position of the wool trade', *JRASE*, 3rd ser., 7, 1896.

Turner, M. E. 'Towards an agricultural prices index for Ireland, 1850–1914', *The Economic and Social Review*, 18, 1987.

'Output and prices in UK agriculture, 1867–1914, and the Great Agricultural Depression reconsidered', *AHR*, 40, 1992.

see also Holderness, B. A. and Turner, M.

Turner, M. E., Beckett, J. V., and Afton, B. *Agricultural Rent in England, 1690–1914*. Cambridge, 1997.

Turner, P. and Wood, R. *P. H. Emerson. Photographer of Norfolk*. London, 1974.

Tyler, C. 'The history of the Agricultural Education Association, 1894–1914', *Agricultural Progress*, 48, 1973.

Unwin, R. *Cottage Plans and Common Sense*. Fabian Society Tract, 109. London, 1902.

Nothing to be Gained by Overcrowding. London, 1912.

Town Planning in Practice; an Introduction to the Art of Designing Cities and Suburbs. London, 1909.

Vamplew, R. 'A grain of truth: the nineteenth-century corn averages', *AHR*, 28, 1980.

Venn, J. A. *The Foundations of Agricultural Economics*. Cambridge, 1933.

Vince, S. W. E. 'The rural population of England and Wales, 1801–1951', PhD thesis, University of London, 1955.

Voller, W. R. *Modern Flour Milling*. 2nd edn, Gloucester, 1892.

Wall, W. Barrow. 'The agriculture of Pembrokeshire', *JRASE*, 2nd ser., 23, 1887.

Wallace, A. R. *Land Nationalisation*. London, 1882.

Walton, J. K. *Lancashire: a Social History, 1558–1939*. Manchester, 1987.

The English Seaside Resort: a Social History, 1750–1914. Leicester, 1983.

'Pedigree cattle and the national herd, *c.* 1750–1950', *AHR*, 34, 1986.

Ward, J. T. 'A study of capital and rental values of agricultural land in England and Wales between 1858 and 1958', PhD thesis, University of London, 1960.

'Farm sale prices over a hundred years', *Estates Gazette*, Centenary Supplement, 3 May 1958.

'West Riding landowners and mining in the nineteenth century', *Yorks. Bulletin of Economic and Social Research*, 15, 1963.

Watson, J. A. Scott. *The History of the Royal Agricultural Society of England, 1839–1939*. London, 1939.

Watson, J. A. Scott and Hobbs, M. E. *Great Farmers*. London, 1937.

Weaver, B. M. Q. 'The history of veterinary anaesthesia', *Veterinary History*, 5, 1988.

Weaver, C. and Ransomes, M. *Ransomes 1789–1989: 200 Years of Excellence*. Ipswich, 1989.

Webb, H. J. *Advanced Agriculture*. London, 1894.

Webber, R. *Covent Garden: Mud Salad Market*. London, 1969.

Webster, C. 'On the farming of Westmorland', *JRASE*, 2nd ser., 4, 1868.

Wells, Roger A. E. 'Social conflict and protest in the English countryside in the early nineteenth century: a rejoinder', *Journal of Peasant Studies*, 8 (4), 1981.

Wells, W. 'On the treatment of the reclaimed bog-land of Whittlesea Mere', *JRASE*, 2nd ser., 6, 1870.

'The drainage of Whittlesea Mere', *JRASE*, 21, 1860.

Welton, T. A. *England's Recent Progress: an Investigation of the Statistics of Migrations, Mortality, etc. in the Twenty Years from 1881 to 1901 as indicating Tendencies towards the Growth or Decay of Particular Communities*. London, 1911.

'On the distribution of population in England and Wales, and its progress in the period of ninety years from 1801 to 1891', *JRSS*, 63, 1900.

Whetham, E. H. *AHEW*, VIII, *1914–1939*. Cambridge, 1978.

'Sectoral advance in English agriculture, 1850–80: a summary', *AHR*, 16, 1968.

'The London milk trade, 1860–1900', *EcHR*, 2nd ser., 33, 1980.

White, J. W. *The Flora of Bristol*. Bristol, 1912.

Whitehead, C. 'The cultivation of hops, fruit and vegetables', *JRASE*, 2nd ser., 14, 1878.

Whitehead, R. A. *Garretts of Leiston*. London, 1964.

Wholmsley, D. 'A landed estate and the railway: Huddersfield, 1844–54', *Journal of Transport History*, 2, 1974.

Wiener, Martin J. *English Culture and the Decline of the Industrial Spirit, 1850–1980*. Cambridge, 1981.

Wild, M. T. 'The Yorkshire wool textile industry', in J. Geraint Jenkins, ed., *The Wool Textile Industry in Great Britain*. London, 1972.

Wilkinson, J. 'The farming of Hampshire', *JRASE*, 22, 1861.

Williams, K. *From Pauperism to Poverty*. London, 1981.

Williams, M. *The Draining of the Somerset Levels*. Cambridge, 1970.

'The enclosure and reclamation of waste land in England and Wales in the eighteenth and nineteenth centuries', *TIBG*, 51, 1970.

Williams, Raymond. *The Country and the City*. London, repr. 1973.

Williamson, Tom and Bellamy, Liz. *Property and Landscape. A Social History of Landownership and the English Countryside*. London, 1987.

Wilmot, S. A. H. 'Agriculture in south-west England in the nineteenth century', in R. J. P. Kain and W. L. D. Ravenhill, eds., *An Historical Atlas of South-West England*, Exeter, 1999.

'Landownership, farm structure, and agrarian change in south-west England, 1800–1900: regional experience and national ideals', PhD thesis, University of Exeter, 1988.

'The Business of Improvement'; Agriculture and Scientific Culture in Britain, c. 1700–c. 1870, Historical Geography Research Series, 24, November, 1990.

Wilson, A. *Forgotten Harvest*. Calne, 1995.

Wilson, D. *Francis Frith's Travels. A Photographic Journey through Victorian Britain*. London, 1985.

Wilson, G. B. *Alcohol and the Nation*. London, 1940.

Wilson, R. 'The British brewing industry since 1750' in L. Richmond and A. Turton, eds., *The Brewing Industry: a Guide to Historical Records*. New York, 1990.

Wingfield-Stratford, E. *The Squire and his Relations*. London, 1956.

Winstanley, M. 'Industrialization and the small farm: family and household economy in nineteenth-century Lancashire', *PP*, 152, 1996.

Winter, G. *A Country Camera, 1844–1914*. Newton Abbot, 1966.

Wiseman, J. *A History of the British Pig*. London, 1986.

Wood, B. A. 'The development of the Nottinghamshire County Council small-holdings estate, 1907–80', *East Midland Geographer*, 8, 1982.

Wood, G. H. 'Rates of wages and hours of labour in various industries in the UK for a series of years, 1907'. Unpublished MS, Library of the Royal Statistical Society, University College, London.

Wood, T. B. *The National Food Supply in Peace and War*. Cambridge, 1917.

Woods, K. S. *The Rural Industries round Oxford: a Survey*. Oxford, 1921.

Rural Crafts of England. A Study of Skilled Workmanship. East Ardsley, Wakefield, 1975.

Woods, R. I. and Smith, C. W. 'The decline of marital fertility in the late nineteenth century: the case of England and Wales', *Population Studies*, 37, 1983.

Woodward, H. B. *The Geology of Soils and Substrata, with Special Reference to Agriculture, Estates and Sanitation*. London, 1912.

Wratislaw, Charles, 'The amount of capital required for the profitable occupation of a mixed arable and pasture farm in a Midland county', *JRASE*, 22, 1861.

Wright, R. P. 'Rotation of farm crops', in R. P. Wright, ed., *The Standard Cyclopaedia of Modern Agriculture and Rural Economy*, x. London, n.d. (c. 1905–14).

Wrigley, E. A. and Schofield, R. *The Population History of England, 1541–1871: A Reconstruction*. 2nd edn, Cambridge, 1989.

[Wynter, Andrew], 'The London commissariat', *Quarterly Review*, 190, 1854.

Yeo, S. and E., eds. *Popular Culture and Class Conflict. Explorations in the History of Labour and Leisure*. Brighton, 1981.

van Zanden, J. L. 'The first "green revolution": the growth of production and productivity in European agriculture, 1870–1914', *EcHR*, 2nd ser., 44, 1991.

PARLIAMENTARY PAPERS AND OTHER OFFICIAL GOVERNMENT
INQUIRIES

Reports spanning a period of years, 1850–1914

BPP, 1854–1914. Annual Trade and Navigation Returns.
BPP, Censuses of England and Wales, General Reports for 1863, 1871, 1881,
    1891, 1901, and 1911 in BPP, 1863, LIII; 1873, LXXI; 1883, LXXX; 1893–4,
    CVI; 1901, XC; 1902, CXVIII, CXIX, CXX, CXXI; 1903, LXXXV, LXXXVI; 1912,
    CXI; 1913, LXXVIII, LXXIX; 1917–18, XXV.
BPP, Agricultural Statistics of Ireland for the Years, 1891, 1901, 1911, 1916 in
    BPP, 1892, LXXXVIII; 1902, CXVI; 1912–13, CVI; 1919, LI.

Annual list in chronological order

BPP, 1839, XIX. First Report of the Commissioners appointed to inquire as to
    the Best Means of Establishing an Efficient Constabulary Force in the
    Counties of England and Wales.
BPP, 1843, XXIV, 24, Census of Ireland for the Year 1841.
BPP, 1843, XII, i. Reports of the Special Assistant Poor Law Commissioners on
    the Employment of Women and Children in Agriculture.
BPP, 1844, XVI. Report of the Commissioners of Inquiry for South Wales.
BPP, 1846, VI, part I. Select Committee of the House of Lords, Report on the
    Burdens of Real Property.
BPP, 1846, (Parts I & II), IX, Select Committee Report on the Game Laws.
BPP, 1852–3, LXXXVIII, Census of Great Britain, 1851. Population Tables.
BPP, 1854–5, VIII, Report from the Select Committee of the House of Lords
    appointed to inquire into the best Mode of obtaining Agricultural Statistics
    from all Parts of the United Kingdom; with the Minutes of Evidence,
    Appendix and Index.
BPP, 1854–5, LIII. Reports by Poor Law Inspectors on Agricultural Statistics
    (England), 1854.
BPP, 1860, XXXIX. Gross Annual Value of Property assessed to the Property and
    Income Tax in England and Wales, for the Year ended 5 April, 1860.
BPP, 1862, LX. Circular sent by Secretary of State to Chairmen of Quarter
    Sessions in England and Wales with Reference to Agricultural Statistics.
BPP, 1867, XVI, Royal Commission on the Employment of Children in Trades
    and Manufactures not regulated by Law, Sixth Report, Appendix
    (Agriculture).
BPP, 1867–8, XVII, Royal Commission on the Employment of Children, Young
    Persons, and Women in Agriculture, First Report, Appendix (Evidence of
    Assistant Commissioners).
BPP, 1868–9, XIII, Royal Commission on the Employment of Children, Young
    Persons and Women in Agriculture, 1867, Second Report, 1869.
BPP, 1870, XIII, Royal Commission on the Employment of Children, Young
    Persons, and Women in Agriculture, Third Report.

BPP, 1870, LXI, Report from the Committee . . . on the Transit of Animals by Sea and Land.

BPP, 1871, LXXI, part II, Census, General Report.

BPP, 1873, XI, Select Committee on Contagious Diseases in Livestock.

BPP, 1874, LXXII, Return of Owners of Land, England and Wales, 1873.

BPP, 1877, IX, Select Committee on Cattle Plague and the Importation of Livestock.

BPP, 1880, XVIII; 1881, XV, XVI, XVII; 1882, XIV, XV, Royal Commission on the Depressed Condition of the Agricultural Interests, 1880–82. *Note esp.* Prelim. Report, 1881, XV; Final Report, 1882, XIV and Reports of the Assistant Commissioners, 1880, XVIII; 1881, XVI, XV; Minutes of Evidence, 1881, XV, XVI; 1882, XIV; Digest, 1881, XVI; 1882, XIV; Appendices, 1881, XVI; 1882, XIV.

BPP, 1881, LXXXIII. Returns showing the Total Quantities of the Various Kinds of Grain and Flour imported into the United Kingdom in 1879 and 1880; Return showing the Total Quantity of Wheat, Barley, and Oats . . . in England and Wales in the Year 1880.

BPP, 1884, XXIX, Royal Commission on Technical Instruction, 2nd Report, vol. I.

BPP, 1884–5, XXII, (The 28th) Annual Report of the Commissioners of the Inland Revenue, 1885.

BPP, 1886, XXI, Royal Commission on Depression in Trade and Industry, First Report.

BPP, 1888, X. Report from the Select Committee on Corn Averages.

BPP, 1888, XXXII. Final Report of the Departmental Committee on Agricultural Dairy Schools (the Paget Committee).

BPP, 1888, XVIIII. Select Committee on Allotments and Small Holdings.

BPP, 1889, XII, Report from the Select Committee on Allotments and Small Holdings.

BPP, 1890, XVII. Select Committee on Allotments and Small Holdings.

BPP, 1890–1, XXXVII–XXXIX. Royal Commission on Market Rights and Tolls.

BPP, 1892, XXVI, Report of the Departmental Committee on the Adulteration of Artificial Manures and Fertilizers and Feeding Stuffs used in Agriculture.

BPP, 1893, XXXIV–XXXVI, Royal Commission on Labour. *Note esp.* C 6894–I, W. E. Bear; C6894-V, A. J. Spencer; 6894-XIV, Wales: D. L. Thomas.

BPP, 1893–7. Reports of the Royal Commission on Agricultural Depression, esp. 1894, XII; XV; XVI and XVII; 1897, XV. *Note esp.* First General Report, 1894, XV; Second Report, 1896, XVI; Final Report, 1897, XV; Reports of the Assistant Commissioners, 1894, XVI (*esp.* Report on Ongar, Braintree, etc. by Pringle); 1895, XVI and XVII; Minutes of Evidence and Appendix, 1894, XVI; 1896, XVII; 1897, XV; Second Report on Depression: Particulars of the Expenditures and Outgoings on Certain Estates in Great Britain and Farm Accounts. 1896, XVI.

BPP, 1895, XL, Royal Commission on Land in Wales and Monmouthshire. Minutes of Evidence, III.

BPP, 1896, XXXIV, Royal Commission on Land in Wales and Monmouthshire, 2nd Report.

BPP, 1896, xcii, Agricultural Returns for Great Britain for 1895.

BPP 1897, xv. Royal Commission on BPP, 1874, lxxii, parts i & ii, Return of Owners of Land, 1872–3.

BPP, 1898, xxxiv, Minutes of Evidence . . . upon the Inland Transit of Cattle.

BPP, 1898, lxxxv. Customs Tariff of the United Kingdom from 1800 to 1897.

BPP, 1900, lxxxii. Report by Mr Wilson Fox on the Wages and Earnings of Agricultural Labourers in the United Kingdom.

BPP, 1903, lxvii. British and Foreign Trade and Industry, Memoranda, Statistical Tables and Charts . . . by the Board of Trade with Reference to . . . Industrial Conditions.

BPP, 1903, lxviii, Return of Food Supplies (Imported) since 1870.

BPP, 1903, lxviii, Report on Wholesale and Retail Prices in the United Kingdom in 1902 with Comparative Statistical Tables for a Series of Years.

BPP, 1905, lxxxiv, Second Series of British and Foreign Trade and Industrial Conditions, Memoranda and Statistical Tables.

BPP, 1905, xcvii, (Second) Report on the Wages, Earnings, and Conditions of Employment of Agricultural Labourers in the United Kingdom, by A. Wilson Fox.

BPP, 1906, xxiv. Minutes of Evidence before the Departmental Committee upon the Fruit Industry of Great Britain.

BPP, 1907, xxxiii, Second Report of the Royal Commission on Canals and Inland Navigation of the UK.

BPP, 1907, xxxiv. Royal Commission on Coast Erosion and the Reclamation of Tidal Lands in the United Kingdom, First Report; Minutes of Evidence, appendix xxiii.

BPP, 1908, xxi. Board of Agriculture, Report of the Departmental Committee on Agricultural Education in England and Wales (The Reay Committee).

BPP, 1908, cxxi, Agricultural Statistics, 1907 . . . Prices and Supplies of Corn, Livestock, and other Agricultural Produce.

BPP, 1909, cii, British and Foreign Trade and Industry, Memoranda and Statistical Tables and Charts (1854–1908).

BPP, 1909, xv, Committee on Combinations in the Meat Trade.

BPP, 1910, lxxxiv. Board of Trade, Report . . . into the Earnings and Hours of Labour of Workpeople in the United Kingdom. v. Agriculture in 1907.

BPP, 1910, xii, Fourth and Final Report of the Royal Commission on Canals and Inland Navigations, Vol. vii.

BPP, 1910, lxxiv. Return of the Average Rate of Weekly Earnings of Agricultural Labourers in the Unions of England and Wales.

BPP, 1910, cix. Census of Production (1907), Final Report and Preliminary Tables, Part iii.

BPP, 1911, xiv. Royal Commission on Coast Erosion and Afforestation, Third (and Final) Report.

BPP, 1911, xxii, Report of the Departmental Committee appointed to inquire into the British Export Trade in Livestock with the Colonies.

BPP, 1912. Department of Technical Instructions. Agricultural Output of Ireland, 1908.

BPP, 1912–13, x, Board of Agriculture and Fisheries, The Agricultural Output of Great Britain in 1908 . . . in Connection with the Census of Production Act, 1906.

BPP 1912–13, xi. Board of Agriculture, Report to the Board of Agriculture and Fisheries of an Enquiry into Agricultural Credit and Agricultural Cooperation in Germany: with some Notes of German Livestock Insurance by J. R. Cahill.

BPP, 1912–13, xxv. Minutes of Evidence of Departmental Committee on British Export Trade in Livestock and Combinations in the Meat Trade.

BPP, 1912–13, xlvii, Report of the Departmental Committee on the Position of Tenant Farmers on the Occasion of any Change in the Ownership of their Holdings.

BPP, 1914–16, lxi, Seventeenth Abstract of Labour Statistics.

BPP, 1916, ix, The Food Supply of the United Kingdom: a Report drawn up by a Committee of the Royal Society at the Request of the President of the Board of Trade.

BPP, 1917–18, xxxvi, Agricultural Statistics, Ireland, 1915: Return of Prices of Crops, Livestock and other Irish Agricultural Products.

BPP, 1918, viii, Report of Engineering Trades (New Industries).

BPP, 1919, viii, Report of the Departmental Committee on Agricultural Machinery (Summaries of Evidence).

BPP, 1919, xxv, Third Interim Report of the Committee on the Production and Distribution of Milk.

BPP 1919, ix. Board of Agriculture and Fisheries. Report on Wages and Conditions of Employment in Agriculture.

BPP, 1927, xxv. Agricultural Output of England and Wales, 1925.

BPP, 1933–4, xxvi, Agricultural Output of England and Wales, 1930–31.

BPP, 1935–6, xiv, Royal Commission on Tithe Rentcharge, Minutes of Evidence.

MAF. *Report of the Committee on Stabilisation of Agricultural Prices*, Economic Series, No. 2, 1925.

MAF. *Report on the Trade in Refrigerated Beef, Mutton and Lamb*, Economic Series, No. 6, HMSO, London, 1925.

MAF. *Report on Wool Marketing in England and Wales*, Economic Series, No. 7, 1926.

MAF. *Report on Fruit Marketing in England and Wales*, Economic Series, No. 15, 1927.

MAF. *Report on the Marketing of Wheat, Barley, and Oats in England and Wales*, Economic Series, No. 18, 1928.

MAF. *Agricultural Output and the Food Supplies of Great Britain*. HMSO, London, 1929.

MAF. *Report on the Marketing of Cattle and Beef in England and Wales*, Economic Series, No. 20, 1929.

MAF. *Report on the Marketing of Dairy Produce in England and Wales*, Part 1, *Cheese*, Economic Series, no. 22, 1930.

MAF. *Report on Markets and Fairs in England and Wales*, Part 1, *General Review*,

Economic Series, No. 13, 1927; Part II, *Midland Markets*, Economic Series, No. 14, 1927; Part III, *Northern Markets*, Economic Series, No. 19, 1928; Part IV, *London Markets*, Economic Series, No. 26, 1930.

MAF. *Index Number of Agricultural Prices*. HMSO, London, 1938.

MAFF. *Animal Health: A Centenary, 1865–1965*. HMSO, London, 1965.

MAFF. *A Century of Agricultural Statistics: Great Britain, 1866–1966*. HMSO, London, 1968.

Irish Parliamentary papers

Saorstat Eireann, Dept of Industry and Commerce, *(Irish) Agricultural Statistics, 1847–1926*. Dublin, 1930.

INDEX

Note: Page references in italics indicate tables, figures and maps; references in bold type indicate major discussion of particular topics.

Compiled by Meg Davies

sales 980
  sheep rearing 407, 468
mining 1140, 1142, 1326, 1549
output *161*, *218*, *1912*
population *1946*, *1949*, *1950*
rent levels 219, *1930*, *1931*, *1933*, *1935*
rural industry 1134
statesmen 408, 727
upland areas 1696, *1697*, 1704
wage rates *1998*, *2000*, *2003*, *2004*, *2007*, *2010*
Cunningham, H. 1407–8
Cunningham, Joseph 1592
currency, and bi-metallism 64–6, 146, 157, 327, 341 n.72
curriers 1124–6, *1124*, *1126*
Curtis, John, *Farm Insects* 551, 552–3, 603
Curtler, W. H. R. 2149
cycling 1192, 1538–9, 1565, 1637, 1670, 1752, 2156
Cyclists' Touring Club 1592, 1670
cymhortha (dependence on neighbours) 22–3

Daily News 57, 1321, 1325
Dairy and Cowshed Orders (1885, 1889) 199
dairy associations 670–2, 979
dairy produce **1097–9**
  consumption 42, 43
  in Golden Age **98–107**
  importance 145
  imports 61, 98, 473–4, 1100, 1218
  industrialised 199, 1061, 1097–8
  Irish 78, 283
  output 77, 78, 94, 99, 107, 117, 264, 290–1
    statistics *232*, *236*, *237*, *265*
  prices 38, 150, 156, 164, 473, 478
  product distribution 298, *299*
  production and distribution companies 476–7
  profitability 475–6, 761
  protectionism 64
  quality 199, 201, 404, 621
  retailing 477
  scientific research 605–6, 621
  world trade 39
  see also butter; cheese; cream; milk
dairy schools 169, 199, 202, 437, 605–6, 641, 647
dairying **472–8**
  and arable farming 463–4, 465–6, 902
  as capital-intensive 189, 787
  and cattle feeding 571, 576, 581–2
  factory 197
  farm size 185, 203, 1836, 1841, *1875–6*
  farmers' wives 764, 1375, 1381
  hygiene 204, 445, 477, 503, 511, 807, 996, 1045, 1047
  labour costs 154, 308

pig rearing 114, 445, 450, 487–8
profitability 788, 1745
  in Golden Age 73, 98–9, 120, 123–4, *136*, 2152
  in Great Depression 159–60, 164, 188, 189–90, 475–8, 777, 791
  record-keeping 789
  in Recovery period 219
  in regions *see individual regions*
  stock levels 95
  studies 27
  technical education 169, 641–4, 647
  technology 22, 99–100, 203, 216–17, 472–3, 1097–8
  see also mechanisation
Danson, J. C. (quoted) 168–9
Darby, H. C. 890
Darley, Gillian 1332–3, 1336
Dartmoor 411, 413, 426, 466, 1257, 1706–7, 1709–10
Darwin, Charles 86–7, 595, 596, 1620–1, 1623
Daubeny, Charles 602, 611, 616, 619
Davaine, Casimar 588
Davis, Kingsley 1259
Davy, Sir Humphry 594, 598
Day, Alice Catherine 1380–1
Day, Clive 1102–3, 1108, 1115–16, 1125, 1128, 1130, 1145, 1151, 1164, 1172
De Vries, Jan 1216
dealers
  in census reports *1966*, *1968*
  farmers as 172, 978
  as mortgage lenders 919–20
  role in marketing 954, 956, 976, 978, 982–4, 1217
  and sales credit 923–4
  women as 1188
Deane, P. 116, 865, 2153
death duties 326, 700, 757, **932–3**, 943–4, 1585
debt
  estate 693, 719, 725, 748, **751–8**, 917
  labour 1327, 1329, 1346–8, 1381, 1385
  personal 754, 756
  smallholder 785
  tenant farmer 124, 769, 916–17
  yeoman farmer 725
  see also credit; mortgages
Defoe, Daniel 367, 370, 493
deforestation 864, 885, 1647–9, 1745
Delamere Forest (Ches.), clearance 80
Delamotte, Philip Henry 1534
demand
  for cheese 475
  farms 117, 210, 223, 770
  fodder 959
  food 13, 25, 43, 47, 72–3, 75–6, 131, 361, 953, 1085–6, 1541